Integrated Arithmetic & Basic Algebra

Second Edition

Bill E. Jordan
Seminole Community College

William P. Palow
Miami-Dade Community College

Addison
Wesley

Boston   San Francisco   New York
London   Toronto   Sydney   Tokyo   Singapore   Madrid
Mexico City   Munich   Paris   Cape Town   Hong Kong   Montreal

Publisher	*Jason A. Jordan*
Acquisitions Editor	*Jennifer Crum*
Editorial Project Manager	*Ruth Berry*
Managing Editor	*Ron Hampton*
Text and Cover Design Supervision	*Susan C. Raymond*
Cover Design	*Judith Arisman*
Production Coordinator	*Sheila Spinney*
Production Services	*Pre-Press Company*
Assistant Editor	*Lauren Morse*
Media Producer	*Lorie Reilly*
Marketing Manager	*Dona Kenly*
Marketing Coordinator	*Jennifer Berman*
Prepress Services Buyer	*Caroline Fell*
First Print Buyer	*Evelyn Beaton*

Cover Image: Full Circle, wall quilt	*Michael James © 1998*
Cover Photo	*David Caras*

LIBRARY OF CONGRESS CATALOGING-IN-PUBLICATION DATA

Jordan, Bill E.
    Integrated arithmetic and basic algebra/Bill E. Jordan, William P. Palow.—2nd ed.
      p. cm.
    Rev. ed. of: Integrated arithmetic and basic algebra. 1996.
    Includes index.
    ISBN 0-201-64203-4 (student ed., pbk.)
    1. Arithmetic. 2. Algebra. I. Palow, William P. II. Jordan, Bill E. Integrated arithmetic and algebra
III. Title.

QA107 .J675 2000
513'.12—dc21
                                                  00-022298

ISBN 0-201-64203-4

2 3 4 5 6 7 8 9 10 DOW 03 02 01

Contents

Chapter 2 Laws of Exponents, Products and Quotients of Integers and Polynomials

Chapter 3 Linear Equations and Inequalities

Chapter 4 Graphing Linear Equations and Inequalities

Chapter 8

Ratios, Percents, and Applications

Chapter 9

Systems of Linear Equations

Chapter 10

Roots and Radicals

Chapter 11

Solving Quadratic Equations

Preface

Purpose and Goals

Both the content and the methods used in the teaching of mathematics have changed dramatically during the past few years. While calculus reform and the incorporation of technology into the teaching of mathematics are catching most of the headlines, there are also substantial changes taking place in entry-level mathematics courses. The *NCTM Standards* and the *AMATYC Standards for Introductory College Mathematics* have had a significant impact on mathematics before calculus. Some of these recommendations are addressed in this text.

Too often, students learn mathematical concepts in isolation, rarely seeing how many concepts are interrelated. Each concept is viewed as unique, and students often fail to see the "big picture." Students often view algebra and arithmetic as two different subjects rather than recognizing algebra as an extension of arithmetic using variables. This text shows how the concepts of arithmetic and algebra are related by presenting the two subjects *at the same time* rather than in isolation from each other.

This text is written for students whose entry-level math skills would place them into arithmetic or prealgebra. The text contains the same material that would normally be found in an arithmetic/prealgebra text *and* a beginning algebra text. The major goals of *Integrated Arithmetic & Basic Algebra* are:

1. to prepare students whose entry level is arithmetic/prealgebra for intermediate algebra in one semester rather than two (though the text is also well-suited for use over two semesters);
2. to show the close relationships between arithmetic and algebra;
3. to present the material in an organized, easy-to-read, and understandable manner;
4. to help students think critically about mathematics;
5. to show how mathematics relates to other fields and to real-life situations; and
6. to have students understand the basic principles of arithmetic and beginning algebra.

Integrated Arithmetic & Basic Algebra accomplishes these goals through the means described in the following sections.

Approach

Integration of Algebra and Arithmetic

As opposed to a traditional text that presents arithmetic in the beginning and algebra at the end, this text interweaves the two whenever possible. Research by Bill Palow, Jack Rothman, and others working independently has shown that there is little, if any, correlation between a prior knowledge of arithmetic and subsequent success in algebra. One implication of this research is that we do not need to teach arithmetic in its entirety before teaching algebra as is traditionally the case. However, students do need to be proficient in arithmetic in order to function in their everyday lives.

The ideas and algorithms shared by arithmetic and algebra are introduced in an arithmetic context and are then developed through the corresponding algebraic concept. For example, operations with rational numbers and rational expressions are discussed together, whereas traditionally, operations with rational numbers are discussed early on and operations with rational expressions are discussed much later. The integrated presentation helps students recognize the natural extension of arithmetic to algebra.

At the college level, students are familiar with arithmetic although they may not be proficient in it. Another benefit of the authors' integrated approach is that presenting an arithmetic concept and the corresponding algebraic concept simultaneously uses students' familiarity with arithmetic to assist them in learning algebra. Perhaps surprisingly, while writing the text we found that we could often use an algebraic concept to introduce or present an arithmetic concept rather than the other way around. At the college level, students are reluctant to accept the same arithmetic taught in the same manner that has been presented to them since elementary school. Our approach makes students more willing to learn arithmetic because they know that they are going to see an immediate need for arithmetic in learning the corresponding algebra. Concepts are spiralled continually throughout the text, meaning that students visit and revisit an idea as it evolves in different contexts.

Concept Versus Algorithm

Success in mathematics and the ability to use mathematics are dependent upon an understanding of why mathematical principles are true—not just on the ability to do an exercise correctly. Too often, students view mathematics as a "bag of tricks" and determine that the way to learn mathematics is to memorize which trick applies to which situation. If students truly understand why a concept is true, they will remember that concept longer and will be able to apply it to other situations. Hence, the concept (the "why") is introduced and developed before the process or algorithm (the "how") for performing the operations presented. This text does not present the algorithm without first explaining its conceptual basis. For example, in Section 2.7 on simplifying quotients raised to negative powers, Example 3 shows students how to apply rules of exponents that they already know in order to simplify a new type of problem (see page 210). Then the text shows the new rule they have just derived. This approach is used in virtually all sections of the text.

Critical Thinking and Problem Solving

This text places strong emphasis on thinking skills and problem solving. We used these important skills from the very beginning and throughout the text. A wide variety of methods used to emphasize thinking and problem-solving skills are highlighted below.

1. Our problem-solving strategy is first outlined in Section 3.1 (see page 239) and expanded in Section 3.5 (see page 276).
2. Almost every exercise set contains writing exercises that are designed to help students formulate and understand mathematical ideas. Usually these cannot be answered by simply looking up the answer in the text, but often require critical thinking about a mathematical concept (see pages 115, 275, and 560).

3. Uncomplicated group projects are also found in almost all exercise sets (see pages 86, 227, and 453). These give students additional opportunities to formulate and understand mathematical ideas and they often require critical thinking. Sometimes projects show how mathematics relates to careers and other fields of study.

4. Almost all exercise sets contain problems that are not just skills-oriented but that also require students to go beyond the skill to an application of that skill. For example, in the sections on products of polynomials, students are asked to represent the area of geometric figures, the lengths of whose sides are polynomials (see page 189).

5. Special attention has been given to the language of mathematics. In many ways, mathematics is like a language. Students translate from English to math early in Sections 1.6–1.8, 2.1 and 2.5. Just as you cannot understand a language without learning its vocabulary, you cannot achieve a meaningful understanding of mathematics without learning the proper mathematical vocabulary and grammar (see pages 136, 164, and 676).

Follow-up Study

At Seminole Community College, a study was done on 934 students who entered prealgebra and whose progress was tracked through basic algebra. It was found that 58.8 percent were successful in prealgebra, and 61.6 percent of those who continued to basic algebra were successful. However, approximately 21 percent of those successful in prealgebra did not continue on to basic algebra. Consequently, only 23.3 percent of those students entering prealgebra successfully completed basic algebra. At Seminole, our Integrated Arithmetic and Algebra course is a combination of prealgebra and basic algebra in one course. A study of 339 students entering Integrated Arithmetic and Algebra showed that 54.9 percent were successful as compared to the 23.3 percent successful in prealgebra and basic algebra taught as separate courses. Similar results were obtained at Miami-Dade Community College.

Features of the Text

Worktext Format

This text introduces a concept, develops and explains the appropriate algorithm, and demonstrates the algorithm with carefully chosen examples. Students are then asked to practice and apply the concept through the algorithm. This is accomplished through the use of more than 1000 practice exercises and additional practice exercises embedded into the flow of the text. In this manner each topic is introduced, illustrated, and then practiced prior to the section exercises, which then provide ample opportunity for students to practice and draw ideas together. The practice exercises and additional practice exercises are written so that the first practice exercise mirrors the "a" part of the preceding example, the second mirrors the "b" part, etc. This structure enables students to get immediate help with any practice exercise (or additional practice exercise) that is troublesome. The answers to the practice and additional practice exercises are located at the bottom of the page for easy reference. Answers to the odd-numbered section exercises, as well as answers to all of the chapter review exercises and chapter tests, are located in the back of the book.

Examples

We have carefully selected more than 500 examples that cover all of the variations of any particular concept that are appropriate to a text at this level. The examples are worked out step-by-step in a two-column format, with the steps on the left and the explanations on the right. In most texts, the explanations to the right of each step represent what was done to get that step. This order is the reverse of the way we think. We do not do something and then wonder why we did it. In this text, the explanation to the right explains what *is to be done* to get to the next step. This is more consistent with the way we think when solving a problem: we determine what has to been done next and then we do it. This sequence is easier to follow and teaches students good problem-solving techniques.

Approach to Technology

The use of technology should be a natural part of mathematics and should be incorporated as such. The amount of calculator use at this level varies immensely among colleges from none to total immersion. We have incorporated some specific uses of calculators in the early chapters where appropriate (see pages 10, 17, and 79), and many exercise sets include problems that are specifically designated as calculator exercises (see pages 12, 123, and 241). Wherever appropriate, a Calculator Exploration Activity has been added to lead students to discover a mathematical property or concept, and these activities are followed with the appropriate justification (see pages 158, 171, and 682). We feel that, on the whole, instructors should incorporate the amount of technology with which they are comfortable. Given the great number of calculators on the market and the tremendous differences in the ways they operate, we were reluctant to specify keystrokes, but did in some cases for some of the most common calculators. We are very strong supporters of the use of graphing calculators in the teaching of mathematics. However, their use is not included in this text because those who support their usage in a course of this level are in the minority.

Exercise Sets

Each exercise set contains an abundant number of exercises written so that consecutive odd- and even-numbered exercises are similar. There are over 5000 exercises in the text. Great care has been taken to ensure that most variations of each concept are covered in the exercises. Almost all exercise sets contain problems that require students to go beyond the basic skills level and apply the basic skill(s) of that section to a situation. Often this is done using geometric concepts or through solving application problems. Almost all exercise sets contain writing exercises that require the student to understand and think about the concept(s) of that section (see pages 115, 560, and 682). Usually the answers to writing exercises cannot be copied directly from the text. Almost all exercise sets also contain group projects and critical-thinking exercises (see pages 86, 227, and 453), many of which can be assigned as writing exercises.

Geometry

Integrated Arithmetic & Basic Algebra emphasizes geometry more than most texts at this level. Some examples of where geometry has been integrated into other topics are listed below:

- In Sections 1.2–1.4, students are asked to apply the order of operations and the skill of substituting for a variable (learned in Section 1.1) to find perimeters, areas, volumes, and surface areas of geometric figures.
- In Sections 1.2–1.4, students are asked to solve real-world application problems involving geometry that do not require the use of variables. These problems involve perimeters, areas, volumes, and surface areas of geometric figures.
- In Section 3.7, students are asked to solve real-world application problems that involve the use of variables. These problems involve perimeter, angle relationships of parallel lines, and angle relationships of triangles.
- In Section 10.7, students are asked to solve right triangles, rectangles, squares, and real-world problems involving the Pythagorean theorem.
- In many sections throughout the text, students are asked to apply geometric formulas in the exercise sets.

Writing Exercises and Study Tips to Reduce Math Anxiety

Writing about things that cause difficulties is an effective way to understand problems and reduce anxiety. We suggest informal writing, such as prewriting and/or journals, to allow students to write about their feelings and difficulties. We have included extensive writing exercises throughout the text to give students the opportunity to write about mathematics. See page 20 of Section 0.3 for an example of an exercise that asks students to explain how they would analyze a mathematical situation.

Since mathematics is learned differently from other subjects, Study Tip boxes have been dispersed throughout Chapters 0, 1, and 2 to assist and encourage students as they attempt to learn mathematics with less anxiety. For example, on page 122, Study Tip 5 points out that although some things in mathematics need to be memorized, the best way to learn mathematics is by understanding it. Another major philosophy of this text—attention to concept and understanding—has been shown to reduce math anxiety by as much as 40 percent.

Chapter Summaries

Each chapter summary contains the important ideas, definitions, algorithms, and procedures from the chapter and serves as a quick overview and review of the chapter (see pages 149–151, 308–310, and 454–455).

Chapter Review Exercises

Each chapter review contains additional exercises that are keyed to the section of the text that explains the concepts and algorithms needed to do these exercises (see pages 151–154 and 456–457). These allow students to test their understanding of the chapter quickly before attempting the chapter test.

Chapter Tests

Each chapter contains a sample test that students can use to evaluate their understanding of the material covered in the chapter. Test questions are presented in a free-response format (see pages 154–155 and 457–458). As previously mentioned, the answers to all the questions on the Chapter Tests are in the back of the book.

Definitions and Algorithms

All definitions, algorithms, and other material of importance is clearly labeled and boxed for easy reference (see pages 74, 403, and 430). The first time a mathematical term is used, it is printed in boldface type.

Be Careful and Note

Throughout the text, students are warned about common errors with the Be Careful! feature (see pages 138, 161, and 434). Paragraphs that begin with Note provide additional information, explanations, or observations about the concept just discussed (see pages 111, 128, and 278).

What's New in the Second Edition

There have been extensive changes made in the second edition, especially in the order of topics and chapter numbers. It has been our experience that most students are comfortable with whole number and decimal arithmetic and simple operations with fractions. Consequently, we feel that most of the material in the first chapter is review, and we have labeled this chapter as Chapter 0. We have also moved the chapter on graphing linear equations (formerly Chapter 9) closer to the front of the book (Chapter 4) so that it immediately follows the chapter on solving linear equations with one unknown. Following is a summary of the major changes in organization.

- Chapter 0 (formerly Chapter 1): We wanted this chapter to be mostly a skills-building chapter for those who need the review. Consequently, we eliminated some of the material that is repeated later in the text. The number line and order relations were eliminated from Section 0.1, the properties of addition and perimeters of geometric figures were eliminated from Section 0.2, subtraction properties was eliminated from Section 0.3, and properties of multiplication and areas were eliminated from Section 0.4. The remaining sections were streamlined and modified for consistency.

- Chapter 1 (formerly Chapter 2): Optional Calculator Exploration Activity boxes were added to Sections 1.6 and 1.7. Former Sections 2.1 and 2.2 have been combined into one section (Section 1.1). Former Sections 2.3–2.4 have been rearranged as the present Sections 1.2–1.4 so that perimeters (Section 1.2), areas (Section 1.3), and volumes including surface areas (Section 1.4) are discussed separately with appropriate applications. The former Section 2.7 and part of 2.9 have been combined into Section 1.7, and the former Section 2.8 and the remainder of Section 2.9 have been combined into Section 1.8.
- Chapter 2 (formerly Chapter 3): Optional Calculator Exploration Activity boxes were added to Sections 2.1, 2.2, 2.5, 2.6, and 2.7. Products of the form $a(bx + c) + d(ex + f)$ were added to Section 2.3. Former Sections 3.5 and 3.6 were combined into Section 2.6, and scientific notation was moved to the end of the chapter.
- Chapter 3 (formerly Chapter 4): New application problems were added to Sections 3.1–3.3, and there are now three sections on application problems.
- Chapter 4 (formerly Chapter 9): Real-world applications have been added to the exercise sets in Sections 4.2, 4.3, and 4.5. The slope-intercept form of a line is now in Section 4.4 (formerly 9.4) rather than in the section on writing equations on lines, and Section 4.6 is a new section on relations and functions.
- Chapter 5 (formerly Chapter 5): The order of topics has been changed considerably. Sections 5.1 and 5.2 have been combined into new Section 5.1. In Section 5.2, factoring by grouping has been added to the material on removing the GCF. Section 5.3 combines the former Section 5.4 with material from the former Section 5.8 on factoring the sum and difference of cubes. Section 5.6 (formerly 5.7) includes factoring using the ac method as an optional topic, and the old Section 5.8 has been eliminated.
- Chapter 6 (formerly Chapter 6): Sections 6.1 and 6.2 have been combined into one section, Section 6.1.
- Chapter 7 (formerly Chapter 7): The order at the beginning of the chapter has been changed. Section 7.1 was formerly Section 7.4, and Section 7.2 is the former Sections 7.1 and 7.2 combined.
- Chapter 8 (formerly Chapter 8): This chapter has undergone major revision both in content and approach. In Section 8.2, proportions that result in quadratics and equivalent ways of writing proportions were added. Section 8.3 has been completely rewritten with a different approach. Section 8.4 presents a different approach to the simple percent situation, sales tax computation, and discount calculations. In Section 8.5, commission calculations have been moved to the front of the section, and interest calculations have been moved to the back. Percent increase and decrease has been moved from the old Section 8.6 to Section 8.5. The mixture, investment, and money problems have been moved to Chapter 3, thus eliminating the old Section 8.6.
- Chapter 10 (formerly Chapter 11): In Section 10.2, a table of powers of 2, 3, 4, 5, 6, and 7 was added. Optional Calculator Exploration Activity boxes were added to Sections 10.2, 10.3, and 10.4, and more problems have been added to the exercise set for 10.2.

Ancillaries

The following ancillaries are available to help both instructors and students use this text effectively:

For the Student

Student's Solutions Manual This supplement contains solutions for the odd-numbered section exercises (including writing exercises as appropriate), as well as solutions for all of the chapter review exercises and practice tests.
ISBN 0-201-64205-0

InterAct Math® Tutorial Software   Available on a dual-platform, Windows/Macintosh CD-ROM, this interactive tutorial software provides algorithmically generated practice exercises that are correlated at the section level to the content of the text. Every exercise in the program is accompanied by an example and a guided solution designed to involve students in the solution process. The software recognizes common student errors and provides appropriate feedback. It also tracks student activity and scores and can generate printed summaries of students' progress. Instructors can use the InterAct Math Plus® course-management software to create, administer, and track on-line tests and monitor students' performance during their practice sessions in InterAct Math. ISBN 0-201-70732-2

InterAct MathXL: *www.mathxl.com*
InterAct MathXL is a Web-based tutorial system that helps students prepare for tests by allowing them to take practice tests and receive a personalized study plan based on their results. Practice tests are correlated directly to the section objectives in the text, and once a student has taken an on-line practice test, the software scores the test and generates a study plan that identifies strengths, pinpoints topics where more review is needed, and links directly to the appropriate section(s) of the InterAct Math® tutorial software for additional practice and review. A course-management feature allows instructors to view students' test results, study plans, and practice work. Students gain access to the InterAct MathXL Web site through a password-protected subscription; subscriptions can either be bundled with new copies of *Integrated Arithmetic & Basic Algebra* or purchased separately.
InterAct MathXL subscription bundled with the text: ISBN 0-201-71431-0
InterAct MathXL subscription purchased separately: ISBN 0-201-71630-5

Videos   Designed specifically for *Integrated Arithmetic & Basic Algebra*, these videos present a series of lectures correlated directly to the chapter content of the text. These videos use exercises taken from the exercise sets in the text itself, and each segment includes a "stop the tape" feature that encourages students to stop the videotape, work through an example, and resume play to watch the video instructor go over the solution.
ISBN 0-201-70488-9

Digital Video Tutor   The videotapes for *Integrated Arithmetic & Basic Algebra* are also available on CD-ROM, making it easy and convenient for students to watch video segments from a computer at home or on campus. The complete, digitized video set, both affordable and portable for students, is ideal for distance learning or supplemental instruction.

AWL Math Tutor Center   The AWL Math Tutor Center is staffed by qualified mathematics instructors who tutor students via toll-free telephone, fax, or e-mail on examples and exercises from their text. The AWL Math Tutor Center is open five days a week, seven hours a day, and is accessed through a registration number that can be bundled with a new textbook or purchased separately.
Tutor Center registration bundled with the text: ISBN 0-201-71432-9
Tutor Center registration purchased separately: ISBN 0-201-44461-5

For the Instructor

Instructor's Solutions Manual   This supplement contains solutions for the even-numbered section exercises in the text (including writing exercises as appropriate), as well as for the group projects.
ISBN 0-201-64208-5

Answer Book   This supplement contains answers to all the section exercises (including writing exercises as appropriate), chapter tests, and chapter review exercises in the text.
ISBN 0-201-64204-2

Printed Test Bank/Instructor's Resource Guide   This supplement contains two multiple-choice tests per chapter; three free-response tests per chapter; chapter overviews and a sample syllabus; TASP and CLAST correlation guides; and teaching tips for using manipulative materials in the classroom.
ISBN 0-201-64207-7

TestGen-EQ with QuizMaster-EQ   Available on a dual-platform, Windows/Macintosh CD-ROM, this fully networkable software enables instructors to create, edit, and administer tests using a computerized test bank of questions organized according to the chapter content of the text. Six question formats are available, and a built-in question editor allows the user to create graphs, import graphics, and insert mathematical symbols and templates, variable numbers, or text. An Export to HTML feature allows practice tests to be posted to the Internet, and instructors can use QuizMaster-EQ to post quizzes to a local computer network so that students can take them on-line. QuizMaster-EQ automatically grades the quizzes, stores results, and lets the instructor view or print a variety of reports for individual students or for an entire class or section.
ISBN 0-201-70730-6

InterAct Math® Plus   This networkable software provides course-management capabilities and on-line test administration for Addison Wesley Longman's InterAct Math® Tutorial Software (see For the Student). InterAct Math Plus enables instructors to create and administer on-line tests, summarize students' results, and monitor students' progress in the tutorial software, creating an invaluable teaching and tracking resource.
Windows instructor package: ISBN 0-201-63555-0
Macintosh instructor package: ISBN 0-201-64805-9

Acknowledgments

The authors would like to express their appreciation to the following very special people at Addison Wesley Longman for their guidance and, most of all, for being so darn nice and cooperative: Ruth Berry (project manager), Ron Hampton (managing editor), Susan Raymond (designer), Lauren Morse (assistant editor), and most of all our editor Jenny Crum (What a sweetheart!).

In addition, we would especially like to thank the following reviewers, whose contributions were invaluable in producing a text of this quality:

Mary Kay Abbey, Montgomery College
Beverly A. Asplen, Seminole Community College
Richelle Blair, Lakeland Community College
Barbara Blass, Oakland Community College-Royal Oak
Laurie Braga, Loyola University Chicago
Debra Bryant, Tennessee Technological University
Allen J. Bryant, Eastern Illinois University
Karen Sue Cain, Eastern Kentucky University
Valerie Chu, Lemoyne-Owen College
D. J. Clark, Portland Community College
James Condor, Manatee Community College
Pat C. Cook, Weatherford College
Arriana Derrick, Midlands Technical College
Laurel Fischer, Seminole Community College
Barb Gentry, Parkland College
Mary Lou Hammond, Spokane Community College
Janet Hansen, Murray State University
Roberta Lacefield, Waycross College
Theodore Lai, Hudson County Community College

Gail Langer, Oakland Community College-Royal Oak
Christy Linsley, Sam Houston State University
Maria Maspons, Miami-Dade Community College
Dennis Missavage, Chattahoochee Technical Institute
Elizabeth Morrison, Valencia Community College-West Campus
Scott Perkins, Lake Sumter Community College
Richard Plagge, Highline Community College
Angelia Reynolds, Gulf Coast Community College
Candido K. Sanchez, Miami-Dade Community College
Joanne Shansky, Milwaukee Area Technical College
Penny Singerland, Mount Hood Community College
Eileen Synott, Mattatuck Community College
Shirley Treadway, Eastern Illinois Community College
Ellyn Webb, Midlands Technical College

We would also like to thank our colleagues at Seminole Community College and Miami-Dade Community College for allowing us to create the Integrated Arithmetic and Algebra courses at our schools and to class test the materials.

This book has been written with the characteristics and needs of the beginning mathematics student in mind. It is based on the research and experience of authors who have each taught for over a quarter of a century and who have received numerous awards for teaching excellence. We have tried to combine our experience with current learning theory and practice to produce a book that we hope will make a difference. It is our belief that the beginning mathematics student needs a fresh way of viewing mathematics so that he/she is not looking at the same old material in the same old way. We hope that your experiences with this text are as pleasant as ours have been.

B. E. Jordan
W. P. Palow

Chapter 0

Basic Ideas

The ideas of mathematics are called **concepts.** The combining of numbers to obtain another number is known as an **operation.** A process for carrying out an operation is an **algorithm.** An **application** is a situation, usually posed in words, in which a concept is recognized and an algorithm is used to calculate a result. Applications are sometimes called **word problems.**

In this chapter we will introduce and explore the concept of an operation before we introduce and explain the algorithm for that operation. By looking at the concept first we seek to promote understanding, help you to remember the mathematics longer, reduce feelings of uneasiness about mathematics, and help in applications. Applications follow after the mastery of the concepts and algorithms for each operation and will be introduced in Chapter 1.

In Chapter 0 we will concentrate on the first step in the process of problem solving, **identification.** Identification means to know what you "see" when you "see" it. To this end we will learn the vocabulary of arithmetic by paying careful attention to the meaning of the words in the language of mathematics. We will also learn to organize our knowledge by **classifying** or putting like things together. Finally, through classifying, we will learn to recognize *when* to add, subtract, multiply, or divide, so we can combine this ability with knowing *how* to carry out these operations.

Chapter 0 covers the operations of addition, subtraction, multiplication, and division of whole numbers and decimals. Fractions are briefly introduced. Word problems are also introduced.

Everything in Chapter 0 is important. It is intended to give you a "jump start" so that algebraic ideas may be introduced in Chapter 1. However, if you feel you have already mastered this material, you may go directly to Chapter 1.

Reading and Writing Numerals

OBJECTIVES

When you complete this section, you will be able to:

a. Read and write numerals that represent whole numbers.

b. Recognize standard notation and expanded notation of a whole number.

c. Recognize sets and set notation.

Introduction

A **number** is a concept or idea that exists only in our minds. A **numeral** is a symbol that represents a number. How many dollar signs are there in the following? $, $, $, $, $. This number of dollar signs, or other things, may be represented in many different ways depending upon the system used for writing numerals. Some of these are: 五 (Chinese), IIIII (Egyptian), Œ (Ionian Greek), and V (Roman numerals).

The development of the Hindu-Arabic system of numerals with a zero gave mathematicians a powerful tool to develop the present-day subject of arithmetic. The Hindu-Arabic enumeration system uses ten symbols called **digits.** They are 0, 1, 2, 3, 4, 5, 6, 7, 8, and 9. The meaning of each digit comes from counting. They tell us "how many." We associate each of these symbols with a corresponding number of objects in a **set** or collection of things. Sets may be designated by writing the **members** or things in the set between two symbols called **braces**, { }. Set A = {a, b, c} has three members and is thus associated with the digit 3.

Did you know that in the Middle Ages, India was called Hindu?

The most important thing the Hindu-Arabic system gives us is the concept of **place value** in which the meaning of a digit depends on its left-right location. As we go to the left, the value of each place is found by multiplying the value of the place before it by ten. The specific location is called the **value place.** The **meaning of a digit** is found by multiplying the digit by its place value. In the numeral 654, the digit 6 is in the 3rd value place which has a place value of 100 and a meaning of 600. The **value of the numeral** is found by adding the individual amounts represented by the digits of the numeral.

Place values allow construction of numerals that represent any number of things. Some of the values of places in our system of numeration are shown in the following table:

—,	—	—	—,	—	—	—,	—	—	—
B	H	T	M	H	T	T	H	T	O
i	u	e	i	u	e	h	u	e	n
l	n	n	l	n	n	o	n	n	e
l	d		l	d		u	d	s	s
i	r	M	i	r	T	s	r		
o	e	i	o	e	h	a	e		
n	d	l	n	d	o	n	d		
s		l	s		u	d	s		
	M	i		T	s	s			
	i	o		h	a				
	l	n		o	n				
	l	s		u	d				
	i			s	s				
	o			a					
	n			n					
	s			d					
				s					

In Europe, the comma is used in the same way that we use a decimal point in the United States.

The place values are separated into periods of three by a comma so that a numeral is easier to read. As shown on page 2, the place values start on the far right with a value of one, or one unit. As we go to the left, each place value is found by multiplying the one before it by ten. Thus, we obtain one, ten, one hundred, one thousand, ten thousand, one hundred thousand, one million, ten million, one hundred million, one billion, and so on. That is,

one	$= 1$
ten	$= 1 \times 10$
one hundred	$= 1 \times 10 \times 10$
one thousand	$= 1 \times 10 \times 10 \times 10$
ten thousand	$= 1 \times 10 \times 10 \times 10 \times 10$
one hundred thousand	$= 1 \times 10 \times 10 \times 10 \times 10 \times 10$
one million	$= 1 \times 10 \times 10 \times 10 \times 10 \times 10 \times 10$
ten million	$= 1 \times 10 \times 10 \times 10 \times 10 \times 10 \times 10 \times 10$
one hundred million	$= 1 \times 10 \times 10 \times 10 \times 10 \times 10 \times 10 \times 10 \times 10$
one billion	$= 1 \times 10 \times 10 \times 10 \times 10 \times 10 \times 10 \times 10 \times 10 \times 10$
and so on.	

Example 1

Give the place value of the underlined digit.

a. 64,3<u>8</u>9   **Solution:**
3rd place, so $1 \cdot 10 \cdot 10 =$ one hundred.

b. 4<u>0</u>,934   **Solution:**
4th place, so $1 \cdot 10 \cdot 10 \cdot 10 =$ one thousand.

c. 9,487,25<u>4</u>   **Solution:**
1st place, so unit, or one.

Practice Exercises

Give the place value of the underlined digit.

1. 98,5<u>4</u>6

2. 3<u>7</u>8,035

3. 956,<u>0</u>43

The digit in a numeral tells us how many of each of the values we have.

Example 2

Give the meaning of the underlined digit in each case.

a. 46,<u>5</u>03   **Solution:**
The 5 is in the 3rd place, so it represents 5 hundreds or $5 \cdot 10 \cdot 10 = 5 \cdot 100 = 500$.

b. 6,0<u>9</u>4,321   **Solution:**
The 9 is in the 5th place, so it represents 9 ten thousands or $9 \cdot 10 \cdot 10 \cdot 10 \cdot 10 = 9 \cdot 10,000 = 90,000$.

Practice Exercises

Give the meaning of the underlined digit in each case.

4. 14<u>3</u>,706

5. 9<u>7</u>,890,462

The numerals that we have used to represent Hindu-Arabic numbers are said to be in **standard notation** or to be **standard numerals**. Thus, 15,486 is in standard notation, or is called a standard numeral.

As mentioned before, the value represented by a numeral is found by adding the individual amounts represented by the various digits. That is, $15,486 = 1 \cdot 10,000 + 5 \cdot 1,000 + 4 \cdot 100 + 8 \cdot 10 + 6 \cdot 1 = 10,000 + 5,000 + 400 + 80 + 6$. A numeral written in this form is said to be in **expanded notation**.

Example 3

Write in expanded notation.

a. 64,253

Solution:
$$64,253 = 6 \cdot 10,000 + 4 \cdot 1,000 + 2 \cdot 100 + 5 \cdot 10 + 3 \cdot 1$$
$$= 60,000 + 4,000 + 200 + 50 + 3$$

b. 205,650

Solution:
$$205,650 = 2 \cdot 100,000 + 0 \cdot 10,000 + 5 \cdot 1,000 + 6 \cdot 100 + 5 \cdot 10 + 0 \cdot 1$$
$$= 200,000 + 5,000 + 600 + 50$$

Practice Exercises

Write in expanded notation.

6. 4819

7. 214,804

If more practice is needed, do the Additional Practice Exercises.

Additional Practice Exercises

Write the following in expanded notation:

a. 56,149

b. 2,830,456

Using the concept of place value, the standard numeral 6,204,893 means: 6 millions and 2 hundred thousands and 0 ten thousands and 4 thousands and 8 hundreds and 9 tens and 3 ones. However, this is not the way the numeral is properly read aloud or written. The numeral is written six million, two hundred four thousand, eight hundred ninety-three.

When writing out numerals, the numerals twenty-one through twenty-nine, thirty-one through thirty-nine, forty-one through forty-nine, and so on to ninety-nine are hyphenated. The numerals ten, twenty, thirty, forty, and so on to ninety are not hyphenated. The following partial list contains almost everything you need to read or write a numeral:

twenty	twenty-one
thirty	thirty-two
forty	forty-three
fifty	fifty-four
sixty	sixty-five
seventy	seventy-six
eighty	eighty-seven
ninety	ninety-eight

The words that we say while reading the numeral for a number are called its **word name.** The word name is a third type of numeral or symbol for the number.

Please recall that the place values are separated into periods or groups of three place values. From right to left these periods are the units period, the thousands period, the millions period, the billions period, and so on. As we read a numeral we say the numeral within a period, then name the period. For example, in the numeral 414,678, we say, "four hundred fourteen," then we say "thousands" because 414 is in the thousands period. We do not name the units period but simply say "six hundred seventy-eight."

Example 4

Give the word name for the standard numeral.

a. 18,704,234   **Solution:**
Eighteen million, seven hundred four
thousand, two hundred thirty-four

b. 4,914,042   **Solution:**
Four million, nine hundred fourteen
thousand, forty-two

Be careful: Whole numbers
do not have the word "and"
in their names.

IMPORTANT: In Example 4a we did not say, "zero ten thousands." We left out any mention of ten thousands. When reading a numeral that has the digit 0, leave out the place value of the 0.

Practice Exercises

Write the word name for each standard numeral.

8. 6,722,414

9. 7,304,056

Given a word name, we will always be able to write the standard numeral.

Example 5

Express each of the following as a standard numeral:

a. Six thousand, nine hundred six   **Solution:**
6906

b. One hundred million,
eighty-six thousand,
three hundred five

Solution:
100,086,305

Practice Exercises

Express each of the following as a standard numeral:

10. Twelve thousand, five hundred seven

11. Eight million, nine hundred thirteen thousand,
seven hundred thirty-two

The numerals we have studied so far represent **whole numbers.**

Whole Numbers

The set {0, 1, 2, 3, . . . } is the set of whole numbers.

The three dots, . . . , mean that the pattern established by the previous numbers continues forever.

Agreement: Beginning with Section 0.2, whenever we say "numeral," we will mean the standard numeral in the Hindu-Arabic enumeration system unless otherwise specified.

Exercise Set 0.1

Write the word name for each of the following standard numerals:

1. 12	**2.** 824	**3.** 905	**4.** 970
5. 7149	**6.** 2013	**7.** 8902	**8.** 5003

---

9. 30,209

10. 402,668

11. 4,078,074

Do not forget: You can check your spelling by looking at the list on page 4.

12. 9,203,441

13. 29,756,011

14. 17,104,869

Write each of the following as a standard numeral:

15. Four hundred seven

16. Eight thousand, nine hundred eighty-one

17. Fourteen thousand, seventy-three

18. One hundred five thousand, two hundred twenty-seven

19. Nine hundred two thousand, four hundred sixty

20. Forty-five million, two hundred thousand, six hundred one

Give the place value of the underlined digit in each of the following:

21. 8<u>6</u>41

22. 9,7<u>6</u>4,534

23. 85,7<u>3</u>6,394

24. 7,563,0<u>8</u>9

25. 246,576,83<u>4</u>

26. <u>3</u>4,255,768

What is the meaning of the designated digit in each of the following numerals?

27. 7 in 60,171

28. 8 in 338,420

29. 0 in 105,268

30. 2 in 281,093

31. 4 in 34,920,385

32. 4 in 10,233,458

Write each of the following in expanded form:

33. 397

34. 818

35. 3033

36. 740,992

37. 409,135

38. 1,017,819

Challenge Exercises:

39. There are five thousand, two hundred eighty feet in one mile. Write this numeral in expanded notation.

40. Some enumeration systems were repetitive. That is, if you wanted to represent three of something, you repeated the symbol for one, three times. If you wanted to represent thirty things, you wrote the symbol for ten, three times, and so on. In ancient Egypt, the symbol for one was I, the symbol for ten was ∩ . How would you write the numeral for twenty-five in the Egyptian system?

41. Some systems of enumeration were multiplicative. That is, if you wanted to represent thirty things, you wrote the symbol for three followed by the symbol for ten. In China and Japan the symbol for three is 三 and the symbol for ten is 十 . Thus, thirty is symbolized 三 十 . If the symbol for eight is 八 and the symbol for six is 六 , how do the Chinese write eighty-six?

Writing Exercises:

42. Suppose that you are in a culture that counts like this: 1, 2, 3, many. (That is, there are no distinct numerals beyond 3.) Being a large landowner, you have many, many sheep. How could you keep track of the sheep so that you would know if any disappeared overnight? (*Hint:* Try matching sheep with objects.)

43. At least two ancient cultures had enumeration systems based on twenty rather than ten. Why do you think that was true?

Section 0.2	# Addition of Whole Numbers

OBJECTIVES

When you complete this section, you will be able to:

a. Correctly give the sum of any two digits from memory.

b. Use the algorithm for addition of whole numbers without regrouping (carrying).

c. Use the algorithm for addition of whole numbers with regrouping (carrying).

d. Use combinations of numbers to add columns of whole numbers.

Introduction   In Section 0.1 we discussed numbers. In this section we begin our study of operations on numbers. The four basic operations of arithmetic are **addition, subtraction, multiplication,** and **division.** For now we will restrict ourselves to exploring concepts and processes of operations. Properties of operations and applications will be discussed in Chapter 1. We begin with addition.

The concept of addition may be thought of as putting two nonoverlapping sets together to form a combined set. The number of elements in the combined set is the sum

Combination   of the numbers of elements in the individual sets.

Example 1

a. Add seven and three.

    7   add   3
        +

Solution:

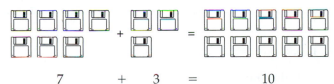

    7   +   3   =   10

If we have a set containing 7 objects and we combine it with a set that contains 3 objects, then we will have a set containing 10 objects.

b. Write a word problem for addition.

Solution:
Fred and Eduardo are roommates at the University of Minnesota. Fred has five textbooks on the bookshelf and Eduardo has four textbooks on the same shelf. How many texts are on the shelf?

In order to talk about several numbers at the same time, we need the idea of a **variable.** A variable is a symbol, usually a letter, which stands for the numbers of a specific set. In this case, the specified set is the set of whole numbers. Any letter may be used as a variable, for example, a, b, c, and so on. We can now offer a formal statement about addition.

DEFINITION The Sum of Whole Numbers

The sum of any two whole numbers, a and b, is the whole number c such that $a + b = c$. The numbers a and b are called **addends** and the number c is called the **sum.**

For example, the sum of 2 and 4 is represented as $2 + 4 = 6$. That is, 2 (addend) + 4 (addend) = 6 (sum).

There are certain sums that we must know and have memorized before we can add anything but the smallest whole numbers. The best way to organize these sums is to put them in a table. This way we can easily find the appropriate sum and look for some very important patterns.

Table of Sums for Two Digits

+	0	1	2	3	4	5	6	7	8	9
0	0	1	2	3	4	5	6	7	8	9
1	1	2	3	4	5	6	7	8	9	10
2	2	3	4	5	6	7	8	9	10	11
3	3	4	5	6	7	8	9	10	11	12
4	4	5	6	7	8	9	10	11	12	13
5	5	6	7	8	9	10	11	12	13	14
6	6	7	8	9	10	11	12	13	14	15
7	7	8	9	10	11	12	13	14	15	16
8	8	9	10	11	12	13	14	15	16	17
9	9	10	11	12	13	14	15	16	17	18

How do we use the table? A line of numbers going left and right is called a **row.** A line of numbers going up and down is called a **column.** The first addend in the sum comes from the first column and the second addend comes from the first row. If we choose 3 in the first column and 5 in the first row, the place where the corresponding column and row intersect is the sum of 3 and 5, which is 8. That is, $3 + 5 = 8$.

After we have mastered the addition facts, we may use them to add any two whole numbers. The process of addition with standard numerals is known as the **addition algorithm.** This process involves lining up the corresponding place values to insure like values are added to like values as digits are added from right to left. This process is shown in the following examples beginning with sums that require no carrying:

Example 2

Add.

a. 154 and 21

Solution:
We line up place values. Units are over units, tens over tens, and so on.

Add units.	*Add tens.*	*Add hundreds.*
↓	↓	↓
154	154	154
+21	+21	+21
5	75	175
4 units and 1 unit is 5 units.	5 tens and 2 tens is 7 tens or 70.	1 hundred and 0 hundred is 1 hundred.

b. 632 and 257

Solution:
We line up place values. Units are over units, tens over tens, and so on.

Add units.	*Add tens.*	*Add hundreds.*
↓	↓	↓
632	632	632
+257	+257	+257
9	89	889
2 units and 7 units is 9 units.	3 tens and 5 tens is 8 tens.	6 hundreds and 2 hundreds is 8 hundreds.

Practice Exercises

Add each of the following:

1. 463 and 16

2. 502 and 456

3. 6034 and 562

Answers:

Practice Exercises 1–3: 1. 479 2. 958 3. 6596

IMPORTANT: Unless a zero lies within the numeral and acts as a place holder, we do not write it until we need to emphasize that it is there, as in Example 3a.

If the sum of the digits in a place is greater than or equal to 10, we **regroup.** To use a simple example, we add 7 and 6. We get more units than we can symbolize in the units place, since the largest single digit we can use is 9. Therefore, we take 3 units from the 6 and add them to the 7. This gives us one group of 10. The remaining 3 units from the 6 are left in the units place and the one group of 10 is "carried" to the tens value place by putting a "1" above the digits already there. This process is called **regrouping** or **carrying** and is illustrated by the following example:

Example 3

Add the following whole numbers:.

a. 48 and 3

Solution:

$$
\begin{array}{r}
\downarrow \\
48 \\
+3 \\
\hline
1
\end{array}
\qquad
\begin{array}{r}
\downarrow \\
1 \\
48 \\
+3 \\
\hline
51
\end{array}
$$

Now 8 ones and 3 ones is 11. Regrouped we get 1 ten and 1 unit.

Put the 1 ten into the tens column as "1." To the unwritten 0 in the tens place, we add 1 ten and 4 tens to get 5 tens.

b. 76 and 58

Solution:

$$
\begin{array}{r}
\downarrow \\
1 \\
76 \\
+58 \\
\hline
4
\end{array}
\qquad
\begin{array}{r}
\downarrow \\
1\,1 \\
76 \\
+58 \\
\hline
34
\end{array}
\qquad
\begin{array}{r}
\downarrow \\
1\,1 \\
76 \\
+58 \\
\hline
134
\end{array}
$$

6 ones and 8 ones is 14 ones. Regroup to 1 ten and 4 ones. Write 4 in the ones place and carry 1 group of ten.

The sum of 1 ten (the carry) and 7 tens and 5 tens is 13 tens. Regroup 10 tens into 1 group of a hundred and carry 1 hundred.

And last, we add the carry to the unwritten 0s in both addends.

c. 706,469 and 514,088

Solution:

$$
\begin{array}{r}
1\ \ 1\,1\ \ \ \\
706{,}469 \\
+514{,}088 \\
\hline
1{,}220{,}557
\end{array}
$$

Starting with the units column, we add each column in turn and carry when necessary.

Practice Exercises

Add each of the following:

4. 25 and 9

5. 6 and 34

6. 68 and 83

7. 99 and 15

8. 140,356 and 92,076

There are techniques that can help in the adding process. One of these is to look for **combinations** of digits that when added give easy sums. We can find various combinations of digits whose sums are 5 or 10 or 15, and so forth. For example, $1 + 4 = 5$, $8 + 2 = 10$, $7 + 3 = 10$, or $7 + 8 = 15$.

Example 4

Add the following:

a. $82 + 73 + 17$

b. 149, 265, 321, and 785

Solution:

The steps are lettered to help you follow the procedure.

1	
82	
73	
+17	
172	

Ones Column
a) Combine 3 and 7 for a partial sum of 10.
b) Add 10 and 2 to get 12.
c) Write 2 for 12 and carry 1 ten.

Tens Column
d) Combine 7 and 8 to get 15.
e) Add 1 and 15 and 1 to get 17.

Solution:

2 2
149
265
321
785
1520

Ones Column
a) 9 + 1 = 10
b) 5 + 5 = 10
10 + 10 = 20
Put 0, carry 2 tens.

Tens Column
c) 4 + 6 = 10
d) 2 + 8 = 10,
10 + 10 + 2 = 22
Put 2, carry 2 hundreds.

Hundreds Column
e) 2 + 1 + 2 = 5
f) 3 + 7 = 10,
5 + 10 = 15
Put 5, carry 1 thousand.

Practice Exercises

Add each of the following columns of numbers (numerals):

9.
```
  39
  82
  46
 +71
```

10.
```
  963
  148
  252
 +377
```

If you cannot find combinations that result in 5, 10, 15, and so forth, add in the usual manner as in Example 3.

The addition of large numbers can be long and tedious. The electronic calculator has given us a quick and easy way to carry out this operation.

OPTIONAL   **CALCULATORS**

Most calculators use what is called algebraic logic. This means that we may enter the numbers and operations on a calculator in the order in which we read the expression. Other calculators, such as the Hewlett-Packard model, use a logic that will not be discussed in this book.

To add numerals on the calculator, do the following:

1. Press the button for each digit of the first addend starting from left to right.

2. Press the operation button $\boxed{+}$.

3. Press the buttons for the next addend.

4. Press $\boxed{+}$.

5. Continue the process of entering the addend and pressing $\boxed{+}$ until you have entered the last addend.

6. Then press $\boxed{=}$ or enter.

The sum should appear in the calculator display.

Let us start by checking the result in Example 3c.

Example C5

Add the following. Keystrokes are separated by commas.

706,469   Press [7], [0], [6], [4], [6], [9], then [+].

+514,088   Press [5], [1], [4], [0], [8], [8], then [=] or [enter].

1,220,557   Sum.

Now we can try a sum with three addends.

Example C6

Add the following:

16,408   Press [1], [6], [4], [0], [8], then [+].

9,862   Press [9], [8], [6], [2], then [+].

+10,343   Press [1], [0], [3], [4], [3], then [=] or [enter].

Practice Exercises

11. Add the following using a calculator:

$$
\begin{array}{r}
4,765 \\
89,230 \\
43,078 \\
47,134 \\
+14,892 \\
\end{array}
$$

Exercise Set 0.2

Add each of the following:

1. $\begin{array}{r} 22 \\ +46 \\ \hline \end{array}$ **2.** $\begin{array}{r} 16 \\ +63 \\ \hline \end{array}$ **3.** $\begin{array}{r} 423 \\ +445 \\ \hline \end{array}$ **4.** $\begin{array}{r} 807 \\ +142 \\ \hline \end{array}$

5. $\begin{array}{r} 903 \\ +27 \\ \hline \end{array}$ **6.** $\begin{array}{r} 607 \\ +89 \\ \hline \end{array}$ **7.** $\begin{array}{r} 86 \\ +924 \\ \hline \end{array}$ **8.** $\begin{array}{r} 62 \\ +419 \\ \hline \end{array}$

9. $\begin{array}{r} 9538 \\ +203 \\ \hline \end{array}$ **10.** $\begin{array}{r} 4847 \\ +608 \\ \hline \end{array}$ **11.** $\begin{array}{r} 4703 \\ +8219 \\ \hline \end{array}$ **12.** $\begin{array}{r} 5237 \\ +3323 \\ \hline \end{array}$

13. $\begin{array}{r} 42,807 \\ +37,155 \\ \hline \end{array}$ **14.** $\begin{array}{r} 29,015 \\ +50,474 \\ \hline \end{array}$ **15.** $\begin{array}{r} 9122 \\ +899 \\ \hline \end{array}$ **16.** $\begin{array}{r} 6458 \\ +7592 \\ \hline \end{array}$

17. $\begin{array}{r} 26 \\ 42 \\ +19 \\ \hline \end{array}$ **18.** $\begin{array}{r} 92 \\ 46 \\ +67 \\ \hline \end{array}$ **19.** $\begin{array}{r} 75 \\ 49 \\ 16 \\ 90 \\ +68 \\ \hline \end{array}$ **20.** $\begin{array}{r} 47 \\ 51 \\ 68 \\ 22 \\ +83 \\ \hline \end{array}$

Answer:

Practice Exercise 11: **11.** 199,099

21. 703
        965
      +444
      ‾‾‾‾

22. 119
        965
      +622
      ‾‾‾‾

23. 599
        290
        656
      +179
      ‾‾‾‾

24. 536
        925
        732
      +357
      ‾‾‾‾

25.   46
        319
       +77
       ‾‾‾

26. 642
         28
      +4118
      ‾‾‾‾‾

27. 4102
         837
        7809
        3281
          56
       +4688
       ‾‾‾‾‾

28. 5049
        2372
         583
        3574
          18
       +7844
       ‾‾‾‾‾

29. 4189
          60
         157
           4
       +2119
       ‾‾‾‾‾

30.   63
        4219
           4
         622
       +8965
       ‾‾‾‾‾

Calculator Exercises: (Optional)

Find the following sums using a calculator:

C1. 81,817 + 48,373 + 2349 + 92,277 + 229,923

C2. 14,778 + 57,375 + 9172 + 40,880 + 600,911

C3. 55,657
         6,990
        21,999
        84,855
       +9,250
       ‾‾‾‾‾‾

C4.   2,349
        10,593
        42,607
        87,074
       +70,663
       ‾‾‾‾‾‾

Challenge Exercise:

31. The algorithm for adding in base five is the same as in base ten, which is our numeral system. For example, $3 + 2 = 10_{five}$, $3 + 3 = 11_{five}$, and $3 + 4 = 12_{five}$. What is the sum of $4 + 4$ in base five? (*Hint:* 12_{five} means one five and two units.)

Writing Exercises:

32. Suppose we did not have a digit zero. How would we write numerals like 204?

33. The Chinese added on an abacus, called a *suan-pan*. This instrument consists of a frame with rods running from one side to the opposite side. Each rod represents a place value. Some people say that an expert on the suan-pan can add as fast as you can say the addends.

Explain how you could add 213 and 152 on the abacus.

Other cultures used the abacus as well. These include the Egyptian, Japanese, Mayan, and Roman civilizations.

The number 135
shown on a suan-pan.

Section 0.3	Subtraction of Whole Numbers

OBJECTIVES

When you complete this section, you will be able to:

a. Use the algorithms for subtraction without regrouping.

b. Use the algorithms for subtraction with regrouping.

c. Check answers to subtraction.

Introduction   In this section we continue our study of the operations on numbers. The operation to be explored in this section is subtraction. Subtraction may be thought of as a taking

Take Away   away of objects from a set or the reduction of the number of objects in a set.

Example 1

a. Subtract three from seven.

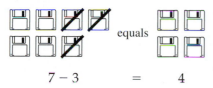

take away

7   −   3

Solution:

equals

7 − 3   =   4

If we take three objects from a set containing seven objects, we have four objects left in the set. The situation is symbolized: $7 - 3 = 4$.

b. Write a subtraction word problem.

Solution:
The math lab at Holly Hills Community College has fifteen computers. Two of them have power supply problems. How many are left in service?

Now that we have an idea of what subtraction is, we offer a definition of subtraction.

DEFINITION: Difference of Whole Numbers

The difference of two whole numbers, *a* and *b*, is the whole number *c* such that $b + c = a$. We call *a* the **minuend**, *b* the **subtrahend**, and *c* the **difference**.

In the sentence $a - b = c$, *b* is subtracted from *a* to get *c*. Thus, $9 - 7 = 2$ becomes

$$
\begin{array}{rl}
9 & \text{Minuend} \\
-7 & \text{Subtrahend} \\
\hline
2 & \text{Difference}
\end{array}
$$

IMPORTANT: Memorize these terms. You will need them later.

If $a - b = c$, then $b + c = a$. In other words, $a - b$ is that number which you add to *b* to get *a*. This discussion also leads to a very important conclusion.

For every subtraction statement, such as $9 - 3 = 6$, $(a - b = c)$, there is a related addition statement, $6 + 3 = 9$, $(c + b = a)$.

If we have mastered the addition facts of Section 0.2, it is unnecessary to memorize a table of subtraction facts. For example, consider $12 - 8$. This may be thought of as,

"What number plus 8 equals 12?" From the addition facts we know that $4 + 8 = 12$. Thus, $12 - 8 = 4$. Since we can apply this logic to any difference, there is no need to memorize a subtraction fact table.

Example 2

Use a related addition statement to find the indicated difference.

a. $9 - 4$

Solution:
The relationship may be thought of as, "What number plus 4 equals 9?" Since $5 + 4 = 9, 9 - 4 = 5$.

b. $15 - 6$

Solution:
The relationship may be thought of as, "What number plus 6 equals 15?" Since $9 + 6 = 15, 15 - 6 = 9$.

Practice Exercises

Find the following differences:

1. $9 - 2$ (What number is added to 2 to get 9?)

2. $8 - 3$ (What number is added to 3 to get 8?)

Although there is only one way of writing subtraction using symbols, there are many ways of saying subtraction in words. The following are some of these ways with the corresponding statement in symbols:

Some Ways to Say Subtraction	Meaning
Find the difference of 18 and 9.	$18 - 9$
Subtract seven from thirteen.	$13 - 7$
Five subtracted from twelve.	$12 - 5$
Find the difference between 14 and 8.	$14 - 8$
Find fifteen minus six.	$15 - 6$

Two ideas are basic to subtraction. The first is that zero subtracted from a number yields the original number, e.g., $5 - 0 = 5$. The second is that any number subtracted from itself yields zero, e.g., $5 - 5 = 0$.

Next, we consider the process or algorithm for subtraction of whole numbers. First, we will show subtraction which requires no **borrowing** or regrouping.

Example 3

Subtract.

a. 5 from 28

Solution:

Line up the place values.
$$
\begin{array}{r} 28 \\ -5 \\ \hline 23 \end{array}
\qquad
\begin{array}{r} 28 \\ -5 \\ \hline 23 \end{array}
$$

8 ones minus 5 ones is 3 ones.

2 tens minus the unwritten 0 tens is 2 tens.

Therefore, 5 subtracted from 28 is 23.

b. 23 from 37

Solution:

Line up the place values.

$$
\begin{array}{r}
\downarrow \\
37 \\
-23 \\
\hline
4
\end{array}
\qquad\qquad
\begin{array}{r}
\downarrow \\
37 \\
-23 \\
\hline
14
\end{array}
$$

7 ones minus 3 ones is 4 ones.   3 tens minus 2 tens is 1 ten.

Therefore, 23 subtracted from 37 is 14.

c. 236 from 7898

Solution:

Line up the place values.

$$
\begin{array}{r}
\downarrow \\
7898 \\
-236 \\
\hline
2
\end{array}
\qquad
\begin{array}{r}
\downarrow \\
7898 \\
-236 \\
\hline
62
\end{array}
\qquad
\begin{array}{r}
\downarrow \\
7898 \\
-236 \\
\hline
662
\end{array}
\qquad
\begin{array}{r}
\downarrow \\
7898 \\
-236 \\
\hline
7662
\end{array}
$$

8 ones minus 6 ones   9 tens minus   8 hundreds minus   7 thousands minus the
                          3 tens   2 hundreds   unwritten 0 thousands

Therefore, 236 subtracted from 7898 is 7662.

IMPORTANT: Place values must be lined up because we may only subtract things that are alike. That is, ones from ones, tens from tens, hundreds from hundreds, and so on.

In summary, to subtract one number from another where no regrouping is necessary, subtract the individual digits of the place values beginning with the units place and continue from right to left.

Practice Exercises

Subtract the following:

3. $24 - 3$ **4.** $77 - 61$ **5.** $279 - 76$ **6.** $982 - 351$

Since there is a related addition statement for each subtraction statement, we can easily check our differences to see if they are correct. That is, if $8 - 5 = 3$, then $3 + 5 = 8$.

Example 4

Subtract 365 from 2599 and check the results.

Solution:

$$
\begin{array}{r}
2599 \\
-365 \\
\hline
+2234 \\
\hline
2599
\end{array}
$$

Add 365 and 2234.

This sum is the same as the minuend. Therefore, the subtraction is correct.

Practice Exercises

Subtract and check the results.

7. $61 - 20$

8. $775 - 34$

We are ready to expand our discussion of subtraction to problems that require regrouping or borrowing to find the differences.

Example 5

Subtract.

a. 36 from 52

    Solution:

$$\begin{array}{r} 52 \\ -36 \\ \hline \end{array}$$

We cannot subtract 6 ones from 2 ones because there is no whole number that you can add to 6 and get 2. Therefore, we regroup by taking 1 ten from the 5 tens which gives us 4 tens and 12 ones.

$$\begin{array}{r} \overset{\scriptstyle 12}{4}2 \\ -36 \\ \hline 6 \end{array}$$

Regrouping allows us to subtract 6 ones from 12 ones to get 6 ones.

$$\begin{array}{r} \overset{\scriptstyle 12}{4}2 \\ -36 \\ \hline 16 \end{array}$$

4 tens minus 3 tens is 1 ten.

Therefore, 36 from 52 is 16.

b. 76 from 494

    Solution:

$$\begin{array}{r} 494 \\ -76 \\ \hline \end{array}$$

There is no whole number that we can add to 6 to get 4. So, we borrow 1 ten from the 9 tens in the next column. This gives us 14 ones and 8 tens.

$$\begin{array}{r} \overset{\scriptstyle 8\,14}{4}94 \\ -76 \\ \hline 8 \end{array}$$

Six ones from 14 ones gives us 8 ones.

$$\begin{array}{r} \overset{\scriptstyle 8\,14}{4}94 \\ -76 \\ \hline 18 \end{array}$$

Seven tens from 8 tens gives us 1 ten.

$$\begin{array}{r} \overset{\scriptstyle 8\,14}{4}94 \\ -76 \\ \hline 418 \end{array}$$

Zero hundreds from 4 hundreds gives us 4 hundreds.

Therefore, 76 from 494 is 418.

Example 6

Find 5003 minus 981.

Solution:

$$\begin{array}{r} 5\,0\,0\,3 \\ -\ 9\,8\,1 \\ \hline 2 \end{array}$$

1 unit from 3 units gives us 2 units.

$$\begin{array}{r} {}^{4}\!{}^{10} \\ 5\,0\,0\,3 \\ -\ 9\,8\,1 \\ \hline 2 \end{array}$$

We cannot take 8 tens from 0 tens, so we need to borrow 1 hundred from the hundreds column. But there are no hundreds, so we must go to the thousands column. From 5 thousands we borrow 1 thousand leaving 4 thousands. We rewrite the 1 thousand we borrowed as 10 hundreds.

$$\begin{array}{r} {}^{9} \\ {}^{4}\,{}^{10}\,{}^{10} \\ 5\,0\,0\,3 \\ -\ 9\,8\,1 \\ \hline 2\,2 \end{array}$$

We still need to borrow from the tens column, so we borrow 1 hundred from the hundreds place leaving 9 hundreds. We rewrite the hundred we borrowed as 10 tens and put it in the tens column. Now, we can take 8 tens from 10 tens and get 2 tens.

$$\begin{array}{r} {}^{9} \\ {}^{4}\,{}^{10}\,{}^{10} \\ 5\,0\,0\,3 \\ -\ 9\,8\,1 \\ \hline 0\,2\,2 \end{array}$$

We take 9 hundreds from 9 hundreds and get 0 hundreds.

$$\begin{array}{r} {}^{9} \\ {}^{4}\,{}^{10}\,{}^{10} \\ 5\,0\,0\,3 \\ -\ 9\,8\,1 \\ \hline 4\,0\,2\,2 \end{array}$$

Finally, we take the unwritten 0 thousands away from 4 thousands and obtain 4 thousands.

Therefore, 5003 minus 981 is 4022.

Practice Exercises

Subtract each of the following. They involve at least one borrowing.

9. $360 - 92$

10. $547 - 277$

11. $1308 - 163$

12. $9007 - 328$

OPTIONAL	**CALCULATORS**

This section is written for calculators that use algebraic logic. Other calculators, such as the Hewlett-Packard model, are not covered in this book.

To subtract on the calculator follow these steps:

1. Press the buttons for each digit of the minuend, starting from the left and going to the right.

2. Press the operations button $\boxed{-}$.

3. Press the buttons for each digit of the subtrahend starting from the left and going to the right.

4. Press the $\boxed{=}$ button or $\boxed{\textbf{enter}}$ button.

The difference between the numerals should appear in the calculator display.

Example C7

Find the difference between 662 and 257. Keystrokes are separated by commas.

Solution: 662   Press [6],[6],[2],[−].

−257   Press [2],[5],[7] [=] or [enter].

405   Difference.

Example C8

Subtract 73,046 from 290,812.

Solution: 290,812   Press [2],[9],[0],[8],[1],[2],[−].

−73,046   Press [7],[3],[0],[4],[6] [=] or [enter].

217,766   Difference.

Practice Exercises

Find the following using a calculator:

13. Find the difference between 486 and 131.

14. Subtract 70,577 from 189,000.

Example 9

Solve the following word problems. Some of them may contain extra information.

a. On Black Monday a certain stock was selling for $15 per share. By Monday of the following week, it was down to $8 per share. How much value did the stock lose in that week?

Solution:
We may think of this problem as asking, "What number must we add to $8 to get $15?" This we recognize from Example 2 is a related addition statement for subtraction. Therefore, $15 − $8 = $7.

b. Professor Jones has 28 students in her Integrated Arithmetic and Algebra class. There are 13 men in the class. How many women are in the class?

Solution:
The class is composed only of men and women. Therefore, the total number of students in the class

(28) minus the number of men (13) equals the unknown number of women. In symbols, 28 − 13 = 15.

c. At the beginning of training camp in August, the Bulldogs pro football team had 60 players report. By September 15, they were down to 44 players on the roster. How many players did they have to take off the roster?

Solution:
Notice that the dates have nothing to do with the answer. They are extra information and should be ignored. The team started with 60 players and had to take away some players to get down to 44. Take away suggests subtraction. Therefore, 60 − 44 = 16.

Practice Exercises

Solve the following word problems:

15. Wolfson Campus has five deans. Three of the deans are women. How many are men?

16. Angela has been working all summer saving to go back to school in the fall. She now has $2,300, but it will cost $5,000 for the academic year. How much must she borrow?

17. One morning the temperature in Calgary, Canada, was 75°F at 8:00 A.M. During the day a cold front came in from the Arctic Circle. By 5:00 P.M., the temperature was 35°F. How much did the temperature drop?

Exercise Set 0.3

Solve each of the following:

1. Find the difference between 71 and 46.

2. Subtract 22 from 47.

3. Eighteen subtracted from 79.

4. Find the difference between 67 and 23.

5. Find 48 minus 15.

6. Take 28 from 79.

Subtract each of the following. (No borrowing.)

7. $78 - 56$

8. $89 - 65$

9. $252 - 41$

10. $665 - 51$

11. $895 - 544$

12. $922 - 911$

13. $1974 - 923$

14. $8707 - 606$

15. $4459 - 3212$

16. $9825 - 7303$

Subtract each of the following. (There is at least one borrow.) Check each difference as in Example 4.

17. $63 - 25$

18. $75 - 38$

19. $394 - 58$

20. $952 - 49$

21. $732 - 279$

22. $655 - 277$

23. $5709 - 934$

24. $2980 - 961$

25. $4102 - 3226$

26. $6028 - 2630$

Subtract each of the following. (You may or may not need to borrow.)

27. $76 - 34$

28. $72 - 31$

29. $85 - 47$

30. $56 - 19$

31. $774 - 666$

32. $927 - 547$

33. $808 - 438$

34. $709 - 256$

35. $3005 - 1132$

36. $8009 - 6679$

Solve the following word problems. Some of them may contain extra information or involve addition as well as subtraction.

37. On payday John Lee had $1500 in his checking account. After paying the bills, he had $230 left. How much did he spend on bills?

38. Freshman Chemistry started the term with 45 students. At the end of the term, 14 people were left in the class. How many students dropped the course?

39. Last year Diana bought a new car. After the appropriate negotiations, she paid $12,000 for the car. Now the car has nagging electrical problems and she wants to trade it in. The dealer says the car is worth $9000. How much value has the car lost since she bought it?

40. In order to get a "B" in Beginning Algebra, Hank needs 480 out of 600 points. He now has 395 points after taking four tests. How much must he get on the fifth and last test to get the "B"?

41. Janet is an airplane pilot. She knows that head winds slow the speed of the aircraft. Her jet flies at 550 miles per hour with no wind. If she encounters a 150-miles-an-hour head wind, how fast is her plane actually going over the ground?

42. Arturo bought his girlfriend a bunch of 25 carnations, some of which were yellow and the others red. He got a special price because there were more red than yellow flowers. If there were 10 yellow carnations, how many red ones were there?

Calculator Exercises: (Optional)

Find the following differences using a calculator:

C1. 58,643 − 35,726

C2. 43,244 − 39,318

C3. 125,100 − 102,415

C4. 800,481 − 681,477

C5. 20,007 − 15,596

C6. 61,000 − 42,221

Challenge Exercises:

43. During the lifetime of a 30-year mortgage, you pay $275,000 in principal and interest. If the principal is $55,000, how much interest is paid?

44. Last year a small business had $90,000 in sales. The expenses for doing business that year were as follows: a) rent—$24,000, b) electricity—$6,000, and c) personnel—$48,000. How much profit was made?

Writing Exercises:

45. Suppose you were offered $20 an hour to teach a 6- or 7-year-old child how to subtract. How would you go about it? Where would you start?

46. Suppose you were offered $30 an hour to teach an adult how to subtract. Where would you start? What would you cover? Would you teach an adult any differently than you would teach a child?

Section 0.4 | Multiplication of Whole Numbers

OBJECTIVES

When you complete this section, you will be able to:

a. Know the multiplication facts through 12 by 12.

b. Multiply in standard notation

Introduction

So far we have studied the addition and subtraction of whole numbers. In this section we consider the operation of multiplication. Multiplication may be indicated in three ways. The first is with a raised dot, $2 \cdot 3 = 6$. The second is with parentheses, $2(3) = 6$. And the third is with a "$\times$," $2 \times 3 = 6$.

Multiplication may be thought of as repeated addition. To multiply two numbers, we use the first number as an addend the number of times specified by the second number.

Example 1

Repeated Addition

a. What is two multiplied by three?

Solution:
Two multiplied by three means three groups of two each.

plus plus equals

$$2 \quad + \quad 2 \quad + \quad 2 \quad = \quad 6$$

Two disks and two disks and two disks equals six disks. That is, three groups of two is the same as one group of six. In symbols, we can say $2 + 2 + 2 = 6$. Therefore, $2 \cdot 3 = 6$.

Did you know that fan is short for fanatic?

b. Write a word problem using multiplication.

Solution:

Snyder is a rock music fan and tapes every concert he can. Today he saw a special price on packages containing five blank tapes each, so he bought three packages. How many tapes did he buy today?

DEFINITION   The Product of Whole Numbers

The **product** of two whole numbers, *a* and *b*, is a whole number, *c*, that is the result of multiplying the **factor** *a* by the **factor** *b*. That is, $a \cdot b = c$.

In a typical multiplication statement, we have (factor) × (factor) = (product). For example, 2 (factor) × 4 (factor) = 8 (product). It is important to know the names of the parts of the statement because later they will be referred to by these names.

Now that we know what multiplication is, we need to memorize some basic products so that we have them at our fingertips when we need them. If you ever forget a product, you can always calculate it yourself by using the concept of repeated addition. For example, $9 \times 8 = 9 + 9 + 9 + 9 + 9 + 9 + 9 + 9 = 72$.

If you can add, you can multiply! Nevertheless, it is much faster to memorize the products from the following table.

Table of Products for Two Numbers

×	0	1	2	3	4	5	6	7	8	9	10	11	12
0	0	0	0	0	0	0	0	0	0	0	0	0	0
1	0	1	2	3	4	5	6	7	8	9	10	11	12
2	0	2	4	6	8	10	12	14	16	18	20	22	24
3	0	3	6	9	12	15	18	21	24	27	30	33	36
4	0	4	8	12	16	20	24	28	32	36	40	44	48
5	0	5	10	15	20	25	30	35	40	45	50	55	60
6	0	6	12	18	24	30	36	42	48	54	60	66	72
7	0	7	14	21	28	35	42	49	56	63	70	77	84
8	0	8	16	24	32	40	48	56	64	72	80	88	96
9	0	9	18	27	36	45	54	63	72	81	90	99	108
10	0	10	20	30	40	50	60	70	80	90	100	110	120
11	0	11	22	33	44	55	66	77	88	99	110	121	132
12	0	12	24	36	48	60	72	84	96	108	120	132	144

This table contains a world of information about multiplication. As in the addition fact table, the lines of numbers going up and down are called columns, and the lines of numbers going right and left are called rows. We find a product by choosing one factor in the first column and the second factor in the first row. Then we find where the row and column meet. For example, choose 4 in the first column and 6 in the first row. Run your finger down the first column until you reach 4. Move your finger to the right until you reach the column under 6 in the first row. You will find the number 24. Therefore, $4 \cdot 6 = 24$.

The table has entries through 12 × 12, which is sufficient for most purposes. The **multiples** of a number are the results of multiplying the number by each of the whole numbers, starting with zero. Therefore, each column shows the multiples of the entry at the top of the column and each row shows the multiples of the number at the beginning of the row. For example, the multiples of 8 are the following: 0, 8, 16, 24, 32, 40, 48, and so on. Thus, we have a handy summary of the first 13 multiples of the numbers zero through twelve.

The process we normally use for multiplication is called the **multiplication algorithm.** As in the addition and subtraction algorithms, we line up the place values in the multiplier and the multiplicand. Starting from the right in the multiplier we multiply each digit of the multiplicand by that digit and write the product directly below the multiplier. Products greater than ten have the tens digit "carried" to the next place value. Since the second digit in the multiplier represents the number of tens, the product of multiplying by that digit is put in the tens column and so on. See the following example:

Example 2

Multiply 607 by 52 using the multiplication algorithm.

Solution:

Read from left to right and down.

607	Factor.	1	Carry 1 ten.
×52	Factor.	607	
4	$2 \cdot 7 = 14$. Write 4, carry 1 ten.	×52	
		14	$2 \cdot 0 = 0, 0 + 10 = 10$. Write 1 in the tens column.
1		1	
607	$2 \cdot 600 = 1200$.	607	$50 \cdot 7 = 350$.
×52	Write 2 in the hundreds place. Write 1 in the thousands place.	×52	Write 5 in tens column. Leave the ones place blank for 0, and carry 300.
1214		1214	
		5	
31	Carry 3 hundreds.	31	
607	$50 \cdot 0 = 0$,	607	$50 \cdot 600 = 3000$.
×52	$0 + 300 = 300$.	×52	Put 0 in the hundreds and 3 in the thousands place.
1214		1214	
35	Put 3 in the hundreds column.	3035	
		31564	Add to get product.

Therefore, 607 multiplied by 52 is 31,564.

Practice Exercises

Multiply in standard notation.

1. 532
      ×48

2. 857
      ×602

Example 3

Solve the following:

a. The maximum seating capacity of a commuter train car is 64 passengers. What is the largest number of seated passengers in a six-car train?

Solution:
We have six groups with 64 members each; therefore, we multiply 64 by 6. Using the multiplication algorithm, we have $6 \cdot 64 = 384$.

Answers:

Practice Exercises 1–2: 1. 25,536 2. 515,914

b. There are 22 rows of 14 acoustic tiles on the ceiling of classroom 2206. What is the total number of tiles in the ceiling?

Solution:

We have 22 groups of 14 members each; therefore, we multiply 22 by 14. Using the multiplication algorithm, we have

$$
\begin{array}{r}
22 \\
\times 14 \\
\hline
88 \\
22 \\
\hline
308 \text{ tiles}
\end{array}
$$

Practice Exercises

Solve the following:

3. A word processor is set for 58 lines per page. If you type five pages, how many lines will there be?

4. The Agricultural and Mechanical University marching band has twelve rows of fourteen musicians each. How many musicians are in the marching band?

OPTIONAL	CALCULATORS

This section is written for calculators that use algebraic logic. This includes most calculators, except Hewlett-Packard calculators. If you enter the numbers and operations as they are read from left to right, then you have a calculator with algebraic logic.

To multiply two numbers on the calculator follow these steps:

For two factors:

1. Press the buttons for each digit of the first factor starting from the left and going to the right.

2. Press the operation button $\boxed{\times}$.

3. Press the button for each digit of the second factor starting from the left and going to the right.

4. Press the $\boxed{=}$ or $\boxed{\textbf{enter}}$ button.

The product should appear in the calculator display.

For more than two factors:

1. Enter each factor.

2. Press the operation button $\boxed{\times}$ after each factor.

3. Proceed until the last factor is entered.

4. Press the $\boxed{=}$ or $\boxed{\textbf{enter}}$ button.

The product should appear in the calculator display.

Example C4

Find the following products using a calculator. Keystrokes are separated by commas.

a. (172)(82)   Keystrokes: `1` , `7` , `2` , `×` , `8` , `2` , `=` or `enter` .

b. (14)(37)(85)   Keystrokes: `1` , `4` , `×` , `3` , `7` , `×` , `8` , `5` , `=` or `enter` .

Practice Exercises

Find the following products using a calculator:

5. (29)(52)

6. (73)(41)(97)

Exercise Set 0.4

Multiply each of the following:

1. 4(17)   **2.** 6(15)   **3.** 5(46)

4. 3(58)   **5.** 2(79)   **6.** 9(87)

7. 6(299)   **8.** 7(541)

Multiply each of the following:

9. 65
       ×8

10. 54
        ×6

11. 27
        ×38

12. 88
        ×42

13. 593
        ×20

14. 336
        ×70

15. 102
        ×86

16. 250
        ×66

17. 508
        ×204

18. 709
        ×306

19. 656
        ×331

20. 591
        ×227

Solve each problem. You may need to do more than just multiply. Be careful! There may be extra information.

21. The new dormitory at University of the North needs to order chairs for 230 rooms. If they need three chairs per room, how many chairs should they order?

22. The main library has 40 tables with 16 chairs each in the undergraduate study area. How many students can be seated at one time?

23. Alice is in the real estate business. Last year she sold eight tract houses for a builder. If her commission was $4200 per house, how much did she make?

24. Including cement, blocks, and labor, it costs $2 a block to build a wall. If the wall is 8 blocks high and 263 blocks long, how many blocks are in the wall?

25. Looking up at a downtown office building, Carlos counted 8 windows for each of the 12 floors of the building. How many windows did he count?

26. The committee for the annual Easter egg hunt ordered 150 dozen eggs for the big event this year. If 800 children show up, will this be enough eggs to make at least one per child?

27. A computer spreadsheet has 15 columns across the paper and 35 rows down the paper. How many possible entries can there be?

28. The northwest section of a large city is laid out in streets and avenues. The city stretches 249 blocks north from Main Street and goes 120 blocks west from Prime Avenue. How many city blocks are there?

Answers:

Practice Exercises 5–6: 5. 1508 6. 290,321

Calculator Exercises: (Optional)

Solve each problem.

C1. The sports section of the local newspaper averages 5 pages on Monday through Saturday and 15 pages on Sunday. How many pages of sports does the paper have in one year given that there are 52 weeks in a year?

C2. The microcomputer lab has 45 computers. If each computer is used for one hour a day by eight stu- dents a day for five days, how many hours of student usage does that make?

C3. A lumber company wants to replant 25 acres of trees. If they can plant 150 rows of 48 trees on each acre, how many trees will they need?

Challenge Exercises:

29. Alpha Club wishes to serve doughnuts at their next meeting. They estimate that each of the 35 expected members will eat two doughnuts. Will five dozen doughnuts be enough?

30. Mario has three 300-calorie workouts per week. Will this be enough to burn off his snack of two scoops of ice cream (180 calories each) and four ounces of potato chips (150 calories per ounce) per day?

31. There are five people in George's family—two adults and three children—ages 3, 7, and 11. The family decided to visit an amusement park and buy the three-day pass. Adult passes are $120 per adult and children's passes are $80 per child if the child is nine years old or younger. How much admission cost will the family have to pay?

32. David is going to have a midnight study break with a snack. He eats six ounces of potato chips at 160 calories per ounce, five cookies at 120 calories each, and drinks two sodas at 350 calories each. How many calories did he consume?

Writing Exercises:

33. The Romans did not have algorithms for multiplica- tion. However, they did multiply numbers. How do you suppose they did it?

34. The modern digital computer and almost all calcu- lators can perform only one operation—addition. Subtraction and multiplication are done through addition. How do you think this is possible?

Section 0.5	# Division of Whole Numbers

O B J E C T I V E S

When you complete this section, you will be able to:

a. Recognize and apply the relationship between multiplication and division.

b. Recognize that division by zero is not possible.

c. Use the algorithm for division of whole numbers without a remainder.

d. Use the algorithm for division of whole numbers with a remainder.

Introduction   We have explored the operations of addition, subtraction, and multiplication of whole numbers in the previous sections of this book. Now we consider the operation of division. Division may be thought of as determining the number of groups of ob-

Repeated Subtraction   jects in a certain number of objects by repeatedly subtracting the number of objects in the group.

Example 1

a. What is 12 divided by 3?

Solution:

One way to think about this question is to say, "How many 3s are in 12, or how many groups of 3 can I take away from one group of 12?" The answer can be found by forming groups of 3 things from 12 things.

We have 12 computer disks. Now we form groups of 3 and take them away from the originals.

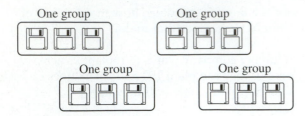

One group   One group

One group   One group

We count 4 groups of 3 each. Therefore, 12 divided by 3 is 4. Notice that there were no disks left over when we formed the groups of 3. That is, the remainder is 0.

b. Write a word problem using division.

Solution:
The judo club wants to transport 30 members to the downtown campus for a tournament. How many cars will they need if each car holds 5 people?

There are several ways we can symbolize division. All of them mean the same but are used in different situations. Consider 15 divided by 3. In each of the four cases,

$$15 \div 3 = 5, \qquad 15/3 = 5, \qquad \frac{15}{3} = 5, \qquad 3\overline{)15}\,^{5}$$

we call 3 the **divisor,** 15 the **dividend,** and 5 the **quotient.** Please memorize these names. We will need them.

There are also many ways that we can express division in words. We have summarized some of these for you. It is very important to recognize a division situation when you see it.

In words		In symbols
How many 3s are in 15?	↔	$15 \div 3$
Divide 15 by 3	↔	$15 \div 3$
15 divided by 3	↔	$15 \div 3$
3 divided into 15	↔	$15 \div 3$
Divide 3 into 15	↔	$15 \div 3$
The quotient of 15 and 3	↔	$15 \div 3$
The quotient of 3 and 15	↔	$3 \div 15$

IMPORTANT: Be careful. A change in the words may not change the meaning of the expression. For example, *15 divided by 3* and *3 divided into 15* mean the same thing. However, a change in the order of the numbers may change the meaning of the expression. For example, the *quotient* of *15 and 3* does not mean the same thing as the *quotient of 3 and 15.* See the preceeding table.

Relationship between multiplication and division

There is a very close relationship between multiplication and division. When we say $3 \cdot 5 = 15$, we mean that there are three 5s in 15. On the other hand, $15 \div 5$ means, "How many 5s are in 15?" We know the answer from the multiplication. There are three 5s in 15. Therefore, $15 \div 5 = 3$.

Each division has a related multiplication:
If $\frac{20}{4} = 5$, then $4 \cdot 5 = 20$. Since $\frac{20}{5} = 4$, then $5 \cdot 4 = 20$.

Each multiplication has two related divisions:
If $3 \cdot 6 = 18$, then both 3 and 6 will divide into 18.
That is, $\frac{18}{3} = 6$ and $\frac{18}{6} = 3$.

A very special case arises if we consider **zero as a divisor.** What does $2 \div 0$ mean? It refers to how many 0s there are in 2. We know how many 1s are in 2: $2 \div 1 = 2$ because $2 \cdot 1 = 2$. But we are not sure how many 0s are in 2. Consider some popular answers.

$$2 \div 0 = 0?$$ The related multiplication is $2 = 0 \cdot 0 = 0$. Is $2 = 0$? No, so $2 \div 0 \neq 0$.

$$2 \div 0 = 2?$$ The related multiplication is $2 = 2 \cdot 0 = 0$. Is $2 = 0$? No, so $2 \div 0 \neq 2$.

Regardless of what number we let equal $2 \div 0$, the related multiplication will always result in the statement $2 = 0$, which is impossible. Therefore, we know that **division by zero is impossible** or division by zero is **undefined**. Now we can make a formal definition of division.

DEFINITION: Division of Whole Numbers

The **quotient** of two whole numbers, *a* and *b* (*b* not zero), is a whole number *c* such that $b \cdot c = a$. The number *a* is called the dividend and the number *b* is called the divisor. Using variables, if $a \div b = c$, then $b \cdot c = a$.

Putting in numbers for the variables means that if $30 \div 6 = 5$, then $6 \cdot 5 = 30$.

The only thing that remains to be learned is how to carry out the process of division. The expression $21 \div 7$ means, "How many 7s are in 21?" We can find the number of 7s in 21 by taking 7 from 21 as many times as we can.

IMPORTANT: Throughout our discussion of division, we will follow the steps by following the numbers of the steps in order.

Example 2

a. Find $21 \div 7$.

Solution:
$21 \div 7$ means, "How many 7s are in 21?"

$$
\begin{array}{r}
21 \\
-7 \\
\hline
14 \\
-7 \\
\hline
7 \\
-7 \\
\hline
0
\end{array}
$$

1. We find the number of 7s in 21 by subtracting 7 from 21 as many times as we can until the difference is less than 7.

2. Since there is nothing left after the third 7 is taken away, we say that the remainder is 0.

Since three 7s were subtracted from 21, $21 \div 7 = 3$, with a remainder of 0.

IMPORTANT: When the remainder is 0, we usually do not write it. Thus, in the above example we would write $21 \div 7 = 3$.

Since there is a related multiplication for every division, we may **check** our division by multiplying the divisor by the quotient to get the dividend. That is,

Check
$21 \div 7 = 3$ because $7 \cdot 3 = 21$.

b. Find $27 \div 6$.

Solution:
$27 \div 6$ means, "How many 6s are in 27?"

$$
\begin{array}{r}
27 \\
-6 \\
\hline
21 \\
-6 \\
\hline
15 \\
-6 \\
\hline
9 \\
-6 \\
\hline
3
\end{array}
$$

1. We find the number of 6s in 27 by subtracting 6 from 27 as many times as we can until the difference is less than 6.

2. Since there is 3 left after the fourth 6 is taken away, we say that the remainder is 3.

Since four 6s were subtracted form 27 and 3 remained, we have $27 \div 6 = 4$ with a remainder of 3 or $27 \div 6 = 4 \, \text{R} \, 3$.

Check
$27 \div 6 = 4 \, \text{R} \, 3$ because $(6 \cdot 4) + 3 = 27$.

Practice Exercises

Find each of the following quotients by repeated subtraction and check using a related multiplication:

1. $18 \div 6$

2. $29 \div 9$

Since multiples are so important to the division process, you should go back and review the multiplication table in Section 0.4.

One process for division uses multiples of the divisor. We talked about multiples in Section 0.4 when we were discussing the patterns from the multiplication table. For example, the multiples of 7 are 7, 14, 21, 28, 35, and so on, and the multiples of 6 are 6, 12, 18, 24, 30, 36, and so on. We will be looking for the largest multiple of the divisor that is less than or equal to the dividend. The quotient is the number you would multiply by the divisor in order to get the multiple previously mentioned.

IMPORTANT: In the examples that follow, the steps are numbered to assist you. Therefore, follow the numbers and not the vertical position of the step.

Example 3

Find the quotient.

a. $21 \div 7$

Solution:

$$7 \overline{)21} \quad \begin{array}{r} 3 \\ \hline \end{array}$$
$$\underline{-21}$$
$$0$$

1. The smallest multiple of 7 less than or equal to 21 is $3 \cdot 7$, so the quotient is 3 because $3 \cdot 7 = 21$.
2. Subtract $3 \cdot 7 = 21$.
3. Remainder.

Check

$7 \cdot 3 = 21$.

b. $27 \div 6$

Solution:

$$6 \overline{)27} \quad \begin{array}{r} 4 \text{ R } 3 \\ \hline \end{array}$$
$$\underline{-24}$$
$$3$$

1. The largest multiple of 6 less than or equal to 27 is $4 \cdot 6 = 24$. So the quotient is 4.
2. Subtract $4 \cdot 6 = 24$.
3. Remainder.

Check

$(6 \cdot 4) + 3 = 24 + 3 = 27$.

Practice Exercises

Find the quotient using multiples and check using a related multiplication.

3. $45 \div 9$

4. $35 \div 8$

Now we look at the division algorithm. Be sure to follow the steps by following the numbers of the steps.

Example 4

Find the quotient of 725 and 5 using the division algorithm.

Solution:

$$5 \overline{)725} \quad \begin{array}{r} 1 \\ \hline \end{array}$$
$$\underline{-5}$$
$$2$$

2. There is one 5 in 7. Put 1 in the hundreds place.
1. How many 5s are there in 7?
3. Multiply 1 times 5.
4. Subtract 5 from 7.

$$5 \overline{)725} \quad \begin{array}{r} 14 \\ \hline \end{array}$$
$$\underline{-5}$$
$$22$$
$$\underline{-20}$$
$$2$$

2. There are four 5s in 22. Put 4 in tens place.
1. Put the 2 in the tens place. This is called **bringing down** 2. How many 5s are in 22?
3. Multiply 4 times 5.
4. Subtract 20 from 22.

$$5 \overline{)725} \quad \begin{array}{r} 145 \\ \hline \end{array}$$
$$\underline{-5}$$
$$22$$
$$\underline{-20}$$
$$25$$
$$\underline{-25}$$
$$0$$

2. There are five 5s in 25. Put 5 in the ones place.
1. Bring down the 5. How many 5s are in 25?
3. Multiply 5 times 5.
4. Subtract 25 from 25. Remainder.

Check

$5 \cdot 145 = 725.$

Therefore, there are 145 fives in 725, or 725 divided into 5 parts has 145 in each part with none left over.

IMPORTANT: In the division algorithm we must be careful to line up corresponding columns that represent the same place values. We thus avoid making careless errors.

Agreement: For the rest of this book, we will use the division algorithm, unless otherwise directed.

We now show a two-digit division.

Example 5

Find the quotient.

a. 12 divided into 456

   Solution:

$$12 \overline{)456}$$

2. There are 0 twelves in 4. Leave the hundreds place blank.
1. How many twelves are in 4?

$$\begin{array}{r} 3 \\ 12 \overline{)456} \\ -36 \\ \hline 9 \end{array}$$

2. There are 3 twelves in 45. Put 3 in the tens place.
1. How many twelves are in 45?
3. Multiply 3 times 12.
4. Subtract 36 from 45.

$$\begin{array}{r} 38 \\ 12 \overline{)456} \\ -36 \\ \hline 96 \\ -96 \\ \hline 0 \end{array}$$

3. There are 8 twelves in 96. Eight is written in the ones place.
1. Put the 6 in the ones place. (Bring down the 6.)
2. How many twelves are in 96?
4. Multiply 12 times 8.
5. Subtract 96 from 96.
6. Remainder.

Therefore, $456 \div 12 = 38.$

   Check

$12 \cdot 38 = 456.$

b. How many 86s are in 3893?

   Solution:

$$86 \overline{)3893}$$

2. There are no 86s in 3. Leave the thousands place blank.
1. How many 86s are in 3?

$$86 \overline{)3893}$$

2. There are no 86s in 38. Leave the hundreds place blank.
1. How many 86s are in 38?

$$\begin{array}{r} 4 \\ 86 \overline{)3893} \\ -344 \\ \hline 45 \end{array}$$

2. 86 is close to 90 and 389 is close to 390, so we guess 4.
1. How many 86s are in 389?
3. Multiply 86 times 4.
4. Subtract 344 from 389.

$$\begin{array}{r} 45 \\ 86 \overline{)3893} \\ -344 \\ \hline 453 \\ -430 \\ \hline 23 \end{array}$$

3. 86 is close to 90 and 453 is close to 450, so we guess 5.
1. Bring down the 3.
2. How many 86s are in 453?
4. Multiply 86 times 5.
5. Subtract 430 from 453.
6. Remainder.

Therefore, $3893 \div 86 = 45 R23.$

   Check

$(86 \cdot 45) + 23 = 3870 + 23 = 3893.$

IMPORTANT: Sometimes we may make a mistake and end with a remainder greater than the divisor. If this happens, just increase the last digit in the quotient until the resulting multiplication yields a remainder less than the divisor.

Practice Exercises

Find each of the quotients:

5. $493 \div 4$ **6.** $873 \div 9$ **7.** $3185 \div 49$ **8.** $1652 \div 15$

Before we consider the solution of word problems, it is worthwhile to review all of the types of wording we can use to express the idea of division as summarized after Example 1.

Example 6

Solve.

a. One version of the game of dominoes requires that all 28 game pieces be drawn at the beginning of play. If three people are playing, how many pieces does each person get?

Solution:
The dominoes are distributed to each player in turn until there are not enough dominoes for each player to receive one. They will go around 9 times with 1 left over. That is, $28 \div 3 = 9$ R 1. So, each player receives 9 pieces and the 1 left over is put aside and not used for play.

b. The area of a standard parking space is 162 square feet (18 feet deep by 9 feet wide). How many of these spaces can we get into a rectangular lot whose area is 6156 square feet (36 feet by 171 feet)?

Solution:
To find the number of parking spaces, divide the area of one space into the total area available. In other words, divide 162 into 6156.

$$
\begin{array}{r}
38 \\
162 \overline{)6156} \\
-486 \\
\hline
1296 \\
-1296 \\
\hline
\end{array}
$$

Therefore, there is room for 38 parking spaces.

Practice Exercises

9. Michael earns $25 each time he mows a lawn. How many lawns would he have to mow to earn the $300 needed to buy a used computer?

10. It is 700 miles from Chicago to Kansas City. If Laura's car gets 35 miles to the gallon on the highway, how many gallons of gasoline would she use on the trip?

OPTIONAL	**CALCULATORS**

This section is written for calculators that use algebraic logic. Most calculators, except the Hewlett-Packard model, use algebraic logic. A calculator has algebraic logic when the steps are entered from left to right as in a mathematical sentence.

To divide one number by another on the calculator follow these steps:

1. Press the buttons for each digit of the dividend starting from the left and going to the right.
2. Press the operation button $\boxed{\div}$.

3. Press the buttons for each digit of the divisor starting from the left and going to the right.

4. Press the $\boxed{=}$ or $\boxed{\textbf{enter}}$ button.

The quotient of the numbers should appear in the calculator display.

Example C7

a. Divide 356 by 4. The keystrokes are separated by commas.

Solution:

$356 \div 4$ Keystrokes: $\boxed{3}$, $\boxed{5}$, $\boxed{6}$, $\boxed{\div}$, $\boxed{4}$, $\boxed{=}$ or $\boxed{\textbf{enter}}$.

The quotient is 89.

Answers:

b. Solve Example 6b with a calculator.

Solution:
Number of spaces: Press [6] , [1] , [5] , [6] , [÷] , [1] , [6] , [2] , [=] or [enter] .
38 will appear.

IMPORTANT: Calculators automatically convert remainders to decimal fractions. Therefore, you will not see anything like $17 \div 5 = 3\ R\ 2$ on a calculator, but you will see a quotient of 3.4. See decimal division in Section 0.8 for further information.

Division on the calculator is checked the same way division by hand is checked. If the product of the quotient and the divisor is the dividend, the division is correct. That is, $30 \div 6 = 5$, if $5 \cdot 6 = 30$.

Practice Exercises (Optional)

Solve each of the following using a calculator:

11. The standard size lot for a house in a certain city is 75 feet by 100 feet. Discounting roads, how many of these lots can a developer get by subdividing 45,000 square feet of land?

12. The displacement or cargo capacity of a certain freighter (ship) is 35,000 tons. Of this 1000 tons is used to transport automobiles. If each automobile weighs 2500 pounds, how many autos can it carry? (one ton = 2000 lb)

The calculator makes the computation a lot easier. Even with a calculator, we still have to know two things. These are **when** to add, subtract, multiply, or divide and **what** to add, subtract, multiply, or divide.

Exercise Set 0.5

Divide by repeated subtraction and check by the related multiplication.

1. $71 \div 9$ **2.** $67 \div 8$ **3.** $86 \div 12$ **4.** $93 \div 11$

Divide by using multiples of the divisor and check by using related multiplication.

5. $57 \div 6$ **6.** $37 \div 4$ **7.** $96 \div 10$ **8.** $72 \div 11$

Divide using the division algorithm and check by using the related multiplication.

9. $4\overline{)72}$ **10.** $7\overline{)91}$

11. $5\overline{)594}$ **12.** $6\overline{)247}$

13. $14\overline{)298}$ **14.** $23\overline{)534}$

15. $67\overline{)979}$ **16.** $93\overline{)788}$

17. $237\overline{)5430}$ **18.** $385\overline{)4964}$

Solve by performing the operation.

19. How many 4s are there in 64?   **20.** Three divides 915 how many times?

21. What is the quotient of 58 and 2?   **22.** What is the quotient of 878 and 9?

23. 228 divided by 12 is what?   **24.** 5 divided into 385 is what?

Answers:

Solve by performing the necessary operation(s). There may be more than one operation needed and there may be more information than needed.

25. Hank wishes to serve the snack at a nursery school. He has only 4-ounce cups. How many children can he serve from a half gallon (64-oz) jug of orange juice?

26. Mass transit wishes to move 30,000 people to and from a downtown parade. With standing room only, one car of the train will hold 200 people. How many train cars do they need to carry the 30,000 people downtown?

27. Four students are going to share equally in the expenses of an automobile trip to New Orleans for the Mardi Gras celebration. If it costs $120 for gasoline, tolls, and parking, how much should each student pay?

28. Five children inherited $12,000 from their mother. The estate is to be shared equally. How much will each child get?

29. Mary is selling flowers for Mother's Day. If she buys one gross (144) of flowers, how many bunches of six flowers each can she make?

30. An automobile dealer has ordered 117 new cars for the beginning of the model year. If the delivery rigs hold nine cars each, how many loads will it take to deliver all of the new cars?

31. Seven students decided to have lunch across the street from campus at the pizza parlor. They shared three pizzas, which along with the drinks cost $35. How much did each student have to pay?

32. The four Diaz brothers have a large paper route of 136 customers. If they split the route equally, how many papers does each brother have to deliver?

33. Winton is a fat cat. If he eats 6 ounces of dry cat food a day, how long will a 2-pound (16 oz/lb) bag last?

34. A suburban congregation of 600 people is planning a new church. If each pew will seat 40 people, how many pews should they order?

Calculator Exercises: (Optional)

Find the following:

C1. $273{,}312 \div 208$

C2. $1{,}544{,}552 \div 697$

C3. In its first year of operation, a large community college generated 36,540 student credit hours. If 12 student credit hours is the same as one full-time student, how many full-time students is this equivalent to?

C4. In the academic year 1987–88, Miami-Dade Community College generated 995,256 credit hours. To be considered as full-time, a student must register for 12 credits. How many full-time students is this equivalent to?

Challenge Exercises:

35. Rosie is making a counted cross-stitch picture having a design which is 110 by 88 stitches. If he is stitching on material which requires 11 stitches per inch, what size frame should he buy?

36. A homeowner wants to fertilize his lawn. His lot measures 120 feet by 110 feet and his house covers an area of 3500 square feet. If a bag of fertilizer covers 8000 square feet, how many bags of fertilizer must the homeowner buy?

Writing Exercises:

37. How are multiplication and division related?

38. How do $0 \div 3$ and $3 \div 0$ differ?

39. How are subtraction, multiplication, and division all related to addition?

40. If the modern computer has only an adder, how can it divide?

Section 0.6	# A Brief Introduction to Fractions

OBJECTIVES

When you complete this section, you will be able to:

a. Write a fraction that represents a given part of a whole (unit).

b. Identify the part of the whole (unit) that is represented by a fraction.

c. Multiply two fractions.

d. Divide two fractions.

e. Add and subtract fractions with the same denominators.

Introduction

In this section we will give a very brief introduction to fractions. We will discuss how to multiply, divide, add, and subtract two fractions. We need just the basic concepts in order to develop additional ideas later in Chapter 0 and in Chapter 2. A thorough discussion of fractions will occur in Chapters 5 and 6.

Writing a fraction that represents a given part of the whole

Fractions can be used to indicate division, but in this section we look at fractions from a different point of view. A **fraction** is a number in the form of $\frac{a}{b}$ with $b \neq 0$. The number a is called the **numerator** and b is the **denominator**. If an object (unit) is divided into a number of equal parts and we select some of those parts, we can represent the part of the object selected by using a fraction. The numerator represents the number of equal parts selected and the denominator represents the number of equal parts into which the object (unit) has been divided.

Example 1

Using a fraction, represent the part of the whole region (unit) that is shaded.

a.
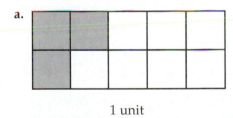
1 unit

Three of the ten equal parts are shaded. Therefore, the numerator is 3 and the denominator is 10. So the fraction is written as $\frac{3}{10}$ and is read as "three tenths."

b.
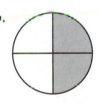
1 unit

Two of the four equal parts are shaded. Therefore, the numerator is 2 and the denominator is 4. So the fraction is written as $\frac{2}{4}$ and is read as "two fourths."

Practice Exercises

Using a fraction, represent the part of the whole region (unit) that is shaded and write the fraction using words.

1.
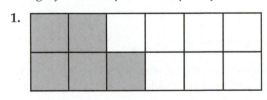
1 unit

2.
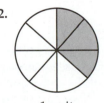
1 unit

The opposite of representing part of a whole by a fraction is identifying the part of the whole that is represented by a fraction.

Answers:

Practice Exercises 1–2: 1. $\frac{5}{12}$, five twelfths 2. $\frac{3}{8}$, three eighths

Example 2

Shade the part of the whole region that is represented by the given fraction.

a. $\frac{1}{4}$

1 unit

The denominator, 4, tells us to divide the region into 4 equal parts. The numerator, 1, tells us to shade any one of the four smaller regions.

Solution:

1 unit

b. $\frac{5}{8}$

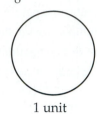

1 unit

The denominator tells us to divide the circle into eight equal regions. The numerator tells us to shade any five of the regions.

Solution:

1 unit

Practice Exercises

Shade the part of the whole that is represented by each of the following fractions:

3. $\frac{7}{15}$

4. $\frac{2}{3}$

Multiplication of fractions can be represented by shading regions. We can represent $\frac{1}{5}$ by dividing a region into five equal parts and shading one of those parts as shown in the following diagram.

Suppose we now wish to further shade $\frac{1}{2}$ of the $\frac{1}{5}$ that is already shaded. We divide each of the five parts into two equal parts and further shade one of the two shaded parts. What part of the total region has been further shaded?

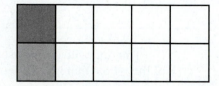

By dividing each of the five equal parts into two equal parts, the entire region has been divided into ten equal parts. Therefore, the one part that was further shaded represents $\frac{1}{10}$ of the entire region. In words, we would say that $\frac{1}{2}$ of $\frac{1}{5}$ is $\frac{1}{10}$. When performing operations with fractions, the word "of" is used to indicate multiplication. Consequently, $\frac{1}{2} \times \frac{1}{5} = \frac{1}{10}$. Note, this is the same as $\frac{1}{2} \times \frac{1}{5} = \frac{1 \cdot 1}{2 \cdot 5} = \frac{1}{10}$. Let us repeat this procedure with a more complicated example.

Example 3

Represent $\frac{2}{3}$ of $\frac{2}{5}$ using shaded regions and give the result as a fractional part of the entire region.

Solution:
The fraction $\frac{2}{5}$ means divide the region (unit) into five equal parts and shade two.

To find $\frac{2}{3}$ of the two parts that are shaded, we will divide each of the 5 parts into three equal parts. Then we further shade two of each of the three parts that were already shaded.

1 unit

Notice that the entire figure is now divided into 15 equal parts of which four have been further shaded. This represents $\frac{4}{15}$ of the entire figure. Therefore, $\frac{2}{3}$ of $\frac{2}{5}$ is $\frac{4}{15}$. As multiplication, $\frac{2}{3} \times \frac{2}{5} = \frac{4}{15}$. Note this can be computed as $\frac{2}{3} \times \frac{2}{5} = \frac{2 \cdot 2}{3 \cdot 5} = \frac{4}{15}$.

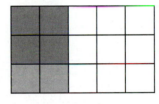

1 unit

Multiplying fractions   Based on the above examples, we are now ready to state the procedure for multiplication of fractions.

Multiplication of Fractions

For all numbers a, b, c, and d with b and $d \neq 0$, $\frac{a}{b} \times \frac{c}{d} = \frac{a \cdot c}{b \cdot d}$. In words, to multiply two fractions, find the product of the numerators divided by the product of the denominators.

Example 4

Find the following products.

a. $\frac{3}{5} \times \frac{2}{7} =$ *Find the product of the numerators divided by the product of the denominators.*

$\frac{3 \cdot 2}{5 \cdot 7} =$ *Multiply.*

$\frac{6}{35}$ *Product.*

b. $\frac{1}{4} \times \frac{3}{8} =$ *Find the product of the numerators divided by the product of the denominators.*

$\frac{1 \cdot 3}{4 \cdot 8} =$ *Multiply.*

$\frac{3}{32}$ *Product.*

c. $8 \times \frac{1}{2} =$ $8 = \frac{8}{1}$.

The product of the numerators divided by the product of the denominators.

$\frac{8}{1} \times \frac{1}{2} =$

$\frac{8 \cdot 1}{1 \cdot 2} =$ *Multiply.*

$\frac{8}{2} =$ $8 \div 2 = 4$.

4 *Product.*

Practice Exercises

Find the following products:

5. $\dfrac{2}{3} \times \dfrac{4}{9}$

6. $\dfrac{1}{4} \times \dfrac{3}{7}$

7. $9 \times \dfrac{1}{3}$

Before we discuss division of fractions, we need to introduce the **multiplicative inverse,** also called the **reciprocal.** Two numbers are multiplicative inverses if their product is 1. For example, the multiplicative inverse (reciprocal) of $\frac{2}{3}$ is $\frac{3}{2}$ since $\frac{2}{3} \cdot \frac{3}{2} = 1$. The multiplicative inverse of 2 is $\frac{1}{2}$ since $2 \cdot \frac{1}{2} = 1$. In general, for a and b with neither a nor b equal to 0, the multiplicative inverse of $\frac{a}{b}$ is $\frac{b}{a}$ since $\frac{a}{b} \cdot \frac{b}{a} = 1$. To find the multiplicative inverse of a number, we interchange the numerator and the denominator. This procedure is sometimes referred to as "inverting the fraction."

Suppose a region is divided into two equal parts so that each part is $\frac{1}{2}$ of the original region as shown in the following diagram:

Now take the $\frac{1}{2}$ that is shaded and *divide* it into two equal parts.

From the figure we can see that one of the two equal shaded parts is $\frac{1}{4}$ of the original region. In symbols, $\frac{1}{2} \div 2 = \frac{1}{4}$. Notice $\frac{1}{2} \times \frac{1}{2} = \frac{1}{4}$ also. Consequently, $\frac{1}{2} \div 2 = \frac{1}{2} \times \frac{1}{2}$. Without shading parts of a figure, we can see that $6 \div 3 = 2$ and $6 \times \frac{1}{3} = 2$, so $6 \div 3 = 6 \times \frac{1}{3}$.

This suggests that dividing by a whole number, except for 0, is the same as multiplying by its multiplicative inverse (reciprocal.)

The preceding suggests the following rule for the division of fractions:

Division of Fractions

For all numbers a, b, c, and d, with b, c, and $d \neq 0$, $\frac{a}{b} \div \frac{c}{d} = \frac{a}{b} \times \frac{d}{c}$. In words, to divide by a number, multiply by its multiplicative inverse (reciprocal).

This rule is usually stated more loosely as "when dividing two fractions, invert the fraction on the right and change the operation to multiplication." It is this form of the rule that we will be using in the following examples.

Answers:

Practice Exercises 5–7: 5. $\dfrac{8}{27}$ 6. $\dfrac{3}{28}$ 7. 3

Example 5

Find the following quotients.

a. $\dfrac{2}{3} \div \dfrac{3}{5} =$ *Invert the fraction on the right and change the operation to multiplication.*

$\dfrac{2}{3} \times \dfrac{5}{3} =$ *Multiply numerator times numerator and denominator times denominator.*

$\dfrac{2 \cdot 5}{3 \cdot 3} =$ *Multiply.*

$\dfrac{10}{9}$ *Quotient.*

b. $\dfrac{3}{7} \div 8$ $8 = \dfrac{8}{1}$.

$\dfrac{3}{7} \div \dfrac{8}{1} =$ *Invert the fraction on the right and change the operation to multiplication.*

$\dfrac{3}{7} \times \dfrac{1}{8} =$ *Multiply numerator times numerator and denominator times denominator.*

$\dfrac{3 \cdot 1}{7 \cdot 8} =$ *Multiply.*

$\dfrac{3}{56} =$ *Quotient.*

Practice Exercises

Find the following quotients.

8. $\dfrac{3}{4} \div \dfrac{2}{7}$

9. $\dfrac{2}{9} \div 5$

Addition of fractions with the same denominators

Shaded regions may also be used to illustrate the addition of fractions with the same denominators. Suppose $\frac{3}{8}$ of a region is shaded.

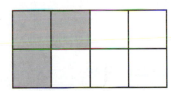

1 unit

Now shade another $\frac{4}{8}$ of the same region.

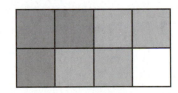

1 unit

Observe that $\frac{7}{8}$ of the region is now shaded. In other words, $\frac{3}{8} + \frac{4}{8} = \frac{3+4}{8} = \frac{7}{8}$. Note that the denominators of the fractions are the same and only the numerators are added. This leads to the following procedure for adding fractions with like denominators:

Addition of Fractions with the Same Denominators

For all numbers a, b, and c, with $c \neq 0$, $\frac{a}{c} + \frac{b}{c} = \frac{a+b}{c}$. In words, to add two or more fractions with the same denominators, add the numerators and place the sum over the same denominator.

Answers:

It can easily be shown that a similar procedure is used to subtract fractions. The procedure is given in the following box:

Subtraction of Fractions with the Same Denominators

For all numbers a, b, and c, with $c \neq 0$, $\dfrac{a}{c} - \dfrac{b}{c} = \dfrac{a-b}{c}$. In words, to subtract two or more fractions with the same denominators, subtract the numerators and place the difference over the same denominator.

Example 6

Find the following sums or differences:

a. $\dfrac{1}{5} + \dfrac{3}{5} =$ *The denominators are the same, so add the numerators and place the sum over the same denominator.*

$\dfrac{1+3}{5} =$ Add 1 and 3.

$\dfrac{4}{5}$ Sum.

b. $\dfrac{7}{9} - \dfrac{2}{9} =$ *The denominators are the same, so subtract the numerators and place the difference over the same denominator.*

$\dfrac{7-2}{9} =$ Subtract 7 and 2.

$\dfrac{5}{9}$ Difference.

c. $\dfrac{2}{13} + \dfrac{4}{13} + \dfrac{1}{13} =$ *The denominators are the same, so add the numerators and place the sum over the same denominator.*

$\dfrac{2+4+1}{13} =$ Add 2, 4, and 1.

$\dfrac{7}{13}$ Sum.

d. $\dfrac{3}{11} + \dfrac{5}{11} - \dfrac{4}{11} =$ *The denominators are the same, so add and subtract the numerators and place the sum and difference over the same denominator.*

$\dfrac{3+5-4}{11} =$ Add 3 and 5, then subtract 4.

$\dfrac{4}{11}$ Answer.

Practice Exercises

Find the following sums:

10. $\dfrac{4}{11} + \dfrac{2}{11}$

11. $\dfrac{11}{15} - \dfrac{4}{15}$

12. $\dfrac{5}{17} + \dfrac{3}{17} + \dfrac{6}{17}$

13. $\dfrac{3}{10} + \dfrac{5}{10} - \dfrac{1}{10}$

Example 7

Solve the following:

a. Each fill-up of a gasoline tank for an outboard motor requires $\frac{1}{2}$ quart of oil. A container has 6 quarts of oil. There is enough oil in the container for how many fill-ups?

Solution:
To find the number of fill-ups, divide the number of quarts of oil in the container by the number of quarts of oil needed per fill-up. Therefore,

$$6 \div \dfrac{1}{2}$$ *Invert $\frac{1}{2}$ and multiply.*

$$6 \times 2 =$$ *Multiply.*

$$12$$ *Therefore, there is enough oil for 12 fill-ups.*

---

b. A recipe for a cake calls for $\frac{2}{3}$ of a cup of chopped nuts. Find the number of cups of nuts needed for 9 such cakes.

Solution:
To find the number of cups of nuts needed for 9 cakes, multiply the number of cups per cake by the number of cakes. Hence,

$$\frac{2}{3} \cdot 9 = \qquad \text{\textit{Think of 9 as }} \frac{9}{1}.$$

$$\frac{2}{3} \cdot \frac{9}{1} = \qquad \text{\textit{Multiply the fractions.}}$$

$$\frac{2 \cdot 9}{3 \cdot 1} = \qquad \text{\textit{Multiply.}}$$

$$\frac{18}{3} = \qquad \text{\textit{Divide.}}$$

$$6 \qquad \text{\textit{Therefore, 6 cups of nuts are required for the 9 cakes.}}$$

c. A certain stock rose $\frac{5}{8}$ of a point on Thursday and $\frac{3}{8}$ of a point on Friday. How much did it gain in two days?

Solution:
To find the total gain, we add the gain on Thursday to the gain on Friday.

$$\frac{5}{8} + \frac{3}{8} = \qquad \text{\textit{Add the numerators and keep the common denominator.}}$$

$$\frac{5+3}{8} = \qquad \text{\textit{Add.}}$$

$$\frac{8}{8} = \qquad \text{\textit{Divide.}}$$

$$1 \qquad \text{\textit{Therefore, the total gain was 1 point.}}$$

d. The charge on a cellular phone was $\frac{11}{15}$ of a full charge. After 3 hours of usage the charge was reduced by $\frac{7}{15}$ of a charge. How much charge was remaining?

Solution:
To find the remaining charge we reduce or subtract the expended charge from the beginning charge.

$$\frac{11}{15} - \frac{7}{15} = \qquad \text{\textit{Subtract the numerators and keep the common denominator.}}$$

$$\frac{11-7}{15} = \qquad \text{\textit{Subtract.}}$$

$$\frac{4}{15} \qquad \text{\textit{Therefore, the remaining charge was }} \frac{4}{15} \text{\textit{ of a full charge.}}$$

Practice Exercises

Solve the following:

14. A baby bottle holds $\frac{1}{8}$ of a quart of formula. If a case of formula contains 16 quarts, how many times can the bottle be filled from a case of formula?

15. A pharmacist fills a prescription for a high blood pressure medication. She uses $\frac{2}{3}$ of the pills in a bottle that is $\frac{2}{5}$ full. What part of a full bottle did she use?

Answers:

16. It took Jennifer $\frac{3}{4}$ of an hour to complete her English homework and $\frac{5}{4}$ of an hour to complete her mathematics homework. How much time did she take to complete these two important sets of homework?

17. At the beginning of a bicycle race Jane had $\frac{15}{17}$ of a bottle of water. After riding the 10-kilometer race she had $\frac{7}{17}$ of a bottle left. How much water did she drink during the race?

Study Tip 1

Motivation—The Key to Success

The attitude with which you approach a course is the single most important factor in determining your success in that course. Begin with a positive attitude and a belief that you can be successful. Mathematics can be learned and understood. We have done our very best to take the mysteries out of mathematics. But for you to reach your goals, you need to make a commitment right now to attend class, to study as we suggest, and to give this course your very best effort. With the proper determination, you can master this course.

Exercise Set 0.6

Using a fraction, represent the part of the whole region that is shaded.

1.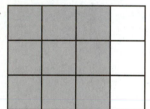

2.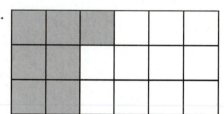

3.

4.

5.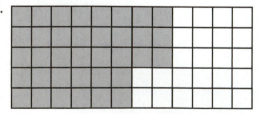

6.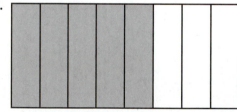

Shade the part of the whole that is represented by the given fraction.

7. $\frac{3}{5}$

8. $\frac{3}{4}$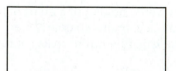

Answers:

9. $\frac{3}{8}$

10. $\frac{2}{4}$

Find the following products:

11. $\frac{1}{3} \times \frac{1}{5}$ 12. $\frac{1}{4} \times \frac{1}{6}$ 13. $\frac{2}{5} \times \frac{4}{9}$

14. $\frac{3}{7} \times \frac{1}{4}$ 15. $\frac{4}{7} \times \frac{3}{5}$ 16. $\frac{5}{8} \times \frac{3}{7}$

17. $\frac{5}{12} \times \frac{7}{17}$ 18. $\frac{4}{13} \times \frac{2}{15}$ 19. $6 \times \frac{1}{3}$

20. $8 \times \frac{1}{4}$ 21. $\frac{1}{5} \times 10$ 22. $\frac{1}{6} \times 12$

Find the following quotients:

23. $3 \div \frac{1}{2}$ 24. $\frac{1}{2} \div 3$ 25. $\frac{3}{5} \div \frac{7}{4}$

26. $\frac{4}{9} \div \frac{3}{2}$ 27. $\frac{1}{2} \div \frac{5}{9}$ 28. $\frac{1}{4} \div \frac{4}{7}$

29. $\frac{7}{11} \div \frac{3}{8}$ 30. $\frac{7}{15} \div \frac{3}{5}$ 31. $\frac{2}{5} \div 3$

32. $\frac{5}{8} \div 6$ 33. $7 \div \frac{3}{4}$ 34. $8 \div \frac{3}{7}$

Find the following sums and differences:

35. $\frac{3}{7} + \frac{2}{7}$ 36. $\frac{2}{15} + \frac{2}{15}$ 37. $\frac{6}{17} + \frac{9}{17}$

38. $\frac{3}{19} + \frac{11}{19}$ 39. $\frac{4}{5} - \frac{2}{5}$ 40. $\frac{7}{9} - \frac{2}{9}$

41. $\frac{9}{13} - \frac{4}{13}$ 42. $\frac{9}{11} - \frac{5}{11}$ 43. $\frac{2}{7} + \frac{1}{7} + \frac{3}{7}$

44. $\frac{4}{23} + \frac{8}{23} + \frac{3}{23}$ 45. $\frac{7}{13} + \frac{4}{13} - \frac{2}{13}$ 46. $\frac{11}{15} + \frac{4}{15} - \frac{7}{15}$

Challenge Exercises:

47. If ten sugar bowls hold $\frac{1}{2}$ cup of sugar each, how much sugar does it take to fill them all?

48. If a tank holds $\frac{1}{4}$ gallon of gasoline, how many times can the tank be filled from a 5-gallon can?

49. A pizza is cut into eight equal pieces. If Sue eats one piece, John eats two pieces, and Ellen eats two pieces, what part of the pizza has been eaten?

50. Allen, Jose, and Ismel take a trip. If Allen has driven $\frac{4}{13}$ of the trip, Jose has driven $\frac{5}{13}$ of the trip, and Ismel has driven $\frac{2}{13}$ of the trip, what part of the trip has been completed?

51. John needs 7 pieces of lumber, each $\frac{7}{12}$ of a foot long. How many feet of lumber does he need?

52. If Candido spends $\frac{11}{8}$ of an hour on his algebra homework each day, how many hours does he spend on his algebra homework per week?

53. A board 12 feet long is to be cut into pieces that are $\frac{2}{3}$ of a foot each. How many pieces will there be?

54. Tia works part-time. She works $\frac{2}{3}$ of a day. How many days will she have to work in order to work the equivalent of 8 full days?

55. If Elena drives to campus, it takes $\frac{13}{15}$ of an hour. If she catches the Metro rail, it takes $\frac{9}{15}$ of an hour to reach campus. How much longer does it take for her to drive?

56. Juan has $\frac{3}{13}$ of the amount on his phone debit card left after talking to his girlfriend. If he started with $\frac{9}{13}$ of the value of his phone card, how much of the card did he spend on the phone call?

Writing Exercises:

57. Name three types of documents in which you have seen fractions used.

58. Write a problem that requires the multiplication of fractions.

59. Write a problem that requires the division of fractions.

60. Write a problem that requires the addition of fractions with the same denominators.

Section 0.7   # Addition and Subtraction of Decimal Numerals

OBJECTIVES

When you complete this section, you will be able to:

a. Recognize, read, and write decimals in standard notation.

b. Determine which decimal is larger or largest.

c. Add decimals without and with carrying.

d. Subtract decimals without and with carrying.

Introduction   In Section 0.6 we introduced and discussed some ideas about fractions. In this section we will be working with special fractions that have only whole numbers in the numerator and only 10, 100, 1000, 10,000, 100,000, and so on as denominators. In order to see where these fractions come from, we need to go back to the idea of place value in the Hindu-Arabic system.

Whole numbers have the following pattern of place values:

$$\leftarrow \quad \frac{\text{one hundred thousand}}{100,000} \quad \frac{\text{ten thousand}}{10,000} \quad \frac{\text{one thousand}}{1000} \quad \frac{\text{one hundred}}{100} \quad \frac{\text{ten}}{10} \quad \frac{\text{one}}{1} \quad \rightarrow$$

If we start on the right and go to the left, we see that the place values increase by a factor of ten as we go from 1 to 10, from 10 to 100, from 100 to 1000, from 1000 to 10,000, and so on. That is, $10 = 10 \cdot 1$, $100 = 10 \cdot 10$, $1000 = 10 \cdot 100$, $10,000 = 10 \cdot 1000$, $100,000 = 10 \cdot 10,000$, and so on. So if we multiply a place value by ten, we get the next place value to the right.

Now we consider the pattern in the opposite direction (from left to right). The opposite operation of multiplication is division, so if we multiplied by ten going in one direction, then we divide by ten going in the other direction. One way to divide by 10 is to multiply by $\frac{1}{10}$. Therefore, if we multiply a place value by $\frac{1}{10}$, we get the next place value to the right. For example, beginning with 100,000 we have the following:

$$\frac{1}{10} \cdot \frac{100{,}000}{1} = \frac{100{,}000}{10} = 10{,}000;$$

$$\frac{1}{10} \cdot \frac{10{,}000}{1} = \frac{10{,}000}{10} = 1{,}000;$$

$$\frac{1}{10} \cdot \frac{1000}{1} = \frac{1000}{10} = 100;$$

$$\frac{1}{10} \cdot \frac{100}{1} = \frac{100}{10} = 10;$$

$$\frac{1}{10} \cdot \frac{10}{1} = \frac{10}{10} = 1;$$

$$\frac{1}{10} \cdot 1 = \frac{1}{10} \cdot \frac{1}{1} = \frac{1}{10} = ?$$

The last division yielded a number, $\frac{1}{10}$, one of those special fractions we talked about earlier. We represent these special numbers with a **decimal point.** The first place value to the right of the ones place is $\frac{1}{10}$, so we represent $\frac{1}{10}$ as .1. Since both the symbol $\frac{1}{10}$ and .1 represent the same number, they are both read as "one tenth."

In some countries, such as France and Germany, a comma is used instead of a decimal point.

Most people prefer to read ".1" as "point one." We strongly advise against this for the beginning student. We want to emphasis the fact that the decimal represents a fraction and that unless we practice saying what the fraction is, the symbols lose their meaning. Therefore, we will use the fractional method for reading decimals through-out the remainder of this chapter. To help avoid confusion we shall call fractions of the form $\frac{1}{10}$ **ratio fractions,** and fractions of the form .1 **decimal fractions.**

Consider the second place value to the right of the decimal point. If we continue our process of multiplying by $\frac{1}{10}$, the value of this next place would be $\frac{1}{10} \cdot \frac{1}{10} = \frac{1}{100}$, symbolized as .01 and read as, "one hundredth." This process of multiplying a place value by $\frac{1}{10}$ is continued forever as we go to the right. We summarize the first few of these decimal fractions next.

Ratio Fraction	Decimal Fraction	Name
$\frac{1}{10}$	.1	one tenth
$\frac{1}{100}$	.01	one hundredth
$\frac{1}{1000}$	.001	one thousandth
$\frac{1}{10{,}000}$	.0001	one ten-thousandth
$\frac{1}{100{,}000}$	.00001	one hundred-thousandth
$\frac{1}{1{,}000{,}000}$	.000001	one millionth

This process may be continued forever.

Please note that the decimal fraction is found by writing the same number of decimal places as we have 0s in the denominator of the corresponding ratio fraction.

From the following chart we read a decimal by starting at the decimal point and going to the right. We use "tenths" for the place value of the first digit to the right of 0, "hundredths" for the place value of the second digit to the right of 0, "thousandths" for the third place value, "ten-thousandths" for the fourth place value, "hundred-thousandths" for the fifth place value, and so on, until we reach the last digit in the decimal fraction. The decimal is read as though it were a whole number, and the place value of the last digit on the right is then given. For example, .25 is read 25 hundredths.

one tenth	*one hundredth*	*one thousandth*	*one ten-thousandth*	*one hundred-thousandth*	*one millionth →*
$\dfrac{1}{10}$	$\dfrac{1}{100}$	$\dfrac{1}{1000}$	$\dfrac{1}{10,000}$	$\dfrac{1}{100,000}$	$\dfrac{1}{1,000,000}$

Example 1

Read as a decimal fraction.

a. .408   **Solution:**
Starting at the decimal point, the place values are tenths, hundredths, thousandths. Thus 8 is in the thousandths place, so we read "four hundred eight thousandths."

b. .0169   **Solution:**
Starting from the decimal point, the place values are tenths, hundredths, thousandths, and ten-thousandths. Thus 9 is in the ten-thousandths place, so we read "one hundred sixty-nine ten-thousandths."

The only time "and" should be used in a numeral is to indicate a decimal point.

Decimal numerals may include whole numbers as well as decimal fractions. We call these numerals **decimal mixed numerals** because they contain both whole numbers and fractions. We read the whole number, then read the decimal point as "and," then read the decimal part of the numeral as in Example 2

Example 2

Read the decimal numeral.

a. 6.485   **Solution:**
We read six (whole number portion) **and** (decimal point) four hundred eighty-five thousandths (decimal portion).

b. 305.14   **Solution:**
We read three hundred five (whole number portion) **and** (decimal point) fourteen hundredths (decimal portion).

Practice Exercises

Read each of the following. Then write the word name.

1. .702   **2.** .4083   **3.** 69.12   **4.** 524.035

If more practice is needed, do the Additional Practice Exercises.

Additional Practice Exercises

Read and write the word name for each:

a. .862   **b.** 2.83   **c.** 68.469

Answers:

As in whole numbers, it is useful to know which decimal number is larger. We can easily tell which number is larger by comparing the digits in corresponding decimal places. In doing this it is convenient to use the symbol > for "greater than" and the symbol < for "less than."

Example 3

Which is the larger or largest number?

a. 38.547 or 206.1

Solution:
We can write each numeral with the same number of digits by writing zeroes both to the left and to the right of the decimal point until each numeral has the same number of place values filled. So 38.547 becomes 038.547 and 206.1 becomes 206.100.

$$0\,3\,8\,.\,5\,4\,7$$
$$|\ |\ |\ |\ |\ |\ |$$
$$2\,0\,6\,.\,1\,0\,0$$

Comparing digits from left to right, 2 is greater than 0, which means that 200 is greater than 0 hundreds. Therefore, 206.1 is greater than 38.547 and is written 206.1 > 38.547.

b. 17.637 or 17.652

Solution:
Since these numerals have the same number of digits both left and right of the decimal point, we go from left to right, checking each place value.

$$1\,7\,.\,6\,3\,7$$
$$|\ |\ |\ |\ |\ |$$
$$1\,7\,.\,6\,5\,2$$

The tens places have the same value (1); the units places have the same value (7); the tenths places have the same value (6). The hundredths places are different. Since 5 is larger than 3, we say 17.652 > 17.637.

c. 2.31, 2.504, or 2.5

Solution:

2.310
2.504
2.500

Line up the place values.

Write as many zeroes as necessary to give each numeral the same number of place values.

$$2\,.\,3\,1\,0$$
$$|\ |\ |\ |\ |$$
$$2\,.\,5\,0\,4$$
$$|\ |\ |\ |\ |$$
$$2\,.\,5\,0\,0$$

From left to right the units digits are the same (2). In the tenths places 5 > 3, so 2.31 is the smallest number. Since the same digit, 5, is in both 2.504 and 2.5, we go to the hundredths digits. The hundredths digits are the same (0) in 2.504 and 2.500, so we go to the thousandths digits. 4 > 0. Therefore, 2.504 is the larger number.

Practice Exercises

Decide which of the following decimals is larger or largest.

5. .38, .308

6. 76.085, 7.9965

7. 554.03, 554.44

8. 7.022, 7.7922, 7.124

If more practice is needed, do the Additional Practice Exercises on page 46.

Additional Practice Exercises

Which is larger?

d. .4157, .4152

e. 67.602, 6.7602

f. 99.056, 99.048

All whole numbers have an understood, but unwritten, decimal point to the right of the numeral. For example, 15 = 15. and so on. Consequently, all whole numbers are also decimals. Therefore, it should not surprise us that the addition of decimals is very similar to the addition of whole numbers and that the properties of addition of whole numbers hold for decimals as well.

We can add only digits that are in the same place value. When adding decimals, we can only add hundreds with hundreds, tens with tens, ones with ones, tenths with tenths, hundredths with hundredths, and so on. Combinations of decimal fractions and/or decimal mixed numerals may be added by putting them in vertical form and lining up the decimal points so that digits with the same place value are in columns. We then add the columns just as with whole numbers and bring the decimal point straight down. This process is summarized below.

Addition Algorithm for Decimals

Decimal fractions and/or decimal mixed numerals are added by the following procedure:

1. Write the numerals with corresponding place values lined up in the same columns. Annex zeroes so that all addends have the same number of decimal places.
2. Add the columns as in whole numbers.
3. Bring the decimal point straight down from the last addend above.

We start our examples with sums of decimal numerals that require no carrying.

Example 4

Add.

a. .24 and .5

Solution:

$$
\begin{array}{r}
.24 \\
+.5 \\
\hline
4
\end{array}
$$
Addend.
Addend.
Sum.

$$
\begin{array}{r}
.24 \\
+.5 \\
\hline
.74
\end{array}
$$
2 tenths and 5 tenths is 7 tenths. Write the decimal point below those in the addends.

4 hundredths and 0 hundredths is 4 hundredths.

b. .16 and .23

Solution:

$$
\begin{array}{r}
.16 \\
+.23 \\
\hline
9
\end{array}
$$
Addend.
Addend.
Sum.

$$
\begin{array}{r}
.16 \\
+.23 \\
\hline
.39
\end{array}
$$
1 tenth and 2 tenths is 3 tenths. Write the decimal point below those in the addends.

6 hundredths and 3 hundredths is 9 hundredths.

Practice Exercises

Add each of the following (no carrying):

9. .48 and .3

10. .25 and .73

11. .26 and .419

12. .061 and .52

Answers:

Additional Practice Exercises d–f: d. .4157 **e.** 67.602 **f.** 99.056 **Practice Exercises 9–12: 9.** .78 **10.** .98 **11.** .679 **12.** .581

Once we have lined up the place values by lining up the decimal points, addition of decimals follows the same pattern as addition of whole numbers. The following examples show situations in which there may or may not be a need to carry.

Example 5

Add.

a. 43.834 and 35.16

Solution:
Line up decimal points and annex zeroes to make the number of digits equal.

$$
\begin{array}{r}
43.834 \\
+35.160 \\
\hline
78.994
\end{array}
$$

b. 89.437 and 62.7381

Solution:
Line up decimal points and write zeroes to make the number of digits equal.

$$
\begin{array}{r}
{\scriptstyle 111\ \ 1} \\
89.4370 \\
+62.7381 \\
\hline
152.1751
\end{array}
$$

c. Add the following: $616.57 + 92.372 + 484.13 + 18.059$.

Solution:
Line up the decimal points to match place values and annex zeroes.

$$
\begin{array}{r}
{\scriptstyle 221\ 21} \\
616.570 \\
092.372 \\
484.130 \\
+018.059 \\
\hline
1211.131
\end{array}
$$

Practice Exercises

Add each of the following:

13. $74.26 + 64.297$

14. $50.076 + 16.77$

15. $6.43 + 8.57 + 47.09$

16. $1.8 + 77.1 + 32.5 + 206.26$

Subtraction of decimals is very much like subtraction of whole numbers. The process for subtraction of decimals is exactly the same as that for subtraction of whole numbers, except that as in addition of decimals, place values must be lined up by lining up the decimal points of the minuend and subtrahend.

Subtraction Algorithm for Decimals

Decimal fractions and/or decimal mixed numerals are subtracted by the following procedure:

1. Write the numerals with corresponding place values lined up in the same columns. Annex zeroes so that the minuend and subtrahend have the same number of decimal places.

2. Subtract the columns as in whole numbers.

3. Bring the decimal point straight down from the subtrahend.

Answers:

The following examples do not require borrowing (regrouping):

Example 6

a. Subtract .3 from .7.

Solution:
Line up decimal points and subtract.

$$
\begin{array}{r}
.7 \\
-.3 \\
\hline
.4
\end{array}
$$

.7   Minuend.

$-.3$ Subtrahend.

.4   Difference.

7 tenths minus 3 tenths equals 4 tenths. Write the decimal point below the decimal point in the subtrahend.

b. Subtract .2 from .35.

Solution:
Line up decimal points. Annex 0 after 2 and subtract.

$$
\begin{array}{r}
.35 \\
-.20 \\
\hline
5
\end{array}
$$

.35   Minuend.

$-.20$ Subtrahend.

5   Difference.

5 hundredths minus 0 hundredths equals 5 hundredths.

$$
\begin{array}{r}
.35 \\
-.20 \\
\hline
.15
\end{array}
$$

3 tenths minus 2 tenths equals 1 tenth. Write the decimal point below the decimal point in the subtrahend.

Once we have lined up the place values by lining up the decimal points, subtraction of decimals follows the same pattern as subtraction of whole numbers. The following examples require borrowing and flow from left to right:

Example 7

a. Subtract 54.32 from 70.86.

Solution:
Line up decimal points and subtract.

$$
\begin{array}{r}
70.86 \\
-54.32 \\
\hline
4
\end{array}
\qquad
\begin{array}{r}
70.86 \\
-54.32 \\
\hline
.54
\end{array}
\qquad
\begin{array}{r}
{}^{6\ 10} \\
\cancel{7}0.86 \\
-54.32 \\
\hline
16.54
\end{array}
$$

b. Subtract 21.43 from 853.02.

Solution:
Line up decimal points and subtract.

$$
\begin{array}{r}
{}^{9} \\
{}^{2\ 10\ 12} \\
85\cancel{3}.\cancel{0}2 \\
-21.43 \\
\hline
9
\end{array}
$$

We cannot take 3 from 2, so we try to borrow from 0. Zero doesn't have anything, so we go to 3. Borrow 1 one from 3 ones. Change the 1 one to 10 tenths. Borrow 1 tenth from the 10 tenths. Change 1 tenth to 10 hundredths and add to 2 hundredths giving 12 hundredths. Three hundredths from 12 hundredths = 9 hundredths.

$$
\begin{array}{r}
{}^{9} \\
{}^{2\ 10\ 12} \\
85\cancel{3}.\cancel{0}2 \\
-21.43 \\
\hline
.59
\end{array}
\qquad
\begin{array}{r}
{}^{9} \\
{}^{2\ 10\ 12} \\
85\cancel{3}.\cancel{0}2 \\
-21.43 \\
\hline
1.59
\end{array}
\qquad
\begin{array}{r}
{}^{9} \\
{}^{2\ 10\ 12} \\
85\cancel{3}.\cancel{0}2 \\
-021.43 \\
\hline
831.59
\end{array}
$$

4 tenths from 9 tenths = 5 tenths.

1 one from 2 ones is 1 one.

2 tens from 5 tens is 3 tens and 0 hundreds from 8 hundreds is 8 hundreds.

c. Subtract .24 from .8.

Solution:
Line up decimal points. Annex 0 after 8 and subtract.

$$
\begin{array}{r}
{}^{7\ 1} \\
.\cancel{8}0 \\
-.24 \\
\hline
6
\end{array}
$$

Minuend

Subtrahend

Difference

Cannot take 4 hundredths from 0 hundredths. Borrow 1 tenth and express as 10 hundredths. Four hundredths from 10 hundredths is 6 hundredths.

$$
\begin{array}{r}
{}^{7\ 1} \\
.\cancel{8}0 \\
-.24 \\
\hline
.5\ 6
\end{array}
$$

Seven tenths minus 2 tenths is 5 tenths. Write the decimal point below the decimal point in the subtrahend.

Practice Exercises

17. .45 − .24

18. .873 − .6

19. 80.80 − 61.23

20. 16 − .0683

21. 22.34 − 15

22. 873.08 − 45.13

When writing a check, we must write both a decimal numeral for the amount to be paid and the word name for the dollar amount to be paid. This way the bank can make sure of the correct amount to be paid. See the following example:

Example 8

Write a check for $85.59.

Solution:

IMPORTANT: The number of pennies is always written as a fraction of a dollar. That is, the 59 cents is written $\frac{59}{100}$ of a dollar.

Example 9

Bryan needed some new clothes because cold weather was coming. He bought a pair of pants for $24.95, a shirt for $14.49, and a pair of shoes for $48.35. If he had a gift certificate for $20.00, how much of his own money did he use?

Solution:

First we add all of the costs.

$$
\begin{array}{ll}
\overset{1\,1\,1}{\$24.95} & \text{Pants.} \\
\$14.49 & \text{Shirt.} \\
+\ \$48.35 & \text{Shoes.} \\
\hline
\$87.79 & \text{Sum.}
\end{array}
$$

Then we subtract the amount of the gift certificate.

$$
\begin{array}{ll}
\$87.79 & \text{Sum.} \\
-\$20.00 & \text{Certificate.} \\
\hline
\$67.79 & \text{Difference.}
\end{array}
$$

Therefore, Bryan would write a check for $67.79, or sixty-seven and $\frac{79}{100}$ dollars for the clothes.

Practice Exercises

23. Write the decimal numeral and the word name as they would appear on a check for the difference of $154.78 and $68.91.

24. Donna went to the discount pharmacy. She bought a toothbrush for $1.49, a deodorant stick for $2.79, a hairbrush for $3.54, a bottle of shampoo for $2.38, and some vitamins for $7.49. At the same time she got a refund for an $8.75 calculator that did not work. How much did she have to pay?

Study Tip 2

Developing a Positive Approach

A positive approach can help ensure success in your math courses. Your approach should include the following:

1. Recognize that your degree of success depends upon you and you alone.
2. Make an all-out effort to do well in the course.
3. Work hard enough to do much better than just pass. Set high but realistic goals for yourself.
4. Make a commitment to overcome *any* setbacks, personal or otherwise, and work hard until the very end of the course.

Commitment and determination can go a long way toward guaranteeing that you will be successful. You **can** do it!

Exercise Set 0.7

Read each of the following, then write the word name:

1. .6

2. .3

3. .84

4. .39

5. .763

6. .499

7. 19.45

8. 93.54

9. 68.607

10. 17.095

Identify the place value of the underlined digit.

11. 8.7$\underline{4}$

12. 9.0$\underline{5}$

13. .0$\underline{4}$7

14. .094$\underline{7}$

15. 12.$\underline{0}$18

16. 34.87$\underline{9}$2

Decide which decimal numeral represents the larger or largest number.

17. .520, 52.0

18. .680, 6.80

19. 18.40, 18043

20. 61.023, 61.000

21. 298.20, 29.80

22. 840.67, 804.67

23. .66499, .66399

24. .89532, .89668

25. 1.01221, 1.011, 1.1022

26. 35,071, 3507, 350.07

Answers:

Add each of the following:

27. .5 + .9

28. .8 + .6

29. .17 + .54

30. .38 + .27

31. 4.37 + 1.69

32. 8.56 + 3.23

33. 78.077 + 41.42

34. 23.98 + 630.03

35. 383.58 + 807.03

36. 573.06 + 12.982

37. 124.97 + 81.3

38. 99.46 + 566.7

Subtract each of the following:

39. .8 − .2

40. .9 − .4

41. .75 − .17

42. .56 − .39

43. 8.63 − 6.06

44. 75.64 − 68.45

45. 48.76 − 13.5

46. 89.95 −14.3

47. 212.16 − 98.204

48. 634.45 − 27.022

49. 20 − .8369

50. 42 − .1895

51. 70.34 − 38

52. 68.95 − 33

Add.

53. 67.46 + 117.85 + 35.19

54. 135.31 + 87.62 + 43.77

55. 346.48 + 73.373 + 686.45 + 52.05

56. 79.626 + 84.84 + 431.28 + 5.422

Write the word name for each of the following money amounts as it would appear on a check:

57. $706.45

58. $908.44

59. $305.02

60. $602.30

Solve each of the following:

61. At one time the city of Houston was owed $4.05 million in unpaid parking tickets and the city of New Orleans was owed $4.24 million in unpaid parking tickets. How much was owed to the two cities in unpaid parking tickets? (Source: *Miami Herald,* Dec. 22, 1989)

62. At the same time the cities of Anaheim and San Diego were owed $1.65 million and $.55 million in unpaid parking tickets. How much was owed to the two cities in unpaid parking tickets? (Source: *Miami Herald,* Dec. 22, 1989)

63. Driving into the parking garage, Diana saw a sign that said "Clearance 7.6 ft." If her van is 6.8 feet tall, how much clearance does she have?

64. Anita and Alice are long-distance truck drivers. They noticed that two overpasses on the interstate highway had clearances of 16.25 feet and 17.50 feet. What was the difference in clearances?

65. On long driving trips Bill likes to stop every couple of hours. On the first part of his trip he drove 129.4 miles, and on the rest of the trip he drove 168.7 miles. How many miles was the total trip?

66. In traveling from New York to San Francisco, Anne paid $350.34 for airfare and $9.45 for the airport shuttle. How much did she pay for transportation?

67. A gallon of gasoline costs $1.24 on the turnpike, and $1.17 off the turnpike. How much cheaper is gasoline off the turnpike?

68. Two college professors were going to the same mathematics meeting in Baltimore. One professor bought her ticket a month ahead of time and paid $249.95, while her colleague bought a ticket at the last minute when the same ticket was reduced to $168.40. What was the difference in fares?

69. Andy was doing some comparison shopping. He found a portable radio with earphones for $34.95 at one store and the same item for $37.98 at a second store. How much did he save by buying the cheaper radio?

70. Mom wants to buy Miguel a video game for his birthday. At the toy store she found one Miguel wants for $33.49, and at the electronics store she saw the same game for $39.99. How much more would she pay if she bought the video game at the electronics store?

71. Robin fills up her gas tank every Friday afternoon after being paid. This week she put in 13.883 gallons

and last week she put in 15.088 gallons. How much gasoline did she use for the two weeks?

72. Jackie is taking her family of two adults and two children to the movies. If adult tickets are $4.50 each and children's tickets are $2.95 each, how much will it cost to take the family to the movies?

Calculator Exercise: (Optional)

C1. John spent $80.36 on air conditioning in May, $122.48 in June, $186.24 in July, $234.89 in August, $167.86 in September, and $89.46 in October. How much did he spend for air conditioning from May through October?

Challenge Exercises:

73. Gasoline sells for 1.05\frac{9}{10}$ per gallon. How much is this in cents?

74. The odometer tells you how far a car has been driven. If the odometer reads 115,436.8 miles at the beginning of a trip and 116,123.5 miles at the end of the trip, how many miles were driven?

Writing Exercises:

75. Statistics say that on the average 1.8 people ride in an automobile. If you own a new car and do not want it to be hit by people opening car doors, which side of a parked car would you park on? Why?

76. The metric system is composed of all decimal numbers. The United States is the only major country in the world that has not adopted the metric system. Why do you suppose that is true?

Section 0.8	**Multiplication and Division of Decimal Numerals**

OBJECTIVES

When you complete this section, you will be able to:

a. Multiply decimals.

b. Divide decimals.

c. Round decimals.

Introduction   As we discussed in the previous section, decimal fractions are a special type of ratio fraction that can easily be written using a decimal point. Recall that $.1 = \frac{1}{10}$; $.01 = \frac{1}{100}$; $.001 = \frac{1}{1000}$; and so on. Since decimal fractions are ratio fractions, it is reasonable to expect that the multiplication of decimal fractions should be based on the multiplication of ratio fractions.

The numerator of a ratio fraction associated with a decimal is a whole number. A process you can use to multiply decimals is to multiply the whole numbers in the numerators, then find the denominator by multiplying the place values of the last digits of the equivalent decimal fractions. For example, if we multiply $(.7)(.31)$ using ratio fractions, we have $(.7)(.31) = \frac{7}{10} \cdot \frac{31}{100} = \frac{7 \cdot 31}{10 \cdot 100} = \frac{217}{1000}$. The 217 comes from multiplying the whole numbers 7 and 31. The thousandths place resulted from multiplying the place value of .7, which is $\frac{1}{10}$, with the place value of the 1 in .31, which is $\frac{1}{100}$. This process of decimal multiplication can be broken down into two parts, as illustrated by Example 1.

Example 1

This example illustrates an interesting fact about the multiplication of decimals. In whole numbers, the product is always larger than any factor. In decimal multiplication, the product can be smaller than either factor. For example, (0.1)(0.2) = 0.02, which is smaller than 0.1 or 0.2.

Multiply.

a. 1.4 by .12

Solution:

$$\begin{array}{r} 14 \\ \times 12 \\ \hline 28 \\ 14 \\ \hline 168 \end{array}$$

First, multiply the numbers, disregarding the decimal point: $14 \cdot 12 = 168$. This is the same as multiplying numerator by numerator of the associated fractions: Second, multiply the place values of the last digits of the associated fractions: $\frac{1}{10} \cdot \frac{1}{100} = \frac{1}{1000}$, or in decimal form, $.1 \cdot 01 = .001$. Eight, the last digit of the product, has to be in the thousandths place, so we have .168. Hence, 1.4(.12) = .168.

b. 1.23 by 8

Solution:

Treat the decimal numerals as whole numbers and multiply.

$$\begin{array}{r} 1\,2 \\ 123 \\ \times 8 \\ \hline 984 \end{array}$$

The place value of the last digit of 1.23 is $\frac{1}{100}$. The place value of 8 is 1. Hence, $\frac{1}{100} \cdot \frac{1}{1} = \frac{1}{100}$, or in decimal form, $.01 \cdot 1 = .01$. Thus, the last digit of our product must be in the hundredths place. Therefore, $8 \cdot 1.23 = 9.84$.

c. 82.47 by .23

Solution:

Carry out the standard process for multiplying whole numbers.

$$\begin{array}{r} 1 \\ 1\,2 \\ 8247 \\ \times 23 \\ \hline 24741 \\ 16494 \\ \hline 189681 \end{array}$$

We have $\frac{1}{100} \cdot \frac{1}{100} = \frac{1}{10,000}$, or in decimal form, $.01 .01 = .0001$. The last digit (1) must be in the ten-thousandths place, so we place the decimal between the 8 and the 9.

Hence, 82.47(.23) = 18.9681.

Before we continue, let us make an observation about multiplying fractions.

$$\text{Consider,} \quad \frac{1}{100} \cdot \frac{1}{1000} = \frac{1}{100,000}$$

In decimals, $(.01)(.001) = .0001$

Number of decimal places, $2 + 3 = 5$

From this observation we see that the sum of the number of decimal places in all of the factors is the number of decimal places in the product. We use this result in the following algorithm:

Multiplication Algorithm for Decimals

To multiply decimal numerals, follow the procedure below:

1. Ignore the decimal points and multiply as in whole numbers.
2. Count the number of decimal places to the right of the decimal point in each factor.
3. Add the number of decimal places in all of the factors.
4. Place the decimal point in the product by counting from right to left the number of decimal places found in step 3. If necessary, annex zeroes.

We now show some examples applying the multiplication algorithm.

Example 2

Multiply using the multiplication algorithm.

a. 1.8 by .9

Solution:
Treat the decimal numerals as whole numbers and multiply.

$$
\begin{array}{r}
1.8 \\
\times.9 \\
\hline
162
\end{array}
$$

1.8   *Factor.*
×.9   *Factor.*
162   *Whole number product.*

We add the number of decimal places in both factors and put the decimal point this number of places to the left in the whole number product, starting with the digit on the right.

$$
\begin{array}{r}
7 \\
1.8 \\
\times.9 \\
\hline
1.62
\end{array}
$$

1.8   *1 decimal place.*
×.9   *1 decimal place.*
1.62   *Decimal product, 2 decimal places.*

1 decimal place + 1 decimal place = 2 decimal places. Therefore, put the decimal point two places to the left, between the 1 and the 6.

b. 2.38 by 6

$$
\begin{array}{r}
24 \\
2.38 \\
\times6 \\
\hline
14.28
\end{array}
$$

2.38   *Factor, 2 decimal places.*
×6   *Factor, 0 decimal places.*
14.28   *Product, 2 + 0 = 2 decimal places*

2 decimal places + 0 decimal places = 2 decimal places. Therefore, put the decimal point two places to the left, between the 4 and the 2.

c. 29.06 by 4.3

Solution:
Treat the decimal numerals as whole numbers and multiply.

$$
\begin{array}{r}
22\ 1 \\
29.06 \\
\times 4.3 \\
\hline
8718 \\
11624 \\
\hline
124.958
\end{array}
$$

29.06   *Factor, 2 decimal places.*
× 4.3   *Factor, 1 decimal place.*
124.958   *Product, 2 decimal places + 1 decimal place = 3 decimal places.*

Practice Exercises

Multiply each of the following using the multiplication algorithm for decimals:

1. 4.3(.7)   **2.** 6.3(4.12)

3. 9(5.03)   **4.** .38(.6)

Calculators: (Optional)

Find the following products using a calculator:

5. .4(61.23)   **6.** .263(55.41)

7. 6.07(19.457)   **8.** 19.06(87.28)

Answers:

Practice Exercises 1–8: 1. 3.01 2. 25.956 3. 45.27 4. .228 5. 24.492 6. 14.57283 7. 118.10399 8. 1663.5568

The process for division of decimals is also based on ratio fractions. We have defined ratio fractions as having meaning only when the denominator is a natural number {1, 2, 3, . . .}. Therefore, the process for division of decimals depends on the divisor being a natural number. Hence, part of the process for division of decimals is to do what is necessary to make the divisor a natural number.

In the next example, please bear in mind that $10 \times .3$ is $10 \times \frac{3}{10} = 3$. Hence, the decimal point appears to have "moved" one place to the right upon multiplication by 10. Multiplication by 100 "moves" the decimal point two places to the right, multiplication by 1,000 "moves" the decimal point three places to the right, and so on. Let us do an example.

Example 3

Find the quotient.

a. $.8 \div .4$

Solution:

$$.8 \div .4 = \frac{.8}{.4} =$$

$$\frac{.8}{.4} \cdot \frac{10}{10} = \frac{.8 \cdot 10}{.4 \cdot 10} =$$ *Multiply by $\frac{10}{10} = 1$ to make the denominator a natural number.*

$$\frac{8}{4} = 2$$

b. $.75 \div .25$

Solution:

$$.75 \div .25 = \frac{.75}{.25}$$

$$\frac{.75}{.25} \cdot \frac{100}{100} = \frac{.75 \cdot 100}{.25 \cdot 100} =$$ *Multiply by $\frac{100}{100} = 1$ to make the denominator a natural number.*

$$\frac{75}{25} = 3$$

c. $1.2 \div .3$

Solution:

$$1.2 \div .3 = \frac{1.2}{.3} =$$

$$\frac{1.2 \cdot 10}{.3 \cdot 10} =$$ *Multiply the numerator and the denominator by 10 to get 3 in the denominator. This is the same as multiplying by $\frac{10}{10}$.*

$$\frac{12}{3} = 4$$

The following illustrates the same procedure using long division.

Example 4

a. $1.2 \div .3$

Solution:
This process is similar to the algorithm for dividing whole numbers.

$$.3\overline{)1.2}$$
$$3\overline{)12.}$$

The denominator (divisor) is placed outside the division sign. The numerator (dividend) is inside the division sign.

Remember, every whole number has a decimal point behind it.

$$3\overline{)12.}^{\;4.}$$
$$\underline{-12}$$
$$0$$

As in the fractional division above, we multiply divisor and dividend by 10 to make the divisor a whole number. The decimal point is now behind the 2 in the dividend. The decimal is brought up into the quotient directly above its place in the dividend.

After the decimal point is placed in the quotient, it is ignored and we proceed as in whole number division.

Therefore, $1.2 \div .3 = 4$.

Suppose the divisor is already a whole number.

b. Divide 8 into 7.2.

Solution:

$$\begin{array}{r} 0.9 \\ 8\overline{)7.2} \\ -72 \\ \hline 0 \end{array}$$

Since the divisor is already a whole number, there is no need to move the decimal to make it a whole number. This means that the decimal point in the dividend stays where it is and the decimal point in the quotient goes right above it.

Proceed as in whole number division.

Practice Exercises

Find the following quotient:

9. $1.5 \div .3$

10. $2.40 \div .80$

11. Divide 9 into 4.5

The algorithm for division of decimals can be summarized in a process for division of decimals. We now state this algorithm.

Division Algorithm for Decimals

To divide decimal numerals:

1. Make the divisor a whole number by "moving the decimal point"; that is, multiply the divisor by whatever it takes to make the divisor a whole number. "Move" the decimal point in the dividend the same number of places it was moved in the divisor. If the divisor is already a whole number, go to step 2.
2. Place the decimal point in the quotient directly above the decimal point in the dividend.
3. Carry out the division as in whole number division.

Example 5

Find the quotient.

a. $21 \div .4$

Solution:

$$.4\overline{)21}$$

$$\begin{array}{r} 52.5 \\ 4\overline{)210.0} \\ -20 \\ \hline 10 \\ -8 \\ \hline 20 \\ -20 \\ \hline 0 \end{array}$$

Multiply the divisor and dividend by 10 to make the divisor a whole number. This will in effect move the decimal point one place to the right in both the divisor and the dividend. Annex 0s as needed to fill in the place values. Write the decimal point in the quotient above the decimal point in the dividend.

Proceed as in whole number division.

Check

$52.5(.4) = 21$

---

b. $13.5 \div .15$

Solution:

$$.15\overline{)13.5}$$

$$\begin{array}{r} 90. \\ 15\overline{)1350.} \\ -135 \\ \hline 0 \end{array}$$

Multiply the divisor and dividend by 100 to make the divisor a whole number. This will in effect move the decimal point two places to the right in both the divisor and the dividend. Annex 0s as needed to fill in the place values. Write the decimal point in the quotient above the decimal point in the dividend.

Proceed as in whole number division.

Check
$(90)(.15) = 13.5.$

c. $.045 \div .5$

Solution:

$$.5\overline{).045}$$

$$\begin{array}{r} .09 \\ 5\overline{).45} \\ -45 \\ \hline 0 \end{array}$$

Multiply the divisor and dividend by 10 to make the divisor a whole number. This will in effect move the decimal point one place to the right in both the divisor and the dividend. Write the decimal point in the quotient above the decimal point in the dividend.

Proceed as in whole number division. Five will not divide into 4, so write 0 in the value place above the 4.

Five divides into 45 nine times. Multiply and subtract.

Check
$(.09)(.5) = .045.$

Practice Exercises

Find the quotient.

12. $48 \div .6$

13. $16.1 \div .23$

14. $.72 \div .8$

As we know from division of whole numbers, the remainder is not always 0. If the remainder is not 0, we usually continue dividing for a given number of decimal places, then round off.

The process of rounding off involves writing a numeral to the nearest desired place value. If we wanted to round .147 to the nearest hundredth, we go to the *thousandths* place, see a 7, and add 1 to the 4 in the *hundredths* place. That is, .147 is .15 to the *nearest hundredth.*

On the other hand, if we wanted to round .147 to the nearest tenth, we would go to the *hundredths* place, see a 4, and drop it and all other digits to the right of it. So .147 to the *nearest tenth* is .1.

Rounding to the Nearest Place Value

To round to the nearest place value do the following:

1. Go to the place value to the right of the place value under consideration.
2. If this digit is 5 or more, add 1 to the digit under consideration and drop all digits that follow the digit under consideration.
3. If this digit is 4 or less, drop it and all that follow it. Keep the digit under consideration as is.

IMPORTANT: The last digit should be in the place value that you are rounding to.

Example 6

Round the indicated number to the indicated place value.

a. .367, nearest tenth

Solution:
The place value to the right of the tenths place has a "6." Since 6 is greater than 5, we add 1 to the 3 in the tenths place and drop all digits which follow the tenths place. Thus, .367 rounded to the nearest tenth is .4.

b. 12.573, nearest hundredth

Solution:
The place value to the right of the hundredths place has a "3." Since 3 is less than 5, we drop the 3 and all digits that follow. Thus, 12.573 rounded to the nearest hundredth is 12.57.

Practice Exercises

Round to the nearest tenth.

15. 15.082

16. 85.349

Round to the nearest hundredth.

17. 65.6805

18. 1.71518

Round to the nearest thousandth.

19. 25.0963

20. .00499

Example 7

Divide and round the answer off to the nearest hundredth.

a. 4.21 by 5

Solution:

$$5\overline{)4.21}$$

$$\begin{array}{r} 0.842 \\ 5\overline{)4.210} \\ -4\ 0 \\ \hline 21 \\ -20 \\ \hline 10 \\ -10 \\ \hline 0 \end{array}$$

The divisor is already a whole number, so write the decimal point in the quotient above the decimal point in the dividend.

The quotient extends to the thousandths place. This is all we need to round off to the nearest hundredth. We round 0.842 to two decimal places by looking at the thousandths place. The thousandths place has a "2" and 2 is less than 5. We drop the "2." Hence, our rounded quotient is .84.

b. .625 by .21

Solution:

$$.21\overline{)\,.625}$$

$$\begin{array}{r} 2.976 \\ 21\overline{)62.500} \\ -42 \\ \hline 205 \\ -189 \\ \hline 160 \\ -147 \\ \hline 130 \\ -126 \\ \hline 4 \end{array}$$

Move the decimal point two places to the right in both divisor and dividend. This is the same as multiplying the divisor and dividend by 100 to make the divisor a whole number. Write the decimal point in the quotient above the decimal point in the dividend. Proceed as in division of whole numbers.

Since we are rounding to the nearest hundredth, we stop at three decimal places in the quotient. This is all we need to round off to the nearest hundredth. We round 2.976 to two decimal places by looking at the thousandths place. The thousandths place has a "6" and 6 is greater than 5. We add 1 to the "7" in the hundredths place and drop digits to the right of the "7." Hence, our rounded quotient is 2.98.

c. 79 by .15

Solution:

$$.15\overline{)79}$$

$$\begin{array}{r} 526.666...\\ .15\overline{)7900.}\\ -75\\ \hline 40\\ -30\\ \hline 100\\ -90\\ \hline 100\\ -90\\ \hline 10 \text{ etc}\end{array}$$

Move the decimal point two places to the right in both divisor and dividend. This is the same as multiplying the divisor and dividend by 100 to make the divisor a whole number. Write the decimal point in the quotient above the decimal point in the dividend. Proceed as in division of whole numbers.

The quotient is a repeating decimal. Since we are rounding to the nearest hundredth, we stop at three decimal places in the quotient. This is all we need to round off to the nearest hundredth. We round 526.666... to two decimal places by looking at the thousandths place. The thousandths place has a "6" and 6 is greater than 5. We add 1 to the "6" in the hundredths place and drop digits to the right of the resulting "7." Hence, our rounded quotient is 526.67.

Note: If we round a number, the original number is not equal to the rounded number. For example, 5.467 rounded to the nearest tenth is 5.5. However. 5.5 ≠ 5.467. Thus, we use a special symbol which means "approximately equal to." That is, 5.5 ≈ 5.467.

Practice Exercises

Divide each of the following. If a quotient goes beyond the hundredths place, round to the nearest hundredth.

21. $21.5 \div 9$

22. $2.4 \div .7$

23. $.234 \div .08$

24. $65 \div .32$

We see applications of decimals everywhere. Our money system is based on the decimal system, and most of our everyday calculations are done in decimals. Most calculators work only with decimals.

Example 8

Solve each of the following.

a. Dad took his family of four out to dinner to celebrate his daughter's thirteenth birthday. If the meal cost $36.92, what was the average cost per person?

Solution:
The average cost per person is like sharing the cost among four people. Therefore, this is a case of division.

$$\begin{array}{r}\$9.23\\ 4\overline{)\$36.92}\end{array}$$ The meals cost $9.23 per person.

b. A package of cookies weighs 1.25 pounds. How much will 8 packages of cookies weigh?

Solution:
Since there are 8 packages with 1.25 pounds in each package, this is a case of multiplication.

$$\begin{array}{r}1.25\\ \times 8\\ \hline 10.00\end{array}$$ The cookies weigh 10 pounds.

Practice Exercises

Solve each of the following. Round to the nearest hundredth where necessary.

25. A shipping container has a 32,580-kilogram capacity. If there are 2.2 pounds per kilogram, what is the capacity in pounds?

26. If a 10-pound bag of potatoes costs $1.49, find, to the nearest cent, how much 1 pound of potatoes costs.

Answers:

Exercise Set 0.8

Multiply with whole number multiplication and place decimal points by multiplication of place values.

1. .5(6)   **2.** .3(8)   **3.** .2(.17)   **4.** .4(.35)

Multiply each of the following using the multiplication algorithm:

5. 7(8.1)   **6.** 5.2(6)   **7.** 3.2(4.4)

8. 6.5(3.3)   **9.** 1.8(.52)   **10.** .15(2.6)

11. 3.9(6)   **12.** 4.8(5)   **13.** 2(6.4)

14. 9(4.1)   **15.** 5.1(8.4)   **16.** 1.2(3.5)

17. 1.5(4.6)   **18.** 3.2(8.3)   **19.** .3(.89)

20. .6(.74)   **21.** 2.59(.15)   **22.** 3.34(.07)

23. 5.27(.603)   **24.** (23.9).834

Divide each of the following using ratio fractions as in Example 3.

25. $2.4 \div .6$ **26.** $5.6 \div .8$

27. $1.05 \div .35$ **28.** $2.50 \div .25$

Round to the indicated value place.

29. .818 nearest hundredth   **30.** .974 nearest hundredth

31. 8.428 nearest tenth   **32.** 2.675 nearest tenth

33. 64.708 nearest one   **34.** 5.607 nearest one

35. .00684 nearest thousandth   **36.** 75.080691 nearest thousandth

37. 4.9899 nearest tenth   **38.** .9999 nearest hundredth

Divide each of the following. If a quotient goes beyond the hundredths place, round to the nearest hundredth.

39. $5.6 \div 8$ **40.** $7.2 \div .9$ **41.** $.50 \div 6$

42. $.101 \div 11$ **43.** $.28 \div .4$ **44.** $.84 \div .8$

45. $20.4 \div .9$ **46.** $35.3 \div .7$ **47.** $97.43 \div .066$

48. $61.16 \div .02$ **49.** $5.23 \div .143$ **50.** $6.81 \div .257$

Solve each of the following. Round to the nearest hundredth where necessary.

51. There are about 1.6 kilometers in 1 mile. If you are traveling at 60 miles per hour, how fast are you going in kilometers per hour?

52. Five people have formed a lottery pool. If they win the $6.7 million jackpot, how much will each person receive?

53. An author has to write 250 pages in 60 days. What is the average number of pages she needs to write each day?

54. An airline pilot has a schedule to meet. At what rate must she fly if she has 4316 miles to travel in 8.3 hours?

55. Arthur Chung has pledged to read 5 books that together have a total of 1089 pages. If he has pledges for 3 cents per page, how much can he raise for the read-a-thon?

56. A student sold 650 copies of the campus humor magazine. If she makes $.48 a copy, how much has she earned for her activities?

57. A railroad club has its own private railroad car. It costs $32,000 to have the local railroad hook up and pull the car to a railroad convention. If there are 28 people going on a trip, what is the cost per person?

58. Carmen wants to drive her car out West on vacation. She estimates that she will drive about 5500 miles. If her car gets 32 miles to the gallon of gasoline, how many gallons will she need?

Calculator Exercises: (Optional)

C1. 32.53(7.41)

C2. 83.32(8.99)

C3. $66.999 \div 546$

C4. $95.348 \div 403$

Challenge Exercises:

59. If the sales tax rate in a certain state is 6 cents on the dollar, how much is the sales tax on $8,400?

60. In an eight-hour period the temperature outside fell from 42°F to 23°F. Assuming that the decline was constant, on the average how many degrees per hour did the temperature fall?

Writing Exercises:

61. Some people say that with decimal fractions we have no need for any other kind of fractions. Do you think that this is true? Why or why not?

62. Why do we line up the decimal points when adding but not when multiplying decimals?

Section 0.9	Linear Measurement in the American and Metric Systems

OBJECTIVES   *When you complete this section, you will be able to:*

a. Convert between different linear units in the American system.

b. Convert between different linear units in the metric system.

Introduction   In this section we will discuss two systems of linear measure, the American and the metric. Linear units are used to measure distances, such as between two points or the perimeter of a figure. Currently, the United States is the only nation in the world using the American system (formerly called the English system until Great Britain switched to the metric system).

In order to measure a distance, we need a unit of measure. The unit of measure is a standardized length that may be used anywhere to express a distance. When expressing a distance, we first give the number of units and then the unit of measure.

Let us devise our own system of measure. Suppose the length of the following line segment is called a *knod*.

————

1 knod

Using the length of the preceding segment as the unit, let us measure the length of the following line segment:

————————————

We will mark off segments each of whose length is 1 knod.

├────┼────┼────┼────┼────┤

1 knod   1 knod   1 knod   1 knod   1 knod

Since there are five segments each of whose length is 1 knod, the length of the entire line segment is 5 knods.

It is possible to have fractional parts of a unit when measuring a distance. Consider the length of the following:

├────┼────┼────┼──┤

1 knod   1 knod   1 knod $\frac{1}{2}$ knod

Therefore, the length of the above line segment is $3\frac{1}{2}$ knods.

The important thing to realize is that we may have any unit of measurement we please as long as everyone has agreed upon it.

American system of linear measure

In the American system of linear measure, the units are inch, foot, yard, rod, furlong, and mile. The rod and furlong are not often used in everyday situations and will not be discussed here. We often abbreviate the names of the units. We abbreviate inch as "in.," foot as "ft," yard as "yd," and mile as "mi." It is possible to measure the length of an object and get different numerical answers because we used different units. In order to change from one unit of measure to another, we must know the conversion factors. The conversion factors for the American system are given in the following table:

Conversion Factors—American System of Linear Measure	
1 ft = 12 in.	1 yd = 3 ft
1 yd = 36 in.	1 mi = 5280 ft
1 mi = 1760 yd	

There are several methods for converting from one unit to another. One method is used extensively in the sciences and is the method that we will use. It is called the **unit-cancellation, unit analysis,** or **factor-label** method. When using this method, we multiply the expression to be converted by the ratio of the conversion factors in which the numerator is the unit into which we wish to convert and the denominator is the unit from which we are converting. For example, if we want to convert a given number of feet to inches, we multiply the given number of feet by the ratio $\frac{12 \text{ in.}}{1 \text{ ft}}$. We then "cancel" the units. The word *cancel* is often used to indicate division, but will not be used as such in this book.

Example 1

Convert the following into the indicated unit:

a. 4 ft = _____ in.   Multiply 4 ft by the ratio $\frac{12 \text{ in.}}{1 \text{ ft}}$.

$4 \text{ ft} \cdot \frac{12 \text{ in.}}{1 \text{ ft}} =$ Divide the ft.

$4 \cancel{\text{ft}} \cdot \frac{12 \text{ in.}}{1 \cancel{\text{ft}}} =$ 12 in. ÷ 1 = 12 in.

$4 \cdot 12 \text{ in.} =$ Multiply.

48 in.   Therefore, 4 ft = 48 in.

c. 2.4 mi = _____ ft   Multiply 2.4 mi by $\frac{5280 \text{ ft}}{1 \text{ mi}}$.

$2.4 \text{ mi} \cdot \frac{5280 \text{ ft}}{1 \text{ mi}} =$ Divide the mi.

$2.4 \cancel{\text{mi}} \cdot \frac{5280 \text{ ft}}{1 \cancel{\text{mi}}} =$ 5280 ft ÷ 1 = 5280 ft

$2.4 \cdot 5280 \text{ ft} =$ Multiply.

12,672 ft   Therefore, 2.4 mi = 12,672 ft.

b. 72 in. = _____ yd   Multiply 72 in. by $\frac{1 \text{ yd}}{36 \text{ in.}}$.

$72 \text{ in.} \cdot \frac{1 \text{ yd}}{36 \text{ in.}} =$ Divide the in.

$72 \cancel{\text{in.}} \cdot \frac{1 \text{ yd}}{36 \cancel{\text{in.}}} =$ Think of 72 as $\frac{72}{1}$.

$\frac{72}{1} \cdot \frac{1 \text{ yd}}{36} =$ Multiply.

$\frac{72 \text{ yd}}{36} =$ Divide 72 by 36. 72 ÷ 36 = 2.

2 yd   Therefore, 72 in. = 2 yd.

Practice Exercises

Convert the following to the indicated unit:

1. 4 yd = _____ ft

2. 48 in. = _____ ft

3. 3 ft = _____ in.

4. 15,840 ft = _____ mi

If more practice is needed, do the Additional Practice Exercises.

Additional Practice Exercises

Convert the following to the indicated unit:

a. 180 in. = _____ yd   **b.** 2 yd = _____ ft   **c.** 1.2 mi = _____ ft   **d.** 36 ft = _____ yd

Conversions within the metric system   One of the greatest advantages of the metric system is that conversions from one linear unit of measure to another are accomplished by using powers of ten. When multiplying or dividing by a power of ten, we simply move the decimal point the appropriate number of places. Study the following for a pattern that gives the number of places the decimal point is moved when multiplying or dividing by a power of 10:

Example 2

Find the following products:

a. $2.763 \times 100 = 276.3$ Decimal point moved right 2 places.

$43.91 \times 1000 = 43,910$ Decimal point moved right 3 places.

$0.0062 \times 10 = 0.062$ Decimal point moved right 1 place.

$0.000123 \times 10,000 = 1.23$ Decimal point moved right 4 places.

b. $467.91 \div 10 = 46.791$ Decimal point moved left 1 place.

$5.78 \div 1000 = 0.00578$ Decimal point moved left 3 places.

$0.0142 \div 100 = 0.000142$ Decimal point moved left 2 places.

$100.93 \div 10{,}000 = 0.010093$ Decimal point moved left 4 places.

Based on the preceding example, we make the following observations:

Multiplication and Division by Powers of Ten

a. When multiplying by a power of 10, move the decimal point to the right the same number of places as there are zeroes in the power of 10.

b. When dividing by a power of 10, move the decimal point to the left the same number of places as there are zeroes in the power of 10.

The basic unit of linear measure in the metric system is the meter, which is abbreviated as "m." The basic unit is then prefixed by one of the following, which tell how many meters or what part of a meter the unit is measuring:

Partial List of Metric Prefixes

kilo (k) = 1000 units

hecto (h) = 100 units

deca (dk or da) = 10 units

deci (d) = $\dfrac{1}{10}$ of a unit

centi (c) = $\dfrac{1}{100}$ of a unit

milli (m) = $\dfrac{1}{1000}$ of a unit

Consequently, "km" means kilometer and equals 1000 meters, "dm" means decimeter and equals $\frac{1}{10}$ of a meter, "mm" means millimeter and means $\frac{1}{1000}$ of a meter. The most commonly used units are the kilometer, which is used in the way mile is used in the American system; the meter, which is used in the way feet or yards are used in the American system; and the centimeter and millimeter, which are used in the way inch is used in the American system.

Conversions within the metric system can be done using the factor-label method, but there is a much easier way. Since the metric system is based on powers of 10, and multiplying or dividing by powers of 10 simply moves the decimal point, all we have to do is determine how many places and in which direction to move the decimal point. One way of doing this is to use the following diagram. The units on the line correspond to place values in our base ten number system.

We locate the unit given and count the number of places, left or right, to the unit into which we are converting. Then we move the decimal point that number of value places and in the same direction.

Example 4

Convert the following into the indicated unit.

a. 3.6 m = _____ cm

Solution:

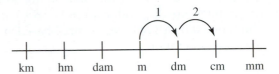

Since cm is two units to the right of m, move the decimal point two places to the right. Therefore, 3.6 m = 360 cm.

b. 7354 dm = _____ hm

Solution:

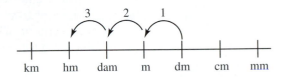

Since hm is three units to the left of dm, move the decimal point three places to the left. Therefore, 7354 dm = 7.354 hm.

c. 0.0178 km = _____ cm

Solution:

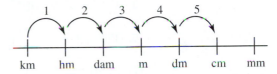

Since cm is five units to the right of km, move the decimal point five places to the right. Therefore, 0.0178 km = 1780 cm. Note: It was necessary to write a 0 after the 8 in order to be able to move the decimal point five places.

d. 9.3 cm = _____ dcm

Solution:

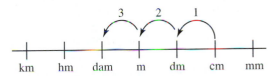

Since dam is three units to the left of cm, move the decimal point three places to the left. Therefore, 9.3 cm = 0.0093 dam. Note: It was necessary to write two 0s before the 9 in order to be able to move the decimal three places.

Practice Exercises

Convert the following to the indicated unit.

5. 76.323 dam = _____ dm

6. 5824 m = _____ hm

7. 2.45 km = _____ dm

8. 2.9 cm = _____ dam

If more practice is needed, do the Additional Practice Exercises.

Additional Practice Exercises

Convert the following to the indicated unit:

e. 4.478 hm = _____ m   **f.** 8345 dm = _____ km   **g.** 6.2 hm = _____ dm   **h.** 0.063 cm = _____ dm

We are discussing only systems of linear measure in this section; therefore, we will not discuss conversions involving mass and volume. However, we will note another advantage of the metric system. Since the metric system was derived from the decimal system, the same system of prefixes is used in measuring mass and volume

Answers:

as in measuring linear units. The only difference is the basic unit. To measure mass, the basic unit is gram. The basic unit to measure volume is liter. To illustrate this, in the metric system an object's weight might be expressed in terms of kilograms (kg) or milligrams (mg). We could measure the volume of a container in kiloliters (kl), decaliters (dal), or milliliters (ml). This means it is not necessary to learn a completely different set of conversion factors for mass and volume like those we have in the American system.

Another example of a decimal-based system is our monetary system. The basic unit is the dollar, $1.00. Going from left to right from the dollar, we have the decadollar or $10.00, the hectodollar or $100.00, and the kilodollar or $1000.00. On the other side, going right from the decimal point, we have the decidollar (dime) or $.10, the centidollar (cent) or $.01, the millidollar (mil) or $.001. All of these decimal-based systems come from the Hindu-Arabaic system of enumeration.

Exercise Set 0.9

Convert the following to the indicated unit:

1. 2 ft = _____ in.

2. 36 in. = _____ ft

3. 9 ft = _____ yd

4. 4 yd = _____ ft

5. 5 mi = _____ ft

6. 10,560 ft = _____ mi

7. 2 mi = _____ yd

8. 3520 yd = _____ mi

9. 60 in. = _____ ft

10. 6 ft = _____ in.

11. 24 ft = _____ yd

12. 15 yd = _____ ft

13. 288 in. = _____ yd

14. 2.5 yd = _____ in.

15. 255 yd = _____ ft

16. 324 in. = _____ yd

17. 5.25 mi = _____ yd

18. 31,680 ft = _____ mi

19. 2.5 yd = _____ in.

20. 7040 yd = _____ mi

Convert the following to the indicated unit:

21. 3 hm = _____ m

22. 500 m = _____ hm

23. 400 mm = _____ dm

24. 15 dm = _____ mm

25. 8 dam = _____ cm

26. 12,000 cm = _____ dam

27. 1.4 km = _____ m

28. 500 m = _____ km

29. .7 m = _____ mm

30. 80 mm = _____ m

31. 458 dm = _____ dam

32. 56 dam = _____ cm

33. 10,000 m = _____ km

34. 7.89 km = _____ m

35. 8.4 dm = _____ m

36. 6.97 hm = _____ km

37. .009 km = _____ m

38. 8.3 m = _____ mm

39. 16.2 dm = _____ hm

40. .005 m = _____ mm

Challenge Exercises:

Convert the following to the indicated unit:

41. 540 in. = _____ yd

42. 243 yd = _____ in.

43. 6.7 mi = _____ ft

44. 18.75 yd = _____ in.

45. 5280 mm = _____ km

46. .0046 hm = _____ cm

47. .0000001 m = _____ mm

48. 8903 km = _____ dm

Writing Exercises:

49. List the conversion factors in the American system for measuring weight.

50. List the conversion factors in the American system for measuring liquid volume.

51. List the conversion factors in the American system for measuring dry volume.

Chapter 0   Summary

Numerals [Section 0.1]
- Digits are the symbols 0, 1, 2, 3, 4, 5, 6, 7, 8, and 9.
- A numeral is a symbol composed of digits used with place value.
- Determining place value involves assigning the digits of a numeral values starting at the right of the numeral with one and going to ten, one hundred, one thousand, ten thousand, one hundred thousand, one million, and so on.
- Place value is determined by the location of the digit within the numeral.

Notation [Section 0.1]
- A numeral is in expanded notation when it is formed by writing the sum of the products of each digit multiplied by its place value. That is, the numeral $abc = 100a + 10b + 1c$.
- Standard notation is the way we normally write numerals.
- The word name of a numeral consists of the words we say when we read aloud a numeral written in standard notation.

Thinking about Addition [Section 0.2]
- Combination: Putting together.

Addition of Whole Numbers [Section 0.2]
- $a + b = c$ a is an addend, b is an addend, and c is the sum of a and b.

Thinking about Subtraction [Section 0.3]
- Take away: Take some from the original.

Difference of Whole Numbers [Section 0.3]
- If $a - b = c$, then $b + c = a$ a is the minuend, b is the subtrahend, and c is the difference.

Thinking about Multiplication [Section 0.4]
- $a \cdot b$ Repeated Addition: The first factor tells how many times the second factor is added.

Product of Whole Numbers [Section 0.4]
- $a \cdot b = c$ a is a factor, b is a factor, and c is the product.

Thinking about Division [Section 0.5]
- $a \div b$ Repeated Subtraction: How many times can we subtract b from a?

Division by Zero Is Impossible [Section 0.5]
- $a \div 0$ is undefined.

Quotient of Whole Numbers [Section 0.5]
- $a \div b = c$, if $b \cdot c = a$ b is the divisor, a is the dividend, and c is the quotient. In division, we say, "How many bs are in a?" Division may be checked with multiplication.

Ratio Fractions [Section 0.6]
- $\frac{a}{b}$ The top number, a, is the numerator and the bottom number, b, is the denominator (not zero). A form of division.

Multiplication of Fractions [Section 0.6]
- $\frac{a}{b} \times \frac{c}{d} = \frac{a \cdot c}{b \cdot d}$ For all numbers a, b, c, d (with b and d not zero), to multiply two fractions, find the product of the numerators divided by the product of the denominators.

Quotient of Ratio Fractions [Section 0.6]
- $\frac{a}{b} \div \frac{c}{d} = \frac{a}{b} \times \frac{d}{c}$ For all numbers a, b, c, and d (with b, c, and d not zero), to divide a fraction by a fraction, multiply by its multiplicative inverse (reciprocal).

Sum of Ratio Fractions with the Same Denominators [Section 0.6]
- $\frac{a}{c} + \frac{b}{c} = \frac{a + b}{c}$ For all numbers a, b, and c, with c not zero, to add two or more fractions with the same denominators, add the numerators and place the sum over the same denominator.

Difference of Ratio Fractions with the Same Denominators [Section 0.6]	• $\frac{a}{c} - \frac{b}{c} = \frac{a-b}{c}$ For c not zero, to subtract two fractions with the same denominator, subtract the numerators and place the difference over the same denominator.
Decimal Fraction [Section 0.7]	• A fraction with a denominator of ten, one hundred, one thousand, and so on, that is written in the form .1, .01, .001, .0001, and so on.
Decimal Point [Section 0.7]	• A symbol used to denote fractions with denominators of ten, one hundred, one thousand, and so on, that is written in decimal form.
Decimal Mixed Numeral [Section 0.7]	• Numerals in decimal notation which include both a whole number and a decimal fraction.
Product of Decimals [Section 0.8]	• $a \cdot b = c$ a is a decimal factor, b is a decimal factor, c is the decimal product.

Multiplication Rule for Decimals [Section 0.8]

• In multiplying decimal numerals,

1. Ignore the decimal points and multiply as in whole numbers.
2. Count the number of decimal value places in each of the factors.
3. Find the sum of the decimal value places in the factors.
4. Place this number of decimal value places in the product counting from right to left.

Quotient of Decimals [Section 0.8]

• $a \div b = c$, if $b \cdot c = a$ b is the divisor, a is the dividend and c is the quotient. Division may be checked with multiplication.

Division Rule for Decimals [Section 0.8]

• In dividing decimal numerals:

1. Make the divisor a whole number by multiplying the divisor and the dividend by the denominator of the divisor. (Make the divisor a whole number by moving the decimal point the required number of places to the right. Then move the decimal point the same number of places to the right in the dividend.) If the divisor is already a whole number, go to step 2.
2. Place the decimal point in the quotient directly above the decimal point in the dividend.
3. Carry out the whole number division.

Rounding to the Nearest Place Value [Section 0.8]

• To round to the nearest place value do the following:

1. Go to the first value place to the right of the place under consideration.
2. If this digit is 5 or more, add 1 to the digit under consideration and drop the digits that follow.
3. If this digit is 4 or less, drop it and keep the digit under consideration.

Conversion Factors— American System [Section 0.9]

• 1 ft = 12 in.
• 1 yd = 3 ft
• 1 yd = 36 in.
• 1 mi = 5,280 ft
• 1 mi = 1760 yd

Conversions Using the Factor-Label Method [Section 0.9]

• To convert from one unit to another using the factor-label method, multiply the expression to be converted by the ratio of the conversion factors in which the numerator is the unit into which we wish to convert and the denominator is the unit from which we are converting.

Metric Prefixes [Section 0.9]
- kilo (k) = 1000 units
- hecto (h) = 100 units
- deca (da or dk) = 10 units
- deci (d) = $\frac{1}{10}$ of a unit
- centi (c) = $\frac{1}{100}$ of a unit
- milli (m) = $\frac{1}{1000}$ of a unit

Converting Within the Metric System [Section 0.9]
- Using the diagram that follows, locate the unit given and count the number of places, left or right, to the unit into which we are converting. Then move the decimal point this number of value places in the same direction.

km   hm   dcm   m   dm   cm   mm

Chapter 0   Review Exercises

Write the word name for each standard numeral. [Section 0.1]

1. 6051

2. 30,459

Write as a standard numeral. [Section 0.1]

3. Thirty-seven thousand, two hundred four

Give the place value of the underlined digit. [Section 0.1]

4. 7̲2,891

What is the meaning of the designated digit in each of the following numerals? [Section 0.1]

5. 4 in 107,461,077

Write in expanded notation. [Section 0.1]

6. 396,071

Insert one of the three symbols (>, =, or <) to make the expression true. [Section 0.1]

7. 177 _____ 717

8. 8208 _____ 8039

Add each of the following. [Section 0.2]

9. 54
     +45

10. 58
      +27

11. 583
      +45

12. 5409
       +747

13. 652,428
       +76,923

Add the following column of figures. [Section 0.2]

14. 5541
       27
       7019
       1510
      +286

Solve the following word problems. [Section 0.2]

15. A football team scored 14 points in the first quarter, 7 points in the second quarter, 0 points in the third quarter, and 21 points in the fourth quarter. How many points did this team score in the four quarters of the game?

16. In 1981, Kay and Dani planted a six-foot-tall oak tree. Since then it has grown thirty-five feet. How tall is it now?

17. Last year the Parks Department planted 18,000 trees. This year they planted 27,000 trees. How many trees have they planted in the last two years?

18. A TV cable is 8 feet long. Twelve more feet of cable are needed to reach the TV. How long must the cable be?

Subtract each of the following. (There will be one or more carries.) [Section 0.3]

19. 729 − 41

20. 623 − 544

21. 82,501 − 7345

Solve the following word problems. Some of them may contain extra information or involve addition as well as subtraction. [Section 0.3]

22. Marti has 350 miles to drive from her hometown to the university she attends. If she has driven 168 miles, how much farther does she have to go?

23. Dr. Lopez is cleaning her office. She has 164 books, but 28 of them will not fit on the shelves. How many will she have left if she gives the 28 extra books to the campus library?

24. Fran and Dan were sitting in the student union comparing how many relatives each has. Fran has 13 first cousins and Dan has 47 first cousins. How many more first cousins does Dan have than Fran?

25. Cesar got his paycheck today. Out of his salary of $875 the following amounts were withheld: $175 for income tax, $70 for FICA, $53 for retirement, and $71 for health insurance. How much does he actually take home?

Multiply each of the following: [Section 0.4]

26. $\begin{array}{r} 65 \\ \times 7 \\ \hline \end{array}$
27. $\begin{array}{r} 76 \\ \times 39 \\ \hline \end{array}$

28. $\begin{array}{r} 306 \\ \times 42 \\ \hline \end{array}$
29. $\begin{array}{r} 671 \\ \times 30 \\ \hline \end{array}$

30. $\begin{array}{r} 415 \\ \times 108 \\ \hline \end{array}$

Solve each of the following exercises. You may need to do more than just multiply. There may be extra information as well. [Section 0.4]

31. As part of her financial aid package, Penny got a job working in the chemistry lab. She washed 15 loads of 24 test tubes in the dishwasher. How many test tubes did she wash?

32. In the Tournament of Roses parade, one band marched 13 members abreast and 15 rows deep. How many band members were there?

33. Juan bought five shirts at $17 each and three pairs of pants at $24 each. How much did he spend on shirts and pants?

Divide by repeated subtraction and check by using the related multiplication. [Section 0.5]

34. $53 \div 8$

Divide using the division algorithm and check by using the related multiplication. [Section 0.5]

35. $625 \div 7$ **36.** $481 \div 13$

Divide using the standard process of long division and check by using the related multiplication. [Section 0.5]

37. $9 \overline{)7119}$ **38.** $18 \overline{)594}$

39. $21 \overline{)9542}$

Solve the following. There may be more than one operation and there may be more information than needed. [Section 0.5]

40. The county will pave a gravel road in front of the property of eight land owners if the land owners will pay the cost of $32,000. If they shared the cost equally, how much would each land owner have to pay?

41. There are 1050 ROTC cadets at Land Grant University. How many classes of 30 students each can be formed?

Multiply. [Section 0.6]   *Divide.* [Section 0.6]

42. $\dfrac{3}{4} \cdot \dfrac{5}{2}$ **43.** $\dfrac{6}{7} \div \dfrac{11}{9}$

Add. [Section 0.6]   *Subtract.* [Section 0.6]

44. $\dfrac{3}{11} + \dfrac{4}{11}$ **45.** $\dfrac{8}{9} - \dfrac{4}{9}$

Add each of the following: [Section 0.7]

46. $1.036 + .093$ **47.** $79.149 + 118$

Subtract each of the following: [Section 0.7]

48. $7.58 - 5.38$ **49.** $201.9 - 55.09$

50. $671.26 - 167.52$

Add the following: [Section 0.7]

51. $16.77 + 371.16 + 58.55 + 21.462$

Solve each of the following. Each exercise may require addition or subtraction or both. [Section 0.7]

52. Jill bought 1.82 pounds of pork chops and 1.99 pounds of spareribs for a family barbecue. How many pounds of pork did she buy?

53. Last month the gas bill included a $10.67 "energy" charge, a $15.32 cost of gas charge, a $6.60 "gas adjustment" charge, and a $1.98 utility tax charge. How much was the total bill?

54. Jose Canseco had a batting average of .269 during the regular baseball season. He had a .357 batting average for the World Series. How much higher was his batting average during the Series?

Multiply with whole number multiplication and place decimal points by multiplication of place values. [Section 0.8]

55. .9(.65)   **56.** .45(2.13)

Multiply each of the following using the multiplication algorithm. [Section 0.8]

57. .04(.85)   **58.** 6.78(5.7)

Divide as in Example 3. [Section 0.8]

59. 3.2 ÷ .8

Round to the indicated place value. [Section 0.8]

60. 38.689 nearest hundredth

Divide each of the following. Round to the nearest hundredth if necessary. [Section 0.8]

61. 2.7 ÷ .9   **62.** 19.3 ÷ .62

63. 78.53 ÷ 2.55

Solve each of the following. If necessary round to the nearest hundredth. [Section 0.8]

64. Little League baseball shoes cost $27.54 per pair. How much would a sponsor have to pay to buy shoes for his team of 12 children?

65. There are 355 milliliters (ml) in 12 ounces (oz) of cola. How many ml are in one oz?

66. Karol sent three pairs of pants and five blouses to the laundry. It cost $1.95 per pair of pants and $2.49 per blouse. How much was her total laundry bill?

Convert the following to the indicated unit using the factor-label method where appropriate: [Section 0.9]

67. 5 ft = _____ in.   **68.** 1.2 yd = _____ ft

69. 90 in. = _____ ft   **70.** 3.7 m = _____ mm

71. 18 km = _____ cm   **72.** .89 hm = _____ dm

73. $\frac{1}{4}$ yd = _____ in.   **74.** 16,896 ft = _____ mi

75. 9215 mm = _____ m   **76.** 43.6 hm = _____ dm

Chapter 0   Test

1. Give the place value of the underlined digit in 9<u>8</u>3,614.

2. Write 82,063 in expanded notation.

3. Add.

728
+674

4. Add.

88,913
+7,108

5.
907
5831
412
6053
+6980

6. Clara collects baseball cards. She has 143 from 1990, 208 from 1991, and 182 from 1992. How many cards does she currently have?

7. Subtract.

11,683
−5,702

8. Subtract.

590,042
−72,672

9. Angela planted 1050 seeds for her biology project. Only 964, however, sprouted. How many seeds did not sprout?

Multiply.

10. 749
        ×8

11. 805
        ×67

12. There are 38 students in each of the 17 classes of Integrated Algebra and Arithmetic. Each student needs a #2 pencil. How many pencils are needed?

Divide, and if necessary, round to the nearest hundredth.

13. 9)855   **14.** 26)966

15. Multiply.   **16.** Subtract.

$\frac{2}{5} \cdot 9$ $\frac{11}{12} - \frac{6}{12}$

17. Which is larger: 70.2259 or 80.967?

18. Add. 68.52 + 469.514

19. Subtract. 60.38 − 45.726

20. Bonnie needs 8.5 pounds of hamburger meat to make hamburgers for her friends after the football game. She already has 2.3 pounds in the freezer. How many more pounds of meat does she need?

Multiply.

21. 8.25(.47)

22. 5.26(.363)

Round to the nearest hundredth.

23. 3.6634

24. 64.999

Divide. Round to the nearest hundredth.

25. 27.283 ÷ 13

26. 843 ÷ .9

Convert the following to the indicated unit:

27. 42 ft = _____ yd

28. 5.5 ft = _____ in.

29. 7.6 hm = _____ cm

30. .09 m = _____ mm

Adding and Subtracting Integers and Polynomials

We begin the chapter by introducing exponents and variables, which may be your first exposure to algebra. Variables allow us to generalize the concepts of arithmetic and then expand on these concepts. We follow this with the order of operations on whole numbers that you will need in order to do the sections on geometry and signed numbers. Operations on signed numbers are spread over Chapters 1 and 2. We next discuss addition and subtraction of signed numbers and then show how these operations are used to combine like terms and polynomials.

Much of arithmetic is the study of performing operations on numbers. Much of algebra is performing these same operations on polynomials.

As you work through the chapter, note how the topics later in the chapter depend upon topics that came earlier in the chapter. This is typical of the structure of mathematics and is the reason you must learn one topic before proceeding to the next. Failure to do this is one of the major reasons some students are not successful in mathematics.

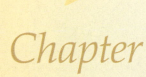

Chapter **1**

| **Section 1.1** | Variables, Exponents, and Order of Operations |

OBJECTIVES

When you complete this section, you will be able to:

a. Identify variables and constants.

b. Read and evaluate expressions raised to powers.

c. Simplify expressions involving more than one operation.

d. Evaluate expressions containing variables when given the value(s) of the variable(s).

Introduction

In this book we will discuss some algebra, much of which is arithmetic from a more general point of view. This more general point of view is made possible by the use of **variables**. You have previously used variables in many formulas for the perimeters, areas, or volumes of several geometric figures. For example, in the formula for the circumference of a circle, $C = 2\pi r$, both C and r are variables. When given a value for r, we can find a value for C. Remember π is a **constant** whose approximate value is 3.14.

DEFINITION Variable and Constant

A *variable* is a symbol, usually a letter, that is used to represent a number. A *constant* is a symbol (a numeral) whose value does not change.

Variables are to mathematics what pronouns are to a language. Pronouns take the place of nouns and variables take the place of numbers.

Constants are used in a variety of ways in mathematics. Often they are given special names depending on how they are used.

Example 1

Identify the constant and the variable(s) in each of the following:

a. $3x$ The constant is 3 and the variable is x.

b. $4xy$ The constant is 4 and the variables are x and y.

c. n This means $1 \cdot n$, so the constant is 1 and the variable is n.

d. 5   The constant is 5 and there is no variable.

Practice Exercises

Identify the constant and the variable(s) in each of the following:

1. $6y$ Constant = _____ , variable = _____ .

2. $5xy$ Constant = _____ , variables = _____ and _____ .

3. z Constant = _____ , variable = _____ .

4. 8   Constant = _____ , variable = _____ .

In arithmetic, multiplication is a short way of performing repeated addition of the same number. For example, $4 \cdot 3$ means $3 + 3 + 3 + 3$. The "4" tells us how many 3s are being added. Likewise, there is a short way of indicating repeated multiplication of the same number using **exponents**. For example, $3 \cdot 3 \cdot 3 \cdot 3$ is written as 3^4, where

---

4 is the **exponent** and 3 is the **base** of the exponent. Since each of the numbers being multiplied is called a **factor,** the exponent tells us how many times the base is used as a factor.

2^3 means $2 \cdot 2 \cdot 2$

In reading expressions that have exponents, the base is read first and then the exponent is read as a power. The preceding expression, 2^3, is read "two to the third power." The power of 2 is most often read as "squared" and the power of 3 is most often read as "cubed." Consequently, 2^3 could also be read as "two cubed." The reasons for reading the second power as squared and the third power as cubed will be made clear after the sections on geometry. There are no special names for powers greater than three.

Since exponents indicate multiplication, constants raised to powers can be evaluated.

Example 2

	Exponential Expression	Read as	Meaning	Value
a.	5^2	5 squared or 5 to the second power	$5 \cdot 5$	25
b.	6^3	6 cubed or 6 to the third power	$6 \cdot 6 \cdot 6$	216
c.	8^5	8 to the fifth power	$8 \cdot 8 \cdot 8 \cdot 8 \cdot 8$	32,768
d.	$\left(\dfrac{2}{3}\right)^4$	Two-thirds to the fourth power.	$\dfrac{2}{3} \cdot \dfrac{2}{3} \cdot \dfrac{2}{3} \cdot \dfrac{2}{3}$	$\dfrac{16}{81}$

Practice Exercises

Complete the following chart:

	Exponential Expression	Read as	Meaning	Value
5.	4^2			
6.	5^3			
7.	3^6			
8.	$\left(\dfrac{1}{2}\right)^5$			

Raising variables to powers has the exact same meaning as raising constants to powers. However, great care must be taken in identifying the base of the exponent.

Example 3

Write each of the following as repeated multiplication:

a. $x^4 = x \cdot x \cdot x \cdot x$

b. $3x^5 = 3 \cdot x \cdot x \cdot x \cdot x \cdot x$ *Note:* 3 is not raised to the fifth power, only the x.

c. $n^3 \cdot y^2 = n \cdot n \cdot n \cdot y \cdot y$

d. $3^3 \cdot x^5 = 3 \cdot 3 \cdot 3 \cdot x \cdot x \cdot x \cdot x \cdot x$

e. $(3a)^3 = (3a)(3a)(3a)$

Example 4

Write each of the following using exponents:

a. $a \cdot a \cdot a \cdot a \cdot a = a^5$

b. $5 \cdot r \cdot r = 5r^2$

c. $n \cdot n \cdot n \cdot p \cdot p \cdot q \cdot q \cdot q = n^3 p^2 q^3$

d. $8 \cdot 8 \cdot 8 \cdot 8 \cdot y \cdot y \cdot y \cdot z \cdot z = 8^4 y^3 z^2$

e. $(8z)(8z)(8z)(8z) = (8z)^4$

Practice Exercises

Write each of the following as repeated multiplication:

9. $z^3 =$ **10.** $4y^3 =$

11. $n^4 y^5 =$ **12.** $7^2 x^6 =$

13. $(2y)^5 =$

Write each of the following using exponents:

14. $a \cdot a \cdot a \cdot a \cdot a =$ **15.** $8 \cdot y \cdot y \cdot y \cdot y \cdot y =$

16. $x \cdot x \cdot x \cdot x \cdot x \cdot y \cdot y \cdot y \cdot z \cdot z =$ **17.** $4 \cdot 4 \cdot 4 \cdot w \cdot w \cdot w \cdot w =$

18. $(6w)(6w)(6w) =$

In mathematics we have the operations of addition, subtraction, multiplication, and division. We just introduced raising to powers, and later we will discuss finding roots. Suppose a problem has more than one operation. Which one do we do first? For example, if we add first, then $2 + 3 \cdot 4 = 5 \cdot 4 = 20$; but if we multiply first, then $2 + 3 \cdot 4 = 2 + 12 = 14$. Both answers cannot be correct, so which one is correct?

Answers:

15. $8y^5$ **16.** $x^5 y^3 z^2$ **17.** $4^3 w^4$ **18.** $(6w)^3$

Practice Exercises 9–18: 9. $z \cdot z \cdot z$ **10.** $4 \cdot y \cdot y \cdot y$ **11.** $n \cdot n \cdot n \cdot n \cdot y \cdot y \cdot y \cdot y \cdot y$ **12.** $7 \cdot 7 \cdot x \cdot x \cdot x \cdot x \cdot x \cdot x$ **13.** $(2y)(2y)(2y)(2y)(2y)$ **14.** a^5

Consider the following situations:

a. You are shopping for some clothes. You buy a shirt for $30 and four pairs of socks for $2 per pair. The amount you would pay is $30 + 4 \cdot 2 = \$38$. Would you add or multiply first?

b. An item is marked three for $12. How much would you pay for two of these items? The amount you would pay is $12 \div 3 \cdot 2 = \$8$. Do you divide or multiply first?

c. Three friends are buying soft drinks for a party. They purchase five cartons at $6 per carton. If they pay equal amounts, how much does each pay? The amount each pays is $5 \cdot 6 \div 3 = \$10$. Would you multiply or divide first?

d. You purchase a jacket for $30 and two shirts marked at $20 each, but with a tag for $5 off on each. You also return a pair of pants originally marked $15 but purchased with a $3 discount. How much money do you owe? You owe $30 + 2(20 - 5) - (15 - 3) = \$48$. In what order would you perform these calculations?

In situation a, you multiply before adding. In situation b, you divide before multiplying, but in situation c you multiply before dividing. In situation d, you simplify inside the parentheses first, then multiply by 2 and finally add and subtract in order from left to right.

It appears that we need some rules as to which order to perform operations in problems that involve more than one operation. Hence we have the following order of operations:

Order of Operations

Order of Operations

If an expression contains more than one operation, the operations are to be performed in the following order:

1. If parentheses or other inclusion symbols (braces or brackets) are present, begin within the innermost and work outward, using the order in steps 3–5 in doing so.

2. If an implied grouping like a fraction bar is present, simplify above and below the fraction bar separately in the order given by steps 3–5.

3. Evaluate all indicated powers.

4. Perform all multiplication or division in the order in which they occur as you work from left to right.

5. Then perform all addition or subtraction in the order in which they occur as you work from left to right.

Simplifying expressions involving more than one operation

Example 5

Simplify the following using the order of operations:

a.
$8 - 6 + 5 =$ Add or subtract left to right, so subtract 6 from 8.
$2 + 5 =$ Add 2 and 5.
7 Answer.

b.
$4 \cdot 5 - 8 =$ Multiply before subtracting, so multiply 4 and 5.
$20 - 8 =$ Subtract 8 from 20.
12 Answer.

c.
$5 + 12 \div 4 =$ Divide before adding, so divide 12 by 4.
$5 + 3 =$ Add 5 and 3.
8 Answer.

d.
$2 + 3 \cdot 4^2 =$ Raise to powers first, so square 4.
$2 + 3 \cdot 16 =$ Multiply before adding, so multiply 3 and 16.
$2 + 48 =$ Add 2 and 48.
50 Answer.

e. $2^3 \cdot 3^2 =$ Raise to powers first.

$8 \cdot 9 =$ Multiply.

72   Answer.

f. $5(6 - 3) + 2 =$ Perform operations inside parentheses first, so subtract 3 from 6.

$5(3) + 2 =$ Multiply before adding, so multiply 5 and 3.

$15 + 2 =$ Add 15 and 2.

17   Answer.

g. $15 - 2(3 + 2^2) =$ Perform operations inside parentheses first, so square 2.

$15 - 2(3 + 4) =$ There is still an operation inside parentheses, so add 3 and 4.

$15 - 2(7) =$ Multiply before adding, so multiply 2 and 7.

$15 - 14 =$ Subtract 14 from 15.

1   Answer.

h. $15 + 3 \cdot 8 \div 12 =$ Multiply 3 and 8.

$15 + 24 \div 12 =$ Divide 24 by 12.

$15 + 2 =$ Add 15 and 2.

17   Answer.

i. $\dfrac{4 + 3 \cdot 2^2}{2 + 3 \cdot 2} =$ Simplify above and below the fraction bar separately. Square 2 in the numerator and multiply in the denominator.

$\dfrac{4 + 3 \cdot 4}{2 + 6} =$ Multiply in the numerator and add in the denominator.

$\dfrac{4 + 12}{8} =$ Add 4 and 12.

$\dfrac{16}{8} =$ Divide 8 into 16.

2   Answer.

Practice Exercises

Simplify the following using the order of operations:

19. $10 - 8 + 5$ **20.** $9 + 3 \cdot 6$ **21.** $12 - 10 \div 5 \cdot 3$ **22.** $5 + 4 \cdot 3^3$

23. $6(12 - 2^2) + 4$ **24.** $15 - 20 \div 4 \cdot 2 - 3$ **25.** $4 + 3(6 + 5 \cdot 3^2) - 10$ **26.** $\dfrac{8 + 15 \cdot 2}{2 \cdot 6 + 7}$

If more practice is needed, do the Additional Practice Exercises.

Additional Practice Exercises

Simplify the following using the order of operations:

a. $6 + 8 - 3$ **b.** $5 + 4(6)$

c. $4 + 2 \cdot 3^3$ **d.** $5^2 \cdot 4^2$

Answers:

e. $7(25 - 15 \div 3) - 8$

f. $6 + 2(9 + 3 \cdot 7)$

g. $24 - 15 \div 3 \cdot 2^2$

h. $\dfrac{6 + 12 \div 3}{2 \cdot 3 - 1}$

OPTIONAL   CALCULATORS

Scientific calculators using algebraic logic have the order of operations built in. Consequently, expressions that do not involve parentheses may be simplified by entering the expression in the order in which it is written. Let us look back at some previous examples and see how they would be done on a scientific calculator. Non-scientific calculators are rarely used on this level and will not be discussed.

Example C1

$4 \cdot 5 - 8 =$
12

Keystrokes: [**4**] [×] [**5**] [−] [**8**] [=]

(Some calculators have an "enter" key instead of =.)

Example C2

$2 + 3 \cdot 4^2 =$
50

Keystrokes: [**2**] [+] [**3**] [×] [**4**] [*x^y*] or [^] [**2**] [=]

(Some calculators have a special key for squaring.)

Example C3

$15 + 3 \cdot 8 \div 12 =$
17

Keystrokes: [**15**] [+] [**3**] [×] [**8**] [÷] [**12**] [=]

If parentheses are involved and your calculator does not have the capability to process expressions in parentheses, it becomes a little tricky. The expression in the parentheses must be evaluated first.

Example C4

$5(6 - 3) + 2 =$
17

Keystrokes: [**6**] [−] [**3**] [=] [×] [**5**] [+] [**2**] [=]

The extra = after the 3 is necessary in order to obtain a value for the expression inside the parentheses before multiplying by 5. If your calculator has parentheses,

the keystrokes are [**5**] [( ] [**6**] [−] [**3**] [ )] [+] [**2**] [=].

Some calculators require a times sign, ×, before the parentheses.

Example C5

$15 + 2(4 + 2^2) =$
31

Keystrokes: [**4**] [+] [**2**] [*x^y*] or [^] [**2**] [=] [×] [**2**] [+] [**15**] [=]

or if your calculator has parentheses: [**15**] [+] [**2**] [( ] [**4**] [+] [**2**] [*x^y*] [**2**] [ )] [=].

Answers:

Additional Practice Exercises e–h: e. 132 f. 66 g. 4 h. 2

Evaluating expressions containing variables   The same order of operations applies when working with expressions containing variables. If a variable and a constant or two variables are written together with no operation sign, the operation is understood to be multiplication. For example, $3x$ means $3 \cdot x$ and xy means $x \cdot y$. We substitute the value(s) for the variable(s) by placing the value in parentheses and simplify by performing the correct order of operations.

Example 6

Let $x = 2$ and $y = 5$ and evaluate.

a. $3x + 4y =$ Substitute 2 for x and 5 for y.

$3(2) + 4(5) =$ Multiply in order from left to right, so multiply 3(2) and 4(5).

$6 + 20 =$ Add 6 and 20.

26   Answer.

b. $2x^2 + 3y^3 =$ Substitute 2 for x and 5 for y.

$2(2)^2 + 3(5)^3$ First raise to powers.

$2(4) + 3(125) =$ Multiply in order from left to right.

$8 + 375 =$ Add.

383   Answer.

c. $5x^2y^2 =$ Substitute 2 for x and 5 for y.

$5(2)^2(5)^2 =$ Raise to powers.

$5 \cdot 4 \cdot 25 =$ Multiply in order from left to right.

$20 \cdot 25 =$ Multiply.

500   Answer.

d. $10x^2 \div y =$ Substitute 2 for x and 5 for y.

$10(2)^2 \div 5 =$ Raise to powers.

$10(4) \div 5 =$ Multiply and divide in order from left to right, so multiply 10 and 4.

$40 \div 5 =$ Divide.

8   Answer.

Practice Exercises

Let $x = 3$ and $y = 4$ and evaluate.

27. $4x + 3y$

28. $2x^2 + 5y$

29. $2x^3y^2$

30. $(8x^3) \div (2y)$

If more practice is needed, do the Additional Practice Exercises.

Additional Practice Exercises

Let $x = 4$ and $y = 5$ and evaluate.

i. $2x^2 + y^2$

j. $3x^2y^2$

k. $3x^3y + 4y^2$

l. $(5x^3) \div (2y)$

Answers:

Practice Exercises 27–30: **27.** 24 **28.** 38 **29.** 864 **30.** 27   *Additional Practice Exercises i–l:* **i.** 57 **j.** 1200 **k.** 1060 **l.** 32

Order of operations occurs in everyday situations.

Example 7

Answer the following:

a. Tara, who has expensive taste, buys a living room suite by paying $500 down and $150 for 18 months. Find the total amount that she pays.

Solution:
The amount that she pays can be represented by $500 + 150 \cdot 18$.

$500 + 150 \cdot 18$	Multiply before adding.
$500 + 2700$	Add.
3200	Therefore, she pays $3200 for the living room suite.

b. Jonathon leased a car with the following terms: he paid $1000 down, $199 per month for 2 years, and 12 cents per mile for all mileage over 24,000. When he returned the car it had 32,000 miles on the odometer. Find the total amount that he paid for the lease.

Solution:
The amount that he paid can be represented by $1000 + 199 \cdot 2 \cdot 12 + .12(32{,}000 - 24{,}000)$. Note that 12 cents was changed to $.12 since all the other amounts were given in dollars.

$1000 + 199 \cdot 2 \cdot 12 + .12(32{,}000 - 24{,}000) =$	Simplify inside the parentheses.
$1000 + 199 \cdot 2 \cdot 12 + .12(8000) =$	Multiply 199 and 2.
$1000 + 398 \cdot 12 + .12(8000) =$	Multiply $398 \cdot 12$.
$1000 + 4776 + .12(8000) =$	Multiply .12 and 8000.
$1000 + 4776 + 960 =$	Add.
6736	Therefore, the lease cost $6736.

Practice Exercises

31. Wolfgang bought a car by paying $1200 down and $350 per month for four years. What is the total amount that he paid for the car?

32. Five people agree to split the cost of some take-out food. Three of the orders cost $4.99 each and two orders cost $6.99 each. They also purchased two 2-liter soft drinks that cost $1.29 each. To the nearest cent (hundredths place), how much did each pay?

Exercise Set 1.1

Write a statement indicating how each of the following would be read and evaluate:

1. 8^2

2. 4^2

3. 5^3

4. 8^3

5. 9^4

6. 2^4

7. 2^5

8. 10^5

9. 10^6

10. 3^4

11. $\left(\dfrac{4}{5}\right)^2$

12. $\left(\dfrac{3}{4}\right)^2$

13. $(1.3)^3$

14. $(2.1)^3$

Write each of the following as multiplication:

15. a^4

16. c^6

17. a^2b^3

18. r^2s^4

19. $5x^3$

20. $3r^5$

21. $(4a)^4$

22. $(5z)^3$

Write each of the following using exponents:

23. $p \cdot p \cdot p \cdot p$

24. $r \cdot r \cdot r \cdot r \cdot r$

25. $b \cdot b \cdot b \cdot b \cdot d \cdot d \cdot d$

26. $t \cdot t \cdot t \cdot t \cdot t \cdot f \cdot f \cdot f \cdot f$

27. $3 \cdot x \cdot x \cdot x \cdot x$

28. $5 \cdot y \cdot y \cdot y \cdot y \cdot y \cdot y \cdot y$

29. $(4r)(4r)(4r)(4r)(4r)(4r)$

30. $(6w)(6w)(6w)(6w)(6w)$

Simplify the following using the order of operations:

31. $7 - 3 + 5$

32. $8 + 6 - 4$

33. $5 - 6 \div 2$

34. $8 + 15 \div 3$

35. $7 + 2 \cdot 6$

36. $10 - 2 \cdot 4$

37. $8 + 3 \cdot 4^2$

38. $25 - 3 \cdot 2^3$

39. $4^2 + 5 \cdot 6$

40. $3^3 + 12 \div 3$

41. $24 - 3(8 - 2)$

42. $25 - 5(12 - 7)$

43. $10 + (13 - 9) \div 2$

44. $15 + (24 - 6) \div 3$

45. $2^3 \cdot 6^2$

46. $2^5 \cdot 4^2$

47. $3^2 \cdot 4^3 \cdot 2^2$

48. $5^2 \cdot 7^2 \cdot 2^3$

49. $\left(\dfrac{1}{2}\right)^3 \cdot \left(\dfrac{2}{3}\right)^2$

50. $\left(\dfrac{4}{3}\right)^2 \cdot \left(\dfrac{1}{4}\right)^3$

51. $2^3 + 3 \cdot 4^2$

52. $3^4 - 4 \cdot 2^3$

53. $26 - 18 \div 3^2$

54. $32 - 24 \div 2^3$

55. $3^4 - 36 \div 3^2$

56. $5^3 - 4 \cdot 3^3$

57. $22 + 18 \div 3 - 12$

58. $31 - 24 \div 6 + 8$

59. $32 - 4^2 + (8 - 3)$

60. $9 - 2^2 + (11 - 7)$

61. $45 \div (18 - 3^2)$

62. $36 \div (22 - 2^2)$

63. $3(11 - 8) \div 3$

64. $4(21 - 6) \div 20$

65. $4(2^4) - 32$

66. $5(3^3) - 45$

67. $6(4^2 - 8) \div 12$

68. $5(6^2 - 28) \div 10$

69. $17 - 9 + 8 \cdot 4 \div 16$

70. $29 - 12 \cdot 5 \div 30$

71. $48 \div 6 \cdot 5 - 16$

72. $64 \div 8 \cdot 4 - 17$

73. $5 + 35 \div 7 \cdot 3 - 16$

74. $14 + 81 \div 9 \cdot 2 - 26$

75. $200 - 84 \div 4 \cdot 2^3 + 29$

76. $225 - 48 \div 8 \cdot 3^2 - 17$

77. $\dfrac{8 + 2 \cdot 3}{7}$

78. $\dfrac{6 + 5 \cdot 4}{13}$

79. $\dfrac{9 - 12 \div 4}{8 \div 4}$

80. $\dfrac{12 - 16 \div 8}{15 \div 3}$

81. $\dfrac{6^2 + 8 \cdot 2^2}{4^2 + 1}$

82. $\dfrac{8^2 + 6 \cdot 4^2}{6^2 + 4 \cdot 11}$

Calculator Exercises:

Evaluate the following using a calculator:

C1. 3^7

C2. 12^3

C3. $(7.32)^2$

C4. $4^4 \cdot 6^5$

C5. $(4.13)(8.95) + (2.6)^2$

C6. $20.48 \div 5.12 - 3.15$

C7. $51.23(6.14^2 - 23.46) - 17.3^2$

C8. $8.1^3 \div 2.7^2 - 4.7 - 3.1^3$

Let x = 2 and y = 3 and evaluate.

83. $3x + 4y$ **84.** $5x + 3y$ **85.** $8x^2$

86. $5y^3$ **87.** x^2y **88.** xy^4

89. $5x^2y^2$ **90.** $7x^3y$ **91.** $3x^2 + 4y^2$

92. $6x^3 - 2y^2$ **93.** $6x^2 \div y$ **94.** $18x^3 \div y^2$

95. $\dfrac{x^2 + 2y^2}{2x^2 + 3}$ **96.** $\dfrac{3x^2 + y^2}{2x^2 - 1}$

Challenge Exercises: (97–100)

Find the value of each of the following when x = $\frac{1}{3}$ and y = $\frac{2}{5}$:

97. xy^2 **98.** x^3y **99.** $9x^2y^2$ **100.** $25x^2y^3$

Calculator Exercises:

Find the value of the following for x = 2.4 and y = 3.6:

C9. $1.7x + 2.6y$ **C10.** $4.2y - 1.2x$ **C11.** $1.6x^2 + 3.7y^3$

C12. $3.3x^3 + 8.7y^2$ **C13.** $6x^3 \div 3y$ **C14.** $1.2x^4 \div .6y^2$

Solve each of the following:

101. Sadie bought a sofa by paying $150 down and $28 per month for 18 months. The total cost to Sadie is $150 + 28(18)$. Find the cost.

102. Diane bought a refrigerator by paying $225 down and $48 per month for 12 months. The total cost to Diane is $225 + 48(12)$. Find the cost.

103. Deon buys a used car by paying $600 down and $190 per month for 24 months. What is the total amount Deon paid for the car?

104. An automobile lease company offers a plan where the customer pays $1000 down and $399 per month for 24 months. Find the total cost of leasing the car.

105. Under the property settlement terms of his divorce, Manuel is to pay his ex-wife $3000 per month for the first four years and then $3500 per month for the next six years. How much will Manual pay in property settlement?

106. Harold buys a lot with the following terms: he is to pay $250 per month for the first five years and $350 per month for the next five years. Find the total amount Harold will pay for the lot.

107. Five fraternity brothers agree to split the cost equally for three pizzas that cost $10.99 each

and five soft drinks that cost 79 cents each. Find the amount (to the nearest cent) that each pays.

108. Six people agree to split the cost equally for some Chinese takeout. If two orders cost $3.99 each, three orders cost $3.50 each, one order costs $4.50, and they also purchase three bottles of soft drinks that cost $1.29 each, find the amount (to the nearest cent) that each pays.

109. Ezra leased a car with the following terms: he agreed to pay $800 down, $299 per month for three years, and 15 cents per mile for all mileage in excess of 45,000. When Ezra returned the car at the end of the lease period, it had 58,000 miles on the odometer. The total cost of the lease can be represented as $800 + 299(3)(12) + .15(58,000 - 45,000)$. Find the total amount Ezra paid for the lease.

110. Candace leased a car with the following terms: she agreed to pay $1500 down, $499 per month for 24 months, and 20 cents per mile for all mileage in excess of 24,000. When she returned the car at the end of the lease period, it had 32,000 miles on the odometer. Find the total amount Candace paid for the lease.

Challenge Exercises (111–114):

Simplify the following using the order of operations:

111. $6^2 + 2^3(4 \cdot 5^2 - 4 \cdot 5) \div 20 + 2$

112. $4^3 + 3 \cdot 5(36 \div 3^2 + 7 \cdot 4) \div 8 - 3^2$

Evaluate each of the following for x = 3, y = 4 *and* z = 6:

113. $x^2y \div 2z + 3y^3 \div xy^2 - z^2$

114. $y^2z^2 \div x^2y \div (2y)(4x^2y)$

Writing Exercises:

115. How do constants and variables differ in meaning?

116. Give three different ways of indicating multiplication.

117. When a number is written with an exponent, what is the meaning of the exponent?

118. Why is the second power often called *squared* and the third power is often called *cubed*?

119. Why is an agreement on order of operations necessary?

Critical-Thinking Exercises:

120. Compare the meaning of multiplication to the meaning of raising a number to a power.

121. Write a word problem that could represent the expression $3 \cdot 2 + 2(20 - 5)$.

Writing Exercises or Group Activities:

If done as a group project, each group should write at least three exercises of each type. Then exchange exercises with another group and solve. If done as writing exercises, each student should both write and solve one of each.

122. Write an application problem similar to Exercises 101–110 that involves at least two operations.

123. Write an application problem similar to Exercises 101–110 that involves at least three operations.

Section 1.2 | # Perimeters of Geometric Figures

OBJECTIVES

When you complete this section, you will be able to:

a. Find the perimeters of geometric figures including triangles, squares, and rectangles.

b. Find the circumference of a circle.

c. Solve application problems involving perimeters of geometric figures.

Introduction   One concept that involves both the evaluation of expressions for specific values of the variable(s) and the order of operations is the evaluation of geometric formulas.

The **perimeter** of a geometric figure is the distance around it. Think of the figure as being made of string. If we cut the string and stretched it out, the total length of the string is the same as the perimeter of the figure.

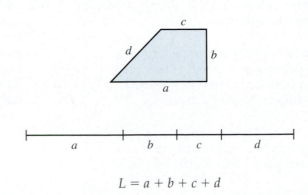

$$L = a + b + c + d$$

Finding the perimeter of geometric figures

Consequently, to find the perimeter of a figure, we find the sum of the lengths of the sides. Since the perimeter is a measure of distance, it is expressed as a linear unit such as inch, foot, mile, centimeter, or meter. If the sides are not given in the same units (e.g., feet, inches), then you must convert to the same unit before doing any calculations.

Be Careful: When substituting into geometric formulas, be sure all the measurements are in the same units. For example, one side cannot be in feet and another in inches.

Perimeter of Geometric Figures

If $a, b, c, \ldots$ are the lengths of the sides of a geometric figure, then the perimeter of the geometric figure is the sum of the lengths of the sides. That is,

$$P = a + b + c + \cdots$$

Example 1

Find the perimeter of each of the following:

a.

To find the perimeter, add the lengths of the sides.
Hence, $P = 3$ in. $+ 5$ in. $+ 4$ in. $+ 4$ in. $= 16$ in.

b.

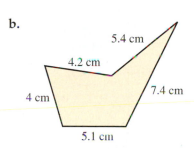

To find the perimeter, add the lengths of the sides. Hence,
$P = 4$ cm $+ 5.1$ cm $+ 7.4$ cm $+ 5.4$ cm $+ 4.2$ cm $= 26.1$ cm.

c. Leave the answer in inches.

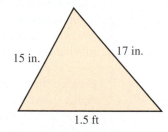

First we need to change all measurements into the same unit.
Let's change 1.5 ft into inches. 1.5 feet $= 1.5(12) = 18$ inches.
To find the perimeter, we add the lengths of all the sides.
Hence, $P = 18$ in. $+ 17$ in. $+ 15$ in. $= 50$ inches.

Practice Exercises

Find the perimeter of each of the following:

1. 9 mm   6 mm   7 mm   7 mm

2. 7.3 in.   5 in.   4.1 in.   3.2 in.   7.6 in.

3. Leave answer in feet.   2 yd   7 ft   8 ft

Answers:

Perimeters of triangles, rectangles, squares, and the circumference of a circle

Some special geometric figures have formulas for finding their perimeter. Following are examples of some common geometric figures, their characteristics, and the formula for calculating the perimeter (circumference, in the case of the circle) of each:

Figure	Characteristics	Formula for Perimeter
Triangle	Three sides of length a, b, and c.	$P = a + b + c$
Rectangle	Four sides, all angles are right angles (90°), opposite sides are equal in length. The two longest sides are usually called the length and the two shortest sides the width.	$P = 2L + 2W$
Square	Four sides of equal length and four right angles. The length of each side is s.	$P = 4s$
Circle	The set of all points in a plane that are an equal distance from a given point in the plane. The given point is called the **center** of the circle and the given distance is the **radius**. The distance around the circle is called the **circumference** rather than the perimeter. A line segment with end points on the circle and passing through the center is called a **diameter**. Hence, $d = 2r$ where d is the diameter and r is the radius.	*Formulas for Circumference* $C = \pi d$ or $C = 2\pi r$ $\pi \approx 3.14$ or $\frac{22}{7}$ The symbol $\approx$ means "is approximately equal to." ***Note:*** Many calculators have a special $\boxed{\pi}$ key. If you use a $\boxed{\pi}$ key, your results may differ slightly from those in the text.

When evaluating geometric formulas, we suggest the following procedure:

Evaluating Geometric Formulas

1. Write the appropriate formula.
2. Substitute the known quantities.
3. Evaluate using the order of operations.

Example 2

Find the perimeter of each of the following:

a.

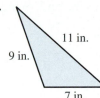

11 in.

9 in.

7 in.

$P = a + b + c$ Formula for the perimeter of a triangle. Substitute for a, b, and c.

$P = 7$ in. $+ 9$ in. $+ 11$ in.   Add.

$P = 27$ in.   Therefore, the perimeter is 27 inches.

b. A rectangle whose length is 2 feet and whose width is 8 inches. Give the answer in inches.

Since the sides are given in different units, we must express both in the same unit. We will use inches in this case. Since 1 foot = 12 inches, 2 feet = 2 · 12 inches = 24 inches.

$P = 2L + 2W$ Formula for the perimeter of a rectangle. Substitute for L and W.

$P = 2(24$ in.$) + 2(8$ in.$)$ Multiply before adding.

$P = 48$ in. $+ 16$ in.   Add.

$P = 64$ in.   Therefore, the perimeter is 64 inches.

c.

3.2 ft

3.2 ft   3.2 ft

3.2 ft

$P = 4s$ Formula for the perimeter of a square. Substitute for s.

$P = 4(3.2$ ft$)$ Multiply.

$P = 12.8$ ft   Therefore, the perimeter is 12.8 feet.

d. Find the circumference of the circle below. Use $\pi \approx 3.14$.

8 ft

$C = \pi d$ Formula for the circumference of a circle. Substitute for π and d.

$C \approx (3.14)(8$ ft$)$ Multiply.

$C \approx 25.12$ ft   Therefore, the circumference is approximately 25.12 feet.

e. Find the circumference of a circle with radius $\frac{9}{5}$ feet. Use $\pi \approx \frac{22}{7}$.

$C = 2\pi r$ Formula for the circumference of a circle. Substitute for π and r.

$C \approx 2\left(\frac{22}{7}\right)\left(\frac{9}{5} \text{ ft}\right)$ Multiply.

$C \approx \frac{396}{35}$ ft   Therefore, the circumference is $\frac{396}{35}$ feet.

Note: In the formula for the circumference of a circle the constant π is used. The ratio of the circumference to the diameter of any circle is π. Since π is a constant, exact answers are often left in terms of π rather than substituting a value and multiplying. The preceding example, Example 2d, could have been done as: $C = \pi d = \pi(8$ ft.$) = 8\pi$ ft and Example 2e would be $C = 2\pi r = 2\pi\left(\frac{9}{5}\right)$ ft $= \frac{18}{5}\pi$ ft.

Practice Exercises

Find the perimeter of each of the following:

4. A triangle whose sides are 2 ft, 18 in., and 15 in.. Give the answer in inches.

5.

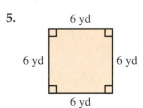

6 yd

6 yd   6 yd

6 yd

6.

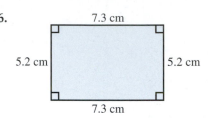

7.3 cm

5.2 cm   5.2 cm

7.3 cm

Find the circumference of the following circles:

7. Diameter of 5.6 yards. Use $\pi \approx 3.14$ and round the answer to the nearest tenth of a yard.

8. Use $\pi \approx \dfrac{22}{7}$.

$\frac{11}{3}$ in.

If more practice is needed, do the Additional Practice Exercises.

Additional Practice Exercises

Find the perimeter of each of the following:

a. A triangle whose sides are 3 inches, 5 inches, and 6 inches.

b.

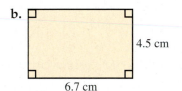

4.5 cm

6.7 cm

c.

6.9 m

6.9 m   6.9 m

6.9 m

Find the circumference of the following circles:

d. Use $\pi \approx 3.14$ and round the answer to the nearest tenth of a meter.

3.4 m

e. Radius of $\frac{5}{3}$ ft. Use $\pi \approx \frac{22}{7}$.

The use of these geometric formulas occurs in many everyday situations. In order to solve these problems, we must determine the appropriate formula, the value(s) of the variable(s), substitute for the variables, and simplify. Often there are additional factors that must be taken into consideration. We illustrate with some examples.

Answers:

d. 10.7 m   e. $\frac{21}{220}$ ft

Practice Exercises 4–8: 4. 57 in.   5. 24 yd   6. 25 cm   7. 17.6 yd   8. $\frac{484}{21}$ in.   **Additional Practice Exercises a–e:** a. 14 in.   b. 22.4 cm   c. 27.6 m

Example 3

Applications involving perimeter

a. An irregularly shaped flower bed has sides whose length are 4 ft, 5 ft, 7 ft, 6 ft, and 8 ft. The flower bed is to be enclosed using landscape timbers. How many feet of timbers are needed?

Solution:
The length of the needed timbers is the same as the perimeter of the flower bed. Since the perimeter is the sum of the lengths of all the sides:

$$P = 4 \text{ ft} + 5 \text{ ft} + 7 \text{ ft} + 6 \text{ ft} + 8 \text{ ft}$$
$$P = 30 \text{ ft} \qquad \text{It would require 30 feet of timbers}$$
to enclose the flower bed.

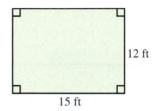

b. A carpenter needs to put molding around the ceiling of a rectangular room whose length is 15 feet and whose width is 12 feet. How many feet of molding does he need? See the figure in the margin.

Solution:
This is the same as finding the perimeter of a rectangle whose length is 15 feet and whose width is 12 feet.

$$P = 2L + 2W \qquad \text{Formula for the perimeter of a rectangle.}$$
$$P = 2(15 \text{ ft}) + 2(12 \text{ ft}) \qquad \text{Substitute for } L \text{ and } W.$$
$$P = 30 \text{ ft} + 24 \text{ ft} \qquad \text{Multiply before adding.}$$
$$P = 54 \text{ ft} \qquad \text{Therefore, the carpenter needs 54 feet of molding.}$$

c. An architect wishes to put a border around a circular fountain whose radius is 3 feet. How much of the border material will she need? See the figure in the margin.

Solution:
This is the same as finding the circumference of a circle whose radius is 3 feet.

$$C = 2\pi r \qquad \text{Formula for the circumference of a circle.}$$
$$C \approx 2(3.14)(3 \text{ ft}) \qquad \text{Substitute for } \pi \text{ and } r.$$
$$C \approx 18.84 \text{ ft} \qquad \text{Therefore, the gardener needs approximately 18.84 feet of border material.}$$

Practice Exercises

Solve each of the following. The figures for some are in the margin.

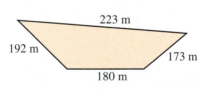

9. A surveyor finds the lengths of the sides of a field with four sides to be 180 meters, 192 meters, 223 meters, and 173 meters. Find the perimeter of the field.

10. A bicycle tire is 26 inches in diameter. How far does the bicycle travel in 10 revolutions of the wheel? Use 3.14 for π.

11. A drainage ditch is dug around a square field each of whose sides is 180 feet long. If Charles is paid $.60 per foot for digging the ditch, how much money does he earn?

Study Tip 4

Reading Mathematics Critically

Mathematics is read differently than other disciplines. In a novel or a story, a whole chapter may contribute only a single idea to the plot. In mathematics, however, every single word or symbol has a precise meaning and each may contribute one or more ideas to the overall subject. Consequently, you must read slowly and carefully so that you understand the meaning of each word and symbol.

You must be able to tell the difference between expressions that look very much alike but have different meanings. For example, "the difference of 8 and 5, written $8 - 5$" is very different from "the difference of 5 and 8, written $5 - 8$."

Read with pencil and paper available. If you do not understand what the author did in getting from one step to the next, take the time to work it out for yourself.

Exercise Set 1.2

Find the perimeter of each of the following:

1.
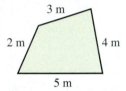
3 m, 2 m, 4 m, 5 m

2.
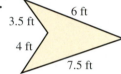
3.5 ft, 6 ft, 4 ft, 7.5 ft

3.
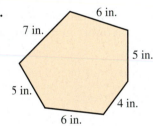
6 in., 7 in., 5 in., 5 in., 6 in., 4 in.

4.
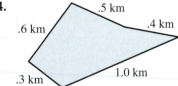
.5 km, .6 km, .4 km, .3 km, 1.0 km

5. Give the answer in inches.

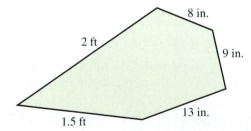
8 in., 2 ft, 9 in., 1.5 ft, 13 in.

6. Give the answer in dm.

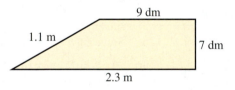
9 dm, 1.1 m, 7 dm, 2.3 m

Find the perimeter of each of the following rectangles:

7.
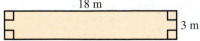
18 m, 3 m

8.
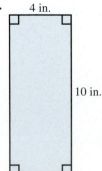
4 in., 10 in.

9.

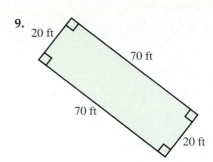

10.

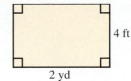

Give answers in terms of feet.

11. Length of 3 yds and width of 5 ft.
Give the answer in feet.

12. Length of 2 ft and width of 19 in.
Give the answer in inches.

Find the perimeter of each of the following squares:

13.

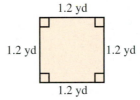

14. Each of whose sides is 9 decimeters long.

15. Each side is 14 cm.

16. Each side is 7 ft.

Find the perimeter of the following triangles:

17.

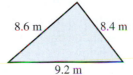

18.

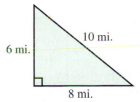

19. Leave the answer in millimeters

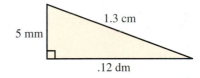

20. Leave answer in inches

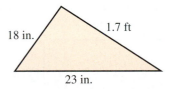

21. A triangle whose sides are 6 inches, 7 inches, and
9 inches in length and whose height to the 6 inches
side is 5 inches.

22. A triangle whose sides are 12 km, 14 km, and 18 km
long, and the height to the 14 km side is 10 km.

Find the exact value of the circumference of each of the following circles. (Leave answers in terms of π.)

23.

24.

Find the circumference of each of the following circles. Use π ≈ 3.14 and round the answers to the nearest tenth of a unit.

25. A circle whose radius is 5.5 feet.

26. A circle with a diameter of 9.8 centimeters.

Find the circumference of each of the following circles. Use $\pi \approx \dfrac{22}{7}$.

27. A circle with a radius of $\dfrac{7}{3}$ in.

28. A circle with a diameter of $\dfrac{3}{5}$ ft.

Challenge Exercises: (29–32)

29. Find the perimeter of the following:

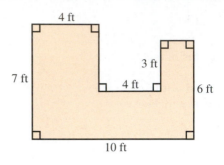

30. Use π ≈ 3.14.

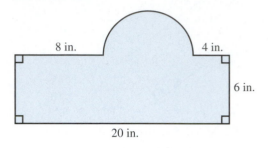

31. Use π ≈ 3.14.

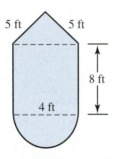

32. Use π ≈ 3.14.

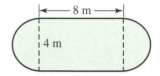

Solve each of the following. If necessary, use π ≈ 3.14 and round your answers to the nearest hundredth of a unit.

33. A carpenter wants to make a square window frame. If the window measures 52 inches on each of the four sides, how much material does he need to make the frame?

34. A rectangular football field measures 120 yards long by 150 feet wide. If the school band is to march all the way around the field, how far will they march?

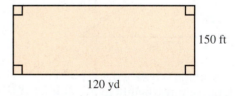

35. A police officer wishes to tape off an irregularly shaped crime scene. From point A to point B is 12 feet, from point B to point C is 15 feet, from point C to point D is 10 feet, from point D to point E is 6 feet, and from point E to point A is 26 feet. How much tape will the officer need to go all the way around the scene?

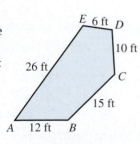

36. Magna has a square garden plot that measures 18 feet on each side. If fencing costs $.50 a linear foot, how much does it cost to fence her garden?

37. A local sandwich shop wishes to put up a sign with a flourescent light border. The shape and size of the sign are given in the diagram below. If the flourescent tubing costs $5.00 per foot, how much does it cost for the border around the sign?

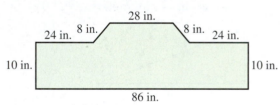

38. A landscape architect is designing a triangular flower bed to be put in the city's new park. If the flower bed is to be 8 yards by 12 yards by 8 yards, how many landscape timbers, each of which is 4 feet long, are needed to go around the flower bed? If the timbers cost $4.95 each, find the cost of the timbers.

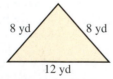

39. A lawn sprinkler rotates on a shaft so that it wets a circular region of the lawn. If the water sprays out a distance of 50 feet from the sprinkler, what is the circumference of the circle of wet grass?

40. Plumbers measure water pipes by their diameter. What is the circumference of a pipe that is 2.5 inches in diameter?

41. The diameter of the bore of the barrels of the largest guns used on a battleship is 16 inches. What is the circumference of the bore of these barrels?

42. The runway at a certain airport is 500 yards long and 60 yards wide. How many runway lights are needed if lights are placed every 20 yards at both edges of the runway?

43. A boat trailer tire has a diameter of 14 inches. How far will the tire roll in five revolutions?

Writing Exercise or Group Project:

If done as a group project, each group should write at least one exercise. The groups should then exchange exercises with other groups and solve the exercise(s) that they receive.

44. Write at least one application problem involving the perimeter of a rectangle.

45. Write at least one application problem involving the circumference of a circle.

Section 1.3	# Areas of Geometric Figures

OBJECTIVES

When you complete this section, you will be able to:

a. Find the areas of rectangles, squares, triangles, parallelograms, trapezoids, and circles.

b. Solve application problems involving area.

Introduction

In the previous section we found the perimeter or distance around a geometric figure. In this section we find the size of the surface of a geometric figure.

Developing the concept of area

The units used for measuring perimeter are linear units that measure distances. Suppose we want to carpet the floor of a room that is 14 feet long and 10 feet wide. We could not measure the amount of carpet needed using linear units since the size of the floor cannot be expressed as a distance. The floor of the room is a surface. The measure of the size of a surface is called its **area**. The units used to measure area are **square units**. A square unit is a square each of whose sides is one linear unit. The following are some examples of square units:

Area units of measure

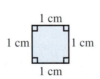

1 square cm (cm^2)

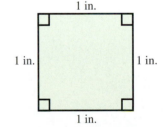

1 square in. (in.^{2})

Instead of writing square cm or square in., we will use the notation cm^2 and in.^{2}. Thus, ft^2 is read "square foot" and m^2 is read "square meter." This is why the exponent "2" is often read as "squared."

In finding the area of a surface, we find the number of square units that are contained within the surface. For example, in the 10 feet by 14 feet room mentioned previously, the area of the floor is the same as the number of square feet of carpet needed to cover the floor. We will develop the formula for the area of a rectangle to reinforce the concept of area and then give the formulas for other geometric figures.

Example 1

Find the area of a rectangle 5 centimeters long and 3 centimeters wide:

Solution:

Area of a rectangle   We begin by drawing the rectangle.

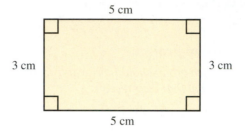

We now draw horizontal and vertical lines one centimeter apart.

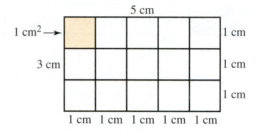

We have created a number of squares within the rectangle. The length of each side of each square is 1 centimeter, so each square is 1 cm^2. Each row contains 5 squares and there are 3 rows of squares. Therefore, there are $5 \cdot 3 = 15$ squares. Consequently, the area of the rectangle is 15 cm^2. Notice that this is the same as multiplying the length (5 cm) times the width (3 cm). Therefore, the area of a rectangle with length L and width W is $A = LW$.

One advantage of the $unit^2$ notation is that we will automatically get the correct unit of measure if we will write the unit as well as the numerical value when performing computations. In the rectangle above, $A = (5 \text{ cm})(3 \text{ cm}) = (5 \cdot 3)(\text{cm} \cdot \text{cm}) = 15 \text{ cm}^2$ (think of cm · cm as cm^2).

In the following chart, we give formulas for the areas of some common geometric figures. As was the case with perimeters, the dimensions must be given in the same units.

Figure	Characteristics	Formula
Rectangle	Four sides, all angles are right angles (90°), opposite sides are equal in length. The two longest sides are usually called the length and the two shortest sides the width.	$A = LW$
Square	Four sides of equal length and four right angles. The length of each side is s.	$A = s^2$

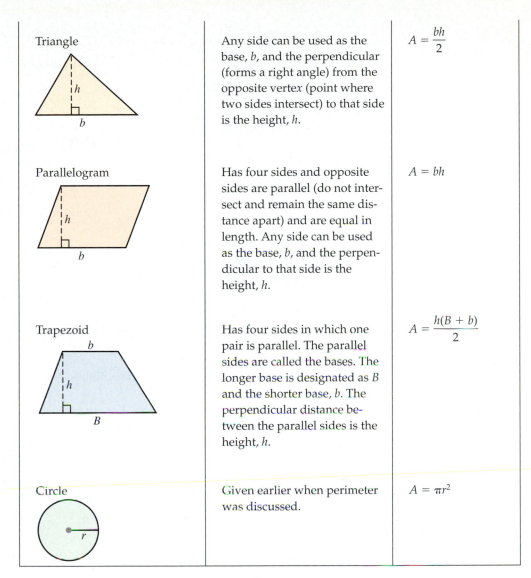

Triangle	Any side can be used as the base, b, and the perpendicular (forms a right angle) from the opposite vertex (point where two sides intersect) to that side is the height, h.	$A = \dfrac{bh}{2}$
Parallelogram	Has four sides and opposite sides are parallel (do not intersect and remain the same distance apart) and are equal in length. Any side can be used as the base, b, and the perpendicular to that side is the height, h.	$A = bh$
Trapezoid	Has four sides in which one pair is parallel. The parallel sides are called the bases. The longer base is designated as B and the shorter base, b. The perpendicular distance between the parallel sides is the height, h.	$A = \dfrac{h(B + b)}{2}$
Circle	Given earlier when perimeter was discussed.	$A = \pi r^2$

We use the same procedure for finding areas that we used in the last section for finding perimeters: 1. write formula, 2. substitute, 3. evaluate.

Example 2

Find the area of each of the following:

a.

$A = LW$ Substitute 4 in. for L and 2 in. for W.

$A = (4 \text{ in.})(2 \text{ in.})$ Multiply.

$A = 8 \text{ in.}^2$ Therefore, the area is 8 square inches.

b.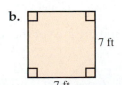

$A = s^2$ Substitute 7 ft for s.

$A = (7 \text{ ft})^2$ Raise to the power.

$A = 49 \text{ ft}^2$ Therefore, the area is 49 square feet.

c.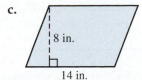

$A = bh$ Substitute for 14 in. for b and 8 in. for h.

$A = (14 \text{ in.})(8 \text{ in.})$ Multiply.

$A = 112 \text{ in.}^2$ Therefore, the area is 112 square inches.

d.

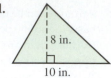

$$A = \frac{bh}{2}$$ Substitute for b and h.

$$A = \frac{(10 \text{ in.})(8 \text{ in.})}{2}$$ Multiply in the numerator.

$$A = \frac{80 \text{ in.}^2}{2}$$ Divide.

$$A = 40 \text{ in.}^2$$ Therefore, the area is 40 square inches.

e.

$$A = \frac{h(B + b)}{2}$$ Substitute for h, B, and b.

$$A = \frac{(3 \text{ in.})(8 \text{ in.} + 6 \text{ in.})}{2}$$ Add inside parentheses.

$$A = \frac{(3 \text{ in.})(14 \text{ in.})}{2}$$ Multiply in the numerator.

$$A = \frac{42 \text{ in.}^2}{2}$$ Divide.

$$A = 21 \text{ in.}^2$$ Therefore, the area is 21 square inches.

f. Circle with diameter of 8.2 inches. Round the answer to the nearest hundredth of an in.^{2}. Use $\pi \approx 3.14$.

Since the diameter is 8.2 inches, the radius is $\dfrac{8.2}{2} = 4.1$ *inches.*

$A = \pi r^2$ Substitute for r.

$A = \pi(4.1 \text{ in.})^2$ Square 4.1 in. and substitute for π.

$A \approx (3.14)(16.81 \text{ in.}^2)$ Multiply.

$A \approx 52.7834 \text{ in.}^2$ Round to the nearest hundredth of an in.^{2}.

$A \approx 52.78 \text{ in.}^2$ Therefore, the area is approximately 52.78 square inches.

Note: The exact answer to Example 2f is $A = \pi r^2 = \pi(4.1)^2 = 16.81\pi \text{ in.}^2$.

Practice Exercises

Find the area of each of the following:

1. Find the area of the following rectangle:

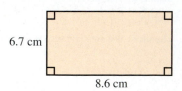

6.7 cm

8.6 cm

2. Find the area of the following square:

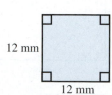

12 mm

12 mm

3. Find the area of the following parallelogram:

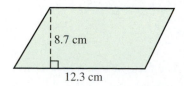

8.7 cm

12.3 cm

4. Find the area of the following triangle:

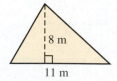

8 m

11 m

5. Find the area of the following trapezoid. Give your answer in square inches.

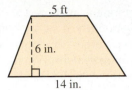

.5 ft

6 in.

14 in.

6. Find the area of the following circle. Use π ≈ 3.14.

10 ft

If more practice is needed, do the Additional Practice Exercises.

Additional Practice Exercises

Find the area of the following rectangle:

a.

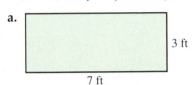

3 ft

7 ft

Find the area of the following parallelogram:

c.

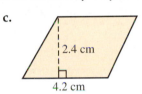

2.4 cm

4.2 cm

Find the area of the following trapezoid:

e.

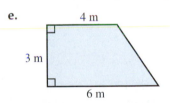

4 m

3 m

6 m

Find the area of the following square:

b.

4.2 cm

Find the area of the following triangle:

d. Base of 4.2 cm and height of 5 cm. Round the answer to the nearest tenth of a cm², if necessary.

Find the area of the following circle:

f. Diameter of 12.8 ft. Round the answer to the nearest hundredth.

Applications problems   There are many applications of areas of geometric figures.

Example 3

a. A piece of wood is in the shape of a rectangle 8 feet long and 6 feet wide. If the wood costs $1.70 per square foot, what is the cost of the piece of wood? See the figure to the right.

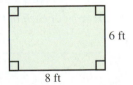

6 ft

8 ft

Strategy: We must first find the number of square feet of wood in the piece. This is the same as the area of a rectangle 6 feet long and 8 feet wide.

Solution: $A = LW$ Substitute for L and W.

$A = (8 \text{ ft})(6 \text{ ft})$ Multiply.

$A = 48 \text{ ft}^2$ Therefore, there are 48 square feet of wood.

Since the wood costs $1.70 per square foot and there are 48 ft², multiply 48 times $1.70. Therefore, cost = (48)(1.70) = $81.60.

Answers:

Practice Exercises 5–6: 5. 60 in.² 6. 78.5 ft²   **Additional Practice Exercises a–f:** a. 21 ft² b. 17.64 cm² c. 10.08 cm² d. 10.5 cm² e. 15 m² f. 128.61 ft²

b. An irrigation pipe 150 feet long is anchored at one end and the other end is mounted on wheels. Consequently, it sweeps out a circle when in operation. Suppose the field is planted in potatoes. The potatoes can be fertilized by putting liquid fertilizer into the irrigation system. When diluted, one gallon of fertilizer will fertilize 900 ft² of potatoes and costs $8.50. How much does it cost to fertilize the field? See the figure in the margin:

Strategy:

We must determine how many gallons of fertilizer are needed to fertilize a circular field 150 feet in radius. Since one gallon will fertilize 900 ft², the number of gallons needed is equal to the area of the field divided by 900. Therefore, we must first determine the area of the field.

Solution: $A = \pi r^2$ Substitute for π and r.

$A \approx (3.14)(150 \text{ ft})^2$ Square 150 feet.

$A \approx (3.14)(22{,}500 \text{ ft}^2)$ Multiply.

$A \approx 70{,}650 \text{ ft}^2$ Therefore, the area of the field is approximately 70,650 square feet.

Since one gallon will fertilize 900 ft², the number of gallons of fertilizer needed is $70650 \div 900 = 78.5$ gallons. Each gallon costs $8.50, so the cost of the fertilizer = $(78.5)(8.50) = \$667.25$.

Practice Exercises

Solve each of the following. The figures are in the margin.

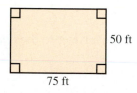

7. How many square feet of sod will it take to cover a rectangular yard 75 feet long and 50 feet wide?

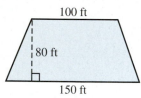

8. A field is in the shape of a trapezoid whose parallel sides are 150 feet and 100 feet long. The height of the trapezoid is 80 feet. A bag of grass seed costs $6.00 and will plant 500 square feet. How much will the seed to plant the field cost?

Exercise Set 1.3

Find the area of each of the following rectangles:

1.

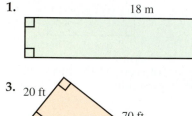

2.

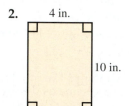

3.

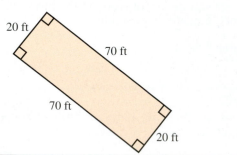

4.

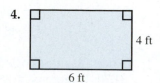

Answers:

5. Length of 3 yds and width of 5 ft. Give answer in square feet.

6. Length of 2 ft and width of 19 in. Give answer in square inches.

Find the area of each of the following squares:

7.

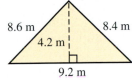

1.2 yd
1.2 yd
1.2 yd
1.2 yd

8. Each of whose sides is 9 decimeters long.

9. Each side is $\frac{5}{7}$ cm.

10. Each side is $\frac{7}{2}$ ft.

Find the area of the following triangles:

11.

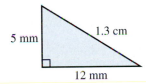

8.6 m
8.4 m
4.2 m
9.2 m

12.

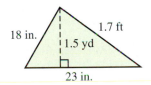

6 mi.
10 mi.
8 mi.

13. Leave the answer in mm².

5 mm
1.3 cm
12 mm

14. Leave answer in in.².

18 in.
1.7 ft
1.5 yd
23 in.

15. A triangle whose sides are 6 inches, 7 inches, and 9 inches in length, and whose height to the 6 inches side is 5 inches.

16. A triangle whose sides are 12 km, 14 km, and 18 km long, and whose height to the 14 km side is 10 km.

Find the exact value of the area of each of the following circles. (Leave answers in terms of π.)

17.

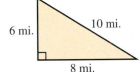

7 m

18.

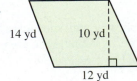

12 ft

Find the area of each of the following circles. Use π ≈ 3.14 and round the answers to the nearest tenth of a unit.

19. A circle with a radius of 5.5 feet.

20. A circle with a diameter of 9.8 centimeters.

Find the area of each of the following circles. Use $\pi \approx \frac{22}{7}$.

21. A circle with a radius of $\frac{7}{3}$ in.

22. A circle with a diameter of $\frac{3}{5}$ ft.

Find the area of each of the following parallelograms:

23.

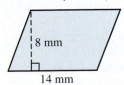

8 mm
14 mm

24.

14 yd
10 yd
12 yd

25.

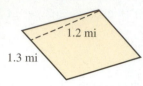

1.2 mi

1.3 mi

26.

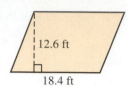

12.6 ft

18.4 ft

27. A parallelogram whose base is 31 mm and whose height is 2.3 cm. Leave answer in cm^2.

28. A parallelogram whose base is 7 ft and whose height is 2 yd. Leave answer in ft^2.

Find the areas of the following trapezoids:

29.

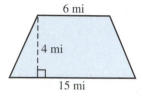

6 mi

4 mi

15 mi

30.

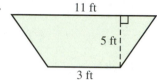

11 ft

5 ft

3 ft

31. Leave answer in square inches.

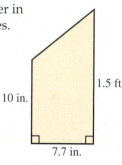

1.5 ft

10 in.

7.7 in.

32. A trapezoid with a large base of 11.3 centimeters, a small base of 9.7 cm, and an altitude of 6.5 cm.

Challenge Exercises: (33–35)

Find the area of each of the following:

33.

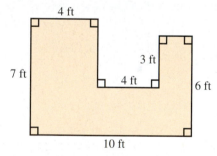

4 ft

7 ft

3 ft

4 ft

6 ft

10 ft

34. Use $\pi = 3.14$.

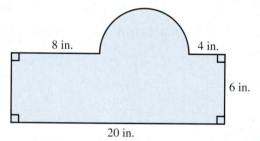

8 in.

4 in.

6 in.

20 in.

35. Find the area inside the square and outside the circle. Use $\pi \approx 3.14$.

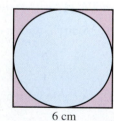

6 cm

Solve the following:

36. The floor of a rectangular shower stall measures 52 inches long and 32 inches wide. What is the area of the floor of the shower stall?

37. A certain type of plant needs 1 square yard of ground to grow in. If a field is 200 yards long and 150 yards wide, how many plants can be grown in the field? If each plant costs $2.95, find the cost of the plants.

38. A circular fountain has a diameter of 10 meters. How many square-shaped tiles are needed to cover the bottom of the fountain if each tile is 1 square decimeter? Assume no waste. If each tile costs $.69, find the cost of the tiles.

39. A university wants to install artificial turf in its football stadium. The football field, including the end zones, is rectangular and measures 150 yards long and 120 yards wide. How many square yards of artificial turf are needed?

40. The airport authority wants to design a triangular caution sign. How many square centimeters of reflective material will be needed for the sign if the height of the triangle is 90 centimeters and the base is 100 centimeters?

41. A business concern wants to build a decorative wall in the shape of a trapezoid that measures 25 feet on the top base and 40 feet on the bottom base. If the wall is 8 feet high, how many 1-foot square ceramic pieces of tile are necessary to cover the wall? Assume no waste. If each piece of tile costs $2.59, find the cost of the tiles.

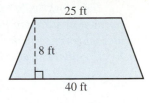

42. Acoustic ceiling tile with an area of 1 square foot costs $.49 a tile. Find the cost of the tile for an acoustical ceiling measuring 20 feet long by 17 feet wide?

43. A family wishes to put an addition on their house. The roof of the addition is rectangular and measures 25 feet long and 12 feet wide. If roofing costs $3.50 a square foot, how much will the material for the new roof cost?

Challenge Exercise:

44. An artist wishes to make one wall of his studio out of glass brick to capture the natural light. The wall measures 26 feet long and 12 feet high. If the glass blocks are 6-inch squares and cost $1.25 each, how much will the blocks for the wall cost?

Writing Exercise or Group Project:

If done as a group project, each group should write at least one exercise. The groups should then exchange exercises with other groups and solve the exercise(s) that they receive.

45. Write at least one application problem involving the area of a rectangle.

46. Write at least one application problem involving the area of a circle.

Group Project:

47. Look around your campus or local community and find examples of at least five of the geometric figures that we studied in this section. Find the area of each.

Section 1.4	**Volumes and Surface Areas of Geometric Figures**

OBJECTIVES   *When you complete this section, you will be able to:*

a. Find the volumes of rectangular solids, cubes, right circular cylinders, and cones.

b. Find the surface area of rectangular solids, cubes, and right circular cylinders.

c. Solve application problems involving volumes and surface areas.

Introduction   In the previous sections we found the perimeter and area of two-dimensional figures. In this section, we concentrate on three-dimensional figures.

Finding volumes of geometric solids   If we wanted to find how much cypress mulch it would take to fill a box 4 feet long, 2 feet wide, and 3 feet high, we could not use linear units since we are not finding a distance, and we could not use square units since we are not finding the size of a surface. In order to find the volume of a solid object, we must use another type of unit called a **cubic unit.**

The word "cube" is often misused in everyday English. Have you ever seen an ice cube that is really a cube?

Imagine a box, including the top, each of whose faces is a square. This type of figure is a **cube.** A cube is a three-dimensional geometric solid with six faces each of which is a square. A cubic unit is a cube each of whose faces is a square whose sides are one unit each. For example, the following figure is a cubic inch (in.³).

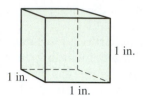

When finding the **volume** of a solid, we are finding the number of cubic units it takes to fill the solid. This may be a little confusing since the volumes we encounter in everyday situations are usually liquid measures of volume like ounces and gallons.

Volume of a rectangular solid

A **rectangular solid** is a geometric figure with six faces, each of which is a rectangle. A box is an example of a rectangular solid. See the following figure:

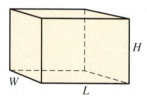

We will develop the formula for the volume of a rectangular solid and then give the formulas for the volumes of other types of solids. Let us find the volume of a rectangular solid that is 3 feet long, 4 feet wide, and 2 feet high. We are finding the number of cubic feet necessary to fill the solid. First we calculate how many cubic feet (ft³) it would take to cover the bottom of the rectangular solid.

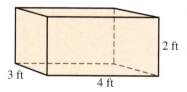

We mark off units of one foot each along the length and width of the base of the solid and draw horizontal lines at these units.

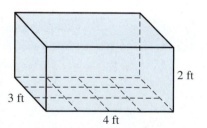

The lines divide the base into 3 · 4 = 12 unit squares each of whose sides is one foot. Since each face of a cube is a square, we can place one cubic foot on top of each of these squares forming one layer of 12 cubic feet on the base.

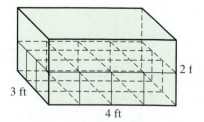

Since the height is 2 feet, it would take another layer of cubes identical to this layer to completely fill the rectangular solid. Therefore, it would take $12 \cdot 2 = 24$ ft³ to fill the rectangular solid. Notice the length (3) times the width (4) gave us the number of cubes in the first layer (12). The height (2) gave us the number of layers. Therefore, the number of cubic feet in the solid is $12 \cdot 2 = 24$ ft³. Therefore, the volume is $3 \cdot 4 \cdot 2 = 24$ ft³. Consequently, the volume of a rectangular solid is $V = LWH$. Following are some common geometric solids and the formulas for their volumes:

Type of Solid	**Characteristics**	**Formula**
Rectangular Solid	Six faces all of which are rectangles.	$V = LWH$
Cube	A rectangular solid in which each face is a square. The intersection of any two faces is called an **edge** and is denoted by e. All edges are of the same length.	$V = e^3$
Right Circular Cylinder	Looks like a tin can. The top and bottom are circles and the "sides" are perpendicular to the top and bottom. The radius of the base is denoted by r and the distance between the top and bottom is the height, h.	$V = \pi r^2 h$ Note that this is the area of the base times the height.
Cone	The bottom, called the **base**, is a circle and the figure comes to a single point at the top, called the vertex. The radius of the base is denoted by r. The perpendicular distance between the vertex and the base is called the *height* and is denoted by h.	$V = \dfrac{1}{3}\pi r^2 h$ or $\dfrac{\pi r^2 h}{3}$

Example 1

Find the volumes of the following geometric solids. If necessary, round answers to the nearest hundredth.

a.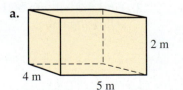

4 m   5 m   2 m

Solution:

$V = LWH$ Formula for the volume of a rectangular solid. Substitute for L, W, and H.

$V = (5\text{ m})(4\text{ m})(2\text{ m})$ Multiply.

$V = 40\text{ m}^3$ Therefore, the volume is 40 m³.

b. Find the volume of the following cube:

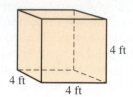

$V = e^3$	Formula for the volume of a cube. Substitute for e.
$V = (4 \text{ ft})^3$	Raise to the power.
$V = 64 \text{ ft}^3$	Therefore, the volume is 64 cubic feet.

Find the volumes of the following right circular cylinders. Use $\pi \approx 3.14$.

c.

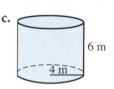

$V = \pi r^2 h$	Substitute for π, r, and h.
$V \approx (3.14)(4 \text{ m})^2(6 \text{ m})$	Raise to powers first.
$V \approx (3.14)(16 \text{ m}^2)(6 \text{ m})$	Multiply.
$V \approx 301.44 \text{ m}^3$	Therefore, the volume is approximately 301.44 cubic meters.

d. Radius of the base is 2.1 meters and the height is 3.2 meters. Round the answer to the nearest hundredth of a cubic meter.

$V = \pi r^2 h$	Substitute for π, r, and h.
$V \approx (3.14)(2.1 \text{ m})^2(3.2 \text{ m})$	Raise to powers first.
$V \approx (3.14)(4.41 \text{ m}^2)(3.2 \text{ m})$	Multiply.
$V \approx 44.31168 \text{ m}^2$	Round to the nearest hundredth.
$V \approx 44.31 \text{ m}^3$	Therefore, the volume is 44.31 cubic meters to the nearest hundredth.

Note: Since π has a constant value, we often leave the results in terms of π rather than substituting and multiplying out. For example, we could have done Examples 1c and d as:

$$V = \pi r^2 h = \pi(4 \text{ m})^2(6 \text{ m}) = \pi(16 \text{ m}^2)(6 \text{ m}) = 96\pi \text{ m}^3.$$
$$V = \pi r^2 h = \pi(2.1 \text{ m})^2(3.2 \text{ m}) = \pi(4.41 \text{ m}^2)(3.2 \text{ m}) = 14.112\pi \text{ m}^3.$$

e. Find the volume of the following cone. Leave answer in terms of π.

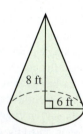

$V = \dfrac{\pi r^2 h}{3}$	Substitute for r and h.
$V = \dfrac{\pi(6 \text{ ft})^2(8 \text{ ft})}{3}$	Raise to powers first.
$V = \dfrac{\pi(36 \text{ ft}^2)(8 \text{ ft})}{3}$	Multiply in the numerator.
$V = \dfrac{\pi(288 \text{ ft}^3)}{3}$	Divide.
$V = 96\pi \text{ ft}^3$	Therefore, the volume is 96π cubic feet.

Practice Exercises

Find the volume of the following geometric solids:

1. Rectangular solid with length of 13 mm, width of 2 mm, and height of 24 mm.

2. Find the volume of the following cube:

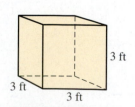

Answers:

3. Right circular cylinder for which radius of the base is $\frac{3}{2}$ cm and the height is $\frac{11}{4}$ cm. Use $\pi \approx \frac{22}{7}$.

4. Find the volume of the following cone: Leave answer in terms of π.

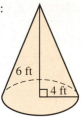

If more practice is needed, do the Additional Practice Exercises.

Additional Practice Exercises

Find the volume of the following rectangular solid:

a. Length of 8 in., width of 4 in., and height of 5 in.

Find the volume of the following cube:

b.

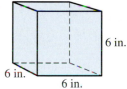

6 in.

6 in.

6 in.

Find the volume of the following right circular cylinder:

c. Radius of the base is 5.1 ft and the height is 2.4 ft. Leave answer in terms of π.

Find the volume of the following cone:

d. Leave answer in terms of π.

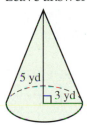

Three-dimensional figures are made up of surfaces that often are of the types that we studied in Section 1.3. For example, a rectangular solid has six surfaces all of which are rectangles and a cube has six surfaces all of which are squares. When we find the areas of all these surfaces, we have found the **surface area** of the solid. Following are some of the solids for which we previously found the volume, along with the formulas for their surface areas.

Type of Solid	Rectangular Solid	Cube	Right Circular Cylinder
	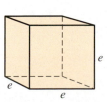	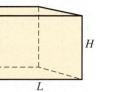	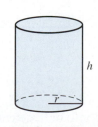
Formula for Surface Area	$SA = 2LW + 2LH + 2WH$	$SA = 6e^2$	$SA = 2\pi r^2 + 2\pi rh$

Example 2

Find the surface area of each of the following:

a.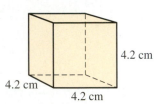

3 in.

8 in.

6 in.

Solution:

$SA = 2LW + 2LH + 2WH$ Substitute.

$SA = 2(8 \text{ in.})(6 \text{ in.}) + 2(8 \text{ in.})(3 \text{ in.}) + 2(6 \text{ in.})(3 \text{ in.})$ Multiply.

$SA = 96 \text{ in.}^2 + 48 \text{ in.}^2 + 36 \text{ in.}^2$ Add.

$SA = 180 \text{ in.}^2$ Answer.

b.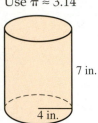

4.2 cm

4.2 cm

4.2 cm

Solution:

$SA = 6e^2$ Substitute.

$SA = 6(4.2 \text{ cm})^2$ Evaluate $(4.2)^2$.

$SA = 6(17.64 \text{ cm}^2)$ Multiply.

$SA = 105.84 \text{ cm}^2$ Answer.

c. Use $\pi \approx 3.14$

7 in.

4 in.

Solution:

$SA = 2\pi r^2 + 2\pi rh$ Substitute.

$SA \approx 2(3.14)(4 \text{ in.})^2 + 2(3.14)(4 \text{ in.})(7 \text{ in.})$ Square 4.

$SA \approx 2(3.14)(16 \text{ in.}^2) + 2(3.14)(4 \text{ in.})(7 \text{ in.})$ Multiply.

$SA \approx 100.48 \text{ in.}^2 + 175.84 \text{ in.}^2$ Add.

$SA \approx 276.32 \text{ in.}^2$ Answer.

Practice Exercises

Find the surface area of each of the following:

5.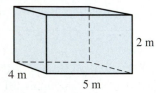

2 m

4 m

5 m

6.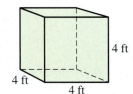

4 ft

4 ft

4 ft

7. Leave answer in terms of π.

6 m

5 m

Exercise Set 1.4

Find the volume and surface area of the following rectangular solids:

1.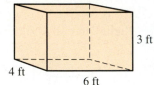

3 ft

4 ft

6 ft

2.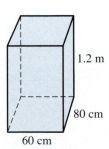

1.2 m

80 cm

60 cm

3. Length of 8 meters, width of 5.2 meters, and height of 6.5 meters.

4. Length of 8 feet, width of 7 feet, and height of 4 yards.

Find the volume and surface area of the following cubes:

5.

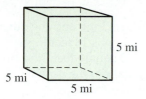

5 mi
5 mi
5 mi

6. Length of each edge is 2.3 kilometers.

7. Cube each of whose edges is $\frac{3}{5}$ cm.

8. Cube each of whose edges is $\frac{4}{7}$ ft.

Find the volume and surface area of the following right circular cylinders. Leave answers in terms of π.

9.

12 cm

5 cm

10.

8 ft

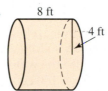

4 ft

11. Radius of the base is 4.2 meters and the height is 6 meters.

12. Radius of the base is 3 centimeters and the height is 8.5 centimeters.

Find the volumes of the following cones. Round the answer(s) to the nearest tenth.

13. Use π ≈ 3.14.

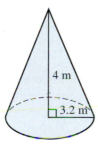

4 m

3.2 m

14. Leave answer in terms of π.

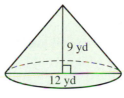

9 yd

12 yd

Challenge Exercises: 15–19

Find the volumes of the following cones:

15. Base radius of $\frac{5}{7}$ in. and height of 14 in. Use $\pi \approx \frac{22}{7}$.

16. Base diameter of 14 ft and height of $\frac{23}{11}$ ft. Use $\pi \approx \frac{22}{7}$.

Find the volume of the following figure. Leave answer in terms of π.

17.

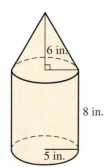

6 in.

8 in.

5 in.

18. A sphere is a solid in the shape of a ball. The formula for the volume of a sphere is $V = \frac{4}{3}\pi r^3$. Find the volume of a sphere whose radius is 6 inches. Leave the answer in terms of π.

19. Find the volume of the following propane tank. Leave the answer in terms of π.

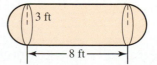

3 ft

8 ft

Answer the following:

20. A cereal box is 11 inches high, 7 inches long, and 2 inches wide. How many cubic inches of cereal can it hold?

21. The body of a dump truck used for hauling dirt and sand is in the shape of a rectangular solid and measures 8 feet by 12 feet by 4 feet. How many cubic yards of sand will it hold? Round answer to the nearest hundredth of a cubic yard.

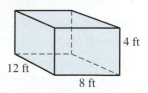

4 ft
12 ft
8 ft

22. A building contractor wishes to put a basement in a new office building on campus. The basement is to measure 12 feet deep, and the building is to measure 90 feet long by 40 feet wide. How many cubic yards of dirt must be removed before the basement can be built?

23. A storage tank is in the shape of a right circular cylinder that has a base diameter of 42 inches and a height of 2 yards. Find the volume in cubic inches and leave the answer in terms of π.

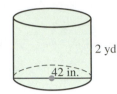

2 yd

42 in.

24. John and Kay wish to fill their swimming pool. Their pool is shaped like a rectangular solid measuring 15 feet wide by 40 feet long with the water at a depth of 6 feet. If water costs $.75 per 100 cubic feet, how much would it cost to fill the pool?

25. Alfredo is estimating the cost of an addition to his home. He will build two rectangular-shaped rooms with a combined area that measures 60 feet by 18 feet. If the concrete slab (foundation) must be 8 inches thick and the concrete costs $27.00 a cubic yard, how much will the concrete for the slab cost?

26. It costs the Department of Transportation $.43 per square foot for the material used in making traffic signs. Each sign is rectangular with a length of 2 feet and a width of 1.5 feet. Find the cost of the material to make 400 such signs.

27. A room is 12 feet long and 18 feet wide. If carpet costs $18.95 per square yard, find the cost of the carpet needed to carpet the floor of the room.

28. A carpenter is building a wooden chest in the shape of a rectangular solid. The chest is to be 2 feet wide, 3 feet long and 1.5 feet high. The wood costs $5.50 per square foot. Find the cost of the wood needed to build the chest.

29. A manufacturer of cardboard boxes has received an order for 500 boxes that are to be 18 inches wide, 24 inches long and 20 inches high. How much

cardboard will she need in order to manufacture these boxes?

30. Two walls of a room are 10 feet long and the other two are 16 feet long. The ceiling is 12 feet high. If a gallon of paint will cover 500 square feet, find (to the nearest hundredth of a gallon) the number of gallons it will take to paint the room.

31. Two walls of a room are 12 feet long and the other two are 15 feet long. The ceiling is 10 feet high. If it costs $.80 per square foot to plaster the walls, find the cost of plastering the walls and ceiling.

32. As part of a mathematics exhibit at the science center, a cube is constructed so that the length of each edge is 15 inches. The material used in making the cube costs $.12 per square inch. Find the cost of the material.

33. A cardboard box is in the shape of a cube with each edge 2 feet long. The box is to be wrapped with wrapping paper that costs $.35 per square foot. Find the cost of the paper.

34. A container of raisins is in the shape of a right circular cyclinder with a radius of 2 inches and height of 6 inches. What is the surface area of the container? Use $\pi \approx 3.14$.

35. A can of premium chunk white chicken is in the shape of a right circular cyclinder and is 4 inches in diameter and 2 inches high. What is the surface area of the can? Use $\pi \approx 3.14$.

36. A cereal box is in the shape of a right circular cyclinder and has a radius of 3 inches and a height of 10 inches. The cardboard used to make the box costs $.0002 per square inch. Find the cost of the material to the nearest cent. Use $\pi \approx 3.14$.

37. A manufacturer of cans receives an order for 10,000 cans with a radius of 5 cm and a height of 15 cm. If the material used in constructing the cans costs $.0002 per cm^2, find the cost of the material used in filling the order. Use $\pi \approx 3.14$.

Challenge Exercise:

38. A gasoline storage tank at a service station is in the shape of a right circular cylinder with a diameter of 6 feet and a height of 12 feet. If there are 231 cubic inches in a gallon and the gasoline costs the service station owner $.89 per gallon, how much will it cost to fill the tank? Use $\pi \approx 3.14$ and round the number of gallons to the nearest whole gallon.

Writing Exercises or Group Projects:

If done as a group project, each group should write at least one exercise of each type. Then exchange exercises with another group and solve. If done as writing exercises, each student should both write and solve one of each type.

39. Write an application problem involving the volume of a rectangular solid.

40. Write an application problem involving the volume of a right circular cyclinder.

41. Write an application problem involving the surface area of a rectangular solid.

42. Write an application problem involving the surface area of a right circular cyclinder.

Group Project:

43. Look around your campus or local community and find examples of at least five of the geometric figures that we studied in this section. Find the volume and surface area (except for a cone) of each.

| Section 1.5 | Introduction to Integers |

OBJECTIVES

When you complete this section, you will be able to:

a. Identify natural numbers, whole numbers, and integers.

b. Graph natural numbers, whole numbers, and integers.

c. Determine order relations for integers.

d. Find the negative (opposite) of an integer.

e. Find the absolute value of an integer.

Introduction

A **set** is any collection of objects and the objects that make up the set are called the **elements** of the set. In mathematics, the sets that we are most often concerned with are sets whose elements are numbers. Sets are usually named by using a capital letter and are indicated by enclosing its elements in braces and seperated by commas. If the elements of a set are listed individually, then the set is said to be listed by **roster.** For example, {a, b, c} is read "The set whose elements are a, b, and c." and the elements are listed by roster. If all the elements of set A are also elements of set B, then set A is a **subset** of set B. For example, the set {1, 2} is a subset of the set {1, 2, 3, 4 }.

Types of numbers

Historically, numbers have been invented as they were needed. Our ancestors probably kept track of "how many" by using tally marks or stones, one for each object being "counted." No doubt the first numbers needed were those that indicated how many. We call these the **natural** or **counting** numbers.

Set of Natural Numbers

$N = \{1, 2, 3, 4, 5, \ldots\}$

If the set of natural numbers is expanded to include 0, we get another set of numbers called the **whole numbers.** Note that the natural numbers are a subset of the whole numbers since all the elements of the set of natural numbers are also elements of the set of whole numbers.

Set of Whole Numbers

$W = \{0, 1, 2, 3, 4, \ldots\}$

If limited to just the whole numbers, we would have no way of representing 10° below zero or a loss of $150. Consequently, **negative numbers** were invented to help us make these representations. If $+10°$ represents 10° above 0° then $-10°$ represents 10° below 0°. Likewise, if $+45$ feet represents 45 feet above sea level, then -45 feet would represent 45 feet below sea level.

Practice Exercises

1. If +3 yards represents a gain of 3 yards, then _____ represents a loss of 3 yards.

2. If −$100 represents a withdrawal of $100, then _____ represents a deposit of $100.

3. If +12° represents a rise of 12° in temperature, then _____ represents a drop of 12° in temperature.

If additional practice is needed, do the Additional Practice Exercises.

Additional Practice Exercises

a. If +23 represents going up 23 floors in an elevator, how would you represent going down 23 floors?

b. If +$250 means receiving $250, what would −$250 represent?

c. If +33 represents going 33 miles north, how would you represent going 33 miles south?

If the set of whole numbers is expanded to include the negatives of the natural numbers, we get a new set of numbers called the **integers.** Note that both the natural numbers and the whole numbers are subsets of the integers.

Set of Integers
$I = \{ \ldots, -3, -2, -1, 0, +1, +2, +3, \ldots \}$

Positive and negative numbers are often referred to as signed numbers, with 0 being neither positive nor negative. Another way of representing the integers is by using the number line. The number line is like a thermometer that is horizontal instead of vertical. To construct a number line, draw a horizontal line and choose an arbitrary point which is labeled "0." Choose a unit of length and mark off the line both to the left and right of 0 in terms of this unit. Label the units to the right of 0 with the positive integers and the units to the left with the negative integers. Positive numbers are usually written without the + sign. For example, +3 is written as 3 and +6 is written as 6. The resulting number line should look something like this.

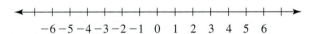

Graphing the integers   In this manner any integer can be paired with a point on the number line. This procedure is called **graphing.**

Example 1

Graph the integers {−4, −2, 0, 3}.

We put a dot on the number line at the location of each of the integers −4, −2, 0, 3.

Answers:

Practice Exercises 1–3: **1.** −3 yds **2.** +$100 **3.** −12°   *Additional Practice Exercises a–c:* **a.** −23 **b.** giving $250 **c.** −33

Practice Exercises

Graph the following sets of integers:

4. {−5, −2, 3, 6}

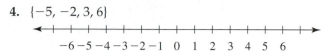

$$-6\ -5\ -4\ -3\ -2\ -1\ \ 0\ \ 1\ \ 2\ \ 3\ \ 4\ \ 5\ \ 6$$

5. {−4, −3, −1, 1, 2, 3}

$$-6\ -5\ -4\ -3\ -2\ -1\ \ 0\ \ 1\ \ 2\ \ 3\ \ 4\ \ 5\ \ 6$$

If more practice is needed, do the Additional Practice Exercises.

Additional Practice Exercises

Draw a number line and graph each of the following:

d. {−3, −1, 0, 2, 4}

e. {−4, −2, 3, 5, 6}

Note: There are points on the number line that do not have integers assigned to them. Consequently, other types of numbers, other than decimals and fractions, will be needed in order to assign a number to every point on the number line. These types of numbers will be discussed later in the text.

The integers possess the property of being **ordered**. By this, we mean that when given two integers, it is possible to determine whether they are equal or which is the larger or smaller of the two.

To show the order relationships of numbers, we use the following symbols:

Order Symbols

$a = b$ Read "*a* is equal to *b*." This means that the number represented by *a* has the same value as the number represented by *b*. For example, $5 = 5$.

$a \neq b$ Read "*a* is not equal to *b*." This means that the number represented by *a* does not have the same value as the number represented by *b*. For example, $-3 \neq 4$.

$a < b$ Read "*a* is less than *b*." This means that the number represented by *a* is to the left of the number represented by *b* on the number line. For example, $2 < 3$ because 2 is to the left of 3 on the number line.

$a \leq b$ Read "*a* is less than or equal to *b*." This means that the number represented by *a* is either less than the number represented by *b* or the number represented by *a* is equal to the number represented by *b*. For example: $3 \leq 4, 5 \leq 5$.

$a > b$ Read "*a* is greater than *b*." This means that the number represented by *a* is to the right of the number represented by *b* on the number line. For example, $5 > 1$ because 5 is to the right of 1 on the number line.

$a \geq b$ Read "*a* is greater than or equal to *b*." This means that the number represented by *a* is either greater than the number represented by *b* or the number represented by *a* is equal to the number represented by *b*. For example: $4 \geq 2, 7 \geq 7$.

Remember, the smaller number is always to the left of the larger on the number line. Or equivalently, the larger is always to the right of the smaller.

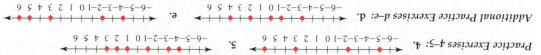

Answers:

Additional Practice Exercises d–e: d.

e.

Practice Exercises 4–5: 4.

5.

Look back at the graph of Example 1.

Example 2

Determining order relations for integers

a. 0 is to the left of 3. Therefore, $0 < 3$ or equivalently $3 > 0$. We could also write $0 \leq 3$ or $3 \geq 0$.

b. -2 is to the left of 0. Therefore, $-2 < 0$ or $0 > -2$.

c. -4 is to the left of -2. Therefore, $-4 < -2$.

Practice Exercises

Insert the proper symbol (=, <, or >) in order to make each of the following true:

6. a. -2 _____ 3   **b.** 4 _____ -5 **c.** -5 _____ -8 **d.** 0 _____ -6

  e. 8 _____ 8   **f.** 4 _____ $2 + 5$ **g.** -9 _____ -5

From the following diagram, you can see that each integer, except 0, can be paired with another integer the same distance from 0, but on the opposite side. Each such integer is called the **opposite** or **negative** of the other. Zero is its own opposite.

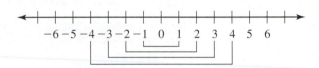

The negative of 3 is -3, the negative of -4 is 4, and so on. The negative of a number is usually indicated by placing a negative sign before the number. Consequently, the negative of -4 is written $-(-4)$ and equals 4 since 4 is the negative or opposite of -4. Likewise, $-(-6) = 6$ and $-(-2) = 2$. This leads us to an important characteristic of signed numbers.

> **Observation**
>
> $-(-x) = x$ for all values of x

Another important concept associated with signed numbers is that of **absolute value**. The absolute value of a number is defined as the distance from 0 to the number. Absolute value is indicated by placing the number whose absolute value you wish to find between vertical bars. Since absolute value is a distance, it can never be negative.

Example 3

Find the indicated absolute values.

a. $|4| = 4$ The distance from 0 to 4 is 4.   **b.** $|-4| = 4$ The distance from 0 to -4 is 4.

c. $|0| = 0$ The distance from 0 to 0 is 0.   **d.** $|-8| = 8$ The distance from 0 to -8 is 8.

e. $|8| = 8$ The distance from 0 to 8 is 8.

Answers:

As stated earlier, the absolute value of a number is never negative. As long as we are finding absolute values of constants, there is no problem because it is obvious whether the constant is positive, negative, or zero. The difficulty occurs when we are asked to represent the absolute values of variables. Since a variable may be replaced by any number, the variable x could represent either a positive number, a negative number, or 0. Consequently, $x < 0$, $x > 0$, or $x = 0$ depending on what we choose to substitute for x. Therefore, x is not necessarily positive nor is $-x$ necessarily negative. If x is positive, then $-x$ is negative; in symbols, if $x > 0$, then $-x < 0$. In words, the opposite of a positive number is a negative number. If x is negative, then $-x$ is positive; that is, if $x < 0$, $-x > 0$. In words, the opposite of a negative number is a positive number. For example, if $x = -3$, then $-x = -(-3) = 3$. So when asked to represent $|x|$, we have to consider three cases; one case when $x > 0$, another when $x < 0$, and another when $x = 0$. The case where $x = 0$ is usually combined with the case where $x > 0$.

Absolute Value of x

a. If $x \geq 0$, $|x| = x$.

b. If $x < 0$, $|x| = -x$.

In words, if x is positive or 0, then x is its own absolute value. If x is negative, the opposite of x is its absolute value.

In words this says that the absolute value of a positive number is the same positive number ($|6| = 6$) and the absolute value of a negative number is the negative or opposite of that negative number ($|-5| = -(-5) = 5$).

Example 4

Find the following using the definition of absolute value:

a. $|2| = 2$ Since $2 \geq 0$, $|x| = x$.

b. $|-8| = -(-8) = 8$ Since $-8 < 0$, $|x| = -x$.

c. $|0| = 0$ Since $0 \geq 0$, $|x| = x$.

d. $|n| = \begin{cases} n \text{ for } n \geq 0 \\ -n \text{ for } n < 0 \end{cases}$ We must include all possibilities since we do not know if n is positive, negative, or zero.

e. $-|10| = -10$ The "−" is outside the absolute value marks, so first find the absolute value of 10 and then take its negative.

f. $-|-15| = -15$ The "−" is outside the absolute value marks, so first find the absolute value of −15 and then take its negative.

g. $|10 - 6| =$ Simplify inside the absolute value marks.
$|-4| =$ Find the absolute value of 4.
4 Answer.

Practice Exercises

Find the value of each of the following:

7. $|9| = $ _____

8. $|-10| = $ _____

9. $|13| = $ _____

10. $|-5| = $ _____

11. $|y| = $ _____

12. $-|18| = $ _____

13. $-|-23| = $ _____

14. $|11 - 4| = $ _____

Exercise Set 1.5

Answer each of the following using signed numbers:

1. If $+8$ represents 8 units to the right, how do you represent 8 units to the left?

2. If -80 feet represents a scuba diver descending 80 feet, how do you represent a scuba diver ascending 50 feet?

3. If $+12$ pounds represents gaining 12 pounds, what would -10 pounds represent?

4. If $-\$25$ represents losing \$25, what would $+\$55$ represent?

5. If 250 miles represents 250 miles east, what would -250 miles represent?

6. If $+5$ represents gaining 5 points on an exam, what would -5 represent?

Draw a number line for each of the following and graph the indicated integers:

7. $\{-5, -4, -1, 0\}$

8. $\{-3, -1, 2, 3\}$

9. $\{-10, -7, -2, 0, 4,\}$

10. $\{-8, -3, 0, 2, 6\}$

Give three examples of each of the following (11–12).

11. Integers that are not natural numbers.

12. Integers that are not whole numbers.

13. Give one example of a whole number that is not a natural number.

*Answer the following as **always true**, **sometimes true**, or **never true**:*

14. A natural number is an integer.

15. An integer is a whole number.

16. A whole number is a natural number.

17. A whole number is an integer.

18. An integer is a natural number.

19. A natural number is a whole number.

Find the negative (opposite) of each of the following:

20. 19

21. 17

22. -12

23. -15

24. $|4|$

25. $|8|$

26. $|-6|$

27. $|-8|$

28. $-|7|$

29. $-|15|$

Insert the proper symbol ($=$, $<$, or $>$) in order to make each of the following true:

30. 8 _____ 10

31. 29 _____ 12

32. 13 _____ -3

33. -10 _____ 8

34. -8 _____ -4

35. -3 _____ -7

36. $|6|$ _____ 6

37. $|-5|$ _____ -5

38. $-|-4|$ _____ 4

39. -7 _____ $-|-7|$

40. $|10|$ _____ $|-10|$

41. $-|-10|$ _____ $-(-10)$

42. $|24 - 18|$ _____ $|8 - 1|$

43. $|32 - 15|$ _____ $|13 - 5|$

44. $-(-5)$ _____ 5

45. 8 _____ $-(-8)$

46. -10 _____ $-(-10)$

Challenge Exercises:

Evaluate each of the following:

47. $2 \, |{-4}| + 3 \, |{-5}| - 2 \cdot 3^2$

48. $4 \, |5| - 2 \, |{-3}| + 3 \cdot 2^3$

49. $5 \, |4^2 + 3 \cdot 2| - 4^2 \cdot 2$

50. $6 \, |5 \cdot 6^2 - 8 \cdot 7| - 8 \cdot 2^3$

Writing Exercises:

51. Give three examples of numbers that are not integers and explain why they are not integers.

52. Does $-(-x)$ always represent a positive number? Why or why not?

53. Why is the absolute value of a number always non-negative?

54. Why is any positive number greater than any negative number?

55. What does $\not\le$ mean?

Group Project:

56. In this section, examples of the occurrence of negative numbers were given, including the withdrawal of money and a drop in temperature. Give two more examples where negative numbers occur other than those given in the text.

Section 1.6	Addition of Integers

OBJECTIVES   *When you complete this section, you will be able to:*

a. Add integers with the same and opposite signs.

b. Translate verbal expressions into mathematical expressions and simplify.

c. Identify properties of addition.

Introduction   The addition of integers occurs in many everyday situations. If the temperature rises 8° from 9:00 to 10:00 A.M. and then rises 5° from 10:00 to 11:00, there has been a total change in temperature of $8° + 5° = 13°$ from 9:00 to 11:00. If a football team gains 8 yards (+8) on first down and loses 3 yards (−3) on second down, then there is a net gain on the two downs of $(+8) + (-3) = 5$ yards. If a gambler loses \$20 (−20) on the first hand and wins \$8 (+8) on the second hand, there is a total loss of $(-\$20) + \$8 = -\$12$ on the two hands. If a scuba diver descends 30 feet (−30) and then descends another 20 feet (−20), then she has descended a total of $(-30) + (-20) = -50$ feet.

OPTIONAL   CALCULATOR EXPLORATION ACTIVITY

Using a calculator, complete the following:

Column A	Column B
$3 + 3 =$	$6 + (-2) =$
$4 + 5 =$	$-6 + 2 =$
$6 + 2 =$	$4 + (-6) =$
$(-3) + (-4) =$	$5 + (-2) =$
$(-5) + (-7) =$	$-8 + 5 =$
$(-2) + (-6) =$	$-4 + 9 =$

(continued)

Based on your answers to Column A, answer the following:

1. When adding two numbers with the same sign, is the absolute value of the answer the sum or difference of the absolute values of the numbers being added?

2. If the signs of the numbers being added are the same, how does the sign of the answer compare with the signs of the numbers being added?

3. Based on your answers to 1 and 2, write a rule for adding two numbers with the same sign.

Based on your answers to Column B answer the following:

4. When adding two numbers with the opposite signs, is the absolute value of the answer the sum or difference of the absolute values of the numbers being added?

5. If the signs of the numbers being added are opposites, how does the sign of the answer compare with the signs of the numbers being added?

6. Based on your answers to 4 and 5, write a rule for adding two numbers with opposite signs.

Following are some justifications for the conclusions reached in questions 1–6:

Adding real numbers   When adding integers, it is convenient to think of them as directed numbers. In the following examples the number line will be used to show the addition of integers with direction to the right as positive and direction to the left as negative. We will always begin at 0. The operation sign, +, is like a verb since it is telling you what action to perform. Think of the action as "and then go." Do not confuse the addition sign with the + sign indicating a positive number.

Example 1

Find the following sums using the number line:

a. $(+3) + (+2)$ (This is usually written as $3 + 2$ since a number written without a sign is assumed to be positive.)

This means begin at 0 and go 3 units to the right (because 3 is positive). Then go 2 more to the right (because 2 is also positive). You should now be at $+5$, showing that $(+3) + (+2) = +5$ or $3 + 2 = 5$. Using the number line it appears as follows:

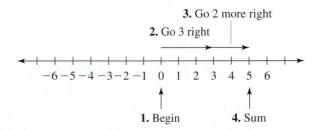

Answers:

Calculator Exploration Activity: **1.** sum **2.** same **3.** Add the absolute values of the numbers being added and give the answer the same sign. **4.** difference **5.** Same as the number with the larger absolute value. **6.** Subtract the absolute values of the numbers being added and give the answer the sign of the number with the larger absolute value.

b. $(+6) + (-2)$ (Usually written as $6 + (-2)$ or $6 - 2$.)

This means begin at 0 and go 6 units to the right. Then go 2 units to the left. You should now be at 4. Therefore, $6 + (-2) = 4$ as the following illustrates:

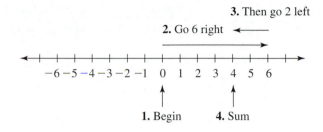

c. $-6 + (+4)$ (Usually written as $-6 + 4$.)

This means begin at 0 and go 6 units to the left. Then go 4 units to the right. You should now be at -2. Therefore, $-6 + 4 = -2$ as the following illustrates:

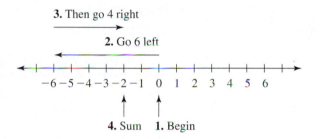

d. $(-4) + (-2)$

This means begin at 0 and go 4 units left. Then go 2 more units left. You should now be at -6. Therefore, $(-4) + (-2) = -6$ as the following illustrates:

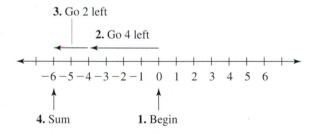

Practice Exercises

Use the number line to find the following sums:

1. $2 + 4$

2. $5 + (-4)$

3. $-6 + 3$

4. $-2 + (-3)$

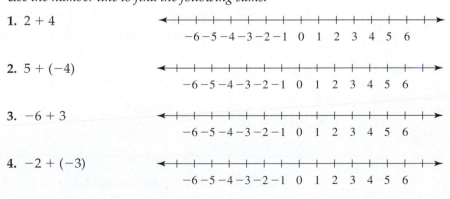

Answers:

Following is a table summarizing the results of Example 1:

Problem	Signs of Addends	Add or Subtract the Absolute Values of the Addends	Sign of the Answer
$(+3) + (+2) = 5$	same (both +)	add	positive—the same as sign of the addends
$(+6) + (-2) = 4$	opposite	subtract	positive—the same as addend with larger absolute value
$-6 + (+4) = -2$	opposite	subtract	negative—the same as addend with larger absolute value
$(-4) + (-2) = -6$	same (both −)	add	negative—the same as sign of the addends

One way to remember whether to add or subtract the absolute values is to remember what is happening on the number line. If you go in the same direction, add the absolute values. If you go in the opposite direction, subtract the absolute values.

Look back at the situations in the introduction and see that the results are consistent. This brings us to a general rule for adding signed numbers.

Adding Signed Numbers

a. To add numbers with the same sign, add the absolute values of the numbers and give the answer the same sign as the numbers being added.

b. To add numbers with opposite signs, subtract the absolute values of the numbers and give the answer the sign of the number with the larger absolute value.

Example 2

Find the following sums:

a. $7 + 5 =$ *Signs are the same, so add the absolute values.*
    12   *Both numbers are positive, so sum is positive.*

b. $6 + (-13) =$ *Signs are opposite, so subtract the absolute values.*
    -7 *−13 has the larger absolute value, so the sum is negative.*

c. $-6 + (-7) =$ *Signs are the same, so add the absolute values.*
    -13 *Both numbers are negative, so the sum is negative.*

d. $-14 + 5 =$ *Signs are opposite, so subtract the absolute values.*
    -9 *−14 has the larger absolute value, so the sum is negative.*

Practice Exercises

Find the following sums:

5. $9 + 7$ **6.** $-12 + 7$ **7.** $-8 + (-5)$ **8.** $9 + (-16)$

If more practice is needed, do the Additional Practice Exercises.

Additional Practice Exercises

Find the following sums:

a. $7 + 12$

b. $-15 + 9$

c. $-12 + (-15)$

d. $13 + (-13)$

If we are adding more than two numbers, apply the order of operations by adding in order from left to right. If parentheses are present, simplify inside the parentheses first. To avoid confusion, brackets are often used to show grouping when parentheses are already present.

Example 3

Find the following sums:

a. $4 + (-8) + 7 =$ First add 4 and −8.

$\quad -4 + 7 =$ Now add −4 and 7.

$\quad 3$ Sum.

b. $-24 + 16 + (-18) =$ First add −24 and 16.

$\quad -8 + (-18) =$ Now add −8 and −18.

$\quad -26$ Sum.

c. $26 + [17 + (-13)] =$ Perform operations inside brackets first, so add 17 and −13.

$\quad 26 + 4 =$ Add 26 and 4.

$\quad 30$ Sum.

d. $(-52 + 34) + [-21 + (-8)] =$ Perform operations inside parentheses first.

$\quad -18 + (-29) =$ Add −18 and −29.

$\quad -47$ Sum.

Practice Exercises

Find the following sums:

9. $-10 + 8 + (-15)$

10. $32 + [-18 + (-21)]$

11. $-9 + [15 + (-6)]$

12. $(-45 + 34) + [36 + (-22)]$

If more practice is needed, do the Additional Practice Exercises.

Additional Practice Exercises

Find the following sums:

e. $-14 + 9 + (-16)$

f. $24 + (-9) + (-22)$

g. $18 + 27 + (-17)$

h. $(-23 + 52) + (-45 + 33)$

Often it is necessary to translate expressions written in English into mathematical expressions. There are many expressions in English that indicate addition. For example, "sum," "more than," and "increased by" all imply addition.

Example 4

Write a numerical expression for each of the following and evaluate:

a. The sum of -7 and -2.

Solution: Sum means add, so this means add -7 and -2.

$-7 + (-2) =$ Signs are the same, so add the absolute values.

-9 Both signs are negative, so the sum is negative.

b. -10 increased by 4

Solution: This implies that you begin with -10 and increase it by 4, which means add 4 to -10.

$-10 + 4 =$ Signs are opposite, so subtract the absolute values.

-6 -10 has the larger absolute value, so the sum is negative.

c. 6 more than -2

Solution: This implies that you begin with -2 and you need 6 more which means add -2 and 6.

$-2 + 6 =$ Signs are opposite, so subtract the absolute values.

4 6 has the larger absolute value, so the sum is positive.

Practice Exercises

Write a numerical expression for each of the following and evaluate:

13. the sum of -12 and -5

14. -8 increased by 14

15. 5 more than -13

16. -6 added to 4

If more practice is needed, do the Additional Practice Exercises.

Additional Practice Exercises

Write a numerical expression for each of the following and evaluate:

i. The sum of 8 and -3.

j. -6 increased by 4.

k. 9 more than -23.

l. -6 added to -5.

One way to remember the commutative property is to think how each day you commute from home to school. You take the same route each way, but travel in the opposite direction. Similarly, the commutative property involves the same numbers (routes), but in the opposite order.

The operation of addition has several properties that are of such importance that they are given names for easy reference. You may have already noticed that the order in which the numbers are added makes no difference. For example, $2 + 4 = 4 + 2$, and $5 + 9 = 9 + 5$. The same is true for the addition of integers. This relationship is known as the **commutative property of addition**. It says that the order in which the numbers are added does not matter since the sum is the same.

Example 5

a. $-4 + 7 = 7 + (-4)$ Simplify each side.

$3 = 3$

b. $-5 + (-8) = -8 + (-5)$ Simplify each side.

$-13 = -13$

Answers:

Additional Practice Exercises i–l: **i.** $8 + (-3) = 5$ **j.** $-6 + 4 = -2$ **k.** $-23 + 9 = -14$ **l.** $-5 + (-6) = -11$

Practice Exercises 13–16: **13.** $-12 + (-5) = -17$ **14.** $-8 + 14 = 6$ **15.** $-13 + 5 = -8$ **16.** $4 + (-6) = -2$

If three or more numbers are added, the manner in which they are grouped makes no difference. For example, $2 + (3 + 4) = (2 + 3) + 4$. So it does not matter which we add first, the "3" and the "4" or the "2" and the "3." This relationship is called the **associative property of addition.** It says the manner in which the numbers are grouped does not matter since the sum is the same.

Example 6

a. $4 + (5 + 8) = (4 + 5) + 8$ Simplify each side by adding inside the parentheses.

$\quad\ \ 4 + 13 = 9 + 8$ Add.

$\quad\quad\quad\ 17 = 17$ Therefore, both sides are equal.

b. $-6 + [(-4) + 9] = [-6 + (-4)] + 9$ Simplify inside brackets first.

$\quad\quad\ \ -6 + 5 = -10 + 9$ Add.

$\quad\quad\quad\quad\ -1 = -1$ Therefore, both sides are equal.

The number **0** is called the **additive identity** or the identity element for addition. Adding 0 to any number gives a result identical to the original number. If the sum of two numbers is 0, the two numbers are **additive inverses** of each other. In Section 1.5 these were called negatives or opposites.

Example 7

a. The additive inverse of 5 is -5 because $5 + (-5) = 0$.

b. The additive inverse of -14 is 14 because $-14 + 14 = 0$.

c. 0 is its own additive inverse because $0 + 0 = 0$.

Following is a summary of the properties of addition:

Properties of Addition

All real numbers a, b, and c satisfy the following properties:

Commutative Property of Addition:
$a + b = b + a$ Addition may be performed in any order.

Associative Property of Addition:
$a + (b + c) = (a + b) + c$ Addition may be grouped in any manner.

Additive Inverse:
Every real number, a, has an additive inverse, $-a$, so that $a + (-a) = (-a) + a = 0$.

Additive Identity:
The real number 0 is called the additive identity because $a + 0 = 0 + a = a$. The sum of any number and 0 is identical to the number.

Example 8

Give the name of the property illustrated by each of the following:

a. $3 + (-5) = -5 + 3$ *Order is different. Therefore, commutative property of addition.*

b. $4 + (-6 + 2) = [4 + (-6)] + 2$ *Order is same and grouping is different.*
Therefore, associative property of addition.

c. $5 + (8 + 4) = 5 + (4 + 8)$ *Grouping is same and order is different. Therefore, commutative property of addition.*

d. $5[6 + (-6)] = 5(0)$ *6 1 (26) 5 0. Therefore, additive inverse property.*

e. $-6 + 0 = -6$ *Additive identity property.*

Practice Exercises

Give the name of the property illustrated by each of the following:

17. $-8 + 2 = 2 + (-8)$ _____

18. $7 + (-6 + 3) = [7 + (-6)] + 3$ _____

19. $4(6 + 3) = 4(3 + 6)$ _____

20. $-9(-3 + 3) = -9(0)$ _____

21. $5 + 0 = 5$ _____

Study Tip 5

Memorizing and Understanding Mathematics

Some things in mathematics must be memorized. Symbols, definitions, rules, and algorithms have to be both memorized and understood. We all forget, so those things that must be memorized also need to be reviewed periodically. We can do this by writing definitions, rules, and so on, on 3×5 cards or 8×5 cards and reading them while we wait for a bus or train or between classes at the college or university. However, you should realize that you cannot memorize everything in a mathematics course. You may be able to memorize enough for one unit exam, but there is always a final exam.

Psychologists tell us that we learn best when what we are learning has meaning for us. That is, we can learn more and keep it longer when we understand the material. The more we review, think about, and see how things fit together, the more meaning these things have for us and the better we can understand and apply them.

In mathematics there is a reason for everything that we do, and we must know and understand that reason. Mathematics is not learned just by doing problems. It is learned by doing the problems and understanding why we did them the way we did. The "why" of mathematics is as important, if not more important, than the "how" of mathematics. In general, mathematics is to be *understood*, not memorized.

Exercise Set 1.6

Find the following sums:

1. $2 + 16$

2. $9 + 5$

3. $-2 + 13$

4. $-5 + 165$

5. $7 + (-14)$

6. $8 + (-18)$

7. $17 + (-8)$

8. $14 + (-5)$

9. $-17 + 6$

10. $-21 + 3$

11. $-7 + (-4)$

12. $-6 + (-3)$

13. $13 + 35$

14. $32 + 43$

15. $-24 + 35$

Answers:

Practice Exercises 17–21: **17.** commutative property of addition **18.** associative property of addition **19.** commutative property of addition **20.** additive inverse **21.** additive identity

16. $-54 + 67$

17. $-25 + 19$

18. $-42 + 37$

19. $63 + (-34)$

20. $72 + (-54)$

21. $4.3 + (-2.5)$

22. $-6.7 + 5.3$

23. $-9.2 + 4.5$

24. $7.6 + (-4.1)$

25. $8 + (-5) + 4$

26. $9 + (-6) + (-6)$

27. $-8 + 10 + (-7)$

28. $-6 + 12 + (-4)$

29. $13 + (-17) + (-21)$

30. $21 + (-16) + (-32)$

31. $-16 + (-23) + (-31)$

32. $-25 + (-32) + (-47)$

33. $[9 + (-3)] + (-8 + 5)$

34. $[7 + (-4)] + (-6 + 9)$

35. $(-13 + 9) + (-21 + 14)$

36. $(-17 + 8) + (-27 + 17)$

37. $[-4 + (-7)] + (-10 + 4)$

38. $[-8 + (-5)] + (-13 + 7)$

39. $[16 + (-32)] + [17 + (-6)]$

40. $[-5 + (-43)] + [-37 + (-21)]$

Challenge Exercises: (41–46)

Find the following sums of fractions:

41. $\dfrac{7}{11} + \dfrac{-3}{11}$

42. $\dfrac{-5}{9} + \dfrac{2}{9}$

43. $\dfrac{-7}{17} + \dfrac{3}{17}$

44. $\dfrac{-8}{19} + \dfrac{5}{19}$

45. $-\dfrac{2}{11} + \left(-\dfrac{5}{11}\right)$

46. $\dfrac{-4}{13} + \left(-\dfrac{2}{13}\right)$

Calculator Exercises:

Find the following sums using a calculator:

C1. $14.62 - 33.21$

C2. $-67.43 + 49.62$

C3. $-896.456 + (-45.7689)$

C4. $-985.35 + (-83678.587)$

C5. $(-765.45 + 65.78) + [473.21 + (-860.32)]$

C6. $[-75849.359 + (-94327.432)] + [79235.937 + (-27807.857)]$

Give the name of the property illustrated by each of the following:

47. $6 + (-2) = -2 + 6$ _____

48. $5 + (6 + 4) = (5 + 6) + 4$ _____

49. $-8 + 8 = 0$ _____

50. $57 + 0 = 57$ _____

51. $7(5 + 3) = 7(3 + 5)$ _____

52. $6[4 + (-2)] = 6[-2 + 4]$ _____

53. $-5(-4 + 4) = -5(0)$ _____

54. $0 + 8 = 8 + 0$ _____

55. $12 + (-4 + 9) = 12 + [9 + (-4)]$ _____

56. $0 + (6 + 4) = 6 + 4$ _____

57. $(3 + 5) + (-7) = 3 + [5 + (-7)]$ _____

58. $(-10 + 10) + 7 = 0 + 7$ _____

Complete each of the following using the given property:

59. $-5 + 6 =$ _____ commutative for addition

60. $-4 + (9 + 12) =$ _____ associative for addition

61. $-4 + (9 + 12) =$ _____ commutative for addition

62. $0 + 34 =$ _____ additive identity

63. $-14 + 14 =$ _____ additive inverse

64. $45 + 0 =$ _____ commutative for addition

65. $7(5 + 8) =$ _____ commutative for addition

Write a numerical expression for each of the following and evaluate:

66. The sum of -9 and 5

67. The sum of -11 and 8

68. The sum of $-3, 6,$ and -7

69. The sum of $-6, -4,$ and 10

70. 6 increased by 9

71. 8 increased by 11

72. -7 increased by 14

73. -12 increased by 7

74. 23 more than 36

75. 35 more than 49

76. 17 more than -30

77. 36 more than -18

Problems 78–85 involve changes in temperature. Write each as an addition problem and solve.

Current temp	Change	Addition	New temperature	Current temp	Change	Addition	New temperature
78. 78°	Drops 15°	_____	_____	**82.** −5°	Rises 15°	_____	_____
79. 92°	Drops 25°	_____	_____	**83.** −12°	Rises 24°	_____	_____
80. 18°	Drops 28°	_____	_____	**84.** −14°	Drops 12°	_____	_____
81. 12°	Drops 19°	_____	_____	**85.** −21°	Drops 7°	_____	_____

Problems 86–89 involve a scuba diver. Express each as an addition problem and solve.

Current depth	Change in depth	Addition	New depth
86. −43 feet	Ascends 27 feet	_____	_____
87. −68 feet	Ascends 39 feet	_____	_____
88. −45 feet	Descends 22 feet	_____	_____
89. −65 feet	Descends 43 feet	_____	_____

Solve each of the following word problems using addition of integers:

Use the following information for Exercises 90–93. In Europe the first floor of a building is called ground floor and what we call the second floor is called the first floor, and so on. Consequently, we could think of the ground floor as being the "0" floor. Suppose a European hotel has a 30th floor and four levels of underground parking garages. Represent each of the following as an addition problem and solve. See the figure in the margin.

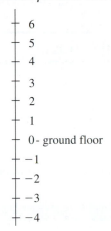

90. If an elevator in a European hotel starts on the 27th floor and goes down 29 levels, where does it stop?

91. Suppose the elevator in Exercise 90 begins on the 16th floor and goes down 19 levels. Where does it stop?

92. Suppose the elevator in Exercise 90 begins on the third parking level below ground and goes up 23 levels. Where does it stop?

93. Suppose the elevator in Exercise 90 begins on the second parking level below ground and goes up 14 levels. Where does it stop?

94. Hernando has a balance of $130 in his checking account. He makes a deposit of $230. What is his new balance?

95. Farshad has a balance of $289 in his checking account. He writes a check for $146. What is his new balance?

96. Sara has a balance of $319 in her checking account. She makes a deposit of $132 and then writes a check for $350. What is her new balance?

97. Mira has a balance of $293 in her checking account. She writes a check for $167 and later makes a deposit of $250. What is her new balance?

98. At the opening bell, a stock was selling for $53\frac{3}{8}$. During the day it gained $3\frac{2}{8}$ points. What was the closing price?

99. At the opening bell, a stock was selling for $37\frac{3}{8}$ and gained $1\frac{7}{8}$ before the final bell. What was the closing price?

100. A hiker begins at the rim of the Grand Canyon and descends 1500 feet into the canyon. He then turns around and ascends 850 feet. How far is the hiker from the rim?

101. The balance in a credit card account is $1396 before a payment of $450 is made. Find the new balance.

102. A group of explorers is located in Death Valley and is 32 feet below sea level. An airplane drops supplies from a height of 1800 feet. How far did the supplies fall?

103. On successive plays a running back gained 5 yards, lost 2 yards, and gained 8 yards. If a net gain of 10 yards is needed to earn a first down, did the team earn a first down? Why?

Challenge Exercises:

Evaluate each of the following:

104. $4^2(-7 + 2 \cdot 5) \div 3 \cdot 4$

105. $(-3 + 6 \cdot 3^2) \div 17 \cdot 6$

106. $3^2 \cdot 4(-12 + 4^2) \div 2^2 \cdot 3^2$

107. $48 \div 2^2 \cdot 3(-18 + 2 \cdot 4^2)$

Writing Exercise:

108. Why is 0 its own additive inverse (opposite)?

Critical Thinking:

109. In your own words, explain why you add the absolute values of two numbers with like signs when finding their sum.

110. In your own words, explain why you subtract the absolute values of two numbers with unlike signs when finding their sum.

Group Projects:

111. In the introduction, several examples were given where the addition of integers occurred in everyday situations. Give three additional examples not found in the text.

112. Interview an engineer, physics or chemistry instructor, or other professional and make a list of at least three instances in which they have used addition of signed numbers.

Section 1.7 | **Subtraction of Integers and Combining Like Terms**

OBJECTIVES   *When you complete this section, you will be able to:*

a. Subtract integers.

b. Combine like terms.

Introduction   In this section, we will see that we already know how to subtract signed numbers.

OPTIONAL | **CALCULATOR EXPLORATION ACTIVITY**

Complete the following using a calculator:

	Column A	Column B
1.	$4 - (+3) =$	$4 + (-3) =$
2.	$-6 - (+5) =$	$-6 + (-5) =$
3.	$7 - (-3) =$	$7 + (+3) =$
4.	$-8 - (-2) =$	$-8 + (+2) =$

By looking at the corresponding lines in Columns A and B, answer the following:

Subtracting a number is the same as _____ its _____.

Following are some real world situations that support the preceding conclusion:

As with the addition of integers, the subtraction of integers occurs in many everyday situations. Look for a pattern in the following examples using temperature:

Example 1

a. If the temperature changes from 50° to 70°, what is the change in temperature? We know the temperature has risen 20°, which can be represented as +20°, but how do we find it mathematically? The change is (the new temperature) − (the old temperature). Therefore, $70° - (+50°) = +20°$ represents the change in temperature. The positive indicates the temperature has risen by 20°. Note that $70° - (+50°) = 70° + (-50°) = 20°$.

b. If the temperature changes from 60° to 50°, what is the change in temperature? We know the temperature has dropped 10°, which can be represented by −10°. The change can be found mathematically by taking (the new temperature) − (the old temperature). Therefore, the change would be $50° - (+60°) = -10°$ where the negative indicates that the temperature has dropped. Note that $50° - (+60°) = 50° + (-60)° = -10°$.

Answers:

Calculator Exploration: **1.** Subtracting a number is the same as adding its negative.

c. If the temperature changes from $-10°$ to $20°$, what is the change in temperature? We know the temperature has risen $30°$, which can be represented by $+30°$. The change can be found mathematically by taking (the new temperature) $-$ (the old temperature). Therefore, the change would be $20° - (-10°) = +30°$ where the positive indicates that the temperature has risen. Note that $20° - (-10°) = 20° + (+10)° = +30°$.

d. Likewise, if the temperature changes from $-20°$ to $-10°$, how much has it changed? We know it has risen $10°$. The change in temperature is again found mathematically by taking (the new temperature) $-$ (the old temperature). Therefore, the change would be $(-10°) - (-20°) = +10°$. The positive indicates the temperature has risen by $10°$. Notice $(-10°) - (-20°) = -10° + (+20°) = +10°$.

Following is a table summarizing the results of Example 1:

Temperature Change	Written as Subtraction	Written as Addition
From $50°$ to $70°$	$70° - (+50°) = 20°$	$70° + (-50°) = 20°$
From $60°$ to $50°$	$50° - (+60°) = -10°$	$50° + (-60°) = -10°$
From $-10°$ to $20°$	$20° - (-10°) = 30°$	$20° + (+10°) = 30°$
From $-20°$ to $-10°$	$-10° - (-20°) = 10°$	$-10° + (+20°) = 10°$

From the preceding examples it seems there is a very close relationship between subtraction and addition since each subtraction problem can be rewritten as addition. You will notice that the procedure is to change the sign of the number being subtracted and change the operation from subtraction to addition. This leads to the following procedure for subtracting integers:

Procedure for Subtracting

$a - b = a + (-b)$.   To subtract a number, add its negative (opposite). Remember, a and b can represent either positive or negative numbers.

Example 2

Rewrite each of the following as addition and evaluate:

a. $7 - 5 =$ $7 - 5$ means $7 - (+5)$, so change the subtraction to addition and $+5$ to -5.

　　$7 + (-5) =$ Add.

　　2 Difference.

b. $-6 - 2 =$ $-6 - 2$ means $-6 - (+2)$, so change the subtraction to addition and $+2$ to -2.

　　$-6 + (-2) =$ Add.

　　-8 Difference.

c. $8 - (-5) =$ Change the subtraction to addition and -5 to $+5$.

　　$8 + 5 =$ It is not necessary to write $+ (+5)$ since 5 means $+5$.

　　13 Difference.

d. $-12 - (-8) =$ Change the subtraction to addition and -8 to $+ 8$.

　　$-12 + 8 =$ Add.

　　-4 Difference.

Practice Exercises

Rewrite each of the following as addition and evaluate:

1. $12 - 8$

2. $-7 - 5$

3. $9 - (-4)$

4. $-14 - (-6)$

If more practice is needed, do the Additional Practice Exercises.

Additional Practice Exercises

Evaluate each of the following:

a. $8 - 13$

b. $-3 - 9$

c. $5 - (-12)$

d. $-16 - (-9)$

Note: Since $a - b = a + (-b)$, we will no longer write two signs when adding or subtracting, except in the case of subtracting a negative number. *We will assume the operation is addition and the sign is the sign of the number following it.* For example, $6 - 4 = 6 + (-4) = 2$ and $-8 - 5 = -8 + (-5) = -13$.

If there are more than two numbers, simplify inside any parentheses or brackets first from the innermost to the outermost, then add or subtract in order from left to right as in the following examples:

Example 3

Evaluate the following. Remember the order of operations.

a. $8 - 10 + 5 =$ Add 8 and -10. $(8 - 10 = 8 + (-10))$

  $-2 + 5 =$ Add -2 and 5. $(-2 + 5 = -2 + (+5))$

  3 Sum.

b. $-14 - 7 + 6 - (-4) =$ Add -14 and -7. $(-14 - 7 = -14 + (-7))$

  $-21 + 6 - (-4) =$ Add -21 and 6. $(-21 + 6 = -21 + (+6))$

  $-15 - (-4) =$ Change $-(-4)$ to $+ 4$.

  $-15 + 4 =$ Add -15 and 4. $(-15 + 4 = -15 + (+4))$

  -11 Sum.

c. $14 - (9 - 21) =$ Simplify inside parentheses first, so add 9 and -21.

  $14 - (-12) =$ Change $-(-12)$ to $+ 12$.

  $14 + 12 =$ Add 14 and 12.

  26 Sum.

d. $(-16 - 8) - (18 - 9) =$ Simplify inside parentheses first.

  $-24 - 9 =$ Add -24 and -9.

  -33 Sum.

Answers:

e. $5 - [7 - (-6 - 5)] =$ Simplify inside parentheses first.

     $5 - [7 - (-11)] =$ Change $-(-11)$ to $+ 11$.

     $5 - [7 + 11] =$ Simplify inside brackets next.

     $5 - 18 =$ Add 5 and -18.

     -13 Sum.

Note: In Example 3e, we used brackets and parentheses to show grouping since it is confusing to have parentheses inside parentheses.

Practice Exercises

Evaluate the following:

5. $-7 + 10 - 8$

6. $10 - (-6) + 4 - 12$

7. $17 - (-5 - 8)$

8. $(8 - 16) - (12 - 21)$

9. $8 - [-9 - (-24 + 18)]$

If more practice is needed, do the Additional Practice Exercises.

Additional Practice Exercises

Simplify the following:

e. $-9 + 7 - 3$

f. $14 - 6 - (-7) + 5$

g. $15 - (-11 + 7)$

h. $(12 - 19) - (-6 + 13)$

i. $5 - [6 - (-15 + 9)]$

Addition and subtraction of integers is used to combine like terms. A **term** is a number, variable, or product and/or quotient of numbers and variables raised to powers. Examples of terms are 5, x, $5x$, $-3x^2$, $2x^3y^2$, $-7x^2y^3z^4$, $\frac{x}{y}$ and $4x^{-2}$. If we had three apples and two bananas and purchased four apples and six bananas we would now have seven apples and eight bananas. We add the number of apples to the number of apples and the number of bananas to the number of bananas, but not the number of apples to the number of bananas. If we represent apples with a and bananas with b, this becomes $3a + 2b + 4a + 6b = 7a + 8b$. Though very simplistic, this is the basic idea behind adding **like terms,** which is one of the topics of this section.

Distributive property   Before we can discuss adding terms, we need another number property. Let us simplify $3(5 + 2)$ in two different ways. If we follow the order of operations, we get $3(5 + 2) = 3(7) = 21$. If we find $3 \cdot 5 + 3 \cdot 2$, we get $15 + 6 = 21$. Therefore, $3(5 + 2) = 3 \cdot 5 + 3 \cdot 2$. This example can be generalized into the **distributive property** stated as follows:

Distributive Property of Multiplication over Addition

For all numbers a, b, and c, $a(b + c) = a \cdot b + a \cdot c$.

Answers:

The distributive property states that it does not matter whether we add inside the parentheses and then multiply or multiply first and then add. We will not discuss multiplication of signed numbers until a later section, so Example 4 will be limited to natural numbers only. The distributive property holds, however, for all numbers.

Example 4

Verify the distributive property for each of the following by evaluating both sides and showing that they are equal:

a. $2(3 + 5) = 2 \cdot 3 + 2 \cdot 5$ *$a(b + c) = ab + ac$. Simplify using the order of operations.*

$2(8) = 6 + 10$

$16 = 16$

b. $5(2 + 8) = 5 \cdot 2 + 5 \cdot 8$ *$a(b + c) = ab + ac$. Simplify using the order of operations.*

$5(10) = 10 + 40$

$50 = 50$

Practice Exercises

Verify the distributive property for each of the following. Rewrite each expression as in Example 1, then evaluate each side and show they are equal.

10. $3(4 + 2)$

11. $5(6 + 3)$

We need to be able to recognize the distributive property when we see it in various forms. Instead of writing it as $a(b + c) = ab + ac$, another form of the distributive property is $(b + c)a = ba + ca$. This equation may be "turned around" and written as $ba + ca = (b + c)a$. For example, $3 \cdot 4 + 5 \cdot 4 = (3 + 5) \cdot 4$. It is this form that we will use in Example 5.

Terms that have the same variables with the same exponents on these variables are called **like terms**. Examples of like terms are $3x$ and $5x$, $6y$ and $9y$, $11z^2$ and $9z^2$, $3x^2y$ and $7x^2y$. The terms $4x$ and $3y$ are not like terms because the variables are not the same. The terms $3x$ and $3x^2$ are not like terms because the exponents on the variables are not the same.

Now we will use the distributive property in the form $ba + ca = (b + c)a$ to add like terms. Study the following examples carefully:

Example 5

Find the following sums:

a. $2 \cdot x + 4 \cdot x =$ *Apply $ba + ca = (b + c)a$.*

$(2 + 4)x =$ *Now add 2 and 4.*

$6x$ *Therefore, $2x + 4x = 6x$.*

b. $5xy - 3xy =$ *Apply $ba + ca = (b + c)a$.*

$(5 - 3)xy =$ *Now add 5 and −3.*

$2xy$ *Therefore, $5xy - 3xy = 2xy$.*

c. $3x^2 + 8x^2 =$ *Apply $ba + ca = (b + c)a$.*

$(3 + 8)x^2 =$ *Now add 3 and 8.*

$11x^2$ *Therefore, $3x^2 + 8x^2 = 11x^2$.*

d. $3x + 7y$ *The distributive property does not apply, so this expression cannot be simplified.*

In Examples 5a–c, we were adding like terms. In each case the variable was placed outside the parentheses with the coefficients inside and the coefficients were then added. This leads to the following procedure for adding like terms:

Answers:

Practice Exercises 10–11: 10. $3(4 + 2) = 3 \cdot 4 + 3 \cdot 2$ $3(6) = 12 + 6$ $18 = 18$ **11.** $5(6 + 3) = 5 \cdot 6 + 5 \cdot 3$ $5(9) = 30 + 15$ $45 = 45$

Addition of Like Terms

To add like terms, add the numerical coefficients and leave the variable portion unchanged.

The preceding rule greatly simplifies the addition of like terms since we no longer have to apply the distributive property each time we add.

Example 6

Simplify the following, if possible, by adding like terms:

a. $6x + 2x =$ These are like terms, so add the coefficients, 6 and 2.

$8x$ Leave the variable portion unchanged.

b. $4x - 7x =$ These are like terms, so add the coefficients, 4 and -7.

$-3x$ Leave the variable portion unchanged.

c. $5x^2 + 4x^2 =$ These are like terms, so add the coefficients, 5 and 4.

$9x^2$ Leave the variable portion unchanged.

d. $6xy - 2xy + 8 =$ $6xy$ and $-2xy$ are like terms, so add them.

$4xy + 8$

e. $5x + 4y - 2x + 6y =$ $5x$ and $-2x$ are like terms and $4y$ and $6y$ are also like terms,

$3x + 10y$ so add each pair.

Note: Some people prefer to use the commutative and associative properties to rearrange the terms so the like terms are together before adding. Example 3e could be rewritten as follows: $5x + 4y - 2x + 6y = 5x - 2x + 4y + 6y = 3x + 10y$.

f. $3x^2 + 4x - 7x^2 + 5 + 3x =$ $3x^2$ and $-7x^2$ are like terms and so are $4x$ and $3x$.

$-4x^2 + 7x + 5$ Add each pair.

As in Example 3e, the terms could be rearranged before adding. So,
$3x^2 + 4x - 7x^2 + 5 + 3x = 3x^2 - 7x^2 + 4x + 3x + 5 = -4x^2 + 7x + 5.$

g. $2x - (-5x) =$ Change to addition.

$2x + 5x =$ Add like terms.

$7x$

h. $3x + 4y - 3x^2$ There are no like terms, so this expression cannot be simplified.

Practice Exercises

Simplify the following, if possible:

12. $7x + 5x$

13. $4xy + 8xy$

14. $6y^2 + 9y^2$

15. $5x - 9x + 3x$

16. $4x^2y - 5 + 8x^2y$

17. $7z + 6x - 3z + 9x$

18. $4x - 5x^2 - 6x + 4 - 3x$

19. $4x - (-7x)$

20. $6a + 7b + 5a^2$

If more practice is needed, do the Additional Practice Exercises.

Additional Practice Exercises

Simplify the following, if possible:

j. $4y + 6y$

k. $5ab + 9ab$

l. $7x - 4x + 5x$

m. $5z^2 + 9z^2$

n. $9x^2y - 5x + 3x^2y$

o. $6y + 2z - 9y - 7z$

p. $7x - 6x^2 - 9x + 6 + 7x^2$

q. $7y - (-5y)$

r. $7r^2 + 3s - 5t$

Often problems present themselves to us in the form of words rather than symbols. When this happens it is necessary to translate from English to mathematics. Remember, "sum" means add and "difference" means subtract.

Example 7

Write each of the following as a subtraction problem and evaluate:

a. The difference of 7 and 12

> **Solution:**
> The difference of 7 and 12 means subtract 12 from 7, which is written as $7 - 12$.
>
> $$7 - 12 = \quad \text{Add 7 and } -12.$$
> $$-5 \quad \text{Therefore, the difference of 7 and 12 is } -5.$$

b. $8a$ less than $2a$

> **Solution:**
> "$8a$ less than" means subtract $8a$ from $2a$, which is written as $2a - 8a$.
>
> $$2a - 8a = \quad \text{Add the coefficients 2 and } -8.$$
> $$-6a \quad \text{Therefore, } 8a \text{ less than } 2a \text{ is } -6a.$$

c. The difference of 4 and -5 added to 10

> **Solution:**
> The difference of 4 and -5 means subtract -5 from 4, which is $4 - (-5)$. This entire difference is added to 10, which is written as $10 + [4 - (-5)]$. The brackets around $4 - (-5)$ are necessary to show that we are adding the difference of 4 and -5 to 10. Without the brackets we would have $10 + 4 - (-5)$, which means we are adding 10 and 4 and then finding the difference between that sum and -5. So we have:
>
> $$10 + [4 - (-5)] = \quad \text{Rewrite } 4 - (-5) \text{ as } 4 + 5.$$
> $$10 + [4 + 5] = \quad \text{Add 4 and 5. Drop the brackets.}$$
> $$10 + 9 = \quad \text{Add 10 and 9.}$$
> $$19 \quad \text{Therefore, the difference of 4 and } -5 \text{ added to 10 is 19.}$$

d. The sum of $-8x$ and $6x$ subtracted from the difference of $-3x$ and $2x$.

> **Solution:**
> The sum of $-8x$ and $6x$ means add $-8x$ and $6x$, which is written as $-8x + 6x$. The difference of $-3x$ and $2x$ means subtract $2x$ from $-3x$ and is written as $-3x - 2x$. We need to subtract $-8x + 6x$ from $-3x - 2x$. This is written as $(-3x - 2x) - (-8x + 6x)$. The parentheses are necessary to show we are subtracting the *sum* of $-8x$ and $6x$ (in parentheses) from the *difference* of $-3x$ and $2x$ (in parentheses). So we have:

Answers:

Additional Practice Exercises j–r: **j.** $10y$ **k.** $14ab$ **l.** $8x$ **m.** $14z^2$ **n.** $12x^2y - 5x$ **o.** $-3y - 5z$ **p.** $x^2 - 2x + 6$ **q.** $12y$ **r.** cannot be simplified

$$(-3x - 2x) - (-8x + 6x) = \qquad \text{Simplify inside the parentheses first.}$$
$$(-5x) - (-2x) = \qquad \text{Rewrite } (-5x) - (-2x) \text{ as } -5x + 2x.$$
$$-5x + 2x = \qquad \text{Add.}$$
$$-3x \qquad \text{Therefore, the sum of } -8x \text{ and } 6x$$

subtracted from the difference of $-3x$ and $2x$ is $-3x$.

Practice Exercises

Write each of the following as a subtraction problem and evaluate:

21. The difference of -5 and -9.

22. $10x$ less than $-3x$.

23. The difference of 3 and -1 added to -3.

24. The sum of $5b$ and $-8b$ subtracted from the difference of $-4b$ and $6b$.

Exercise Set 1.7

Evaluate each of the following:

1. $9 - 3$

2. $11 - 7$

3. $6 - 12$

4. $9 - 15$

5. $8 - (-4)$

6. $6 - (-7)$

7. $-5 - (-9)$

8. $-7 - (-8)$

9. $-14 - (-6)$

10. $-9 - (-3)$

11. $7.5 - 9.2$

12. $4.6 - 10.3$

13. $-7 + 4 - 6$

14. $-5 + 9 - 8$

15. $-13 - 6 + 7$

16. $-17 - 7 + 4$

17. $21 - (-8) - 4$

18. $14 - 6 - 9$

19. $10.6 - 5.1 - 3.3$

20. $11.4 - 2.1 + 6.3$

21. $8 - (6 - 2)$

22. $12 - (7 - 5)$

23. $9 - (5 - 8)$

24. $7 - (3 - 8)$

25. $(16 - 12) - (8 - 5)$

26. $(26 - 17) - (7 - 3)$

27. $(14 - 23) - (31 - 26)$

28. $(16 - 28) - (34 - 27)$

29. $(-9 + 24) - (-19 - 5)$

30. $(-14 + 29) - (-24 - 9)$

31. $8 - [3 - (6 - 9)]$

32. $6 - [4 - (3 - 9)]$

33. $-13 + [24 - (-9 + 3)]$

34. $-25 + [37 - (-8 + 7)]$

35. $[(33 - 24) - 12] - 23$

36. $[(46 - 39) - 32] - 44$

Challenge Exercises: (37–46)

37. $\dfrac{-11}{8} + \dfrac{3}{8}$

38. $-\dfrac{8}{6} + \dfrac{7}{6}$

39. $-\dfrac{15}{20} - \dfrac{8}{20}$

40. $-\dfrac{14}{21} - \dfrac{12}{21}$

41. $\dfrac{11}{17} - \dfrac{18}{17}$

42. $\dfrac{6}{13} - \dfrac{10}{13}$

43. $\dfrac{20}{24} - \left(-\dfrac{9}{24}\right)$

44. $\dfrac{11}{19} - \left(-\dfrac{3}{19}\right)$

45. $\left(\dfrac{7}{13} - \dfrac{5}{13}\right) - \left(\dfrac{10}{13} - \dfrac{2}{13}\right)$

46. $\left(\dfrac{10}{17} - \dfrac{14}{17}\right) - \left(\dfrac{21}{17} - \dfrac{12}{17}\right)$

Calculator Exercises:

Find the following differences, using a calculator when necessary:

C1. $7564.968 - (-8573.3547)$

C2. $6503.954 - (-2759.8735)$

C3. $56.45 - [45.93 - (45.98 - 65.43)]$

C4. $87.4 - [8.94 - (5.7 - 87.64)]$

Simplify the following by adding like terms:

47. $4x + 7x$

48. $6x + 9x$

49. $-3y + 9y$

50. $-7y + 4y$

51. $9ab - 2ab$

52. $2xy - 7xy$

53. $-4x^2 - 3x^2$

54. $-5y^2 - 4y^2$

55. $12z - 8z + 3$

56. $15z - 9z + 7$

57. $-8y + 2y - 3$

58. $-10a - 3a + 6$

59. $3x - 8x - 6x$

60. $4z - 5z - 2z$

61. $-4r + 7r - 8r$

62. $-7b - 2b - 3b$

63. $5xy + 6x - 9xy + 2x$

64. $9xy + 5x - 3xy + 8x$

65. $3x^2 - 2y^2 - 6x^2 - 5x^2$

66. $4x^2 + 5y^2 - 6x^2 + 3y^2$

67. $9z - 5 + 3x + 3z - 2x - 7$

68. $8y - 4k + 2 - 4y + 3 - 9k$

69. $9b^2 - 8b + 2 - 2b^2 - 2b + 10$

70. $4d^2 - 4 + 4d + 4d^2 - 9 - 3d$

71. $m^2r + 3r - 5mr^2 + 3r + 5mr^2$

72. $n^2t - 8t + 6tn^2 - 2t - n^2t$

Write each of the following as a subtraction problem and evaluate. Remember, the difference of a and b means subtract b from a, which is written a − b.

73. The difference of 7 and -5

74. The difference of 9 and -5

75. The difference of $-10a$ and $6a$

76. The difference of $-14a$ and $9a$

77. 5 less than 8

78. 7 less than 12

79. $8m$ less than $-5m$

80. $13n$ less than $-8n$

81. 6 less 9

82. 7 less 12

83. -13 subtracted from 9

84. -21 subtracted from 11

85. $32r$ subtracted from $26r$

86. $29r$ subtracted from $21r$

87. The difference of 6 and -5 added to 14

88. The difference of 8 and -3 added to 23

89. $15ab$ added to the difference of $-4ab$ and $-6ab$

90. $7mn$ added to the difference of $9mn$ and $-5mn$

91. The sum of -5 and 6 subtracted from the difference of 9 and -4

92. The sum of 7 and -3 subtracted from the sum of -7 and 5

93. The difference of $-5y$ and $-2y$ added to the difference of $7y$ and $-3y$

94. The difference of $4y$ and $9y$ added to the difference of $-5y$ and $8y$

Problems 95–98 involve a scuba diver. Write as subtraction and complete the table.

	Previous Depth	Present Depth	Subtraction	Change in Depth
95.	−18 ft	−62 ft	_____	_____
96.	−22 ft	−37 ft	_____	_____
97.	−34 ft	−15 ft	_____	_____
98.	−84 ft	−49 ft	_____	_____

Write each as subtraction and answer the following:

99. A submarine is submerged at a depth of 145 feet. An airplane searching for the sub flies over at a height of 350 feet. What is the distance between them?

100. If the submarine in Exercise 99 is submerged at a depth of 245 feet and the airplane is at an altitude of 435 feet, what is the distance between them?

101. If a company loses $5000 in its first year of operation and has a profit of $12,000 the following year, how much greater was the company's earnings the second year?

102. If a company has a profit of $6500 dollars one year and a loss of $7300 the next, how much less did the company earn the second year than it did the first?

103. A mine shaft began at 500 feet above sea level and ended at 300 feet below sea level. Suppose a miner is 250 feet above sea level and descends to 125 feet below sea level. What is his change in altitude? Remember, change = new altitude − old altitude.

104. Suppose the miner in Exercise 103 is 210 feet below sea level and ascends to 109 feet below sea level. What is his change in altitude?

105. An airplane is flying over Death Valley at a height of 2500 feet when it flies over a hiker whose elevation is 187 feet below sea level. What is the distance between the airplane and the hiker?

106. For the first half of the fiscal year, a company had a profit of $123,000. During the second half of the year, it had a loss of $105,000. Find the difference in the company's earnings for the first and second halves of the year.

107. A gauge initially had a reading of 17 psi and later had a reading of −6 psi. By how much did the reading change?

108. On the same day in February, the low temperature in East Yellowstone, Montana was −36°F and the low temperature in Orlando, Florida was 52°F. How much higher was the low temperature in Orlando than in East Yellowstone?

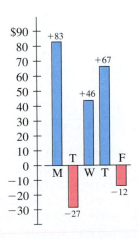

John conducts a small part-time business from his home. The following graph shows the daily profit or loss for his business. Use the graph to answer Exercises 109–110.

109. What was the total profit/loss for the week?

110. How much more did the company earn on Monday than on Tuesday?

Challenge Exercises (111–120):

Evaluate the following:

111. $5^2[6^2 \cdot 2 - (-4)] \div 19$

112. $2^4[48 - (-12) - 32] \div 7 \cdot 4$

113. $5^2\{6 - 2[10 - 3(-9 + 12)]\}$

114. $2^3\{24 - 3[12 - 2(-7 + 11)]\}$

Simplify the following:

115. $\dfrac{4}{5}x^2y + \dfrac{3}{7}xy^2 - \dfrac{2}{5}x^2y + \dfrac{2}{7}xy^2$

118. $\dfrac{4}{17}a^4b^3 - \dfrac{5}{19}a^2b - \dfrac{8}{17}a^4b^3 - \dfrac{3}{19}a^2b$

116. $\dfrac{7}{9}ab^2 + \dfrac{3}{11}a^2b^2 - \dfrac{5}{9}ab^2 - \dfrac{5}{11}a^2b^2$

119. $\dfrac{6}{23}x^3y - \dfrac{16}{29}xy^2 - \dfrac{17}{23}x^3y + \dfrac{7}{29}xy^2$

117. $\dfrac{3}{13}m^3n^2 - \dfrac{5}{9}mn^2 - \dfrac{7}{13}m^3n^2 - \dfrac{2}{9}mn^2$

120. $\dfrac{16}{17}ab^2 + \dfrac{5}{11}a^2b - \dfrac{9}{17}ab^2 - \dfrac{7}{11}a^2b$

Writing Exercises:

121. Is subtraction commutative? Why or why not? (*Hint:* Try some examples.)

122. Is subtraction associative? Why or why not? (*Hint:* Try some examples.)

Writing Exercise or Group Project:

If done as a group project, each group should write at least two exercises. Then exchange with another group and solve. If done as a writing exercise, each student should write and solve one exercise.

123. Write an application problem involving the subtraction of integers, at least one of which is negative.

Section 1.8   Polynomial Definitions and Combining Polynomials

OBJECTIVES

When you complete this section, you will be able to:

a. Identify terms.

b. Identify the numerical coefficient and the variable(s) of a term.

c. Find the degree of a term.

d. Identify types of polynomials.

e. Write polynomials in descending order and find their degrees.

f. Evaluate polynomials for a specified value of the variable.

Introduction

In arithmetic, operations are performed on numbers. In algebra, operations are performed on algebraic expressions. An **algebraic expression** is any constant, variable, or combination of constants and variables using the operations of addition, subtraction, multiplication, division, raising to powers, or taking of roots (which will be discussed later). Examples of algebraic expressions are:

$$5, \; -3, \; x, \; 3x, \; x^3, \; -4x^4y, \; 4x^2 - 5x + 2, \; \dfrac{2x}{y}, \; \dfrac{2x-5}{3x+5}, \; \dfrac{2x^3 - 4y^2}{4y^4 + 3x^2}$$

Identifying terms

We defined "term" in Section 1.7, but will repeat the definition. A **term** is a number, variable, or product and/or quotient of numbers and variables raised to powers. Examples of terms are 5, x, $5x$, $-3x^2$, $2x^3y^2$, $-7x^2y^3z^4$, $\dfrac{x}{y}$ and $4x^{-2}$. An expression like $3x + 2$ is not a term, since addition or subtraction is not allowed in a term. This is actually the sum of two terms. We have not used 0 or negative integers as powers, but we will do so in Chapter 2 when we show that $x^0 = 1$ if $x \neq 0$. We mention this now because we will need it when we discuss the degree of a polynomial later in this section.

Definition of a polynomial

A **polynomial** is the sum or difference of a **finite** number of terms in which there is no variable as a divisor and the exponents on the variables are whole numbers. A set is finite if it is possible to represent the number of elements in the set with a whole number. Examples of polynomials are x, $2x + 3$, $5x^2 + 2x - 4$, $2x^3 + 2x^2 - 4x + 6$, $\dfrac{2}{3}x^2 + \dfrac{4}{5}x - 4$, and 3.

Expressions like $3x^{-1} + 2$ and $\frac{3}{x} + 6$ are not polynomials because negative exponents on variables or division by variables is not allowed in a polynomial.

Identifying types of polynomials

Polynomials can also be classified by the number of terms they contain as illustrated in the following table:

Name	Number of Terms	Examples	Terms
Monomial	One term	4 x $-5x^2$	4 x $-5x^2$
Binomial	Two terms	$3x - 2$ $4x^3 - 3x$	$3x, -2$ $4x^3, -3x$
Trinomial	Three terms	$y^2 - 3y + 5$ $6x^3 + 2x^2 - 4x$	$y^2, -3y, 5$ $6x^3, 2x^2, -4x$

Note: Polynomials with more than three terms are not given special names. You will be asked to suggest names for some of these in the writing exercises.

Example 1

Classify each of the following as a monomial, binomial, trinomial, or none of these:

a. $2x + 4$ *There are two terms, 2x and 4. Therefore, this is a binomial.*

b. $3x^3$ *There is only one term. Therefore, this is a monomial.*

c. $2y^3 + 4y - 2$ *There are three terms, $2y^3$, 4y, and -2. Therefore, this is a trinomial.*

d. $6x^2 - 4x^4 + 5x - 4$ *There are four terms. Therefore, this is none of these.*

e. -4 *There is one term. Therefore, this is a monomial.*

f. $\dfrac{5}{y} - 10$ *We have division by a variable. Therefore, this is not a polynomial.*

Practice Exercises

Classify each of the following as monomial, binomial, trinomial, or none of these:

1. $2y^2 + 4$ **2.** $8x^4$

3. $-3x^2 - 6x^5 - 7x$ **4.** $2x^4 - 2x - x^2 + 9$

5. 7 **6.** $\dfrac{3}{x} + 5$

Coefficients, variables, and degree of monomials

When writing a term, the constant usually written before the variable(s) is called the **numerical coefficient** or simply the coefficient. For example, in the term $3x$, the numerical coefficient is 3. If a monomial has only one variable, the exponent on the variable is called the **degree** of the monomial. If there is more than one variable, the degree of the monomial is the sum of all the exponents on all the variables.

Answers:
Practice Exercises 1–6: **1.** binomial **2.** monomial **3.** trinomial **4.** none of these **5.** monomial **6.** none of these (not a polynomial)

Example 2

Give the coefficient, variable, and degree of each of the following monomials:

	Monomial	Coefficient	Variable(s)	Degree
a.	$4x^3$	4	x	3
b.	$-8z^6$	-8	z	6
c.	y	$1\ (y = 1y^1)$	y	1
d.	$2m^3n^4$	2	m, n	$3 + 4 = 7$
e.	-3	-3	none	0 since $-3 = -3x^0$

Practice Exercises

> **Be careful!** In Example 2e the degree of -3 is 0 since $-3 = -3x^0$. It is often mistakenly given as degree of 1. Remember, the degree of a term is the exponent of the variable, so $-3x$ has a degree of 1. Thus, -3 and $-3x$ could not have the same degree.

Give the coefficient, variable(s), and degree of each of the following.

7. $-8x^3$ coefficient = _____ variable(s) = _____ degree = _____

8. y coefficient = _____ variable(s) = _____ degree = _____

9. $2x^2y^4$ coefficient = _____ variable(s) = _____ degree = _____

10. -5 coefficient = _____ variable(s) = _____ degree = _____

Writing polynomials in descending order and finding their degrees

    A polynomial of one variable is written in **descending order** when the term of highest degree is written first, the term of next highest degree second, and so on. For example, $3x^2 - 2x + 4$ is in descending order, but $2x - 3x^3 - 5$ is not because the term of highest degree is not written first. The **degree of a polynomial** is the same as the degree of the term of the polynomial with the highest degree. See the chart in Example 3.

Example 3

Write each polynomial in descending order and find the degree of the polynomial:

Polynomial	Descending Order	Terms	Degree of Terms	Degree of Polynomial
$3 + 4x$	$4x + 3$	$4x, 3$	$1, 0$	1
$2x - 4 + x^2$	$x^2 + 2x - 4$	$x^2, 2x, -4$	$2, 1, 0$	2
$4x - 5x^3 - 7x^4 + 2$	$-7x^4 - 5x^3 + 4x + 2$	$-7x^4, -5x^3, 4x, 2$	$4, 3, 1, 0$	4

Practice Exercises

Write each of the following polynomials in descending order and give its degree:

Polynomial	Descending Order	Degree
11. $-5 + 3x$	_____	_____
12. $-5 + 2x^2 + 8x$	_____	_____
13. $6x - 7x^4 - 8x^2 + x^3$	_____	_____

Answers:

Practice Exercises 7–13: **7.** $-8; x; 3$ **8.** $1; y; 1$ **9.** $2; x$ and $y; 6$ **10.** $-5;$ none; 0 **11.** $3x - 5$, degree 1 **12.** $2x^2 + 8x - 5$, degree 2 **13.** $-7x^4 + x^3 - 8x^2 + 6x$, degree 4

If more practice is needed, do the Additional Practice Exercises.

Additional Practice Exercises

Write each of the following in descending order and give its degree:

a. $3 - 7x$

b. $7 - 3x + 5x^2$

c. $5y - 4y^2 + 6y^4 - y^3$

Evaluating polynomials

Since polynomials contain variables and variables represent numbers, it is possible to evaluate a polynomial when given a value for its variable. Remember to keep the order of operations in mind when evaluating the polynomial. For convenience in evaluating, polynomials are often written using a special notation. The polynomial is given a name, usually a letter, and then its variable is specified in parentheses. For example, we might write $P(x) = x^2 + 2x - 3$ where P is the name we have given the polynomial and x is the variable. When we wish to assign the variable a value, the value replaces the variable inside the parentheses and is substituted for the variable throughout the expression. If we wish to evaluate $P(x) = x^2 + 2x - 3$ for $x = 3$, we write $P(3) = 3^2 + 2(3) - 3$ and then simplify. So, $P(3) = 9 + 6 - 3 = 12$. $P(x)$ is read "P of x" and means the value of P for a specific value of x. Consequently, $P(3) = 12$ means that the polynomial named P has a value of 12 when x has a value of 3.

Example 4

Evaluate the following polynomials for the given value of the variable:

a. $P(x) = 2x + 4; x = 3$ — Substitute 3 for x.

$P(3) = 2(3) + 4$ — Multiply 2 and 3.

$P(3) = 6 + 4$ — Add 6 and 4.

$P(3) = 10$ — Therefore, the value of P when x is 3 is 10.

b. $P(t) = t^2 + 3t - 4; t = 2$ — Substitute 2 for t.

$P(2) = (2)^2 - 3(2) - 4$ — Raise to powers first, so square 2.

$P(2) = 4 - 3(2) - 4$ — Multiply 3 and 2.

$P(2) = 4 - 6 - 4$ — Add in order left to right.

$P(2) = -2 - 4$ — Add.

$P(2) = -6$ — Therefore, the value of P when t is 2 is -6.

Practice Exercises

Evaluate each of the following polynomials for the given value of the variable:

14. $P(x) = 2x + 5; x = 2$

15. $P(y) = y^2 + 3y - 6; y = 4$

If more practice is needed, do the Additional Practice Exercises.

Additional Practice Exercises

Evaluate each of the following for the given value of the variable:

d. $P(x) = 3x - 7; x = 2$

e. $P(t) = t^2 + 4t - 4; t = 6$

Evaluations of polynomials occur in many situations in the sciences and in the real world.

Example 5

Solve the following:

A particle moves along a straight line such that the distance from the starting point is given by $s(t) = t^2 + 3t + 2$ where $s(t)$ is the distance after t seconds. Find the location of the particle after 3 seconds by finding $s(3)$.

$$s(t) = t^2 + 3t + 2 \qquad \text{To find } s(3) \text{ replace } t \text{ with 3.}$$
$$s(3) = 3^2 + 3(3) + 2 \qquad \text{Square 3.}$$
$$s(3) = 9 + 3(3) + 2 \qquad \text{Multiply 3(3).}$$
$$s(3) = 9 + 9 + 2 \qquad \text{Add in order left to right.}$$
$$s(3) = 20 \qquad \text{Therefore, the particle is 20 units from the starting point after 3 seconds.}$$

Practice Exercise

16. If x units of a product are sold, the revenue is given by $R(x) = 1000x + .2x^2$. Find the revenue when 25 units are sold by finding $R(25)$.

The polynomial plays a central role in elementary algebra. A great deal of the remainder of this book will be spent performing operations on polynomials.

Since polynomials are made of terms, adding or subtracting polynomials is very similar to adding like terms. Parentheses are used to show which polynomials are being added or subtracted, so we must remove the parentheses and combine like terms. If there is no sign preceding the parentheses, there is an understood positive sign and we do not change any signs. We found the negative of an integer by changing its sign. In the same manner we will find the negative of a polynomial by changing *all* of its signs.

Example 6

Remove the parentheses from each of the following:

a. $(2x - 3) = 2x - 3$ There is an understood + sign before the parentheses, so do not change any signs.

b. $-(3x - 6) = -3x + 6$ There is a − sign before the parentheses, so change all the signs of the polynomial.

c. $(-4x + 2) = -4x + 2$ There is an understood + sign before the parentheses, so do not change any signs.

d. $-(2x^2 + 4x - 5) =$
$-2x^2 - 4x + 5$ There is a − sign before the parentheses, so change all the signs of the polynomial.

To add or subtract polynomials, we remove the parentheses and add like terms.

Example 7

Find the following sums or differences:

a. $(2x + 4) + (3x - 6) =$ Remove the parentheses. Do not change signs.
$2x + 4 + 3x - 6 =$ Add like terms.
$5x - 2$ Sum.

Answer:

b. $(2x^2 - 6x + 4) + (x^2 - 3x - 7) =$ Remove parentheses.
   $2x^2 - 6x + 4 + x^2 - 3x - 7 =$ Add like terms.
   $3x^2 - 9x - 3$ Sum.

c. $(2x^2 - 6x + 3) - (x^2 - 2x + 7) =$ Remove parentheses and change all the signs of the second polynomial.
   $2x^2 - 6x + 3 - x^2 + 2x - 7 =$ Add like terms.
   $x^2 - 4x - 4$ Difference.

d. $(x^2 + 6x - 5) - (-2x^2 + 7x - 3) =$ Remove parentheses and change all the signs of the second polynomial.
   $x^2 + 6x - 5 + 2x^2 - 7x + 3 =$ Add like terms.
   $3x^2 - x - 2$ Difference.

Practice Exercises

Add or subtract the following polynomials:

17. $(4x + 6) + (3x + 5)$

18. $(x^2 - 5x + 4) + (2x^2 - 3x - 6)$

19. $(3x^2 - 7x - 4) - (x^2 + 5x - 6)$

20. $(2x^2 + 4x - 7) - (-3x^2 - 8x + 3)$

If more practice is needed, do the Additional Practice Exercises.

Additional Practice Exercises

Add or subtract the following polynomials:

f. $(5x - 3) + (-3x + 6)$

g. $(x^2 - 4x + 2) + (x^2 + 6x - 5)$

h. $(x^2 + x - 7) - (x^2 - 8x - 8)$

i. $(3x^2 + 2x - 9) - (-x^2 + 2x + 3)$

As was the case with integers, addition and subtraction of polynonials is often expressed in words.

Example 8

Write each of the following as addition and/or subtraction and simplify:

a. Find the sum of $2x^2 + 3x - 6$ and $4x^2 - 5x + 2$.

Solution:
Since sum means add, this is written as
$(2x^2 + 3x - 6) + (4x^2 - 5x + 2)$.

$(2x^2 + 3x - 6) + (4x^2 - 5x + 2) =$ Remove the parentheses.
$2x^2 + 3x - 6 + 4x^2 - 5x + 2 =$ Add like terms.
$6x^2 - 2x - 4$ Sum.

b. Subtract $3y^2 - 4y + 3$ from $y^2 - 5y + 7$.

Solution:
This implies that we begin with $y^2 - 5y + 7$ and subtract $3y^2 - 4y + 3$ from it. This is written as
$(y^2 - 5y + 7) - (3y^2 - 4y + 3)$.

$(y^2 - 5y + 7) - (3y^2 - 4y + 3) =$ Remove parentheses.
$y^2 - 5y + 7 - 3y^2 + 4y - 3 =$ Add like terms.
$-2y^2 - y + 4$ Difference.

c. From the sum of $4x + 5$ and $2x - 8$ subtract $7x - 1$.

Solution:
This means that we first add $4x + 5$ and $2x - 8$ and from this sum subtract $7x - 1$.
This is written as $[(4x + 5) + (2x - 8)] - (7x - 1)$.

$[(4x + 5) + (2x - 8)] - (7x - 1) =$ Remove the parentheses inside the brackets.

$[4x + 5 + 2x - 8] - (7x - 1) =$ Add like terms inside the brackets.

$[6x - 3] - (7x - 1) =$ Remove the parentheses and brackets.

$6x - 3 - 7x + 1 =$ Add like terms.

$-x - 2$ Answer.

Practice Exercises

Write each of the following as addition and/or subtraction and simplify:

21. Find the sum of $5z^2 - 7z + 4$ and $2z^2 + 4z - 6$.

22. Subtract $4a^2 - 5a - 5$ from $2a^2 + 7a - 8$.

23. From the sum of $5a + 2$ and $-2a - 5$ subtract $8a - 1$.

Optional: *Adding and Subtracting Polynomials Vertically*

     Addition and subtraction of polynomials may be done vertically as well as horizontally. The procedure is similar to the way whole numbers and decimals are added. We place digits with the same place value underneath each other and add. In adding polynomials vertically, we put them in descending order, then place like terms underneath each other and add. Of course, there is no regrouping (carrying) when adding polynomials.

Example 9

Add the following vertically:

a. 374 and 68

Solution:
Place the digits with the same place value underneath each other and add.

$$\begin{array}{r} 374 \\ 68 \\ \hline 442 \end{array}$$

b. Add $13p^2 + 3p - 4$ and $7p^2 - 4p - 4$.

Solution:
Place like terms underneath each other and add.

$$\begin{array}{r} 13p^2 + 3p - 4 \\ 7p^2 - 4p - 4 \\ \hline 20p^2 - p - 8 \end{array}$$

c. Add $9w^2 + 12w - 14$ and $-5w^2 + 12$.

Solution:
Place like terms underneath each other and add.

$$\begin{array}{r} 9w^2 + 12w - 14 \\ -5w^2 \qquad + 12 \\ \hline 4w^2 + 12w - 2 \end{array}$$

d. Add $3x^3 - 5x + 7$ and $2x^2 + 4x - 9$.

Solution:
Place like terms underneath each other and add.

$$\begin{array}{r} 3x^3 \qquad - 5x + 7 \\ 2x^2 + 4x - 9 \\ \hline 3x^3 + 2x^2 - x - 2 \end{array}$$

Remember, when subtracting polynomials we add the negative of the polynomial being subtracted. Consequently, all its signs are changed. When subtracting vertically, the polynomial being subtracted is the one on the bottom, so we change the signs of the polynomial on the bottom and add.

Example 10

Subtract the following vertically:

a. Subtract $3y^2 - 8y - 10$ from $5y^2 - 9y + 3$.

Solution:
Written horizontally, this is $(5y^2 - 9y + 3) - (3y^2 - 8y - 10)$, so it is necessary to change the signs of $3y^2 - 8y - 10$. Written vertically, it appears as follows:

Subtract:

$$5y^2 - 9y + 3$$
$$3y^2 - 8y - 10$$

Change the signs of the polynomial on the bottom since it is the polynomial being subtracted and change the operation to addition.

Add:

$$\begin{array}{r} 5y^2 - 9y + 3 \\ -3y^2 + 8y + 10 \\ \hline 2y^2 - y + 13 \end{array}$$

b. Find the difference of $6x^2 - 11x + 9$ and $3x^2 + 6$.

Solution:
Just like with whole numbers, difference means subtract the second expression from the first.

Subtract:

$$6x^2 - 11x + 9$$
$$3x^2 + 6$$

Change the signs of the bottom polynomial and add.

Add:

$$\begin{array}{r} 6x^2 - 11x + 9 \\ -3x^2 - 6 \\ \hline 3x^2 - 11x + 3 \end{array}$$

c. Subtract $4x^2 - 6x - 8$ from $3x^3 - 7x^2 + 5$. (This is the same as saying "Find the difference of $3x^3 - 7x^2 + 5$ and $4x^2 - 6x - 8$.")

Solution:
Subtract:

$$3x^3 - 7x^2 + 5$$
$$ 4x^2 - 6x - 8$$

Change the signs of the bottom polynomial and add.

Add:

$$\begin{array}{r} 3x^3 - 7x^2 + 5 \\ - 4x^2 + 6x + 8 \\ \hline 3x^3 - 11x^2 + 6x + 13 \end{array}$$

Practice Exercises

Add or subtract the following vertically:

24. Add $5x^2 + 7x - 3$ and $3x^2 - 8x + 5$.

25. Add $10x^2 + 9x - 6$ and $-4x^2 - 9$.

26. Add $3x^3 - 4x + 5$ and $2x^2 + 6x - 9$.

27. Subtract $6x^2 + 7x - 9$ from $2x^2 + 5x + 7$.

28. Find the difference of $8y^2 - 9y + 3$ and $5y^2 + 9$.

29. Subtract $7x^2 - 8x + 2$ from $8x^3 - 9x^2 + 4$.

Study Tip 6

Making Sure: Confidence Building

We feel good about what we have done when we know that it is correct. We can tell if our work is accurate by checking our answers. Sometimes the answer is in the back of the book, but most of the time we need to make our own check. The check should be different from and shorter than reworking the exercise. For example, the check for subtraction is addition: $8 - 5 = 3$, if $3 + 5 = 8$. Also, the check for division is multiplication: $54 \div 9 = 6$, if $6 \cdot 9 = 54$. Always check your work!

Exercise Set 1.8

Classify each of the following as a monomial, binomial, trinomial, *or none of these:*

1. $2x^2$ _____

2. $3y^3$ _____

3. $x^2 - 3x + 4$ _____

4. $x^2 - 5x + 2$ _____

5. $5 + 8x$ _____

6. $3 - 7x$ _____

7. $2x - 2x^2 + 4$ _____

8. $4 - 3x + 5x^2$ _____

9. $\dfrac{2x^2}{y^2} - 4x$ _____

10. $\dfrac{3y^3}{2x^2} - 2y$ _____

11. 5 _____

12. -7 _____

13. $x^3 + 3x - 4x^2 - 3$ _____

14. $x^4 - 3x + 2x^3 + 4$ _____

Give the coefficient, variable(s), and degree of each of the following:

	Coefficient	Variable(s)	Degree
15. $3x$	_____	_____	_____
16. $2y$	_____	_____	_____
17. x^2	_____	_____	_____
18. y^4	_____	_____	_____
19. $-5x$	_____	_____	_____

Answers:

Practice Exercises 24–29: **24.** $8x^2 - x + 2$ **25.** $6x^2 + 9x - 15$ **26.** $3x^3 + 2x^2 + 2x - 4$ **27.** $-4x^2 - 2x + 16$ **28.** $3y^2 - 9y - 6$ **29.** $8x^3 - 16x^2 + 8x + 2$

	Coefficient	Variable(s)	Degree
20. $-7z$	_____	_____	_____
21. $4x^3$	_____	_____	_____
22. $7y^5$	_____	_____	_____
23. $-x^3$	_____	_____	_____
24. $-y^5$	_____	_____	_____
25. $10x^2y^4$	_____	_____	_____
26. $12x^5y^2$	_____	_____	_____
27. $-13x^3z^3$	_____	_____	_____
28. $-16y^4z^6$	_____	_____	_____

Write each of the following in descending order and give the degree of the polynomial:

	Descending Order	Degree
29. $4 - 5x$	_____	_____
30. $5 + 2y$	_____	_____
31. $3x - 4 + 2x^2$	_____	_____
32. $5x - 6 + 4x^2$	_____	_____
33. $3x - 4x^2 - x^4 + x^3$	_____	_____
34. $2y + 3y^3 - y^4 + 5$	_____	_____
35. $x + 3x^3$	_____	_____
36. $4 + 3x^2$	_____	_____

Evaluate each of the following for the indicated value of the variable:

37. $P(x) = 3x - 9; x = 3$ **38.** $P(y) = 4y - 3; y = 2$

39. $P(x) = 2x^2 - 6; x = 4$ **40.** $P(x) = 2x^2 - 5; x = 3$

41. $P(x) = x^2 + 3x - 5; x = 1$ **42.** $P(x) = x^2 + 4x - 3; x = 5$

43. $P(s) = 2s^2 + 5s - 7; s = 4$ **44.** $P(s) = 3s^2 + 6s - 12; s = 5$

45. $P(x) = x^3 + 2x^2 + 3x - 3; x = 2$ **46.** $P(x) = x^3 + 4x^2 + 2x + 5; x = 3$

Calculator Exercises:

C1. Use a calculator to evaluate $P(x) = x^3 - 5.1x^2 + 4.3x - 6.7$ for $x = 4.7$.

C2. Use a calculator to evaluate $P(x) = 3.2x^4 - 6.5x^3 + 8.5x - 9.8$ for $x = 2.3$.

If an object is dropped from a resting position, the distance (in feet) it falls is given by s(t) = 16t². *Use this to answer Exercises 47–48.*

47. While hunting in rural Georgia, a hunter discovers an old homesite with an open well. In order to estimate the depth of the well, he drops a stone into it and observes that it takes $t = 2.5$ seconds for the stone to hit the water. Find the depth of the well by finding $s(2.5)$

48. If a baseball is dropped from the top of the Sears Tower, find how far it has fallen after 3 seconds by finding $s(3)$.

A particle moves along a line such that the distance from the starting point is given by s(t) = t³ + 2t² + 3t + 5 *with t measured in seconds. Use this to answer Exercises 49–50.*

49. Find the location of the particle after 2 seconds by finding $s(2)$.

50. Find the location of the particle after 5 seconds by finding $s(5)$.

A furniture manufacturing company finds that the cost of manufacturing x sofas is given by C(x) = 3500 + 23x + .4x². *Use this to answer Exercises 51–52.*

51. Find the cost of producing 100 sofas by finding $C(100)$.

52. Find the cost of producing 500 sofas by finding $C(500)$.

A farmer has 200 feet of fencing and wishes to enclose a rectangular field. If the length of the field is x, then the area is A(x) = 100x − x². *Use this to answer Exercises 53–54.*

53. Find the area when the length is 50 feet by finding $A(50)$.

54. Find the area when the length is 30 feet by finding $A(30)$.

Find the following sums and differences of polynomials:

55. $(6x - 5y) + (3x + 2y)$

56. $(3x + 8y) + (7x - 2y)$

57. $(5x - 4) - (3x - 6)$

58. $(4y + 7) - (6y - 2)$

59. $(2x^2 - 7x + 9) + (3x^2 + 2x + 5)$

60. $(3x^2 - 5x + 4) + (4x^2 - 5x + 8)$

61. $(2m^2c^2 + 5mc - 9c^2 + 5) + (3m^2c^2 - mc - 2)$

62. $(3q^2d^2 + 6qd - 4d^2 + 4) + (3q^2d^2 - qd - 3)$

63. $(4r^2 + 3r - 5) - (9r^2 - 3r + 5)$

64. $(2k^2 + 9k - 5) - (9k^2 - 9k + 5)$

65. $(8rp^2 - 5rp) - (r^3 + 4rp) + (4rp - 7rp^2)$

66. $(-5m^2t + 9) - (4mt^2 + 9m - 9) + (6mt^2 + 5m^2t)$

67. $(4h^2 + 7h - 4) - (3h^2 - 4h - 5) + (-11h - 10)$

68. $(3p^3 - 9p + 7) + (3p^2 - 5p + 6) - (8p^2 - 3)$

Represent each of the following using addition and/or subtraction and simplify:

69. Add $7x^2 + 9x - 1$ and $4x^2 - 3x + 4$.

70. Add $4x^2 + 9x - 2$ and $9x^2 - 3x - 7$.

71. Subtract $6x^2 + 7x - 3$ from $3x^2 - 3x + 1$.

72. Subtract $2z^2 + z - 3$ from $4z^2 + 7z - 4$.

73. Find the sum of $3x + 13$ and $5x - 8$.

74. Find the sum of $5x + 16$ and $7x - 3$.

75. Find the difference of $4x - 7$ and $2x - 12$.

76. Find the difference of $5y + 7$ and $2y - 6$.

77. Find the sum of $x^2 - 3x + 4$ and $2x^2 + 4x - 6$.

78. Find the sum of $y^2 - 7y + 3$ and $3y^2 + 6y - 8$.

79. Find the difference of $z^2 + 7z - 9$ and $z^2 - 5z + 2$.

80. Find the difference of $w^2 - 8w + 4$ and $w^2 - 7w - 9$.

81. From the sum of $2x^2 - 7x - 2$ and $x^2 + 5$ subtract $2x^2 + 4x - 3$.

82. From the sum of $3y^2 + 8y - 6$ and $2y^2 - 9$ subtract $4y^2 - 7y - 4$.

83. To the difference of $3x - 7$ and $4x + 2$ add $5x - 9$.

84. To the difference of $4x + 2$ and $6x - 3$ add $9x + 5$.

Recall that the perimeter of a triangle is found by adding the lengths of all three sides. For example, if the sides of a triangle are x inches, 4x inches, and 2x inches, the perimeter is x + 4x + 2x = 7x inches. Find the following:

85. Find the perimeter of a triangle whose sides are $2x$ cm, $5x$ cm, and x cm.

86. Find the perimeter of a triangle whose sides are x ft, $6x$ ft, and $4x$ ft.

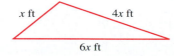

87. The lengths of the sides of a triangular sign are x inches, $(2x - 3)$ inches, and $(5x + 4)$ inches. What is the perimeter of the sign?

88. The lengths of the sides of a triangular piece of land are x feet, $(3x - 2)$ feet, and $(7x + 1)$ feet. How many feet of fencing would it take to enclose it?

A quadrilateral is a figure with four sides. The perimeter of a quadrilateral is found by adding the lengths of all four sides. For example, if the sides are 2x ft, 3x ft, x ft and 2x ft, the perimeter is 2x + 3x + x + 2x = 8x ft

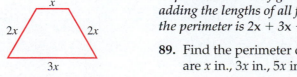

89. Find the perimeter of a quadrilateral whose sides are x in., $3x$ in., $5x$ in., and $7x$ in.

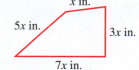

90. Find the perimeter of a quadrilateral whose sides are x m, $4x$ m, $2x$ m, and $8x$ m.

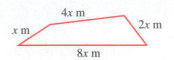

91. If the dimensions of a yard are x feet, $(2x + 3)$ feet, $(x + 4)$ feet, and $(3x + 2)$ feet, how many feet of fencing would it take to enclose it?

92. If the dimensions of a patio are x feet, $(3x - 1)$ feet, $(4x + 2)$ feet, and $(5x - 7)$ feet, how many feet of screen would it take to enclose it? Assume that each piece of screen will reach to the ceiling.

Represent each of the following using variables. Make a drawing for Exercises 96–98 similar to those in Exercises 93 and 95.

93. John is building a piece of furniture. He needs a board that is x feet long and another that is $(2x - 2)$ feet long. What is the total length of the two boards needed?

$$\underset{x \text{ ft} \qquad\qquad (2x-2) \text{ ft}}{\rule{1pt}{0pt}\rule{300pt}{0.5pt}}$$

94. If John has x dollars and Mary has $(3x - 5)$ dollars, how much money do they have combined?

95. Christine is making a shelf for a closet and has a board $(2x - 5)$ feet long. She has to cut off a piece $(x - 3)$ feet long. How long is the remaining piece?

$$\overset{\longleftarrow\quad (2x-5)\text{ ft}\quad\longrightarrow}{\underset{(x-3)\text{ ft}}{\rule{300pt}{0.5pt}}}$$

96. A carpenter is installing baseboard and has a board $(3y + 2)$ feet long. He has to cut off a piece $(y + 4)$ feet long. How long is the remaining piece?

97. If Chan climbs $(3x + 3)$ meters and then climbs $(2x - 5)$ meters more, how high has he climbed altogether?

98. If Yolanda has driven $(5x - 6)$ miles and then drives $(3x + 2)$ miles farther, what is the total distance she has driven?

Challenge Exercises:

99. $\left(\dfrac{5}{7}x + \dfrac{3}{11}y\right) + \left(\dfrac{1}{7}x - \dfrac{4}{11}y\right)$

100. $\left(\dfrac{4}{13}a - \dfrac{3}{7}b\right) + \left(\dfrac{5}{13}a - \dfrac{2}{7}b\right)$

101. $\left(\dfrac{9}{13}r - \dfrac{6}{17}m\right) - \left(\dfrac{7}{13}r + \dfrac{10}{17}m\right)$

102. $\left(\dfrac{3}{7}m - \dfrac{4}{5}n\right) - \left(\dfrac{5}{7}m - \dfrac{1}{5}n\right)$

103. $\left(\dfrac{2}{3}a^2 - \dfrac{3}{5}a + \dfrac{4}{15}\right) + \left(-\dfrac{1}{3}a^2 + \dfrac{1}{5}a + \dfrac{3}{15}\right)$

104. $\left(\dfrac{4}{9}x^2 + \dfrac{2}{11}x - \dfrac{9}{19}\right) + \left(\dfrac{4}{9}x^2 - \dfrac{7}{11}x + \dfrac{4}{19}\right)$

105. $\left(\dfrac{8}{23}y^2 - \dfrac{8}{19}y + \dfrac{6}{13}\right) - \left(\dfrac{12}{23}y^2 - \dfrac{7}{19}y + \dfrac{2}{13}\right)$

106. $\left(\dfrac{10}{17}y^2 + \dfrac{16}{29}y - \dfrac{5}{31}\right) - \left(\dfrac{6}{17}y^2 + \dfrac{23}{29}y + \dfrac{16}{31}\right)$

Writing Exercises:

107. How does an algebraic expression differ from a polynomial?

108. Give three examples of algebraic expressions that are not polynomials.

109. If special names were given to polynomials of more than three terms, what suggestions would you have for the name of a polynomial of four terms? Five terms?

110. How do you find the degree of a monomial? A polynomial with two or more terms?

111. What is meant by writing a polynomial in descending order?

112. Why does 2^4y^2 have degree 2 and x^4y^2 have degree 6?

113. How does a term differ from a monomial?

Critical-Thinking Exercises:

114. Are all terms monomials? Explain. Are all monomials terms? Explain.

115. Do the following: 1. Choose any number; 2. Triple the number; 3. Add 1; 4. Subtract the original number; 5. Repeat steps 3 and 4; 6. Repeat steps 3 and 4 again; 7. The result is always 3; Write a paragraph explaining why.

116. Write a "riddle" similar to the one in Exercise 115.

Chapter 1   Summary

Definition of Natural Number Exponents: [Section 1.1]
- a^n means multiply n factors of a.

Definition of Variables and Constants: [Section 1.1]
- A variable is a symbol, usually a letter, that is used to represent a number.
- A constant is a symbol, a number, whose value does not change.

Order of Operations: [Section 1.1]
- If an expression contains more than one operation, operations are to be performed in the following order:

  1. If parentheses or other inclusion symbols (braces or brackets) are present, begin within the innermost and work outward, using the order in steps 3–5 in doing so.
  2. If an implied grouping like a fraction bar is present, simplify above and below the fraction bar separately in the order given by steps 3–5.
  3. Evaluate all indicated powers.
  4. Perform all multiplication or division in the order in which they occur as you work from left to right.
  5. Then perform all additions or subtractions in the order in which they occur as you work from left to right.

Perimeter: [Section 1.2]
- The perimeter of a geometric figure is the distance around the figure. The distance around a circle is the circumference. Perimeter and circumference are expressed in linear units.

Perimeter Formulas: [Section 1.2]

Figure	Formula
Triangle	$P = a + b + c$
Rectangle	$P = 2L + 2W$
Square	$P = 4s$
Circle	$C = 2\pi r$ or $C = \pi d$

Area: [Section 1.3] • The area is the measure of the size of a surface and is expressed in square units.

Area Formulas: [Section 1.3]

Figure	Formula
Rectangle	$A = LW$
Square	$A = s^2$
Triangle	$A = \dfrac{bh}{2}$
Parallelogram	$A = bh$
Trapezoid	$A = \dfrac{h(B + b)}{2}$
Circle	$A = \pi r^2$

Volume: [Section 1.4] • The volume of a geometric solid is a measure of how much it will hold and is expressed in cubic units.

Volume Formulas: [Section 1.4]

Solid	Formula
Rectangular Solid	$V = LWH$
Cube	$V = e^3$
Right Circular Cylinder	$V = \pi r^2 h$
Cone	$V = \dfrac{\pi r^2 h}{3}$

Surface Area: [Section 1.4] • The surface area of a three-dimensional object is the sum of the areas of all its surfaces and is expressed in square units.

Surface Area Formulas: [Section 1.4]

Solid	Formula
Rectangular Solid	$SA = 2LW + 2LH + 2WH$
Cube	$SA = 6e^2$
Right Circular Cylinder	$SA = 2\pi r^2 + 2\pi rh$

Types of Numbers: [Section 1.5]
- Natural Numbers = $\{1, 2, 3, \ldots\}$
- Whole Numbers = $\{0, 1, 2, 3, \ldots\}$
- Integers = $\{\ldots, -3, -2, -1, 0, 1, 2, 3, \ldots\}$

Order Symbols: [Section 1.5]
- $=$ equals
- $\neq$ does not equal
- $<$ is less than
- $\leq$ is less than or equal to
- $>$ is greater than
- $\geq$ is greater than or equal to

Negative of a Negative: [Section 1.5]
- $-(-x) = x$ for all x.

Absolute Value: [Section 1.5]
- $|x| = x$ if $x \geq 0$ and $|x| = -x$ if $x < 0$.

Rules for Addition of Integers: [Section 1.6]
- To add integers with the same sign, add their absolute values and give the answer the same sign as the numbers being added.
- To add integers with opposite signs, subtract their absolute values and give the answer the sign of the number with the larger absolute value.

Properties of Addition: [Section 1.6]
- Commutative:
  $a + b = b + a$ Addition may be performed in any order.
- Associative:
  $a + (b + c) = (a + b) + c$ Addition may be grouped in any manner.

- Inverse: For every integer a, there exists an integer $-a$ such that $a + (-a) = -a + a = 0$.
- Identity: The integer 0 is the additive identity since $0 + a = a + 0 = a$.

Rule for Subtraction: [Section 1.7]
- $a - b = a + (-b)$ To subtract a number, add its inverse (opposite).

Definition of Term: [Section 1.7, 1.8]
- A term is a number, variable, or product and/or quotient of numbers and variables raised to powers.

Definition of Like Terms: [Section 1.7, 1.8]
- Like terms are terms that have the same variables with the same exponents on these variables.

Distributive Property of Multiplication over Addition: [Section 1.7]
- For all numbers a, b, and c, $a(b + c) = ab + ac$.

Addition of Like Terms: [Section 1.7]
- To add like terms, add the numerical coefficients and leave the variable portion unchanged.

Numerical Coefficient: [Section 1.8]
- The numerical coefficient of a term is the number factor of the term.

Degree of a Term: [Section 1.8]
- The degree of a term is the exponent on the variable if there is only one variable and is the sum of the exponents on the variables if there is more than one variable.

Definition of Polynomial: [Section 1.8]
- A polynomial is the sum of a finite number of terms that do not have a variable as a divisor and the exponents on the variables are whole numbers.

Types of Polynomials: [Section 1.8]
- A monomial is a polynomial with one term.
- A binomial is a polynomial with two terms.
- A trinomial is a polynomial with three terms.

Descending Order: [Section 1.8]
- A polynomial is written in descending order if the term of highest degree is written first, the term of second highest degree is written second, and so on.

Degree of a Polynomial: [Section 1.8]
- The degree of a polynomial is the same as the degree of the term of the polynomial with the highest degree.

Addition of Polynomials: [Section 1.8]
- To add polynomials, add the like terms and leave the sum in descending order.

Subtraction of Polynomials: [Section 1.8]
- To subtract a polynomial, add the negative of the polynomial.

Chapter 1   Review Exercises

Write each of the following using exponents: [Section 1.1]

1. $9 \cdot 9 \cdot 9 \cdot 9 \cdot 9$

2. $(-2)(-2)(-2)(-2)$

3. $3 \cdot 3 \cdot 3 \cdot 5 \cdot 5 \cdot 5 \cdot 5$

4. $12 \cdot 9 \cdot 9 \cdot 9 \cdot 2 \cdot 2$

5. $3 \cdot 3 \cdot 3 \cdot a \cdot a \cdot a$

6. $7 \cdot 7 \cdot 7 \cdot r \cdot r \cdot r \cdot r \cdot s \cdot s$

Evaluate each of the following: [Section 1.1]

7. $2^2 \cdot 3^4$

8. $5 \cdot 2^4$

9. $3^2 \cdot 5^2 \cdot 2^2$

10. $10^2 \cdot 2^2 \cdot 5^3$

Find the value of each of the following when $x = 3$ and $y = 2$: [Section 1.1]

11. $x^3 y$

12. $x^2 y^2$

13. $5x^2 y$

14. $4xy^5$

15. Find the area of a square each of whose sides is 13 inches long. [Section 1.1]

16. Find the volume of a cube each of whose edges is 6 centimeters. [Section 1.1]

Evaluate each of the following using the order of operations: [Section 1.1]

17. $8 - 2 \cdot 3$

18. $2^4 + 24 \div 6$

19. $29 - 4(7 - 2)$

20. $5 + 2 \cdot 3^2$

21. $4^2 - 24 \div 2^2$

22. $64 \div (3^3 - 11)$

23. $8 + 48 \div 8 \cdot 3 + 5$

24. $12 + 8 \cdot 9 \div 3 - 9$

Find the value of each of the following when x = 6 and y = 3: [Section 1.1]

25. $5x - 2y^2$

26. $2x^2 \div 3y$

27. $5xy^2$

Find the perimeter of each of the following: [Section 1.2]

28.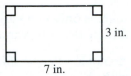
7 in.
3 in.

29.
9 ft
9 ft

30.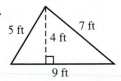
5 ft   7 ft
4 ft
9 ft

31. Find the circumference. Use $\pi \approx 3.14$.

5 m

Find the areas of each of the following: [Section 1.3]

32.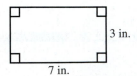
7 in.
3 in.

33.
9 ft
9 ft

34.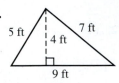
5 ft   7 ft
4 ft
9 ft

35. Find the area. Use $\pi \approx 3.14$.

5 m

36.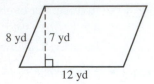
8 yd   7 yd
12 yd

37.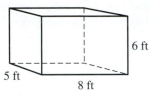
9 cm
6 cm
18 cm

Find the volume and surface area of each of the following: [Section 1.4]

38.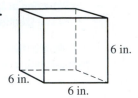
6 ft
5 ft   8 ft

39.
6 in.
6 in.   6 in.

40. Leave answer in terms of π.

7 m
5 m

41. Find the volume only. Leave answer in terms of π.

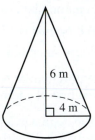
6 m
4 m

Answer the following: [Sections 1.2–1.4]

42. A rancher is planning to fence in a holding pen for his cattle. If the pen is to be in the shape of a square 250 feet on each side, how much fencing will he need?

43. An irrigation system consists of a pipe that is anchored on one end to remain stationary and the other end is on wheels. When the system is turned on, the pipe rotates about the stationary end and the end on wheels sweeps out a circle. If the pipe is 150 feet long, how far do the wheels travel in one rotation?

44. A rectangular bedspread is 120 inches long and 90 inches wide. Pat is going to sew a lace border around the bedspread. If the lace costs $1.50 per yard, how much will the lace cost?

45. Find the cost of the carpet needed to cover the floor of a room that is 14 yards long and 10 yards wide if the carpet costs $13.00 per square yard?

46. A window is in the shape of a square each of whose sides is 52 inches long. What is the area of the window?

47. A sign for an advertisement is in the shape of a trapezoid. If the parallel sides are 10 inches and 6 inches long and the height of the sign is 8 inches, find the area of the sign.

48. Find the area watered in one revolution by the irrigation system in Exercise 43.

49. A tabletop is in the shape of a triangle two of whose sides are 17 inches long and the third of which is 16 inches long. The height to the 16-inch side is 15 inches. Find the area of the tabletop.

50. A box of rice is in the shape of a rectangular solid that is six inches long, one inch wide, and eight inches high. a) What is the volume of the box? b) How much cardboard was used in constructing the box?

51. A can of soup is in the shape of a right circular cylinder whose diameter is three inches and whose height is five inches. a) What is the volume of the can? b) How much steel was used in constructing the can? Use $\pi \approx 3.14$.

52. A drinking cup is in the shape of a right circular cylinder whose radius is two inches and whose height is six inches. What is the volume of the cup?

Answer each of the following using a signed number (integer): [Section 1.5]

53. If + $25 represents winning $25, how would you represent losing $50?

54. If −5 feet represents the depth of a hole, how would you represent the height of a pile of dirt 7 feet high?

Give the negative (opposite) of each of the following: [Section 1.5]

55. 23

56. −35

57. $|7|$

58. $|-32|$

59. $-|64|$

60. $-|-86|$

Insert the proper symbol (=,<, or >) to make each of the following true: [Section 1.5]

61. 5 _____ 9

62. −28 _____ 9

63. −8 _____ −10

64. $|12|$ _____ $|-12|$

65. $-|-16|$ _____ $|16|$

66. $|-32|$ _____ $-|32|$

Find the following sums and/or differences: [Sections 1.6 and 1.7]

67. 19 − 13

68. 9 − (−6)

69. −16 − (−9)

70. 8 − 12 + 17

71. −24 − (−33) + 14

72. (−9 + 5) + (5 − 8)

73. (−53 − 37) − (32 − 27)

74. (−13 − 32) − (41 − 56)

Give the name of the property illustrated by each of the following: [Section 1.6]

75. 9 + 12 = 12 + 9

76. 6 + (5 + 1) = (6 + 5) + 1

77. 6 + (5 + 1) = 6 + (1 + 5)

78. (3 + 8) + 5 = (8 + 3) + 5

79. 9 + (−4 + 4) = 9 + 0

80. 0 + (−3 + 2) = −3 + 2

Write an expression for each of the following and evaluate: [Sections 1.6 and 1.7]

81. The sum of −9 and 6

82. The difference of 15 and −8

83. 16 more than −5

84. −6 less than 3

85. −9 increased by 8

86. −3 decreased by −7

87. The sum of 18 and −8 added to the sum of −5 and 13

88. The difference of −5 and 9 added to the difference of −9 and −2

89. The difference of 4 and −5 subtracted from the sum of 14 and −8

90. The product of 5 and the sum of −6 and 10

Simplify the following by adding like terms: [Section 1.7]

91. $2x - 5x$

92. $8x + 4y - 6x - 3y$

93. $-13w^3 + 7u^2 - 5 + 8w^3 - 13u^2 + 8$

94. $14x^2y - 23xy^2 - 19x^2y + 29xy^2 + 4$

Give the coefficient, variable(s), and degree of each of the following: [Section 1.8]

	Coefficient	Variable(s)	Degree
95. $9x^2$	_____	_____	_____
96. $-13a^6$	_____	_____	_____
97. $-9x^5y^4$	_____	_____	_____

Write each of the following in descending order and classify each as a monomial, binomial, trinomial, or none of these. Give the degree of those that are polynomials. [Section 1.8]

98. $4x^3 + 6x$

99. $7y^2 - 8y + 2$

100. $5x^7$

101. $9x^4 - 7x^5 + 2x$

102. $2a^3 - 5a^2 + 14a^6 + 19$

103. $\dfrac{3x^6}{y} + 9x - 8$

Evaluate each of the following for the indicated value of the variable: [Section 1.8]

104. $P(x) = 3x^2 + 2x - 8; x = 1$

105. $P(t) = t^2 + 6t - 9; t = 3$

106. $P(a) = 2a^3 + 5a; a = 2$

107. $P(b) = 3b^4 + 5b - 5; b = 3$

Find the following sums and/or differences of polynomials. Leave the answers in descending order. [Section 1.8]

108. $(4x + 3) + (6x - 8)$

109. $(7a + 3b) - (2a - 8b)$

110. $(u^4 - 2u^3 + u) + (4u^4 - 3u^3 + u)$

111. $(4a^2 - 7a + 2) - (2a^2 + 7a - 3)$

112. $(z^4 - 3z^3 + 6z) + (2 - 4z + 4z^3)$

113. $(x^4 - 1 - 2x^2 + 3x^3) - (3x^3 - 4x^4 + 2x^2)$

114. $(3z^2 + 5z - 1) - (7z^2 - 7z + 8) + (5z^2 - 6z + 1)$

Write each of the following as addition and/or subtraction and simplify: [Section 1.8]

115. Find the sum of $5x^2 - 9x + 5$ and $6x^2 - 2x - 1$.

116. Find the sum of $3x^3 + 4x^2 - 4$ and $2x^2 - 4x - 7$.

117. Find the difference of $4x^2 + 7x - 2$ and $6x^2 - 6x + 3$.

118. Subtract $3u^4 - 3u^3 + 4u^2 - 7$ from $u^4 + u^2 - 3u + 1$.

119. From the sum of $-x^3 - 4x^2 + 2x + 2$ and $2x^3 - 3x^2 - 2x + 1$, subtract $3x^3 + 6x^2 - 2x - 2$.

120. Find the perimeter of a triangle whose sides are $3x - 5$ ft, $4x + 1$ ft, and $2x - 2$ ft.

121. If a board is $(2x + 6)$ feet long and a piece $(x - 5)$ feet long is cut off, what is the length of the remaining piece?

122. If John climbs $(x + 5)$ feet and then climbs $(3x - 8)$ feet farther, what is the total distance he has climbed?

123. If Clem has $(4x - 6)$ dollars and spends $(2x + 3)$ dollars, how much money does he have left?

Chapter 1   Test

Write each of the following using exponents:

1. $6 \cdot 6 \cdot 6 \cdot 6 \cdot 5 \cdot 5$

2. $4 \cdot 4 \cdot x \cdot x \cdot x \cdot x \cdot y \cdot y$

3. $3 \cdot x \cdot x \cdot x \cdot x \cdot x$

4. $(4a)(4a)(4a)$

Evaluate each of the following:

5. $36 - 2 \cdot 3^2 \div 6$

6. $5(8^2 - 16) \div 12 + 3$

7. $7 + 81 \div 3 \cdot 3 - 18$

8. $\dfrac{8^2 + 6 \cdot 4^2}{6^2 + 4 \cdot 11}$

If x = 4 and y = 2, find the value of the following:

9. $8xy \div 2y^3$

10. $\dfrac{2x^2 + y^2}{2y^2 + 1}$

Find the perimeter and area of the following:

11.

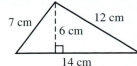

7 cm 12 cm

6 cm

14 cm

12.

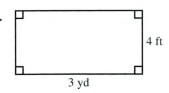

4 ft

3 yd

Find the area of each of the following:

13.

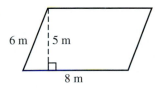

6 m 5 m

8 m

14.

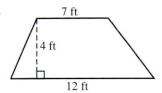

7 ft

4 ft

12 ft

15. Find the circumference and area of the given circle. Use $\pi \approx 3.14$.

14 m

Find the volume and surface area of the following:

16.

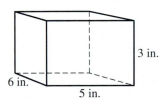

3 in.

6 in.

5 in.

17.

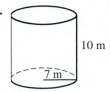

10 m

7 m

18. Magna plans to fence in a square garden plot that measures 18 feet on each side. If fencing costs $.50 per linear foot, how much will it cost to fence in the plot?

19. A rectangular wall is 15 feet long and 12 feet high. How much will it cost to paint the wall if the paint costs $.40 per square foot?

20. A tank is in the shape of a rectangular solid with a length of 6 feet, a width of 4 feet, and a height of 3 feet. The tank is to be filled with water that weighs 62.5 pounds per cubic foot. Find the weight of the water in the tank.

Insert the proper symbol (<, >, or =) in order to make each of the following true:

21. -3 _____ $|-6|$

22. $-|-10|$ _____ $|10|$

Find the following sums or differences:

23. $-16 + 5$

24. $-9 + (-6) + 8$

25. $(-11 + 5) - (7 - 12)$

26. $-12 - [22 - (-8 - 2)]$

Name the property illustrated by each of the following:

27. $(-5 + 7) + 4 = [7 + (-5)] + 4$

28. $8[4 + (-4)] = 8(0)$

Write an expression for each of the following and evaluate:

29. The sum of -7 and 11

30. The difference of 9 and -6 added to -12

31. The sum of -3 and 9 subtracted from the sum of 4 and -13

32. The difference of $x^2 - 6x + 2$ and $2x^2 + 5x - 3$

33. The sum of $3x + 6$ and $2x - 3$ subtracted from $4x + 5$

34. A board is $(3x + 6)$ meters long and a piece $(2x - 2)$ meters long is cut off. How long is the remaining piece?

35. Answer the following for the polynomial
$3x^2 + 4x^3 + 6x$:

a. What is the numerical coefficient of the second term?

b. What is the degree of the polynomial?

c. Write the polynomial in descending order.

d. Is the polynomial a *monomial, binomial, trinomial,* or *none of these*?

36. Evaluate $P(x) = 3x^2 - 2x + 8$ for $x = 2$.

Simplify the following:

37. $9xy + 6yz - 4xy - 10yz$

38. $(2x^2 - 5x + 7) + (3x^2 + 2x - 11)$

39. $(6n^3 + 7n^2 - 8) - (9n^2 - 7n - 6)$

40. Is $-x$ always a negative number? Explain.

In this chapter we will continue our discussion of operations on integers and polynomials. In Chapter 1 we found the sums and differences of integers and polynomials. In this chapter we will find the products and quotients of integers, the products of polynomials, the quotients of monomials, and the quotients of polynomials and monomials. We will not discuss the division of polynomials by polynomials other than monomials until Chapter 6.

Laws of Exponents, Products and Quotients of Integers and Polynomials

Chapter 2

<table>
<tr><td>**Section 2.1**</td><td># Multiplication of Integers</td></tr>
</table>

OBJECTIVES

When you complete this section, you will be able to:

a. Multiply signed numbers.

b. Raise signed numbers to powers.

c. Identify properties of multiplication.

Introduction

In Sections 1.6 and 1.7 we learned to add and subtract signed numbers. In this section we will use our knowledge of how signed numbers are added to develop rules for multiplication of signed numbers.

OPTIONAL   **CALCULATOR EXPLORATION ACTIVITY**

Using a calculator, complete the following.

Column A	Column B
$3(3) =$	$-3(3) =$
$3(2) =$	$-3(2) =$
$3(1) =$	$-3(1) =$
$3(0) =$	$-3(0) =$
$3(-1) =$	$-3(-1) =$
$3(-2) =$	$-3(-2) =$
$3(-3) =$	$-3(-3) =$

Based on your answers to column A, answer the following:

1. Is the product of two positive numbers positive or negative?

2. Is the product of a positive number and a negative number positive or negative?

Based on your answers to column B, answer the following:

3. Is the product of a negative number and a positive number positive or negative?

4. Is the product of two negative numbers positive or negative?

5. What is the product of any number and 0?

Conclusions based on patterns can often be misleading, so we now give a mathematical basis for the conclusions reached in questions 1–4 above.

Recall that multiplication is a shortcut for repeatedly adding the same number. For example, $3(5) = 5 + 5 + 5$. The first number, 3, tells us how many of the second number, 5, are being added. Likewise, $4(6) = 6 + 6 + 6 + 6$. In the following examples, we will develop rules for multiplying a positive number by a positive number, a positive number by a negative number, a negative number by a positive number, and a negative number by a negative number.

Example 1

Finding products of signed numbers

The multiplication of a positive number by a positive number is already familiar to you from arithmetic.

a. $3 \cdot 5 = 5 + 5 + 5 = 15$ Positive 3 times positive 5 equals positive 15.

b. $4 \cdot 6 = 6 + 6 + 6 + 6 = 24$ Positive 4 times positive 6 equals positive 24.

It is apparent from the preceding example, and your prior knowledge of arithmetic, that a positive number multiplied by a positive number results in a positive number. Example 2 illustrates the product of a positive number and a negative number.

Example 2

Find the following products.

a. $+3(-5) = (-5) + (-5) + (-5) =$ Definition of multiplication.

   -15 Therefore, $+3$ times -5 equals -15.

b. $+4(-6) = (-6) + (-6) + (-6) + (-6) =$ Definition of multiplication.

   -24 Therefore, $+4$ times -6 equals -24.

The preceding example implies that a positive number multiplied by a negative number results in a negative number.

In order to discuss a negative number multiplied by a positive number, we need the **commutative property for multiplication.** We know from arithmetic that $3(4) = 4(3)$ and $6(2) = 2(6)$. These products suggest that the order in which two numbers are multiplied makes no difference. We can generalize this to the statement of the commutative property for multiplication.

Commutative Property for Multiplication

For any two integers a and b, $a \cdot b = b \cdot a$.

Example 3

Find the following products:

a. $(-3)(3) =$ Apply the commutative property.

   $(3)(-3) =$ Multiply.

   -9 Therefore, -3 times $+3$ equals -9.

b. $(-6)(4) =$ Apply the commutative property.

   $(4)(-6)$ Multiply.

   -24 Therefore, -6 times $+4$ equals -24.

From Example 3, we can conclude that a negative number multiplied by a positive number results in a negative number. Unfortunately, we cannot use the definition of multiplication to develop a rule for multiplying a negative number by a negative number. For example, $(-3)(-5)$ cannot mean add -3, -5s.

Example 4

Look for a pattern in the following:

$$(3)(-4) = -12$$
$$(2)(-4) = -8$$
$$(1)(-4) = -4$$
$$(0)(-4) = 0$$

$\}$ Increase of 4
$\}$ Increase of 4
$\}$ Increase of 4

Notice that the number on the left is decreasing by one and the product is increasing by 4 in each case. Let's continue the pattern.

$$(3)(-4) = -12$$
$$(2)(-4) = -8$$
$$(1)(-4) = -4$$
$$(0)(-4) = 0$$
$$(-1)(-4) = 4 \;\Big\}\; \text{Increase of 4}$$
$$(-2)(-4) = 8 \;\Big\}\; \text{Increase of 4}$$

The preceding pattern implies that a negative number multiplied by a negative number results in a positive number.

Looking back at Examples 1 and 4, we notice that if the signs are both positive or both negative, the product is positive. From Examples 2 and 3, we see that if the signs are different (one positive and one negative), the product is negative. Based on these examples, we make the following rules for multiplying signed numbers:

Products of Signed Numbers

The product of two numbers with the same (like) signs is positive. The product of two numbers with different (unlike) signs is negative.

Using symbols, $(+)(+) = +$, $(-)(-) = +$, $(+)(-) = -$, and $(-)(+) = -$.

Example 5

Find the following products:

a. $4 \cdot 7 = 28$ Same signs. Therefore, the product is positive.

b. $(-5)(5) = -25$ Different signs. Therefore, the product is negative.

c. $(-6)(-7) = 42$ Same signs. Therefore, the product is positive.

d. $(5)(-8) = -40$ Different signs. Therefore, the product is negative.

Practice Exercises

Find the following products:

1. $6 \cdot 3$ **2.** $(-8)(4)$

3. $(-8)(-3)$ **4.** $(7)(-6)$

If more practice is needed, do the Additional Practice Exercises.

Additional Practice Exercises

Find the following products:

a. $5 \cdot 3$ **b.** $(-6)(-4)$

c. $(8)(-3)$ **d.** $7(-8)$

Answers:

If more than two numbers are multiplied, remember to multiply in order from left to right unless the numbers are grouped using the associative property.

Example 6

Find the following products:

a. $4(-3)(5) =$ Multiply 4 and -3.

$(-12)(5) =$ Multiply -12 and 5.

-60 Product.

b. $(-5)(-2)(4) =$ Multiply -5 and -2.

$(10)(4) =$ Multiply 10 and 4.

40 Product.

c. $6(-4)(-5)(-3) =$ Multiply 6 and -4.

$(-24)(-5)(-3) =$ Multiply -24 and -5.

$(120)(-3) =$ Multiply 120 and -3.

-360 Product.

d. $(-2)(-3)(-5)(-1) =$ Multiply -2 and -3.

$6(-5)(-1) =$ Multiply 6 and -5.

$(-30)(-1) =$ Multiply -30 and -1.

30 Product.

Be Careful *Remember that we multiply in order from left to right. A common error when multiplying signed numbers is to "distribute" the first factor over the remaining factors. For instance, in Example 6a,*
$4(-3)(5) \neq (-12)(20).$

By looking closely at Example 6, you may discover an easy way to determine the sign of the product in advance.

Products with Negative Numbers

If there is an even number of negative signs, the product is positive. If there is an odd number of negative signs, the product is negative.

Remember, the even numbers are $2, 4, 6, \ldots$ and the odd numbers are $1, 3, 5, \ldots$.

Practice Exercises

Find the following products:

5. $(6)(4)(-3)$

6. $(5)(-4)(-3)$

7. $(-3)(-4)(6)(-2)$

8. $(-3)(-1)(-4)(-5)$

If more practice is needed, do the Additional Practice Exercises.

Additional Practice Exercises

Find the following products:

e. $(4)(-3)(6)$

f. $(-2)(7)(-3)$

g. $(-3)(-2)(-5)(4)$

h. $(-3)(-1)(-3)(-5)$

Answers:

Practice Exercises 5–8: 5. -72 6. 60 7. -144 8. 60   *Additional Practice Exercises e–h:* e. -72 f. 42 g. -120 h. 45

Now that we know how to multiply signed numbers, we can raise signed numbers to powers. Recall from Section 1.1 that an exponent tells us how many times the base is used as a factor.

OPTIONAL   CALCULATOR EXPLORATION ACTIVITY

Use a calculator to evaluate the following. If your calculator does not have parentheses, be very careful.

Column A	**Column B**	**Column C***
$(-5)^2 =$	$(-5)^3 =$	$-5^2 =$
$(-3)^4 =$	$(-3)^5 =$	$-3^4 =$
$(-2)^6 =$	$(-2)^7 =$	$-2^5 =$
$(-1)^8 =$	$(-1)^9 =$	$-1^9 =$

1. Based on your answers to column A, is the result of raising a negative number to an even power positive or negative?

2. Based on your answers to column B, is the result of raising a negative number to an odd power positive or negative?

3. Based on your answers to column C, is the negative of a positive number raised to any positive integer power positive or negative?

We will now confirm these results mathematically.

*Some calculators without parentheses give incorrect answers to some of these.

In Example 7, look for a relationship between the exponent and the sign of the answer.

Example 7

Evaluate the following:

a. $(-3)^2 = (-3)(-3) = 9$

b. $(-5)^3 = (-5)(-5)(-5) = (25)(-5) = -125$

c. $(-1)^4 = (-1)(-1)(-1)(-1) = 1(-1)(-1) = (-1)(-1) = 1$

d. $(-2)^5 = (-2)(-2)(-2)(-2)(-2) = 4(-2)(-2)(-2) = (-8)(-2)(-2) = (16)(-2) = -32$

Since the exponent indicates how many times the base is used as a factor, the procedure for finding the sign of a product applies to finding the sign of a negative number raised to a power.

Be Careful *There is a great deal of difference between $(-3)^2$ and -3^2. The exponent applies only to the symbol immediately preceding it, so the "$-$" is not part of the base unless we use parentheses. Consequently, $(-3)^2 = (-3)(-3) = 9$. But in the expression -3^2, which is the negative of 3^2, the base of the exponent is 3 and not -3. You may think of -3^2 as $-1 \cdot 3^2$, so the meaning is $-1 \cdot 3 \cdot 3 = -3 \cdot 3 = -9$. The only way to raise a negative number to a power is to put the negative number inside parentheses and the power outside the parentheses, as in $(-3)^2$.*

Raising a Negative Number to a Power

If a negative number is raised to an even power, the answer is positive. If a negative number is raised to an odd power, the answer is negative.

Example 8

Evaluate the following:

a. $(-2)^2 = (-2)(-2) = 4$

b. $-2^2 = -1 \cdot 2 \cdot 2 = -2 \cdot 2 = -4$

c. $-8^2 = -1 \cdot 8 \cdot 8 = -8 \cdot 8 = -64$

d. $(-8)^2 = (-8)(-8) = 64$

e. $(-3)^3 = (-3)(-3)(-3) = 9(-3) = -27$

f. $-3^3 = -1 \cdot 3^3 = -1 \cdot 3 \cdot 3 \cdot 3 = -3 \cdot 3 \cdot 3 = -9 \cdot 3 = -27$

Practice Exercises

Evaluate the following:

9. $(-4)^2$ **10.** $(-2)^3$ **11.** $(-4)^4$

12. -2^4 **13.** $(-5)^2$ **14.** -5^2

If more practice is needed, do the Additional Practice Exercises.

Additional Practice Exercises

Evaluate the following:

i. $(-6)^2$ **j.** $(-4)^3$ **k.** $(-3)^4$

l. -2^5 **m.** $(-9)^2$ **n.** -4^2

In problems involving both multiplication and powers, remember to raise to powers first.

Example 9

Find the following products:

a. $(-2)^2(3)^3 =$ Square -2 and cube 3.
$(4)(27) =$ Multiply 4 and 27.
108 Product.

b. $(-3)^3(-1)^2 =$ Cube -3 and square -1.
$(-27)(1) =$ Multiply -27 and 1.
-27 Product.

c. $(-5)^2(-4)^3 =$ Square -5 and cube -4.
$(25)(-64) =$ Multiply 25 and -64.
-1600 Product.

Answers:

Practice Exercises

Find the following products:

15. $(-3)^2(2)^2$

16. $(-1)^3(-2)^3$

17. $(-4)^2(-5)^3$

If more practice is needed, do the Additional Practice Exercises.

Additional Practice Exercises

Find the following products:

o. $(-2)^2(4)^2$

p. $(-4)^2(-5)^2$

q. $(4)^2(-3)^3$

Earlier, the commutative property of multiplication was mentioned. Like addition, multiplication has a number of properties that are summarized in the following box:

Properties of Multiplication

Commutative
$ab = ba$ Multiplication may be performed in any order.

Associative
$a(bc) = (ab)c$ Multiplication may be grouped in any manner.

Identity
The number 1 is the identity for multiplication, since $a \cdot 1 = 1 \cdot a = a$, for any number a.

Inverse
For any number $a \neq 0$, there exists a number $\frac{1}{a}$ such that $a \cdot \frac{1}{a} = 1$.

Multiplication by 0
$a \cdot 0 = 0 \cdot a = 0$ for all numbers a.

Example 10

Identify the property illustrated by each of the following:

a. $(-3)(2) = (2)(-3)$ Order is different. Therefore, commutative of multiplication.

b. $2[(-3)(-5)] = [2(-3)](-5)$ Grouping is different. Therefore, associative of multiplication.

c. $2(-3 + 5) = (-3 + 5)2$ Order is different. Therefore, commutative of multiplication.

d. $(-4)1 = -4$ Identity for multiplication.

e. $5 \cdot \dfrac{1}{5} = 1$ Inverse for multiplication.

f. $5 \cdot 0 = 0$ Multiplication by 0.

Practice Exercises

Identify the property illustrated by each of the following:

18. $(5)(-6) = (-6)(5)$

19. $-3[2(-4)] = [(-3)2](-4)$

20. $-3(5 + 6) = (5 + 6)(-3)$

21. $1(-7) = -7$

22. $6 \cdot \dfrac{1}{6} = 1$

23. $8 \cdot 0 = 0$

Exercise Set 2.1

Find the following products:

1. $6(-4)$

2. $(-12)(-8)$

3. $(-6)(4)$

4. $(14)(-5)$

5. $(-6)(-4)$

6. $(-8)(-9)$

7. $(-7)(3)$

8. $(-5)(8)$

9. $(-11)(-4)$

10. $(12)(-3)$

11. $(-2.5)(-1.4)$

12. $(-6.2)(4.3)$

13. $(-1.6)(5.7)$

14. $(-2.1)(-7.6)$

15. $(-1)(3)(-5)$

16. $(-4)(-1)(6)$

17. $(4)(-3)(4)$

18. $(5)(-4)(6)$

19. $(-5)(-3)(-6)$

20. $(-4)(-7)(-10)$

21. $(-2)(4)(-3)(-5)$

22. $(-6)(4)(-3)(-1)$

23. $(-2)(-4)(-5)(-3)$

24. $(-2)(5)(-8)(-2)$

25. -9^2

26. $(-9)^2$

27. $(-11)^2$

28. -11^2

29. -3^4

30. $(-3)^4$

31. $(3)^2(-2)^3$

32. $(-4)^2(-2)^3$

33. $(-2)^2(3)^3$

34. $(-4)^2(3)^2$

35. $(-1)^5(-5)^3$

36. $(-6)^2(-2)^3$

Challenge Exercises: (37–44)

37. $\left(\dfrac{2}{3}\right)\left(-\dfrac{4}{5}\right)$

38. $\left(-\dfrac{3}{4}\right)\left(\dfrac{5}{7}\right)$

39. $\left(-\dfrac{8}{15}\right)\left(-\dfrac{7}{5}\right)$

40. $\left(-\dfrac{2}{3}\right)\left(-\dfrac{2}{5}\right)$

41. $\left(\dfrac{3}{4}\right)\left(-\dfrac{1}{5}\right)\left(-\dfrac{3}{2}\right)$

42. $\left(-\dfrac{1}{3}\right)\left(-\dfrac{2}{5}\right)\left(\dfrac{4}{7}\right)$

43. $\left(-\dfrac{1}{4}\right)\left(-\dfrac{5}{2}\right)\left(-\dfrac{9}{8}\right)$

44. $\left(-\dfrac{5}{2}\right)\left(-\dfrac{3}{8}\right)\left(-\dfrac{5}{18}\right)$

Calculator Exercises:

Find the following products using a calculator. The type of calculator you use determines how you will enter a negative number. On many calculators, you first enter the number and then press the $\boxed{+/-}$ key. This key changes the sign of the number currently showing on the display. Do not attempt to use the subtraction key. On most graphing calculators, there is a special key indicated by $\boxed{(-)}$ that is used to indicate a negative number.

C1. $(-5.74)(8.65)$

C2. $(98.67)(-6.45)$

C3. $(-89.54)(-67.34)$

C4. $(-8.546)(-6.768)$

Answers:

Evaluate each of the following for x = −2 and y = −4:

45. $3x$

46. $4y$

47. $-5x$

48. $-2y$

49. $6x^2$

50. $3y^2$

51. $-5x^3$

52. $-4y^3$

53. x^2y^3

54. x^2y^2

55. x^3y^2

56. x^3y^3

Write an expression for each of the following and evaluate:

57. The product of 5 and −3 increased by 6

58. The product of −4 and 7 increased by 9

59. 5 decreased by the product of 4 and −7

60. −3 decreased by the product of −5 and 3

61. The product of 3 and the square of −4

62. The product of −5 and the square of 5

63. The product of −6 and −3 decreased by 8

64. The product of −3 and −5 decreased by 4

65. The product of 4 and −2 added to the product of −5 and 4

66. The product of 8 and −3 added to the product of −5 and 7

Identify the property illustrated by each of the following:

67. $(-2)[(3)(-4)] = [(-2)(3)])(-4)$

68. $[(-4)(9)](-5) = (-4)[(9)(-5)]$

69. $3(-4 + 6) = (-4 + 6)(3)$

70. $(-4)\,(5 - 7) = (5 - 7)(-4)$

71. $(-9) \cdot 1 = -9$

72. $1 \cdot (-6) = -6$

73. $(-4)(-7) + 2 = (-7)(-4) + 2$

74. $5 + (-5)(8) = 5 + (8)(-5)$

75. $-3 \cdot \dfrac{-1}{3} = 1$

76. $5 \cdot \dfrac{1}{5} = 1$

77. $0 \cdot 7 = 0$

78. $9 \cdot 0 = 0$

Complete each of the following using the given property of multiplication:

79. $3[(-9)(-7)] = $ _____   Associative property

80. $6[(-2)(3)] = $ _____   Associative property

81. $3(-7) = $ _____   Commutative property

82. $(-5)(8) = $ _____   Commutative property

83. $4(-9 + 2) = $ _____   Distributive property

84. $5(-2 + 9) = $ _____   Distributive property

85. $4(-9 + 2) = $ _____   Commutative property

86. $5(-2 + 9) = $ _____   Commutative property

87. $(-2) \cdot 1 = $ _____   Identity for multiplication

88. $(-5) \cdot 1 = $ _____   Identity for multiplication

89. $8 \cdot \dfrac{1}{8} = $ _____   Inverse for multiplication

90. $4 \cdot \dfrac{1}{4} = $ _____   Inverse for multiplication

91. $2 \cdot 0 = $ _____   Multiplication property of 0

92. $0 \cdot 5 = $ _____   Multiplication property of 0

Challenge Exercises: (93–96)

Evaluate each of the following:

93. $-2 \cdot 3^2 + 5 \cdot (-4)^2 \div (5)(2)$

94. $4 \cdot (-6)^2 \div (3)(2)^3$

95. $-4[-5 - (-6)]^2 + (-7)$

96. $-6 - 5[-3^2 - (-6)]^2$

Answer the following:

97. When a cold front moved through, the temperature dropped at an average rate of 4° per hour. Using a signed number, find the change in temperature after three hours.

98. On a sunny day in Florida, the temperature rose at an average rate of 3.5° per hour from 10:00 A.M. to 2:00 P.M. Using a signed number, find the change in temperature from 10:00 A.M. to 2:00 P.M.

99. On a steep mountain road, the elevation dropped 7 feet per 100 feet. Using a signed number, find the change in elevation after 1000 feet.

100. On a steep roof, there is a vertical drop of 2 feet for every horizontal change of 10 feet. Using a signed number, find the vertical change for a horizontal change of 25 feet.

101. On Black Monday, a certain stock dropped at an average rate of $\frac{3}{4}$ points per hour. Using a signed number, find the drop after eight hours.

102. A parachutist is descending at a rate of 8 feet per second. Using a signed number, find the number of feet that she descends in 15 seconds.

Writing Exercises:

103. If a product has 16 positive factors and 23 negative factors, is the product positive or negative? Why?

104. Is $(-1)^{48}$ positive or negative? Why? Is $(-1)^{97}$ positive or negative? Why?

Critical Thinking Exercises:

105. Why is a product that contains an even number of negative signs positive?

106. Why is a product that contains an odd number of negative signs negative?

107. Compare/contrast the commutative and associative properties of multiplication.

108. Why does 0 not have a multiplicative inverse?

109. Why is $-x^2$ always nonpositive?

Group Project:

110. Write an application exercise similar to Exercises 97–102. Exchange your exercise with another group and solve their exercise.

Section 2.2	Multiplication Laws of Exponents

OBJECTIVES

When you complete this section, you will be able to:

a. Simplify expressions of the form $a^m \cdot a^n$.

b. Simplify expressions of the form $(ab)^n$.

c. Simplify expressions of the form $(a^m)^n$.

d. Simplify expressions involving two or more of the preceding forms a through c.

Introduction   Previously, we raised whole numbers and integers to powers. In this section, we will develop rules to simplify products involving exponents whose bases are the same, to raise a number to a power to another power, and to raise the product of two numbers to a power. All of these rules depend on the definition of a positive integer exponent. Remember, if n is a positive integer, a^n means the product of n factors of a.

OPTIONAL   CALCULATOR EXPLORATION ACTIVITY 1

Use a calculator to evaluate each of the following columns:

Column A	Column B
1. $2^2 \cdot 2^3 =$	$2^5 =$
2. $3^3 \cdot 3^4 =$	$3^7 =$
3. $4^2 \cdot 4^4 =$	$4^6 =$

By looking at the corresponding lines in columns A and B, answer the following:

1. If two exponential expressions with the same base are multiplied, leave the

_____ unchanged and _____ the exponents.

We now give a mathematical justification for the preceding observation.

Simplifying expressions of the form $a^m a^n$

By the definition of exponents $(a^2)(a^3) = (a \cdot a)(a \cdot a \cdot a) = a \cdot a \cdot a \cdot a \cdot a = a^5$.

2 factors of a + 3 factors of a = 5 factors of a

Also, $(a^4)(a^3) = (a \cdot a \cdot a \cdot a)(a \cdot a \cdot a) = a \cdot a \cdot a \cdot a \cdot a \cdot a \cdot a = a^7$

4 factors of a + 3 factors of a = 7 factors of a

Based on the two preceding examples, we generalize to the first law of exponents for multiplication.

First Law of Exponents: Product Rule

$a^m \cdot a^n = a^{m+n}$ To multiply two expressions with the same base, leave the base unchanged and add the exponents.

The preceding property is easily extended to more than two expressions. For example, $x^3 \cdot x^4 \cdot x^2 = x^{3+4+2} = x^9$. Remember, if no exponent is given, the exponent is understood to be 1.

Example 1

Find the following products. Leave the answers in exponential form.

a. $x^3 \cdot x^5 =$ Apply $a^m \cdot a^n = a^{m+n}$.

$x^{3+5} =$ Add the exponents.

x^8 Therefore, $x^3 \cdot x^5 = x^8$.

b. $2^4 \cdot 2^2 =$ Apply $a^m \cdot a^n = a^{m+n}$.

$2^{4+2} =$ Add the exponents. Note: The base remains 2.

2^6 Therefore, $2^4 \cdot 2^2 = 2^6$.

c. $3 \cdot 3^4 =$ Apply $a^m \cdot a^n = a^{m+n}$.

$3^{1+4} =$ Add the exponents. Remember, 3 means 3^1.

3^5 Therefore, $3 \cdot 3^4 = 3^5$.

d. $a^5 \cdot a^4 \cdot a =$ Apply $a^m \cdot a^n = a^{m+n}$.

$a^{5+4+1} =$ Add the exponents.

a^{10} Therefore, $a^5 \cdot a^4 \cdot a = a^{10}$.

If the terms have numerical coefficients, we use the commutative and associative properties to rearrange the order of the factors and multiply the coefficients and variable factors with the same bases separately.

Answers:

Example 2

Find the following products:

Be Careful *When multiplying exponential expressions with constant bases as in parts b and c above, a common error is to multiply the bases as well as add the powers. A product of the type $3^2 \cdot 3^4$ means multiply 2 factors of 3 by 4 factors of 3 for a total of 6 factors of 3, which is written as 3^6. A common error is to write the product $3^2 \cdot 3^4$ as 9^6, which means multiply 6 factors of 9, when in fact, 9 is not used as a factor at all.*

a. $(2x^2)(3x^3)$ Rearrange the order of the factors.

$(2 \cdot 3)(x^2 \cdot x^3)$ Multiply 2 and 3. Apply $a^m \cdot a^n = a^{m+n}$.

$6x^{2+3}$ Add the exponents.

$6x^5$ Therefore, $(2x^2)(3x^3) = 6x^5$.

b. $(-4x^4y^2)(6x^3y^5)$ Rearrange the order of the factors.

$(-4 \cdot 6)(x^4 \cdot x^3)(y^2 \cdot y^5)$ Multiply -4 and 6. Apply $a^m \cdot a^n = a^{m+n}$.

$-24x^{4+3}\,y^{2+5}$ Add the exponents.

$-24x^7y^7$ Therefore, $(-4x^4y^2)(6x^3y^5) = -24x^7y^7$.

c. $(3^2 \cdot 4^4)(3^5 \cdot 4^2)$ Rearrange the order of the factors.

$(3^2 \cdot 3^5)(4^4 \cdot 4^2)$ Apply $a^m \cdot a^n = a^{m+n}$.

$3^{2+5} \cdot 4^{4+2}$ Add the exponents.

$3^7 \cdot 4^6$ Therefore, $(3^2 \cdot 4^4)(3^5 \cdot 4^2) = 3^7 \cdot 4^6$.

Practice Exercises

Find the following products:

1. $x^2 \cdot x^4$

2. $3^2 \cdot 3^6$

3. $b^4 \cdot b^3 \cdot b^5$

4. $(4x^4)(3x^5)$

5. $(5y^2z^3)(-6y^3z^5)$

6. $(2^4 \cdot 5^3)(2^2 \cdot 5^4)$

If more practice is needed, do the Additional Practice Exercises.

Additional Practice Exercises

Find the following products:

a. $z^4 \cdot z^3$

b. $2^4 \cdot 2^2$

c. $c^3 \cdot c^2 \cdot c^5$

d. $(5x^3)(2x^5)$

e. $(-4w^3z)(3w^5z^3)$

f. $(3^4 \cdot 6^3)(3^3 \cdot 6^4)$

OPTIONAL   **CALCULATOR EXPLORATION ACTIVITY 2**

Evaluate each of the following using a calculator. Be careful if your calculator does not have parentheses.

Column A	Column B
1. $(2 \cdot 3)^3 =$	$2^3 \cdot 3^3 =$
2. $(3 \cdot 5)^3 =$	$3^3 \cdot 5^3 =$
3. $(3 \cdot 4)^2 =$	$3^2 \cdot 4^2 =$

By looking at corresponding lines of columns A and B, answer the following:

2. Raising a product to a power is the same as raising each _____ to the _____.

We now provide a mathematical justification for the preceding observation.

Answers:

Simplifying expressions of the form $(ab)^n$

In an expression of the form $(ab)^3$, the base of the exponent 3 is the product ab. Consequently, $(ab)^3 = (ab)(ab)(ab)$. Using the commutative and associative properties of multiplication, $(ab)(ab)(ab) = (a \cdot a \cdot a)(b \cdot b \cdot b) = a^3b^3$. Therefore, $(ab)^3 = a^3b^3$. Likewise, $(2y)^4 = (2y)(2y)(2y)(2y) = (2 \cdot 2 \cdot 2 \cdot 2)(y \cdot y \cdot y \cdot y) = 2^4y^4 = 16y^4$. These examples suggest the following generalization.

Second Law of Exponents: Power of a Product

$(a \cdot b)^n = a^nb^n$. To raise a product to a power, raise each factor to the power.

The preceding property can be extended to include more than two factors. For example, $(2ab)^4 = (2ab)(2ab)(2ab)(2ab) = (2 \cdot 2 \cdot 2 \cdot 2)(a \cdot a \cdot a \cdot a)(b \cdot b \cdot b \cdot b) = 2^4a^4b^4 = 16a^4b^4$.

Example 3

Find the following products:

a. $(xy)^2 = x^2y^2$

b. $(3y)^4 = 3^4y^4 = 81y^4$ *Do not forget to raise 3 to the fourth power!*

c. $(-4z)^2 = (-4)^2z^2 = 16z^2$

d. $(abc)^5 = a^5b^5c^5$

e. $(-4ab)^3 = (-4)^3a^3b^3 = -64a^3b^3$

Practice Exercises

Find the following products:

7. $(ab)^5$ 8. $(5x)^3$

9. $(-6x)^2$ 10. $(xyz)^6$

11. $(3ab)^4$

If more practice is needed, do the Additional Practice Exercises.

Additional Practice Exercises

Find the following products:

g. $(ef)^7$ h. $(4w)^3$

i. $(-5y)^4$ j. $(rst)^8$

k. $(-5st)^3$

Answers:

Practice Exercises 7–11: 7. a^5b^5 8. $125x^3$ 9. $36x^2$ 10. $x^6y^6z^6$ 11. $81a^4b^4$ **Additional Practice Exercises g–k:** g. e^7f^7 h. $64w^3$ i. $625y^4$ j. $r^8s^8t^8$ k. $-125s^3t^3$

OPTIONAL	CALCULATOR EXPLORATION ACTIVITY 3

Evaluate each of the following columns using a calculator:

Column A   **Column B**

1. $(2^2)^3$ $2^6 =$

2. $(3^4)^2 =$ $3^8 =$

3. $(2^3)^3 =$ $2^9 =$

By comparing the corresponding lines in columns A and B, answer the following:

3. To raise a number to a power to another power, leave the _____ unchanged and

_____ the exponents.

Following is justification for the preceding observation:

Simplifying expressions of the form $(a^m)^n$

In the expression $(a^2)^3$, the base of the exponent 3 is a^2. Consequently, $(a^2)^3 = a^2 \cdot a^2 \cdot a^2$. From the Product Rule, we know $a^2 \cdot a^2 \cdot a^2 = a^{2+2+2} = a^6$. Since multiplication is a shortcut for repeated additions of the same number, $2 + 2 + 2 = 3 \cdot 2$. Therefore, $(a^2)^3 = a^{(3)(2)} = a^6$. In the same manner, $(b^3)^4 = b^3 \cdot b^3 \cdot b^3 \cdot b^3 = b^{3+3+3+3} = b^{(4)(3)} = b^{12}$. These examples can be generalized into the third law of exponents.

Third Law of Exponents: Power to a Power

$(a^m)^n = a^{mn}$.   To raise a number to a power to another power, leave the base unchanged and multiply the powers.

Example 4

Simplify the following:

a. $(x^2)^4 =$ Apply $(a^m)^n = a^{mn}$.

  $x^{(2)(4)} =$ Multiply the exponents.

  x^8 Therefore, $(x^2)^4 = x^8 \cdot$

b. $(2^3)^4 =$ Apply $(a^m)^n = a^{mn}$.

  $2^{(3)(4)} =$ Multiply the exponents.

  2^{12} Therefore, $(2^3)^4 = 2^{12}$.

c. $(y^4)^2 =$ Apply $(a^m)^n = a^{mn}$.

  $y^{(4)(2)} =$ Multiply the exponents.

  y^8 Therefore, $(y^4)^2 = y^8$.

Practice Exercises

Simplify the following:

12. $(a^7)^2$ **13.** $(4^3)^5$ **14.** $(z^4)^5$

If more practice is needed, do the Additional Practice Exercises.

Additional Practice Exercises

Simplify the following:

l. $(q^2)^5$ **m.** $(5^4)^4$ **n.** $(w^3)^6$

Simplifying expressions involving more than one law of exponents

Often it is necessary to simplify expressions that involve more than one of the laws of exponents. Be careful in distinguishing which rule applies to each situation.

Example 5

Simplify each of the following. Leave answers in exponential form.

a. $(x^2y^3)^3 =$ Apply $(ab)^n = a^nb^n$.

$(x^2)^3(y^3)^3 =$ Apply $(a^m)^n = a^{mn}$.

x^6y^9 Answer.

b. $(2x^3y)^4 =$ Apply $(ab)^n = a^nb^n$.

$2^4(x^3)^4(y^1)^4=$ Apply definition of exponents and $(a^m)^n = a^{mn}$.

$16x^{12}y^4$ Answer.

c. $(2xy)^2(3xy)^3 =$ Apply $(ab)^n = a^nb^n$.

$2^2 \cdot x^2 \cdot y^2 \cdot 3^3 \cdot x^3 \cdot y^3 =$ Regroup and apply the definition of exponent.

$(4 \cdot 27)(x^2 \cdot x^3)(y^2 \cdot y^3) =$ Apply $a^n \cdot a^m = a^{m+n}$.

$108x^5y^5$ Answer.

d. $(x^3y^2)^2(x^4y^3)^2=$ Apply $(ab)^n = a^nb^n$.

$(x^3)^2(y^2)^2(x^4)^2(y^3)^2=$ Apply $(a^m)^n = a^{mn}$.

$x^6y^4 \cdot x^8y^6 =$ Regroup.

$(x^6 \cdot x^8)(y^4 \cdot y^6) =$ Apply $a^m \cdot a^n = a^{m+n}$.

$x^{14}y^{10}$ Answer.

Practice Exercises

Simplify the following:

15. $(x^4y^2)^4$ **16.** $(-4a^3b)^2$ **17.** $(3ab)^3(2ab)^4$ **18.** $(a^3b^4)^2(a^2b^4)^3$

If more practice is needed, do the Additional Practice Exercises.

Additional Practice Exercises

Simplify the following:

o. $(w^4z^3)^2$ **p.** $(-2x^4y^4)^4$ **q.** $(cd)^4(cd)^3$ **r.** $(xy^5)^3(x^3y^5)^4$

Sometimes it is necessary to add like terms after applying one or more laws of exponents. Remember, when adding like terms we add the coefficients only. *We do not add the exponents.* Also remember to follow the order of operations.

Answers:

p. $16x^{16}y^{16}$ **q.** c^7d^7 **r.** $x^{15}y^{35}$

Additional Practice Exercises o–r: **o.** w^8z^6 **Practice Exercises 15–18:** **15.** $x^{16}y^8$ **16.** $16a^6b^2$ **17.** $432a^7b^7$ **18.** $a^{12}b^{20}$

Example 6

Simplify the following:

a. $(2x)(4x) + (6x)(2x) =$ Multiply before adding. Use $a^m \cdot a^n = a^{m+n}$.

$8x^2 + 12x^2 =$ Add like terms. Add coefficients only.

$20x^2$ Answer.

b. $(2x)^3 + (3x)(-2x^2) =$ Simplify $(2x)^3$ and $(3x)(-2x^2)$.

$8x^3 - 6x^3 =$ Add like terms. Add coefficients only.

$2x^3$ Answer.

c. $(-3x^2y)(2xy) + (4xy^2)(3x^2) =$ Multiply before adding. Use $a^m \cdot a^n = a^{m+n}$.

$-6x^3y^2 + 12x^3y^2 =$ Add like terms. Add coefficients only.

$6x^3y^2$ Answer.

Practice Exercises

Simplify the following:

19. $(4y)(2y) + (3y)(-5y)$ **20.** $(3r)^4 + (4r^2)(-10r^2)$ **21.** $(5a^2b^2)(-2ab^2) + (4ab^3)(3a^2b)$

If more practice is needed, do the Additional Practice Exercises.

Additional Practice Exercises

Simplify the following:

s. $(-2x)(4x) + (4x)(3x)$ **t.** $(4x)^2 + (5x)(-2x)$ **u.** $(6pq)(4p^2q) + (5p^2q^2)(3p)$

Study Tip 7

Preparing For Class

How well you prepare for class determines how much you will get from that class. Organize your day and set aside at least one hour each day to study mathematics. It is best to study in concentrated short intervals of approximately thirty minutes, with five to ten minute breaks in between. The following are some things you should do:

1. Read your class notes as soon after class as possible, preferably the same day. Highlight important formulas, statements, and so on.
2. Write any definitions, rules, formulas, or important statements on your review cards. If your instructor went over these things in class, they are important. Check the chapter summary.
3. Read the textbook slowly and carefully with pencil and paper available. Mark down things that you do not understand and come back to them after you have read all of the material. Check the material that was previously covered for something similar.

(continued)

Answers:

Practice Exercise 19-21: **19.** $-7y^2$ **20.** $41r^4$ **21.** $2a^3b^4$ Additional Practice Exercises s-u: **s.** $4x^2$ **t.** $6x^2$ **u.** $39p^3q^2$

Study Tip 7 *(continued)*

4. If necessary, use other textbooks and study guides. Many texts have supplemental study guides that may be purchased in the college bookstore. Many math learning centers, or the library, have texts that may be borrowed. Your instructor may also have additional texts or materials.

5. Do your homework as soon as possible after class is over—most definitely before the next class meeting. Do not skip steps. The reason for a step can often be found in something that you learned previously. Putting each step down on paper will help reinforce the principles being covered and help you remember them. Careless errors often occur when steps are skipped. Review your homework just prior to attending the next class meeting. Make a list of things that you are unsure of and ask your instructor about them.

6. Preview the text to be covered in class the next day by reading the text with pencil in hand. Mark those parts you are unsure of and make a list of questions to ask your instructor if his or her explanation does not fully answer your specific question.

7. If your instructor does not have time in class to answer all of your questions, make an appointment to see him or her during office hours.

Exercise Set 2.2

Simplify the following using the laws of exponents:

1. $r^2 \cdot r^6$

2. $s^3 \cdot s^5$

3. $4^3 \cdot 4^4$

4. $5^4 \cdot 5^5$

5. $x^3 \cdot x^4 \cdot x^2$

6. $y^3 \cdot y^6 \cdot y^5$

7. $(4x^5)(5x^4)$

8. $(3y^5)(6y^6)$

9. $(-7a^4)(-3a^4)$

10. $(-6b^6)(5b^6)$

11. $(x^5y^3)(x^3y^4)$

12. $(a^3b^6)(a^7b^3)$

13. $(4a^5b^4)(2a^5b^7)$

14. $(5y^6z^2)(7y^5z^5)$

15. $(ab)^5$

16. $(xy)^7$

17. $(2y)^4$

18. $(3a)^5$

19. $(-4x)^2$

20. $(-7x)^2$

21. $-(5x)^2$

22. $-(8b)^4$

23. $-(-3c)^3$

24. $(-5a)^3$

25. $(-3b)^3$

26. $(x^4)^5$

27. $(y^3)^6$

28. $(3^4)^2$

29. $(4^5)^3$

30. $(ab)^5(ab)^4$

31. $(xy)^2(xy)^5$

32. $(2d)^4(2d)^5$

33. $(4x)^5(4x)^3$

34. $(a^3b^3)^5$

35. $(c^4d^3)^6$

36. $(3^4x^5)^3$

37. $(4^5a^3)^4$

38. $-(2^2a^4)^3$

39. $-(4^4b^5)^5$

40. $(-5^3x^2)^4$

41. $(-6^2b^3)^3$

42. $(2x^3y^4)^3(3xy^3)^2$

43. $(3a^5b^2)^3(2ab^3)^2$

44. $(3^2a^4)^3(3^4a^5)^3$

45. $(4^4b^4)^3(4^2b^4)^4$

46. $(x^2y^3z^2)^3(x^4y^4z^2)^3$

47. $(a^3b^2c^3)^2(a^2b^3c^4)^4$

Simplify the following:

48. $(5y)(4y) + (3y)(-6y)$

49. $(6a)(-2a) + (7a)(4a)$

50. $(2z)^3 + (3z)(-5z^2)$

51. $(3x)^3 + (-4x)(6x^2)$

52. $(-2n)^4 + (2n^2)^2$

53. $(2x)^6 + (5x^3)^2$

54. $(3x^2)(-2x) + (-3x)(-5x^2)$

55. $(-6a^3)(2a) - (-4a^2)(-3a^2)$

56. $(2x^2y)^2 - (3x^3y)(-4xy)$

57. $(4m^2n^3)^2 + (-3mn^5)(6m^3n)$

Challenge Exercises: (58–67)

Simplify the following:

58. $x^{2n} \cdot x^{3n}$

59. $a^{5x} \cdot a^{3x}$

60. $x^{2m+1} \cdot x^{3m-2}$

61. $a^{3b-2} \cdot a^{2b-1}$

62. $(x^a y^b)(x^{2a} y^{3b})$

63. $(2^{3x} a^{2y} b^z)(2^{4x} a^{3y} b^{2z})$

64. $(a^m b^n)^p$

65. $(2x^a y^b)^c$

66. $(x^{n-1})^2$

67. $(b^{2n-1})^3$

Writing Exercises:

68. Are $(-2)^2$ and -2^2 equal? Explain why or why not.

69. Are $(-2)^3$ and -2^3 equal? Are their meanings the same? Explain why or why not.

70. Are 3^2 and $3 \cdot 2$ equal? Explain why or why not.

71. If we simplify $(2x^3)(3x^3)$, we multiply the coefficients and add the exponents. If we simplify $2x^3 + 3x^3$, we add the coefficients only and do not change the exponents. Why?

Section 2.3	Products of Polynomials

OBJECTIVES   *When you complete this section, you will be able to:*

a. Find the product of two or more monomials.

b. Find the product of a monomial and a polynomial.

c. Find the product of two polynomials.

Introduction   The laws of exponents developed in the previous section are used extensively in finding the products of polynomials. In showing that $a^m \cdot a^n = a^{m+n}$ and in subsequent examples, we were multiplying monomials by monomials. For example, $(2x^3 y^2)(3x^3 y^4) = 6x^6 y^6$ is the product of two monomials.

Product of a monomial and a polynomial   In multiplying a monomial and a polynomial, we use the distributive property, which you will recall is $a(b + c) = ab + ac$. The distributive property is easily extended to include multiplication of a monomial over polynomials of more than two terms. For example, $a(b + c + d) = ab + ac + ad$ and $a(b + c + d + e) = ab + ac + ad + ae$, and so on.

Example 1

Find the following products:

a. $2(x + 4) =$ Apply the distributive property.

  $2 \cdot x + 2 \cdot 4 =$ Multiply 2 and 4.

  $2x + 8$ These are not like terms. Therefore, this is the product.

Be Careful You cannot add unlike terms! A common mistake is to find the product and then "add" the results by adding the coefficients and exponents. For example, $2x(x^2 + 3x) = 2x^3 + 6x^2 \neq 8x^5$.

b. $2x^2(x - 5) =$ Apply the distributive property.

  $2x^2 \cdot x - 2x^2 \cdot 5 =$ $2x^2 \cdot x = 2x^3$ and $2x^2 \cdot 5 = 10x^2$.

  $2x^3 - 10x^2$ These are not like terms. Therefore, this is the product.

c. $3x^2 y(2xy^2 + 4x^3 y^2) =$ Apply the distributive property.

  $(3x^2 y)(2xy^2) + (3x^2 y)(4x^3 y^2) =$ Apply $a^m \cdot a^n = a^{m+n}$.

  $6x^3 y^3 + 12x^5 y^3$ These are not like terms. Therefore, this is the product.

d. $-2x^2(3x^2 - 4x + 6) =$ Apply the distributive property.

$(-2x^2)(3x^2) + (2x^2)(4x) - (2x^2)(6) =$ Apply $a^m \cdot a^n = a^{m+n}$.

$-6x^4 + 8x^3 - 12x^2$ These are not like terms. Therefore, this is the product.

Practice Exercises

Find the following products:

1. $3(x + 5)$ **2.** $3x^3(x - 6)$

3. $2ab^2(4a^2b - 3a^2b^2)$ **4.** $-2y^2(4y^3 - 3y - 5)$

If more practice is needed, do the Additional Practice Exercises.

Additional Practice Exercises

Find the following products:

a. $4(a + 2)$ **b.** $a(3a - 6)$

c. $3yz^2(3yz - 4y^3z^2)$ **d.** $-4a^2(3a^3 - 5a^2 + 4)$

It is often necessary to combine multiples of polynomials as in the following:

Example 2

Simplify the following:

a. $2(x - 4) + 3(x + 2) =$ Apply the distributive property.

$2x - 8 + 3x + 6 =$ Add like terms.

$5x - 2$ Therefore, $2(x - 4) + 3(x + 2) = 5x - 2$.

b. $3(2a^2 - 4a + 5) + 4(a^2 + 2a - 3) =$ Apply the distributive property.

$6a^2 - 12a + 15 + 4a^2 + 8a - 12 =$ Add like terms.

$10a^2 - 4a + 3$ Therefore, $3(2a^2 - 4a + 5) + 4(a^2 + 2a - 3)$

$= 10a^2 - 4a + 3$.

c. $4(2y - 5) - 2(3y - 6) =$ Apply the distributive property.

$8y - 20 - 6y + 12 =$ Add like terms.

$2y - 8$ $4(2y - 5) - 2(3y - 6) = 2y - 8$.

d. $-3(b^2 + 3b - 2) - 2(2b^2 - 4b - 5) =$ Apply the distributive property.

$-3b^2 - 9b + 6 - 4b^2 + 8b + 10 =$ Add like terms.

$-7b^2 - b + 16.$ Therefore, $-3(b^2 + 3b - 2) - 2(2b^2 - 4b - 5) =$

$-7b^2 - b + 16$.

Practice Exercises

Simplify the following:

5. $2(r - 5) + 3(2r + 1)$

6. $4(2z^2 - 3z - 2) + 2(3z^2 + 4z - 5)$

7. $-3(2r - 3) - 4(3r - 2)$

8. $4(2p^2 + 3p - 6) - 3(4p^2 - 6p - 2)$

If more practice is needed, do the Additional Practice Exercises.

Additional Practice Exercises

Simplify the following:

e. $3(x - 1) + 2(x + 3)$

f. $5(x^2 - 3x + 2) + 2(x^2 + 2x - 3)$

g. $6(2x - 1) - 3(x + 2)$

h. $-3(x^2 - 6x + 2) - (3x^2 + 4x - 1)$

In multiplying a binomial and a polynomial of two or more terms, we need to multiply each term of the polynomial by each term of the binomial. The easiest way to do this is to use the distributive property twice. We distribute each term of the binomial over the polynomial as illustrated by the following examples:

Example 3

Find the following products.

a.

$(x + 2)(x + 3) =$	Rewrite using the distributive property.
$x(x + 3) + 2(x + 3) =$	Apply the distributive property.
$x^2 + 3x + 2x + 6 =$	Combine like terms.
$x^2 + 5x + 6$	Product.

b.

$(2x - 4)(3x - 2) =$	Rewrite using the distributive property.
$2x(3x - 2) - 4(3x - 2) =$	Apply the distributive property.
$6x^2 - 4x - 12x + 8 =$	Combine like terms.
$6x^2 - 16x + 8$	Product.

c.

$(2a - 3)(2a^2 - 4a + 5) =$	Rewrite using distributive property.
$2a(2a^2 - 4a + 5) - 3(2a^2 - 4a + 5) =$	Apply the distributive property.
$4a^3 - 8a^2 + 10a - 6a^2 + 12a - 15 =$	Combine like terms.
$4a^3 - 14a^2 + 22a - 15$	Product.

d.

$(a + 5)^2 =$	Rewrite as a product.
$(a + 5)(a + 5) =$	Rewrite using the distributive property.
$a(a + 5) + 5(a + 5) =$	Apply the distributive property.
$a^2 + 5a + 5a + 25 =$	Combine like terms.
$a^2 + 10a + 25$	Product.

Be Careful Notice that $(a + 5)^2 \neq a^2 + 5^2$. Do not confuse $(a + b)^2$ with $(a \cdot b)^2$.

Practice Exercises

Find the following products:

9. $(y + 4)(y + 6)$ **10.** $(3x - 2)(4x - 1)$ **11.** $(3a + 2)(a^2 - 3a + 2)$ **12.** $(x - 4)^2$

If more practice is needed, do the Additional Practice Exercises.

Additional Practice Exercises

Find the following products:

i. $(b + 4)(b + 1)$ **j.** $(2a - 4)(4a + 3)$ **k.** $(2b + 3)(b^2 + 4b - 1)$ **l.** $(a - 6)^2$

| OPTIONAL | **VERTICAL MULTIPLICATION OF POLYNOMIALS** |

Multiplication of polynomials can also be done vertically in much the same manner that whole numbers are multiplied. Since any two terms can be multiplied, it is not necessary to align like terms before multiplying as was necessary with addition. When multiplying whole numbers, it is necessary to line up digits with the same place value before adding to get a final answer. *When multiplying polynomials vertically, it is necessary to align like terms before adding.* Study the following examples:

Example 4

a. Find the product of 46 and 67 vertically.

$$
\begin{array}{r}
46 \\
\underline{67} \\
322 \\
\underline{276} \\
3082
\end{array}
$$

Product of 7 and 46.
Product of 60 and 46. Align place values.
Product.

b. Find the product of $3x + 2$ and $2x - 7$ vertically.

$$
\begin{array}{r}
3x + 2 \\
\underline{2x - 7} \\
-21x - 14 \\
\underline{6x^2 + 4x} \\
6x^2 - 17x - 14
\end{array}
$$

Product of $3x + 2$ and -7.
Product of $3x + 2$ and $2x$. Align like terms and add.
Product.

c. Find the product of $2x^2 - 4x + 2$ and $3x + 4$ vertically.

$$
\begin{array}{r}
2x^2 - 4x + 2 \\
\underline{3x + 4} \\
8x^2 - 16x + 8 \\
\underline{6x^3 - 12x^2 + 6x} \\
6x^3 - 4x^2 - 10x + 8
\end{array}
$$

Product of 4 and $2x^2 - 4x + 2$.
Product of $3x$ and $2x^2 - 4x + 2$. Align like terms and add.
Product.

Practice Exercises

Find the following products:

13. $3x - 5$
   $\underline{2x + 3}$

14. $3x^2 - 5x + 2$
   $\underline{\hspace{1.5cm} 3x - 1}$

If more practice is needed, do the Additional Practice Exercises.

Additional Practice Exercises

Find the following products vertically.

m. $(2x - 1)(3x - 2)$

n. $(4x^2 - 5x + 1)(x - 4)$

If the dimensions of a geometric figure are given in terms of variables, it is possible to represent the area in terms of these same variables. Remember, the formula for the area of a rectangle is $A = LW$ and the formula for the area of a triangle is $A = \frac{bh}{2}$.

Example 5

Write an expression for the area of each of the following using the given dimensions:

a.   A rectangle with $L = 2x - 3$ and $W = 3x - 4$

$A = LW$	Substitute for L and W.
$A = (2x - 3)(3x - 4)$	Rewrite using the distributive property.
$A = 2x(3x - 4) - 3(3x - 4)$	Apply the distributive property.
$A = 6x^2 - 8x - 9x + 12$	Add like terms.
$A = 6x^2 - 17x + 12$	Therefore, the area is represented as $6x^2 - 17x + 12$.

b.   A triangle with $b = 4x$ and $h = 5x$

$A = \dfrac{bh}{2}$	Substitute for b and h.
$A = \dfrac{(4x)(5x)}{2}$	Multiply $4x$ and $5x$.
$A = \dfrac{20x^2}{2}$	Divide 2 into 20.
$A = 10x^2$	Therefore, the area is represented as $10x^2$.

Practice Exercises

Write an expression for the area of each of the following using the given dimensions:

15.   A rectangle with $L = x - 3$ and $W = 2x + 2$

16.   A triangle with $b = 5x$ and $h = 6x$

If more practice is needed, do the Additional Practice Exercises.

Additional Practice Exercises

Write an expression for the area of each of the following using the given dimensions:

o. Rectangle with $L = x - 5$ and $W = x + 3$

p. Triangle with $b = 7a$ and $h = 4a$

Exercise Set 2.3

Find the products of the following monomials:

1. $(7ax^2)(-a^2x)$

2. $(5rq^3)(-3r^4q)$

3. $(6fkb^5)(-7f)(-f^5k)$

4. $(-5ab^2c)(2b)(-3ab^3c^2)$

5. $(-3p^3r)(4p^2qr^4)(-2q^4r^3)$

6. $(5p^4r^2)(-4p^3q^2)(-q^3r^4)$

Calculator Exercises:

Find the following products using a calculator as needed:

C1. $(-6.78x^3y^5)(7.93x^6y^8)$

C2. $(8.52a^6b^2)(-9.97a^2b)$

C3. $(7.68x^4y^5)(-4.5x^7y^4)$

C4. $(-67.4m^4n^8)(-56.9m^2n^8)$

Find the products of the following monomials and polynomials:

7. $2(x + 3)$

8. $5(y - 6)$

9. $-4(2x - 3)$

10. $-3(4y - 2)$

11. $3(2x^2 - 6x + 5)$

12. $4(3a^2 + 2a - 5)$

13. $-2(6b^2 - 2b + 4)$

14. $-5(2a^2 + 5a - 7)$

15. $2b(2b^2 - 4b + 5)$

16. $4x(5x^2 - 6x - 2)$

17. $2h^3(h^2 + 8h - 5)$

18. $3y^2(y^2 + 5y - 9)$

19. $2b^4(b^2 + 7b - 6)$

20. $2b^5(b^2 + 6b - 5)$

21. $-2r^3(3r^3 - 4r^2 + 2r - 3)$

22. $-3s^2(4s^4 - 5s^2 + 6s - 5)$

23. $5x^2y(2x^2y^2 - 4xy^3)$

24. $3ab^2(6a^3b - 2a^2b^2)$

25. $-4p^2q^3(3p^4q - p^2q^2 + q)$

26. $-3u^2v^2(4u + 3u^4v^2 - u^2v^4)$

Calculator Exercises:

Find the following products using a calculator as needed:

C5. $4.3x^3(5.3x^2 - 7.8x - 7.2)$

C6. $-6.4y^2(-4.1y^2 - 9.3y + 9.2)$

Simplify the following:

27. $2(2x - 3) + 3(x + 4)$

28. $4(3x + 1) + 2(2x - 4)$

29. $3(6a - 4) - 2(3a + 2)$

30. $5(2b - 6) - 4(3b + 2)$

31. $4(2y - 3) + 2(-3y + 4)$

32. $3(6x - 2) + 4(-5x + 3)$

33. $-3(7a - 2b) + 2(5a - 6b)$

34. $-4(2c - 3d) + 3(4c + 2d)$

35. $-2(4x - 4) - 3(2x + 5)$

36. $-6(2x - 2) - 2(3x - 5)$

37. $3(4c^2 - 3c + 2) + 4(3c^2 + 2c - 3)$

38. $6(2m^2 - 3m - 4) + 3(3m^2 - 6m - 8)$

Answers:

39. $5(3r^2 + 2r - 5) - 2(4r^2 - 3r - 6)$

40. $8(3a^2 - 2a + 1) - 6(5a^2 - 4a - 2)$

41. $-2(4a^2 - 3a + 1) - 3(2a^2 - 5a - 2)$

42. $-3(2r^2 - 3r - 4) - 2(3r^2 + 4r - 5)$

Challenge Exercises: (43–44)

Simplify the following:

43. $3(a^2 + 2a - 6) + 2[3(2a^2 - 3a + 2) - 2(a^2 - 4a - 5)]$

44. $4(2b^2 - 3b + 5) - 3[2(b^2 + 5b - 6) - 3(2b^2 - 3b + 1)]$

Find the products of the following polynomials:

45. $(a + 4)(a + 5)$

46. $(q + 7)(q + 2)$

47. $(z + 5)(z - 8)$

48. $(w - 1)(w + 5)$

49. $(x + 4)(x - 4)$

50. $(x - 5)(x + 5)$

51. $(2x - y)(2x + y)$

52. $(x + 4y)(x - 2y)$

53. $(x + y)(x - 2y)$

54. $(r - d)(r + 3d)$

55. $(4w + 3)(5w - 8)$

56. $(2c - 1)(9c + 5)$

57. $(3z - 4)(2z - 6)$

58. $(7a - 5)(2a - 3)$

59. $(2a - b)(3a + 4b)$

60. $(4x - 3y)(2x + 5y)$

61. $(x - 1)(x^2 + x + 3)$

62. $(x - 2)(x^2 + x - 4)$

63. $(x + 3)(2x^2 - 4x + 3)$

64. $(x + 4)(3x^2 - 2x + 5)$

65. $(x + 2)(x^2 - 2x + 4)$

66. $(x - 3)(x^2 + 3x + 9)$

67. $(2x - 3)(3x^2 + 4x - 3)$

68. $(3x - 4)(3x^2 - 5x - 2)$

Calculator Exercises:

Find the following products using a calculator:

C7. $(5.3x - 4.8)(7.1x + 3.6)$

C8. $(6.2x + 3.8)(2.6x - 7.2)$

Challenge Exercises: (69–76)

69. $(3a^2 + 2b^2)(4a^2 - b^2)$

70. $(4x^2 + 3y^2)(2x^2 - 3y^2)$

71. $(2a^2 + 3)(2a^2 + 4a - 5)$

72. $(3x^2 - 1)(2x^2 - 3x + 2)$

73. $(a^2 + 3a + 2)(a^2 - 4a + 3)$

74. $(x^2 - 2x + 5)(2x^2 + 3x - 2)$

75. $(x^n + 2)(x^n - 3)$

76. $(2x^a - 3)(3x^a + 4)$

Find the following products:

77. $2x + 4$
$\underline{x + 3}$

78. $3x + 2$
$\underline{x + 5}$

79. $3x - 5$
$\underline{2x - 5}$

80. $4x - 3$
$\underline{3x - 2}$

81. $a^2 - 5a + 6$
$\underline{2a + 4}$

82. $b^2 - 4b + 7$
$\underline{3b + 5}$

83. $2y^2 + 3y - 2$
$\underline{4y - 2}$

84. $3t^2 - 5t + 4$
$\underline{5t - 1}$

Recall that the formula for the area of a rectangle is A = LW. Write an expression for the area of each of the following rectangles with the given length and width. For example, if L = 2x and W = x − 3, then A = 2x(x − 3) = 2x² − 6x.

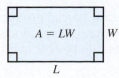

85. $L = x$, $W = 2x$

86. $L = 2x$, $W = 3x$

87. $L = 4x$, $W = x + 6$

88. $L = 5x$, $W = 3x - 5$

89. $L = x + 2$, $W = 2x - 3$

90. $L = x - 3$, $W = 3x + 1$

91. $L = x^2 - x + 2$, $W = 3x + 2$

92. $L = x^2 + 2x + 9$, $W = 4x - 3$

Recall that the formula for the area of a triangle is $A = \frac{bh}{2}$. Write an expression for the area of each of the following triangles.
For example, if b = 3a and h = 6a, then $A = \frac{3a \cdot 6a}{2} = \frac{18a^2}{2} = 9a^2$.

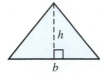

93. $b = 2y, h = 4y$

95. $b = 6z, h = 3z$

94. $b = 4x, h = 6x$

96. $b = 4a, h = 5a$

97. Fran, who is an architect, has a strange habit. In all the houses that she designs, the width of the living room is 8 feet less than twice the length. If y represents the length of a living room, how would you represent the amount of carpet needed to carpet the living room?

98. The Martinez family wants to sod their backyard. They found the yard's length is 3 feet more than twice its width. If x represents the width of their backyard, how would you represent the amount of sod they need to sod their backyard?

99. A rectangular pasture is $(3x + 5)$ feet long and $(3x - 2)$ feet wide. How many square feet of land are in the pasture?

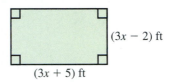

100. A rectangular bathroom wall is $(2x + 8)$ feet long and $(3x - 9)$ feet high. How many square feet of tile would it take to cover it?

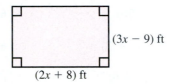

Challenge Exercises:

101. One brand of pile carpet sells for $12.50 per square yard. How would you represent the cost to carpet a room $(x + 3)$ yards long and x yards wide?

102. A certain type of sod sells for $.50 per square foot. How would you represent the cost to sod a rectangular yard which is $(2x + 4)$ feet long and $(x + 2)$ feet wide?

Writing Exercises:

103. Is $(a + 3)^2 = a^2 + 3^2$? Why or why not?

104. Is the product of a monomial and a trinomial always a trinomial? Explain.

105. Can the product of two binomials be a binomial? Explain.

Section 2.4	**Special Products**

OBJECTIVES

When you complete this section, you will be able to:

a. Find the product of two binomials.

b. Find products of the form $(a + b)(a - b)$.

c. Square binomials.

d. Optional: Recognize products of the form $(a + b)(a^2 - ab + b^2)$ and $(a - b)(a^2 + ab + b^2)$.

Introduction   The abilities of multiplying binomials quickly and recognizing special products will be very important in Chapter 5. In the previous section, we found the products of binomials by applying the distributive property twice. We need to be able to multiply binomials more quickly and easily. Consider the following:

$$(x + 3)(2x + 4) = x(2x + 4) + 3(2x + 4) = 2x^2 + 4x + 6x + 12$$

Multiplying binomials using FOIL

The term, $2x^2$, is the product of the <u>F</u>irst terms of the two binomials, x and $2x$. The term, $4x$, is the product of the <u>O</u>utside terms (terms farthest apart), x and 4. To find the term, $6x$, multiply the <u>I</u>nside terms (terms closest together), 3 and $2x$. The term, 12, is the product of the <u>L</u>ast terms of the two binomials, 3 and 4. The underlined letters spell the word FOIL, which is an easy way to remember which terms to multiply and allows us to multiply binomials mentally. The only thing left to do is add like terms and get the final product of $2x^2 + 10x + 12$. The following diagram illustrates the use of FOIL:

First   Last   F   O   I   L

$$(x + 3)(2x + 4) = 2x^2 + 4x + 6x + 12$$

Inside

Outside

Now combine the like terms $4x$ and $6x$ to get the product $2x^2 + 10x + 12$.

Note: If you have a really good imagination, the above diagram resembles a face.

Example 1

Find the following products using FOIL:

a.

Outside   Last

$$(x - 4)(x + 2) = x^2 + 2x - 4x - 8 = x^2 - 2x - 8$$

First   Inside

b. $(2x - 5)(2x + 3) =$ Apply FOIL.
$2x \cdot 2x + 2x \cdot 3 - 5 \cdot 2x - 5 \cdot 3 =$ Multiply.
$4x^2 + 6x - 10x - 15 =$ Add like terms.
$4x^2 - 4x - 15$ Product.

c. $(4a - b)(3a - 2b) =$ Apply FOIL.
$4a \cdot 3a - 4a \cdot 2b - b \cdot 3a + b \cdot 2b =$ Multiply.
$12a^2 - 8ab - 3ab + 2b^2 =$ Add like terms.
$12a^2 - 11ab + 2b^2$ Product.

d. $(x + 2y)(3a + b) =$ Apply FOIL.
$3ax + bx + 6ay + 2by$ Product, since there are no like terms.

Practice Exercises

Find the following products using FOIL:

1. $(q + 2)(q - 4)$ **2.** $(3a - 2)(5a - 3)$ **3.** $(2w + 5t)(3w - 2t)$ **4.** $(3a + 2b)(4c + 5d)$

If more practice is needed, do the Additional Practice Exercises.

Additional Practice Exercises

Find the following products using FOIL:

a. $(r + 5)(r - 3)$ **b.** $(5x - 2)(3x + 4)$ **c.** $(4y + 3z)(2y + 7z)$ **d.** $(2a + 4b)(3c - 5d)$

Products of the form $(a + b)(a - b)$

Products of the form $(a + b)(a - b)$ are of particular interest, since the answer is a binomial instead of the usual trinomial. Notice the factors are the sum and difference of the exact same two terms. Products of this form are often referred to as **conjugate pairs.** As you will see, the sum of the outer and inner products is always zero, which leaves the difference of two squares.

Answers:

Additional Practice Exercises a–d: a. $r^2 + 2r - 15$ **b.** $15x^2 + 14x - 8$ **c.** $8y^2 + 34yz + 21z^2$ **d.** $6ac - 10ad + 12bc - 20bd$
Practice Exercises 1–4: 1. $q^2 - 2q - 8$ **2.** $15a^2 - 19a + 6$ **3.** $6w^2 + 11tw - 10t^2$ **4.** $12ac + 15ad + 8bc + 10bd$

Example 2

Find the following product.

$(a + b)(a - b) =$ Apply FOIL.

$a^2 - ab + ab - b^2 =$ Combine like terms.

$a^2 - b^2$ Sum of outer and inner products is 0. The product is the difference of squares.

This example leads to the following observation:

Products of the Form $(a + b)(a - b)$ (Conjugate Pairs)

The product of two binomials that are the sum and difference of the same terms results in a binomial that is a difference of squares. In symbols, $(a + b)(a - b) = a^2 - b^2$.

From the preceding example, we see that in the product of conjugate pairs the outside and inside products always have a sum of zero. Therefore, it is not necessary to compute them.

Example 3

Find the following products:

a. $(3x - 4)(3x + 4) =$ Form of $(a + b)(a - b)$.

   $9x^2 - 16$ Multiply first by first and last by last.

b. $(3b + 4c)(3b - 4c) =$ Form of $(a + b)(a - b)$.

   $9b^2 - 16c^2$ Multiply first by first and last by last.

Note that in each case the answer is the difference of squares.

Practice Exercises

Find the following products:

5. $(x + 6)(x - 6)$

6. $(2a - 6b)(2a + 6b)$

If more practice is needed, do the Additional Practice Exercises.

Additional Practice Exercises

Find the following products:

e. $(a + 1)(a - 1)$

f. $(2p + 5q)(2p - 5q)$

Another special product is the square of a binomial. In the following illustration, the outside and inside products are always the same. This suggests a quick way of squaring a binomial mentally.

Answers:

Example 4

Square the following binomials:

a. $(a + b)^2 =$ Apply the definition of exponent.

$(a + b)(a + b) =$ Apply FOIL.

$a^2 + ab + ab + b^2 =$ Add like terms. Note that the outside and inside products are equal.

$a^2 + 2ab + b^2$ Product. Note that the middle term is twice the product ab.

b. $(a - b)^2 =$ Apply the definition of exponent.

$(a - b)(a - b) =$ Apply FOIL.

$a^2 - ab - ab + b^2 =$ Add like terms. Note that the outside and inside products are equal.

$a^2 - 2ab + b^2$ Product. Note that the middle term is twice the product $-ab$.

This suggests the following procedure for squaring a binomial quickly:

Square of a Binomial

To square a binomial:

1. Square the first term.
2. Add two times the product of the two terms.
3. Add the square of the last term.

In symbols, $(a + b)^2 = a^2 + 2ab + b^2$ and $(a - b)^2 = a^2 - 2ab + b^2$.

This procedure is diagrammed as follows for the square of the sum of two terms:

First term squared   Last term squared

$$(a + b)^2 = a^2 + 2ab + b^2$$

Two times the product of the terms

Example 5

Square the following binomials:

a. $(x + 3)^2 =$ Apply $(a + b)^2 = a^2 + 2ab + b^2$.

$x^2 + 2(x)(3) + 3^2 =$ Multiply.

$x^2 + 6x + 9$ Answer.

b. $(y - 4)^2$ Apply $(a - b)^2 = a^2 - 2ab + b^2$.

$y^2 - 2(y)(4) + (-4)^2$ Multiply.

$y^2 - 8y + 16$ Answer.

c. $(w - 2q)^2$ Apply $(a - b)^2 = a^2 - 2ab + b^2$.

$w^2 - 2(w)(2q) + (-2q)^2$ Multiply.

$w^2 - 4wq + 4q^2$ Answer.

d. $(2a + 3b)^2$ Apply $(a + b)^2 = a^2 + 2ab + b^2$.

$(2a)^2 + 2(2a)(3b) + (3b)^2$ Multiply and raise to powers.

$4a^2 + 12ab + 9b^2$ Answer.

Practice Exercises

Square the following binomials:

7. $(a + 2)^2$ **8.** $(b - 5)^2$ **9.** $(c + 3d)^2$ **10.** $(3x - 2y)^2$

Answers:

Practice Exercises 7–10: 7. $a^2 + 4a + 4$ 8. $b^2 - 10b + 25$ 9. $c^2 + 6cd + 9d^2$ 10. $9x^2 - 12xy + 4y^2$

If more practice is needed, do the Additional Practice Exercises.

Additional Practice Exercises

Square the following:

g. $(q + 5)^2$

h. $(r - 2)^2$

i. $(c - 2d)^2$

j. $(4a + 2b)^2$

OPTIONAL	**PRODUCTS THAT RESULT IN THE SUM OR DIFFERENCE OF CUBES**

Another special product is that of a binomial (two terms) multiplied by a trinomial (three terms) where the trinomial has the following properties: 1. The first term of the trinomial is the square of the first term of the binomial. 2. The second term of the trinomial is the negative of the product of the terms of the binomial. 3. The third term of the trinomial is the square of the last term of the binomial. The following illustrates the form of this product:

Square of first term   Square of last term
↓   ↓

$$(a + b)(a^2 - ab + b^2)$$

↑

Negative of the product of the two terms

A product of this form always results in the sum of two cubes if the binomial is a sum. The product will be the difference of two cubes if the binomial is a difference. The terms of the product are the cubes of the terms of the binomial.

Example 6

Find the following products:

a. $(a + b)(a^2 - ab + b^2) =$ Rewrite using distributive property.

$a(a^2 - ab + b^2) + b(a^2 - ab + b^2) =$ Apply distributive property.

$a^3 - a^2b + ab^2 + a^2b - ab^2 + b^3 =$ Add like terms.

$a^3 + b^3$ Sum of the cubes of the terms of the binomial.

b. $(x - y)(x^2 + xy + y^2) =$ Rewrite using distributive property.

$x(x^2 + xy + y^2) - y(x^2 + xy + y^2) =$ Apply distributive property.

$x^3 + x^2y + xy^2 - x^2y - xy^2 - y^3 =$ Add like terms.

$x^3 - y^3$ Difference of the cubes of the terms of the binomial.

c. $(x + 2)(x^2 - 2x + 4) =$ Form of $(a + b)(a^2 - ab + b^2)$.

$x^3 + 2^3 =$ Sum of the cubes of the terms of the binomial.

$x^3 + 8$ Answer.

d. $(2x - 3y)(4x^2 + 6xy + 9y^2) =$ Form of $(a - b)(a^2 + ab + b^2)$.

$(2x)^3 - (3y)^3 =$ Difference of cubes. Apply $(ab)^n = a^n b^n$.

$8x^3 - 27y^3$ Answer.

Answers:

Additional Practice Exercises 8–j: **g.** $q^2 + 10q + 25$ **h.** $r^2 - 4r + 4$ **i.** $c^2 - 4cd + 4d^2$ **j.** $16a^2 + 16ab + 4b^2$

Practice Exercises

Find the following products:

11. $(c + d)(c^2 - cd + d^2)$

12. $(w - z)(w^2 + wz + z^2)$

13. $(y + 3)(y^2 - 3y + 9)$

14. $(3a - b)(9a^2 + 3ab + b^2)$

If more practice is needed, do the Additional Practice Exercises.

Additional Practice Exercises:

Find the following products.

k. $(r + s)(r^2 - rs + s^2)$

l. $(a - b)(a^2 + ab + b^2)$

m. $(a + 4)(a^2 - 4a + 16)$

n. $(2a + 3b)(4a^2 - 6ab + 9b^2)$

Study Tip 8

Making Effective Use of Class Time

You need to get as much out of your time in class as possible. It is here that you have the benefit of having an expert on the subject at your disposal. The following are some things you should do in order to get the most out of class time:

1. Attend all class meetings. Be on time so that you do not miss anything. You will not have anyone to explain the material to you. If you do miss a class, be sure to find out what the assignment was and complete it *before* the next class meeting.
2. Participate in class. Do not just listen to the instructor. Think along with the instructor and anticipate the next step when he or she is doing an example. Answer questions that the instructor asks. Ask questions when you do not understand what the teacher has said or why he or she has said it.
3. Make an effort to understand as much as possible before you leave class.
4. Sit as close to the front of the classroom as possible so you can see and hear everything. You will also find yourself more involved with classroom activities.
5. Keep a mathematics notebook that has your class notes in one section and your homework in another. When taking notes, leave extra space so you will have room to add additional comments as you review your notes.
6. If you do not understand an example, ask the instructor to explain the first step that you do not understand.
7. Get help outside of class either from the instructor, a classmate, or, if available, the math learning center. Forming a study group with others in the class is an excellent idea.
8. Try tape-recording the class and use the recording to supplement your notes. Listen to the recording as you review your notes. Don't forget to ask the instructor's permission to record the class.

Exercise Set 2.4

Find the product of the following binomials using FOIL:

1. $(x + 3)(x + 2)$ **2.** $(y + 4)(y + 5)$ **3.** $(a + 3)(a - 4)$ **4.** $(b - 4)(b + 6)$

5. $(c - 5)(c - 2)$ **6.** $(q - 5)(q - 1)$ **7.** $(a + 3b)(a + 4b)$ **8.** $(t + 5s)(t + 2s)$

9. $(r - 5s)(r + 3s)$ **10.** $(x + 4y)(x - 6y)$ **11.** $(2a + 3b)(3a + 2b)$ **12.** $(3s + 5t)(2s + 3t)$

13. $(3x - 5y)(2x + 3y)$ **14.** $(4x + 3y)(5x - 2y)$ **15.** $(2x - 5y)(3x - 2y)$ **16.** $(3a - 5b)(2a - 7b)$

Calculator Exercises:

Find the following products using a calculator:

C1. $(3.1x + 6.3y)(8.2x - 5.2y)$ **C2.** $(8.3x - 3.1y)(9.5x + 4.8y)$

Find the following products of the form (a + b)(a − b):

17. $(x + 3)(x - 3)$ **18.** $(x + 5)(x - 5)$ **19.** $(p - q)(p + q)$

20. $(r - s)(r + s)$ **21.** $(a + 4b)(a - 4b)$ **22.** $(w + 3q)(w - 3q)$

23. $(6a - b)(6a + b)$ **24.** $(3x + y)(3x - y)$ **25.** $(4a + 5)(4a - 5)$

26. $(5a + 2)(5a - 2)$ **27.** $(2x + 5y)(2x - 5y)$ **28.** $(3x + 5y)(3x - 5y)$

Calculator Exercises:

Find the following products using a calculator:

C3. $(7.2a + 6.7)(7.2a - 6.7)$ **C4.** $(2.8x - 8.3)(2.8x + 8.3)$

Square the following binomials:

29. $(x + 2)^2$ **30.** $(y + 6)^2$ **31.** $(a - 4)^2$

32. $(b - 7)^2$ **33.** $(r + s)^2$ **34.** $(p + q)^2$

35. $(t - s)^2$ **36.** $(r - s)^2$ **37.** $(2x + 3)^2$

38. $(3x + 4)^2$ **39.** $(3a - 1)^2$ **40.** $(5a - 2)^2$

41. $(2x + 5y)^2$ **42.** $(3a + 4b)^2$ **43.** $(4x - 3y)^2$

44. $(5x - 6y)^2$ **45.** $(5t^2 - w)^2$ **46.** $(b + 2z^2)^2$

Calculator Exercises:

Square the following binomials using a calculator:

C5. $(4.7x + 3.7y)^2$ **C6.** $(5.7a - 7.4b)^2$

Find the following products:

47. $(2a + 5b)(3a - 2b)$ **48.** $(4x - 3y)(2x + 5y)$

49. $(4a - 3b)(4a + 3b)$ **50.** $(3s + 5)(3s - 5)$

51. $(3a + 5)^2$ **52.** $(2a - 7)^2$

53. $(2 - 3b)(4 - 3b)$

54. $(3 + 5c)(3 + 2c)$

55. $(6 + y)(6 - y)$

56. $(7 + x)(7 - x)$

57. $(4 + 3w)^2$

58. $(5 + 4c)^2$

59. $(4c + 7d)(4c - 7d)$

60. $(5x - 8y)(5x + 8y)$

Write an expression or equation representing each of the following and simplify when possible:

61. The square of the sum of 3 and x

64. 5 less than the square of y

62. The square of the difference of y and 4

65. The square of the difference of x and 5 is equal to 25.

63. 3 more than the square of x

66. The square of the sum of x and 7 is equal to 16.

Recall that the formula for the area of a rectangle is $A = LW$. Represent the area of the following rectangles in exercises 67–70. For example, if $L = x + 2$ and $W = x - 5$, then $A = (x + 2)(x - 5) = x^2 - 3x - 10$.

67. $L = x + 5$, $W = x - 8$

68. $L = y - 7$, $W = y + 9$

69. $L = 4a - 5$, $W = 3a + 9$

70. $L = 5x - 6$, $W = x + 5$

Recall that the area of a square is $A = s^2$, where s is the length of a side. For example, the area of a square each of whose sides is $x + 7$ is $A = (x + 7)^2 = x^2 + 14x + 49$. Represent the area of each of the squares in exercises 71–74:

71. $s = x + 8$

72. $s = y - 9$

73. $s = 3a - 5b$

74. $s = 4y + 3z$

Optional Exercises: (75–82)

Find the following products:

75. $(p + q)(p^2 - pq + q^2)$

76. $(t - s)(t^2 + ts + s^2)$

77. $(x + 3)(x^2 - 3x + 9)$

78. $(a + 5)(a^2 - 5a + 25)$

79. $(b - 4)(b^2 + 4b + 16)$

80. $(r - 2)(r^2 + 2r + 4)$

81. $(3a + 2b)(9a^2 - 6ab + 4b^2)$

82. $(2z - 4w)(4z^2 + 8wz + 16w^2)$

Writing Exercises:

83. In finding products of the form $(a + b)(a - b)$, there is no middle term. Why?

85. By assigning values to a and b, show that $(a + b)^2 \neq a^2 + b^2$. In general, show that $(a + b)^n \neq a^n + b^n$ for $n = 2, 3,$ and 5.

84. When squaring a binomial, why is it necessary to multiply the product of the two terms by 2 to get the middle term?

Group Project:

86. Using the formulas for the area of a rectangle ($A = LW$) and the area of a square ($A = s^2$), construct a geometric figure that illustrates that $(a + b)^2 = a^2 + 2ab + b^2$. (*Hint:* Begin with a square each of whose sides is of length $a + b$.)

Section 2.5 | Division of and Order of Operations with Integers

OBJECTIVES

When you complete this section, you will be able to:

a. Divide integers.

b. Evaluate expressions containing integers that involve combinations of addition, subtraction, multiplication, division, and raising to powers.

c. Evaluate expressions containing variables when given the value(s) of the variables.

Introduction

In Section 2.1, we discussed the multiplication of integers and found that $(+)(+) = +$, $(+)(-) = -$, $(-)(+) = -$, and $(-)(-) = +$. In other words, we found that the product of two numbers with the same sign is positive, and the product of two numbers with opposite signs is negative. In this section, we will develop similar rules for division, but first we need to define division.

We know that $\frac{24}{3} = 8$, because $3 \cdot 8 = 24$. Likewise, $\frac{12}{4} = 3$ because $4 \cdot 3 = 12$. Division can always be checked by multiplication. Generalizing on these examples, we state the following definition of division:

DEFINITION Division

For all numbers a, b, and c with $b \neq 0$, if $\frac{a}{b} = c$, then $b \cdot c = a$.

In the above definition, $b \neq 0$. In other words, you cannot divide by 0. This is because of the definition of division. If $\frac{a}{0} = c$, then $0 \cdot c = a$, which is impossible since 0 times any real number is 0. For example, if $\frac{3}{0} =$ _____ then $0 \cdot$ _____ $= 3$, which is impossible since 0 times any number is 0. We would say that $\frac{3}{0}$ is undefined.

OPTIONAL | CALCULATOR EXPLORATION ACTIVITY

Evaluate each of the following columns using a calculator:

Column A	Column B
$\frac{12}{6} =$	$\frac{-12}{6} =$
$\frac{18}{2} =$	$\frac{18}{-2} =$
$\frac{-16}{-4} =$	$\frac{-16}{4} =$
$\frac{-21}{-3} =$	$\frac{21}{-3} =$

1. Based on your answers to Column A, is the quotient of two numbers with the same sign positive or negative?

2. Based on your answers to Column B, is the quotient of two numbers with different signs positive or negative?

We now give justification for the preceding observations.

Answers:

Calculator Exploration Activity: 1. positive 2. negative

Dividing a positive number by a positive number

From our prior knowledge of arithmetic, we know that a positive number divided by a positive number is a positive number. For example, $\frac{18}{6} = 3$ and $\frac{28}{7} = 4$. Stated in symbols, $\frac{(+)}{(+)} = (+)$, because $(+)(+) = (+)$.

Using the definition of division just stated, we will develop rules for dividing a positive number by a negative number, a negative number by a positive number, and a negative number by another negative number.

Example 1

Dividing a positive number by a negative number

a. Find $\frac{12}{-3}$.

If $\frac{12}{-3} = $ _____, then $-3 \cdot$ _____ $= 12$. Since $(-3)\underline{\,(-4)\,} = 12$, the blank would be filled by -4. Therefore, $\frac{12}{-3} = -4$. Check: $(-3)(-4) = 12$.

b. Find $\frac{32}{-4}$.

If $\frac{32}{-4}$, then $-4 \cdot$ _____ $= 32$. Since $(-4)\underline{\,(-8)\,} = 32$, the blank would be filled by -8. Therefore, $\frac{32}{-4} = -8$. Check: $(-4)(-8) = 32$.

From Example 1, we can conclude that a positive number divided by a negative number results in a negative number. Stated in symbols, $\frac{(+)}{(-)} = (-)$, because $(-)(-) = (+)$.

Example 2

Dividing a negative number by a positive number

a. Find $\frac{-18}{3}$.

If $\frac{-18}{3} = $ _____, then $3 \cdot$ _____ $= -18$. Since $(3)\underline{\,(-6)\,} = -18$, the blank would be filled with -6. Therefore, $\frac{-18}{3} = -6$. Check: $(3)(-6) = -18$.

b. Find $\frac{-12}{4}$.

If $\frac{-12}{4} = $ _____, then $4 \cdot$ _____ $= -12$. Since $(4)\underline{\,(-3)\,} = -12$, the blank would be filled with -3. Therefore, $\frac{-12}{4} = -3$. Check: $(4)(-3) = -12$.

From Example 2, we can conclude that a negative number divided by a positive number results in a negative number. Written symbolically, $\frac{(-)}{(+)} = (-)$, because $(+)(-) = (-)$.

Example 3

Dividing a negative number by a negative number

a. Find $\frac{-8}{-2}$.

If $\frac{-8}{-2} = $ _____, then $-2 \cdot$ _____ $= -8$. Since $(-2)\underline{\,(4)\,} = -8$, the blank would be filled by 4. Therefore, $\frac{-8}{-2} = 4$. Check: $(-2)(4) = -8$.

b. Find $\frac{-27}{-3}$.

If $\frac{-27}{-3} = $ _____, then $-3 \cdot$ _____ $= -27$. Since $(-3)\underline{\,(9)\,} = -27$, the blank would be filled by 9. Therefore, $\frac{-27}{-3} = 9$. Check: $(-3)(9) = -27$.

From Example 3, we can conclude that a negative number divided by a negative number results in a positive number. Stated symbolically, $\frac{(-)}{(-)} = (+)$, because $(-)(+) = (-)$.

Note that the rules for division of signed numbers are exactly the same as the rules for multiplication as summarized in the following box:

> ### Division of Signed Numbers
>
> The quotient of two numbers with the same sign is positive and the quotient of two numbers with opposite signs is negative. In symbols: $\frac{(+)}{(+)} = (+)$, $\frac{(+)}{(-)} = (-)$, $\frac{(-)}{(+)} = (-)$, and $\frac{(-)}{(-)} = (+)$.

Example 4

Find the following quotients:

a. $\dfrac{10}{-2} = -5$ A positive divided by a negative is a negative.

b. $\dfrac{-14}{-7} = 2$ A negative divided by a negative is a positive.

c. $\dfrac{24}{8} = 3$ A positive divided by a positive is a positive.

d. $\dfrac{-26}{13} = -2$ A negative divided by a positive is a negative.

Practice Exercises

Find the following quotients:

1. $\dfrac{24}{-3}$ **2.** $\dfrac{-21}{-7}$ **3.** $\dfrac{4}{0}$ **4.** $\dfrac{-36}{4}$

If more practice is needed, do the Additional Practice Exercises.

Additional Practice Exercises

Find the following quotients:

a. $\dfrac{-16}{8}$ **b.** $\dfrac{-24}{-6}$ **c.** $\dfrac{15}{0}$ **d.** $\dfrac{54}{-9}$

When translating from English to mathematics, one way of indicating division is by using the word "quotient." When using "quotient," look for the words "of" and "and." If division is indicated by using a fraction, the expression following "of" is the numerator and the expression following "and" is the denominator.

Example 5

Write an expression for each of the following and simplify:

a. The quotient of 24 and -6 increased by 7

Solution:
The quotient of 24 and -6 means $\frac{24}{-6}$. The quotient is increased by 7, which means add 7 to the quotient. Consequently, the quotient of 24 and -6 increased by 7 is expressed as:

$\dfrac{24}{-6} + 7 =$ Divide before adding.

$-4 + 7 =$ Add.

3 Answer.

b. The quotient of -15 and -3 subtracted from the quotient of 36 and -9

$$\frac{36}{-9} - \frac{-15}{-3} = \qquad \text{Find each quotient.}$$

$$-4 - 5 = \qquad \text{Add.}$$

$$-9 \qquad \text{Answer.}$$

Solution:
The quotient of -15 and -3 means $\frac{-15}{-3}$ and the quotient of 36 and -9 means $\frac{36}{-9}$. The first quotient is subtracted from the second, which is expressed as:

Practice Exercises

Write an expression for each of the following and simplify:

5. The quotient of -35 and 7 decreased by 8

6. The quotient of 48 and -12 subtracted from the quotient of -56 and 8

Because integers had not been discussed in Section 1.1 when we did order of operations, we were limited to operations with whole numbers only. Now that we know how to perform operations on integers, we will revisit the order of operations. We repeat the order of operations.

Order of Operations

If an expression contains more than one operation, operations are to be performed in the following order:

1. If parentheses or other inclusion symbols (braces or brackets) are present, begin within the innermost and work outward, using the order in steps 3–5 in doing so.
2. If a fraction bar is present, simplify above and below the fraction bar separately in the order given in steps 3–5.
3. Evaluate all indicated powers.
4. Perform all multiplication or division in the order in which they occur as you work from left to right.
5. Then perform all addition or subtraction in the order in which they occur as you work from left to right.

Example 6

Evaluate the following using the order of operations agreement:

a. $-5 - 6 + 8 =$ Add and subtract in order from left to right.

$-11 + 8 =$ Continue adding.

-3 Answer.

b. $8 - 15 \div (-3) =$ Divide before adding. $-15 \div (-3) = 5$.

$8 + 5 =$ Add.

13 Answer.

c. $-7 - 3(-2)^2 =$ Raise to powers first. $(-2)^2 = 4$.

$-7 - 3 \cdot 4 =$ Multiplication before subtraction.

$-7 - 12 =$ Add.

-19 Answer.

d. $6 - (12 - 18) \div 3 =$ Add numbers inside parentheses first.

$6 - (-6) \div 3 =$ Division before subtraction.

$6 - (-2) =$ $-(-2) = 2$.

$6 + 2 =$ Add.

8 Answer.

e. $-3 + 5 \cdot 6^2 \div (-9)(-2) + 6 =$ Raise to powers first.

$-3 + 5 \cdot 36 \div (-9)(-2) + 6 =$ Multiply 5 and 36.

$-3 + 180 \div (-9)(-2) + 6 =$ Divide 180 by -9.

$-3 + (-20)(-2) + 6 =$ Multiply -20 and -2.

$-3 + 40 + 6 =$ Add -3 and 40.

$37 + 6 =$ Add 37 and 6.

43 Answer.

f. $-3(5^2 - 4) \div (-7) =$ Evaluate power inside parentheses first.

$-3(25 - 4) \div (-7) =$ Add inside parentheses.

$-3(21) \div (-7) =$ Multiply before dividing in order from left to right.

$-63 \div (-7) =$ Divide.

9 Answer.

g. $\dfrac{4 - 6^2}{4^2 - 8} =$ Simplify the numerator and the denominator separately.

$\dfrac{4 - 36}{16 - 8} =$ Add in the numerator and the denominator.

$\dfrac{-32}{8} =$ Divide.

-4 Answer.

h. $\dfrac{-5(-2) - [(-4)(-5) + 6^2]}{(-2)^2(6) - 3} =$ Evaluate power inside the brackets in the numerator and the power in the denominator first.

$\dfrac{-5(-2) - [(-4)(-5) + 36]}{(4)(6) - 3} =$ Multiply inside the brackets in the numerator and multiply in the denominator.

$\dfrac{-5(-2) - [20 + 36]}{24 - 3} =$ Add inside the brackets in the numerator and add in the denominator.

$\dfrac{-5(-2) - 56}{21} =$ Multiply in the numerator.

$\dfrac{10 - 56}{21} =$ Add in the numerator.

$\dfrac{-46}{21} = -\dfrac{46}{21}$ Answer.

Practice Exercises

Evaluate the following using the order of operations:

7. $-12 + 5 - 8$

8. $-9 - 18 \div (-9)$

9. $6 - (-3)(-4)^2$

10. $8 - (5 - 13) \div (-4)$

11. $6(7^2 - 13) \div (-9)$

12. $7 - 3 \cdot 8^2 \div (-8)(-3) + 5$

13. $\dfrac{6 - 6^2}{5^2 - 8}$

14. $\dfrac{4(-2) - [(-5)(3) - 21]}{(-3)^2(-2) + 15}$

If more practice is needed, do the Additional Practice Exercises.

Additional Practice Exercises

Evaluate the following using the order of operations agreement:

e. $9 - 12 - 8$

f. $-6 - (-8) \div 2$

g. $9 - 4(-3)^3$

h. $-7 - (7 - 19) \div (-3)$

i. $-2(8^2 - 22) \div (-14)$

j. $-4 + 2 \cdot 4^3 \div (-16) - 7$

k. $\dfrac{3^3 - 2^4}{4^2 - 9}$

l. $\dfrac{4(-3) - [(-8)(3) - 4]}{(-4)^2(-1) - 9}$

In Sections 1.1 and 1.2, variables were introduced and we evaluated expressions containing variables. The values used for the variables were whole numbers, since operations with integers had not been discussed. We can now use integer values as illustrated by the following examples:

Example 7

Evaluate each of the following for $x = -2$ *,* $y = -3$ *, and* $z = 6$ *:*

a.

$5x^2 - 2y =$	Substitute -2 for x and -3 for y.
$5(-2)^2 - 2(-3) =$	First raise to powers.
$5(4) - 2(-3) =$	Multiply in order from left to right.
$20 + 6 =$	Add.
26	Answer.

b.

$-3xy^2z =$	Substitute for x, y, and z.
$-3(-2)(-3)^2(6) =$	First raise to powers.
$-3(-2)(9)(6) =$	Multiply in order from left to right.
$6 \cdot 9 \cdot 6 =$	Continue multiplying.
$54 \cdot 6 =$	Continue multiplying.
324	Answer.

c.

$x - y(x^4 - 2z) =$	Substitute for x, y, and z.
$-2 - (-3)[(-2)^4 - 2 \cdot 6] =$	Raise to powers inside the parentheses.
$-2 - (-3)(16 - 2 \cdot 6) =$	Multiply inside the parentheses.
$-2 - (-3)(16 - 12) =$	Add inside the parentheses.
$-2 - (-3)(4) =$	Multiply before adding.
$-2 - (-12) =$	$-(-12) = 12$.
$-2 + 12 =$	Add.
10	Answer.

Answers:

Practice Exercises 11–14: 11. -24 12. -60 13. $\dfrac{-30}{17} = \dfrac{30}{-17}$ 14. $\dfrac{-28}{3} = \dfrac{28}{-3}$

Additional Practice Exercises e–l: e. -11 f. -2 g. 117 h. -11 i. 6 j. -19 k. $\dfrac{11}{7}$ l. $\dfrac{16}{-25} = \dfrac{-16}{25}$

d. $\dfrac{2x^2 - 3y}{2z^2 - 3x^4} =$ Substitute for x, y, and z.

$\dfrac{2(-2)^2 - 3(-3)}{2(6)^2 - 3(-2)^4} =$ First raise to powers.

$\dfrac{2 \cdot 4 - 3(-3)}{2 \cdot 36 - 3 \cdot 16} =$ Multiply in order, left to right.

$\dfrac{8 + 9}{72 - 48} =$ Add.

$\dfrac{17}{24}$ Answer.

Practice Exercises

Evaluate each of the following for x $= -3$, y $= 2$, *and* z $= -4$:

15. $3x^2 - 4z$ **16.** $-2xy^2z^2$ **17.** $-2y - x(3y^2 - 2z^2)$ **18.** $\dfrac{4y - 5x^2}{3y^3 - z^2}$

If more practice is needed, do the Additional Practice Exercises.

Additional Practice Exercises

Evaluate each of the following for x $= -4$, y $= 3$, *and* z $= -1$:

m. $4z^2 - 2x^2$ **n.** $-5x^2yz^2$ **o.** $-3y + 2x(2z^2 - 2x^2)$ **p.** $\dfrac{3y^2 - 5z}{3x^2 - 5y}$

Study Tip 9

Preparing for a Test

Proper test preparation is one of the most important factors in determining how well you do in a course. The following suggestions should help:

1. Set your goal high. Try for a score of 100% rather than just a passing grade. The higher you set your goal, the more complete your preparation and the higher you should score.
2. Avoid developing a mental block. Thoroughly prepare for the test. Inadequate preparation causes a loss of confidence that can produce a mental block. So be prepared.
3. Begin your test preparation early. Do not wait until the night before the test to begin studying. Begin your preparation at least a week in advance and study at least an hour a day for the exam.
4. Be organized: make a list of specific topics to be covered on the test. Find and solve specific problems for each topic. Be sure to include all types of problems that could be contained within each topic.
5. Start at the beginning of the material to be tested and work through each section in turn. Master each section before going to the next.
6. Practice by doing the chapter review in your textbook. Do all of the exercises if you can. If you can't do an exercise, go back to the indicated section and study the examples. If you still can't do an exercise, ask your instructor.

Answers:

o. 231 p. $\dfrac{33}{32}$

Practice Exercises 15–18: 15. 43 16. 384 17. -64 18. $-\dfrac{37}{8}$ **Additional Practice Exercises m–p:** m. -28 n. -240

7. In your review, answer each problem, confirm that the answer is correct (check your work!), and examine your understanding of the problem. Do not allow yourself to get "stuck." If you can't do an exercise after ten minutes, go to the next exercise.
8. Review both your notes and the text to clear up any questions that you might have. Think about the material as you review it. How does it relate to previous material? How do different parts of the material relate to other parts?
9. Be able to distinguish between the different types of problems that might be on the test.
10. Try to find or construct a practice test. The practice test might be one of your teacher's previous tests, tests from the text, or tests from a study guide.

Exercise Set 2.5

Find the following quotients:

1. $\dfrac{36}{-9}$

2. $\dfrac{-36}{-6}$

3. $\dfrac{-42}{7}$

4. $\dfrac{63}{9}$

5. $\dfrac{-48}{-8}$

6. $\dfrac{56}{-7}$

7. $\dfrac{6}{0}$

8. $\dfrac{-8}{0}$

9. $\dfrac{144}{-18}$

10. $\dfrac{-392}{98}$

Challenge Exercises: (11–14)

11. $\dfrac{3}{4} \div \left(-\dfrac{5}{3}\right)$

12. $-\dfrac{2}{3} \div \left(-\dfrac{5}{4}\right)$

13. $-\dfrac{3}{5} \div \dfrac{5}{4}$

14. $\left(-\dfrac{7}{3}\right) \div \left(-\dfrac{9}{2}\right)$

Write an expression for each of the following and simplify:

15. The quotient of 12 and −4 increased by 8

16. The quotient of −15 and 5 increased by 9

17. The quotient of −18 and −9 decreased by −6

18. The quotient of −24 and −4 decreased by −9

19. 12 less the quotient of 36 and −9

20. −5 less the quotient of 45 and −5

21. The quotient of 48 and −16 subtracted from 7

22. The quotient of −28 and −7 subtracted from −8

23. The quotient of −9 and 3 added to the quotient of 14 and −7

24. The quotient of −24 and 6 added to the quotient of −32 and 8

25. The quotient of −36 and 18 subtracted from the quotient of 48 and −8

26. The quotient of −42 and 6 subtracted from the quotient of 54 and −9

Answer the following:

27. When a cold front moved through an area, the temperature dropped 28° in four hours. Using a signed number, find the average drop per hour.

28. Paula went on a diet and lost ten pounds in four weeks. Using a signed number, find the average number of pounds per week that Paula lost.

29. On a steep mountain road the elevation dropped at the rate of 9 feet per 150 feet. Using signed numbers, find the rate at which the road dropped per foot.

30. A football team lost 12 yards in three plays. Using a signed number, find the average loss per play.

31. A parachutist descends 54 feet in six seconds. Using a signed number, find the rate at which she is descending per second.

32. A stock dropped 5 points in four hours. Using a signed number, find the average drop per hour.

Evaluate the following:

33. $-3 + 7 - 6$

34. $-7 + 9 - 5$

35. $12 - 5 + 15$

36. $14 - 9 + 6$

37. $8 - 4 \cdot 6$

38. $5 - 7 \cdot 3$

39. $-5 + 4(-7)$

40. $-3 + 7(-6)$

41. $8 - 12 \div 3$

42. $6 - 18 \div 3$

43. $9 - 21 \div (-7)$

44. $4 - 15 \div (-5)$

45. $10 - 2(-4)^2$

46. $14 - 4(-2)^2$

47. $9 - (8 - 16) \div 4$

48. $5 - (3 - 12) \div 3$

49. $-8 - (15 - 5) \div (-5)$

50. $-4 - (13 - 7) \div (-3)$

51. $6 + 4 \cdot 5^2 \div 25(-2) + 3$

52. $3 + 3 \cdot 8^2 \div 32(-3) + 5$

53. $-5 - 2 \cdot 6^2 \div (-36)(-4) + 8$

54. $-9 - 6 \cdot 4^2 \div (-16)(-5) + 7$

55. $4(6^2 - 4) \div (-8)$

56. $6(4^2 + 12) \div (-7)$

57. $-3^2(6 - 3^2)$

58. $-4^2(8 - 4^2)$

59. $-36 \div (4^2 - 7)$

60. $-48 \div (5^2 - 9)$

61. $\dfrac{7^2 - 13}{5^2 - 7}$

62. $\dfrac{9^2 - 17}{6^2 - 4}$

63. $\dfrac{2(-3)^2 - 6}{-3(-2)^2 + 6}$

64. $\dfrac{2(-3)^2 + 6}{3(-3)^2 - 3}$

65. $\dfrac{2(4^2) - 7^2}{5^2 - 3(-3)^2}$

66. $\dfrac{3(-2)^2 - 5^2}{9^2 - 4(-5)^2}$

67. $\dfrac{(-3)(-5) - [5(-4) + 7^2]}{(-3)^2(4) - 7}$

68. $\dfrac{(-2)(-6) - [6(-3) + 5^2]}{(-4)^2(3) - 21}$

Calculator Exercises:

Evaluate each of the following using a calculator. If necessary, round off to the nearest hundredth.

C1. $7.02 + (5.72)(9.43)$

C2. $9.40 + (7.34)(6.92)$

C3. $18.2 - 208.32 \div 16.8$

C4. $14.9 - 325.26 \div (-23.4)$

C5. $(56.2)^2 + (-82.4)^2$

C6. $(-12.4)^3 - (58.9)^2$

C7. $68.23 - 34.7(56.1^2 - 67.2) \div 39.1$

C8. $95.35 + 62.5(9.8^2 - 68.9) \div 62.8$

C9. $\dfrac{(13.1)^2 + (32.1)(45.3)^2}{(12.6)^3 - (76.3)(18.5)^2}$

C10. $\dfrac{(5.7)^4 - (19.7)^2(96.3)}{(26.7)^2 + (14.6)^2(31.1)^2}$

Evaluate each of the following for x = − 2, y = 3, and z = − 3:

69. $4x^2 - 2y$

70. $5z^2 - 3x$

71. $-3y^3 + 4x$

72. $-5x^2 + 3z$

73. $5x^2 - 3y^2$

74. $2y^2 - 3x^2$

75. $3x^2y$

76. $4y^2z$

77. $-5xz^2$

78. $-6yz^2$

79. $3x^2y^2z$

80. $2xy^2z^2$

81. $3y - 2(3x + z)$

82. $3x - 4(2y - x)$

83. $-3x + z(3x - 4y)$

84. $-3z + x(4y - 3x)$

85. $-y^2 - (2x^2 - 3z)$

86. $-x^2 - (3z^2 - 4x)$

87. $2x^3y - x^2(3z^2 - 4x^2)$

88. $3y^3x - z^2(5x^2 - y^2)$

89. $\dfrac{2x + 3y}{3y + 2z}$

90. $\dfrac{3x + 3z}{4y + 2x}$

91. $\dfrac{x^2 - y^2}{y^2 + z^2}$

92. $\dfrac{y^2 + x^2}{x^2 - z^2}$

93. $\dfrac{2x^3 - 3y^2}{3x^3 - 2z^2}$

94. $\dfrac{4x^3 - 2y^2}{5x^3 - 3z^3}$

Answer the following:

95. The owner of a produce stand sold 55 pounds of tomatoes at a profit of 15 cents per pound. As the tomatoes began to age, she sold 22 pounds at a loss of 8 cents per pound. What was her net profit/loss on the sale of the tomatoes?

96. The produce stand owner in Exercise 95 also sold 20 pounds of onions at a profit of 12 cents per pound and later sold 8 pounds at a loss of 15 cents per pound. What was her net profit/loss on the sale of the onions?

97. During a recent price war among the airlines, 30 seats on one flight were sold at a profit of $40 per seat and 60 seats were sold at a loss of $25 per seat. Find the net profit/loss for this flight.

98. During the same price war as in Exercise 97, another flight sold 45 seats at a profit of $25 per seat and 50 seats at a loss of $25 per seat. Find the net profit/loss for this flight.

99. Ms. Jones recently sold some of her stock. She made a profit of $30 per share on 45 shares, had a loss of $12 per share on 20 shares, and a profit of $8 per share on 36 shares. What was her net profit/loss on the sale of the stock?

100. Mr. Lopez sold 20 shares of stock at a loss of $12 per share, 32 shares at a profit of $20 per share, and 40 shares at a loss of $7 per share. What was his net profit/loss from the sale of the stock?

Writing Exercises:

101. Why is the quotient of a negative number and a positive number equal to a negative number?

102. Why is the quotient of a negative number and a negative number equal to a positive number?

103. Why is $\frac{2}{0}$ undefined? (Do not say because you cannot divide by 0!)

104. Is division commutative? Why or why not?

105. What is the error in the following? $8 - 3(2^2 + 4) = 8 - 3(4 + 4) = 8 - 3(8) = 5(8) = 40$. Rework the problem correctly.

106. What is the error in the following? $5 - 2 \cdot 4^2 = 5 - 8^2 = 5 - 64 = -59$. Rework the problem correctly.

Group Project:

107. Write a problem involving the order of operations on integers similar to 95–100. Exchange your problem with another group and solve the problem that your group receives.

Section 2.6   # Quotient Rule and Integer Exponents

OBJECTIVES

When you complete this section, you will be able to:

a. Simplify exponential expressions using the property, $\frac{a^m}{a^n} = a^{m-n}$.

b. Simplify expressions with zero and negative integer exponents.

Introduction   In Section 2.2, we developed laws of exponents that involved products of expressions with exponents. For reference, these were: (1) $a^m \cdot a^n = a^{m+n}$, (2) $(a^m)^n = a^{mn}$, and (3) $(ab)^n = a^n b^n$. In this section we will develop similar laws for quotients of expressions with exponents.

OPTIONAL	CALCULATOR EXPLORATION ACTIVITY 1

Evaluate each of the following columns using a calculator.

Column A   **Column B**

1. $\dfrac{2^5}{2^2} =$ $2^3 =$

2. $\dfrac{3^7}{3^4} =$ $3^3 =$

3. $\dfrac{5^6}{5^4} =$ $5^2 =$

By looking at the corresponding lines of columns A and B, answer the following:

1. To divide two numbers with the same base, leave the _____ unchanged and _____ the exponents.

Following is justification for the preceding observation:

Developing $\dfrac{a^m}{a^n} = a^{m-n}$

Remember, the exponent indicates how many times the base is to be used as a factor. Another fact that we will need is that any number (other than 0) divided by itself is equal to 1. We also need the procedure for multiplying fractions. Recall that we multiply numerator times numerator and denominator times denominator. For example, $\dfrac{2}{3} \cdot \dfrac{5}{7} = \dfrac{2 \cdot 5}{3 \cdot 7} = \dfrac{10}{21}$.

Example 1

Simplify the following using the definition of exponents. Leave the answer in exponential form.

a. $\dfrac{2^6}{2^3} =$ Rewrite using the definition of exponents.

$\dfrac{2 \cdot 2 \cdot 2 \cdot 2 \cdot 2 \cdot 2}{2 \cdot 2 \cdot 2} =$ Rewrite as the multiplication of fractions.

$\dfrac{2}{2} \cdot \dfrac{2}{2} \cdot \dfrac{2}{2} \cdot \dfrac{2}{1} \cdot \dfrac{2}{1} \cdot \dfrac{2}{1} =$ $\dfrac{2}{2} = 1$ and $\dfrac{2}{1} = 2$.

$1 \cdot 1 \cdot 1 \cdot 2 \cdot 2 \cdot 2 =$ Apply the identity for multiplication.

$2 \cdot 2 \cdot 2 =$ Rewrite using the definition of exponents.

2^3 Answer.

Notice that three of the six factors of 2 in the numerator were divided by the three factors of 2 in the denominator, leaving $6 - 3 = 3$ factors of 2 in the numerator. Therefore, $\dfrac{2^6}{2^3} = 2^{6-3} = 2^3$. Notice that the base remained as 2.

b. $\dfrac{x^5}{x^2} =$ Rewrite using the definition of exponents.

$\dfrac{x \cdot x \cdot x \cdot x \cdot x}{x \cdot x} =$ Rewrite as the multiplication of fractions.

$\dfrac{x}{x} \cdot \dfrac{x}{x} \cdot \dfrac{x}{1} \cdot \dfrac{x}{1} \cdot \dfrac{x}{1} =$ $\dfrac{x}{x} = 1$, if $x \neq 0$ and $\dfrac{x}{1} = x$.

$1 \cdot 1 \cdot x \cdot x \cdot x =$ Apply the identity for multiplication.

$x \cdot x \cdot x =$ Rewrite using the definition of exponents.

x^3 Answer.

Again, notice that two of the factors of x in the numerator were divided by the two factors of x in the denominator, leaving $5 - 2 = 3$ factors of x in the numerator. Therefore, $\frac{x^5}{x^2} = x^{5-2} = x^3$.

Note that in each of the preceding examples, the exponent in the numerator was larger than the exponent in the denominator. Based on these examples, we generalize to the following law of exponents for division of expressions with the same base but different exponents:

Be Careful A common error is to divide the bases as well as subtract the exponents. The base does not change when using this law of exponents. For example, $\frac{3^5}{3^3} \neq 1^{5-3}$. Using the Quotient Rule, $\frac{3^5}{3^3} = 3^{5-3} = 3^2$.

Fourth Law of Exponents: Quotient Rule

For any two positive integers m and n, with $m > n$ and $a \neq 0$, $\frac{a^m}{a^n} = a^{m-n}$. To divide two numbers with the same base, subtract the bottom exponent from the top exponent, leaving the base unchanged.

Example 2

Simplify the following using the Quotient Rule. Leave the quotients in exponential form. All variables represent nonzero quantities.

a. $\frac{3^5}{3^2} =$ Apply $\frac{a^m}{a^n} = a^{m-n}$.

$3^{5-2} =$ Subtract the exponents.

3^3 Quotient.

b. $\frac{5^6}{5^2} =$ Apply $\frac{a^m}{a^n} = a^{m-n}$.

$5^{6-2} =$ Subtract the exponents.

5^4 Quotient.

c. $\frac{z^7}{z^4} =$ Apply $\frac{a^m}{a^n} = a^{m-n}$.

$z^{7-4} =$ Subtract the exponents.

z^3 Quotient.

Practice Exercises

Simplify the following using the Quotient Rule. Leave the answers in exponential form. All variables represent nonzero quantities.

1. $\frac{4^5}{4^3} =$ **2.** $\frac{6^8}{6^4} =$ **3.** $\frac{x^6}{x^3} =$

If more practice is needed, do the Additional Practice Exercises.

Additional Practice Exercises

Simplify the following using the Quotient Rule. Leave the answers in exponential form. All variables represent nonzero quantities.

a. $\frac{5^6}{5^4}$ **b.** $\frac{9^7}{9^4}$ **c.** $\frac{r^5}{r^2}$

All the exponents that we have discussed to this point have been positive integers. If the restriction that $m > n$ is removed from the Quotient Rule, it would be possible to have zero or negative exponents. Using the Quotient Rule, $\frac{2^3}{2^3} = 2^{3-3} = 2^0$. What does 2^0 mean? It surely cannot mean use 2 as a factor 0 times. Example 3 will give meaning to 0 as an exponent.

Answers:

Example 3

Simplify the following:

a. $\frac{2^4}{2^4} = \frac{2 \cdot 2 \cdot 2 \cdot 2}{2 \cdot 2 \cdot 2 \cdot 2} = \frac{2}{2} \cdot \frac{2}{2} \cdot \frac{2}{2} \cdot \frac{2}{2} = 1 \cdot 1 \cdot 1 \cdot 1 = 1$: However, if we use the Quotient Rule $\frac{2^4}{2^4} = 2^{4-4} = 2^0$. Since $\frac{2^4}{2^4} = 1$ and $\frac{2^4}{2^4} = 2^0$, we conclude that $2^0 = 1$.

b. $\frac{x^3}{x^3} = \frac{x \cdot x \cdot x}{x \cdot x \cdot x} = \frac{x}{x} \cdot \frac{x}{x} \cdot \frac{x}{x} = 1 \cdot 1 \cdot 1 = 1$. If we use the Quotient Rule, $\frac{x^3}{x^3} = x^{3-3} = x^0$. Since $\frac{x^3}{x^3} = 1$ and $\frac{x^3}{x^3} = x^0$, we conclude that $x^0 = 1$ for $x \neq 0$.

Defining 0 exponents   Based on Example 3, we make the following definition for 0 exponents:

> **DEFINITION   0 Exponents**
>
> For all $a \neq 0$, $a^0 = 1$. Therefore, any nonzero number raised to the 0 power is equal to 1.

In using 0 as an exponent, great care must be taken in determining the base of the exponent, as illustrated by the following. Remember, the base is the symbol immediately preceding the exponent unless parentheses are used, in which case the base is everything inside the parentheses.

Example 4

Evaluate each of the following. Assume all variables have nonzero values.

a. $5^0 = 1$ Definition of 0 exponents.

b. $5x^0 = 5 \cdot 1 = 5$ The base of the 0 exponent is x only, not 5x. Remember, x fi 0.

c. $(5x)^0 = 1$ The base of the 0 exponent is 5x.

d. $-3y^0 + 7x^0 =$ The bases of the 0 exponents are y and x only. Apply the definition of 0 exponents.

   $-3 \cdot 1 + 7 \cdot 1 =$ Apply the identity for multiplication.

   $-3 + 7 =$ Add -3 and 7.

   4 Answer.

e. $(4x)^0 - 6x^0 =$ The bases of the 0 exponents are 4x and x. Apply the definition of 0 exponents.

   $1 - 6 \cdot 1 =$ Apply the identity for multiplication.

   $1 - 6 =$ Add 1 and -6.

   -5 Answer.

Practice Exercises

Evaluate each of the following. Assume x and z $\neq$ 0.

4. $8^0 =$

5. $8x^0 =$

6. $(8x)^0 =$

7. $4z^0 - 9z^0 =$

8. $6x^0 - (-3x)^0 =$

If more practice is needed, do the Additional Practice Exercises.

Additional Practice Exercises

Evaluate each of the following:

d. 10^0

e. $10y^0$

f. $(10x)^0$

g. $3a^0 - 7a^0$

h. $(-5x)^0 - 8x^0$

Defining negative exponents

Now we will consider negative exponents.

| OPTIONAL | CALCULATOR EXPLORATION ACTIVITY 2 |

Evaluate each of the following columns using a calculator:

Column A   **Column B**

1. $2^{-3} =$ $\dfrac{1}{2^3} =$

2. $5^{-2} =$ $\dfrac{1}{5^2} =$

3. $3^{-4} =$ $\dfrac{1}{3^4} =$

By looking at the corresponding lines of columns A and B, answer the following:

2. Raising a number to a negative power is the same as _____

_____.

Following is justification for the previous observation:

Using the Quotient Rule, $\dfrac{3^2}{3^5} = 3^{2-5} = 3^{-3}$ What does 3^{-3} mean? It cannot mean use 3 as a factor -3 times. The following examples will give meaning to these types of expressions:

Example 5

Simplify the following:

a. $\dfrac{3^2}{3^4} = \dfrac{3 \cdot 3}{3 \cdot 3 \cdot 3 \cdot 3} = \dfrac{3}{3} \cdot \dfrac{3}{3} \cdot \dfrac{1}{3} \cdot \dfrac{1}{3} = 1 \cdot 1 \cdot \dfrac{1}{3} \cdot \dfrac{1}{3} = \dfrac{1}{3^2}$. If we use the Quotient Rule, we get $\dfrac{3^2}{3^4} = 3^{2-4} = 3^{-2}$. Since $\dfrac{3^2}{3^4} = \dfrac{1}{3^2}$ and $\dfrac{3^2}{3^4} = 3^{-2}$, then $3^{-2} = \dfrac{1}{3^2}$.

b. $\dfrac{x^3}{x^4} = \dfrac{x \cdot x \cdot x}{x \cdot x \cdot x \cdot x} = \dfrac{x}{x} \cdot \dfrac{x}{x} \cdot \dfrac{x}{x} \cdot \dfrac{1}{x} = 1 \cdot 1 \cdot 1 \cdot \dfrac{1}{x} = \dfrac{1}{x}$. If we use the Quotient Rule, we get $\dfrac{x^3}{x^4} = x^{3-4} = x^{-1}$. Since $\dfrac{x^3}{x^4} = \dfrac{1}{x}$ and $\dfrac{x^3}{x^4} = x^{-1}$, then $x^{-1} = \dfrac{1}{x}$.

From the preceding example, we generalize to the following definition for negative integer exponents.

DEFINITION   Negative Exponents
For all $x \neq 0$, $x^{-n} = \dfrac{1}{x^n}$.

It is interesting to note that $x^n \cdot x^{-n} = x^{n+(-n)} = x^0 = 1$, $x \neq 0$. This means that x^n and x^{-n} are multiplicative inverses (reciprocals). For example, $2^1 = 2$ and $2^{-1} = \frac{1}{2}$ so $2 \cdot 2^{-1} = 2 \cdot \frac{1}{2} = 1$. Another way of stating it is that x^n denotes multiplication and x^{-n} denotes division.

Example 6

Rewrite each of the following with positive exponents only. Assume all variables represent nonzero quantities.

a. $4^{-3} = \frac{1}{4^3} = \frac{1}{64}$ Definition of negative exponents and $4^3 = 64$.

b. $x^{-5} = \frac{1}{x^5}$ Definition of negative exponents.

c. $3a^{-4} = 3 \cdot \frac{1}{a^4} = \frac{3}{a^4}$ The base of the exponent is a only, not $3a$.

d. $(4x)^{-2} = \frac{1}{(4x)^2} =$ The base of the exponent is $(4x)$.

$\frac{1}{4^2 x^2} = \frac{1}{16x^2}$ Apply $(ab)^n = a^n b^n$ and $4^2 = 16$.

> *Be Careful* A common mistake is to confuse negative exponents with negative numbers. The value of $3^{-2} \neq -9$, and also $3^{-2} \neq (-2)(3)$. The value of $3^{-2} = \frac{1}{3^2} = \frac{1}{9}$. Another common error occurs in exercises like Example 6c, $3a^{-4}$. This is often mistakenly written as $\frac{1}{3a^4}$.

Practice Exercises

Rewrite each of the following with positive exponents only. Assume all variables represent nonzero quantities.

9. $5^{-3} =$ **10.** $y^{-6} =$ **11.** $6y^{-4}$ **12.** $(7z)^{-3}$

If you need more practice, do the Additional Practice Exercises.

Additional Practice Exercises

Rewrite each of the following with positive exponents only:

i. 8^{-3} **j.** r^{-5} **k.** $-3a^{-5}$ **l.** $(4x)^{-2}$

Look for a pattern in simplifying an expression of the form $\frac{1}{x^n}$.

| OPTIONAL | CALCULATOR EXPLORATION ACTIVITY 3 |

Evaluate each of the following columns using a calculator:

Column A   **Column B**

1. $\frac{1}{2^{-3}} =$ $2^3 =$

2. $\frac{1}{5^{-2}} =$ $5^2 =$

3. $\frac{1}{3^{-4}} =$ $3^4 =$

By looking at the corresponding lines of columns A and B, answer the following:

3. One divided by a number raised to a negative power is the same as _____

_____.

Following is justification for the previous observation:

Example 7

a. In the expression $(2^{-3})^{-1}$, the base of the exponent -1 is (2^{-3}). Therefore, by the definition of a negative exponent, $(2^{-3})^{-1} = \frac{1}{2^{-3}}$. However, using the $(a^m)^n = a^{mn}$, which can be shown to also apply to negative exponents, $(2^{-3})^{-1} = 2^{(-3)(-1)} = 2^3$. Since $(2^{-3})^{-1} = \frac{1}{2^{-3}}$ and $(2^{-3})^{-1} = 2^3$, then $\frac{1}{2^{-3}} = 2^3$.

b. In the expression $(x^{-4})^{-1}$, the base of the exponent -1 is (x^{-4}). Therefore, by the definition of a negative exponent, $(x^{-4})^{-1} = \frac{1}{x^{-4}}$. But, if we use the *Power to a Power law of exponents*, $(x^{-4})^{-1} = x^4$. Therefore, $\frac{1}{x^{-4}} = x^4$ since they both equal $(x^{-4})^{-1}$.

From the preceding examples, we make the following observation:

Observation

For all $x \neq 0$, $\frac{1}{x^{-n}} = x^n$.

Example 8

Write the following with positive exponents only. Assume all variables represent nonzero quantities.

a. $\dfrac{1}{2^{-4}} = 2^4$ Apply $\dfrac{1}{x^{-n}} = x^n$.

b. $\dfrac{1}{x^{-6}} = x^6$ Apply $\dfrac{1}{x^{-n}} = x^n$.

c. $\dfrac{2}{x^{-2}} = 2x^2$ Apply $\dfrac{1}{x^{-n}} = x^n$.

Practice Exercises

Write the following with positive exponents only. Assume all variables represent nonzero quantities.

13. $\dfrac{1}{5^{-7}} =$

14. $\dfrac{1}{x^{-8}} =$

15. $\dfrac{4}{y^{-3}} =$

Since 0 and negative exponents now have meaning, it is no longer necessary to restrict $m > n$ in the Quotient Rule. Therefore, $\frac{a^m}{a^n} = a^{m-n}$ for all integer values of m and n.

Generalized Quotient Rule

For all integer values of m and n, $\frac{a^m}{a^n} = a^{m-n}$ for $a \neq 0$.

It can be shown that all the laws of exponents hold for zero and negative exponents. The following examples make that assumption. Remember, $a^m \cdot a^n = a^{m+n}$, $(ab)^n = a^n b^n$, and $(a^m)^n = a^{mn}$.

Example 9

Simplify each of the following. Leave answers in exponential form with positive exponents only. Assume all variables represent nonzero quantities.

a. $3^{-3}3^5 =$ Apply $a^m a^n = a^{m+n}$.

$3^{-3+5} =$ Add exponents.

3^2 Answer.

b. $a^4 a^{-6} =$ Apply $a^m a^n = a^{m+n}$.

$a^{4+(-6)} =$ Add exponents.

$a^{-2} =$ Apply $a^{-n} = \dfrac{1}{a^n}$.

$\dfrac{1}{a^2}$ Answer.

c. $\dfrac{x^{-4}}{x^2} =$ Apply $\dfrac{a^m}{a^n} = a^{m-n}$.

$x^{-4-2} =$ Subtract exponents.

$x^{-6} =$ Apply $a^{-n} = \dfrac{1}{a^n}$.

$\dfrac{1}{x^6}$ Answer.

Example 9c could also be done as follows: $\dfrac{x^{-4}}{x^2} = x^{-4} \cdot \dfrac{1}{x^2} = \dfrac{1}{x^4} \cdot \dfrac{1}{x^2} = \dfrac{1}{x^6}$

d. $\dfrac{a^{-3}}{a^{-5}} =$ Apply $\dfrac{a^m}{a^n} = a^{m-n}$.

$a^{-3-(-5)} =$ $-(-x) = x$.

$a^{-3+5} =$ Add the exponents.

a^2 Answer.

e. $(w^{-3})^2 =$ Apply $(a^m)^n = a^{mn}$.

$w^{(-3)(2)} =$ Multiply the exponents.

$w^{-6} =$ Apply $a^{-n} = \dfrac{1}{a^n}$.

$\dfrac{1}{w^6}$ Answer.

Practice Exercises

Simplify each of the following. Leave answers in exponential form with positive exponents only. Assume all variables represent nonzero quantities.

16. $4^{-5}4^7 =$

17. $w^3 w^{-5} =$

18. $\dfrac{b^{-4}}{b^2} =$

19. $\dfrac{h^4}{h^{-3}} =$

20. $(a^{-4})^{-2} =$

If more practice is needed, do the Additional Practice Exercises.

Additional Practice Exercises

Simplify each of the following. Leave answers in exponential form with positive exponents only. Assume all variables represent nonzero quantities.

m. $2^{-5} \cdot 2^3$

n. $r^{-2}r^5$

o. $\dfrac{z^{-3}}{z^3}$

p. $\dfrac{r^2}{r^{-5}}$

q. $(a^{-4})^2$

Exercise Set 2.6

Simplify each of the following. Leave answers in exponential form with positive exponents only. Assume all variables represent nonzero quantities.

1. $\dfrac{2^6}{2^2}$

2. $\dfrac{3^8}{3^3}$

3. $\dfrac{c^5}{c}$

4. $\dfrac{d^7}{d}$

5. $\dfrac{z^9}{z^5}$

6. $\dfrac{a^8}{a^6}$

7. 2^{-4}

8. 3^{-2}

9. a^{-6}

10. b^{-7}

11. $\dfrac{1}{5^{-3}}$

12. $\dfrac{1}{8^{-4}}$

13. $\dfrac{1}{y^{-5}}$

14. $\dfrac{1}{z^{-9}}$

15. $\dfrac{x^2}{x^5}$

16. $\dfrac{y^4}{y^5}$

17. $\dfrac{5^3}{5^6}$

18. $\dfrac{4^2}{4^6}$

19. 8^0

20. 6^0

21. $3a^0$

22. $7z^0$

23. $(-4s)^0$

24. $(-7h)^0$

25. $(3x)^0 + (2y)^0$

26. $(6w)^0 - (2z)^0$

27. $3x^0 - 4z^0$

28. $2x^{-3}$

29. $5a^{-2}$

30. $-7b^{-4}$

31. $-9c^{-6}$

32. $(2x)^{-3}$

33. $(5a)^{-2}$

34. $8q^0 - 2q^0$

35. $2^{-3} \cdot 2^5$

36. $4^5 \cdot 4^{-2}$

37. $5^{-6} \cdot 5^4$

38. $6^3 \cdot 6^{-7}$

39. $x^{-3}x^{-2}$

40. $y^{-4}y^{-5}$

41. $z^{-4}z^2$

42. q^5q^{-2}

43. $\dfrac{3^{-2}}{3^2}$

44. $\dfrac{2^{-4}}{2^5}$

45. $\dfrac{y^{-4}}{y^5}$

46. $\dfrac{4^4}{4^{-5}}$

47. $\dfrac{6^5}{6^{-2}}$

48. $\dfrac{x^4}{x^{-3}}$

49. $\dfrac{r^2}{r^{-5}}$

50. $\dfrac{5^{-3}}{5^{-5}}$

51. $\dfrac{6^{-2}}{6^{-3}}$

52. $\dfrac{x^{-6}}{x^{-3}}$

53. $\dfrac{z^{-8}}{z^{-4}}$

54. $(5^2)^5$

55. $(8^3)^6$

56. $(p^4)^4$

57. $(r^5)^4$

58. $(4^5)^{-4}$

59. $(6^3)^{-5}$

60. $(t^6)^{-2}$

61. $(y^7)^{-3}$

62. $(6^{-1})^4$

63. $(8^{-3})^6$

64. $(b^4)^{-2}$

65. $(z^3)^{-5}$

66. $(6^{-2})^{-5}$

67. $(9^{-5})^{-2}$

68. $(a^{-2})^{-4}$

Answers:

Challenge Exercises:

Simplify each of the following. Leave answers with positive exponents only. Assume all variables represent nonzero quantities.

69. $(2x^{-3}y^2)^2(-3x^3y^{-3})^3$

70. $(-4a^2b^{-3})^3(5a^{-3}b^{-3})^2$

71. $(2m^{-4}n^2)^{-2}(-2m^2n^{-4})^3$

72. $(8c^{-5}d^3)^{-1}(8c^3d^{-5})^2$

73. $\dfrac{(4a^{-4}b^2)^3}{(2a^3b^{-4})^3}$

74. $\dfrac{(6m^{-5}n^2)^2}{(3m^2n^{-6})^2}$

75. $\dfrac{x^{3a}}{x^a}$

76. $\dfrac{y^{4b}}{y^{2b}}$

77. $\dfrac{a^{3m-1}}{a^{2m-2}}$

78. $\dfrac{b^{3n+6}}{b^{2n-4}}$

Writing Exercises:

79. How does 4^{-3} differ from $(-3)(4)$?

81. What is wrong with the following:
$\dfrac{4^5}{4^3} = 1^{5-3} = 1^2 = 1$?

80. Why is $3^{-3} \neq -27$?

Critical Thinking Exercises:

82. Is it possible to raise a positive number to a power and get a negative answer? Why or why not?

83. We know that $x^0 = 1$ if $x \neq 0$. Why can $x \neq 0$?

Section 2.7   # Power Rule for Quotients and Using Combined Laws of Exponents

OBJECTIVES

When you complete this section, you will be able to:

a. Simplify exponential expressions using the property $\left(\dfrac{a}{b}\right)^n = \dfrac{a^n}{b^n}$.

b. Simplify expressions that involve integer exponents using more than one law of exponents.

Introduction   One of the laws of exponents for products is $(ab)^n = a^n b^n$. This means that to raise a product to a power, you raise each of the factors to the power. There is a similar property for quotients.

OPTIONAL CALCULATOR EXPLORATION ACTIVITY

Evaluate each of the columns using a calculator.

Column A	Column B
1. $\left(\dfrac{2}{5}\right)^2 =$	$\dfrac{2^2}{5^2} =$
2. $\left(\dfrac{3}{4}\right)^3 =$	$\dfrac{3^3}{4^3} =$
3. $\left(\dfrac{3}{2}\right)^4 =$	$\dfrac{3^4}{2^4} =$

By comparing corresponding lines in columns A and B, answer the following:

1. Raising a fraction to a power is the same as _____.

Following is justification for the above observation:

Developing $\left(\dfrac{a}{b}\right)^n = \dfrac{a^n}{b^n}$ Remember: to multiply fractions, you multiply numerator times numerator and divide by denominator times denominator. In general, $\dfrac{a}{b} \cdot \dfrac{c}{d} = \dfrac{a \cdot c}{b \cdot d}$.

Example 1

Simplify each of the following using the definition of exponents. Leave the answer in exponential form. Assume all variables represent nonzero quantities.

a. $\left(\dfrac{2}{3}\right)^3 =$ Rewrite using the definition of exponents.

$\dfrac{2}{3} \cdot \dfrac{2}{3} \cdot \dfrac{2}{3} =$ Multiply the fractions.

$\dfrac{2 \cdot 2 \cdot 2}{3 \cdot 3 \cdot 3} =$ Rewrite using the definition of exponents.

$\dfrac{2^3}{3^3}$ Answer.

Notice that both the numerator and the denominator are raised to the third power.

b. $\left(\dfrac{x}{y}\right)^4 =$ Rewrite using the definition of exponents.

$\dfrac{x}{y} \cdot \dfrac{x}{y} \cdot \dfrac{x}{y} \cdot \dfrac{x}{y} =$ Multiply the fractions.

$\dfrac{x \cdot x \cdot x \cdot x}{y \cdot y \cdot y \cdot y} =$ Rewrite using the definition of exponents.

$\dfrac{x^4}{y^4}$ Answer.

Notice that both the numerator and the denominator were raised to the fourth power.

Based on the preceding example, we generalize to the following law for raising a quotient of a power:

Fifth Law of Exponents: Power Rule for Quotients

For any positive integer n and $b \neq 0$, $\left(\dfrac{a}{b}\right)^n = \dfrac{a^n}{b^n}$. To raise a quotient to a power, raise both the numerator and the denominator to the power.

Example 2

Simplify the following using the Power Rule for Quotients. Leave the answer in exponential form. Assume all variables represent nonzero quantities.

a. $\left(\dfrac{2}{5}\right)^4 = \dfrac{2^4}{5^4}$ Applying $\left(\dfrac{a}{b}\right)^n = \dfrac{a^n}{b^n}$.

b. $\left(\dfrac{a}{b}\right)^5 = \dfrac{a^5}{b^5}$ Applying $\left(\dfrac{a}{b}\right)^n = \dfrac{a^n}{b^n}$.

c. $\left(\dfrac{3}{y}\right)^2 = \dfrac{3^2}{y^2}$ Applying $\left(\dfrac{a}{b}\right)^n = \dfrac{a^n}{b^n}$.

Practice Exercises

Simplify the following using the Power Rule for Quotients. Leave the answer in exponential form. Assume all variables represent nonzero quantities.

1. $\left(\dfrac{5}{3}\right)^6$

2. $\left(\dfrac{r}{s}\right)^8$

3. $\left(\dfrac{z}{5}\right)^3$

Answers:

Practice Exercises 1–3: 1. $\dfrac{5^6}{3^6}$ 2. $\dfrac{r^8}{s^8}$ 3. $\dfrac{z^3}{5^3}$

Calculator Exploration Activity: 1. raising the numerator and denominator to the power.

Another interesting property occurs when a quotient is raised to a negative power. We need to recall the rule for dividing fractions, which tells us to multiply by the reciprocal of the fraction on the right and change the operation to multiplication. For example, $\frac{1}{2} \div \frac{3}{5} = \frac{1}{2} \cdot \frac{5}{3}$.

Example 3

Simplify the following leaving the quotient with positive exponents only:

a. $\left(\frac{a}{b}\right)^{-3} =$ Apply $\left(\frac{a}{b}\right)^n = \frac{a^n}{b^n}$.

$\frac{a^{-3}}{b^{-3}} =$ Rewrite as multiplication.

$\frac{a^{-3}}{1} \cdot \frac{1}{b^{-3}} =$ Apply $a^{-n} = \frac{1}{a^n}$ and $\frac{1}{a^{-n}} = a^n$.

$\frac{1}{a^3} \cdot b^3 =$ Multiply.

$\frac{b^3}{a^3} =$ Apply $\frac{a^n}{b^n} = \left(\frac{a}{b}\right)^n$.

$\left(\frac{b}{a}\right)^3$ Answer.

Therefore, $\left(\frac{a}{b}\right)^{-3} = \left(\frac{b}{a}\right)^3$. Notice the fraction has been inverted and the exponent has been changed to positive. Therefore, we have the following rule:

> ### Quotients Raised to Negative Powers
> For $a \neq 0$ and $b \neq 0$, $\left(\frac{a}{b}\right)^{-n} = \left(\frac{b}{a}\right)^n$. To raise a fraction to a negative power, invert the fraction and change the power to positive.

Example 4

Simplify the following, leaving answers with positive exponents only. Assume that all variables represent nonzero quantities.

a. $\left(\frac{3}{5}\right)^{-2} =$ Apply $\left(\frac{a}{b}\right)^{-n} = \left(\frac{b}{a}\right)^n$.

$\left(\frac{5}{3}\right)^2 =$ Apply $\left(\frac{a}{b}\right)^n = \frac{a^n}{b^n}$.

$\frac{5^2}{3^3} =$ Simplify the powers.

$\frac{25}{9}$ Answer.

b. $\left(\frac{x}{y}\right)^{-3} =$ Apply $\left(\frac{a}{b}\right)^{-n} = \left(\frac{b}{a}\right)^n$.

$\left(\frac{y}{x}\right)^3 =$ Apply $\left(\frac{a}{b}\right)^n = \frac{a^n}{b^n}$.

$\frac{y^3}{x^3}$ Answer.

Practice Exercises

Simplify the following, leaving answers with positive exponents only. Assume that all variables represent nonzero quantities.

4. $\left(\frac{3}{5}\right)^{-3} =$

5. $\left(\frac{a}{b}\right)^{-5} =$

Often it is necessary to use two or more of the laws of exponents to simplify an expression. For reference, a list of these laws and the properties of integer exponents follows:

Laws of Exponents

For any integers m and n and any real numbers a and b:

1. $a^m a^n = a^{m+n}$ Product Rule
2. $(ab)^n = a^n b^n$ Power Rule for Products
3. $(a^m)^n = a^{mn}$ Power to a Power
4. $\dfrac{a^m}{a^n} = a^{m-n}, a \neq 0$ Quotient Rule
5. $\left(\dfrac{a}{b}\right)^n = \dfrac{a^n}{b^n}, b \neq 0$ Power Rule for Quotients
6. $\left(\dfrac{a}{b}\right)^{-n} = \left(\dfrac{b}{a}\right)^n$ if $a, b, \neq 0$ Quotients to Negative Powers
7. $a^0 = 1$ if $a \neq 0$ Definition of Zero Exponents
8. $a^{-n} = \dfrac{1}{a^n}$ if $a \neq 0$ Definition of Negative Exponents
9. $\dfrac{1}{a^{-n}} = a^n$ if $a \neq 0$ Definition of Negative Exponents

Simplifying expressions using more than one law of exponents

Example 5

Simplify each of the following. Leave answers in exponential form with positive exponents only. Assume all variables represent nonzero quantities.

a. $\dfrac{x^2 x^4}{x^3} =$ Apply $a^m a^n = a^{m+n}$ in the numerator.

$\dfrac{x^6}{x^3} =$ Apply $\dfrac{a^m}{a^n} = a^{m-n}$.

$x^{6-3} =$ Subtract the exponents.

x^3 Answer.

b. $(3x^{-2}y^3)^{-3} =$ Apply $(ab)^n = a^n b^n$.

$(3^{-3})(x^{-2})^{-3}(y^3)^{-3} =$ Apply $a^{-n} = \dfrac{1}{a^n}$ and $(a^m)^n = a^{mn}$.

$\dfrac{1}{3^3}x^6 y^{-9} =$ Apply $a^{-n} = \dfrac{1}{a^n}$ and $3^3 = 27$.

$\dfrac{1}{27}x^6 \cdot \dfrac{1}{y^9} =$ Multiply the fractions.

$\dfrac{x^6}{27y^9}$ Answer.

c. $\dfrac{(4x^{-3})^2}{x^2 x^{-4}} =$ Apply $(a^m)^n = a^{mn}$ in the numerator and $a^m a^n = a^{m+n}$ in the denominator.

$\dfrac{4^2 x^{-6}}{x^{-2}} =$ Apply $\dfrac{a^m}{a^n} = a^{m-n}$ and $4^2 = 16$.

$16x^{-6-(-2)} =$ $-(-2) = +2$.

$16x^{-6+2} =$ Add exponents.

$16x^{-4} =$ Apply $x^{-n} = \dfrac{1}{x^n}$.

$\dfrac{16}{x^4}$ Answer.

d. $\left(\dfrac{x^4}{y^2}\right)^3 = $ Apply $\left(\dfrac{a}{b}\right)^n = \dfrac{a^n}{b^n}$.

$\dfrac{(x^4)^3}{(y^2)^3} = $ Apply $(a^m)^n = a^{mn}$.

$\dfrac{x^{12}}{y^6}$ Answer.

e. $\dfrac{(x^{-2})^3(x^3)^4}{(x^{-3})^3} = $ Apply $(a^m)^n = a^{mn}$.

$\dfrac{x^{-6}x^{12}}{x^{-9}} = $ Apply $a^m a^n = a^{m+n}$ in the numerator.

$\dfrac{x^6}{x^{-9}} = $ Apply $\dfrac{a^m}{a^n} = a^{m-n}$.

$x^{6-(-9)} = $ $-(-9) = +9$.

$x^{6+9} = $ Add exponents.

x^{15} Answer.

Practice Exercises

Simplify each of the following. Leave answers in exponential form with positive exponents only. Assume all variables represent nonzero quantities.

6. $\dfrac{x^3 x^5}{x^4}$

7. $(2a^{-3}b^2)^{-4}$

8. $\dfrac{(3x^4)^{-2}}{x^{-4}x^3}$

9. $\left(\dfrac{a^4}{b^2}\right)^4$

10. $\dfrac{(a^3)^{-3}(a^4)^{-2}}{(a^{-4})^2}$

If more practice is needed, do the Additional Practice Exercises.

Additional Practice Exercises

Simplify each of the following. Leave answers in exponential form with positive exponents only. Assume all variables represent nonzero quantities.

a. $\dfrac{x^5 x^3}{x^6}$

b. $(4p^4 q^{-3})^{-2}$

c. $(4z^{-4})^{-3}$

d. $\dfrac{(3x^{-1})^4}{x^{-3}x^{-2}}$

e. $\left(\dfrac{p^3}{q^2}\right)^5$

f. $\dfrac{(b^{-1})^4(b^{-3})^{-2}}{(b^3)^{-1}}$

Answers:

Practice Exercises 6–10: 6. x^4 7. $\dfrac{16b^8}{a^{12}}$ 8. $\dfrac{9x^2}{a^6}$ 9. $\dfrac{a^{16}}{b^8}$ 10. $\dfrac{1}{a^6}$ **Additional Practice Exercises a–e:** a. x^2 b. $\dfrac{q^6}{16p^8}$ c. $\dfrac{z^{12}}{64}$ d. $81x$ e. $\dfrac{p^{15}}{q^{10}}$ f. b^5

Study Tip 10 *(continued)*

1. Arrive early so you will be ready when the test is passed out.
2. As soon as you get the test, write down any formulas or definitions that you might need.
3. Read the test over from front to back. Mark the questions that you know you can answer with a check mark. Mark the ones you are unsure about with a question mark and the ones you know you cannot answer with an "x."
4. Answer the questions in this order:
   a) Those you know how to do, starting with the ones you think are easiest.
   b) Those you are unsure about.
   c) Those you initially thought you did not know how to do. Sometimes doing the ones that you know reminds you of how to do others.
5. Estimate the amount of time needed for each question by dividing the number of minutes allowed for the test by the number of items on the test. If you are spending more than this amount of time on an item, go to the next item and come back to it later if you have time.
6. Read the directions to each question carefully. Underline the key words, such as *not equal.*
7. On an open-ended test, write down the information given, what you are asked to find, and any relevant formulas, definitions, or theorems. Sometimes an estimate of the answer will give you a clue about how to solve the problem.
8. On an open-ended test, show all your work in a neat and organized manner. Box in your answers.
9. Check all your answers, if time permits.
10. If you are taking a multiple-choice test that has a penalty for wrong answers and you are able to narrow the answer to two choices—guess. If there is no penalty, guess at all answers and leave nothing blank.
11. Take all of the time permitted for the test. Never worry about being the last one to leave a test. Generally the first students to leave a test are either the ones who know everything—or those who know very little.

Remember, the idea in taking a test is to show what you know, so attempt what you know first.

Exercise Set 2.7

Write each of the following without exponents and evaluate:

1. $\left(\dfrac{2}{3}\right)^2$

2. $\left(\dfrac{3}{4}\right)^3$

3. $\left(\dfrac{3}{5}\right)^3$

4. $\left(\dfrac{4}{3}\right)^2$

5. $\left(\dfrac{5}{2}\right)^3$

6. $\left(\dfrac{2}{5}\right)^{-2}$

7. $\left(\dfrac{3}{2}\right)^{-2}$

8. $\left(\dfrac{5}{4}\right)^{-3}$

9. $\left(\dfrac{3}{7}\right)^{-4}$

Write each of the following with positive exponents only. Assume all variables represent nonzero quantities.

10. $\left(\dfrac{r}{s}\right)^2$

11. $\left(\dfrac{a}{b}\right)^3$

12. $\left(\dfrac{x}{z}\right)^{-4}$

13. $\left(\dfrac{p}{q}\right)^{-4}$

14. $\left(\dfrac{3a}{b}\right)^4$

15. $\left(\dfrac{2x}{y}\right)^3$

16. $\left(\dfrac{3x}{y}\right)^{-4}$

17. $\left(\dfrac{p}{3q}\right)^{-3}$

18. $(r^{-2}s^2)^3$

19. $(a^3b^{-5})^4$

20. $(z^{-2}w^3)^{-4}$

21. $(x^4y^{-2})^{-6}$

22. $(2x^3)^{-2}$

23. $(4x^2)^{-3}$

24. $(-4x^{-3})^{-4}$

25. $(-2y^{-5})^{-2}$

26. $(2x^{-3}y^{-2})^3$

27. $(5a^{-2}b^{-5})^2$

28. $(3x^3y^{-4})^{-4}$

29. $(2w^3z^{-2})^{-2}$

30. $\left(\dfrac{x^3}{y^5}\right)^3$

31. $\left(\dfrac{w^3}{z^6}\right)^3$

32. $\left(\dfrac{a^5}{b^3}\right)^{-5}$

33. $\left(\dfrac{x^4}{z^6}\right)^{-5}$

34. $\dfrac{(2x^2)(3x^7)}{(x^3)(x^3)}$

35. $\dfrac{(4w^3)(6w^4)}{(w^2)(w^2)}$

36. $\dfrac{(-5a^2)(3a^5)}{(a^6)(a^4)}$

37. $\dfrac{(x^4)(3x^6)}{(5x^8)(2x^3)}$

38. $\dfrac{(7x^3y^2)(6x^4y^4)}{(5x^2y^3)(x^5y^5)}$

39. $\dfrac{(x^4)^2}{(x^2)^3}$

40. $\dfrac{(y^3)^4}{(y^5)^2}$

41. $\dfrac{(a^{-3})^3}{(a^3)^2}$

42. $\dfrac{(z^{-2})^4}{(z^5)^2}$

43. $\dfrac{(a^{-2})^{-3}}{(a^{-4})^2}$

44. $\dfrac{(p^{-4})^{-2}}{(p^{-5})^3}$

45. $\dfrac{(2y^2)^4}{(y^4)^4}$

46. $\dfrac{(3q^3)^3}{(q^2)^4}$

47. $\dfrac{(x^2)^5}{x^2x^4}$

48. $\dfrac{(y^4)^3}{y^3y^2}$

49. $\dfrac{(3x^4)^0}{(2x^2)^2}$

50. $\dfrac{(5c^3)^2}{(8c^6)^0}$

51. $\dfrac{(7^0g^4)^3}{(2g^2)^4}$

52. $\dfrac{(4a^4)^3}{(5^0a^2)^4}$

53. $\dfrac{(x^{-2})^4}{x^3x^4}$

54. $\dfrac{(z^{-1})^5}{z^5z^2}$

55. $\dfrac{(q^2)^3(q^4)^2}{(q^3)^5}$

56. $\dfrac{(p^4)^2(p^5)^3}{(p^2)^7}$

57. $\dfrac{(4a^{-4})^3(2a^2)^{-3}}{(2a^{-2})^3}$

58. $\dfrac{(3b^5)^{-3}(2b^3)^{-3}}{(36b^4)^{-2}}$

59. $\dfrac{(2x^{-3})^3}{x^2x}$

60. $\dfrac{(2y^{-4})^4}{y^5y^2}$

61. $\dfrac{(-3a^{-3})^4}{(a^{-3})^5}$

Challenge Exercises:

Simplify the following: Leave answers with positive exponents only. Assume all variables represent nonzero quantities.

62. $\dfrac{(2x^{-3}y^2)^4(3xy^{-3})^{-2}}{(3x^{-4}y^{-2})^{-4}}$

63. $\dfrac{(4a^{-3}b^{-2})^{-2}(6a^3b^{-5})^3}{(2a^{-2}b^{-2})^{-4}}$

64. $\dfrac{(x^m)^n}{(x^p)^q}$

65. $(a^mb^n)^q$

66. $\dfrac{(x^ay^b)^c}{(x^my^n)^d}$

Section 2.8	**Division of Polynomials by Monomials**

OBJECTIVES

When you complete this section, you will be able to:

a. Divide monomials by monomials.

b. Divide polynomials of more than one term by monomials.

Introduction   Recall from Section 1.8 that a monomial is a polynomial with only one term. Examples are 4, $5x^2$, $6xy$, $-4z^2y^3$, and $8p^3q^2r^5$. Remember, the exponent on any variable must be a whole number. This means the exponent on a variable cannot be negative, and there cannot be a variable in the denominator. Recall also that the number in front of the variable is called the coefficient. In order to divide a monomial by another monomial, we will use the Quotient Rule discussed in Section 2.6. The division of polynomials by polynomials other than monomials will be discussed in Section 6.7.

Example 1

Dividing a monomial by a monomial

Find the quotients of the following monomials. Leave answers with positive exponents only. Assume all variables represent nonzero quantities only.

a. $\dfrac{4x^4}{2x^3} =$ Divide the coefficients and variables separately.

$\dfrac{4}{2} \cdot \dfrac{x^4}{x^3} =$ Divide the coefficients and apply $\dfrac{a^m}{a^n} = a^{m-n}$ to the variables.

$2x$ Answer.

b. $\dfrac{-12x^5y^3}{3x^2y^2} =$ Divide the coefficients and like variables separately.

$\dfrac{-12}{3} \cdot \dfrac{x^5}{x^2} \cdot \dfrac{y^3}{y^2} =$ Divide the coefficients and apply $\dfrac{a^m}{a^n} = a^{m-n}$ to the variables.

$-4x^3y$ Answer.

c. $\dfrac{6x^3}{-2x^5} =$ Divide the coefficients and the variables separately.

$\dfrac{6}{-2} \cdot \dfrac{x^3}{x^5} =$ Divide the coefficients and apply $\dfrac{a^m}{a^n} = a^{m-n}$ to the variables.

$-3x^{-2} =$ Apply $x^{-n} = \dfrac{1}{x^n}$.

$-3 \cdot \dfrac{1}{x^2} =$ Multiply.

$\dfrac{-3}{x^2} = -\dfrac{3}{x^2}$ Answer.

Note: The answer to Example 1c is not a monomial since there is division by a variable. Consequently, the quotient of two monomials may not be a monomial, just as the quotient of two integers may not be an integer.

d. $\dfrac{24a^4b^3z^6}{8a^2b^3z^8} =$ Divide the coefficients and like variables separately.

$\dfrac{24}{8} \cdot \dfrac{a^4}{a^2} \cdot \dfrac{b^3}{b^3} \cdot \dfrac{z^6}{z^8} =$ Divide the coefficients and apply $\dfrac{a^m}{a^n} = a^{m-n}$ to the variables.

$3a^2b^0z^{-2} =$ Apply $x^0 = 1$ and $x^{-n} = \dfrac{1}{x^n}$.

$3a^2 \cdot 1 \cdot \dfrac{1}{z^2} =$ Multiply.

$\dfrac{3a^2}{z^2}$ Answer.

e. $\dfrac{(4p^2q^3)^2}{8p^3q^4} =$ Apply $(ab)^n = a^nb^n$ in the numerator.

$\dfrac{(4)^2(p^2)^2(q^3)^2}{8p^3q^4} =$ Apply $(a^m)^n$ in the numerator.

$\dfrac{16p^4q^6}{8p^3q^4} =$ Divide the coefficients and apply $\dfrac{a^m}{a^n} = a^{m-n}$ to the variables.

$2pq^2$ Answer.

Practice Exercises

Find the quotients of the following monomials. Leave answers with positive exponents only. Assume all variables represent nonzero quantities only.

1. $\dfrac{8y^6}{2y^3}$

2. $\dfrac{-18x^8y^7}{6x^3y^2}$

3. $\dfrac{16p^3}{-4p^5}$

4. $\dfrac{-15x^3y^5z^2}{3x^3y^3z^2}$

5. $\dfrac{(6x^3y^4)^2}{9x^5y^4}$

If more practice is needed, do the Additional Practice Exercises.

Additional Practice Exercises

Find the quotients of the following monomials. Leave answers with positive exponents only. Assume all variables represent nonzero quantities.

a. $\dfrac{18a^6}{9a^4}$

b. $\dfrac{-24p^6q^6}{6p^4q^5}$

c. $\dfrac{-32r^5}{-4r^7}$

d. $\dfrac{-12a^4b^5c^3}{2a^3b^5c^6}$

e. $\dfrac{(4a^4b^4)^3}{8a^6b^8}$

Now we will divide polynomials by monomials. From Section 0.6 we know how to add fractions with a common denominator, $\dfrac{a}{c} + \dfrac{b}{c} = \dfrac{a+b}{c}$. If we "turn this expression around," it gives us the method for dividing a polynomial by a monomial, $\dfrac{a+b}{c} = \dfrac{a}{c} + \dfrac{b}{c}$. We summarize in the following rule:

Dividing polynomials by monomials

Division of a Polynomial by a Monomial

$\dfrac{a+b}{c} = \dfrac{a}{c} + \dfrac{b}{c}$. To divide a polynomial by a monomial, divide each term of the polynomial by the monomial.

Example 2

Find the following quotients of polynomials and monomials. Assume all variables represent nonzero quantities only.

a. $\dfrac{4x+6}{2} =$ Divide each term of the polynomial by the monomial.

$\dfrac{4x}{2} + \dfrac{6}{2} =$ Divide the coefficients.

$2x + 3$ Quotient.

b. $\dfrac{6r^4 + 9r^5}{3r^2} =$ Divide each term of the polynomial by the monomial.

$\dfrac{6r^4}{3r^2} + \dfrac{9r^5}{3r^2} =$ Divide the coefficients and apply $\dfrac{a^m}{a^n} = a^{m-n}$ to the variables.

$2r^2 + 3r^3$ Quotient.

Answers:

Practice Exercises 1–5: 1. $4y^3$ 2. $-3x^5y^5$ 3. $-\dfrac{4}{p^2}$ 4. $-5y^2$ 5. $4xy^4$ **Additional Practice Exercises a–e:** a. $2a^2$

b. $-4p^2q$ c. $\dfrac{8}{r^2}$ d. $-\dfrac{6c^3}{c^3}$ e. $8a^6b^4$

c. $\dfrac{8y^4 - 12y^2}{4y^3} =$ Divide each term of the polynomial by the monomial.

$\dfrac{8y^4}{4y^3} - \dfrac{12y^2}{4y^3} =$ Divide the coefficients and apply $\dfrac{a^m}{a^n} = a^{m-n}$ to the variables.

$2y - 3y^{-1} =$ Apply $\dfrac{a^{-n}}{n} = \dfrac{1}{a^n}$.

$2y - \dfrac{3}{y}$ Quotient.

Note: The answer to Example 2c is not a polynomial. Therefore, the quotient of a polynomial with a monomial need not be a polynomial. Also, a common error is to write the answer $2y - \dfrac{3}{y}$ as $\dfrac{2y - 3}{y}$.

d. $\dfrac{10n^5 + 15n^4 - 3n^3}{5n^2} =$ Divide each term of the polynomial by the monomial.

$\dfrac{10n^5}{5n^2} + \dfrac{15n^4}{5n^2} - \dfrac{3n^3}{5n^2} =$ Divide the coefficients and apply $\dfrac{a^m}{a^n} = a^{m-n}$ to the variables.

$2n^3 + 3n^2 - \dfrac{3n}{5}$ Answer.

e. $\dfrac{12x^3y^4 - 16x^4y^3 + 3x^2y^5}{4x^3y^4}$ Divide each term of the polynomial by the monomial.

$\dfrac{12x^3y^4}{4x^3y^4} - \dfrac{16x^4y^3}{4x^3y^4} + \dfrac{3x^2y^5}{4x^3y^4} =$ Divide the coefficients and apply $\dfrac{a^m}{a^n} = a^{m-n}$ to the variables.

$3x^0y^0 - 4xy^{-1} + \dfrac{3}{4}x^{-1}y =$ Apply $x^0 = 1$ and $x^{-n} = \dfrac{1}{x^n}$.

$3 - \dfrac{4x}{y} + \dfrac{3y}{4x}$ Answer.

Practice Exercises

Find the following quotients of polynomials and monomials. Assume all variables represent nonzero quantities only.

6. $\dfrac{6x + 12}{2}$

7. $\dfrac{10a^5 + 25a^3}{5a^2}$

8. $\dfrac{12r^4 - 20r^2}{4r^3}$

9. $\dfrac{24p^6 - 12p^4 + 5p^7}{6p^3}$

10. $\dfrac{21p^2q^4 - 14p^5q^3 + 6p^3q^7}{7p^3q^5}$

If more practice is needed, do the Additional Practice Exercises.

Additional Practice Exercises

Find the following quotients of polynomials and monomials:

f. $\dfrac{8x - 12}{4}$

g. $\dfrac{21x^6 + 15x^3}{3x^2}$

h. $\dfrac{15b^5 - 25b^2}{5b^4}$

i. $\dfrac{8x^6 + 16x^5 - 24x^3}{8x^3}$

j. $\dfrac{9r^4s^5 - 15r^2s^3 + 21r^5s^4}{3r^3s^4}$

Answers:

Practice Exercises 6–10: 6. $3x + 6$ 7. $2a^3 + 5a$ 8. $3r - \dfrac{5}{r}$ 9. $4p^3 - 2p + \dfrac{5p^4}{6}$ 10. $3q^{-1} - 2p^2 + \dfrac{6q^2}{7}$

Additional Practice Exercises f–j: f. $2x - 3$ g. $7x^4 + 5x$ h. $3b - \dfrac{5}{b^2}$ i. $x^3 + 2x^2 - 3$ j. $3rs - \dfrac{5}{rs} + 7r^2$

Exercise Set 2.8

Find the quotients of the following monomials:

1. $\dfrac{5x^2}{x}$

2. $\dfrac{6y^3}{y}$

3. $\dfrac{-3z^4}{z^2}$

4. $\dfrac{-5a^5}{a^3}$

5. $\dfrac{6x^4}{3x^2}$

6. $\dfrac{14b^5}{b^2}$

7. $\dfrac{-15a^4}{3a^7}$

8. $\dfrac{-24w^5}{8w^9}$

9. $\dfrac{18n^6}{-3n^4}$

10. $\dfrac{28u^8}{-4u^2}$

11. $\dfrac{30c^4}{15c^4}$

12. $\dfrac{32w^6}{16w^6}$

13. $\dfrac{x^4y^5}{x^2y^3}$

14. $\dfrac{a^6b^4}{a^4b^3}$

15. $\dfrac{r^3s^6}{-r^5s^2}$

16. $\dfrac{-q^5r^2}{qr^5}$

17. $\dfrac{18x^5y^3}{6x^3y}$

18. $\dfrac{28y^6z^4}{7y^3z^2}$

19. $\dfrac{-32x^3y^5}{8x^6y^2}$

20. $\dfrac{-16a^2b^6}{8a^4b^3}$

21. $\dfrac{x^3y^6z^4}{x^2y^4z^6}$

22. $\dfrac{a^5b^3c^8}{a^3b^7c^4}$

23. $\dfrac{48x^4y^6z^8}{-12x^6y^3z^{10}}$

24. $\dfrac{36a^3b^5c^2}{-18a^5b^2c}$

25. $\dfrac{-28m^2n^5p^3}{-14m^2n^3p^6}$

26. $\dfrac{-32x^4y^2z^6}{-8x^4y^6z^3}$

27. $\dfrac{(6x^3y^5)^2}{4x^4y^8}$

28. $\dfrac{(8a^4b^3)^2}{16a^6b^4}$

29. $\dfrac{-(5x^5y^2)^2}{5x^{10}y^6}$

Find the following quotients of polynomials and monomials:

30. $\dfrac{2x+8}{2}$

31. $\dfrac{3x+9}{3}$

32. $\dfrac{x^2+x}{x}$

33. $\dfrac{a^2-a}{a}$

34. $\dfrac{2x+4}{x}$

35. $\dfrac{3y+7}{y}$

36. $\dfrac{4x^2-2x}{2x}$

37. $\dfrac{12z^2-6z}{6z}$

38. $\dfrac{12x^3+8x}{4x^2}$

39. $\dfrac{24y^3+16y}{8y^2}$

40. $\dfrac{6x^3y^4-12x^4y^5}{3x^2y^2}$

41. $\dfrac{20a^5b^3-10a^3b^6}{5a^2b^2}$

42. $\dfrac{12x^2y^4+24x^5y^2}{6x^4y^3}$

43. $\dfrac{18mn^4+27m^4n^2}{9m^2n^3}$

44. $\dfrac{4x^2-6x+2}{2}$

45. $\dfrac{6x^2+9x-3}{3}$

46. $\dfrac{x^3-x^2-x}{x}$

47. $\dfrac{y^4-y^3-y^2}{y}$

48. $\dfrac{9x^4-3x^3+6x^2}{3x}$

49. $\dfrac{15m^4-5m^3+3m^2}{5m}$

50. $\dfrac{6x^3-12x^2+2x}{3x^2}$

51. $\dfrac{12y^4-16y^3+8y}{4y^2}$

52. $\dfrac{21x^2y^4-12x^3y^5+18x^5y^3}{3xy^2}$

53. $\dfrac{24m^3n^2-16m^4n^3+32m^5n^4}{8m^2n^2}$

54. $\dfrac{6x^3-4x}{3x^2}$

55. $\dfrac{12b^4-3b^2}{4b^3}$

56. $\dfrac{5a^6b^2-15a^3b^5}{5a^5b^3}$

57. $\dfrac{24mn^5 + 18m^4n^2}{6m^3n^4}$

58. $\dfrac{16x^5 + 8x^4 - 32x^3}{8x^4}$

59. $\dfrac{28r^7 - 14r^5 + 21r^3}{7r^5}$

60. $\dfrac{36m^4n^5 - 27m^3n^3 + 4m^5n^4}{9m^3n^4}$

61. $\dfrac{32p^3q^5 - 3p^6q^4 + 20p^4q^8}{4p^4q^6}$

Write an expression for each of the following and simplify:

62. Find the quotient of $18x^6y^9$ and $9x^4y^3$.

63. Find the quotient of $14m^4n^8$ and $7m^3n^6$.

64. Divide $-24a^3b^2$ by $-12a^5b$.

65. Divide $-32m^4n^2$ by $-16m^7n$.

66. Find the quotient of $32x^5 - 36x^4$ and $4x^2$.

67. Find the quotient of $16a^7 - 24a^5$ and $8a^3$.

68. Divide $16x^4 - 32x^2 + 24$ by $8x^2$.

69. Divide $12y^5 - 16y^3 - 8y$ by $4y^3$.

70. Divide the sum of $6x^2y^3$ and $-2x^2y^3$ by $2xy$.

71. Divide the sum of $8x^3y^3$ and $-2x^3y^3$ by $3xy$.

72. Divide the sum of $2x^4 - 7x^3 + 3x^2$ and $4x^4 - 2x^3 - 6x^2$ by $-3x^2$.

73. Divide the sum of $4y^5 - 6y^3 + y^2$ and $2y^5 - 4y^3 + 7y^2$ by $2y^2$.

Writing Exercises:

74. Is the quotient of a monomial with a monomial always a monomial? Why or why not? If not, give examples.

75. Is the quotient of a polynomial with a monomial always a polynomial? Why or why not? If not, give examples.

76. Does the quotient of a trinomial with a monomial always have three terms?

77. What is wrong with the following?

$\dfrac{4x^2 + 5x + 6}{6} = \dfrac{4x^2 + 5x + \not{6}}{\not{6}} = 4x^2 + 5x + 1.$

Critical Thinking Exercises:

78. Without using a calculator, evaluate $\dfrac{x^{1000} - x^{999}}{x^{999}}$ when $x = 1000$.

79. The quotient of a polynomial and $3xy$ is $2x^2y - 6x^3$. What is the polynomial? Why?

80. If $\dfrac{12x^a - 18x^b + 9x^c + 6x^d}{3x} = 4x^3 - 6x^2 + 3x + 2$, find a, b, c, and d.

81. If the area of a rectangle is $8x^3 + 12x^2 + 4$ and the width is x^2, find the length.

Section 2.9	An Application of Exponents: Scientific Notation

OBJECTIVES   *When you complete this section you will be able to:*

a. Change numbers written in scientific notation to numbers in standard notation.

b. Write numbers in scientific notation.

c. Multiply and divide very large and/or very small numbers using scientific notation.

Introduction   A certain computer can perform 15,000 computations per second. Therefore, in one day it can perform $(15,000)(60)(60)(24)$ computations. When this calculation was performed on one calculator, the result was 1.296E9. What does this mean?

Very large and very small numbers are used frequently in both the sciences and in everyday situations to refer to such things as the population of the United States, the national debt, and the gross national product. Frequently (as in the case in the previous paragraph), these numbers have so many digits that a calculator cannot display all of them. In such cases, we use an alternative method of writing extremely large and small numbers called **scientific notation.**

> **DEFINITION Scientific Notation**
>
> A number is written in scientific notation if it is in the form of $a \times 10^n$ where $1 \le a < 10$ and n is an integer.

To say $1 \le a < 10$ means that there is one nonzero digit to the left of the decimal point. Remember that the integers are $\{\ldots -3, -2, -1, 0, 1, 2, 3, \ldots\}$. Examples of numbers written in scientific notation are:

$$2.3 \times 10^4, 4.06 \times 10^7, 9.23 \times 10^{-5}, \text{ and } 5.34 \times 10^{-3}.$$

In performing the calculation $(15,000)(60)(60)(24)$, the calculator gave the answer in scientific notation with the number following the "E" representing the exponent of 10. Consequently, this computer can perforn 1.296×10^9 calculations per day. What does this number represent?

Before we write numbers in scientific notation, we need to make some observations regarding powers of 10 and how multiplying and dividing by powers of 10 affects the movement of the decimal point. Study the table below:

Power of 10	Value
10^0	1
10^1	10
10^2	100
10^3	1000
10^4	10,000

Observe that the exponent of 10 is the same as the number of zeros following the 1 in the value of the power of 10. Consequently, 10^7 equals 1 followed by seven 0s or 10,000,000. Study the following table:

Power of 10	Value
10^{-1}	$\dfrac{1}{10} = .1$
10^{-2}	$\dfrac{1}{10^2} = \dfrac{1}{100} = .01$
10^{-3}	$\dfrac{1}{10^3} = \dfrac{1}{1000} = .001$
10^{-4}	$\dfrac{1}{10^4} = \dfrac{1}{10,000} = .0001$

Notice the total number of decimal places is the same as the absolute value of the exponent of 10. Consequently, 10^{-6} has a total of 6 decimal places and equals .000001.

Study the following examples and look for a relationship between the exponent of 10 and the movement of the decimal point. When graphing integers on the number line, movement to the right is positive and movement to the left is negative. We will use the same idea in moving the decimal point.

Example 1

Find the following products.

a. 3.2×10^3 $10^3 = 1000$. Write vertically and multiply.

$$
\begin{array}{r}
3.2 \\
\times\ 1000 \\
\hline
3200
\end{array}
$$

Place three zeroes behind 32 and mark off one decimal place. Notice that the decimal has been moved three places to the right, which is the same as the exponent of 10.

b. 4.67×10^5 $10^5 = 100,000$. Write vertically and multiply.

$$
\begin{array}{r}
4.67 \\
\times\ 100000 \\
\hline
467000
\end{array}
$$

Place five zeroes behind 467 and mark off two decimal places. Notice that the decimal has been moved five places to the right, which is the same as the exponent of 10.

c. 2.452×10^{-4} $10^{-4} = .0001.$ Write vertically and multiply.

$$
\begin{array}{r}
2.452 \\
\times \ .0001 \\
\hline
.0002452
\end{array}
$$

$1 \times 2452 = 2452.$ Mark off seven decimal places. Notice that the decimal has been moved four places to the left, which is the same as the absolute value of the exponent of 10.

d. 9.6×10^{-2} $10^{-2} = .01.$ Write vertically and multiply.

$$
\begin{array}{r}
9.6 \\
\times \ .01 \\
\hline
.096
\end{array}
$$

$1 \times 96 = 96.$ Mark off two decimal places. Notice that the decimal has been moved two places to the left, which is the same as the absolute value of the exponent of 10.

Based on the results of Example 1, we make the following observation:

Multiplying by Powers of 10

To find a product of the form $a \times 10^n$, move the decimal n places to the right if n is positive, or move the decimal $|n|$ places to the left if n is negative.

An easy way to remember this rule is that a positive exponent on 10 makes the product larger, so move the decimal point to the right. A negative exponent makes the product smaller, so move the decimal point to the left.

Since all the numbers in Example 1 were given in scientific notation, the preceding rule is used to change a number from scientific notation into standard notation.

Example 2

Change each of the following from scientific notation into standard notation:

a. 6.89×10^4 Since 4 is positive, make the product larger by moving the decimal point four places to the right.

68,900   Standard notation.

b. 3.01×10^{-3} Since -3 is negative, make the product smaller by moving the decimal point three places to the left.

.00301   Standard notation.

Practice Exercises

Change each of the following from scientific notation to standard notation:

1. 9.42×10^5

2. 1.72×10^{-5}

3. 7.82×10^2

4. 2.05×10^{-6}

Now that we know how to multiply numbers by powers of 10 and change numbers from scientific notation to standard notation, we are ready to write numbers in scientific notation. Remember that in order for a number to be written in scientific notation, $a \times 10^n$, the "a" must have one nonzero digit to the left of the decimal. Therefore, our first task is to properly place the decimal. Our second task is to determine the exponent of 10 so that the number written in scientific notation is equal to the original number.

Example 3

Write the following numbers in scientific notation:

a. 91,000   **Solution:**
Since we need one nonzero digit to the left of the decimal, place the decimal between 9 and 1. Therefore, 91,000 written in scientific notation is of the form 9.1×10^n. Since $9.1 < 91,000$, we need to multiply 9.1 by a power of 10 that will make the product larger by moving the decimal point four places to the right. Therefore, $n = 4$. Consequently, $91,000 = 9.1 \times 10^4$.

b. .000091   **Solution:**
Again, the decimal is placed between 9 and 1. Therefore, .000091 written in scientific notation is of the form 9.1×10^n. Since $9.1 > .000091$, we need to multiply 9.1 by a power of 10 that will make the product smaller by moving the decimal point five places to the left. Therefore, $n = -5$. Consequently, $.000091 = 9.1 \times 10^{-5}$.

c. 43,600,000   **Solution:**
Since we need one nonzero digit to the left of the decimal, place the decimal between 4 and 3. Therefore, 43,600,000 written in scientific notation is of the form 4.36×10^n. Since $4.36 < 43,600,000$, we need to multiply 4.36 by a power of 10 that will make the product larger by moving the decimal point seven places to the right. Therefore, $n = 7$. Consequently, $43,600,000 = 4.36 \times 10^7$.

d. .00361   **Solution:**
Since we need one nonzero digit to the left of the decimal, place the decimal between 3 and 6. Therefore, .00361 written in scientific notation is of the form 3.61×10^n. Since $3.61 > .00361$, we need to multiply 3.61 by a power of 10 that will make the product smaller by moving the decimal point three places to the left. Therefore, $n = -3$. Consequently, $.00361 = 3.61 \times 10^{-3}$.

Practice Exercises

Write the following numbers in scientific notation:

5. 470,000

6. .00000056

7. 5630

8. .000972

There is frequently a need for scientific notation in the various branches of the natural sciences, such as chemistry, physics, biology, and astronomy.

Example 4

a. The average distance between Earth and the sun is 93,000,000 miles. Express this distance using scientific notation.

Solution:
The decimal is placed between 9 and 3. So $93,000,000 = 9.3 \times 10^n$. If $9.3 \times 10^n = 93,000,000$, the decimal must be moved seven places to the right. Therefore, $n = 7$. Consequently, the average distance between the earth and sun is 9.3×10^7 miles.

b. Einstein stated that the energy, E, equivalent to the mass, m, can be calculated by the equation $E = mc^2$. According to this equation, 9.0×10^7 joules of energy are equivalent to 1.0×10^{-6} grams of mass. Write the number of grams of mass in standard form and give the fractional part of a gram that it represents.

Answers:

A joule is a unit of measure of energy. 4.184 J = 1 calorie. Hence, 4.184 J is the amount of heat required to raise the temperature of 1 gram of water 1 degree Celsius.

Solution:

Since the exponent of 10 is negative, move the decimal six places to the left. Therefore, 1.0×10^{-6} grams = .000001 grams. As a fraction, $.000001 = \frac{1}{1,000,000}$. Hence, according to Einstein's equation, one one-millionth of a gram of mass is converted into 90,000,000 joules of energy. This accounts for the tremendous amount of energy released during an atomic reaction.

Practice Exercises

9. The age of Earth is estimated at about 45,000,000,000 years. Express the age of Earth in scientific notation.

10. The radius of an atom of a certain substance is 1.42×10^{-7} millimeters. Write the radius of this atom in standard form.

In order to perform operations on numbers written in scientific notation, we need to recall two laws of exponents: $a^m \cdot a^n = a^{m+n}$ and $\frac{a^m}{a^n} = a^{m-n}$. The method that we use to multiply and divide numbers written in scientific notation is very much like the method used to multiply and divide monomials. We will refer to the "a" of $a \times 10^n$ as the coefficient and the "10^n" as the exponential part. Consider the following:

Monomials	**Scientific Notation**
$(3.2x^2)(1.6x^4)$	$(3.2 \times 10^2)(1.6 \times 10^4)$
Regroup the factors so the coefficients and variable factors are together.	Regroup the factors so the coefficients and exponential factors are together.
$(3.2 \cdot 1.6)(x^2 \cdot x^4)$	$(3.2 \cdot 1.6) \times (10^2 \cdot 10^4)$
Multiply the coefficients and the variable factors.	Multiply the coefficients and the exponential factors.
$5.12x^6$	5.12×10^6
Therefore, $(3.2x^2)(1.6x^4) = 5.12x^6$.	Therefore, $(3.2 \times 10^2)(1.6 \times 10^4) = 5.12 \times 10^6$.
$\dfrac{2.6x^5}{1.3x^3}$	$\dfrac{2.6 \times 10^5}{1.3 \times 10^3}$
We think of this as coefficient divided by coefficient, and variable factors divided by variable factors.	We think of this as coefficient divided by coefficient, and exponential factors divided by exponential factors.
$\dfrac{2.6}{1.3} \cdot \dfrac{x^5}{x^3}$	$\dfrac{2.6}{1.3} \times \dfrac{10^5}{10^3}$
Divide the coefficients and the variables separately.	Divide the coefficients and the exponentials separately.
$2x^2$	2×10^2
Therefore, $\dfrac{2.6x^5}{1.3x^3} = 2x^2$.	Therefore, $\dfrac{2.6 \times 10^5}{1.3 \times 10^3} = 2 \times 10^2$.

Example 5

Simplify the following using scientific notation. Leave answers in standard notation.

a. $(4.1 \times 10^3)(2.3 \times 10^{-6})$ Regroup the factors so the coefficients and exponential parts are together.

$(4.1 \cdot 2.3) \times (10^3 \cdot 10^{-6})$ Multiply coefficients and $10^3 \cdot 10^{-6} = 10^{3+(-6)} = 10^{-3}$.

9.43×10^{-3} Move the decimal three places to the left.

$.00943$ Product.

b. $\dfrac{7.44 \times 10^{-2}}{3.1 \times 10^4}$ Divide coefficient by coefficient and exponential part by exponential part.

$\dfrac{7.44}{3.1} \times \dfrac{10^{-2}}{10^4}$ Divide coefficients and $\dfrac{10^{-2}}{10^4} = 10^{-2-4} = 10^{-6}$.

2.4×10^{-6} Move the decimal six places to the left.

$.0000024$ Quotient.

c. $(4100)(56,000)$ Rewrite each number in scientific notation.

$(4.1 \times 10^3)(5.6 \times 10^4)$ Regroup the factors so the coefficients and exponential parts are together.

$(4.1 \cdot 5.6) \times (10^3 \cdot 10^4)$ Multiply coefficients and $10^3 \cdot 10^4 = 10^{3+4} = 10^7$.

22.96×10^7 Move the decimal seven places to the right.

$229,600,000$ Product.

d. $\dfrac{(120,000)(.0018)}{3600}$ Rewrite each number in scientific notation.

$\dfrac{(1.2 \times 10^5)(1.8 \times 10^{-3})}{3.6 \times 10^3}$ Regroup in the numerator.

$\dfrac{(1.2 \cdot 1.8) \times (10^5 \cdot 10^{-3})}{3.6 \times 10^3}$ Multiply coefficients and $10^5 \cdot 10^{-3} = 10^{5-3} = 10^2$.

$\dfrac{2.16 \times 10^2}{3.6 \times 10^3}$ Divide coefficient by coefficient and exponential part by exponential part.

$\dfrac{2.16}{3.6} \times \dfrac{10^2}{10^3}$ Divide. $\dfrac{10^2}{10^3} = 10^{2-3} = 10^{-1}$.

$.6 \times 10^{-1}$ Move the decimal one place to the left.

$.06$ Answer.

Practice Exercises

Simplify each of the following using scientific notation. Leave answers in standard notation.

11. $(4.4 \times 10^4)(2.3 \times 10^3)$ **12.** $\dfrac{9.3 \times 10^7}{3.0 \times 10^3}$ **13.** $(5300)(.000025)$ **14.** $\dfrac{.00345}{15000}$

Computations involving scientific notation often occur in application problems.

Example 6

Solve the following using scientific notation. Leave the answers in standard notation.

a. A certain type of computer can perform a calculation in .000003 of a second. How long would it take this computer to perform 5 billion (5,000,000,000) calculations?

Solution:
The time required to perform 5,000,000,000 calculations is equal to:

$(5,000,000,000)(.000003)$ Write each in scientific notation.

$(5 \times 10^9)(3 \times 10^{-6})$ Regroup the factors.

$(5 \cdot 3) \times (10^9 \cdot 10^{-6})$ Multiply coefficients and $10^9 \cdot 10^{-6} = 10^{9-6} = 10^3$.

15×10^3 Move the decimal three places to the right.

15000 Therefore, it takes the computer 15000 seconds to perform 5 billion computations.

b. The solubility constant of barium sulfate is 1.5×10^{-9} and the solubility constant of silver bromide is 5×10^{-13}. How many times greater is the solubility constant of barium sulfate than that of silver bromide?

Solution:
The number of times greater the solubility constant of barium sulfate is than that of silver bromide is equal to the solubility constant of barium sulfate divided by the solubility constant of silver bromide.

$\dfrac{1.5 \times 10^{-9}}{5 \times 10^{-13}}$ Divide coefficient by coefficient and exponential part by exponential part.

$\dfrac{1.5}{5} \times \dfrac{10^{-9}}{10^{-13}}$ Divide. $\dfrac{10^{-9}}{10^{-13}} = 10^{-9-(-13)} = 10^{-9+13} = 10^{4}$.

$.3 \times 10^{4}$ Move the decimal four places to the right.

3000 Therefore, the solubility constant of barium sulfate is 3000 times greater than the solubility constant of silver bromide.

Practice Exercises

Solve the following using scientific notation. Leave the answers in standard notation.

15. The mass of a helium atom is 6.65×10^{-24} grams. Find the mass of a sample of helium that contains 3.0×10^{26} atoms.

16. The approximate distance from Earth to the planet Saturn is 7.942×10^{8} miles. How many hours would it take a spacecraft traveling at 20,000 miles per hour to reach Saturn?

Exercise Set 2.9

Write the following in standard notation:

1. 3.5×10^{4} **2.** 4.7×10^{2} **3.** 9.5×10^{-3} **4.** 6.3×10^{-5}

5. 4.79×10^{6} **6.** 3.07×10^{8} **7.** 9.24×10^{-6} **8.** 2.19×10^{-2}

9. 1×10^{2} **10.** 7×10^{5} **11.** 1×10^{-8} **12.** 9×10^{-4}

Write the following in scientific notation:

13. 7600   **14.** 83000   **15.** .00035   **16.** .0026

17. 857000   **18.** 13400000   **19.** .000000498   **20.** .000913

21. 600000   **22.** 10000000   **23.** .00000001   **24.** .00006

Perform the following operations using scientific notation. Leave answers in standard notation.

25. $(1.2 \times 10^{3})(2.5 \times 10^{5})$ **26.** $(3.4 \times 10^{4})(1.7 \times 10^{2})$ **27.** $(4.3 \times 10^{6})(3.2 \times 10^{-4})$

28. $(8.3 \times 10^{7})(5.1 \times 10^{-5})$ **29.** $(3.12 \times 10^{-6})(4.23 \times 10^{3})$ **30.** $(7.21 \times 10^{-8})(4.82 \times 10^{4})$

31. $(6.28 \times 10^{-3})(4.21 \times 10^{-5})$ **32.** $(3.04 \times 10^{-2})(2.5 \times 10^{-4})$ **33.** $\dfrac{4.2 \times 10^{7}}{2.1 \times 10^{4}}$

34. $\dfrac{3.9 \times 10^{6}}{1.3 \times 10^{2}}$ **35.** $\dfrac{3.6 \times 10^{-3}}{2.4 \times 10^{4}}$ **36.** $\dfrac{4.5 \times 10^{-5}}{1.8 \times 10^{2}}$

37. $\dfrac{1.8 \times 10^{3}}{1.5 \times 10^{-3}}$ **38.** $\dfrac{1.065 \times 10^{5}}{4.26 \times 10^{-5}}$ **39.** $\dfrac{1.28 \times 10^{-3}}{2.56 \times 10^{-5}}$

Answers:

40. $\dfrac{7.29 \times 10^{-5}}{2.43 \times 10^{-2}}$

41. $\dfrac{(1.25 \times 10^{-5})(5 \times 10^{7})}{2.5 \times 10^{-2}}$

42. $\dfrac{(3.6 \times 10^{6})(4.8 \times 10^{-4})}{1.44 \times 10^{-3}}$

43. $\dfrac{(5.4 \times 10^{-3})(7.2 \times 10^{6})}{(2.7 \times 10^{-2})(1.6 \times 10^{-1})}$

44. $\dfrac{(7.4 \times 10^{5})(1.0 \times 10^{-3})}{(1.5 \times 10^{-2})(2.5 \times 10^{8})}$

Perform the following operations using scientific notation. Leave answers in standard notation.

45. (55000)(2300)

46. (350000)(2600000)

47. (5600)(.000045)

48. (7200000)(.00036)

49. (.0000025)(.000000036)

50. (.00000034)(.0031)

51. $\dfrac{345000000}{15000}$

52. $\dfrac{1280000}{1600}$

53. $\dfrac{1800}{72000000}$

54. $\dfrac{22500}{112500000}$

55. $\dfrac{37400}{.0017}$

56. $\dfrac{9240000}{.0000021}$

57. $\dfrac{.000612}{.0034}$

58. $\dfrac{.0084}{.00000056}$

59. $\dfrac{(120000)(.0018)}{3600}$

60. $\dfrac{(24000)(.00000028)}{560000}$

61. $\dfrac{(8000)(.000252)}{(.00063)(400)}$

62. $\dfrac{(112000)(.0015)}{(.00002)(1400)}$

63. The astronomical unit is 150,000,000 kilometers. Write the astronomical unit in scientific notation.

64. The net assets of a company are $1,250,000,000. Write the assets in scientific notation.

65. The density of the element mercury is 13,600 kilograms per cubic meter. Express the density of mercury in scientific notation.

66. The coefficient of linear expansion of aluminum is .000024. Express the linear expansion coefficient of aluminum in scientific notation.

67. The wavelength of an X ray is .0000013 meters. Express the wavelength of the X ray in scientific notation.

68. The solubility product constant of lead sulfate is .000000013. Express the solubility constant of lead sulfate in scientific notation.

69. The speed of light is approximately 2.997925×10^{8} meters per second. Express the speed of light in standard notation.

70. The half-life of a radioactive substance is the time required for one-half of the amount present to decay. The half-life of actinium is 7.04×10^{8} years. Express the half-life of actinium in standard notation.

71. The density of gold is 1.93×10^{4} kilograms per cubic meter. Express the density of gold in standard notation.

72. The density of hydrogen is 8.99×10^{-2} kilograms per cubic meter. Express the density of hydrogen in standard notation.

73. The mass of a hydrogen atom is 1.673×10^{-24} grams. Find the mass of the atoms in a sample of hydrogen that contains 1,000,000,000 hydrogen atoms.

74. If a computer can execute a command in .00005 seconds, how long will it take the computer to execute 5 million commands?

75. The distance from Earth to the nearest star, Alpha Centauri, is approximately 25,200,000,000,000 miles. A light-year is the distance that light can travel in one year and is approximately 6,000,000,000,000 miles. Find the distance to Alpha Centauri to the nearest tenth of a light-year.

76. Density is defined as $\frac{\text{mass}}{\text{volume}}$. If the mass of the earth is 5.98×10^{27} grams and the volume of Earth is 1.08×10^{27} cubic centimeters, find the density of Earth to the nearest tenth of a gram per cubic centimeter.

77. In chemistry the number of moles (a measure of the amount of a substance) is equal to $\frac{\text{number of molecoles of the substance}}{6.02 \times 10^{23}}$. Find the number of moles of a particular substance if 9.03×10^{23} molecules are present.

Challenge Exercises:

78. The mass of a hydrogen atom is 1.673×10^{-24} grams. How many atoms are in a 10-gram sample of hydrogen?

79. In Exercise 75, the distance from Earth to Alpha Centauri was given as approximately 25,200,000,000,000 miles. Spacecraft travel at ap- proximately 25,000 miles per hour. Approximately how many years would it take a spacecraft to reach Alpha Centauri? Is this a problem for manned space travel?

80. The mass of a helium atom is 6.65×10^{-24} grams. How many atoms are in a 4-gram sample of helium?

Writing Exercise:

81. Why is scientific notation preferable when performing operations on extremely large or extremely small numbers?

Group Project:

82. Interview some science instructors, engineers, astronomers, or others, and find at least five examples of ways they have used scientific notation.

Chapter 2   Summary

Rules for Multiplying Signed Numbers: [Section 2.1]
- The product of two numbers with the same sign is positive: $(+)(+) = (+)$ and $(-)(-) = (+)$
- The product of two numbers with opposite signs is negative: $(+)(-) = (-)$ and $(-)(+) = (-)$

Signed Numbers to Powers: [Section 2.1]
- A negative number raised to an even power is positive.
- A negative number raised to an odd power is negative.
- The base of $(-a)^n$ is $-a$, so $(-a)^n$ means multiply n, $-a$'s.
- The base of $-a^n$ is a, so $-a^n$ means $-1 \cdot a^n$.

Properties of Multiplication: [Section 2.1]
- Commutative:
- $ab = ba$ Multiplication may be performed in any order.
- Associative:
  $a(bc) = (ab)c$ Factors may be grouped in any manner.
- Identity:
  The number 1 is the identity for multiplication, since $a \cdot 1 = 1 \cdot a = a$.
- Inverse:
  For any number $a \neq 0$, there exists a number $\frac{1}{a}$ such that $a \cdot \frac{1}{a} = 1$.

Multiplication Laws of Exponents: [Section 2.2]
- Product Rule: $a^m \cdot a^n = a^{m+n}$
- Power of a Product: $(ab)^n = a^n b^n$
- Power to a Power: $(a^m)^n = a^{mn}$

Products of Polynomials: [Section 2.3]
- Monomial by monomial: Multiply coefficients and multiply variables with the same base using the Product Rule.
- Polynomials by monomials: Use the distributive property to multiply each term of the polynomial by the monomial. If necessary, add like terms.

- Polynomial by polynomial: Use the distributive property to multiply the second polynomial by each term of the first polynomial.

$$(a + b)(c + d + e) = a(c + d + e) + b(c + d + e)$$

Special Products: [Section 2.4]

- Binomial by binomial—FOIL
- Sum and difference of the same terms: Multiply the first and last only. The product is the difference of squares.

$$(a + b)(a - b) = a^2 - b^2$$

- Square of a binomial: Square the first term, add twice the product of the terms, add the square of the last term.

$$(a + b)^2 = a^2 + 2ab + b^2$$

$$(a - b)^2 = a^2 - 2ab + b^2$$

- Products resulting in the sum of cubes: $(a + b)(a^2 - ab + b^2) = a^3 + b^3$
- Products resulting in the difference of cubes: $(a - b)(a^2 + ab + b^2) = a^3 - b^3$

Definition of Division: [Section 2.5]

- If $\frac{a}{b} = c$, then $bc = a$, $b \neq 0$.

Rules for Dividing Signed Numbers: [Section 2.5]

- The quotient of two numbers with the same sign is positive. $\frac{(+)}{(+)} = (+)$ and $\frac{(-)}{(-)} = (+)$

- The quotient of two numbers with opposite signs is negative. $\frac{(+)}{(-)} = (-)$ and $\frac{(-)}{(+)} = (-)$

Order of Operations on Integers: [Section 2.5]

- The order of operations on integers is the same as with whole numbers. If needed, see the Chapter 1 summary.

Laws of Exponents for Quotients: [Sections 2.6 and 2.7]

- Quotient Rule: $\frac{a^m}{a^n} = a^{m-n}$, $c \neq 0$
- Power Rule for Quotients: $\left(\frac{a}{b}\right)^n = \frac{a^n}{b^n}$, $b \neq 0$

Integer Exponents: [Sections 2.6 and 2.7]

1. $x^0 = 1$, if $x \neq 0$

2. $a^{-n} = \frac{1}{a^n}$, $a \neq 0$

3. $\frac{1}{a^{-n}} = a^n$, $a \neq 0$

4. $\left(\frac{a}{b}\right)^{-n} = \left(\frac{b}{a}\right)^n$, $a, b \neq 0$

Division of a Polynomial by a Monomial: [Section 2.8]

- To divide a polynomial by a monomial, divide each term of the polynomial by the monomial. In symbols, $\frac{a + b}{c} = \frac{a}{c} + \frac{b}{c}$, $c \neq 0$.

Scientific Notation: [Section 2.9]

1. A number is written in scientific notation if it is in the form $a \times 10^n$ with $1 \leq a < 10$ and n an integer.

2. To change a number from scientific notation to standard notation, move the decimal n places to the right if n is positive and $|n|$ places to the left if n is negative.

3. To write a number written in standard notation in scientific notation, place the decimal so there is one nonzero digit to the left of the decimal. Then determine n by the number of places the decimal would have to be moved so the number in scientific notation will be equal to the number in standard notation.

4. To multiply two numbers in scientific notation, multiply the coefficients and the exponential parts separately.

5. To divide two numbers in scientific notation, divide the coefficients and exponential parts separately.

Chapter 2   Review Exercises

Find the following products: [Section 2.1]

1. $(-4)(2)$

2. $(-5)(-3)$

3. $(4)(-6)(3)$

4. $(-4)(5)(-3)(-2)$

5. $(-3)^2(4)^2$

6. $(-1)^3(-4)^3$

Evaluate each of the following for x = −3 and y = −2: [Section 2.1]

7. $5x$

8. $-3y^2$

9. x^2y^3

10. $-4xy^3$

Write an expression for each of the following and evaluate: [Section 2.1]

11. The product of −4 and 8 increased by 7

12. −6 decreased by the product of −5 and 5

13. The sum of the product of −6 and 3 and the product of 6 and −4

Give the name of the property illustrated by each of the following: [Section 2.1]

14. $(-4)(5) = (5)(-4)$

15. $(-2 \cdot 5)(6) = -2(5 \cdot 6)$

16. $(-5)(-3)(7) = (-3)(-5)(7)$

17. $-5(4 + 7) = (-5)(4) + (-5)(7)$

18. $9 \cdot \dfrac{1}{9} = 1$

19. $1 \cdot 13 = 13$

Simplify the following using the multiplication laws of exponents: [Section 2.2]

20. $w^5 \cdot w^8$

21. $4^3 \cdot 4^5 \cdot 4^6$

22. $(8x^3)(-5x^7)$

23. $(6a^4b^6)(-5a^2b^6)$

24. $(xy)^7$

25. $(5d)^3$

26. $(-2ab)^4$

27. $(xy)^5(xy)^4$

28. $(4x)^3(-2x)^5$

29. $(x^5)^4$

30. $(4^3)^8$

31. $(a^4b^6)^4$

32. $(4a^3b^7)^4$

33. $(a^4b^3)^3(a^5b^2)^5$

34. $(2m^3n^3)^3(3m^2n^5)^2$

Find the products of the following polynomials: [Sections 2.3 and 2.4]

35. $(-3x^4y^3)(-5x^6y^2)$

36. $6y(3y^2 - 2)$

37. $-5x^2(4x^2 - 5)$

38. $7c^3(3c^2 - 7c + 2)$

39. $-2x^2y^3(3x^4y - 8xy^5 + 4x^3)$

40. $2(3x - 4) + 3(2x + 7)$

41. $3(5x - 5) - 4(3x - 6)$

42. $-2(2a^2 + 4a - 5) + 3(3a^2 - 5a + 1)$

43. $5(b^2 + 2b - 1) - 3(2b^2 - 3b - 6)$

44. $(x + 2)(x + 5)$

45. $(4m + 5)(3m - 6)$

46. $(5m - 6n)(4m + 3n)$

47. $(x - 9)(x + 9)$

48. $(5x - 6)(5x + 6)$

49. $(2s + 5t)(2s - 5t)$

50. $(m + n)^2$

51. $(3a - 2)^2$

52. $(6a - 2b)^2$

53. $(a + 4)(a^2 - 5a + 4)$

54. $(2m - 5)(3m^2 + 2m - 4)$

55. $(4x - 2y)(3x^2 - 5xy + 2y^2)$

Write an expression for each of the following and simplify: [Section 2.3]

56. Find the sum of two times $3x + 7$ and five times $-2x + 1$.

57. Find the difference of three times $4m - 2$ and three times $2m + 5$.

Find the following products: (Optional Exercises 58–59): [Section 2.3]

58. $(x - 4)(x^2 + 4x + 16)$

59. $(3x + 5)(9x^2 - 15x + 25)$

60. Find the area of the rectangle with length $5a - 7$ and width $3a + 2$. *[Section 2.3]*

61. Find the area of the triangle with base $5x$ and height $4x$. *[Section 2.3]*

62. Find the area of the square each of whose sides is $3z + 8$. *[Section 2.4]*

Write an expression for each of the following and simplify. [Section 2.4]

63. The square of the quantity 4 less than the product of 6 and x

64. The square of the quantity 2 more than the product of 3 and z

Find the following quotients: [Section 2.5]

65. $\dfrac{48}{-6}$

66. $\dfrac{-28}{-14}$

67. $\dfrac{-18}{2}$

68. $\dfrac{16}{13 - 17}$

Simplify the following using the Order of Operations: [Section 2.5]

69. $\dfrac{23 - 35}{-6}$

70. $\dfrac{-120}{(15)(-4)}$

71. $\dfrac{(-16)(6)}{(-8)(3)}$

72. $-6 + 13 - 9$

73. $8 - 4 \cdot 5$

74. $17 + 36 \div (-4)$

75. $14 - 3(-5)^2$

76. $7 + 2(8) \div (-4)$

77. $6 - 2(9 + 4)$

78. $15 + 4 \cdot 6^2 \div (-9)(3) - 7$

79. $-4(6^2 - 2^2) \div (-8)$

80. $\dfrac{(-3)(-6) + 6}{(-6)(3) + 6}$

81. $\dfrac{2 \cdot 4^2 + 16}{-3 \cdot 2^2 + 4}$

82. $\dfrac{8^2 - 4^2}{-2(3^2 - 1)}$

83. $\dfrac{(-4)(5) + [6(-4) + 6^2]}{(-3)^2(4) - 28}$

Find the value of each of the following for $x = -3$, $y = 2$, and $z = -2$: [Section 2.5]

84. $3x^2 - 4z$

85. $-4x^3y^2$

86. $3xy^2z$

87. $3x - 4(y - 3z)$

88. $-y^2 - (2x^2 - 3z^2)$

89. $\dfrac{3x - 2y}{4y - 2z}$

90. $\dfrac{2x^2 - 2z^3}{5x^2 + 3z^2}$

Simplify each of the following. Leave answers in exponential form with positive exponents only. Assume all variables represent nonzero quantities. [Section 2.6]

91. $\dfrac{2^8}{2^3}$

92. $\dfrac{m^8}{m^6}$

93. 3^{-4}

94. x^{-6}

95. $\dfrac{5^4}{5^6}$

96. $3x^0 + (5x)^0$

97. $5^{-5} \cdot 5^4$

98. z^4z^{-2}

99. $x^{-4}x^{-2}$

100. $\dfrac{5^{-4}}{5^2}$

101. $\dfrac{a^{-3}}{a^{-6}}$

102. $\dfrac{7^3}{7^{-7}}$

Write each of the following without exponents and evaluate: [Section 2.7]

103. $\left(\dfrac{3}{5}\right)^2$

104. $\left(\dfrac{3}{2}\right)^{-2}$

Write each of the following in simplest form with positive exponents only: [Section 2.7]

105. $(3^{-5})^2$

106. $(r^{-5})^{-4}$

107. $(x^{-4}y^4)^{-3}$

108. $(4x^{-2})^{-3}$

109. $(5a^4)^{-3}$

110. $(2x^{-4}y^6)^{-3}$

111. $\left(\dfrac{m^4}{n^3}\right)^{-2}$

112. $\dfrac{(m^{-4})^{-5}}{(m^3)^{-4}}$

113. $\dfrac{-24n^6}{-3n^3}$

114. $\dfrac{m^4n^6}{m^3n^2}$

115. $\dfrac{-48m^8n^5}{-8m^{10}n^5}$

116. $\dfrac{(-6x^3)(-5x^5)}{(2x^2)(-3x^4)}$

117. $\dfrac{(3m^{-4}n^3)(-8m^3n^{-6})}{(4m^5n^{-3})(-2m^{-3}n^6)}$

118. $\dfrac{(m^3)^4(m^{-6})^2}{(m^8)^{-2}(m^{-3})^{-3}}$

Find the following quotients of polynomials by monomials:
[Section 2.8]

119. $\dfrac{9x - 18}{9}$

120. $\dfrac{5y^3 + 3y}{y}$

121. $\dfrac{28a^2 - 16a}{4a}$

122. $\dfrac{36m^7n^3 + 18m^4n^7}{9m^5n^5}$

123. $\dfrac{12x^4 - 9x^3 - 21x^2}{3x^2}$

124. $\dfrac{8m^5n^3 - 32m^2n^7 + 24m^6n^4}{8m^6n^5}$

Write an expression for each of the following and simplify:
[Section 2.8]

125. The quotient of $48m^6n^3$ and $-16m^4n^5$

126. The quotient of $24a^6 - 16a^4$ and $8a^3$

127. Divide the sum of $4x^4 - 2x^3 - 3x^2$ and $2x^4 - 7x^3 + 2x^2$ by $3x^2$.

128. The quotient of $28x^6 - 40x^3$ and $4x^4$

129. Divide the sum of $5x^2 - 7x + 5$ and $3x^2 + 3x - 9$ by $2x$.

Write the following numbers in standard notation:
[Section 2.9]

130. 4.6×10^5

131. 3.06×10^6

132. 7.03×10^{-5}

133. 7.123×10^{-3}

Write the following numbers in scientific notation:
[Section 2.8]

134. 97,000,000,000

135. 46,700

136. .00000478

137. .000307

Simplify each of the following using scientific notation. Leave answers in standard form. [Section 2.8]

138. $(4.5 \times 10^4)(3.1 \times 10^{-7})$

139. $(5.07 \times 10^{-5})(4.06 \times 10^{-2})$

140. $\dfrac{4.8 \times 10^4}{1.6 \times 10^6}$

141. $\dfrac{3.9 \times 10^{-5}}{1.3 \times 10^3}$

142. $(4600)(.0000012)$

143. $(.000047)(.00000034)$

144. $\dfrac{48000}{.00096}$

145. $\dfrac{.00033}{.00000022}$

Chapter 2   Test

Find the following products:

1. $(-3)(5)(-4)$

2. $(-2)^3(3)^2$

3. $(3x^4y^3)(-2xy^2)$

4. $(3x^2y)(-4x^2y)^2$

5. $2x^2(3x - 4)$

6. $p^2q(2pq^3 - 4p^2q - 2)$

7. $4(2x - 3) - 5(x - 4)$

8. $(3p + 4)(2p - 1)$

9. $(3a - 4b)(2a - b)$

10. $(3y - 1)(y^2 + 2y + 5)$

11. $(5x - 3y)(5x + 3y)$

12. $(2a - 7)^2$

Simplify the following. If the exercise involves exponents, leave the answer with positive exponents only.

13. $\dfrac{-24 - 8}{8}$

14. $\dfrac{4 \cdot 5^2 - 25}{(-3)(-4) + 13}$

15. $(4m^3n^5)(-6m^{-6}n^2)$

16. $8x^0 - (3x)^0$

17. $\dfrac{36x^4y^3z^5}{-18x^7y^3z^2}$

18. $(3x^{-5}y^4)^{-3}$

19. $\dfrac{(4a^6b^{-3})^2}{(2a^3b^4)^3}$

20. $\dfrac{(6a^{-2}b^{-4})(-8a^{-5}b^8)}{(4a^{-3}b^4)(-6a^6b^{-6})}$

21. $\dfrac{8m^6n^3 - 32m^4n^5 + 20m^5n^2}{4m^3n^2}$ **22.** $\dfrac{8x^3 - 10x^2 + 6x}{2x^2}$

Evaluate each of the following:

23. $-8 - (12 - 36) \div (-4)$ **24.** $\dfrac{4(-2)^3 + 2}{8^2 + 2(-3)^3}$

25. Find the value of $2y - (3x^2 - 3y) \div x^3$, if $x = -2$ and $y = -4$.

Write an expression for each of the following and simplify:

26. The product of 7 and -3 subtracted from the quotient of 12 and -4

27. The product of $x + 3$ and $2x - 5$ added to $3x^2 - 7x + 5$

28. Write each of the following in scientific notation:

a) 4,790,000,000   b) .0000000749

29. Evaluate $(43,000,000)(.000025)$ using scientific notation. Leave the answer in standard notation.

30. Evaluate $\dfrac{97020000}{.00231}$ using scientific notation. Leave the answer in scientific notation.

O ne of the primary uses of the addition, subtraction, multiplication, and division of signed numbers is in solving linear equations and inequalities. The notions of equations and inequalities were previously mentioned in Chapter 1. In this chapter, we will expand on these notions and use them to solve various types of real-world and not-so-real-world application problems.

Linear Equations and Inequalities

Chapter

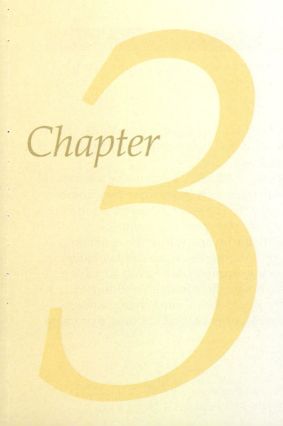

Addition Property of Equality

O B J E C T I V E S

When you complete this section, you will be able to:

a. Solve linear equations using the Addition Property of Equality.

b. Translate mathematics expressions into English and English expressions into mathematics.

c. Solve various types of real-world problems.

Introduction

An **equation** is a mathematical sentence indicating that two expressions are equal. For example, both $3 + 2 = 5$ and $x + 4 = 9$ are equations. A **solution** of an equation is any value of the variable that makes the equation true. For example, $x = 5$ is a solution of $x + 4 = 9$, because if we replace x with 5, we get $5 + 4 = 9$, which is a true statement. When the solutions are written as a set, we refer to it as the **solution set.** So, the solution set of $x + 4 = 9$ is $\{5\}$.

Finding the solution(s) of an equation is called **solving the equation.** In this and the following sections, we will discuss techniques that allow us to solve **linear equations.** A linear equation is any equation that can be put in the form of $ax + b = c$, where a, b, and c are constants and $a \neq 0$.

Suppose we begin with an equation and add the same number to both sides of the equation. Is the resulting sentence still an equation? Consider the following:

$$4 + 5 = 9 \qquad \text{Add 6 to both sides.}$$
$$4 + 5 + 6 \overset{?}{=} 9 + 6 \qquad \text{Simplify both sides.}$$
$$15 = 15 \qquad \text{Adding 6 to both sides of the equation did not change the equality.}$$

$$4 + 5 = 9 \qquad \text{Add } -8 \text{ to both sides.}$$
$$4 + 5 - 8 \overset{?}{=} 9 - 8 \qquad \text{Simplify both sides.}$$
$$1 = 1 \qquad \text{Adding } -8 \text{ to both sides of the equation did not change the equality.}$$

The preceding examples lead us to the following generalization:

Addition Property of Equality

If $a = b$, then $a + c = b + c$ for any number c. In words, if the same number is added to both sides of an equation, the result is still an equation.

Equations that have the same solutions are called **equivalent equations.** If the Addition Property of Equality is applied to an equation containing variables, the resulting equation is equivalent to the original equation.

Solving equations using the Addition Property of Equality requires the use of two principles. The first is the Additive Inverse Property, which states that every number, a, has an additive inverse, $-a$, such that $a + (-a) = 0$. The second property is the Additive Property of Zero, which states that $0 + a = a + 0 = a$ for any number a.

The goal of equation solving is to get the variable on one side of the equation and a constant on the other. By doing so, we will get the simplest equivalent equation in the form of (variable) = (constant). In the following examples, we have a variable and a constant on the same side of the equation. We isolate the variable by adding the additive inverse of the constant to both sides of the equation and simplifying. After solving an equation, we can verify that our solution is correct by substituting that value into the original equation and simplifying.

Example 1

Solve the following:

a. $y - 3 = 9$ Since 3 is the additive inverse of -3, add 3 to both sides of the equation.

$y - 3 + 3 = 9 + 3$ Simplify both sides of the equation.

$y + 0 = 12$ $y + 0 = y$.

$y = 12$ Therefore, $y = 12$ is the solution and $\{12\}$ is the solution set.

CHECK:

$y - 3 = 9$ Substitute 12 for y.

$12 - 3 = 9$ Simplify the left side of the equation.

$9 = 9$ Therefore, 12 is the correct solution.

b. $6 = 8 + z$ Since -8 is the additive inverse of 8, add -8 to both sides of the equation.

$-8 + 6 = -8 + 8 + z$ Simplify both sides of the equation.

$-2 = 0 + z$ $0 + z = z$.

$-2 = z$, or $z = -2$ Therefore, -2 is the solution and $\{-2\}$ is the solution set.

CHECK:

$6 = 8 + z$ Substitute -2 for z.

$6 = 8 + (-2)$ Simplify the right side of the equation.

$6 = 6$ Therefore, -2 is the correct solution.

Note: We will no longer write the solution sets of equations except in the cases where there are no solutions. If there are no solutions, then the solution set is the empty set indicated by $\emptyset$.

Practice Exercises

Solve the following:

1. $v - 8 = 7$

2. $r + 3 = 4$

If more practice is needed, do the Additional Practice Exercises.

Additional Practice Exercises

Solve the following:

a. $x - 3 = 6$

b. $r + 4 = -1$

It is often necessary to simplify one or both sides of an equation before applying the Addition Property of Equality. The following steps are suggested to do so:

Solving Linear Equations Using the Addition Property of Equality

1. If necessary, remove any inclusion symbols (parentheses, brackets, braces). This will often involve the distributive property.
2. If necessary, simplify each side of the equation by adding like terms.
3. If necessary, use the Addition Property of Equality to get the variable on one side and the constant on the other.
4. Check the answer in the original equation.

This procedure will be modified in Section 3.3 when we discuss the most general types of linear equations.

Example 2

Solve the following:

a. $t + 8 - 6 = 11 + 4$ — Simplify both sides of the equation by adding like terms.

$t + 2 = 15$ — Since -2 is the additive inverse of 2, add -2 to both sides of the equation.

$t + 2 + (-2) = 15 + (-2)$ — Simplify both sides of the equation.

$t = 13$ — Therefore, 13 is the solution.

CHECK:

$t + 8 - 6 = 11 + 4$ — Substitute 13 for t.

$13 + 8 - 6 = 11 + 4$ — Simplify both sides of the equation.

$15 = 15$ — Therefore, $t = 13$ is the correct solution.

b. $6a + 4 - 5a - 2 = -6$ — Simplify the left side of the equation.

$a + 2 = -6$ — Since -2 is the additive inverse of 2, add -2 to both sides of the equation.

$a + 2 + (-2) = -6 + (-2)$ — Simplify both sides of the equation.

$a = -8$ — Therefore, the solution is $a = -8$.

The check is left as an exercise for the student.

c. $6.3u + 5.4 - 5.3u = 20.6$ — Simplify the left side of the equation.

$u + 5.4 = 20.6$ — Since -5.4 is the additive inverse of 5.4, add -5.4 to both sides of the equation.

$u + 5.4 + (-5.4) = 20.6 + (-5.4)$ — Simplify both sides of the equation.

$u = 15.2$ — Therefore, $u = 15.2$ is the solution.

CHECK:

$6.3u + 5.4 - 5.3u = 20.6$ — Substitute 15.2 for u.

$6.3(15.2) + 5.4 - 5.3(15.2) = 20.6$ — Multiply first.

$95.76 + 5.4 - 80.56 = 20.6$ — Add.

$20.6 = 20.6$ — Therefore, 15.2 is the correct solution.

Note: In Example 2a, we added -2 to both sides of the equation. Rather than adding -2, we could have subtracted 2 since subtraction is the same as the addition of the negative. Recall from Chapter 1 that $a - b = a + (-b)$ and $a + (-b) = a - b$. Likewise, in Example 2c we could have subtracted 5.4 from both sides of the equation instead of adding -5.4.

d. $3x - 2(x + 4) = -5$ Apply the distributive property on the left side of the equation.

$3x - 2x - 8 = -5$ Simplify the left side.

$x - 8 = -5$ Since 8 is the additive inverse of -8, add 8 to both sides of the equation.

$x - 8 + 8 = -5 + 8$ Simplify both sides of the equation.

$x = 3$ Therefore, $x = 3$ is the solution.

CHECK:

The check is left as an exercise for the student.

e. $4 = 3(3x - 2) - 4(2x - 1)$ Apply the distributive property on the right side of the equation.

$4 = 9x - 6 - 8x + 4$ Simplify the right side.

$4 = x - 2$ Since 2 is the additive inverse of -2, add 2 to both sides of the equation.

$4 + 2 = x - 2 + 2$ Simplify both sides of the equation.

$6 = x$, or $x = 6$ Therefore, $x = 6$ is the solution.

CHECK:

The check is left as an exercise for the student.

Practice Exercises

Solve using the Addition Property:

3. $3 + t - 6 = 8 - 10$

4. $8b + 5 - 7b + 1 = -4$

5. $-5.7a + 4.6 + 6.7a = 16.5 - 9.2$

6. $6(x - 2) - 5x = -10$

7. $3 = 2(4x + 5) - 7(x + 2)$

If you need more practice, do the Additional Practice Exercises.

Additional Practice Exercises

Solve using the addition property:

c. $12 + s - 1 = 5$

d. $4b + 7 - 3b - 2 = -3$

e. $5.2x + 7.4 - 4.2x = 11.6$

f. $4(x + 3) - 3x = 7$

g. $6 = 5(x - 2) - 2(2x - 4)$

Translating from mathematics to English   In problem solving, we must be able to translate from English to mathematics. We will start by translating from mathematics to English and then translate from English to mathematics. The examples that follow, although they may have few applications in the real world, will serve to help you begin to think mathematically. Real-world examples will follow.

Answers:

Practice Exercises 3–7: 3. $t = 1$ 4. $b = -10$ 5. $a = 2.7$ 6. $x = 2$ 7. $x = 7$ Additional Practice Exercises c–g: c. $s = -6$ d. $b = -8$ e. $x = 4.2$ f. $x = -5$ 8. $x = 8$

Example 3

Translate each of the following from mathematics to English. There is more than one translation.

Mathematics	English
a. $t + 8 = 12$	Some number plus eight is twelve, or
	Eight added to some number is twelve, or
	The sum of some number and eight is twelve, or
	Eight more than some number is twelve, or
	A number increased by eight is twelve.
b. $y - 5 = 11$	Some number minus five is eleven, or
	Five subtracted from some number is eleven, or
	The difference of some number and five is eleven, or
	Five less than some number is eleven, or
	A number decreased by five equals eleven, or
	Some number less five is eleven.
c. $17 = u + 6$	Seventeen is some number plus six, or
	Seventeen equals six added to some number, or
	Seventeen equals a number increased by six, or
	Seventeen is six more than some number, or
	Seventeen is the sum of some number and six.
d. $4 = x - 21$	Four equals some number minus twenty-one, or
	Four is twenty-one subtracted from some number, or
	Four is some number decreased by twenty-one, or
	Four is twenty-one less than some number, or
	Four equals the difference of some number and twenty-one.

Practice Exercises

Translate from mathematical sentences into English sentences. There is more than one possible answer.

8. $r + 7 = 9$ **9.** $t - 4 = 17$ **10.** $15 = x + 6$ **11.** $28 = y - 19$

Example 4

Translating from English to mathematics.

Translate from English to mathematics and solve. Use x as the variable in each exercise.

English	Mathematics
a. A number plus one equals four.	$x + 1 = 4$
	$x + 1 - 1 = 4 - 1$
	$x = 3$
b. A number minus ten is fifteen.	$x - 10 = 15$
	$x - 10 + 10 = 15 + 10$
	$x = 25$
c. The sum of twelve and some number is twenty.	$12 + x = 20$
	$12 - 12 + x = 20 - 12$
	$x = 8$

Answers:

Practice Exercises 8–11: **8.** Seven plus some number is nine. **9.** Four subtracted from a number is seventeen. **10.** Fifteen is the sum of a number and six. **11.** Twenty-eight is the difference of some number and nineteen.

d. The difference of some number and thirteen is six.

$$x - 13 = 6$$
$$x - 13 + 13 = 6 + 13$$
$$x = 19$$

e. Some number added to eighteen is twenty-nine.

$$18 + x = 29$$
$$18 - 18 + x = 29 - 18$$
$$x = 11$$

Practice Exercises

Translate from English to mathematical sentences and solve. Use x as the variable.

12. The sum of a number and eleven is twenty-nine.

13. The difference of a number and six is fourteen.

14. A number increased by four is ten.

15. Eight subtracted from some number is twenty-three.

The following real-world problems may be solved without using algebra if you really think about them. Later on, however, the situations will be far too complicated to solve without the techniques of algebra. Consequently, even though you may be able to do these problems without algebra, use algebraic techniques as practice for the more complicated problems in later sections of this chapter. The procedure we will use in this section is outlined in the box that follows. As the problems become more complicated, we will refine the procedure.

Procedure for Solving Real-World Problems with One Variable

1. Identify the unknown and represent it with a variable.
2. Write a word equation relating the known and unknown quantities.
3. Write an algebraic equation using the word equation as a guide.
4. Solve the algebraic equation.
5. Check the solution in the wording of the original problem.

Example 5

Let x represent the unknown in each of the following. Write an equation and solve.

a. John scored ten points more on his algebra test than Mike. If John scored 93, what was Mike's score?

Solution:
1. Identify the unknown and represent it with a variable. The unknown is Mike's score, so let x represent Mike's score.

2. Write a word equation relating the known and unknown quantities.

Since John scored ten points more than Mike,

Mike's score	plus	ten points	equals	John's score

Answers:

Practice Exercises 12–15: **12.** $x + 11 = 29, x = 18$ **13.** $x - 6 = 14, x = 20$ **14.** $x + 4 = 10, x = 6$ **15.** $x - 8 = 23, x = 31$

3. Write an algebraic equation using the word equation as a guide.

Mike's score	plus	ten points	equals	John's score
x	$+$	10	$=$	93

4. Solve the equation.

$x + 10 = 93$ Subtract 10 from both sides of the equation (add -10).

$x + 10 - 10 = 93 - 10$ Simplify both sides.

$x = 83$ Therefore, Mike's score was 83.

5. Check the solution against the wording of the original problem.

Is John's score ten points more than Mike's score? Since 93 is 10 more than 83, the answer is yes and our solution is correct.

Note: Other correct equations are $93 - x = 10$ and $x = 93 - 10$.

b. The limit on the number of crappie you can catch in Florida is 50 per person per day. Alice has already caught 32. How many more can she catch until she has her limit?

Solution:

1. Identify the unknown and represent it with a variable. The unknown is how many more crappie Alice can catch, so let n represent the number of additional crappie she can catch.

2. Write a word equation relating the known and unknown quantities.

Since she has already caught 32 and she can catch 50,

The number Alice has already caught	plus	Additional number she can catch	equals	Fifty

3. Write an algebraic equation using the word equation as a guide.

The number Alice has already caught	plus	Additional number she can catch	equals	Fifty
32	$+$	n	$=$	50

4. Solve the equation.

$32 + n = 50$ Subtract 32 from both sides of the equation (add -32).

$32 - 32 + n = 50 - 32$ Simplify both sides.

$n = 18$ Therefore, Alice can catch 18 more.

5. Check the solution against the wording of the original problem.

Does the number that she has already caught (32) plus the additional number she can catch (18) equal the limit (50)? Yes. Therefore, the solution is correct.

Note: Other correct equations are $x = 50 - 32$ and $50 - x = 32$.

Practice Exercises

Let x represent the unknown in each of the following. Write an equation using the given word equation as a guide and solve.

16. The cost of a car is $2500 less than the cost of a mini-van. If the cost of the car is $18,350, find the cost of the minivan.

$$\boxed{\text{cost of minivan}} - \boxed{2500} = \boxed{\text{cost of car}}$$

17. Jenny got married six years ago. If she is presently 28 years old, how old was she when she got married?

$$\boxed{\text{age when she got married}} + \boxed{6} = \boxed{\text{present age}}$$

Exercise Set 3.1

Solve each of the following:

1. $t + 5 = 13$

2. $z + 9 = 12$

3. $r + 8 = 2$

4. $s + 9 = 4$

5. $a - 3 = 5$

6. $b - 6 = 2$

7. $x - 6.3 = 4.2$

8. $y - 5.7 = 3.6$

9. $m + 7.3 = 6.1$

10. $v + 8.9 = 4.3$

11. $7 + w - 1 = 15$

12. $16 + u - 3 = 23$

13. $18 - 15 + x = 10$

14. $20 - 12 + y = 17$

15. $z - 5 - 4 = -8 + 2$

16. $n - 5 - 3 = -10 + 3$

17. $5t + 7 - 4t = 1 + 5$

18. $3z - 5 - 2z = 1 - 6$

19. $6.1x - 1.3 - 5.1x = 6.8$

20. $4.3x - 3.7 - 3.3x = 8.1$

21. $7 + 5 = 6r + 8 - 5r$

22. $3 + 6 = 7a + 4 - 6a$

23. $-4 - 6 = -8x - 10 + 9x$

24. $-8 - 4 = -3x - 12 + 4x$

25. $4(x + 3) - 3x = 10$

26. $3(x + 5) - 2x = 12$

27. $7x - 3(2x + 4) = -6$

28. $11x - 5(2x + 1) = -2$

29. $16x - 5(3x - 2) - 7 = 5$

30. $17x - 4(4x - 3) - 8 = 6$

31. $3(3.4x - 1.2) - 9.2x = -6.2$

32. $4(2.4x - 1.6) - 8.6x = -3.3$

33. $-2(4.7y - 5.4) + 10.4y = 3.1$

34. $-3(1.6b - 2.4) + 5.8b = -6.2$

35. $6(t + 3) - 5(t + 4) = 10$

36. $8(w - 5) - 7(w - 4) = 15$

37. $3(3x - 5) - 2(4x - 3) = -5$

38. $5(5x - 3) - 6(4x - 2) = -3$

39. $-4(5x - 4) + 7(3x - 2) - 6 = 8 - 2$

40. $-6(4x - 2) + 5(5x - 3) - 9 = 9 - 3$

Challenge Exercises: (41–42)

41. $3.2(1.5x - 3.4) - 3.8x + 5.32 = -6.33$

42. $4.6(2.5x + 1.4) - 10.5x - 7.34 = -8.21$

Translate the following mathematical sentences into English in at least two ways:

43. $x - 5 = 19$

44. $22 - t = 11$

45. $13 + y = 7$

46. $25 = u + 8$

47. $17 = w - 6$

48. $44 = v + 9$

Translate the following English sentences into mathematical sentences and solve. Use x as the variable.

49. Some number minus two is equal to nine.

50. Twelve increased by some number is nineteen.

Answers:

Practice Exercises 16–17: **16.** $x - 2500 = 18,350, x = \$20,850$ **17.** $x + 6 = 28, x = 22$ years old

51. Fourteen subtracted from some number is eight.

52. Six subtracted from some number is nineteen.

53. The difference of a number and two is fifteen.

54. A number decreased by seven is thirty-two.

55. The sum of five and some number is thirteen.

56. Five more than some number is negative two.

57. Eight less than some number is negative ten.

58. The sum of a number and negative three is negative seven.

59. The difference of some number and two is negative eight.

Let x represent the unknown in each of the following. Write an equation and solve. A typical word equation that can be expressed as an equation of the form studied in this section is given to encourage you to solve the problem by using an equation rather than arithmetic.

60. On a trip, Francine drove 75 miles more than Hector drove. If Francine drove 260 miles, how far did Hector drive?

$$\boxed{\text{number of miles Hector drove}} + 75 = \boxed{\text{number of miles Francine drove}}$$

61. Shasta and Charlene shared the driving responsibilities on a trip. If the total trip was 550 miles and Shasta drove 275 miles, how far did Charlene drive?

$$\boxed{\text{number of miles Shasta drove}} + \boxed{\text{number of miles Charlene drove}} = \boxed{\text{total miles}}$$

62. Horace needs $465 to buy a new canoe. If he has already saved $312, how much more does he need?

$$\boxed{\text{amount Horace has}} + \boxed{\text{amount Horace needs}} = \boxed{\text{cost of canoe}}$$

63. Jenny and Bryan are going to chip in to buy their parents an anniversary gift. If the gift costs $160 and Bryan pays $75, how much does Jenny pay?

$$\boxed{\text{amount Jenny pays}} + \boxed{\text{amount Bryan pays}} = \boxed{\text{cost of gift}}$$

64. John and Jim shared expenses on a trip from Orlando, Florida to Atlanta, Georgia. John paid $30 less than Jim paid. If Jim paid $243, how much did John pay?

$$\boxed{\text{amount John paid}} + \boxed{30} = \boxed{\text{amount Jim paid}}$$

65. In a recent football game, the Dolphins scored 8 points fewer than the Bills scored. If the Dolphins scored 24 points, how many points did the Bills score?

$$\boxed{\text{number of points Bills scored}} - 8 = \boxed{\text{number of points Dolphins scored}}$$

66. A mixture contains 23 more gallons of water than of alcohol. If the mixture contains 47 gallons of water, how many gallons of alcohol does it contain?

$$\boxed{\text{number of gallons of alcohol}} + 23 = \boxed{\text{number of gallons of water}}$$

67. Connie needs a total of 800 points in order to get an A in her English I class. If she already has 632 points, how many more does she need?

$$\boxed{\text{number of points Connie has}} + \boxed{\text{number of points Connie needs}} = \boxed{\text{number of points needed for an A}}$$

68. The Orlando Magic scored 13 more points than the New Jersey Nets scored in a recent game. If the Nets scored 92 points, how many did the Magic score?

$$\boxed{\text{number of points Nets scored}} + 13 = \boxed{\text{number of points Magic scored}}$$

69. Pamika plans to buy a new TV. The TV costs $46 less at a wholesale store than at an electronics store. If the price of the TV is $326 at the electronics store, what is the price at the wholesale store?

$$\boxed{\text{price at wholesale store}} + 46 = \boxed{\text{price at electronics store}}$$

70. The selling price of a computer is $1195. If the markup is $210, find the store's cost.

$$\boxed{\text{store's cost}} + \boxed{\text{markup}} = \boxed{\text{selling price}}$$

71. Francine went on a diet and lost 23 pounds. If she weighed 118 pounds after the diet, what did she weigh before the diet?

$$\boxed{\text{weight before diet}} - \boxed{\text{weight loss}} = \boxed{\text{weight after diet}}$$

72. Mount Rainier in Washington state is 7726 feet higher than Mount Mitchell in North Carolina. If Mount Mitchell is 6684 feet high, find the height of Mount Rainier.

$$\boxed{\text{height of Mount Rainier}} - 7726 = \boxed{\text{height of Mount Mitchell}}$$

73. A local department store received $18,000 more from the sale of men's shirts than from the sale of men's pants. If they received $40,000 for the sale of shirts, how much did they receive from the sale of men's pants?

$$\boxed{\text{amount from sale of pants}} + 18,000 = \boxed{\text{amount from sale of shirts}}$$

74. A board is cut into two pieces. The longer piece is 5 feet in length and is 3 feet longer than the shorter piece. Find the length of the shorter piece.

$$\boxed{\text{length of shorter piece}} + 3 = \boxed{\text{length of longer piece}}$$

75. The perimeter of a triangle is 18 inches. If two of the sides have lengths of 6 inches and 8 inches, what is the length of the third side? ($P = a + b + c$)

76. If the perimeter of a triangle is 24 centimeters and two of the sides have lengths of 8 centimeters and 6 centimeters, what is the length of the third side? ($P = a + b + c$)

Writing Exercises:

77. What does it mean to say that a number "solves an equation"?

78. Why do we not need a "Subtraction Property of Equality"?

Writing Exercise or Group Project:

79. Write two application problems similar to 60−76 that require the use of the Addition Property of Equality. Exchange with another group or person and solve.

| Section 3.2 | Multiplication Property of Equality |

OBJECTIVES

When you complete this section, you will be able to:

a. Solve equations using the Multiplication Property of Equality.

b. Solve equations that require simplifications using the Multiplication Property of Equality.

c. Solve real-world problems.

Introduction   The Addition Property of Equality allows us to solve one type of equation. It does not, however, allow us to solve equations like $3x = 6$. If we subtracted 3 from both sides, we would get $3x - 3 = 3$, a more complicated equation than the one we began with. We need another technique to be able to solve this type of equation.

In Section 3.1, we found that we could add the same number to both sides of an equation and the equality was preserved. Would the equality be preserved if we multiply both sides of an equation by the same nonzero number?

$$6 = 6 \qquad \text{Multiply both sides by 3.}$$
$$3(6) \overset{?}{=} 3(6) \qquad \text{Simplify both sides.}$$
$$18 = 18 \qquad \text{Therefore, the equality was preserved when both sides were multiplied by 3.}$$

$$4 + 3 = 7 \qquad \text{Multiply both sides by 2.}$$
$$2(4 + 3) \overset{?}{=} 2(7) \qquad \text{Apply the distributive property.}$$
$$2(4) + 2(3) \overset{?}{=} 2(7) \qquad \text{Multiply before adding.}$$
$$8 + 6 \overset{?}{=} 14 \qquad \text{Add.}$$
$$14 = 14 \qquad \text{Therefore, the equality was preserved when both sides were multiplied by 2.}$$

$$2 \cdot 9 = 18 \qquad \text{Multiply both sides by } \tfrac{1}{6}.$$
$$\frac{1}{6} \cdot (2 \cdot 9) = \frac{1}{6} \cdot 18 \qquad 2 \cdot 9 = 18 \text{ and } \tfrac{1}{6} \cdot 18 = \tfrac{18}{6}.$$
$$\frac{18}{6} = \frac{18}{6} \qquad \text{Divide.}$$
$$3 = 3 \qquad \text{Therefore, the equality was preserved when both sides were divided by 6.}$$

Note: Multiplying by $\frac{1}{6}$ is the same as dividing by 6.

From the arithmetic equations above, it appears that if we multiply (or divide) both sides of an equation by any nonzero number the equality is preserved. We formally state this as the Multiplication Property of Equality.

Multiplication Property of Equality

If $a = b$, and $c \neq 0$, then $a \cdot c = b \cdot c$. In words, both sides of an equation may be multiplied (or divided) by any nonzero number and the result will still be an equation.

As with the Addition Property of Equality, when we apply the Multiplication Property of Equality to an equation with variables, the resulting equation is equivalent to the original.

To solve equations using the Multiplication Property of Equality, we need to recall two properties from Section 2.1. The first is the multiplicative inverse, also called the **reciprocal.** Two numbers are multiplicative inverses if their product is 1. For example, the multiplicative inverse (reciprocal) of a is $\frac{1}{a}$, because $\frac{1}{a}(a) = \frac{a}{a} = 1$. The multiplicative inverse of $\frac{a}{b}$ is $\frac{b}{a}$, because $\frac{a}{b} \cdot \frac{b}{a} = \frac{ab}{ba} = 1$. For example, the reciprocal of $\frac{3}{4}$

is $\frac{4}{3}$, because $\frac{3}{4} \cdot \frac{4}{3} = \frac{3 \cdot 4}{4 \cdot 3} = \frac{12}{12} = 1$. The second is the multiplication property of 1, which states that $1 \cdot x = x \cdot 1 = x$.

Suppose we are asked to solve the equation $3u = 15$. As in Section 3.1, the goal in solving this equation is to get the variable with a coefficient of 1 on one side of the equation and the constant on the other. This is done by multiplying both sides of the equation by the reciprocal of the coefficient of the variable.

Example 1

Solve the following.

a. $3u = 15$ Since $\frac{1}{3}$ is the reciprocal of 3, multiply both sides by $\frac{1}{3}$.

$\frac{1}{3} \cdot 3u = \frac{1}{3} \cdot 15$ $\frac{1}{3} \cdot 3 = 1$ and $\frac{1}{3} \cdot 15 = \frac{15}{3} = 5$.

$1(u) = 5$ $1 \cdot u = u$.

$u = 5$ Therefore, $u = 5$ is the solution.

CHECK:

$3u = 15$ Substitute 5 for u.

$3(5) = 15$ Multiply.

$15 = 15$ Therefore, 5 is the solution.

Note: Division by 3 is the same as multiplication by $\frac{1}{3}$. Consequently, we could have done Example 2a as follows:

$3u = 5$ Divide both sides by 3.

$\dfrac{3u}{3} = \dfrac{15}{3}$ $\frac{3}{3} = 1$ and $\frac{15}{3} = 5$.

$u = 5$ Therefore, 5 is the solution.

This is the method that we will use in future examples, when appropriate.

b. $-.6v = 1.8$ Divide both sides by $-.6$.

$\dfrac{-.6v}{-.6} = \dfrac{1.8}{-.6}$ $\frac{-.6}{-.6} = 1$ and $\frac{1.8}{-.6} = -3$.

$1 \cdot v = -3$ $1 \cdot v = v$.

$v = -3$ Therefore, $v = -3$ is the solution.

CHECK:

$-.6v = 1.8$ Substitute -3 for v.

$-.6(-3) = 1.8$ Multiply.

$1.8 = 1.8$ Therefore, -3 is the solution.

c. $\dfrac{t}{-4} = 7$ Rewrite as a product.

$\dfrac{1}{-4}t = 7$ Since -4 is the reciprocal of $\frac{1}{-4}$, multiply both sides by -4.

$-4 \cdot \dfrac{1}{-4}t = -4 \cdot 7$ $-4 \cdot \frac{1}{-4} = 1$ and $-4 \cdot 7 = -28$.

$1 \cdot t = -28$ $1 \cdot t = t$.

$t = -28$ Therefore, -28 is the solution.

CHECK:

$\dfrac{t}{-4} = 7$ Substitute -28 for t.

$\dfrac{-28}{-4} = 7$ Divide.

$7 = 7$ Therefore, -28 is the correct solution.

d. $\dfrac{3}{2}z = 60$ Since $\dfrac{2}{3}$ is the reciprocal of $\dfrac{3}{2}$, multiply both sides by $\dfrac{2}{3}$.

$\dfrac{2}{3} \cdot \dfrac{3}{2}z = \dfrac{2}{3} \cdot 60$ $\dfrac{2}{3} \cdot \dfrac{3}{2} = 1$ and $\dfrac{2}{3} \cdot 60 = \dfrac{2}{3} \cdot \dfrac{60}{1} = \dfrac{120}{3} = 40.$

$1 \cdot z = 40$ $1 \cdot z = z.$

$z = 40$ Therefore, $z = 40$ is the solution.

CHECK:

$\dfrac{3}{2}z = 60$ Substitute 40 for z.

$\dfrac{3}{2}(40) = 60$ Multiply.

$\dfrac{120}{2} = 60$ Divide.

$60 = 60$ Therefore, 40 is the solution.

e. $-x = 5$ $-x = -1 \cdot x.$ Multiply both sides by the reciprocal of -1, which is -1.

$-1(-x) = -1 \cdot 5$ Simplify both sides.

$x = -5$ Therefore, $x = -5$ is the solution.

CHECK:

$-x = 5$ Substitute -5 for x.

$-(-5) = 5$ Apply the double negative rule.

$5 = 5$ Therefore, -5 is the correct solution.

Practice Exercises

Solve the following:

1. $5v = -20$ **2.** $-.5y = 32.5$ **3.** $\dfrac{w}{2} = 6$

4. $-\dfrac{7}{9}u = 21$ **5.** $3 = -t$

If you need more practice, do the Additional Practice Exercises.

Additional Practice Exercises:

Solve the following.

a. $42 = 6n$ **b.** $.7p = -6.3$ **c.** $\dfrac{x}{-3} = 3$

d. $\dfrac{8}{5}s = -32$ **e.** $-x = 4$

When applying the Addition Property of Equality in the previous section, it was often necessary to simplify one or both sides of the equation. The same is true when the Multiplication Property of Equality is involved.

Answers:

Example 2

Solve each of the following. If the equation involves decimals, leave the answer as a decimal. Otherwise, when necessary, leave the answer as a fraction.

a. $8x + 4x = -48$ Simplify the left side of the equation.

$12x = -48$ Divide both sides by 12.

$\dfrac{12x}{12} = \dfrac{-48}{12}$ $\dfrac{12}{12} = 1$ and $\dfrac{-48}{12} = -4$.

$1 \cdot x = -4$ $1 \cdot x = x$.

$x = -4$ Therefore, $x = -4$ is the solution.

CHECK:

$8x + 4x = -48$ Substitute -4 for x.

$8(-4) + 4(-4) = -48$ Multiply.

$-32 - 16 = -48$ Add.

$-48 = -48$ Therefore, -4 is the solution.

b. $5y = 9 + 15$ Simplify the right side of the equation.

$5y = 24$ Divide both sides by 5.

$\dfrac{5y}{5} = \dfrac{24}{5}$ $\dfrac{5}{5} = 1$.

$1 \cdot y = \dfrac{24}{5}$ $1 \cdot y = y$.

$y = \dfrac{24}{5}$ or 2.8 Therefore, the solution is $y = \dfrac{24}{5}$.

CHECK:

The check is left as an exercise for the student.

c. $2.4z - 3.6z = 4.8$ Simplify the left side of the equation.

$-1.2z = 4.8$ Divide both sides by -1.2.

$\dfrac{-1.2z}{-1.2} = \dfrac{4.8}{-1.2}$ $\dfrac{-1.2}{-1.2} = 1$ and $\dfrac{4.8}{-1.2} = -4$.

$1 \cdot z = -4$ $1 - z = z$.

$z = -4$ Therefore, $t = -4$ is the solution.

CHECK:

The check is left as an exercise for the student.

Practice Exercises

Solve each of the following. If the equation involves decimals, leave the answer as a decimal. Otherwise, when necessary, leave the answer as a fraction.

6. $16t - 7t = 45$ **7.** $10q = 35 - 15$ **8.** $3.3 = 5.8s - 5.5s$

If you need more practice, do the Additional Practice Exercises.

Additional Practice Exercises

Solve using the multiplication property. If the equation involves decimals, leave the answer as a decimal. Otherwise, leave the answer as a fraction.

f. $2x + 4x = 24$ **g.** $2t = -22 + 36$ **h.** $5.7a - 2.3a = -10.2$

As in the previous section, we will translate from mathematics to English and from English to mathematics.

Example 3

Translate each of the following from mathematics to English. There may be more than one translation into English.

Mathematics	English
a. $2y = 8$	Twice a number is eight.
	Two times a number equals eight.
	The product of two and a number is eight.
b. $\dfrac{x}{4} = 5$	The quotient of some number and four is five.
	Some number divided by four is five.
c. $32 = -8t$	Thirty-two is negative eight times a number.
	Thirty-two is the product of negative eight and a number.
d. $.6s = 3$	Six-tenths of a number is three.
	The product of six-tenths and a number is three.
	Six-tenths times some number is three.

Practice Exercises

Translate from mathematical sentences to English sentences. There is more than one possible translation.

9. $9x = 45$ **10.** $\dfrac{v}{8} = 2$

11. $33 = -3w$ **12.** $.5z = -13$

If more practice is needed, do the Additional Practice Exercises.

Additional Practice Exercises

Translate from mathematical sentences to English sentences. There is more than one possible translation.

i. $7y = 28$ **j.** $\dfrac{x}{3} = -4$

k. $15 = -2x$ **l.** $.4t = 16$

Answers:

k. Fifteen is -2 times some number. **l.** Four-tenths of a number is sixteen. **Additional Practice Exercises f–l: i.** Seven times a number is twenty-eight. **j.** The quotient of a number and three is negative four. **10.** The quotient of a number and eight is two. **11.** Thirty-three is negative three times a number. **12.** Five-tenths of some number is negative thirteen. **Additional Practice Exercises f–l: f.** $x = 4$ **g.** $t = 7$ **h.** $a = -3$ **Practice Exercises 9–12: 9.** The product of nine and a number is forty-five.

Example 4

Translate from English to mathematics and solve. Let x be the variable in each case.

English	**Mathematics**

a. The product of four and a number is twenty.

$$4x = 20$$
$$\frac{4x}{4} = \frac{20}{4}$$
$$x = 5$$

b. The quotient of some number and three is negative six.

$$\frac{x}{3} = -6$$
$$3 \cdot \frac{x}{3} = 3(-6)$$
$$x = -18$$

c. Five and six-tenths equals negative ten times a number.

$$5.6 = -10x$$
$$\frac{5.6}{-10} = \frac{-10x}{-10}$$
$$-.56 = x$$

d. Eight-tenths of a number equals negative two and four-tenths.

$$.8x = -2.4$$
$$\frac{.8x}{.8} = \frac{-2.4}{.8}$$
$$x = -3$$

Practice Exercises

Translate from English to mathematical sentences and solve. Use x as the variable.

13. Five times a number is forty.

14. The quotient of some number and negative two is eight.

15. Negative four and two-tenths is negative seven times some number.

16. Seven-tenths of a number is twenty-one.

If you need more practice, do the Additional Practice Exercises.

Additional Practice Exercises

Translate from English to mathematical sentences and solve. Use x as the variable.

m. The product of six and a number is twelve.

n. The quotient of some number and negative three is five.

o. Three and five-tenths equals seven-tenths of a number.

As in the previous section, most of the following real-world problems can be solved without the use of algebra. However, we ask that you continue to follow the format used in Section 3.1. Sometimes it is necessary to convert a percent into a decimal. Recall that to do this you drop the percent sign and move the decimal point two places to the left. For example, 23% = .23 and 6% = .06.

Example 5

a. Jude earns $52.00 for working eight hours. What is his hourly wage?

Solution:

1. Identify the unknown and represent it with a variable. The unknown is Jude's hourly wage, so let w represent his hourly wage.

2. Write a word equation.

| number of hours Jude worked | times | his hourly wage | equals | total earnings |

3. Write an algebraic equation using the word equation as a guide.

| number of hours Jude worked | times | his hourly wage | equals | total earnings |
| 8 | · | w | = | 52 |

4. Solve the equation.

$8w = 52$ Divide both sides by 8.

$\dfrac{8w}{8} = \dfrac{52}{8}$ Simplify both sides.

$w = \$6.50$ Therefore, Jude earns $6.50 per hour.

5. Check the solution against the wording of the original problem.

If Jude worked eight hours at $6.50 per hour, his total earnings would be 8($6.50), or $52.00. Therefore, our solution is correct.

b. Sondra works as a salesperson. Her commission is 7% of her total sales. If she earned $245 one week, what were her total sales for that week?

Solution:

1. Identify the unknown and represent it with a variable. The unknown is Sondra's total sales, so let s represent her sales for the week.

2. Write a word equation.

| rate of commission | times | sales for the week | equals | earnings for the week |

3. Write an algebraic equation using the word equation as a guide. Remember, 7% = .07.

| rate of commission | times | sales for the week | equals | earnings for the week |
| .07 | · | s | = | 245 |

4. Solve the equation.

$.07s = 245$ Divide both sides by .07.

$\dfrac{.07s}{.07} = \dfrac{245}{.07}$ Simplify both sides.

$s = 3500$ Therefore, Sondra's sales for the week were $3500.

5. Check the solution against the wording of the original problem.

Sondra's earnings are equal to (.07)($3500), or $245. Therefore, our solution is correct.

c. A carpenter is installing wood for a hardwood floor. He knows that the area of the rectangular room is 252 square feet and the length of the room is 18 feet. What is the width of the room?

Solution:
This question is quite different from the preceding problems 5a and 5b in that we can apply a formula that will serve as our equation. Consequently, all we need to do is write down the formula, substitute for the unknowns, and solve the resulting equation. In this case, we know the area and the length of the rectangle and are asked to find the width.

$A = LW$ Formula for the area of a rectangle. Substitute for A and L.

$252 = 18W$ Divide both sides of the equation by 18.

$\dfrac{252}{18} = \dfrac{18W}{18}$ Simplify both sides.

$14 \text{ ft} = W$ Therefore, the width is 14 feet.

CHECK:

$A = LW$ Substitute for A, L, and W.

$252 = 18 \cdot 14$ Multiply 18 and 14.

$252 = 252$ Therefore, the solution is correct.

Practice Exercises

Solve the following:

17. Hernando drove 230 miles in five hours. What is his rate in miles per hour? (rate · time = distance)

18. A bank pays 6% interest per year on savings accounts. If Sally received $1080 interest on her savings account last year, how much does she have in savings?

19. A sign company is to construct a triangular sign whose area is 72 square feet. If the base of the sign is 12 feet, what is the height? $\left(A = \dfrac{bh}{2} \right)$

Exercise Set 3.2

Solve the following:

1. $7u = 56$

2. $5v = 70$

3. $-3n = 36$

4. $-62 = 2m$

5. $42 = -14x$

6. $-12y = 96$

7. $-64 = -16p$

8. $-75 = -15q$

9. $7u = 5.6$

10. $6v = 4.8$

11. $.3x = 15$

12. $.8y = 24$

13. $-.1t = -11$

14. $-.6m = -4.2$

15. $\dfrac{n}{2} = 8$

16. $\dfrac{r}{5} = 3$

17. $\dfrac{q}{-3} = 6$

18. $\dfrac{w}{-4} = 6$

19. $\dfrac{8}{7}z = -56$

20. $\dfrac{8}{9}q = -72$

21. $-\dfrac{2}{3}w = -6$

22. $-\dfrac{3}{4}x = -6$

23. $\dfrac{8m}{5} = -120$

24. $-91 = \dfrac{7n}{13}$

25. $\dfrac{2x}{3} = 6$

26. $\dfrac{3x}{5} = 10$

27. $4a = -16 + 8$

Answers:

28. $6b = -36 + 24$

29. $6y = 55 - 13$

30. $-4z = -39 + 15$

31. $3x + 5x = 24$

32. $4x + 7x = -22$

33. $9x - 3x = -24$

34. $11x - 4x = -28$

35. $4x - 7x = 15 - 6$

36. $5x - 9x = 12 - 8$

37. $-2.9u + 3.6u = 6.3$

38. $-4.6z + 2.2z = 4.8$

39. $3.4x + 5.1x = 30 - 4.5$

Translate the following mathematical sentences into English sentences in at least two ways:

40. $7t = 46$

41. $62 = 4z$

42. $-14 = 2h$

43. $-39 = 3k$

44. $5.6 = 7x$

45. $2.7y = 5.6$

Translate from English to mathematical sentences and solve. Use x as the variable.

46. Eight times a number is sixty-four.

47. Five times a number is thirty-five.

48. The product of six and some number is negative forty-eight.

49. The product of four and some number is negative twenty-eight.

50. Six-tenths of a number is twelve.

51. Four-tenths of a number equals two.

52. The quotient of a number and two is six.

53. The quotient of a number and four is five.

54. Some number divided by five is negative two.

55. Some number divided by negative three is eight.

56. The product of two and some number is divided by five. The result is ten. What is the number?

Solve each of the following:

57. John is to construct a rectangular sign that is 12 feet long and has an area of 96 ft². How wide is the sign? ($A = LW$)

58. Fred is buying carpet for his rectangular living room. He arrives at the carpet store and can remember only that the length of the room is 18 feet and the area is 216 square feet. Find the width of the room. ($A = LW$)

59. Paul is buying a triangular piece of land whose area is 1500 yd² and whose base is 100 yards. What is the height?

60. A triangular traffic sign has an area of 6 ft² and a height of 3 feet. What is the base of the sign?

61. Leslie has a piece of apple pie à la mode. The piece of pie has three times as many calories as the low-calorie ice cream served with it. If the pie has 450 calories, how many calories does the ice cream have?

62. Bill spent $14.50 for four fishing lures, all of which had the same price. To the nearest cent, how much did each cost?

63. Catalina worked 12 hours last week and earned $102. What is her hourly wage?

64. Susan is building a bookcase. How many 3.5-foot-long shelves can she cut from a board that is 14 feet long?

65. During a particularly boring speech, two-fifths of the audience left. If forty-six people left, how many people were originally in the audience?

66. Two-thirds of a class passed the last exam. If 18 people passed, how many students are in the class?

67. On her morning walk, Frankie passes a tree located at a point that is $\frac{1}{6}$ of the total distance she plans to walk from her house. If she has walked $\frac{2}{3}$ of a mile, how far does she plan to walk?

68. After a stock split, Harry will have $1\frac{1}{4}$ times as many shares of stock as he had before the split. If he has 225 shares after the split, how many shares did he have before the split?

69. Three-sevenths of the employees of a company earn at least $28,000 per year. If 123 people earn at least $28,000 per year, how many employees does the company have?

70. Katrina works as a salesperson and is paid a 6% commission on her sales. If she receives a $9000 commission on a sale, what was the amount of the sale?

71. The state sales tax in Florida is 6%. If the sales tax on a car purchased in Florida is $1050, what was the selling price of the car?

72. Teschima is renting a TV for $3.50 per week. If a new TV costs $280, how many weeks will it take for her rental fee to equal the cost of a new TV?

73. A wholesale club charges an annual fee of $80 that entitles its members to a 16% discount. How much would a member have to purchase per year for the savings to equal the amount of the membership fee?

74. A sign is in the shape of an isosceles triangle whose base is 16 inches long. If the area of the sign is 48 square inches, what is the height of the sign? ($A = \frac{1}{2} bh$)

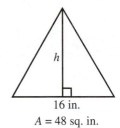

16 in.

$A = 48$ sq. in.

Writing Exercise:

75. Why do we not need a "Division Property of Equality"?

Writing Exercise or Group Project:

76. Write two application problems similar to exercises 57–74 that involve the use of the Multiplication Property of Equality. Exchange with another group or person and solve.

Section 3.3

Combining Properties in Solving Linear Equations

OBJECTIVES

When you complete this section, you will be able to:

a. Solve linear equations using both the Addition and Multiplication Properties of Equality.

b. Solve linear equations that require simplification of one or both sides using the Addition and/or Multiplication Properties of Equality.

Introduction   In this section, we will combine the Addition and Multiplication Properties of Equality to solve linear equations in one variable. For example, suppose we were asked to solve $3x + 9 = -6$. In order to solve for x, we must get rid of both 3 and 9. Which one do we get rid of first? The procedure for solving such equations is outlined in the following box:

Solving Linear Equations

1. If necessary, simplify both sides of the equation as much as possible. This could involve using the distributive, commutative, and associative properties and combining like terms.

2. If necessary, use the Addition Property of Equality to get all the terms with variables on one side of the equation and all the constant terms on the other.

3. If necessary, use the Multiplication Property of Equality to eliminate any coefficient on the variable other than one.

4. Check the answer in the original equation.

Up to this point, all the equations that we have solved in this chapter have had all the variables on one side. If, after simplifying both sides of the equation, there are terms with variables on both sides, use the Addition Property of Equality to get the terms with the variable on one side of the equation and the constants on the other. We still use the principle of the additive inverse to accomplish this. For example, the additive inverse of $2x$ is $-2x$, since $2x + (-2x) = 0$. In solving equations like $3x + 4 = -5x + 28$, we can eliminate any one of the four terms by adding its additive inverse to both sides of the equation. In Example 1b–e we will first eliminate a term with a variable and then a constant. When an equation has variable terms on both sides of the equation we will usually eliminate the variable term with the smaller coefficient in order to avoid negative coefficients in the next-to-last step, though this is not necessary.

Example 1

Solve the following:

a. $3x + 9 = -6$ Subtract 9 from both sides of the equation.

$3x + 9 - 9 = -6 - 9$ Simplify both sides.

$3x = -15$ Divide both sides by 3.

$\dfrac{3x}{3} = \dfrac{-15}{3}$ Simplify both sides.

$x = -5$ Therefore, the solution is $x = -5$.

CHECK:

$3x + 9 = -6$ Substitute -5 for x.

$3(-5) + 9 = -6$ Multiply before adding.

$-15 + 9 = -6$ Add.

$-6 = -6$ Therefore, -5 is the correct solution.

b. $3x + 4 = -5x + 28$ We could eliminate any of the four terms.
We will eliminate $-5x$ by adding its inverse, $5x$, to both sides of the equation.

$3x + 5x + 4 = -5x + 5x + 28$ Simplify both sides.

$8x + 4 = 28$ Subtract 4 from both sides.

$8x + 4 - 4 = 28 - 4$ Simplify both sides.

$8x = 24$ Divide both sides by 8.

$\dfrac{8x}{8} = \dfrac{24}{8}$ Simplify both sides.

$x = 3$ Therefore, $x = 3$ is the solution.

CHECK:

$3x + 4 = -5x + 28$ Substitute 3 for x.

$3(3) + 4 = -5(3) + 28$ Multiply before adding.

$9 + 4 = -15 + 28$ Add.

$13 = 13$ Therefore, 3 is the solution.

c. $12z - 3.8 = 5z + .4$ We could eliminate any of the four terms.
We will eliminate $5z$ by adding its inverse, $-5z$, to both sides of the equation.

$12z - 5z - 3.8 = 5z - 5z + .4$ Simplify both sides.

$7z - 3.8 = .4$ Add 3.8 to both sides.

$7z - 3.8 + 3.8 = .4 + 3.8$ Simplify both sides.

$7z = 4.2$ Divide both sides by 7.

$\dfrac{7z}{7} = \dfrac{4.2}{7}$ Simplify both sides.

$z = .6$ Therefore, $z = .6$ is the solution.

CHECK:

$12z - 3.8 = 5z + .4$	Substitute .6 for z.
$12(.6) - 3.8 = 5(.6) + .4$	Multiply before adding.
$7.2 - 3.8 = 3.0 + .4$	Add.
$3.4 = 3.4$	Therefore, .6 is the solution.

d.

$9y - 5 - 8y = 5y + 10 - 3y$	Simplify both sides of the equation by adding like terms.
$y - 5 = 2y + 10$	Subtract y from both sides.
$y - y - 5 = 2y - y + 10$	Simplify both sides.
$-5 = y + 10$	Subtract 10 from both sides.
$-5 - 10 = y + 10 - 10$	Simplify both sides.
$-15 = y$	Therefore, $y = -15$ is the solution.

The check is left as an exercise for the student.

e.

$7(r - 1) + 10 = 5(2 + r)$	Apply the distributive property.
$7r - 7 + 10 = 10 + 5r$	Simplify both sides of the equation by adding like terms.
$7r + 3 = 10 + 5r$	To eliminate $5r$, subtract $5r$ from both sides.
$7r - 5r + 3 = 10 + 5r - 5r$	Simplify both sides.
$2r + 3 = 10$	Subtract 3 from both sides.
$2r + 3 - 3 = 10 - 3$	Simplify both sides.
$2r = 7$	Divide both sides by 2.
$\dfrac{2r}{2} = \dfrac{7}{2}$	Simplify both sides.
$r = \dfrac{7}{2}$ or 3.5	Therefore, $r = \dfrac{7}{2}$ is the solution.

The check is left as an exercise for the student.

Practice Exercises

Solve each of the following:

1. $4x - 7 = 9$

2. $2r - 4 = 11 - 3r$

3. $.7 - t = 3t - 2.1$

4. $9 - 6r - 3 - 7r = 2r + 2 - 11r$

5. $4(r + 2) = 3(r + 4) + 6$

If you need more practice, do the Additional Practice Exercises.

Additional Practice Exercises

Solve the following:

a. $5x + 2 = 12$

b. $4x - 15 = 5 - 6x$

c. $2.3 - .4w = .5w - 2.2$

d. $3z + 5 + 5z = 2z - 1$

e. $6(p - 4) - 3p = 18 - 3p$

Note: The more complicated equations involving fractions are discussed in Chapter 7, when solving equations with rational expressions is discussed.

In solving equations, there are two situations that require special attention. These are **identities** and **contradictions**. An *identity* is an equation that is true for all values of the variable for which the equation is defined. When solving an equation that is an identity, we arrive at a statement that is obviously true, such as $5 = 5$. Since any number solves the identity, we will denote the solutions as "all real numbers." A *contradiction* is an equation that has no solution. When solving an equation that is a contradiction, we arrive at a statement that is obviously false, such as $3 = 10$. The fact that there is no solution is usually indicated by using the symbol $\varnothing$, which stands for the empty set, the set that has no elements.

Example 2

Determine whether each equation is an identity or a contradiction and indicate the solutions.

a. $3(x + 2) - 4 = x + 2x + 2$ Simplify each side of the equation.

$3x + 6 - 4 = 3x + 2$ Add like terms.

$3x + 2 = 3x + 2$ Subtract $3x$ from both sides.

$3x - 3x + 2 = 3x - 3x + 2$ Simplify both sides.

$2 = 2$ This is obviously a true statement. Therefore, this is an identity and the solutions are all real numbers.

b. $4x - 3(x + 5) = 2(2x - 3) - 3(x + 5)$ Simplify both sides of the equation.

$4x - 3x - 15 = 4x - 6 - 3x - 15$ Add like terms.

$x - 15 = x - 21$ Subtract x from both sides.

$x - x - 15 = x - x - 21$ Add like terms.

$-15 = -21$ This is obviously a false statement. Therefore, this is a contradiction and there are no solutions. Equivalently, the solution set is $\varnothing$.

Practice Exercises

Determine whether each equation is an identity or a contradiction and indicate the solution or solution set.

6. $2(2x + 4) - 11 = 4(x - 2) + 5$

7. $4(x - 1) + 2(x - 3) = 4(x - 2) + 2x$

We continue to practice translating mathematical sentences into English and English sentences into mathematics.

Example 3

Translate each of the following mathematical sentences into English sentences. There may be more than one translation.

Mathematics	**English**
a. $2x + 3 = 5$	Three more than twice a number equals five.
	Twice a number plus three is five.
	The sum of twice a number and three is five.
	Twice a number increased by three is five.
	Three more than the product of two and a number is five.

Answers:

Practice Exercises 6–7: **6.** Identity, all real numbers **7.** Contradiction, no solutions, or $\varnothing$.

b. $4u - 12 = 10u$ — Four times a number minus twelve equals ten times the number.

Four times a number decreased by twelve is ten times the number.

The difference between four times a number and twelve equals ten times the number.

Twelve less than four times a number is ten times that number.

c. $8(10 - t) = 96$ — Eight times the difference of ten and a number is ninety-six.

The product of eight and the difference of ten and some number is ninety-six.

The difference of ten and a number multiplied by eight is ninety-six.

d. $1.5 = .75v - .5v$ — One and five-tenths equals seventy-five hundredths of a number decreased by five-tenths of that number.

One and five-tenths is the difference of seventy-five hundredths of a number and five-tenths of that number.

One and five-tenths equals seventy-five hundredths of a number minus five-tenths of that number.

Practice Exercises

Translate each of the following mathematical sentences into English sentences. There is more than one translation.

8. $3w - 7 = 4$ **9.** $5x + 6 = 7x$ **10.** $12(3 - m) = 48$ **11.** $x + .06x = 24$

If more practice is needed, do the Additional Practice Exercises.

Additional Practice Exercises:

Translate each of the following mathematical sentences into English sentences. There may be more than one translation.

f. $15 - 5z = 10$ **g.** $5x + 6 = 7x$ **h.** $26 = 2(a - 4)$ **i.** $.5u - 12 = .4$

Example 4

Translate the following English sentences into mathematical sentences and solve. Let x be the variable in each case.

English	**Mathematics**
a. Six more than four times a number equals ten.	$4x + 6 = 10$
	$4x + 6 - 6 = 10 - 6$
	$4x = 4$
	$x = 1$

Note: The numbers are not always written in the same order in mathematics as in English. "Five subtracted from three times a number" is translated "$3x - 5$."

b. Five subtracted from three times a number equals four times that number.

$$3x - 5 = 4x$$
$$3x - 3x - 5 = 4x - 3x$$
$$-5 = x$$

c. Twice the sum of some number and five is thirty.

$$2(x + 5) = 30$$
$$2x + 2(5) = 30$$
$$2x + 10 = 30$$
$$2x + 10 - 10 = 30 - 10$$
$$2x = 20$$
$$x = 10$$

d. The total of a number, seven-tenths of the number, and negative two-tenths of the number equals six-tenths.

$$x + .7x - .2x = .6$$
$$1.7x - .2x$$
$$1.5x = .6$$
$$x = .4$$

Practice Exercises

Translate the following English sentences into mathematical sentences and solve. Use x as the variable.

12. Eight less than three times a number is sixteen.

13. Eighteen decreased by five times a number is equal to four times the number.

14. Four times the difference of a number and five is thirty.

15. Twenty-one equals a number decreased by three-tenths of the number.

If you need more practice, do the Additional Practice Exercises.

Additional Practice Exercises:

Translate the following English sentences into mathematical sentences and solve.

j. Five more than three times a number is negative four.

k. Six times the difference of a number and one is twenty-four.

l. The product of eight and a number decreased by six is fourteen.

m. The sum of six hundredths of a number and the number is seven and forty-two hundredths.

Follow the same technique as in the previous two sections to solve the following:

Example 5

Solve the following:

a. An appliance repairman charges a service charge of $25 and a base rate of $18 per hour. If the charges for repairing a refrigerator are $79 (excluding parts), how many hours did the job take?

Solution:
1. Identify the unknown and represent it with a variable. The unknown is the number of hours it took to repair the refrigerator, so let h represent the number of hours it took to repair the refrigerator.

Answers:

Additional Practice Exercises j–m: j. $3x + 5 = -4, x = -3$ k. $6(x - 1) = 24, x = 5$ l. $8(x - 6) = 14, x = 14 \frac{4}{31}$ or 7.75 m. $.06x + x = 7.42, x = 7$

Practice Exercises 12–15: 12. $3x - 8 = 16, x = 8$ 13. $18 - 5x = 4x, x = 2$ 14. $4(x - 5) = 30, x = 30 \frac{2}{25}$ or 12.5 15. $21 = x - .3x, x = 30$

2. Write the word equation.

| service charge | plus | charge per hour | times | number of hours | equals | total charges |

3. Write the algebraic equation.

| service charge | plus | charge per hour | times | number of hours | equals | total charges |

$$25 \quad + \quad 18 \quad \cdot \quad h \quad = \quad 79$$

4. Solve the equation.

$25 + 18h = 79$	Subtract 25 from both sides.
$25 - 25 + 18h = 79 - 25$	Simplify both sides.
$18h = 54$	Divide both sides by 18.
$\dfrac{18h}{18} = \dfrac{54}{18}$	Simplify both sides.
$h = 3$	Therefore, it took 3 hours.

5. Check the answer against the wording of the problem.

If the repairman works three hours, his charges would be $3(18) = \$54$, and $\$54 + \$25 = \$79$. Therefore, the answer is correct.

b. If Pascual had $10 more than he presently has, then twice that amount would be enough to purchase a compact disc player that sells for $280. How much does he presently have?

Solution:

1. Identify the unknown and represent it with a variable. The unknown is the amount of money Pascual presently has, so let a represent the amount he presently has. Then $a + 10$ represents the amount he would have if he had $10 more.

2. Write a word equation.

| two | times | $10 more than he presently has | equals | 280 |

3. Write the algebraic equation.

| two | times | $10 more than he presently has | equals | 280 |

$$2 \quad \cdot \quad (a + 10) \quad = \quad 280$$

4. Solve the equation.

$2(a + 10) = 280$	Apply the distributive property.
$2a + 20 = 280$	Subtract 20 from both sides.
$2a + 20 - 20 = 280 - 20$	Simplify both sides.
$2a = 260$	Divide both sides by 2.
$\dfrac{2a}{2} = \dfrac{260}{2}$	Simplify both sides.
$a = 130$	Therefore, Pascual presently has $130.

5. Check the answer against the wording of the problem.

If Pascual has $130, then he would have $140 if he had $10 more. Since $2(\$140) = \$280$, the answer is correct.

c. A rancher needs to fence in a corral whose perimeter is 2000 feet and whose length is 600 feet. How much fencing does she need for the width? ($P = 2L + 2W$)

Solution:

This question is quite different from the preceding 5a and 5b in that we can apply a formula that will serve as our equation. Consequently, all we need to do is write the formula for the perimeter of a rectangle, substitute for the variables, and solve the resulting equation. We know that $P = 2000$ feet and $L = 600$ feet.

$P = 2L + 2W$	Substitute for P and L.
$2000 = 2(600) + 2W$	Simplify the left side of the equation.
$2000 = 1200 + 2W$	Subtract 1200 from both sides.
$2000 - 1200 = 1200 - 1200 + 2W$	Simplify both sides.
$800 = 2W$	Divide both sides by 2.
$\dfrac{800}{2} = \dfrac{2W}{2}$	Simplify both sides.
$400 = W$	Therefore, the width is 400 feet.

CHECK:

$P = 2L + 2W$	Substitute for P, L, and W.
$2000 = 2(600) + 2(400)$	Multiply on the right.
$2000 = 1200 + 800$	Add on the right.
$2000 = 2000$	Therefore, the solution is correct.

Practice Exercises

Solve the following:

16. A TV repair service charged $22 per hour plus $43 for parts. If the bill for repairing a TV was $76, how many hours did the repair take?

17. Carla earned $390 last week for 40 hours of work. Included in the $390 was a $50 bonus. What was her hourly wage?

18. A box manufacturer receives an order for some boxes, each of which has a surface area of 292 in.², and a length of 8 inches, and a width of 6 inches. What is the height of the boxes? ($SA = 2LW + 2LH + 2WH$.)

Exercise Set 3.3

Solve the following:

1. $3x + 5 = 11$

2. $5x + 7 = -8$

3. $-7x - 6 = 8$

4. $-3x + 5 = -10$

5. $17 = 4a - 3$

6. $-8 = 6x + 4$

7. $5x - 8 = 4x - 5$

8. $9y + 4 = 8y + 1$

9. $10 - 3u = 12 - 4u$

10. $3r + 5 = 2r + 4$

11. $7 - 9v = v - 3$

12. $4w + 8 = 15 - 6w$

13. $3n - 6 = 2n - 6$

14. $2m + 13 = 4 - 8m$

15. $9p + 3.4 = 8p + 1.4$

16. $.4q + .7 = .8 - .6q$

17. $3v + 7 + 2v = 18 - 1$

18. $6w + 4 - 8w = -14 + 46$

19. $4k + 5 = 7k - 7 + 9k$

20. $-11b - 5 + 5b = 23 - 10$

21. $3a - 4a + 9 = -12a + 18 - 6a$

22. $17z - 11z - 17 = -4z + 13 - 5z$

23. $2.3y - 1.6 - .8y = 4.4$

24. $4.7 - 5.2c - .3c = -6.3$

25. $n + 5 = 3(n + 7)$

26. $2t - 9 = 3(t - 2)$

27. $2(5r - 2) = 9r + 1$

28. $2(w - 4) = 5w - 14$

29. $6u - 17 = 4(u + 3) - 3$

30. $2v + 9 = 11 + 7(v - 1)$

31. $14(w - 2) + 13 = 4w + 5$

32. $2x + 2(3x - 4) = -23 + 5x$

33. $2y + 13(y - 1) = 5y + 12$

34. $8(2z - 3) - z = 5z - 34$

35. $.2(8x - 90) + .9x = .1x + 6$

36. $1.4u - .5(44 - 6u) = 66 + 2.4u$

37. $.3n + .2(4n - 1) = .4 + .5n$

38. $.3z + .7(2z + 1) = 2.9 - .5z$

39. $7(p - 3) = 3(1 - p) - 4$

40. $5(3q - 2) + 20 = 5(q - 4)$

41. $15(n + 2) = 5(n + 4) - 10$

42. $8(m - 5) = 7(m - 5) - 9$

43. $5(u - 2) + 7(3 - u) = 9u$

44. $5(h + 2) - 7(h - 1) = -h$

45. $4(b - 3) + 2(b + 2) = 2(b - 2)$

46. $3(v - 8) + 2(v + 5) = 7(v - 4)$

47. $5(a - 1) - 9(a - 2) = -3(2a + 1) - 2$

48. $4(k + 4) + 13(k - 1) = -3(k - 3) + 14$

49. $9 - .2(v - 8) = .3(6v - 8)$

50. $.5(t - 6) + .4(1 - t) = -3.4$

Determine whether the following are identities or contradictions and indicate the solutions.

51. $2x + 4 = 2(x + 1) + 2$

52. $3(x + 1) - 2 = 3x + 5$

53. $3(2x - 5) - 4x = 2(x - 3) - 9$

54. $2(4x - 7) + 10 = 2(x - 5) + 2(3x + 3)$

55. $4(2x - 1) - 3(x + 5) = 5(x - 2) + 7$

56. $3(x - 7) - 4(x - 3) = 5 - (x + 3)$

Challenge Exercises: (57–58)

Solve the following:

57. $.4(x + 2) = .3(x - 5) - .5(x - 3) + 2$

58. $1.8(x - 2) + 5(x - 5) + .8 = -.4(8x + 7)$

Translate each of the following mathematical sentences into English sentences in at least two ways.

59. $4w - 3 = 7$

60. $5x + 8 = 19$

61. $6v - 9 = 3v$

62. $16n - 8 = 6n$

63. $7y + 2.8 = 6.3$

64. $q + .05q = 45$

65. $12(y - 1) = 9$

66. $4(t + 3) = 36$

Translate the following English sentences into mathematical sentences and solve. Use x as the variable.

67. Eight less than three times a number equals thirteen.

68. One is five more than twice a number.

69. The difference of five times a number and six is nine.

70. The sum of three times a number and seven is four.

71. Zero is equal to nine times the sum of a number and three.

72. The product of four and the difference of a number and five is three.

Solve the following:

75. A plumber charges $30 for a service call and $45 per hour base rate. Her charge for a repair job was $210, excluding parts. How many hours did the plumber work?

76. An appliance repairman charges $25 dollars for a service call plus his hourly wage. If a repair job costs $109 (excluding parts) and took three hours, what is his hourly wage?

77. An auto repair service charges $35 per hour plus parts. If the parts to repair a transmission cost $160 and the total bill was $335, how many hours did the job take?

78. A TV repair shop charges an hourly rate plus the cost of parts. Find the hourly rate for a job that took one and one-half hours if the parts cost $65 and the bill was $113.

79. If Yo Chen had $15 more than she presently has, then with three times that amount she could buy a camera which costs $525. How much does she presently have?

80. Five times $20 less than Bryan presently has is enough to buy a new bike that costs $300. How much does Bryan presently have?

81. Frank earned $395 last week for working 40 hours. Included in the $395 was a bonus of $25. Find his hourly wage.

82. Susie earned $277 for working 28 hours. Included in the $277 was a $60 bonus. Find her hourly wage.

83. A particular copy machine requires 15 cents for the first copy and 4 cents for each additional copy. How many copies can be made for $4.95?

84. Joan uses a coupon in the paper for $25 off the cost of repairs at Joe's Garage. If her bill was $147.50, including $60 for parts and three hours of labor, how much did Joe charge per hour for labor?

85. A local department store offers $15 off on purchases totaling $100 or more. Ahmed purchases four sports

73. Four times the difference of three times a number and two is sixteen.

74. The product of seven and four more than a number is sixty-three.

coats, each costing the same, and a silk tie that costs $28. Find the cost of one sports coat if the bill was $355.

86. Hernando earns $1\frac{1}{4}$ times his hourly wage on all hours worked in excess of 40 hours per week. Last week he worked 46 hours and earned $391.40. What is his hourly wage?

87. Various long-distance phone companies have different plans. Company A has a $5.95 charge per month plus 10¢ per minute with 30 free minutes per month. Company B charges $4.95 per month and 8¢ per minute with no free minutes. Find the number of minutes per month so that the charges would be the same for Company A as for Company B.

88. A country club offers two types of memberships. Plan A charges $500 per year plus $15 per round of golf, and Plan B charges $675 per year plus $10 per round of golf. How many rounds of golf would have to be played per year for the changes under Plans A and B to be equal?

89. Saline solution has 1 part salt to 8 parts water by weight. Find the number of ounces of salt and the number of ounces of water in 108 ounces of the solution.

90. A rectangular patio has a perimeter of 74 feet and a length of 22 feet. What is the width? ($P = 2L + 2W$)

91. A rectangular parking lot has a perimeter of 2000 yards and a width of 450 yards. What is the length? ($P = 2L + 2W$)

92. A storage tank is in the shape of a rectangular solid. The surface area is 236 ft^2, the length is 8 feet, and the width is 6 feet. What is the height? ($SA = 2LW + 2LH + 2WH$)

93. A shipping container is in the shape of a rectangular solid and has a surface area of 376 ft^2. If the container is 10 feet long and 6 feet high, find the width. ($SA = 2LW + 2LH + 2WH$)

Challenge Exercises: (94–95)

94. The area of a trapezoid is 76 ft², with a height of 12 feet. One base has length of 12 feet. What is the length of the other base? $(A = \frac{1}{2}h(B + b))$

95. The surface area of a right circular cylinder is 48π in.² and the radius of the base is 3 inches. Find the height.

Writing Exercises:

96. When solving an equation of the form $ax + b = c$, why do we eliminate b before eliminating a?

97. How does simplifying the expression $2(x - 3) + 4(x - 6)$ differ from solving the equation $2(x - 3) + 4(x - 6) = 0$?

Writing Exercise or Group Project:

If done as a group project, write two application problems, exchange with another group, and then solve.

98. Write and solve an application problem whose equation will be in the form $ax + b = c$.

| Section 3.4 | Solving Linear Inequalities |

OBJECTIVES

When you complete this section, you will be able to:

a. Solve linear inequalities using the Addition Property of Inequality and graph the solutions.

b. Solve linear inequalities using the Multiplication Property of Inequality and graph the solutions.

c. Solve linear inequalities using both the Addition and Multiplication Properties of Inequality and graph the solutions.

Introduction

In Sections 3.1–3.3, we learned to solve linear equations. In this section, we will solve linear inequalities. Before we discuss solving linear inequalities, we will expand on the concept of inequality and how inequalities can be represented on the number line.

The number line represents the integers and all of the numbers between the integers. The numbers between the integers are the **rational numbers** (to be discussed in Chapter 6) and the **irrational numbers** (to be discussed in Chapter 10).

Choose a number, say 3. We put a dot on 3 on the number line. Consider some unknown number x. There are three possibilities for the position of x in relation to 3 on the number line.

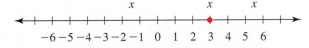

1. x is to the right of 3, so $x > 3$, or

2. x is right on top of 3, so $x = 3$, or

3. x is to the left of 3, so $x < 3$.

The number line has been divided into three parts, $x > 3$, $x = 3$, and $x < 3$. Expressions like $x > 3$ and $x < 3$ are called **inequalities.** Those numbers that make an inequality true are **solutions of the inequality.** One convenient method of representing the solutions of an inequality is by graphing the solutions on the number line. The expression $x \geq 3$ means $x > 3$, or $x = 3$. The graph of $x = 3$ consists of a dot on 3, and the graph of $x > 3$ consists of everything to the right of 3. So the graph of $x \geq 3$ is a dot on 3 with everything to the right shaded. An open dot is used to indicate that the point is not a part of the solution. So the graph of $x > 3$ is an open dot on 3 with everything to the right shaded.

Example 1

Graph each of the following on the number line.

a. $x > 5$

Since any number greater than 5 is to the right of 5 on the number line, we put an open dot on 5 (to show that 5 is not a solution) and shade the number line to the right of 5. The shaded region represents all values of x such that $x > 5$, and the arrow indicates that it extends to the right forever.

b. $x < 5$

The condition $x < 5$ is drawn in much the same way as for $x > 5$. The numbers less than 5 are to the left of 5 on the number line. We put an open dot on 5 (to show that 5 is not a solution) and then shade to the left of 5 on the number line. The shaded region extends forever, as indicated by the arrow.

c. $x \geq -4$

The numbers greater than -4 are to the right of -4 on the number line. We put a solid dot on -4 (to indicate that -4 is a solution) and shade the number line to the right of -4. The arrow indicates the region extends forever.

Practice Exercises

Graph the following on the number line:

1. $x > 3$

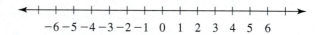

2. $x < -1$

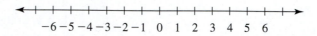

3. $x \leq -3$

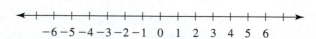

4. $x \geq 1$

Inequalities of the form $a < x < b$ are called **compound inequalities**. The compound inequality $2 < x < 5$ is read "two is less than x and x is less than 5." On the number line, 2 is to the left of x and x is to the left of 5. Consequently, x is any number between 2 and 5. The graph of this compound inequality is shown as follows:

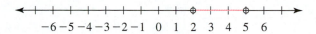

The open dots indicate that neither 2 nor 5 are solutions.

Example 2

Graph the following on the number line:

a. $0 \le x < 5$

Put a solid dot on 0 (to indicate that 0 is a solution) and an open dot on 5 (to indicate that 5 is not a solution). Since 0 is to the left of x and x is to the left of 5, x can be any number between 0 and 5. Therefore, shade the region between 0 and 5.

b. $-5 \le x \le 4$

Put a solid dot on both -5 and 4 to indicate that they are solutions. Since -5 is to the left of x and x is to the left of 4, x can be any number between -5 and 4. Therefore, shade the region between -5 and 4.

Practice Exercises

Graph the following:

5. $-5 < x < 0$

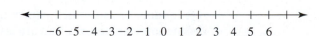

6. $3 < x < 6$

Now we are ready to discuss solving linear inequalities. To solve linear equations, we established the Addition and Multiplication Properties of Equality. To solve linear inequalities, we will need to establish similar properties for inequalities.

Example 3

a. $3 < 8$ Add 5 to both sides of the inequality.

$3 + 5 \overset{?}{<} 8 + 5$ Simplify both sides.

$8 < 13$ This is a true statement, so adding 5 to both sides of the inequality had no effect on the order symbol.

---

b. $9 > 7$

$9 - 3 \overset{?}{>} 7 - 3$ Subtract 3 from both sides of the inequality. (Same as adding -3.)

Simplify both sides.

$6 > 4$ This is a true statement, so subtracting 3 from both sides of the inequality had no effect on the order symbol.

c. $-12 < -2$

$-12 + 8 \overset{?}{<} -2 + 8$ Add 8 to both sides of the inequality.

Simplify both sides.

$-4 < 6$ This is a true statement, so adding 8 to both sides of the inequality had no effect on the order symbol.

Example 3 leads us to the following statement of the Addition Property of Inequality:

Addition Property of Inequality

For all real numbers a, b, and c, if $a < b$, then $a + c < b + c$ and if $a > b$, then $a + c > b + c$. In words, any real number may be added to (or subtracted from) both sides of an inequality without affecting the order.

In other words, adding the same number to both sides of an inequality results in an equivalent inequality. The Addition Property of Inequality is also true for the inequalities "$\geq$" and "$\leq$."

As in solving equations with the Addition Property of Equality, we wish to get the variable by itself on one side of the inequality and the constant on the other side. It is often easier to understand the solutions of an inequality if we graph them, so we will graph the solutions of each of the following examples:

Example 4

Solve the following and graph the solutions:

a. $x - 6 < 3$ Add 6 to both sides. Adding the same quantity to both sides of an inequality does not change the order symbol.

$x - 6 + 6 < 3 + 6$ Simplify both sides.

$x < 9$ Therefore, the solutions are all numbers less than 9.

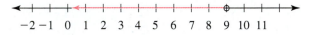

$$-2 \; -1 \; 0 \; 1 \; 2 \; 3 \; 4 \; 5 \; 6 \; 7 \; 8 \; 9 \; 10 \; 11$$

Unfortunately, there is no easy way to check the solution because there are an infinite number of values of x which make the inequality true. However, we may choose some values of x that are less than 9 and some that are greater than 9 and substitute them into the original inequality to get an indication of whether our solution is correct.

Let $x = 8$. From $x - 6 < 3$, we have $8 - 6 < 3$ or $2 < 3$. This is a true statement so, $x = 8$ is a solution and $8 < 9$.

On the other hand, let $x = 10$. Now we have $10 - 6 < 3$, or $4 < 3$. This not true, so $x = 10$ is not a solution and $10 > 9$.

Therefore, values less than 9 seem to make the inequality true and values greater than 9 seem to make it false. Hence, the solution is reasonable.

b. $2(y - 3) - y \geq -3$ Apply the distributive property.

$2y - 6 - y \geq -3$ Simplify the left side.

$y - 6 \geq -3$ Add 6 to both sides. Adding the same quantity to both sides of an inequality does not change the order symbol.

$y - 6 + 6 \geq -3 + 6$ Simplify both sides.

$y \geq 3$ Therefore, the solutions are all numbers greater than or equal to 3.

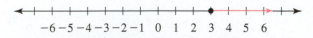

$$-6 \; -5 \; -4 \; -3 \; -2 \; -1 \; 0 \; 1 \; 2 \; 3 \; 4 \; 5 \; 6$$

c. $-2(2x - 4) + 5(x + 3) > 11$ Apply the distributive property.

$-4x + 8 + 5x + 15 > 11$ Simplify the left side of the inequality.

$x + 23 > 11$ Subtract 23 from both sides. Subtracting the same quantity from both sides of an inequality does not change the order symbol.

$x + 23 - 23 > 11 - 23$ Simplify both sides.

$x > -12$ Therefore, the solutions are all numbers greater than -12.

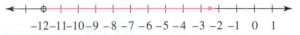

$-12\,-11\,-10\,-9\,-8\,-7\,-6\,-5\,-4\,-3\,-2\,-1\,0\,1$

Practice Exercises

Solve the following inequalities and graph the solutions:

7. $t + 3 > 1$ **8.** $3(s - 2) - 2s \leq 3$ **9.** $5(2t - 3) - 3(3t - 4) > 2$

If you need more practice, do the Additional Practice Exercises.

Additional Practice Exercises

Solve the following inequalities and graph the solutions:

a. $a - 8 + 2 < 0$ **b.** $6(u + 2) - 5u < 9$ **c.** $4(4z - 2) - 5(3z - 1) \leq 4$

The Addition Property of Inequality cannot be used to solve all linear inequalities, just as the Addition Property of Equality alone will not solve all linear equations. In order to solve an inequality like $3x \leq 9$, we need to divide both sides by 3. What effect, if any, will this have on the order symbol? We investigate the effect of multiplication and division in the following examples:

Example 5

a. $4 < 9$ Multiply both sides of the inequality by 2.

$2(4) \overset{?}{<} 2(9)$ Simplify both sides.

$8 < 18$ This is a true statement, so multiplying both sides by 2 had no effect on the order symbol.

b. $5 < 7$ Multiply both sides of the inequality by -3.

$(-3)5 \overset{?}{<} (-3)7$ Simplify both sides.

$-15 < -21$ This is a false statement. In order to make it true, the $<$ symbol must be changed to a $>$ symbol.

$-15 > -21$

c. $6 > -1$ Multiply both sides of the inequality by 4.

$4(6) \overset{?}{>} 4(-1)$ Simplify both sides.

$24 > -4$ This is a true statement, so multiplying both sides by 4 had no effect on the order symbol.

d. $8 \geq -9$ Multiply both sides of the inequality by -5.

$8(-5) \overset{?}{\geq} -9(-5)$ Simplify both sides.

$-40 \geq 45$ This is a false statement. In order to make it true, the $\geq$ symbol must be changed to a $\leq$ symbol.

$-40 \leq 45$

Answers:

c. $z \leq 7$

Additional Practice Exercises a–c: **a.** $a < 6$

b. $u > -3$

9. $t < 5$

8. $s \leq 9$

Practice Exercises 7–9: 7. $t > -2$

Example 5 suggests the following: If we multiply both sides of an inequality by a positive number, as in Examples 5a and 5c, the inequality symbol remains unchanged. However, if we multiply an inequality by a negative number, as in Examples 5b and 5d, the inequality symbol changes (reverses). In other words, what was greater became less and what was less became greater. When the order symbol is unchanged, we sometimes say the sense of the inequality was not changed, and when the order symbol is changed, we say the sense of the inequality is reversed. These results are formally stated in the following properties:

Multiplication Properties of Inequality

Positive Factor:

For all real numbers a, b, and c, if $a < b$ and $c > 0$, then $a \cdot c < b \cdot c$, and if $a > b$, then $a \cdot c > b \cdot c$. In words, if both sides of an inequality are multiplied (or divided) by a positive number, then the order symbol (sense) is unchanged.

Negative Factor:

For all real numbers a, b, and c, if $a < b$ and $c < 0$, then $a \cdot c > b \cdot c$, and if $a > b$ and $c < 0$, then $a \cdot c < b \cdot c$. In words, if both sides of an inequality are multiplied (or divided) by a negative number, then the order symbol (sense) must be reversed.

---

*The important thing to remember is that the direction of the inequality changes when we **multiply (or divide)** both sides of the **inequality** by a **negative number**. Otherwise, solving linear inequalities is **exactly the same** as solving linear equations.*

Just as in the addition properties, the multiplication properties are true for the inequalities "$\geq$" and "$\leq$," as well as for "$>$" and "$<$."

Example 6

Solve and graph the solution of each of the following:

a. $4x \geq 20$ Divide both sides of the inequality by 4. **Dividing** by a **positive** number **does not** change the direction of the inequality symbol.

$\dfrac{4x}{4} \geq \dfrac{20}{4}$ Simplify both sides.

$x \geq 5$ Therefore, the solutions are all numbers greater than or equal to 5.

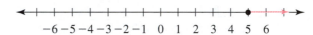

b. $-6y > 18$ Divide both sides of the inequality by -6. **Dividing** by a **negative** number **reverses** the direction of the order symbol.

$\dfrac{-6y}{-6} < \dfrac{18}{-6}$ Simplify both sides.

$y < -3$ Therefore, the solutions are all numbers less than -3.

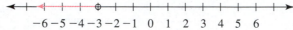

c. $\dfrac{x}{-2} \geq 1$ Multiply both sides of the inequality by -2. **Multiplying** by a **negative** number **reverses** the direction of the order symbol.

$-2 \cdot \dfrac{x}{-2} \leq -2 \cdot 1$ Simplify both sides.

$x \leq -2$ Therefore, the solutions are all numbers less than or equal to -2.

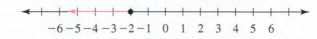

Practice Exercises

Solve and graph each of the following:

10. $2z > 4$ **11.** $-5n < 35$ **12.** $\dfrac{u}{-3} > 2$

If more practice is needed, do the Additional Practice Exercises

Additional Practice Exercises

Solve and graph each of the following:

d. $12 \geq 4v$ **e.** $-3r \geq 27$ **f.** $\dfrac{x}{-4} \leq -3$

Up to this point, we have discussed linear inequalities that require the use of only one of the properties of inequality. The following examples require both the Addition and Multiplication Properties of Inequality:

Example 7

Solve each of the following and graph the solutions:

a. $3x + 5 > x - 3$ Subtract x from both sides of the inequality. **Subtracting** from both sides **does not** affect the order symbol.

$3x - x + 5 > x - x - 3$ Simplify both sides.

$2x + 5 > -3$ Subtract 5 from both sides. **Subtracting** from both sides **does not** affect the order symbol.

$2x + 5 - 5 > -3 - 5$ Simplify both sides.

$2x > -8$ Divide both sides by 2. **Dividing** by a **positive** number **does not** affect the order symbol.

$\dfrac{2x}{2} > \dfrac{-8}{2}$ Simplify both sides.

$x > -4$ Therefore, the solutions are all numbers greater than -4.

←—+—+—+—⊕—+—+—+—+—+—+—+▸+—+—+—+—→
−6 −5 −4 −3 −2 −1 0 1 2 3 4 5 6

b. $16x + 8 - 4x > -16 + 10x$ Simplify both sides of the inequality.

$12x + 8 > -16 + 10x$ Subtract $12x$ from both sides. **Subtracting** from both sides **does not** affect the order symbol.

$12x - 12x + 8 > -16 + 10x - 12x$ Simplify both sides.

$8 > -16 - 2x$ Add 16 to both sides. **Adding** to both sides **does not** affect the order symbol.

$8 + 16 > -16 + 16 - 2x$ Simplify both sides.

$24 > -2x$ Divide both sides by -2. **Dividing** by a **negative** number **reverses** the order symbol.

$\dfrac{24}{-2} < \dfrac{-2x}{-2}$ Simplify both sides.

$-12 < x$ or $x > -12$ Therefore, the solutions are all numbers greater than -12.

←—+—⊕—+—+—+—+—+—+—+—+▸+—+—+—+—→
−12 −11 −10 −9 −8 −7 −6 −5 −4 −3 −2 −1 0 1

Answers:

$$c. \quad -2(y - 5) + 3y \geq 5(y + 6) + y$$

Distribute -2 and 5.

$$-2y + 10 + 3y \geq 5y + 30 + y$$

Simplify both sides of the inequality.

$$10 + y \geq 6y + 30$$

Subtract $6y$ from both sides. **Subtracting** from both sides **does not** affect the order symbol.

$$10 + y - 6y \geq 6y - 6y + 30$$

Simplify both sides.

$$10 - 5y \geq 30$$

Subtract 10 from both sides. **Subtracting** from both sides **does not** affect the order symbol.

$$10 - 10 - 5y \geq 30 - 10$$

Simplify both sides.

$$-5y \geq 20$$

Divide both sides by -5. **Dividing** by a **negative** number **reverses** the order symbol.

$$\frac{-5y}{-5} \leq \frac{20}{-5}$$

Simplify both sides.

$$y \leq -4$$

Therefore, the solutions are all numbers less than or equal to -4.

Practice Exercises

Solve the following and graph the solutions:

13. $5x - 3 > x - 15$

14. $4 - 3x \leq -8 + x$

15. $-4t + 7 + 16t > 10t + 12$

16. $20r - 5(6 + 2r) \leq 4r + 4(r - 7)$

If you need more practice, do the Additional Practice Exercises.

Additional Practice Exercises

Solve the following:

g. $2t - 6 > 14 - 3t$

h. $6 - 2x \leq -4 + 3x$

i. $8z - 3 + 2z \leq 10z - 10 - 7z$

j. $7(-8 + u) - 2u \geq -14 + 3(2u - 15)$

Compound inequalities are solved using the same techniques as for other linear inequalities. The variable is located between the inequality symbols. Our task is to eliminate the constants so the variable will be isolated between the inequality symbols. It is important to remember that whatever you do to the middle part to isolate the variable must be done to all three parts of the inequality.

Example 8

Solve each of the following inequalities:

a. $-3 \leq x + 2 \leq 3$

Subtract 2 from all three expressions.
Subtracting does not affect the order symbol.

$$-3 - 2 \leq x + 2 - 2 \leq 3 - 2$$

Simplify all three expressions.

$$-5 \leq x \leq 1$$

Therefore, the solutions are all numbers greater than or equal to -5 and less than or equal to 1.

Answers:

$\text{h. } x \geq 2 \quad \text{i. } z \leq -1 \quad \text{j. } u \leq 3$

Practice Exercises 13–16: $13. \ x > -3 \quad 14. \ x \geq 3 \quad 15. \ t > 2.5 \quad 16. \ t \leq 1$ **Additional Practice Exercises 8–j:** $8. \ t > 4$

b. $1 < 6x + 7 \le 31$

Subtract 7 from all three expressions. **Subtracting does not** affect the order symbol.

$1 - 7 < 6x + 7 - 7 \le 31 - 7$ Simplify all expressions.

$-6 < 6x \le 24$

Divide each expression by 6. **Dividing** by a **positive** number **does not** affect the order symbols.

$\dfrac{-6}{6} < \dfrac{6x}{6} \le \dfrac{24}{6}$ Simplify all expressions.

$-1 < x \le 4$

Therefore, the solutions are all numbers greater than -1 and less than or equal to 4.

c. $0 \le -1.5y - 3 < 4.5$

Add 3 to all three expressions. **Adding does not** affect the order symbols.

$0 + 3 \le -1.5y - 3 + 3 < 4.5 + 3$ Simplify all expressions.

$3 \le -1.5y < 7.5$

Divide each expression by -1.5. **Dividing** by a **negative** number **reverses** the order symbols.

$\dfrac{3}{-1.5} \ge \dfrac{-1.5y}{-1.5} > \dfrac{7.5}{-1.5}$ Simplify all expressions.

$-2 \ge y > -5$ or $-5 < y \le -2$

Therefore, the solutions are all numbers greater than -5 and less than or equal to -2.

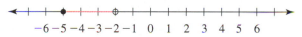

Practice Exercises

Solve the following inequalities:

17. $-4 \le x - 3 \le 2$ **18.** $-13 \le -4t - 5 < 15$ **19.** $3 < .2w + 3.6 < 3.6$

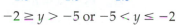

If more practice is needed, do the Additional Practice Exercises.

Additional Practice Exercises

Solve the following inequalities:

k. $3 < x + 5 < 7$ **l.** $-8 < 3v + 4 \le 19$ **m.** $-2 < .5x - 2.5 < -1$

Problems are often posed in terms of inequalities as well as in terms of equations. Therefore, we need to be able to translate these relationships from mathematics to English as well.

Example 9

Translate each of the following from mathematics to English. There can be more than one translation.

Mathematics	English
a. $4y > 12$	Four times a number is greater than twelve.
	The product of four and a number is more than twelve.

Answers:

b. $2x - 5 \le 20$ Twice a number, decreased by five, is less than or equal to twenty.
Twice a number, decreased by five, is at most twenty.
The difference of twice a number and five is less than or equal to twenty.
Five less than twice a number is at most 20.

c. $3(u + 6) < -9$ Three times the sum of a number and 6 is less than negative nine.
The product of three and the sum of a number and six is less than negative nine.

d. $5(2v - 4) + 3 \ge 8$ Five times the difference of twice a number and four increased by three is greater than or equal to eight.
The product of five and the difference of two times a number and four increased by three is at least eight.

Practice Exercises

Translate from mathematical sentences to English sentences:

20. $8x < 4$

21. $15 > 3t + 6$

22. $2(x - 4) > 3$

23. $4(3x + 2) - 5 \le 7$

Example 10

Translate from English to mathematics and solve. There is only one mathematical translation for each English sentence. Use x as the variable.

English	**Mathematics**
a. Six times a number is greater than eighteen.	$6x > 18$
	$x > 3$
b. Five more than negative three times a number is less than seventeen.	$-3x + 5 < 17$
	$-3x + 5 - 5 < 17 - 5$
	$-3x < 12$
	$\dfrac{-3x}{-3} > \dfrac{12}{-3}$
	$x > -4$
c. Seven times the difference of a number and four is at most thirty-five.	$7(x - 4) \le 35$
	$7x - 28 \le 35$
	$7x - 28 + 28 \le 35 + 28$
	$7x \le 63$
	$\dfrac{7x}{7} \le \dfrac{63}{7}$
	$x \le 9$

Answers:

Practice Exercises 20–23: (*There may be more than one English sentence for each exercise.*) **20.** Eight times a number is less than four. **21.** Fifteen is greater than six more than three times a number. **22.** Two times the difference of some number and four is greater than three. **23.** The product of four and the sum of three times some number and two, decreased by five, is at most seven.

d. Three times the difference of a number and
four, increased by five is at least eleven.

$$3(x - 4) + 5 \geq 11$$
$$3x - 12 + 5 \geq 11$$
$$3x - 7 \geq 11$$
$$3x - 7 + 7 \geq 11 + 7$$
$$3x \geq 18$$
$$\frac{3x}{3} \geq \frac{18}{3}$$
$$x \geq 6$$

Practice Exercises

Translate from English to mathematical sentences and solve. Use x as the variable.

24. The sum of twice a number and nine is less than
seventeen.

25. The difference of twelve and four times a number is
at most twenty.

26. The difference of seven times a number and fourteen is at least zero.

27. Sixty-three is greater than three times the sum of
four times a number and one.

Exercise Set 3.4

Graph each of the following:

1. $x > -2$

2. $x < -3$

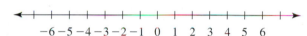

3. $x > 4$

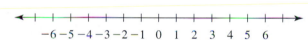

4. $x > -6$

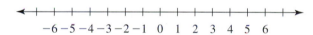

5. $x \geq 3$

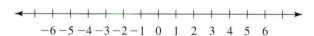

6. $-2 \leq x$

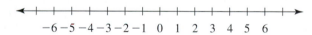

7. $1 < x < 6$

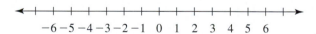

8. $-3 < x < 2$

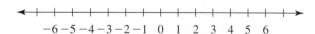

9. $-5 < x < 3$

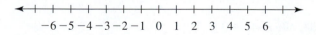

10. $-6 < x < -1$

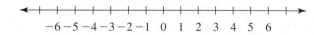

11. $-4 \leq x < -3$

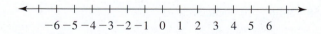

12. $2 < x \leq 6$

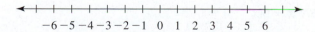

Solve by using the Addition Property of Inequality and graph the solutions.

13. $q + 6 < 4$

14. $v - 8 < 5$

15. $6p - 5 - 5p \leq -3$

16. $4w - 2 - 3w \geq 6$

17. $3p - 6 - 2p < -8$

18. $5q + 30 - 4q \leq 14$

19. $3(s + 2) - 2s \geq 7$

20. $4(s - 2) - 3s \leq -3$

21. $4(2x - 5) - 7x + 16 > -2$

22. $2(3y - 5) - 5y + 6 < 2$

23. $8(2u - 3) - 5(3u + 4) > -3$

24. $7(3x - 6) - 4(5x - 6) > -10$

25. $-3(9t - 3) + 4(7t - 2) \leq 3 - 5$

26. $-4(8t + 5) + 3(11t + 5) \leq 4 - 7$

Solve by using the Multiplication Property of Inequality.

27. $3y < 15$

28. $4t \geq 32$

29. $-14u > 42$

30. $16w \leq -64$

31. $-2a < -14$

32. $-6n \geq -5.4$

33. $\dfrac{x}{-3} < 5$

34. $\dfrac{u}{-2} \geq 4$

35. $\dfrac{r}{-1} > -2$

36. $\dfrac{b}{-6} < -1$

Solve by using the Addition and Multiplication Properties of Inequality.

37. $2q - 16 > 8$

38. $3v - 13 < 6$

39. $-4a + 5 \leq -7$

40. $-5r + 2 \geq -8$

41. $8p - 15 \geq 5p$

42. $5v + 4 \leq 2v$

43. $7w + 6 > 4w - 18$

44. $11z - 6 > 4z + 8$

45. $3a - 1.6 \leq .8 + a$

46. $4b - 2.4 \geq 9b + .6$

47. $12 - .3t > -1.1t + 36$

48. $6.8 - 7r \geq 3r - .2$

49. $10s - 2 - 12s < 8s + 2 - 6s$

50. $7z + 16 - 4z > 10 - z + 6$

51. $12p - 9p + 12 \geq 7p - 15 + 5p$

52. $15q + 9 - 6q > 4q + 27 + 11q$

53. $7(w - 1) \geq -12$

54. $9(y + 4) \leq 0$

55. $5(2m - 3) > m + 21$

56. $4(3n + 6) < 9n - 3$

57. $x + 6(x - 3) \leq -25$

58. $2(4w - 5) - 3w < 18$

59. $7u - 10(2u + 3) \geq 22$

60. $12(t - 5) - 6t > 3t$

61. $2(1 - 4u) < -3u + 7$

62. $9v - 24 \geq 6(5 - 3v)$

63. $7(n - 2) > 3(n + 4) - 26$

64. $5(m + 4) < 28 - 4(m + 2)$

65. $8(3z - 2) + 15 \le 11z + 2(10 - 4z)$

66. $10s - 3(4 - 5s) > s + 6(4 + s)$

67. $3(r + .9) + .2r > 3r + 1.7$

68. $.7(t - 3) + .8 < 1.5 + .3t$

69. $-2 < x - 6 < 4$

70. $6 < y + 7 < 12$

71. $-6 \le 2y \le 8$

72. $-4 \le 4y \le 12$

73. $-11 \le 4x - 3 < 9$

74. $-3 < 5y - 8 \le 12$

75. $3 < -2y + 5 \le 7$

76. $7 \le -3y - 2 < 16$

77. $-3.5 < 2t - 1.5 < 8.5$

78. $-1 \le 10s - 3 \le 5$

Challenge Exercises: (79–80)

79. $6(3y - 5) + 4(2y - 7) \ge 8y - 13 + 9y$

80. $-5(3 - 2x) + 9(3x - 6) < 7(4x - 1) + 1$

Translate the following mathematical sentences into English sentences in at least two ways:

81. $5x \ge 10$

82. $9 > 2y$

83. $3s - 5 > 2$

84. $15 + 4z < 12$

85. $28 \le 4(t - 2)$

86. $31 \ge 9(2 - u)$

Translate the following English sentences into mathematical sentences and solve. Use x as the variable.

87. Six times some number is less than eighteen.

88. Thirty-six is at least negative three times a number.

89. Nineteen decreased by four times a number is greater than eighteen.

90. Eight more than five times a number is at most twelve.

91. Three times the sum of twice a number and seven is at least negative three.

92. Seventeen is greater than or equal to twice the difference of a number and five.

93. Four times a number decreased by six is less than eighteen.

94. Twenty-two is less than or equal to five times a number, plus seven.

95. Four times the difference of a number and three is at most eight.

96. Three times the sum of a number and two is at most nine.

97. Three more than two times the difference of a number and one is at most five.

98. Three less than four times the sum of a number and one is at most eleven.

Writing Exercises:

99. List at least five situations where we use inequalities but do not specifically state the inequality. For example, "You must be twenty-one to vote," means your age must be greater than or equal to twenty-one years.

100. Compare solving a linear equation with solving a linear inequality. How are they similar? How do they differ?

101. How does the solution(s) of a linear equation compare with the solution(s) of a linear inequality?

Critical-Thinking Exercise:

102. Recall that for any two numbers x and a, either $x < a$, $x = a$, or $x > a$. With these in mind, answer the following questions:

   a. What does "not greater than" mean?

   b. What does "not equal to" mean?

   c. What does "not less than or equal to" mean?

   d. What does "at least" mean?

| Section 3.5 | Traditional Application Problems |

OBJECTIVES

When you complete this section, you will be able to:

a. Solve real-world problems that are general in nature.

b. Solve real-world problems involving consecutive , consecutive odd, and consecutive even integers.

c. Solve real-world problems involving distance, time, and rate.

Introduction

This is the first of two sections of application problems of the type traditionally found in algebra books. You will probably encounter few such problems in your everyday life. Their main purpose is to help you think mathematically and to see the power of algebra in problem solving.

In previous sections of this chapter, we solved real-world problems that involved only one unknown. In this section, we will solve several types of application problems, many of which involve two or more unknowns. If there are two or more unknowns, we modify our procedure slightly.

Procedure for Solving Real-World Problems

1. Identify the unknown(s) and represent one of them with a variable. It is *usually* best to let the variable represent the smaller or smallest quantity.

2. Represent all other unknowns in the problem in terms of this variable.

3. Write a word equation that gives the relationship between the known and unknown quantities.

4. Write an algebraic equation using the word equation as a guide.

5. Solve the algebraic equation and answer the question.

6. Be sure that you have answered the question that was asked.

7. Check the solution against the wording of the original problem.

If there are two unknowns in a real-world problem, there must be two conditions in the problem: one that gives the relationship between the unknowns and one that gives the information necessary to write the equation.

Often there are special approaches that are used for particular types of real-world problems. In solving application problems, it is often necessary to change percents into decimals—something that you have no doubt done before. You will recall that the traditional procedure is to drop the % sign and move the decimal two places to the left. For example, 23% = .23, 6% = .06, and 182% = 1.82. A thorough discussion of percents is found in Section 8.3.

General Problems

Example 1

---

a. Last week, Alena earned $23 more than George did at their respective part-time jobs. Together, they earned $145. How much did Alena earn?

Solution:

1. Identify the unknown(s). We are asked to find how much Alena earned. However, we also do not know how much George earned. George earned less than Alena, so:

Let n represent the amount that George earned.

2. Represent all other unknowns in terms of the same variable. What condition in the problem gives the relationship between the amounts that George and Alena earned? We are told that Alena earned $23 more than George earned. Since we let n represent George's total earnings, 23 more than n is $n + 23$, so:

$n + 23$ represents the amount that Alena earned.

3. Write a word equation. What condition gives us the information needed to write the equation? We are told that together they earned $145. Consequently,

$$\boxed{\text{amount Alena earned}} + \boxed{\text{amount George earned}} = \boxed{\text{amount earned together}}$$

4. Write an algebraic equation using the "word" equation as a guide.

$$\boxed{\text{amount Alena earned}} + \boxed{\text{amount George earned}} = \boxed{\text{amount earned together}}$$
$$(n + 23) \qquad + \qquad n \qquad = \qquad 145$$

5. Solve the equation.

$(n + 23) + n = 145$	Remove the parentheses.
$n + 23 + n = 145$	Add like terms.
$2n + 23 = 145$	Subtract 23 from both sides.
$2n = 122$	Divide both sides by 2.
$n = 61$	Solution of the equation.

6. Have we answered the question asked? We were asked to find the amount that Alena earned, but n represents the amount that George earned. The amount that Alena earned is $n + 23 = 61 + 23 = \$84$.

7. Check the answer against the wording of the problem. Is the amount that Alena earned $23 more than George earned? Yes, since $84 is $23 more than $61. Is the total of their earnings $145? Yes, since $61 + $84 = $145. Therefore, the answers are correct.

b. Pablo works as a salesperson in a department store and is paid $175 per week plus 6% of his sales. Last week he earned $505. What was the amount of his sales?

1. Identify the unknown(s). We are asked to find the amount of his sales. Let x represent the amount of Pablo's sales.

2. Represent all other unknowns in terms of the same variable. There are no other unknowns in this problem.

3. Write a word equation. What condition gives us the information needed to write the equation? We know that Pablo earns $175 per week plus 6% of his sales. Consequently,

$$\boxed{175} + \boxed{6\% \text{ of his sales}} = \boxed{\text{amount earned}}$$

4. Write an algebraic equation using the word equation as a guide. In order to represent 6% of his sales, we need to change 6% to .06. When using percents, "of" means "to multiply." Consequently, since x represents the amount of Pablo's sales, 6% of his sales is $.06x$.

$$\boxed{175} + \boxed{6\% \text{ of his sales}} = \boxed{\text{amount earned}}$$
$$175 + \qquad .06x \qquad = \qquad 505$$

5. Solve the equation.

$$175 + .06x = 505$$ Subtract 175 from each side of the equation.

$$175 - 175 + .06x = 505 - 175$$ Simplify each side of the equation.

$$.06x = 330$$ Divide both sides by .06.

$$\frac{.06x}{.06} = \frac{330}{.06}$$ Simplify both sides of the equation.

$$x = 5500$$ Solution of the equation.

6. Have we answered the question that was asked? We were asked to find the amount of Pablo's sales, and x represented the amount of his sales. Therefore, we have answered the question.

7. Check the answer against the wording of the problem. If the amount of his sales is $5500, then he earns $.06(5500) = \$330$. The amount he earned for the week is $\$330 + \$175 = \$505$. Therefore, our solution is correct.

Practice Exercises

1. Jude purchased a saw and a drill. If the total cost was $238, and the drill cost $54 less than the saw, find the cost of the saw.

2. Felicia is paid $215 per month plus 8% of her sales. Last month she earned $1655. What was the amount of her sales?

Consecutive Integer Problems

Consecutive integers are integers that follow one another in order. For example, 1, 2, 3, 4, and so on, are consecutive integers. To get the next largest consecutive integer, we add 1 to the previous integer. For example, if we begin with 1, then $1 + 1 = 2, 2 + 1 = 3$, $3 + 1 = 4$, and so on. Therefore, if we let $n =$ some integer, then the second integer is $n + 1$, the third is $(n + 1) + 1 = n + 2$, the fourth is $(n + 2) + 1 = n + 2 + 1 = n + 3$, and so on.

Consecutive odd integers are every other integer starting with an odd integer. For example, 1, 3, 5, 7, and so on, are consecutive odd integers. To get the next largest consecutive odd integer, we add 2 to the previous odd integer. Therefore, if we let $n =$ the first odd integer, then the second odd integer is $n + 2$, the third odd integer is $(n + 2) + 2 = n + 4$, and the fourth odd integer is $(n + 4) + 2 = n + 6$, and so on.

Consecutive even integers are every other integer starting with an even integer. For example, 2, 4, 6, 8, and so on, are consecutive even integers. Again, to get the next largest consecutive even integer, we add 2 to the previous even integer. Hence, if we let $n =$ the first even integer, then $n + 2$ is the second even integer, $n + 4$ is the third even integer, $n + 6$ is the fourth even integer, and so on.

Let us put all three types of integers together for comparison.

Integer	Consecutive	Consecutive Odd	Consecutive Even
First	n	n	n
Second	$n + 1$	$n + 2$	$n + 2$
Third	$n + 2$	$n + 4$	$n + 4$
Fourth	$n + 3$	$n + 6$	$n + 6$
and so on	and so on	and so on	and so on

Note: Do not be confused because both consecutive odd and consecutive even integers are represented in the same way in the preceding chart. The difference is in whether the first integer is odd or even. If the first integer is odd, then that integer plus 2 is also odd. Likewise for even integers.

Example 2

---

a. The sum of three consecutive odd integers is -9. Find the integers.

Solution:

1. Identify the unknown(s). We are asked to find three consecutive odd integers.

   Let n represent the (smallest) of the three consecutive odd integers.

2. Represent the other unknown(s) in terms of the same variable. What condition in the problem describes the relationship between the unknowns? The integers are all both consecutive and odd. Consequently,

   $n + 2$ represents the second integer and
   $n + 4$ represents the third.

3. Write a word equation. What condition gives the information needed to write the equation? We are told that the sum of the three consecutive odd integers is -9. Thus,

   $$\boxed{\text{first integer}} + \boxed{\text{second integer}} + \boxed{\text{third integer}} = \boxed{-9}$$

4. Write an algebraic equation using the word equation as a guide.

   $$\boxed{\text{first integer}} + \boxed{\text{second integer}} + \boxed{\text{third integer}} = \boxed{-9}$$
   $$n \quad + \quad (n+2) \quad + \quad (n+4) \quad = -9$$

5. Solve the equation.

$n + (n + 2) + (n + 4) = -9$	Remove the parentheses.
$n + n + 2 + n + 4 = -9$	Add like terms.
$3n + 6 = -9$	Subtract 6 from both sides of the equation.
$3n = -15$	Divide both sides by 3.
$n = -5$	Solution of the equation.

6. Have we answered the question that was asked? No, since we were asked to find three consecutive odd integers. Therefore, the complete solution is $n = -5$, $n + 2 = -3$, and $n + 4 = -1$.

7. Check the answers against the wording of the original problem. Are -1, -3, and -5 consecutive odd integers? Yes. Is the sum of -1, -3, and -5 equal to -9? Yes. Therefore, we have the correct solution.

b. Three times the larger of two consecutive even integers is 10 less than five times the smaller. Find the larger integer.

Solution:

1. Identify the unknown(s). We are looking for two consecutive even integers. Let n represent the first (smaller) of the two consecutive integers.

2. Represent all other variables in terms of the same variable. What condition in the problem gives the information needed to represent the other unknown in terms of n? We know the integers are consecutive even integers. Therefore,

   $n + 2$ represents the second (larger) of the two consecutive even integers.

3. Write a word equation. What condition gives us the information needed to write the equation? We know that three times the larger is 10 less than five times the smaller. Thus,

   $$\boxed{\text{three times the larger integer}} = \boxed{\text{10 less than five times the smaller integer}}$$

4. Write an algebraic equation using the word equation as a guide.

three times the larger integer	=	10 less than five times the smaller integer
$3(n + 2)$	=	$5n - 10$

5. Solve the equation.

$3(n + 2) = 5n - 10$	Distribute the 3 on the left side of the equation.
$3n + 6 = 5n - 10$	Subtract $3n$ from both sides.
$3n - 3n + 6 = 5n - 3n - 10$	Simplify both sides.
$6 = 2n - 10$	Add 10 to both sides.
$16 = 2n$	Divide both sides by 2.
$8 = n$	Solution of the equation.
or $n = 8$	

6. Have we answered the question that was asked? No: n represents the smaller integer, but we were asked to find the larger integer. The larger integer is represented as $n + 2$, so the larger integer is $8 + 2 = 10$.

7. Check the solution against the wording of the problem. Are 8 and 10 consecutive even integers? The answer is yes. Is three times the larger 10 less than five times the smaller? Three times the larger is $3(10) = 30$, and five times the smaller is $5(8) = 40$, and 30 is 10 less than 40. Thus we have the correct solution.

Practice Exercises

3. Find three consecutive even integers whose sum is 54.

4. Three consecutive odd integers are such that three times the sum of the first and second is seven more than the third. Find the smallest integer.

Distance, Rate, and Time Problems

If an object is traveling at a constant rate (speed), the distance traveled depends upon how long it has been traveling (time). This relationship is given by the equation $d = rt$, where d represents the distance traveled, r is the rate (speed), and t is the time. For example, if a truck travels at a constant rate of 50 miles per hour for five hours, the distance it has traveled is $d = (50)(5) = 250$ miles.

Even though the problems in this section may be just as easily solved using the procedure previously used, we suggest that you use the procedure that follows as preparation for the problems in Chapter 7.

Solving Distance, Rate, and Time Problems

1. Make a chart with columns for the moving objects, the distance, the rate, and the time.

2. Fill in one column with known numerical values.

3. Assign a variable to one of the unknowns and represent any other unknowns in terms of this variable.

4. Fill in the remaining column from the first two, using the appropriate relationships between d, r, and t. In this section, $d = rt$.

5. Write the equation, using the information in the last column that was filled in. It is sometimes helpful to draw a diagram.

6. Solve the equation.

7. Be sure that you have answered the question that was asked.

8. Check the solution against the wording of the original problem.

Example 3

a. The distance between Atlanta, Georgia, and Charleston, West Virginia, is 500 miles. A truck leaves Atlanta traveling toward Charleston at an average rate of 47 miles per hour. At the same time, a bus leaves Charleston traveling toward Atlanta at an average rate of 53 miles per hour. Assuming they are traveling on the same route, how long will it take until they meet?

Solution:

1. Draw a chart and label the columns.

Moving Objects	d	r	t
Truck			
Bus			

2. Fill in one column with known numerical values. We know that the rate at which the truck is traveling is 47 miles per hour and the rate at which the bus is traveling is 53 miles per hour.

Moving Objects	d	r	t
Truck		47	
Bus		53	

3. Assign a variable to one of the unknowns and represent any other unknowns in terms of this variable. We are looking for the number of hours it will take until they meet. Since they left at the same time, the truck and car will be traveling for the same number of hours. Let t represent the number of hours for each. Fill in the t column.

Moving Objects	d	r	t
Truck		47	t
Bus		53	t

4. Fill in the remaining column, using the formula $d = rt$. Consequently, the distance the truck travels is its rate (47) times its time (t), or $d = 47t$. Likewise, for the bus, $d = 53t$.

Moving Objects	d	r	t
Truck	$47t$	47	t
Bus	$53t$	53	t

5. Write the equation using the information from the last column filled in. Since the last column filled in was the distance column, the equation must involve the distances traveled by the truck and the bus. A diagram might be helpful.

Truck: $d = 47t$ Bus: $d = 53t$

Atlanta ——————————————— Charleston
←———— Total distance = 500 miles ————→

From the diagram, we can see that:

distance the truck traveled	+	distance the bus traveled	=	distance between the cities
$47t$	+	$53t$	=	500

6. Solve the equation.

$$47t + 53t = 500 \qquad \text{Add like terms.}$$
$$100t = 500 \qquad \text{Divide both sides of the equation by 100.}$$
$$t = 5 \qquad \text{Solution of the equation.}$$

7. Have we answered the question that was asked? We are looking for the number of hours until the truck and the bus meet; t represents the number of hours that each will be traveling until they meet. Consequently, we have answered the question.

8. Check the solution against the wording of the original problem.

The total distance traveled by the truck in five hours is $5(47) = 235$ miles. The total distance the bus has traveled in five hours is $5(53) = 265$ miles. The total miles traveled by both in five hours, therefore, is $235 + 265 = 500$ miles, the distance between the two cities. Our solution is correct.

b. John leaves his home in Lake City, Florida, traveling north on I-75 at an average rate of 45 miles per hour. Two hours later, his wife, Jan, leaves home and takes the same route traveling at an average rate of 60 miles per hour. How many hours will it take Jan to catch up to John?

Solution:

1. Draw a chart and label the columns.

Moving Objects	d	r	t
John			
Jan			

2. Fill in one column with known numerical values. We know that John's rate is 45 miles per hour and Jan's rate is 60 miles per hour.

Moving Objects	d	r	t
John		45	
Jan		60	

3. Assign a variable to one of the unknowns and represent any other unknowns in terms of this variable. We are looking for the number of hours it will take Jan to catch up to John, but John left two hours before Jan. If we let t represent John's time, then $t - 2$ will represent Jan's time. Fill in the t column.

Moving Objects	d	r	t
John		45	t
Jan		60	$t - 2$

4. Fill in the remaining column using the formula $d = rt$. The distance that John travels is his rate (45) times his time (t), or $d = 45t$. Likewise, for Jan, $d = 60(t - 2)$.

Moving Objects	d	r	t
John	$45t$	45	t
Jan	$60(t - 2)$	60	$t - 2$

5. Write the equation, using the information from the last column filled in. Since the last column filled in was the distance column, the equation must involve the distances traveled by John and Jan. A diagram might be helpful.

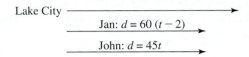

From the diagram we can see that:

distance John traveled	=	distance Jan traveled
$45t$	=	$60(t - 2)$

6. Solve the equation.

$$45t = 60(t - 2)$$ Distribute on the right side of the equation.
$$45t = 60t - 120$$ Subtract 60t from both sides.
$$-15t = -120$$ Divide both sides by -15.
$$t = 8$$ Solution of the equation.

7. Have we answered the question that was asked? We are looking for the number of hours until Jan catches up to John; t represents the number of hours that John has been traveling. The number of hours that Jan has been traveling is represented by $t - 2$, or $8 - 2 = 6$. Therefore, it will take Jan six hours to catch up to John.

8. Check the solution against the wording of the original problem.

The total distance traveled by John in 8 hours is $8(45) = 360$ miles. The total distance Jan has traveled in 6 hours is $6(60) = 360$ miles. The total miles traveled by each is 360 miles. Thus, our solution is correct.

Practice Exercises

5. A car traveling at an average rate of 44 miles per hour leaves Jacksonville, Florida, traveling toward Jackson, Mississippi. At the same time, a bus leaves Jackson traveling toward Jacksonville at an average rate of 36 miles per hour. Assuming they travel the same route and the distance between Jacksonville and Jackson is 600 miles, how long will it take until they meet?

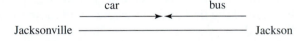

6. A recreational vehicle leaves Tallahassee, Florida, heading west on I-10 for Houston, Texas, at an average speed of 35 miles per hour. Four hours later, a truck also leaves Tallahassee on I-10 for Houston at an average rate of 55 miles per hour. How many hours will it take the truck to catch up to recreational vehicle?

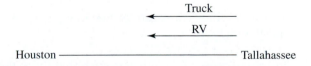

Exercise Set 3.5

Solve the following general applications problems:

1. A two-by-four that is 8 feet long is cut so that one piece is 2 feet longer than the other. Find the length of each piece.

2. Rick and Hilda were recently married. At the wedding, there were 25 more guests of the bride than of

the groom. If 157 people attended, how many were guests of the groom?

3. Last year, Laura jogged 262 miles more than Pat jogged. If the total number of miles they jogged was 1308, how many miles did Pat jog?

4. The mass transit authority operates two types of buses. A smaller bus that carries 24 fewer passengers than a larger bus is used for short distances. If the total number of passengers that can be carried by the two types of buses is 112, how many passengers can the smaller bus carry?

5. Last week, Richard earned $43 less than Patricia earned. If their combined salary for the week was $557, how much did each earn?

6. Tara purchased a rocking chair and a bench at a flea market. The total cost for the two items was $111.00. If the bench cost $15 more than the rocking chair, find the cost of each.

7. Marlene is doing her income tax return. She knows that she spent a total of $594 on two business trips last year. If one trip cost $136 more than the other, find the cost of each trip.

8. The Garcia family owns a van and a car. The car gets twice as many miles per gallon of gasoline as the van gets. If the combined mileage of the two is 54 miles per gallon, how many miles per gallon does the van get?

9. It costs $5 a day extra for after-school care at a nursery school for preschool children. If the total cost for a five-day week of nursery school including after school care is $85, what is the charge per day for the nursery school without after-school care?

10. It costs $35 per credit hour plus a one-time lab fee of $15 to register for courses at Mountain High Community College. If Zane received a $250.00 student loan, what is the maximum number of credit hours that he can pay for from the loan?

11. Yolanda's job as a salesclerk pays her $25 per day plus 3% commission on all her sales. If her total wages for Monday were $70, find the amount of her sales.

12. A performer charges $1000 plus 10% of the gate receipts. She earned $16,000 for her last performance. What were the gate receipts?

13. François received a raise of 4% of her present salary. If her new salary is $35,980, what is her present salary?

14. The sales tax in Orlando, Florida, is 6%. If a college cafeteria in Orlando wants the cost of the lunch special to be $4.24 including tax, what should they charge for the lunch special?

Solve the following consecutive integer problems:

15. The sum of two consecutive odd integers is 24. Find the two integers.

16. The sum of two consecutive even integers is 98. Find the integers.

17. The sum of three consecutive integers is 75. Find the integers.

18. The sum of three consecutive even integers is 132. Find the integers.

19. Find three consecutive integers such that twice the smallest is 8 less than the largest.

20. Find three consecutive even integers such that three times the smallest is 12 more than the largest.

21. Find three consecutive even integers such that the sum of the first and the second is 16 more than the third.

22. Find three consecutive odd integers such that three times the third is 17 more than the sum of the first and second.

23. Find three consecutive even integers such that twice the sum of the first and the second is 18 more than the third.

24. Find three consecutive even integers such that five times the second equals twice the sum of the first and the third.

Solve the following distance, rate, and time problems:

25. The distance between Charlotte, North Carolina, and Buffalo, New York, is 700 miles. Ralph leaves Charlotte traveling toward Buffalo at an average rate of 56 miles per hour. At the same time, Charles leaves Buffalo traveling toward Charlotte at an average rate of 44 miles per hour. Assuming they are traveling on the same route, how long will it take until they meet?

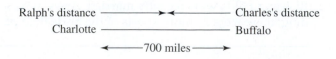

26. A commercial passenger jet leaves Miami, Florida, traveling to Reno, Nevada, at an average speed of 560 miles per hour. At the same time, a private jet leaves Reno traveling toward Miami at an average rate of 440 miles per hour. How long will it take until they meet if the distance from Miami to Reno is 3000 miles?

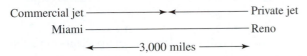

27. Frank leaves Memphis, Tennessee, traveling toward Albuquerque, New Mexico, at an average speed of 60 miles per hour. Two hours later, Marsha leaves Albuquerque traveling toward Memphis at an average speed of 50 miles per hour. If the distance from Memphis to Albuquerque is 1000 miles, how long after Marsha leaves will it take them to meet?

28. The distance between Norfolk, Virginia, and Buffalo, New York, is 595 miles. Grandma Rosie leaves Norfolk traveling toward Buffalo at an average speed of 35 miles per hour. Two hours later, Grandpa Julius leaves Buffalo traveling toward Norfolk at an average speed of 40 miles per hour. How long after Grandpa Julius leaves will it take them to meet?

29. Tameka leaves Memphis, Tennessee, on I-40 traveling west at an average speed of 30 miles per hour. Three hours later, Akiva also leaves Memphis traveling the same route at an average speed of 40 miles per hour. How long will it take Akiva to catch up to Tameka?

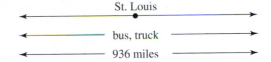

30. Polly leaves Savannah, Georgia, on I-95 traveling north at an average speed of 43 miles per hour. Two hours later, Eric also leaves Savannah traveling

north on I-95 but at an average speed of 52 miles per hour. How long will it take Eric to catch up to Polly?

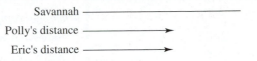

31. Pia leaves Kansas City, Missouri, traveling west on I-70 at an average speed of 42 miles per hour. At the same time, Alma leaves Kansas City traveling east on I-70 at an average speed of 36 miles per hour. How long will it take until they are 468 miles apart?

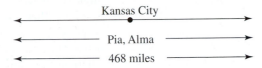

32. A truck leaves St. Louis, Missouri, traveling north on I-55 at an average rate of 62 miles per hour. At the same time, a bus leaves St. Louis traveling south on I-55 at an average rate of 55 miles per hour. How long will it take until they are 936 miles apart?

33. Harold leaves Denver, Colorado, traveling east on I-70 at an average speed of 44 miles per hour. Two hours later, Selma leaves Denver traveling west on I-70 at an average rate of 48 miles per hour. How long after Harold leaves will it take until they are 364 miles apart?

34. A motorcycle leaves Colorado Springs, Colorado, traveling north on I-25 at an average speed of 47 miles per hour. One hour later, a car leaves Colorado Springs traveling south on I-25 at an average speed of 54 miles per hour. How long after the motorcycle leaves will it take until they are 552 miles apart?

Challenge Exercises:

35. An architect wishes to build a fountain that has four pipes. If the pipes were arranged in order from the shortest to the longest, each pipe would be 3 feet longer than the previous one. If she has 54 feet of pipe, how long should each pipe be?

36. Jim and Amy planned their family so that the two children were born four years apart. If the sum of the ages of the children is 22 years, how old is each child?

Writing Exercises:

37. How would you go about teaching a friend to solve application problems?

Writing Exercises or Group Projects:

If done as a group project, have each group write two exercises of each type, exchange them with another group, and then solve them.

38. Write and solve an application problem involving consecutive integers.

39. Write and solve an application problem involving distance, rate, and time.

Other Types of Traditional Application Problems

OBJECTIVES

When you complete this section, you will be able to:

a. Solve real-world problems involving money.

b. Solve real-world problems involving investments.

c. Solve real-world problems involving mixtures.

Introduction   In this section, we continue our discussion of traditional types of application problems. All of the problems in this section can be solved using a chart that is modified slightly (mostly headings) for each type of problem. There is an extra step involved in these problems that we did not do previously: we will need to calculate the value (in the case of money) or the amount of pure substance (in the case of mixtures) before writing the equation. Another approach to the problems in this section is given in Chapter 9, where systems of equations, the preferred method of some, are used.

A situation commonly encountered in application problems is when the total amount is known and parts of the total need to be represented using a variable.

Example 1

Represent each of the following:

a. If a board is 12 feet long, and a piece x feet long is cut off, represent the length of the remaining piece in terms of x.

Solution:
The length of the other piece is the total length of the board minus the length of the piece cut off. Thus, the length of the remaining piece is $(12 - x)$ feet.

b. A paint contractor buys 36 gallons of paint, some of which is acrylic and some of which is oil-based. If he buys x gallons of acrylic, represent the number of gallons of oil-based paint in terms of x.

Solution:
The number of gallons of oil-based paint is the total number of gallons purchased minus the number of gallons of acrylic paint. Therefore, the number of gallons of oil-based paint is $(36 - x)$ gallons.

Practice Exercises

Represent each of the following:

1. Sharla has a piece of ribbon 65 inches long and cuts off x inches to wrap a package. How long is the remaining piece?

2. There is a total of 40 marbles in a jar, some black and the remainder white. If there are x black marbles in the jar, represent the number of white marbles in terms of x.

Money Problems

We will learn to solve two types of problems involving money. One type of problem will involve the purchase of two or more items at different costs. We will be given the total number of items purchased and the total cost, and will then be asked to find how many items of each type were purchased. The second type is similar to the first, but involves coins. We suggest the following procedure to solve both types of problems:

Solving Problems Involving Money

1. Draw a chart with four columns and a row for each item purchased. Label the columns Type of Item, Price per Unit, Number of Units, and Total Cost. For the second type of problem, the columns should be labeled Type of Coin, Value of Coin, Number of Coins, and Total Value. Fill in the Type of Item (or Type of Coin) column first.

2. Fill in the Price per Unit (or Value of Coin) column. These values will usually be given, but may involve a variable.

3. Fill in the Number of Units (or Number of Coins) column. This will usually involve a variable. It is usually necessary to assign the variable to the number of units of one item and express the number of units of the other item in terms of this variable.

4. Fill in the Total Value (or Total Cost) column. The total cost of value is the product of the price per unit and the number of units.

5. Write the word equation. This will almost always be in a form such that the sum of the values of the individual items equals the total value of all the items.

6. Write the algebraic equation using the word equation as a guide.

7. Solve the equation.

8. Be sure you answered the question that was asked.

9. Check the solution against the wording of the original equation.

Example 2

a. A paint contractor paid $372 for 24 gallons of paint to paint the inside and outside of a house. If the paint for the inside costs $12 per gallon and the paint for the outside costs $18 per gallon, how many gallons of each did he buy?

Solution:

1. Draw a chart and label the columns. Fill in the Type of Item (in this case, paint) column.

Type of Paint	Price per Gallon	Number of Gallons	Total Cost
Inside			
Outside			

2. Fill in the Price per Unit (in this case, gallon) column.

Type of Paint	Price per Gallon	Number of Gallons	Total Cost
Inside	12		
Outside	18		

3. Fill in the Number of Units (in this case, gallons) column. Since we are looking for the number of gallons of each type of paint, we must assign a variable. There is no particular advantage (or disadvantage) to letting our variable represent either, so let n represent the number of gallons of inside paint. Since there is a total of 24 gallons, $24 - n$ represents the number of gallons of outside paint.

Type of Paint	Price per Gallon	Number of Gallons	Total Cost
Inside	12	n	
Outside	18	$24 - n$	

4. Fill in the Total Cost column. The total cost is the price per gallon times the number of gallons. Therefore, $12n$ represents the cost of the inside paint in dollars and $18(24 - n)$ represents the cost of the outside paint in dollars.

Type of Paint	Price per Gallon	Number of Gallons	Total Cost
Inside	12	n	$12n$
Outside	18	$24 - n$	$18(24 - n)$

5. Write a word equation.

$$\boxed{\text{cost of inside paint}} + \boxed{\text{cost of outside paint}} = \boxed{\text{total cost of the paint}}$$

6. Write an algebraic equation using the word equation as a guide.

$$\boxed{\text{cost of inside paint}} + \boxed{\text{cost of outside paint}} = \boxed{\text{total cost of the paint}}$$

$$12n \qquad + \qquad 18(24 - n) \qquad = \qquad 372$$

7. Solve the algebraic equation.

$$12n + 18(24 - n) = 372 \qquad \text{Distribute 18.}$$
$$12n + 432 - 18n = 372 \qquad \text{Add like terms.}$$
$$-6n + 432 = 372 \qquad \text{Subtract 432 from both sides.}$$
$$-6n = -60 \qquad \text{Divide both sides by } -6.$$
$$n = 10 \qquad \text{Solution of the equation.}$$

8. Be sure that you answered the question that was asked. We were asked to find the number of gallons of each type of paint. Since n represents the number of gallons of inside paint, we still need to find the number of gallons of outside paint. Since $24 - n$ represents the number of gallons of outside paint, we therefore have $24 - 10 = 14$ gallons of outside paint.

9. Check the solutions against the wording of the original problem. Do we have a total of 24 gallons of paint? Since $10 + 14 = 24$, the answer is yes. Is the total value of the paint $372? Ten gallons of inside paint is worth $10(12) = \$120$ and 14 gallons of outside paint is worth $18(14) = \$252$. The total value of the paint is $\$120 + \$252 = \$372$, so our solutions are correct.

b. A collection of 23 coins is made up of nickels and dimes and is worth $1.90. How many of each type of coin are there?

Solution:
This is a special type of money problem that involvs coins. At first glance, this problem may seem very different from the preceding problem, but they are in reality essentially the same. We will simply change the headings on our columns to Type of Coin, Value of Coin, Number of Coins, and Total Value.

1. Draw the chart and label the columns. Fill in the type of coin column.

Type of Coin	Value of Coin	Number of Coins	Total Value
Nickel			
Dime			

2. Fill in the Value of Coin column expressing the value in cents.

Type of Coin	Value of Coin	Number of Coins	Total Value
Nickel	5		
Dime	10		

3. Fill in the Number of Coins column. Since we are looking for the number of nickels and dimes, we need to assign a variable to one of them. Let n represent the number of nickels. Since we have a total of 23 coins, $23 - n$ represents the number of dimes.

Type of Coin	Value of Coin	Number of Coins	Total Value
Nickel	5	n	
Dime	10	$23 - n$	

4. Fill in the Total Value column. The total value of each type of coin is the value of the coin times the number of coins. Therefore, $5n$ represents the value of the nickels in cents and $10(23 - n)$ represents the value of the dimes in cents.

Type of Coin	Value of Coin	Number of Coins	Total Value
Nickel	5	n	$5n$
Dime	10	$23 - n$	$10(23 - n)$

5. Write a word equation.

$$\boxed{\text{value of nickels}} + \boxed{\text{value of dimes}} = \boxed{\text{total value of the collection}}$$

6. Write an algebraic equation using the word equation as a guide.

$$\boxed{\text{value of nickels}} + \boxed{\text{value of dimes}} = \boxed{\text{total value of the collection}}$$

$$5n \qquad + \qquad 10(23 - n) \qquad = \qquad 190$$

Note: Since the value of the nickels and the value of the dimes were given in cents, it was necessary to change $1.90 into 190¢.

7. Solve the algebraic equation.

$5n + 10(23 - n) = 190$	Distribute 10.
$5n + 230 - 10n = 190$	Add like terms.
$-5n + 230 = 190$	Subtract 230 from both sides.
$-5n = -40$	Divide both sides by -5.
$n = 8$	Solution of the equation.

8. Be sure that you answered the question that was asked. We were asked to find the number of each type of coin. Since n represents the number of nickels, we still need to find the number of dimes. Since $23 - n$ represents the number of dimes, we therefore have $23 - 8 = 15$ dimes.

9. Check the solutions against the wording of the original problem. Do we have a total of 23 coins? Since $8 + 15 = 23$, the answer is yes. Is the total value of the collection $1.90? Eight nickels are worth $8(.05) = \$.40$ and 15 dimes are worth $15(.10) = \$1.50$. The total value of the collection is $\$.40 + \$1.50 = \$1.90$, so our solutions are correct.

Practice Exercises

3. A landscape company pays $7 each for rose bushes and $5 each for azaleas. If they paid $82 for 14 plants, how many of each did they buy?

4. A cashier has a total of 30 nickels and quarters in the cash register. If the total value of the coins is $3.90, how many of each type does he have?

Answers:

Investment Problems

In doing investment problems, it is necessary to change percents to decimals, as in Section 3.5. Remember to drop the % sign and move the decimal two places to the left. For example, 12% = .12 and 6% = .06.

   Investment problems are special types of *money problems*. Consequently, the procedure is virtually the same except the column headings are changed to Type of Investment, Interest Rate (expressed as a decimal), Amount Invested, and Interest Earned. The amount of simple interest earned in a year is found by multiplying the annual interest rate expressed as a decimal times the amount invested. For example, the interest earned on $2000 invested at 5% interest for one year is (.05)(2000) = $100. Also, the total interest earned is the sum of the interests earned on each investment.

Example 3

A professional organization has its cash in two investments. Part of the money is in an account paying 8% interest annually and the remainder is in an account paying 14% interest annually. The amount in the account paying 14% interest is $7000 more than the amount in the account paying 8% interest. If the combined interest earned in one year from the two accounts is $4940, find the amount in each account.

Solution:
1. Draw and label a chart.

Type of Investment	Interest Rate	Amount Invested	Interest Earned
8%			
14%			

2. Fill in the Interest Rate column by converting the percents to decimals.

Type of Investment	Interest Rate	Amount Invested	Interest Earned
8%	.08		
14%	.14		

3. Fill in the Amount Invested column. We know that there is $7000 more invested at 14% than at 8%. So if we let x represent the amount invested at 8%, then $x + 7000$ represents the amount invested at 14%.

Type of Investment	Interest Rate	Amount Invested	Interest Earned
8%	.08	x	
14%	.14	$x + 7000$	

4. Fill in the Interest Earned column. Remember, the interest earned is the interest rate times the amount invested.

Type of Investment	Interest Rate	Amount Invested	Interest Earned
8%	.08	x	$.08x$
14%	.14	$x + 7000$	$.14(x + 7000)$

5. Write a word equation. We know that the total interest earned on the two investments is $4940. Therefore,

$$\boxed{\text{interest earned at 8\%}} + \boxed{\text{interest earned at 14\%}} = \boxed{\text{total interest earned}}$$

6. Write the algebraic equation using the word equation as a guide.

$$\boxed{\text{interest earned at 8\%}} + \boxed{\text{interest earned at 14\%}} = \boxed{\text{total interest earned}}$$

$$.08x \quad + \quad .14(x + 7000) \quad = \quad 4940$$

7. Solve the equation.

$$.08x + .14(x + 7000) = 4940 \qquad \text{Distribute the .14.}$$
$$.08x + .14x + 980 = 4940 \qquad \text{Add like terms.}$$
$$.22x + 980 = 4940 \qquad \text{Subtract 980 from both sides.}$$
$$.22x = 3960 \qquad \text{Divide both sides by .22.}$$
$$x = 18,000 \qquad \text{Solution of equation.}$$

8. Have we answered the question that was asked? Partially. Since x represented the amount invested at 8%, we still need to find the amount invested at 14%. Since $x + 7000$ represents the amount invested at 14%, we therefore have $18,000 + \$7000 = \$25,000$ invested at 14%.

9. Check the solutions against the wording of the original problem. The interest earned on $18,000 at 8% is $(.08)(18,000) = \$1440$ and the interest earned on $25,000 at 14% is $(.14)(25000) = \$3500$. So the total interest on the two investments is $\$1440 + \$3500 = \$4940$, which is what the combined interest is supposed to be. So our solutions are correct.

There are several variations of investment problems.

Practice Exercise

5. A college support fund has its money in two investments. The first investment pays 18% interest annually and is $20,000 more than the second investment, which pays 12% interest annually. If the total interest earned from the two investments is $18,600 per year, how much is in the first investment?

Mixture Problems

Mixture problems are very much like money and investment problems. We will consider two types of mixture problems: liquid and dry. We will learn to solve dry mixture problems first. To do so, we will again use a chart, but change the headings to Type of Ingredient, Price per Unit, Number of Units, and Total Value. We will also need an extra line at the bottom for the mixture. The key to solving dry mixture problems is that the sum of the values of each of the ingredients must equal the value of the mixture.

Example 4

How many pounds of peppermint candy worth $1.80 per pound must be mixed with 15 pounds of butterscotch worth $2.30 a pound to get a mixture worth $2.10 per pound?

Solution:
1. Draw the chart, label the columns, and fill in the types of candy.

Type of Candy	Price per Pound	Number of Pounds	Total Value
Peppermint			
Butterscotch			
Mixture			

2. Fill in the Price per Pound column.

Type of Candy	Price per Pound	Number of Pounds	Total Value
Peppermint	1.80		
Butterscotch	2.30		
Mixture	2.10		

3. Fill in the Number of Pounds column. Since we are looking for the number of pounds of peppermint candy, let x represent the number of pounds of peppermint candy. Also, we know that we have 15 pounds of butterscotch. The number of pounds of candy in the mixture is the number of pounds of peppermint plus the number of pounds of butterscotch. Therefore, the number of pounds of candy in the mixture is $x + 15$.

Type of Candy	Price per Pound	Number of Pounds	Total Value
Peppermint	1.80	x	
Butterscotch	2.30	15	
Mixture	2.10	$x + 15$	

4. The total value of each type of candy is the price per pound times the number of pounds. Use this fact to fill in the Total Value column.

Type of Candy	Price per Pound	Number of Pounds	Total Value
Peppermint	1.80	x	$1.80x$
Butterscotch	2.30	15	$(2.30)(15) = 34.50$
Mixture	2.10	$x + 15$	$2.10(x + 15)$

5. Write the word equation.

$$\boxed{\text{value of peppermint}} + \boxed{\text{value of butterscotch}} = \boxed{\text{value of mixture}}$$

6. Write the algebraic equation using the word equation as a guide.

$$\boxed{\text{value of peppermint}} + \boxed{\text{value of butterscotch}} = \boxed{\text{value of mixture}}$$
$$1.80x \qquad + \qquad 34.50 \qquad = \qquad 2.10(x + 15)$$

7. Solve the equation.

$1.80x + 34.50 = 2.10(x + 15)$	Distribute 2.10.
$1.80x + 34.50 = 2.10x + 31.50$	Subtract $1.80x$ from both sides.
$34.50 = .3x + 31.50$	Subtract 31.50 from both sides.
$3 = .3x$	Divide both sides by .3.
$10 = x$	Solution of the equation.

8. Have we answered the question that was asked? Yes. We were looking for the number of pounds of peppermint candy and x represents the number of pounds of peppermint candy.

9. Check the solution against the wording of the original problem. The value of 10 pounds of peppermint candy is $1.80(10) = \$18.00$, and from the preceding chart we know the value of 15 pounds of butterscotch is $34.50. Consequently, $18.00 + $34.50 = $52.50. The mixture sells for $2.10 per pound and we have $10 + 15 = 25$ pounds. Therefore, the value of the mixture is $2.10(25) = \$52.50$, the same as the sum of the values of the peppermint and butterscotch. Our solution is correct.

Other variations of this type of problem include cases where the number of units and the price per unit of each of the ingredients are known and we are asked to find the price per unit of the mixture. Another type gives the price per unit of each of the ingredients and the price per unit and number of units in the mixture and asks us to find the number of units of each ingredient. The approach is the same in all cases.

Liquid mixture problems are done in exactly the same way as dry mixture problems, except that we deal with the amount of pure substance in each ingredient and in the mixture instead of with values. Therefore, we will not repeat the procedure, but will illustrate with an example. The key to doing this type of problem is that the amount of pure substance in a solution (alloy, etc.) is the percent that is pure times the number of units (gallons, pounds, etc.) of the solution. For example, if we have 8 gallons of a solution that is 20% alcohol, the amount of pure alcohol in the solution is $(.20)(8) = 1.6$ gallons.

Example 5

How much of a solution that is 30% alcohol must be added to 55 liters of a solution that is 60% alcohol to obtain a solution that is 40% alcohol?

Solution:

1. Draw a chart similar to the one for dry mixtures. Note that the headings are different. Fill in the types of solutions.

Type of Solution	Part Pure Alcohol	Volume	Amount Pure Alcohol
30%			
60%			
40% (mixture)			

2. Fill in the Part Pure Alcohol column by converting the percents to decimals.

Type of Solution	Part Pure Alcohol	Volume	Amount Pure Alcohol
30%	.30		
60%	.60		
40% (mixture)	.40		

3. Fill in the Volume column. Since we are looking for the amount (volume) of 30% alcohol, let x represent the volume of 30% alcohol. We know that we have 55 liters of 60% alcohol. Consequently, when the two are mixed we will have $(x + 55)$ liters of the 40% mixture.

Type of Solution	Part Pure Alcohol	Volume	Amount Pure Alcohol
30%	.30	x	
60%	.60	55	
40% (mixture)	.40	$x + 55$	

4. Fill in the Amount Pure Alcohol column. Remember, the amount of pure alcohol in each solution equals the part pure alcohol times the volume.

Type of Solution	Part Pure Alcohol	Volume	Amount Pure Alcohol
30%	.30	x	$.30x$
60%	.60	55	$(.60)(55) = 33$
40% (mixture)	.40	$x + 55$	$.40(x + 55)$

5. Write the word equation.

amount pure alcohol in 30% solution	+	amount pure alcohol in 60% solution	=

amount pure alcohol in the 40% mixture

6. Write the algebraic equation using the word equation as a guide.

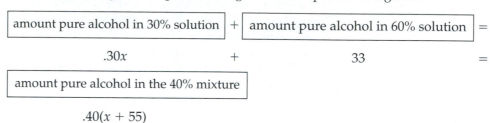

amount pure alcohol in 30% solution	+	amount pure alcohol in 60% solution	=

$$.30x \qquad\qquad + \qquad\qquad 33 \qquad\qquad =$$

amount pure alcohol in the 40% mixture

$$.40(x + 55)$$

7. Solve the equation.

$.30x + 33 = .40(x + 55)$	Distribute .40 on the right side.
$.30x + 33 = .40x + 22$	Subtract .30x from both sides.
$33 = .10x + 22$	Subtract 22 from both sides.
$11 = .10x$	Divide both sides by .10.
$110 = x$	Solution of the equation.

8. Have we answered the question that was asked? Yes, since x represents the amount of the 30% solution and that is what we were asked to find.

9. Check the solution against the wording of the original problem. In 110 liters of 30% solution there would be $.30(110) = 33$ liters of pure alcohol. From the preceding chart, we know that in 55 liters of 60% solution there are 33 liters of pure alcohol. So there is a total of $33 + 33 = 66$ liters of pure alcohol in the two that are mixed. We have a total of $110 + 55 = 165$ liters of the 40% mixture. In this mixture there is $.40(165) = 66$ liters of pure alcohol, the same as the amount of pure alcohol in the two that are mixed to get the 40% solution. Thus our solution is correct.

As with dry mixture problems, there are several variations of this type of problem.

Practice Exercises

6. How many pounds of coffee from the Dominican Republic that sells for $3.00 per pound should be mixed with 9 pounds of coffee from Colombia that sells for $4.50 per pound, in order to get a mixture that sells for $3.90 per pound?

7. How much of a solution that is 20% baking soda must be added to 60 ounces of a solution that is 5% baking soda to make a solution that is 10% baking soda?

Exercise Set 3.6

Represent each of the following:

1. Sally is taking a trip of 400 miles. If she has traveled x miles, represent the remainder of the trip in terms of x.

2. Francine walks 5 miles every morning. If she has already walked x miles, represent the distance remaining in terms of x.

Answers:

3. There is a total of 24 nickels and dimes in a box. If x of these are nickels, represent the number of dimes in terms of x.

4. A seamstress has a box containing 50 buttons, some of which are brass and the remainder of which are glass. If there are x brass buttons, represent the number of glass buttons in terms of x.

5. A nursery has a total of 150 azalea and camellia plants. If there are x azaleas, represent the number of camellias in terms of x.

6. A class has 35 students. If x of the students are girls, represent the number of boys in terms of x.

Solve the following money problems:

7. An electrical contractor paid $1040 for 36 light fixtures to be installed in a hotel lobby. If the first type of fixture costs $25 each and the second type of fixture costs $32 each, how many of each kind did she buy?

8. A locksmith installed inside and outside locks in a new classroom building. The locksmith paid $1216 for 80 locks. If inside locks cost $14 each and outside locks cost $26 each, how many of each kind of lock did he buy?

9. The managers of an office building have decided to reduce the cost of air conditioning by installing overhead fans. They paid $5225 for 105 fans, some of which measure 36 inches and the remainder of which measure 54 inches. If the 36-inch fans cost $45 each and the 54-inch fans cost $65 each, how many of each kind of fan did they buy?

10. An accounting firm purchased a supply of computer disks. They paid $725 for 60 boxes of disks. If the double density disks cost $8 a box and the high density disks cost $15 a box, how many of each kind did they buy?

11. Winston paid $7.46 for 36 stamps. If some of the stamps cost 18¢ each and the remainder cost 25¢ each, how many of each kind did he buy?

12. Gayle paid $17.40 for 42 pens at the campus bookstore. If she bought some pens costing $.35 each and others costing $.50 each, how many of each kind did she buy?

13. The owners of an exterminator business have replaced their fleet of 27 cars and trucks by purchas-

Solve the following investment problems:

21. When he retired, Professor Gomez received a lump sum in cash for his unused sick days. He invested part of the money in bonds that paid 6% interest annually. He invested $16,000 more in certificates of deposit at 10% annual interest than he invested in bonds. The total interest earned per year from the two investments was $3200. How much money did he invest at each rate?

ing new cars for $12,000 each and new trucks for $9500 each. If they paid a total of $294,000, how many of each type of vehicle did they buy?

14. A building contractor has paid $2088 for 54 doors. If inside doors cost $35 each and outside doors cost $46 each, how many of each kind of door did he buy?

15. At the end of his shift, a cashier has only $5 and $20 bills in his drawer. If he has 48 bills worth a total of $900, how many of each kind of bill does he have?

16. A credit union carries only $20 and $50 traveler's checks. A customer buys $1500 worth of traveler's checks. If he bought a total of 39 checks, how many $50 checks did he buy?

17. A child goes to a bank to deposit her savings. She has $4.00 worth of nickels and dimes. If she has five more nickels than dimes, how many of each type of coin does she have?

18. A certain vending machine accepts only dimes and quarters. If the person who services the machine collects 52 coins with a total value of $8.95, how many of each kind of coin was in the machine?

19. A parking meter will accept dimes and quarters only. At the end of the day, the meter attendants collected 55 coins whose total value was $10.75. How many coins of each type were there?

20. A vending machine will accept nickels and dimes only. The attendant checks the machine and finds a total of 42 coins whose value is $3.40. How many coins of each type were there?

22. A petroleum geologist received a $50,000 bonus for discovering a new oil deposit. She invested part of the money in bonds that paid 15% interest annually and put the rest in a life insurance policy that paid 8% interest annually. How much did she put into the life insurance policy if she expects to earn $5400 total from both investments each year?

23. A professional football player received a $200,000 bonus for signing his contract. He invested part of the money in home mortgages that yield 12% interest annually and he put the rest in a business that was yielding a 20% profit annually. How much did he invest in the business if he expects to earn a total of $27,200 per year from both investments?

Solve the following mixture problems:

25. How many ounces of pipe tobacco that sells for $2.50 per ounce must be mixed with 6 ounces that sells for $3.50 per ounce if the mixture is to sell for $3.10 per ounce?

26. How many pounds of candy that sells for $2.75 per pound must be mixed with 10 pounds that sells for $3.50 per pound to get a mixture that sells for $3.25 per pound?

27. If 8 pounds of almonds that sell for $2.25 per pound are mixed with 4 pounds of cashews that sell for $4.50 per pound, what should be the selling price of the mixture?

28. If 12 ounces of white chocolate that sells for $3.75 per ounce is mixed with 6 ounces of dark chocolate that sells for $4.50 per ounce, what should be the selling price of the mixture?

29. A specialty store owner mixes peanuts worth $1.80 per pound with cashew nuts worth $5.40 per pound. How many pounds of each type of nut is needed to make a 12-pound mixture worth $3.00 a pound?

30. A coffee company wishes to make a special blend of coffee by mixing coffee beans that cost $3.20 a pound with coffee beans that cost $6.20 per pound.

Solve the following exercises. Types of problems are mixed.

37. How many gallons of pure water must be added to 10 gallons of a solution that is 25% soap to make a solution that is 5% soap?

38. How many pounds of an alloy that is 30% tin must be added to 40 pounds of an alloy that is 15% tin to make an alloy that is 25% tin?

39. Professor Matas bought materials for a workshop for elementary school teachers. She bought number stickers for $.49 each and strings of beads for $1.19 a string. She bought 10 more number stickers than strings of beads. If she paid a total of $30.10, how many of each item did she buy?

40. Jeannette Castillo won $20,000 as part of the state lottery. After her celebration party, she put part of the money into bonds paying 7% interest. The rest,

24. A retired couple sold their home. They invested part of the money in home mortgages at 15% interest annually. They invested $20,000 less than this amount into certificates of deposit at 9% annual interest. How much money did they invest if they received $15,000 per year interest from the two investments?

How many pounds of each type of coffee bean is needed to make 40 pounds of coffee that costs $3.95 per pound?

31. How many milliliters of a 40% solution of battery acid must be added to 30 milliliters of an 8% solution to make a 20% solution of acid?

32. How many ounces of a 10% baking soda solution must be added to 40 ounces of a 2% baking soda solution to make a 5% baking soda solution?

33. How many tons of an alloy that is 5% nickel must be added to 15 tons of an alloy that is 20% nickel to make an alloy that is 10% nickel?

34. How many cups of a party mix that is 60% peanuts must be added to 5 cups of a party mix that is 30% peanuts to make a party mix that is 45% peanuts?

35. How many liters of pure baking soda must be added to 60 liters of a solution that is 10% baking soda to get a solution that is 15% baking soda?

36. How much pure water must be added to 20 liters of a solution that is 15% salt to make a solution that is 8% salt?

which was $10,000 less than the first part, she invested in a stock that was paying 20% in dividends (such as interest). How much did she put in each investment if she earned a total of $2050 per year?

41. How much pure alcohol must be added to 50 gallons of gasohol that is 10% alcohol to make gasohol that is 20% alcohol?

42. Mr. Jacob bought two prescription medications. He paid $.83 a tablet for the first medication and $1.15 a tablet for the second medication. Altogether, there were 86 tablets for which he paid $79.70. How many of the more expensive tablets did he buy?

Challenge Exercises:

43. A vending machine accepts nickels, dimes, and quarters. At the end of the day, there were three more dimes than nickels and four fewer quarters than nickels. If the total value of the coins was $4.10, find the number of each type of coin.

44. A banker has three investments that pay 6%, 5%, and 8% each. She has twice as much invested at 5% as at 6% and $5000 more invested at 8% than at 6%. If the total interest earned per year from the three investments is $2800, find the amount invested at each rate.

Writing Exercises or Group Projects:

If done as a group project, each group should write two exercises of each type, exchange them with another group, and solve them.

45. Write and solve an application problem involving money.

46. Write and solve an application problem involving mixtures.

47. Write and solve an application problem involving investments.

Section 3.7	**Geometric Application Problems**

OBJECTIVE

When you complete this section, you will be able to solve real-world problems involving geometric concepts.

Introduction   Linear equations are often found in geometric applications. There are essentially two types of problems. The first type are those for which known formulas (Section 1.4) give us a relationship between the unknowns of the problem. When solving this type of problem, we are either given a value for the variable(s) in the formula or we have to represent the unknowns in the formula in terms of variables. We then substitute into the formula and solve the resulting equation. Example 1 will be of this type. We will discuss the second type after Example 1. We do not have to look for conditions within the problem to give us the equation since it will be some known formula. We suggest the following steps in solving both types of problems:

Solving Geometric Application Problems

1. Identify the unknown(s) and represent one of them with a variable.
2. Represent all other variables in the problem in terms of the same variable.
3. Write the formula. A sketch might be helpful.
4. Substitute into the formula.
5. Solve the resulting equation.
6. Be sure that you have answered the question that was asked.
7. Check the solution against the wording of the original problem.

Example 1

a. An urban area has a large park in the shape of a rectangle. The park is 15 city blocks longer than it is wide. If the perimeter is 50 city blocks, find the length and width of the park.

Solution:
1. Identify the unknown(s) and represent one of them with a variable. We are looking for the length and width. Since the width is the smaller:

   Let x represent the width.

2. Represent all other unknowns in the problem in terms of the same variable. Since the length is 15 more than the width and x represents the width:

   $x + 15$ represents the length.

3. Write the formula. The formula needed is the formula for the perimeter of a rectangle. Note that a word equation is not necessary.

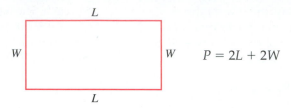

4. Substitute into the formula. We know from the problem that $P = 50$. We have also represented the width with x and the length with $x + 15$. Therefore:

$$50 = 2(x + 15) + 2x$$

5. Solve the equation.

$50 = 2(x + 15) + 2x$ Distribute 2 on the right side.

$50 = 2x + 30 + 2x$ Add like terms.

$50 = 4x + 30$ Subtract 30 from both sides.

$20 = 4x$ Divide both sides by 4.

$5 = x$ Solution of the equation.

6. Have we answered the question that was asked? Partially. We were asked to find both the length and width. Since x represents the width, we have found the width only. The length is represented by $x + 15$, so the length is $5 + 15 = 20$ blocks.

7. **Check:** Is the length 15 blocks more than the width? Yes, since 20 is 15 more than 5. Is the perimeter 50 blocks? Yes, since $2(20) + 2(5) = 50$. So, our solution is correct.

b. The length of the longest side of a triangle is 1 less than twice the length of the shortest side. The length of the third side is 2 more than the length of the shortest side. If the perimeter of the triangle is 21 inches, find the length of the sides of the triangle.

Solution:
1. Identify the unknown(s) and represent one of them with a variable. We are looking for the lengths of all three sides of the triangle. Let x represent the shortest side.

2. Represent all other unknowns in terms of the same variable. We know that the length of the longest side is 1 less than twice the length of the shortest side, so:

$2x - 1$ represents the length of the longest side.

We also know that the length of the third side is 2 more than the length of the shortest side, so:

$x + 2$ represents the length of the third side.

3. Write the formula. In this case, we need the formula for the perimeter of a triangle.

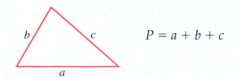

4. Substitute into the formula. We know that $P = 21$ inches and that the sides are represented by $x, 2x - 1$, and $x + 2$. Therefore:

$$21 = x + (2x - 1) + (x + 2)$$

5. Solve the equation.

$21 = (x) + (2x - 1) + (x + 2)$	Remove the parentheses.
$21 = x + 2x - 1 + x + 2$	Add like terms.
$21 = 4x + 1$	Subtract 1 from both sides.
$20 = 4x$	Divide both sides by 4.
$5 = x$	Solution of the equation.

6. Have we answered the question that was asked? Partially. We were asked to find the lengths of all three sides. The length of the longest side is $2x - 1 = 2 \cdot 5 - 1 = 10 - 1 = 9$ inches, and the length of the third side is $x + 2 = 5 + 2 = 7$ inches.

7. **Check:** Is the longest side 1 less than twice the shortest side? Yes, since $9 = 2 \cdot 5 - 1$. Is the third side 2 more than the shortest side? Yes, since $7 = 5 + 2$. Is the perimeter 21 inches? Yes, since $5 + 9 + 7 = 21$. Therefore, our solutions are correct.

Practice Exercises

1. A rectangular rug on the living room floor is 4 feet longer than it is wide. If the distance around the rug is 40 feet, find the dimensions (length and width) of the rug.

2. The length of the longest side of a triangle is twice the length of the shortest side. The length of the middle-sized side is 5 more than the length of the shortest side. Find the lengths of the sides of the triangle if the perimeter is 33 feet.

The second type of geometric application problem is that for which we have to represent the unknown(s) in terms of a variable and then write an equation expressing the relationship between the unknowns using some known geometric relationship. Example 2 will be of this type. These relationships often involve angles. An angle is usually denoted by using the symbol $\angle$. The size of an angle is measured in degrees and is called the measure of the angle. The measure of $\angle A$ is denoted as $m\angle A$. Two angles with the same number of degrees (measure) are called **congruent** angles and the symbol for congruent is $\cong$. These known geometric relationships may come from, but are not limited to, the following:

1. The sum of the measures of all the angles of any triangle is $180°$.

2. Two angles are **supplementary** if the sum of their measures is $180°$. Each angle is the supplement of the other.

3. Two angles are **complementary** if the sum of their measures is $90°$. Each angle is the complement of the other.

4. Angle relationships involving parallel lines. Two or more lines are parallel if they lie in the same plane and do not intersect. Any line intersecting two or more lines is called a **transversal.** In the figure that follows, L_1 and L_2 are parallel and T is a transversal. The angle relationships are summarized as follows:

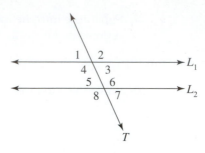

a. **Alternate interior angles** are congruent. The two pairs of alternate interior angles are $\angle 4$ and $\angle 6$, and $\angle 3$ and $\angle 5$. Therefore, $m\angle 4 = m\angle 6$ and $m\angle 3 = m\angle 5$.

b. **Alternate exterior angles** are congruent. The two pairs of alternate exterior angles are $\angle 1$ and $\angle 7$, and $\angle 2$ and $\angle 8$. Therefore, $m\angle 1 = m\angle 7$ and $m\angle 2 = m\angle 8$.

c. **Corresponding angles** are congruent. The four pairs of corresponding angles are $\angle 1$ and $\angle 5$, $\angle 2$ and $\angle 6$, $\angle 3$ and $\angle 7$, and $\angle 4$ and $\angle 8$. Therefore, $m\angle 1 = m\angle 5$, $m\angle 2 = m\angle 6$, $m\angle 3 = m\angle 7$, and $m\angle 4 = m\angle 8$.

d. **Interior angles** on the same side of the transversal are supplementary. The interior angles on the same side of the transversal are $\angle 3$ and $\angle 6$ and also $\angle 4$ and $\angle 5$. Therefore, $m\angle 3 + m\angle 6 = 180°$ and $m\angle 4 + m\angle 5 = 180°$.

Our procedure for solving will be exactly the same except for writing the equation that in this case comes from a known geometric fact rather than a known formula.

Example 2

a. In a triangle, the measure of the largest angle is twice the measure of the smallest angle. If the measure of the middle-sized angle is 12 more than the measure of the smallest angle, find the largest angle.

Solution:

1. Identify the unknown(s) and represent one of them with a variable. We are looking for the measure of each of the three angles of the triangle.

   Let t represent the measure of the smallest angle.

2. Represent all other unknowns in the problem in terms of the same variable. We know that the measure of the largest angle is twice the measure of the smallest, so:

   $2t$ represents the measure of the largest angle.

   We also know that the measure of the middle-sized angle is 12 more than the measure of the smallest angle, so:

   $t + 12$ represents the measure of the middle-sized angle.

3. What is the known fact that will help us write an equation? We know that the sum of the measures of all the angles of a triangle is 180°.

$$m\angle 1 + m\angle 2 + m\angle 3 = 180°$$

4. Substitute into the equation. We have represented the angles by t, $2t$, and $t + 12$, so:

   $$t + 2t + (t + 12) = 180°$$

5. Solve the equation.

$t + 2t + (t + 12) = 180$	Remove parentheses.
$t + 2t + t + 12 = 180$	Simplify the left side of the equation.
$4t + 12 = 180$	Subtract 12 from both sides.
$4t + 12 - 12 = 180 - 12$	Simplify both sides.
$4t = 168$	Divide both sides by 4.
$\dfrac{4t}{4} = \dfrac{168}{4}$	Simplify both sides.
$t = 42°$	Solution of the equation.

6. Have we answered the question that was asked? No, the largest angle is $2t = 2(42°) = 84°$.

7. Check: The largest angle is 84°. Is this twice the smallest? The smallest angle is 42° and $84° = 2(42°)$. So, the answer is yes. Is the middle angle 12° more than the smallest? The middle angle is $t + 12 = 42° + 12° = 54°$, and 54° is 12° more than 42°. Is the sum of the three angles 180°? Since $42° + 54° + 84° = 180°$, yes. Therefore, we have the correct solution.

b. Given the following figure with L_1 parallel to L_2, find the value of x and the $m\angle A$ and $m\angle B$.

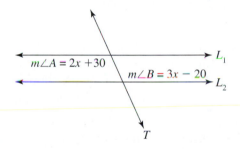

Solution:

1. Identify the unknown(s) and represent one of them with a variable. Since we are given a labeled figure, this step is not necessary in this case.

2. Represent all other unknowns in terms of the same variable. Again, this is not necessary.

3. What is the known geometric relationship? $\angle A$ and $\angle B$ are alternate interior angles of two parallel lines. Therefore, their measures are equal.

4. Write the equation.

$$m\angle A = m\angle B \text{ so:}$$
$$2x + 30 = 3x - 20$$

5. Solve the equation.

$2x + 30 = 3x - 20$	Subtract $2x$ from both sides.
$30 = x - 20$	Add 20 to both sides.
$50 = x$	Therefore, $x = 50$.

6. Have we answered the question that was asked? Partially. We still need to find $m\angle A$ and $m\angle B$. $m\angle A = 2x + 30 = 2 \cdot 50 + 30 = 100 + 30 = 130°$. $m\angle B = 3x - 20 = 3 \cdot 50 - 20 = 150 - 20 = 130°$.

7. Check: Are the two angles congruent? Yes, since both have measures of 130°.

Practice Exercises

3. The measure of the smallest angle of a triangle is 18° less than the measure of the middle angle. If the measure of the largest angle is four times the measure of the smallest angle, find all three angles.

4. The measure of an angle is 30° less than twice the measure of its supplement. Find the measure of each of the angles.

Exercise Set 3.7

1. The width of a rectangular piece of cloth is 1 foot shorter than the length. If the perimeter is 10 feet, how long is the cloth?

2. A mathematics classroom at Urban College is rectangular in shape and measures 6 meters longer than it is wide. If the perimeter is 44 meters, what are its length and width?

3. A playing field for soccer is rectangular in shape and is 40 yards longer than it is wide. If the perimeter is 220 yards, what are its length and width?

4. A rectangular office building is four times as long as it is wide. If the perimeter is 900 feet, what are the length and width of the building?

5. A rectangular parking space has a length that is 2 feet less than three times the width. If the perimeter is 60 feet, what are the dimensions of the parking space?

6. The length of a rectangular sign is 4 inches less than twice the width. The perimeter of the sign is 28 inches. Find the length and the width.

7. The length of the longest side of a triangle is twice the length of the shortest side and the length of the third side is 2 inches more than the shortest side. If the perimeter is 18 inches, find the length of each of the three sides of the triangle.

8. The length of the longest side of a triangle is three times the length of the shortest and the length of the third side is 4 feet more than the length of the shortest. If the perimeter is 19 feet, find the length of each of the three sides of the triangle.

9. Two sides of a triangular flower bed are equal in length and the third side is 4 feet less than twice the length of the equal sides. If the perimeter of the flower bed is 28 feet, find the length of each of the three sides of the flower bed.

10. Two sides of a triangular sign are equal in length and the length of the third side is 3 inches less than twice the length of the equal sides. If the perimeter of the sign is 21 inches, find the length of each of the three sides of the sign.

11. One angle of a triangle measures 90°. One of the remaining two angles is twice as large as the other. Find the measure of the remaining two angles.

12. Two of the angles of a triangle are equal in measure. The measure of the remaining angle is twice the sum of the measures of the two equal angles. Find the measures of the three angles.

13. The measure of the first angle of a triangle is three times the measure of the second angle and the measure of the third angle is 5 more than the measure of the first angle. Find the measures of the three angles.

14. The measure of the smallest angle of a triangle is .3 of the measure of the largest angle and the measure of the middle angle is .5 of the measure of the largest angle. Find all three angles.

15. The measure of the complement of an angle is 60° less than twice the measure of the angle. Find the measures of both angles.

16. The measure of the supplement of an angle is 10° more than four times the measure of the angle. Find the measures of both angles.

17. The measure of the supplement of an angle is 20° more than three times the measure of the angle. Find the measure of both angles.

18. The measure of the complement of an angle is 15° less than four times the measure of the angle. Find the measure of both angles.

19. Given the following figure where L_1 and L_2 are parallel, find the value of x and the measures of $\angle A$ and $\angle B$.

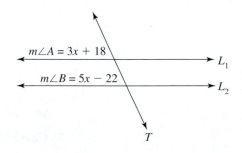

$m\angle A = 3x + 18$
$m\angle B = 5x - 22$

20. Given the following figure where L_1 and L_2 are parallel, find the value of x and the measures of $\angle A$ and $\angle B$.

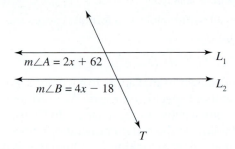

$m\angle A = 2x + 62$

$m\angle B = 4x - 18$

L_1

L_2

T

21. Given the following figure where L_1 and L_2 are parallel, find the value of x and the measures of $\angle A$ and $\angle B$.

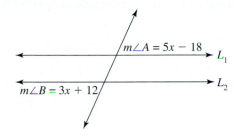

$m\angle A = 5x - 18$

L_1

$m\angle B = 3x + 12$

L_2

22. Given the following figure where L_1 and L_2 are parallel, find the value of x and the measures of $\angle A$ and $\angle B$.

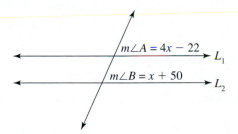

$m\angle A = 4x - 22$

L_1

$m\angle B = x + 50$

L_2

23. Given the following figure where L_1 and L_2 are parallel, find the value of x and the measures of $\angle A$ and $\angle B$.

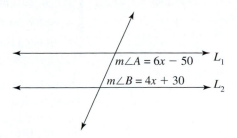

$m\angle A = 6x - 50$

L_1

$m\angle B = 4x + 30$

L_2

24. Given the following figure where L_1 and L_2 are parallel, find the value of x and the measures of $\angle A$ and $\angle B$.

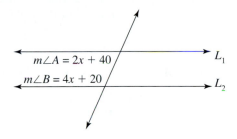

$m\angle A = 2x + 40$

L_1

$m\angle B = 4x + 20$

L_2

Challenge Exercise:

25. The following type of problem often appears on standardized tests. Find the value of x in the figure that follows:

$37°$

$28°$

$x°$

Writing Exercises or Group Projects:

If done as a group project, each group should write two exercises of each type, exchange them with another group, and then solve them.

26. Write an application problem involving the perimeter of a triangle.

27. Write an application problem involving angle relationships.

Section 3.8	Using and Solving Formulas

OBJECTIVES

When you complete this section, you will be able to:

a. Solve for the unknown when given all of the values in a formula except one.

b. Solve a formula for a given variable.

Introduction

Formulas are equations that express relationships that apply to a whole class of problems. They are like ready-made equations without specific numbers. We used several geometric formulas in Chapter 1 and in previous sections of this chapter. In Chapter 1, we simply substituted the given values and followed the order of operations since the formula was already solved for the quantity that we were looking for. Often that is not the case, and we have to use the techniques for solving linear equations to find the desired value.

Solving for the unknown when given all of the values in a formula except one

In using a formula, we must know the values of all of the variables except the one for which we are solving. We substitute the known values for the variables and then solve for the unknown variable using the techniques from the previous sections of this chapter.

Example 1

Solve each of the following for the indicated variable:

a. The formula for distance is $d = rt$. If $r = 30$ miles per hour and $t = 5$ hours, find d.

Solution:

$d = rt$	Replace r with 30 and t with 5.
$d = 30 \cdot 5$	Multiply.
$d = 150$	Therefore, the distance is 150 miles.

c. The formula for finding the circumference (distance around) of a circle is $C = 2\pi r$. If $C = 9.42$ ft, find r. Use $\pi \approx 3.14$.

Solution:

$C = 2\pi r$	Replace C with 9.42 and π with 3.14.
$9.42 \approx 2(3.14)r$	Simplify the right side of the equation.
$9.42 \approx 6.28r$	Divide both sides by 6.28.
$\dfrac{9.42}{6.28} \approx \dfrac{6.28r}{6.28}$	Simplify both sides.
$1.5 \approx r$	Therefore, the radius is approximately 1.5 feet.

b. The formula for the perimeter of a rectangle is $P = 2L + 2W$. If $P = 20$ in. and $L = 6$ in., find W.

Solution:

$P = 2L + 2W$	Replace P with 20 and L with 6.
$20 = 2(6) + 2W$	Simplify the right side of the equation.
$20 = 12 + 2W$	Subtract 12 from both sides.
$8 = 2W$	Divide both sides by 2.
$4 = W$	Therefore, the width is 4 inches.

d. The formula for the area of a triangle is $A = \dfrac{bh}{2}$. If $A = 60$ in.^{2} and $b = 15$ in., find h.

Solution:

$A = \dfrac{bh}{2}$	Replace A with 60 and b with 15.
$60 = \dfrac{15h}{2}$	Multiply both sides of the equation by $\dfrac{2}{15}$.
$\dfrac{2}{15}(60) = \dfrac{2}{15} \cdot \dfrac{15h}{2}$	Simplify both sides. $\frac{2}{15}(60) = \frac{2}{15} \cdot \frac{60}{1} = \frac{120}{15} = 8.$
$8 = 1h$ or $h = 8$	Therefore, $h = 8$ inches.

Practice Exercises

Solve the following:

1. The formula for the volume of a box is $V = LWH$. If $L = 10$ feet, $H = 6$ feet, and $V = 180$ cubic feet, find W.

2. In physics, the formula for work is $W = Fd$, where W represents work in foot pounds, F represents force in pounds, and d represents distance in feet. If $F = 40$ pounds and $d = 2$ feet, find W.

3. In electricity, the formula for power is $P = iV$, where P represents power in watts, i represents current in amps, and V represents voltage. If $P = 140$ watts and $i = 10$ amps, find V.

4. The formula for rate is $r = \frac{d}{t}$, where d is the distance and t is the time. Find the distance if the rate is 45 miles per hour and the time is four hours.

If more practice is needed, do the Additional Practice Exercises.

Additional Practice Exercises

Solve the following:

a. The formula for interest is $I = prt$. If $I = \$67.50$, $p = \$1500$, and $t = .5$, find r.

b. The formula for the area of a rectangle is $A = LW$. If $W = 3.5$ m and $A = 21$ m^2, find L.

c. The formula for the area of a circle is $A = \pi r^2$. If $r = 2$ m, find A. Use $\pi \approx 3.14$.

d. The formula for the area of a trapezoid is $A = \frac{h(B + b)}{2}$. If $A = 18$ ft^2, $h = 4$ ft, and $B = 6$ ft, find b.

Each variable in a formula may be solved in terms of the other variables of that formula. If we have several problems in which we are asked to find the same variable, we may want to solve the formula for that variable before substituting values for the other variables. This way we do not have to solve several similar equations for the same variable several times.

Example 2

Solve the following formulas for the indicated variable:

a. $d = rt$, for t

Solution:
We treat all the other variables as if they were numbers and solve.

$d = rt$ ⟶ Divide both sides of the equation by r.

$\dfrac{d}{r} = \dfrac{rt}{r}$ ⟶ Simplify the right side. $\left(\dfrac{r}{r} = 1\right)$

$\dfrac{d}{r} = t$ ⟶ Therefore, $t = \dfrac{d}{r}$.

b. $A = \dfrac{1}{2}bh$, for h

$A = \dfrac{1}{2}bh$ ⟶ Multiply both sides of the equation by 2.

$2A = 2\left(\dfrac{1}{2}\right)bh$ ⟶ Simplify both sides.

$2A = bh$ ⟶ Divide both sides by b. $\left(\dfrac{b}{b} = 1\right)$

$\dfrac{2A}{b} = \dfrac{bh}{b}$ ⟶ Simplify both sides.

$\dfrac{2A}{b} = h$ ⟶ Therefore, $h = \dfrac{2A}{b}$.

c. Solve $P = 2L + 2W$ for W.

$P = 2L + 2W$ ⟶ Subtract $2L$ from both sides of the equation.

$P - 2L = 2L - 2L + 2W$ ⟶ Simplify the right side.

$P - 2L = 2W$ ⟶ Divide both sides by 2.

$\dfrac{P - 2L}{2} = \dfrac{2W}{2}$ ⟶ Simplify the right side.

$\dfrac{P - 2L}{2} = W$ ⟶ Therefore, $W = \dfrac{P - 2L}{2}$.

d. Solve $3x + 4y = 5$ for y.

$3x + 4y = 5$ ⟶ Subtract $3x$ from both sides of the equation.

$3x - 3x + 4y = 5 - 3x$ ⟶ Simplify the left side.

$4y = 5 - 3x$ ⟶ Divide both sides by 4.

$\dfrac{4y}{4} = \dfrac{5 - 3x}{4}$ ⟶ Simplify the left side.

$y = \dfrac{5 - 3x}{4}$ ⟶ Therefore, $y = \dfrac{5 - 3x}{4}$.

Answers:

Practice Exercises

Solve each of the following formulas for the given variable:

5. $W = Fd$ for d

6. $E = \dfrac{1}{2}mv^2$ for m

7. Solve $P = 2L + 2W$ for L.

8. $2x + 3y = 9$ for y

If more practice is needed, do the Additional Practice Exercises.

Additional Practice Exercises

Solve each of the following for the variable indicated:

e. $A = LW$, for L.

f. $P = \dfrac{W}{t}$, for t. (Power)

g. $V = \pi r^2 h$, for h. (Volume of a right circular cylinder)

h. $4x + 2y = 7$, for y

Exercise Set 3.8

Solve the formula for the indicated variable with the given information:

1. $A = \dfrac{bh}{2}$ for A with $b = 2$ mi and $h = 8$ mi

2. $A = \dfrac{bh}{2}$ for b with $A = 30$ in.^{2} and $h = 10$ in.

3. $A = \dfrac{bh}{2}$ for h with $A = 14$ ft^2 and $b = 7$ ft

4. $d = rt$ for d with $r = 60$ miles per hour and $t = 3$ hours

5. $d = rt$ for t with $r = 50$ km per hour and $d = 75$ km

6. $d = rt$ for r with $t = 6$ hours and $d = 720$ km

7. $C = 2\pi r$ for r with $C = 25.12$ in. and $\pi = 3.14$

8. $C = 2\pi r$ for C with $r = 5\pi$ m and $\pi = 3.14$

9. $P = 2W + 2L$ for P with $W = 6$ in. and $L = 15$ in.

10. $P = 2W + 2L$ for W with $L = 7$ cm and $P = 24$ cm

In the following exercises we will omit units, since many of them are unfamiliar to us: Solve each formula for the indicated variable with the given value(s).

11. $A = \dfrac{1}{2}(B + b)h$ for A with $b = 5$, $B = 7$, and $h = 4$

12. $A = \dfrac{1}{2}(B + b)h$ for h with $B = 6$, $b = 8$, and $A = 35$

13. $V = iR$ for V with $i = 3$ and $R = 5$

(Ohm's law in electricity)

14. $V = iR$ for R with $V = 120$ and $i = 3$

15. $P = iV$ for P with $i = .5$ and $V = 120$

(Electricity: power formula)

16. $P = iV$ for i with $P = 150$ and $V = 120$

17. $W = mg$ for W with $m = 82$ and $g = 32$

(Physics: formula for weight)

---

18. $W = mg$ for m with $W = 170$ and $g = 32$

19. $P = \dfrac{F}{A}$ for P with $F = 180$ and $A = 15$

(Physics: pressure on an area)

20. $P = \dfrac{F}{A}$ for F with $P = 50$ and $A = 8$

Solve each formula for the indicated variable.

21. $K = C + 273$ for C.

22. $K - 273 = C$ for K

(Conversion formulas for Kelvin and Celsius temperatures)

23. $P = kT$ for T

24. $P = kT$ for k

(Chemistry: gas law)

25. $F = ma$ for m

26. $F = ma$ for a

(Physics: law of motion)

27. $v = gt$ for t

28. $v = gt$ for g

(Physics: velocity of falling object)

29. $PV = kT$ for V

30. $PV = kT$ for T

(Chemistry: universal gas law)

31. $E = mc^2$ for m

32. $E = mc^2$ for c^2

(Nuclear Physics: conversion from mass to energy)

33. $I = prt$ for r

34. $V = LWH$ for W

(Finance: formula for calculating interest)

35. $a^2 + b^2 = c^2$ for a^2

36. $a^2 + b^2 = c^2$ for b^2

(Geometry: Pythagorean theorem for right triangles)

37. $d = \dfrac{1}{2}gt^2$ for g

38. $d = \dfrac{1}{2}gt^2$ for t^2

(Physics: distance traveled in free fall)

39. $D = \dfrac{M}{V}$ for M

40. $D = \dfrac{M}{V}$ for V

Solve each of the following for y:

41. $2x + y = 3$

42. $3x + y = 5$

43. $3x + 2y = 6$

44. $4x + 3y = 8$

45. $6x - 3y = 7$

46. $4x - 2y = 7$

Challenge Exercises:

Solve the given formula for the indicated variable.

47. $A = \dfrac{1}{2}(B + b)h$ for B.

48. $A = \dfrac{1}{2}(B + b)h$ for b.

49. $F = k\dfrac{q_1 q_2}{r^2}$ for q_1

50. $F = \dfrac{9}{5}C + 32$ for C.

Chapter 3   Summary

Definition Linear Equations **[Section 3.1]**	• A linear equation is any equation that can be put in the form $ax + b = c$, where only one variable is raised to the first power and a, b, and c are constants.
Addition Property of Equality **[Section 3.1]**	• If $a = b$, then $a + c = b + c$ for any number c. In words, we may add any number to both sides of an equation to get an equation with the same solution(s).
Multiplication Property of Equality **[Section 3.2]**	• If $a \neq b$, and $c - 0$, then $a \cdot c = b \cdot c$. In words, we may multiply both sides of an equation by a nonzero number to get another equation with the same solution(s).

Procedure: Solving Linear Equations [Section 3.3]

1. If necessary, simplify both sides of the equation as much as possible. This could involve using the distributive, commutative, and associative properties and adding like terms.

2. If necessary, use the Addition Property of Equality to get all the terms with variables on one side of the equation and all the constant terms on the other.

3. If necessary, use the Multiplication Property of Equality to eliminate any coefficient on the variable.

4. Check the answer in the original equation.

Identities and Contradictions **[Section 3.3]**	• An **identity** is an equation that is true for all values of the variable for which the equation is defined. When solving an identity, we arrive at a statement that is obviously true, such as $10 = 10$. The solutions of an identity are "all real numbers." • A **contradiction** is an equation that has no solution. When solving a contradiction, we arrive at a statement that is obviously false, such as $5 = 10$. The solution set is indicated by 0.

Procedure: Graphing Inequalities [Section 3.4]

• Equalities are graphed by placing a dot on the number line at the appropriate point.

• Inequalities are graphed by:

   a. placing an open dot on the number line at the appropriate place and shading to the left for less than ($<$) or shading to the right for greater than ($>$).

   b. placing a shaded dot on the number line at the appropriate place and shading to the left for less than or equal to ($\leq$) or shading to the right for greater than or equal to ($\geq$).

Addition Properties of Inequalities [Section 3.4]

a. If $a < b$, then $a + c < b + c$ for any c.

b. If $a > b$, then $a + c > b + c$ for any c.

• If we add any number to both sides of an inequality, we get an inequality with the same solutions.

Multiplication Properties of Inequalities [Section 3.4]

a. If $a < b$ and $c > 0$, then $a \cdot c < b \cdot c$.

b. If $a > b$ and $c > 0$, then $a \cdot c > b \cdot c$.

• If we multiply both sides of the inequality by a *positive number*, we get another inequality with the same solutions.

a. If $a < b$ and $c < 0$, then $a \cdot c > b \cdot c$.

b. If $a > b$ and $c < 0$, then $a \cdot c < b \cdot c$.

• If we multiply both sides of an inequality by *a negative number*, we must reverse the direction of the inequality symbol.

Problem-Solving Procedures [Section 3.5]

Procedure: Solving Real-World Problems

1. Identify the unknown(s) and represent one of them with a variable. It is *usually* best to let the variable represent the smaller or smallest quantity.

2. Represent all other unknowns in the problem in terms of this variable.

3. Write a word equation.

4. Write an algebraic equation using the word equation as a guide.

5. Solve the algebraic equation.

6. Be sure that you have answered the question that was asked.

7. Check the solution against the wording of the original problem.

Procedure: Solving Distance, Rate, and Time Problems

1. Make a chart with columns for the moving objects, the distance, the rate, and the time.

2. Fill in one column with known numerical values.

3. Assign a variable to one of the unknowns and represent any other unknowns in terms of this variable. Fill in another column with those expressions containing variables.

4. Fill in the remaining column from the first two, using the appropriate relationships between d, r, and t. In this section, $d = rt$.

5. Write the equation using the information in the last column that was filled in. It is sometimes helpful to draw a diagram.

6. Solve the equation.

7. Be sure that you have answered the question that was asked.

8. Check the solution against the wording of the original problem.

Money, Investment, and Mixture Problems [Section 3.6]

Procedure: Solving Money Problems

1. Draw a chart with four columns and a row for each item purchased. Label the columns Type of Item, Price per Unit, Number of Units, and Total Cost. Fill in the Type of Item column first.

2. Fill in the Price per Unit column. These values will usually be given, but may involve a variable.

3. Fill in the Number of Units column. This will usually involve a variable. It is usually necessary to assign the variable to the number of units of one item and express the number of units of the other item in terms of this variable.

4. Fill in the Total Value column. The total value is the product of the price per unit and the number of units.

5. Write the word equation. This will almost always be in a form such that the sum of the values of the individual items equals the total value of all the items.

6. Write the algebraic equation using the word equation as a guide.

7. Solve the equation.

8. Be sure you answered the question that was asked.

9. Check the solution against the wording of the original equation.

- For coin problems, change the headings to Type of Coin, Value of Coin, Number of Coins, and Total Value.

- For investment problems, change the headings to Type of Investment, Interest Rate (expressed as a decimal), Amount Invested, and Interest Earned.

- For mixture problems, change the headings to Type of Ingredient, Price per Unit, Number of Units, and Total Value. Also, add an extra row for the mixture.

Geometric Applications Problems
[Section 3.7]

- Know the formulas for the perimeters, areas, and volumes of common geometric figures.

- The sum of all three angles of any triangle is 180°.

- Two angles are *supplementary* if the sum of their measures is 180°.

- Two angles are *complementary* if the sum of their measures is 90°.

- If two lines are parallel, then a) pairs of alternate interior angles are congruent, b) pairs of alternate exterior angles are congruent, and c) pairs of corresponding angles are congruent, d) interior angles on the same side of the transversal are supplementary.

Chapter 3   Review Exercises

Solve using the Addition Property of Equality. [Section 3.1]

1. $x - 6 = -2$

2. $x + \dfrac{2}{3} = \dfrac{5}{6}$

3. $6v - 2 = 5v$

4. $10 + z = 7$

5. $5z - 28 - 4z = 12$

6. $6 + 2w + 3 = w$

7. $4u + 13 = 3u + 10$

8. $11 + 15b = 14b + 18$

9. $-4a + 14 + 9a = 5a + 2 - a$

10. $7z - 6 + 2z = 6z - 1 + 2z$

Translate the mathematical sentence into an English sentence in two ways. [Section 3.1]

11. $2x - 5 = 14$

12. $5 + x = -2$

Translate the English sentence into a mathematical sentence and solve. Use x as the variable. [Section 3.1]

13. Twelve more than some number is negative two.

14. A number decreased by four is six.

Solve the following:

15. John and Shelley share the cost of a TV. If the TV costs $360 and John pays $220, how much does Shelley pay?

16. Harold bought a suit for $230 after a discount of $50. What was the price of the suit before the discount?

Find the reciprocal of each of the following: [Section 3.2]

17. $\dfrac{-2}{7}$

18. .4

Solve using the Multiplication Property of Equality. [Section 3.2]

19. $5x = -40$

20. $35 = -7y$

21. $\dfrac{3}{5}t = 9$

22. $\dfrac{x}{3} = -4$

Translate the mathematical sentence into an English sentence. [Section 3.2]

23. $14y = -28$

24. $\dfrac{2}{3}x = -6$

Translate the English sentence into a mathematical sentence and solve. Use x as the variable. [Section 3.2]

25. The product of four and some number is negative twelve.

26. A number divided by negative two is three.

Solve the following: [Section 3.2]

27. A TV costs three times as much as a VCR. If the TV costs $468, how much does the VCR cost?

28. The area of a parallelogram is 84 m² and the base is 14 m. What is the height? ($A = bh$)

Solve the following equations: [Section 3.3]

29. $5w - 13 = 7w + 15$

30. $11 - 4z = 7 + 44$

Solve the following.

31. The cost of a TV is three times the cost of a VCR. If the TV costs $495, find the cost of the VCR.

32. Hans is renting a sofa for $12.75 per month. If a new sofa costs $306, how many months will it take for the rental fees to equal the cost of a new sofa?

33. A high school boasts that $\frac{3}{5}$ of its teachers have a master's degree. If 45 teachers have a master's degree, how many teachers does the school have?

34. The value of a piece of property increased by 2% last year, which represented an increase of $3600. What was the value of the property before the 2% increase in value?

35. The perimeter of a rectangle is 42 cm and the length is 12 cm. What is the width?

36. The surface area of a rectangular solid is 236 in.^{2}, the length is 6 inches, and the width 5 inches. What is the height?

Simplify, then solve. [Section 3.3]

37. $5t - 11 - 9t + 6 = 19 + 8t$

38. $4 - 5u + 12 + 16u = -6u - 9$

39. $33 - 2(7r + 9) = -9 - 6r$

40. $19 + 24s = 8 - 3(5 - 9s)$

41. $.6(8 - 5v) + .4 = v + 1.2$

42. $3.6 - 4(.7p - 15) = 2.2p - 1.4$

43. $5(q - 3) - 7(11 - 2q) = 3$

44. $8(10 - 4a) + 15 = -3(a + 7)$

Determine whether the following are identities or contradictions. Give the solutions of each. [Section 3.3]

45. $2(x - 3) + 4x = 6(x - 2)$

46. $3(2x - 1) - 4x = 2(x - 4) + 5$

Translate the mathematical sentence into an English sentence. [Section 3.3]

47. $3x - 2 = 5$ 48. $4(2x - 9) = 22$

Translate the English sentence into a mathematical sentence and solve. Use x as the variable. [Section 3.3]

49. The sum of four times a number and three is eleven.

50. Fifty-four is equal to six times the difference of three times a number and nine.

Solve each of the following: [Section 3.3]

51. The charges for repairing a transmission were $255 including parts. If the mechanic worked for four hours and the parts cost $75, how much does the mechanic charge per hour?

52. Rachel paid $37.10 for a blouse, including 6% for taxes. What was the cost of the blouse?

Graph each of the following on the number line: [Section 3.4]

53. $v \geq 0$ 54. $w < 4$

55. $-1 \leq x < 5$ 56. $-3 \leq y \leq 2$

Solve each of the following inequalities by using the Addition Property of Inequality. [Section 3.4]

57. $10p - 8 < 9p + 7$ 58. $6 - 4q \leq -5q + 18$

59. $1.6x - 4 > 12 - .4x$ 60. $9.2 - 15y \geq -14y + 7$

61. $14z + 9 - 9z < 6z + 21$

62. $10 - s + 14 > 7 - 2s - 3$

Translate the mathematical sentence into an English sentence in two ways. [Section 3.4]

63. $19 \leq t + 9$

Translate the English sentence into a mathematical sentence. [Section 3.4]

64. Eleven is greater than or equal to some number increased by six.

Solve by using the Multiplication Property of Inequality. [Section 3.4]

65. $6t \leq -42$ 66. $-8u < 48$

Solve by using the Addition and Multiplication Properties of Inequalities. [Section 3.4]

67. $9 - 4x > 6x - 11$

68. $33y + 6 \geq 17y + 54$

69. $4z + 32 - 5z < 6z - 13 + 2z$

70. $21 - 8r - 7 > 9r - 1 - 2r$

71. $5(2u - 2) - 6u \leq 14$

72. $7v > 8(3 - v) - 9$

73. $6(2s + 4) - 9 \geq -3(4s + 3)$

74. $3(t - 4) + 6(2t + 3) \leq 6t - 39$

75. $-3 < 2x + 5 < 7$

76. $6 \leq 3x - 9 \leq 12$

77. $-6 \leq -4x + 2 \leq 10$

78. $-5 < -3x + 4 \leq 13$

Solve each of the following: [Section 3.5]

79. In a large lecture class, there were 53 more men than women. If there were 329 students in the class, how many women were in the class?

80. The perimeter of a triangle is 16 feet and the length of two of the sides are 5 feet and 4 feet. What is the length of the third side?

81. This year the legislature increased tuition for college students by .1 of last year's amount. If tuition is now $35 per credit hour, what was it last year? (Round to the nearest cent.)

82. The sum of three consecutive integers is 33. Find the integers.

83. Three consecutive odd integers are such that the sum of four times the first and the second is three times the third. Find the integers.

Solve each of the following: [Section 3.6]

84. An office purchased 50 reams of paper for which they paid $260. There were two types of paper. The lighter weight paper cost $4.00 per ream and the heavier paper cost $7.00 per ream. How many reams of each type of paper did they purchase?

85. At the end of each day, David puts all the nickels and dimes that he has in his pockets in a jar. At the end of the week, he has a total of 32 coins whose total value is $2.20. How many of each type of coin does he have?

86. Amy matched five of the six numbers in the Florida lotto, and after the celebration party, had $6500 left,

which she decided to invest. She invested part of the money in municipal bonds paying 6% per year and the remainder in certificates of deposit paying 5% per year. How much did she invest in each if her total income from the two investments is $360 per year?

87. A chemist needs 500 ml of a solution that is 40% sulfuric acid, but all that she can find in the lab are solutions that are 25% sulfuric acid and 50% sulfuric acid. How much of each should she mix?

88. A supermarket sells a mixture of blueberries and strawberries. How many pounds of blueberries that sell for $2.00 per pound must be mixed with 12 pounds of strawberries that sell for $3.00 per pound if the mixture sells for $2.60 per pound?

Solve the following: [Section 3.7]

89. In a suburban area, a residential lot is typically 25 feet longer than it is wide. If the perimeter of the lot is 350 feet, what are the dimensions of the lot?

90. In a triangle, the middle angle is 45° and the largest angle is 15° more than twice the smallest angle. Find the smallest and largest angles.

91. The measure of an angle is 16° more than the measure of its complement. Find the measure of each.

92. Given that L_1 is parallel to L_2 in the following figure, find x.

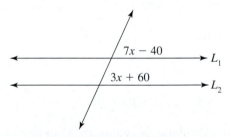

Solve the formula for the indicated variable with the given information. [Section 3.8]

93. $C = 2\pi r$ for C if $r = 3$. Use $\pi \approx 3.14$.

Solve the formula for the indicated variable. [Section 3.8]

94. $ax + b = c$ for x

Chapter 3   Test

Solve the following equations:

1. $3u + 4 = 2u - 9$

2. $30 - 5v - 19 = 8 - 6v$

3. $-7k = 42$

4. $.8m = -.54$

5. $6v - 13 = 8v + 7$

6. $5t + 7 - 9t = 4t - 30$

7. $13 - .5(y + 4) = 3.5y - 9$

8. $-3(2a - 6) + 15 = 4(a + 11) - 5a$

Solve the following inequalities:

9. $12p - 3 > 1 + 11p$

10. $14 - 2.3q \leq 6 - 1.3q$

11. $5u - 10 - 3u \geq u + 7$

12. $14x - 7 + 2x > 9x + 42$

13. $10(y - 5) \leq 12y - 38$

14. $23 - 5(4z - 1) < 7(6 - 3z) + 2z$

15. $-3 < 2x + 1 < 7$

Translate the mathematical sentences into English sentences.

16. $\dfrac{x}{3} + 4 = 6$

17. $2(s - 5) = 3s$

Translate the English sentence into a mathematical sentence.

18. Three times the difference of a number and six is one more than the number.

Solve each of the following:

19. A drywall hanger knows that the area of a wall is 128 ft² and the length of the wall is 16 feet. What is the height of the wall?

20. An animal trap for catching small animals is in the shape of a rectangular solid with a surface area of 1384 in.². If the length is 26 inches and the width is 12 inches, what is the height?

21. Ericka receives $150 plus a 4% commission on all her sales. If she earned $360 last weekend, what were her sales?

22. Oscar paid $124 for four pairs of shorts that all cost the same. If he had a coupon for $20 off, what was the cost of each pair of shorts?

23. A certain wine is 12% alcohol by volume. If a bottle contains 750 milliliters of wine, how many milliliters of alcohol does the wine contain?

24. The women's basketball team at Central College outscored their opponents by 143 points last season. If the total number of points scored by both Central and their opponents when they played each other was 4921, how many points did Central score?

25. The sum of three consecutive integers is equal to four times the second integer. Find the integers.

26. A building contractor purchased a total of 25 light bulbs, some of which were incandescent and some of which were fluorescent, at a total cost of $24. If the incandescent bulbs cost $.60 each and the fluorescent bulbs cost $1.50 each, how many of each did he buy?

27. A long-distance runner left his house running at an average rate of 8 miles per hour. Fifteen minutes (one-fourth of an hour) later, his son left his house on his bike traveling on the same route at an average rate of 12 miles per hour. How long will it take the son to catch up with his father?

28. A television picture tube is 9 cm longer than it is wide. If the perimeter of the tube is 142 cm, what are the length and width of the tube?

Solve the following for the indicated variable:

29. $P = 2W + 2L$ for W

30. $A = \dfrac{1}{2}(B + b)h$ for B

Graphing Linear Equations and Inequalities

In Chapter 3, we solved linear equations and inequalities that contained only one variable. In Chapter 5.8, we will solve quadratic equations with one variable by factoring. In this chapter, we will learn to solve linear equations and inequalities that contain two variables.

The solution of a linear equation with one variable is usually a single number, but sometimes there is no solution or there may be an infinite number of solutions. In order to solve equations with two variables, we will first need to define what is meant by the solution of an equation with two variables. We will also need a method of denoting solutions of equations with two variables that allows us to specify a value of each variable in the equation.

In Chapter 1, we learned to graph whole numbers and integers on the number line. In this chapter, we will learn to graph the solutions of equations with two variables. Since a solution will consist of a value for each variable, the solution cannot be graphed on a number line. Hence, graphing a solution of an equation with two variables will require the use of something other than a number line. The graph of the solutions of an equation with two variables will be quite different from the graph of a set of integers.

In Chapter 3, we learned to graph the solutions of linear inequalities with one unknown. In this chapter, we will learn to graph the solutions of linear inequalities with two variables. There will be similarities between the graphs of linear inequalities with two variables and the graphs of linear inequalities with one variable in that both require the shading of regions.

We end the chapter by studying some characteristics of the graphs of linear equations with two variables and by writing equations with two variables that satisfy certain given conditions.

Section 4.1	Ordered Pairs and Solutions of Linear Equations with Two Variables

OBJECTIVES

When you complete this section, you will be able to:

a. Verify solutions of equations with two variables.

b. Write solutions of linear equations with two variables as ordered pairs.

c. Determine if a given ordered pair is a solution of a given equation.

d. Find the missing number in an ordered pair for a given linear equation.

e. Plot points on the rectangular coordinate system.

f. Represent solutions of everyday situations as ordered pairs.

Introduction

In this chapter, we will discuss **linear equations with two variables**. A linear equation with two variables is any equation of the form $ax + by = c$, or any equation that can be put into that form, where a and b are constants and not equal to 0 at the same time.

In Chapter 3, we solved linear equations with one variable. The solution of a linear equation with one variable is usually a single number. The solution can be verified by replacing the variable with the value and simplifying. For example, the solution of $3x + 7 = 19$ is $x = 4$. We can verify that $x = 4$ is a solution of $3x + 7 = 19$ by replacing x with 4 and simplifying. So $3(4) + 7 = 12 + 7 = 19$. Therefore, the solution is correct.

Verifying solutions of linear equations in two variables

Any solution of a linear equation with two variables must have a value for each variable. Hence, any solution has two values. We check a solution of a linear equation with two variables in much the same way that we check a solution of a linear equation with one variable. We replace each variable with its value and simplify.

Example 1

Verify that the following are solutions of the given equations:

a. $x + 2y = 6; x = 4, y = 1$ Substitute 4 for x and 1 for y.

$4 + 2 \cdot 1 = 6$ Multiply 2 and 1.

$4 + 2 = 6$ Add 4 and 2.

$6 = 6$ Therefore, the solutions are correct.

b. $x + 2y = 6; x = -2, y = 4$ Substitute -2 for x and 4 for y.

$-2 + 2(4) = 6$ Multiply 2 and 4.

$-2 + 8 = 6$ Add -2 and 8.

$6 = 6$ Therefore, the solutions are correct.

Notice two things from Example 1. First, the equation in part a is the same as the equation in part b, but the solutions are different. Consequently, a linear equation in two variables has more than one solution. In fact, it has an infinite number of solutions since we can assign any value to either variable and find the corresponding value for the remaining variable. Second, it is cumbersome to write the solutions in the form given in Example 1. We would have to write "two of the solutions to the linear equation $x + 2y = 6$ are $x = 4$ and $y = 1$, and $x = -2$ and $y = 4$." For this reason, we introduce the notion of an **ordered pair.**

An ordered pair consists of two numbers enclosed in parentheses and separated by a comma. They are called ordered pairs because one variable is assigned to the first number of the pair and a different variable is assigned to the second. Hence, the numbers are written in a specific order determined by the variables.

Writing solutions of linear equations with two variables as ordered pairs

If a set of ordered pairs is in the form (x,y), the first number in the pair is the x-value and the second is the y-value. So, in the ordered pair $(-4,6)$, $x = -4$ and $y = 6$, and in the ordered pair $(2,-3)$, $x = 2$ and $y = -3$. In Example 1, the solution

$x = 4$ and $y = 1$ is represented as the ordered pair $(4,1)$, and the solution $x = -2$ and $y = 4$ is represented as $(-2,4)$. This leads us to the following definition:

Determining if a given ordered pair is a solution of a given equation

Solutions of Linear Equations of Two Variables

If the solutions of $ax + by = c$ are ordered pairs in the form (x,y), the ordered pair (u,v) is a solution if the replacement of x with u and y with v results in a true statement.

Example 2

Determine whether the given ordered pair is a solution of the given equation.

a. $3x - 4y = 9; (7, 3)$ Substitute 7 for x and 3 for y.

$3(7) - 4(3) = 9$ Multiply before adding.

$21 - 12 = 9$ Add 21 and -12.

$9 = 9$ Therefore, (7,3) is a solution.

b. $2x + 3y = 6; (-3, 0)$ Substitute -3 for x and 0 for y.

$2(-3) + 3(0) = 6$ Multiply before adding.

$-6 + 0 = 6$ Add -6 and 0.

$-6 \neq 6$ Therefore, $(-3,0)$ is not a solution.

c. $4x - y = 4; (-1, -8)$ Substitute -1 for x and -8 for y.

$4(-1) - (-8) = 4$ Multiply before adding and $-(-8) = 8$.

$-4 + 8 = 4$ Add -4 and 8.

$4 = 4$ Therefore, $(-1,-8)$ is a solution.

Practice Exercises

Determine whether the given ordered pair is a solution of the given equation. Assume the ordered pairs are in the form (x,y).

1. $2x - 4y = 10; (1, -2)$ **2.** $3x + y = 8; (-3, 1)$ **3.** $3x - y = 7; (2, 1)$

If more practice is needed, do the Additional Practice Exercises.

Additional Practice Exercises

Determine whether the given ordered pair is a solution of the given equation. Assume the ordered pairs are in the form (x,y).

a. $x + 3y = 7; (1, 2)$ **b.** $2x - y = 8; (-4, -2)$ **c.** $3x + 5y = 10; (0, 2)$

Finding the missing member of an ordered pair for a given equation

Suppose we have an equation and know one member of an ordered pair. We can find the other member by substituting the known value for the variable it represents and solving for the other variable.

Example 3

Use the given equation to find the missing member of each of the following ordered pairs. Assume the ordered pair is in the form (x,y).

a. $y = 2x - 6$; $(0, \_)$, $(\_, 0)$, $(5, \_)$, $(-3, \_)$

Solution:

Since the ordered pairs are in the form of (x,y), the ordered pair $(0, \_)$ means $x = 0$ and we need to find y.

$$y = 2(0) - 6 \qquad \text{Substitute 0 for } x.\ 2(0) = 0.$$
$$y = 0 - 6 \qquad \text{Add 0 and } -6.$$
$$y = -6 \qquad 0 - 6 = -6.\ \text{Therefore, the ordered pair is } (0, -6).$$

The ordered pair $(\_, 0)$ means $y = 0$ and we need to find x.

$$0 = 2x - 6 \qquad \text{Substitute 0 for } y.\ \text{Add 6 to both sides.}$$
$$0 + 6 = 2x - 6 + 6 \qquad \text{Simplify both sides.}$$
$$6 = 2x \qquad \text{Divide both sides by 2.}$$
$$\frac{6}{2} = \frac{2x}{2} \qquad \text{Simplify both sides.}$$
$$3 = x \qquad \text{Therefore, the ordered pair is } (3,0).$$

In the ordered pair $(5, \_)$, $x = 5$ and we need to find y. Substitute 5 for x and simplify.

$$y = 2(5) - 6$$
$$y = 10 - 6$$
$$y = 4$$

Therefore, the ordered pair is $(5, 4)$.

In the ordered pair $(-3, \_)$, $x = -3$ and we need to find y. Substitute -3 for x and simplify.

$$y = 2(-3) - 6$$
$$y = -6 - 6$$
$$y = -12$$

Therefore, the ordered pair is $(-3, -12)$.

b. $2x - 3y = 6$; $(0, \_)$, $(\_, 0)$, $(6, \_)$, $(\_, -4)$

Solution:

Since the ordered pair is in the form (x,y), the ordered pair $(0, \_)$ means $x = 0$ and we need to find y.

$$2(0) - 3y = 6 \qquad \text{Substitute 0 for } x.\ 2(0) = 0.$$
$$0 - 3y = 6 \qquad 0 - 3y = -3y.$$
$$-3y = 6 \qquad \text{Divide both sides by } -3.$$
$$y = -2 \qquad \text{Therefore, the ordered pair is } (0, -2).$$

The ordered pair $(\_, 0)$ means $y = 0$ and we need to find x.

$$2x - 3(0) = 6 \qquad \text{Substitute 0 for } y.\ 3(0) = 0.$$
$$2x - 0 = 6 \qquad 2x - 0 = 2x.$$
$$2x = 6 \qquad \text{Divide both sides by 2.}$$
$$x = 3 \qquad \text{Therefore, the ordered pair is } (3, 0).$$

In the ordered pair $(6, \_)$, $x = 6$ and we need to find y. Substitute 6 for x and solve for y.

$$2(6) - 3y = 6$$
$$12 - 3y = 6$$
$$-3y = -6$$
$$y = 2$$

Therefore, the ordered pair is $(6, 2)$.

In the ordered pair $(\_, -4)$, $y = -4$ and we need to find x. Substitute -4 for y and solve for x.

$$2x - 3(-4) = 6$$
$$2x + 12 = 6$$
$$2x = -6$$
$$x = -3$$

Therefore, the ordered pair is $(-3, -4)$.

Use the given equation to find the missing member of each of the following ordered pairs. Assume the ordered pair is in the form (x, y).

c. $x = 2$; $(\_, 0)$, $(\_, -2)$, $(\_, 4)$

Solution:
Notice that there is no "y" term in this equation. We can think of the equation as $x + 0y = 2$. For the ordered pair $(\_, 0)$, $y = 0$ and we need to find x.

$$x + 0(0) = 2 \qquad \text{Substitute 0 for } y. \ 0(0) = 0.$$
$$x + 0 = 2 \qquad x + 0 = x.$$
$$x = 2 \qquad \text{Therefore, the ordered pair is } (2, 0).$$

In the ordered pair $(\_, -2)$, $y = -2$ and we need to find x. Substitute -2 for y and simplify.

$$x + 0(-2) = 2$$
$$x + 0 = 2$$
$$x = 2$$

Therefore, the ordered pair is $(2, -2)$.

In the ordered pair $(\_, 4)$, $y = 4$ and we need to find x. Substitute 4 for y and simplify.

$$x + 0(4) = 2$$
$$x + 0 = 2$$
$$x = 2$$

Therefore, the ordered pair is $(2, 4)$.

Note: In Example 2c, it soon becomes clear that $x = 2$ for any value of y since $0 \cdot y = 0$ for all values of y. When one variable is missing from a linear equation, the remaining variable must always equal the given constant. For example, if $y = -2$, then all ordered pairs solving this equation must have a y-value of -2. The x-value may be anything. Some solutions are $(0, -2)$, $(-3, -2)$, and $(5, -2)$. In general, the solutions are represented by $(x, -2)$ where x can have any value.

Practice Exercises

Use the given equation to find the missing member of each of the following ordered pairs. Assume the ordered pair is in the form (x, y).

4. $y = 3x + 12$; $(0, \_)$, $(\_, 0)$, $(-3, \_)$ **5.** $3x + 4y = 12$; $(0, \_)$, $(\_, 0)$, $(-4, \_)$ **6.** $y = -3$; $(0, \_)$, $(5, \_)$, $(-5, \_)$

If more practice is needed, do the Additional Practice Exercises.

Additional Practice Exercises

Use the given equation to find the missing member of each of the following ordered pairs. Assume the ordered pair is in the form (x, y).

d. $y = -3x + 9$; (0, _) (_, 0),(4, _)   **e.** $x - 3y = 9$; (0, _),(_, 0),(_, -2)   **f.** $x = -4$; (_, 0),(_, -2),(_, 3)

In Chapter 3, we solved linear equations with one variable. It is possible to graph the solution of a linear equation on the number line. For example, the solution of $2x + 7 = 5$ is $x = -1$. This solution is graphed on the number line by placing a dot at -1 as follows:

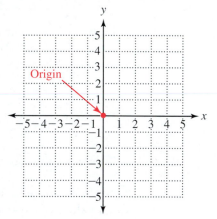

A solution of $2x + y = 8$ is (2, 4). Since a solution to a linear equation in two variables is an ordered pair, we cannot graph a solution on the number line. Since there are two variables, we need two number lines—one for each variable. Suppose the ordered pairs are in the form of (x, y). We draw a horizontal number line and label it the x-line. At the 0 point of the x-line, we draw a vertical number line and put its 0 point at the point of intersection of the two lines. We call this the y-line. The x-line is called the **x-axis** and the y-line is called the **y-axis**. The point of intersection of the two axes is called the **origin.** On the x-axis, we label the points to the right of the origin with positive numbers and those to the left with negative. On the y-axis, we number the points above the origin with positive numbers and those below with negative. See the following figure:

Historical note: The invention of the rectangular coordinate system is attributed to René Descartes. One story is that the inspiration for its development came from watching a fly crawling on a ceiling near the corner of the room. Descartes observed that he could give the location of the fly by using only two numbers that gave the perpendicular distance from each of the walls to the fly. Imagine that! A fly is responsible for one of the most important inventions in the history of mathematics!

This is known as the rectangular or Cartesian (in honor of its inventor, René Descartes) coordinate system. The x- and y-axes divide the coordinate system into four regions called **quadrants.** The quadrants are numbered counter-clockwise with the first quadrant being the upper right quadrant. See the following figure:

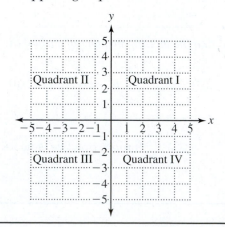

Answers:

Additional Practice Exercises *d–f*: **d.** (0, 9), (3, 0), (4, −3) **e.** (0, −3), (9, 0), (3, −2) **f.** (−4, 0), (−4, −2), (−4, 3)

Plotting points in the rectangular
coordinate system

Since the *x*-axis is horizontal, the *x*-value of the ordered pair tells us how far and in which direction to go horizontally from the origin. Go to the right if *x* is positive and to the left if *x* is negative. Since the *y*-axis is vertical, the *y*-value tells us how far and in which direction to go vertically. Go up if *y* is positive and down if *y* is negative. For example, to plot the ordered pair (4, 5), we begin at the origin and go 4 units to the right and then we go 5 units up as illustrated in the following graph:

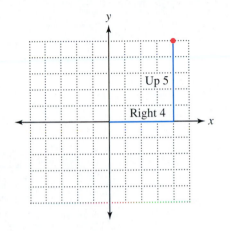

In an ordered pair, the first number of the ordered pair is called the **abscissa** and the second number is called the **ordinate**. Together, they are called the **coordinates** of the point.

Example 4

Plot the points represented by the following ordered pairs on the rectangular coordinate system.

a. (3, 5)

b. (−4, 3)

c. (−3, −4)

d. (3, −5)

e. (0, 2)

f. (4, 0)

Solution:
We first give a description of how the ordered pairs are plotted and then the points are shown on the rectangular coordinate system.

a. To plot the point (3, 5), begin at the origin. Go 3 units to the right and 5 units up.

b. To plot the point (−4, 3), begin at the origin. Go 4 units to the left and 3 units up.

c. To plot the point (−3, −4), begin at the origin. Go 3 units to the left and 4 units down.

d. To plot the point (3, −5), begin at the origin. Go 3 units to the right and 5 units down.

e. To plot the point (0, 2), do not go right or left any units. Go up 2 units from the origin.

f. To plot the point (4, 0), begin at the origin. Go 4 units to the right and stay there. Do not go up or down any units.

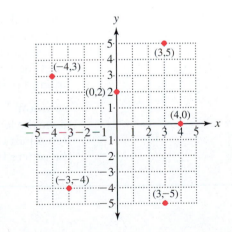

Practice Exercises

Plot the points represented by the following ordered pairs on the rectangular coordinate system:

7. $(3, 4)$

8. $(-2, 5)$

9. $(-4, -1)$

10. $(4, -3)$

11. $(-6, 0)$

12. $(0, -6)$

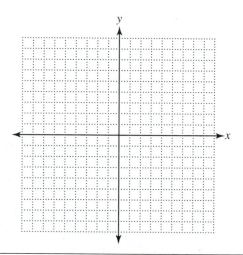

Writing solutions to everyday situations in the form of ordered pairs

Ordered pairs can occur in many types of situations. Suppose a car is traveling at a constant rate of 60 miles per hour. At the end of one hour the car has traveled $60 \cdot 1 = 60$ miles, at the end of two hours the car has traveled $60 \cdot 2 = 120$ miles, and so forth. The distance the car travels is $d = 60t$ where t is the number of hours. If we let the ordered pairs be of the form (t, d), then $(1, 60)$ means that after one hour the car traveled 60 miles and $(3, 180)$ means that after three hours the car traveled 180 miles.

Example 5

a. The length of a rectangle remains 5 feet while the width is allowed to vary. If W represents the width and A represents the area, then $A = 5W$. Represent the area for the given widths using ordered pairs in the form (W, A). Use $W = 2$ ft, 4 ft, and 9 ft. (The formula for the area of a rectangle is $A = LW$, so $A = 5W$.) What does the ordered pair $(8, 40)$ represent?

Solution:
Since the length has a constant value of 5 feet, the area is $A = 5W$. Since the ordered pairs are in the form (W, A) and $W = 2$ ft, 4 ft, and 9 ft, we need to complete the ordered pairs $(2,)$, $(4,)$, and $(9,)$.

Answers:

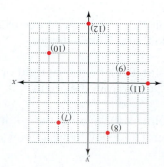

Practice Exercises 7–12:

$W = 2$ ft	$W = 4$ ft	$W = 9$ ft
$A = 5 \cdot 2 = 10$ ft^2	$A = 5 \cdot 4 = 20$ ft^2	$A = 5 \cdot 9 = 45$ ft^2

Therefore, the ordered pair is (2, 10). This means when the width is 2 ft, the area is 10 ft^2.

Therefore, the ordered pair is (4, 20). This means when the width is 4 ft, the area is 20 ft^2.

Therefore, the ordered pair is (9, 45). This means when the width is 9 ft, the area is 45 ft^2.

The ordered pair (8, 40) means that when the width is 8 ft, the area is 40 ft^2.

b. The distance a free-falling object, like a rock, falls is given by $s = 16t^2$, where s represents the distance in feet and t represents the elapsed time in seconds. Represent the distances for each of the given times using ordered pairs of the form (t, s) for $t = 1, 4,$ and 6 seconds. What does the ordered pair (3, 144) represent?

Solution:
Since $t = 1, 4,$ and 6 and the ordered pairs are of the form (t, s), we need to complete the ordered pairs (1,), (4,), and (6,).

$t = 1$ second	$t = 4$ seconds	$t = 6$ seconds
$s = 16(1)^2 = 16 \cdot 1 =$	$s = 16(4)^2 = 16 \cdot 16 =$	$s = 16(6)^2 = 16 \cdot 36 =$
16 ft	256 ft	576 ft

Therefore, the ordered pair is (1, 16). This means the object will travel 16 feet in 1 second.

Therefore, the ordered pair is (4, 256). This means the object will travel 256 feet in 4 seconds.

Therefore, the ordered pair is (6, 576). This means the object will travel 576 feet in 6 seconds.

The ordered pair (3, 144) means that the object will fall 144 feet in 3 seconds.

Practice Exercises

13. In the formula $p = br$, p stands for percentage, b for base, and r for rate. If the rate is fixed at 8% (.08), the equation is $p = .08b$. Represent the percentages for each of the given bases using ordered pairs of the form (b, p) for $b = \$200, \$500,$ and \$1000. What does the ordered pair (600, 48) represent?

14. The formula for the area of a circle is $A = \pi r^2$ where A represents the area and r represents the radius. Represent the areas of the following circles using ordered pairs of the form (r, A) for $r = 1$ cm, 3 cm, and 5 cm. Leave A in terms of π. (Do not substitute for π.) What does the ordered pair $(4, 16\pi)$ mean?

Exercise Set 4.1

Determine whether the given ordered pair is a solution of the given equation. Assume the ordered pairs are of the form (x,y).

1. $y = 3x + 2; (1, 5)$

2. $y = 2x + 4; (-1, 2)$

3. $y = -4x + 5; (-1, 1)$

4. $y = -2x - 4; (2, 0)$

5. $2x + y = 4; (0, 4)$

6. $x + 3y = 9; (9, 0)$

7. $x - 4y = 8; (0, 8)$

8. $3x - y = 12; (-4, 0)$

9. $2x + 3y = 12; (3, 2)$

10. $2x + 5y = 10; (-5, 4)$

11. $3x - 5y = 15; (-5, -6)$

12. $4x - 3y = 12; (-3, -8)$

13. $4x - 5y = -20; (0, -4)$

14. $7x - 3y = -21; (-3, 0)$

15. $2x + 3y = 6; (-\frac{3}{2}, 3)$

16. $3x + 4y = 8; (-\frac{4}{3}, 3)$

Use the given equation to find the missing member of each of the given ordered pairs. Assume the ordered pairs are in the form (x,y).

17. $y = 2x + 3; (0, \_),(\_, 0),(3, \_),(\_, 7)$

18. $y = 3x - 5; (0, \_),(\_, 0),(-2, \_),(\_, 4)$

Answers:

19. $y = -2x - 4$; $(0, \_),(\_, 0),(-3, \_),(\_, -8)$

20. $y = -3x - 5$; $(0, \_),(\_, 0),(-2, \_),(\_, 4)$

21. $2x - 3y = 12$; $(0, \_),(\_, 0),(-6, \_),(\_, -6)$

22. $3x - 2y = 18$; $(0, \_),(\_, 0),(4, \_),(\_, 6)$

23. $4x + 5y = 20$; $(0, \_),(\_, 0),(-10, \_),(\_, 8)$

24. $6x + 5y = 30$; $(0, \_),(\_, 0),(10, \_),(\_, 12)$

25. $2x + 3y = -6$; $(0, \_),(\_, 0),(9, \_),(\_, -4)$

26. $3x + 5y = -30$; $(0, \_),(\_, 0),(-10, \_),(\_, 9)$

27. $4x - 3y = -12$; $(0, \_),(\_, 0),(6, \_),(\_, -12)$

28. $5x - 4y = -20$ $(0, \_),(\_, 0),(12, \_),(\_, -20)$

29. $x = 4$; $(\_, 0),(\_, 3),(\_, -2),(\_, -4)$

30. $y = -3$; $(0, \_),(2, \_),(-2, \_),(-4, \_)$

31. $y = 4$; $(2, \_),(3, \_),(-4, \_),(-1, \_)$

32. $x = -1$; $(\_, 2),(\_, 3),(\_, -3),(\_, -5)$

Plot the points represented by the following ordered pairs on the rectangular coordinate system. Assume each unit represents 1.

33. $(3, 5),(-3, 1),(0, 5),(2, -4)$

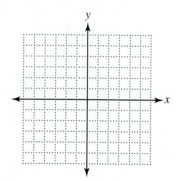

34. $(-4, 0),(4, -2),(-3, 4),(0, 4)$

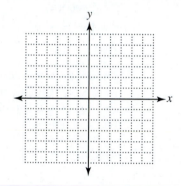

35. $(3, 7),(-2, -4),(0, 2),(4, -5)$

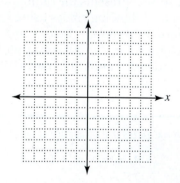

36. $(6, -2),(-4, 5),(2, -1),(3, 4)$

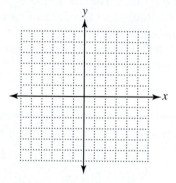

37. $(3, 0), (-2, 5), (3, -6), (-4, -2)$

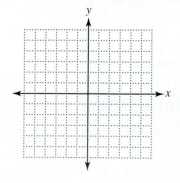

38. $(-4, 0), (0, 3), (-1, -5), (4, -6)$

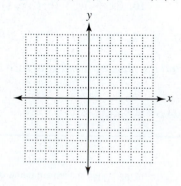

On the following grid some points have been plotted. Give the coordinates of each.

39.

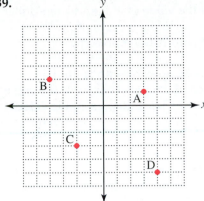

40.

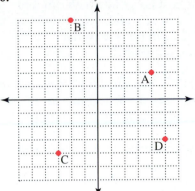

41.

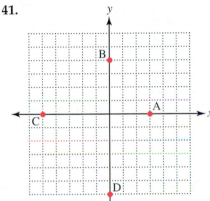

42.

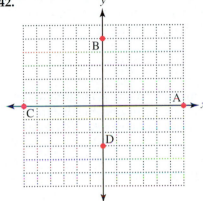

43.

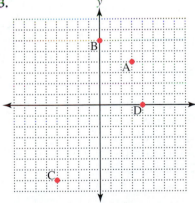

44.

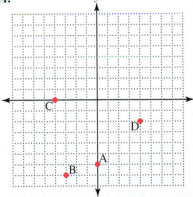

Represent each of the following as ordered pairs:

45. A motorcycle is traveling at a constant speed of 40 miles per hour. The formula for the distance traveled at the end of t hours is $d = 40t$. a) Represent the distance traveled after 1, 2, and 5 hours as ordered pairs in the form (t, d). b) What does the ordered pair (8, 320) represent?

46. A boat is traveling at a constant speed of 25 miles per hour. The formula for the distance traveled at the end of t hours is $d = 25t$. a) Represent the distance traveled at the end of 1, 3, and 6 hours as ordered pairs in the form (t, d). b) What does the ordered pair (5, 125) represent?

47. The width of a rectangle remains constant at 6 m and the length is allowed to vary. The formula for the area is $A = L(6) = 6L$. a) Represent the area for $L = 2$ m, 4 m, and 5 m as ordered pairs in the form (L, A). b) What does the ordered pair (9, 54) represent?

48. The length of a rectangle remains constant at 5 feet and the width is allowed to vary. The formula for the perimeter is $P = 2(5) + 2W = 10 + 2W$. a) Represent the perimeter for $W = 2$ ft, 3 ft, and 6 ft as ordered pairs in the form (W, P). b) What does the ordered pair (5, 20) represent?

49. A particular brand of candy bar costs 60 cents each. If n represents the number of candy bars purchased, the formula for the cost is $c = 60n$ with c in cents. **a.** Represent the cost of 1, 3, and 7 candy bars, as ordered pairs in the form (n, c). **b.** What does the ordered pair (4, 240) represent?

50. A type of bolt costs 35 cents each. If n represents the number of bolts purchased, the formula for the cost is $c = 35n$ with c in cents. **a.** Represent the cost of 5, 10, and 15 bolts as ordered pairs in the form (n, c). **b.** What does the ordered pair (4, 140) represent?

51. If v represents the value of a collection of five-dollar bills and n represents the number of bills, then $v = 5n$. **a.** Represent the value of 2, 5, and 8 five-dollar bills as ordered pairs in the form (n, v). **b.** What does the ordered pair (10, 50) represent?

52. If v represents the value of a collection of ten-dollar bills and n represents the number of bills, then $v = 10n$. **a.** Represent the value of 3, 5, and 7 ten-dollar bills as ordered pairs in the form (n, v). **b.** What does the ordered pair (8, 80) represent?

53. If an object is thrown upward with an initial velocity of 100 feet per second, then its velocity at the end of t seconds is $v = 100 - 32t$. **a.** Represent the velocity after 1, 3, and 6 seconds as ordered pairs in the form (t, v). **b.** What does the ordered pair (2, 36) represent?

54. If an object is thrown upward with an initial velocity of 150 feet per second, then its velocity at the end of t seconds is $v = 150 - 32t$. **a.** Represent the velocity after 2, 3, and 5 seconds as ordered pairs in the form (t, v). **b.** What does the ordered pair (6, −42) represent?

Challenge Exercises:

55. A particle is moving along a line according to the formula $s = 6t^2$ where s represents the distance in inches from the starting point after t seconds. **a.** Represent the distance after 1, 3, and 5 seconds as ordered pairs in the form (t, s). **b.** What does the ordered pair (2, 24) represent?

56. A particle is moving along a line according to the formula $s = 8t^2$ where s represents the distance in feet from the starting point after t seconds. **a.** Represent the distance after 2, 4, and 6 seconds as ordered pairs in the form (t, s). **b.** What does the ordered pair (3, 72) represent?

57. If an object is thrown upward with an initial velocity of 150 feet per second from a height of 200 feet, the height of the object, s, in feet after t seconds is given by the formula $s = -16t^2 + 150t + 200$. **a.** Represent the height of the object after 1, 2, and 4 seconds as ordered pairs of the form (t, s). **b.** What does the ordered pair (3, 506) represent?

58. If an object is thrown upward with an initial velocity of 100 feet per second from a height of 125 feet, the height of the object, s, in feet after t seconds is given by the formula $s = -16t^2 + 100t + 125$. **a.** Represent the height of the object after 2, 4, and 5 seconds as ordered pairs of the form (t, s). **b.** What does the ordered pair (3, 281) represent?

59. A culture of bacteria is growing according to the formula $N = 100(2^t)$ where N is the number present after t hours. **a.** Represent N after 1, 2, and 3 hours as ordered pairs of the form (t, N). **b.** What does the ordered pair (4, 1600) represent?

60. A culture of bacteria is growing according to the formula $N = 150(3^t)$ where N is the number present after t hours. **a.** Represent N after 1, 2, and 3 hours as ordered pairs of the form (t, N). **b.** What does the ordered pair (4, 12,150) represent?

Writing Exercises:

61. A linear equation with two unknowns may be written in the form $ax + by = c$. Solutions of such equations can be represented as ordered pairs of the form (x, y). A linear equation with three unknowns may be written in the form $ax + by + cz = d$. How would you represent solutions to these equations?

62. Write a problem similar to the preceding problems 45–54 whose solutions may be represented as ordered pairs. Find three solutions and represent them as ordered pairs.

63. Why does a linear equation with two unknowns have more that one solution? Why does it have an infinite number of solutions?

64. If the graph of an ordered pair is in the first quadrant, both the x- and y-values of the ordered pair are positive. Describe the x- and y-values of an ordered pair whose graph is in the second quadrant. Why? Third quadrant. Why? Fourth quadrant. Why?

| Section 4.2 | Graphing Linear Equations with Two Variables |

OBJECTIVES

When you complete this section, you will be able to:

Graph linear equations with two variables on the rectangular coordinate system.

Introduction   Using the rectangular coordinate system, it is possible to plot solutions of equations with two variables. The plot of the solutions is called the **graph** of the equation. Since there is an infinite number of solutions, it is not possible to plot them all. All the graphs in this section will be of linear equations (consequently, the graphs are straight lines) that we will graph using the following procedure:

Graphing Linear Equations

To graph a linear equation, we perform the following:

a. Determine three ordered pairs that solve the equation by letting x have a value and finding y or letting y have a value and finding x. (Only two points are required. The third point is used as a check.)

b. Plot the ordered pairs.

c. Draw the line that contains the three points.

A straight line is determined by two points. However, it is a good idea to find a third point as a check, since a straight line cannot always be drawn through any three distinct points. For example, a single straight line cannot be drawn through the following three points:

Example 1

Draw the graph of each of the following equations.

a. $y = 2x + 1$

Solution:
We need to find some ordered pairs from the solution set. We can let either variable have any value we choose and then solve for the other variable. Since we have to plot the resulting ordered pairs, it is best to use values small in absolute value. This equation is solved for y, so it would be easier to assign values to x and then find y.

Let $x = 0$.	Let $x = 2$.	Let $x = -3$.
$y = 2(0) + 1$	$y = 2(2) + 1$	$y = 2(-3) + 1$
$y = 0 + 1$	$y = 4 + 1$	$y = -6 + 1$
$y = 1$	$y = 5$	$y = -5$

Therefore, the ordered pair is $(0, 1)$.

Therefore, the ordered pair is $(2, 5)$.

Therefore, the ordered pair is $(-3, -5)$.

Note: One easy way of keeping track of the ordered pairs is to put them in a table. For example, the preceding ordered pairs could be put in a table as follows:

x	y
0	1
2	5
−3	−5

Now plot the points on the rectangular coordinate system. It appears that the three points all lie on a line, so draw a line through the three points.

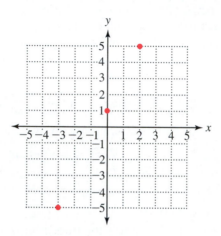
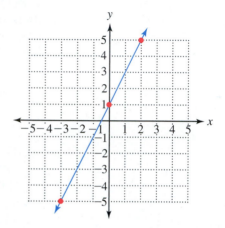

Note: There are a number of things we need to point out at this time. First, it can be shown that the graph of any linear equation with two variables ($ax + by = c$) is a straight line. That is why they are called *"linear."* Second, the coordinates of any point on the line solve the equation, and the graph of any ordered pair that solves the equation will be on the line. Third, the line continues in both directions indefinitely. This is usually indicated by putting arrow heads on each end of the line.

b. $3x − 2y = 6$

Solution:
We find ordered pairs that solve the equation by letting either variable have a value and solving for the other variable.

Let $x = 0$.

$3(0) − 2y = 6$
$0 − 2y = 6$
$-2y = 6$
$y = -3$

Therefore, the ordered pair is $(0, -3)$.

Let $y = 0$.

$3x − 2(0) = 6$
$3x − 0 = 6$
$3x = 6$
$x = 2$

Therefore, the ordered pair is $(2, 0)$.

Let $x = -2$.

$3(-2) − 2y = 6$
$-6 − 2y = 6$
$-2y = 12$
$y = -6$

Therefore, the ordered pair is $(-2, -6)$

In a table, the ordered pairs would appear as follows:

x	y
0	−3
2	0
−2	−6

Plot the ordered pairs on the rectangular coordinate system and draw a line through them.

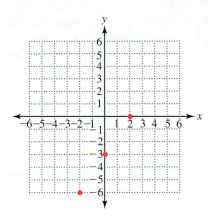
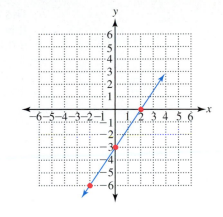

Practice Exercises

Draw the graph of each of the following equations. Assume each unit on the coordinate system represents 1.

1. $y = x + 3$

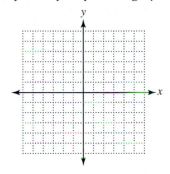

2. $2x + y = 4$

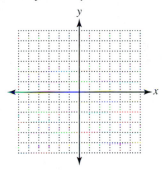

If more practice is needed, do the Additional Practice Exercises.

Additional Practice Exercises

Draw the graph of each of the following equations. Each unit on the coordinate system represents 1.

a. $y = 3x - 2$

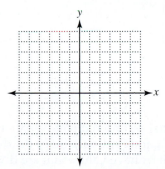

b. $4x + 2y = 8$

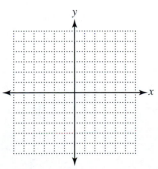

Answers:

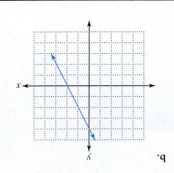
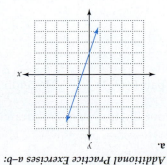
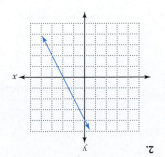
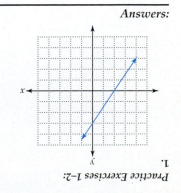

The standard form of a linear equation is $ax + by = c$. If one of the terms is missing, it is still a linear equation but may require some special considerations.

If the equation is of the form $y = ax$ (or can be put into that form), then the equation is solved for y. The easiest way to find ordered pairs is to assign values to x and find y. Note that if $x = 0$, then $y = 0$, so $(0,0)$ (which is the origin) is a point on the line. Consequently, all lines whose equations are of the form $y = ax$ pass through the origin.

Example 2

Graph the following equations. Assume each unit is 1.

a. $y = 2x$ **Solution:**

If we let $x = 0$, then $y = 0$, so one ordered pair is $(0, 0)$ Find two other points by assigning any value to x and finding y.

Let $x = 2$.	Let $x = -3$.
$y = 2(2)$	$y = 2(-3)$
$y = 4$	$y = -6$

Therefore, the ordered pair is $(2, 4)$.   Therefore, the ordered pair is $(-3, -6)$.

Plot the points on the coordinate system and draw a line through them.

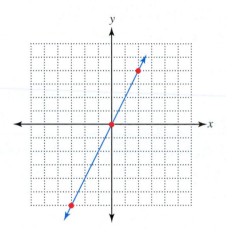

b. $y = \dfrac{1}{3}x$ **Solution:**

If we let $x = 0$, we get $y = 0$, so one ordered pair is $(0,0)$. Find two more points by assigning values to x and finding y. Since the denominator is 3, it would be a good idea to let x have values that are divisible by 3. Otherwise we will get fractions that are more difficult to plot.

Let $x = 3$.	Let $x = 6$.
$y = \dfrac{1}{3} \cdot 3 = \dfrac{3}{3} = 1$	$y = \dfrac{1}{3} \cdot 6 = \dfrac{6}{3} = 2$

Therefore, the ordered pair is $(3, 1)$.   Therefore, the ordered pair is $(6, 2)$.

Plot the points on the coordinate system and draw a line through them.

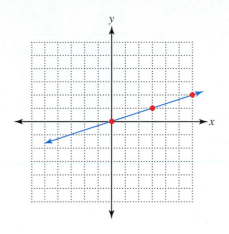

Practice Exercises

Graph the following equations. Assume each unit is 1.

3. $y = 3x$

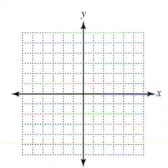

4. $y = \frac{1}{5}x$

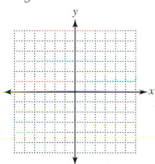

Graphing horizontal and vertical lines   The other special cases of $ax + by = c$ occur if $a = 0$ or $b = 0$. If $b = 0$, the resulting equation is of the form $x = c$ and if $a = 0$, then $y = c$ where c is any constant. We can think of an equation of the form $x = c$ to be $x + 0y = c$. This means that regardless of the value for y, $x = c$. Therefore, all solutions to the equation $x = c$ are of the form (c, y) where y can have any value. Similarly, we can think of an equation of the form $y = c$ to be $0x + y = c$. This means that $y = c$ for all values of x. Therefore, all solutions of the equation $y = c$ are of the form (x, c) where x can have any value.

Answers:

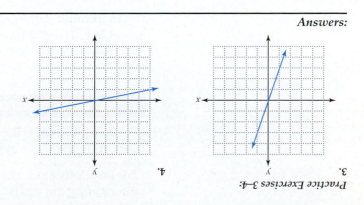

Example 3

Graph the following equations. Assume each unit is 1.

a. $x = 3$ **Solution:**

Think of $x = 3$ as $x + 0y = 3$. Therefore, $x = 3$ for all values of y. So, all solutions of $x = 3$ are of the form $(3, y)$ where y can have any value. Thus, some solutions are $(3, 0)$, $(3, -2)$, and $(3, 4)$.

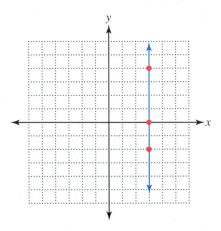

b. $y = -2$ **Solution:**

Think of $y = -2$ as $0x + y = -2$, This means that $y = -2$ for all values of x. Thus, all solutions of $y = -2$ are of the form $(x, -2)$ where x can have any value. Some solutions are $(0, -2)$, $(3, -2)$ and $(-2, -2)$. Plot the ordered pairs and draw a line through the points.

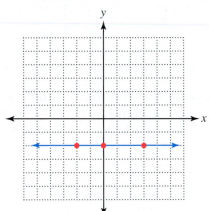

In Example 3, the graph of $x = 3$ is a vertical line and the graph of $y = -2$ is a horizontal line. The graphs of any two ordered pairs with the same x-value will always lie on the same vertical line, and the graphs of any two ordered pairs with the same y-value will always be on the same horizontal line. This leads to the following observation:

Vertical and Horizontal Lines

For any constant k, the graph of an equation of the form $x = k$ is a vertical line intersecting the x-axis at k.

For any constant k, the graph of an equation of the form $y = k$ is a horizontal line intersecting the y-axis at k.

Practice Exercises

Graph the following equations. Assume each unit is 1.

5. $x = -1$

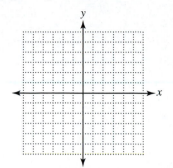

6. $y = 4$

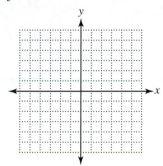

Note: There are a couple of observations we need to make regarding vertical and horizontal lines. If two or more points lie on the same vertical line, they have the same x-values. Also, if two or more points have the same x-values, they lie on the same vertical line. Likewise, if two or more points lie on the same horizontal line, they have the same y-values. Also, if two or more points have the same y-value, they lie on the same horizontal line.

Exercise Set 4.2

Complete the following ordered pairs and use them to draw the graph of the equation:

1. $y = 2x - 3$
   $(0,\ )\ (2,\ ),\ (-2,\ )$

2. $y = 3x - 1$
   $(0,\ )\ (1,\ )\ (-2,\ )$

3. $x + y = 5$
   $(0,\ )\ (\ ,0)\ (2,\ )$

4. $x - y = 3$
   $(0,\ )\ (\ ,0)\ (-2,\ )$

5. $2x + y = 6$
   $(0,\ )\ (\ ,0)\ (\ ,4)$

6. $x - 3y = 9$
   $(0,\ )\ (\ ,0)\ (\ ,-2)$

Find three ordered pairs that solve the equation and draw the graph of each of the following:

7. $y = x + 2$

8. $y = x - 4$

9. $y = 2x + 4$

10. $y = 3x - 5$

11. $y = -2x + 1$

12. $y = -3x + 5$

13. $x + y = 3$

14. $x + y = 6$

15. $x + 3y = 9$

16. $x + 4y = 8$

17. $x - 2y = 8$

18. $x - 3y = 9$

19. $4x - y = 2$

20. $x - 4y = 6$

21. $2x + 3y = 6$

22. $3x + 2y = 6$

23. $2x - 5y = 10$

24. $5x - 2y = 10$

25. $2x + 6y = 12$

26. $6x - 2y = 12$

Answers:

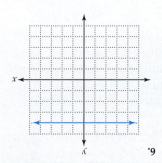

6.

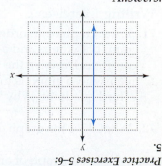

5.

Practice Exercises 5–6:

27. $3x - 4y = 12$ **28.** $5x - 3y = 15$ **29.** $2x + 5y = 5$ **30.** $2x + 7y = 7$

31. $3x - 5y = 10$ **32.** $4x - 3y = 8$ **33.** $4x - 3y = 10$ **34.** $3x - 4y = 14$

35. $y = x$ **36.** $y = 4x$ **37.** $y = -3x$ **38.** $y = -4x$

39. $2x + y = 0$ **40.** $3x - y = 0$ **41.** $-2x + y = 0$ **42.** $-4x - y = 0$

43. $y = \frac{1}{4}x$ **44.** $y = \frac{1}{2}x$ **45.** $y = -\frac{1}{3}x$ **46.** $y = -\frac{1}{5}x$

47. $y = \frac{2}{3}x$ **48.** $y = \frac{3}{4}x$ **49.** $x = 2$ **50.** $x = 5$

51. $y = 3$ **52.** $y = 7$ **53.** $x = -4$ **54.** $x = -2$

55. $y = -4$ **56.** $y = -7$ **57.** $y = -9$ **58.** $x + 5 = 0$

59. $x - 6 = 0$ **60.** $y - 7 = 0$ **61.** $y + 1 = 0$

Write an equation representing each of the following and then graph the equation.

62. The sum of the x- and y-coordinates is 8.

63. The difference of the x- and y-coordinates is 4.

64. The sum of three times the x-coordinate and two times the y-coordinate is 12.

65. The difference of three times the x-coordinate and four times the y-coordinate is 12.

66. The y-coordinate is 5 less than twice the x-coordinate.

67. The y-coordinate is 2 more than three times the x-coordinate.

68. Two times the y-coordinate is equal to 6 less than the x-coordinate.

Challenge Exercises: (69–70)

69. A rock is dropped from the top of a tall building. If y represents the velocity and x represents the time in seconds after the object is dropped, then $y = -32x$.
  a. Find the velocity after 0 seconds, 2 seconds, and 4 seconds. Write the results as ordered pairs.
  b. Draw the graph for values of x such that $0 \leq x \leq 5$. You will need to change the units on the y-axis.
  c. Use the graph to estimate the number of seconds after the object is thrown when the velocity is -96 feet per second.

70. A stone is thrown upward from the top of a tall building with an initial velocity of 150 feet per second. If y represents the velocity and x the number of seconds after the object is thrown, then $y = -32x + 150$.
  a. Find the velocity after 1 second, 2 seconds, and 3 seconds. Write the results as ordered pairs.
  b. Draw the graph.
  c. Use the graph to estimate the number of seconds after the object is thrown when the velocity is -10 feet per second.

Writing Exercises:

71. If the graph of an ordered pair is in the first quadrant, both the x- and y-values of the ordered pair are positive. Describe the x- and y-values of an ordered pair whose graph is in the second quadrant. Why? Third quadrant. Why? Fourth quadrant. Why?

72. What does the graph of an equation represent?

73. Why is the graph of an equation in the form $x = k$ a vertical line?

74. Why is the graph of an equation in the form $y = k$ a horizontal line?

75. Since one and only one line can be drawn through any two distinct points, why is it a good idea to always find three points when graphing a line?

76. What is the equation of the x-axis?

| **Section 4.3** | Graphing Linear Equations Using Intercepts |

OBJECTIVES

When you complete this section, you will be able to:

a. Find the *x*- and *y*-intercepts from the graph of a linear equation with two variables.

b. Graph a linear equation with two variables by finding the *x*- and *y*-intercepts.

c. Graph vertical and horizontal lines.

Introduction   In the previous section, we graphed linear equations by finding three ordered pairs that were solutions of the equations, plotting the ordered pairs, and drawing a line through the points. In this section, we will continue that process by finding special points called the *intercepts.*

Finding the *x*- and *y*-intercepts   The point where a graph intersects the *x*-axis is the **x-intercept,** and the point where the graph intersects the *y*-axis is the **y-intercept.** See the following figure:

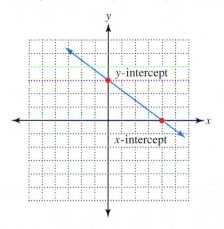

Notice the *y*-value of the *x*-intercept is 0, and the *x*-value of the *y*-intercept is 0. Any point on the *x*-axis has a *y*-value of 0 and any point on the *y*-axis has an *x*-value of 0. Therefore, the coordinates of the *x*-intercept are of the form $(a, 0)$ and the coordinates of the *y*-intercept are of the form $(0, b)$.

Example 1

Find the x- and y-intercepts of the following graphs. Assume each unit is one.

a.

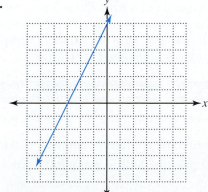

b.

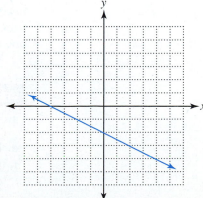

Solution:

The graph crosses the *x*-axis at -3, so the *x*-intercept is $(-3, 0)$. The graph crosses the *y*-axis at 6, so the *y*-intercept is $(0, 6)$.

The graph crosses the *x*-axis at -4, so the *x*-intercept is $(-4, 0)$. The graph crosses the *y*-axis at -2, so the *y*-intercept is $(0, -2)$.

Note: Often the intercepts are not written as ordered pairs. Sometimes only the value at which the line crosses the axis is given. In the preceding Example 1a, we could say the x-intercept is −3 and the y-intercept is 6.

Practice Exercises

Find the x- *and* y-*intercepts of the following graphs.*

1.

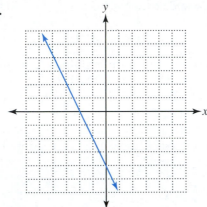

2.

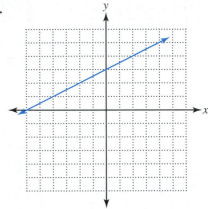

If you need more practice, do the Additional Practice Exercises.

Additional Practice Exercises

Find the x- *and* y-*intercepts of the following graphs:*

a.

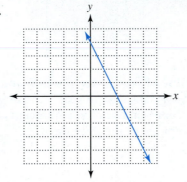

b.

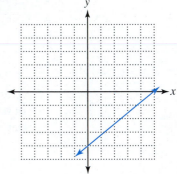

Since only one straight line may be drawn through any two distinct points, we need only two ordered pairs from the solution set to draw the graph of a linear equation. Usually, the two easiest ordered pairs to find are the x- and y-intercepts, since one of the coordinates is 0. However, find a third point as a check.

The procedure is summarized as follows:

Graphing Linear Equations Using the Intercept Method

To graph a linear equation using the intercept method, we perform the following:

a. Let $x = 0$ to find the y-intercept and let $y = 0$ to find the x-intercept.

b. Find a third point by letting x or y have any value and solve for the remaining variable.

c. Plot the ordered pairs.

d. Draw a line through the three points.

Answers:

Practice Exercises 1–2: 1. (−2, 0), (0, −4) 2. (−6, 0), (0, 3)   *Additional Practice Exercises a–b:* a. (2, 0), (0, 4) b. (5, 0), (0, −4)

Example 2

Draw the graph of each of the following by finding the x- and y-intercepts. Find a third point as a check. Assume each unit on the coordinate system represents one.

a. $2x + y = 6$

Solution:

To find the x-intercept, let $y = 0$.	To find the y-intercept, let $x = 0$.	To find another point, let $x = -1$.
$2x + 0 = 6$	$2(0) + y = 6$	$2(-1) + y = 6$
$2x = 6$	$0 + y = 6$	$-2 + y = 6$
$x = 3$	$y = 6$	$y = 8$

Therefore, the x-intercept is $(3, 0)$.

Therefore, the y-intercept is $(0, 6)$.

Therefore, the ordered pair is $(-1, 8)$.

In a table, the ordered pairs appear as follows:

x	y
3	0
0	6
−1	8

Plot the points on the rectangular coordinate system and draw a line through them.

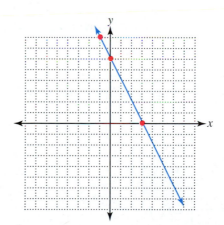

b. $2x + 3y = 9$

Solution:

To find the x-intercept, let $y = 0$.	To find the y-intercept, let $x = 0$.	To find a third point, let $x = -3$.
$2x + 3(0) = 9$	$2(0) + 3y = 9$	$2(-3) + 3y = 9$
$2x + 0 = 9$	$0 + 3y = 9$	$-6 + 3y = 9$
$2x = 9$	$3y = 9$	$3y = 15$
$x = \dfrac{9}{2}$	$y = 3$	$y = 5$

Therefore, the x-intercept is $\left(\dfrac{9}{2}, 0\right)$.

Therefore, the y-intercept is $(0, 3)$.

Therefore, the ordered pair is $(-3, 5)$.

Plot the points on the rectangular coordinate system and draw a line through them.

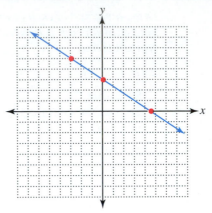

Note: In Example 3b, the *x*-intercept has a fractional value. In plotting the correspond-ing point, it was necessary to estimate its location. This can be a real problem if the lo-cation of the fraction is difficult to estimate. For example, $(\frac{3}{11}, \frac{12}{5})$ would be difficult to plot accurately. If you can find an *x*-value that gives an integer value for *y*, then adding any integer multiple of the coefficient of *y* to that *x*-value will give other *x*-values that result in integer values for *y*. In other words, if you have an *x*-value that results in an integer *y*-value, then (*x*-value) + (multiple of *y*-coefficient) = (another *x*-value that gives an integer *y*-value). For example, if we have $2x + 3y = 12$ and the *y*-intercept is (0,4). So we have a point with an integer *x*-value of 0 and the coefficient of *y* in the equation is 3. Therefore, *x* = 0 + (any integer multiple of 3) will give another *x*-value that results in an integer value for *y*. Hence, *x* = 3, 6, 9, −3, −6, −9, and so on, will give integer values for *y*. The opposite applies for *y*-values. For the equation $2x + 5y = 8$, the *x*-intercept is (4,0) and the coefficient of *x* = 2. Hence *y* = 0 + (any multiple of 2) gives another *y*-value that results in an integer value for *x*. Hence *y* = 2, 4, 6, −2, −4, −6, and so on will give integer values for *x*.

Practice Exercises

Draw the graph of each of the following by finding the x- and y-intercepts. Find a third point as a check. Assume each unit on the coordinate system represents 1.

3. $x + 4y = 8$

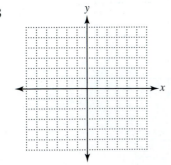

4. $2x - 5y = 15$

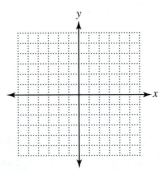

Answers:

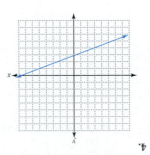

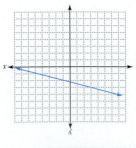

4.

3.

We graphed vertical and horizontal lines in Section 4.2, but now we have another way of approaching these graphs. Since a vertical line has an equation of the form $x = k$, we can think of a vertical line as having an x-intercept at k and no y-intercept (hence it is parallel to the y-axis). Likewise since a horizontal line has an equation of the form $y = k$, we can think of a horizontal line as having a y-intercept of k and no x-intercept (hence it is parallel to the x-axis).

Example 3

Graph each of the following:

a. $x = 4$ **Solution:**
The graph has an x-intercept at $x = 4$ and no y-intercept. Therefore, it is a vertical line.

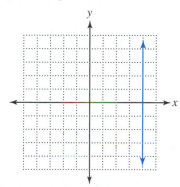

b. $y = -3$ **Solution:**
The graph has a y-intercept of -3 and no x-intercept. Therefore, it is a horizontal line.

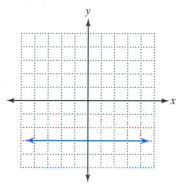

Practice Exercises

Draw the graph of each of the following:

5. $x = -2$

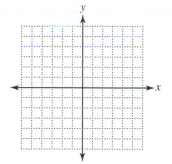

6. $y = 4$

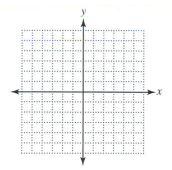

Some linear equations are easier graphed by methods other than using the intercepts. For example, equations that are solved for y such as $y = \frac{2}{3}x + 5$, are easier to graph by finding three ordered pairs that solve the equation or by using a method to be discussed in Section 4.5.

Answers:

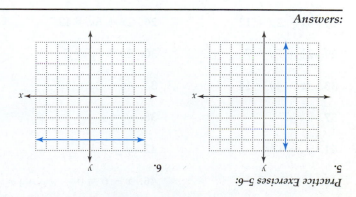

6.

5.

Practice Exercises 5–6:

Exercise Set 4.3

Identify the x- and y-intercepts of the following:

1.

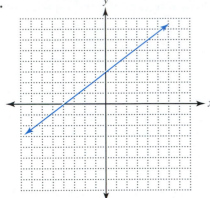

2.

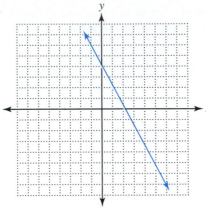

3.

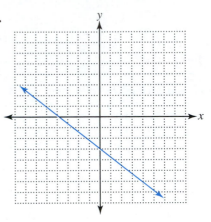

4.

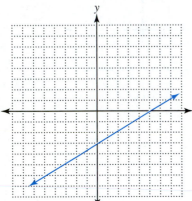

5.

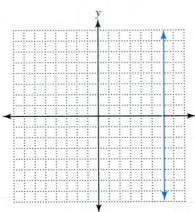

6.

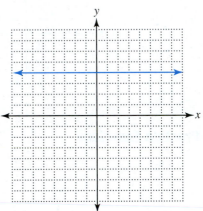

Draw the graph of the lines whose intercepts are given as follows:

7. $(3, 0)$ $(0, 4)$

8. $(2, 0)$ $(0, 5)$

9. $(-3, 0)$ $(0, 4)$

10. $(-5, 0)$ $(0, 2)$

11. $(-4, 0)$ $(0, -3)$

12. $(-7, 0)$ $(0, 5)$

Graph each of the following using the intercept method:

13. $x + y = 6$ **14.** $x + y = 3$ **15.** $x + 4y = 8$ **16.** $x + 3y = 9$

17. $x - 3y = 9$ **18.** $x - 2y = 8$ **19.** $x - 4y = 6$ **20.** $4x - y = 2$

21. $3x + 2y = 6$ **22.** $2x + 3y = 6$ **23.** $5x - 2y = 10$ **24.** $2x - 5y = 10$

25. $6x - 2y = 12$ **26.** $2x + 6y = 12$ **27.** $5x - 3y = 15$ **28.** $3x - 4y = 12$

29. $2x + 7y = 7$ **30.** $2x + 5y = 5$ **31.** $4x - 3y = 8$ **32.** $3x - 5y = 10$

33. $3x - 4y = 14$ **34.** $4x - 3y = 10$ **35.** $x = 5$ **36.** $x = 2$

37. $y = 7$ **38.** $y = 3$ **39.** $x = -2$ **40.** $x = -4$

41. $y = -7$ **42.** $y = -4$ **43.** $x + 5 = 0$ **44.** $y + 9 = 0$

45. $y - 7 = 0$ **46.** $x - 6 = 0$ **47.** $y + 1 = 0$

Write an equation for each of the following:

48. The Maitland Public Library was having a plant sale. If Katrina bought x plants for $2 each and y plants for $3 each, write an equation that shows the possible combinations of $2 and $3 plants if the total cost was $24.

49. If x represents the length of a rectangle and y represents the width, write an equation that shows the possible combinations of lengths and widths if the perimeter is 32 feet.

50. If there are x $5 bills and y $10 bills, write an equation that shows the possible combinations of $5 and $10 bills if the total value of the bills is $450.

51. John receives $6 per hour for regular time worked and $9 per hour for overtime. If John works x hours of regular time and y hours of overtime, write an equation that shows the possible combinations of regular hours and overtime hours if John received $420.

52. Abdul took a trip. For the first part of the trip, he averaged 50 miles per hour and for the second part he averaged 60 miles per hour. If Abdul drove x hours at 50 miles per hour and y hours at 60 miles per hour, write an equation that shows the possible combinations of hours if the trip was 500 miles long.

53. In basketball, a field goal counts 2 points and a free throw counts 1 point. If Jamal scored x field goals and y free throws, write an equation that shows the possible combinations of field goals and free throws if he scored 27 points.

Challenge Exercises (54–59)

54. If the graph of $ax + by = 15$ has an x-intercept of $(5, 0)$ and a y-intercept of $(0, 3)$, find a and b.

55. If the graph of $ax + by = -8$ has an x-intercept of $(4, 0)$ and a y-intercept of $(0, -2)$, find a and b.

Find the intercepts of the following:

56.

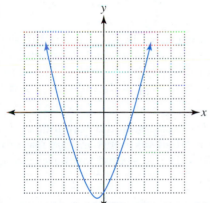

57.

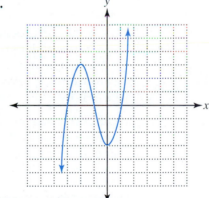

58.

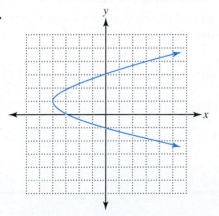

59.

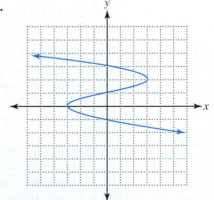

Writing Exercise:

60. When graphing an equation of the form $ax + by = c$, why do we usually find the x- and y-intercepts as two of the points?

Slope of a Line

OBJECTIVES   *When you complete this section, you will be able to:*

a. Find the slope of a line by using the slope formula.

b. Find the slope of a line from its graph.

c. Graph a line when given a point on the line and the slope of the line.

Introduction   Thus far, we have been graphing lines by finding at least two points on the line and then drawing a line through the points. In this section, we will learn another method.
Consider the graphs of the following lines, all of which contain the point (3, 3):

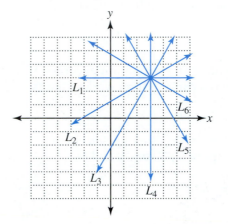

All these lines pass through the same point yet there are substantial differences. L_1 is horizontal, L_4 is vertical, L_5 slants downward (a "falling" line) from left to right, and L_2 slants upward (a "rising" line) from left to right. Also, both L_2 and L_3 slant upward from left to right, but L_3 is steeper than L_2. One way of describing these differences is by using **slope.**

Note: You may not be familiar with the use of subscripts. In the notations L_1, L_2, L_3, and so on, the Ls denote lines. The numbers are called *subscripts* and are used to denote different lines. Hence, L_1 stands for the first line, L_2 stands for the second line, and so on.

Slope of a Line

The slope of a line is the ratio of the change in x to the change in x as you go from one point on a line to any other point on the line.

An informal way of thinking of slope is that it is the ratio of the rise in a line to the run of the line between any two points on the line as illustrated by the following figure: as such it can be thought of as the rate of the change in y with respect to x.

This informal way of thinking of slope is consistent with the formal definition. The rise is a vertical change and, since the y-axis is vertical, any vertical change is a change in y. Also, the run is a horizontal change and, since the x-axis is horizontal, any horizontal change is a change in x.

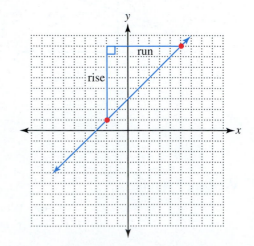

Derivation of the slope formula

As is often the case in mathematics, the definition gives us no indication of how to actually perform the task. Suppose we want to find the slope of the line that contains the points (2, 3) and (7, 9). Plot the points and draw the line containing these points. Then go from point (2, 3) to (7, 9) by first going vertically up from (2, 3) and then horizontally to (7, 9) as illustrated below.

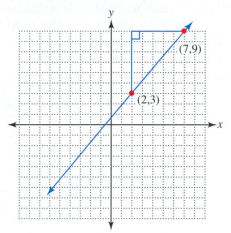

Since any two points on the same vertical line have the same *x*-value and any two points on the same horizontal line have the same *y*-value, the point of intersection of the horizontal and vertical lines is (2, 9).

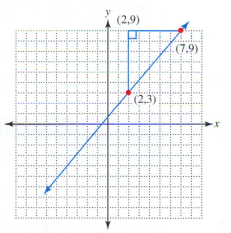

The rise is the change in *y* as we go from (2, 3) to (2, 9). The *y*-value changes from 3 to 9, which is a total change of 6. This change in *y* (rise) can be found by finding the difference of the *y*-values (9 − 3 = 6). The run is the change in *x* as we go from (2, 9) to (7, 9). The *x*-value changes from 2 to 7, which is a change of 5. This change in *x* (run) can be found by finding the difference of the *x*-values (7 − 2 = 5).

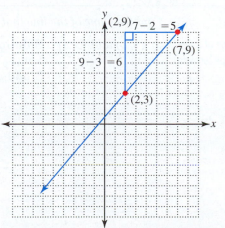

Therefore, the slope is $\frac{\text{rise}}{\text{run}} = \frac{\text{change in } y}{\text{change in } x} = \frac{9-3}{7-2} = \frac{6}{5}$.

Instead of going through all of this each time we want to find the slope of a line, we use any line of the form $ax + by = c$ and any two points on the line. We will denote the points as (x_1, y_1) and (x_2, y_2) and derive a formula for slope. Subscripts are used to indicate that both points have x- and y-values, but these values may be different. The letter m is customarily used for slope. Consider the following figure:

*It is not really known why the letter **m** is used to denote slope. One belief is that it comes from the French word "monter" which means "to climb."*

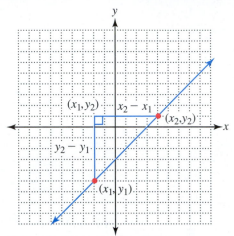

Since any two points on the same vertical line have the same x-value and any two points on the same horizontal line have the same y-value, the coordinates of the point of intersection of the vertical and horizontal line segments from (x_1, y_1) to (x_2, y_2) are (x_1, y_2). See the preceding figure. As in the previous example, the change in y is the difference in the y-values. Thus, the change in y is $y_2 - y_1$. The change in x is the difference in the x-values. Thus, the change in x is $x_2 - x_1$. Therefore, the formula for the slope of a line containing the points (x_1, y_1) and (x_2, y_2) is $m = \frac{y_2 - y_1}{x_2 - x_1}$.

Slope Formula

The slope of a line containing the points whose coordinates are (x_1, y_1) and (x_2, y_2) is $m = \frac{y_2 - y_1}{x_2 - x_1}$.

Note: The slope of the line is the difference of the y-values divided by the difference of the x-values. This difference can be found in either order. Consequently, the slope could also be $m = \frac{y_1 - y_2}{x_1 - x_2}$. However, we must be very careful. Whichever y-value point is used first in the numerator, that x-value point must also be used first in the denominator. Also, very simple reduction of fractions is needed in the remainder of this chapter. A thorough discussion of this topic is in Chapter 6.

Example 1

Find the slope of the line that contains each of the following pairs of points:

Finding the slope of a line using the slope formula

a. $(1, 3)$ and $(5, 9)$ **Solution:**

Let $(x_1, y_1) = (1, 3)$ and $(x_2, y_2) = (5, 9)$. Hence, $x_1 = 1, y_1 = 3, x_2 = 5$, and $y_2 = 9$. Substitute these values into the slope formula.

$$m = \frac{y_2 - y_1}{x_2 - x_1}$$ Substitute for y_2, y_1, x_2, and x_1.

$$m = \frac{9 - 3}{5 - 1}$$ Simplify the numerator and the denominator.

$$m = \frac{6}{4}$$ Reduce to lowest terms.

$$m = \frac{3}{2}$$ Therefore, the slope is $\frac{3}{2}$.

Alternate Solution:
If we let $(x_1, y_1) = (5, 9)$ and $(x_2, y_2) = (1, 3)$, then $m = \frac{3-9}{1-5} = \frac{-6}{-4} = \frac{3}{2}$. Notice that we get the same result.

b. $(-2, 3)$ and $(4, -2)$ **Solution:**
Let $(x_1, y_1) = (-2, 3)$ and $(x_2, y_2) = (4, -2)$. Therefore, $x_1 = -2$, $y_1 = 3$, $x_2 = 4$, and $y_2 = -2$. Substitute these values into the slope formula.

$$m = \frac{y_2 - y_1}{x_2 - x_1} \qquad \text{Substitute for } y_2, y_1, x_2, \text{ and } x_1.$$

$$m = \frac{-2 - 3}{4 - (-2)} \qquad 4 - (-2) = 4 + 2.$$

$$m = \frac{-2 - 3}{4 + 2} \qquad \text{Simplify the numerator and the denominator.}$$

$$m = \frac{-5}{6} = -\frac{5}{6} \qquad \text{Therefore, the slope is } -\frac{5}{6}.$$

c. $(3, 2)$ and $(-1, 2)$ **Solution:**
Let $(x_1, y_1) = (3, 2)$ and $(x_2, y_2) = (-1, 2)$. Therefore, $x_1 = 3$, $y_1 = 2$, $x_2 = -1$, and $y_2 = 2$. Substitute these values into the slope formula.

$$m = \frac{y_2 - y_1}{x_2 - x_1} \qquad \text{Substitute for } y_2, y_1, x_2, \text{ and } x_1.$$

$$m = \frac{2 - 2}{-1 - 3} \qquad \text{Simplify the numerator and denominator.}$$

$$m = \frac{0}{-4} \qquad \text{Simplify.}$$

$$m = 0 \qquad \text{Therefore, the slope is 0.}$$

Note: The graph of the line containing $(3, 2)$ and $(-1, 2)$ is horizontal.

d. $(3, 6)$ and $(3, 2)$ **Solution:**
Let $(x_1, y_1) = (3, 6)$ and $(x_2, y_2) = (3, 2)$. Therefore, $x_1 = 3$, $y_1 = 6$, $x_2 = 3$, and $y_2 = 2$. Substitute these values into the slope formula.

$$m = \frac{y_2 - y_1}{x_2 - x_1} \qquad \text{Substitute for } y_2, y_1, x_2, \text{ and } x_1.$$

$$m = \frac{2 - 6}{3 - 3} \qquad \text{Simplify the numerator and the denominator.}$$

$$m = \frac{-4}{0} \qquad \text{Division by 0 is undefined. Therefore, the slope of the line through } (3, 6) \text{ and } (3, 2) \text{ is undefined.}$$

Note: The graph of the line containing $(3, 6)$ and $(3, 2)$ is vertical.

Practice Exercises

Find the slope of the line which contains each of the following pairs of points:

1. $(2, 4), (5, 8)$ **2.** $(-2, 3), (5, -2)$ **3.** $(-3, 4), (-1, 4)$ **4.** $(3, -5), (3, -2)$

If more practice is needed, do the Additional Practice Exercises.

Additional Practice Exercises

Find the slope of the line which contains each of the following pairs of points:

a. $(4, 7)$ and $(6, 3)$ **b.** $(-1, 4)$ and $(4, -3)$ **c.** $(2, 2)$ and $(-4, 2)$ **d.** $(-5, -2)$ and $(-5, -4)$

Finding the slope from a graph   Sometimes it is necessary to find the slope from a graph. To do so, locate two points on the line and use their coordinates in the slope formula.

Example 2

Find the slope of the following lines:

a.

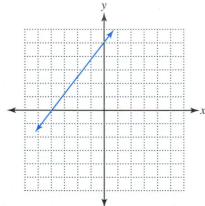

b.

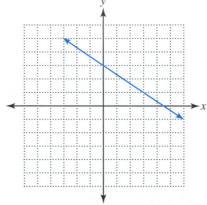

Solution:
We need two points on the line. A good place to check is the intercepts. It appears that the x-intercept is $(-4, 0)$ and the y-intercept is $(0, 5)$.

$$m = \frac{y_2 - y_1}{x_2 - x_1} \qquad \text{Substitute.}$$

$$m = \frac{5 - 0}{0 - (-4)} \qquad \text{Simplify.}$$

$$m = \frac{5}{4}$$

Solution:
We need two points. The y-intercept is $(0, 3)$ but the x-intercept does not appear to be an integer. Look for another point. The graph seems to go through $(3, 1)$

$$m = \frac{y_2 - y_1}{x_2 - x_1} \qquad \text{Substitute.}$$

$$m = \frac{1 - 3}{3 - 0} \qquad \text{Simplify.}$$

$$m = -\frac{2}{3}$$

Practice Exercises

Find the slopes of the following lines:

5.

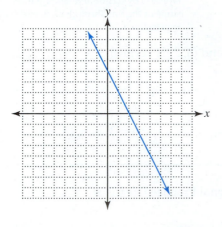

6.

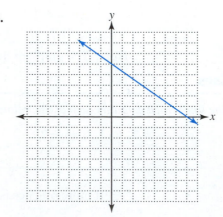

Answers:

Graphing a line given a point on the line and the slope

Using slope, we now have another method of graphing a line. Remember, the slope represents the change in y (vertical change) divided by the change in x (horizontal change). If we know a point on the line and the slope, we can find the graph of the line by using the following procedure:

Graphing a Line Given a Point and the Slope

1. Plot the given point.

2. From the given point, we can find another point by going vertically (up or down) the number of units given by the numerator of the slope and horizontally (right or left) the number of units given by the denominator.

3. Draw a line through these two points.

Example 3

Draw the graph of each of the following lines containing the given point and with the given slope:

a. $(2, 3)$ with $m = \dfrac{3}{4}$

Solution:

Plot the point with coordinates $(2, 3)$. To locate another point on the line, from $(2, 3)$ go up 3 units and then go to the right 4 units and plot another point. Draw the line through these two points.

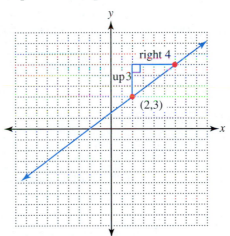

b. $(-2, 2)$ and $m = 3$

Solution:

Plot the point with coordinates $(-2, 2)$. Think of 3 as $\frac{3}{1}$. Therefore, from $(-2, 2)$ go 3 units up and then go 1 unit to the right and plot another point. Draw the line through these two points.

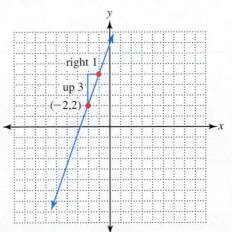

c. $(-3, -1)$ and $m = -\dfrac{2}{3}$

Solution:

Plot the point with coordinates $(-3, -1)$. The slope $-\dfrac{2}{3}$ may be thought of as $\dfrac{-2}{3}$, or $\dfrac{2}{-3}$. We will use $\dfrac{-2}{3}$. Therefore, from $(-3, -1)$ go 2 units down , then go 3 units to the right and plot another point. Draw the line through these two points.

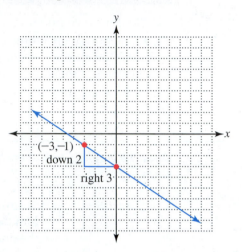

Practice Exercises

Draw the graph of each of the following lines containing the given point with the given slope:

7. $(3, 4)$, $m = \dfrac{2}{3}$

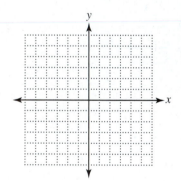

8. $(-4, 2)$, m $= -2$

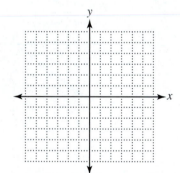

9. $(-5, 4)$, $m = -\dfrac{5}{3}$

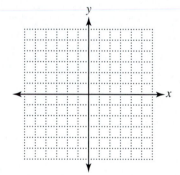

If more practice is needed, do the Additional Practice Exercises.

Answers:

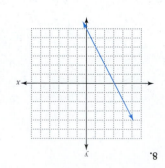

9.

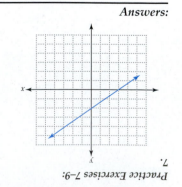

8.

7.

Additional Practice Exercises

Draw the graph of each of the following lines containing the given point with the given slope:

e. $(-1,4)$; $m = \dfrac{1}{3}$ **f.** $(3,-4)$; $m = 2$ **g.** $(-3,5)$; $m = -\dfrac{2}{3}$

We now make two observations about the slope of a line. Since a positive slope means go up from a given point and then go to the right, a line with *positive slope* rises as we go from left to right. Since a line with a *negative slope* can mean go down from a given point and then go to the right, a line with negative slope drops as we go from left to right. See the following figures:

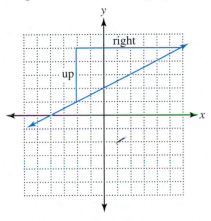

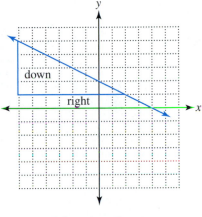

Positive slope   Negative slope

The greater the absolute value of the slope, the greater the vertical change is in relation to the horizontal change. Therefore, the greater the absolute value of the slope, the steeper the line. See the following figures:

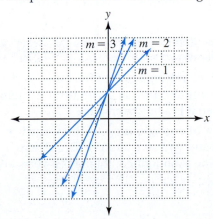

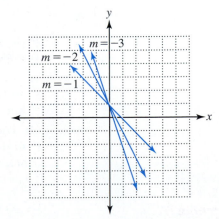

Answers:

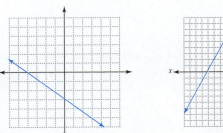

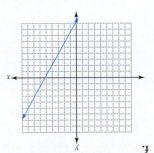

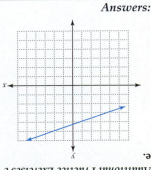

g.   **f.**   **e.**

Additional Practice Exercises e–g:

Recall that the graphs of equations of the form $y = k$ are horizontal lines and the graphs of equations of the form $x = k$ are vertical lines. To find the slopes of these lines, we will locate two points on the line and use the slope formula.

Example 4

a. Find the slope of the graph of $x = 3$.

Solution:
All points on the graph of $x = 3$ have an x-value of 3 and y can have any value. So two points on the graph are $(3, 0)$ and $(3, 1)$. Substitute these values into the slope formula.

$$m = \frac{y_2 - y_1}{x_2 - x_1} \quad \text{Substitute.}$$

$$m = \frac{1 - 0}{3 - 3} \quad \text{Simplify.}$$
Division by 0 is undefined. Therefore, there is no number that represents the slope
$$m = \frac{1}{0}$$
of the graph of $x = 3$. We say the slope is undefined.

Note that there is also no y-intercept.

b. Find the slope of the graph of $y = 2$.

Solution:
All points on the graph of $y = 2$ have a y-value of 2 and x can have any value. Two points on the graph are $(0, 2)$ and $(3, 2)$. Use these points in the slope formula.

$$m = \frac{y_2 - y_1}{x_2 - x_1} \quad \text{Substitute.}$$

$$m = \frac{2 - 2}{3 - 0} \quad \text{Simplify.}$$

$$m = \frac{0}{3} \quad \text{Simplify.}$$

$$m = 0 \quad \text{Therefore, the slope is 0.}$$

Note that there is no x-intercept.

Since the equation of a vertical line is in the form of $x = k$, x remains constant. Since the slope is the change in y divided by the change in x and x is a constant, there is no change in x. Consequently, the denominator is 0, which makes the slope of the vertical line undefined. Similarly, a horizontal line has an equation of the form $y = k$, which means that y is constant. Therefore, there is no change in y so the numerator is 0. This means that the slope of the horizontal line is always 0.

From the preceding, we make the following generalizations:

Slopes of Vertical and Horizontal Lines

a. The graph of any equation of the form $y = b$ is a horizontal line whose slope is **0** and whose y-intercept is b. There is no x-intercept (unless $b = 0$). If the slope of a line is 0, then it is a horizontal line.

b. The graph of any equation of the form $x = a$ is a vertical line whose slope is **undefined** and whose x-intercept is a. There is no y-intercept (unless $a = 0$). If the slope of a line is undefined, then it is a vertical line.

Example 5

Find the slope of each of the following.

a. $x = 4$

Solution:
The graph of $x = 4$ is a vertical line whose slope is undefined and there is no y-intercept.

b. $y = -3$

Solution:
The graph of $y = -3$ is a horizontal line whose slope is 0 and the y-intercept is -3.

Draw the graph of each of the following lines containing the given point and with the given slope.

c. $(3, -2)$ with $m = 0$

Solution:
Since the slope is 0, this is a horizontal line through the point $(3, -2)$.

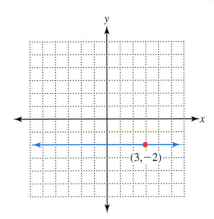

d. $(-5, 2)$ with undefined slope.

Solution:
Since the slope is undefined, the graph is a vertical line through $(-5, 2)$

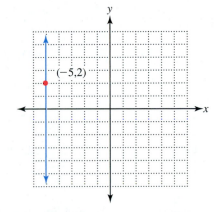

Practice Exercises

Find the slope of each of the following:

10. $x = -2$

11. $y = 7$

Draw the graph of each of the following lines containing the given point and with the given slope:

12. $(-2, 4)$ with undefined slope

13. $(2, 3)$ with $m = 0$

If more practice is needed, do the Additional Practice Exercises.

Additional Practice Exercises

Find the slope of each of the following:

h. $y = -4$

i. $x = 5$

Draw the graph of each of the following lines containing the given point and with the given slope:

j. $(4, 5)$ with $m = 0$

k. $(-4, 2)$ with undefined slope.

Answers:

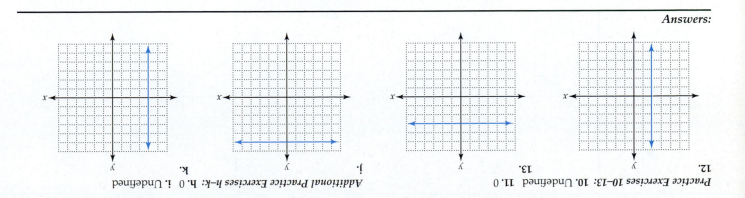

Practice Exercises 10–13: 10. Undefined **11.** 0
12. 13.

Additional Practice Exercises h–k: h. 0 **i.** Undefined
j. k.

Exercise Set 4.4

Find the slope of the line that contains each of the following pairs of points:

1. $(2, 4), (4, 5)$

2. $(1, 5), (3, 8)$

3. $(3, 1), (1, 5)$

4. $(3, 6), (1, 3)$

5. $(-1, 4), (2, -2)$

6. $(-3, 2), (1, -6)$

7. $(3, -2), (-3, 2)$

8. $(5, -4), (1, -2)$

9. $(-4, -2), (-8, 2)$

10. $(-5, -2), (-1, -8)$

11. $(0, 3), (4, 6)$

12. $(6, 0), (4, -2)$

13. $(-2, -4), (6, 0)$

14. $(-7, -2), (-2, 0)$

15. $(3, 5), (-1, 5)$

16. $(4, 8), (-3, 8)$

17. $(2, 6), (2, 3)$

18. $(-3, 4), (-3, -1)$

19. $(3, 6), (-4, 6)$

20. $(5, -2), (5, 4)$

Find the slope of each of the following lines from the graph:

21.

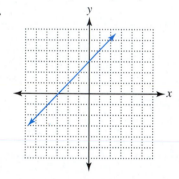

22.

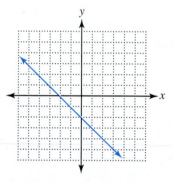

23.

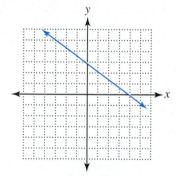

24.

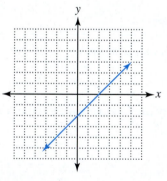

25.

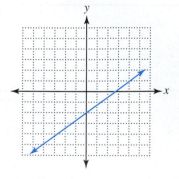

26.

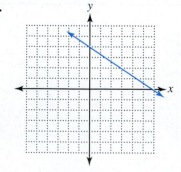

27.

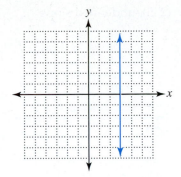

28.

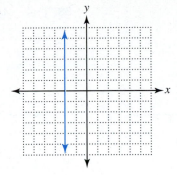

29.

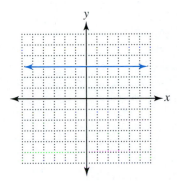

30.

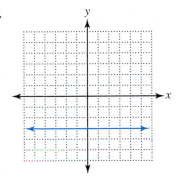

Draw the graph of each of the following lines containing the given point and with the given slope:

31. $(-3, 2)$, $m = \dfrac{3}{4}$

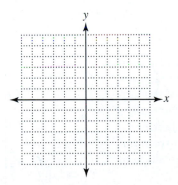

32. $(4, -3)$, $m = \dfrac{3}{2}$

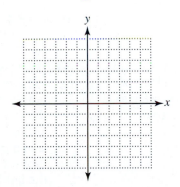

33. $(-1, -3)$, $m = -\dfrac{5}{3}$

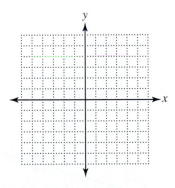

34. $(-3, -5)$, $m = -\dfrac{4}{3}$

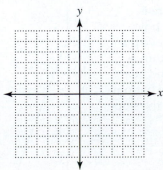

35. $(0, 2)$, $m = 3$

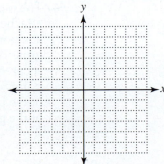

36. $(4, 0)$, $m = 4$

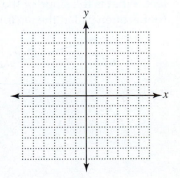

37. $(-1, 3), m = -2$

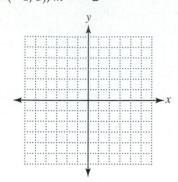

38. $(-4, -1), m = -4$

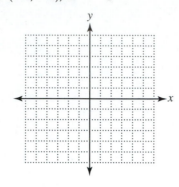

39. $(2, 5),$ undefined slope

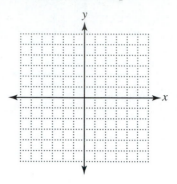

40. $(-2, 4),$ undefined slope

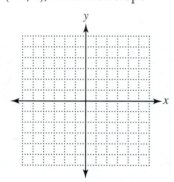

41. $(-2, 6), m = 0$

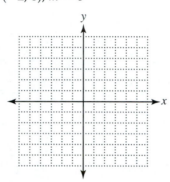

42. $(5, -2), m = 0$

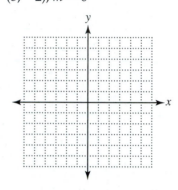

Find the slope of the graph of each of the following:

43. $x = 5$ **44.** $x = 8$ **45.** $x = -1$ **46.** $x = -6$

47. $y = 4$ **48.** $y = 7$ **49.** $y = -4$ **50.** $y = -8$

Answer the following:

51. A mountain road rises 30 feet vertically over a horizontal distance of 500 feet. a) What is the slope of the road? b) When the slope of a road is written as a percent, it is called the *grade.* What is the grade?

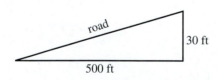

52. The roof of a house rises 6 feet over a span of 20 feet. What is the slope of the roof? (The slope of a roof is often called the *pitch.*)

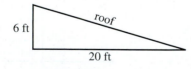

53. A carpenter is working on the outside of a house using a ladder. If the foot of the ladder is 9 feet from the wall and the top of the ladder is at a point 24 feet from the ground, what is the slope of the ladder?

54. The top of a guy wire is attached to a utility pole at a point 30 feet above the ground and the bottom is secured to the ground at a point 8 feet from the bottom of the pole. What is the slope of the guy wire?

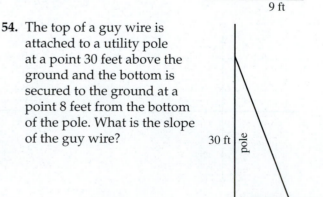

Writing Exercises:

55. How does the graph of a line whose slope is $\frac{2}{3}$ compare with the graph of a line whose slope is $-\frac{2}{3}$?

56. How does the graph of a line whose slope is $\frac{5}{2}$ compare with the graph of a line whose slope is $\frac{1}{3}$?

57. Why is the slope of a vertical line undefined?

58. Why is the slope of a horizontal line 0?

59. Why does a line with positive slope rise from left to right?

60. Why does a line with negative slope fall from left to right?

Section 4.5	Slope-Intercept Form of a Line

OBJECTIVES

When you complete this section, you will be able to:

a. Write an equation in slope-intercept form and find the slope and y-intercept from the equation.

b. Graph a line using the slope-intercept form.

c. Write the equation of a line given the slope and y-intercept.

d. Determine if two lines are parallel or perpendicular.

Introduction In the last section, we found the slope of a line containing two points by using the slope formula. In this section, we learn a method for finding the slope of a line when given the equation of the line.

If we are given the equation of a line and asked to find the slope, one method would be to find two points on the line and use the slope formula. The intercepts are usually the two easiest points to find.

Example 1

Find the slope of the line represented by each of the following equations:

a. $3x - 2y = 6$

Solution:
Find the x- and y-intercepts by letting $y = 0$ and $x = 0$, respectively.

x-intercept. Let $y = 0$.

$$3x - 2(0) = 6$$
$$3x = 6$$
$$x = 2$$

Therefore, the x-intercept is (2,0).

y-intercept. Let $x = 0$.

$$3(0) - 2y = 6$$
$$0 - 2y = 6$$
$$y = -3$$

Therefore, the y-intercept is $(0, -3)$.

Find the slope using the points whose coordinates are $(2, 0)$ and $(0, -3)$.

$$m = \frac{y_2 - y_1}{x_2 - x_1} \quad \text{Substitute for } x_1, y_1, x_2, \text{ and } y_2.$$

$$m = \frac{-3 - 0}{0 - 2} \quad \text{Simplify.}$$

$$m = \frac{-3}{-2} = \frac{3}{2} \quad \text{Therefore, the slope is } \frac{3}{2}.$$

b. $3x + 5y = 15$

Solution:
Find the x- and y-intercepts.

x-intercept. Let $y = 0$. y-intercept. Let $x = 0$.

$$3x + 5(0) = 15 \qquad\qquad\qquad 3(0) + 5y = 15$$
$$3x + 0 = 15 \qquad\qquad\qquad\quad 0 + 5y = 15$$
$$3x = 15 \qquad\qquad\qquad\qquad\quad 5y = 15$$
$$x = 5 \qquad\qquad\qquad\qquad\qquad y = 3$$

Therefore, the x-intercept is $(5, 0)$.   Therefore, the y-intercept is $(0, 3)$.

Find the slope using the points whose coordinates are $(5, 0)$ and $(0, 3)$.

$$m = \frac{y_2 - y_1}{x_2 - x_1} \qquad \text{Substitute.}$$

$$m = \frac{3 - 0}{0 - 5} \qquad \text{Simplify the numerator and denominator.}$$

$$m = \frac{3}{-5} = -\frac{3}{5} \qquad \text{Slope.}$$

Finding the slope of a line from the equation of the line
If the x and/or y-intercepts are fractions, this method can be somewhat complicated. To show another method of finding the slope of a line when the equation is given, we will solve each of the equations from Example 3 for y and make an observation.

Example 1a   Example 1b

$$3x - 2y = 6 \qquad\qquad\qquad\qquad 3x + 5y = 15$$
$$3x - 3x - 2y = -3x + 6 \qquad\qquad 3x - 3x + 5y = -3x + 15$$
$$-2y = -3x + 6 \qquad\qquad\qquad\quad 5y = -3x + 15$$
$$\frac{-2y}{2} = \frac{-3x + 6}{-2} \qquad\qquad\qquad \frac{5y}{5} = \frac{-3x + 15}{5}$$
$$y = \frac{-3x}{-2} + \frac{6}{-2} \qquad\qquad\qquad y = \frac{-3x}{5} + \frac{15}{5}$$
$$y = \frac{3}{2}x - 3 \qquad\qquad\qquad\qquad y = -\frac{3}{5}x + 3$$

In Example 1a, we found the slope of $3x - 2y = 6$ to be $\frac{3}{2}$. When we solved $3x - 2y = 6$ for y, the coefficient of the x-term is also $\frac{3}{2}$. In Example 1b, we found the slope of $3x + 5y = 15$ to be $-\frac{3}{5}$. When we solved $3x + 5y = 15$ for y, we found the coefficient of the x-term is also $-\frac{3}{5}$. This is no coincidence. It can be shown that if a linear equation with two variables is solved for y, the graph of the equation is a line whose slope is the coefficient of x. Further note that the y-intercept of Example 1a is -3, which is the constant when $3x - 2y = 6$ was solved for y giving $y = \frac{3}{2}x - 3$. Likewise, for Example 1b the y-intercept is 5, which is the constant when the equation was solved for y giving $y = -\frac{3}{5}x + 5$. Consequently, when a linear equation is solved for y, we can read both the slope and the y-intercept from the equation.

Slope-Intercept Form of a Line

When any linear equation is solved for y, the resulting equation is of the form $y = mx + b$ where m is the slope and b is the y-intercept of the graph of the equation. This is called the slope-intercept form of a line.

Example 2

Find the slope and y-intercept of the line represented by each of the following equations.

a. $4x + y = 5$

Solution:
Solve the equation for y.

$$4x + y = 5$$

Subtract $4x$ from both sides of the equation.

$$4x - 4x + y = -4x + 5$$

Add like terms.

$$0 + y = -4x + 5$$

$0 + y = y$.

$$y = -4x + 5$$

Therefore, the slope is -4 and the y-intercept is 5.

b. $5x - 2y = 7$

Solution:
Solve the equation for y.

$$5x - 2y = 7$$

Subtract $5x$ from both sides of the equation.

$$5x - 5x - 2y = -5x + 7$$

Add like terms.

$$0 - 2y = -5x + 7$$

$0 - 2y = -2y$.

$$-2y = -5x + 7$$

Divide both sides by -2.

$$\frac{-2y}{-2} = \frac{-5x + 7}{-2}$$

Simplify both sides.

$$y = \frac{5}{2}x - \frac{7}{2}$$

Therefore, the slope is $\frac{5}{2}$ and the y-intercept is $-\frac{7}{2}$.

Practice Exercises

Find the slope and y-intercept of the line represented by each of the following equations:

1. $x + 2y = 6$

2. $6x - 5y = 11$

If more practice is needed, do the Additional Practice Exercises.

Additional Practice Exercises

Find the slope and y-intercept of the line represented by each of the following equations:

a. $5x + y = 9$

b. $7x - 3y = 8$

Graphing a line using the slope-intercept method

The slope-intercept form of a linear equation gives us yet another method of graphing linear equations. The y-intercept is a point on the graph. By knowing a point and the slope we can draw the graph using the technique given earlier in Section 4.3.

Example 3

Graph the following using the slope and y-intercept:

a. $y = \frac{2}{3}x + 2$

Solution:
From the equation, we have the slope is $\frac{2}{3}$ and the y-intercept is 2. Plot the y-intercept and from that point go up 2 and 3 to the right and plot another point.

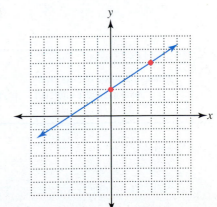

b. $3x + 4y = 8$ **Solution:**

First we must get the line into slope-intercept form by solving for y.

$$3x + 4y = 8 \qquad \text{Subtract } 3x \text{ from both sides of the equation.}$$

$$4y = -3x + 8 \qquad \text{Divide both sides by 4.}$$

$$y = \frac{-3x + 8}{4} \qquad \text{Simplify the right side.}$$

$$y = -\frac{3}{4}x + 2 \qquad \text{Therefore, the slope is } -\frac{3}{4} \text{ and the } y\text{-intercept is 2.}$$

Plot the y-intercept. From that point, go down 3 and to the right 4 and plot another point.

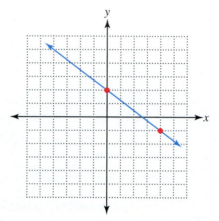

Practice Exercises

Graph each of the following using the slope-intercept method:

3. $y = \frac{3}{5}x - 1$

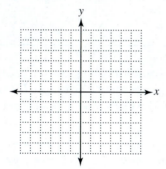

4. $3x + 2y = 6$

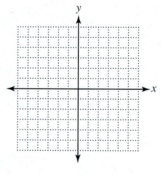

Writing equations of lines given the slope and y-intercept   If we know the slope and y-intercept of a line, then we can use the slope-intercept form to write the equation of the line.

Answers:

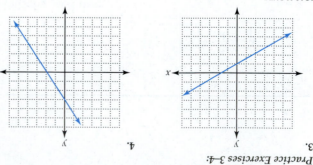

Practice Exercises 3–4:

3.

4.

Example 4

Write the equations of the following lines. Leave the answers in slope-intercept form.

a. Slope of 3 and y-intercept of 4.

Solution:

$y = mx + b$ Substitute 3 for m and 4 for b.

$y = 3x + 4$ Therefore, the equation of the line with slope of 3 and y-intercept of $(0,4)$ is $y = 3x + 4$.

b. The slope is $\frac{3}{4}$ and the line contains $(0, \frac{2}{3})$.

Solution:

The point $(0, \frac{2}{3})$, is the y-intercept, so $b = \frac{2}{3}$.

$y = mx + b$ Substitute $\frac{3}{4}$ for m and $\frac{2}{3}$ for b.

$y = \frac{3}{4}x + \frac{2}{3}$ Therefore, the equation of the line with slope of $\frac{3}{4}$ and containing $(0,\frac{2}{3})$ is $y = \frac{3}{4}x + \frac{2}{3}$.

Practice Exercises

Write the equations of the following lines. Leave the answers in slope-intercept form.

5. The slope is $\frac{5}{4}$ and the y-intercept is -3.

6. The slope is $-\frac{3}{5}$ and the line contains $(0, -\frac{3}{4})$

Parallel and perpendicular lines

Two lines are **parallel** if they are in the same plane and do not intersect. In the figure that follows, L_1 is the graph of a line containing $(2, 3)$ and with a slope of $\frac{1}{2}$, and L_2 is the graph of a line containing $(2, -2)$ also with a slope of $\frac{1}{2}$.

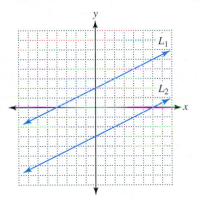

Both lines have slope of $\frac{1}{2}$ and the graphs appear to be parallel lines. It can be shown that two lines are parallel if and only if they have equal slopes. This means that if two lines are parallel, then they have equal slopes. It also means that if two lines have equal slopes, then they are parallel.

Two lines are perpendicular if they intersect at right angles. In the figure that follows, L_1 is the graph of a line containing $(2, 3)$ with a slope of $\frac{2}{3}$, and L_2 is the graph of a line containing $(2, 3)$ and with a slope of $-\frac{3}{2}$.

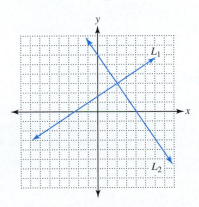

The two lines appear to be perpendicular. That is, they meet at a 90°, or right angle. Notice the product of their slopes is -1. That is, $\left(\frac{2}{3}\right)\left(-\frac{3}{2}\right) = -1$. It can be shown that two lines are perpendicular if and only if the product of their slopes is -1. This means that if two lines are perpendicular, the product of their slopes is -1. It also means that if the product of the slopes of two lines is -1, the two lines are perpendicular. If m_1 represents the slope of a line and m_2 represents the slope of any line perpendicular to m_1, then $m_1 m_2 = -1$. Solving for m_1, we get $m_1 = -\frac{1}{m_2}$. This means that m_1 is the negative of the reciprocal of m_2. Thus, if a line has a slope of 3, then any line perpendicular to it has a slope of $-\frac{1}{3}$. If a line has a slope of $-\frac{3}{4}$, then any line perpendicular to it has a slope of $\frac{4}{3}$.

Parallel and Perpendicular Lines

Two nonvertical lines are **parallel** if and only if they have **equal slopes**.

Two lines are **perpendicular** if and only if the product of their slopes is -1, provided neither is vertical or horizontal. In other words, two nonvertical lines are perpendicular if the slopes are negatives and reciprocals. Any vertical line is perpendicular to any horizontal line and vice versa.

From the preceding, we see that if an equation is given in $ax + by = c$ form, the slope is $-\frac{a}{b}$. For example, if $2x + 3y = 5$, then $a = 2$, $b = 3$ and $m = -\frac{2}{3}$, and if $4x - 5y = 7$, then $a = 4$, $b = -5$ and $m = -\frac{4}{-5} = \frac{4}{5}$.

Example 5

Determine whether the lines determined by the following pairs of points are parallel, perpendicular, or neither:

a. L_1: $(-3, 2)$ and $(1, 5)$, L_2: $(2, -1)$ and $(-2, -4)$

b. L_1: $(3, -1)$ and $(1, 3)$, L_2: $(-1, -3)$ and $(1, -2)$

Solution:
In order to determine if the lines are parallel or perpendicular, we need the slopes of the lines.

Slope of L_1

$$m = \frac{y_2 - y_1}{x_2 - x_1}$$

$$m = \frac{5 - 2}{1 - (-3)}$$

$$m = \frac{5 - 2}{1 + 3}$$

$$m = \frac{3}{4}$$

Slope of L_2

$$m = \frac{y_2 - y_1}{x_2 - x_1}$$

$$m = \frac{-4 - (-1)}{-2 - 2}$$

$$m = \frac{-4 + 1}{-2 - 2}$$

$$m = \frac{-3}{-4}$$

$$m = \frac{3}{4}$$

Since the slopes are equal, L_1 and L_2 are parallel.

Solution:
In order to determine if the lines are parallel or perpendicular, we need the slopes of the lines.

Slope of L_1

$$m = \frac{y_2 - y_1}{x_2 - x_1}$$

$$m = \frac{3 - (-1)}{1 - 3}$$

$$m = \frac{3 + 1}{1 - 3}$$

$$m = \frac{4}{-2} = -2$$

Slope of L_2

$$m = \frac{y_2 - y_1}{x_2 - x_1}$$

$$m = \frac{-2 - (-3)}{1 - (-1)}$$

$$m = \frac{-2 + 3}{1 + 1}$$

$$m = \frac{1}{2}$$

Since $(-2)\left(\frac{1}{2}\right) = -1$, L_1 and L_2 are perpendicular.

Often all we need from an equation is the slope and not the y-intercept. If we write the general equation of a linear equation ($ax + by = c$) in slope-intercept form, we will get a formula for slope.

$ax + by = c$	Subtract ax from both sides.
$by = -ax + c$	Divide both sides by b.
$\dfrac{by}{b} = \dfrac{-ax + c}{b}$	Simplify both sides.
$y = -\dfrac{a}{b}x + \dfrac{c}{b}$	Therefore, the slope is $-\dfrac{a}{b}$.

c. L_1: $2x + 5y = 10$ and L_2: $5x - 2y = 4$

Solution:
In order to determine if the lines are parallel or perpendicular, we need to know the slopes.

The slope of $L_1 = -\frac{a}{b} = -\frac{2}{5}$ and the slope of

$L_2 = -\frac{a}{b} = -\frac{5}{-2} = \frac{5}{2}$.

Since $\left(-\frac{2}{5}\right)\left(\frac{5}{2}\right) = -1$, L_1 and L_2 are perpendicular.

d. L_1: $3x + 4y = 8$ and L_2: $4x + 3y = 6$

Solution:
In order to determine if the two lines are parallel or perpendicular, we need the slopes.

The slope of $L_1 = -\frac{a}{b} = -\frac{3}{4}$ and the slope of

$L_2 = -\frac{a}{b} = -\frac{4}{3}$.

Since the slopes are neither equal nor is their product -1, L_1 and L_2 are neither parallel nor perpendicular.

Practice Exercises

Determine whether the lines containing the following pairs of points are parallel, perpendicular or neither:

7. L_1: $(4, 5)$ and $(-1, 2)$, L_2: $(-2, -1)$ and $(3, 2)$

8. L_1: $(5, -2)$ and $(3, 1)$, L_2: $(4, 6)$ and $(1, 4)$

Determine whether the graphs of the lines with the given equations are parallel, perpendicular, or neither.

9. L_1: $x + 3y = 6$ and L_2: $3x + y = 4$

10. L_1: $2x - 7y = 5$ and L_2: $2x - 7y = -7$

If more practice is needed, do the Additional Practice Exercises.

Additional Practice Exercises

Determine whether the following pairs of lines are parallel, perpendicular or neither:

d. L_1: $(1, 1)$ & $(2, 3)$, L_2: $(-3, 2)$ & $(-4, 0)$

e. L_1: $(-2, 2)$ & $(1, 3)$, L_2: $(2, -1)$ & $(3, 2)$

f. L_1: $2x + y = 4$, L_2: $2x + y = 6$

g. L_1: $3x - 2y = 4$, L_2: $2x + 3y = 4$

Exercise Set 4.5

Find the slope and the y-intercept of the line represented by each of the following equations. Graph each equation using the slope-intercept method.

1. $y = -2x + 9$

2. $y = -3x + 8$

3. $3x - y = 7$

4. $2x - y = 11$

5. $2x + 3y = 9$

6. $3x + 2y = 12$

7. $5x - 3y = 15$

8. $2x - 5y = 10$

9. $2x + y = 0$

10. $3x - y = 0$

11. $3x + 2y = 5$

12. $5x - 2y = -3$

Find the slope of each of the following:

13. $x = 8$

14. $x = 5$

15. $x = -6$

16. $x = -1$

17. $y = 7$

18. $y = 4$

19. $y = -8$

20. $y = -4$

Write the equations of the following lines. Leave the answers in $y = mx + b$ *form.*

21. Slope is 2 and y-intercept is 4

22. Slope is -5 and y-intercept is -2

23. Slope is $\frac{3}{5}$ and y-intercept is -1

24. Slope is $\frac{5}{2}$ and y-intercept is 4

25. Slope is $-\frac{2}{3}$ and y-intercept is $\frac{4}{5}$

26. Slope is $-\frac{6}{5}$ and y-intercept is $-\frac{5}{4}$

Determine whether the following lines containing the given pairs of points are parallel, perpendicular, or neither:

27. L_1: $(2, -4)$ and $(4, 2)$,
L_2: $(2, -5)$ and $(4, 1)$

28. L_1: $(1, -3)$ and $(4, 0)$,
L_2: $(-1, 4)$ and $(1, 10)$

29. L_1: $(4, -3)$ and $(-1, -1)$,
L_2: $(-2, 2)$ and $(-4, -3)$

30. L_1: $(-4, 5)$ and $(2, 3)$,
L_2: $(-1, -2)$ and $(0, 1)$

31. L_1: $(-6, 2)$ and $(-2, 0)$,
L_2: $(8, -3)$ and $(10, 1)$

32. L_1: $(-5, -1)$ and $(7, 7)$,
L_2: $(1, 5)$ and $(-2, 3)$

33. L_1: $(3, -7)$ and $(-1, -1)$,
L_2: $(8, -2)$ and $(2, 2)$

34. L_1: $(4, -5)$ and $(5, -1)$,
L_2: $(3, -4)$ and $(7, -3)$

35. L_1: $(2, 4)$ and $(2, -2)$,
L_2: $(-4, 1)$ and $(-4, 6)$

36. L_1: $(3, 5)$ and $(-2, 5)$,
L_2: $(-1, 2)$ and $(4, 2)$

37. L_1: $(-2, 3)$ and $(4, 3)$,
L_2: $(6, 1)$ and $(6, 7)$

38. L_1: $(-4, 1)$ and $(-4, 6)$,
L_2: $(-3, 5)$ and $(3, 5)$

Determine whether the graphs of the lines with the given equations are parallel, perpendicular, or neither:

39. L_1: $4x + y = 7$,
L_2: $4x + y = -3$

40. L_1: $3x - y = 5$,
L_2: $3x - y = 9$

41. L_1: $4x + 3y = 6$,
L_2: $3x - 4y = 12$

42. L_1: $5x - 3y = 11$,
L_2: $3x + 5y = 8$

43. L_1: $4x + 6y = 5$,
L_2: $8x + 12y = 9$

44. L_1: $5x + 10y = 15$,
L_2: $4x - 2y = 5$

45. L_1: $x = 4$,
L_2: $x = -1$

46. L_1: $y = 5$,
L_2: $y = -5$

47. L_1: $x = 6$,
L_2: $y = 2$

48. L_1: $y = -5$,
L_2: $x = 3$

Challenge Exercises:

49. A lawn-maintenance worker earns $6.00 per hour.

   a. If y represents the worker's earnings for working x hours, write an equation representing the worker's earnings.

   b. Graph the equation.

   c. What is the relationship between the slope of the line and the worker's hourly rate of pay?

50. A local nursery sells only azalea plants. There is an initial start-up cost of $360 per day and a cost of $3.00 per azalea sold per day.

   a. If y represents the cost per day and x the number of azaleas sold per day, write an equation for the daily costs of the nursery.

   b. Graph the equation. (Adjust scale on y-axis.)

   c. What is the relationship between the slope of the line and the cost per azalea?

51. A salesperson at a paint store earns $100 per week plus a commission of $2.00 on every gallon of paint he or she sells.

   a. If y represents the salesperson's salary for selling x gallons of paint, write an equation that represents the salesperson's salary.

   b. Graph the equation. (Adjust scale on y-axis.)

   c. What is the relationship between the slope of the line and the commission per gallon of paint sold?

52. A piece of machinery costs $3000 and depreciates at the rate of $600 per year.

   a. If y represents the value of the machinery after x years, write an equation representing the value of the machinery.

   b. Graph the equation. (Adjust scale on y-axis.)

   c. What is the relationship between the slope of the line and the per-year depreciation?

Writing Exercises:

53. Give the equation of a line in $ax + by = c$ form that has **a.** positive slope, and **b.** negative slope.

54. Given the equations $y = -2x - 3$, $y = -\frac{1}{2}x - 3$, $y = -.75x - 3$ and $y = x - 3$, which equation has the steepest graph? Why?

55. The standard equation of a line is $ax + by = c$. What is the relationship between a and b if the line has a positive slope? a negative slope?

56. Why must parallel lines have equal slopes?

57. Are the lines whose equations are $y = 2x + 3$ and $y = -2x + 3$ perpendicular? Why or why not?

58. Can two perpendicular lines both have negative slopes? Why or why not?

59. How can you tell if two lines are parallel without graphing them?

Point-Slope Form of a Line

OBJECTIVES

When you complete this section, you will be able to:

a. Write the equation of a line that contains a given point and has a given slope.

b. Write the equation of a line that contains two given points.

c. Write the equation of a line that contains a given point and is parallel or perpendicular to a line whose equation is given.

Introduction   If we are given a point on a line and the slope of the line, we can draw a line that contains the given point and that has the given slope. Also, if we are given two points, we can draw the line containing the two points. How do we find the equations of these lines? These are the questions we will answer in this section.

Suppose we want to write the equation of a line that contains (2, 3) and has a slope of 2. Let us draw the graph.

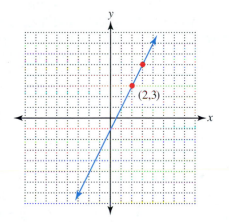

The slope of the line is the ratio of the change in y to the change in x. In this case, this ratio is $\frac{2}{1}$. This does not mean that we must go up 2 units and to the right 1 unit in order to get from one point of the line to another. We could go up 4 and to the right 2. We could even go fractional parts of a unit as long as the ratio of the change in y to the change in x is 2. That is, since $\frac{4}{2} = \frac{-10}{-5} = \frac{2}{1}$, we could use any of these ratios to indicate a slope of 2. The ratio is the same for any two points on the line.

If we let (x, y) represent any other point on the line, then the slope from (2, 3) to (x, y) is 2. If we go from (2, 3) to (x, y) by going vertically and then horizontally, the point of intersection of the vertical and horizontal line segments is $(2, y)$. The change in y is $y - 3$ and the change in x is $x - 2$. See the following figure:

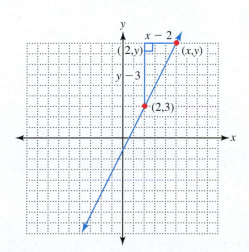

To write the equation of the line, use the slope formula and let $(x_1, y_1) = (2, 3)$ and $(x_2, y_2) = (x, y)$. You will not be asked to write equations using this procedure, but we need this procedure to understand where the formula comes from.

$$m = \frac{y_2 - y_1}{x_2 - x_1}$$

Substitute for m, x_1, y_1, x_2, and y_2.

$$2 = \frac{y - 3}{x - 2}$$

Multiply both sides by $x - 2$.

$$\frac{y-3}{x-2}(x-2) = \frac{y-3}{1} \cdot \frac{1}{x-2} \cdot \frac{x-2}{1} =$$

$$2(x - 2) = \frac{y - 3}{x - 2}(x - 2)$$

$$(y - 3) \cdot \frac{x-2}{x-2} = (y - 3) \cdot 1 = y - 3.$$

$$2(x - 2) = y - 3$$

Apply the distributive property.

$$2x - 4 = y - 3$$

Add 4 to both sides.

$$2x = y + 1$$

Subtract y from both sides.

$$2x - y = 1$$

Therefore, the equation of the line through $(2,3)$ with slope of 2 is $2x - y = 1$.

Writing the equation of a line given a point on the line and the slope of the line

Now let us use the preceding procedure to develop a formula for the equation of a line when we are given a point on the line and the slope of the line. Let (x_1, y_1) be the coordinates of the given point on the line and m be the slope of the line. Let (x, y) be any other point of the line. The slope between (x_1, y_1) and (x, y) is m. Substitute into the slope formula with $(x_2, y_2) = (x, y)$. Consequently,

$$m = \frac{y - y_1}{x - x_1}$$

Multiply both sides by $x - x_1$.

$$\frac{y-y_1}{x-x_1}(x-x_1) = \frac{y-y_1}{1} \cdot \frac{1}{x-x_1} \cdot \frac{x-x_1}{1} =$$

$$m(x - x_1) = \frac{y - y_1}{x - x_1}(x - x_1)$$

$$(y - y_1) \cdot \frac{x-x_1}{x-x_1} = (y - y_1) \cdot 1 = y - y_1.$$

$$m(x - x_1) = y - y_1$$

Rewrite as $y - y_1 = m(x - x_1)$.

$$y - y_1 = m(x - x_1)$$

Therefore, the equation of a line containing the point (x_1, y_1) and with slope of m is $y - y_1 = m(x - x_1)$.

We now state this more formally as the point-slope formula.

Be Careful In using the point-slope formula, we substitute the coordinates of the given point for x_1 and y_1 and the value of the slope for m. The variables are x and y, so they are not replaced.

Point-Slope Formula

The equation of the line containing the given point (x_1, y_1) and with the given slope of m is $y - y_1 = m(x - x_1)$.

Example 1

Write the equation of each of the following lines that contain the given point and have the given slope in ax + by = c form with a and b integers:

a. $(3, -4)$ and $m = \dfrac{2}{3}$

Solution:
$(x_1, y_1) = (3, -4)$. Therefore, $x_1 = 3$ and $y_1 = -4$.

$$y - y_1 = m(x - x_1)$$

Substitute for x_1, y_1, and m.

$$y - (-4) = \frac{2}{3}(x - 3)$$

$y - (-4) = y + 4$ and multiply both sides by 3.

$$3(y + 4) = 3 \cdot \frac{2}{3}(x - 3)$$

Multiply 3 and $\frac{2}{3}$. $(3 \cdot \frac{2}{3} = \frac{3}{1} \cdot \frac{2}{3} = \frac{6}{3} = 2)$

$$3(y + 4) = 2(x - 3)$$

Distribute on both sides.

$$3y + 12 = 2x - 6$$

Subtract 12 from both sides.

$$3y + 12 - 12 = 2x - 6 - 12$$

Simplify both sides.

$$3y = 2x - 18$$

Subtract $2x$ from both sides.

$$3y - 2x = 2x - 2x - 18$$

$$-2x + 3y = -18$$

$$-1(-2x + 3y) = -1(-18)$$

$$2x - 3y = 18$$

Write in $ax + by = c$ form. It is preferred that the coefficient of x be positive, so multiply both sides by -1. Simplify both sides. Therefore, the equation of the line containing $(3, -4)$ with slope of $\frac{2}{3}$ is $2x - 3y = 18$.

b. $(-2, 4)$ and $m = -\frac{3}{4}$

Solution:

$(x_1, y_1) = (-2, 4)$ so $x_1 = -2$ and $y_1 = 4$.

$$y - y_1 = m(x - x_1)$$

Substitute for x_1, y_1, and m.

$$y - 4 = -\frac{3}{4}(x - (-2))$$

$x - (-2) = x + 2$ Multiply both sides by 4.

$$4(y - 4) = 4\left(-\frac{3}{4}\right)(x + 2)$$

$4\left(-\frac{3}{4}\right) = \left(\frac{4}{1}\right)\left(-\frac{3}{4}\right) = -\frac{12}{4} = -3$

$$4(y - 4) = -3(x + 2)$$

Apply the distributive property.

$$4y - 16 = -3x - 6$$

Add 16 to both sides.

$$4y = -3x + 10$$

Add $3x$ to both sides. Write the x-term first.

$$3x + 4y = 10$$

Therefore, the equation of the line containing $(-2, 4)$ and with slope of $-\frac{3}{4}$ is $3x + 4y = 10$.

Alternative Method:

It is also possible to write the equations of a line when given a point and the slope using the slope-intercept formula, $y = mx + b$. Since we are given a point and the slope, we know a value for x, y, and m. We substitute these values into $y = mx + b$ and solve for b. We illustrate this procedure using Example 1a, where we were given the point $(3, -4)$ (so $x = 3$ and $y = -4$). We were also given that $m = \frac{2}{3}$.

$$y = mx + b$$

Substitute 3 for x, -4 for y, and $\frac{2}{3}$ for m.

$$-4 = \frac{2}{3}(3) + b$$

$\frac{2}{3}(3) = \frac{2}{3} \cdot \frac{3}{1} = \frac{6}{3} = 2$

$$-4 = 2 + b$$

Subtract 2 from both sides of the equation.

$$-6 = b$$

Substitute $\frac{2}{3}$ for m and -6 for b in $y = mx + b$.

$$y = \frac{2}{3}x - 6$$

Equation in slope-intercept form.

Practice Exercises

Write the equation of each of the following lines that contain the given point and have the given slope in **ax + by = c** *form with* **a** *and* **b** *integers.*

1. $(-1, 3)$ and $m = \frac{3}{5}$

2. $(4, -2)$ and $m = -\frac{3}{2}$

If more practice is needed, do the Additional Practice Exercises.

Additional Practice Exercises

Write the equations of the following lines in ax + by = c *form with* a *and* b *integers.*

a. $(1, -3)$ and $m = \dfrac{1}{4}$

b. $(-3, 5)$ and $m = -3$

Writing the equation of a line containing two given points   The point-slope formula requires that we know a point on the line and the slope of the line. If we know the coordinates of two points on the line, we do not know the slope, but we can find it. Consequently, we can use the point-slope formula to write the equation of a line if we know two points on the line.

Example 2

Find the equation of the line that contains the following pairs of points in ax + by = c *form with* a *and* b *integers:*

a. $(2, 4)$ and $(-1, 2)$

Solution:
Using the slope formula with $(x_1, y_1) = (2, 4)$ and $(x_2, y_2) = (-1, 2)$, find the slope of the line.

$$m = \frac{y_2 - y_1}{x_2 - x_1} \qquad \text{Substitute for } x_1, y_1, x_2, \text{ and } y_2.$$

$$m = \frac{2 - 4}{-1 - 2} \qquad \text{Simplify.}$$

$$m = \frac{-2}{-3} = \frac{2}{3} \qquad \text{Therefore, the slope is } \frac{2}{3}.$$

Since the slope between any two points on the line is the same, we may use either of the given points in the point-slope formula. Use $(2, 4)$ as (x_1, y_1).

$$y - y_1 = m(x - x_1) \qquad \text{Substitute for } m, x_1, \text{ and } y_1.$$

$$y - 4 = \frac{2}{3}(x - 2) \qquad \text{Multiply both sides by 3.}$$

$$3(y - 4) = 3 \cdot \frac{2}{3}(x - 2) \qquad \text{Multiply 3 and } \frac{2}{3}.$$

$$3(y - 4) = 2(x - 2) \qquad \begin{array}{l}\text{Apply the distributive property}\\ \text{on both sides.}\end{array}$$

$$3y - 12 = 2x - 4 \qquad \text{Add 12 to both sides.}$$

$$3y = 2x + 8 \qquad \text{Subtract } 2x \text{ from both sides.}$$

$$3y - 2x = 8 \qquad \begin{array}{l}\text{Rewrite in } ax + by = c \text{ form by}\\ \text{putting } -2x \text{ first.}\end{array}$$

$$-2x + 3y = 8 \qquad \text{Multiply both sides by } -1.$$

$$2x - 3y = -8 \qquad \begin{array}{l}\text{Therefore, the equation of the}\\ \text{line containing } (2,4) \text{ and } (-1,2) \text{ is}\\ 2x - 3y = -8.\end{array}$$

b. $(-3, -3)$ and $(-1, 1)$

Solution:
Find the slope of the line using the slope formula with $(x_1, y_1) = (-3, -3)$ and $(x_2, y_2) = (-1, 1)$.

Answers:

Additional Practice Exercises a–b: **a.** $x - 4y = 13$ **b.** $3x + y = -4$

$$m = \frac{y_2 - y_1}{x_2 - x_1} \qquad \text{Substitute for } x_1, y_1, x_2, \text{ and } y_2.$$

$$m = \frac{1 - (-3)}{-1 - (-3)} \qquad \text{Simplify the numerator and the denominator.}$$

$$m = \frac{1 + 3}{-1 + 3} \qquad \text{Continue simplifying.}$$

$$m = \frac{4}{2} = 2 \qquad \text{Therefore, the slope is 2.}$$

Since the slope between any two points on the line is the same, we may use either of the given points in the point-slope formula. Use $(-1, 1)$.

$$y - y_1 = m(x - x_1) \qquad \text{Substitute for } m, x_1, \text{ and } y_1.$$
$$y - 1 = 2(x - (-1)) \qquad x - (-1) = x + 1.$$
$$y - 1 = 2(x + 1) \qquad \text{Distribute on the right side.}$$
$$y - 1 = 2x + 2 \qquad \text{Add 1 to both sides.}$$
$$y = 2x + 3 \qquad \text{Subtract } 2x \text{ from both sides.}$$
$$-2x + y = 3 \qquad \text{Multiply both sides by } -1.$$
$$2x - y = -3 \qquad \begin{array}{l}\text{Therefore, the equation of the line}\\ \text{containing } (-3,-3) \text{ and } (-1,1) \text{ is}\\ 2x - y = -3.\end{array}$$

Practice Exercises

Find the equation of the line that contains the following pairs of points in $ax + by = c$ *form with* a *and* b *integers:*

3. $(4, 3)$ and $(-1, -7)$

4. $(-2, 5)$ and $(-4, 2)$

If more practice is needed, do the Additional Practice Exercises.

Additional Practice Exercises

Find the equation of the line that contains the following pairs of points. Leave the answers in $ax + by = c$ *form.*

c. $(5, 2)$ and $(3, 8)$

d. $(0, 3)$ and $(-2, 4)$

Writing the equations of
vertical and horizontal lines

Recall that the graph of a linear equation with two variables of the form $x = k$ is a vertical line. Thus all the coordinates of the points on a vertical line have the same x-value. To write the equation of a vertical line, all we need to know is the x-coordinate of one point on the line. The equation is $x =$ (the known x-coordinate). Recall also that the slope of a vertical line is undefined.

You will also recall that the graph of a linear equation with two variables of the form $y = k$ is a horizontal line. Thus all the points on a horizontal line have the same y-coordinate. To write the equation of a horizontal line, all we need to know is the y-coordinate of one point on the line. The equation is $y =$ (the known y-coordinate). Also recall that the slope of a horizontal line is 0.

Example 3

Write the equation of each of the following lines:

a. Containing $(3, 5)$ and vertical.

Solution:
Since the line is vertical, the equation is of the form $x = k$. We know the x-coordinate of one point on the line is 3. Therefore, the equation is $x = 3$.

b. Containing $(-2, 4)$ with undefined slope.

Solution:
Since the slope is undefined, we know the line is vertical. Therefore, the equation is of the form $x = k$. We know the x-coordinate of one point on the line is -2. Therefore, the equation is $x = -2$.

c. Containing $(3, 5)$ and horizontal.

Solution:
Since the line is horizontal, the equation of the line is in the form of $y = k$. We know the y-coordinate of one point on the line is 5. Therefore, the equation is $y = 5$.

d. Containing $(-3, -2)$ and $m = 0$.

Solution:
Since $m = 0$, we know the line is horizontal. Therefore, the equation is of the form $y = k$. We know the y-coordinate of one point on the line is -2. Therefore, the equation of the line is $y = -2$.

Practice Exercises

Write the equation of each of the following lines:

5. Containing $(-3, 6)$ and vertical

6. Containing $(2, 1)$ with $m = 0$

7. Containing $(4, -1)$ and horizontal

8. Containing $(-3, 2)$ with undefined slope

If more practice is needed, do the Additional Practice Exercises.

Additional Practice Exercises

Write the equation of each of the following lines:

e. Vertical containing $(2, 7)$

f. Horizontal containing $(-5, 7)$

g. Containing $(-3, 2)$ with undefined slope.

h. Containing $(5, -3)$ with $m = 0$

Writing the equation of a line containing a given point and parallel or perpendicular to a given line
    If we know a point on the line and the slope of the line, then we can write the equation of the line. Suppose we are given the equation of a line and the coordinates of a point not on the line. Since parallel lines have equal slopes, we can write the equation of a line containing the given point and parallel to the given line by using the slope of the given line. Similarly, we can write the equation of a line containing the given point and perpendicular to the given line by using the negative of the reciprocal of the slope of the given line. Recall that the slope of a line in $ax + by = c$ form is $M = -\dfrac{a}{b}$.

Example 4

Write the equations of the following lines:

a. Containing $(-2, 3)$ and parallel to the graph of $3x + 5y = 8$.

Solution:
To write the equation of a line, we need a point on the line and the slope of the line. We know that the point whose coordinates are $(-2, 3)$ is on the line. Since parallel

Answers:

Practice Exercises 5–8: 5. $x = -3$ 6. $y = 1$ 7. $y = -1$ 8. $x = -3$ **Additional Practice Exercises e–h:** e. $x = 2$

f. $y = 7$ g. $x = -3$ h. $y = -3$

lines have equal slope, the slope of the line we are looking for is equal to the slope of the line whose equation is $3x + 5y = 8$. The slope of $3x + 5y = 8$ is $\frac{-a}{b} = -\frac{3}{5}$.

The slope of any line parallel to this line is also $-\frac{3}{5}$. Therefore, we are looking for the equation of a line containing $(-2, 3)$ and with slope of $-\frac{3}{5}$. Since we know a point and the slope, use the point-slope formula.

$y - y_1 = m(x - x_1)$	Substitute for m, x_1, and y_1.
$y - 3 = -\dfrac{3}{5}(x - (-2))$	$x - (-2) = x + 2$ and multiply both sides by 5.
$5(y - 3) = 5\left(-\dfrac{3}{5}\right)(x + 2)$	Multiply 5 and $-\dfrac{3}{5}$.
$5(y - 3) = -3(x + 2)$	Distribute on both sides.
$5y - 15 = -3x - 6$	Add 15 to both sides.
$5y - 15 + 15 = -3x - 6 + 15$	Simplify both sides.
$5y = -3x + 9$	Add $3x$ to both sides.
$3x + 5y = -3x + 3x + 9$	Simplify the right side.
$3x + 5y = 9$	Therefore, the equation of the line containing $(-2,3)$ and parallel to the graph of $3x + 5y = 8$ is $3x + 5y = 9$.

b. Containing $(4, -3)$ and perpendicular to the graph of $2x + 5y = 7$.

Solution:
To write the equation of a line, we need a point on the line and the slope of the line. We know the point whose coordinates are $(4, -3)$ is on the line. Since the line we are looking for is perpendicular to the graph of $2x + 5y = 7$, its slope is the negative of the reciprocal of the slope of $2x + 5y = 7$. The slope of $2x + 5y = 7$ is $\frac{-a}{b} = -\frac{2}{5}$.

The slope of any line perpendicular to the given line is the negative reciprocal of $-\frac{2}{5}$. Therefore, the slope of any line perpendicular to $2x + 5y = 7$ is $\frac{5}{2}$. Hence, we are looking for the equation of a line containing $(4, -3)$ with slope of $\frac{5}{2}$. Use the point-slope formula.

$y - y_1 = m(x - x_1)$	Substitute for m, x_1, and y_1.
$y - (-3) = \dfrac{5}{2}(x - 4)$	$y - (-3) = y + 3$ and multiply both sides by 2.
$2(y + 3) = 2\left(\dfrac{5}{2}\right)(x - 4)$	Multiply 2 and $\dfrac{5}{2}$.
$2(y + 3) = 5(x - 4)$	Distribute on both sides.
$2y + 6 = 5x - 20$	Subtract $5x$ from both sides.
$-5x + 2y + 6 = 5x - 5x - 20$	Simplify both sides.
$-5x + 2y + 6 = -20$	Subtract 6 from both sides.
$-5x + 2y + 6 - 6 = -20 - 6$	Simplify both sides.
$-5x + 2y = -26$	Multiply both sides by -1.
$-1(-5x + 2y) = -1(-26)$	Simplify both sides.
$5x - 2y = 26$	Therefore, the equation of the line containing $(4,-3)$ and perpendicular to the graph of $2x + 5y = 7$ is $5x - 2y = 26$.

Practice Exercises

Write the equations of the following lines. Leave answers in the form of $ax + by = c$.

9. Containing $(-1, 2)$ and parallel to the graph of $3x - 6y = 8$

10. Containing $(-3, -4)$ and perpendicular to $4x + 2y = 7$

Answers:

Frequently, real-world situations can be modeled using linear equations with two variables.

Example 5

a. A salesperson works for a salary plus commission. If x represents the value of the merchandise sold and y represents the salary earned per month, then $(15{,}000, 3150)$ indicates that, for sales of $15,000, the salary was $3150 and $(10{,}000, 2150)$ means that on $10,000 worth of sales, the salary was $2150.

1. Write a linear equation that gives the salary, y, in terms of the sales, x.

Solution:
We are being asked to write the equation of a line that contains the points $(15{,}000, 3150)$ and $(10{,}000, 2150)$ and to leave the answer in the form $y = mx + b$. We will use the point-slope formula, so we need the slope.

$$m = \frac{y_2 - y_1}{x_2 - x_1} \qquad \text{Substitute.}$$

$$m = \frac{3150 - 2150}{15{,}000 - 10{,}000} \qquad \text{Simplify.}$$

$$m = \frac{1000}{5000} \qquad \text{Reduce.}$$

$$m = \frac{1}{5}$$

Now use either of the points and the point-slope formula. We will use $(10{,}000, 2150)$.

$$y - y_1 = m(x - x_1) \qquad \text{Substitute.}$$

$$y - 2150 = \frac{1}{5}(x - 10{,}000) \qquad \text{Distribute } \tfrac{1}{5}.$$

$$y - 2150 = \frac{1}{5}x - 2000 \qquad \text{Add 2150 to both sides.}$$

$$y = \frac{1}{5}x + 150. \qquad \text{Therefore, the salary, } y, \text{ in terms of the sales, } x, \text{ is } y = \tfrac{1}{5}x + 150.$$

2. Use the equation found in part 1 to find the salary for sales of $18,000.

Solution:
Since x represents the sales, we replace x with 18,000 and find y, the salary.

$$y = \frac{1}{5}x + 150 \qquad \text{Substitute 18,000 for } x.$$

$$y = \frac{1}{5}(18{,}000) + 150 \qquad \text{Multiply } \tfrac{1}{5} \text{ and 18,000.}$$

$$y = 3600 + 150 \qquad \text{Add.}$$

$$y = 3750 \qquad \text{Therefore, the salesperson earns } \$3750 \text{ for sales of } \$18{,}000.$$

3. Use the equation found in part 1 to find the total value of the sales in a month when the salary was $1150.

Solution:
Since y represents the salary, we will replace y with $1150 and find x, the sales.

$$y = \frac{1}{5}x + 150 \qquad \text{Substitute 1150 for } y.$$

$$1150 = \frac{1}{5}x + 150 \qquad \text{Subtract 150 from both sides.}$$

$$1000 = \frac{1}{5}x \qquad \text{Multiply both sides by 5.}$$

$$5000 = x \qquad \begin{array}{l}\text{Therefore, the salesperson's sales were}\\ \text{\$5000 in order to earn \$1150 in a month.}\end{array}$$

b. A college buys a mower for \$10,000. After six years, it has depreciated to a value of \$4000. Assuming the mower depreciates the same each year (called linear depreciation), answer the following:

   1. Write a linear equation giving the value of the mower, y, in terms of the number of years after it was purchased, x. Leave the answer in $y = mx + b$ form.

   Solution:
   Since x represents the number of years after it was purchased and y represents the value of the mower, the fact that the mower cost \$10,000 can be represented as the ordered pair $(0, 10000)$ and the fact that after six years the mower was worth \$4000 can be written as $(6, 4000)$. We need the equation of a line containing these two points and leave the answer in $y = mx + b$ form. First find the slope.

$$m = \frac{y_2 - y_1}{x_2 - x_1} \qquad \text{Substitute.}$$

$$m = \frac{4000 - 10{,}000}{6 - 0} \qquad \text{Simplify.}$$

$$m = \frac{-6000}{6} \qquad \text{Divide.}$$

$$m = -1000$$

   Now use the slope-intercept formula and $(0, 10{,}000)$.

$$y = mx + b \qquad \text{Substitute. } -1000 \text{ for } m \text{ and } 10{,}000 \text{ for } b.$$

$$y = -1000x + 10{,}000 \qquad \begin{array}{l}\text{Therefore, the value of the mower, } y, \text{ in}\\ \text{terms of the number of years after it was}\\ \text{purchased, } x, \text{ is } y = -1000x + 10{,}000.\end{array}$$

   2. Use the equation found in part 1 to find the value of the mower eight years after it was purchased.

   Solution:
   Since x represents the number of years after it was purchased, replace x with 8 and find y.

$$y = -1000x + 10{,}000 \qquad \text{Substitute 8 for } x.$$

$$y = -1000(8) + 10{,}000 \qquad \text{Multiply } -1000 \text{ and 8.}$$

$$y = -8000 + 10{,}000 \qquad \text{Add.}$$

$$y = 2000 \qquad \begin{array}{l}\text{Therefore, eight years after it was pur-}\\ \text{chased, the mower will be worth \$2000.}\end{array}$$

   3. How many years after it was purchased will the mower have no value?

   Solution:
   Since y represents the value of the mower, we are being asked to find x when $y = 0$.

$$y = -1000x + 10{,}000 \qquad \text{Substitute 0 for } y.$$

$$0 = -1000x + 10{,}000 \qquad \text{Add } 1000x \text{ to both sides of the equation.}$$

$$1000x = 10{,}000 \qquad \text{Divide both sides of the equation by 1000.}$$

$$x = 10 \qquad \begin{array}{l}\text{Therefore, the mower will have no value}\\ \text{after ten years.}\end{array}$$

Practice Exercises

Answer the following:

11. John is an appliance salesperson and earns a salary of $100 per week plus $7.50 for each appliance sold.

   a. Write a linear equation that gives his weekly earnings, y, in terms of the number of appliances sold, x.

   b. Use the equation found in part a to find John's earnings during a week in which he sold 15 appliances.

   c. How many appliances would he have to sell in order to make $400 in one week?

12. The Jones family purchased a lot for $20,000. After five years, the value of the lot had increased to $30,000. Assuming the lot increases by the same amount each year, answer the following:

   a. Write a linear equation that gives the value of the lot, y, in terms of the number of years after it was purchased, x.

   b. Use the equation found in part a to find the value of the lot nine years after it was purchased.

   c. How many years after it was purchased will the lot have a value of $50,000?

Exercise Set 4.6

Write the equation of each of the following lines that contain the given point and have the given slope in $ax + by = c$ *form with* a *and* b *integers:*

1. $(3, 1)$ and $m = 2$

2. $(5, 2)$ and $m = 3$

3. $(-2, 4)$ and $m = -2$

4. $(4, -2)$ and $m = -3$

5. $(-3, 5)$ and $m = \dfrac{1}{4}$

6. $(4, -5)$ and $m = \dfrac{1}{2}$

7. $(2, -4)$ and $m = -\dfrac{1}{3}$

8. $(-2, -1)$ and $m = -\dfrac{1}{5}$

9. $(5, 1)$ and $m = \dfrac{4}{3}$

10. $(6, -2)$ and $m = \dfrac{5}{2}$

11. $(-2, -5)$ and $m = -\dfrac{2}{3}$

12. $(4, -1)$ and $m = -\dfrac{5}{3}$

13. $(3, 4)$ and vertical

14. $(-2, 5)$ and vertical

15. $(2, 3)$ and horizontal

16. $(-1, -7)$ and horizontal

17. $(5, 2)$ and undefined slope

18. $(-6, 2)$ and undefined slope

19. $(5, 4)$ and $m = 0$

20. $(-6, -1)$ and $m = 0$

Write the equation of the line that contains the following pairs of points in $ax + by = c$ *form with* a *and* b *integers:*

21. $(4, -3)$ and $(-1, 7)$

22. $(2, -5)$ and $(4, -2)$

23. $(1, 5)$ and $(-3, 1)$

24. $(4, -3)$ and $(-2, -1)$

25. $(-1, -5)$ and $(-4, 1)$

26. $(5, -2)$ and $(-3, 4)$

27. $(6, 0)$ and $(-3, 6)$

28. $(0, 4)$ and $(3, 5)$

29. $(5, -3)$ and $(5, 1)$

30. $(-2, 4)$ and $(-2, -1)$

31. $(5, -4)$ and $(-1, -4)$

32. $(7, 2)$ and $(-3, 2)$

Write the equations of the following lines:

33. Containing $(3, -1)$ and parallel to the graph of $2x + 4y = 10$

34. Containing $(-4, 6)$ and parallel to the graph of $3x + 5y = 4$

35. Containing $(-3, -4)$ and parallel to the graph of $4x + 8y = 9$

36. Containing $(-4, 2)$ and parallel to the graph of $3x + 9y = 10$

37. Containing $(4, -1)$ and perpendicular to the graph of $x + 4y = 7$

38. Containing $(-5, 0)$ and perpendicular to the graph of $3x - y = 6$

Answers:

c. 15 years

Practice Exercises 11–12: 11. a. $y = 7.5x + 100$ **b.** $212.50 **c.** 40 appliances **12. a.** $y = 2000x + 20,000$ **b.** $38,000

39. Containing $(-5, 3)$ and perpendicular to the graph of $2x + 6y = 7$

40. Containing $(4, -1)$ and perpendicular to the graph of $8x + 4y = 15$

41. Containing $(6, -3)$ and parallel to the graph of $x = -5$

42. Containing $(7, -3)$ and parallel to the graph of $y = 2$

43. Containing $(4, -4)$ and perpendicular to the graph of $x = 6$

44. Containing $(-2, 3)$ and perpendicular to the graph of $y = 3$

Answer the following:

45. The cost to drive a rental car depends on the number of miles driven. The ordered pair (100,40) indicates that it costs $40 to drive 100 miles and the ordered pair (200, 55) indicates that it costs $55 to drive 200 miles.

  a. Write a linear equation giving the cost, y, in terms of the number of miles, x.

  b. Using the equation found in part a, find how much it costs to drive 500 miles.

  c. How many miles would you have to drive if the cost is $62.50?

46. A salesperson works for a fixed salary plus a commission on sales. The ordered pair (1000, 250) indicates earnings of $250 for $1000 in sales and the ordered pair (3000, 450) indicates earnings of $450 for $3000 in sales.

  a. Write a linear equation giving the salary, y, in terms of the value of merchandise sold, x.

  b. Use the equation in part a to find the salary for sales of $4000.

  c. Find the total sales for a week in which the salesperson earned $750.

47. The cost to ride a taxi depends on the number of miles that you ride. The ordered pair (2, 2.80) means that a ride of 2 miles costs $2.80 and (6, 6) means that a ride of 6 miles costs $6.

  a. Write a linear equation that gives the cost, y, in terms of the number of miles, x.

  b. Use the equation found in part a to find the cost of a taxi ride of 8 miles.

  c. Find the length of a taxi ride that costs $5.20.

48. The cost of an item depends on the number produced. The ordered pair (2, 80) indicates that when two items are produced the cost per item is $80. The ordered pair (3, 70) indicates that when three items are produced the cost per item is $70.

  a. Write a linear equation giving the cost per item, y, in terms of the number of items produced, x.

  b. Use the equation in part a to find the cost per item if five items are produced.

  c. Use the equation in part a to find the number of items produced if the cost per item is $40.

49. A farmer purchases a tractor for $50,000. After five years, the tractor has depreciated to a value of $25,000. Assuming linear depreciation, answer the following:

  a. Write a linear equation that gives the value of the tractor, y, in terms of the number of years after it was purchased, x.

  b. Use the equation found in part a to find the value of the tractor after eight years.

  c. Use the equation found in part a to find the number of years until the tractor has no value.

50. A basketball trading card cost $1.50 new, and after four years has a value of $2.70. Assuming the card increases in value the same amount each year, answer the following:

  a. Write a linear equation that gives the value of the card, y, in terms of the number of years after it was purchased, x.

  b. Use the equation found in part a to find the value of the card ten years after it was purchased.

  c. Use the equation found in part a to find the number of years until the card is worth $9.00.

51. Sally is a salesperson at a model home center and her salary is $200 per month plus a 5% commission on her sales.

  a. Write a linear equation that gives her salary, y, in terms of her sales, x.

  b. Use the equation found in part a to find how much she will earn during a month in which her sales are $250,000.

  c. Use the equation found in part a to find her sales for a month in which she earned $8950.

52. The cost of renting a car is $30 per day plus $.15 per mile.

  a. Write a linear equation giving the cost, y, in terms of the number of miles driven, x, for a one-day rental.

  b. Use the equation found in part a to find the cost if the car is driven 300 miles in one day.

  c. Use the equation found in part a to find how many miles the car was driven in one day if the cost was $52.50.

Section 4.7	Graphing Linear Inequalities with Two Variables

OBJECTIVE   *When you complete this section, you will be able to:*

Graph linear inequalities with two unknowns.

In Section 4.2, we graphed linear equations. In this section we will graph **linear inequalities**. A linear inequality with two variables is in the form of $ax + by > c$, $ax + by \geq c$, $ax + by < c$, $ax + by \leq c$, or can be put into one of these forms. Just as any ordered pair of numbers that makes a linear equation true is a solution to the equation, any ordered pair that makes a linear inequality true is a solution of the inequality.

Example 1

Determine if the given ordered pair is a solution of the given inequality.

a. $y > 3x - 1$; $(2, 7)$

  Solution:

  $y > 3x - 1$ Substitute 2 for x and 7 for y.
  $7 > 3(2) - 1$ Simplify the right side of the equation.
  $7 > 5$ This is true, so $(2, 7)$ is a solution.

b. $3x + 2y \geq 6$; $(-4,5)$

  Solution:

  $3x + 2y \geq 6$; $(-4,5)$ Substitute -4 for x and 5 for y.
  $3(-4) + 2(5) \geq 6$ Multiply before adding.
  $-12 + 10 \geq 6$ Add.
  $-2 \geq 6$ This is a false statement, so $(-4, 5)$ is not a solution.

In Section 3.4, we solved and graphed the solutions of linear inequalities with one unknown. For example, the solution of $2x - 3 > 5$ is $x > 4$. To graph $x > 4$, we first graph $x = 4$ with an open dot to indicate that 4 is not a part of the solution. The graph of $x > 4$ is the set of all points of the line on one side of 4 (either to the right or left of 4). To determine which side, use a test value for x, say 5. Substitute 5 for x and we get $5 \geq 4$, which is a true statement. Since 5 is to the right of 4, the graph of $x > 4$ is all the points to the right of 4. See the following graph:

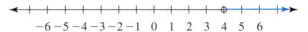

$$-6\ -5\ -4\ -3\ -2\ -1\ \ 0\ \ 1\ \ 2\ \ 3\ \ 4\ \ 5\ \ 6$$

To graph $x \geq 4$, we would use a closed dot, rather than an open dot, to indicate that 4 is a part of the solution.

The graph of the solutions of a linear inequality with one variable of the form $x > a$ or $x < a$ is all the points of the number line that are on one side of $x = a$ or the other. Similarly, the graph of the solutions of a linear inequality with two variables is all the points of the rectangular coordinate system on one side of the graph of $ax + by = c$ or the other. In other words, all the points on the graph of $ax + by = c$ make the equation true. All other points in the coordinate system make either $ax + by > c$ or $ax + by < c$ true. We indicate the solutions by shading all the points on the side of the line that solve the inequality.

To graph the solutions of a linear inequality with two variables, we follow a procedure similar to graphing the solutions of a linear inequality with one variable. The procedure is outlined as follows:

Graphing linear inequalities
with two unknowns

Graphing Linear Inequalities with Two Variables

To graph a linear inequality with two variables:

1. Graph the equality $ax + by = c$. If the line is part of the solution ($\leq$ or $\geq$), draw a solid line. If the line is not part of the solution ($<$ or $>$), draw a dashed line.

2. Pick a test point not on the line. If the coordinates of the test point solve the inequality, then all points on the same side of the line as the test point solve the inequality. If the coordinates of the test point do not solve the inequality, then all the points on the other side of the line from the test point solve the inequality.

3. Shade the region on the side of the line that contains the solutions of the inequality.

Example 2

Graph the following linear inequalities with two variables:

a. $2x + y > 8$

Solution:
Draw the graph of $2x + y = 8$ with a dashed line, since the points on the line do not solve the inequality. The x-intercept is $(4, 0)$ and the y-intercept is $(0, 8)$ and we use the point $(2, 4)$ as a check. Remember, to find the x-intercept, let $y = 0$ and to find the y-intercept, let $x = 0$.

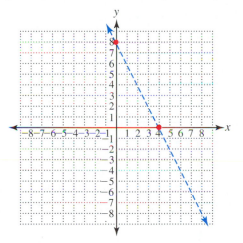

Pick a test point. The origin, $(0, 0)$, is often a convenient test point. Substitute $(0, 0)$ into the inequality. $2(0) + 0 > 8, 0 + 0 > 8, 0 > 8$. This is a false statement, so $(0, 0)$ is not a solution of $2x + y > 8$. Therefore, the solutions are on the opposite side of the line from $(0, 0)$. Shade the region on the opposite side of the line from $(0, 0)$. You may want to test a point on the shaded side just to be sure.

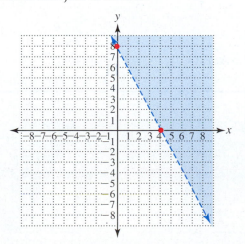

b. $3x - y \leq 6$

Solution:
Draw the graph of $3x - y = 6$ as a solid line, since the points on the line satisfy the inequality less than or *equal to*. Use the x-intercept $(2, 0)$ and the y-intercept $(0, -6)$ and another point as a check.

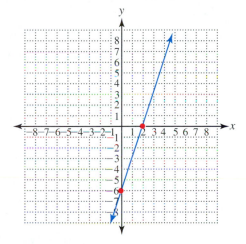

Pick a test point. Again, $(0, 0)$ is a convenient choice. Substitute $(0, 0)$ into the inequality. $3(0) - 0 \leq 6$, $0 - 0 \leq 6, 0 \leq 6$. This is a true statement. Therefore, $(0, 0)$ solves the inequality. Consequently, all points on the same side of the line as $(0, 0)$ solve the inequality. Shade the region on the same side of the line as $(0, 0)$.

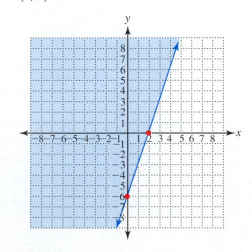

Practice Exercises

1. Following is the graph of $2x + y = 6$ as a dashed line. Shade the side that solves $2x + y < 6$.

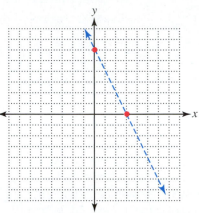

Graph the following linear inequalities with two variables:

2. $3x - 4y \geq 12$

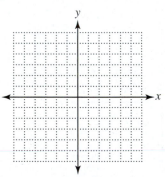

3. $5x - 2y < 10$

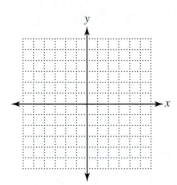

If more practice is needed, do the Additional Practice Exercises.

Answers:

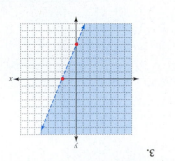

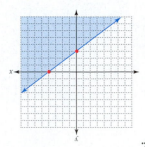

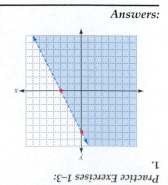

3.   **2.**   **1.**

Practice Exercises 1–3:

Additional Practice Exercises

Graph the following linear inequalities:

a. Following is the graph of $x + 2y = 4$ as a dashed line. Shade the side that solves $x + 2y < 4$.

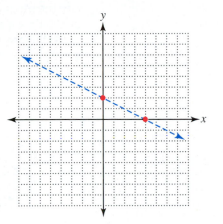

b. $3x - y \geq 4$

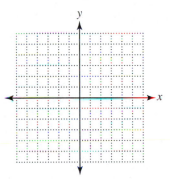

c. $4x + y > 8$

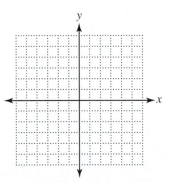

If the line passes through the origin, we cannot use the origin as a test point, since the test point must lie in a region on one side of the line. In this case, choose any point you wish as long as it is clearly not on the line.

Answers:

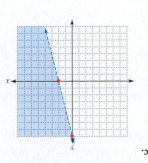
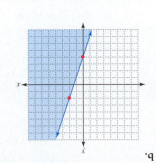
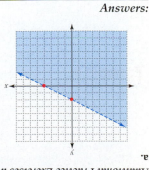

c.   b.   a.

Additional Practice Exercises a–c:

Example 3

Graph the following inequality:

a. $y \leq 3x$ **Solution:**

Draw the graph of $y = 3x$ as a solid line, since the points on the line satisfy the inequality less than or *equal to*. Since both the x- and y-intercepts are $(0, 0)$, we need another point other than the intercepts in order to draw the graph. Choose a value for x and solve for y. Let $x = 3$, then $y = 3(3) = 9$. Therefore, another ordered pair is $(3, 9)$.

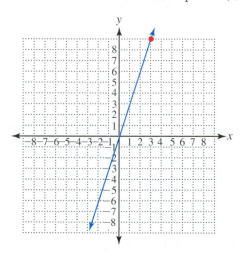

Since $(0, 0)$ lies on the line, we cannot use it as the test point. Choose any other point that clearly is not on the line, say $(2, 0)$. Substitute $(2, 0)$ into the inequality. $0 \leq 3(2), 0 \leq 6$. This is true. Therefore, all the solutions are on the same side of the line as $(2, 0)$. Shade this region.

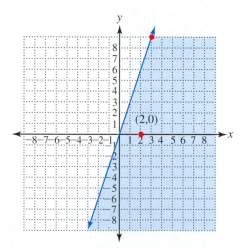

In the case of vertical and horizontal lines, it is not necessary to use a test point. The situation is very much like graphing the solutions to a linear inequality with one variable.

Example 4

Graph the following linear inequalities **with two variables:**

a. $x > -2$ **Solution:**

Draw the graph of $x = -2$ as a dashed line, since the coordinates of the points on the line do not satisfy the inequality. The graph of $x = -2$ is a vertical line intersecting the x-axis at -2.

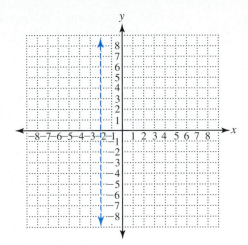

The solutions to $x > -2$ consist of all points whose x-values are greater than -2. Just as on the number line, these points lie to the right of the graph of $x = -2$. Therefore, shade the region to the right of $x = -2$.

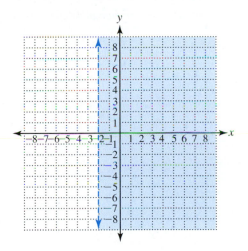

b. $y \leq 3$

Solution:
Draw the graph of $y = 3$ as a solid line, since the coordinates of the points on the line satisfy the inequality. The graph of $y = 3$ is a horizontal line that intersects the y-axis at 3.

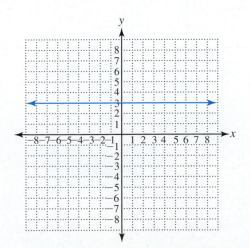

The solutions to $y \leq 3$ consist of all points whose y-values are less than 3. These points lie below the line $y = 3$. Therefore, shade the region below the line.

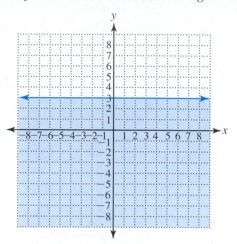

Practice Exercises

Graph the following linear inequalities with two variables. Assume each unit on the coordinate system is 1.

3. $y < -3x$ **4.** $x < -3$ **5.** $y \geq 2$

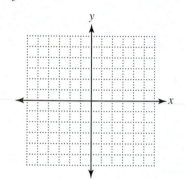   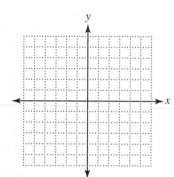   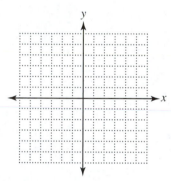

If more practice is needed, do the Additional Practice Exercises.

Answers:

   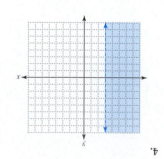   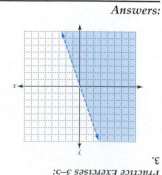

5.   **4.**   **3.**

Practice Exercises 3–5:

Additional Practice Exercises

Graph the following inequalities with two variables:

c. $y \geq x$

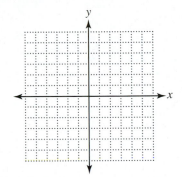

d. $x \geq 4$

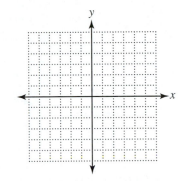

e. $y < -1$

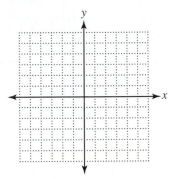

Exercise Set 4.7

Determine whether the given point is a solution of the given inequality.

1. $y > 3x - 2$; $(1, 0)$

2. $y > -2x + 3$; $(2, 0)$

3. $2x + 3y \leq 6$; $(-2, 5)$

4. $3x - 4y \leq 8$; $(-2, -3)$

5. $3x - 2y \geq 12$; $(2, -3)$

6. $4x - 3y \leq 8$; $(5, 4)$

For exercises 7–10, a dashed line has been graphed. Shade the side of the line that is the solution of the given inequality.

7. $x + y > 6$

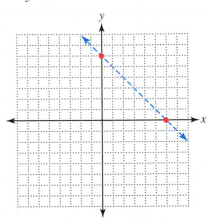

8. $x - y < 4$

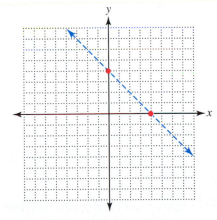

Answers:

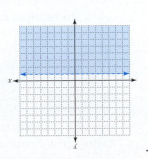

e.

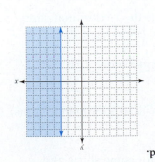

d.

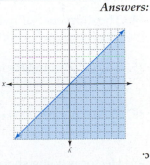

c.

Additional Practice Exercises c–e:

9. $3x + y < -6$

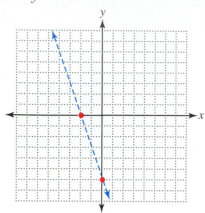

10. $x - 4y > 8$

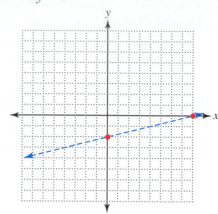

Graph the following linear inequalities with two variables. Assume each unit on the coordinate system represents 1.

11. $y < x + 2$

12. $y > x - 6$

13. $y \geq 2x + 3$

14. $y \geq 3x - 4$

15. $y < -2x + 5$

16. $y \geq -3x + 2$

17. $y > \dfrac{1}{3}x + 2$

18. $y \leq \dfrac{2}{5}x - 3$

19. $2x + y > 6$

20. $3x - y \leq 9$

21. $x + 2y \leq 4$

22. $x + 4y < 8$

23. $2x - 3y > 6$

24. $5x - 2y \leq 15$

25. $3x + 4y \leq 12$

26. $y \leq 2x$

27. $y > x$

28. $y < -4x$

29. $y \geq -x$

30. $y > -2x$

31. $y \leq -4x$

32. $x \geq 2$

33. $x < 4$

34. $y > 3$

35. $y \leq 4$

36. $x + 1 \geq 0$

37. $x + 5 < 0$

38. $y + 2 \leq 0$

39. $y + 6 > 0$

40. $y \leq 0$

Write an inequality representing each of the following and graph:

41. The sum of twice the *x*-coordinate and four times the *y*-coordinate is less than 12.

42. The sum of three times the *x*-coordinate and four times the *y*-coordinate is greater than or equal to 12.

43. The difference of the *x*-coordinate and two times the *y*-coordinate is greater than or equal to 8.

44. The difference of two times the *x*-coordinate and five times the *y*-coordinate is less than 10.

45. The *y*-coordinate is less than 5 more than three times the *x*-coordinate.

46. The *y*-coordinate is greater than 3 more than four times the *x*-coordinate.

47. Hector is going shopping for some shirts and shorts. The shirts cost $20 each and the shorts cost $25 each. If he buys *x* shirts and *y* shorts, write an inequality that shows the possible combinations of shirts and shorts if Hector can spend no more than $150.

48. Tickets to a high school basketball game cost $3 for students and $5 for adults. If *x* student tickets and *y* adult tickets are sold, write an inequality that shows the possible combinations of student and adult tickets if the receipts from ticket sales must be at least $500.

49. An electronics company assembles TVs and VCRs. It takes three hours to assemble a TV and two hours to assemble a VCR. If *x* TVs and *y* VCRs are assembled per day, write an inequality that shows the possible combinations of TVs and VCRs if the company must assemble at least 500 units per day to keep up with demand.

Challenge Exercises: (50–53)

50. A nut company uses *x* ounces of peanuts in a regular mix and *y* ounces of peanuts in a special party mix. The company has at most 500 ounces of peanuts in stock.

   a. Write an inequality describing the situation.

   b. Graph the inequality.

   c. Is (100, 300) a solution of this inequality?

51. A chemical company needs *x* liters of sulfuric acid to make chemical A and *y* liters to make chemical B. The company has at most 200 liters of sulfuric acid in stock.

   a. Write an inequality describing the situation.

   b. Graph the situation.

   c. Is (150, 100) a solution of the inequality?

52. A battery company sells two types of batteries. It makes a profit of $10 on battery A and a profit of $15 on battery B. The company must have sales of at least $3000 per week to break even.

   a. Let *x* represent the number of battery As sold per week and *y* represent the number of battery Bs sold each week. Write an inequality describing the situation.

   b. Graph the inequality.

   c. Is (200, 100) a solution of the inequality?

53. A small company is in the business of manufacturing lawn chairs. It takes two hours of labor to make a regular chair and three hours of labor to make a lounge chair. There is a maximum of 84 hours per day of labor available.

   a. Let *x* represent the number of regular chairs produced and *y* represent the number of lounge chairs produced. Write an inequality describing the situation.

   b. Graph the inequality.

   c. Is (20, 20) a solution of the inequality?

Writing Exercises:

54. How is graphing a linear inequality with two variables like graphing a linear inequality with one variable? How does it differ?

55. If your test point is on one side of the boundary line and it fails to solve the inequality, why are the solutions on the other side of the line?

56. In choosing a test point, why can't you choose a point on the line?

Section 4.8   # Relations and Functions

OBJECTIVES   *When you complete this section you will be able to:*

   a. Recognize functions when given as a set of ordered pairs, as a graph, or as an equation.

   b. Determine the domain and range of relations and functions.

   c. Use functional notation.

Introduction   In Section 3.1, a set was defined as any collection of objects. The objects that make up the set are called the **elements** of the set. In algebra, these elements are often numbers. Recall that sets are indicated by using braces, { }. In this section, we will discuss sets whose elements are ordered pairs of numbers. In an ordered pair, the first number is often called the first component and the second number the second component. For example, in the ordered pair (2, −3) the first component is 2 and the second is −3. With this in mind we have the following definitions:

DEFINITION   Relation

A **relation** is any set of ordered pairs. The set of all first components of the ordered pairs is the **domain** of the relation and the set of all second components is called the **range.**

If the ordered pairs are given in the form (x,y), the domain is the set of all x-values and the range is the set of all y-values.

Example 1

Give the domain and range of each of the following relations:

a. $\{(1, 2), (-3, 4), (6, 3), (-4, -5)\}$

Solution:

The domain is the set of all first components, which is $\{1, -3, 6, -4\}$.

The range is the set of all second components, which is $\{2, 4, 3, -5\}$.

b. $\{(-3, 2), (-1, -4), (2, -4), (2, 6)\}$

Solution:

The domain is the set of all first components, which is $\{-3, -1, 2\}$.

The range is the set of all second components, which is $\{2, -4, 6\}$.

Note: In Example 1b, two ordered pairs have the first component 2 and two ordered pairs have the same second component -4, but these are listed only once in the domain and range respectively.

A **function** is a special type of relation.

DEFINITION **Function**

A **function** is a relation in which every first component (element from the domain) is paired with exactly one second component (element from the range).

Another way of thinking of a function is that no two ordered pairs can have the same first components and different second components. More specifically, if the ordered pairs are in the form (x,y), then a function is a correspondence that assigns exactly one value of y to each value of x. As such, y is said to be a function of x. The variable x is the **independent variable** and y is the **dependent variable.** Since a function is a relation, the domain of the function is the set of all first components and the range is the set of all second components.

Example 2

Recognizing functions as sets of ordered pairs.

Determine which of the following relations are functions and give the domain and range of each:

a. $\{(-3, 4), (-1, -3), (0, 2), (3, 5)\}$

Solution:

Since no two ordered pairs have the same first component and different second components, this is a function. The domain is $\{-3, -1, 0, 3\}$ and the range is $\{4, -3, 2, 5\}$.

b. $\{(-4, 2), (-2, 4), (1, -7), (3, 4)\}$

Solution:

Since no two ordered pairs have the same first component and different second components, this is a function. Notice that the ordered pairs $(-2, 4)$ and $(3, 4)$ have the same second component paired with different first components. This does not violate the definition of a function. The domain is $\{-4, -2, 1, 3\}$ and the range is $\{2, 4, -7\}$.

c. {(−5, 3), (3, 6), (−4, −2), (3, 1), (5, 3)}

Solution:
This is not a function since the first component 3 is paired with two different second components, 6 and 1. The domain is {−5, 3, −4, 5} and the range is {3, 6, −2, 1}.

Practice Exercises

Determine which of the following relations are functions and give the domain and range of each:

1. {(−5, −3), (−2, −1), (3, 5), (4, −6)}   **2.** {(3, −2), (−4, 6), (3, 2), (5, 7), (−3,3)}   **3.** {(−5, 2), (6, −4), (−4, 5), (5, 2)}

Recognizing graphs of functions

The relation in Example 2c was not a function. Let's take a look at the graph of this relation.

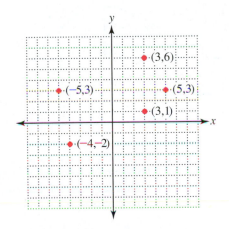

The two ordered pairs that prevented this relation from being a function were (3, 6) and (3, 1) because the first component, 3, was paired with two second components, 6 and 1. We know from Section 4.2 that the graph of any linear equation of the form $x = k$ is a vertical line. Consequently, if two ordered pairs have the same x-value, then their graph must be on the same vertical line. You will note that the graphs of (3, 6) and (3, 1) are on the same vertical line as demonstrated in the following graph:

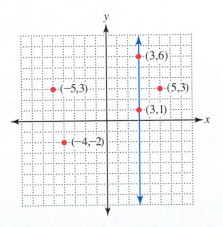

This observation leads to the following graphical test for a function:

Vertical Line Test

If any vertical line intersects the graph of a relation in more than one point, then that relation is not a function.

An equivalent way of stating the preceding test is that no vertical line may intersect the graph of a function in more than one point.

Although we have not graphed anything except points and straight lines in this book, this test can be applied to any graph.

Example 3

Determine whether each of the following is the graph of a function:

a.
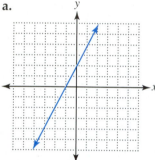

b.
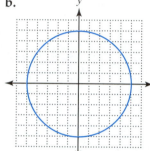

c.
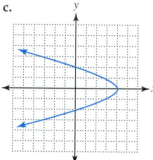

d.
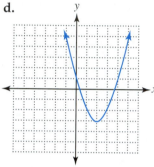

Solution:

The graphs of *a* and *d* are the graphs of functions, but the graphs of *b* and *c* are not, since each may be intersected by a vertical line in more than one point as illustrated by the following:

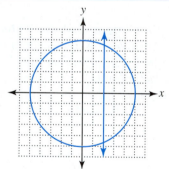

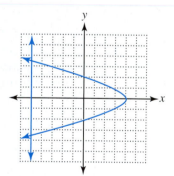

Practice Exercises

Determine whether each of the following is the graph of a function:

4.
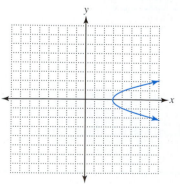

5.
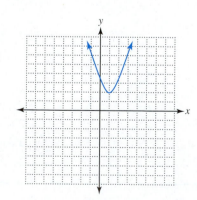

6.

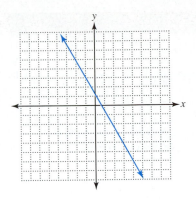

7.

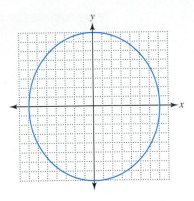

Recognizing functions as equations   Rather than as sets of ordered pairs or graphs, functions are usually given in the form of equations. If the equation defines y as a function of x, then for each value assigned to x the equation must give exactly one value for y.

Example 4

Determine whether the following equations define y *as a function of* x:

a. $y = 3x + 5$

b. $y^2 = x - 3$

Solution:
For each value of x, we find the corresponding value of y by multiplying the value of x by 3 and adding 5. This results in exactly one value for y. Therefore, y is a function of x. For example, if $x = 2$, then $y = 3 \cdot 2 + 5 = 6 + 5 = 11$ and only 11. This corresponds to the ordered pair (2, 11) and this is the only ordered pair whose x-value is 2.

Solution:
Let x have a value, say 4. Then $y^2 = 4 - 3 = 1$. This is asking, "What number squared is equal to 1?" Since $1^2 = 1$ and $(-1)^2 = 1$, there are two numbers whose square is 1. Thus $y = 1$ and $y = -1$. Since there are two values for y for some values of x, this is not a function.

Practice Exercises

Determine whether the following equations define y *as a function of* x.

8. $y = x^2$

9. $y^2 = x + 2$ (*Hint:* Let $x = -1$.)

If more practice is needed, do the Additional Practice Exercises.

Additional Practice Exercises

Determine whether the following equations define y *as a function of* x.

a. $y = 2x - 3$

b. $y^2 = x - 4$ (*Hint:* Let $x = 5$.)

Using functional notation   If y is a function of x, it is convenient to name the function using a special notation called **functional notation** where y is replaced with a symbol of the form $f(x)$ that is read "the value of f at x," or more commonly, "f of x." For example, if $y = 2x + 3$, then y is a function of x. Replacing y with $f(x)$, we have $f(x) = 2x + 3$. Function names are usually lowercase letters, so any lowercase letter may be used as a function name. So $g(x) = 2x + 3$ is the same function.

> ## f(x) Notation
>
> When using notation of the form $f(x)$:
>
> **a.** f is the name of the function. (Frequently other symbols, especially other lower-case letters, are used instead of f.)
>
> **b.** x is a value from the domain.
>
> **c.** $f(x)$ is the value of the function in the range that corresponds with the value of x from the domain. It is a y-value, so the results can be written as the ordered pair $(x, f(x))$.

Note: $f(x)$ does not mean $f \cdot (x)$. It means the value of f at x.

When finding the value of $f(x)$ for a specific value of x, we are **evaluating the function.** This is accomplished by substituting the value for x and evaluating the expression using the order of operations. For example, if $f(x) = 2x + 3$ and we wish to find the value of the function when $x = 3$, then we are finding $f(3)$. To do this, replace x with 3 in the equation of the function and evaluate. So, $f(3) = 2 \cdot 3 + 3 = 6 + 3 = 9$. This can be written as the ordered pair $(3, 9)$.

Actually, any letter or symbol can be used for both the name of the function and the independent variable. For example, $f(x) = 3x + 5$, $g(t) = 3t + 5$, and $h(r) = 3r + 5$ all represent the same function, since they all represent the same set of ordered pairs.

Example 5

Given $f(x) = 3x - 4$, *find each of the following. Represent each result as an ordered pair.*

a. $f(2)$

Solution:
To find $f(2)$, replace x with 2 and simplify.

$$f(2) = 3 \cdot 2 - 4 \quad \text{Multiply.}$$
$$= 6 - 4 \quad \text{Add.}$$
$$= 2 \quad \text{Therefore, } f(2) = 2.$$

As an ordered pair, this is $(2, 2)$.

b. $f(-3)$

Solution:
To find $f(-3)$, replace x with -3 and simplify.

$$f(-3) = 3(-3) - 4 \quad \text{Multiply.}$$
$$= -9 - 4 \quad \text{Add.}$$
$$= -13 \quad \text{Therefore, } f(-3) = -13.$$

As an ordered pair, this is $(-3, -13)$.

c. $f(.5)$

Solution:
To find $f(.5)$, replace x with .5 and simplify.

$$f(.5) = 3(.5) - 4 \quad \text{Multiply.}$$
$$= 1.5 - 4 \quad \text{Add.}$$
$$= -2.5 \quad \text{Therefore, } f(.5) = -2.5.$$

As an ordered pair, this is $(.5, -2.5)$.

d. $f(a)$

Solution:
To find $f(a)$, replace x with a.

$$f(a) = 2a - 4 \quad \text{Since this cannot be simplified, } f(a) = 2a - 4.$$

As an ordered pair, this is $(a, 2a - 4)$.

Practice Exercises

Given $g(x) = 4x - 2$, *find the following. Represent each result as an ordered pair.*

10. $g(2)$

11. $g(-4)$

12. $g\left(\dfrac{1}{2}\right)$

13. $g(c)$

Answers:

Practice Exercises 10–13: **10.** 6, (2, 6) **11.** −18, (−4, −18) **12.** 0, $\left(\dfrac{1}{2}, 0\right)$ **13.** $4c − 2$, $(c, 4c − 2)$

If more practice is needed, do the Additional Practice Exercises.

Additional Practice Exercises

Given h(t) = 2t + 5, *find the following:*

c. $h(4)$

d. $h(-1)$

e. $h(1.3)$

f. $h(b)$

Exercise Set 4.8

Determine which of the following relations are functions and give the domain and range of each:

1. $\{(-6, 2), (-3, -2), (5, -7), (7, 9)\}$

2. $\{(-8, -5), (-3, 6), (0, 7), (2, 3)\}$

3. $\{(-4, 2), (-1, 3), (3, 2), (5, 1)\}$

4. $\{(7, -3), (4, 7), (-5, 2), (3, -3)\}$

5. $\{(-8, 2), (-6, 3), (-8, -3), (5, 1)\}$

6. $\{(-5, 1), (3, -2), (6, 2), (-5, 4)\}$

Determine whether each of the following is the graph of a function:

7.

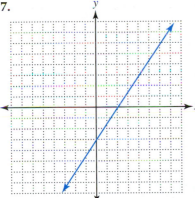

8.

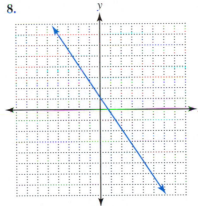

9.

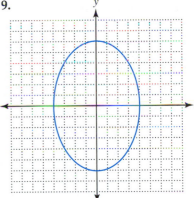

10.

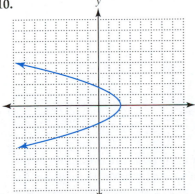

11.

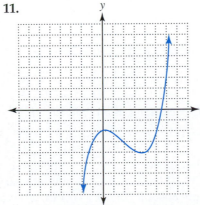

12.

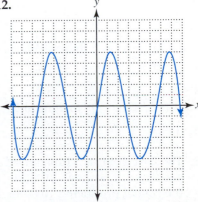

13.

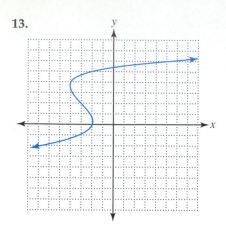

14.

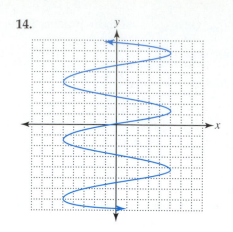

Determine whether each of the following define y *as a function of* x:

15. $y = -2x + 3$

16. $y = 6x - 5$

17. $y = 2x^2$

18. $y = -3x^2 + 2$

19. $y^2 = x - 3$ (*Hint:* Let $x = 4$.)

20. $y^2 = x - 2$ (*Hint:* Let $x = 3$.)

21. $y^2 = 2x + 5$ (*Hint:* Let $x = 2$.)

22. $y^2 = 4x + 1$ (*Hint:* Let $x = 2$.)

23. $y^2 = 16 - x^2$ (*Hint:* Let $x = 0$.)

24. $y^2 = 25 - x^2$ (*Hint:* Let $x = 0$.)

25. $y = x^3$

26. $y = x^3 - 1$

27. $y = |2x + 3|$

28. $y = |x + 7|$

29. $y < x - 6$

30. $y \geq 2x + 3$

31. $y = (x - 4)^2$

32. $y = (x + 5)^2$

Given f(x) = 3x + 5, *find the following. Write each result as an ordered pair.*

33. $f(0)$

34. $f(4)$

35. $f(a)$

Given g(z) = z² − 2, *find the following. Write each result as an ordered pair.*

36. $g(3)$

37. $g(-1)$

38. $g(a)$

Given h(t) = 16t², *find the following. Write each result as an ordered pair.*

39. $h(1)$

40. $h(-2)$

41. $h(a)$

Given r(a) = a³ + 2, *find the following. Write each result as an ordered pair.*

42. $r(0)$

43. $r(-2)$

44. $r(b)$

Given f(x) = |2x − 3|, *find the following. Write each result as an ordered pair.*

45. $f(-2)$

46. $f(4)$

47. $f(z)$

48. The cost, $C(x)$, as a function of the number of items produced, x, is given by $C(x) = 20x + 150$. a) Find the cost of producing 10 items. b) Find the cost of producing 15 items. c) What does $C(30) = 750$ mean?

49. For a rental car, the cost, $C(x)$, as a function of the number of miles driven, x, is given by $C(x) = .12x + 22$. a) Find the cost of driving 100 miles. b) Find the cost of driving 350 miles. c) What does $C(250) = 52$ mean?

50. The value, $V(x)$, of a car as a function of the number of years after it was purchased, x, is given by $V(x) = -1500x + 18,000$. a) Find the value of the car after three years. b) Find the value of the car after six years. c) What does $V(10) = 3000$ mean?

51. The monthly salary, $S(x)$, of a salesperson as a function of her sales, x, is given by $S(x) = .10x + 175$. a) Find her salary for sales of $15,000. B) Find her salary for sales of $10,000. c) What does $s(20,000) = 2175$ mean?

Challenge Exercises: (52–55)

Determine whether the following define y *as a function of* x:

52. $y = |x + 3|$

53. $|y| = 2x - 1$

54. $x = |y| + 3$

55. $|x| = y + 7$

Writing Exercises:

56. How do functions and relations differ?

58. If $y^2 = 2x + 3$, then y is not a function of x. Why not?

57. Why does the vertical line test for a function work?

Chapter 4   Summary

Linear Equations with Two Variables: [Section 4.1]
- A linear equation with two variables is any equation of the form $ax + by = c$, or any equation that can be put into that form, with a and b both not equal to zero at the same time.

Ordered Pairs: [Section 4.1]
- An ordered pair is a pair of numbers enclosed in parentheses, separated by a comma, and with a variable assigned to each number. The first number is the **abscissa** and the second is the **ordinate.**

Solutions of Linear Equations with Two Variables: [Section 4.1]
- If the solutions of $Ax + By = C$ are ordered pairs in the form (x,y), the ordered pair (u,v) is a solution if the replacement of x with u and y with v into the given equation results in a true statement.

Finding a Missing Member of an Ordered Pair: [Section 4.1]
- If an equation and one member of an ordered pair are given, find the other member of the ordered pair by substituting the given member for the variable it represents and solving the equation for the other variable.

Rectangular (Cartesian) Coordinate System: [Section 4.1]
- The rectangular coordinate system is made up of a horizontal and a vertical number line that intersect at the zero point of each. The point of intersection is called the **origin.** The horizontal number line is the **x-axis** and the vertical number line is the **y-axis.** The four regions that the x- and y-axis divide the coordinate plane into are called **quadrants.** The quadrants are numbered counter-clockwise beginning with the top right.

Plotting Points on the Rectangular Coordinate System: [Section 4.1]
- When plotting points, ordered pairs are assumed to be in the form (x, y). To plot a point, begin at the origin and go to the left or right the number of units given by the first number of the ordered pair. From that point, go up or down the number of units given by the second number of the ordered pair.

Graphing Linear Equations with Two Variables: [Section 4.2]
- To graph a linear equation with two variables, find at least two ordered pairs (three is recommended) that are solutions of the equation. Plot the points on the rectangular coordinate system and draw a straight line through them.

Graphing Lines that Contain the Origin: [Section 4.2]
- The graph of any equation of the form $y = ax$ contains the origin. Since both intercepts are zero, find at least one other point by assigning values to x and solving for y.

X- and Y-Intercepts: [Section 4.3]
- The point(s) where a graph crosses the x-axis is (are) called the x-intercept(s). To find the x-intercept(s), let $y = 0$ and solve for x.
- The point(s) where a graph crosses the y-axis is (are) called the y-intercept(s). To find the y-intercept(s), let $x = 0$ and solve for y.

Vertical and Horizontal Lines: [Section 4.3 and 4.5]
- The graph of any equation of the form $x = h$, where h is a constant, is a vertical line.
- The x-values of all points on the vertical line are h.
- The graph of any equation of the form $y = k$, where k is a constant, is a horizontal line.
- The y-values of all points on the horizontal line are k.

Definition of Slope: [Section 4.4]	• The **slope** of a line is the ratio of the change in y to the change in x between any two points on the line. Alternately, the slope is the ratio of the rise over the run between any two points on the line.
Slope Formula: [Section 4.4]	• The slope of the line containing the two points (x_1, y_1) and (x_2, y_2) is $m = \frac{y_2 - y_1}{x_2 - x_1}$.
Graphing a Line Given a Point and the Slope: [Section 4.4]	1. Plot the given point.
	2. From the given point, we can find another point by going vertically (up or down) the number of units given by the numerator and horizontally (right or left) the number of units given by the denominator.
	3. Draw a line through these two points.
Observations about Slope: [Section 4.4]	1. A line with positive slope rises from left to right.
	2. A line with negative slope falls from left to right.
	3. The greater the absolute value of the slope, the steeper the line.
Slopes of Vertical and Horizontal Lines: [Section 4.4]	• The slope of a vertical line is **undefined.** The slope of a horizontal line is **0.**
Determining Slope and *Y*-intercept from the Equation: [Section 4.5]	• To find the slope of the graph of a line from its equation, solve the equation for y. The resulting equation is in the form $y = mx + b$ where m is the slope and b is the y-intercept.
Parallel and Perpendicular Lines: [Section 4.5]	• Two lines are **parallel** if and only if their slopes are equal. Two lines are **perpendicular** if and only if the product of their slopes is **−1.**
Writing Equations of Lines: [Section 4.6 and 4.5]	1. To write the equation of a line given a point (x_1, y_1) on the line and the slope m of the line, use the point-slope formula $y - y_1 = m(x - x_1)$. Substitute for m, x_1, and y_1. Never substitute for x and y.
	2. To write the equation of a line that contains two given points: a) Find the slope of the line. b) Substitute into the point-slope formula using the slope that you found and either of the two given points as (x_1, y_1).
	3. To write the equation of a vertical line, set x equal to the x-value of any point on the line.
	4. To write the equation of a horizontal line, set y equal to the y-value of any point on the line.
	5. To write the equation of a line containing a given point and parallel to a given line, determine the slope of the line. The slope of the line you are looking for is equal to the slope of the given line. Substitute the slope and the coordinates of the given point into the point-slope formula.
	6. To write the equation of a line containing a given point and perpendicular to a given line, determine the slope of the line. The slope of the line you are looking for is equal to the negative reciprocal of the slope of the given line. Substitute the slope and the coordinates of the given point into the point-slope formula.
	7. To write the equation of a line given the slope and y-intercept, substitute for the slope and the y-intercept into the slope-intercept formula, $y = mx + b$. [Section 4.5]
Graphing Linear Inequalities with Two Variables: [Section 4.7]	• To graph a linear inequality with two variables:
	1. Graph the equality $ax + by = c$. If the line is part of the solution ($\leq$ or $\geq$), draw a solid line. If the line is not part of the solution ($<$ or $>$), draw a dashed line.
	2. Pick a test point not on the line. If the coordinates of the test point solve the inequality, then all points on the same side of the line as the test point solve the inequality. If the coordinates of the test point do not solve the inequality, then all the points on the other side of the line from the test point solve the inequality.

3. Shade the region on the side of the line that contains the solutions of the inequality.

Relations: [Section 4.8] • A **relation** is any set of ordered pairs. In algebra, relations are usually given in the form of equations, tables, or graphs.

Functions: [Section 4.8] • A **function** is a relation in which each first component (element from the domain) of the ordered pair is paired with exactly one second component (element from the range). In other words, no two ordered pairs can have the same first components and different second components.

Vertical Line Test: [Section 4.8] • If any vertical line intersects the graph of a relation in more than one point, the graph does not represent a function.

Determining if an Equation Represents a Function: [Section 4.8] • An equation represents a function if each value of the independent variable (often x) results in exactly one value of the dependent variable (often y).

$f(x)$ Notation: [Section 4.8] • When using $f(x)$ notation:

    a. f is the name of the function.

    b. x is a value from the domain.

    c. $f(x)$ is the y-value from the range that is paired with x from the domain.

Chapter 4   Review Exercises

Determine whether the given ordered pair is a solution of the given equation. Assume the ordered pairs are in the form of (x,y). [Section 4.1]

1. $y = 4x + 5$; $(-2,-3)$ **2.** $y = -3x + 5$; $(2,1)$

3. $4x + 5y = 20$; $(10,-4)$ **4.** $7x - 2y = 14$; $(4,7)$

5. $x = 3$; $(3,5)$ **6.** $y = -5$; $(-5,5)$

Use the given equation to find the missing member of each of the following ordered pairs. Assume the ordered pairs are in the form (x,y). [Section 4.1]

7. $y = -3x + 3$; $(0, \_)$, $(\_, 0)$, $(3, \_)$, $(\_, -3)$

8. $y = 4x + 2$; $(0, \_)$, $(\_, 0)$, $(-5, \_)$, $(\_, -6)$

9. $3x + 2y = 12$; $(0, \_)$, $(\_, 0)$, $(-4, \_)$, $(\_, 3)$

10. $5x - 3y = 15$; $(0, \_)$, $(\_, 0)$, $(6, \_)$, $(\_, 5)$

11. A motorcycle is traveling at a constant speed of 60 miles per hour. The formula for the distance traveled after t hours is $d = 60t$. Represent the distance traveled after 2, 4, and 7 hours as ordered pairs in the form (t, d). What does the ordered pair (3, 180) represent?

12. An accountant is paid $25 per hour. If p represents her pay and n represents the number of hours worked, represent the accountant's pay for 2, 5, and 7 hours work as ordered pairs in the form of (n, p). What does the ordered pair (4, 100) represent?

Plot the points represented by the following ordered pairs on the rectangular coordinate system. Assume each unit represents 1. [Section 4.2]

13. $(0, 2)$, $(-3, 4)$, $(0, 1)$

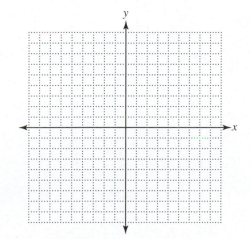

14. $(5, -2), (4, -2), (-6, 7)$

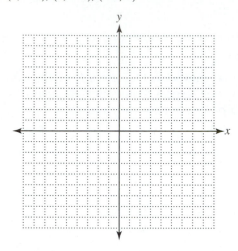

Draw the graph of each of the following equations. Assume each unit represents 1. [Section 4.2]

15. $y = 3x - 6$

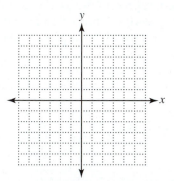

16. $y = -2x - 1$

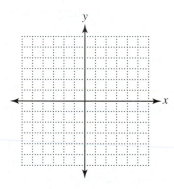

17. $3x + y = 9$

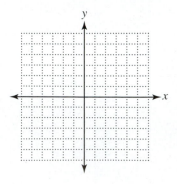

18. $4x - y = 8$

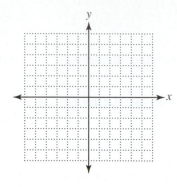

19. $2x + 3y = -6$

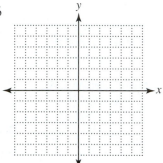

20. $2x - 6y = 12$

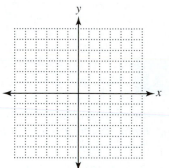

21. $y = -5x$

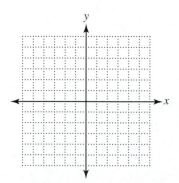

22. $y = 3x$

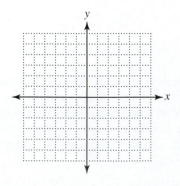

23. $x = 7$

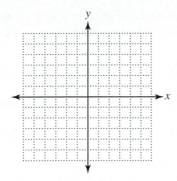

24. $y = -6$

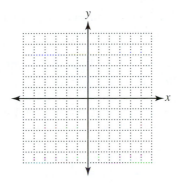

25. Write an equation for the following: The sum of twice the *x*-coordinate and four times the *y*-coordinate is 16.

26. Francine works two part-time jobs. She receives $5 per hour when working at the supermarket and $7 per hour when working at the drug store. If *x* represents the number of hours that she works at the supermarket and *y* the number of hours that she works at the drug store, write an equation showing the possible combinations of hours if she earned $145 last week.

Find the slope of the line that contains each of the following pairs of points: [Section 4.4]

27. $(4, 2)$ and $(5, 4)$ **28.** $(6, 3)$ and $(3, 1)$

29. $(4, -1)$ and $(-2, 2)$ **30.** $(-2, -7)$ and $(0, -2)$

31. $(4, 2)$ and $(4, -2)$ **32.** $(2, -5)$ and $(-4, -5)$

Draw the graph of each of the following lines containing the given point and with the given slope. Assume each unit represents 1. [Section 4.4]

33. $(2, -3)$, $m = \dfrac{4}{3}$

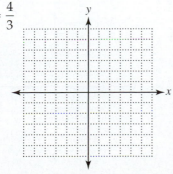

34. $(-3, 4)$, $m = -\dfrac{2}{3}$

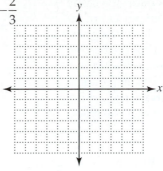

35. $(3, -1)$, $m = 3$

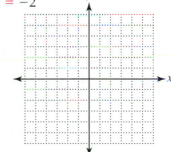

36. $(-1, -4)$, $m = -2$

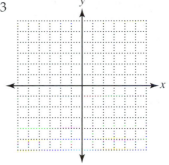

37. $(5, -2)$, undefined slope

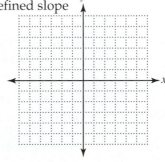

38. $(-2, 4)$, $m = 0$

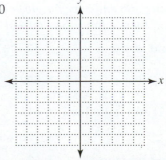

39. The operator of a crane lowers an object onto a loading dock at a point 30 feet from the base of the crane. If the tip of the crane is 20 feet directly above the loading dock, what is the slope of the crane?

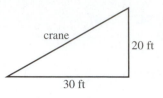

40. A road drops 8 feet vertically for every 100 feet horizontally. What is the slope of the road?

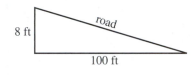

Find the slope and y-intercept of the line represented by each of the following equations: [Section 4.5]

41. $x + 2y = 8$

42. $5x - 2y = 15$

43. $x = 5$

44. $y = -3$

Determine whether the following lines are parallel, perpendicular, or neither. [Section 4.5]

45. L_1: $(4, -1)$ and $(10, 1)$, L_2: $(-4, 2)$ and $(2, 4)$

46. L_1: $(2, -2)$ and $(-3, -4)$, L_2: $(-1, -1)$ and $(-3, 4)$

47. L_1: $(-7, 3)$ and $(-1, -1)$, L_2: $(-2, 8)$ and $(2, 2)$

48. L_1: $3x + 4y = 8$, L_2: $4x - 3y = 9$

49. L_1: $4x + 6y = 5$, L_2: $2x + 3y = 6$

50. L_1: $5x + 4y = 12$, L_2: $4x + 5y = 15$

51. L_1: $x = 3$, L_2: $x = -\frac{1}{3}$

52. L_1: $y = 4$, L_2: $x = 5$

Write the equation of each of the following lines. Leave the answer in slope-intercept form. [Section 4.5]

53. $m = -3$ and y-intercept is $(0, \frac{5}{6})$

54. $m = \frac{6}{5}$ and y-intercept is $(0, 4)$

55. Horizontal line with y-intercept $(0, -3)$

Write the equation of each of the following lines that contain the given point and have the given slope. Leave answers in ax + by = c form. [Section 4.6]

56. $(1, 3)$ and $m = 3$

57. $(-2, 5)$ and $m = -2$

58. $(4, -1)$ and $m = -\frac{5}{2}$

59. $(-5, 4)$ and $m = \frac{4}{3}$

60. $(-7, -1)$ and undefined slope

61. $(3, -5)$ and $m = 0$

62. $(6, 2)$ and horizontal

63. $(-8, 5)$ and vertical

Write the equation of the line which contains the following pairs of points. Leave answers in ax + by = c form. [Section 4.6]

64. $(-3, 4)$ and $(7, -1)$

65. $(5, 1)$ and $(1, -3)$

66. $(0, 6)$ and $(6, -3)$

67. $(4, 0)$ and $(5, 3)$

68. $(-3, 5)$ and $(2, 5)$

69. $(2, 6)$ and $(2, -3)$

Write the equation of each of the following lines. Leave answers in ax + by = c form. [Section 4.6]

70. Contains $(6, -4)$ and perpendicular to the graph of $5x - 3y = 15$

71. Contains $(-4, 3)$ and parallel to the graph of $9x + 3y = 11$

72. Contains $(-5, 4)$ and parallel to the graph of $4x + 6y = 18$

73. Contains $(-1, -1)$ and perpendicular to the graph of $4x - y = 5$

74. Contains $(2, -5)$ and parallel to the graph of $x = 6$

75. Contains $(4, -1)$ and parallel to the graph of $y = 3$

76. Contains $(5, 3)$ and perpendicular to the graph of $x = -1$

Answer the following: [Section 4.6]

77. The fixed cost of operating a sandwich shop is $250 per day. In addition, each sandwich they sell costs the shop an average of $1.50.

    a. Write a linear equation that gives the daily operating costs, y, in terms of the number of sandwiches sold, x.

    b. Use the equation found in part a to find the cost for a day when 300 sandwiches were sold.

    c. Using the equation found in part a, find the number of sandwiches sold on a day when the costs were $587.50.

Graph the following linear inequalities with two variables. Assume each unit represents 1. [Section 4.7]

78. $y < x + 5$

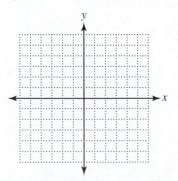

79. $y \geq \dfrac{5}{2}x - 3$

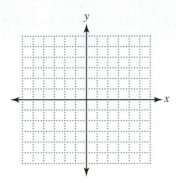

80. $2x - 5y \geq 10$

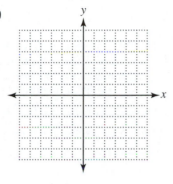

81. $y < -5x$

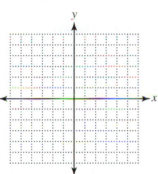

82. $x \geq -6$

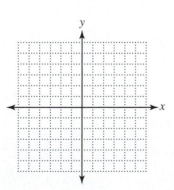

83. $y \leq 8$

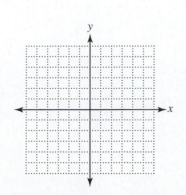

84. Write an inequality for the following. The difference of three times the y-coordinate and two times the x-coordinate is at least 8.

85. Sharon goes shopping for some slacks and blouses. The slacks cost $18 each and the blouses cost $15 each. If x represents the number of pairs of slacks and y the number of blouses, write an inequality that shows the possible combinations of slacks and blouses if Sharon can spend no more than $180.

Determine whether the following relations are functions and give the domain and range of each: [Section 4.8]

86. $\{(-3, 2), (-1, 4), (0, 7), (-1, 5)\}$

87. $\{(-4, 2), (-1, 4), (3, 2), (6, 5)\}$

Determine whether each of the following is the graph of a function: [Section 4.8]

88.

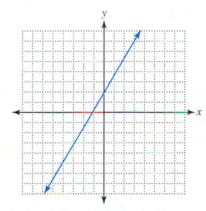

89.

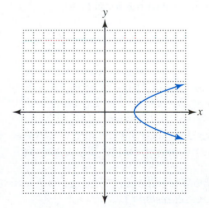

90.

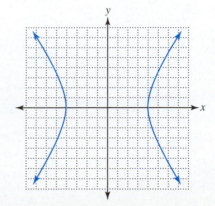

91.

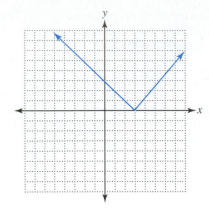

Determine whether each of the following define y *as a function of* x: *[Section 4.8]*

92. $y = -2x + 1$

93. $y = |2x + 6|$

94. $y^2 = x + 6$
    (*Hint:* Let $x = -2$.)

95. $x^2 = y + 5$

96. Given $f(x) = 2x^2 + 3$, find each of the following. Represent each result as an ordered pair.

  a. $f(2)$

  b. $f(-3)$

  c. $f(a)$

97. Given $f(x) = x^3 - 3x^2 + 2$, find each of the following. Represent each result as an ordered pair.

  a. $g(-2)$

  b. $g(0)$

  c. $g(b)$

98. A company purchases a piece of machinery for $30,000. The value, $V(x)$, of the machinery as a function of the number of years after it was purchased, x, is given by $-1800x + 30,000$.

  a. Find the value of the machinery after four years.

  b. Find the value of the machinery after ten years.

  c. What does $V(8) = 15,600$ mean?

Chapter 4   Test

1. If the solutions of $5x - 3y = 9$ are in the form of (x,y), is the ordered pair $(-3, -8)$ a solution?

2. Use the given equation to find the missing member of each of the given ordered pairs. Assume the ordered pairs are in the form (x,y).

   $3x + 4y = -24$; $(0, \_)$, $(\_, 0)$, $(-4, \_)$, $(\_, 3)$

3. A particular type of pipe costs $2.00 per foot. The cost, C, of L feet of pipe is represented as $C = 2L$. Represent the cost of 1 ft, 2 ft, and 4 ft of this pipe as ordered pairs in the form of (L, C) where L is the length of the piece of pipe and C is the cost. What does the ordered pair $(10, 20)$ represent?

Graph each of the following. Assume each unit on the coordinate system represents 1.

4. $y = 2x - 6$

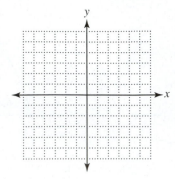

5. $3x + 5y = -15$

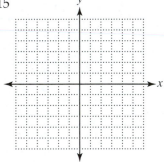

6. $3x + 4y = -6$

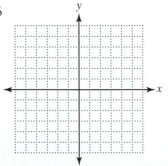

7. $x = -5$

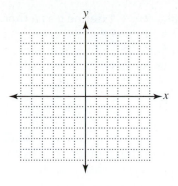

8. A hardware store sells hammers for $9 each and shovels for $20 each. If x represents the number of hammers and y the number of shovels sold, write an equation showing all possible combinations of hammers and shovels if the hardware store received $480 for the sale of hammers and shovels.

Graph each of the following:

9. $y \geq -\dfrac{2}{3}x + 4$

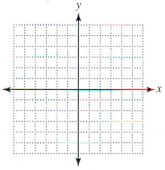

10. $4x - 5y > 20$

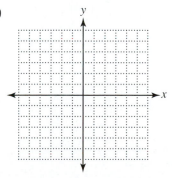

11. $y \leq -6$

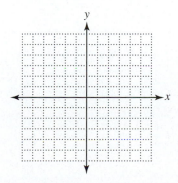

12. Containing $(-1, 4)$ with slope $= \dfrac{3}{5}$.

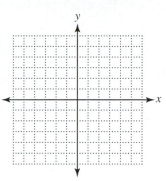

13. Containing $(4, 0)$ with undefined slope.

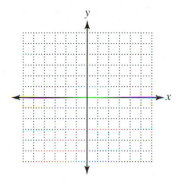

Find the slope of the lines that contain the following pairs of points:

14. $(3, -2)$ and $(-4, 5)$

15. $(4, -2)$ and $(-2, -2)$

Find the slope of the lines with the given equations.

16. $x - 4y = 7$

17. $5x - 2y = 3$

18. $x = -3$

Determine whether the following pairs of lines are parallel, perpendicular, or neither parallel nor perpendicular:

19. L_1 contains $(4, -5)$ and $(-1, -3)$; L_2 contains $(-7, 2)$ and $(-3, 12)$.

20. L_1 is the graph of $4x - 7y = 5$ and L_2 is the graph of $8x - 14y = 11$.

Write the equation of each of the following lines. Leave answers in $ax + by = c$ *form.*

21. Contains $(4, -2)$ with $m = -\dfrac{5}{2}$

22. Contains $(6, -1)$ and $(2, -3)$

23. Contains $(-2, 3)$ and parallel to the graph of $x - 3y = 6$

24. Contains $(5, -6)$ and is horizontal

25. How does the graph of a line whose slope is -2 compare with the graph of a line whose slope is 2?

26. Does the following set of ordered pairs represent a function? Give the domain and range.

$\{(-4, 7), (8, -2), (6, -4), (8, 3)\}$

27. Does the following graph represent the graph of a function?

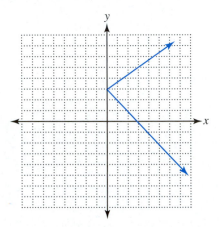

28. Does $y = 2x^2 + 1$ define y as a function of x?

29. Given $f(x) = x^2 - 3x + 2$, find each of the following. Write the results as ordered pairs.

   a. $f(2)$

   b. $f(-1)$

30. The cost, $C(x)$, of operating a small business as a function of the number of days it operates, x, is given by $C(x) = 500x + 2000$.

   a. What is the cost of operating five days?

   b. What does $C(10) = 7000$ mean?

In this chapter, we continue to explore the close relationship between arithmetic and algebra. First, we define and find factors and divisors of natural numbers, a topic from arithmetic. Then we extend these arithmetic concepts into algebra. In arithmetic, there are a couple of procedures for writing a number in terms of its prime factors. In algebra, however, we factor polynomials according to their type.

Factors, Divisors, and Factoring

The prerequisite skills for this chapter are in Chapter 2, where you learned the product and quotient laws of exponents and the techniques for multiplying polynomials. As you will see, factoring a polynomial is the opposite of finding the product of polynomials. For that reason you might want to review Chapter 2 prior to beginning Chapter 5.

A good knowledge of factoring is crucial for many topics to be discussed in later chapters. In Chapter 6, for example, we must be able to factor in order to reduce and find the products or quotients of rational expressions. In Chapter 7, you must be able to factor in order to find the sum or differences of rational expressions.

Chapter

5

Prime Factorization and Greatest Common Factor

OBJECTIVES

When you complete this section, you will be able to:

a. Find all natural number factors of a given number.

b. Recognize prime and composite numbers.

c. Determine if a number is divisible by 2, 3, or 5.

d. Find the prime factorization of composite numbers.

e. Find the greatest common factor for two or more natural numbers.

f. Find the greatest common factor for two or more monomials that contain variables.

Introduction

In this section we will limit ourselves to natural numbers. We are familiar with finding the products of numbers like $5 \cdot 6 = 30$. Recall that the number 30 is called the product of 5 and 6 and that 5 and 6 are called factors of 30. The expression $5 \cdot 6$ is called a **factorization** of 30.

Factorization of a Number

A factorization of a number, *c*, consists of writing *c* as the product of two or more numbers. For example, if $ab = c$, then ab is a factorization of *c*. *a* and *b* are factors of *c*.

A number may have more than one factorization, and consequently, more than two factors. Let us consider all possible factorizations of several numbers by using the following table:

Number	Factorization(s)	Factors	Number of Factors
1	$1 \cdot 1$	1	1
2	$1 \cdot 2$	1, 2	2
3	$1 \cdot 3$	1, 3	2
4	$1 \cdot 4, 2 \cdot 2$	1, 2, 4	3
5	$1 \cdot 5$	1, 5	2
6	$1 \cdot 6, 2 \cdot 3$	1, 2, 3, 6	4
7	$1 \cdot 7$	1, 7	2
8	$1 \cdot 8, 2 \cdot 4$	1, 2, 4, 8	4
9	$1 \cdot 9, 3 \cdot 3$	1, 3, 9	3
10	$1 \cdot 10, 2 \cdot 5$	1, 2, 5, 10	4
11	$1 \cdot 11$	1, 11	2
12	$1 \cdot 12, 2 \cdot 6, 3 \cdot 4, 2 \cdot 2 \cdot 3$	1, 2, 3, 4, 6, 12	6
13	$1 \cdot 13$	1, 13	2

and so on.

Looking through the preceding table we observe some patterns.

1. Some numbers like 2, 3, 5, 7, 11, and 13, have only two factors, the factors being 1 and the number itself.

2. Some numbers like 4, 6, 8, 9, 10, and 12, have more than two factors.

3. There is only one number with only one factor, the number 1.

4. All of the numbers have 1 and the number itself as factors.

Prime and composite numbers These observations lead to the following definitions:

Prime Number, Composite Number, or Neither

A **prime number** is a natural number with two, and only two, factors. These factors are 1 and the number itself.

A **composite number** is a natural number with more than two factors. These factors are 1, the number itself, and at least one other number.

The natural number 1 is **neither** prime nor composite.

Example 1

Classify each of the following as prime or composite:

a. 2   Prime, since 2 has only two factors: 1 and 2.

b. 11   Prime, since 11 has only two factors: 1 and 11.

c. 4   Composite, since 4 has three factors: 1, 2, and 4.

d. 19   Prime, since 19 has only two factors: 1 and 19.

e. 12   Composite, since 12 has six factors: 1, 2, 3, 4, 6 and 12.

Practice Exercises

Classify each of the following as prime or composite:

1. 7   **2.** 8   **3.** 10

4. 13   **5.** 41   **6.** 57

The number 20 is divisible by 5, since $20 \div 5 = 4$ with 0 as the remainder. In general, the number a is **divisible** by the number b if the remainder is 0 when a is divided by b. As mentioned before, there is an important relationship between multiplication and division. Since $20 \div 4 = 5$, then $4 \cdot 5 = 20$. This means 4 is a factor of 20. Furthermore, since $20 \div 5 = 4$, then $5 \cdot 4 = 20$. This means 5 is also a factor of 20. From these examples, we suspect that if a number is divisible by a second number, then the second number is a factor of the first. Furthermore, if a nonzero number is a factor of another number, then that number is divisible by the factor.

Divisor and Divisibility

If the natural number a is a **divisor** of a natural number c, then a is a nonzero factor of c. Furthermore, if the number a is a nonzero factor of the number c, then a is a divisor of c. The number c is said to be **divisible** by the number a.

This states that nonzero factors and divisors are the same thing. So, if 18 is divisible by 3, then 3 is a factor of 18. Conversely, if 3 is a factor of 18, since $3 \neq 0$, then 3 is a divisor of 18. Consequently, a prime number can be thought of as a number that is divisible only by itself and 1.

Answers:

Alternative Definition of a Prime Number

A natural number is prime if it is divisible only by itself and 1. The number 1 is neither prime nor composite.

Example 2

Find the divisors (factors) of each of the following:

Number	Answer
a. 14	Since $1 \cdot 14 = 14$ and $2 \cdot 7 = 14$, the divisors (factors) of 14 are 1, 2, 7, and 14.
b. 20	Since $1 \cdot 20 = 20$, $2 \cdot 10 = 20$, and $4 \cdot 5 = 20$, the divisors (factors) of 20 are 1, 2, 4, 5, 10, and 20.
c. 5	Since $1 \cdot 5 = 5$, the divisors (factors) of 5 are 1 and 5.
d. 48	Since $1 \cdot 48 = 48$, $2 \cdot 24 = 48$, $3 \cdot 16 = 48$, $4 \cdot 12 = 48$, and $6 \cdot 8 = 48$, the divisors (factors) of 48 are 1, 2, 3, 4, 6, 8, 12, 16, 24, and 48.

Practice Exercises

Find the divisors (factors) of each of the following:

7. 9   **8.** 28   **9.** 19   **10.** 96

Divisibility by 2, 3, and 5   When the numbers are somewhat larger, as in Practice Exercise 10, it is sometimes difficult to determine if the number is divisible by something other than one and itself. There are some easy ways of determining if a number is divisible by 2, 3, or 5. They are summarized in the following rules:

Rules of Divisibility for 2, 3, and 5

1. A natural number is divisible by 2 if the last digit of the number is 0, 2, 4, 6, or 8. In other words, if the number is even then it divisible by 2.
2. A natural number is divisible by 3 if the sum of the digits of the number is divisible by 3.
3. A natural number is divisible by 5 if the last digit of the number is 0 or 5.

Example 3

Determine if each of the following is divisible by 2, 3, 5 or none of these:

a. 20   The last digit is 0, so 20 is divisible by both 2 and 5. The sum of the digits is 2 ($2 + 0 = 2$) and 2 is not divisible by 3, so 20 is not divisible by 3.

b. 39   39 is not even, so it is not divisible by 2. The last digit is not 0 or 5, so 39 is not divisible by 5. The sum of the digits is 12 ($3 + 9 = 12$) and 12 is divisible by 3, so 39 is divisible by 3. ($39 \div 3 = 13$)

c. 60   60 is even, so it is divisible by 2. The last digit is 0, so 60 is divisible by 5. The sum of the digits is 6 and 6 is divisible by 3, so 60 is divisible by 3.

d. 73   73 is not even, so it is not divisible by 2. The last digit is not 0 or 5, so 73 is not divisible by 5. The sum of the digits is 10 and 10 is not divisible by 3, so 73 is not divisible by 3. Does this mean 73 is prime?

Practice Exercises

Determine if each of the following is divisible by 2, 3, 5 or none of these:

11. 40   **12.** 87   **13.** 120   **14.** 91

Finding prime factorizations

It is possible to write any composite number as the product of prime numbers. This procedure is referred to as **prime factorization.** There is one and only one prime factorization of any composite number. For example,

Number	Prime Factorization
6	$2 \cdot 3$ (Remember, $2 \cdot 3 = 3 \cdot 2$. Why?)
9	$3 \cdot 3 = 3^2$
12	$2 \cdot 6 = 2 \cdot 2 \cdot 3 = 2^2 \cdot 3$

If the numbers are small, finding the prime factorization is relatively easy. For larger numbers, the following procedure is helpful in finding the prime factorization:

Procedure for Finding a Prime Factorization

a. Divide the composite number by the smallest prime number by which it is divisible.

b. Then divide the resulting quotient by the smallest prime number by which it is divisible.

c. Continue this process until the quotient is prime.

d. The prime factorization will be the product of all the divisors and the last quotient.

To assist us, the first few prime numbers are $\{2, 3, 5, 7, 11, 13, 17, 19, 23, 29, \ldots\}$. We illustrate the procedure with some examples.

Example 4

Find the prime factorization for each of the following:

a. 24   2 is the smallest prime number that divides evenly into 24.

$2\overline{)24}$; 12 12 is also divisible by 2.

$2\overline{)12}$; 6 6 is also divisible by 2.
$2\overline{)24}$

$2\overline{)6}$; 3 3 is prime.
$2\overline{)12}$
$2\overline{)24}$

Therefore, $24 = 2 \cdot 2 \cdot 2 \cdot 3 = 2^3 \cdot 3$, which is the product of all the divisors and the last quotient and is the prime factorization of 24.

b. 36   2 is the smallest prime number that divides evenly into 36.

$$2\overline{)36} = 18$$

18 is also divisible by 2.

$$2\overline{)18} = 9$$
$$2\overline{)36}$$

3 is the smallest prime that divides into 9 evenly.

$$3\overline{)9} = 3$$
$$2\overline{)18}$$
$$2\overline{)36}$$

3 is prime.

Therefore, $36 = 2 \cdot 2 \cdot 3 \cdot 3 = 2^2 \cdot 3^2$, which is the product of all the divisors and the last quotient and is the prime factorization of 36.

c. 105   3 is the smallest prime that divides evenly into 105.

$$3\overline{)105} = 35$$

5 is the smallest prime that divides evenly into 35.

$$5\overline{)35} = 7$$
$$3\overline{)105}$$

7 is prime.

Therefore, $105 = 3 \cdot 5 \cdot 7$ is the prime factorization of 105.

d. 89   *89 is not divisible by 2, 3, 5, 7, 11, 13, and so on. We suspect that 89 might be prime, but how do we know? Do we have to try dividing 89 by every prime number until we get to 89? Fortunately, we do not.* **We can stop looking for prime factors when we arrive at a prime number whose square is greater than the number we are trying to factor.** *In this case, $11^2 > 89$, so there will be no prime factors greater than 11. Therefore, we conclude that 89 is prime.*

Observation

If no prime factor has been found that is smaller than a prime number whose square is greater than the number we are attempting to factor, then the number is prime.

The preceding observation can save us time when attempting to factor numbers that are prime. For example, if we were attempting to factor 163, we would not try any prime factors greater than 13, since 13 is prime and $13^2 = 169$, which is greater than 163.

Practice Exercises

Find the prime factorization of each of the following:

15. 80   **16.** 84   **17.** 90   **18.** 73

If more practice is needed, do the Additional Practice Exercises.

Find the prime factorization for each of the following:

a. 40   **b.** 54   **c.** 150   **d.** 108

Prime factorizations have many uses, one of which is finding the **greatest common factor** of two or more natural numbers or algebraic expressions.

First we need to know what is meant by the greatest common factor. The factors of 12 are {1, 2, 3, 4, 6, 12} and the factors of 18 are {1, 2, 3, 6, 9, 18}. Therefore, the factors that 12 and 18 have in common are {1, 2, 3, 6}. The largest factor that 12 and 18 have in common is 6. Therefore, 6 is the greatest common factor for 12 and 18. *Remember, a number is divisible by all its factors.* So, the greatest common factor for a group of numbers is the largest number that will divide evenly into each number of the group. For 12 and 18, 6 is the greatest common factor, so 6 is the largest number that will divide evenly into both.

Finding the greatest common factor of two or more natural numbers

Listing the factors and picking out the greatest common factor works for small numbers, but it would certainly be difficult to use this procedure for a pair of large numbers like 108 and 144. We will find the greatest common factor for groups of numbers using the procedure outlined as follows:

Finding the Greatest Common Factor (GCF)

1. Write each number as the product of its prime factors.
2. Select those factors that are common in each prime factorization.
3. For each factor that is common, select the smallest exponent that appears in any of the prime factorizations.
4. The greatest common factor is the product of the factors found in step 3.

We demonstrate this procedure with some examples.

Example 5

Find the greatest common factor (GCF) for each of the following using prime factorization:

a. 12 and 18 (We already know the answer is 6 from the introduction.)

$12 = 2^2 \cdot 3$ Prime factorization.

$18 = 2 \cdot 3^2$ Prime factorization.

The factors that are common are 2 and 3. The smallest power of 2 that appears in either factorization is 2^1 in $2 \cdot 3^2$ and the smallest power of 3 that appears in either factorization is 3^1 in $2^2 \cdot 3$. Therefore, the GCF is $2^1 \cdot 3^1 = 6$. Consequently, 6 is the largest number that will divide evenly into 12 and 18.

b. 60 and 72

$60 = 2^2 \cdot 3 \cdot 5$ Prime factorization.

$72 = 2^3 \cdot 3^2$ Prime factorization.

The factors that are common are 2 and 3 (5 is not a factor of 72). The smallest power of 2 that appears in either factorization is 2^2 in $2^2 \cdot 3 \cdot 5$ and the smallest power of 3 is 3^1 in $2^2 \cdot 3 \cdot 5$. Therefore, the GCF is $2^2 \cdot 3 = 12$. This means that 12 is the largest number that will divide into 60 and 72 evenly.

Answers:

Additional Practice Exercises *a–d:* **a.** $2^3 \cdot 5$ **b.** $2 \cdot 3^3$ **c.** $2 \cdot 3 \cdot 5^2$ **d.** $2^2 \cdot 3^3$

c. 6 and 35

$6 = 2 \cdot 3$ Prime factorization.

$35 = 5 \cdot 7$ Prime factorization.

There are no prime factors in common, so the greatest common factor is 1. Remember, 1 is a factor of every number. In other words, the largest number that will divide evenly into 6 and 35 is 1.

d. 270 and 315

$270 = 2 \cdot 3^3 \cdot 5$ Prime factorization.

$315 = 3^2 \cdot 5 \cdot 7$ Prime factorization.

The factors that are common are 3 and 5 (2 is not a factor of 315 and 7 is not a factor of 270). The smallest power of 3 appearing in either factorization is 3^2 in $3^2 \cdot 5 \cdot 7$ and the smallest power of 5 is 5^1 in both prime factorizations. Therefore, the greatest common factor is $3^2 \cdot 5 = 45$. This means 45 is the largest number that will divide into 270 and 315 evenly.

e. 54, 120 and 126

$54 = 2 \cdot 3^3$ Prime factorization.

$126 = 2 \cdot 3^2 \cdot 7$ Prime factorization.

$120 = 2^3 \cdot 3 \cdot 5$ Prime factorization.

The factors that are common are 2 and 3. The smallest power of 2 that appears in any factorization is 2^1, which occurs in $2 \cdot 3^3$ and $2 \cdot 3^2 \cdot 7$. The smallest power of 3 is 3^1 in $2^3 \cdot 3 \cdot 5$. Therefore, the greatest common factor is $2 \cdot 3 = 6$. Consequently, 6 is the largest number that will divide into 54, 126, and 120 evenly.

Why does this procedure work? Since we are looking for the greatest *common* factor, we select only those factors that are *common*. Why the smallest exponent? The exponent indicates how many times the base is used as a *factor*. We are looking for the greatest number of factors that are common. Hence the greatest number of times the *factor* is *common* is the smallest power of that *factor*.

Practice Exercises

Find the greatest common factor for each of the following:

19. 24 and 30   **20.** 54 and 90   **21.** 15 and 77

22. 600 and 180   **23.** 56, 112 and 140

If more practice is needed, do the Additional Practice Exercises.

Additional Practice Exercises

Find the greatest common factor for each of the following:

e. 55 and 21   **f.** 60 and 140   **g.** 12, 28, and 42

Finding the greatest common
factor of two or more monomials
that contain variables

The procedure for finding the GCF of a monomial containing variables is exactly the same, except usually easier, since we do not have to factor the variables.

Answers:

Practice Exercises 19–23: **19.** 6 **20.** 18 **21.** 1 **22.** 60 **23.** 28   Additional Practice Exercises e–g: **e.** 1 **f.** 20 **g.** 2

Example 6

Find the GCF for each of the following:

a. a^2b^3 and a^4b^2 The factors a and b are common. The smallest power of a is a^2 and the smallest power of b is b^2. Therefore, the GCF is a^2b^2. This means a^2b^2 is the monomial of greatest degree that divides evenly into a^2b^3 and a^4b^2.

b. x^4yz^3 and $x^3y^3z^3$ The factors x, y, and z are common. The smallest power of x is x^3, the smallest power of y is y^1, and the smallest power of z is z^3. Therefore, the GCF is x^3yz^3. This means x^3yz^3 is the monomial of greatest degree that divides evenly into x^4yz^3 and $x^3y^3z^3$.

c. a^2b^4 and c^3d^2 There are no factors in common, therefore the GCF is 1.

d. $6c^2d^3$ and $8cd^5$ We need to consider the coefficients and the variables separately. The GCF of 6 and 8 is 2. The variables c and d are common, the smallest power of c is c^1 and the smallest power of d is d^3. Therefore, the GCF is $2cd^3$. So, $2cd^3$ is the monomial of greatest degree that will divide evenly into $6c^2d^3$ and $8cd^5$.

e. $12a^3b^2$ and $18b^4d$ The GCF of 12 and 18 is 6. The only variable in common is b and the smallest power of b is b^2. Therefore, the GCF is $6b^2$.

f. $8x^2yz^4$, $12x^4y^2z^3$, and $14x^2y^3$ The GCF of 8, 12, and 14 is 2. The only variables in common are x and y and the smallest powers are x^2 and y^1, respectively. Therefore, the GCF is $2x^2y$.

Practice Exercises

Find the GCF of each of the following:

24. c^4d^2 and c^3d^5

25. xy^3 and c^2d

26. $18a^3b^4$ and $24a^2b^5$

27. $15x^2y^3$ and $18x^3z^2$

28. $15x^4y^2z$, $9x^2z^3$, and $18xy^3z^2$

If more practice is needed, do the Additional Practice Exercises.

Additional Practice Exercises

Find the GCF of each of the following:

h. p^3q^2 and pq^4

i. $m^3n^2p^3$ and $a^4b^3c^5$

j. $6a^4b^2$ and $9ab^2$

k. $24m^4n^3$ and $36n^2p^3$

l. $10x^3yz^2$, $15y^3z^4$ and $25x^2y^4z^5$

Exercise Set 5.1

Classify each of the following as prime or composite:

1. 9
2. 15
3. 17
4. 23

5. 28
6. 35
7. 51
8. 69

9. 71
10. 83

Answers:

Practice Exercises 24–28: 24. c^3d^2 25. 1 26. $6a^2b^4$ 27. $3x^2$ 28. $3xz$ **Additional Practice Exercises h–l:** h. pq^2 i. 1 j. $3ab^2$ k. $12n^2$ l. $5yz^2$

Determine if each of the following is divisible by 2, 3, 5, or none of these:

11. 30   **12.** 25   **13.** 57   **14.** 51

15. 48   **16.** 96   **17.** 97   **18.** 113

19. 145   **20.** 185   **21.** 732   **22.** 822

23. 1620   **24.** 5340

Find the prime factorization of each of the following:

25. 14   **26.** 15   **27.** 46   **28.** 58

29. 28   **30.** 20   **31.** 29   **32.** 31

33. 45   **34.** 63   **35.** 100   **36.** 225

37. 42   **38.** 30   **39.** 154   **40.** 232

41. 180   **42.** 126   **43.** 315   **44.** 525

45. 756   **46.** 360

Challenge Exercises: (47–50)

List all the natural number factors of the following:

47. 108   **48.** 144

Find the prime factorizations of each of the following:

49. 1512   **50.** 7056

Find the greatest common factor (GCF) of each of the following:

51. 6 and 9   **52.** 8 and 12   **53.** 65 and 35

54. 22 and 55   **55.** 33 and 10   **56.** 65 and 28

57. 12 and 45   **58.** 68 and 76   **59.** 90 and 140

60. 60 and 140   **61.** 180 and 270   **62.** 165 and 175

63. 6, 13, and 21   **64.** 14, 35, and 49   **65.** 8, 16, and 24

66. 12, 30, and 42   **67.** 16, 32, and 56   **68.** 24, 48, and 60

69. 48, 96, and 120   **70.** 72, 108, and 180   **71.** mn^3 and m^2n

72. a^4b and ab^2 **73.** c^3d^2 and cd^3 **74.** pq^4 and p^3q^4

75. a^3b^5 and a^3b^2 **76.** m^4n^6 and m^2n^6 **77.** j^3k^4 and a^2b^3

78. r^3s^6 and s^3t **79.** $a^3b^3c^3$ and a^2bc^2 **80.** $x^4y^3z^2$ and xy^4z^3

81. pq^2r^4 and p^2r^3

82. a^4b^2c and a^2c^3

83. $18pq^4$ and $30p^2q$

84. $16c^5d$ and $24cd^3$

85. $27s^3t^3$ and $36s^2t^5$

86. $12p^2q^3$ and $24p^5q^4$

87. $9a^2b^4c^2, 15a^3bc^2, 12ab^3c^2$

88. $5x^3y^6z, 10xy^2z, 20x^2y^2z$

89. $8a^4b^3c^2, 16a^6b^3c^2, 4a^2b^5c^5$

90. $12r^5s^4t^2, 24r^3s^4t^3, 28r^2s^4t^5$

91. $24p^2q^4, 33q^2r^3, 42p^3r^4$

92. $24a^4c^2, 40a^5b^2, 56b^4c^2$

93. Juan has 3 nickels and 5 dimes and Fred has 2 nickels and 8 dimes. How many of each type of coin do they have in common?

94. Danielle has 4 dimes and 6 quarters and Sandy has 2 dimes and 4 quarters. How many of each type of coin do they have in common?

95. Frank has 6 New York Yankees baseball cards and 8 Atlanta Braves baseball cards. Bill has 4 New York Yankees baseball cards and 10 Atlanta Braves baseball cards. How many cards of the same teams do they have in common?

96. Murlene has 3 yellow blouses and 5 red blouses. Hazel has 6 yellow blouses and 2 red blouses. How many blouses of the same color do they have in common?

97. Bill and Don went fishing. Bill caught 5 redfish and 8 trout, while Don caught 3 redfish and 12 trout. How many fish of the same species do they have in common?

Challenge Exercises: (98–101)

Find the GCF for the following:

98. 108 and 144

99. 240 and 384

100. $48a^3b^2(x + 2)^3, 64a^4b(x + 2)^2$

101. $36x^5y^7(a - 3)^4, 54x^4y^9(a - 3)^7$

Writing Exercises:

102. What is the difference between a prime and a composite number?

103. Why is 1 not considered to be a prime number?

104. How can you determine if a number is divisible by 4? by 6?

105. Under what circumstances would a number be a common divisor of two other numbers?

106. A friend of yours is having trouble finding the GCF of two large numbers. Describe how you would help him.

107. Following is an example of finding the GCF of 108 and 144 using a different procedure. Study the example and write the procedure in words.

$$
\begin{array}{r|l r}
2 & 108 & 144 \\ \hline
2 & 54 & 72 \\ \hline
3 & 27 & 36 \\ \hline
3 & 9 & 12 \\ \hline
& 3 & 4
\end{array}
$$

Therefore, the GCF is $2 \cdot 2 \cdot 3 \cdot 3 = 2^2 \cdot 3^2 = 36$.

Factoring Polynomials with Common Factors and by Grouping

OBJECTIVES

When you complete this section, you will be able to:

a. Factor a polynomial by removing the greatest common factor.

b. Factor by grouping.

Introduction

In Section 5.1, we learned how to write a composite number like 6 and a monomial like $2x^2$ as the product of prime factors. This section is the first of six in which we will learn to write composite polynomials (other than monomials) as a product of **prime polynomials.** A prime polynomial, like a prime number, is a polynomial that can be written only as a product of itself and 1. For example, $x + 4$ is prime, but $x^2 - 4$ is not since $x^2 - 4 = (x + 2)(x - 2)$.

When asked to "factor a polynomial," what are we actually being asked to do? We know that when two or more numbers are multiplied, each is called a factor of the product. So, factoring a polynomial means to write the polynomial as the product of two or more prime polynomials in the same way factoring a composite number means to write the composite number as the product of two or more prime numbers. In Chapter 2, we multiplied polynomials. Factoring is "undoing the multiplication." Consider the following:

Composite Number	Factors	Composite Polynomial	Factors
$15 = 3 \cdot 5$	$3, 5$	$x^2 - 4 = (x + 2)(x - 2)$	$x + 2, x - 2$
$22 = 2 \cdot 11$	$2, 11$	$2x^2 + 4x = 2x(x + 2)$	$2, x, x + 2$

Removing the greatest common factor

The distributive property in the form $ab + ac = a(b + c)$ provides the technique for factoring by **removing the greatest common factor.** In the binomial $ab + ac$, a is a factor of the first term, ab, and a is also a factor of the second term, ac. Therefore, a is a factor common to both terms. The common factor (a in this case) is written outside the parentheses and the polynomial inside the parentheses is determined by one of two methods.

1. We determine the term that must be multiplied by the common factor in order to give the corresponding term of the polynomial being factored. In the preceding example, $ab + ac = a(\_ + \_)$. The first blank is filled by the term whose product with a is ab. This is b. The second blank is the term whose product with a is ac. This is c. Consequently, $ab + ac = a(b + c)$. This process is sometimes referred to as **inspection.**

2. Find the quotient of each of the terms of the polynomial with the common factor. Using division, divide the first term, ab, by the common factor a, $\frac{ab}{a} = b$. Therefore, the first term inside the parentheses is b. Divide the second term, ac, by the common factor a, $\frac{ac}{a} = c$. Therefore, the second term inside the parentheses is c. So, $ab + ac = a(b + c)$. This division process works because of the definition of division. The common factor is a and the first term of the polynomial to be factored is ab. But $\frac{ab}{a} = b$ (the first term inside the parentheses). By the definition of division, a (the common factor) $\cdot b$ (the first term inside the parentheses) $= ab$ (the first term of the polynomial being factored).

Any factorization may be checked by finding the product of the factors. If the product is the same as your original polynomial, then your factorization is correct. For example, $2x + 2y = 2(x + y)$ is correct since $2(x + y) = 2x + 2y$. However, we must be careful that the factor that we have removed is the greatest common factor. If the polynomial inside the parentheses still has a common factor, then we have not removed the greatest common factor. Try again!

When adding like terms, we were really just removing a common factor and adding the numerical coefficients that remained inside the parentheses. For example, $4x + 7x$ has a common factor of x. Rewrite $4x + 7x$ as $(4 + 7)x$ by removing the common factor x. Then add 4 and 7 and get $11x$.

In the examples and exercises that follow, you will need to remember the Product Rule for exponents ($a^m \cdot a^n = a^{m+n}$) and the Quotient Rule for exponents $\frac{a^m}{a^n} = a^{m-n}$.

Example 1

Factor each of the following by removing the greatest common factor:

a. $bx + bz$ The greatest common factor of bx and bz is b. Write b outside the parentheses.

$b(\_ + \_)$ The first blank represents a term whose product with b is bx. The term is x. In expressions that are more complicated, we may need to divide each term of the polynomial by the common factor. If we divide the first term of the polynomial, bx, and by the common factor, b, we get the first term inside the parentheses, x. That is, $\frac{bx}{b} = x$.

$b(x + \_)$ The second blank must be a term whose product with b is bz. By inspection or division, the blank represents z. So we now have:

$b(x + z)$ Therefore, $bx + bz = b(x + z)$ where the first factor is b and the second factor is $x + z$.

CHECK: $b(x + z) = bx + bz$.   Therefore, the factorization is correct.

b. $4x + 2$ The GCF of $4x$ and 2 is 2. Write 2 outside the parentheses.

$2(\_ + \_)$ The first blank represents a term whose product with 2 is $4x$. Since $2 \cdot 2x = 4x$, the blank represents $2x$. Using division, $\frac{4x}{2} = 2x$.

$2(2x + \_)$ The second blank represents a term whose product with 2 is 2. Since $2 \cdot 1 = 2$, the blank represents 1. Using division, $\frac{2}{2} = 1$.

$2(2x + 1)$ Therefore, $4x + 2 = 2(2x + 1)$ where the first factor is 2 and the second is $2x + 1$.

CHECK: $2(2x + 1) = 4x + 2$.   Therefore, the factorization is correct.

c. $8x^2 + 12x^3$ The greatest common factor of $8x^2$ and $12x^3$ is $4x^2$. Write $4x^2$ outside the parentheses.

$4x^2(\_ + \_)$ The first blank is a term whose product with $4x^2$ is $8x^2$. Since $4x^2 \cdot 2 = 8x^2$, the blank represents 2. Using division, $\frac{8x^2}{4x^2} = 2$.

$4x^2(2 + \_)$ The second blank is a term whose product with $4x^2$ is $12x^3$. Since $4x^2 \cdot 3x = 12x^3$, the second blank represents $3x$. Using division, $\frac{12x^3}{4x^2} = 3x$.

$4x^2(2 + 3x)$ Therefore, $8x^2 + 12x^3 = 4x^2(2 + 3x)$ where the first factor is $4x^2$ and the second is $2 + 3x$.

CHECK: $4x^2(2 + 3x) = 8x^2 + 12x^3$.   Therefore, the factorization is correct.

d. $12x^2y^3 - 18xy^4$ The GCF of $12x^2y^3$ and $18xy^4$ is $6xy^3$. Write $6xy^3$ outside the parentheses.

$6xy^3(\_ + \_)$ The first blank represents $2x$, since $6xy^3 \cdot 2x = 12x^2y^3$ or $\frac{12x^2y^3}{6xy^3} = 2x$.

$6xy^3(2x + \_)$ The second blank represents $-3y$, since $6xy^3(-3y) = -18xy^4$ or $\frac{-18xy^4}{6xy^3} = -3y$.

$6xy^3(2x - 3y)$ Therefore, $12x^2y^3 - 18xy^4 = 6xy^3(2x - 3y)$ where the first factor is $6xy^3$ and the second is $2x - 3y$.

CHECK: $6xy^3(2x - 3y) = 12x^2y^3 - 18xy^4$.   Therefore, the factorization is correct.

e. $10x^2y^3 - 5x^3y - 20x^2y^2$ The GCF is $5x^2y$. Write $5x^2y$ outside the parentheses.

$5x^2y(\_ + \_ + \_)$ The first term inside the parentheses is $2y^2$, the second $-x$, and the third $-4y$. This gives us:

$5x^2y(2y^2 - x - 4y)$ Therefore, $10x^2y^3 - 5x^3y - 20x^2y^2 = 5x^2y(2y^2 - x - 4y)$.

CHECK: $5x^2y(2y^2 - x - 4y) = 10x^2y^3 - 5x^3y - 20x^2y^2$.   Therefore, the factorization is correct.

Practice Exercises

Factor each of the following by removing the greatest common factor:

1. $pq + pr$

2. $6a + 9$

3. $8y^3 + 16y^2$

4. $10a^3b^2 - 15a^2b^2$

5. $8a^2b^4 - 16ab^2 - 12a^2b^3$

If more practice is needed, do the Additional Practice Exercises.

Additional Practice Exercises

Factor each of the following by removing the greatest common factor:

a. $mn + mp$

b. $8p - 6$

c. $12p^4 + 18p^3$

d. $6c^2d^2 - 9c^3d^4$

e. $8a^2 - 12a + 4$

f. $9x^2y^4 - 15xy^3 - 18x^3y^4$

We usually want the coefficient of the first term inside the parentheses to be positive, so it is sometimes necessary to remove a negative common factor as illustrated by the following examples:

Example 2

Factor the following by removing a negative common factor:

a. $-2a + 4b$ Since we want the first coefficient inside the parentheses to be positive, the GCF is -2.

$-2(\_ + \_)$ The first term is a, since $\dfrac{-2a}{-2} = a$, and the second term is $-2b$, since $\dfrac{4b}{-2} = -2b$.

$-2(a - 2b)$ Therefore, $-2a + 4b = -2(a - 2b)$.

CHECK: $-2(a - 2b) = -2a + 4b$.   Therefore, the factorization is correct.

b. $-3x^3 + 9x^2 - 3x$ The GCF is $-3x$. Write $-3x$ outside the parentheses.

$-3x(\_ + \_ + \_)$ The first term is x^2, since $\dfrac{-3x^3}{-3x} = x^2$, the second term is $-3x$, since $\dfrac{9x^2}{-3x} = -3x$, and the third term is 1, since $\dfrac{-3x}{-3x} = 1$.

$-3x(x^2 - 3x + 1)$ Therefore, $-3x^3 + 9x^2 - 3x = -3x(x^2 - 3x + 1)$.

CHECK: $-3x(x^2 - 3x + 1) = -3x^3 + 9x^2 - 3x$.   Therefore, the factorization is correct.

c. $-a + b$ We want the first term inside the parentheses to be positive, so factor out -1. We do not write -1, just the "$-$."

$-(\_ + \_)$ The first term is a, since $\dfrac{-a}{-1} = a$ and the second is $-b$, since $\dfrac{b}{-1} = -b$.

$-(a - b)$ Therefore, $-a + b = -(a - b)$.

CHECK: $-(a - b) = -a + b$.   Therefore, the factorization is correct.

Answers:

f. $3xy^3(3xy - 5 - 6x^2y)$
Additional Practice Exercises a–f: a. $m(n + p)$ **b.** $2(4p - 3)$ **c.** $6p^3(2p + 3)$ **d.** $3c^2d^2(2 - 3cd^2)$ **e.** $4(2a^2 - 3a + 1)$
Practice Exercises 1–5: 1. $p(q + r)$ **2.** $3(2a + 3)$ **3.** $8y^2(y + 2)$ **4.** $5a^2b^2(2a - 3)$ **5.** $4ab^2(2ab^2 - 4 - 3ab)$

Practice Exercises

Factor the following by removing a negative common factor:

6. $-9x + 12y$

7. $-4a^4 + 2a^3 - 6a^2$

8. $-x + y$

If more practice is needed, do the Additional Practice Exercises.

Additional Practice Exercises

Factor the following by removing a negative common factor:

g. $-10p + 15q$

h. $-8p^3 + 4p^2 - 16p$

i. $-r + s$

In the previous examples and practice exercises, the common factor has been a monomial, but that is not always the case. The common factor can also be any polynomial. In the following examples, the common factor is a binomial:

Example 3

Factor the following by removing the common binomial factor:

a. $a(c + d) + b(c + d)$ The factors of the first term are a and $(c + d)$, and the factors of the second term are b and $(c + d)$. So, the common factor is $(c + d)$. Write $(c + d)$ outside the parentheses. To show the entire quantity $(c + d)$ is a factor, put it in parentheses.

$(c + d)(\_ + \_)$ The first blank is the term whose product with $(c + d)$ is $a(c + d)$, which is a.

$(c + d)(a + \_)$ The second blank is the term whose product with $(c + d)$ is $b(c + d)$ and is b.

$(c + d)(a + b)$ Therefore, $a(c + d) + b(c + d) = (c + d)(a + b)$. The check is more difficult in this case. Compare to the distributive property $x(y + z)$ where x is $(c + d)$, y is a, and z is b. Hence, $(c + d)(a + b) = (c + d)a + (c + d)b = a(c + d) + b(c + d)$ by the commutative property of multiplication.

b. $x(x + 2) - 5(x + 2)$ The common factor is $(x + 2)$.

$(x + 2)(\_ + \_)$ The first blank is the term whose product with $(x + 2)$ is $x(x + 2)$. This is x.

$(x + 2)(x + \_)$ The second blank is the term whose product with $(x + 2)$ is $-5(x + 2)$. This is -5.

$(x + 2)(x - 5)$ Therefore, $x(x + 2) - 5(x + 2) = (x + 2)(x - 5)$.

Practice Exercises

Factor the following by removing the common binomial factor:

9. $x(y - z) + w(y - z)$

10. $y(y + 4) - 5(y + 4)$

If more practice is needed, do the Additional Practice Exercises.

Answers:

Additional Practice Exercises

Factor the following by removing the common binomial factor:

j. $r(s + t) + p(s + t)$

k. $z(z + 5) - 6(z + 5)$

Factoring by grouping   We will now apply removing both common monomial and binomial factors to factor polynomials with four terms. We will group terms and remove a common factor from each group. The results will have a common binomial factor that we will then remove. Study the following examples carefully:

Example 4

Factor the following:

a. $ab + ac + db + dc$ Group the first two terms together and remove the common factor of a. Group the third and fourth terms together and remove a common factor of d.

$a(b + c) + d(b + c)$ We now have a common factor of $(b + c)$, which we will remove.

$(b + c)(a + d)$ Factorization.

CHECK: $(b + c)(a + d) = ab + bd + ac + cd$, which is correct, except for the order of the terms. What property states that the order of the terms does not matter?

b. $xy + xz + 3y + 3z$ Group the first two terms together and remove the common factor of x. Group the third and fourth terms together and remove the common factor of 3.

$x(y + z) + 3(y + z)$ Now remove the common factor of $y + z$.

$(y + z)(x + 3)$ Factorization.

CHECK: $(y + z)(x + 3) = xy + 3y + xz + 3z$, which is correct.

c. $4x - xy + 20 - 5y$ Remove the common factor of x from the first two terms and the common factor of 5 from the remaining terms.

$x(4 - y) + 5(4 - y)$ Now remove the common factor of $4 - y$.

$(4 - y)(x + 5)$ Factorization.

CHECK: $(4 - y)(x + 5) = 4x + 20 - xy - 5y$, which is correct.

Note: When factoring by grouping, it is often possible to group the terms in more than one way. The previous Example 1b could have been done as follows:

$xy + xz + 3y + 3z$ Change the order of the terms to $xy + 3y + xz + 3z$.

$xy + 3y + xz + 3z$ Group the first two terms together and remove the common factor of y. Group the third and fourth terms together and remove the common factor of z.

$y(x + 3) + z(x + 3)$ Now remove the common factor of $x + 3$.

$(x + 3)(y + z)$ Factorization.

Note the answer is the same except for the order of the factors.

Practice Exercises

Factor the following:

11. $ab - ac + xb - xc$

12. $xy + xz + 7y + 7z$

13. $3y + xy + 12 + 4x$

Answers:

12. $(y + x)(z + 7)$ **13.** $(3 + x)(y + 4)$

Additional Practice Exercises j–k: j. $(s + t)(r + p)$ **k.** $(z + 5)(z - 6)$ **Practice Exercises 11–13: 11.** $(b - c)(a + x)$

If more practice is needed, do the Additional Practice Exercises.

Additional Practice Exercises

Factor the following:

l. $mn + mp + qn + qp$

m. $xc - xf + 5c - 5f$

n. $5a + ab + 15 + 3b$

The goal of this type of factoring is to have the exact same polynomial inside the parentheses after we have grouped the terms and removed the common factors from each group. Consequently, it is sometimes necessary to remove a negative factor in order to make the signs of the binomial factors agree.

Example 5

Factor the following:

a. $ax + az - bx - bz$ Group the first two terms and remove the common factor of a.

$a(x + z) - bx - bz$ Since we want $x + z$ inside both sets of parentheses, factor $-b$ from the remaining terms. Be careful! This will change the signs of both terms inside the parentheses.

$a(x + z) - b(x + z)$ Now remove the common factor of $x + z$.

$(x + z)(a - b)$ Factorization.

CHECK: $(x + z)(a - b) = ax - bx + az - bz$, which is correct.

b. $3x - xr - 21 + 7r$ Remove the common factor of x from the first two terms.

$x(3 - r) - 21 + 7r$ We want $3 - r$ inside both sets of parentheses, so remove the factor of -7 from the remaining two terms.

$x(3 - r) - 7(3 - r)$ Now remove the common factor of $(3 - r)$.

$(3 - r)(x - 7)$ Factorization.

CHECK: $(3 - r)(x - 7) = 3x - 21 - rx + 7r$, which is correct.

c. $3x - xy - 3 + y$ Group the first two terms and remove the common factor of x.

$x(3 - y) - 3 + y$ Notice the two remaining terms are the same as the terms inside the parentheses except the signs are opposite. Factor -1 from the last two terms.

$x(3 - y) - 1(3 - y)$ Now remove the common factor of $3 - y$.

$(3 - y)(x - 1)$ Factorization.

CHECK: $(3 - y)(x - 1) = 3x - 3 - xy + y$, which is correct.

Practice Exercises

Factor the following:

14. $xy + hy - xc - hc$

15. $4a - ab - 16 + 4b$

16. $xy - y - x + 1$

If more practice is needed, do the Additional Practice Exercises.

Additional Practice Exercises

Factor the following:

o. $nm + nw - xm - xw$

p. $4x - xr - 20 + 5r$

q. $xe - e - x + 1$

Answers:

Exercise Set 5.2

Factor the following by removing the greatest common factor:

1. $cd + cf$

2. $gh + gk$

3. $rs - rt$

4. $xy - xz$

5. $3x + 9$

6. $4x + 8$

7. $8x - 12$

8. $15x - 10$

9. $x^2 + x$

10. $y^2 + y$

11. $r^4 - r^2$

12. $q^3 - q^2$

13. $7a^3 + 14a^2$

14. $5r^4 + 10r^2$

15. $18p^3 - 9p^5$

16. $16r^3 - 8r^4$

17. $15x^3 + 5x$

18. $12d^4 + 6d$

19. $22r^5 - 11r^2$

20. $26t^4 - 13t$

21. $12c^3 + 8c^2$

22. $14b^5 + 8b^3$

23. $18z^2 - 12z^6$

24. $15w^4 - 10w^6$

25. $x^2y^3 + x^2y^2$

26. $r^3s^4 + r^4s^3$

27. $u^4v^2 - u^3v^3$

28. $p^5q^2 - p^2q^4$

29. $c^3d^4 + cd^2$

30. $a^4b^2 + a^3b$

31. $x^2y^2 - x^3y^4$

32. $a^3b^2 - a^5b^5$

33. $18a^2b^4 + 12a^3b^2$

34. $18r^3s^2 + 27r^5s$

35. $9xy^3 - 12x^2y^2$

36. $24fg^4 - 18f^4g^3$

37. $9x^2y^4 + 3xy^3$

38. $8y^3z^2 + 4y^2z$

39. $9x^2 - 12x + 6$

40. $12x^2 - 8x + 12$

41. $15x^4 - 10x^3 - 20x^2$

42. $24y^5 - 16y^4 - 32y^2$

43. $6a^4 + 12a^2 - 24$

44. $9s^5 + 18s^3 - 27$

45. $22x^3 - 33x^2 - 11x$

46. $21a^4 - 28a^2 - 7a$

47. $16c^3d^2 - 24c^2d^4 + 36cd^2$

48. $16r^2s^4 - 8r^3s^4 + 12rs^3$

49. $14x^3y^3 - 21x^2y^4 - 7xy^2$

50. $25y^4z^4 - 30y^3z^2 - 5y^2z^2$

Factor the following by removing a negative common factor:

51. $-3m + 6n$

52. $-4a + 8b$

53. $-16c + 8d$

54. $-20p + 15q$

55. $-14x + 7$

56. $-18a + 9$

57. $-4x^2 + 8x - 16$

58. $-6x^2 + 18x - 24$

59. $-10x^3 - 15x^2 + 25x$

60. $-8a^3 - 24a^2 + 16a$

Factor the following by removing the common binomial factor:

61. $a(m + n) + b(m + n)$

62. $x(p + q) + y(p + q)$

63. $a(c + 2) - b(c + 2)$

64. $r(u + 5) - s(u + 5)$

65. $t(t - 3) + 6(t - 3)$

66. $y(y - 3) + 3(y - 3)$

67. $a(a - 6) - 7(a - 6)$

68. $b(b - 5) - 4(b - 5)$

69. What are the factors in the expression $(2x + 3)(3x - 5)$?

70. What are the factors in the expression $(5x - 2)(2x + 7)$?

Challenge Exercises: (71–74)

Remove the greatest common factor from each of the following:

71. $360x^6y^5 - 540x^8y^4$

72. $756a^9b^6 - 504a^7b^9$

73. $3(a + b)^3 + 9(a + b)^2$

74. $4(x + y)^4 - 12(x + y)^2$

Factor each of the following by grouping:

75. $xz + xw + yz + yw$

76. $ac + ad + bc + bd$

77. $mn - 4m + 3n - 12$

78. $dc - 2d + 5c - 10$

79. $ab + 5a + 15 + 3b$

80. $mn + 6n + 24 + 4m$

81. $x^2 - xy + 5x - 5y$

82. $a^2 - ab + 3a - 3b$

83. $ab - 3a + b - 3$

84. $xy - 5x + y - 5$

85. $xm + xn - my - ny$

86. $cd + cf - de - ef$

87. $mn - 6n - 3m + 18$

88. $rs - 7s - 4r + 28$

89. $cd - ce - 3d + 3e$

90. $pr - pq - 7r + 7q$

91. $2x^2 - xy + 6x - 3y$

92. $3r^2 - rs + 12r - 4s$

93. $rs + 4r - s - 4$

94. $yz + 5y - z - 5$

95. $cd - 3c - d + 3$

96. $ab - 7a - b + 7$

97. $10a + 4ab + 15 + 6b$

98. $12x + 9bx + 8 + 6b$

99. $8xz - 4xw - 2yz + yw$

100. $15ac - 5ad - 3bc + bd$

101. $6ac + 4ad - 9bc - 6bd$

102. $6xz + 9xw - 8yz - 12yw$

103. $8x^2 - 2xy + 12x - 3y$

104. $6a^2 - 9ay + 10a - 15y$

Writing Exercises:

105. $3a^2b^2(6b + 4a - 8) = 18a^2b^3 + 12a^3b^2 - 24a^2b^2$. However, $3a^2b^2(6b + 4a - 8)$ is not the correct prime factorization of $18a^2b^3 + 12a^3b^2 - 24a^2b^2$. Why not?

106. How can we tell if we have removed the greatest common factor? (i.e., What will happen if the factor we remove is a factor of the polynomial but not the greatest common factor?)

107. After factoring a polynomial, how can you determine if your factorization is correct?

Section 5.3

Factoring General Trinomials with Leading Coefficients of One

OBJECTIVE

When you complete this section, you will be able to:

Factor a general trinomial whose leading coefficient is one.

Introduction   When most people think of factoring, it is the factoring of trinomials that usually comes to mind.

We will consider two types of trinomials: those with positive last terms and those with negative last terms. We will first look at the products of binomials that result in trinomials whose last terms are positive.

Example 1

Find the following products:

a. $(x + 3)(x + 7) = x^2 + 7x + 3x + 21$ Apply FOIL.

$= x^2 + 10x + 21$ Add like terms.

Therefore, $x^2 + 10x + 21 = (x + 3)(x + 7)$ in factored form.

b. $(x - 3)(x - 7) = x^2 - 7x - 3x + 21$ Apply FOIL.

$\qquad\qquad\qquad = x^2 - 10x + 21$ Add like terms.

Therefore, $x^2 - 10x + 21 = (x - 3)(x - 7)$ in factored form.

From Example 1, we observe that the signs of the last terms of the trinomials are positive. Further observe that the signs of the second terms of both binomials are the same as the sign of the second term of the trinomial.

From Example 1a,

$$x^2 + 10x + 21 = (x + 3)(x + 7)$$

Last sign positive.

$$x^2 + 10x + 21 = (x + 3)(x + 7)$$

If the last sign is positive, these three signs are the same.

From Example 1b,

$$x^2 - 10x + 21 = (x - 3)(x - 7)$$

Last sign positive.

$$x^2 - 10x + 21 = (x - 3)(x - 7)$$

If the last sign is positive, these three signs are the same.

When multiplying binomials whose second signs are the same, both the inside and outside products have the same sign. Consequently, we add the absolute values of the outside and inside products in order to find the absolute value of the coefficient of the middle term of the trinomial. This will be helpful when deciding which factors of the last term to try when factoring the trinomial.

$$(x + 3)(x + 7) = x^2 + 7x + 3x + 21 = x^2 + 10x + 21$$

Same sign, so add the absolute values of 7 and 3 to get 10, which is the absolute value of the coefficient of the middle term.

$$(x - 3)(x - 7) = x^2 - 7x - 3x + 21 = x^2 - 10x + 21$$

Same sign, so add the absolute values of 7 and 3 to get 10, which is the absolute value of the coefficient of the middle term.

Consider the following. Remember, we are doing FOIL in reverse and if the last sign of the trinomial is positive, both binomial factors have the same sign as the sign of the middle term of the trinomial.

Example 2

Factor the following:

a. $x^2 + 4x + 3 =$ The sign of the last term is $+$ and the sign of the middle term is $+$. Therefore, both of the signs of the binomials will be $+$.

$(\_ + \_)(\_ + \_) =$ The first blank in each binomial must be filled by terms whose product is x^2. The only reasonable factorization of x^2 is $x \cdot x$.

$(x + \_)(x + \_) =$ The second blank in each binomial must be filled by terms whose product is 3 but which will also give the sum of the outside and inside products of $4x$. The only factors of 3 are $1 \cdot 3$.

$(x + 1)(x + 3)$ The outside product is $3x$ and the inside product is x. The sum is $4x$, which is what we wanted.

x

$3x$

$4x$

$(x + 1)(x + 3)$ Factorization.

CHECK: $(x + 1)(x + 3) = x^2 + 3x + x + 3 = x^2 + 4x + 3$, which is the original trinomial. Therefore, this factorization is correct.

Note: Since multiplication is commutative, $(x + 3)(x + 1)$ is also correct.

b. $x^2 - 7x + 10 =$ The sign of the last term of the trinomial is + and the sign of the middle term is −. Therefore, both of the signs of the binomials are −.

$(\_ - \_)(\_ - \_) =$ The first blanks of each binomial must be filled with terms whose product is x^2. The only reasonable factorization of x^2 is $x \cdot x$.

$(x - \_)(x - \_) =$ The second blanks must be filled by terms whose product is 10 and which give the sum of the outside and inside products of $-7x$. The possible factors of 10 are $1 \cdot 10$ and $2 \cdot 5$. Try 2 and 5.

$(x - 2)(x - 5)$ The outside product is $-5x$ and the inside product is $-2x$. The sum is $-7x$, which is what we wanted.

$(x - 2)(x - 5)$ Factorization.

CHECK : $(x - 2)(x - 5) = x^2 - 5x - 2x + 10 = x^2 - 7x + 10$, which is the original trinomial. Therefore, this factorization is correct.

c. $a^2 - 6ab + 8b^2 =$ The sign of the last term is + and the sign of the middle term is −. Therefore, both of the signs of the binomials are −.

$(\_ - \_)(\_ - \_) =$ The first blanks must be filled by terms whose product is a^2. The only factorization of a^2 is $a \cdot a$.

$(a - \_)(a - \_) =$ The second blanks must be filled with terms whose product is $8b^2$ and give a sum of the outside and inside products of $-6ab$. The choices of factors of $8b^2$ are $b \cdot 8b$ or $2b \cdot 4b$. Try $2b$ and $4b$.

$(a - 2b)(a - 4b)$ The outside product is $-4ab$ and the inside product is $-2ab$. The sum is $-6ab$, which is what we wanted.

$(a - 2b)(a - 4b)$ Factorization.

CHECK: $(a - 2b)(a - 4b) = a^2 - 4ab - 2ab + 8b^2 = a^2 - 6ab + 8b^2$. Therefore, this factorization is correct.

Practice Exercises

Factor the following:

1. $a^2 + 8a + 7$ 2. $b^2 - 9b + 14$ 3. $y^2 - 6yz + 5z^2$

If more practice is needed, do the Additional Practice Exercises.

Additional Practice Exercises

Factor the following:

a. $b^2 + 9b + 8$ b. $y^2 - 9yz + 14z^2$ c. $x^2 + 8xy + 15y^2$

We will now look at products of binomials that result in trinomials whose last terms are negative.

Example 3

Find the following products:

a. $(x + 3)(x - 7) = x^2 - 7x + 3x - 21$ Apply FOIL.

$= x^2 - 4x - 21$ Add like terms.

Therefore, $x^2 - 4x - 21 = (x + 3)(x - 7)$ in factored form.

b. $(x - 3)(x + 7) = x^2 + 7x - 3x - 21$ Apply FOIL.

$= x^2 + 4x - 21$ Add like terms.

Therefore, $x^2 + 4x - 21 = (x - 3)(x + 7)$ in factored form.

Looking at Example 3, we notice the sign of the last term of the trinomial in both examples is $-$. In each case, one binomial factor has a $+$ sign and the other a $-$ sign. In other words, the signs are opposites.

From Example 3a,

$$x^2 - 4x - 21 = (x + 3)(x - 7)$$
$\uparrow$

Last sign negative.

$$x^2 - 4x - 21 = (x + 3)(x - 7)$$
$\qquad\qquad\qquad\qquad \uparrow \qquad \uparrow$

If the last sign is negative, these signs are opposites.

From Example 3b,

$$x^2 + 4x - 21 = (x - 3)(x + 7)$$
$\uparrow$

Last sign negative.

$$x^2 + 4x - 21 = (x - 3)(x + 7)$$
$\qquad\qquad\qquad\qquad \uparrow \qquad \uparrow$

If the last sign is negative, these signs are opposites.

Since the signs of the binomials are opposites, the outside and inside products are opposite in sign. This means the absolute value of the coefficient of the middle term of the trinomial is found by subtracting the absolute values of the coefficients of the outside and inside products of the binomials. This will be helpful when deciding which factors of the last term to try when factoring the trinomial.

From Example 3a,

$$(x + 3)(x - 7) = x^2 - 7x + 3x - 21 = x^2 - 4x - 21$$
$\qquad\qquad\qquad\qquad \uparrow \qquad \uparrow$

These are opposite in sign, so subtract the absolute values of 7 and 3 to get 4, which is the absolute value of the coefficient of the middle term of the trinomial.

From Example 3b,

$$(x - 3)(x + 7) = x^2 + 7x - 3x - 21 = x^2 + 4x - 21$$
$\qquad\qquad\qquad\qquad \uparrow \qquad \uparrow$

These are opposite in sign, so subtract the absolute values of 7 and 3 to get 4, which is the absolute value of the coefficient of the middle term of the trinomial.

When factoring a trinomial whose last term is negative, we will not be concerned with the signs initially. We will first find binomial factors that give the correct first and last terms of the trinomial and such that the difference of the outside and inside products of the binomials gives the middle term, disregarding the sign, of the trinomial. When this is accomplished, we will decide which factor has the $+$ and which has the $-$. This will become clearer by doing some examples.

Example 4

Factor the following:

a. $x^2 - x - 6$ The sign of the last term is $-$. Therefore, the signs of the binomials are opposites.

(_ _)(_ _)   The first blanks must be filled by terms whose product is x^2. The only choice is x and x. Notice we did not put in any signs, since we will determine them last.

$(x \ \_)(x \ \_)$ The second blanks must be filled in by terms whose product is 6 and such that the difference of the outside and inside products is 1. The possible factorizations of 6 are $1 \cdot 6$ and $2 \cdot 3$. Try 1 and 6.

The outside product is $6x$ and the inside product is x. But $6x - x = 5x$ and the middle term of the trinomial is x, disregarding the sign. Therefore, 1 and 6 are not the correct choices. Try 2 and 3.

The outside product is $3x$ and the inside product is $2x$ and $3x - 2x = x$, which is the middle term of the trinomial, except for the sign.

Since the middle term of the trinomial is $-x$, we want $-3x$ and $+2x$ as the outside and inside products. Therefore, use -3 and $+2$.

$(x + 2)(x - 3)$ Factorization.

CHECK: $(x + 2)(x - 3) = x^2 - 3x + 2x - 6 = x^2 - x - 6$. Therefore, this factorization is correct.

b. $x^2 + 4xy - 12y^2$ The last sign of the trinomial is $-$, so the binomial factors will have opposite signs.

(_ _)(_ _)   The first blanks must be filled by terms whose product is x^2. Again, the only reasonable choice is x and x.

$(x \ \_)(x \ \_)$ The second blanks must be filled with terms whose product is $12y^2$ and the difference of the outside and inside products is $4xy$. The possible factorizations of $12y^2$ are $y \cdot 12y$, $2y \cdot 6y$, and $3y \cdot 4y$. Try y and $12y$.

The outside product is $12xy$ and the inside product is xy. However, $12xy - xy = 11xy$ and the middle term is $4xy$. Therefore, this is not the correct choice of factors of $12y^2$. Try $2y$ and $6y$.

The outside product is $6xy$ and the inside product is $2xy$. Also, $6xy - 2xy = 4xy$, which is the correct middle term.

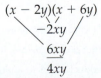

Since the middle term is $+4xy$, we need $+6xy$ and $-2xy$ as the outside and inside products. Therefore, use $+6y$ and $-2y$.

$(x - 2y)(x + 6y)$ Factorization.

CHECK: $(x - 2y)(x + 6y) = x^2 + 6xy - 2xy - 12y^2 = x^2 + 4xy - 12y^2$. Therefore, this factorization is correct.

c. $x^2 + 3x - 8$ The last sign is $-$, so the signs of the binomials are opposites.

$(\_\_\_)(\_\_\_)$ The first blanks must be filled in by terms whose product is x^2. This is, again, x and x.

$(x\_\_)(x\_\_)$ The second blanks must be filled by terms whose product is 8 and the difference in the outside and inside products is $3x$. The possible factors of 8 are $1 \cdot 8$ and $2 \cdot 4$. Try 1 and 8.

$(x\quad1)(x\quad8)$ The outside product is $8x$ and the inside product is x, but $8x - x = 7x$, which is not the correct middle term. Try 2 and 4.

$(x\quad2)(x\quad4)$ The outside product is $4x$ and the inside product is $2x$, but $4x - 2x = 2x$, which is not the correct middle term. Since we have tried all possible combinations of factors of 8 and none work, this trinomial is **prime**.

Practice Exercises

Factor the following. If the polynomial cannot be factored, write prime.

4. $x^2 + 3x - 10$ **5.** $a^2 - 2ab - 15b^2$ **6.** $b^2 + 4b - 10$

If more practice is needed, do the Additional Practice Exercises.

Additional Practice Exercises

Factor the following. If the polynomial cannot be factored, write prime.

d. $x^2 + 5x - 14$ **e.** $x^2 - 5x - 6$ **f.** $a^2 + 3a - 12$

Following is a summary of the techniques discussed in this section:

Summary

1. If the last sign of the trinomial is positive, both binomial factors will have the same sign as the middle sign of the trinomial. Also, the absolute values of the coefficients of the outside and inside products of the binomial factors must have a sum equal to the absolute value of the coefficient of the middle term of the trinomial being factored.

2. If the last term of the trinomial is negative, the signs of the binomial factors will be opposites, Also, the difference of the absolute values of the coefficients of the outside and inside products of the binomial factors must equal the absolute value of the coefficient of the middle term of the trinomial being factored.

We will do one of each type in the following example:

Example 5

Factor the following trinomials:

a. $x^2 - 11x + 24 =$ The sign of the last term is $+$ and the sign of the second term is $-$. Therefore, the signs of both binomial factors are $-$.

$(\_-\_)(\_-\_) =$ The first blanks must be filled with terms whose product is x^2. As before, the only reasonable choice is x and x.

$(x-\_)(x-\_) =$ The second blanks must be filled with terms whose product is 24 and which give the sum of the outside and inside products of -11. The possible factors of 24 are $1 \cdot 24, 2 \cdot 12, 3 \cdot 8$, and $4 \cdot 6$. Try 3 and 8.

$(x-3)(x-8)$ Factorization.

CHECK: $(x-3)(x-8) = x^2 - 8x - 3x + 24 = x^2 - 11x + 24$, which is the original trinomial. Therefore, this factorization is correct.

Answers:

Additional Practice Exercises d–f: d. $(x+7)(x-2)$ **e.** $(x-6)(x+1)$ **f.** Prime

Practice Exercises 4–6: 4. $(x+5)(x-2)$ **5.** $(a-5b)(a+3b)$ **6.** Prime

b. $b^2 - b - 20$ The sign of the last term is $-$. Therefore, the signs of the binomials are opposites.

$(\underline{\ }\ \underline{\ })(\underline{\ }\ \underline{\ })$ The first blanks must be filled by terms whose product is b^2. The only choice is b and b.

$(b\ \ \underline{\ })(b\ \ \underline{\ })$ The second blanks must be filled with terms whose product is 24 and the difference of the outside and inside products is 1. The possible factors of 20 are $1 \cdot 20$, $2 \cdot 10$, and $4 \cdot 5$. Try 2 and 10.

$(b\quad 2)(b\quad 10)$
$2b$
$10b$
$8b$
The outside product is $10b$ and the inside product is $2b$. However, $10b - 2b = 8b$ and the middle term of the trinomial is $-b$. Therefore, this is not the correct choice of factors of 10. Try 4 and 5.

$(b\quad 4)(b\quad 5)$
$4b$
$5b$
b
The outside product is $5b$ and the inside product is $4b$. Also, $5b - 4b = b$, which is the correct middle term, except for the sign.

$(b + 4)(b - 5)$
$4b$
$-5b$
$-b$
To get $-b$, we need $-5b$ and $+4b$ as the inside and outside products. Therefore, use -5 and $+4$.

$(b + 4)(b - 5)$ Factorization.

CHECK: $(b + 4)(b - 5) = b^2 - 5b + 4b - 20 = b^2 - b - 20$, which is the original trinomial. Therefore, this factorization is correct.

Practice Exercises

Factor the following:

7. $a^2 + 13a + 40$

8. $r^2 - 7r - 18$

9. $m^2 + 2m - 24$

10. $x^2 + 16x + 48$

If more practice is needed, do the Additional Practice Exercises.

Additional Practice Exercises

Factor the following:

g. $x^2 - x - 42$

h. $c^2 - 12c + 32$

As in the previous sections on factoring, if a common factor is present, we must first remove it and try to factor the polynomial inside the parentheses.

Example 6

Factor the following completely:

a. $3x^2 + 12x - 63 =$ Remove the GCF of 3.

$3(x^2 + 4x - 21) =$ The last sign of the trinomial is $-$, so the signs of the binomial factors will be opposites.

$3(\_\ \_)(\_\ \_) =$ The first blanks must be filled with terms whose product is x^2. Again, this is x and x.

$3(x\ \_)(x\ \_) =$ The second blanks must be filled with terms whose product is 21 and the difference of the outside and inside products is $4x$. Try 3 and 7.

$3(x\ \ 3)(x\ \ 7) =$ The outside product is $7x$ and the inside product is $3x$ and $7x - 3x = 4x$, which is the desired middle term. Since we want $+4x$, we need $+7x$ and $-3x$ as the outside and inside products. Therefore, use -3 and $+7$.

$3(x - 3)(x + 7)$ Factorization.

CHECK: $3(x - 3)(x + 7) = 3(x^2 + 7x - 3x - 21) = 3(x^2 + 4x - 21) = 3x^2 + 12x - 63$. Therefore, this factorization is correct.

b. $a^3b - 9a^2b^2 + 8ab^3$ Remove the GCF of ab.

$ab(a^2 - 9ab + 8b^2)$ The last sign of the trinomial is $+$ and the middle sign is $-$. Therefore, both signs of the binomials are $-$.

$ab(\_ - \_)(\_ - \_) =$ The first blanks are filled by a and a.

$ab(a - \_)(a - \_) =$ The second blanks must be filled with terms whose product is $8b^2$ and which give outside and inside products whose sum is $-9ab$. Try b and $8b$.

$ab(a - b)(a - 8b)$ Factorization.

CHECK: $ab(a - b)(a - 8b) = ab(a^2 - 8ab - ab + 8b^2) = ab(a^2 - 9ab + 8b^2) = a^3b - 9a^2b^2 + 8ab^3$. Therefore, this factorization is correct.

Practice Exercises

Factor the following:

11. $2x^2 - 18x + 36$

12. $y^3z - 15y^2z^2 - 16yz^3$

If more practice is needed, do the Additional Practice Exercises.

Additional Practice Exercises

Factor the following:

i. $4x^2 - 40x + 36$

j. $m^3n - m^2n^2 - 12mn^3$

Exercise Set 5.3

Factor the following, if possible. If the polynomial is not factorable, write prime.

1. $x^2 + 3x + 2$

2. $x^2 + 6x + 5$

3. $x^2 - 4x + 3$

4. $x^2 - 8x + 7$

5. $a^2 + 5a + 6$

6. $a^2 + 7a + 10$

7. $b^2 - 6b + 8$

8. $c^2 - 9c + 14$

9. $y^2 - 4y - 5$

10. $z^2 - 2z - 3$

11. $a^2 + 2a - 35$

12. $b^2 + 4b - 21$

13. $x^2 + 10x + 16$

14. $z^2 + 11z + 18$

15. $r^2 + r - 72$

16. $s^2 + s - 20$

17. $a^2 - 12a + 27$

18. $c^2 - 13c + 40$

19. $x^2 + 4x + 8$

20. $a^2 - 6a + 8$

21. $x^2 - 12x + 35$

22. $n^2 - 10n + 24$

23. $y^2 + y - 42$

24. $x^2 + x - 72$

25. $z^2 + 15z + 36$

26. $b^2 + 11b + 24$

27. $a^2 + 4ab + 3b^2$

28. $c^2 + 12cd + 11d^2$

29. $r^2 - 8rs + 7s^2$

30. $q^2 - 6qr + 5r^2$

31. $c^2 + 7cd - 18d^2$

32. $a^2 + 10ab - 24b^2$

33. $r^2 - 7rs - 44s^2$

34. $a^2 - ab - 72b^2$

35. $y^2 + 3yz - 40z^2$

36. $m^2 + 11mn - 42n^2$

37. $a^2 - 3ab - 54b^2$

38. $x^2 - 4xy - 45y^2$

39. $r^2 - 15rs + 36s^2$

40. $a^2 - 6ab - 27b^2$

Factor the following completely:

41. $3a^2 + 15a + 18$

42. $4x^2 + 12x + 8$

43. $5z^2 - 25z + 30$

44. $6a^2 - 30a + 36$

45. $x^3 - 3x^2 - 28x$

46. $y^3 - y^2 - 20y$

47. $y^4 - 20y^3 + 75y^2$

48. $c^4 - 10c^3 + 21c^2$

49. $3a^3 - 3a^2 - 12a$

50. $4x^3 - 4x^2 - 80x$

51. $x^3y + 2x^2y - 80xy$

52. $a^3b + 14a^2b - 51ab$

53. $x^2y^2 + 9xy^2 + 20y^2$

54. $y^4z^3 + 13y^3z^3 + 30y^2z^3$

55. $4r^2s + 20rs - 56s$

56. $3a^2b + 6ab - 45b$

57. $3a^2b - 3ab - 84b$

58. $2x^2y - 10xy - 100y$

Challenge Exercises: (59–67)

Factor the following:

59. $x^2 - 3x - 108$

60. $x^2 + 7x - 144$

61. $y^2 + 24y + 128$

62. $6 - x - x^2$

63. $24 + 2x - x^2$

64. $10 + 7x + x^2$

65. $18 + 9a + a^2$

66. $10 - 3x - x^2$

67. $16 - 6b - b^2$

Find all integer values of k *such that each of the following is factorable over the integers:*

68. $x^2 + kx + 6$

69. $x^2 + kx + 12$

Recall that the formula for the area of a rectangle is A = LW. *If the following trinomials represent the areas of rectangles, find binomials that could represent the length and width.*

70. $A = x^2 + 19x + 48$

71. $A = x^2 + 12x + 32$

72. $A = y^2 + 3y - 54$

The volume of a rectangular solid is V = LWH. *If each of the following trinomials represents the volume of a rectangular solid, find monomials and/or binomials that could represent the length, width, and height:*

73. $V = x^3 + 6x^2 - 40x$

74. $V = 4x^3 - 12x^2 - 112x$

Writing Exercises:

75. If the sign of the last term of a trinomial is positive, why must both signs of the second terms of the binomial factors be the same?

76. Given $f(x) = (x + 4)(x + 2)$ and $g(x) = x^2 + 6x + 8$;

    a. Find $f(-2)$ and $g(-2)$.

    b. What is the relationship between $f(-2)$ and $g(-2)$? Why?

| Section 5.4 | Factoring General Trinomials with First Coefficient Other Than One |

O B J E C T I V E   *When you complete this section, you will be able to:*

Factor trinomials whose first coefficient is a natural number other than one.

Introduction   In the previous section, all of the trinomials that we factored had a coefficient of one on the first term. The factorization of these trinomials was made easier since all we needed to do was find the factors of the last term that had either a sum or difference that equaled the coefficient of the middle term. The same trial and error procedure used in the last section applies to trinomials whose first coefficient is not one. However, these are more difficult because the location of the terms in the binomial gives different outside and inside products while using the exact same numbers. In other words, the number of possibilities is greatly increased as the following illustrates:

Example 1

Factor the following. If the trinomial cannot be factored, write **prime.**

a. $3x^2 + 5x + 2$ The last term is $+$ and the middle term is $+$, so both signs of the binomial factors are $+$.

$(\_ + \_)(\_ + \_)$ The first blanks must be filled by terms whose product is $3x^2$. Since 3 is prime, the only choices are $3x$ and x.

$(3x + \_)(x + \_)$ The second blanks must be filled by terms whose product is 2 and the sum of the outside and inside products is $5x$. The only factors of 2 are 2 and 1. Try 1 in the first blank and 2 in the second blank

$(3x + 1)(x + 2)$
x
$6x$
$7x$
    The outside product is $6x$ and the inside product is x. However, $6x + x = 7x$ and we needed $5x$. Therefore, this factorization is incorrect. Try 2 in the first blank and 1 in the second.

$(3x + 2)(x + 1)$
$2x$
$3x$
$5x$
    The outside product is $3x$ and the inside product is $2x$, and $3x + 2x = 5x$, which is the correct middle term.

$(3x + 2)(x + 1)$ Factorization.

CHECK: $(3x + 2)(x + 1) = 3x^2 + 3x + 2x + 2 = 3x^2 + 5x + 2$. Therefore, this factorization is correct.

b. $5x^2 + 8x - 4$ The last sign is $-$. Therefore, the signs of the binomial are different, so we will be looking for the difference of the outside and inside products.

$(\_ \ \_)(\_ \ \_)$ The first blanks must be filled by terms whose product is $5x^2$. These are $5x$ and x.

$(5x \ \_)(x \ \_)$ The second blanks must be filled by terms whose product is 4 and the difference of the outside and inside products is $8x$. The possible factorizations of 4 are $1 \cdot 4$ and $2 \cdot 2$. Try 1 in the first blank and 4 in the second.

$(5x \ 1)(x \ 4)$
x
$20x$
$19x$
    The outside product is $20x$ and the inside product is x, and $20x - x = 19x$. Since we wanted $8x$, this is the wrong choice. Try 4 in the first blank and 1 in the second.

$(5x \ 4)(x \ 1)$
$4x$
$5x$
x
    The outside product is $5x$ and the inside product is $4x$, and $5x - 4x = x$. Therefore, this choice is also incorrect. This means that 1 and 4 will not work, so try 2 and 2.

$(5x \ 2)(x \ 2)$
$2x$
$10x$
$8x$
    The outside product is $10x$ and the inside product is $2x$, and $10x - 2x = 8x$, which is what we want.

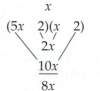
$(5x - 2)(x + 2)$
$-2x$
$10x$
$8x$
    Since the middle term is $+8x$, we want $+10x$ and $-2x$ for the outside and inside products. Hence, -2 goes in the first blank and 2 goes in the second.

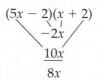
$(5x - 2)(x + 2)$ Factorization.

CHECK: $(5x - 2)(x + 2) = 5x^2 + 10x - 2x - 4 = 5x^2 + 8x - 4$. Therefore, this factorization is correct.

c. $6a^2 - 11a - 10$ | The last sign is $-$, so the binomial factors are different in sign.

$(\_\ \_)(\_\ \_)$ | The first blanks must be filled by terms whose product is $6a^2$. This could be $6a \cdot a$ or $3a \cdot 2a$. Try $3a \cdot 2a$.

$(3a\ \_)(2a\ \_)$ | The second blanks must be filled by terms whose product is 10 and such that the difference of the outside and inside products is $11a$. The possible factors of 10 are $1 \cdot 10$ or $2 \cdot 5$. Try 2 and 5 with 2 in the first blank and 5 in the second.

This gives an outside product of $15a$ and an inside product of $4a$. Also, $15a - 4a = 11a$, which is what we want.

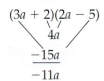

Since the middle term is $-11a$, we need $-15a$ and $+4a$. Put $+$ in the first factor and $-$ in the second.

$(3a + 2)(2a - 5)$ | Factorization.

CHECK: $(3a + 2)(2a - 5) = 6a^2 - 15a + 4a - 10 = 6a^2 - 11a - 10$. Therefore, this factorization is correct.

Note: In any factorization, the first thing we always look for is a greatest common factor. If there is no greatest common factor other than 1 (or once it has been removed), no factor may contain a common factor. This fact may greatly reduce the number of trial and error attempts.

e. $3x^2 + 14x + 10$ | The last sign is $+$ and the middle sign is $+$, so both signs of the binomial factors are $+$. The only choice for the first terms are $3x$ and x.

$(3x + \_)(x + \_)$ | The choices for the second blanks are 1 and 10 or 2 and 5. Try 2 in the first blank and 5 in the second.

$(3x + 2)(x + 5)$ | CHECK: $(3x + 2)(x + 5) = 3x^2 + 15x + 2x + 10 = 3x^2 + 17x + 10$, which is not correct. Try 5 in the first factor and 2 in the second.

$(3x + 5)(x + 2)$ | CHECK: $(3x + 5)(x + 2) = 3x^2 + 6x + 5x + 10 = 3x^2 + 11x + 10$, which is not correct. Since 2 and 5 will not work, try 1 and 10 with 1 in the first factor and 10 in the second.

$(3x + 1)(x + 10)$ | CHECK: $(3x + 1)(x + 10) = 3x^2 + 30x + x + 10 = 3x^2 + 31x + 10$, which is incorrect. Try 10 in the first factor and 1 in the second.

$(3x + 10)(x + 1)$ | CHECK: $(3x + 10)(x + 1) = 3x^2 + 3x + 10x + 10 = 3x^2 + 13x + 10$ which is incorrect. We have tried all possible combinations of the factors of $3x^2$ and 10, and none of them work. Since this exhausts all the possibilities, this polynomial is **prime (cannot be factored).**

Note: If the factors were $(3x - 2)(x + 5)$, we would get the correct middle term, but the last term would be -10 instead of $+10$.

d. $4a^2 + 11ab + 6b^2$ | The last sign is $+$ and the middle sign is $+$, so both signs of the binomial factors will be $+$. The first terms of the binomial factors must have a product of $4a^2$. The possibilities are $4a \cdot a$ and $2a \cdot 2a$. Let's try $2a$ and $2a$ and then show why this choice is impossible.

$(2a + \_)(2a + \_)$ | The blanks must be filled by terms whose product is $6b^2$ and such that the sum (since the signs are the same) of the outside and inside products is $11ab$. The possibilities are $6b \cdot b$ and $3b \cdot 2b$. Try $6b$ and b.

$(2a + 6b)(2a + b)$ | This cannot work, since the first factor has a common factor of 2. It would not help to put $6b$ in the second factor and b in the first since we would then have a common factor of 2 in the second factor. Try $3b$ and $2b$.

$(2a + 3b)(2a + 2b)$ | The second factor has a common factor of 2, so this cannot work. The same situation would occur if $2b$ were put in the first factor and $3b$ in the second. Therefore, the first terms cannot be $2a$ and $2a$, so try $4a$ and a.

$(4a + \_)(a + \_)$ | Again the possible terms for the second blanks are $6b$ and b or $3b$ and $2b$. Try $6b$ in the first blank and $2b$ in the second.

$(4a + 6b)(a + b)$ | The first factor has a common factor of 2, so this cannot be correct. Put $6b$ in the second factor and b in the first.

$(4a + b)(a + 6b)$ | Neither factor has a common factor, but the outside product is $24ab$ and the inside product is ab. Since $24ab + ab = 25ab$ and the middle term is $11ab$, this is incorrect. Try $3b$ in the first factor and $2b$ in the second.

$(4a + 3b)(a + 2b)$ | Factorization. The outside product is $8ab$ and the inside product is $3ab$, and $8ab + 3ab = 11ab$, which is what we need. Notice we could not have put $2b$ in the first factor and $3b$ in the second. Why?

$(4a + 3b)(a + 2b)$ | Factorization.

CHECK: $(4a + 3b)(a + 2b) = 4a^2 + 8ab + 3ab + 6b^2 = 4a^2 + 11ab + 6b^2$. Therefore, this factorization is correct.

Note: After doing a few of these, you should be able to do most, if not all, of the above trial and error steps mentally.

Alternative Method—The "ac" Method

There is another method of factoring trinomials that is often easier than the trial and error method presented earlier. This is called the "ac" method.

Suppose we find the product of $(mx + n)(px + q)$

$(mx + n)(px + q)$	Apply FOIL.
$mpx^2 + mqx + npx + nq$	Remove the common factor of x from the middle two terms.
$mpx^2 + (mq + np)x + nq$	

We now make some observations. If we multiply the coefficient of the x^2 term (which is mp) with the constant term (which is nq), we get $mpnq$. If we multiply the terms of the coefficient of the x term (which are mq and np), we get $mqnp$, which, except for the order of the factors, is the same thing we got when multiplying the coefficient of the x^2 term and the constant term. This means that the coefficient of the x term must be the sum of two factors of the product of the coefficient of the x^2 term and the constant term. We summarize the procedure in the following box:

Factoring Trinomials Using the "ac" Method

Given a trinomial in the form of $ax^2 + bx + c$:

1. Find the product of a and c.

2. Find two factors of ac whose sum is b.

3. Rewrite bx as the sum of two terms whose coefficients are the numbers found in step 2.

4. Factor the resulting polynomial by grouping.

The procedure is essentially the same if the trinomial is of the form $ax^2 + bxy + cy^2$. We illustrate the procedure using the trinomials from Example 1.

Example 2

Factor each of the following using the "ac" method of factoring:

a. $3x^2 + 5x + 2 =$

The product of a and c is $3 \cdot 2 = 6$, so we need two factors of 6 whose sum is b, which is 5. Six can be factored as $1 \cdot 6$ or $(-1)(-6)$, $2 \cdot 3$ or $(-2)(-3)$. Since $2 + 3 = 5$ (which is b), we rewrite $5x$ as $2x + 3x$.

$3x^2 + 2x + 3x + 2 =$ Factor by grouping. Remove the common factor of x from the first two terms and write $3x + 2$ as $1 \cdot (3x + 2)$.

$x(3x + 2) + 1 \cdot (3x + 2) =$ Remove the common factor of $3x + 2$.

$(3x + 2)(x + 1)$ Factorization.

b. $5x^2 + 8x - 4 =$

$ac = (5)(-4) = -20$ and $b = 8$, so we need two factors of -20 whose sum is 8. We can factor -20 as $(-1)(20)$ or $(1)(-20)$, $(-2)(10)$ or $(2)(-10)$, $(-4)(5)$ or $(4)(-5)$. Since $-2 + 10 = 8$, we rewrite $8x$ as $-2x + 10x$.

$5x^2 - 2x + 10x - 4 =$ Factor by grouping.

$x(5x - 2) + 2(5x - 2) =$ Remove the common factor of $5x - 2$.

$(5x - 2)(x + 2)$ Factorization.

c. $6a^2 - 11a - 10 =$

$ac = 6(-10) = -60$ and $b = -11$, so we need two factors of -60 whose sum is -11. We can factor -60 as $(-1)(60)$ or $(1)(-60)$, $(-2)(30)$ or $(2)(-30)$, $(-3)(20)$ or $(3)(-20)$, $(-4)(15)$ or $(4)(-15)$, $(-5)(12)$ or $(5)(-12)$, $(-6)(10)$ or $(6)(-10)$. Since $4 - 15 = -11$, we write $-11a$ as $4a - 15a$.

$6a^2 + 4a - 15a - 10 =$ Factor by grouping.

$2a(3a + 2) - 5(3a + 2) =$ Remove the common factor of $3a + 2$.

$(3a + 2)(2a - 5)$ Factorization.

d. $4a^2 + 11ab + 6b^2 =$

$ac = 4 \cdot 6 = 24$ and $b = 11$, so we need two factors of 24 whose sum is 11. We can factor 24 as $1 \cdot 24$ or $(-1)(-24)$, $2 \cdot 12$ or $(-2)(-12)$, $3 \cdot 8$ or $(-3)(-8)$, $4 \cdot 6$ or $(-4)(-6)$. Since $3 + 8 = 11$, we rewrite $11ab$ as $3ab + 8ab$.

$4a^2 + 3ab + 8ab + 6b^2 =$ Factor by grouping.

$a(4a + 3b) + 2b(4a + 3b) =$ Remove the common factor of $4a + 3b$.

$(4a + 3b)(a + 2b)$ Factorization.

e. $3x^2 + 14x + 10$

$ac = 3 \cdot 10 = 30$ and $b = 14$, so we need two factors of 30 whose sum is 14. We can factor 30 as $1 \cdot 30$ or $(-1)(-30)$, $2 \cdot 15$ or $(-2)(-15)$, $3 \cdot 10$ or $(-3)(-10)$, $5 \cdot 6$ or $(-5)(-6)$. None of these factors have a sum of 30, so this trinomial is prime.

Each of these methods has its advantages and disadvantages. If you use the "ac" method, you can often save yourself a lot of work by looking at the signs of the original polynomial and use that information as a guide in choosing the factors. For example, in Example 2d, both factors must be positive.

Practice Exercises

Factor the following. If the trinomial will not factor, write prime.

1. $2x^2 - 5x + 3$

2. $4x^2 + 17x - 15$

3. $6x^2 + x - 12$

4. $6c^2 - 13cd + 6d^2$

5. $2x^2 + 5x + 6$

If more practice is needed, do the Additional Practice Exercises.

Additional Practice Exercises

Factor the following. If the trinomial will not factor, write prime.

a. $5x^2 - 9x + 4$

b. $7a^2 + 40a - 12$

c. $4b^2 + 5b - 6$

d. $6c^2 - 19cd + 15d^2$

e. $4x^2 - 7x - 3$

As in all cases of factoring, the first thing we should look for is a common factor, and if there is one, we should remove it.

Example 3

Factor the following. If the expression will not factor, write prime.

a. $18x^2 - 36x + 16$ Remove the common factor of 2.

$2(9x^2 - 18x + 8)$ Factor the trinomial inside the parentheses.

$2(3x - 2)(3x - 4)$ Factorization.

CHECK: $2(3x - 2)(3x - 4) = 2(9x^2 - 18x + 8) = 18x^2 - 36x + 16$. Therefore, this factorization is correct.

b. $12a^3 + 27a^2 - 27a$ Remove the common factor of $3a$.

$3a(4a^2 + 9a - 9)$ Factor the trinomial inside the parentheses.

$3a(4a - 3)(a + 3)$ Factorization.

CHECK: $3a(4a - 3)(a + 3) = 3a(4a^2 + 9a - 9) = 12a^3 + 27a^2 - 27a$. Therefore, this factorization is correct.

Practice Exercises

Factor the following completely. If the expression will not factor, write prime.

6. $9b^2 + 30b - 24$

7. $12cd^2 - 38cd + 20c$

If more practice is needed, do the Additional Practice Exercises.

Additional Practice Exercises

f. $24a^2 + 52a + 24$

g. $16a^2b + 36ab - 10b$

Exercise Set 5.4

Factor the following completely. If the polynomial will not factor, write prime.

1. $2a^2 + 7a + 5$ **2.** $3a^2 - 22a + 7$ **3.** $2a^2 + 5a - 3$

4. $5b^2 + 3b - 2$ **5.** $7m^2 - 34m - 5$ **6.** $5y^2 - 2y - 3$

7. $3x^2 - 4x + 7$ **8.** $5c^2 - 24c + 5$ **9.** $5c^2 - 2cd + 7d^2$

10. $7x^2 - 20xy - 3y^2$ **11.** $11x^2 + 16xy + 5y^2$ **12.** $3c^2 - 10cd + 7d^2$

13. $2x^2 + 13x + 10$ **14.** $3x^2 - 11x + 6$ **15.** $5x^2 + 6x - 8$

16. $7a^2 - 9a - 10$ **17.** $4x^2 + 8x + 3$ **18.** $6x^2 + 13x + 5$

19. $10x^2 - 33x - 7$ **20.** $21x^2 + 32x - 5$ **21.** $2x^2 + 11x + 12$

22. $3x^2 + 26x + 20$ **23.** $3a^2 - 14a + 16$ **24.** $2t^2 - 13t + 18$

25. $5c^2 + 11c - 12$ **26.** $3z^2 - 13z - 30$ **27.** $2b^2 + 13b - 24$

28. $5r^2 - 36r - 32$ **29.** $6x^2 + 17x + 5$ **30.** $6x^2 + 7x + 5$

31. $8x^2 - 22x + 5$ **32.** $8x^2 - 18x + 7$ **33.** $24a^2 - 14a - 3$

34. $24c^2 - 13c - 7$ **35.** $16n^2 - 2n - 5$ **36.** $18m^2 + 9m - 5$

37. $6x^2 + 19x + 15$ **38.** $8x^2 + 18x + 9$ **39.** $12c^2 - 17c + 6$

40. $24n^2 - 42n + 15$ **41.** $18a^2 + 9a - 35$ **42.** $16d^2 - 14d - 15$

43. $8w^2 - 6w - 27$ **44.** $15t^2 + 7t - 30$ **45.** $20a^2 - 9a + 20$

46. $18y^2 + 3y - 28$ **47.** $6a^2 + 19a - 15$ **48.** $12c^2 - 8c - 15$

49. $6a^2 + 29ab + 35b^2$ **50.** $8a^2 + 34ab + 21b^2$ **51.** $18c^2 - 9cd - 20d^2$

52. $24r^2 - 25rs - 25s^2$ **53.** $12a^2 + 7ab - 12b^2$ **54.** $18x^2 - 9xy - 20y^2$

Answers:

Practice Exercises 6–7: 6. $3(3b - 2)(b + 4)$ 7. $2c(3d - 2)(2d - 5)$
Additional Practice Exercises f–g: f. $4(2a + 3)(3a + 2)$ g. $2b(4a - 1)(2a + 5)$

55. $12f^2 - 32fg + 21g^2$

56. $16a^2 + 48ab + 35b^2$

57. $6a^2 + 3a - 18$

58. $6r^2 - 26r - 20$

59. $4x^2y - 13xy + 3y$

60. $3a^2b^2 - 10a^2b + 8a^2$

61. $6x^3 - 21x^2 + 15x$

62. $2x^3 - 14x^2 + 20x$

63. $18x^3 - 3x^2 - 36x$

Challenge Exercises: (64–74)

Factor the following. If the polynomial will not factor, write prime.

64. $32x^2 - 134x - 45$

65. $36x^2 + 29x - 20$

66. $24x^2 - 55x - 24$

67. $(x + 2)^2 - (x + 2) - 6$

68. $(x - 3)^2 - 2(x - 3) - 8$

69. $6 + a - 2a^2$

70. $8 + 10a + 3a^2$

71. $12 + b - 6b^2$

72. $10 + 11a - 6a^2$

73. $18 + 18n - 8n^2$

74. $40 - 33x - 18x^2$

Determine all integer values of k *so that the following will be factorable over the integers:*

75. $2x^2 + kx + 5$

76. $3x^2 + kx + 8$

Writing Exercises:

77. How would you explain to a classmate why $x^2 + 9x + 15$ is prime?

78. Explain why $12x^2 - x + 6$ cannot be factored correctly as $(3x + 2)(4x - 3)$.

79. Explain why $6x^2 + 21x + 15$ cannot be factored correctly as $(3x + 3)(2x + 5)$.

80. After factoring a trinomial, the results are $(2x - 3)(5x + 7)$. What is the trinomial? Why?

Section 5.5   # Factoring Binomials

OBJECTIVES

When you complete this section, you will be able to:

a. Factor binomials that are the difference of two squares.

b. Factor the difference of squares where one or more of the terms is a binomial.

c. Factor the sum and difference of cubes.

Introduction   Recall that the product of two binomials that differed only by the signs of the second terms always resulted in a binomial that was the difference of two squares (Section 2.4). For example, $(a + b)(a - b) = a^2 - b^2$. In this case we are multiplying $(a + b)$ and $(a - b)$, so they are factors of $a^2 - b^2$. Consequently, the difference of squares, $a^2 - b^2$, factors into the sum, $(a + b)$, and difference, $(a - b)$, of the expressions being squared. Thus, in factored form $a^2 - b^2 = (a + b)(a - b)$.

Factoring the Difference of Squares

A binomial that is the difference of squares factors into two binomials. One of the factors is the sum of the expressions being squared and the other factor is the difference of the expressions being squared. Using variables, $a^2 - b^2 = (a + b)(a - b)$. Expressions of the form $a + b$ and $a - b$ are often called **conjugate pairs.**

Note: $a^2 - b^2$ could be thought of as $a^2 + 0ab - b^2$ and factored using the methods of Section 5.3. However, it is quicker and easier to factor if we recognize it as the difference of squares and factor using the preceding procedure.

To assist you in recognizing perfect squares, refer to Table 1 in the Appendix. Each entry in the "n^2" column is the square of the corresponding entry in the "n" column. Consequently, each entry in the n^2 column is a perfect square. You should be able to recognize all perfect squares up to 225 from memory.

Example 1

Factor the following difference of squares:

a. $x^2 - y^2 =$ The first term is the square of x and the second term is the square of y. Hence, the correct factorization is the product of the sum and difference of x and y.

$(x + y)(x - y)$ Factorization.

CHECK: $(x + y)(x - y) = x^2 - y^2$. Therefore, the factorization is correct.

Note: According to the commutative property of multiplication, the factors may be written in either order. So, $(x - y)(x + y)$ is also correct. Also, we could think of $x^2 - y^2$ as $x^2 + 0xy - y^2$ and factor using the methods of Section 5.3.

b. $a^2 - 9 =$ a^2 is the square of a and 9 is the square of 3.

$a^2 - 3^2 =$ Therefore, the correct factorization is the product of the sum and difference of a and 3.

$(a + 3)(a - 3)$ Factorization.

CHECK: $(a + 3)(a - 3) = a^2 - 9$. Therefore, the factorization is correct.

> *Be Careful* $x^2 + y^2$ is the sum of squares and cannot be factored with real coefficients. One common error is $x^2 + y^2 = (x + y)^2$. This is incorrect since $(x + y)^2 = x^2 + 2xy + y^2$. There is an extra $2xy$ when it is multiplied out. Another common error is $x^2 + y^2 = (x + y)(x + y)$. This is the same as $(x + y)^2$. Do not confuse $(ab)^n$ with $(a + b)^n$, since $(ab)^n = a^n b^n$, but $(a + b)^n \neq a^n + b^n$.

c. $y^2 - 4z^2 =$ y^2 is the square of y and $4z^2$ is the square of $2z$.

$y^2 - (2z)^2 =$ Therefore, the correct factorization is the product of the sum and difference of y and $2z$.

$(y + 2z)(y - 2z)$ Factorization.

CHECK: $(y + 2z)(y - 2z) = y^2 - 4z^2$. Therefore, the factorization is correct.

d. $16m^2 - 9n^2 =$ $16m^2$ is the square of $4m$ and $9n^2$ is the square of $3n$.

$(4m)^2 - (3n)^2 =$ Therefore, the correct factorization is the product of the sum and difference of $4m$ and $3n$.

$(4m + 3n)(4m - 3n)$ Factorization.

CHECK: $(4m + 3n)(4m - 3n) = 16m^2 - 9n^2$. Therefore, the factorization is correct.

Practice Exercises

Factor the following:

1. $a^2 - b^2$ **2.** $c^2 + 16$ **3.** $x^2 - 9y^2$ **4.** $25m^2 - 4n^2$

If more practice is needed, do the Additional Practice Exercises.

Additional Practice

Factor the following:

a. $p^2 - q^2$ **b.** $n^2 + 25$ **c.** $9c^2 - d^2$ **d.** $4a^2 - 9b^2$

Answers:

Additional Practice Exercises *a–d:* a. $(p + q)(p - q)$ **b.** prime **c.** $(3c + d)(3c - d)$ **d.** $(2a + 3b)(2a - 3b)$

Practice Exercises 1–4: 1. $(a + b)(a - b)$ **2.** prime **3.** $(x + 3y)(x - 3y)$ **4.** $(5m + 2n)(5m - 2n)$

Sometimes binomials are the differences of squares, even though they do not appear to be. Also, any factorization is not complete until each of the factors is prime. For example, $12 = 2 \cdot 6$, but the factorization is not complete since 6 is not prime. We will also need to recall that $(a^m)^n = a^{mn}$.

Example 2

Factor the following completely:

a. $x^4 - y^4 =$ x^4 is the square of x^2 and y^4 is the square of y^2.

$(x^2)^2 - (y^2)^2 =$ The factorization is the product of the sum and the difference of x^2 and y^2.

$(x^2 + y^2)(x^2 - y^2)$ $x^2 - y^2$ is still the difference of squares and must be factored again.

$(x^2 + y^2)(x + y)(x - y)$ Complete factorization.

CHECK: $(x^2 + y^2)(x + y)(x - y) = (x^2 + y^2)(x^2 - y^2) = x^4 - y^4$. Therefore, the factorization is correct.

b. $a^4 + b^4 = (a^2)^2 + (b^2)^2$ This is the sum of squares and cannot be factored. It is prime.

c. $a^4 - 81 =$ a^4 is the square of a^2 and 81 is the square of 9.

$(a^2)^2 - 9^2 =$ The factorization is the product of the sum and the difference of a^2 and 9.

$(a^2 + 9)(a^2 - 9)$ $a^2 - 9$ is also the difference of squares and must be factored.

$(a^2 + 9)(a + 3)(a - 3)$ Complete factorization.

CHECK: $(a^2 + 9)(a + 3)(a - 3) = (a^2 + 9)(a^2 - 9) = a^4 - 81$. Therefore, the factorization is correct.

Practice Exercises

Factor the following. If the polynomial cannot be factored, write prime.

5. $p^4 - q^4$ **6.** $r^4 + s^4$ **7.** $x^4 - 16$

In Section 5.2, we found that we could remove a common binomial factor, as in $x(x + 2) - 3(x + 2) = (x + 2)(x - 3)$. In factoring the difference of squares, it is also possible for one or more of the expressions being squared to be a binomial. Study the following examples:

Example 3

Factor the following:

a. $(a + b)^2 - 4 =$ $(a + b)^2$ is the square of $(a + b)$ and 4 is the square of 2.

$(a + b)^2 - 2^2 =$ The factorization is the product of the sum and difference of $(a + b)$ and 2.

$[(a + b) + 2][(a + b) - 2)]$ Remove the parentheses.

$(a + b + 2)(a + b - 2)$ Factorization.

b. $9 - (x - y)^2 =$ 9 is the square of 3 and $(x - y)^2$ is the square of $(x - y)$.

$3^2 - (x - y)^2 =$ The factorization is the product of the sum and difference of of 3 and $(x - y)$.

$[3 + (x - y)][3 - (x - y)]$ Remove the parentheses.

$(3 + x - y)(3 - x + y)$ Factorization.

Practice Exercises

Factor the following:

8. $(a + b)^2 - 25$ **9.** $4 - (m - n)^2$

If more practice is needed, do the Additional Practice Exercises.

Answers:

Practice Exercises 5–9: 5. $(p^2 + q^2)(p + q)(p - q)$ **6.** prime **7.** $(x^2 + 4)(x + 2)(x - 2)$ **8.** $(a + b + 5)(a + b - 5)$ **9.** $(2 + m - n)(2 - m + n)$

Additional Practice Exercises

Factor the following:

e. $(p - q)^2 - 9$

f. $36 - (c + d)^2$

In some cases, there may be a common factor that must be removed before factoring the difference of squares.

Example 4

Factor the following completely:

a. $18x^2 - 8y^2 =$ First remove the common factor of 2.

$2(9x^2 - 4y^2) =$ $9x^2$ is the square of $3x$ and $4y^2$ is the square of $2y$.

$2[(3x)^2 - (2y)^2] =$ Factor the difference of squares as the product of the sum and difference of $3x$ and $2y$.

$2(3x + 2y)(3x - 2y)$ Complete factorization.

b. $25x^2 - 100 =$ First remove the common factor of 25.

$25(x^2 - 4) =$ Factor the difference of squares.

$25(x + 2)(x - 2)$ Complete factorization.

Note: In Example 4b, $25x^2 - 100 = (5x)^2 - 10^2$, so it is the difference of squares. If we had factored as $(5x + 10)(5x - 10)$, the factorization would not have been complete since the factors are not prime. Each has a common factor of 5. Remember, always remove the GCF first.

Practice Exercises

Factor the following completely:

10. $12x^2 - 27y^2$

11. $16a^2 - 64b^2$

OPTIONAL   **FACTORING THE SUM AND DIFFERENCE OF CUBES**

Factoring the sum and difference of cubes

Previously, we learned that products of the form $(a + b)(a^2 - ab + b^2) = a^3 + b^3$, which is the sum of cubes (Section 2.4). We also found that $(a - b)(a^2 + ab + b^2) = a^3 - b^3$, which is the difference of cubes. From these products, we can factor the sum and difference of cubes as follows:

Factoring the Sum and Difference of Cubes

$a^3 + b^3 = (a + b)(a^2 - ab + b^2)$

$a^3 - b^3 = (a - b)(a^2 + ab + b^2)$

Let's make some observations about the factored forms. Notice that there is a binomial factor and a trinomial factor. The first term of the binomial factor is the expression whose cube is the first term of the sum or difference of cubes. The second term of the binomial is the expression whose cube is the second term of the sum or difference of

---

cubes. The sign of the binomial is the same as the sign of the sum or difference of cubes. The first term of the trinomial is the square of the first term of the binomial. The second term of the trinomial is the negative of the product of the two terms of the binomial. The third term of the trinomial is the square of the second term of the binomial.

Square of first term   Square of second term
$$a^3 + b^3 = (a + b)(a^2 - ab + b^2)$$
Negative of the product of the two terms

The preceding illustration is for the sum of cubes, but the rule is exactly the same for the difference of cubes.

Example 5

Factor the following:

a. $x^3 - 27 =$ Rewrite as $x^3 - 3^3$. Therefore, this is the difference of cubes. The first term of the binomial is the expression whose cube is x^3, so it is x. The sign of the binomial is $-$, since this is a difference of cubes. The second term of the binomial is the expression whose cube is 27. Since $3^3 = 27$, the second term is 3.

$(x - 3)(\_\ \_\ \_) =$ The first term of the trinomial is the square of the first term of the binomial. Therefore, the first term of the trinomial is x^2.

$(x - 3)(x^2\ \_\ \_) =$ The second term of the trinomial is the negative of the product of the terms of the binomial, which gives $-(-3x) = 3x$.

$(x - 3)(x^2 + 3x\ \_) =$ The last term of the trinomial is the square of the second term of the binomial. Since $3^2 = 9$, the last term of the trinomial is 9.

$(x - 3)(x^2 + 3x + 9)$ Factorization. Check by multiplying.

b. $8x^3 + 125 =$ Rewrite as $(2x)^3 + 5^3$. Therefore, this is the sum of cubes. The first term of the binomial is the expression whose cube is $8x^3$. Therefore, the first term is $2x$. The second term is the expression whose cube is 125. Therefore, the second term is 5. The sign is $+$, since this is the sum of cubes.

$(2x + 5)(\_\ \_\ \_) =$ The first term of the trinomial is the square of the first term of the binomial. Since $(2x)^2 = 4x^2$, the first term of the trinomial is $4x^2$.

$(2x + 5)(4x^2\_\ \_) =$ The second term of the trinomial is the negative of the product of the terms of the binomial. Since $-(2x)(5) = -10x$, the second term of the trinomial is $-10x$.

$(2x + 5)(4x^2 - 10x\ \_) =$ The last term of the trinomial is the square of the second term of the binomial. Since $5^2 = 25$, the last term of the trinomial is 25.

$(2x + 5)(4x^2 - 10x + 25)$ Factorization. Check by multiplying.

Practice Exercises

Factor each of the following:

12. $a^3 - 8$

13. $27x^3 + 64$

If more practice is needed, do the Additional Practice Exercises.

Additional Practice Exercises

Factor the following:

g. $y^3 - 1$

h. $8c^3 + 27$

Answers:

Exercise Set 5.5

Factor the following differences of squares. If the polynomial cannot be factored, write prime.

1. $m^2 - n^2$ **2.** $r^2 - s^2$ **3.** $x^2 - 4$ **4.** $p^2 - 1$

5. $r^2 - 64$ **6.** $a^2 - 81$ **7.** $m^2 + n^2$ **8.** $r^2 + s^2$

9. $c^2 - 25d^2$ **10.** $t^2 - 25s^2$ **11.** $a^2 - 100b^2$ **12.** $y^2 - 64z^2$

13. $4x^2 - 25y^2$ **14.** $9a^2 - 16b^2$ **15.** $49p^2 - 81q^2$ **16.** $25c^2 - 9d^2$

17. $9x^2 + 4y^2$ **18.** $36a^2 + 25b^2$ **19.** $121a^2 - 49b^2$ **20.** $169f^2 - 36g^2$

Factor the following completely:

21. $a^4 - b^4$ **22.** $r^4 - t^4$ **23.** $x^4 - 1$

24. $y^4 - 81$ **25.** $16x^4 - 81$ **26.** $256r^4 - 1$

Factor the following completely:

27. $(x + y)^2 - 9$ **28.** $(p + q)^2 - 25$ **29.** $36 - (x - y)^2$

30. $49 - (r - s)^2$ **31.** $4(x - y)^2 - 9$ **32.** $16(a - b)^2 - 25$

Factor the following completely:

33. $3x^2 - 75$ **34.** $5y^2 - 20$ **35.** $16x^2 - 36y^2$ **36.** $50x^2 - 32y^2$

37. $27x^2 - 48$ **38.** $20c^2 - 125$ **39.** $9x^2 - 81$ **40.** $4x^2 - 36$

41. $3x^4 - 48$ **42.** $2y^4 - 162$

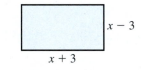

Problems 43 and 44 require the use of the formula for the area of a rectangle, which is A = LW. *For example, if* A = $x^2 - 9$, *then* A = $(x + 3)(x - 3)$, *so L could be* x + 3 *and W could be* x − 3. *See the figure in the margin.*

43. If the area of a rectangle is represented by $9x^2 - 49$, how can the length and width be represented?

44. If the area of a rectangle is represented by $16z^2 - 81$, how can the length and width be represented?

Challenge Exercises: (45–56)

Factor the following completely:

45. $196x^2 - 625$ **46.** $324a^2 - 400$ **47.** $256c^2 - 361d^2$ **48.** $441r^2 - 576s^2$

49. $338x^2 - 288$ **50.** $720a^2 - 125$ **51.** $x^8 - y^8$ **52.** $a^6 - b^6$

53. $81(x + 2)^2 - 16y^2$ **54.** $36(3x - 5)^2 - 25y^2$ **55.** $x^{2n} - y^{2n}$ **56.** $x^{4n} - 81$

Factor the following sum or difference of cubes (optional):

57. $x^3 + y^3$ **58.** $r^3 - s^3$ **59.** $c^3 + 8$

60. $d^3 + 27$ **61.** $n^3 - 64$ **62.** $r^3 - 125$

63. $m^3 + 1$ **64.** $b^3 - 1$ **65.** $8x^3 + y^3$

66. $64a^3 + b^3$ **67.** $27a^3 - 8$ **68.** $125c^3 - 64$

69. $8x^3 + 125y^3$ **70.** $64r^3 - 216s^3$

Writing Exercises:

71. What is wrong with the following factorization? $16x^2 - 144 = (4x + 12)(4x - 12)$. What is the correct factorization?

What is wrong with the following factorizations? Refactor each correctly.

72. $8x^3 + 64 = (2x + 4)(4x^2 - 8x + 16)$

73. $x^3 - 64 = (x - 4)(x^2 - 8x + 16)$

74. Given $f(x) = x^2 - 16$ and $g(x) = (x + 4)(x - 4)$.

  a. Find $f(2)$ and $g(2)$.

  b. What is the relationship between $f(2)$ and $g(2)$? Why?

Section 5.6	Factoring Perfect Square Trinomials

OBJECTIVE — *When you complete this section, you will be able to:*

Recognize and factor perfect square trinomials.

Introduction

Natural numbers that result from squaring integers are called **perfect squares.** The first few perfect squares are 1, 4, 9, 16, 25, 36, 49, 64, 81, 100, and so on, because $1^2 = 1$, $2^2 = 4$, $3^2 = 9$, and so on. In this section, we will be factoring **perfect square trinomials.** A perfect square trinomial results from the square of a binomial. For example, $(a + b)^2 = a^2 + 2ab + b^2$. Since $a^2 + 2ab + b^2$ is the result of squaring the binomial $a + b$, it is a perfect square trinomial.

How do we recognize a perfect square trinomial? The answer is in knowing the shortcut method of squaring a binomial (Section 2.4). Recall that the procedure for squaring a binomial is:

  1. Square the first term of the binomial.

  2. Multiply the product of the two terms of the binomial by two.

  3. Square the last term of the binomial.

The procedure is diagrammed as follows for $(a + b)^2$.

First term squared   Last term squared
$$(a + b)^2 = a^2 + 2ab + b^2$$
Two times the product of the terms

For example, $(2x + 3)^2 = (2x)^2 + 2(2x)(3) + 3^2 = 4x^2 + 12x + 9$. Since $4x^2 + 12x + 9$ is the square of $2x + 3$, it is a perfect square trinomial.

Reversing the procedure, we see that a perfect square trinomial must have the following characteristics:

Characteristics of a Perfect Square Trinomial

  1. The first and last terms of the trinomial must be perfect squares.

  2. The middle term (ignore the sign for now) must be two times the product of the terms whose squares give the first and last terms of the trinomial.

From the preceding example, we know that $a^2 + 2ab + b^2$ is a perfect square trinomial. The following diagram illustrates the characteristics of a perfect square trinomial using $a^2 + 2ab + b^2$:

The square of a   The square of b
$$a^2 + 2ab + b^2$$
Two times a times b

We illustrate with the following examples:

Example 1

Determine if the following are perfect square trinomials:

a. $x^2 - 6x + 9$ The first term is the square of x and the last term is the square of 3. The middle
$\quad$ ↑ $\quad$ ↑ $\quad$ ↑   term, ignoring the sign, is $2(x)(3)$. Therefore, this is a perfect square trinomial.
$(x)^2$ $2\cdot3\cdot x$ 3^2

b. $4x^2 - 20xy + 25y^2$ The first term is the square of $2x$ and the last term is the square of $5y$.
$\quad$ ↑ $\quad\quad$ ↑ $\quad\quad$ ↑   The middle term, ignoring the sign, is $2(2x)(5y)$. Therefore, this is a
$(2x)^2$ $2\cdot2x\cdot5y$ $(5y)^2$ perfect square trinomial.

c. $9a^2 - 12ay + 8y^2$ The last term, $8y^2$, is not the square of a term. Therefore, this is not a perfect
$\quad$ square trinomial.

d. $9p^2 + 12p + 16$ The first term is the square of $3p$ and the last term is the square of 4. However,
$\quad$ the middle term is not equal to $2(3p)(4)$. Therefore, this is not a perfect square
$\quad$ trinomial.

Practice Exercises

Determine if the following are perfect square trinomials:

1. $y^2 - 8y + 16$

2. $16a^2 + 24ab + 9b^2$

3. $25x^2 + 20x + 32$

4. $9a^2 - 15a + 25$

Now that we can recognize a perfect square trinomial, how do we find the binomial whose square results in the perfect square trinomial? Again, the answer lies in the procedure for squaring the binomial. In the introduction, we found that $4x^2 + 12x + 9$ resulted from squaring $2x + 3$. Notice that the first term of the binomial $2x$ is the term whose square gives the first term of the trinomial $4x^2$. The second term of the binomial 3 is the term whose square gives the last term of the trinomial 9. The sign between the terms of the binomial is the same as the sign of the second term of the perfect square trinomial. Based on the preceding, the procedure for factoring a perfect square trinomial is:

Factoring a Perfect Square Trinomial

1. Determine if the trinomial is a perfect square trinomial.

2. If it is, write a binomial squared. The first term of the binomial is the term whose square gives the first term of the trinomial. The second term of the binomial is the term whose square gives the last term of the trinomial.

3. The sign between the terms of the binomial is the same as the sign of the middle term of the trinomial.

From previous examples, we know the $a^2 + 2ab + b^2 = (a + b)^2$. So we will illustrate the procedure using $a^2 + 2ab + b^2$.

$$a^2 + 2ab + b^2 = (a\ \ \ )$$

$(a)^2$ First term is a.

$$a^2 + 2ab + b^2 = (a\ \ \ b)$$

$(b)^2$ Second term is b.

$$a^2 + 2ab + b^2 = (a + b)$$

These two signs are the same.

Therefore, $a^2 + 2ab + b^2 = (a + b)^2$.

Note: Perfect square trinomials can also be factored using the methods of Sections 5.3 and 5.4. However, it is usually easier and quicker if you recognize that it is a perfect square trinomial and factor accordingly. Also, in Chapter 7, it will be necessary to find the least common denominator of rational expressions. In order to do so, perfect square trinomials must be written as the square of a binomial.

Example 2

Determine whether each of the following is a perfect square trinomial. If it is, factor it as the square of a binomial.

a. $x^2 - 6x + 9 =$

From Example 1a, we know that this is a perfect square trinomial.

$(x\ \ \ )^2 =$ x^2 is the square of x, so the first term of the binomial is x.

$(x\ \ \ 3)^2 =$ 9 is the square of 3, so the second term of the binomial is 3.

$(x - 3)^2$ The sign of the middle term of the trinomial is $-$, so the sign of the binomial is $-$.

CHECK: $(x - 3)^2 = (x)^2 - 2(x)(3) + 3^2 = x^2 - 6x + 9$. Therefore, the factorization is correct.

b. $x^2 - 10x + 25 =$

The first term is the square of x, the last term is the square of 5, and the middle term is $2(x)(5)$. Therefore, it is a perfect square trinomial.

$(x\ \ \ )^2$ x^2 is the square of x, so the first term of the binomial is x.

$(x\ \ \ 5)^2$ 25 is the square of 5, so the second term of the binomial is 5.

$(x - 5)^2$ The sign of the middle term is $-$, so the sign of the binomial is $-$.

CHECK: $(x - 5)^2 = (x)^2 + 2(x)(-5) + (-5)^2 = x^2 - 10x + 25$.

c. $4x^2 + 20xy + 25y^2 =$

From Example 1b, we know that this is a perfect square trinomial.

$(2x\ \ \ )^2$ $4x^2$ is the square of $2x$, so the first term of the binomial is $2x$.

$(2x\ \ \ 5y)^2$ $25y^2$ is the square of $5y$, so the second term of the binomial is $5y$.

$(2x + 5y)^2$ The sign of the middle term is $+$, so the sign of the binomial is $+$.

CHECK: $(2x + 5y)^2 = (2x)^2 + 2(2x)(5y) + (5y)^2 = 4x^2 + 20xy + 25y^2$. Therefore, the factorization is correct.

d. $9p^2 + 24p + 16 =$

The first term is the square of $3p$, the last term is the square of 4, and the middle term is $2(3p)(4)$. Therefore, this is a perfect square trinomial.

$(3p + 4)^2$ $9p^2$ is the square of $3p$, 16 is the square of 4, and the sign of the middle term is $+$. So, $9p^2 + 24p + 16 = (3p + 4)^2$.

CHECK: The check is left as an exercise for the student.

Practice Exercises

Determine whether each of the following is a perfect square trinomial. If it is, factor it as the square of a binomial.

5. $y^2 - 8y + 16$ **6.** $16a^2 + 24ab + 9b^2$ **7.** $a^2 - 16a + 64$ **8.** $4x^2 + 20xy + 25y^2$

If more practice is needed, do the Additional Practice Exercises.

Additional Practice Exercises

Factor the following if it is a perfect square trinomial:

a. $a^2 - 4a + 4$ **b.** $9a^2 - 6ab + b^2$

Answers:

Practice Exercises 5–8: 5. $(y - 4)^2$ **6.** $(4a + 3b)^2$ **7.** $(a - 8)^2$ **8.** $(2x + 5y)^2$ **Additional Practice Exercises a–b: a.** $(a - 2)^2$ **b.** $(3a - b)^2$

As in the previous section, if there is a GCF, remove it and then see if the resulting polynomial can be factored further.

Example 3

Factor the following:

a. $3x^2 - 30x + 75 =$ First remove the GCF of 3.

$3(x^2 - 10x + 25) =$ The resulting trinomial is a perfect square.

$3(x - 5)^2$ The first term is the square of x, the last term is the square of 5, and the sign of the second term is $-$. Check by multiplying.

b. $x^3y + 8x^2y + 16xy =$ Remove the GCF of xy.

$xy(x^2 + 8x + 16) =$ The resulting trinomial is a perfect square.

$xy(x + 4)^2$ x^2 is the square of x, 16 is the square of 4, and the sign of the second term is $+$. Check by multiplying.

Practice Exercises

Factor the following:

9. $5x^2 - 60x + 180$

10. $x^4y^2 - 14x^3y^2 + 49x^2y^2$

If more practice is needed, do the Additional Practice Exercises.

Additional Practice Exercises

Factor the following:

c. $8x^2 - 16x + 8$

d. $x^3 - 18x^2 + 81x$

Exercise Set 5.6

Factor the following, if the trinomial is a perfect square. If it is not a perfect square, write **prime**.

1. $x^2 + 4x + 4$ **2.** $a^2 - 6a + 9$ **3.** $x^2 + 2x + 1$

4. $b^2 - 2b + 1$ **5.** $x^2 + 12x + 36$ **6.** $c^2 - 14c + 49$

7. $y^2 + 20y + 100$ **8.** $r^2 + 22r + 121$ **9.** $25x^2 - 10x + 1$

10. $36x^2 + 12x + 1$ **11.** $a^2 - 2ab + b^2$ **12.** $c^2 + 2cd + d^2$

13. $4a^2 + 4ab + b^2$ **14.** $16a^2 - 8ab + b^2$ **15.** $16c^2 - 24c + 9$

16. $25x^2 + 20x + 4$ **17.** $9x^2 + 30x + 25$ **18.** $4x^2 + 28x + 49$

19. $16x^2 - 8xy + y^2$ **20.** $49x^2 + 14xy + y^2$ **21.** $25x^2 + 40xy + 16y^2$

22. $36x^2 - 84xy + 49y^2$ **23.** $4a^2 - 14ab + 49b^2$ **24.** $25c^2 + 35cd + 49d^2$

25. $8x^2 - 24x + 18$ **26.** $27x^2 - 36x + 12$ **27.** $2x^2 + 28x + 98$

28. $3x^2 - 54x + 243$ **29.** $4x^2 + 80x + 400$ **30.** $5x^2 - 60x + 180$

31. $25x^4 - 30x^3 + 9x^2$ **32.** $4x^3y + 36x^2y + 81xy$

33. $50x^3y + 80x^2y^2 + 32xy^3$ **34.** $81x^4 - 36x^3y + 4x^2y^2$

Answers:

Practice Exercises 9–10: **9.** $5(x-6)^2$ **10.** $x^2y^2(x-7)^2$ Additional Practice Exercises c–d: **c.** $8(x-1)^2$ **d.** $x(x-9)^2$

Problems 35 and 36 require the use of the formula for the area of a square, which is A = s², where s is the length of each side of the square. For example, if A = x² + 2x + 1, then A = (x + 1)². This means that each side of the square is (x + 1). See the figure in the margin.

$x + 1$

$x + 1$

35. If $x^2 - 14x + 49$ represents the area of a square, what is the length of each side?

36. If $4a^2 + 36a + 81$ represents the area of a square, what is the length of each side?

Challenge Exercises:

Factor the following:

37. $81x^2 - 90xy + 25y^2$

38. $121x^2 - 286xy + 169y^2$

39. $100a^4 + 300a^2b^2 + 225b^4$

40. $144x^4 + 192x^2y^2 + 64y^4$

Find the value of **k** *so that each of the following will be a perfect square trinomial:*

41. $x^2 + 14x + k$

42. $4x^2 + kx + 25$

43. $kx^2 + 40x + 16$

44. $9a^2 + 30ab + kb^2$

Writing Exercise:

45. If the middle term in a perfect square trinomial is negative, why is the middle term of the binomial whose square equals the trinomial also negative?

46. Given $f(x) = 4x^2 + 12x + 9$ and $g(x) = (2x + 3)^2$,
   a. Find $f(1)$ and $g(1)$.
   b. What is the relationship between $f(1)$ and $g(1)$? Why?

Section 5.7   # Mixed Factoring

OBJECTIVES

When you complete this section, you will be able to:

a. Recognize and factor polynomials that have common factors, that are the difference of squares, or that are trinomials.

b. Recognize and factor polynomials by grouping or (optional) that are the sum or difference of cubes.

Introduction

In Sections 5.2–5.6, we concentrated on a particular type of factoring in each section. This made matters somewhat easier, since we knew which form to look for. When the different types of factoring are mixed, we must first recognize the form of the polynomial in order to know which procedure to use. The following check list is recommended:

Procedure for Complete Factoring

A. First, if there is a greatest common factor other than 1, remove it.

B. How many terms does the polynomial have?

   1. If the polynomial has two terms:

      a. Is it the difference of squares? If so, factor using $a^2 - b^2 = (a + b)(a - b)$.

      b. Is it the sum or difference of cubes? If so, factor using
      $a^3 + b^3 = (a + b)(a^2 - ab + b^2)$ or $a^3 - b^3 = (a - b)(a^2 + ab + b^2)$.

   2. If the polynomial has three terms:

      a. If necessary, put the trinomial in descending order.   *(continued)*

Procedure for Complete Factoring *(continued)*

    b. Is it a perfect square trinomial? If so, factor using $a^2 + 2ab + b^2 = (a + b)^2$ or $a^2 - 2ab + b^2 = (a - b)^2$.

    c. If the trinomial is not a perfect square trinomial, factor using trial and error. Check the factorization using FOIL.

  3. If the polynomial has more than three terms, try factoring by grouping.

C. The factorization is not complete unless each factor is prime. Check each factor and factor any that are not prime.

We will use the preceding procedure in the following examples:

Example 1

Factor each of the following completely:

a. $x^4 - 10x^3 + 25x^2$ Remove the GCF of x^2.

    $x^2(x^2 - 10x + 25)$ The trinomial is a perfect square.

    $x^2(x - 5)^2$ Factorization is complete.

    CHECK: $x^2(x - 5)^2 = x^2(x^2 - 10x + 25) = x^4 - 10x^3 + 25x^2$. Therefore, the factorization is correct.

b. $25x^2 - 81$ There are no common factors.

    $(5x)^2 - 9^2$ The binomial is the difference of squares.

    $(5x + 9)(5x - 9)$ Each factor is prime, so the factorization is complete.

    CHECK: $(5x + 9)(5x - 9) = 25x^2 - 81$. Therefore, the factorization is correct.

c. $8x^2 + 2x - 21$ There are no common factors. The trinomial is not a perfect square, so factor using trial and error.

    $(4x + 7)(2x - 3)$ Each factor is prime, so the factorization is complete.

    CHECK: $(4x + 7)(2x - 3) = 8x^2 - 12x + 14x - 21 = 8x^2 + 2x - 21$. Therefore, the factorization is correct.

d. $16a^4 - 81$ There are no common factors.

    $(4a^2)^2 - 9^2$ The binomial is the difference of squares.

    $(4a^2 + 9)(4a^2 - 9)$ $4a^2 + 9$ is prime, but $4a^2 - 9$ is the difference of squares.

    $(4a^2 + 9)(2a + 3)(2a - 3)$ Each factor is prime, so the factorization is complete.

    CHECK: $(4a^2 + 9)(2a + 3)(2a - 3) = (4a^2 + 9)(4a^2 - 9) = 16a^4 - 81$. Therefore, the factorization is correct.

e. $12bc - 9b + 20c - 15$ There are no common factors. There are more than three terms, so factor by grouping. Remove the factor of $3b$ from the first two terms and 5 from the remaining terms.

    $3b(4c - 3) + 5(4c - 3)$ Remove the greatest common factor of $4c - 3$.

    $(4c - 3)(3b + 5)$ Each factor is prime, so the factorization is complete.

    CHECK: $(4c - 3)(3b + 5) = 12bc + 20c - 9b - 15$. Therefore, the factorization is correct.

f. $12x^2y - 22xy - 70y$ Remove the greatest common factor of $2y$.

    $2y(6x^2 - 11x - 35)$ The trinomial factor is not a perfect square, so factor by trial and error.

    $2y(2x - 7)(3x + 5)$ Each factor is prime, so the factorization is complete.

    CHECK: $2y(2x - 7)(3x + 5) = 2y(6x^2 - 11x - 35) = 12x^2y - 22xy - 70y$. Therefore, the factorization is correct.

g. $6x^2 + 4x + 5$ There is no common factor. The trinomial is not a perfect square. Attempts to factor by trial and error will show that the trinomial is prime.

Practice Exercises

Factor the following completely. If the polynomial will not factor, write **prime**.

1. $4x^4 + 12x^3 + 9x^2$ **2.** $81a^2 - 49$ **3.** $9x^2 + 6x - 35$ **4.** $16c^4 - d^4$

5. $6xy - 8x - 21y + 28$ **6.** $48x^2y - 20xy - 8y$ **7.** $4x^2 - x + 6$

The ability to factor quickly and accurately is one of the most important skills in algebra. This skill will be used extensively in Chapters 6 and 7.

Answers:

Practice Exercises 1–7: 1. $x^2(2x + 3)^2$ **2.** $(9a + 7)(9a - 7)$ **3.** $(3x - 5)(3x + 7)$ **4.** $(4c^2 + d^2)(2c + d)(2c - d)$ **5.** $(3y - 4)(2x - 7)$ **6.** $4y(4x + 1)(3x - 2)$ **7.** prime

Exercise Set 5.7

Factor each of the following completely. If the polynomial will not factor, write prime.

1. $16 - 8x^2$

2. $12 - 6y^2$

3. $c^2 + 12c + 27$

4. $d^2 + 13d + 40$

5. $3a^3 - 9a^2 - 54a$

6. $2x^3 + 8x^2 - 64x$

7. $x^2 - 100$

8. $a^2 - 144$

9. $c^2 - 18c - 81$

10. $r^2 + 22r + 121$

11. $ax + bx + 2a + 2b$

12. $bx - 2x + by - 2y$

13. $6a^2 - 48a + 72$

14. $5r^2 + 75r + 180$

15. $r^2 + r - 72$

16. $b^2 - b - 56$

17. $m^3n^2 + m^2n^4$

18. $a^4b^3 + a^2b$

19. $12a^4 - 75a^2$

20. $27x^4 - 48x^2$

21. $9a^2 + 30ab + 25b^2$

22. $4p^2 - 28pq + 49q^2$

23. $16r^2 + 40r - 96$

24. $30a^2 - 87a + 30$

25. $a^2b^2 - 64$

26. $p^2q^2 - 81$

27. $5n^2 + 22n + 8$

28. $3m^2 + 16m + 16$

29. $8c^2 + 4cd + d^2$

30. $10b^2 - 5bc + c^2$

31. $x(y + z) + w(y + z)$

32. $r(s - 3) - t(s - 3)$

33. $(a - b)^2 - 36$

34. $(r - s)^2 - 169$

35. $10xy - 4y - 15x + 6$

36. $10xy + 15y - 8x - 12$

37. $a^4 - 4b^2$

38. $25y^4 - 16$

39. $x^2 + 4$

40. $a^2 + 25$

41. $16b^4 - 1$

42. $n^4 - 625$

43. $6x^2 + 31x + 5$

44. $2d^2 + d - 10$

45. $6x^2 + 9xz - 2xy - 3yz$

46. $6a^2 + 4ac - 9ab - 6bc$

47. $ab + 3a - b - 3$

48. $xy - xz - y + z$

49. $16u^3 + 46u^2v - 6uv^2$

50. $18y^3 + 24y^2z - 10yz^2$

51. $2a^2b - 5a^2b^2 + 7ab^2$

52. $x^3y - 3x^2y^2 + 5xy^3$

53. $8c^2 + 33cd + 4d^2$

54. $7r^2 - 50rs + 7s^2$

55. $9ax + 3bx - 9a - 3b$

56. $8ac + 12ad - 6bc - 9bd$

57. $6x^2 - x - 15$

58. $8x^2 - 14x - 15$

59. $12ac - 9ad + 8bc - 6bd$

60. $8mx - 10nx - 12my + 15ny$

61. $36x^2 - 144y^2$

62. $144x^2 - 9y^2$

63. $x^2 + 2x - 120$

64. $x^2 + 18x + 80$

65. $36a^4b^3 - 39a^3b^4 - 12a^2b^5$

66. $20x^5y - 42x^4y^2 + 18x^3y^4$

Factor each of the following completely: (optional)

67. $8z^3 + 125$

68. $64c^3 + 27$

69. $27c^3 - 8d^3$

70. $125t^3 - 64d^3$

71. $24c^3 - 3d^3$

72. $4t^3 + 500$

Challenge Exercises: 73–78 (do not involve optional type of factoring)

73. $12x^2 - 23x - 24$

74. $20x^2 - 39x + 18$

75. $81a^4 - 625b^4$

76. $256r^4 - 81s^4$

77. $x^2 + 10x + 25 - y^2$

78. $a^2 - 12ab + 36b^2 - 9$

Writing Exercise:

79. Why is it important to first remove any GCF when factoring?

Group Project:

80. Write two polynomials that would be factored by each of the following techniques:

     a. Removing the GCF.

     b. Difference of squares.

     c. Sum or difference of cubes.

     d. Perfect square trinomial.

     e. Trial and error using FOIL.

     f. Grouping

Section 5.8 | # Solving Quadratic Equations by Factoring

OBJECTIVES

When you complete this section, you will be able to:

a. Solve quadratic equations by factoring.

b. Solve application problems using quadratic equations.

Introduction   We learned how to solve linear equations in Chapter 3 by using the Addition and Multiplication Properties of Equality. In this section, we will learn to solve a different type of equation called **quadratic equations.** A quadratic equation is any equation of the form $ax^2 + bx + c = 0$ with $a \neq 0$, or any equation that can be put into that form. Solving quadratic equations involves the techniques of solving linear equations and one additional property, the zero product property.

Zero Product Property

If $ab = 0$, then $a = 0$, or $b = 0$, or $a = b = 0$. In words, if the product of two factors is zero, then one or both of the factors is zero.

This property is easily extended to three or more factors. For example, if $abc = 0$, then at least one of the following is true: $a = 0$, $b = 0$, or $c = 0$. When this property is applied to the solutions of equations, it means find the value(s) of the variable that make each factor equal to 0. For example, if $(x + 3)(x - 2) = 0$, then either $x + 3 = 0$ or $x - 2 = 0$, since one of the factors must be 0.

Example 1

Solve the following using the zero product property:

a. $x(3x + 2) = 0$ Set each factor equal to 0.

     $x = 0$ or $3x + 2 = 0$ Solve the linear equation.

         $3x = -2$

         $x = -\dfrac{2}{3}$ Therefore, the solutions are 0 and $-\dfrac{2}{3}$.

b. $(a + 3)(a - 1) = 0$ Set each factor equal to 0.

     $a + 3 = 0$ or $a - 1 = 0$ Solve each linear equation.

     $a = -3$ $a = 1$ Therefore, the solutions are -3 and 1.

c. $(b - 4)(2b + 5)(3b - 4) = 0$ Set each factor equal to 0.

     $b - 4 = 0$ or $2b + 5 = 0$ or $3b - 4 = 0$ Solve each linear equation.

     $b = 4,$ $2b = -5$ $3b = 4$

             $b = -\dfrac{5}{2}$ $b = \dfrac{4}{3}$ The solutions are 4, $-\dfrac{5}{2}$ and $\dfrac{4}{3}$.

Practice Exercises

Solve the following:

1. $b(3b + 2) = 0$

2. $(n + 4)(3n - 5) = 0$

3. $(y - 3)(7y + 2)(5y - 3) = 0$

Suppose we were to multiply the factors of Example 1b. We get:

$$(a + 3)(a - 1) = 0$$
$$a^2 + 2a - 3 = 0$$

The result is a quadratic equation for which $(a + 3)(a - 1)$ is the factored form. Therefore, factoring and the zero factor property give us a method for solving some quadratic equations.

Solving quadratic equations

Solving Quadratic Equations

1. If necessary, rewrite the equation in the form $ax^2 + bx + c = 0$ or $0 = ax^2 + bx + c$.
2. Factor $ax^2 + bx + c$.
3. Set each factor equal to 0. (zero product property)
4. Solve the resulting linear equations.
5. Check the solutions.

Example 2

Solve the following.

a. $2x^2 + 6x = 0$ Factor the left side of the equation.

$2x(x + 3) = 0$ Set each factor equal to 0.

$2x = 0, x + 3 = 0$ Solve the linear equations.

$x = 0, \quad x = -3$ Therefore, the solutions are 0 and -3.

CHECK: $x = 0$ $x = -3$

Substitute 0 for x.   Substitute -3 for x.

$2(0)^2 + 6(0) = 0$ $2(-3)^2 + 6(-3) = 0$

$2 \cdot 0 + 0 = 0$ $2(9) - 18 = 0$

$0 + 0 = 0$ $18 - 18 = 0$

$0 = 0$ $0 = 0$

Therefore, 0 and -3 are the correct solutions.

c. $c^2 - 16 = 0$ Factor the left side of the equation.

$(c + 4)(c - 4) = 0$ Set each factor equal to 0.

$c + 4 = 0, c - 4 = 0$ Solve each linear equation.

$c = -4 \quad c = 4$ Therefore, the solutions are 4 and -4.

CHECK: The check is left as an exercise for the student.

b. $x^2 - 2x - 15 = 0$ Factor the left side of the equation.

$(x - 5)(x + 3) = 0$ Set each factor equal to 0.

$x - 5 = 0, x + 3 = 0$ Solve each linear equation.

$x = 5 \quad\quad x = -3$ Therefore, the solutions are 5 and -3.

CHECK: $x = 5$ $x = -3$

Substitute 5 for x.   Substitute -3 for x.

$5^2 - 2(5) - 15 = 0$ $(-3)^2 - 2(-3) - 15 = 0$

$25 - 10 - 15 = 0$ $9 + 6 - 15 = 0$

$0 = 0$ $0 = 0$

Therefore, 5 and -3 are the correct solutions.

d. $6x^2 - 10 = 11x$ The equation must first be written in the form of $ax^2 + bx + c = 0$. Subtract $11x$ from both sides and write the left side in descending order.

$6x^2 - 11x - 10 = 0$ Factor.

$(2x - 5)(3x + 2) = 0$ Set each factor equal to 0.

$2x - 5 = 0, 3x + 2 = 0$ Solve each linear equation.

$2x = 5, \quad\quad 3x = -2$ Therefore, the solutions are $\frac{5}{2}$ and $-\frac{2}{3}$.

$$x = \frac{5}{2} \quad\quad x = -\frac{2}{3}$$

CHECK: The check is left as an exercise for the student.

e. $x(x + 10) = 24$ Simplify the left side of the equation.

    $x^2 + 10x = 24$ Subtract 24 from both sides.

    $x^2 + 10x - 24 = 0$ Factor.

    $(x + 12)(x - 2) = 0$ Set each factor equal to 0.

    $x + 12 = 0$ $x - 2 = 0$ Solve each equation.

    $x = -12,$ $x = 2$ Therefore, the solutions are -12 and 2.

    CHECK: The check is left as an exercise for the student.

Practice Exercises

Solve the following:

4. $2r^2 - 6r = 0$ **5.** $z^2 - 4z - 21 = 0$

6. $m^2 - 36 = 0$ **7.** $8x^2 - 15 = 14x$

8. $x(x + 3) = 40$

If more practice is needed, do the Additional Practice Exercises.

Additional Practice Exercises

Solve the following:

a. $4c^2 - 8c = 0$ **b.** $a^2 + 9a + 14 = 0$

c. $b^2 - 1 = 0$ **d.** $6x^2 - 3 = 7x$

e. $x(x + 2) = 48$

Applications with quadratic equations

     Application problems often result in quadratic equations. Care must be taken since some solutions to the equations may not be realistic solutions to the original problem. For example, if x represents the length of the side of a rectangle, then x cannot be negative. However, the equation may have a negative solution. We will use the same technique developed in Chapter 3 to solve the following problems, but will not elaborate as much.

Example 3

a. Find two consecutive positive even integers whose product is 80.

Solution:
Let x = the smaller of the two consecutive even integers. Then,
$x + 2$ = the larger of the two consecutive even integers.
Since *product* means *multiply*:

smaller integer	times	larger integer	equals	80
x	$\cdot$	$(x + 2)$	$=$	80

Answers:

Additional Practice Exercises *a–e:* **a.** $c = 0, 2$ **b.** $a = -7, -2$ **c.** $b = 1, -1$ **d.** $x = -\dfrac{1}{3}, \dfrac{3}{2}$ **e.** $x = 6, -8$

Practice Exercises *4–8:* **4.** $r = 0, 3$ **5.** $z = -3, 7$ **6.** $m = 6, -6$ **7.** $x = -\dfrac{5}{7}, \dfrac{2}{7}$ **8.** $x = 5, -8$

$$x(x + 2) = 80 \qquad \text{Simplify the left side of the equation.}$$
$$x^2 + 2x = 80 \qquad \text{Subtract 80 from both sides.}$$
$$x^2 + 2x - 80 = 0 \qquad \text{Factor.}$$
$$(x + 10)(x - 8) = 0 \qquad \text{Set each factor equal to 0.}$$
$$x + 10 = 0, x - 8 = 0 \qquad \text{Solve each equation.}$$
$$x = -10 \qquad x = 8 \qquad \text{Possible solutions are } -10 \text{ and } 8.$$

Since we were asked for consecutive *positive* even integers, we discard -10. If 8 is the smaller of the two consecutive positive even integers, then $x + 2 = 10$ is the larger. Consequently, the integers are 8 and 10.

Check:
Are 8 and 10 consecutive positive even integers? Yes. Is the product of 8 and 10 equal to 80? Yes. Then 8 and 10 are the correct solutions.

b. The length of a rectangular screen is 3 feet less than twice the width. Find the dimensions of the screen if the area is 54 square feet.

Solution:
Let x = the width. Then, $2x - 3$ = the length. (The length is 3 less than twice the width.)

The formula for the area of a rectangle is $A = LW$.

$$A = LW \qquad \text{Substitute for } A, L, \text{ and } W.$$
$$54 = (2x - 3)(x) \qquad \text{Simplify the right side of the equation.}$$
$$54 = 2x^2 - 3x \qquad \text{Subtract 54 from both sides.}$$
$$0 = 2x^2 - 3x - 54 \qquad \text{Factor.}$$
$$0 = (2x + 9)(x - 6) \qquad \text{Set each factor equal to 0.}$$
$$2x + 9 = 0, x - 6 = 0 \qquad \text{Solve each equation.}$$
$$2x = -9, \qquad x = 6$$
$$x = -\frac{9}{2}$$

Since the width of a rectangle cannot be negative, we discard $-\frac{9}{2}$. Since we let x represent the width, the width is 6 feet. The length is represented by $2x - 3$, so the length is $2(6) - 3 = 12 - 3 = 9$ feet. The dimensions of the screen are 6 feet wide and 9 feet long.

Check:
Is the length, which is 9, 3 less than twice the width, which is 6? Yes. Is the area 54 square feet? Since $A = LW = 9 \cdot 6 = 54$ square feet, yes. Therefore, our solution is correct.

c. If a ball is thrown vertically upward from ground level with an initial velocity of 80 feet per second, the equation giving its height above the ground is $h = 80t - 16t^2$ where h is the height of the ball in feet and t is the number of seconds after the ball was thrown. a. Find the number of seconds until the ball returns to the ground. b. Find the number of seconds until the ball is 96 feet above the ground.

Solution:
Part a:
When the ball returns to ground level, its height above the ground is 0, so substitute 0 for h and solve for t.

$$0 = 80t - 16t^2 \qquad \text{Remove the common factor of } 16t.$$
$$0 = 16t(5 - t) \qquad \text{Set each factor equal to 0.}$$
$$16t = 0, 5 - t = 0 \qquad \text{Solve each equation.}$$
$$t = 0, \qquad 5 = t \qquad \begin{array}{l} t = 0 \text{ indicates the ball was at ground} \\ \text{level when it was thrown. Therefore,} \\ \text{the ball returns to ground level after 5} \\ \text{seconds.} \end{array}$$

Check:
After 5 seconds, $h = 80(5) - 16(5)^2 = 80(5) - 16(25) = 400 - 400 = 0$, which indicates the ball is on the ground.

Part b:
If the ball is 96 feet above the ground, $h = 96$, so substitute 96 for h and solve for t.

$$96 = 80t - 16t^2 \qquad \begin{array}{l} \text{Add } 16t^2 \text{ and } -80t \text{ to both sides of} \\ \text{the equation.} \end{array}$$
$$16t^2 - 80t + 96 = 0 \qquad \text{Remove the common factor of 16.}$$
$$16(t^2 - 5t + 6) = 0 \qquad \text{Factor } t^2 - 5t + 6.$$
$$16(t - 2)(t - 3) = 0 \qquad \text{Set } t - 2 \text{ and } t - 3 \text{ equal to 0. } (16 \neq 0)$$
$$t - 2 = 0, t - 3 = 0 \qquad \text{Solve each equation.}$$
$$t = 2, \qquad t = 3 \qquad \begin{array}{l} \text{Therefore, the ball is 96 feet above the} \\ \text{ground after 2 seconds and again af-} \\ \text{ter 3 seconds.} \end{array}$$

Check: The check is left as an exercise for the student.

d. The formula for finding the sum of n consecutive positive integers beginning with one is $S = \frac{n(n+1)}{2}$ where S is the sum and n is the number of positive integers. For example, in the sum $1 + 2 + 3 + 4 + 5 + 6 + 7 + 8 + 9 + 10$, $n = 10$, since we are summing the first 10 positive integers. The sum is $S = \frac{(10)(10+1)}{2} = \frac{(10)(11)}{2} = 55$. How many integers would it take for the sum to be 36?

Solution:

Since the sum is 36, substitute 36 for S and solve for n.

$36 = \dfrac{n(n+1)}{2}$ Multiply both sides of the equation by 2.

$2(36) = 2 \cdot \dfrac{n(n+1)}{2}$ Simplify both sides.

$72 = n^2 + n$ Subtract 72 from both sides.

$0 = n^2 + n - 72$ Factor.

$0 = (n + 9)(n - 8)$ Set each factor equal to 0.

$n + 9 = 0, n - 8 = 0$ Solve each equation.

$n = -9, \qquad n = 8$ Since we are looking for positive integers, 8 is the only possible answer.

Check:

Substitute 8 for n and find S.

$S = \frac{8(8+1)}{2} = \frac{8(9)}{2} = \frac{72}{2} = 36$, which is what it is supposed to be. Therefore, our answer is correct.

e. A business executive of a large corporation has to visit every office in her region (except her home office) twice per year. If V represents the total number of visits and n is the number of offices, then $V = n(n - 1)$. How many offices are in her region if she makes 42 visits per year?

Since she makes 42 visits per year, substitute 42 for V and solve for n.

$42 = n(n - 1)$ Simplify the right side of the equation.

$42 = n^2 - n$ Subtract 42 from both sides.

$0 = n^2 - n - 42$ Factor.

$0 = (n - 7)(n + 6)$ Set each factor equal to 0.

$n - 7 = 0, n + 6 = 0$ Solve each equation.

$n = 7, \qquad n = -6$ Since there cannot be a negative number of offices, there are 7 offices in her region.

Check: The check is left as an exercise for the student.

Practice Exercises

Solve each of the following:

9. Find two consecutive integers so that twice the square of the larger is 57 more than three times the smaller.

10. The length of a rectangular rug is 5 feet less than twice the width. If the area of the rug is 88 square feet, find the dimensions of the rug.

11. If an object is launched vertically upward from ground level with an initial velocity of 128 feet per second, the height of the object above the ground is given by $h = 128t - 16t^2$ where h is the height of the object and t is the number of seconds after it is launched. **a.** Find how long it will take the object to return to the ground. **b.** After how many seconds will the object be 240 feet above the ground?

12. Using the formula from Example 3d, find how many positive integers beginning with 1 it would take for the sum to be 45.

13. Using the formula from Example 3e, find how many offices are in a region if the executive has to make 90 visits per year.

Equations of the form $ax^2 + bx + c = 0$ are called quadratic equations. Functions of the form $f(x) = ax^2 + bx + c$, accordingly, are called **quadratic functions.** Quadratic functions are evaluated in the same manner as any other functions.

Answers:

Example 4

Given f(x) = x² + x − 6, *find the following:*

a. $f(0)$

Solution:
To find $f(0)$, substitute 0 for x and simplify.

$f(x) = x^2 + x - 6$ Substitute 0 for x.
$f(0) = 0^2 + 0 - 6$ Simplify.
$f(0) = -6$ Therefore $f(0) = -6$. Remember, this means the point $(0,-6)$ lies on the graph of $y = x^2 + x - 6$.

b. $f(-3)$

Solution:
To find $f(-3)$, substitute -3 for x and simplify.

$f(x) = x^2 + x - 6$ Substitute -3 for x.
$f(-3) = (-3)^2 + (-3) - 6$ Raise to powers.
$f(-3) = 9 - 3 - 6$ Add.
$f(-3) = 0$ Therefore, $f(-3) = 0$. This means the point $(-3,0)$ lies on the graph of $y = x^2 + x - 6$.

c. $f(4)$

Solution:
To find $f(4)$, substitute 4 for x and simplify.

$f(x) = x^2 + x - 6$ Substitute 4 for x.
$f(4) = 4^2 + 4 - 6$ Raise to powers.
$f(4) = 16 + 4 - 6$ Add.
$f(4) = 14$ Therefore, $f(4) = 14$, which means the point $(4,14)$ lies on the graph of $y = x^2 + x - 6$.

Practice Exercises

Given f(x) = x² + 2x − 8, *find the following:*

14. $f(0)$ **15.** $f(-4)$ **16.** $f(3)$

Exercise Set 5.8

Solve the following using the zero product property:

1. $(t - 6)(t + 4) = 0$ **2.** $(u + 3)(u - 8) = 0$ **3.** $(2v - 10)(v - 9) = 0$

4. $(w + 2)(5w + 15) = 0$ **5.** $r(5r - 2) = 0$ **6.** $s(3s - 5) = 0$

7. $6z(7z + 11) = 0$ **8.** $13p(8p - 15) = 0$ **9.** $(a - 3)(a + 2)(a - 5) = 0$

10. $(b + 4)(b - 6)(b - 1) = 0$ **11.** $(q + 7)(3q - 5)(4q - 1) = 0$ **12.** $(2x - 9)(6x + 2)(13x - 5) = 0$

13. $(2y - 4)^2 = 0$ **14.** $(5z + 3)^2 = 0$

Solve the following:

15. $x^2 - x - 6 = 0$ **16.** $2d^2 + 7d - 4 = 0$ **17.** $4m^2 - 23m + 15 = 0$

18. $n^2 + 10n + 24 = 0$ **19.** $5s^2 - 10s = 0$ **20.** $21k - 7k^2 = 0$

21. $3r^2 - 2r = 0$ **22.** $4h - 9h^2 = 0$ **23.** $v^2 - 4 = 0$

24. $9 - u^2 = 0$ **25.** $3w^2 - 75 = 0$ **26.** $72 - 2x^2 = 0$

27. $t^2 + 12 = 7t$

28. $a^2 - 45 = 4a$

29. $3y^2 - 8 = 10y$

30. $24b^2 - 15 = -2b$

31. $c(c - 12) = -36$

32. $d(d + 10) = -25$

33. $4y(y + 5) = -25$

34. $3x(3x - 8) = -16$

Solve each of the following:

35. Find two consecutive positive even integers whose product is 168.

36. Find two consecutive negative odd integers whose product is 143.

37. Find two consecutive negative even integers such that the sum of their squares is 52.

38. Find two consecutive negative odd integers such that the difference of four times the square of the smaller and twice the square of the larger is 34.

39. If the sum of an integer squared and 12 times the integer is -32, find the integer.

40. If the difference of an integer squared and twice the integer is 63, find the integer.

41. If the difference of five times the square of an integer and eight times the integer is 21, find the integer.

42. If the sum of three times the square of an integer and 13 times the integer is 30, find the integer.

43. An American flag is 4 feet longer than it is wide. Find the dimensions of the flag if the area is 32 square feet. (Recall that $A = LW$.)

44. A rectangular patio is 5 yards shorter than it is long. Find the dimensions of the patio if the area of the patio is 50 square yards. (Recall that $A = LW$.)

45. The length of the cover of a textbook is 3 inches more than its width. Find the dimensions of the cover of the book if the area of the cover is 88 square inches. (Recall that $A = LW$.)

46. The side of a rectangular box is such that the length is 7 centimeters less than twice the width. Find the dimensions of the side of the box if its area is 130 square centimeters. (Recall that $A = LW$.)

47. A search party is looking for a lost person known to be in a triangular region bounded by three highways. If one of the highways is used as the base of the region, the height to that base is 4 miles less than the base. If the area of the region is 6 square miles, find the base and height. (Recall that $A = \frac{bh}{2}$.)

48. The height of a triangular piece of plywood is two feet more than the base. If the area is 24 square feet, find the base and the height. (Recall that $A = \frac{bh}{2}$.)

49. A projectile is launched vertically upward from ground level with an initial velocity of 112 feet per second. The equation giving its height above the ground is $h = 112t - 16t^2$ where h is the height in feet and t is the number of seconds after it was launched. **a.** Find the number of seconds until the projectile returns to the ground. **b.** Find the number of seconds until the projectile is 192 feet above the ground.

50. A rock is thrown vertically upward from ground level with an initial velocity of 96 feet per second. The equation giving its height above the ground is $h = 96t - 16t^2$ where h is the height in feet and t is the number of seconds after it was thrown. **a.** Find the number of seconds until the rock returns to the ground. **b.** Find the number of seconds until the rock is 80 feet above the ground.

51. If a heavy object is dropped from the top of a building that is 256 feet high, the equation giving the height of the object is $h = 256 - 16t^2$ where h is the height above the ground and t is the number of seconds after the object was dropped. How long will it take the object to reach the ground?

52. If a heavy object is dropped from the top of a cliff that is 144 feet high, the equation giving the height of the object is $h = 144 - 16t^2$ where h is the height above the ground and t is the number of seconds after the object was dropped. How long will it take the object to reach the ground?

In Exercises 53–56, use the formula $S = \frac{n(n + 1)}{2}$. *See Example 3d.*

53. How many positive integers, beginning with 1, will it take for the sum to be 28?

54. How many positive integers, beginning with 1, will it take for the sum to be 45?

55. Logs are stacked in a triangular-shaped pile with one log on the top row, two logs on the second row, three logs on the third row, and so on. If there is a total of 21 logs in the stack, how many logs are on the bottom row?

56. Fence posts in a lumber yard are stacked in a triangular-shaped pile with one post on the top row, two posts on the second row, three posts on the third row, and so on. If there is a total of 78 posts in the pile, find the number of posts on the bottom row.

In Exercises 57–58, use the formula V = n(n − 1). See Example 3e.

57. How many offices are in the region if the executive makes 72 visits per year?

58. How many offices are in the region if the executive makes 56 visits per year?

Answer the following:

59. Given $f(x) = x^2 + 2x - 8$, find the following:

    a. $f(0)$ **b.** $f(2)$ **c.** $f(-3)$

60. Given $f(x) = x^2 - 9x + 18$, find the following:

    a. $f(0)$ **b.** $f(6)$ **c.** $f(-3)$

Challenge Exercises:

Solve each of the following:

61. $x^4 - 13x^3 + 36x^2 = 0$

62. $3a^3 - 147a = 0$

63. $5t^4 + 30t^3 = 0$

64. $32x^3 - 144x^2 + 162x = 0$

Writing Exercises:

65. Compare the methods of solving linear equations and quadratic equations.

66. When solving a quadratic equation, why must one side of the equation be equal to 0?

67. Suppose we are solving a quadratic equation by factoring and we arrive at $4(x - 2)(x + 5) = 0$. In order to solve this equation, we set the factors $x - 2$ and $x + 5 = 0$. Why do we not also set 4 equal to 0? What happens to the 4?

Writing Exercise or Group Project:

If done as a group project, each group should write two exercises of each type, exchange with another group, and then solve them.

68. Write and solve an applications problem involving consecutive integers.

69. Write and solve an applications problem involving the area of a rectangular-shaped region.

70. Given $f(x) = x^2 - 2x - 3$. (Remember, $y = f(x)$.)

    a. Let $f(x) = 0$ and solve the resulting equation for x. That is, solve $0 = x^2 - 2x - 3$.

    b. Following is the graph of $f(x) = x^2 - 2x - 3$. What are the x-intercepts?

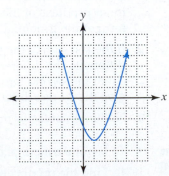

    c. What is the relationship between the solutions of the equation $0 = x^2 - 2x - 3$ found in part a to the x-intercepts found in part b? Why?

Chapter 5   Summary

Definitions: [Section 5.1]

- **Factor:** The number a is a factor of the number b if b can be written as the product of a and some other number.

- **Divisible:** The number b is divisible by the number a if a divides evenly into b. That is, if a is divided into b, the remainder is 0.

- **Prime:** A natural number is prime if it is greater than one and its only factors are one and itself. Equivalently, a number is prime if it is greater than one and divisible only by one and itself.

- **Composite:** A natural number, other than one, is composite if it is not prime. Equivalently, a number is composite if it has at least one factor other than one and itself.

Prime Factorization: [Section 5.1]

- A number is written in prime factorization form if it is written as the product of its prime factors.

Greatest Common Factor: [Section 5.1]

- The greatest common factor of two or more numbers is the largest number that is a factor of all the numbers.

- The greatest common factor of two or more monomials (other than a number) is the monomial of greatest degree that is a factor of each of the monomials.

Finding the Greatest Common Factor: [Section 5.1]

1. Write each number or monomial as the product of its prime factors.
2. Select those factors that are common in each prime factorization.
3. For each factor that is common, select the smallest exponent that occurs in any of the prime factorizations.
4. The greatest common factor is the product of each factor that is common with the smallest power in any of the prime factorizations.

Removing the Greatest Common Factor: [Section 5.2]

- Determine the greatest common factor of the terms of the polynomial. Use the distributive property to write the polynomial as the product of the greatest common factor and another polynomial. The terms of the polynomial within the parentheses may be determined either by inspection or by division.

Factoring by Grouping: [Section 5.2]

- If the polynomial has more than three terms, factor by grouping. The goal is to get the same binomial as a common factor and then remove it.

Factoring a General Trinomial: [Sections 5.3 and 5.4]

1. If the last sign of the trinomial is positive, both signs of the binomial factors will be the same as the sign of the middle term of the trinomial. The sum of the outer and inner products of the binomials (ignoring the signs) will be equal to the second term of the trinomial (ignoring the sign).

2. If the last sign of the trinomial is negative, the signs of the binomial factors will be different. The difference of the outer and inner products (ignoring the signs) will equal the second term of the trinomial (ignoring the sign). Find the terms which give the correct difference, then determine the signs.

3. No binomial factor may contain a common factor.

Factoring a Trinomial Using the "ac" Method [Section 5.4]

- To factor a trinomial of the form $ax^2 + bx + c$:
  1. Find the product of a and c.
  2. Find two factors of ac whose sum is b.
  3. Rewrite bx as the sum of two terms whose coefficients are the numbers found in step 2.
  4. Factor the resulting polynomial by grouping.

Difference of Squares: [Section 5.5]

- A binomial that is the difference of squares factors into the sum and difference of the expressions being squared. $a^2 - b^2 = (a + b)(a - b)$

Factoring the Sum or Difference of Cubes: [Section 5.5: optional]

- The sum or difference of cubes factors into a binomial and a trinomial. The terms of the binomial are the expressions whose cubes give the sum or difference of cubes. The sign of the binomial is the same as the sign of the sum or difference of cubes. The first term of the trinomial is the square of the first term of the binomial. The second term of the trinomial is the negative of the product of the terms of the binomial. The third term of the trinomial is the square of the second term of the binomial.

  $a^3 + b^3 = (a + b)(a^2 - ab + b^2)$ and $a^3 - b^3 = (a - b)(a^2 + ab + b^2)$

Perfect Square Trinomial: [Section 5.6]

- A trinomial is a perfect square trinomial if the first and last terms are squares of expressions and the middle term is two times the product of the expressions being squared. The binomial whose square gives the perfect square trinomial has a first term whose square gives the first term of the trinomial, has a second term whose square gives the last term of the trinomial, and has the same middle sign as the second sign of the trinomial.

  $a^2 + 2ab + b^2 = (a + b)^2$ and $a^2 - 2ab + b^2 = (a - b)^2$

Factoring a Polynomial: [Section 5.7]

A. First, remove all common factors.

B. How many terms does the polynomial have?

  1. If the polynomial has two terms:

    a. Is it the difference of squares? If so, factor using the form $a^2 - b^2 = (a + b)(a - b)$.

    b. Is it the sum or difference of cubes? If so, factor using the forms $a^3 + b^3 = (a + b)(a^2 - ab + b^2)$ or $a^3 - b^3 = (a - b)(a^2 + ab + b^2)$.

  2. If the polynomial has three terms:

    a. Is it a perfect square trinomial? If so, factor using the forms $a^2 + 2ab + b^2 = (a + b)^2$ or $a^2 - 2ab + b^2 = (a - b)^2$.

    b. If the trinomial is not a perfect square, factor using trial and error. Check the factorization using FOIL.

  3. If the polynomial has more than three terms, factor by grouping.

C. The factorization is not complete unless each factor is prime. Check each factor, and if necessary, factor any that are not prime until all factors are prime.

Solving Quadratic Equations: [Section 5.8]

1. The zero product property: If $ab = 0$, then $a = 0$, $b = 0$, or $a = b = 0$.

2. Using the zero product property to solve quadratic equations.

    a. If necessary, rewrite the equation in the form $ax^2 + bx + c = 0$.

    b. Factor $ax^2 + bx + c$.

    c. Set each of the factors equal to 0.

    d. Solve the resulting equations.

    e. Check the solutions.

3. Solving application problems involving quadratic equations.

    a. Read the problem and assign a variable to an unknown.

    b. Represent any other unknown in terms of the variable.

    c. Write the equation using information from the problem or other known facts.

    d. Solve the equation.

    e. See if all solutions of the equation make sense when compared with the original problem.

    f. Check all solutions when compared with the original wording of the problem.

Chapter 5 Review Exercises

Classify each of the following as prime or composite:
[Section 5.1]

1. 87

2. 43

Determine the prime factorization of each of the following:
[Section 5.1]

3. 60

4. 308

Find the greatest common factor of each of the following:
[Section 5.1]

5. 126 and 75

6. 54 and 140

7. 28, 112, and 294

8. 48, 42, and 180

9. m^2n^4 and m^3n

10. a^3b^4 and b^2c^5

11. $18r^5s^3t^2$ and $24r^2s^2t^3$

12. $15a^3b^2$, $10ab^3$, and $20a^3b^2c$

Factor the following by removing the greatest common factor: [Section 5.2]

13. $ax + ay$

14. $6x^2 + 3$

15. $18x^2y - 24xy$

16. $r^4s^3 + r^6s^4$

17. $18x^2 - 24x + 12$

18. $8a^2b^4 - 12a^3b^5 + 18ab^3$

19. $r(s - 2) - 5(s - 2)$

20. $m(n - p) - q(n - p)$

21. Factor $-15x^3 + 5x^2 - 25x$ by removing a negative common factor.

Completely factor the following by grouping: [Section 5.2]

22. $ax + bx + ay + by$

23. $rs - 8r + 3s - 24$

24. $ab - 6a - 3b + 18$

25. $xy - 7y - 4x + 28$

26. $3ab - 3a - b + 1$

27. $6x^2 + 15x - 4xy - 10y$

Completely factor the following trinomials with leading co-efficients of one: [Section 5.3]

28. $x^2 + 12x + 11$

29. $x^2 - 12x + 35$

30. $x^2 + 6x - 7$

31. $m^2 - 3m - 10$

32. $x^2 - 11x + 24$

33. $r^2 + rs - 72s^2$

34. $a^2 + 25ab - 54b^2$

35. $m^2 - 19mn - 42n^2$

36. $2x^2 + 24x + 70$

37. $y^3 + 7y^2z - 44yz^2$

Completely factor the following trinomials with leading co-efficient other than one: [Section 5.4]

38. $5x^2 - 7x + 2$

39. $3z^2 - 2z - 5$

40. $5x^2 + 18x - 8$

41. $3a^2 + 26a + 16$

42. $8a^2 + 13ab - 7b^2$

43. $18a^2 + a - 5$

44. $35x^2 + 29xy + 6y^2$

45. $20a^2 + 9ab - 18b^2$

46. $20a^2 - 66a - 14$

47. $36c^3 - 51c^2d + 18cd^2$

Completely factor the following differences of squares:
[Section 5.5]

48. $g^2 - h^2$

49. $r^2 - 121$

50. $25x^2 - 16$

51. $49a^2 - 64b^2$

52. $6x^2 - 24$

53. $64c^2 - 16d^2$

54. $(m + n)^2 - 81$

55. $64 - (c - d)^2$

56. $r^2 + s^2$

57. $m^4 - n^4$

Completely factor the following sums or differences of cubes: [Section 5.5: optional]

58. $m^3 + n^3$

59. $r^3 - 1$

60. $a^3 + 125$

61. $64x^3 + 27y^3$

Completely factor the following perfect square trinomials: [Section 5.6]

62. $r^2 + 18r + 81$

63. $x^2 - 24x + 144$

64. $x^2 + 8x + 64$

65. $49a^2 - 14a + 1$

66. $49a^2 + 28ab + 4b^2$

67. $25c^2 + 30cd + 9d^2$

68. $18p^2 + 24p + 32$

69. $4x^2 - 40x + 100$

Completely factor the following polynomials that involve all types of factoring: [Section 5.7]

70. $26a^2 + 13a$

71. $81x^2 - y^2$

72. $2x^2 - 6x - 80$

73. $x^2 - 2x - 63$

74. $22a^2b - 11a^3b^2 + 33a^2b$

75. $9m^2 - 48m + 48$

76. $3xy - 27y - 4x + 36$

77. $4(a + b)^2 - 49$

78. $8x^2 + 28x - 56$

79. $49a^2 - 28ab + 4b^2$

80. $6x^2 + 21x - 45$

81. $6x(y - 5) + 2(y - 5)$

82. $27x^2y - 75y$

83. $121a^2 - 169$

84. $81x^2 - 72xy + 16y^2$

85. $9x^2 + 81$

86. $8x^2 + 24x - 4xy - 12y$

87. $r^2 + 6r - 40$

88. $x^4 + 81$

89. $y^4 - 625$

90. $3x^3y^2 - 12x^2y^3 - 135xy^4$

91. $8x^2 - 15x + 7$

Solve the following equations: [Section 5.8]

92. $x^2 + 5x = 0$

93. $x^2 = 6x$

94. $4x^2 - 81 = 0$

95. $9x^2 = 64$

96. $x^2 - 2x - 35 = 0$

97. $y^2 - 4y - 21 = 0$

98. $2x^2 - 4 = 7x$

99. $3x^2 - 8 = -10x$

Solve the following application problems: [Section 5.8]

100. The length of a rectangle is 5 less than twice the width. If the area of the rectangle is 25 square centimeters, find the dimensions of the rectangle.

101. Find two consecutive odd integers whose product is 63.

102. If 2 is added to the length of each side of a square, the area is increased by 28 square yards. Find the length of a side of the original square.

103. The sum of the square of an integer and three times the integer is equal to 28. Find the integer(s).

Chapter 5   Test

1. Classify each of the following as prime or composite:

  a. 111   **b.** 73

2. Find the prime factorization of 252.

Find the greatest common factor for each of the following:

  3. 72 and 96   **4.** $48x^3y^5$ and $60x^2yz^3$

Factor each of the following. If the polynomial cannot be factored, write **prime**.

  5. $x^3 + x^2$ **6.** $m^2 - 18m + 81$

  7. $m^2 - 3mn - 40n^2$ **8.** $49a^2 - 9b^2$

  9. $14x^3y - 28x^2y^2 + 35x^4y^3$ **10.** $9y^2 - 24y + 16$

  11. $16a^2 + 25b^2$ **12.** $r(s + 5) - 7(s + 5)$

13. $(x - y)^2 - 100$

14. $3q^2 + 6q - 105$

15. $3a^2 - 13a - 30$

16. $8a^2 - 34ab + 35b^2$

17. $27x^2 - 12y^2$

18. $6ab - 4a + 15b - 10$

19. $5a^2 + 50ab + 125b^2$

20. $x^4 - 625$

21. $20a^2 - 7ab - 6b^2$

22. The area of a rectangle is represented by $x^2 - 2x - 24$. How may the length and width be represented as binomials?

Solve the following:

23. $6x^2 + 4x = 0$

24. $6x^2 - 19x + 10 = 0$

25. The length of a rectangle is 4 less than three times the width. Find the dimensions of the rectangle if the area is 32 square inches.

This chapter and Chapter 7 show the closest relationship between arithmetic and algebra of any chapters in this book. Polynomials are to elementary algebra as natural numbers are to arithmetic. In arithmetic, we learn to add, subtract, multiply, and divide natural numbers. Thus far we have learned to add, subtract, and multiply polynomials. In this chapter, we will learn to divide them as well. The next step in arithmetic is to make fractions from the natural numbers and learn to add, subtract, multiply, and divide these fractions. The corresponding step in algebra is to make rational expressions from polynomials and learn to add, subtract, multiply, and divide these rational expressions. This is what we will be doing in Chapters 6 and 7.

Multiplication and Division of Rational Numbers and Expressions

Chapter

6

Section 6.1	Fractions and Rational Numbers

OBJECTIVES

When you complete this section, you will be able to:

a. Write a fraction that represents a given part of a whole (unit).

b. Identify the part of a whole (unit) that is represented by a fraction.

c. Represent what portion a given group of objects is of a total group of objects using a fraction.

d. Identify the number of objects in a group of objects that is represented by a given fraction.

e. Represent a rational number as division and change rational numbers into decimals.

f. Change an improper fraction into a mixed number and a mixed number into an improper fraction.

g. Graph rational numbers.

Introduction   A brief introduction to fractions was given in Chapter 0. In this section, we are going to expand on the concept of what a fraction is and what it represents. Some of the material is a repeat of the material in Section 0.6, but in an expanded form.

A fraction is a number of the form $\frac{a}{b}$ with $b \neq 0$. The number a is called the numerator and b is called the denominator. If the numerator and denominator of a fraction are both integers and the denominator is not 0, or if a number can be put into that form, the fraction is called a **rational number.**

Rational Number

A number is rational if it can be written in the form $\frac{a}{b}$ with a and b integers and $b \neq 0$.

There is a distinction between a fraction and a rational number. The expression $\frac{\pi}{3}$ is a fraction, but it is not a rational number, since π is not an integer. Natural numbers, whole numbers, and integers are also rational numbers, since any of these can be written as a fraction with the denominator of 1. For example, the natural number $5 = \frac{5}{1}$ and is therefore, a rational number.

Writing a fraction that represents a given part of the whole   If an object is divided into a number of equal parts and we select some of those parts, we can represent the part of the whole object (unit) selected by using a fraction. The numerator represents the number of equal parts selected and the denominator represents the number of equal parts into which the whole object (unit) has been divided.

Example 1

Using a fraction, represent the part of the whole region that is shaded.

a.

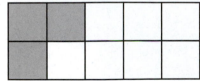

1 unit

b.

1 unit

Three of the ten equal parts are shaded. The numerator is the number of equal parts shaded and the denominator is the number of equal parts into which the figure is divided. Therefore, the fraction is written as $\frac{3}{10}$ and is read as "three tenths."

Two of the four equal parts are shaded. Therefore, the numerator is 2 and the denominator is 4. So, the fraction is written as $\frac{2}{4}$ and is read as "two fourths."

Practice Exercises

Using a fraction, represent the part of the whole region (unit) that is shaded and write the name of the fraction in words.

1.

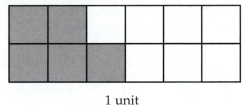

1 unit

2.

1 unit

This concept can also be applied to situations in which the shading of figures is not involved. If a pizza is cut into eight equal pieces and five of these pieces have been eaten, then $\frac{5}{8}$ of the pizza has been eaten.

Example 2

Represent each of the following using a fraction:

a. For a fishing tournament, the lake has been divided into 15 regions of equal size. To prevent overcrowding of any one region, contestants are allowed to fish four of the regions that are determined by a drawing. What part of the lake is each person allowed to fish?

Solution:
Since the lake has been divided into 15 equal regions, the denominator is 15. The number of regions each person is allowed to fish is the numerator, 4. Therefore, the part of the lake each person is allowed to fish is $\frac{4}{15}$.

b. The Ecology Club sponsored a cleanup of a small stream. The stream was divided into 14 sections of equal size. If the ΔΔΔ sorority cleaned three of these sections, what part of the stream did they clean?

Solution:
Since the stream was divided into 14 equal sections, the denominator is 14. The number of sections ΔΔΔ cleaned, 3, is the numerator. Therefore, ΔΔΔ cleaned $\frac{3}{14}$ of the stream.

Practice Exercises

Represent each of the following using a fraction:

3. A developer has divided a piece of land into 50 lots of equal size. Thus far, houses have been built on 27 of these lots. What part of the land has been built on?

4. An algebra class has 11 males and 15 females. Females make up what fractional part of the class?

Identifying the part of the whole that is represented by a fraction   In the previous exercises, we represented part of a whole by a fraction. Now we will identify the part of the whole that is represented by a fraction.

Example 3

Shade the part of the whole region that is represented by the given fraction.

a. $\dfrac{1}{4}$

The denominator tells us to divide the region into four equal parts. The numerator tells us to shade any one of the four smaller regions.

Solution:

b. $\dfrac{5}{8}$

The denominator tells us to divide the circle into eight equal regions and the numerator tells us to shade any five of the regions.

Solution:

Practice Exercises

Shade the part of the whole that is represented by each of the following fractions:

5. $\dfrac{7}{15}$

6. $\dfrac{2}{3}$

The fractions that we have discussed thus far are called **proper fractions.** A fraction is proper if the numerator is less than the denominator A fraction whose numerator is greater than or equal to its denominator is called **improper.** Examples of improper fractions are $\frac{5}{5}, \frac{7}{3}$ or $\frac{11}{8}$. A **mixed number** represents a quantity greater than one and has a whole number part and a fractional part. Examples of mixed numbers are $2\frac{2}{3}, 3\frac{4}{7}$ and $1\frac{3}{5}$. Improper fractions and mixed numbers may also be represented by shaded regions as the following illustrates.

Example 4

Represent each of the following by an improper fraction and a mixed number:

a.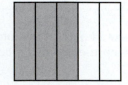

Solution:
Each figure is divided into five equal parts. Since one whole figure is shaded and three of the five equal parts of the other figure are shaded, we represent this with the mixed number $1\frac{3}{5}$. Since each figure is divided into five equal parts and eight of these regions are shaded, we can represent this with the improper fraction $\frac{8}{5}$.

Answers:

b.

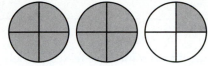

c.

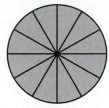

Solution:
Two complete circles and one of the four equal sections of the third circle are shaded. This is represented by the mixed number $2\frac{1}{4}$. Since each figure is divided into four equal parts and 9 equal parts are shaded, this is represented by the improper fraction $\frac{9}{4}$.

This figure is divided into eight equal regions and all eight are shaded. Therefore, this represents the improper fraction $\frac{8}{8}$.

This example cannot be represented by a mixed number, since its value is not greater than one.

Practice Exercises

Represent each of the following as an improper fraction and as a mixed number, if possible:

7.

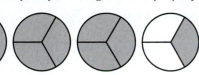

8.

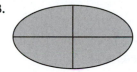

Using a fraction to represent part of a group of objects   Another way a fraction may be used is to represent a part of a group of objects. In this case, the denominator represents the total number of objects in the group and the numerator represents the number of objects selected from the group.

Example 5

Write a fraction that represents the part of the group of figures that is shaded.

a.

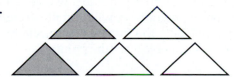

There are five figures in the group, so the denominator is 5. Two of the figures are shaded, so the numerator is 2. Therefore, the fraction that represents the part of the total number of figures that are shaded is $\frac{2}{5}$.

b.

There are ten figures in the group, so the denominator is 10. Six of the figures are shaded, so the numerator is 6. Therefore, the fraction that represents the part of the total number of figures that are shaded is $\frac{6}{10}$.

Practice Exercises

Write a fraction that represents the part of the figures that are shaded.

9.

10.

As before, this situation can be illustrated without using shaded figures. For example, if Maria has $13 dollars and loans Harold $10, then she has loaned Harold $\frac{10}{13}$ of her money. If there are ten computer disks in a box and Michelle uses three full boxes and seven disks from a fourth box, then she has used $3\frac{7}{10}$ or $\frac{37}{10}$ boxes of disks.

Answers:

Example 6

Represent each of the following by a fraction:

a. To build a workbench, Tom bought 50 nails and he used 43. What part of the total number of nails purchased did he use?

Solution:
The denominator is the total number of nails purchased, 50. The numerator is the number he used, 43. Therefore, Tom used $\frac{43}{50}$ of the nails he purchased.

b. For target practice, Bryan took three boxes of shells that contained 25 shells each. He shot two full boxes and 18 shells from the third box. Of the total number of boxes that he took, what part did he shoot?

Solution:
As a mixed number, we would represent the number of boxes that he shot as $2\frac{18}{25}$. The two represents the two full boxes and the $\frac{18}{25}$ represents the fractional part of the third box that he shot. To represent this as an improper fraction, we think of each box as being divided into 25 equal parts. The total number of shells that he shot is $25 + 25 + 18 = 68$. Therefore, the part of the boxes that he shot is $\frac{68}{25}$.

Practice Exercises

Represent each of the following by a fraction:

11. In order to carpet the living room, Manuel purchased 130 square yards of carpet. He actually used only 123 square feet. What part of the carpet purchased did he use?

12. Picture hangers come in boxes containing ten in each box. An art gallery recently used five full boxes and three from another box. Represent the number of boxes used as a mixed number and as an improper fraction.

Identifying the number of objects in a group of objects that is represented by a fraction   We were previously using fractions to represent what part a given number of objects is of a total group of objects. The opposite is to identify the number of objects in a group of objects that is represented by a fraction. As before, we will first illustrate by shading figures. In this case, the numerator represents the number of objects to be shaded and the denominator represents the number of objects in the group.

Example 7

Shade the number of objects represented by each of the following fractions:

a. $\frac{4}{7}$

Solution:
Since there are seven triangles, $\frac{4}{7}$ represents four of the seven. Therefore, we would shade any four of the seven triangular regions.

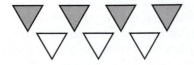

b. $\frac{2}{5}$

Solution:
Since there are five figures, $\frac{2}{5}$ represents two of the five figures. Therefore, we shade two of the five.

Practice Exercises

Shade the number of figures represented by each of the following:

13. $\dfrac{2}{9}$

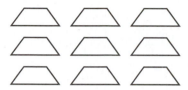

14. $\dfrac{3}{6}$

As in previous cases, we do not have to rely on shaded figures for this interpretation of a fraction as illustrated by the following examples:

Example 8

a. If a club has ten members and $\dfrac{3}{10}$ of the membership are put on a committee, how many people are on the committee?

Solution:
The denominator represents the number of people in the club. The numerator represents the number of people put on the committee. Hence, there are three people on the committee.

b. Bill and Sandy went fishing and caught 23 fish. If Sandy caught $\dfrac{13}{23}$ of the fish, how many did she catch?

Solution:
The denominator, 23, represents the total number of fish caught. The numerator, 13, represents the number caught by Sandy. Therefore, Sandy caught 13 fish.

Practice Exercises

15. A jar contains 35 marbles of which $\dfrac{12}{35}$ are black. How many black marbles are in the jar?

16. Jose spent $54 on groceries. He spent $\dfrac{13}{54}$ of the money for meat. How much did he spend for meat?

Representing rational numbers as division and converting rational numbers into decimals

Any rational number represents division where the numerator is divided by the denominator. For example, $\dfrac{3}{4} = 3 \div 4$. In general, $\dfrac{a}{b} = a \div b$. This interpretation allows us to write a rational number as a decimal and to change an improper fraction to a mixed number. Division of decimals and rounding off were discussed in Chapter 0. You may need to review these procedures before continuing.

Example 9

Change the following rational numbers into decimals. If necessary, round the answer to the nearest hundredth.

a. $\dfrac{3}{4}$

Solution:
$\dfrac{3}{4}$ means $3 \div 4$. Since $4 > 3$, it is necessary to put a decimal point and some zeroes after the 3.

$$
\begin{array}{r}
.75 \\
4\overline{)3.00} \\
-2\,8 \\
\hline
20 \\
-20 \\
\hline
0
\end{array}
$$

Therefore, $\dfrac{3}{4} = .75$

Answers:

b. $\dfrac{9}{16}$

c. $\dfrac{12}{5}$

Solution:

$\frac{9}{16}$ means 9 ÷ 16. As in Example 9a, we will need to put a decimal point after the 9 followed by one or more zeroes.

$$
\begin{array}{r}
.562 = .56 \\
16\overline{)9.000} \\
-8\,0 \\
\hline
1\,00 \\
-96 \\
\hline
40 \\
32 \\
\hline
\end{array}
$$

In order to round off to the nearest hundredth, we need an answer to the nearest thousandth.

Solution:

$\frac{12}{5}$ means 12 ÷ 5. Since 5 will not divide into 12 evenly, put a decimal point after the 12 followed by one or more zeroes.

$$
\begin{array}{r}
2.4 \\
5\overline{)12.00} \\
-10 \\
\hline
2\,0 \\
-2\,0 \\
\hline
0 \\
\end{array}
$$

d. $\dfrac{24}{7}$

Solution:

$\frac{24}{7}$ means 24 ÷ 7. As in Example 1c, place a decimal point and as many zeroes as needed after 24.

$$
\begin{array}{r}
3.428 = 3.43 \\
7\overline{)24.000} \\
-21 \\
\hline
30 \\
-28 \\
\hline
20 \\
-14 \\
\hline
60 \\
-56 \\
\hline
\end{array}
$$

In order to round to the nearest hundredth, we need the answer to the nearest thousandth.

Practice Exercises

Write each of the following as a decimal. If necessary, round off to the nearest hundredth.

17. $\dfrac{3}{8}$ **18.** $\dfrac{12}{17}$ **19.** $\dfrac{15}{6}$ **20.** $\dfrac{29}{9}$

Remember from Chapter 0 that the remainder can be put over the divisor if you want a fractional answer instead of a decimal. If the dividend is larger than the divisor, this procedure results in a mixed number. Therefore, this gives us a procedure for changing improper fractions into mixed numbers.

Example 10

Change the following improper fractions into mixed numbers:

a. $\dfrac{11}{4}$

$$
\begin{array}{r}
2 \\
4\overline{)11} \\
-8 \\
\hline
3 \\
\end{array}
$$

Solution:

$\frac{11}{4}$ means 11 ÷ 4. We will divide 4 into 11 and put the remainder over 4.

3 is the remainder, therefore $\frac{11}{4} = 2\frac{3}{4}$.

b. $\dfrac{69}{5}$

$$\begin{array}{r} 13 \\ 5\overline{)69} \\ -5 \\ \hline 19 \\ -15 \\ \hline 4 \end{array}$$

Solution:

$\dfrac{69}{5}$ means $69 \div 5$. We will divide 5 into 69 and put the remainder over 5.

The remainder is 4. Therefore, $\dfrac{69}{5} = 13\dfrac{4}{5}$.

Practice Exercises

Change each of the following improper fractions into a mixed number:

21. $\dfrac{23}{6}$

22. $\dfrac{30}{13}$

The definition of division provides us with an easy method for converting mixed numbers into improper fractions. In Example 2a, the improper fraction $\dfrac{11}{4}$ was converted into the mixed number $2\dfrac{3}{4}$ by division. The procedure for checking the answer to a division problem is to multiply the divisor, 4, by the quotient, 2, and add the remainder, 3. This results in the dividend ($4 \cdot 2 + 3 = 11$). Note if $\dfrac{11}{4}$ is written as the mixed number $2\dfrac{3}{4}$, the divisor, 4, is the denominator of the fractional part. The quotient, 2, is the whole part of the mixed number, and the remainder, 3, is the numerator of the fractional part. Therefore, to change a mixed number to an improper fraction, we multiply the denominator of the fractional part by the whole part and add the numerator of the fractional part. This result is then written as the numerator and the denominator is the denominator of the fractional part. In symbols, $2\dfrac{3}{4} = \dfrac{4 \cdot 2 + 3}{4} = \dfrac{11}{4}$.

Example 11

Change the following mixed numbers into improper fractions:

a. $4\dfrac{2}{3}$

b. $5\dfrac{7}{8}$

Solution:

Multiply the denominator, 3, by the whole part, 4, and add the numerator, 2. Put the result over the denominator, 3. Therefore, $4\dfrac{2}{3} = \dfrac{3 \cdot 4 + 2}{3} = \dfrac{14}{3}$.

Solution:

Multiply the denominator, 8, by the whole part, 5, and add the numerator, 7. Therefore, $5\dfrac{7}{8} = \dfrac{8 \cdot 5 + 7}{8} = \dfrac{47}{8}$.

Practice Exercises

Change each of the following mixed numbers into improper fractions:

23. $6\dfrac{3}{4}$

24. $11\dfrac{4}{9}$

Using fractions to indicate distance If the units of the number line are divided into equal parts, then we can locate points on the number line whose distance from 0 is represented by a rational number. Remember, this procedure is called graphing and was first discussed in Chapter 1 when we graphed integers.

Answers:

Practice Exercises 21–24: 21. $3\dfrac{5}{6}$ 22. $2\dfrac{4}{13}$ 23. $\dfrac{27}{4}$ 24. $\dfrac{103}{9}$

Example 12

Graph the following rational numbers:

a. $\dfrac{3}{4}$

b. $-\dfrac{5}{2}$

Solution:
The denominator, 4, means divide each unit into four equal parts. Since it is positive, go to the right of 0. The numerator, 3, means put the dot on the third division point.

Solution:
The denominator, 2, means divide each unit into two equal parts. The negative means go to the left of 0, and the numerator, 5, means put the dot on the fifth division point.

If we have several rational numbers to graph whose denominators are different, we will estimate the location of the point corresponding to the rational number.

Example 13

Graph the following rational numbers on the number line:

a. $\left\{-\dfrac{7}{4}, -\dfrac{4}{5}, 0, \dfrac{1}{2}, \dfrac{7}{4}, 2\right\}$

b. $\left\{-\dfrac{10}{3}, -\dfrac{5}{2}, -\dfrac{1}{4}, 1, \dfrac{5}{3}, \dfrac{13}{4}\right\}$

Solution:
Since the denominators are not all the same, we will not be able to divide each unit evenly. Consequently, we will estimate the location of each point.

Solution:
As in Example 5a, we will estimate the location of each point.

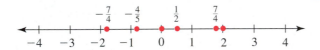

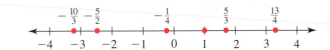

Practice Exercises

Graph the following rational numbers on the number line:

25. $\left\{-\dfrac{14}{4}, -\dfrac{5}{3}, -\dfrac{1}{2}, \dfrac{3}{4}, \dfrac{11}{6}\right\}$

26. $\left\{-\dfrac{11}{3}, -\dfrac{11}{5}, -1, \dfrac{3}{8}, \dfrac{4}{3}, \dfrac{13}{4}\right\}$

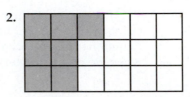

Exercise Set 6.1

Using a fraction, represent the part of the whole region that is shaded.

1.

2.

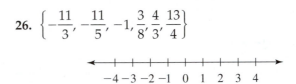

3.

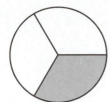

4.

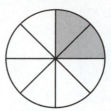

5.

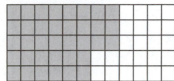

6.

Represent each of the following using a fraction:

7. A deck of cards has 52 cards. In a certain card game, each player receives seven cards. What fractional part of the whole deck does each player receive?

8. A pizza has been cut into 16 pieces of equal size. If Alto eats 3 pieces, what fractional part of the whole pizza has he eaten?

9. In a 10-kilometer race, Aretha has completed 7 kilometers. What fractional part of the race has she completed? What part does she have remaining?

10. A company of soldiers goes on a 20-mile march. After four hours, they have marched nine miles. What fractional part of the total march have they completed? What part do they have remaining?

11. Ray went skeet shooting and shot at 50 clay birds of which he hit 47. What fractional part of the birds did he hit? What part did he miss?

12. Donnie has 75 crab traps and has checked 53 of them. What fractional part of his traps has he checked? What part remains to be checked?

Shade the part of the whole that is represented by the given fraction.

13. $\dfrac{3}{5}$

14. $\dfrac{3}{4}$

15. $\dfrac{3}{8}$

16. $\dfrac{2}{3}$

Represent each of the following by an improper fraction and a mixed number, if possible:

17.

18.

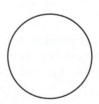

19.

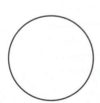

20.

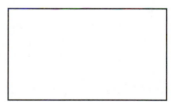

21. 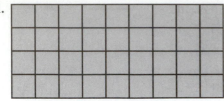   **22.**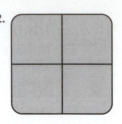

Write a fraction that represents the part of the total number of figures that is shaded.

23. 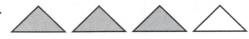   **24.**

25. 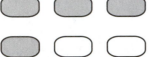   **26.**

27. 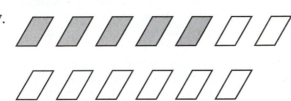   **28.**

Represent each of the following as a fraction:

29. For a fishing trip, John bought a package of plastic worms that contained 20 worms. He used seven of them. What fractional part of the package of worms did he use?

30. Karen received a basket for Easter that contained 25 chocolate eggs. She gave 12 of them to her friends. What fractional part of the eggs did she give away?

31. Armando and Claudia purchased 400 feet of fencing to fence in their yard. They used 357 feet. What part of the fencing purchased did they use?

32. Pat and Allen purchased 50 feet of landscape timbers to build a flower box. They used 47 feet. What portion of the landscape timbers purchased did they use?

33. The Salvation Army set a goal of $5000 for a fund-raising function but raised only $3837. What part of their goal did they achieve?

34. The First Baptist Church hoped to raise $750 for the retirement center but raised only $673. What part of their goal did they achieve?

35. Phi Theta Kappa needed $450 to send their president to the state convention and raised $481. Use a fraction to represent the amount of money they raised in comparison to what they needed.

36. In order to get an NBA franchise, the Orlando Magic needed to sell 10,000 season tickets before their first year of competition. They actually sold 10,562. Use a fraction to represent the number sold in comparison to the number they needed to sell.

37. For a cookout, the Indian Y-Guides purchased ten dozen eggs. They cooked nine dozen eggs and seven single eggs from the remaining dozen. Represent the part of the eggs they cooked as an improper fraction and as a mixed number. Give the answer in terms of dozens.

38. For a party, the Johnsons purchased 12 cartons of soft drinks that contained eight cans each. After the party, they found that nine full cartons plus three cans had been drunk. Represent the part of the soft drinks purchased that had been drunk as an improper fraction and as a mixed number. Give the answer in terms of cartons.

Shade the number of objects represented by each of the following fractions:

39. $\dfrac{6}{8}$ 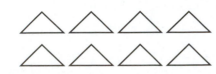   **40.** $\dfrac{4}{9}$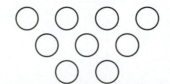

41. $\dfrac{5}{12}$

42. $\dfrac{7}{10}$

43. $\dfrac{6}{8}$

44. $\dfrac{5}{16}$

Answer the following:

45. In an algebra class with 23 students, $\dfrac{10}{23}$ of the class made the grade of C on the Chapter 2 test. How many students made a C on the test?

46. The U.S. Senate has 100 members. On an environmental bill, $\dfrac{79}{100}$ of the Senate voted in favor of the bill. How many senators voted against the bill?

47. There are 13 slices of bacon in a one-pound package. If $\dfrac{6}{13}$ of the package was cooked for breakfast, how many slices were not cooked?

48. There are 128 bricks in a fireplace. If $\dfrac{23}{128}$ of the bricks are red, how many red bricks are in the fireplace?

49. A college campus has 53 buildings. If $\dfrac{12}{53}$ of the buildings are dormitories, how many of the buildings are dormitories?

50. A lot had 11 trees on it. In order to build a house on the lot, $\dfrac{5}{11}$ of the trees had to be cut down. How many trees were cut down?

Change the following fractions into decimals. If necessary, round answers to the nearest hundredth.

51. $\dfrac{1}{4}$

52. $\dfrac{3}{6}$

53. $\dfrac{7}{8}$

54. $\dfrac{5}{8}$

55. $\dfrac{9}{16}$

56. $\dfrac{10}{17}$

57. $\dfrac{2}{3}$

58. $\dfrac{4}{9}$

59. $\dfrac{9}{4}$

60. $\dfrac{13}{4}$

61. $\dfrac{22}{5}$

62. $\dfrac{27}{5}$

63. $\dfrac{33}{25}$

64. $\dfrac{27}{20}$

65. $\dfrac{31}{14}$

66. $\dfrac{32}{15}$

Change the following improper fractions into mixed numbers:

67. $\dfrac{13}{5}$

68. $\dfrac{19}{6}$

69. $\dfrac{36}{7}$

70. $\dfrac{44}{9}$

71. $\dfrac{47}{13}$

72. $\dfrac{57}{22}$

73. $\dfrac{63}{29}$

74. $\dfrac{71}{34}$

Change the following mixed numbers into improper fractions:

75. $4\dfrac{2}{5}$

76. $3\dfrac{3}{4}$

77. $5\dfrac{4}{9}$

78. $7\dfrac{3}{8}$

79. $6\dfrac{5}{12}$

80. $2\dfrac{8}{15}$

81. $12\dfrac{5}{6}$

82. $14\dfrac{7}{8}$

83. $14\dfrac{21}{23}$

84. $18\dfrac{31}{40}$

Graph the following rational numbers. Estimate as needed.

85. $\left\{-2\dfrac{3}{4}, -1\dfrac{1}{2}, -\dfrac{2}{3}, \dfrac{3}{5}, 2\right\}$

<---+---+---+---+---+---+---+---+---+--->
$\quad$ −4 −3 −2 −1 $\;$ 0 $\;$ 1 $\;$ 2 $\;$ 3 $\;$ 4

86. $\left\{-3\dfrac{2}{5}, -3, -2\dfrac{1}{2}, -\dfrac{2}{3}, 3\dfrac{3}{4}\right\}$

<---+---+---+---+---+---+---+---+---+--->
$\quad$ −4 −3 −2 −1 $\;$ 0 $\;$ 1 $\;$ 2 $\;$ 3 $\;$ 4

87. $\left\{-4, -2\dfrac{2}{5}, -\dfrac{3}{4}, 1\dfrac{4}{5}, 3\dfrac{5}{6}\right\}$

<---+---+---+---+---+---+---+---+---+--->
$\quad$ −4 −3 −2 −1 $\;$ 0 $\;$ 1 $\;$ 2 $\;$ 3 $\;$ 4

88. $\left\{-1\dfrac{1}{3}, -1, \dfrac{3}{5}, 2\dfrac{1}{2}, 3\dfrac{1}{4}\right\}$

<---+---+---+---+---+---+---+---+---+--->
$\quad$ −4 −3 −2 −1 $\;$ 0 $\;$ 1 $\;$ 2 $\;$ 3 $\;$ 4

89. $\left\{-4, -3\dfrac{3}{4}, -\dfrac{1}{2}, 2, 3\dfrac{2}{5}\right\}$

<---+---+---+---+---+---+---+---+---+--->
$\quad$ −4 −3 −2 −1 $\;$ 0 $\;$ 1 $\;$ 2 $\;$ 3 $\;$ 4

Writing Assignments:

90. Using the concept of shading, explain why $\dfrac{6}{8} = \dfrac{3}{4}$.

91. In this section, we have given two interpretations of fractions [part of a whole (unit) and part of a given number of objects]. Give at least one more interpretation.

92. Write a word problem involving a fraction that represents a part of a whole.

93. Write a word problem involving a fraction that represents a part of a given number of objects.

94. How do fractions and rational numbers differ?

95. List all the different interpretations of fractions illustrated in this section.

Section 6.2 | **Reducing Rational Numbers and Rational Expressions**

OBJECTIVES   *When you complete this section, you will be able to:*

a. Reduce rational numbers to lowest terms.

b. Simplify rational expressions that consist of the quotients of monomials.

Introduction   In Section 6.1, you probably noticed that some of the shaded regions could have been represented by fractions other than the ones given. In Example 1b of Section 6.1, a circle is divided into four equal regions of which two were shaded. The following circle is shown below for your convenience:

1 unit

$\quad$ The fraction that represents the shaded part of the circle is $\dfrac{2}{4}$. Notice that $\dfrac{2}{4}$ is also $\dfrac{1}{2}$ of the circle. Consequently, $\dfrac{2}{4}$ and $\dfrac{1}{2}$ represent the same part of the circle. Thus, $\dfrac{2}{4} = \dfrac{1}{2}$. Such fractions are said to be **equivalent.**

DEFINITION   Equivalent Fractions

Two or more fractions are equivalent if they represent the same quantity.

Look at the following rectangles where each rectangle represents one unit:

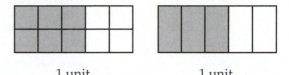

1 unit   1 unit

The fraction representing the shaded part of the first rectangle is $\frac{6}{10}$, and the fraction representing the shaded part of the second rectangle is $\frac{3}{5}$. Since the rectangles are identical in size and the exact same amount is shaded, we can see that $\frac{6}{10} = \frac{3}{5}$. In other words, $\frac{6}{10}$ and $\frac{3}{5}$ are equivalent fractions.

We can change $\frac{6}{10}$ into $\frac{3}{5}$ using the following procedure. Write the numerator and the denominator of   in terms of their prime factors. Then divide the numerator and the denominator by the factors common to both. Remember that any number (except 0) divided by itself equals one and that the product of one and any number is that number. So, $\frac{6}{10} = \frac{2 \cdot 3}{2 \cdot 5} = \frac{\cancel{2} \cdot 3}{\cancel{2} \cdot 5} = \frac{1 \cdot 3}{1 \cdot 5} = \frac{3}{5}$. This procedure is called **reducing a fraction to lowest terms.**

DEFINITION   Lowest Terms

A fraction is reduced to lowest terms if the numerator and the denominator have no common factors other than one.

Reducing fractions to lowest terms

The procedure for reducing a fraction to lowest terms was given in the previous discussion and is summarized as follows:

Reducing a Fraction to Lowest Terms

1. Write the numerator and the denominator as the product of their prime factors. (It is best not to use exponents in doing so.)

2. Divide the numerator and the denominator by the factors common to both.

3. Multiply any remaining factors.

The same procedure applies whether the fraction is proper or improper. We illustrate this procedure with some examples.

Example 1

Reduce the following fractions to lowest terms:

a. $\dfrac{8}{12} =$ Write the numerator and the denominator as the product of prime factors.

$\dfrac{2 \cdot 2 \cdot 2}{2 \cdot 2 \cdot 3} =$ Divide the numerator and the denominator by the factors that are common.

$\dfrac{\cancel{2} \cdot \cancel{2} \cdot 2}{\cancel{2} \cdot \cancel{2} \cdot 3} =$ Each division leaves a quotient of 1 that is not written since the product of 1 and any number is that number.

$\dfrac{2}{3}$ Therefore, $\frac{8}{12} = \frac{2}{3}$.

Note: Some texts refer to this procedure as "canceling." The authors of this text prefer not to use the term *canceling* since it does not adequately convey the fact that we can divide only by common factors. Using canceling could lead to serious problems in the next section when reducing rational expressions, since parts of terms are often mistakenly "canceled" with parts of other terms.

b. $\dfrac{21}{126} =$ Write the numerator and the denominator as the product of prime factors.

$\dfrac{3 \cdot 7}{2 \cdot 3 \cdot 3 \cdot 7} =$ Divide the numerator and the denominator by the factors that are common.

$\dfrac{\cancel{3} \cdot \cancel{7}}{2 \cdot 3 \cdot \cancel{3} \cdot \cancel{7}} =$ We divided by all the factors that were in the numerator. Each time we divide, there is a quotient of 1 left. In this case, we must write the 1 since it is the only factor left in the numerator.

$\dfrac{1}{6}$ Therefore, $\dfrac{21}{126} = \dfrac{1}{6}$.

c. $\dfrac{60}{198} =$ Write the numerator and the denominator as the product of prime factors.

$\dfrac{2 \cdot 2 \cdot 3 \cdot 5}{2 \cdot 3 \cdot 3 \cdot 11} =$ Divide the numerator and the denominator by the factors that are common.

$\dfrac{\cancel{2} \cdot 2 \cdot \cancel{3} \cdot 5}{\cancel{2} \cdot \cancel{3} \cdot 3 \cdot 11} =$ Multiply the remaining factors.

$\dfrac{10}{33}$ Therefore, $\dfrac{60}{198} = \dfrac{10}{33}$.

d. $\dfrac{54}{45} =$ Write the numerator and the denominator as the product of prime factors.

$\dfrac{2 \cdot 3 \cdot 3 \cdot 3}{3 \cdot 3 \cdot 5} =$ Divide the numerator and the denominator by the factors that are common.

$\dfrac{2 \cdot \cancel{3} \cdot \cancel{3} \cdot 3}{\cancel{3} \cdot \cancel{3} \cdot 5} =$ Multiply the remaining factors.

$\dfrac{6}{5}$ Therefore, $\dfrac{54}{45} = \dfrac{6}{5}$.

Practice Exercises

Reduce the following fractions to lowest terms:

1. $\dfrac{18}{24}$ **2.** $\dfrac{15}{60}$

3. $\dfrac{180}{378}$ **4.** $\dfrac{30}{12}$

If more practice is needed, do the Additional Practice Exercises.

Additional Practice Exercises

Reduce the following fractions to lowest terms:

a. $\dfrac{15}{40}$ **b.** $\dfrac{14}{56}$

c. $\dfrac{105}{315}$ **d.** $\dfrac{32}{18}$

Answers:

Practice Exercises 1–4: 1. $\frac{3}{4}$ 2. $\frac{1}{4}$ 3. $\frac{10}{21}$ 4. $\frac{5}{2}$ *Additional Practice Exercises a–d:* a. $\frac{3}{8}$ b. $\frac{1}{4}$ c. $\frac{1}{3}$ d. $\frac{16}{9}$

In algebra, the numerators and the denominators of fractions are polynomials. Such fractions are called **rational expressions.**

> **DEFINITION**   **Rational Expression**
>
> A rational expression is an algebraic expression of the form $\frac{P}{Q}$ where P and Q are polynomials and $Q \neq 0$.

In this section, we will limit ourselves to rational expressions whose numerators and denominators are monomials. We have already discussed this type of rational expression in Chapter 2, where we discussed the division laws of exponents. In this section, we will not be using the laws of exponents but will be reducing these rational expressions using the same procedure we used to reduce rational numbers. Compare the following reduction of a rational number with the reduction of a rational expression where the instructions on the right apply to both exercises:

$$\frac{45}{54} \qquad \frac{a^3 b^4}{a^5 b^3}$$ Write the numerator and denominator as prime factors.

$$\frac{3 \cdot 3 \cdot 5}{2 \cdot 3 \cdot 3 \cdot 3} \qquad \frac{a \cdot a \cdot a \cdot b \cdot b \cdot b \cdot b}{a \cdot a \cdot a \cdot a \cdot a \cdot b \cdot b \cdot b}$$ Divide by the common factors.

$$\frac{\cancel{3} \cdot \cancel{3} \cdot 5}{2 \cdot \cancel{3} \cdot \cancel{3} \cdot 3} \qquad \frac{\cancel{a} \cdot \cancel{a} \cdot \cancel{a} \cdot \cancel{b} \cdot \cancel{b} \cdot \cancel{b} \cdot b}{\cancel{a} \cdot \cancel{a} \cdot \cancel{a} \cdot a \cdot a \cdot \cancel{b} \cdot \cancel{b} \cdot \cancel{b}}$$ Multiply the remaining factors.

$$\frac{5}{6} \qquad \frac{b}{a^2}$$ Reduced forms.

Example 2

Reduce the following rational expressions to lowest terms:

a. $\dfrac{x^2 y^4}{x^5 y^2} =$ Write the numerator and the denominator in terms of prime factors.

$$\frac{x \cdot x \cdot y \cdot y \cdot y \cdot y}{x \cdot x \cdot x \cdot x \cdot x \cdot y \cdot y} =$$ Divide by factors common to the numerator and the denominator.

$$\frac{\cancel{x} \cdot \cancel{x} \cdot \cancel{y} \cdot \cancel{y} \cdot y \cdot y}{\cancel{x} \cdot \cancel{x} \cdot x \cdot x \cdot x \cdot \cancel{y} \cdot \cancel{y}} =$$ Multiply the remaining factors.

$$\frac{y^2}{x^3}$$ Therefore, $\frac{x^2 y^4}{x^5 y^2} = \frac{y^2}{x^3}$.

b. $\dfrac{8x^2 y^3}{12x^4 y} =$ Write the numerator and the denominator in terms of prime factors.

$$\frac{2 \cdot 2 \cdot 2 \cdot x \cdot x \cdot y \cdot y \cdot y}{2 \cdot 2 \cdot 3 \cdot x \cdot x \cdot x \cdot x \cdot y} =$$ Divide by factors common to the numerator and the denominator.

$$\frac{\cancel{2} \cdot \cancel{2} \cdot 2 \cdot \cancel{x} \cdot \cancel{x} \cdot \cancel{y} \cdot y \cdot y}{\cancel{2} \cdot \cancel{2} \cdot 3 \cdot \cancel{x} \cdot \cancel{x} \cdot x \cdot x \cdot \cancel{y}} =$$ Multiply the remaining factors.

$$\frac{2y^2}{3x^2}$$ Therefore, $\frac{8x^2 y^3}{12x^4 y} = \frac{2y^2}{3x^2}$.

Note: Many times reducing these types of rational expressions is easier than reducing arithmetic fractions because the factorization is usually easier.

c. $\dfrac{20 a^3 b^2}{24 ab} =$ Write the numerator and the denominator in terms of prime factors.

$$\frac{2 \cdot 2 \cdot 5 \cdot a \cdot a \cdot a \cdot b \cdot b}{2 \cdot 2 \cdot 2 \cdot 3 \cdot a \cdot b} =$$ Divide by factors common to the numerator and the denominator.

$$\frac{\cancel{2} \cdot \cancel{2} \cdot 5 \cdot \cancel{a} \cdot a \cdot a \cdot \cancel{b} \cdot b}{\cancel{2} \cdot \cancel{2} \cdot 2 \cdot 3 \cdot \cancel{a} \cdot \cancel{b}} =$$ Multiply the remaining factors.

$$\frac{5 a^2 b}{6}$$ Therefore, $\frac{20 a^3 b^2}{24 ab} = \frac{5 a^2 b}{6}$.

Practice Exercises

Reduce the following rational expressions to lowest terms:

5. $\dfrac{m^3n^6}{m^4n^2}$

6. $\dfrac{18x^4y^2}{21x^2y^3}$

7. $\dfrac{-27a^5b^2c^3}{36a^2bc^4}$

If more practice is needed, do the Additional Practice Exercises.

Additional Practice Exercises

Reduce the following rational expressions to lowest terms:

e. $\dfrac{a^2b^5}{a^4b}$

f. $\dfrac{14cd^3}{28c^3d^2}$

g. $\dfrac{12r^4s^4w^5}{-18r^2s^6w^3}$

The procedure of dividing the numerator and the denominator by factors common to both will be used often in the remainder of this chapter. Consequently, it is very important you understand this procedure.

Exercise Set 6.2

Reduce the following fractions to lowest terms:

1. $\dfrac{3}{18}$

2. $\dfrac{4}{16}$

3. $\dfrac{12}{8}$

4. $\dfrac{18}{12}$

5. $\dfrac{40}{84}$

6. $\dfrac{16}{56}$

7. $\dfrac{36}{54}$

8. $\dfrac{32}{72}$

9. $\dfrac{90}{60}$

10. $\dfrac{100}{80}$

11. $\dfrac{72}{54}$

12. $\dfrac{84}{60}$

13. $\dfrac{72}{108}$

14. $\dfrac{120}{144}$

15. $\dfrac{120}{96}$

16. $\dfrac{72}{124}$

17. $\dfrac{210}{110}$

18. $\dfrac{300}{108}$

19. $\dfrac{90}{315}$

20. $\dfrac{180}{210}$

Reduce the following rational expressions:

21. $\dfrac{a^2b^3}{a^4b}$

22. $\dfrac{c^3d}{c^2d^4}$

23. $\dfrac{-r^3s^2}{r^2s^2}$

24. $\dfrac{-q^4r^3}{q^4r}$

25. $\dfrac{a^3b^4}{a^4b^6}$

26. $\dfrac{u^2v^5}{u^3v^6}$

27. $\dfrac{12x^5y^3}{3x^2y^2}$

28. $\dfrac{18x^5y^7}{6x^3y^2}$

29. $\dfrac{-16x^4y^6}{8x^3y^4}$

30. $\dfrac{-24p^6q^6}{6p^4q^5}$

31. $\dfrac{-28m^3n^5}{16m^7n^6}$

32. $\dfrac{32c^2d^6}{-24c^4d^7}$

33. $\dfrac{a^5b^3c^6}{a^3b^6c^4}$ **34.** $\dfrac{x^3y^6z^4}{x^2y^4z^6}$ **35.** $\dfrac{-32m^2n^5p^3}{-18m^2n^3p^4}$ **36.** $\dfrac{-27x^3y^4z^2}{6x^3y^2z^3}$

37. $\dfrac{15r^3s^5w}{-25r^4s^6w}$ **38.** $\dfrac{36qr^4s^7}{-27qr^5s^9}$ **39.** $\dfrac{-32x^4y^3z^2}{16x^2y^2z^2}$ **40.** $\dfrac{42w^3x^4y^3}{-14w^2x^3y}$

Writing Exercises:

41. Why is it important to reduce fractions to lowest terms?

Group Project:

42. Find three examples of an "everyday life" situation in which a fraction occurs that is not reduced to lowest terms.

Section 6.3   # Further Reduction of Rational Expressions

OBJECTIVE   *When you complete this section, you will be able to:*

Reduce rational expressions whose numerators and denominators are not monomials.

Introduction   In the previous section, we reduced rational expressions whose numerators and denominators were monomials. The procedure we used was to factor the numerator and the denominator into prime factors and then divide by the factors that were common to both the numerator and the denominator. If the numerator and the denominator are polynomials other than monomials, we will use the same procedure.

Before we discuss reducing rational expressions, we need to find the value(s) of the variable(s) for which a rational expression is defined. We know that division by 0 is not defined. For example, if $\frac{4}{0} = x$, then $0 \cdot x = 4$, which is impossible since 0 times anything is 0. Consequently, the denominator cannot have the value of 0. This means that a rational expression is defined for all value(s) of the variable(s) for which the denominator does not equal 0. Those values can be found by setting the denominator equal to 0 and solving the resulting equation. The variable(s) cannot equal any resulting value(s) since they make the denominator equal to 0.

Example 1

Find the value(s) of the variable(s) for which the following are defined:

a. $\dfrac{x}{x + 3}$

b. $\dfrac{x}{2x + 3}$

Solution:

$\frac{x}{x+3}$ is defined for all values of x except those that make the denominator equal to 0. To find the value(s) that make the denominator equal to 0, set the denominator equal to 0 and solve the equation.

$x + 3 = 0$ Subtract 3 from both sides of the equation.

$x + 3 - 3 = 0 - 3$ Simplify both sides.

$x + 0 = -3$ Additive identity.

$x = -3$ Solution of the equation.

Since the denominator is equal to 0 for $x = -3$, $x \neq -3$. So the values of x for which the expression is defined are all numbers except 3.

Solution:

$\frac{x}{2x+3}$ is defined for all values of x except those that make the denominator equal to 0. To find the value(s) that make the denominator equal to 0, set the denominator equal to 0 and solve the equation.

$2x + 3 = 0$ Subtract 3 from both sides of the equation.

$2x = -3$ Divide both sides by 2.

$x = -\dfrac{3}{2}$ Solution of the equation.

Since the denominator is equal to 0 for $x = -\frac{3}{2}$, $x \neq -\frac{3}{2}$. So the values of x for which the expression is defined are all numbers except $-\frac{3}{2}$.

c. $\dfrac{a + b}{a - b}$

d. $\dfrac{x - 5}{x^2 - 2x - 24}$

Solution:

$\dfrac{a + b}{a - b}$ is defined for all values of a and b except those that make the denominator equal to 0. To find those values, set the denominator equal to 0 and solve the equation.

$a - b = 0$	Add b to both sides of the equation.
$a - b + b = 0 + b$	Simplify both sides.
$a + 0 = b$	Additive identity.
$a = b$	Solution of the equation.

Since the denominator is equal to 0 when $a = b$, $a \neq b$. For example, if $a = 2$, b can have any value except 2, and so on.

Solution:

$\dfrac{x - 5}{x^2 - 2x - 24}$ is defined for all values of x except those that make the denominator equal to 0. To find these values, set the denominator equal to 0 and solve for x.

$x^2 - 2x - 24 = 0$	Factor.
$(x + 4)(x - 6) = 0$	Set each factor equal to 0.
$x + 4 = 0, x - 6 = 0$	Solve each equation.
$x = -4, \quad x = 6$	Solutions of the equation.

Since the denominator is equal to 0 when $x = -4$ or $x = 6$, $x \neq -4$ and $x \neq 6$. So the values of x for which the expression is defined are all numbers except -4 and 6.

Practice Exercises

Find the values of the variable(s) for which the following are defined:

1. $\dfrac{x - 5}{x + 4}$

2. $\dfrac{4x}{3x - 2}$

3. $\dfrac{c - d}{2c + d}$

4. $\dfrac{c - 1}{c^2 - 4c - 21}$

If more practice is needed, do the Additional Practice Exercises.

Additional Practice Exercises

Find the values of the variable(s) for which the following are defined:

a. $\dfrac{x + 6}{x + 5}$

b. $\dfrac{4a}{5a + 3}$

c. $\dfrac{2a}{a + b}$

d. $\dfrac{m - 6}{m^2 + 6m + 8}$

We know that any number, other than 0, divided by itself is equal to 1. For example, $\frac{5}{5} = 1$ and $\frac{-4}{-4} = 1$. We used this fact in the last section to reduce rational numbers. We used an extension of this fact when reducing rational expressions with monomial numerators and denominators by indicating that a variable divided by itself also has a value of 1 providing the variable does not equal 0. For example, $\frac{x}{x} = 1$, if $x \neq 0$. We are going to extend this fact again in this section by saying any polynomial divided by itself is equal to 1, providing the polynomial is not equal to 0. For example, $\frac{x + 3}{x + 3} = 1$, if $x + 3 \neq 0$. Since $x + 3$ represents a number for any value of x, we are simply saying that any nonzero number divided by itself is 1.

The procedure for reducing rational expressions whose numerators and/or denominators are not monomials is exactly the same as for those whose numerators and denominators are monomials. Compare the following:

$$\frac{8x^2y^3}{12x^4y} \qquad\qquad \frac{x^2 + 3x + 2}{x^2 - 2x - 3}$$

Write the numerator and the denominator in terms of prime factors.

$$\frac{2 \cdot 2 \cdot 2 \cdot x \cdot x \cdot y \cdot y \cdot y}{2 \cdot 2 \cdot 3 \cdot x \cdot x \cdot x \cdot x \cdot y} \qquad \frac{(x + 1)(x + 2)}{(x + 1)(x - 3)}$$

Divide by factors common to the numerator and the denominator.

$$\frac{\cancel{2} \cdot \cancel{2} \cdot 2 \cdot \cancel{x} \cdot \cancel{x} \cdot \cancel{y} \cdot y \cdot y}{\cancel{2} \cdot \cancel{2} \cdot 3 \cdot \cancel{x} \cdot \cancel{x} \cdot x \cdot x \cdot \cancel{y}} \qquad \frac{\cancel{(x + 1)}(x + 2)}{\cancel{(x + 1)}(x - 3)}$$

Multiply the remaining factors.

$$\frac{2y^2}{3x^2} \qquad\qquad \frac{x + 2}{x - 3}$$

Reduce to lowest terms.

Example 2

Reduce the following rational expressions. Assume the variables cannot equal any value for which the denominator is equal to 0.

a. $\dfrac{5(x + 1)}{8(x + 1)} =$
The numerator and the denominator are already factored, so divide the numerator and the denominator by the common factor of $x + 1$.

$\dfrac{5\cancel{(x + 1)}}{8\cancel{(x + 1)}} = \qquad \dfrac{x + 1}{x + 1} = 1 \text{ if } x \neq -1.$

$\dfrac{5}{8}$
Answer if $x \neq -1$.

b. $\dfrac{4x + 16}{3x + 12} =$
Factor the numerator and the denominator.

$\dfrac{4(x + 4)}{3(x + 4)} =$
Divide the numerator and the denominator by the common factor of $x + 4$.

$\dfrac{4\cancel{(x + 4)}}{3\cancel{(x + 4)}} = \qquad \dfrac{x + 4}{x + 4} = 1, \text{ if } x \neq -4.$

$\dfrac{4}{3}$
Answer if $x \neq -4$.

c. $\dfrac{x^2 - 9}{x^2 - x - 12} =$
Factor the numerator and the denominator.

$\dfrac{(x + 3)(x - 3)}{(x + 3)(x - 4)} =$
Divide the numerator and the denominator by the common factor of $x + 3$.

$\dfrac{\cancel{(x + 3)}(x - 3)}{\cancel{(x + 3)}(x - 4)} = \qquad \dfrac{x + 3}{x + 3} = 1, \text{ if } x \neq -3.$

$\dfrac{x - 3}{x - 4}$
Answer if $x \neq 4$ or $x \neq -3$.

d. $\dfrac{x + 4}{x^2 - 16} =$
Factor the denominator. The numerator is prime.

$\dfrac{x + 4}{(x + 4)(x - 4)} =$
Divide the numerator and the denominator by the common factor of $x + 4$.

$\dfrac{\cancel{x + 4}}{\cancel{(x + 4)}(x - 4)} = \qquad \dfrac{x + 4}{x + 4} = 1, \text{ if } x \neq -4.$

$\dfrac{1}{x - 4}$
Answer if $x \neq 4$ or -4. Remember, each time we divide by a common factor, there is a factor of 1 left. We do not write the 1 unless it is the only factor remaining in the numerator.

Be Careful A common error is to divide by terms instead of factors. Consider the following: $\frac{6 + 3}{3} \neq \frac{6 + \cancel{3}}{\cancel{3}} \neq 6 + 1 = 7$. If we follow the order of operations, $\frac{6 + 3}{3} = \frac{9}{3} = 3$. But $7 \neq 3$. We have two different answers to the same question. Since we followed the order of operations in the second case, we know that 3 is the correct answer. What is wrong with the first case? We have divided by a *term* and not a *factor*. Remember, terms are separated by $+$ and $-$ signs and factor implies multiplication. Therefore, before you can reduce a rational expression, you must first write the numerator and the denominator in *factored* form. Then you divide by the factors that are common to both. In the following, both the correct and the incorrect methods are shown:

CORRECT

$$\frac{x^2 - 6x + 8}{x^2 + 2x - 8} = \frac{(x - 4)(x - 2)}{(x - 2)(x + 4)} = \frac{(x - 4)\cancel{(x - 2)}}{\cancel{(x - 2)}(x + 4)} = \frac{x - 4}{x + 4}$$

Since $x - 2$ is a factor common to both the numerator and the denominator, we can divide both the numerator and the denominator by $x - 2$.

INCORRECT

$$\frac{x^2 - 6x + 8}{x^2 + 2x - 8} = \frac{\cancel{x^2} - 6x + \cancel{8}}{\cancel{x^2} + 2x - \cancel{8}} = \frac{1 - 6x + 1}{1 - 2x - 1} = \frac{2 - 6x}{+ 2x}$$

x^2 and 8 are *terms* common to the numerator and the denominator. They are not *factors*. **You cannot divide by terms!**

Practice Exercises

Reduce the following rational expressions. Assume the variables cannot equal any value that makes the denominator equal to 0.

5. $\dfrac{4(x + y)}{7(x + y)}$

6. $\dfrac{3x + 6}{5x + 10}$

7. $\dfrac{x^2 - 4}{x^2 + 9x + 14}$

8. $\dfrac{x - 3}{x^2 - 7x + 12}$

If more practice is needed, do the Additional Practice Exercises.

Additional Practice Exercises

Reduce the following rational expressions to lowest terms:

e. $\dfrac{2(y + 6)}{5(y + 6)}$

f. $\dfrac{12x + 4y}{21x - 7y}$

g. $\dfrac{x^2 - 2x - 15}{x^2 + 7x + 12}$

h. $\dfrac{x + 1}{x^2 + 4x + 3}$

There is one instance in reducing rational expressions that requires special attention. This occurs if the numerator and the denominator are exactly the same except for the signs of the terms. For example, $\frac{a - b}{b - a}$. The numerator has $+a$ and $-b$ while the denominator has $-a$ and $+b$, so the signs of the terms are opposites. In order to reduce this type of fraction, we need to recall a technique that was discussed earlier where we factored out -1 when factoring by grouping.

Example 3

Factor -1 from each of the following and rewrite the expression inside the parentheses so the term with the positive coefficient is first:

a. $b - a =$ Removing the factor of -1 changes the signs of all the terms inside the parentheses.

$-1(-b + a) =$ Check using the distributive property. Hint: $-(-b + a) = -1(-b + a)$.

$-1(a - b)$ By the commutative property, $-b + a = a - b$.

b. $3 - 2x =$ Removing the factor of -1 changes the signs of all the terms inside the parentheses.

$-1(-3 + 2x) =$ Check using the distributive property.

$-1(2x - 3)$ By the commutative property.

Answers:

Practice Exercises

Remove -1 *from each of the following and rewrite the expression inside the parentheses so the term with the positive coefficient is first:*

9. $4 - a$

10. $3y - 2$

We will now apply this technique to reducing rational expressions that contain factors in the numerator and the denominator with opposite signs. We can remove the -1 from either factor, but it is usually easier if we remove -1 from the factor in the numerator.

Example 4

Reduce the following rational expressions to lowest terms:

a. $\dfrac{a - b}{b - a} =$ The numerator and the denominator are the same except the signs of the terms are opposites. Remove -1 from the numerator.

$\dfrac{-1(-a + b)}{b - a} =$ Apply the commutative property in the numerator.

$\dfrac{-1(b - a)}{b - a} =$ Divide by the common factor of $b - a$.

$\dfrac{-1(\cancel{b - a})}{\cancel{b - a}} =$ Anything (except 0) divided by itself equals 1.

$\dfrac{-1}{1} =$ $\dfrac{-1}{1} = -1$.

-1 Answer.

b. $\dfrac{x^2 - 1}{1 - x} =$ Factor the numerator.

$\dfrac{(x - 1)(x + 1)}{1 - x} =$ $x - 1$ and $1 - x$ are opposite in signs. Remove -1 from $1 - x$ in the denominator.

$\dfrac{(x - 1)(x + 1)}{-1(-1 + x)} =$ Apply the commutative property to $-1 + x$.

$\dfrac{(x - 1)(x + 1)}{-1(x - 1)} =$ Divide by the common factor of $x - 1$.

$\dfrac{(\cancel{x - 1})(x + 1)}{-1(\cancel{x - 1})} =$ 1 is left in both the numerator and the denominator where the common factors were divided.

$\dfrac{1(x + 1)}{-1} =$ $\dfrac{1}{-1} = -1$.

$-1(x + 1) =$ Distribute the -1.

$-x - 1$ Answer.

> **Be Careful** Rational expressions of the form $\dfrac{4 + x}{4 - x}$ are often mistaken for the previous type. In this rational expression, not all the terms in the numerator and the denominator are opposites in sign. This expression cannot be reduced because $\dfrac{4 + x}{4 - x} = \dfrac{4 + x}{-(x - 4)}$, so there is no common factor.

Note: The answer $-(x + 1)$ is usually acceptable and sometimes preferred. This problem could also have been done by removing a factor of -1 from $(x - 1)$ in the numerator.

Practice Exercises

Reduce the following rational expressions to lowest terms:

11. $\dfrac{c - d}{d - c}$

12. $\dfrac{x^2 - 4}{2 - x}$

If more practice is needed, do the Additional Practice Exercises.

Additional Practice Exercises

Reduce the following rational expressions to lowest terms:

i. $\dfrac{r - s}{s - r}$

j. $\dfrac{x^2 - 9}{3 - x}$

We defined functions in Section 4.8 and discussed quadratic functions in Section 5.8. Following is the definition of another type of function:

Rational Function

A function of the form $f(x) = \dfrac{p(x)}{q(x)}$, $q(x) \neq 0$, where $p(x)$ and $q(x)$ are polynomials is called a rational function.

Rational functions are evaluated in the same manner as any other function.

Example 6

Given $f(x) = \dfrac{x + 4}{x^2 - 16}$, *find the following:*

a. $f(0)$

Solution:
To find $f(0)$, replace x with 0 and simplify.

$f(x) = \dfrac{x + 4}{x^2 - 16}$ Replace x with 0.

$f(0) = \dfrac{0 + 4}{0^2 - 16}$ Simplify.

$f(0) = \dfrac{4}{-16}$ Reduce to lowest terms.

$f(0) = -\dfrac{1}{4}$ Therefore, $f(0) = -\dfrac{1}{4}$.

b. $f(2)$

Solution:
To find $f(2)$, replace x with 2 and simplify.

$f(x) = \dfrac{x + 4}{x^2 - 16}$ Replace x with 2.

$f(2) = \dfrac{2 + 4}{2^2 - 16}$ Simplify by adding in the numerator and raising to a power in the denominator.

$f(2) = \dfrac{6}{4 - 16}$ Add in the denominator.

$f(2) = \dfrac{6}{-12}$ Reduce to lowest terms.

$f(2) = -\dfrac{1}{2}$ Therefore, $f(2) = -\dfrac{1}{2}$.

Answers:

Practice Exercises 11–12: **11.** -1 **12.** $-x - 2$ *Additional Practice Exercises i–j:* **i.** -1 **j.** $-x - 3$

c. $f(4)$

Solution:

To find $f(4)$, replace x with 4 and simplify.

$$f(x) = \frac{x + 4}{x^2 - 16}$$ Replace x with 4.

$$f(4) = \frac{4 + 4}{4^2 - 16}$$ Simplify by adding in the numerator and raising to a power in the denominator.

$$f(4) = \frac{8}{16 - 16}$$ Add in the denominator.

$$f(4) = \frac{8}{0}$$ Since division by 0 is undefined, $f(4)$ does not exist.

$f(4)$ is undefined.

Practice Exercises

Given $f(x) = \frac{x - 2}{x^2 + x - 6}$, find the following:

13. $f(0)$ 　　　　　　　**14.** $f(4)$ 　　　　　　　**15.** $f(-3)$

Rational functions play a very important role in upper level mathematics.

Exercise Set 6.3

Find the value(s) of the variable(s) for which the following are defined:

1. $\dfrac{5}{x + 5}$ 　　　**2.** $\dfrac{9}{a - 6}$ 　　　**3.** $\dfrac{r + 3}{r - 8}$ 　　　**4.** $\dfrac{y + 3}{y + 10}$

5. $\dfrac{x + 3}{3x + 12}$ 　　　**6.** $\dfrac{y - 5}{4y - 8}$ 　　　**7.** $\dfrac{6a}{3a - 4}$ 　　　**8.** $\dfrac{5b}{4b - 5}$

9. $\dfrac{8x}{x + 2y}$ 　　　**10.** $\dfrac{7y}{x - 3y}$ 　　　**11.** $\dfrac{5}{(x - 7)(x + 3)}$ 　　　**12.** $\dfrac{m}{(m + 5)(m - 3)}$

13. $\dfrac{x - 3}{x^2 - 3x - 10}$ 　　　**14.** $\dfrac{x + 2}{x^2 + x - 12}$ 　　　**15.** $\dfrac{r + 3}{2r^2 - 7r - 15}$ 　　　**16.** $\dfrac{n + 7}{3n^2 + 2n - 8}$

Reduce the following rational expressions to lowest terms:

17. $\dfrac{4(x + 3)}{9(x + 3)}$ 　　　**18.** $\dfrac{3(y - 2)}{5(y - 2)}$ 　　　**19.** $\dfrac{-5(2a - 7)}{6(2a - 7)}$

20. $\dfrac{-8(3b + 4)}{9(3b + 4)}$ 　　　**21.** $\dfrac{4(x - 10)}{6(x - 10)}$ 　　　**22.** $\dfrac{6(x + 5)}{8(x + 5)}$

23. $\dfrac{2x + 12}{3x + 18}$ 　　　**24.** $\dfrac{3x + 6}{5x + 10}$ 　　　**25.** $\dfrac{8x + 12y}{20x + 30y}$

26. $\dfrac{24a - 16b}{36a - 24b}$ 　　　**27.** $\dfrac{-3x - 15}{4x + 20}$ 　　　**28.** $\dfrac{-5x - 25}{6x + 30}$

Answers:

29. $\dfrac{x^2 + 3x + 2}{x^2 - 2x - 3}$

30. $\dfrac{x^2 - 4x + 3}{x^2 - 5x + 6}$

31. $\dfrac{a^2 + 3a - 10}{a^2 + 8a + 15}$

32. $\dfrac{c^2 + 3c - 18}{c^2 - c - 6}$

33. $\dfrac{x - 6}{x^2 - 4x - 12}$

34. $\dfrac{x - 2}{x^2 - 10x + 16}$

35. $\dfrac{x^2 - 9}{x^2 - 2x - 15}$

36. $\dfrac{x^2 - 25}{x^2 - x - 30}$

37. $\dfrac{4x^2 - 9y^2}{6x^2 - xy - 15y^2}$

38. $\dfrac{9x^2 - 25y^2}{6x^2 - 5xy - 25y^2}$

39. $\dfrac{4a^2 - 4ab - 3b^2}{6a^2 - ab - 2b^2}$

40. $\dfrac{6a^2 + 13ab + 6b^2}{3a^2 - 7ab - 6b^2}$

Challenge Exercises (41–44)

41. $\dfrac{ac - ad + bc - bd}{ax + ay + bx + by}$

42. $\dfrac{px + py + qx + qy}{mx - nx + my - ny}$

43. $\dfrac{xy - 4x + 3y - 12}{xy - 6x + 3y - 18}$

44. $\dfrac{ab + 5a - 3b - 15}{bc - 2b + 5c - 10}$

Factor -1 from each of the following and rewrite the expression inside the parentheses so the term with the positive coefficient is first:

45. $y - x$

46. $s - r$

47. $2d - c$

48. $3y - x$

49. $3r - 2s$

50. $4a - 3b$

Reduce the following rational expressions to lowest terms:

51. $\dfrac{m - n}{n - m}$

52. $\dfrac{q - p}{p - q}$

53. $\dfrac{2b - a}{a - 2b}$

54. $\dfrac{3m - n}{n - 3m}$

55. $\dfrac{a^2 - 25}{5 - a}$

56. $\dfrac{c^2 - 49}{7 - c}$

57. $\dfrac{x^2 - 3x - 10}{5 - x}$

58. $\dfrac{x^2 + 4x - 21}{3 - x}$

59. $\dfrac{4 - x}{x^2 - 9x + 20}$

Answer the following:

60. Given $f(x) = \dfrac{x + 7}{x^2 + 5x - 14}$, find the following:
   a. $f(0)$ b. $f(3)$ c. $f(2)$

61. Given $f(x) = \dfrac{x - 6}{x^2 - 8x + 12}$, find the following:
   a. $f(0)$ b. $f(-1)$ c. $f(2)$

Writing Exercises:

62. What is wrong with the following?
$$\frac{3x + 4}{2} = \frac{3x + \overset{2}{\cancel{4}}}{\cancel{2}} = 3x + 2$$

63. How does a term differ from a factor?

64. Are $a - b$ and $b - a$ always equal? If not, under what circumstances will $a - b$ and $b - a$ be equal?

65. Why does $\dfrac{b - a}{a - b} = -1$?

Group Project:

66. Given $f(x) = \dfrac{x - 4}{x^2 - 7x + 12}$ and $g(x) = \dfrac{1}{x - 3}$.
   a. Find $f(2)$ and $g(2)$.
   b. How are $f(2)$ and $g(2)$ related? Why?
   c. Find $f(-2)$ and $g(-2)$.
   d. How are $f(-2)$ and $g(-2)$ related? Why?
   e. Find $f(4)$ and $g(4)$.
   f. Does $f(4) = g(4)$? Why or why not?

| **Section 6.4** | **Multiplication of Rational Numbers and Expressions** |

OBJECTIVES

When you complete this section, you will be able to:

a. Multiply two or more fractions.

b. Multiply fractions and mixed numbers.

c. Multiply whole numbers and fractions or whole numbers and mixed numbers.

Introduction   In Chapter 0, we introduced the procedure for multiplying fractions but did so on a very elementary level. We needed this procedure to develop properties of exponents in Chapter 2. In Chapter 2, you were not required to multiply fractions, only to recall the procedure. In this section, we will expand on the concepts developed in Section 0.6.

In Section 6.1, we used shaded regions of figures to represent a fractional part of the entire region (unit). We can represent $\frac{1}{5}$ by dividing a region into five equal parts and shading one of those parts as the following shows:

1 unit

Suppose we now wish to further shade $\frac{1}{2}$ of the $\frac{1}{5}$ that is already shaded. We divide everything into two equal parts with a horizontal line segment. The one part that was shaded has now been divided into two equal parts. So, $\frac{1}{2}$ of $\frac{1}{5}$ is one of these two equal parts which we will shade further. What part of the total region has been further shaded?

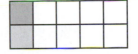

1 unit

By dividing each of the five equal parts into two equal parts, the entire region (unit) has been divided into ten equal parts. Therefore, the one part that was further shaded represents $\frac{1}{10}$ of the entire region. We would say that $\frac{1}{2}$ of $\frac{1}{5}$ is $\frac{1}{10}$. When performing operations on fractions, the word "of" is used to indicate multiplication. Consequently, $\frac{1}{2} \cdot \frac{1}{5} = \frac{1}{10}$. Note that this is the same as $\frac{1}{2} \cdot \frac{1}{5} = \frac{1 \cdot 1}{2 \cdot 5} = \frac{1}{10}$. Let us repeat this procedure with a more complicated example and see if it still works.

Example 1

Represent $\frac{2}{3}$ of $\frac{2}{5}$ using shaded regions and give the result as a fractional part of the entire region.

Solution:
The fraction $\frac{2}{5}$ means divide the region into five equal parts and shade two of them.

To find $\frac{2}{3}$ of the two parts that are shaded, we will divide each of the five equal parts into three equal parts. Each of the parts that were shaded has now been divided into three equal parts. To indicate $\frac{2}{3}$ of each of these shaded parts, further shade two of these three equal parts.

You will notice the entire figure is now divided into 15 equal parts of which four have been further shaded. This represents $\frac{4}{15}$ of the entire figure. Therefore, $\frac{2}{3}$ of $\frac{2}{5}$ is $\frac{4}{15}$. As multiplication, $\frac{2}{3} \cdot \frac{2}{5} = \frac{4}{15}$. Note that this is the same as $\frac{2}{3} \cdot \frac{2}{5} = \frac{2 \cdot 2}{3 \cdot 5} = \frac{4}{15}$.

Multiplying fractions   Based on the preceding examples, we are now ready to state the procedure for multiplication of fractions.

Multiplication of Fractions

The product of two fractions is found by computing the product of the numerators and dividing by the product of the denominators. In symbols, $\frac{a}{b} \cdot \frac{c}{d} = \frac{a \cdot c}{b \cdot d}$. In other words, numerator times numerator divided by denominator times denominator.

The preceding procedure is easily extended to three or more fractions as follows:

$$\frac{a}{b} \cdot \frac{c}{d} \cdot \frac{e}{f} = \frac{a \cdot c \cdot e}{b \cdot d \cdot f} \quad \text{and} \quad \frac{a}{b} \cdot \frac{c}{d} \cdot \frac{e}{f} \cdot \frac{g}{h} = \frac{a \cdot c \cdot e \cdot g}{b \cdot d \cdot f \cdot h}.$$

Example 2

Find the following products:

a. $\dfrac{3}{5} \cdot \dfrac{2}{7} =$ Find the product of the numerators divided by the product of the denominators.

$\dfrac{3 \cdot 2}{5 \cdot 7} =$ Perform the multiplications.

$\dfrac{6}{35}$ Product.

b. $\dfrac{1}{4} \cdot \dfrac{3}{8} \cdot \dfrac{5}{7} =$ Find the product of the numerators divided by the product of the denominators.

$\dfrac{1 \cdot 3 \cdot 5}{4 \cdot 8 \cdot 7} =$ Perform the multiplications.

$\dfrac{15}{224}$ Product.

Practice Exercises

Find the following products:

1. $\dfrac{2}{3} \cdot \dfrac{4}{9}$

2. $\dfrac{1}{4} \cdot \dfrac{3}{2} \cdot \dfrac{5}{8}$

After finding the product of two or more fractions, it is often necessary to reduce to lowest terms. Also, you will need to recall the rules of signs for multiplication. If there is an even number of negative signs, the product is positive. If there is an odd number of negative signs, the product is negative.

Example 3

Find the following products. If necessary, reduce to lowest terms.

a. $\dfrac{2}{5} \cdot \left(-\dfrac{3}{8}\right) =$ Find the product of the numerators divided by the product of the denominators. The product of a positive and a negative is negative.

$-\dfrac{2 \cdot 3}{5 \cdot 8} =$ Write the numerator and the denominator in terms of prime factors.

$-\dfrac{2 \cdot 3}{5 \cdot 2 \cdot 2 \cdot 2} =$ Divide by the common factor of 2.

$-\dfrac{\cancel{2} \cdot 3}{5 \cdot \cancel{2} \cdot 2 \cdot 2} =$ Multiply the remaining factors.

$-\dfrac{3}{20}$ Product.

b. $\dfrac{3}{15} \cdot \dfrac{10}{9} =$ Find the product of the numerators divided by the product of the denominators.

$\dfrac{3 \cdot 10}{15 \cdot 9} =$ Write the numerator and the denominator in terms of prime factors.

$\dfrac{3 \cdot 2 \cdot 5}{3 \cdot 5 \cdot 3 \cdot 3} =$ Divide by the common factors of 3 and 5.

$\dfrac{\cancel{3} \cdot 2 \cdot \cancel{5}}{\cancel{3} \cdot \cancel{5} \cdot 3 \cdot 3} =$ Multiply the remaining factors.

$\dfrac{2}{9}$ Product.

---

c. $-\dfrac{4}{5} \cdot \dfrac{15}{14} \cdot \left(-\dfrac{7}{10}\right) =$

Find the product of the numerators divided by the product of the denominators. The product of two negatives is positive.

$\dfrac{4 \cdot 15 \cdot 7}{5 \cdot 14 \cdot 10} =$

Write the numerator and the denominator in terms of prime factors.

$\dfrac{2 \cdot 2 \cdot 3 \cdot 5 \cdot 7}{5 \cdot 2 \cdot 7 \cdot 2 \cdot 5} =$

Divide by the common factors of 2, 2, 5, and 7.

$\dfrac{\cancel{2} \cdot \cancel{2} \cdot 3 \cdot \cancel{5} \cdot \cancel{7}}{\cancel{5} \cdot \cancel{2} \cdot \cancel{7} \cdot \cancel{2} \cdot 5} =$

Multiply the remaining factors.

$\dfrac{3}{5}$

Product.

Note: You may reduce before actually multiplying. This is permitted because each numerator becomes a

factor of the numerator of the product and likewise for the denominators. Therefore, we may divide by factors common to any numerator and denominator before we multiply. Example 3b could be done as follows.

$\dfrac{3}{15} \cdot \dfrac{10}{9} =$

Factor the numerators and the denominators into primes.

$\dfrac{3}{3 \cdot 5} \cdot \dfrac{2 \cdot 5}{3 \cdot 3} =$

Divide by the common factors of 3 and 5.

$\dfrac{\cancel{3}}{\cancel{3} \cdot \cancel{5}} \cdot \dfrac{2 \cdot \cancel{5}}{3 \cdot 3} =$

Multiply the remaining numerators and the denominators.

$\dfrac{2}{3 \cdot 3} =$

Perform the multiplication.

$\dfrac{2}{9}$

Product.

This technique is more like the technique we will be using in the next section on multiplication of rational expressions. For that reason, we will use this method throughout the remainder of this section.

Practice Exercises

Find the following products. If necessary, reduce to lowest terms.

3. $\dfrac{3}{4} \cdot \dfrac{10}{21}$

4. $\dfrac{5}{8} \cdot \left(-\dfrac{14}{25}\right)$

5. $\dfrac{6}{7} \cdot \left(-\dfrac{21}{8}\right) \cdot \left(-\dfrac{4}{9}\right)$

If more practice is needed, do the Additional Practice Exercises.

Additional Practice Exercises

Find the following products. If necessary, reduce to lowest terms.

a. $\dfrac{2}{9} \cdot \dfrac{15}{6}$

b. $-\dfrac{3}{8} \cdot \dfrac{4}{9}$

c. $\dfrac{10}{3} \cdot \left(-\dfrac{15}{16}\right) \cdot \dfrac{12}{5}$

Multiplying mixed numbers

Since mixed numbers can be converted into improper fractions, the same technique used for multiplying fractions can be used to multiply mixed numbers. Likewise, since whole numbers may be thought of as fractions whose denominator is 1, this technique applies to products involving fractions and whole numbers also.

Example 4

Find the products of the following numbers. Leave answers reduced to lowest terms.

a. $-2\dfrac{2}{7} \cdot 3\dfrac{1}{2} =$ Change $2\dfrac{2}{7}$ and $3\dfrac{1}{2}$ into improper fractions.

$-\dfrac{16}{7} \cdot \dfrac{7}{2} =$ Write the numerators and the denominators in terms of prime factors.

$-\dfrac{2 \cdot 2 \cdot 2 \cdot 2}{7} \cdot \dfrac{7}{2} =$ Divide by the common factors of 2 and 7.

$-\dfrac{\cancel{2} \cdot 2 \cdot 2 \cdot 2}{\cancel{7}} \cdot \dfrac{\cancel{7}}{\cancel{2}} =$ Multiply the remaining factors.

$-\dfrac{2 \cdot 2 \cdot 2}{1} =$ Perform the multiplication.

-8 Product.

b. $12 \cdot \dfrac{5}{16} =$ $12 = \dfrac{12}{1}$.

$\dfrac{12}{1} \cdot \dfrac{5}{16} =$ Write the numerators and the denominators in terms of prime factors.

$\dfrac{2 \cdot 2 \cdot 3}{1} \cdot \dfrac{5}{2 \cdot 2 \cdot 2 \cdot 2} =$ Divide by the common factors of 2 (twice).

$\dfrac{\cancel{2} \cdot \cancel{2} \cdot 3}{1} \cdot \dfrac{5}{\cancel{2} \cdot \cancel{2} \cdot 2 \cdot 2} =$ Multiply the remaining factors.

$\dfrac{3 \cdot 5}{1 \cdot 2 \cdot 2} =$ Perform the multiplications.

$\dfrac{15}{4}$ Product.

Practice Exercises

Find the products of the following numbers. Leave answers reduced to lowest terms.

6. $-3\dfrac{1}{3} \cdot 6\dfrac{3}{5}$

7. $8 \cdot \dfrac{5}{6}$

If more practice is needed, do the Additional Practice Exercises.

Additional Practice Exercises

Find the products of the following mixed numbers. Leave answers reduced to lowest terms.

d. $\dfrac{2}{3} \cdot 4\dfrac{1}{2}$

e. $-18 \cdot \left(-1\dfrac{5}{6}\right)$

Multiplying fractions and whole numbers   If the numerators and the denominators of the rational expressions are monomials, the product is found in much the same way we multiply rational numbers. Compare the following. The instructions on the right apply to both exercises. Assume all variables have nonzero values.

Rational Number	**Rational Expression**	
$\dfrac{3}{4} \cdot \dfrac{10}{21}$	$\dfrac{a^2}{y^2} \cdot \dfrac{y^4}{a^5}$	Write the numerators and the denominators in terms of prime factors.
$\dfrac{3}{2 \cdot 2} \cdot \dfrac{2 \cdot 5}{3 \cdot 7}$	$\dfrac{a \cdot a}{y \cdot y} \cdot \dfrac{y \cdot y \cdot y \cdot y}{a \cdot a \cdot a \cdot a \cdot a}$	Divide by the common factors.
$\dfrac{\cancel{3}}{\cancel{2} \cdot 2} \cdot \dfrac{\cancel{2} \cdot 5}{\cancel{3} \cdot 7}$	$\dfrac{\cancel{a} \cdot \cancel{a}}{\cancel{y} \cdot \cancel{y}} \cdot \dfrac{\cancel{y} \cdot \cancel{y} \cdot y \cdot y}{\cancel{a} \cdot \cancel{a} \cdot a \cdot a \cdot a}$	Multiply the remaining factors.
$\dfrac{5}{14}$	$\dfrac{y^2}{a^3}$	Product.

Example 5

Find the following products. Leave answers reduced to lowest terms and assume all variables have nonzero values.

a. $\dfrac{a^2}{y^2} \cdot \dfrac{y^3}{a^3} =$ Write the numerators and the denominators in terms of prime factors.

$\dfrac{a \cdot a}{y \cdot y} \cdot \dfrac{y \cdot y \cdot y}{a \cdot a \cdot a} =$ Divide by common factors.

$\dfrac{\cancel{a} \cdot \cancel{a}}{\cancel{y} \cdot \cancel{y}} \cdot \dfrac{\cancel{y} \cdot \cancel{y} \cdot y}{\cancel{a} \cdot \cancel{a} \cdot a} =$ Multiply remaining factors.

$\dfrac{y}{a}$ Product.

b. $\dfrac{3mn^2}{5m^3n} \cdot \dfrac{10m^2n}{9m^2n^3} =$ Write the numerators and the denominators in terms of prime factors.

$\dfrac{3 \cdot m \cdot n \cdot n}{5 \cdot m \cdot m \cdot m \cdot n} \cdot \dfrac{2 \cdot 5 \cdot m \cdot m \cdot n}{3 \cdot 3 \cdot m \cdot m \cdot n \cdot n \cdot n} =$ Divide by common factors.

$\dfrac{\cancel{3} \cdot \cancel{m} \cdot \cancel{n} \cdot \cancel{n}}{\cancel{5} \cdot \cancel{m} \cdot \cancel{m} \cdot \cancel{m} \cdot \cancel{n}} \cdot \dfrac{2 \cdot \cancel{5} \cdot \cancel{m} \cdot \cancel{m} \cdot \cancel{n}}{\cancel{3} \cdot 3 \cdot m \cdot m \cdot \cancel{n} \cdot \cancel{n} \cdot n} =$ Multiply remaining factors.

$\dfrac{2}{3m^2n}$ Product.

c. $\dfrac{3xy^2}{4a^2b^2} \cdot \dfrac{2ab}{9x^3y^3} =$ Write the numerators and the denominators in terms of prime factors.

$\dfrac{3 \cdot x \cdot y \cdot y}{2 \cdot 2 \cdot a \cdot a \cdot b \cdot b} \cdot \dfrac{2 \cdot a \cdot b}{3 \cdot 3 \cdot x \cdot x \cdot x \cdot y \cdot y \cdot y} =$ Divide by common factors.

$\dfrac{\cancel{3} \cdot \cancel{x} \cdot \cancel{y} \cdot \cancel{y}}{\cancel{2} \cdot 2 \cdot \cancel{a} \cdot a \cdot \cancel{b} \cdot b} \cdot \dfrac{\cancel{2} \cdot \cancel{a} \cdot \cancel{b}}{3 \cdot \cancel{3} \cdot \cancel{x} \cdot x \cdot x \cdot \cancel{y} \cdot \cancel{y} \cdot y} =$ Multiply remaining factors. Remember, each time we divide, there is a factor of 1 left that usually is not written. Since 1 is all that is left in the numerator, it must be written.

$\dfrac{1}{6abx^2y}$ Product.

Note: The products in Example 1 can also be found by using the laws of exponents developed in Chapter 2. Example 6b can also be done as follows:

$\dfrac{3mn^2}{5m^3n} \cdot \dfrac{10m^2n}{9m^2n^3} =$ Multiply the fractions.

$\dfrac{3mn^2 \cdot 10m^2n}{5m^3n \cdot 9m^2n^3} =$ Multiply the coefficients and apply $a^m \cdot a^n = a^{m+n}$.

$\dfrac{30m^3n^3}{45m^5n^4} =$ Apply $\dfrac{a^m}{a^n} = a^{m-n}$.

$\dfrac{30}{45}m^{-2}n^{-1} =$ Reduce the coefficient and apply $a^{-n} = \dfrac{1}{a^n}$.

$\dfrac{2}{3} \cdot \dfrac{1}{m^2} \cdot \dfrac{1}{n} =$ Multiply the fractions.

$\dfrac{2}{3m^2n}$ Product.

Practice Exercises

Find the following products. Leave answers reduced to lowest terms and assume all variables have nonzero values.

8. $\dfrac{x^3}{y^2} \cdot \dfrac{y^4}{x}$

9. $\dfrac{8a^3b}{7a^3b} \cdot \dfrac{21ab^2}{4a^2b}$

10. $\dfrac{2a^2b}{15xy^3} \cdot \dfrac{5xy}{4a^3b}$

If more practice is needed, do the Additional Practice Exercises.

Additional Practice Exercises

Find the following products. Leave answers reduced to lowest terms.

f. $\dfrac{s^3}{r^2} \cdot \dfrac{r^4}{s^5}$

g. $\dfrac{6r^3s}{5r^2s^2} \cdot \dfrac{15rs^2}{2r^3s^2}$

h. $\dfrac{3rs^2}{14p^2q^3} \cdot \dfrac{7pq^2}{6r^2s^2}$

Application problems often require the use of fractions.

Example 6

Answer the following:

a. A bookcase has six shelves. If each shelf is $2\frac{2}{3}$ feet long, what is the total length of the six shelves?

Solution:
Since there are six shelves, each of which is $2\frac{2}{3}$ feet long, the total length of the six shelves is the product of 6 and $2\frac{2}{3}$.

$6 \cdot 2\dfrac{2}{3}$ Change $2\frac{2}{3}$ into an improper fraction.

$6 \cdot \dfrac{8}{3}$ $6 = \dfrac{6}{1}$ and write 6 and 8 in terms of prime factors.

$\dfrac{2 \cdot 3}{1} \cdot \dfrac{2 \cdot 2 \cdot 2}{3}$ Divide by the common factor of 3.

$\dfrac{2 \cdot \cancel{3}}{1} \cdot \dfrac{2 \cdot 2 \cdot 2}{\cancel{3}}$ Multiply the remaining factors.

16 Therefore, the total length of the six shelves is 16 feet.

b. Find the area of a triangle with a base of $3\frac{1}{5}$ inches if the altitude to that base is $3\frac{3}{4}$ inches. The formula for the area of a triangle is $A = \frac{1}{2}bh$.

Solution:

$A = \dfrac{1}{2}bh$ Substitute into the formula.

$A = \dfrac{1}{2} \cdot 3\dfrac{1}{5} \cdot 3\dfrac{3}{4}$ Change the mixed numbers to improper fractions.

$A = \dfrac{1}{2} \cdot \dfrac{16}{5} \cdot \dfrac{15}{4}$ Write in terms of prime factors.

$A = \dfrac{1}{2} \cdot \dfrac{2 \cdot 2 \cdot 2 \cdot 2}{5} \cdot \dfrac{3 \cdot 5}{2 \cdot 2}$ Divide by the common factors.

$A = \dfrac{1}{\cancel{2}} \cdot \dfrac{\cancel{2} \cdot \cancel{2} \cdot \cancel{2} \cdot 2}{\cancel{5}} \cdot \dfrac{3 \cdot \cancel{5}}{\cancel{2} \cdot \cancel{2}}$ Multiply the remaining factors.

$A = 6 \text{ in.}^2$ Therefore, the area is 6 square inches.

Practice Exercises

Answer the following:

11. A beaker is $\frac{3}{4}$ full of sulfuric acid. If a student uses $\frac{2}{3}$ of the contents of the beaker in an experiment, what fractional part of a full beaker does the student use?

12. Find the area of a rectangle whose length is 12 meters and whose width is $4\frac{1}{4}$ meters. ($A = LW$)

Answers:

Exercise Set 6.4

Find the following products. Leave answers reduced to lowest terms and assume all variables have nonzero values.

1. $\dfrac{2}{5} \cdot \dfrac{3}{7}$

2. $\dfrac{1}{2} \cdot \dfrac{3}{10}$

3. $\dfrac{5}{8} \cdot \left(-\dfrac{9}{11}\right)$

4. $\dfrac{7}{12} \cdot \left(-\dfrac{5}{6}\right)$

5. $\dfrac{21}{8} \cdot \dfrac{16}{3}$

6. $\dfrac{18}{5} \cdot \dfrac{5}{3}$

7. $-\dfrac{3}{2} \cdot \left(-\dfrac{8}{9}\right)$

8. $-\dfrac{3}{8} \cdot \left(-\dfrac{7}{12}\right)$

9. $\dfrac{5}{8} \cdot \dfrac{32}{15}$

10. $\dfrac{3}{5} \cdot \dfrac{10}{9}$

11. $\dfrac{10}{3} \cdot \dfrac{6}{5}$

12. $\dfrac{8}{7} \cdot \dfrac{6}{5}$

13. $-\dfrac{2}{3} \cdot 1\dfrac{2}{3}$

14. $\dfrac{2}{5} \cdot \left(-2\dfrac{2}{3}\right)$

15. $1\dfrac{7}{8} \cdot \dfrac{4}{5}$

16. $4\dfrac{4}{5} \cdot \dfrac{3}{4}$

17. $-2\dfrac{1}{2} \cdot \left(-\dfrac{11}{15}\right)$

18. $-2\dfrac{1}{7} \cdot \left(-\dfrac{21}{10}\right)$

19. $5\dfrac{1}{2} \cdot \left(-1\dfrac{5}{11}\right)$

20. $-2\dfrac{2}{5} \cdot 1\dfrac{5}{12}$

21. $2\dfrac{2}{3} \cdot 3\dfrac{3}{5}$

22. $2\dfrac{1}{2} \cdot 1\dfrac{9}{21}$

23. $-10 \cdot \dfrac{5}{8}$

24. $16 \cdot \left(-\dfrac{5}{6}\right)$

25. $-24 \cdot \left(-2\dfrac{5}{6}\right)$

26. $-18 \cdot \left(-3\dfrac{2}{9}\right)$

27. $4\dfrac{1}{3} \cdot 9$

28. $4\dfrac{3}{4} \cdot 12$

29. $-\dfrac{3}{7} \cdot \dfrac{14}{5}$

30. $\dfrac{5}{11} \cdot \left(-\dfrac{22}{7}\right)$

31. $4\dfrac{1}{2} \cdot 3\dfrac{1}{3}$

32. $3\dfrac{2}{3} \cdot 1\dfrac{5}{22}$

33. $-2\dfrac{1}{4} \cdot \left(-1\dfrac{1}{3}\right)$

34. $-2\dfrac{2}{5} \cdot \left(-2\dfrac{1}{12}\right)$

35. $15 \cdot \dfrac{9}{5}$

36. $28 \cdot \dfrac{11}{7}$

37. $\dfrac{7}{18} \cdot \left(-2\dfrac{1}{4}\right)$

38. $-\dfrac{8}{15} \cdot 3\dfrac{1}{3}$

39. $\dfrac{7}{13} \cdot \left(-\dfrac{26}{21}\right) \cdot \dfrac{3}{4}$

40. $\dfrac{6}{15} \cdot \dfrac{25}{8} \cdot \left(-\dfrac{9}{10}\right)$

41. $5\dfrac{3}{5} \cdot \left(-1\dfrac{11}{14}\right) \cdot \left(-1\dfrac{3}{10}\right)$

42. $-6\dfrac{3}{8} \cdot 2\dfrac{2}{15} \cdot \left(-1\dfrac{3}{17}\right)$

43. $3\dfrac{4}{7} \cdot \dfrac{9}{10} \cdot 14$

44. $3\dfrac{5}{9} \cdot \dfrac{15}{16} \cdot 12$

45. $\dfrac{r^3}{s^3} \cdot \dfrac{s^5}{r^6}$

46. $\dfrac{a^5}{b^3} \cdot \dfrac{b^2}{a^3}$

47. $\dfrac{x^2}{y^3} \cdot \dfrac{y^5}{x}$

48. $\dfrac{a^4}{b^2} \cdot \dfrac{b^4}{a}$

49. $\dfrac{m^2}{n^2} \cdot \dfrac{n^2}{m^4}$

50. $\dfrac{c^3}{d^5} \cdot \dfrac{d^2}{c^3}$

51. $\dfrac{x^2 y^2}{a^2 b^3} \cdot \dfrac{a^4 b}{x y^4}$

52. $\dfrac{r^4 s^2}{t^3 u^3} \cdot \dfrac{t^5 u}{r^6 s}$

53. $\dfrac{x^3 y^2}{x y^3} \cdot \dfrac{x^2 y^4}{x^3 y^5}$

54. $\dfrac{m^2 n^2}{m^3 n^2} \cdot \dfrac{m n^3}{m^2 n}$

55. $\dfrac{r^4 s}{r^3 s^2} \cdot \dfrac{r^2 s^3}{r^4 s^3}$

56. $\dfrac{a^2 b^3}{a^3 b^2} \cdot \dfrac{a^2 b}{a^3 b^3}$

57. $\dfrac{6x^2}{5y^3} \cdot \dfrac{10y^5}{8x}$

58. $\dfrac{12a^4}{7b^5} \cdot \dfrac{14b^2}{16a^2}$

59. $\dfrac{2x^2 y^4}{3w^4 z^2} \cdot \dfrac{15w^3 z^4}{3x^5 y^2}$

60. $\dfrac{21a^4 b^3}{12c^2 d^6} \cdot \dfrac{9c^4 d^2}{14a^2 b^5}$

61. $\dfrac{8x^2 y^4}{12x^3 y^2} \cdot \dfrac{4x^3 y^3}{6x y^3}$

62. $\dfrac{10a^2 b^2}{18a^2 b^3} \cdot \dfrac{12a^3 b}{15a^2 b^4}$

63. $\dfrac{24a^2 bc^3}{16ab^3 c^2} \cdot \dfrac{32a^3 b^2 c^3}{18a^4 b^3 c^3}$

64. $\dfrac{30x^3 y^2 z}{42x^2 y z^2} \cdot \dfrac{15x^2 y^2 z^2}{25x^4 y z}$

Answer the following:

65. The latest creation of a fashion designer of women's apparel requires $4\frac{1}{3}$ square yards of material per dress. How many square yards of material would it take to make 24 of these dresses?

66. A tailor needs $3\frac{3}{4}$ square yards of material to make a suit. How many square yards of material would be needed to make 16 suits?

67. A wiring harness for a boat trailer requires $40\frac{1}{3}$ yards of wire. How many yards of wire would the manufacturer of this trailer need for 36 trailers?

68. It takes $4\frac{5}{12}$ feet of ribbon to wrap a birthday package. How many feet of ribbon would it take to wrap 18 of these packages?

69. A chemist has a beaker that is $\frac{2}{3}$ full of hydrochloric acid. In an experiment, she uses $\frac{1}{3}$ of the acid in the beaker. What fractional part of a full beaker did she use?

70. A pharmacist fills a prescription for a medication for high blood pressure. He uses $\frac{3}{4}$ of a bottle that is $\frac{2}{3}$ full. What part of a full bottle does it take to fill this prescription?

71. A board is $3\frac{3}{8}$ feet long. A carpenter needs $\frac{2}{3}$ of the board to make a shelf. How long is the piece he needs?

72. A tree is $26\frac{2}{3}$ feet high. A woodpecker has drilled a hole in the tree $\frac{3}{4}$ of the way to the top. How far is the hole above the ground?

Recall from Chapter 1 that the formula for the area of a rectangle is A = LW where A represents the area, L represents the length, and W represents the width. If the length of a rectangle is $2\frac{3}{4}$ inches and the width is $3\frac{2}{3}$ inches, then $A = 2\frac{3}{4} \cdot 3\frac{2}{3} = \frac{11}{4} \cdot \frac{11}{3} = \frac{121}{12}$ square inches. Find the area of each of the following rectangles:

73. $L = 2\frac{1}{2}$ cm and $W = 3\frac{3}{4}$ cm

74. $L = 3\frac{3}{5}$ m and $W = 4\frac{2}{3}$ m

75. $L = 12$ ft and $W = 3\frac{1}{6}$ ft

76. $L = 8\frac{2}{3}$ yd and $W = 6$ yd

Recall that the formula for the area of a triangle is $A = \frac{1}{2}$ bh. where A is the area, b is the base, and h is the height. If the base of a triangle is $\frac{3}{4}$ feet and the height is $\frac{3}{5}$ feet, then $A = \frac{1}{2} \cdot \frac{3}{4} \cdot \frac{3}{5} = \frac{9}{40}$ square feet. Find the areas of the following triangles:

77. $b = \frac{7}{8}$ m and $h = \frac{9}{10}$ m

78. $b = \frac{7}{12}$ yd and $h = \frac{5}{6}$ yd

79. $b = 2\frac{1}{6}$ ft and $h = 3\frac{1}{4}$ ft

80. $b = 4\frac{3}{10}$ m and $h = 5\frac{3}{5}$ m

Recall that the formula for the volume of a rectangular box is V = LWH where L is the length, W is the width, and H is the height. If the length is $2\frac{1}{4}$ inches, the width is $3\frac{1}{5}$ inches, and the height is $1\frac{3}{8}$ inches, then $V = 2\frac{1}{4} \cdot 3\frac{1}{5} \cdot 1\frac{3}{8} = \frac{9}{4} \cdot \frac{16}{5} \cdot \frac{11}{8} = \frac{99}{10}$ cubic inches. Find the volumes of the following rectangular boxes:

81. $L = 2\frac{3}{4}$ cm, $W = 3\frac{3}{11}$ cm, and $H = 1\frac{2}{9}$ cm

82. $L = 3\frac{3}{5}$ in., $W = 2\frac{2}{9}$ in., and $H = 2\frac{1}{10}$ in.

83. $L = 3\frac{1}{5}$ yd, $W = 10$ yd, and $H = 2\frac{1}{4}$ yd

84. $L = 2\frac{5}{8}$ m, $W = 16$ m, and $H = 4\frac{2}{3}$ m

Challenge Exercises:

Recall that the formula for finding the area of a trapezoid is $A = \frac{1}{2}h(B + b)$ where A represents the area, h represents the height, B represents the longer of the two parallel sides, and b represents the shorter of the two parallel sides. See the figure in the margin. Find the areas of the following trapezoids:

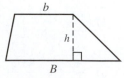

85. $h = 1\frac{2}{3}$ in., $B = 8$ in., $b = 4$ in.

86. $h = 2\frac{2}{5}$ cm, $B = 8$ cm, $b = 7$ cm

Writing Exercise:

87. In finding the product of fractions, why are we allowed to divide by factors common to the numerators and the denominators before multiplying?

Section 6.5	Further Multiplication of Rational Expressions

OBJECTIVES

When you complete this section, you will be able to:

a. Multiply rational expressions whose numerators and denominators are monomials.

b. Multiply rational expressions whose numerators and denominators are not monomials.

c. Multiply rational expressions that contain factors of the form $a - b$ and $b - a$.

Introduction

In Section 6.4, we multiplied rational numbers and rational expressions with monomial numerators and denominators using the following procedure:

1. Factor the numerators and the denominators into prime factors.
2. Divide the numerators and the denominators by factors common to both.
3. Multiply the remaining factors in the numerators and divide this by the product of the remaining factors in the denominators.

Products of rational expressions whose numerators and denominators are rational expressions

This is the same technique we will use to multiply rational expressions whose numerators and/or denominators are not integers or monomials. The only difference is the manner in which we factor the numerators and the denominators, since they are polynomials instead of integers or monomials. Compare the following. Assume the variables cannot have values that make any denominator equal to 0.

Monomials

$$\frac{a^2b^2}{a^3b^3} \cdot \frac{a^2b^4}{a^3b}$$

$$\frac{a \cdot a \cdot b \cdot b}{a \cdot a \cdot a \cdot b \cdot b \cdot b} \cdot \frac{a \cdot a \cdot b \cdot b \cdot b \cdot b}{a \cdot a \cdot a \cdot b}$$

$$\frac{\cancel{a} \cdot \cancel{a} \cdot \cancel{b} \cdot \cancel{b}}{\cancel{a} \cdot \cancel{a} \cdot \cancel{a} \cdot \cancel{b} \cdot \cancel{b} \cdot \cancel{b}} \cdot \frac{\cancel{a} \cdot \cancel{a} \cdot \cancel{b} \cdot \cancel{b} \cdot b \cdot b}{\cancel{a} \cdot a \cdot a \cdot \cancel{b}}$$

$$\frac{b^2}{a^2}$$

Nonmonomials

$$\frac{2x - 6}{x^2 + x - 2} \cdot \frac{x^2 + 3x - 4}{x^2 - 9}$$

$$\frac{2(x - 3)}{(x + 2)(x - 1)} \cdot \frac{(x + 4)(x - 1)}{(x + 3)(x - 3)}$$

$$\frac{2\cancel{(x - 3)}}{(x + 2)\cancel{(x - 1)}} \cdot \frac{(x + 4)\cancel{(x - 1)}}{(x + 3)\cancel{(x - 3)}}$$

$$\frac{2(x + 4)}{(x + 2)(x + 3)}$$

Factor the numerators and denominators.
Divide by the common factors.
Multiply the remaining factors.
Product.

Example 1

Find the following products. Leave answers reduced to lowest terms.

a. $\dfrac{a + b}{2x^2} \cdot \dfrac{4x}{(a + b)^2} =$ Factor the numerators and the denominators.

$\dfrac{a + b}{2 \cdot x \cdot x} \cdot \dfrac{2 \cdot 2 \cdot x}{(a + b)(a + b)} =$ Divide by the common factors.

$\dfrac{\cancel{a + b}}{2 \cdot \cancel{x} \cdot x} \cdot \dfrac{\cancel{2} \cdot 2 \cdot \cancel{x}}{\cancel{(a + b)}(a + b)} =$ Multiply the remaining factors.

$\dfrac{2}{x(a + b)}$ Product.

b. $\dfrac{2x - 6}{3y + 12} \cdot \dfrac{2y + 8}{5x - 15} =$ Factor the numerators and the denominators.

$\dfrac{2(x - 3)}{3(y + 4)} \cdot \dfrac{2(y + 4)}{5(x - 3)} =$ Divide by the common factors.

$\dfrac{2\cancel{(x - 3)}}{3\cancel{(y + 4)}} \cdot \dfrac{2\cancel{(y + 4)}}{5\cancel{(x - 3)}} =$ Multiply the remaining factors.

$\dfrac{2 \cdot 2}{3 \cdot 5} =$ Perform the multiplications.

$\dfrac{4}{15}$ Product.

$$\text{c. } \frac{x^2 - x - 6}{x^2 + 2x - 15} \cdot \frac{x^2 + 4x - 12}{x^2 - 4} =$$

Factor the numerators and the denominators.

$$\frac{(x + 2)(x - 3)}{(x + 5)(x - 3)} \cdot \frac{(x - 2)(x + 6)}{(x + 2)(x - 2)} =$$

Divide by the common factors.

$$\frac{(x + 2)(x - 3)}{(x + 5)(x - 3)} \cdot \frac{(x - 2)(x + 6)}{(x + 2)(x - 2)} =$$

Multiply the remaining factors.

$$\frac{x + 6}{x + 5}$$

Product.

Practice Exercises

Find the following products. Leave answers reduced to lowest terms.

1. $\dfrac{3x^2 y}{(w + z)^2} \cdot \dfrac{w + z}{4xy^3}$

2. $\dfrac{3a + 12}{5b - 30} \cdot \dfrac{b - 6}{4a + 16}$

3. $\dfrac{a^2 + 6a + 8}{a^2 + 5a - 6} \cdot \dfrac{a^2 + 3a - 18}{a^2 + a - 12}$

If more practice is needed, do the Additional Practice Exercises.

Additional Practice Exercises

Find the following products. Leave answers reduced to lowest terms.

a. $\dfrac{5rs}{(r + s)^2} \cdot \dfrac{2(r + s)}{7r^2 s}$

b. $\dfrac{4a - 12}{5b + 20} \cdot \dfrac{2b + 8}{3a - 9}$

c. $\dfrac{a^2 + a - 2}{a^2 - 3a - 10} \cdot \dfrac{a^2 - a - 20}{a^2 + 3a - 4}$

Products of rational expressions containing factors of the form $a - b$ and $b - a$

In Section 6.3, we reduced rational expressions and sometimes found it necessary to remove "-1" from a factor of the numerator or the denominator. This occurred if the terms of a factor in the numerator and the denominator were exactly the same except that the signs were opposites. For example, $\frac{a - b}{b - a}$. The same situation may occur when multiplying rational expressions.

Example 2

Find the following products. Reduce answers to lowest terms.

a. $\dfrac{4 - a}{5} \cdot \dfrac{8}{a - 4} =$

Remove a "-1" from $4 - a$.

$$\frac{-1(-4 + a)}{5} \cdot \frac{8}{a - 4} =$$

Commutative property, $-4 + a = a - 4$.

$$\frac{-1(a - 4)}{5} \cdot \frac{8}{a - 4} =$$

Divide by the common factors.

$$\frac{-1(a - 4)}{5} \cdot \frac{8}{a - 4} =$$

Multiply the remaining factors.

$$\frac{-8}{5} \text{ or } -\frac{8}{5}$$

Product. $\frac{-8}{5} = -\frac{8}{5}$, since a negative divided by a positive is negative.

Answers:

Practice Exercises 1–3: 1. $\frac{3x}{4y^2(w + z)}$ 2. $\frac{3}{20}$ 3. $\frac{a + 2}{a - 1}$ Additional Practice Exercises a–c: a. $\frac{10}{7r(r + s)}$ b. $\frac{8}{15}$ c. 1

b. $\dfrac{3-a}{a^2+3a+2}\cdot\dfrac{a^2-4}{a^2+2a-15}=$

Factor everything that is not prime. $3-a$ and $a-3$ are the same except the signs are opposites.

$\dfrac{3-a}{(a+2)(a+1)}\cdot\dfrac{(a+2)(a-2)}{(a-3)(a+5)}=$

Factor a "-1" from $3-a$.

$\dfrac{-1(-3+a)}{(a+2)(a+1)}\cdot\dfrac{(a+2)(a-2)}{(a-3)(a+5)}=$

Rewrite $-3+a$ as $a-3$.

$\dfrac{-1(a-3)}{(a+2)(a+1)}\cdot\dfrac{(a+2)(a-2)}{(a-3)(a+5)}=$

Divide by the common factors.

$\dfrac{-1(a-3)}{(a+2)(a+1)}\cdot\dfrac{(a+2)(a-2)}{(a-3)(a+5)}=$

Multiply the remaining factors.

$\dfrac{-1(a-2)}{(a+1)(a+5)}$

Product.

Note: The answer to Example 2b could be written as $-\dfrac{a-2}{(a+1)(a+5)}$ or $\dfrac{2-a}{(a+1)(a+5)}$. Why?

Practice Exercises

Find the following products. Leave answers reduced to lowest terms.

4. $\dfrac{7-x}{a}\cdot\dfrac{b}{x-7}$

5. $\dfrac{y^2+3y-10}{y^2-25}\cdot\dfrac{y+4}{2-y}$

If more practice is needed, do the Additional Practice Exercises.

Additional Practice Exercises

Find the following products. Leave answers reduced to lowest terms.

d. $\dfrac{rs}{b-2}\cdot\dfrac{2-b}{2a}$

e. $\dfrac{m^2-7m+12}{m^2-2m-8}\cdot\dfrac{m-6}{3-m}$

Exercise Set 6.5

Find the following products. Leave answers reduced to lowest terms.

1. $\dfrac{(x+y)^2}{6a}\cdot\dfrac{8a^2}{x+y}$

2. $\dfrac{12x}{(a+b)^2}\cdot\dfrac{a+b}{16x^2}$

3. $\dfrac{3x^2y^4}{10(r+s)}\cdot\dfrac{5(r+s)^2}{12xy^2}$

4. $\dfrac{6(p+q)^2}{5rs}\cdot\dfrac{15r^3s^4}{18(p+q)}$

5. $\dfrac{3x+18}{7x-7}\cdot\dfrac{2x-2}{5x+30}$

6. $\dfrac{2r-8}{3r+6}\cdot\dfrac{5r+10}{3r-12}$

7. $\dfrac{8x+12}{18x-24}\cdot\dfrac{9x-12}{10x+15}$

8. $\dfrac{24x+16}{30x-18}\cdot\dfrac{45x-27}{18x+12}$

9. $\dfrac{4a-16}{3a-3}\cdot\dfrac{a^2-1}{a^2-16}$

10. $\dfrac{x^2-9}{x+3}\cdot\dfrac{x-5}{x^2-25}$

11. $\dfrac{4x^2-25}{3x-4}\cdot\dfrac{9x^2-16}{2x+5}$

12. $\dfrac{16x^2-9}{4x+3}\cdot\dfrac{25x^2-1}{5x+1}$

13. $\dfrac{a-6}{7} \cdot \dfrac{9}{6-a}$

14. $\dfrac{b-4}{5} \cdot \dfrac{3}{4-b}$

15. $\dfrac{3x-4}{14} \cdot \dfrac{21}{4-3x}$

16. $\dfrac{5-2b}{18} \cdot \dfrac{12}{2b-5}$

17. $\dfrac{4x^2y^3}{2x-6} \cdot \dfrac{12-4x}{6x^3y}$

18. $\dfrac{6b^3c^3}{70-10b} \cdot \dfrac{5b-35}{8b^2c^5}$

19. $\dfrac{x-3}{x^2+x-2} \cdot \dfrac{x+2}{x^2+x-12}$

20. $\dfrac{a-4}{a^2-2a-3} \cdot \dfrac{a+1}{a^2-2a-8}$

21. $\dfrac{x^2-4}{x^2-3x-10} \cdot \dfrac{x^2-8x+15}{x^2-9}$

22. $\dfrac{x^2+3x-10}{x^2-16} \cdot \dfrac{x^2-9x+20}{x^2-25}$

23. $\dfrac{x^2-2x-15}{x^2+x-30} \cdot \dfrac{x^2+5x-6}{x^2+7x+12}$

24. $\dfrac{a^2-2a-24}{a^2+9a+20} \cdot \dfrac{a^2-3a-10}{a^2-4a-12}$

25. $\dfrac{x^2+2x-15}{x^2-9x+18} \cdot \dfrac{x^2-4x-12}{x^2+3x-10}$

26. $\dfrac{y^2-7y+10}{y^2+4y-12} \cdot \dfrac{y^2+10y+24}{y^2-2y-15}$

27. $\dfrac{6x^2+x-12}{2x^2-5x-12} \cdot \dfrac{3x^2-14x+8}{9x^2-18x+8}$

28. $\dfrac{2x^2+3x-20}{4x^2+11x-3} \cdot \dfrac{4x^2-9x+2}{2x^2-9x+10}$

29. $\dfrac{3x^2-5x-2}{4x^2-11x+6} \cdot \dfrac{4x^2+5x-6}{3x^2-8x-3}$

30. $\dfrac{8x^2+10x-3}{3x^2+4x-4} \cdot \dfrac{x^2+6x+8}{2x^2+11x+12}$

31. $\dfrac{b^2-16}{b^2+b-12} \cdot \dfrac{b^2+2b-15}{4-b}$

32. $\dfrac{y^2-y-20}{y^2-2y-24} \cdot \dfrac{y+3}{5-y}$

33. $\dfrac{2x^2-11x+12}{2x^2+11x-21} \cdot \dfrac{3x^2+20x-7}{4-x}$

34. $\dfrac{3x^2-17x+10}{3x^2+7x-6} \cdot \dfrac{2x^2+x-15}{5-x}$

Challenge Exercises:

Find the following products. Reduce answers to lowest terms.

35. $\dfrac{ac+3a+2c+6}{ad+a+2d+2} \cdot \dfrac{ad-5a+2d-10}{bc+3b-4c-12}$

36. $\dfrac{mn+2m+4n+8}{np-3n+2p-6} \cdot \dfrac{pq+5p-3q-15}{mq+4m+4q+16}$

37. $\dfrac{ac-ad+bc-bd}{am+an-bm-bn} \cdot \dfrac{ac+ad-bc-bd}{ac+ad+bc+bd}$

38. $\dfrac{xw-xz-yw+yz}{xa-xb+ya-yb} \cdot \dfrac{xw+xz+yw+yz}{xw-xz-yw-yz}$

39. $\dfrac{a^3-b^3}{a^2+ab+b^2} \cdot \dfrac{2a^2+ab-b^2}{a^2-b^2}$

40. $\dfrac{3x^2+10x-8}{x^2-4} \cdot \dfrac{x^3-8}{x^2+2x+4}$

Writing Exercises:

41. What is wrong with the following? Rework the problem correctly.

$\dfrac{a^2+b^2}{a^2-b^2} \cdot \dfrac{a^2+2ab-3b^2}{a^2+ab-6b^2} = \dfrac{(a+b)(a+b)}{(a+b)(a-b)} \cdot \dfrac{(a-b)(a+3b)}{(a+3b)(a-2b)} =$

$\dfrac{\cancel{(a+b)}(a+b)}{\cancel{(a+b)}(a-b)} \cdot \dfrac{(a-b)(a+3b)}{(a+3b)(a-2b)} = \dfrac{a+b}{a-2b}$

42. What is wrong with the following? Rework the problem correctly.

$\dfrac{a^2+b^2}{a^2-b^2} \cdot \dfrac{a^2+2ab-3b^2}{a^2+ab-6b^2} = \dfrac{\cancel{a^2}+\cancel{b^2}}{\cancel{a^2}-\cancel{b^2}} \cdot \dfrac{\cancel{a^2}+2ab-\cancel{3b^2}}{\cancel{a^2}+ab-\cancel{6b^2}} =$

$\dfrac{2ab}{ab-2}$

Division of Rational Numbers and Expressions

OBJECTIVES

When you complete this section, you will be able to:

a. Divide rational numbers.

b. Divide rational expressions.

Introduction   In Sections 6.4 and 6.5, we multiplied rational numbers and rational expressions. In this section, we will learn how to divide rational numbers and rational expressions. Before we discuss division of rational numbers and expressions, we need to review the multiplicative inverse, also called the reciprocal. Two numbers are multiplicative inverses if their product is 1. For example, $\frac{3}{4}$ and $\frac{4}{3}$ are multiplicative inverses (reciprocals) since $\frac{3}{4} \cdot \frac{4}{3} = 1$. In general, for a and $b \neq 0$, the multiplicative inverse of $\frac{a}{b}$ is $\frac{b}{a}$, since $\frac{a}{b} \cdot \frac{b}{a} = 1$. This procedure is sometimes referred to as "inverting the fraction." To find the multiplicative inverse of a number (other than 0), we interchange the numerator and the denominator. Remember, 0 does not have a multiplicative inverse.

Suppose a region is divided into two equal parts so that each part is $\frac{1}{2}$ of the original region as shown in the following figure:

Now divide each of the two equal regions into two equal parts by drawing a horizontal line segment through the figure. Now further shade one of the two parts that was already shaded.

From the figure, we can see that this one part that we further shaded is $\frac{1}{4}$ of the original region. We have taken $\frac{1}{2}$ of the region and divided it by 2. This resulted in $\frac{1}{4}$ of the original region. In symbols, $\frac{1}{2} \div 2 = \frac{1}{4}$. Notice that $\frac{1}{2} \cdot \frac{1}{2} = \frac{1}{4}$ also. Consequently, $\frac{1}{2} \div 2 = \frac{1}{2} \cdot \frac{1}{2}$. Without shading parts of a figure, we can see that $6 \div 3 = 2$ and $6 \cdot \frac{1}{3} = 2$ also. Thus, $6 \div 3 = 6 \cdot \frac{1}{3}$.

This suggests that dividing by a whole number, except for 0, is the same as multiplying by its multiplicative inverse (reciprocal).

To show that the same procedure applies if we are dividing by a fraction is a little more difficult. Remember, any fraction represents division and division may be represented as a fraction. For example, $\frac{2}{3} = 2 \div 3$, and $5 \div 8 = \frac{5}{8}$. Suppose we have $\frac{2}{3} \div \frac{7}{5}$.

Since division may be represented as a fraction, $\frac{2}{3} \div \frac{7}{5}$ may be represented as $\dfrac{\frac{2}{3}}{\frac{7}{5}}$.

In Sections 6.2 and 6.3, we reduced fractions by dividing the numerator and the denominator by factors common to both. We are now going to do something similar by multiplying the numerator and the denominator by the same number. This is permissible since we are actually multiplying by 1 (any number, except 0, divided by itself equals 1). You will not be asked to actually do this in this section but simply to understand that it can be done. This procedure will be discussed in great detail in Chapter 7. Study the following:

$$\frac{2}{3} \div \frac{7}{5} =$$ Rewrite as a fraction.

$$\frac{\frac{2}{3}}{\frac{7}{5}} =$$ Multiply the numerator and the denominator by $\frac{5}{7}$.

$$\frac{\frac{2}{3} \cdot \frac{5}{7}}{\frac{7}{5} \cdot \frac{5}{7}} =$$ $\frac{7}{5} \cdot \frac{5}{7} = 1$.

$$\frac{\frac{2}{3} \cdot \frac{5}{7}}{1} =$$ Divide by 1.

$$\frac{2}{3} \cdot \frac{5}{7} =$$ Therefore, $\frac{2}{3} \div \frac{7}{5} = \frac{2}{3} \cdot \frac{5}{7}$. Notice the division by $\frac{7}{5}$ was changed into multiplication by $\frac{5}{7}$.

$$\frac{2 \cdot 5}{3 \cdot 7} =$$ Multiplication of fractions.

$$\frac{10}{21}$$ Quotient.

The preceding two cases suggest the following rule for the division of rational numbers:

Division of Rational Numbers

For all integers a, b, c, and d with b, c, and $d \neq 0, \frac{a}{b} \div \frac{c}{d} = \frac{a}{b} \cdot \frac{d}{c}$. In words, to divide by a number, multiply by its multiplicative inverse (reciprocal).

This rule is usually stated more loosely as "when dividing rational numbers, invert the one on the right and change the operation to multiplication." It is this form of the rule that we will use in the following examples.

We summarize the procedure for finding quotients of rational expressions as follows:

Quotients of Rational Expressions

To find the quotient of two rational expressions:

1. Invert the rational expression on the right and change the operation to multiplication.

2. Factor all numerators and denominators completely.

3. Divide by all factors common to the numerators and the denominators.

4. Find the product of all remaining factors.

Example 1

Find the following quotients. Leave answers reduced to lowest terms.

a. $\dfrac{2}{3} \div \dfrac{5}{3} =$ Invert the number on the right and change the operation to multiplication.

$\dfrac{2}{3} \cdot \dfrac{3}{5} =$ Divide by the common factor of 3.

$\dfrac{2}{\cancel{3}} \cdot \dfrac{\cancel{3}}{5} =$ Multiply the remaining factors.

$\dfrac{2}{5}$ Quotient.

b. $-\dfrac{5}{8} \div 3\dfrac{3}{4} =$ Change the mixed number to an improper fraction.

$-\dfrac{5}{8} \div \dfrac{15}{4} =$ Invert the number on the right and change the operation to multiplication.

$-\dfrac{5}{8} \cdot \dfrac{4}{15} =$ Write the numerators and the denominators in terms of prime factors.

$-\dfrac{5}{2 \cdot 2 \cdot 2} \cdot \dfrac{2 \cdot 2}{3 \cdot 5} =$ Divide by the common factors.

$-\dfrac{\cancel{5}}{\cancel{2} \cdot \cancel{2} \cdot 2} \cdot \dfrac{\cancel{2} \cdot \cancel{2}}{3 \cdot \cancel{5}} =$ Multiply the remaining factors.

$-\dfrac{1}{6}$ Quotient.

c. $4\dfrac{2}{5} \div 11 =$ Change the mixed number to an improper fraction and write the whole number 11 as the fraction $\dfrac{11}{1}$.

$\dfrac{22}{5} \div \dfrac{11}{1} =$ Invert the number on the right and change the operation to multiplication.

$\dfrac{22}{5} \cdot \dfrac{1}{11} =$ Write the numerators and the denominators in terms of prime factors.

$\dfrac{2 \cdot 11}{5} \cdot \dfrac{1}{11} =$ Divide by the common factor of 11.

$\dfrac{2 \cdot \cancel{11}}{5} \cdot \dfrac{1}{\cancel{11}} =$ Multiply the remaining factors.

$\dfrac{2}{5}$ Quotient.

d. $-3\dfrac{1}{3} \div \left(-2\dfrac{1}{7}\right) =$ Change the mixed numbers into improper fractions.

$-\dfrac{10}{3} \div \left(-\dfrac{15}{7}\right) =$ Invert the fraction on the right and change the operation to multiplication.

$-\dfrac{10}{3} \cdot \left(-\dfrac{7}{15}\right) =$ Write the numerators and the denominators in terms of prime factors.

$-\dfrac{2 \cdot 5}{3} \cdot \left(-\dfrac{7}{3 \cdot 5}\right) =$ Divide by the common factor of 5.

$-\dfrac{2 \cdot \cancel{5}}{3} \cdot \left(-\dfrac{7}{3 \cdot \cancel{5}}\right) =$ Multiply the remaining factors.

$\dfrac{14}{9}$ Quotient.

Practice Exercises

Find the following quotients. Leave answers reduced to lowest terms.

1. $\dfrac{3}{5} \div \dfrac{7}{5}$

2. $\dfrac{10}{7} \div \left(-2\dfrac{1}{2}\right)$

3. $3\dfrac{3}{7} \div 8$

4. $-5\dfrac{1}{3} \div \left(-2\dfrac{2}{5}\right)$

If more practice is needed, do the Additional Practice Exercises.

Additional Practice Exercises

Find the following quotients. Leave answers reduced to lowest terms.

a. $-\dfrac{5}{8} \div \dfrac{3}{8}$

b. $\dfrac{8}{5} \div 2\dfrac{2}{3}$

c. $3\dfrac{3}{5} \div (-9)$

d. $-2\dfrac{1}{3} \div \left(-2\dfrac{4}{5}\right)$

Answers:

Practice Exercises 1–4: 1. $\dfrac{3}{7}$ 2. $-\dfrac{4}{7}$ 3. $\dfrac{3}{7}$ 4. $\dfrac{20}{9}$ Additional Practice Exercises a–d: a. $-\dfrac{5}{3}$ b. $\dfrac{3}{5}$ c. $-\dfrac{2}{5}$ d. $\dfrac{5}{6}$

Division of rational expressions is done in exactly the same manner as division of rational numbers. We will invert the rational expression on the right and change the operation to multiplication.

Example 2

Find the following quotients. Leave answers reduced to lowest terms.

a. $\dfrac{a^2b^3}{xy^2} \div \dfrac{ab^4}{x^2y^3} =$

Invert the rational expression on the right and change the operation to multiplication.

$\dfrac{a^2b^3}{xy^2} \cdot \dfrac{x^2y^3}{ab^4} =$

Write the numerators and the denominators in terms of prime factors.

$\dfrac{a \cdot a \cdot b \cdot b \cdot b}{x \cdot y \cdot y} \cdot \dfrac{x \cdot x \cdot y \cdot y \cdot y}{a \cdot b \cdot b \cdot b \cdot b} =$

Divide by the factors common to the numerator and the denominator.

$\dfrac{\cancel{a} \cdot a \cdot \cancel{b} \cdot \cancel{b} \cdot \cancel{b}}{\cancel{x} \cdot \cancel{y} \cdot \cancel{y}} \cdot \dfrac{x \cdot x \cdot \cancel{y} \cdot \cancel{y} \cdot y}{\cancel{a} \cdot \cancel{b} \cdot \cancel{b} \cdot \cancel{b} \cdot b} =$

Multiply the remaining factors.

$\dfrac{axy}{b}$

Quotient.

b. $\dfrac{4x + 8}{6} \div \dfrac{5x + 10}{9} =$

Invert the rational expression on the right and change the operation to multiplication.

$\dfrac{4x + 8}{6} \cdot \dfrac{9}{5x + 10} =$

Factor the numerators and the denominators into prime factors.

$\dfrac{2 \cdot 2(x + 2)}{2 \cdot 3} \cdot \dfrac{3 \cdot 3}{5(x + 2)} =$

Divide by the factors common to the numerator and the denominator.

$\dfrac{\cancel{2} \cdot 2(\cancel{x + 2})}{\cancel{2} \cdot \cancel{3}} \cdot \dfrac{\cancel{3} \cdot 3}{5(\cancel{x + 2})} =$

Multiply the remaining factors.

$\dfrac{6}{5}$

Quotient.

c. $\dfrac{2x^2 + 7x + 3}{x^2 - 9} \div \dfrac{2x^2 + 11x + 5}{x^2 - 3x} =$

Invert the rational expression on the right and multiply.

$\dfrac{2x^2 + 7x + 3}{x^2 - 9} \cdot \dfrac{x^2 - 3x}{2x^2 + 11x + 5} =$

Factor the numerators and the denominators.

$\dfrac{(2x + 1)(x + 3)}{(x + 3)(x - 3)} \cdot \dfrac{x(x - 3)}{(2x + 1)(x + 5)} =$

Divide by common factors.

$\dfrac{(\cancel{2x + 1})(\cancel{x + 3})}{(\cancel{x + 3})(\cancel{x - 3})} \cdot \dfrac{x(\cancel{x - 3})}{(\cancel{2x + 1})(x + 5)} =$

Multiply the remaining factors.

$\dfrac{x}{x + 5}$

Quotient.

Practice Exercises

Find the following quotients. Leave answers reduced to lowest terms.

5. $\dfrac{r^3s^2}{p^2q^2} \div \dfrac{rs^3}{p^4q}$

6. $\dfrac{8x}{2x + 6} \div \dfrac{6}{4x + 12}$

7. $\dfrac{2x^2 + 7x + 6}{x^2 - 4} \div \dfrac{2x^2 - 3x - 9}{x^2 - 2x}$

If more practice is needed, do the Additional Practice Exercises.

Additional Practice Exercises

Find the following quotients. Leave answers reduced to lowest terms.

e. $\dfrac{m^4 n^2}{a^3 b^2} \div \dfrac{m^2 n^5}{a b^4}$

f. $\dfrac{10}{5x + 20} \div \dfrac{6}{8x + 32}$

g. $\dfrac{x^2 - x - 6}{x^2 - 16} \div \dfrac{3x^2 - 8x - 3}{x^2 + 4x}$

Exercise Set 6.6

Find the following quotients of fractions. Leave answers reduced to lowest terms.

1. $\dfrac{1}{4} \div \dfrac{4}{7}$

2. $\dfrac{1}{5} \div \dfrac{5}{8}$

3. $\dfrac{3}{7} \div \left(-\dfrac{5}{7}\right)$

4. $-\dfrac{4}{9} \div \dfrac{7}{9}$

5. $\dfrac{2}{3} \div \dfrac{5}{9}$

6. $\dfrac{3}{4} \div \dfrac{7}{12}$

7. $-\dfrac{5}{8} \div \left(-\dfrac{11}{32}\right)$

8. $-\dfrac{8}{15} \div \left(-\dfrac{3}{5}\right)$

9. $-\dfrac{18}{25} \div \dfrac{9}{20}$

10. $\dfrac{28}{45} \div \left(-\dfrac{24}{25}\right)$

11. $\dfrac{8}{9} \div \dfrac{2}{3}$

12. $\dfrac{14}{15} \div \dfrac{7}{3}$

13. $6 \div \dfrac{2}{3}$

14. $12 \div \dfrac{3}{4}$

15. $-72 \div \dfrac{8}{3}$

16. $-144 \div \dfrac{16}{9}$

17. $24 \div \left(-\dfrac{15}{7}\right)$

18. $36 \div \left(-\dfrac{24}{5}\right)$

19. $\dfrac{10}{9} \div 5$

20. $\dfrac{2}{7} \div 4$

21. $-\dfrac{28}{9} \div (-20)$

22. $-\dfrac{30}{11} \div (-24)$

23. $2\dfrac{2}{5} \div \dfrac{4}{3}$

24. $3\dfrac{1}{3} \div \dfrac{3}{5}$

25. $-1\dfrac{2}{3} \div \dfrac{15}{16}$

26. $-2\dfrac{4}{5} \div \dfrac{7}{10}$

27. $3\dfrac{1}{2} \div 3\dfrac{3}{8}$

28. $4\dfrac{2}{3} \div 1\dfrac{2}{5}$

29. $1\dfrac{7}{8} \div \left(-4\dfrac{1}{6}\right)$

30. $2\dfrac{4}{7} \div \left(-2\dfrac{5}{8}\right)$

31. $15 \div 1\dfrac{3}{7}$

32. $24 \div 2\dfrac{5}{8}$

33. $-6\dfrac{2}{3} \div 8$

34. $2\dfrac{8}{11} \div (-12)$

35. $4\dfrac{4}{5} \div 6$

36. $3\dfrac{1}{7} \div 11$

Solve the following:

37. If $2\dfrac{3}{4}$ ounces of silver cost \$44, what is the cost of one ounce?

38. If a developer paid \$286,000 for $4\dfrac{2}{5}$ acres of land, how much did she pay for 1 acre?

39. How many $\dfrac{1}{4}$-acre lots can a developer get from a $3\dfrac{1}{2}$-acre tract of land?

40. How many servings of $\dfrac{2}{3}$ pound each can be gotten from a 12-pound bag of dog food?

41. A bag of cookies that is $\dfrac{2}{3}$ full is divided equally among four people. What part of a full bag did each receive?

42. It takes 10 gallons of water to fill an aquarium $\dfrac{2}{5}$ full. How many gallons of water does the aquarium hold?

Answers:

Additional Practice Exercises e–g: e. $\dfrac{a^2 m^2}{b^3 n^3}$ f. $\dfrac{8}{3}$ g. $\dfrac{x(x + 2)}{(x - 4)(3x + 1)}$

Find the following quotients of rational expressions. Leave answers reduced to lowest terms.

43. $\dfrac{a}{b^4} \div \dfrac{a^3}{b^2}$

44. $\dfrac{x^3}{y^3} \div \dfrac{x^5}{y^2}$

45. $\dfrac{8a^2}{9b^3} \div \dfrac{4a^3}{3b^2}$

46. $\dfrac{12c^2}{7d^3} \div \dfrac{8c^4}{21d^4}$

47. $\dfrac{a^3b^2}{c^4d^3} \div \dfrac{ab^5}{c^2d}$

48. $\dfrac{m^4n^6}{p^2q^3} \div \dfrac{m^6n^3}{p^4q^4}$

49. $\dfrac{8x^7g^3}{15x^2g^4} \div \dfrac{3xg^2}{4x^2g^3}$

50. $\dfrac{10c^5g^5}{12c^2g^4} \div \dfrac{5cg^2}{4c^2g^3}$

51. $\dfrac{5x + 15}{8} \div \dfrac{7x + 21}{6}$

52. $\dfrac{3x - 18}{8} \div \dfrac{4x + 24}{12}$

53. $\dfrac{4a - 8}{15a} \div \dfrac{3a - 6}{5a^2}$

54. $\dfrac{6b + 24}{7b^2} \div \dfrac{7b + 28}{14b^4}$

55. $\dfrac{4p + 12q}{3p - 6q} \div \dfrac{5p + 15q}{6p - 12q}$

56. $\dfrac{5r + 10s}{3r - 12s} \div \dfrac{6r + 12s}{4r - 16s}$

57. $\dfrac{10x + 15}{18x - 6} \div \dfrac{20x + 30}{9x - 3}$

58. $\dfrac{12y + 8}{4y - 10} \div \dfrac{18y + 12}{8y - 20}$

59. $\dfrac{a - b}{6a - 9b} \div \dfrac{b - a}{4a - 6b}$

60. $\dfrac{x - y}{12x - 3y} \div \dfrac{y - x}{8x - 2y}$

61. $\dfrac{h}{h^2 + 13h + 42} \div \dfrac{4h^2 + 28h}{4h + 24}$

62. $\dfrac{b}{b^2 + 11b + 30} \div \dfrac{6b + 30}{6b^2 + 36b}$

63. $\dfrac{d^2 - d - 12}{2d^2} \div \dfrac{3d^2 + 13d + 12}{d}$

64. $\dfrac{y^2 + y - 20}{7y^2} \div \dfrac{3y^2 + 19y + 20}{y}$

65. $\dfrac{m^2 - 25}{2m - 12} \div \dfrac{3m + 15}{m^2 - 36}$

66. $\dfrac{h^2 - 36}{3h - 21} \div \dfrac{4h + 24}{h^2 - 49}$

67. $\dfrac{x + 8}{x - 8} \div \dfrac{x^2 + 16x + 64}{x^2 - 16x + 64}$

68. $\dfrac{w + 3}{w - 3} \div \dfrac{w^2 + 6w + 9}{w^2 - 6w + 9}$

69. $\dfrac{a^2 - 5a + 6}{a^2 - 9a + 18} \div \dfrac{a^2 - 6a + 8}{a^2 - 9a + 20}$

70. $\dfrac{y^2 + 3y - 40}{y^2 + 3y - 18} \div \dfrac{y^2 + 2y - 48}{y^2 + 2y - 35}$

71. $\dfrac{2x^2 - 7x - 4}{3x^2 - 14x + 8} \div \dfrac{2x^2 + 7x + 3}{3x^2 - 8x + 4}$

72. $\dfrac{8x^2 + 6x - 9}{8x^2 + 14x - 15} \div \dfrac{2x^2 + 11x + 12}{4x^2 + 19x + 12}$

Challenge Exercises:

Find the following quotients:

73. $\dfrac{ab + 3a + 2b + 6}{bc + 4b + 3c + 12} \div \dfrac{ac - 3a + 2c - 6}{bc + 4b - 4c - 16}$

74. $\dfrac{xy - 3x + 4y - 12}{xy + 6x - 3y - 18} \div \dfrac{xy + 5x + 4y + 20}{xy + 5x - 3y - 15}$

Writing Exercises:

75. When performing the division $\dfrac{a}{b} \div \dfrac{c}{d}$, why must b, c, and $d \neq 0$?

76. Write and solve an application problem involving rational numbers.

| Section 6.7 | Division of Polynomials (Long Division) |

O B J E C T I V E   *When you complete this section, you will be able to:*

Divide two polynomials using long division.

Introduction   In Section 6.2, we reduced fractions by factoring the numerator and the denominator and then dividing by the factors that were common to both. This is one method by which we can divide polynomials, but it has some limitations. What if the numerator and the denominator have no common factors? Can we still divide the polynomials? We can, and the method by which we perform this type of division is the topic of this section.

In Chapter 0, we discussed dividing whole numbers using long division. The procedure for dividing polynomials follows almost the exact same procedure. (Except it is often easier!) We will divide two whole numbers and two polynomials side by side and note the similarities in the procedure.

Divide 1058 by 23.

$$23\overline{)1058}$$

Divide $2x^2 + 5x - 12$ by $x + 4$.

$$x + 4\overline{)2x^2 + 5x - 12}$$

1. 23 will not go into 10, so divide 23 into 105. We guess 4. Put the 4 above the 5.

$$\begin{array}{r} 4 \\ 23\overline{)1058} \end{array}$$

1. Divide x into $2x^2$. There is no guessing as in long division of whole numbers. $\frac{2x^2}{x} = 2x$. Put the $2x$ above the $2x^2$.

$$\begin{array}{r} 2x \\ x + 4\overline{)2x^2 + 5x - 12} \end{array}$$

2. Multiply 4 and 23 and put the product beneath 105.

$$\begin{array}{r} 4 \\ 23\overline{)1058} \\ 92 \end{array}$$

2. Multiply $2x$ and $x + 4$ and put the product beneath the like terms of $2x^2 + 5x - 12$.

$$\begin{array}{r} 2x \\ x + 4\overline{)2x^2 + 5x - 12} \\ 2x^2 + 8x \end{array}$$

3. Subtract 92 from 105.

$$\begin{array}{r} 4 \\ 23\overline{)1058} \\ -92 \\ \hline 13 \end{array}$$

3. Subtract $2x^2 + 8x$ from $2x^2 + 5x$ by changing the signs of the bottom polynomial to $-2x^2 - 8x$ and adding.

$$\begin{array}{r} 2x \\ x + 4\overline{)2x^2 + 5x - 12} \\ -2x^2 - 8x \\ \hline -3x \end{array}$$

4. Bring down the 8.

$$\begin{array}{r} 4 \\ 23\overline{)1058} \\ -92 \\ \hline 138 \end{array}$$

4. Bring down the -12.

$$\begin{array}{r} 2x \\ x + 4\overline{)2x^2 + 5x - 12} \\ -2x^2 - 8x \\ \hline -3x - 12 \end{array}$$

5. Divide 23 into 138 and put the quotient over the 8.

$$\begin{array}{r} 46 \\ 23\overline{)1058} \\ -92 \\ \hline 138 \end{array}$$

5. Divide x into $-3x$ and put the quotient (which is -3) over the $5x$.

$$\begin{array}{r} 2x - 3 \\ x + 4\overline{)2x^2 + 5x - 12} \\ -2x^2 - 8x \\ \hline -3x - 12 \end{array}$$

6. Multiply 6 and 23 and put the product under 138.

$$23\overline{)1058}$$
$$\underline{-92}$$
$$138$$
$$\underline{138}$$
(quotient 46)

6. Multiply -3 and $x + 4$ and put the product (which is $-3x - 12$) under $-3x - 12$.

$$x + 4\overline{)2x^2 + 5x - 12}$$
(quotient $2x - 3$)
$$\underline{-2x^2 - 8x}$$
$$-3x - 12$$
$$\underline{-3x - 12}$$

7. Subtract 138 from 138.

$$23\overline{)1058}$$
(quotient 46)
$$\underline{-92}$$
$$138$$
$$\underline{-138}$$
$$0$$

7. Subtract $-3x - 12$ from $-3x - 12$ by changing signs of the bottom polynomial to $3x + 12$ and adding.

$$x + 4\overline{)2x^2 + 5x - 12}$$
(quotient $2x - 3$)
$$\underline{-2x^2 - 8x}$$
$$-3x - 12$$
$$\underline{+3x + 12}$$
$$0$$

8. Remainder of 0. Therefore, 23 divides into 1058 evenly.

Check:
$(23)(46) = 1058$, so 46 is the correct answer.

8. Remainder of 0. Therefore, $x + 4$ divides into $2x^2 + 5x - 12$ evenly.

Check:
$(x + 4)(2x - 3) = 2x^2 + 5x - 12$, so $2x - 3$ is the correct answer.

As you can see, the procedure is almost exactly the same in both cases. We illustrate the procedure for dividing polynomials with some more examples.

Example 1

Find the following quotients using long division:

a. $\dfrac{2x^2 + x + 15}{x + 3}$

Solution:

$\dfrac{2x^2 + x - 15}{x + 3}$ means $(2x^2 + x - 15) \div (x + 3)$, which is written as $x + 3\overline{)2x^2 + x - 15}$ using long division.

$x + 3\overline{)2x^2 + x - 15}$ (quotient $2x$) $\dfrac{2x^2}{x} = 2x$ and put the $2x$ over $2x^2$.

$x + 3\overline{)2x^2 + x - 15}$ (quotient $2x$), $2x^2 + 6x$ $2x(x + 3) = 2x^2 + 6x$ and put $2x^2 + 6x$ underneath the like terms of $2x^2 + x + 15$.

$x + 3\overline{)2x^2 + x - 15}$ (quotient $2x$), $\underline{-2x^2 - 6x}$, $-5x$ Subtract $2x^2 + 6x$ by changing all the signs and adding.

$x + 3\overline{)2x^2 + x - 15}$ (quotient $2x$), $\underline{-2x^2 - 6x}$, $-5x - 15$ Bring down -15.

$x + 3\overline{)2x^2 + x - 15}$ (quotient $2x - 5$), $\underline{-2x^2 - 6x}$, $-5x - 15$ $\dfrac{-5x}{x} = -5$ and put the -5 over the $+x$.

$x + 3\overline{)2x^2 + x - 15}$ (quotient $2x - 5$) $-5(x + 3) = -5x - 15$ and put $-5x - 15$ under the other $-5x - 15$.
$\underline{2x^2 - 6x}$
$-5x - 15$
$\underline{-5x - 15}$

$x + 3\overline{)2x^2 + x - 15}$ (quotient $2x - 5$)   Subtract $-5x - 15$ by changing signs and adding.
$\underline{2x^2 - 6x}$
$-5x - 15$
$\underline{+5x + 15}$
0

The remainder is 0. Therefore, $x + 3$ divides into $2x^2 + x - 15$ evenly. Hence $x + 3$ is a factor of $2x^2 + x - 15$.

Check:
$(x + 3)(2x - 5) = 2x^2 + x - 15$, so $2x - 5$ is the correct answer.

b. $\dfrac{6x^2 + 11x - 10}{3x - 2}$

Solution:

Using long division, this is written as

$3x - 2\overline{)6x^2 + 11x - 10}.$

$\begin{array}{r} 2x \\ 3x - 2\overline{)6x^2 + 11x - 10} \end{array}$ $\dfrac{6x^2}{3x} = 2x$ and put the $2x$ over $6x^2$.

$\begin{array}{r} 2x \\ 3x - 2\overline{)6x^2 + 11x - 10} \\ 6x^2 - \ 4x \end{array}$ $2x(3x - 2) = 6x^2 - 4x.$

$\begin{array}{r} 2x \\ 3x - 2\overline{)6x^2 + 11x - 10} \\ -6x^2 + \ 4x \\ \hline 15x \end{array}$ Subtract $6x^2 - 4x$ by changing signs and adding.

$\begin{array}{r} 2x \\ 3x - 2\overline{)6x^2 + 11x - 10} \\ -6x^2 + \ 4x \\ \hline 15x - 10 \end{array}$ Bring down the -10.

$\begin{array}{r} 2x + \ 5 \\ 3x - 2\overline{)6x^2 + 11x - 10} \\ -6x^2 + \ 4x \\ \hline 15x - 10 \end{array}$ $\dfrac{15x}{3x} = 5$ and put the 5 over the $11x$.

$\begin{array}{r} 2x + \ 5 \\ 3x - 2\overline{)6x^2 + 11x - 10} \\ -6x^2 + \ 4x \\ \hline 15x - 10 \\ 15x - 10 \end{array}$ $5(3x - 2) = 15x - 10.$

$\begin{array}{r} 2x + \ 5 \\ 3x - 2\overline{)6x^2 + 11x - 10} \\ -6x^2 + \ 4x \\ \hline 15x - 10 \\ -15x + 10 \\ \hline 0 \end{array}$ Subtract $15x - 10$ by changing signs and adding.

The remainder is 0. Therefore, $3x - 2$ divides into $6x^2 + 11x - 10$ evenly. Hence $3x - 2$ is a factor of $6x^2 + 11x - 10$.

Check:

$(3x - 2)(2x + 5) = 6x^2 + 11x - 10$, so $2x + 5$ is the correct answer.

Practice Exercises

Find the following quotients using long division:

1. $\dfrac{x^2 - 2x - 24}{x + 4}$

2. $\dfrac{8x^2 - 6x - 9}{2x - 3}$

If more practice is needed, do the Additional Practice Exercises.

Additional Practice Exercises

Find the following quotients using long division:

a. $\dfrac{3x^2 + 13x - 30}{x + 6}$

b. $\dfrac{12x^2 + 2x - 2}{3x - 1}$

Actually, all the examples we have done to this point could have been done by factoring the numerator and the denominator and dividing by common factors as though we were reducing rational expressions. In the introductory comments, we posed the question as to whether we could divide two polynomials if there were no common factors. We see that we can by using long division. If the polynomial that is the divisor is not a factor of the dividend, there will be a remainder just as in the division of whole numbers. For example, 3 is not a factor of 11, so $11 \div 3 = 3$ with a remainder of 2. Recall that the remainder is usually written as the numerator with the divisor as the denominator. Consequently, $11 \div 3 = 3\frac{2}{3}$.

Answers:

Practice Exercises 1–2: **1.** $x - 6$ **2.** $4x + 3$ *Additional Practice Exercises a–b:* **a.** $3x - 5$ **b.** $4x + 2$

Before performing long division, it is necessary to do two things:

1. Write the divisor and the dividend in descending powers of the variable.

2. If there is a missing power of the variable, leave a space for each term with a missing power. This is usually done by inserting 0 times the variable raised to the power that is missing. The 0 serves as a "place holder" much as it does in writing a number like 407.

Note: In the following example, we will not show all the steps, only the actual division. If you get stuck, look back at Example 1.

Example 2

Find the following quotients using long division:

a. $\dfrac{5x^2 + x^3 - 12 + 2x}{x + 2}$

Solution: First, we need to rewrite the numerator in descending order, then rewrite as long division.

$$\frac{5x^2 + x^3 - 12 + 2x}{x + 2} = \frac{x^3 + 5x^2 + 2x - 12}{x + 2} = x + 2 \overline{)x^3 + 5x^2 + 2x - 12}$$

Divide as in Example 1.

$$
\begin{array}{r}
x^2 + 3x - 4 \\
x + 2 \overline{)x^3 + 5x^2 + 2x - 12} \\
\underline{-x^3 - 2x^2} \\
3x^2 + 2x \\
\underline{-3x^2 - 6x} \\
-4x - 12 \\
\underline{+4x + 8} \\
-4
\end{array}
$$

Since x will not divide into -4, -4 is the remainder. Hence $x + 2$ is not a factor of $x^3 + 5x^2 + 2x - 12$. Remember, the remainder is written over the divisor. Therefore,

$$\frac{x^3 + 5x^2 + 2x - 12}{x + 2} = x^2 + 3x - 4 + \frac{-4}{x + 2} \text{ is the solution.}$$

Remember, in checking a division problem that has a remainder, we multiply the divisor and the quotient, then add the remainder. The result is equal to the dividend. Earlier, we stated that $11 \div 3 = 3$ with a remainder of 2. To check, we have $3 \cdot 3 + 2 = 9 + 2 = 11$. If the preceding division is correct, then $(x + 2)(x^2 + 3x - 4) + (-4) = x^3 + 5x^2 + 2x - 12$.

Check:
$(x + 2)(x^2 + 3x - 4) + (-4) = x(x^2 + 3x - 4) + 2(x^2 + 3x - 4) - 4 = x^3 + 3x^2 - 4x + 2x^2 + 6x - 8 - 4 = x^3 + 5x^2 + 2x - 12$. Therefore, the answer is correct.

b. $\dfrac{8x^3 - 22x + 8}{2x - 3}$

Solution: The x^2 term is missing, so insert $0x^2$ between $8x^3$ and $-22x$ and divide as before.

$$
\begin{array}{r}
4x^2 + 6x - 2 \\
2x - 3 \overline{)8x^3 + 0x^2 - 22x + 8} \\
\underline{-8x^3 + 12x^2} \\
12x^2 - 22x \\
\underline{-12x^2 + 18x} \\
-4x + 8 \\
\underline{+4x - 6} \\
2
\end{array}
$$

Therefore, $\dfrac{8x^3 - 22x + 8}{2x - 3} = 4x^2 + 6x - 2 + \dfrac{2}{2x - 3}.$

Check:
The check is left as an exercise for the student.

Practice Exercises

Use long division to find the following quotients:

3. $\dfrac{16x - 12x^2 + 3x^3 - 12}{x - 2}$

4. $\dfrac{8x^3 + 12x^2 - 6}{2x - 1}$

If more practice is needed, do the Additional Practice Exercises.

Additional Practice Exercises

Find the following quotients using long division:

c. $\dfrac{-11x^2 + 3x^3 - 14x - 10}{x - 4}$

d. $\dfrac{2x^3 - 14x + 5}{2x - 4}$

Exercise Set 6.7

Find the following quotients using long division:

1. $\dfrac{x^2 - 2x - 15}{x + 3}$

2. $\dfrac{x^2 + 2x - 8}{x + 4}$

3. $\dfrac{x^2 - 4x - 21}{x - 7}$

4. $\dfrac{x^2 - 2x - 24}{x - 6}$

5. $\dfrac{2x^2 - x - 10}{2x - 5}$

6. $\dfrac{3x^2 - 22x + 24}{3x - 4}$

7. $\dfrac{6x^2 + 7x - 20}{3x - 4}$

8. $\dfrac{8x^2 + 6x - 9}{4x - 3}$

9. $\dfrac{4x^2 - 25}{2x + 5}$

10. $\dfrac{9x^2 - 49}{3x + 7}$

11. $\dfrac{2x^2 + 4x - 28}{2x - 6}$

12. $\dfrac{3x^2 - 13x - 25}{3x + 5}$

13. $\dfrac{12x^2 - 5x + 6}{4x - 3}$

14. $\dfrac{12x^2 + 22x + 14}{3x + 7}$

15. $\dfrac{6x^2 + 13x - 16}{2x + 7}$

16. $\dfrac{15x^2 + 29x - 10}{5x - 2}$

17. $\dfrac{2x^3 - 14x^2 + 27x - 12}{x - 4}$

18. $\dfrac{3x^3 + 22x^2 + 33x - 10}{x + 5}$

19. $\dfrac{4x^3 - 16x + 6x^2 + 6}{2x - 1}$

20. $\dfrac{6x^3 + 25x - 10x^2 - 14}{3x - 2}$

21. $\dfrac{6x^3 + 11x^2 - 8x + 9}{2x + 5}$

22. $\dfrac{12x^3 - 17x^2 + 12x - 8}{3x - 2}$

23. $\dfrac{3x^3 - 73x - 10}{x - 5}$

24. $\dfrac{2x^3 - 35x - 12}{x + 4}$

25. $\dfrac{-2x^2 + 4x^3 + 18}{2x - 3}$

26. $\dfrac{22x^2 + 12x^3 - 2}{3x + 1}$

27. $\dfrac{16x^3 - x + 10}{4x + 3}$

28. $\dfrac{4x^3 - 42x + 23}{2x - 6}$

29. $\dfrac{x^3 - 8}{x - 2}$

30. $\dfrac{x^3 + 27}{x + 3}$

31. $\dfrac{x^4 - 16}{x - 2}$

32. $\dfrac{x^4 - 81}{x - 3}$

33. $\dfrac{x^3 + 9}{x + 3}$

34. $\dfrac{x^3 - 12}{x - 2}$

Answers:

Additional Practice Exercises c–d: c. $3x^2 + x - 10 + \dfrac{-50}{x-4}$ d. $x^2 + 2x - 3 + \dfrac{-7}{2x-4}$

Practice Exercises 3–4: 3. $3x^2 - 6x + 4 + \dfrac{-4}{x-2}$ 4. $4x^2 + 8x + 4 + \dfrac{-2}{2x-1}$

Challenge Exercises:

35. Find the value of k, so $x - 6$ will divide evenly into $2x^2 - 15x + k$.

36. Find the value of k, so $x + 4$ will divide evenly into $3x^2 + 6x + k$.

37. Find the value of k, so $x - 2$ will divide evenly into $4x^2 + kx - 6$.

38. Find the value of k, so $x + 3$ will divide evenly into $3x^2 + kx + 6$.

Writing Exercises:

39. Is $x + 4$ a factor of $2x^3 + 5x^2 - 11x + 4$? Why or why not?

40. Is $x - 3$ a factor of $3x^3 - 7x^2 - 8x + 6$? Why or why not?

Group Project:

41. Find three second-degree polynomials that are divisible by $x + 3$.

42. Find three second-degree polynomials that leave a remainder of 3 when divided by $x - 2$.

Chapter 6   Summary

Interpretations of Fractions: [Section 6.1]	**a.** A fraction may represent a given part of a whole (unit). The denominator represents the number of equal parts into which the unit is divided, and the numerator represents the number of equal parts selected.
	b. A fraction may represent the part a given number of objects is of a total number of objects. The numerator represents the number of objects selected, and the denominator represents the total number of objects in the group.
	c. A fraction may represent division in which the numerator is divided by the denominator. This allows us to represent a fraction as a decimal.
Proper and Improper Fractions: [Section 6.1]	**a.** A fraction is **proper** if the numerator is less than the denominator.
	b. A fraction is **improper** if the numerator is greater than or equal to the denominator.
Mixed Number: [Section 6.1]	• A number is mixed if it is greater than one and has a whole part and a fractional part.
Changing a Rational Number into a Decimal: [Section 6.1]	• To change a rational number into a decimal, divide the denominator into the numerator, and if necessary, round the quotient to the indicated number of decimal places.
Changing a Mixed Number into an Improper Fraction and Vice Versa: [Section 6.1]	**a.** To change a mixed number into an improper fraction, multiply the denominator of the fractional part by the whole part and add the numerator of the fractional part. This is the numerator of the improper fraction, and the denominator is the denominator of the fractional part.
	b. To change an improper fraction into a mixed number, divide the denominator into the numerator. The number of times the denominator divides into the numerator is the whole part, and the number left over is the numerator of the fractional part.
Graphing Rational Numbers: [Section 6.1]	• Rational numbers are graphed on the number line by putting a dot at the location of the point that is the indicated distance and direction from 0. The denominator represents the number of equal parts into which the units are divided, and the numerator represents the number of equal parts from 0 to the location.
Equivalent Fractions: [Section 6.2]	• Two or more fractions are equivalent if they represent the same quantity.
Lowest Terms: [Section 6.2]	• A fraction is reduced to lowest terms if the numerator and the denominator have no common factors other than 1.

Reducing a Fraction to Lowest Terms: [Section 6.2]	**a.** Write the numerator and the denominator as the product of their prime factors. **b.** Divide the numerator and the denominator by the common factors.

Rational Expressions: [Sections 6.2 and 6.3]

a. A rational expression is an algebraic expression of the form $\frac{P}{Q}$ where P and Q are polynomials and $Q \neq 0$.

b. To find the value(s) for which a rational expression is undefined, set the denominator equal to 0 and solve the resulting equation.

Simplifying Rational Expressions: [Sections 6.2 and 6.3]

• A rational expression is simplified (reduced to lowest terms) if the numerator and the denominator have no common factors other than 1. The procedure for simplifying rational expressions is exactly the same as reducing fractions to lowest terms.

Multiplication of Rational Numbers: [Section 6.4]

a. To multiply two or more rational numbers, multiply numerator times numerator and denominator times denominator. In symbols, $\frac{a}{b} \cdot \frac{c}{d} = \frac{a \cdot c}{b \cdot d}$.

b. To multiply a fraction and mixed number or two or more mixed numbers, change the mixed numbers into improper fractions and multiply as with any other fractions.

c. To multiply a fraction and a whole number, think of the whole number as a fraction whose denominator is 1 and multiply as with any other fractions.

d. All products should be reduced to lowest terms by using the procedure for reducing fractions to lowest terms. You may divide by factors common to the numerator and the denominator before multiplying.

Multiplication of Rational Expressions: [Section 6.5]

a. Rational expressions are multiplied in exactly the same manner as rational numbers. That is, multiply numerator times numerator and put the product over denominator times denominator.

b. If the product of two or more rational expressions contains factors that differ in sign only, remove a factor of -1 from one of the factors. For example, $b - a = -1(a - b)$.

Division of Rational Numbers and Expressions: [Section 6.6]

• To divide rational numbers or expressions, invert the expression on the right and change the operation to multiplication.

Division of Polynomials (Long Division): [Section 6.7]

• Long division of polynomials is done using the same algorithm as division of whole numbers, except like terms are aligned instead of digits with the same place value. The steps are:

  1. Divide the first term of the divisor into the first term of the dividend (after both have been put into descending order and missing powers are inserted as $0 \cdot$ variable raised to the missing power).

  2. Multiply the entire divisor by the first quotient (the result of the first division).

  3. Align like terms and subtract the expression found in step 2.

  4. Bring down the next term of the dividend.

  5. Continue steps 1–4 until all terms of the dividend have been used. The remainder (if any) will have a lower degree than the divisor.

Chapter 6   Review Exercises

Using a fraction, represent the part of the whole region that is shaded. [Section 6.1]

1.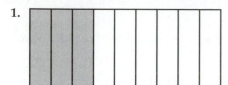

2.

Shade the part of the whole region that is represented by the given fraction. [Section 6.1]

3. $\frac{5}{7}$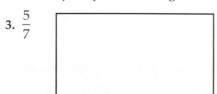

4. $\frac{3}{8}$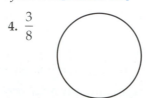

Represent each of the following with an improper fraction and a mixed number, if possible: [Section 6.1]

5.

6.

Write a fraction that represents the part of the total number of figures that is shaded: [Section 6.1]

7.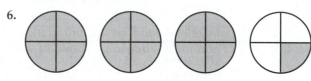

8.

Represent each of the following using a fraction: [Section 6.1]

9. The Kelleys decided to build a patio out of flagstone. They purchased 120 stones and used 113. What fractional part of the stones purchased did they use?

10. The biology department received 250 frogs for a lab exercise. Of the 250 received, 197 were used. What fractional part of the frogs were used in the lab exercise?

11. Kim bought a pack of notebook paper containing 500 sheets. She used 387 sheets in her Calculus II class. What fractional part of the pack of paper did she use in her Calculus II class?

12. In a deck of 52 cards, there are 12 face cards. What fractional part of the deck is made up of face cards?

13. Joe has 45 problems for homework in his arithmetic and algebra class. He completed 28 of them before taking a break. What fractional part of his homework has he completed? What fractional part does he still have to do?

14. A chemistry lab assistant was setting up the lab for an experiment. She needed 250 milliliters of sulfuric acid, but could find only 225 milliliters. What fractional part of what she needed could she find? What fractional part was missing?

15. The VFW set a goal of $600 in a fund-raising campaign to benefit the children's home. They were able to raise $573. What fractional part of their goal were they able to achieve?

16. The student government association requested $1000 to hold a film festival. The college administration agreed to budget $750 for the event. What fractional part of the amount they requested did the student government association receive?

17. A particular type of fishhook is packaged in blister packs containing 25 each. A charter boat captain used 8 of these packages and 12 hooks from the ninth package in one week. Represent the number of packages he used as a mixed numeral and as an improper fraction.

18. Cereal may be purchased in an assortment pack containing ten individual boxes of cereal. Last month, the Lockwood family ate three complete assortment packs and three boxes from the fourth. Represent the number of assortment packs they ate as a mixed numeral and as an improper fraction.

19. In a Calculus I class of 27 students, $\frac{12}{27}$ of the class are female. How many females are in the class?

20. For a probability experiment, a bowl contained 20 balls. If $\frac{9}{20}$ of the balls were blue, how many blue balls were in the bowl?

Change the following fractions into decimals. If necessary, round the answer to the nearest hundredth. [Section 6.1]

21. $\frac{5}{16}$
22. $\frac{7}{9}$
23. $\frac{11}{13}$
24. $\frac{12}{19}$

25. $\frac{11}{5}$
26. $\frac{17}{4}$
27. $\frac{17}{12}$
28. $\frac{23}{11}$

Convert the following improper fractions into mixed numbers: [Section 6.1]

29. $\frac{17}{8}$
30. $\frac{35}{6}$
31. $\frac{53}{21}$
32. $\frac{81}{33}$

Change the following mixed numbers into improper fractions: [Section 6.1]

33. $5\frac{4}{7}$
34. $6\frac{7}{9}$
35. $3\frac{7}{18}$
36. $8\frac{5}{23}$

Represent the distance from 0 to the dot using a rational number. [Section 6.1]

37.
38.

Graph the following rational numbers. Estimate as needed. [Section 6.1]

39. $\left\{-\frac{5}{3}, -\frac{2}{5}, 0, \frac{3}{4}, \frac{9}{4}\right\}$

40. $\left\{-3\frac{1}{2}, -1\frac{1}{6}, -\frac{1}{3}, 1\frac{4}{5}, 3\frac{1}{2}\right\}$

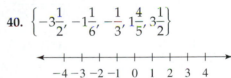

Reduce the following fractions to lowest terms: [Section 6.2]

41. $\frac{36}{42}$
42. $\frac{64}{72}$
43. $\frac{120}{124}$
44. $\frac{84}{96}$

Reduce the following rational expressions to lowest terms. [Section 6.2]

45. $\frac{x^3 y}{x^2 y^3}$
46. $\frac{m^2 n^4}{m^4 n^5}$
47. $\frac{r^3 s^5}{rs^3}$

48. $\frac{32x^2 y^5}{8xy^3}$
49. $\frac{-42p^6 q^2}{27p^3 q^4}$
50. $-\frac{18m^3 n^2}{54m^5 n^5}$

Find the value(s) of the variable(s) for which the following are defined: [Section 6.3]

51. $\frac{x}{x + 9}$
52. $\frac{x + 3}{(x + 7)(x + 2)}$
53. $\frac{x + 4}{x - 5y}$
54. $\frac{3a - 4}{a^2 + 4a - 12}$

Reduce the following rational expressions to lowest terms: [Section 6.3]

55. $\frac{7(2x + 9)}{5(2x + 9)}$
56. $\frac{4a - 20}{7a - 35}$
57. $\frac{a^2 + 4a - 21}{a^2 + 9a + 14}$

58. $\dfrac{2x - 3}{6x^2 - x - 12}$

59. $\dfrac{25x^2 - 9}{10x^2 + x - 3}$

60. $\dfrac{2c + 3d}{8c^2 + 26cd + 21d^2}$

61. $\dfrac{3x - 4y}{4y - 3x}$

62. $\dfrac{5 - 2x}{4x^2 - 25}$

Find the following products of rational numbers. Leave answers reduced to lowest terms. [Section 6.4]

63. $\dfrac{18}{7} \cdot \dfrac{7}{12}$

64. $\dfrac{21}{10} \cdot \left(-\dfrac{15}{14}\right)$

65. $-\dfrac{5}{16} \cdot 1\dfrac{13}{15}$

66. $4\dfrac{4}{5} \cdot 3\dfrac{1}{3}$

67. $-32 \cdot \left(-2\dfrac{3}{8}\right)$

68. $\dfrac{9}{16} \cdot \dfrac{28}{15} \cdot \dfrac{10}{7}$

69. A worker works for $6\dfrac{1}{2}$ hours and is paid \$6.80 per hour. How much did the worker earn?

70. A rope is $5\dfrac{1}{4}$ feet long. If $\dfrac{2}{3}$ of the rope is cut off, how long is the piece that was cut off?

71. Find the area of a rectangle whose length is $2\dfrac{1}{4}$ feet and whose width is $1\dfrac{1}{3}$ feet.

72. Find the area of a triangle whose base is $3\dfrac{3}{5}$ inches and whose height is $4\dfrac{2}{3}$ inches.

Find the following products. Leave answers reduced to lowest terms. [Section 6.5]

73. $\dfrac{x^5}{y^3} \cdot \dfrac{y^2}{x^2}$

74. $\dfrac{28m^2n^3}{15p^2q^6} \cdot \dfrac{25p^4q^3}{14mn}$

75. $\dfrac{m - n}{32m^3n} \cdot \dfrac{36mn^3}{(m - n)^2}$

76. $\dfrac{14x - 21}{30x - 40} \cdot \dfrac{15x - 20}{8x - 12}$

77. $\dfrac{25x^2 - 16}{16x + 24} \cdot \dfrac{12x + 18}{5x - 4}$

78. $\dfrac{2x - 7}{12} \cdot \dfrac{10}{7 - 2x}$

79. $\dfrac{x^2 - 16}{x^2 + 6x + 8} \cdot \dfrac{x^2 - 3x - 10}{x^2 - 25}$

80. $\dfrac{8x^2 + 2x - 15}{6x^2 + x - 12} \cdot \dfrac{3x^2 - 13x + 12}{3x^2 - 7x - 6}$

81. $\dfrac{y^2 - 9}{y^2 - 3y - 18} \cdot \dfrac{y^2 - 4y - 12}{3 - y}$

Find the following quotients. Leave answers reduced to lowest terms. [Section 6.6]

82. $\dfrac{5}{8} \div \dfrac{15}{24}$

83. $-\dfrac{32}{25} \div \dfrac{28}{35}$

84. $-144 \div \dfrac{9}{16}$

85. $-\dfrac{32}{19} \div (-24)$

86. $4\dfrac{1}{5} \div \left(-\dfrac{7}{14}\right)$

87. $-12 \div 3\dfrac{3}{8}$

88. If $3\dfrac{2}{5}$ ounces of gold cost \$1224, how much does one ounce cost?

89. If a car used $12\dfrac{8}{10}$ gallons of gas on a trip of 416 miles, what was the average miles per gallon for the trip?

Find the following quotients of rational expressions. Leave answers reduced to lowest terms. [Section 6.6]

90. $\dfrac{39a^2b^4}{27x^3y^2} \div \dfrac{26a^3b}{28xy^4}$

91. $\dfrac{8y - 20}{9} \div \dfrac{6y - 15}{10}$

92. $\dfrac{21a + 7b}{16a - 24b} \div \dfrac{42a + 14b}{24a - 36b}$

93. $\dfrac{x - y}{8x - 4y} \div \dfrac{y - x}{12x - 6y}$

94. $\dfrac{z^2}{z^2 - 2z - 8} \div \dfrac{9z^3 + 3z^4}{z^2 + 5z + 6}$

95. $\dfrac{4p^2 - 9}{24p + 28} \div \dfrac{10p - 15}{36p^2 - 49}$

96. $\dfrac{16p^2 - 8pq - 3q^2}{8p^2 + 22pq + 5q^2} \div \dfrac{8p^2 - 10pq + 3q^2}{10p^2 + pq - 3q^2}$

Find the following quotients using long division: [Section 6.7]

97. $\dfrac{12x^2 + 7x - 10}{3x - 2}$

98. $\dfrac{15x^2 - 29x - 16}{5x + 2}$

99. $\dfrac{6a^3 - a^2 - 9a + 7}{3a + 4}$

100. $\dfrac{16z^2 - 16z + 12}{4z - 3}$

101. $\dfrac{-27x - 2x^2 + 8x^3 + 18}{2x - 3}$

102. $\dfrac{3x^3 - 24x - 7}{x - 3}$

Chapter 6   Test

Represent each of the following using a fraction:

1. Freda, who works in an electronics lab, has 84 connections to solder. If she has completed soldering 23 connections, what fractional part has she completed? What fractional part remains to be soldered?

2. At Rick's Seafood Restaurant, oysters are served on trays that contain 12 oysters each. If Natasha eats all the oysters on two trays and five from a third tray, represent the number of trays of oysters she ate as: 1) a mixed number, and 2) an improper fraction.

3. Represent the shaded part of the region using a fraction.

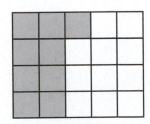

4. Write $\frac{5}{16}$ as a decimal rounded to the nearest hundredth.

5. Change $\frac{29}{13}$ to a mixed number.

6. Change $6\frac{4}{7}$ to an improper fraction.

7. Graph $\left\{ -\frac{5}{2}, -\frac{5}{4}, -\frac{1}{3}, 1, \frac{6}{5} \right\}$. Estimate as needed.

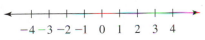

$-4\ -3\ -2\ -1\ \ 0\ \ 1\ \ 2\ \ 3\ \ 4$

8. Reduce $\frac{72}{84}$ to lowest terms.

Find the value(s) for which the following are defined:

9. $\dfrac{4a}{3a - 12}$

10. $\dfrac{6b}{b^2 - 2b - 8}$

Simplify the following by reducing to lowest terms:

11. $\dfrac{42a^3b^4}{16ab^6}$

12. $\dfrac{6x^2 + 11x - 10}{4x^2 + 4x - 15}$

13. It took $1\frac{3}{4}$ gallons of paint to paint the trim on a house. How many gallons of paint would it take to paint six such houses?

14. If $2\frac{1}{2}$ pounds of coffee cost \$11.25, what is the cost of 1 pound?

Find the following products and quotients:

15. $\dfrac{18}{25} \cdot \dfrac{35}{24}$

16. $6\dfrac{2}{3} \cdot 1\dfrac{7}{8}$

17. $2\dfrac{4}{7} \div 27$

18. $\dfrac{27a^2b^4}{14x^4y} \cdot \dfrac{35xy^3}{18a^5b^2}$

19. $\dfrac{7-b}{42} \cdot \dfrac{26}{b-7}$

20. $\dfrac{a^2 + 4a - 21}{a^2 + 3a - 28} \div \dfrac{a^2 + 3a - 18}{a^2 + 8a + 12}$

21. $\dfrac{16y^2 - 25}{6y^2 - 17y - 14} \cdot \dfrac{3y^2 + 2y}{8y^2 - 2y - 15}$

22. Find $\dfrac{12x^2 - 11x - 12}{3x - 5}$ using long division.

C hapter 7 is a continuation of the study of rational numbers and rational expressions begun in Chapter 6. In Chapter 6, we reduced rational numbers and rational expressions to lowest terms, multiplied and divided rational numbers and expressions, and divided polynomials using long division.

We begin Chapter 7 with the addition of rational numbers and fractions with like denominators, followed by two sections that cover the skills that are necessary to be able to add and subtract rational numbers and expressions with unlike denominators. We then add and subtract rational numbers and expressions and simplify complex fractions. We end the chapter by solving equations that contain rational numbers and expressions and then use this skill to solve application problems.

Addition and Subtraction of Rational Numbers and Expressions

Chapter

7

Section 7.1	Addition and Subtraction of Rational Numbers and Expressions with Like Denominators

OBJECTIVES

When you complete this section, you will be able to:

a. Add and subtract fractions with common denominators.

b. Add and subtract mixed numbers whose fractional parts have common denominators.

c. Add and subtract rational expressions with common denominators.

Introduction

This is the first of two sections in which we will be adding fractions and rational expressions. The technique used in adding fractions depends upon the denominators. In this section, the denominators will be the same number or expression, whereas in Section 7.4, the denominators will be different. When fractions have the same denominators, we say the fractions have a **common denominator.** We begin with a graphical illustration of $\frac{4}{15} + \frac{3}{15}$. The figure is divided into 15 equal regions of which 4 are shaded. Therefore, $\frac{4}{15}$ of the unit is shaded.

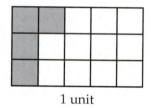

1 unit

If we shade 3 more regions, this shading represents $\frac{3}{15}$ of the unit. What part of the unit is now shaded?

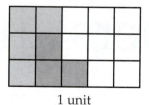

1 unit

Since 7 of the 15 regions are now shaded, $\frac{7}{15}$ of the region is shaded. Therefore, $\frac{4}{15} + \frac{3}{15} = \frac{7}{15}$. In this case, the fractions we added had common denominators. Therefore, to add fractions with common denominators we add only the numerators and leave the denominators unchanged. This leads to the following procedure for adding fractions with common denominators:

Addition of Fractions with Common Denominators

To add fractions with common denominators, add the numerators and put the sum over the common denominator. In symbols, $\frac{a}{c} + \frac{b}{c} = \frac{a+b}{c}$. Reduce if possible.

This principle is easily extended to three or more fractions: $\frac{a}{d} + \frac{b}{d} + \frac{c}{d} = \frac{a+b+c}{d}$, and so on.

Example 1

Find the following sums:

a. $\dfrac{2}{7} + \dfrac{4}{7} =$ Put the sum of the numerators over the common denominator.

$\dfrac{2+4}{7} =$ Add the numerators.

$\dfrac{6}{7}$ Sum.

b. $\dfrac{1}{11} + \dfrac{6}{11} + \dfrac{3}{11} =$ Put the sum of the numerators over the common denominator.

$\dfrac{1+6+3}{11} =$ Add the numerators.

$\dfrac{10}{11}$ Sum.

Subtracting fractions with common denominators

Before we subtract fractions, we need to look at equivalent forms of negative fractions. By using the rules of signs for division of signed numbers, we see that $-\dfrac{a}{b} = \dfrac{-a}{b}$. That is, the negative of a fraction is the negative of the numerator divided by the same denominator. For example $-\dfrac{6}{2} = \dfrac{-6}{2}$, since they both equal -3. Consequently, we may change subtraction of fractions into addition as follows. The difference $\dfrac{a}{c} - \dfrac{b}{c}$ becomes the sum $\dfrac{a}{c} + \dfrac{-b}{c} = \dfrac{a + (-b)}{c} = \dfrac{a-b}{c}$. We summarize as follows:

> ## Subtraction of Fractions with Common Denominators
>
> To subtract fractions with common denominators, put the difference of the numerators over the common denominator. In symbols, $\dfrac{a}{c} - \dfrac{b}{c} = \dfrac{a-b}{c}$. Reduce if possible.

Example 2

Find the following differences:

a. $\dfrac{4}{5} - \dfrac{2}{5} =$ Put the difference of the numerators over the common denominator.

$\dfrac{4-2}{5} =$ $4 - 2 = 2$

$\dfrac{2}{5}$ Difference.

b. $\dfrac{11}{9} - \dfrac{-5}{9} =$ Put the difference of the numerators over the common denominator.

$\dfrac{11 - (-5)}{9} =$ $-(-5) = +5$.

$\dfrac{11 + 5}{9} =$ Add the numerators.

$\dfrac{16}{9}$ Difference.

c. $\dfrac{1}{5} - \dfrac{3}{5} =$ Put the difference of the numerators over the common denominator.

$\dfrac{1-3}{5} =$ $1 - 3 = -2$.

$\dfrac{-2}{5} =$ Apply equivalent form of negative fractions.

$-\dfrac{2}{5}$ Difference.

Practice Exercises

Find the following sums or differences:

1. $\dfrac{1}{5} + \dfrac{3}{5}$

2. $\dfrac{11}{15} + \dfrac{2}{15}$

3. $\dfrac{8}{17} + \dfrac{7}{17} + \dfrac{4}{17}$

4. $\dfrac{9}{13} - \dfrac{5}{13}$

5. $\dfrac{4}{9} - \dfrac{-1}{9}$

6. $\dfrac{3}{11} - \dfrac{9}{11}$

If more practice is needed, do the Additional Practice Exercises.

Additional Practice Exercises

Find the following sums or differences:

a. $\dfrac{1}{3} + \dfrac{4}{3}$

b. $\dfrac{5}{17} + \dfrac{7}{17}$

c. $\dfrac{3}{5} + \dfrac{2}{5} + \dfrac{1}{5}$

d. $\dfrac{11}{13} - \dfrac{5}{13}$

e. $\dfrac{7}{11} - \dfrac{-3}{11}$

f. $\dfrac{1}{7} - \dfrac{6}{7}$

When adding or subtracting fractions, it is often necessary to reduce the answer to lowest terms as in the following:

Example 3

Find the following sums or differences. Leave answers reduced to lowest terms.

a. $\dfrac{5}{9} + \dfrac{7}{9} =$ Put the sum of the numerators over the common denominator.

$\dfrac{5+7}{9} =$ $5 + 7 = 12.$

$\dfrac{12}{9} =$ Factor the numerator and the denominator into prime factors.

$\dfrac{2 \cdot 2 \cdot 3}{3 \cdot 3} =$ Divide by the common factor 3.

$\dfrac{2 \cdot 2 \cdot \cancel{3}}{3 \cdot \cancel{3}} =$ Multiply the remaining factors.

$\dfrac{4}{3}$ Sum.

b. $\dfrac{11}{15} - \dfrac{2}{15} =$ Put the difference of the numerators over the common denominator.

$\dfrac{11-2}{15} =$ $11 - 2 = 9.$

$\dfrac{9}{15} =$ Factor the numerator and the denominator into prime factors.

$\dfrac{3 \cdot 3}{3 \cdot 5} =$ Divide by the common factor 3.

$\dfrac{\cancel{3} \cdot 3}{\cancel{3} \cdot 5} =$ Multiply the remaining factors.

$\dfrac{3}{5}$ Difference.

Note: In the following examples, we will assume you remember how to reduce fractions and will omit the steps showing the prime factorizations and divisions by the common factors.

Answers:

Practice Exercises 1–6: 1. $\frac{4}{5}$ 2. $\frac{13}{15}$ 3. $\frac{19}{17}$ 4. $\frac{4}{13}$ 5. $\frac{5}{9}$ 6. $-\frac{6}{11}$

Additional Practice Exercises a–f: a. $\frac{5}{3}$ b. $\frac{12}{17}$ c. $\frac{6}{5}$

d. $\frac{6}{13}$ e. $\frac{10}{11}$ f. $-\frac{5}{7}$

c. $\dfrac{11}{18} - \dfrac{7}{18} + \dfrac{5}{18} =$ Put the numerators over the common denominator.

$\dfrac{11 - 7 + 5}{18} =$ $11 - 7 + 5 = 9$.

$\dfrac{9}{18} =$ Reduce to lowest terms.

$\dfrac{1}{2}$ Answer.

d. $\dfrac{23}{28} - \dfrac{5}{28} - \dfrac{11}{28} =$ Put the numerators over the common denominator.

$\dfrac{23 - 5 - 11}{28} =$ $23 - 5 - 11 = 7$.

$\dfrac{7}{28} =$ Reduce to lowest terms.

$\dfrac{1}{4}$ Answer.

Practice Exercises

Find the following sums and/or differences. Leave answers reduced to lowest terms.

7. $\dfrac{19}{24} + \dfrac{7}{24}$

8. $\dfrac{11}{18} - \dfrac{7}{18}$

9. $\dfrac{5}{12} + \dfrac{11}{12} - \dfrac{7}{12}$

10. $\dfrac{29}{36} - \dfrac{7}{36} - \dfrac{13}{36}$

If more practice is needed, do the Additional Practice Exercises.

Additional Practice Exercises

Find the following sums and/or differences. Leave answers reduced to lowest terms.

g. $\dfrac{13}{16} + \dfrac{5}{16}$

h. $\dfrac{9}{14} - \dfrac{3}{14}$

i. $\dfrac{35}{48} - \dfrac{25}{48} + \dfrac{5}{48}$

j. $\dfrac{25}{27} - \dfrac{8}{27} - \dfrac{2}{27}$

Adding mixed numbers with common denominators   Mixed numbers may be added or subtracted in two different ways. We may combine the fractional parts and whole parts separately or we may convert the mixed numbers to improper fractions and then combine. Converting to improper fractions avoids many of the problems associated with adding and subtracting mixed numbers and is more like the procedure used to combine algebraic fractions. For these reasons, it is the only method we will discuss and we will leave the answers as improper fractions.

Example 4

Add or subtract the following mixed numbers:

a. $2\dfrac{3}{4} + 5\dfrac{1}{4} =$ Convert to improper fractions.

$\dfrac{11}{4} + \dfrac{21}{4} =$ Put the sum of the numerators over the common denominator.

$\dfrac{11 + 21}{4} =$ $11 + 21 = 32$.

$\dfrac{32}{4} =$ $32 \div 4 = 8$.

8 Sum.

b. $4\dfrac{3}{8} - 6\dfrac{1}{8} =$ Convert to improper fractions.

$\dfrac{35}{8} - \dfrac{49}{8} =$ Put the difference of the numerators over the common denominator.

$\dfrac{35 - 49}{8} =$ $35 - 49 = -14$.

$\dfrac{-14}{8} =$ Reduce to lowest terms.

$-\dfrac{7}{4}$ Difference.

Answers:

Practice Exercises 7–10: 7. $\dfrac{13}{12}$ 8. $\dfrac{2}{9}$ 9. $\dfrac{3}{4}$ 10. $\dfrac{1}{4}$ **Additional Practice Exercises 8–j:** g. $\dfrac{9}{8}$ h. $\dfrac{3}{7}$ i. $\dfrac{5}{16}$ j. $\dfrac{5}{9}$

$$\textbf{c. } 6\frac{1}{12} - 2\frac{11}{12} + 3\frac{7}{12} = \qquad \text{Convert to improper fractions.}$$

$$\frac{73}{12} - \frac{35}{12} + \frac{43}{12} = \qquad \text{Combine fractions.}$$

$$\frac{73 - 35 + 43}{12} = \qquad 73 - 35 + 43 = 81.$$

$$\frac{81}{12} = \qquad \text{Reduce to lowest terms.}$$

$$\frac{27}{4} \qquad \text{Answer.}$$

Practice Exercises

Add or subtract the following mixed numbers:

11. $4\frac{1}{6} + 7\frac{5}{6}$ **12.** $6\frac{7}{8} - 9\frac{3}{8}$ **13.** $3\frac{17}{30} + 6\frac{9}{30} - 2\frac{11}{30}$

If more practice is needed, do the Additional Practice Exercises.

Additional Practice Exercises

Add or subtract the following mixed numbers:

k. $2\frac{1}{5} + 5\frac{3}{5}$ **l.** $3\frac{7}{10} - 6\frac{3}{10}$ **m.** $2\frac{11}{21} - 3\frac{5}{21} + 4\frac{1}{21}$

The addition of rational expressions with common denominators is done exactly the same way as the addition of fractions. We will first do some examples in which the denominators are monomials. Compare the following:

$$\frac{4}{13} + \frac{6}{13} \qquad \frac{4}{x} + \frac{6}{x} \qquad \text{Put the sum of the numerators over the common denominator.}$$

$$\frac{4+6}{13} \qquad \frac{4+6}{x} \qquad \text{Add the numerators.}$$

$$\frac{10}{13} \qquad \frac{10}{x} \qquad \text{Sum}$$

Example 5

Find the sums of the following rational expressions. Leave answers reduced to lowest terms.

a. $\dfrac{3}{a} + \dfrac{6}{a} =$ Put the sum of the numerators over the common denominator.

$\dfrac{3+6}{a}$ Add the numerators.

$\dfrac{9}{a}$ Sum.

b. $\dfrac{m}{p} + \dfrac{n}{p} =$ Put the sum of the numerators over the common denominator.

$\dfrac{m+n}{p}$ Sum

Answers:

Practice Exercises 11–13: 11. 12   12. $-\frac{5}{2}$ 13. $\frac{7}{2}$ **Additional Practice Exercises k–m:** k. $\frac{39}{5}$ l. $-\frac{5}{13}$ m. $\frac{10}{3}$

c. $\dfrac{3}{x^2} + \dfrac{y}{x^2} =$ Put the sum of the numerators over the common denominator.

$\dfrac{3+y}{x^2}$ Sum.

d. $\dfrac{3x}{10y} - \dfrac{7x}{10y} =$ Put the difference of the numerators over the common denominator.

$\dfrac{3x-7x}{10y} =$ $3x - 7x = -4x.$

$\dfrac{-4x}{10y} =$ Reduce to lowest terms.

$-\dfrac{2x}{5y}$ Difference.

Practice Exercises

Find the sums or differences of the following rational expressions. Leave answers reduced to lowest terms.

14. $\dfrac{6}{b} + \dfrac{4}{b}$

15. $\dfrac{r}{t} + \dfrac{s}{t}$

16. $\dfrac{m}{n} - \dfrac{6}{n}$

17. $\dfrac{3x}{16z} + \dfrac{5x}{16z}$

If more practice is needed, do the Additional Practice Exercises.

Additional Practice Exercises

Find the sums of the following rational expressions. Leave answers reduced to lowest terms.

n. $\dfrac{8}{y} - \dfrac{3}{y}$

o. $\dfrac{x}{z} - \dfrac{y}{z}$

p. $\dfrac{v}{u^3} + \dfrac{w}{u^3}$

q. $\dfrac{5a}{14b} - \dfrac{11a}{14b}$

If the common denominators are binomials or trinomials, it is often necessary to factor the numerator of the sum or difference and then reduce.

Example 6

Find the sums or differences of the following rational expressions. Leave answers reduced to lowest terms. Assume the denominator does not equal 0.

a. $\dfrac{3}{a+2} + \dfrac{5}{a+2} =$ Put the sum of the numerators over the common denominator.

$\dfrac{3+5}{a+2} =$ $3 + 5 = 8.$

$\dfrac{8}{a+2}$ Sum.

b. $\dfrac{c}{c+4} - \dfrac{2}{c+4} =$ Put the difference of the numerators over the common denominator.

$\dfrac{c-2}{c+4}$ Difference.

c. $\dfrac{2y}{y-3} - \dfrac{6}{y-3} =$ Put the difference of the numerators over the common denominator.

$\dfrac{2y-6}{y-3} =$ Factor the numerator.

$\dfrac{2(y-3)}{y-3} =$ Divide by the common factor $y - 3$.

2 Difference.

d. $\dfrac{x^2}{x^2+4x-21} + \dfrac{2x-15}{x^2+4x-21} =$ Put the sum of the numerators over the common denominator.

$\dfrac{(x^2)+(2x-15)}{x^2+4x-21} =$ Remove the parentheses.

$\dfrac{x^2+2x-15}{x^2+4x-21} =$ Factor the numerator and the denominator.

$\dfrac{(x+5)(x-3)}{(x+7)(x-3)} =$ Divide by the common factor $x - 3$.

$\dfrac{x+5}{x+7}$ Sum.

Answers:

Practice Exercises 14–17: 14. $\dfrac{10}{b}$ **15.** $\dfrac{r+s}{t}$ **16.** $\dfrac{m-6}{n}$ **17.** $\dfrac{x}{z}$ **Additional Practice Exercises n–q: n.** $\dfrac{5}{y}$ **o.** $\dfrac{x-y}{z}$ **p.** $\dfrac{v+w}{u^3}$ **q.** $-\dfrac{3a}{7b}$

Be Careful *In exercises like Example 6e, where we are finding the difference of rational expressions, a common error is to forget to change all the signs of the numerator of the second fraction. To avoid making this type of error, you should write the step with the difference of the numerators in parentheses.*

e. $\dfrac{x^2 + 5x}{x^2 + 7x + 12} - \dfrac{8x + 28}{x^2 + 7x + 12} =$ Put the difference of the numerators over the common denominator.

$\dfrac{(x^2 + 5x) - (8x + 28)}{x^2 + 7x + 12} =$ Remove the parentheses.

$\dfrac{x^2 + 5x - 8x - 28}{x^2 + 7x + 12} =$ Add like terms.

$\dfrac{x^2 - 3x - 28}{x^2 + 7x + 12} =$ Factor the numerator and the denominator.

$\dfrac{(x - 7)(x + 4)}{(x + 4)(x + 3)} =$ Divide by the common factor of $x + 4$.

$\dfrac{x - 7}{x + 3}$ Difference.

Practice Exercises

Find the following sums or differences of rational expressions. Leave answers reduced to lowest terms.

18. $\dfrac{6}{a - 5} + \dfrac{2}{a - 5}$ **19.** $\dfrac{3a}{a + 2} - \dfrac{a}{a + 2}$ **20.** $\dfrac{x}{x - 4} - \dfrac{4}{x - 4}$

21. $\dfrac{x^2}{x^2 + 3x - 18} + \dfrac{5x - 6}{x^2 + 3x - 18}$ **22.** $\dfrac{x^2 - 2x}{x^2 - 5x + 6} - \dfrac{5x - 10}{x^2 - 5x + 6}$

If more practice is needed, do the Additional Practice Exercises.

Additional Practice Exercises

Find the following sums or differences of rational expressions. Leave answers reduced to lowest terms.

r. $\dfrac{12}{2x + 1} - \dfrac{5}{2x + 1}$ **s.** $\dfrac{y}{3y + 4} + \dfrac{3y}{3y + 4}$ **t.** $\dfrac{6x}{2x + 5} + \dfrac{15}{2x + 5}$

u. $\dfrac{2x^2}{x^2 - 9} + \dfrac{7x + 3}{x^2 - 9}$ **v.** $\dfrac{x^2}{x^2 - x - 6} - \dfrac{2x + 8}{x^2 - x - 6}$

There is one type of problem that requires special attention, so we will discuss it separately. Earlier in this section, we discussed equivalent forms of negative fractions and we indicated that $-\dfrac{a}{b} = \dfrac{-a}{b}$. It is also true that $\dfrac{a}{-b} = \dfrac{-a}{b}$, since both are equal to $-\dfrac{a}{b}$. For example, $\dfrac{4}{-7} = \dfrac{-4}{7}$ and $\dfrac{5}{-(a - b)} = \dfrac{-5}{a - b}$. In general a fraction has three signs, any two of which may be changed without changing the value of the fraction. For example, $+\dfrac{+2}{+3} = +\dfrac{-2}{-3} = -\dfrac{-2}{+3} = -\dfrac{+2}{-3}$.

Answers:

Practice Exercises 18–22: 18. $\dfrac{8}{a - 5}$ 19. $\dfrac{2a}{a + 2}$ 20. 1   21. $\dfrac{x - 1}{x - 3}$ 22. $\dfrac{x - 5}{x - 3}$ **Additional Practice Exercises r–v:** r. $\dfrac{7}{2x + 1}$ s. $\dfrac{4y}{3y + 4}$ t. 3   u. $\dfrac{2x + 3}{x - 3}$ v. $\dfrac{x - 4}{x - 3}$

Example 7

Find the following sums or differences. Leave the answers reduced to lowest terms.

a. $\dfrac{5}{x-2} + \dfrac{3}{2-x} =$ $\qquad$ *Rewrite $2 - x$ as $-(x - 2)$.*

$\dfrac{5}{x-2} + \dfrac{3}{-(x-2)} =$ $\qquad$ *Rewrite $\dfrac{3}{-(x-2)}$ as $\dfrac{-3}{x-2}$.*

$\dfrac{5}{x-2} + \dfrac{-3}{x-2} =$ $\qquad$ *Find the sum.*

$\dfrac{5 + (-3)}{x-2} =$ $\qquad$ *$5 + (-3) = 2$.*

$\dfrac{2}{x-2}$ $\qquad$ *Sum.*

b. $\dfrac{a+b}{a-b} - \dfrac{2a-3b}{b-a} =$ $\qquad$ *Rewrite $b - a$ as $-(a - b)$.*

$\dfrac{a+b}{a-b} - \dfrac{2a-3b}{-(a-b)} =$ $\qquad$ *Rewrite $\dfrac{2a-3b}{-(a-b)}$ as $\dfrac{-(2a-3b)}{a-b}$.*

$\dfrac{a+b}{a-b} - \dfrac{-(2a-3b)}{a-b} =$ $\qquad$ *Remove the parentheses.*

$\dfrac{a+b}{a-b} - \dfrac{-2a+3b}{a-b} =$ $\qquad$ *Find the difference.*

$\dfrac{(a+b) - (-2a+3b)}{a-b} =$ $\qquad$ *Remove the parentheses.*

$\dfrac{a+b+2a-3b}{a-b} =$ $\qquad$ *Add like terms.*

$\dfrac{3a-2b}{a-b}$ $\qquad$ *Answer.*

Practice Exercises

Find the following sums or differences. Leave answers reduced to lowest terms.

23. $\dfrac{8}{y-3} + \dfrac{5}{3-y}$

24. $\dfrac{3x+2y}{x-y} - \dfrac{x-5y}{y-x}$

If more practice is needed, do the Additional Practice Exercises.

Additional Practice Exercises

Find the following sums or differences. Leave answers reduced to lowest terms.

w. $\dfrac{-5}{a-4} + \dfrac{2}{4-a}$

x. $\dfrac{2b-3c}{b-c} - \dfrac{b-2c}{c-b}$

Exercise Set 7.1

Find the following sums or differences:

1. $\dfrac{2}{7} + \dfrac{3}{7}$

2. $\dfrac{3}{8} + \dfrac{1}{8}$

3. $\dfrac{7}{18} + \dfrac{5}{18}$

4. $\dfrac{4}{25} + \dfrac{9}{25}$

5. $\dfrac{3}{16} + \dfrac{1}{16} + \dfrac{5}{16}$

6. $\dfrac{6}{29} + \dfrac{12}{29} + \dfrac{3}{29}$

7. $\dfrac{3}{4} - \dfrac{2}{4}$

8. $\dfrac{8}{9} - \dfrac{7}{9}$

9. $\dfrac{10}{17} - \dfrac{-5}{17}$

10. $\dfrac{9}{22} - \dfrac{-9}{22}$

11. $\dfrac{7}{11} - \dfrac{9}{11}$

12. $\dfrac{5}{12} - \dfrac{10}{12}$

Answers:

Practice Exercises 23–24: **23.** $\dfrac{3}{y-3}$ **24.** $\dfrac{4x-3y}{x-y}$ **Additional Practice Exercises w–x:** **w.** $\dfrac{-7}{a-4}$ **x.** $\dfrac{3b-5c}{b-c}$

Find the following sums or differences. Leave the answers reduced to the lowest terms.

13. $\dfrac{5}{14} + \dfrac{9}{14}$

14. $\dfrac{7}{24} + \dfrac{11}{24}$

15. $\dfrac{9}{10} - \dfrac{3}{10}$

16. $\dfrac{17}{25} - \dfrac{12}{25}$

17. $\dfrac{8}{15} + \dfrac{3}{15} - \dfrac{6}{15}$

18. $\dfrac{6}{21} - \dfrac{8}{21} + \dfrac{16}{21}$

19. $\dfrac{17}{18} - \dfrac{6}{18} - \dfrac{5}{18}$

20. $-\dfrac{9}{32} - \dfrac{-4}{32} + \dfrac{29}{32}$

Add or subtract the following mixed numbers:

21. $8\dfrac{2}{7} + 4\dfrac{3}{7}$

22. $3\dfrac{9}{10} + 6\dfrac{3}{10}$

23. $10\dfrac{7}{9} - 5\dfrac{4}{9}$

24. $7\dfrac{3}{14} - 12\dfrac{10}{14}$

25. $8\dfrac{10}{27} + 9\dfrac{16}{27} - 11\dfrac{5}{27}$

26. $13\dfrac{24}{33} - 9\dfrac{10}{33} + 4\dfrac{8}{33}$

Find the sums or differences of the following rational expressions. Leave your answers reduced to lowest terms.

27. $\dfrac{a}{b} + \dfrac{c}{b}$

28. $\dfrac{x}{t} + \dfrac{y}{t}$

29. $\dfrac{u}{v} - \dfrac{5}{v}$

30. $\dfrac{8}{w} - \dfrac{z}{w}$

31. $\dfrac{11r}{21u} + \dfrac{4m}{21u}$

32. $\dfrac{13p}{30n} + \dfrac{7q}{30n}$

33. $\dfrac{7}{2y} + \dfrac{3}{2y} - \dfrac{5}{2y}$

34. $\dfrac{5}{2x} + \dfrac{7}{2x} - \dfrac{4}{2x}$

35. $\dfrac{6x}{7y^2} - \dfrac{2x}{7y^2} - \dfrac{5x}{7y^2}$

36. $\dfrac{9a}{11y^2} - \dfrac{6a}{11y^2} + \dfrac{2a}{11y^2}$

37. $\dfrac{6}{n+5} + \dfrac{3}{n+5}$

38. $\dfrac{4}{2-m} + \dfrac{7}{2-m}$

39. $\dfrac{9}{t-8} - \dfrac{7}{t-8}$

40. $\dfrac{8}{4-z} - \dfrac{12}{4-z}$

41. $\dfrac{5c}{c+2} - \dfrac{2c}{c+2}$

42. $\dfrac{5z}{6-b} - \dfrac{7z}{6-b}$

43. $\dfrac{v}{v-6} - \dfrac{11}{v-6}$

44. $\dfrac{w}{7+w} - \dfrac{2}{7+w}$

45. $\dfrac{3a}{a+2} + \dfrac{5a}{a+2} - \dfrac{2a}{a+2}$

46. $\dfrac{8x}{x+3} - \dfrac{2x}{x+3} - \dfrac{4x}{x+3}$

47. $\dfrac{4a-2b}{2a-3b} - \dfrac{2a+b}{2a-3b} + \dfrac{4a-6b}{2a-3b}$

48. $\dfrac{3x-2y}{3x+4y} + \dfrac{5x+5y}{3x+4y} - \dfrac{2x-5y}{3x+4y}$

49. $\dfrac{x^2}{x^2-x-12} + \dfrac{5x+6}{x^2-x-12}$

50. $\dfrac{y^2}{y^2-5y+4} + \dfrac{11y+30}{y^2-5y+4}$

51. $\dfrac{t^2-4t}{t^2+10t+21} - \dfrac{-6t+3}{t^2+10t+21}$

52. $\dfrac{v^2+6}{v^2-3v-10} - \dfrac{2v+14}{v^2-3v-10}$

53. $\dfrac{u^2}{u^2-9} + \dfrac{2u-15}{u^2-9}$

54. $\dfrac{w^2+3w}{w^2+4w+4} - \dfrac{5w+8}{w^2+4w+4}$

55. $\dfrac{x^2-2x}{x^2-12x+36} + \dfrac{-5x+6}{x^2-12x+36}$

56. $\dfrac{q^2-3}{q^2-49} - \dfrac{5q+11}{q^2-49}$

57. $\dfrac{2x^2+x-3}{x^2+6x+5} + \dfrac{x^2-2x+4}{x^2+6x+5} - \dfrac{2x^2+x+4}{x^2+6x+5}$

58. $\dfrac{3x^2-2x+3}{x^2+x-12} - \dfrac{x^2+x-2}{x^2+x-12} - \dfrac{x^2+4x-7}{x^2+x-12}$

Find the sums or differences. Leave the answers reduced to lowest terms.

59. $\dfrac{3}{7} + \dfrac{4}{-7}$

60. $\dfrac{13}{15} + \dfrac{8}{-15}$

61. $\dfrac{-9}{17} - \dfrac{8}{-17}$

62. $\dfrac{19}{22} - \dfrac{2}{-22}$

63. $\dfrac{3}{t-4} + \dfrac{11}{4-t}$

64. $\dfrac{-5}{7-t} + \dfrac{8}{t-7}$

65. $\dfrac{v}{u-v} - \dfrac{3v}{v-u}$

66. $\dfrac{2x}{x-y} + \dfrac{-3y}{y-x}$

67. $\dfrac{2m-n}{m-n} - \dfrac{m-3}{n-m}$

68. $\dfrac{p-5q}{p-q} - \dfrac{4p-6q}{q-p}$

69. $\dfrac{2a-3b}{3a-b} + \dfrac{3a+2b}{b-3a}$

70. $\dfrac{4x+3y}{2x-5y} + \dfrac{2x-y}{5y-2x}$

Solve the following:

71. A triangular flower bed has sides whose lengths are $3\frac{3}{8}$ ft, $2\frac{2}{8}$ ft and $4\frac{5}{8}$ ft. How much border material will it take to enclose the bed?

72. Fred is wrapping a triangular package whose sides are $8\frac{6}{16}$ in., $7\frac{3}{16}$ in., and $5\frac{7}{16}$ in. How long would a piece of ribbon need to be in order to wrap around the package?

73. A rectangular picture is $6\frac{3}{4}$ in. long and $4\frac{1}{4}$ in. wide. What is the perimeter of the picture?

74. A rectangular traffic sign is $18\frac{5}{8}$ in. long and $12\frac{3}{8}$ in. wide. What is the perimeter of the sign?

Find the perimeters of the following triangles whose sides have the given lengths:

75. $\dfrac{5}{24a}$ ft, $\dfrac{7}{24a}$ ft, $\dfrac{11}{24a}$ ft

76. $\dfrac{7}{28x}$ yd, $\dfrac{9}{28x}$ yd, $\dfrac{8}{28x}$ yd

77. $\dfrac{5a}{2a+3b}$ dm, $\dfrac{3b}{2a+3b}$ dm, $\dfrac{2a+4b}{2a+3b}$ dm

78. $\dfrac{3m}{2m+n}$ ft, $\dfrac{m+2n}{2m+n}$ ft, $\dfrac{2m-n}{2m+n}$ ft

79. $\dfrac{2x^2-3x}{x^2-2x-15}$ cm, $\dfrac{-x^2+x-4}{x^2-2x-15}$ cm, $\dfrac{x-8}{x^2-2x-15}$ cm

80. $\dfrac{3x^2+2}{x^2-3x-10}$ in., $\dfrac{-x^2+4x+1}{x^2-3x-10}$ in., $\dfrac{-x^2+2x+5}{x^2-3x-10}$ in.

Find the perimeters of the following rectangles, given their lengths and widths.

81. $L = \dfrac{6}{x}$ m, $W = \dfrac{4}{x}$ m

82. $L = \dfrac{8}{3y}$ m, $W = \dfrac{5}{3y}$ m

83. $L = \dfrac{a+2b}{2a-3b}$ ft, $W = \dfrac{2a-b}{2a-3b}$ ft

84. $L = \dfrac{2x-4y}{3x-5y}$ cm, $W = \dfrac{x+5y}{3x-5y}$ cm

85. $L = \dfrac{4x^2+8x+3}{8x^2+2x-3}$ m, $W = \dfrac{4x^2+10x+6}{8x^2+2x-3}$ m

86. $L = \dfrac{x^2+4x+3}{2x^2-5x-12}$ in., $W = \dfrac{x^2+3x+3}{2x^2-5x-12}$ in.

Challenge Exercises:

Find the following sums or differences:

87. $\dfrac{x^2+2x}{x^3+2x^2-15x} - \dfrac{15}{x^3+2x^2-15x}$

88. $\dfrac{y^2+9y}{y^3-36y} + \dfrac{18}{y^3-36y}$

89. $\dfrac{t^3-4t^2}{t^3-2t^2+t} - \dfrac{-3t}{t^3-2t^2+t}$

Writing Exercises:

90. When adding or subtracting rational numbers or rational expressions that have the same denominator, why do we add or subtract the numerators but we do not add or subtract the denominators?

91. In this section, we began with a graphical illustration for the addition of rational numbers. Make your own illustration for the addition of fractions and explain how it works.

92. Show a graphical illustration for the subtraction of fractions and explain how it works.

93. What is wrong with the following? $\dfrac{4x-1}{x+6} - \dfrac{2x-3}{x+6} = \dfrac{4x-1-2x-3}{x+6} = \dfrac{x-4}{x+6}$. What is the correct answer?

Least Common Multiple and Equivalent Rational Expressions

OBJECTIVES

When you complete this section, you will be able to:

a. Find the least common multiple of two or more integers.

b. Find the least common multiple of two or more polynomials.

c. Change a fraction into an equivalent fraction with a specified denominator.

d. Change a rational expression into an equivalent rational expression with a specified denominator.

Introduction

Previously, we discussed the greatest common factor of two or more integers. One interpretation of the greatest common factor is that it is the largest integer that will divide evenly into all the integers for which it is the greatest common factor. In this section, we discuss the **least common multiple** of two or more integers. One interpretation of the least common multiple is that it is the smallest positive integer that is divisible by all the integers for which it is the least common multiple.

Before discussing the technique for finding the least common multiple, let us discuss the idea of multiples. To find the multiples of a number, multiply the number by 1, 2, 3, 4, and so on. For example, the multiples of 3 are $3 \cdot 1 = 3, 3 \cdot 2 = 6, 3 \cdot 3 = 9, 3 \cdot 4 = 12$, and so on.

The multiples of 3 are {3, 6, 9, 12, 15, 18, 21, 24, . . .}. The multiples of 4 are {4, 8, 12, 16, 20, 24, . . .}. Looking at the multiples of 3 and 4, we see that 12 and 24 are multiples of both. Consequently, we call 12 and 24 common multiples of 3 and 4. Other common multiples are 36, 48, 60, and so on. Of all the common multiples of 3 and 4, the smallest is 12. Therefore, 12 is the least common multiple of 3 and 4. Notice that each multiple of 3 is divisible by 3 and each multiple of 4 is divisible by 4. Consequently, the common multiples of 3 and 4 are divisible by both 3 and 4. Numbers that are divisible by 3 and 4 have 3 and 4 as factors.

Least Common Multiple (LCM)

The integer a is the least common multiple of the integers b and c, if a is the smallest positive integer that is a multiple of both b and c. Equivalently, a is the least common multiple of b and c, if a is the smallest positive integer that is divisible by both b and c.

The preceding definition is easily extended to more than two integers.

The technique we use for finding the LCM depends upon finding the prime factorizations of the numbers for which we wish to find the LCM. Prime factorization was discussed in Section 5.1. You may wish to review this technique before continuing.

Finding the LCM of two or more integers

Finding the Least Common Multiple (LCM)

Procedure for finding the LCM of two or more numbers.

1. Write each number as the product of its prime factors.

2. Select every factor appearing in any prime factorization. If a factor appears in more than one of the prime factorizations, select the factor with the largest exponent.

3. The LCM is the product of the factors selected in step 2.

This technique for finding the LCM works because any factor of a number is also a factor of any multiple of that number. By taking the largest exponent of every factor that appears in any prime factorization, we are assuring ourselves that their product (the LCM) contains every factor of each number and only those numbers that are factors.

Example 1

Find the LCM of the following:

a. 3 and 7

Solution:
Since 3 and 7 are both prime numbers, the different factors that appear are 3 and 7. The largest exponent on each is 1. Consequently, the LCM is their product: $3 \cdot 7 = 21$. Note that 21 is divisible by both 3 and 7.

b. 108 and 144

Solution:

$$108 = 2^2 \cdot 3^3 \qquad \text{Prime factorization.}$$
$$144 = 2^4 \cdot 3^2 \qquad \text{Prime factorization.}$$

The different factors that appear in either factorization are 2 and 3. The largest exponent on 2 in either factorization is 4 in 2^4, and the largest exponent on 3 in either factorization is 3 in 3^3. Consequently, the LCM is $2^4 \cdot 3^3 = 16 \cdot 27 = 432$. Note that 432 is divisible by both 108 and 144.

c. 90 and 189

Solution:

$$90 = 2 \cdot 3^2 \cdot 5 \qquad \text{Prime factorization.}$$
$$189 = 3^3 \cdot 7 \qquad \text{Prime factorization.}$$

The different factors that appear in either factorization are 2, 3, 5, and 7. The largest exponent on 2 is 1 in 2, the largest exponent of 3 is 3 in 3^3, the largest exponent of 5 is 1 in 5, and the largest exponent of 7 is 1 in 7. Consequently, the LCM is $2 \cdot 3^3 \cdot 5 \cdot 7 = 2 \cdot 27 \cdot 5 \cdot 7 = 54 \cdot 5 \cdot 7 = 270 \cdot 7 = 1890$. Verify that 1890 is divisible by both 90 and 189.

d. 50, 45, and 42

Solution:

$$50 = 2 \cdot 5^2 \qquad \text{Prime factorization.}$$
$$45 = 3^2 \cdot 5 \qquad \text{Prime factorization.}$$
$$42 = 2 \cdot 3 \cdot 7 \qquad \text{Prime factorization.}$$

The different factors that appear in the factorizations are 2, 3, 5, and 7. The largest exponent of 2 is 1 in 2, the largest exponent of 3 is 2 in 3^2, the largest exponent of 5 is 2 in 5^2, and the largest exponent of 7 is 1 in 7. Consequently, the LCM is $2 \cdot 3^2 \cdot 5^2 \cdot 7 = 2 \cdot 9 \cdot 25 \cdot 7 = 18 \cdot 25 \cdot 7 = 450 \cdot 7 = 3150$. Verify that 3150 is divisible by 50, 45, and 42.

Practice Exercises

Find the LCM of the following:

1. 7 and 11
2. 48 and 72
3. 60 and 126
4. 28, 30, and 33

Answers:

Practice Exercises 1–4: 1. $7 \cdot 11 = 77$ 2. $2^4 \cdot 3^2 = 144$ 3. $2^2 \cdot 3^2 \cdot 5 \cdot 7 = 1260$ 4. $2^2 \cdot 3 \cdot 5 \cdot 7 \cdot 11 = 4620$

If more practice is needed, do the Additional Practice Exercises.

Additional Practice Exercises

Find the LCM of the following:

a. 3 and 5   **b.** 225 and 375   **c.** 90 and 168   **d.** 20, 63, and 70

Finding the LCM of two or more monomials

The procedure for finding the LCM of two or more monomials involving variables is exactly the same as for finding the LCM from prime factorizations. Many times it is actually easier, since we do not have to find the prime factorizations for the variable factors.

Example 2

Find the LCM of the following:

a. a^2b^3 and a^4b

Solution:
The different factors that appear are a and b. The largest exponent of a is 4 in a^4, and the largest exponent of b is 3 in b^3. Consequently, the LCM is a^4b^3. Note that $\frac{a^4b^3}{a^2b^3} = a^2$ and $\frac{a^4b^3}{a^4b} = b^2$. Consequently, a^4b^3 is divisible by both a^2b^3 and a^4b.

b. x^3y^2z and x^2y^4w

Solution:
The different factors that appear are x, y, z, and w. The largest exponent on x is 3 in x^3, the largest exponent of y is 4 in y^4, the largest exponent of z is 1 in z, and the largest exponent of w is 1 in w. Consequently, the LCM is x^3y^4zw. Note that x^3y^4zw is divisible by both x^3y^2z and x^2y^4w.

If the monomial expressions have numerical coefficients, we think of the exercise in two parts. We find the prime factorizations of the coefficients and use the same method as before.

Example 3

Find the LCM of the following:

a. $12r^3s^5$ and $18rs^2$

Solution:
We find the prime factorizations of the coefficients and use the same method as before.

$$12 = 2^2 \cdot 3r^3s^5 \qquad \text{Prime factorization of 12.}$$
$$18 = 2 \cdot 3^2rs^2 \qquad \text{Prime factorization of 18.}$$

The different factors that appear are 2, 3, r, and s. The largest exponent on 2 is 2 in 2^2, the largest exponent on 3 is 2 in 3^2, the largest exponent on r is 3 in r^3, and the largest exponent on s is 5 in s^5. Therefore, the LCM is $2^2 \cdot 3^2r^2s^5 = 36r^3s^5$. Note that $36r^3s^5$ is divisible by $12r^3s^5$ and $18rs^2$.

b. $6a^5b^2c^2$, $9a^2bc^2$, and $10a^2b^3c^2$

Solution:
We find the prime factorizations of the coefficients and use the same method as before.

$$6 = 2 \cdot 3a^5b^2c^2 \qquad \text{Prime factorization of 6.}$$
$$9 = 3^2a^2bc^2 \qquad \text{Prime factorization of 9.}$$
$$10 = 2 \cdot 5a^2b^3c^2 \qquad \text{Prime factorization of 10.}$$

The different factors that appear are 2, 3, 5, a, b, and c. The largest exponent on 2 is 1 in 2, the largest exponent on 3 is 2 in 3^2, the largest exponent on 5 is 1 in 5, the largest exponent on a is 5 in a^5, the largest exponent on b is 3 in b^3, and the largest exponent on c is 2 in c^2. Therefore, the LCM is $2 \cdot 3^2 \cdot 5a^5b^3c^2 = 90a^5b^3c^2$. Note that $90a^5b^3c^2$ is divisible by $6a^5b^2c^2$, $9a^2bc^2$, and $10a^2b^3c^2$.

Answers:

Additional Practice Exercises *a–d:* a. $3 \cdot 5 = 15$ **b.** $3^2 \cdot 5^3 = 1125$ **c.** $2^3 \cdot 3^2 \cdot 5 \cdot 7 = 2520$ **d.** $2^2 \cdot 3^2 \cdot 5 \cdot 7 = 1260$

Practice Exercises

Find the LCM of the following:

5. x^2y^5 and x^4y

6. r^4st^3 and $r^2s^3u^2$

7. $15a^2b^4$ and $12a^3b^4$

8. $8x^2yz^3$, $12x^4y^2z$ and $15x^4y^2z^2$

If more practice is needed, do the Additional Practice Exercises.

Additional Practice Exercises

Find the LCM of the following:

e. a^3b^2 and a^2b^4

f. $m^2n^4o^2$ and $m^5n^3o^4$

g. $14r^6s^2$ and $20r^2s^2$

h. $4m^3n^3$, $6m^2n^4$, and $15m^3n^4$

In order to find the LCM of two or more polynomials that are not monomials, the procedure is still the same. However, we usually leave the LCM in factored form rather than multiply it out.

Example 4

Find the LCM of the following:

a. $x + 3$ and $x - 2$

Solution:
Since $x + 3$ and $x - 2$ are both prime polynomials, the LCM is their product $(x + 3)(x - 2)$. Note that $(x + 3)(x - 2)$ is divisible by both $x + 3$ and $x - 2$.

b. $y - 6$ and $y^2 + 5y$

Solution:

$$y - 6 \qquad\qquad y - 6 \text{ is a prime polynomial.}$$
$$y^2 + 5y = y(y + 5) \qquad \text{Prime factorization.}$$

The different factors that appear in either prime factorization are y, $y - 6$, and $y + 5$. The largest exponent of each is 1. Therefore, the LCM is $y(y - 6)(y + 5)$. Note that $y(y - 6)(y + 5)$ is divisible by $y - 6$ and $y^2 + 5y$.

c. $z + 4$ and $z^2 + 6z + 8$

Solution:

$$z + 4 \qquad\qquad z + 4 \text{ is a prime polynomial.}$$
$$z^2 + 6z + 8 = (z + 4)(z + 2) \qquad \text{Prime factorization.}$$

The different factors that appear are $z + 4$ and $z + 2$. The largest exponent on $z + 4$ is 1 in $z + 4$, and the largest exponent on $z + 2$ is 1 in $z + 2$. Therefore, the LCM is $(z + 4)(z + 2)$. Note that $(z + 4)(z + 2)$ is divisible by $z + 4$ and $z^2 + 6z + 8$.

Answers:

Practice Exercises 5–8: 5. x^4y^5 **6.** $r^4s^3t^3u^2$ **7.** $60a^3b^4$ **8.** $120x^4y^2z^3$ **Additional Practice Exercises e–h: e.** a^3b^4 **f.** $m^5n^4o^4$ **g.** $140r^6s^2$ **h.** $60m^3n^4$

d. $x^2 + 2x - 15$ and $x^2 - 5x + 6$

Solution:

$$x^2 + 2x - 15 = (x - 3)(x + 5) \qquad \text{Prime factorization.}$$
$$x^2 - 5x + 6 = (x - 3)(x - 2) \qquad \text{Prime factorization.}$$

The different factors that appear are $x - 3$, $x + 5$, and $x - 2$. The largest exponent on $x - 3$ is 1 in $x - 3$, the largest exponent on $x + 5$ is 1 in $x + 5$, and the largest exponent on $x - 2$ is 1 in $x - 2$. Therefore, the LCM is $(x - 3)(x + 5)(x - 2)$. Note that $(x - 3)(x + 5)(x - 2)$ is divisible by $x^2 + 2x - 15$ and $x^2 - 5x + 6$.

e. $x^2 + 5x + 4$ and $x^2 + 2x + 1$

Solution:

$$x^2 + 5x + 4 = (x + 1)(x + 4) \qquad \text{Prime factorization.}$$
$$x^2 + 2x + 1 = (x + 1)^2 \qquad \text{Prime factorization.}$$

The different factors that appear are $x + 1$ and $x + 4$. The largest exponent on $x + 1$ is 2 in $(x + 1)^2$, and the largest exponent on $x + 4$ is 1 in $x + 4$. Therefore, the LCM is $(x + 1)^2(x + 4)$. Note that $(x + 1)^2(x + 4)$ is divisible by $x^2 + 5x + 4$ and $x^2 + 2x + 1$.

f. $4r^2 - 9$, $6r^2 + 7r - 3$, and $6r^2 - 11r + 3$

Solution:

$$4r^2 - 9 = (2r + 3)(2r - 3) \qquad \text{Prime factorization.}$$
$$6r^2 + 7r - 3 = (2r + 3)(3r - 1) \qquad \text{Prime factorization.}$$
$$6r^2 - 11r + 3 = (3r - 1)(2r - 3) \qquad \text{Prime factorization.}$$

The different factors that appear are $2r + 3$, $2r - 3$, and $3r - 1$. Since the largest exponent on each is 1, the LCM is $(2r + 3)(2r - 3)(3r - 1)$. Note that $(2r + 3)(2r - 3)(3r - 1)$ is divisible by $4r^2 - 9$, $6r^2 + 7r - 3$, and $6r^2 - 11r + 3$.

Practice Exercises

Find the LCM of the following:

9. $a + 4$ and $a - 3$

10. $b + 3$ and $b^2 + 3b$

11. $b - 4$ and $b^2 - 2b - 8$

12. $y^2 - 7y + 10$ and $y^2 + 3y - 10$

13. $x^2 + 4x + 4$ and $x^2 - 3x - 10$

14. $4x^2 - 25$, $8x^2 + 14x - 15$, and $8x^2 - 26x + 15$

If more practice is needed, do the Additional Practice Exercises.

Additional Practice Exercises

Find the LCM of the following:

i. $b - 5$ and $b + 2$

j. $m - 3$ and $m^2 - 7m + 12$

k. $c^2 + 3c - 18$ and $c^2 + c - 12$

l. $4x^2 - 12x + 9$ and $8x^2 - 18x + 9$

m. $9x^2 - 4$, $6x^2 + x - 2$, and $6x^2 - 7x + 2$

In Section 6.2, we introduced the idea of equivalent fractions. Look at the following rectangles where each rectangle represents one unit:

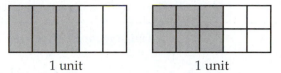

1 unit   1 unit

The fraction representing the shaded part of the first rectangle is $\frac{3}{5}$. Divide the unit horizontally into halves in order to get the second rectangle. The fraction representing the shaded part of the second rectangle is $\frac{6}{10}$. The shaded area is the same before and after the division. Therefore, only the name representing the part of the unit that is shaded has changed. Hence, we see that $\frac{6}{10} = \frac{3}{5}$. In other words, $\frac{6}{10}$ and $\frac{3}{5}$ are equivalent fractions.

From the example, we have $\frac{6}{10} = \frac{3}{5}$. Notice that the product of the numerator of the first fraction with the denominator of the second is equal to the product of the denominator of the first fraction and the numerator of the second. That is, $6 \cdot 5 = 10 \cdot 3$. This observation leads to a more formal definition of equivalent fractions than the one given earlier in Section 6.2.

DEFINITION   Equivalent Fractions

Two or more fractions are equivalent if they represent the same quantity. Equivalently, $\frac{a}{b} = \frac{c}{d}$ if $a \cdot d = b \cdot c$.

The preceding procedure is often referred to as **cross multiplication** because the lines of multiplication "cross" each other.

$$\frac{a}{b} \times \frac{c}{d}$$

$$a \cdot d = b \cdot c$$

In reducing fractions to lowest terms, we divide the numerators and the denominators by all common factors common to both to get equivalent fractions reduced to lowest terms. For example, $\frac{6}{10} = \frac{2 \cdot 3}{2 \cdot 5} = \frac{3}{5}$. In this section, we are going to do the opposite and generate equivalent fractions by multiplying the numerator and the denominator of a fraction by the same number. This is permissible because any nonzero number divided by itself is 1, and any number multiplied by 1 is equivalent to the original number. Since we are multiplying the numerator and the denominator by the same number, we are multiplying the fraction by 1. This is called the **fundamental property of fractions.**

Fundamental Property of Fractions

If $\frac{a}{b}$ is a fraction and c is any number except 0, then $\frac{a}{b} = \frac{a}{b} \cdot \frac{c}{c} = \frac{ac}{bc}$. In words, we can multiply both the numerator and the denominator of a fraction by any number except 0 and get a fraction equivalent to the original fraction.

Note: If we cross multiply $\frac{a}{b} = \frac{ac}{bc}$, we get $abc = bac$. By what properties of multiplication does $abc = bac$?

Changing a fraction into an equivalent fraction with a specified denominator   In the examples that follow, we will use the fundamental property of fractions to create equivalent fractions with a given denominator. We must decide by what we need to multiply the denominator of the first fraction in order to get the denominator of the second fraction. Then we will multiply the numerator and the denominator of the first fraction by this number to get an equivalent fraction with the denominator of the second fraction. Since the purpose of these exercises is to aid you in a later section in adding fractions with unlike denominators, we will limit our discussion to finding missing numerators only.

Example 5

Find the missing numerator so that the fractions are equal.

a. $\dfrac{1}{3} = \dfrac{?}{6}$ Since $2 \cdot 3 = 6$, multiply $\dfrac{1}{3}$ by $\dfrac{2}{2}$.

$\dfrac{2}{2} \cdot \dfrac{1}{3} = \dfrac{2}{6}$ Therefore, $\dfrac{1}{3} = \dfrac{2}{6}$.

CHECK: $1 \cdot 6 = 3 \cdot 2$, so the answer is correct.

b. $\dfrac{3}{4} = \dfrac{?}{12}$ Since $3 \cdot 4 = 12$, multiply $\dfrac{3}{4}$ by $\dfrac{3}{3}$.

$\dfrac{3}{3} \cdot \dfrac{3}{4} = \dfrac{9}{12}$ Therefore, $\dfrac{3}{4} = \dfrac{9}{12}$.

CHECK: $3 \cdot 12 = 4 \cdot 9$, so the answer is correct.

Note: The number that you must multiply the first fraction by to get an equivalent fraction with the given denominator can be found by dividing the denominator of the first fraction into the denominator of the new fraction. In Example 1b, $12 \div 4 = 3$. Therefore, we multiply $\dfrac{3}{4}$ by $\dfrac{3}{3}$.

c. $\dfrac{7}{12} = \dfrac{?}{108}$ It may not be obvious what we have to multiply 12 by in order to get 108. Since $108 \div 12 = 9$, multiply $\dfrac{7}{12}$ by $\dfrac{9}{9}$.

$\dfrac{9}{9} \cdot \dfrac{7}{12} = \dfrac{63}{108}$ Therefore, $\dfrac{7}{12} = \dfrac{63}{108}$.

CHECK: Does $7 \cdot 108 = 12 \cdot 63$?

d. $\dfrac{11}{8} = \dfrac{?}{96}$ It may not be obvious what we have to multiply 8 by in order to get 96. Since $96 \div 8 = 12$, multiply $\dfrac{11}{8}$ by $\dfrac{12}{12}$.

$\dfrac{12}{12} \cdot \dfrac{11}{8} = \dfrac{132}{96}$ Therefore, $\dfrac{11}{8} = \dfrac{132}{96}$.

CHECK: Does $11 \cdot 96 = 8 \cdot 132$?

Practice Exercises

Find the missing numerator so that the two fractions are equal.

15. $\dfrac{1}{5} = \dfrac{?}{30}$ **16.** $\dfrac{2}{3} = \dfrac{?}{18}$ **17.** $\dfrac{5}{16} = \dfrac{?}{144}$ **18.** $\dfrac{13}{9} = \dfrac{?}{126}$

If more practice is needed, do the Additional Practice Exercises.

Additional Practice Exercises

Find the missing numerator so that the two fractions are equal.

n. $\dfrac{1}{6} = \dfrac{?}{24}$ **o.** $\dfrac{3}{7} = \dfrac{?}{35}$ **p.** $\dfrac{5}{6} = \dfrac{?}{96}$ **q.** $\dfrac{17}{8} = \dfrac{?}{120}$

Answers:

Practice Exercises 15–18: **15.** 6 **16.** 12 **17.** 45 **18.** 182   Additional Practice Exercises n–q: **n.** 4 **o.** 15 **p.** 80 **q.** 255

The procedure is exactly the same if we have rational expressions instead of rational numbers. In the exercises that follow, assume all variables in the denominator have nonzero values.

Example 6

Find the missing numerator so that the rational expressions are equal. Assume all variables in the denominator have nonzero values.

a. $\dfrac{x}{y} = \dfrac{?}{yz}$ Since $y \cdot z = yz$, multiply $\dfrac{x}{y}$ by $\dfrac{z}{z}$.

$\dfrac{z}{z} \cdot \dfrac{x}{y} = \dfrac{xz}{yz}$ Therefore, $\dfrac{x}{y} = \dfrac{xz}{yz}$.

CHECK: $xyz = yxz$, so the answer is correct.

b. $\dfrac{3}{5x} = \dfrac{?}{15x^3}$ Since $3x^2 \cdot 5x = 15x^2$, multiply $\dfrac{3}{5x}$ by $\dfrac{3x^2}{3x^2}$.

$\dfrac{3x^2}{3x^2} \cdot \dfrac{3}{5x} = \dfrac{9x^2}{15x^3}$ Therefore, $\dfrac{3}{5x} = \dfrac{9x^2}{15x^3}$.

CHECK: $3 \cdot 15x^3 = 5x \cdot 9x^2$, since both equal $45x^3$. Therefore, the answer is correct.

c. $\dfrac{12x^2}{7y^2} = \dfrac{?}{28y^4}$ Since $4y^2 \cdot 7y^2 = 28y^4$, multiply $\dfrac{12x^2}{7y^2}$ by $\dfrac{4y^2}{4y^2}$.

$\dfrac{4y^2}{4y^2} \cdot \dfrac{12x^2}{7y^2} = \dfrac{48x^2y^2}{28y^4}$ Therefore, $\dfrac{12x^2}{7y^2} = \dfrac{48x^2y^2}{28y^4}$.

CHECK: Does $12x^2 \cdot 28y^4 = 7y^2 \cdot 48x^2y^2$?

Note: In Example 6c, it may not be obvious that we need to multiply $7y^2$ by $4y^2$ in order to get $28y^4$. We can use the same procedure we used in Example 5. Since $28y^4 \div 7y^2 = 4y^2$, we need to multiply by $4y^2$.

d. $\dfrac{8}{5x^2y} = \dfrac{?}{30x^3y^4}$ Since $30x^3y^4 \div 5x^2y = 6xy^3$, we need to multiply by $6xy^3$.

$\dfrac{6xy^3}{6xy^3} \cdot \dfrac{8}{5x^2y} = \dfrac{48xy^3}{30x^3y^4}$ Therefore, $\dfrac{8}{5x^2y} = \dfrac{48xy^3}{30x^3y^4}$.

CHECK: Does $8 \cdot 30x^3y^4 = 5x^2y \cdot 48xy^3$?

Practice Exercises

Find the missing numerator so that the rational expressions are equal. Assume all variables in the denominator have nonzero values.

19. $\dfrac{a}{b} = \dfrac{?}{bd}$ **20.** $\dfrac{5}{3a} = \dfrac{?}{18a^4}$ **21.** $\dfrac{8a^3}{5b^2} = \dfrac{?}{30b^5}$ **22.** $\dfrac{6}{11a^2b^3} = \dfrac{?}{55a^3b^3c}$

If more practice is needed, do the Additional Practice Exercises.

Additional Practice Exercises

Find the missing numerator so that the rational expressions are equal.

r. $\dfrac{m}{n} = \dfrac{?}{np}$ **s.** $\dfrac{3}{8x} = \dfrac{?}{48x^5}$ **t.** $\dfrac{7y^2}{4z^3} = \dfrac{?}{24z^6}$

Answers:

Practice Exercises 19–22: **19.** ad **20.** $30a^3$ **21.** $48a^3b^3$ **22.** $30ac$ **Additional Practice Exercises r–t:** **r.** mp **s.** $18x^4$ **t.** $42y^2z^3$

If the denominators of the rational expressions are polynomials other than monomials, we will need to factor those denominators that are not prime. By doing so, we can determine the expression we need to multiply the first rational expression by so that we can convert it into a rational expression with the same denominator as the second.

Example 7

Find the missing numerator so that the rational expressions will be equal.

a. $\dfrac{5a}{2a + 8} = \dfrac{?}{6a + 24}$ Factor the denominators.

$\dfrac{5a}{2(a + 4)} = \dfrac{?}{6(a + 4)}$ Since $3 \cdot 2(a + 4) = 6(a + 4)$, multiply $\dfrac{5a}{2(a + 4)}$ by $\dfrac{3}{3}$.

$\dfrac{3}{3} \cdot \dfrac{5a}{2(a + 4)} = \dfrac{15a}{6(a + 4)}$ Therefore, $\dfrac{5a}{2a + 8} = \dfrac{15a}{6a + 24}$.

Note: The checks can be done in the usual manner but may often be very difficult. For that reason, equivalent rational expressions are usually not checked.

b. $\dfrac{7}{x^2 - 2x} = \dfrac{?}{x(x - 2)(x + 3)}$ Factor $x^2 - 2x$.

$\dfrac{7}{x(x - 2)} = \dfrac{?}{x(x - 2)(x + 3)}$ The factor missing is $x + 3$, so we multiply by $\dfrac{x + 3}{x + 3}$.

$\dfrac{x + 3}{x + 3} \cdot \dfrac{7}{x(x - 2)} = \dfrac{7(x + 3)}{x(x - 2)(x + 3)}$ Multiply

$= \dfrac{7x + 21}{x(x - 2)(x + 3)}$ Therefore, $\dfrac{7}{x^2 - 2x} = \dfrac{7x + 21}{x(x - 2)(x + 3)}$.

c. $\dfrac{7x}{x^2 + 2x - 8} = \dfrac{?}{(x - 5)(x - 2)(x + 4)}$ Factor $x^2 + 2x - 8$.

$\dfrac{7x}{(x - 2)(x + 4)} = \dfrac{?}{(x - 5)(x - 2)(x + 4)}$ Since the denominator of the first fraction is missing the factor $x - 5$, we need to multiply by $\dfrac{x - 5}{x - 5}$.

$\dfrac{x - 5}{x - 5} \cdot \dfrac{7x}{x^2 + 2x - 8} = \dfrac{7x(x - 5)}{(x - 5)(x - 2)(x + 4)}$ Multiply.

$= \dfrac{7x^2 - 35x}{(x - 5)(x - 2)(x + 4)}$ Therefore, $\dfrac{7x}{x^2 + 2x - 8} = \dfrac{7x^2 - 35x}{(x - 5)(x - 2)(x + 4)}$.

Practice Exercises

Find the missing numerator so that the rational expressions are equal.

23. $\dfrac{7x}{3x - 6} = \dfrac{?}{9x - 18}$

24. $\dfrac{2a}{b^2 - 3b} = \dfrac{?}{b(b - 3)(b + 2)}$

25. $\dfrac{3x}{x^2 - 8x + 15} = \dfrac{?}{(x + 4)(x - 3)(x - 5)}$

If more practice is needed, do the Additional Practice Exercises.

Additional Practice Exercises

Find the missing numerator so that the rational expressions are equal.

u. $\dfrac{9}{4a + 8} = \dfrac{?}{12a + 24}$

v. $\dfrac{4z}{m^2 + 5m} = \dfrac{?}{m(m + 4)(m + 5)}$

w. $\dfrac{6x}{x^2 + x - 12} = \dfrac{?}{(x + 6)(x + 4)(x - 3)}$

Exercise Set 7.2

Find the LCM of the following whole numbers:

1. 3 and 5

2. 2 and 7

3. 9 and 11

4. 13 and 15

5. 25 and 30

6. 42 and 60

7. 36 and 60

8. 35 and 75

9. 70 and 154

10. 30 and 105

11. 28, 63, and 42

12. 36, 60, and 72

Find the LCM of the following monomials:

13. xy and x^3y^3

14. r^3s^5 and rs

15. u^6v and u^2v^4

16. t^3w^8 and t^5w

17. a^2b and $a^3b^6c^4$

18. m^8np^2 and m^9p^2

19. x^2y^6z and $x^4y^4z^3$

20. v^7st^6 and v^8s^3t

21. $3u^2v$ and $9u^3v^5$

22. $5w^4x^6$ and $15w^8x$

23. $24y^3z^4$ and $36y^5z^2$

24. $32r^2w^7$ and $48r^3w^6$

25. $15m^5np^9$ and $45m^3p^7$

26. $18a^3b^3$ and $54ab^4c^3$

27. $10r^4s^2t^2$ and $12r^4s^8t^3$

28. $21x^5yz^7$ and $27x^2y^6z^8$

29. $9u^3v$, $36u^2v^2$, and $27u^4v^6$

30. $18w^3x^2$, $42w^6x$, and $30w^5x^5$

31. $21r^4s^9t^5$, $14r^4s^3t$, and $28r^2s^8t^6$

32. $60a^2b^5c^2$, $12a^2b^8c^3$, and $48ab^5c^4$

Find the LCM of the following polynomials:

33. $y + 4$ and $y + 1$

34. $t - 5$ and $t - 2$

35. $w + 3$ and $w^2 - 6w$

36. $u - 7$ and $u^2 + 5u$

37. $v - 5$ and $v^2 - 2v - 15$

38. $r + 9$ and $r^2 + 8r - 9$

39. $x^2 - x - 12$ and $x^2 + 8x + 15$

40. $y^2 - 8y + 7$ and $y^2 + y - 2$

41. $z^2 - 4z - 5$ and $z^2 + 2z + 1$

42. $b^2 - 6b + 9$ and $b^2 - 8b + 15$

43. $t^2 - 9$, $t^2 - 2t - 15$, and $t^2 - 7t + 12$

44. $9y^2 - 16$, $3y^2 + 7y + 4$, $3y^2 - 2y - 8$

Challenge Exercises:

Find the LCM of the following:

45. 66, 132, and 33

46. $105x^2yz^3$, $75x^4y^5z$, and $195x^3y^6z^7$

47. $3y^2 - 75$ and $6y^2 - 60y + 150$

48. $24a^2 + 24ab - 90b^2$, $24a^2 - 4ab - 48b^2$

Answers:

Find the missing numerator so that the fractions will be equal.

49. $\dfrac{3}{4} = \dfrac{?}{28}$

50. $\dfrac{6}{7} = \dfrac{?}{35}$

51. $\dfrac{5}{8} = \dfrac{?}{56}$

52. $\dfrac{1}{9} = \dfrac{?}{99}$

53. $1 = \dfrac{?}{5}$

54. $2 = \dfrac{?}{4}$

55. $\dfrac{15}{2} = \dfrac{?}{8}$

56. $\dfrac{11}{3} = \dfrac{?}{21}$

57. $\dfrac{9}{5} = \dfrac{?}{135}$

58. $\dfrac{17}{9} = \dfrac{?}{117}$

Find the missing numerator so that the rational expressions are equal.

59. $\dfrac{c}{e} = \dfrac{?}{ef}$

60. $\dfrac{a}{b} = \dfrac{?}{cb}$

61. $\dfrac{4}{5x} = \dfrac{?}{30x}$

62. $\dfrac{3}{7y^2} = \dfrac{?}{42y^2}$

63. $\dfrac{5t}{8z} = \dfrac{?}{56z^3}$

64. $\dfrac{8u}{9v} = \dfrac{?}{63v^5}$

65. $\dfrac{9r}{11s^2} = \dfrac{?}{121s^5}$

66. $\dfrac{4m}{13n^3} = \dfrac{?}{91n^7}$

67. $\dfrac{9p}{4q^4} = \dfrac{?}{28q^6}$

68. $\dfrac{17w}{5x^5} = \dfrac{?}{40x^8}$

69. $\dfrac{12y^2}{7z^4} = \dfrac{?}{63z^9}$

70. $\dfrac{15a^3}{8b^2} = \dfrac{?}{72b^6}$

71. $\dfrac{9t}{14x^2y} = \dfrac{?}{42x^5y^3}$

72. $\dfrac{5r^2}{16ps^4} = \dfrac{?}{64p^4s^7}$

Find the missing numerator so that the rational expressions are equal.

73. $\dfrac{y}{y+3} = \dfrac{?}{5y+15}$

74. $\dfrac{z}{2-z} = \dfrac{?}{8-4z}$

75. $\dfrac{3t}{2t-10} = \dfrac{?}{6t-30}$

76. $\dfrac{5a}{3a+12} = \dfrac{?}{12a+48}$

77. $\dfrac{4c}{5c+10} = \dfrac{?}{20c+40}$

78. $\dfrac{6h}{14-7h} = \dfrac{?}{42-21h}$

79. $\dfrac{11}{r-1} = \dfrac{?}{r^2+2r-3}$

80. $\dfrac{2u}{4+u} = \dfrac{?}{12-u-u^2}$

81. $\dfrac{5v}{v^2-5v} = \dfrac{?}{v(v-5)(v+2)}$

82. $\dfrac{7x}{x^2+8x} = \dfrac{?}{x(x-1)(x+8)}$

83. $\dfrac{y-2}{y^2-10y+21} = \dfrac{?}{(y+4)(y-7)(y-3)}$

84. $\dfrac{u+5}{u^2+6u-16} = \dfrac{?}{(u+2)(u-2)(u+8)}$

85. $\dfrac{w-1}{w^2-16} = \dfrac{?}{(w+4)(w-4)(w+3)}$

86. $\dfrac{z+8}{z^2-25} = \dfrac{?}{(z+5)(z-3)(z-5)}$

87. $\dfrac{n-3}{n^2+6n+9} = \dfrac{?}{(n+3)(n-9)(n+3)}$

88. $\dfrac{m+5}{m^2-8m+16} = \dfrac{?}{(m-4)(m+8)(m-4)}$

Challenge Exercises:

Find the missing numerator so that the rational expressions are equal.

89. $\dfrac{t}{x^2+4x+4} = \dfrac{?}{(x+2)^3}$

90. $\dfrac{4z}{y^2-6y+9} = \dfrac{?}{y^3-6y^2+9y}$

Writing Exercises:

91. In this section, we have developed the LCM for both arithmetic and algebra. How do you think we will use the LCM in arithmetic? How do you think we will use the LCM in algebra?

92. Why is the LCM divisible by each of the terms for which it is the LCM?

93. In choosing the factors for the LCM, why do we select only factors that appear in the prime factorizations of the terms whose LCM we are finding? What would happen if we selected a factor not found in the prime factorizations?

94. In selecting the factors for the LCM, why is the largest exponent on each factor in the prime factorizations selected?

95. In selecting the factors for the LCM, why must each different factor that appears in any prime factorization be selected?

96. In creating a fraction that is equivalent to a fraction with a given denominator, why is it necessary to multiply both the numerator and the denominator by the *same* expression?

97. Explain how you would find the missing denominator in the following: $\frac{6}{5xy^2} = \frac{42x^3y^2}{?}$

98. Explain how you would show that $\frac{3}{4}$ is equal to $\frac{6}{8}$ by using shading in a diagram.

Section 7.3	The Least Common Denominator of Fractions and Rational Expressions

OBJECTIVES

When you complete this section, you will be able to:

a. Find the least common denominator for two or more fractions and convert each fraction into an equivalent fraction with the least common denominator as its denominator.

b. Find the least common denominator for two or more rational expressions and convert each rational expression into an equivalent rational expression with the least common denominator as its denominator.

Introduction   In Section 7.2, we discussed how to find the least common multiple (LCM) of two or more natural numbers or polynomials. In this section, we will use this concept to find the **least common denominator** of two or more fractions. The least common denominator (LCD) of two or more fractions is the least common multiple of the denominators. You may need to review this procedure before continuing. After finding the LCD, we will write fractions as equivalent fractions with the LCD as their denominators.

Finding the LCD of fractions and   Finding the least common denominator and writing fractions with the least common
writing fractions with the LCD   denominator as their denominators will be very important in the next section.

Example 1

Find the least common denominator for the following. Then write each fraction as an equivalent fraction with the LCD as its denominator.

a. $\frac{1}{3}$ and $\frac{1}{5}$ **Solution:**

Since 3 and 5 are both prime, their LCM is $3 \cdot 5 = 15$. Therefore, the LCD for $\frac{1}{3}$ and $\frac{1}{5}$ is 15. Since $3 \cdot 5 = 15$, we multiply $\frac{1}{3}$ by $\frac{5}{5}$. Since $5 \cdot 3 = 15$, we multiply $\frac{1}{5}$ by $\frac{3}{3}$.

$$\frac{1}{3} = \frac{5}{5} \cdot \frac{1}{3} = \frac{5}{15}, \text{ and}$$

$$\frac{1}{5} = \frac{3}{3} \cdot \frac{1}{5} = \frac{3}{15}.$$

b. $\dfrac{3}{10}$ and $\dfrac{5}{21}$

Solution:

$$10 = 2 \cdot 5$$
$$21 = 3 \cdot 7$$

Prime factorization of the denominators.

Therefore, by choosing the highest power of each factor, the LCM of 10 and 21 is $2 \cdot 3 \cdot 5 \cdot 7 = 210$. So the LCD of $\dfrac{3}{10}$ and $\dfrac{5}{21}$ is 210. Notice the LCD is the same as $10 \cdot 21$. This is because 10 and 21 have no factors in common. Since $10 \cdot 21 = 210$, we multiply $\dfrac{3}{10}$ by $\dfrac{21}{21}$. Since $21 \cdot 10 = 210$, we multiply $\dfrac{5}{21}$ by $\dfrac{10}{10}$.

$$\frac{3}{10} = \frac{21}{21} \cdot \frac{3}{10} = \frac{63}{210}.$$

$$\frac{5}{21} = \frac{10}{10} \cdot \frac{5}{21} = \frac{50}{210}.$$

c. $\dfrac{5}{6}$ and $\dfrac{5}{12}$

Solution:

$$6 = 2 \cdot 3$$
$$12 = 2^2 \cdot 3$$

Prime factorization of the denominators.

By choosing the highest power of each factor, the LCM is $2^2 \cdot 3 = 12$. So the LCD is also 12. Note that 12 is a multiple of 6. If one denominator is a multiple of the other, the larger number is the LCD. Since $6 \cdot 2 = 12$, multiply $\dfrac{5}{6}$ by $\dfrac{2}{2}$. We do not need to multiply $\dfrac{5}{12}$ by anything since it already has 12 as its denominator.

$$\frac{5}{6} = \frac{2}{2} \cdot \frac{5}{6} = \frac{10}{12}, \text{ and}$$

$$\frac{5}{12} = \frac{5}{12}.$$

d. $\dfrac{7}{12}$ and $\dfrac{11}{18}$

Solution:

$$12 = 2^2 \cdot 3$$
$$18 = 2 \cdot 3^2$$

Prime factorization of the denominators.

Therefore, the LCM of 12 and 18 is $2^2 \cdot 3^2 = 36$. So the LCD of $\dfrac{7}{12}$ and $\dfrac{11}{18}$ is also 36. Since $3 \cdot 12 = 36$, we multiply $\dfrac{7}{12}$ by $\dfrac{3}{3}$. Since $2 \cdot 18 = 36$, we multiply $\dfrac{11}{18}$ by $\dfrac{2}{2}$.

$$\frac{7}{12} = \frac{3}{3} \cdot \frac{7}{12} = \frac{21}{36}.$$

$$\frac{11}{18} = \frac{2}{2} \cdot \frac{11}{18} = \frac{22}{36}.$$

e. $\dfrac{4}{15}, \dfrac{5}{12},$ and $\dfrac{7}{10}$

Solution:

$$15 = 3 \cdot 5$$
$$12 = 2^2 \cdot 3$$
$$10 = 2 \cdot 5$$

Prime factorization of the denominators.

Therefore, the LCM of 15, 12, and 10 is $2^2 \cdot 3 \cdot 5 = 60$. So the LCD of $\dfrac{4}{15}, \dfrac{5}{12},$ and $\dfrac{7}{10}$ is 60. Since $15 \cdot 4 = 60$, we multiply $\dfrac{4}{15}$ by $\dfrac{4}{4}$. Since $5 \cdot 12 = 60$, we multiply $\dfrac{5}{12}$ by $\dfrac{5}{5}$. Since $10 \cdot 6 = 60$, we multiply $\dfrac{7}{10}$ by $\dfrac{6}{6}$.

$$\frac{4}{15} = \frac{4}{4} \cdot \frac{4}{15} = \frac{16}{60}.$$

$$\frac{5}{12} = \frac{5}{5} \cdot \frac{5}{12} = \frac{25}{60}.$$

$$\frac{7}{10} = \frac{6}{6} \cdot \frac{7}{10} = \frac{42}{60}.$$

Practice Exercises

Find the least common denominator for the following. Then write each fraction as an equivalent fraction with the LCD as its denominator.

1. $\dfrac{1}{5}$ and $\dfrac{1}{7}$

2. $\dfrac{5}{14}$ and $\dfrac{8}{15}$

3. $\dfrac{2}{9}$ and $\dfrac{7}{18}$

4. $\dfrac{9}{20}$ and $\dfrac{7}{50}$

5. $\dfrac{3}{4}, \dfrac{5}{6},$ and $\dfrac{3}{8}$

If more practice is needed, do the Additional Practice Exercises.

Answers:

Practice Exercises 1–5: (The answers are in the same order as the questions.) **1.** LCD = 35 $\dfrac{7}{35}, \dfrac{5}{35}$ **2.** LCD = 210 $\dfrac{75}{210}, \dfrac{112}{210}$ **3.** LCD = 18 $\dfrac{4}{18}, \dfrac{7}{18}$ **4.** LCD = 100 $\dfrac{45}{100}, \dfrac{14}{100}$ **5.** LCD = 24 $\dfrac{18}{24}, \dfrac{20}{24}, \dfrac{9}{24}$

Additional Practice Exercises

Find the least common denominator for the following. Then write each fraction as an equivalent fraction with the LCD as its denominator.

a. $\dfrac{1}{3}$ and $\dfrac{1}{7}$

b. $\dfrac{1}{22}$ and $\dfrac{1}{6}$

c. $\dfrac{2}{3}$ and $\dfrac{4}{15}$

d. $\dfrac{7}{36}$ and $\dfrac{11}{24}$

e. $\dfrac{1}{6}, \dfrac{4}{9},$ and $\dfrac{3}{10}$

Finding the LCD of rational expressions and writing each with the LCD

Finding the least common denominator of rational expressions follows the same procedure as finding the least common denominator of fractions.

Example 2

Find the least common denominator for the following. Then write each rational expression as an equal rational expression with the LCD as its denominator.

a. $\dfrac{2}{a}$ and $\dfrac{3}{b}$

Solution:

Since the denominators are prime, the LCM is their product ab. So the LCD is ab. Since $a \cdot b = ab$, we multiply $\frac{2}{a}$ by $\frac{b}{b}$. Since $b \cdot a = ab$, we multiply $\frac{3}{b}$ by $\frac{a}{a}$:

$$\frac{2}{a} = \frac{b}{b} \cdot \frac{2}{a} = \frac{2b}{ab}$$

$$\frac{3}{b} = \frac{a}{a} \cdot \frac{3}{b} = \frac{3a}{ab}.$$

b. $\dfrac{b}{a^2d}$ and $\dfrac{c}{ad^2}$

Solution:

The LCM of a^2d and ad^2 is a^2d^2. Therefore, the LCD is also a^2d^2. Since $d \cdot a^2d = a^2d^2$, we multiply $\frac{b}{a^2d}$ by $\frac{d}{d}$. Since $a \cdot ad^2 = a^2d^2$, we multiply $\frac{c}{ad^2}$ by $\frac{a}{a}$.

$$\frac{b}{a^2d} = \frac{b}{a^2d} \cdot \frac{d}{d} = \frac{bd}{a^2d^2}$$

$$\frac{c}{ad^2} = \frac{c}{ad^2} \cdot \frac{a}{a} = \frac{ac}{a^2d^2}.$$

c. $\dfrac{5}{10x^2y^3}$ and $\dfrac{3}{14x^2y^2}$

Solution:

$$10x^2y^3 = 2 \cdot 5x^2y^3 \qquad \text{Prime factorization of the denominators.}$$
$$14x^2y^2 = 2 \cdot 7x^2y^2 \qquad \text{Prime factorization of the denominators.}$$

Therefore, the LCM of $10x^2y^3$ and $14x^2y^2$ is $2 \cdot 5 \cdot 7x^2y^3 = 70x^2y^3$. Consequently, the LCD is $70x^2y^3$. To write $\frac{5}{10x^2y^3}$ as a fraction with a denominator of $70x^2y^3$, we multiply by $\frac{7}{7}$, and to write $\frac{3}{14x^2y^2}$ as a fraction with a denominator of $70x^2y^3$, we multiply by $\frac{5y}{5y}$.

$$\frac{5}{10x^2y^3} = \frac{7}{7} \cdot \frac{5}{10x^2y^3} = \frac{35}{70x^2y^3}.$$

$$\frac{3}{14x^2y^2} = \frac{5y}{5y} \cdot \frac{3}{14x^2y^2} = \frac{15y}{70x^2y^3}.$$

Answers:

Additional Practice Exercises a–e: **a.** LCD = 21 $\frac{7}{21}, \frac{3}{21}$ **b.** LCD = 66 $\frac{3}{66}, \frac{11}{66}$ **c.** LCD = 15 $\frac{10}{15}, \frac{4}{15}$ **d.** LCD = 72 $\frac{14}{72}, \frac{33}{72}$ **e.** LCD = 90, $\frac{15}{90}, \frac{40}{90}, \frac{27}{90}$

Practice Exercises

Find the least common denominator for each of the following. Then write each rational expression as an equivalent rational expression with the LCD as its denominator.

6. $\dfrac{7}{m}$ and $\dfrac{3}{n}$

7. $\dfrac{m}{c^3 d}$ and $\dfrac{n}{c^2 d^2}$

8. $\dfrac{8}{15m^3 n^3}$ and $\dfrac{7}{18m^2 n}$

If more practice is needed, do the Additional Practice Exercises.

Additional Practice Exercises

Find the least common denominator for each of the following. Then write each rational expression as an equivalent rational expression with the LCD as its denominator.

f. $\dfrac{5}{x}$ and $\dfrac{6}{y}$

g. $\dfrac{5}{r^4 s^2}$ and $\dfrac{2}{r^3 s^3}$

h. $\dfrac{3t}{21m^2 n^5}$ and $\dfrac{8s}{12mn^2}$

If the denominators are polynomials other than monomials, we still use the same procedure. It is a little more complicated, since the polynomials must first be factored, if possible, and the factors are now binomials instead of monomials.

Example 3

Find the least common denominator of each of the following. Then write each rational expression as an equivalent rational expression with the LCD as its denominator.

a. $\dfrac{4}{x-3}$ and $\dfrac{7}{x+4}$ **Solution:**

Since $x - 3$ and $x + 4$ are prime, their LCM is $(x - 3)(x + 4)$. Therefore, the LCD is $(x - 3)(x + 4)$. To write $\dfrac{4}{x-3}$ as a fraction with a denominator of $(x - 3)(x + 4)$, we multiply it by $\dfrac{x+4}{x+4}$. To write $\dfrac{7}{x+4}$ as a fraction with a denominator of $(x - 3)(x + 4)$, we multiply it by $\dfrac{x-3}{x-3}$.

$$\frac{4}{x-3} = \frac{x+4}{x+4} \cdot \frac{4}{x-3} = \frac{4(x+4)}{(x+4)(x-3)} = \frac{4x+16}{(x+4)(x-3)}.$$

$$\frac{7}{x+4} = \frac{x-3}{x-3} \cdot \frac{7}{x+4} = \frac{7(x-3)}{(x-3)(x+4)} = \frac{7x-21}{(x-3)(x+4)}.$$

Note: It is customary to find the product of the numerators but not the denominators. You will see the reason for this in the next section.

b. $\dfrac{x}{2x+6}$ and $\dfrac{3x}{6x+18}$ **Solution:**

$$2x + 6 = 2(x + 3) \qquad \text{Prime factorization of the denominators.}$$

$$6x + 18 = 2 \cdot 3(x + 3) \qquad \text{Prime factorization of the denominators.}$$

Therefore, the LCM of $2x + 6$ and $6x + 18$ is $2 \cdot 3(x + 3) = 6(x + 3)$. So the LCD is $6(x + 3)$. To write $\dfrac{x}{2(x+3)}$ as a fraction with a denominator of $6(x + 3)$, we multiply it by $\frac{3}{3}$, since $3 \cdot 2(x + 3) = 6(x + 3)$. The rational expression $\frac{3x}{6x + 18} = \frac{3x}{6(x + 3)}$, so it already has the LCD as its denominator.

$$\frac{x}{2x + 6} = \frac{x}{2(x + 3)} = \frac{3}{3} \cdot \frac{x}{2(x + 3)} = \frac{3x}{6(x + 3)}.$$

c. $\dfrac{x + 2}{x^2 + 3x - 4}$ and $\dfrac{x - 3}{x^2 + 6x + 8}$

Solution: $\quad x^2 + 3x - 4 = (x + 4)(x - 1)$ Prime factorization of the denominators.

$\qquad\qquad\qquad x^2 + 6x + 8 = (x + 4)(x + 2)$ Prime factorization of the denominators.

Therefore, the LCM of $x^2 + 3x - 4$ and $x^2 + 6x + 8$ is $(x + 4)(x - 1)(x + 2)$. So the LCD is $(x + 4)(x - 1)(x + 2)$. Since $\frac{x + 2}{x^2 + 3x - 4} = \frac{x + 2}{(x + 4)(x - 1)}$, we need to multiply it by $\frac{x + 2}{x + 2}$. Since $\frac{x - 3}{x^2 + 6x + 8} = \frac{x - 3}{(x + 4)(x + 2)}$, we need to multiply it by $\frac{x - 1}{x - 1}$.

$$\frac{x + 2}{x^2 + 3x - 4} = \frac{x + 2}{(x + 4)(x - 1)} = \frac{x + 2}{x + 2} \cdot \frac{x + 2}{(x + 4)(x - 1)}$$

$$= \frac{(x + 2)(x + 2)}{(x + 2)(x + 4)(x - 1)}$$

$$= \frac{x^2 + 4x + 4}{(x + 2)(x + 4)(x - 1)}.$$

$$\frac{x - 3}{x^2 + 6x + 8} = \frac{x - 3}{(x + 4)(x + 2)} = \frac{x - 1}{x - 1} \cdot \frac{x - 3}{(x + 4)(x + 2)}$$

$$= \frac{(x - 1)(x - 3)}{(x - 1)(x + 4)(x + 2)}$$

$$= \frac{x^2 - 4x + 3}{(x - 1)(x + 4)(x + 2)}.$$

Practice Exercises

Find the least common denominator of the following. Then write each rational expression as an equal rational expression with the LCD as its denominator.

9. $\dfrac{a}{a - 6}$ and $\dfrac{a}{a + 4}$

10. $\dfrac{3y}{4y - 20}$ and $\dfrac{6y}{6y - 30}$

11. $\dfrac{b - 4}{b^2 - 2b - 35}$ and $\dfrac{b + 3}{b^2 + 7b + 10}$

If more practice is needed, do the Additional Practice Exercises.

Additional Practice Exercises

Find the least common denominator of the following. Then write each rational expression as an equivalent rational expression with the LCD.

i. $\dfrac{x}{x + 1}$ and $\dfrac{y}{x - 2}$

j. $\dfrac{x + 2}{6x - 12}$ and $\dfrac{x - 3}{8x - 16}$

k. $\dfrac{y + 9}{y^2 + 2y - 24}$ and $\dfrac{y - 3}{y^2 - 16}$

Exercise Set 7.3

Find the least common denominator for the following. Then write each fraction as an equivalent fraction with the LCD as its denominator.

1. $\dfrac{1}{2}$ and $\dfrac{1}{7}$

2. $\dfrac{1}{11}$ and $\dfrac{1}{13}$

3. $\dfrac{2}{3}$ and $\dfrac{3}{5}$

4. $\dfrac{4}{17}$ and $\dfrac{8}{19}$

5. $\dfrac{5}{6}$ and $\dfrac{4}{5}$

6. $\dfrac{10}{11}$ and $\dfrac{7}{10}$

7. $\dfrac{2}{5}$ and $\dfrac{7}{15}$

8. $\dfrac{3}{7}$ and $\dfrac{3}{28}$

9. $\dfrac{3}{8}$ and $\dfrac{4}{9}$

10. $\dfrac{5}{16}$ and $\dfrac{11}{15}$

11. $\dfrac{1}{10}$ and $\dfrac{1}{15}$

12. $\dfrac{5}{24}$ and $\dfrac{6}{25}$

13. $\dfrac{11}{40}$ and $\dfrac{13}{30}$

14. $\dfrac{1}{18}$ and $\dfrac{1}{24}$

15. $\dfrac{9}{16}$ and $\dfrac{11}{21}$

16. $\dfrac{41}{75}$ and $\dfrac{61}{100}$

17. $\dfrac{1}{6}, \dfrac{3}{8},$ and $\dfrac{9}{10}$

18. $\dfrac{4}{9}, \dfrac{5}{12},$ and $\dfrac{7}{18}$

19. $\dfrac{11}{12}, \dfrac{5}{18},$ and $\dfrac{17}{24}$

20. $\dfrac{4}{7}, \dfrac{3}{14},$ and $\dfrac{2}{21}$

Find the least common denominator for the following. Then write each rational expression as an equivalent expression with the LCD as its denominator.

21. $\dfrac{2}{n}$ and $\dfrac{3}{m}$

22. $\dfrac{5}{r}$ and $\dfrac{6}{t}$

23. $\dfrac{a}{b}$ and $\dfrac{c}{d}$

24. $\dfrac{x}{y}$ and $\dfrac{u}{v}$

25. $\dfrac{6}{ab}$ and $\dfrac{5}{bc}$

26. $\dfrac{8}{xy}$ and $\dfrac{10}{yz}$

27. $\dfrac{r}{stu}$ and $\dfrac{n}{mtu}$

28. $\dfrac{a}{pqr}$ and $\dfrac{b}{rtq}$

29. $\dfrac{x}{wv^2}$ and $\dfrac{y}{sv}$

30. $\dfrac{m}{p^2q}$ and $\dfrac{n}{pz}$

31. $\dfrac{2t}{a^2b^5}$ and $\dfrac{3z}{a^3b^3}$

32. $\dfrac{6u}{h^4k^2}$ and $\dfrac{4v}{h^3k^7}$

33. $\dfrac{5a}{x^2yz^4}$ and $\dfrac{7d}{xy^5z^3}$

34. $\dfrac{8p}{r^7s^3t^2}$ and $\dfrac{9q}{r^2s^2t^4}$

35. $\dfrac{7}{6u^2v^4}$ and $\dfrac{3}{8uv^3}$

36. $\dfrac{2}{9a^4b^3}$ and $\dfrac{5}{12ab^2}$

37. $\dfrac{4}{15p^6q^8}$ and $\dfrac{9}{20p^4q^5}$

38. $\dfrac{11}{12r^6t^3}$ and $\dfrac{13}{18r^4t^5}$

39. $\dfrac{r}{7x^2y}$ and $\dfrac{t}{14xy^2}$

40. $\dfrac{p}{9mn^3}$ and $\dfrac{q}{27m^3n}$

41. $\dfrac{3a}{10w^6z^7}$ and $\dfrac{6b}{25w^2z^6}$

42. $\dfrac{5r}{24ab^6}$ and $\dfrac{3s}{16a^2b^9}$

Find the least common denominator for the following. Then write each rational expression as an equivalent expression with the LCD as its denominator.

43. $\dfrac{3}{u+8}$ and $\dfrac{9}{u-10}$

44. $\dfrac{5}{4-v}$ and $\dfrac{6}{2-v}$

45. $\dfrac{x}{x+7}$ and $\dfrac{x}{x-5}$

46. $\dfrac{y}{y-3}$ and $\dfrac{y}{y-4}$

47. $\dfrac{t}{3t-9}$ and $\dfrac{5t}{4t-12}$

48. $\dfrac{3z}{5z-10}$ and $\dfrac{z}{7z-14}$

49. $\dfrac{6y}{9y+18}$ and $\dfrac{8y}{15y+30}$

50. $\dfrac{11z}{8x-32}$ and $\dfrac{17x}{12x-48}$

51. $\dfrac{3a}{a^2-1}$ and $\dfrac{a}{a^2+3a-4}$

52. $\dfrac{4b}{b^2-9}$ and $\dfrac{2b}{b^2+b-6}$

53. $\dfrac{v-5}{v^2-2v-3}$ and $\dfrac{3+v}{v^2-5v+6}$

54. $\dfrac{w+6}{w^2+6w+5}$ and $\dfrac{2-w}{w^2+w-20}$

55. $\dfrac{m+1}{m^2+8m+16}$ and $\dfrac{m-4}{m^2+5m+4}$

56. $\dfrac{n-3}{n^2-10n+25}$ and $\dfrac{n-1}{n^2-2n-15}$

57. $\dfrac{x+5}{x^3+x^2-6x}$ and $\dfrac{x-3}{x^4+7x^3+12x^2}$

58. $\dfrac{y-2}{y^5-16y^3}$ and $\dfrac{y-1}{y^3+5y^2+4y}$

Writing Exercises:

59. How is finding the least common denominator of rational numbers like finding the least common denominator of rational expressions? How is it different?

60. What is the relationship between prime factors and the LCD?

Section 7.4	**Addition and Subtraction of Rational Numbers and Expressions with Unlike Denominators**

OBJECTIVES   *When you complete this section, you will be able to:*

a. Add fractions with unlike denominators.

b. Add mixed numbers whose fractional parts have unlike denominators.

c. Add rational expressions with unlike denominators.

Introduction   In Section 7.,1, we combined fractions and rational expressions with common denominators. If the denominators are not common, we have to perform an intermediate step before we can combine the expressions. Suppose we want to find $\frac{1}{2} + \frac{1}{3}$ using shaded regions. We first shade $\frac{1}{2}$ the region.

1 unit

Then we shade $\frac{1}{3}$ of the region.

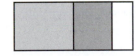

1 unit

What part of the total region is shaded? At this moment, we cannot answer that question. If we go back and divide the region into six equal parts, the $\frac{1}{2}$ becomes $\frac{3}{6}$ and the $\frac{1}{3}$ becomes $\frac{2}{6}$.

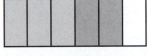

1 unit

Therefore, $\frac{1}{2} + \frac{1}{3} = \frac{3}{6} + \frac{2}{6} = \frac{5}{6}$. Note that 6 is the least common denominator of 2 and 3. What we have done is changed $\frac{1}{2}$ and $\frac{1}{3}$ into equivalent fractions with the LCD as their denominators, and then added as in Section 7.1.

Adding fractions with unlike denominators

Adding Fractions with Unlike Denominators

To add fractions with unlike denominators, change each fraction into an equivalent fraction with the least common denominator as its denominator. Then add the fractions using the techniques of Section 7.1.

Since subtraction of fractions may be expressed in terms of addition, the same procedure applies to subtraction as well. If whole numbers or mixed numbers are involved, first convert them to improper fractions.

Example 1

Find the following sums or differences. Leave answers reduced to lowest terms.

a. $\dfrac{2}{3} + \dfrac{3}{4} =$ The LCD of 3 and 4 is 12. Therefore, multiply $\dfrac{2}{3}$ by $\dfrac{4}{4}$ and multiply $\dfrac{3}{4}$ by $\dfrac{3}{3}$.

$\dfrac{4}{4} \cdot \dfrac{2}{3} + \dfrac{3}{3} \cdot \dfrac{3}{4} =$ Multiply the fractions.

$\dfrac{8}{12} + \dfrac{9}{12} =$ Add the fractions.

$\dfrac{17}{12}$ Sum.

b. $3 - \dfrac{3}{5} =$ Think of 3 as $\dfrac{3}{1}$. The LCD of 1 and 5 is 5. Therefore, multiply 3 by $\dfrac{5}{5}$. Since $\dfrac{3}{5}$ already has the LCD, we do not multiply it by anything. Multiply the fractions.

$\dfrac{5}{5} \cdot \dfrac{3}{1} - \dfrac{3}{5} =$

$\dfrac{15}{5} - \dfrac{3}{5} =$ Subtract the fractions.

$\dfrac{12}{5}$ Difference.

c. $\dfrac{5}{12} + \dfrac{7}{30} =$ The LCD is 60. Therefore, multiply $\dfrac{5}{12}$ by $\dfrac{5}{5}$ and multiply $\dfrac{7}{30}$ by $\dfrac{2}{2}$.

$\dfrac{5}{5} \cdot \dfrac{5}{12} + \dfrac{2}{2} \cdot \dfrac{7}{30} =$ Multiply the fractions.

$\dfrac{25}{60} + \dfrac{14}{60} =$ Add the fractions.

$\dfrac{39}{60} =$ Reduce to lowest terms.

$\dfrac{13}{20}$ Sum.

d. $\dfrac{3}{10} + \dfrac{7}{15} - \dfrac{9}{25} =$ The LCD is 150. Therefore, multiply $\dfrac{3}{10}$ by $\dfrac{15}{15}$, multiply $\dfrac{7}{15}$ by $\dfrac{10}{10}$, and multiply $\dfrac{9}{25}$ by $\dfrac{6}{6}$.

$\dfrac{15}{15} \cdot \dfrac{3}{10} + \dfrac{10}{10} \cdot \dfrac{7}{15} - \dfrac{6}{6} \cdot \dfrac{9}{25} =$ Multiply the fractions.

$\dfrac{45}{150} + \dfrac{70}{150} - \dfrac{54}{150} =$ Add and subtract the fractions.

$\dfrac{61}{150}$ Answer.

e. $3\dfrac{3}{8} - 1\dfrac{1}{2} =$ Change to improper fractions.

$\dfrac{27}{8} - \dfrac{3}{2} =$ The LCD is 8. Therefore, multiply $\dfrac{3}{2}$ by $\dfrac{4}{4}$.

$\dfrac{27}{8} - \dfrac{4}{4} \cdot \dfrac{3}{2} =$ Multiply the fractions.

$\dfrac{27}{8} - \dfrac{12}{8} =$ Subtract the fractions.

$\dfrac{15}{8}$ Difference.

f. $4 - 5\dfrac{7}{12} + 2\dfrac{3}{8} =$ Convert to improper fractions. The LCD is 24. Therefore, multiply $\dfrac{4}{1}$ by $\dfrac{24}{24}$, multiply $\dfrac{67}{12}$ by $\dfrac{2}{2}$, and multiply $\dfrac{19}{8}$ by $\dfrac{3}{3}$.

$\dfrac{4}{1} - \dfrac{67}{12} + \dfrac{19}{8} =$

$\dfrac{24}{24} \cdot \dfrac{4}{1} - \dfrac{2}{2} \cdot \dfrac{67}{12} + \dfrac{3}{3} \cdot \dfrac{19}{8} =$ Multiply the fractions.

$\dfrac{96}{24} - \dfrac{134}{24} + \dfrac{57}{24} =$ Add and subtract the fractions.

$\dfrac{19}{24}$ Answer.

Practice Exercises

Find the following sums or differences. Leave answers reduced to lowest terms.

1. $\dfrac{3}{5} + \dfrac{5}{6}$

2. $\dfrac{5}{9} + 4$

3. $\dfrac{9}{14} - \dfrac{1}{4}$

4. $4\dfrac{5}{6} - 6\dfrac{1}{8}$

5. $\dfrac{2}{3} + \dfrac{5}{8} - \dfrac{7}{9}$

6. $\dfrac{7}{10} - 6 + 5\dfrac{7}{12}$

If more practice is needed, do the Additional Practice Exercises.

Additional Practice Exercises

Find the following sums or differences. Leave answers reduced to lowest terms.

a. $\dfrac{1}{6} + \dfrac{5}{8}$

b. $6 - \dfrac{3}{4}$

c. $\dfrac{5}{6} - \dfrac{4}{9}$

d. $4\dfrac{1}{3} - 6\dfrac{3}{8}$

e. $\dfrac{2}{3} + \dfrac{3}{5} - \dfrac{7}{12}$

f. $7 - 4\dfrac{5}{12} - \dfrac{7}{10}$

Combining rational expressions with unlike denominators

In the following examples with rational expressions, the denominators are monomials that have variable factors. You will notice the procedure is the same as with ordinary fractions.

Example 2

Find the following sums or differences. Leave the answers reduced to lowest terms.

a. $\dfrac{2}{a} + \dfrac{3}{b} =$

The LCD is ab. Therefore, multiply $\dfrac{2}{a}$ by $\dfrac{b}{b}$ and multiply $\dfrac{3}{b}$ by $\dfrac{a}{a}$.

$\dfrac{b}{b} \cdot \dfrac{2}{a} + \dfrac{a}{a} \cdot \dfrac{3}{b} =$ Multiply the rational expressions.

$\dfrac{2b}{ab} + \dfrac{3a}{ab} =$ Add the rational expressions.

$\dfrac{2b + 3a}{ab}$ Sum.

b. $\dfrac{3}{4a} - \dfrac{5}{3a} =$

The LCD is $12a$. Therefore, multiply $\dfrac{3}{4a}$ by $\dfrac{3}{3}$ and multiply $\dfrac{5}{3a}$ by $\dfrac{4}{4}$.

$\dfrac{3}{3} \cdot \dfrac{3}{4a} - \dfrac{4}{4} \cdot \dfrac{5}{3a} =$ Multiply the rational expressions.

$\dfrac{9}{12a} - \dfrac{20}{12a} =$ Subtract the rational expressions.

$-\dfrac{11}{12a}$ Difference.

c. $\dfrac{2x + 4}{4x} + \dfrac{3x - 2}{6x} =$

The LCD is $12x$. Therefore, multiply $\dfrac{2x + 4}{4x}$ by $\dfrac{3}{3}$ and $\dfrac{3x - 2}{6x}$ by $\dfrac{2}{2}$.

$\dfrac{3}{3} \cdot \dfrac{2x + 4}{4x} + \dfrac{2}{2} \cdot \dfrac{3x - 2}{6x} =$ Multiply the rational expressions.

$\dfrac{6x + 12}{12x} + \dfrac{6x - 4}{12x} =$ Add the rational expressions.

$\dfrac{(6x + 12) + (6x - 4)}{12x} =$ Remove the parentheses.

$\dfrac{6x + 12 + 6x - 4}{12x} =$ Add like terms.

$\dfrac{12x + 8}{12x} =$ Factor.

$\dfrac{4(3x + 2)}{4 \cdot 3x} =$ Divide by the common factor 4.

$\dfrac{3x + 2}{3x}$ Sum.

d. $\dfrac{a - 3}{3a^2} - \dfrac{2a - 3}{2a} =$

The LCD is $6a^2$. Therefore, multiply $\dfrac{a - 3}{3a^2}$ by $\dfrac{2}{2}$ and multiply $\dfrac{2a - 3}{2a}$ by $\dfrac{3a}{3a}$.

$\dfrac{2}{2} \cdot \dfrac{a - 3}{3a^2} - \dfrac{3a}{3a} \cdot \dfrac{2a - 3}{2a} =$ Multiply the rational expressions.

$\dfrac{2a - 6}{6a^2} - \dfrac{6a^2 - 9a}{6a^2} =$ Subtract the rational expressions.

$\dfrac{(2a - 6) - (6a^2 - 9a)}{6a^2} =$ Remove the parentheses.

$\dfrac{2a - 6 - 6a^2 + 9a}{6a^2} =$ Add like terms.

$\dfrac{-6a^2 + 11a - 6}{6a^2}$ Difference.

Practice Exercises

Find the following sums or differences. Leave answers reduced to lowest terms.

7. $\dfrac{4}{x} + \dfrac{6}{y}$

8. $\dfrac{7}{6a} - \dfrac{5}{8a}$

9. $\dfrac{b - 5}{5b} + \dfrac{b + 6}{6b}$

10. $\dfrac{3x + 7}{8xy^2} - \dfrac{3x - 4}{6x^2y}$

Answers:

Additional Practice Exercises a–f: a. $\dfrac{19}{24}$ **b.** $\dfrac{21}{4}$ **c.** $\dfrac{7}{18}$ **d.** $-\dfrac{49}{24}$ **e.** $\dfrac{41}{60}$ **f.** $\dfrac{113}{60}$ **Practice Exercises 7–10: 7.** $\dfrac{4y + 6x}{xy}$ **8.** $\dfrac{13}{24a}$ **9.** $\dfrac{11}{30}$ **10.** $\dfrac{9x^2 + 21x - 12xy + 16y}{24x^2y^2}$

If more practice is needed, do the Additional Practice Exercises.

Additional Practice Exercises

Find the following sums or differences. Leave answers reduced to lowest terms.

g. $\dfrac{9}{r} - \dfrac{4}{s}$ **h.** $\dfrac{2}{5d} + \dfrac{3}{10d}$ **i.** $\dfrac{2x - 3}{5x} + \dfrac{3x + 5}{3x}$ **j.** $\dfrac{a + 4}{3a^2b} - \dfrac{2a - 3}{6ab^2}$

If the denominators are binomials or trinomials, it is often necessary to factor the denominators to find the LCD. If possible, factor the numerator also to see if any of the rational expressions will reduce before carrying out any operations. Sometimes we have to factor the numerator of the sum or difference, and then reduce the rational expression. For this reason, we usually leave the LCD in factored form.

Example 3

Find the following sums or differences. Leave answers reduced to lowest terms.

a.

$\dfrac{5}{x - 2} + \dfrac{3}{x + 4} =$ The LCD is $(x - 2)(x + 4)$. Therefore, multiply $\dfrac{5}{x - 2}$ by $\dfrac{x + 4}{x + 4}$ and multiply $\dfrac{3}{x + 4}$ by $\dfrac{x - 2}{x - 2}$.

$\dfrac{x + 4}{x + 4} \cdot \dfrac{5}{x - 2} + \dfrac{x - 2}{x - 2} \cdot \dfrac{3}{x + 4} =$ Multiply the rational expressions.

$\dfrac{5x + 20}{(x + 4)(x - 2)} + \dfrac{3x - 6}{(x + 4)(x - 2)} =$ Add the rational expressions.

$\dfrac{(5x + 20) + (3x - 6)}{(x + 4)(x - 2)} =$ Remove the parentheses.

$\dfrac{5x + 20 + 3x - 6}{(x + 4)(x - 2)} =$ Add like terms.

$\dfrac{8x + 14}{(x + 4)(x - 2)}$ Sum.

Note: In the preceding example, we could factor the numerator and get $\dfrac{2(4x + 7)}{(x + 4)(x - 2)}$, but the expression will not reduce. Consequently, we leave the answer as is.

b.

$\dfrac{x + 2}{4x - 8} - \dfrac{x + 4}{6x - 12} =$ Factor the denominators.

$\dfrac{x + 2}{4(x - 2)} - \dfrac{x + 4}{6(x - 2)} =$ The LCD is $12(x - 2)$. Therefore, multiply $\dfrac{x + 2}{4(x - 2)}$ by $\dfrac{3}{3}$ and $\dfrac{x + 4}{6(x - 2)}$ by $\dfrac{2}{2}$.

$\dfrac{3}{3} \cdot \dfrac{x + 2}{4(x - 2)} - \dfrac{2}{2} \cdot \dfrac{x + 4}{6(x - 2)} =$ Multiply the rational expressions.

$\dfrac{3x + 6}{12(x - 2)} - \dfrac{2x + 8}{12(x - 2)} =$ Subtract the rational expressions.

$\dfrac{(3x + 6) - (2x + 8)}{12(x - 2)} =$ Remove the parentheses.

$\dfrac{3x + 6 - 2x - 8}{12(x - 2)} =$ Add like terms.

$\dfrac{x - 2}{12(x - 2)} =$ Divide by the common factor $x - 2$.

$\dfrac{1}{12}$ Difference.

c.

$\dfrac{4c - 1}{c^2 - 10c + 25} - \dfrac{4}{c - 5} =$ Factor the denominators.

$\dfrac{4c - 1}{(c - 5)^2} - \dfrac{4}{c - 5} =$ The LCD is $(c - 5)^2$. Therefore, multiply $\dfrac{4}{c - 5}$ by $\dfrac{c - 5}{c - 5}$.

$\dfrac{4c - 1}{(c - 5)^2} - \dfrac{c - 5}{c - 5} \cdot \dfrac{4}{c - 5} =$ Multiply the rational expressions.

$\dfrac{4c - 1}{(c - 5)^2} - \dfrac{4c - 20}{(c - 5)^2} =$ Subtract the rational expressions.

$\dfrac{(4c - 1) - (4c - 20)}{(c - 5)^2} =$ Remove the parentheses.

$\dfrac{4c - 1 - 4c + 20}{(c - 5)^2} =$ Add like terms.

$\dfrac{19}{(c - 5)^2}$ Difference.

d. $\dfrac{2x - 3}{3x^2 + 14x + 8} + \dfrac{3x - 1}{3x^2 - 13x - 10} =$ Factor the denominators.

$\dfrac{2x - 3}{(3x + 2)(x + 4)} + \dfrac{3x - 1}{(3x + 2)(x - 5)} =$ The LCD is $(3x + 2)(x + 4)(x - 5)$. Multiply $\dfrac{2x - 3}{(3x + 2)(x + 4)}$ by $\dfrac{x - 5}{x - 5}$ and $\dfrac{3x - 1}{(3x + 2)(x - 5)}$ by $\dfrac{x + 4}{x + 4}$.

$\dfrac{x - 5}{x - 5} \cdot \dfrac{2x - 3}{(3x + 2)(x + 4)} + \dfrac{x + 4}{x + 4} \cdot \dfrac{3x - 1}{(3x + 2)(x - 5)} =$ Multiply.

$\dfrac{2x^2 - 13x + 15}{(3x + 2)(x + 4)(x - 5)} + \dfrac{3x^2 + 11x - 4}{(3x + 2)(x + 4)(x - 5)} =$ Add.

$\dfrac{(2x^2 - 13x + 15) + (3x^2 + 11x - 4)}{(3x + 2)(x + 4)(x - 5)} =$ Remove the parentheses.

$\dfrac{2x^2 - 13x + 15 + 3x^2 + 11x - 4}{(3x + 2)(x + 4)(x - 5)} =$ Add like terms.

$\dfrac{5x^2 - 2x + 11}{(3x + 2)(x + 4)(x - 5)}$ Sum.

Practice Exercises

Find the following sums or differences. Leave answers reduced to lowest terms.

11. $\dfrac{2x}{x + 3} + \dfrac{4}{x - 5}$

12. $\dfrac{x + 6}{4x + 16} - \dfrac{x + 7}{6x + 24}$

13. $\dfrac{2x + 3}{x^2 + 6x + 9} - \dfrac{x + 4}{x + 3}$

14. $\dfrac{x + 2}{2x^2 + 5x - 3} + \dfrac{2x - 3}{2x^2 - 5x + 2}$

If more practice is needed, do the Additional Practice Exercises.

Additional Practice Exercises

Find the following sums or differences. Leave answers reduced to lowest terms.

k. $\dfrac{7}{2x + 1} + \dfrac{3}{3x - 2}$

l. $\dfrac{x - 2}{6x - 30} - \dfrac{x - 1}{8x - 40}$

m. $\dfrac{2y}{y^2 - 4y + 4} - \dfrac{4y + 3}{y - 2}$

n. $\dfrac{x - 4}{x^2 - x - 20} + \dfrac{2x - 1}{x^2 + 7x + 12}$

Exercise Set 7.4

Find the following sums or differences. Leave the answers as fractions reduced to lowest terms.

1. $\dfrac{4}{7} + \dfrac{1}{3}$

2. $\dfrac{8}{9} - \dfrac{3}{4}$

3. $\dfrac{3}{10} + \dfrac{2}{5}$

4. $\dfrac{5}{8} + \dfrac{3}{4}$

5. $\dfrac{7}{6} - \dfrac{2}{3}$

6. $\dfrac{5}{18} - \dfrac{1}{2}$

7. $\dfrac{5}{8} - \dfrac{5}{6}$

8. $\dfrac{9}{11} + \dfrac{4}{9}$

9. $2 + \dfrac{5}{9}$

10. $4 - \dfrac{5}{3}$

11. $\dfrac{13}{4} - 6$

12. $\dfrac{12}{7} - 2$

13. $\dfrac{1}{4} + \dfrac{5}{6}$

14. $\dfrac{11}{24} + \dfrac{13}{24}$

15. $\dfrac{8}{9} - \dfrac{11}{12}$

16. $\dfrac{14}{15} - \dfrac{7}{10}$

17. $\dfrac{7}{12} + \dfrac{1}{9} - \dfrac{5}{6}$

18. $\dfrac{7}{10} + \dfrac{4}{15} - \dfrac{9}{30}$

19. $\dfrac{7}{12} - \dfrac{11}{18} + \dfrac{1}{3}$

20. $\dfrac{7}{12} - \dfrac{5}{8} + \dfrac{1}{6}$

21. $9\dfrac{7}{8} - 3\dfrac{1}{4}$

22. $10\dfrac{5}{12} + 1\dfrac{1}{6}$

23. $7\dfrac{5}{8} + 4\dfrac{2}{12}$

24. $23\dfrac{1}{4} + 34\dfrac{2}{3}$

25. $14\dfrac{1}{9} - 8\dfrac{5}{12}$

26. $9\dfrac{9}{10} - 7\dfrac{6}{15}$

27. $9\dfrac{3}{8} - 7 + \dfrac{11}{16}$

28. $\dfrac{5}{6} + 4 - 3\dfrac{5}{9}$

29. $\dfrac{3}{4} - 4 + 2\dfrac{3}{10}$

30. $-5 + 9\dfrac{7}{18} + \dfrac{11}{24}$

Find the following sums or differences. Leave the answers as rational expressions reduced to lowest terms.

31. $\dfrac{2}{a} + \dfrac{a}{b}$

32. $\dfrac{6}{r} + \dfrac{7}{t}$

33. $\dfrac{5}{8u} - \dfrac{7}{12u}$

34. $\dfrac{6}{15v} - \dfrac{4}{9v}$

35. $\dfrac{3z}{10x} + \dfrac{5z}{4x}$

36. $\dfrac{10w}{21y} + \dfrac{7w}{18y}$

37. $\dfrac{m+8}{2m} - \dfrac{m-7}{3m}$

38. $\dfrac{n-9}{7n} - \dfrac{n-2}{4n}$

39. $\dfrac{3p+1}{10p^3q^2} + \dfrac{9p-2}{6p^2q}$

40. $\dfrac{5t-8}{16s^5t^2} + \dfrac{2t-7}{12s^2t^3}$

41. $\dfrac{4}{k} - \dfrac{6}{k+2}$

42. $\dfrac{5}{j-5} - \dfrac{9}{j}$

43. $\dfrac{4}{w-3} + \dfrac{5}{w+7}$

44. $\dfrac{6}{c+4} + \dfrac{11}{c-1}$

45. $\dfrac{7a}{a-3} - \dfrac{3}{3-a}$

46. $\dfrac{9b}{b-8} - \dfrac{2}{8-b}$

47. $\dfrac{p+10}{p+7} + \dfrac{-3}{p+7}$

48. $\dfrac{q+1}{q-2} + \dfrac{3}{q-2}$

49. $\dfrac{x-4}{3x+9} - \dfrac{x+5}{6x+18}$

50. $\dfrac{y-3}{2y-8} - \dfrac{y-7}{6y-24}$

51. $\dfrac{2t}{t^2-8t+16} + \dfrac{6}{t-4}$

52. $\dfrac{4}{s^2+14s+49} + \dfrac{5s}{s+7}$

53. $\dfrac{u^2-3u}{u^2-6u+9} + \dfrac{8}{u-3}$

54. $\dfrac{v^2+4v}{v^2+8v+16} + \dfrac{3}{v+4}$

55. $\dfrac{2h^2+10h}{h^2-25} - \dfrac{h}{h+5}$

56. $\dfrac{21k+k^2}{k^2-49} - \dfrac{2k}{k-7}$

57. $\dfrac{x+1}{x^2-x-6} + \dfrac{x-5}{x^2+6x+8}$

58. $\dfrac{y+3}{y^2-3y+2} + \dfrac{2y+4}{y^2+y-2}$

59. $\dfrac{2v+5}{v^2-16} - \dfrac{v-9}{v^2-v-12}$

60. $\dfrac{4t}{t^2-9} - \dfrac{3t-2}{t^2-8t+15}$

61. $\dfrac{z-3}{z^2+6z+9} + \dfrac{z+1}{z^2+z-6}$

62. $\dfrac{7w-2}{w^2+2w-24} + \dfrac{9w-8}{3w^2-16w+16}$

Challenge Exercises: (63–64)

63. $\dfrac{x+4}{x^3-36x} + \dfrac{1}{x^2+3x-18}$

64. $\dfrac{y+5}{4y^2+25y+25} - \dfrac{y-1}{3y^2-48y+192}$

Answer the following:

65. Margot walked $2\frac{1}{3}$ miles on Monday, $3\frac{3}{4}$ miles on Tuesday, $1\frac{2}{3}$ miles on Wednesday, $2\frac{3}{8}$ miles on Thursday, $4\frac{1}{4}$ miles on Friday, $2\frac{5}{8}$ miles on Saturday, and did not walk on Sunday because of rain. What was her total miles walked for the week?

66. John jogged $4\frac{1}{2}$ miles on Monday, $2\frac{3}{4}$ miles on Tuesday, $3\frac{2}{3}$ miles on Wednesday, $4\frac{1}{4}$ miles on Thursday, $5\frac{1}{3}$ miles on Friday, $4\frac{2}{3}$ miles on Saturday, and $3\frac{1}{4}$ miles on Sunday. What was his total miles jogged for the week?

67. A room is $16\frac{3}{4}$ feet long and $12\frac{5}{6}$ feet wide. A carpenter is putting molding around the ceiling. If the molding costs \$0.12 per foot, find the cost of the molding.

68. A triangular garden has sides that are $16\frac{5}{8}$ feet, $18\frac{2}{3}$ feet, and $14\frac{5}{6}$ feet in length. How much will it cost to fence in the garden if it costs \$1.60 per foot?

Find the perimeter of each of the following triangles the lengths of whose sides are given:

69. $\frac{1}{a}$ ft, $\frac{1}{a^2}$ ft, $\frac{2}{a}$ ft

70. $\frac{3}{b^2}$ m, $\frac{4}{b}$ m, $\frac{1}{b^2}$ m

71. $\frac{2}{x}$ in., $\frac{3}{y}$ in., $\frac{4}{z}$ in.

72. $\frac{3}{a}$ yd, $\frac{2}{b}$ yd, $\frac{5}{c}$ yd

73. $\frac{4}{a+b}$ ft, $\frac{3}{c+d}$ ft, $\frac{6}{a+b}$ ft

74. $\frac{5}{x+y}$ cm, $\frac{7}{w+z}$ cm, $\frac{4}{w+z}$ cm

Find the perimeters of the following rectangles whose lengths and widths are given:

75. $L = \frac{3}{x}$ ft, $W = \frac{4}{y}$ ft

76. $L = \frac{5}{a}$ yd, $W = \frac{2}{b}$ yd

77. $L = \frac{x}{x+y}$ cm, $W = \frac{y}{2x-y}$ cm

78. $L = \frac{3a}{2a+b}$ mm, $W = \frac{2b}{a-3b}$ mm

79. $L = \frac{a+5}{a^2-2a-24}$ in., $W = \frac{a-2}{a^2+6a+8}$ in.

80. $L = \frac{x-4}{x^2+2x-15}$ cm, $W = \frac{x+2}{x^2-5x+6}$ cm

Writing Exercises:

81. Why is it necessary to rewrite each rational number or rational expression with the LCD as its denominator before adding or subtracting?

83. How are adding and subtracting rational expressions like adding and subtracting rational numbers? How are they different?

82. How are adding and subtracting rational numbers like adding and subtracting whole numbers? How are they different?

Section 7.5	**Complex Fractions**

OBJECTIVES   *When you complete this section, you will be able to:*

  a. Simplify complex fractions whose numerators and/or denominators are rational numbers.

  b. Simplify complex fractions whose numerators and/or denominators are rational expressions.

Introduction   One interpretation of a fraction is that it represents division. For example, $\frac{a}{b} = a \div b$. This interpretation gives us one method of simplifying **complex fractions.** A complex fraction is any fraction whose numerator and/or denominator is also a fraction. Examples of complex fractions are:

$$\frac{\frac{1}{2}}{3}, \quad \frac{\frac{3}{4}}{\frac{5}{8}}, \quad \frac{\frac{2}{3}+\frac{3}{4}}{\frac{4}{5}+\frac{2}{7}}, \quad \frac{\frac{1}{x}}{\frac{x}{y}}, \quad \text{and} \quad \frac{\frac{3}{x}+x}{\frac{x^2-6}{x}}.$$

To simplify a complex fraction means to write it as an ordinary fraction in the form of $\frac{a}{b}$ where a and b are integers and $b \neq 0$. We will first use the division interpretation of a fraction and then show another, sometimes easier, technique. In simplifying complex fractions using the division interpretation of a fraction, we use the following procedure:

> ## Simplifying Complex Fractions Using Division
>
> 1. If necessary, rewrite the numerator and the denominator as single fractions by performing whatever operations are indicated in either.
> 2. Rewrite the complex fraction as the division of the numerator by the denominator.
> 3. Perform the division by inverting the fraction on the right and multiplying.
> 4. If necessary, reduce to lowest terms.

Example 1

Simplify the following complex fractions using division:

a. $\dfrac{\dfrac{2}{3}}{\dfrac{5}{6}} =$ Rewrite as division using $\dfrac{a}{b} = a \div b$.

$\dfrac{2}{3} \div \dfrac{5}{6} =$ Invert $\dfrac{5}{6}$ and change division to multiplication.

$\dfrac{2}{3} \cdot \dfrac{6}{5} =$ Write 6 in terms of prime factors.

$\dfrac{2}{3} \cdot \dfrac{2 \cdot 3}{5} =$ Divide by the common factor of 3.

$\dfrac{2}{\cancel{3}} \cdot \dfrac{2 \cdot \cancel{3}}{5}$ Multiply the results.

$\dfrac{4}{5}$ Quotient.

Note: If the numerator and/or the denominator is the sum or difference of two or more fractions, we must write the numerator and the denominator as a single fraction before we can simplify.

b. $\dfrac{\dfrac{2}{3} + \dfrac{3}{4}}{\dfrac{3}{8} + 1} =$ The LCD in the numerator is 12 and the LCD of the denominator is 8. Change the fractions in the numerator into equivalent fractions whose denominators are 12, and change the fractions in the denominator into equivalent fractions whose denominators are 8.

$\dfrac{\dfrac{4}{4} \cdot \dfrac{2}{3} + \dfrac{3}{3} \cdot \dfrac{3}{4}}{\dfrac{3}{8} + \dfrac{8}{8} \cdot \dfrac{1}{1}} =$ Multiply the fractions.

$\dfrac{\dfrac{8}{12} + \dfrac{9}{12}}{\dfrac{3}{8} + \dfrac{8}{8}} =$ Add the fractions in the numerator and the denominator.

$\dfrac{\dfrac{17}{12}}{\dfrac{11}{8}} =$ Rewrite as division.

$\dfrac{17}{12} \div \dfrac{11}{8} =$ Rewrite as multiplication.

$\dfrac{17}{12} \cdot \dfrac{8}{11} =$ Write 12 and 8 in prime factors.

$\dfrac{17}{2 \cdot 2 \cdot 3} \cdot \dfrac{2 \cdot 2 \cdot 2}{11}$ Divide by the common factors.

$\dfrac{17}{\cancel{2} \cdot \cancel{2} \cdot 3} \cdot \dfrac{\cancel{2} \cdot \cancel{2} \cdot 2}{11}$ Multiply the remaining fractions.

$\dfrac{34}{33}$ Quotient.

The method used in Example 1b is quite involved and has a lot of places where errors may occur. Consequently, many people prefer a second method in which we multiply the numerator and the denominator of the complex fraction by the LCD of all the fractions that appear in the numerator and/or the denominator. The reason we multiply by the LCD of all the fractions is that each denominator divides evenly into the LCD. Therefore, we eliminate all fractions from the numerator and the denominator of the complex fraction. This results in an ordinary fraction. The procedure is summarized as follows:

> ## Simplifying Complex Fractions Using the LCD
>
> 1. Find the LCD of all fractions that occur in the numerator and/or the denominator.
> 2. Multiply both the numerator and the denominator by the LCD.
> 3. Simplify the results.

We demonstrate this method by reworking the examples from Example 1 plus one other case.

Example 2

Simplify the following complex fractions using multiplication:

a. $\dfrac{\frac{2}{3}}{\frac{5}{6}} =$ The LCD for $\frac{2}{3}$ and $\frac{5}{6}$ is 6. Therefore, multiply the numerator and the denominator by 6.

$\dfrac{6\left(\frac{2}{3}\right)}{6\left(\frac{5}{6}\right)} =$ Perform the multiplications.

$\dfrac{4}{5}$ Simplified form.

b. $\dfrac{\frac{2}{3} + \frac{3}{4}}{\frac{3}{8} + 1} =$ The LCD is 24. Therefore, multiply the numerator and the denominator by 24.

$\dfrac{24\left(\frac{2}{3} + \frac{3}{4}\right)}{24\left(\frac{3}{8} + 1\right)} =$ Apply the distributive property.

$\dfrac{24\left(\frac{2}{3}\right) + 24\left(\frac{3}{4}\right)}{24\left(\frac{3}{8}\right) + 24 \cdot 1} =$ Perform the multiplications.

$\dfrac{16 + 18}{9 + 24} =$ Perform the additions.

$\dfrac{34}{33}$ Simplified form.

c. $\dfrac{\frac{6}{8}}{3} =$ Since $6 = \frac{6}{1}$, the LCD of 6 and $\frac{8}{3}$ is 3. Therefore, multiply the numerator and the denominator by 3.

$\dfrac{3\left(\frac{6}{1}\right)}{3\left(\frac{8}{3}\right)} =$ Perform the multiplications.

$\dfrac{18}{8} =$ Reduce to lowest terms.

$\dfrac{9}{4}$ Simplified form.

Practice Exercises

Simplify the following complex fractions using both methods:

1. $\dfrac{\frac{3}{8}}{\frac{5}{6}}$

2. $\dfrac{\frac{8}{9}}{12}$

3. $\dfrac{\frac{4}{3} - \frac{3}{8}}{1 + \frac{7}{6}}$

4. $\dfrac{2 + \frac{3}{4}}{5 - \frac{5}{12}}$

If more practice is needed, do the Additional Practice Exercises.

Additional Practice Exercises

Simplify the following complex fractions using both methods:

a. $\dfrac{\frac{4}{15}}{\frac{2}{5}}$

b. $\dfrac{\frac{8}{4}}{7}$

c. $\dfrac{3 + \frac{2}{3}}{\frac{1}{4} + \frac{5}{12}}$

d. $\dfrac{\frac{7}{10} - 4}{3 - \frac{8}{15}}$

Since the second method in which we use the LCD is always at least as easy as the first using the division interpretation, it is the only method we will use for complex fractions whose numerators and/or denominators contain rational expressions.

Example 3

Simplify the following complex fractions:

a. $\dfrac{\frac{a}{b}}{\frac{c}{d}} =$ The LCD for $\frac{a}{b}$ and $\frac{c}{d}$ is bd. Therefore, multiply the numerator and the denominator by bd.

$\dfrac{bd \cdot \frac{a}{b}}{bd \cdot \frac{c}{d}} =$ Perform the multiplications.

$\dfrac{ad}{bc}$ Simplified form.

b. $\dfrac{\frac{a^2b}{c^2}}{\frac{ab^3}{c^3}} =$ The LCD for $\frac{a^2b}{c^2}$ and $\frac{ab^3}{c^3}$ is c^3. Therefore, multiply the numerator and the denominator by c^3.

$\dfrac{c^3\left(\frac{a^2b}{c^2}\right)}{c^3\left(\frac{ab^3}{c^3}\right)} =$ Perform the multiplications.

$\dfrac{a^2bc}{ab^3} =$ Reduce to lowest terms.

$\dfrac{ac}{b^2}$ Simplified form.

c. $\dfrac{1 - \frac{9}{x^2}}{1 + \frac{3}{x}} =$ The least common denominator for all the fractions in the numerator and the denominator is x^2. Therefore, multiply the numerator and the denominator by x^2.

$\dfrac{x^2\left(1 - \frac{9}{x^2}\right)}{x^2\left(1 + \frac{3}{x}\right)} =$ Apply the distributive property.

$\dfrac{x^2 \cdot 1 - x^2\left(\frac{9}{x^2}\right)}{x^2 \cdot 1 + x^2\left(\frac{3}{x}\right)} =$ Perform the multiplications.

$\dfrac{x^2 - 9}{x^2 + 3x} =$ Factor the numerator and the denominator.

$\dfrac{(x + 3)(x - 3)}{x(x + 3)} =$ Divide by the common factor $(x + 3)$.

$\dfrac{x - 3}{x}$ Simplified form.

d. $\dfrac{1 - \frac{2}{x} - \frac{15}{x^2}}{1 + \frac{7}{x} + \frac{12}{x^2}} =$ The LCD of all the fractions in the numerator and the denominator is x^2. Therefore, multiply the numerator and the denominator by x^2.

$\dfrac{x^2\left(1 - \frac{2}{x} - \frac{15}{x^2}\right)}{x^2\left(1 + \frac{7}{x} + \frac{12}{x^2}\right)} =$ Apply the distributive property.

$\dfrac{x^2 \cdot 1 - x^2\left(\frac{2}{x}\right) - x^2\left(\frac{15}{x^2}\right)}{x^2 \cdot 1 + x^2\left(\frac{7}{x}\right) + x^2\left(\frac{12}{x^2}\right)} =$ Perform the multiplications.

$\dfrac{x^2 - 2x - 15}{x^2 + 7x + 12} =$ Factor the numerator and the denominator.

$\dfrac{(x - 5)(x + 3)}{(x + 4)(x + 3)} =$ Divide by the common factor $(x + 3)$.

$\dfrac{x - 5}{x + 4}$ Simplified form.

e. $\dfrac{x + \frac{9}{x - 6}}{1 + \frac{3}{x - 6}} =$ The LCD of all the fractions in the numerator and the denominator is $x - 6$. Therefore, multiply the numerator and the denominator by $x - 6$.

$\dfrac{(x - 6)\left(x + \frac{9}{x - 6}\right)}{(x - 6)\left(1 + \frac{3}{x - 6}\right)} =$ Apply the distributive property.

$\dfrac{(x - 6)x + (x - 6)\left(\frac{9}{x - 6}\right)}{(x - 6) \cdot 1 + (x - 6)\left(\frac{3}{x - 6}\right)} =$ Perform the multiplications.

$\dfrac{x^2 - 6x + 9}{x - 6 + 3} =$ Factor the numerator and combine like terms in the denominator.

$\dfrac{(x - 3)^2}{x - 3} =$ Divide by the common factor $x - 3$.

$x - 3$ Answer.

Practice Exercises

Simplify the following complex fractions:

5. $\dfrac{\frac{m}{n}}{\frac{p}{q}}$

6. $\dfrac{\frac{x^3y^2}{z}}{\frac{x^2y^4}{z^3}}$

7. $\dfrac{1 - \frac{4}{x^2}}{1 + \frac{2}{x}}$

Answers:

8. $\dfrac{1 - \dfrac{4}{x} - \dfrac{12}{x^2}}{1 + \dfrac{6}{x} + \dfrac{8}{x^2}}$

9. $\dfrac{x + \dfrac{16}{x - 8}}{1 + \dfrac{4}{x - 8}}$

If more practice is needed, do the Additional Practice Exercises.

Additional Practice Exercises

Simplify the following complex fractions:

e. $\dfrac{\dfrac{r}{s}}{\dfrac{t}{u}}$

f. $\dfrac{\dfrac{4cd^3}{3a}}{\dfrac{2c^4d}{6a^2}}$

g. $\dfrac{1 - \dfrac{25}{a^2}}{1 - \dfrac{5}{a}}$

h. $\dfrac{1 - \dfrac{3}{x} - \dfrac{28}{x^2}}{1 - \dfrac{1}{x} - \dfrac{20}{x^2}}$

i. $\dfrac{4x + \dfrac{9}{x - 3}}{2 + \dfrac{3}{x - 3}}$

Exercise Set 7.5

Simplify the following complex fractions:

1. $\dfrac{\dfrac{2}{3}}{\dfrac{5}{6}}$

2. $\dfrac{\dfrac{3}{8}}{\dfrac{5}{4}}$

3. $\dfrac{\dfrac{7}{5}}{\dfrac{9}{10}}$

4. $\dfrac{\dfrac{3}{7}}{\dfrac{11}{14}}$

5. $\dfrac{\dfrac{3}{4}}{\dfrac{1}{6}}$

6. $\dfrac{\dfrac{5}{9}}{\dfrac{4}{15}}$

7. $\dfrac{\dfrac{9}{21}}{\dfrac{3}{14}}$

8. $\dfrac{\dfrac{11}{12}}{\dfrac{3}{14}}$

9. $\dfrac{\dfrac{1}{8}}{2}$

10. $\dfrac{\dfrac{3}{1}}{\dfrac{1}{5}}$

11. $\dfrac{\dfrac{6}{7}}{10}$

12. $\dfrac{\dfrac{15}{8}}{20}$

13. $\dfrac{14}{\dfrac{7}{2}}$

14. $\dfrac{9}{\dfrac{12}{5}}$

15. $\dfrac{\dfrac{4}{9} - \dfrac{1}{3}}{\dfrac{7}{12} - \dfrac{5}{18}}$

16. $\dfrac{\dfrac{3}{4} + \dfrac{1}{12}}{\dfrac{5}{8} - \dfrac{7}{20}}$

17. $\dfrac{\dfrac{6}{5} - \dfrac{7}{6}}{\dfrac{1}{12} + \dfrac{2}{15}}$

18. $\dfrac{\dfrac{11}{14} + \dfrac{3}{2}}{\dfrac{13}{21} - \dfrac{1}{7}}$

19. $\dfrac{8 - \dfrac{9}{4}}{\dfrac{7}{8} + \dfrac{9}{10}}$

20. $\dfrac{\dfrac{7}{9} + \dfrac{5}{12}}{3 - \dfrac{7}{3}}$

21. $\dfrac{\dfrac{7}{6} - \dfrac{1}{3}}{1 - \dfrac{3}{2}}$

22. $\dfrac{6 - \dfrac{2}{9}}{\dfrac{8}{9} + \dfrac{3}{8}}$

23. $\dfrac{5 + \dfrac{7}{2}}{1 + \dfrac{4}{11}}$

24. $\dfrac{9 - \dfrac{5}{4}}{2 + \dfrac{7}{6}}$

Answers:

Practice Exercises 8–9: 8. $\dfrac{x - 9}{x + 4}$ 9. $x - 4$ Additional Practice Exercises e–i: e. $\dfrac{ru}{st}$ f. $\dfrac{4ad^2}{c^3}$ g. $\dfrac{a + 5}{a}$ h. $\dfrac{x - 7}{x - 5}$ i. $2x - 3$

25. $\dfrac{\dfrac{7}{9}+9}{3-\dfrac{4}{3}}$

26. $\dfrac{\dfrac{12}{5}-6}{8+\dfrac{7}{10}}$

27. $\dfrac{\dfrac{m}{8}}{\dfrac{n}{12}}$

28. $\dfrac{\dfrac{p}{15}}{\dfrac{q}{10}}$

29. $\dfrac{\dfrac{5}{d}}{\dfrac{3}{f}}$

30. $\dfrac{\dfrac{7}{g}}{\dfrac{2}{h}}$

31. $\dfrac{\dfrac{x}{y}}{\dfrac{r}{t}}$

32. $\dfrac{\dfrac{u}{v}}{\dfrac{w}{z}}$

33. $\dfrac{\dfrac{g}{h}}{4}$

34. $\dfrac{\dfrac{6}{r}}{s}$

35. $\dfrac{\dfrac{a}{b}}{c}$

36. $\dfrac{\dfrac{m}{n}}{p}$

37. $\dfrac{\dfrac{a}{x^2}}{\dfrac{b}{x}}$

38. $\dfrac{\dfrac{p}{y}}{\dfrac{q}{y^3}}$

39. $\dfrac{\dfrac{u}{v^3}}{\dfrac{w}{v^4}}$

40. $\dfrac{\dfrac{r}{s^5}}{\dfrac{t}{s^2}}$

41. $\dfrac{\dfrac{ac}{b^4}}{\dfrac{ac}{b^3}}$

42. $\dfrac{\dfrac{rs}{t^2}}{\dfrac{rs}{t^5}}$

43. $\dfrac{\dfrac{u^7 v^2}{w^5}}{\dfrac{u^3 v^4}{w}}$

44. $\dfrac{\dfrac{m^6 p^3}{n^7}}{\dfrac{m^5 p^3}{n^4}}$

45. $\dfrac{\dfrac{2}{z}}{1-\dfrac{z}{2}}$

46. $\dfrac{\dfrac{s}{3}}{4-\dfrac{1}{s}}$

47. $\dfrac{1+\dfrac{x}{2}}{1-\dfrac{x}{2}}$

48. $\dfrac{y-\dfrac{y}{3}}{5+\dfrac{y}{3}}$

49. $\dfrac{\dfrac{1}{t}-1}{\dfrac{1}{t^3}-1}$

50. $\dfrac{\dfrac{6}{s^4}-1}{\dfrac{2}{s}+1}$

51. $\dfrac{1+\dfrac{1}{v}-\dfrac{2}{v^2}}{1-\dfrac{4}{v}+\dfrac{3}{v^2}}$

52. $\dfrac{2+\dfrac{13}{w}+\dfrac{6}{w^2}}{1+\dfrac{11}{w}+\dfrac{30}{w^2}}$

53. $\dfrac{\dfrac{1}{u}-\dfrac{2}{u^2}}{1+\dfrac{4}{u}-\dfrac{12}{u^2}}$

54. $\dfrac{\dfrac{4}{p}+\dfrac{4}{p^2}}{3-\dfrac{2}{p}-\dfrac{5}{p^2}}$

55. $\dfrac{\dfrac{2}{x}-\dfrac{7}{x^2}-\dfrac{30}{x^3}}{2+\dfrac{11}{x}+\dfrac{15}{x^2}}$

56. $\dfrac{1-\dfrac{3}{y}-\dfrac{4}{y^2}}{\dfrac{3}{y}-\dfrac{13}{y^2}+\dfrac{4}{y^3}}$

57. $\dfrac{5+\dfrac{4}{r+8}}{r+8}$

58. $\dfrac{\dfrac{7}{s-15}-6}{s-15}$

59. $\dfrac{t+\dfrac{9}{t-7}}{11-\dfrac{3}{t-7}}$

60. $\dfrac{\dfrac{2}{q+4}-q}{\dfrac{13}{q+4}+5}$

Challenge Exercises:

61. $\dfrac{\dfrac{2a}{a+6}+\dfrac{1}{a}}{9a-\dfrac{3}{a+6}}$

62. $\dfrac{\dfrac{8}{b}-\dfrac{7b}{b-1}}{\dfrac{4}{b-1}+\dfrac{21}{b}}$

Writing Exercises:

63. In simplifying complex fractions, why do we multiply the numerator and the denominator by the LCD of all the denominators that occur in the numerator and the denominator?

64. Consider the complex fraction $\dfrac{\dfrac{2}{3}+\dfrac{5}{6}}{\dfrac{1}{4}-\dfrac{1}{3}}$. The LCD of the numerator is 6 and the LCD of the denominator

is 12. Why don't we multiply the numerator by 6 and the denominator by 12? That is, what is wrong

with $\dfrac{6\left(\frac{2}{3} + \frac{5}{6}\right)}{12\left(\frac{1}{4} - \frac{1}{3}\right)}$?

65. Explain why $\dfrac{a^{-2} + b^{-2}}{a^{-1} + b^{-1}} \neq \dfrac{a + b}{a^2 + b^2}$.

Section 7.6

Solving Equations Containing Rational Numbers and Expressions

OBJECTIVES

When you complete this section, you will be able to:

a. Solve equations that contain rational numbers.

b. Solve equations that contain rational expressions.

Introduction   The basis for solving equations that contain fractions or rational expressions is the Multiplication Property of Equality introduced in Chapter 3. In words, it states that we may multiply both sides of an equation by any nonzero number and the result is an equation with the same solution(s) as the original equation. In symbols, if $a = b$ and $c \neq 0$, then $ac = bc$.

Solving equations that contain rational numbers   When solving equations that contain fractions or rational expressions, we will multiply both sides of the equation by the LCD of the fractions or rational expressions. Since all the denominators divide evenly into the LCD, multiplying both sides of the equation by the LCD will eliminate all fractions. The resulting equation is then solved using the same methods used in Chapter 3. We illustrate with some examples.

Example 1

Solve the following equations:

a. $\dfrac{x}{3} - \dfrac{x}{5} = 2$ Multiply both sides of the equation by the LCD of 15.

$15\left(\dfrac{x}{3} - \dfrac{x}{5}\right) = 15 \cdot 2$ Distribute on the left, multiply on the right.

$15\left(\dfrac{x}{3}\right) - 15\left(\dfrac{x}{5}\right) = 30$ Perform the multiplications.

$5x - 3x = 30$ Add like terms.

$2x = 30$ Divide both sides by 2.

$\dfrac{2x}{2} = \dfrac{30}{2}$ Perform the divisions.

$x = 15$ Solution.

CHECK:

Substitute 15 for x.

$\dfrac{15}{3} - \dfrac{15}{5} = 2$

$5 - 3 = 2$

$2 = 2$ Therefore, the solution is correct.

b. $\dfrac{2}{3}a - \dfrac{1}{6}a = \dfrac{8}{9}$ Multiply both sides of the equation by the LCD of 18.

$18\left(\dfrac{2}{3}a - \dfrac{1}{6}a\right) = 18\left(\dfrac{8}{9}\right)$ Distribute on the left, multiply on the right.

$18\left(\dfrac{2}{3}a\right) - 18\left(\dfrac{1}{6}a\right) = 16$ Perform the multiplications.

$12a - 3a = 16$ Combine like terms.

$9a = 16$ Divide both sides by 9.

$a = \dfrac{16}{9}$ Answer.

CHECK:

Substitute $\dfrac{16}{9}$ for a.

$\dfrac{2}{3} \cdot \dfrac{16}{9} - \dfrac{1}{6} \cdot \dfrac{16}{9} = \dfrac{8}{9}$ Perform the multiplications.

$\dfrac{32}{27} - \dfrac{16}{54} = \dfrac{8}{9}$ Reduce $\dfrac{16}{54}$ to lowest terms.

$\dfrac{32}{27} - \dfrac{8}{27} = \dfrac{8}{9}$ Add the fractions on the left side.

$\dfrac{24}{27} = \dfrac{8}{9}$ Reduce the left side to lowest terms.

$\dfrac{8}{9} = \dfrac{8}{9}$ Therefore, the answer is correct.

c. $\dfrac{y+2}{4} = y + 5$ | Multiply both sides of the equation by the LCD of 4.

$4\left(\dfrac{y+2}{4}\right) = 4(y+5)$ | Perform the multiplications.

$y + 2 = 4y + 20$ | Subtract 4y from both sides.

$y + 2 - 4y = 4y + 20 - 4y$ | Combine like terms.

$-3y + 2 = 20$ | Subtract 2 from both sides.

$-3y + 2 - 2 = 20 - 2$ | Perform the subtractions.

$-3y = 18$ | Divide both sides by −3.

$\dfrac{-3y}{-3} = \dfrac{18}{-3}$ | Perform the divisions.

$y = -6$ | Solution.

CHECK:

Substitute −6 for y.

$\dfrac{-6+2}{4} = -6 + 5$ | Simplify both sides.

$\dfrac{-4}{4} = -1$

$-1 = -1$ | Therefore, the answer is correct.

d. $\dfrac{2x-3}{5} - \dfrac{3x+2}{3} = -\dfrac{2}{3}$ | Multiply both sides of the equation by the LCD of 15.

$15\left(\dfrac{2x-3}{5} - \dfrac{3x+2}{3}\right) = 15\left(-\dfrac{2}{3}\right)$ | Distribute on the left and multiply on the right.

$15\left(\dfrac{2x-3}{5}\right) - 15\left(\dfrac{3x+2}{3}\right) = -10$ | Multiply on the left.

$3(2x-3) - 5(3x+2) = -10$ | Distribute on the left.

$6x - 9 - 15x - 10 = -10$ | Combine like terms.

$-9x - 19 = -10$ | Add 19 to both sides.

$-9x - 19 + 19 = -10 + 19$ | Perform the additions.

$-9x = 9$ | Divide both sides by −9.

$\dfrac{-9x}{-9} = \dfrac{9}{-9}$ | Perform the divisions.

$x = -1$ | Solution.

CHECK:

The check is left as an exercise for the student.

Practice Exercises

Solve the following equations and check your answer:

1. $\dfrac{x}{3} - \dfrac{x}{7} = 4$

2. $\dfrac{3}{5}x + \dfrac{3}{10}x = \dfrac{4}{5}$

3. $\dfrac{2x-5}{5} = x + 2$

4. $\dfrac{2z-6}{5} - \dfrac{2z+3}{2} = \dfrac{3}{10}$

If more practice is needed, do the Additional Practice Exercises.

Additional Practice Exercises

Solve the following equations:

a. $\dfrac{a}{9} - \dfrac{a}{4} = \dfrac{5}{6}$

b. $\dfrac{1}{4}x = \dfrac{7}{10}x + \dfrac{1}{2}$

c. $\dfrac{3x+2}{4} = 2x - 2$

d. $\dfrac{3r+1}{4} = \dfrac{13}{12} + \dfrac{2r+5}{6}$

Solving equations that contain rational expressions

If the denominator(s) contains variables, the procedure is essentially the same with one exception. Since division by 0 is impossible, we must eliminate any value of the variable that makes the denominator equal to zero. It is possible to solve an equation correctly and get a number that does not solve the equation because it makes one or more of the denominators equal to 0. Such numbers are called **apparent** or **extraneous** solutions. For this reason, it is mandatory that all answers to equations that contain variables in the denominator be checked in order to determine if they are solutions.

Answers:

Practice Exercises 1–4: 1. $x = 21$ 2. $x = \dfrac{8}{9}$ 3. $x = -5$ 4. $z = -5$ *Additional Practice Exercises a–d:* a. $a = -6$ b. $x = -\dfrac{10}{9}$ c. $x = 2$ d. $r = 4$

Example 2

Solve the following equations:

a. $\dfrac{5}{x} - \dfrac{3}{4} = \dfrac{1}{2}$ Multiply both sides of the equation by the LCD, $4x$.

$4x\left(\dfrac{5}{x} - \dfrac{3}{4}\right) = 4x\left(\dfrac{1}{2}\right)$ Distribute on the left, multiply on the right.

$4x\left(\dfrac{5}{x}\right) - 4x\left(\dfrac{3}{4}\right) = 2x$ Perform the multiplications.

$20 - 3x = 2x$ Add $3x$ to both sides.

$20 = 5x$ Divide both sides by 5.

$4 = x$ Possible solution.

CHECK:

Substitute 4 for x.

$\dfrac{5}{4} - \dfrac{3}{4} = \dfrac{1}{2}$

$\dfrac{2}{4} = \dfrac{1}{2}$

$\dfrac{1}{2} = \dfrac{1}{2}$ Therefore, 4 is the solution.

b. $\dfrac{3y}{y-4} - \dfrac{12}{y-4} = 6$ Multiply both sides of the equation by the LCD, $y - 4$.

$(y-4)\left(\dfrac{3y}{y-4} - \dfrac{12}{y-4}\right) = (y-4)(6)$ Distribute on the left, multiply on the right.

$(y-4)\left(\dfrac{3y}{y-4}\right) - (y-4)\left(\dfrac{12}{y-4}\right) = 6y - 24$

   Multiply on the left.

$3y - 12 = 6y - 24$ Subtract $3y$ from both sides.

$-12 = 3y - 24$ Add 24 to both sides.

$12 = 3y$ Divide both sides by 3.

$4 = y$ Possible solution.

CHECK:

Substitute 4 for y.

$\dfrac{12}{4-4} - \dfrac{12}{4-4} = 6$ Since division by 0 is undefined, there is no solution to this equation. The

$\dfrac{12}{0} - \dfrac{12}{0} = 6$ symbol $\varnothing$ is used to show that there is no solution. It is the symbol for the empty set, which means there is nothing in the set. Hence, no solution.

c. $\dfrac{x}{x+4} = \dfrac{1}{x-2}$

Rather than multiply both sides of the equation by the LCD, a quicker way is to cross multiply. This procedure was discussed in Section 7.2, where it was shown that if $\frac{a}{b} = \frac{c}{d}$ then $ad = bc$. It can be used only when the equation is one fraction equal to another fraction.

$\dfrac{x}{x+4} = \dfrac{1}{x-2}$ Cross multiply.

$x(x-2) = (x+4)(1)$ Apply the distributive property.

$x^2 - 2x = x + 4$ Subtract x from both sides of the equation.

$x^2 - 3x = 4$ Subtract 4 from both sides.

$x^2 - 3x - 4 = 0$ Factor.

$(x-4)(x+1) = 0$ Set each factor equal to 0.

$x - 4 = 0 \qquad x + 1 = 0$ Solve each equation.

$x = 4 \qquad\qquad x = -1$ Therefore, 4 and -1 are possible solutions.

CHECK:

$x = 4 \qquad\qquad\qquad x = -1$

Replace x with 4.   Replace x with -1

$\dfrac{4}{4+4} = \dfrac{1}{4-2} \qquad \dfrac{-1}{-1+4} = \dfrac{1}{-1-2}$

$\dfrac{4}{8} = \dfrac{1}{2} \qquad\qquad \dfrac{-1}{3} = \dfrac{1}{-3}$

$\dfrac{1}{2} = \dfrac{1}{2} \qquad\qquad \dfrac{1}{3} = -\dfrac{1}{3}$ Therefore, both answers are solutions.

d. $\dfrac{3}{2a+2} - \dfrac{4}{3a+3} = \dfrac{5}{6}$ Factor the denominators.

$\dfrac{3}{2(a+1)} - \dfrac{4}{3(a+1)} = \dfrac{5}{6}$ Multiply both sides of the equation by the LCD of $6(a+1)$.

$6(a+1)\left(\dfrac{3}{2(a+1)} - \dfrac{4}{3(a+1)}\right) = 6(a+1)\left(\dfrac{5}{6}\right)$

   Distribute on left, multiply on right.

$6(a+1)\left(\dfrac{3}{2(a+1)}\right) - 6(a+1)\left(\dfrac{4}{3(a+1)}\right) = 5(a+1)$

   Multiply.

$3 \cdot 3 - 2 \cdot 4 = 5a + 5$ Simplify the left side.

$9 - 8 = 5a + 5$ Continue simplifying.

$1 = 5a + 5$ Subtract 5 from both sides.

$-4 = 5a$ Divide both sides by 5.

$-\dfrac{4}{5} = a$ Possible solution.

CHECK:

As you can imagine, the check to this example is very difficult and is omitted. However, if an equation containing rational expressions is solved correctly, the extraneous solutions always make the denominator equal to 0, as in Example 2b. Often it is advisable to see if the possible solution(s) make a denominator equal to 0 rather than doing a complete check. If so, eliminate that value(s). Since $-\frac{4}{5}$ does not make any denominator equal to 0, it is at least a possible solution if not a correct one.

e. $\dfrac{4}{x^2 - x - 6} - \dfrac{2}{x^2 + 3x + 2} = \dfrac{4}{x^2 - 2x - 3}$

Factor the denominators.

$$\dfrac{4}{(x - 3)(x + 2)} - \dfrac{2}{(x + 2)(x + 1)} = \dfrac{4}{(x - 3)(x + 1)}$$

Multiply both sides of the equation by the LCD of $(x - 3)(x + 2)(x + 1)$.

$$(x - 3)(x + 2)(x + 1)\left(\dfrac{4}{(x - 3)(x + 2)} - \dfrac{2}{(x + 2)(x + 1)}\right) = (x - 3)(x + 2)(x + 1)\left(\dfrac{4}{(x - 3)(x + 1)}\right)$$

Distribute on the left side, multiply on the right.

$$(x - 3)(x + 2)(x + 1)\left(\dfrac{4}{(x - 3)(x + 2)}\right) - (x - 3)(x + 2)(x + 1)\left(\dfrac{2}{(x + 2)(x + 1)}\right) = (x + 2)4$$

Perform the multiplications.

$$(x + 1)4 - (x - 3)2 = 4x + 8$$

Simplify the left side.

$$4x + 4 - 2x + 6 = 4x + 8$$

Add like terms on the left side.

$$2x + 10 = 4x + 8$$

Subtract $2x$ from both sides.

$$10 = 2x + 8$$

Subtract 8 from both sides.

$$2 = 2x$$

Divide both sides by 2.

$$1 = x$$

Possible solution.

CHECK:

Note that 1 does not make any denominator equal to 0 and is therefore likely correct. The check is left as an exercise for the student.

Practice Exercises

Solve the following equations:

5. $\dfrac{5}{3z} - \dfrac{1}{2z} = \dfrac{7}{6}$

6. $\dfrac{x}{x + 5} = 4 - \dfrac{5}{x + 5}$

7. $\dfrac{x}{2x + 1} = \dfrac{-4}{x - 3}$

8. $\dfrac{y}{3y + 6} + \dfrac{3}{4y + 8} = \dfrac{3}{4}$

9. $\dfrac{4}{x^2 - 5x + 6} - \dfrac{6}{x^2 + 2x - 8} = \dfrac{3}{x^2 + x - 12}$

If more practice is needed, do the following Additional Practice Exercises.

Additional Practice Exercises

Solve the following equations:

e. $\dfrac{3}{x} - \dfrac{3}{2} = \dfrac{6}{x}$

f. $\dfrac{b}{b - 6} - 5 = \dfrac{6}{b - 6}$

g. $\dfrac{x}{x + 4} = \dfrac{4}{x - 2}$

h. $\dfrac{6}{5m - 10} - \dfrac{4}{2m - 4} = \dfrac{1}{5}$

i. $\dfrac{5}{x^2 - 1} + \dfrac{2}{x^2 + 3x + 2} = \dfrac{7}{x^2 + x - 2}$

Exercise Set 7.6

Solve the following equations:

1. $\dfrac{x}{3} - \dfrac{1}{4} = \dfrac{5}{12}$

2. $\dfrac{y}{6} + \dfrac{5}{2} = 2$

3. $\dfrac{t}{10} + \dfrac{3}{5} = \dfrac{11}{10}$

4. $\dfrac{v}{2} - \dfrac{1}{3} = 14$

5. $\dfrac{1}{3}u - \dfrac{1}{9}u = \dfrac{4}{3}$

6. $\dfrac{1}{7}w + \dfrac{1}{2}w = 3$

7. $\dfrac{2m}{3} - \dfrac{4m}{5} = 1$

8. $\dfrac{3p}{8} + \dfrac{p}{4} = 5$

9. $\dfrac{3}{4}x - \dfrac{1}{3}x = \dfrac{5}{6}$

10. $\dfrac{5}{18}q + \dfrac{2}{9}q = \dfrac{1}{3}$

11. $\dfrac{1}{4}t + \dfrac{1}{3}t + \dfrac{5}{6} = \dfrac{15}{4}$

12. $\dfrac{2}{3}v + \dfrac{1}{6}v - \dfrac{1}{4} = \dfrac{7}{12}$

13. $\dfrac{3x}{4} - \dfrac{2}{5} = \dfrac{3x}{10} + \dfrac{1}{2}$

14. $\dfrac{y}{3} - \dfrac{1}{4} = \dfrac{5y}{12} + \dfrac{1}{6}$

15. $\dfrac{y-3}{4} = y - 1$

16. $t + 5 = \dfrac{t+8}{7}$

17. $\dfrac{3x+6}{2} = x + 2$

18. $2a - 3 = \dfrac{5a+6}{4}$

19. $\dfrac{4v+7}{15} = \dfrac{v+1}{5} + \dfrac{1}{3}$

20. $\dfrac{2y-1}{5} - 1 = \dfrac{y-2}{2}$

21. $\dfrac{2q+1}{4} - \dfrac{3q-5}{16} = \dfrac{7}{8}$

22. $\dfrac{p-1}{2} - \dfrac{p+2}{6} = \dfrac{-7}{12}$

Solve the following equations:

23. $\dfrac{6}{m} + \dfrac{1}{2} = \dfrac{3}{4}$

24. $\dfrac{5}{8} - \dfrac{4}{n} = \dfrac{7}{12}$

25. $\dfrac{5}{r} + \dfrac{3}{r} = 12$

26. $\dfrac{5}{t} - \dfrac{1}{3} = \dfrac{6}{t}$

27. $\dfrac{4}{3u} - \dfrac{1}{2u} = \dfrac{5}{6}$

28. $\dfrac{2}{v} - \dfrac{4}{3v} = \dfrac{3}{5}$

29. $\dfrac{5}{4w} - \dfrac{3}{5w} = \dfrac{-13}{40}$

30. $\dfrac{7}{2z} + \dfrac{11}{6z} = \dfrac{16}{9}$

31. $\dfrac{7}{x-3} = 8 - \dfrac{1}{x-3}$

32. $\dfrac{5}{y+2} + 9 = -\dfrac{4}{y+2}$

33. $\dfrac{t}{t+4} = 3 - \dfrac{12}{t+4}$

34. $\dfrac{8}{z-5} - 2 = \dfrac{z}{z-5}$

35. $\dfrac{4a}{a+6} = \dfrac{10}{a+6} + 2$

36. $\dfrac{5b}{b-8} - 4 = \dfrac{3}{b-8}$

37. $\dfrac{c}{10-c} = \dfrac{1}{c+2}$

38. $\dfrac{x}{2-3x} = \dfrac{1}{3x+2}$

39. $\dfrac{3y}{y-1} = \dfrac{-2}{y+1}$

40. $\dfrac{2n}{n+1} = \dfrac{-5}{n+3}$

41. $\dfrac{m}{m-1} = \dfrac{3m}{4m-3}$

42. $\dfrac{5p}{p+4} = \dfrac{p}{p-1}$

43. $\dfrac{2y}{3y-6} + \dfrac{3}{4y-8} = \dfrac{1}{4}$

44. $\dfrac{2y}{3y+6} - \dfrac{3}{4y+8} = \dfrac{1}{4}$

45. $\dfrac{x}{x-2} - \dfrac{4}{x-1} = \dfrac{2}{x^2-3x+2}$

46. $\dfrac{20z}{z-2} + \dfrac{3}{2z-1} = \dfrac{6}{(z-2)(2z-1)}$

47. $\dfrac{6}{t^2+t-12} = \dfrac{4}{t^2-t-6} + \dfrac{1}{t^2+6t+8}$

48. $\dfrac{7}{v^2-6v+5} - \dfrac{2}{v^2-4v-5} = \dfrac{3}{v^2-1}$

49. $\dfrac{5}{w^2+2w-8} + \dfrac{3}{w^2+w-6} = \dfrac{4}{w^2+7w+12}$

50. $\dfrac{8}{a^2+4a-12} - \dfrac{5}{a^2+a-6} = \dfrac{2}{a^2+9a+18}$

Challenge Exercises:

Solve the following equations:

51. $\dfrac{1}{b^2-b-2} = \dfrac{b}{b^2+2b-8} - \dfrac{1}{b^2+5b+4}$

52. $\dfrac{p+1}{p^2+8p+15} - \dfrac{2}{p^2+7p+10} = \dfrac{1}{p^2+5p+6}$

53. $\dfrac{3}{x^2+2x-24} + \dfrac{x-5}{x^2-16} = \dfrac{x}{x^2+10x+24}$

Writing Exercises:

54. In solving equations that involve rational numbers or rational expressions, why do we multiply both sides of the equation by the LCD?

solving $\frac{y}{3y+5} + \frac{3}{4y+8} = \frac{3}{4}$ different from adding

$\frac{y}{3y+6} + \frac{3}{4y+8} + \frac{3}{4}$?

55. Explain the difference between solving an equation involving rational expressions and adding rational expressions. For example, how is

56. In Example 2c, we solved $\frac{x}{x+4} = \frac{1}{x-2}$ by cross multiplying. Show that this is the same as multiplying both sides by the LCD.

Section 7.7 | # Application with Rational Expressions

O B J E C T I V E S

When you complete this section, you will be able to:

a. Solve application problems involving number properties.

b. Solve application problems involving work.

c. Solve application problems involving distance, rate, and time.

Introduction

Application problems were first discussed in detail in Section 3.5. You may wish to review the procedure and suggested approach before doing this section. Our purpose in considering application problems is to help you to think more mathematically, to better understand the structure of mathematics, to help you translate English into mathematics, and to show the power of mathematics in problem solving. Another purpose of these problems is to serve as a beginning for the more realistic problems that some of you will encounter in more advanced courses.

One of the difficulties in doing applications problems is knowing where to start. All applications problems are not done the same way. There are approaches that are unique to specific types of problems. For this reason, we will discuss number problems, work problems, and distance, rate, time problems separately. We will emphasize the approach needed to do each type of problem. The first type we will discuss is problems involving numbers.

Number problems

Number problems are usually fairly straightforward. You do not have to know special formulas or other facts. All the information needed is contained in the problem. However, there are a few things you need to remember.

a. When working with consecutive integer problems, we let x represent the smallest of the integers, $x + 1$ represents the next largest, $x + 2$ represents the next largest, and so on.

b. When working with consecutive even or consecutive odd integers, we let x represent the smallest even or odd integer, $x + 2$ represents the next largest, $x + 4$ represents the next largest, and so on.

c. The reciprocal of a number is found by putting 1 over the number. For example, the reciprocal of x is $\frac{1}{x}$, $x \neq 0$.

d. In a fraction, the numerator is the number on top and the denominator is the number on bottom.

We suggest the following procedure:

Solving Number Problems

1. Read the problem and identify the unknown(s).
2. Assign a variable to an unknown.

3. If there is more than one unknown, represent the other unknowns in terms of the variable assigned in step 2.
4. Write the equation using information given in the problem.
5. Solve the equation.
6. Check the solution using the wording of the original problem.

Example 1

a. The denominator of a fraction is 3 more than the numerator. If 4 is added to the numerator and subtracted from the denominator, the resulting fraction is $\frac{9}{4}$. Find the original fraction.

Solution:
Let x = the numerator. Then,

$x + 3$ = the denominator. (The denominator is 3 more than the numerator.)

Therefore, the original fraction is $\frac{x}{x+3}$. Since 4 is added to the numerator, the numerator of the new fraction is $x + 4$. Since 4 is also subtracted from the denominator, the denominator of the new fraction is $x + 3 - 4$. Consequently, the new fraction is $\frac{x+4}{x+3-4}$. Since the new fraction is equal to $\frac{9}{4}$, the equation is:

$$\frac{x+4}{x+3-4} = \frac{9}{4} \qquad \text{Simplify the denominator on the left.}$$

$$\frac{x+4}{x-1} = \frac{9}{4} \qquad \text{Cross multiply.}$$

$$4(x+4) = (x-1)9 \qquad \begin{array}{l}\text{Apply the distributive}\\ \text{property.}\end{array}$$

$$4x + 16 = 9x - 9 \qquad \begin{array}{l}\text{Subtract } 9x \text{ from both sides}\\ \text{of the equation.}\end{array}$$

$$-5x + 16 = -9 \qquad \text{Subtract 16 from both sides.}$$

$$-5x = -25 \qquad \text{Divide both sides by } -5.$$

$$x = 5 \qquad \text{Therefore, the numerator is 5.}$$

$$x + 3 = 5 + 3 = 8 \qquad \begin{array}{l}\text{Therefore, the denominator}\\ \text{is 8.}\end{array}$$

Since the numerator is 5 and the denominator is 8, the original fraction is $\frac{5}{8}$.

Check:
If 4 is added to the numerator of $\frac{5}{8}$, the numerator becomes 9. If 4 is subtracted from the denominator, the denominator becomes 4. The new fraction is $\frac{9}{4}$, which is what it is supposed to be. Therefore, our solution is correct.

b. One number is 4 less than the other. If $\frac{1}{2}$ of the larger number is 3 more than $\frac{1}{3}$ of the smaller, find the numbers.

Solution:
Let x = the larger number. Then,

$x - 4$ = the smaller number. (The smaller number is 4 less than the larger.)

We represent $\frac{1}{2}$ of the larger number by $\frac{1}{2}x$ and $\frac{1}{3}$ of the smaller number by $\frac{1}{3}(x - 4)$. Since $\frac{1}{2}$ of the larger number is 3 more than $\frac{1}{3}$ of the smaller, the equation is:

$$\frac{1}{2}x = \frac{1}{3}(x - 4) + 3 \qquad \begin{array}{l}\text{Multiply both sides}\\ \text{of the equation by 6.}\end{array}$$

$$6\left(\frac{1}{2}x\right) = 6\left(\frac{1}{3}(x - 4) + 3\right) \qquad \begin{array}{l}\text{Multiply on the left,}\\ \text{distribute on the}\\ \text{right.}\end{array}$$

$$3x = 6 \cdot \frac{1}{3}(x - 4) + 6(3) \qquad \begin{array}{l}\text{Multiply on the}\\ \text{right side.}\end{array}$$

$$3x = 2(x - 4) + 18 \qquad \text{Distribute.}$$

$$3x = 2x - 8 + 18 \qquad \text{Combine like terms.}$$

$$3x = 2x + 10 \qquad \begin{array}{l}\text{Subtract } 2x \text{ from}\\ \text{both sides.}\end{array}$$

$$x = 10 \qquad \begin{array}{l}\text{Therefore, the larger}\\ \text{number is 10.}\end{array}$$

$$x - 4 = 10 - 4 = 6 \qquad \begin{array}{l}\text{Therefore, the}\\ \text{smaller number is 6.}\end{array}$$

Check:
Since 10 and 6 differ by 4, and $\frac{1}{2}$ of 10 is 5, which is 3 more than $\frac{1}{3}$ of 6 (which is 2), the solutions are correct.

Note: Since you should now be familiar with the techniques in equation solving, not as much detail will be given in the explanations to the right of each step. Also, some of the easier steps will be omitted.

c. For two consecutive even integers, the reciprocal of the smaller is added to three times the reciprocal of the larger. If the resulting sum is $\frac{13}{24}$, find the two consecutive even integers.

Solution:
Let x = the smaller of the two consecutive even integers. Then,

$x + 2$ = the larger of the two consecutive even integers.

The reciprocal of the smaller is $\frac{1}{x}$ and the reciprocal of the larger is $\frac{1}{x+2}$. Three times the reciprocal of the larger is $3\left(\frac{1}{x+2}\right) = \frac{3}{x+2}$. Since *sum* means add and the sum is $\frac{13}{24}$, the equation is:

$$\frac{1}{x} + \frac{3}{x+2} = \frac{13}{24}$$

Multiply both sides of the equation by the LCD $24x(x+2)$.

$$24x(x+2)\left(\frac{1}{x} + \frac{3}{x+2}\right) = 24x(x+2)\left(\frac{13}{24}\right)$$

$$24x(x+2)\left(\frac{1}{x}\right) + 24x(x+2)\left(\frac{3}{x+2}\right) = 13x(x+2)$$ *Simplify both sides.*

$$24(x+2) + 24x(3) = 13x^2 + 26x$$ *Perform the multiplications.*

$$24x + 48 + 72x = 13x^2 + 26x$$ *Simplify the left side.*

$$96x + 48 = 13x^2 + 26x$$ *Continue simplifying.*

$$48 = 13x^2 - 70x$$ *Subtract 96x from both sides.*

$$0 = 13x^2 - 70x - 48$$ *Subtract 48 from both sides.*

$$0 = (13x + 8)(x - 6)$$ *Factor the right side.*

$$13x + 8 = 0, \text{ or } x - 6 = 0$$ *Set each factor equal to 0.*

$$13x = -8 \qquad x = 6$$ *Solve each equation.*

$$x = -\frac{8}{13}$$

The only acceptable answer is 6, since $\frac{-8}{13}$ is not an integer.

If $x = 6$, then $x + 2 = 8$, so the two consecutive even integers are 6 and 8.

Check:
The check is left as an exercise for the student.

Practice Exercises

1. The denominator of a fraction is 2 more than the numerator. If 3 is added to the numerator and subtracted from the denominator, the result is equal to $\frac{9}{5}$. Find the original fraction.

2. Two numbers differ by 2. If $\frac{1}{4}$ of the larger number is 4 less than $\frac{1}{5}$ the smaller number, find the numbers.

3. The sum of a fraction and twice its reciprocal is $3\frac{14}{15}$. Find the fraction. (*Hint:* Let x = the fraction.)

Work problems   The next type of problem to be discussed is the "work" problem. In this type of problem, a task can be performed in a certain amount of time by one individual or thing, and the same task can be performed by another individual or thing in a different amount of time. We are then asked how long it would take for the two working together to perform the same task. There are many variations to this problem. The key to solving this problem is to represent the part of the task that can be done in one unit of time. For example, if it takes five hours to do a job, then $\frac{1}{5}$ is the part of the job that can be done in one hour. If it takes x hours to do a job, then $\frac{1}{x}$ is the part of the job that can be done in one hour.

Solving Work Problems

1. Assign a variable to the unknown.
2. Represent the part of the job that can be done in one unit of time by each individual working alone, and by both working together.
3. The equation is the sum of the parts done by each individual made equal to the part done while working together.
4. Solve the equation.
5. See if the results make sense when compared to the words of the original problem.

Example 2

a. If Mike can paint his room in ten hours and Susan can paint the same room in seven hours, how long would it take them to paint the room working together?

Solution:
Let x = the number of hours needed to paint the room working together. Then,

$\frac{1}{10}$ = the part of the room Mike can paint in one hour working alone.

$\frac{1}{7}$ = the part of the room Susan can paint in one hour working alone.

$\frac{1}{x}$ = the part of the room they can paint in one hour working together.

(part Mike paints) + (part Susan paints) = (part they paint together). So the equation is:

$$\frac{1}{10} + \frac{1}{7} = \frac{1}{x}$$
Multiply both sides of the equation by $70x$.

$$70x\left(\frac{1}{10} + \frac{1}{7}\right) = 70x\left(\frac{1}{x}\right)$$
Distribute on the left, multiply on the right.

$$70x\left(\frac{1}{10}\right) + 70x\left(\frac{1}{7}\right) = 70$$
Perform the multiplications on the left.

$$7x + 10x = 70$$
Combine like terms.

$$17x = 70$$
Divide both sides by 17.

$$x = \frac{70}{17} \text{ hours}$$
Therefore, it takes $\frac{70}{17}$ hours working together.

Does the preceding answer seem reasonable? If both worked at the same rate as Mike, it would take half as long as Mike took, which would be five hours. If both worked at the same rate as Susan, it would take three and one-half hours. The amount of time working together should be between three and one-half and five hours. Is $\frac{70}{17}$ between three and one-half and five?

b. Sal and Melissa work in an electronics factory. Together they can install the picture tubes in 30 televisions in eight hours. If Sal can install the picture tubes in 30 televisions in 14 hours, how long would it take Melissa to install 30 picture tubes working alone?

Solution:
Let x = the number of hours for Melissa to install 30 tubes working alone. Then,

$\frac{1}{x}$ = the part of the job Melissa can complete in one hour.

$\frac{1}{14}$ = the part of the job Sal can complete in one hour.

$\frac{1}{8}$ = the part of the job they can complete in one hour working together.

(part Sal does) + (part Melissa does) = (part they do together). So the equation is:

$$\frac{1}{14} + \frac{1}{x} = \frac{1}{8}$$
Multiply both sides of the equation by $56x$.

$$56x\left(\frac{1}{14} + \frac{1}{x}\right) = 56x\left(\frac{1}{8}\right)$$
Simplify both sides.

$$56x\left(\frac{1}{14}\right) + 56x\left(\frac{1}{x}\right) = 7x$$
Continue simplifying.

$$4x + 56 = 7x$$
Subtract $4x$ from both sides.

$$56 = 3x$$
Divide both sides by 3.

$$\frac{56}{3} = x$$
Therefore, it takes Melissa $18\frac{2}{3}$ hours working alone. Is this answer reasonable?

Practice Exercises

Solve the following:

4. If one pipe can fill a tank in 12 hours and a second pipe can fill a tank in 8 hours, how long would it take to fill the tank with both pipes open?

5. John can clean the pool in six hours. If he and Johanna work together, they can clean the pool in four hours. How long would it take Johanna to clean the pool working alone?

Distance, time, and rate problems

The distance traveled depends upon the rate (speed) and the time traveled according to the formula $d = rt$. For example, if an object moves at the rate of 40 miles per hour for five hours, then $d = 40 \cdot 5 = 200$ miles. Equivalent forms of this formula are $r = \frac{d}{t}$ and $t = \frac{d}{r}$. For example, if an object travels 300 miles in six hours, then $r = \frac{300}{6} = 50$ mph. If an object travels 240 miles at a rate of 60 miles per hour, then $t = \frac{240}{60} = 4$ hours.

The most efficient way of approaching this type of problem is by using a table. The things that are moving are put in a column on the left of the table and the distance, rate, and time are placed in three columns. We will always know the values of both distances, rates, or times. So these columns will be filled in with numbers. We will also know a relationship between two of the remaining quantities. For example, if we know the values of the two distances, then we will also know a relationship between the rates or the times. Suppose we know a relationship between the rates. We let the variable represent one of the rates and express the other rate in terms of the variable. We then use the relationships $d = rt$, $r = \frac{d}{t}$, and $t = \frac{d}{r}$ to fill in the remaining column and write the equation using the information from the last column filled in. This procedure is summarized as follows:

Solving Distance, Rate, and Time Problems

1. Make a chart with columns for the moving things, the distance, the rate, and the time.
2. Fill in one column with known numerical values.
3. Assign a variable and fill in another column with expressions containing variables.
4. Fill in the remaining column from the first two using the appropriate relationships between d, r, and t.
5. Write the equation using the information in the last column filled in.
6. Solve the equation.
7. Check the solution against the wording of the original problem.

Example 3

Solve the following:

a. Tom and John Truong live in cities that are 80 miles apart. Both leave home at 8:00 A.M. traveling toward each other riding bicycles. Tom's speed is two-thirds of John's. Find John's speed if they meet at 10:00 A.M.

Solution:
Consider the following chart:

	d	r	t
Tom			
John			

Since each rider left at 8:00 A.M. and they met at 10:00 A.M., we know that both times are two hours. Put 2 in the time column for each.

	d	r	t
Tom			2
John			2

We also know that Tom's speed is two-thirds of John's. Let John's speed be x, then Tom's speed is $\frac{2}{3}x$. Put these in the table.

	d	r	t
Tom		$\frac{2}{3}x$	2
John		x	2

The rate and time columns have been filled. Since $d = rt$, fill in the d column by finding the products of the expressions in the r and t columns. Therefore, Tom's distance is $\frac{2}{3}x(2) = \frac{4}{3}x$ and John's distance is $2x$.

	d	r	t
Tom	$\frac{4}{3}x$	$\frac{2}{3}x$	2
John	$2x$	x	2

The last column filled in was the distance column, so we write our equation using some known fact(s) about distance. Sometimes it is helpful to draw a diagram.

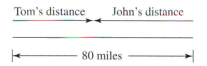

Tom's distance   John's distance

|←——————— 80 miles ———————→|

From the diagram, we see that Tom's distance plus John's distance equals 80. Therefore, the equation is:

$$\frac{4}{3}x + 2x = 80 \qquad \text{Multiply both sides of the equation by 3.}$$

$$3\left(\frac{4}{3}x + 2x\right) = 3 \cdot 80 \qquad \text{Simplify each side.}$$

$$3\left(\frac{4}{3}x\right) + 3 \cdot 2x = 240 \qquad \text{Continue simplifying.}$$

$$4x + 6x = 240 \qquad \text{Add like terms.}$$

$$10x = 240 \qquad \text{Divide both sides by 10.}$$

$$x = 24 \qquad \text{Therefore, John's rate is 24 mph.}$$

Check:
If John's rate is 24 mph, he will travel 48 miles in two hours. Since Tom's rate is two-thirds of John's, he travels at 16 mph. In two hours, Tom will travel 32 miles. Together they have traveled $48 + 32 = 80$ miles, which is the distance between the cities. Therefore, our solution is correct.

b. An airplane flew 450 miles from City A to City B in $1\frac{1}{2}$ hours less than it took to fly 1200 miles from City C to City D. If the airplane flew at the same rate on both trips, find the rate of the airplane.

Solution:
Draw and label the chart.

	d	r	t
A to B			
C to D			

We know the distance from City A to City B is 450 miles and the distance from City C to City D is 1200 miles. We are looking for the rate of the plane. Since it was the same on both trips, let x = the rate of the plane. Enter these values in the table.

	d	r	t
A to B	450	x	
C to D	1200	x	

The d and r columns are filled in. Since $t = \frac{d}{r}$, fill in the t column by dividing the expressions in the d column by the expressions in the r column. Hence, the time from A to B $= \frac{450}{x}$ and the time from C to D $= \frac{1200}{x}$. Enter these values in the table.

	d	r	t
A to B	450	x	$\frac{450}{x}$
C to D	1200	x	$\frac{1200}{x}$

The last column filled is the time column. Therefore, our equation involves time. We know the trip from A to B took $1\frac{1}{2}$ hours less than the time from C to D. So our equation is of the form (time from A to B) = (time from C to D) $-1\frac{1}{2}$. Therefore, the equation is:

$$\frac{450}{x} = \frac{1200}{x} - \frac{3}{2}$$ 　Multiply both sides of the equation by $2x$.

$$2x\left(\frac{450}{x}\right) = 2x\left(\frac{1200}{x} - \frac{3}{2}\right)$$ 　Simplify both sides.

$$2 \cdot 450 = 2x\left(\frac{1200}{x}\right) - 2x\left(\frac{3}{2}\right)$$ 　Continue simplifying.

$$900 = 2(1200) - 3x$$ 　Continue simplifying.

$$900 = 2400 - 3x$$ 　Subtract 2400 from both sides.

$$-1500 = -3x$$ 　Divide both sides by $-3x$.

$$500 = x$$ 　Therefore, the rate of the plane is 500 mph.

Check:
If the plane flies at a rate of 500 mph, the trip of 450 miles would take $\frac{450}{500} = \frac{9}{10}$ of an hour and the trip of 1200 miles would take $\frac{1200}{500} = \frac{12}{5}$ of an hour. The difference in times is $\frac{12}{5} - \frac{9}{10} = \frac{24}{10} - \frac{9}{10} = \frac{15}{10} = \frac{3}{2}$ hours. Therefore, our solution is correct.

c. A river has a current of 5 mph. A boat can make a trip of 20 miles upstream in the same time it can make a trip of 30 miles downstream. Find the rate of the boat in still water.

Solution:
Draw and label the chart.

	d	r	t
upstream			
downstream			

We know the distance upstream is 20 miles and the distance downstream is 30 miles. Let x = the speed of the boat in still water. Since the rate of the current is 5 mph, the current will be slowing the rate of the boat upstream by 5 mph. Therefore, the rate of the boat upstream is $x - 5$. When the boat is going downstream, the current is assisting the boat by 5 mph. Therefore, the rate of the boat downstream is $x + 5$. Enter these values in the table.

	d	r	t
upstream	20	$x - 5$	
downstream	30	$x + 5$	

Now we fill in the column for t. Since $t = \frac{d}{r}$, divide the expressions in the d column by the expressions in the r column. Therefore, the time going upstream is $\frac{20}{x-5}$ and the time going downstream is $\frac{30}{x+5}$. Enter these values in the table.

	d	r	t
upstream	20	$x-5$	$\frac{20}{x-5}$
downstream	30	$x+5$	$\frac{30}{x+5}$

The last column filled was the t column. Therefore, our equation involves time. We know the time going upstream is equal to the time going downstream. So our equation is:

$$\frac{20}{x-5} = \frac{30}{x+5}$$ Cross multiply.

$$20(x+5) = 30(x-5)$$ Simplify both sides.

$$20x + 100 = 30x - 150$$ Subtract $20x$ from both sides.

$$100 = 10x - 150$$ Add 150 to both sides.

$$250 = 10x$$ Divide both sides by 10.

$$25 = x$$ Therefore, the boat travels at 25 mph in still water.

Check:
If the boat travels at the rate of 25 mph in still water, then its speed against the current is 20 mph. In order to go 20 miles upstream, it would take one hour. The speed of the boat with the current is 30 mph. In order to go 30 miles downstream, it would also take one hour. The time going upstream equals the time going downstream. Therefore, our solution is correct.

Practice Exercises

Solve the following:

6. Ellie and Frank live in cities that are 140 miles apart. Both leave home at 7 A.M. traveling by bicycle toward each other. What is Frank's rate if they meet at 11 A.M. and Ellie's rate is three-fourths of Frank's?

7. A plane flew 600 miles in $1\frac{3}{4}$ hours less time than it flew 1125 miles. If the rate is the same for both trips, find the rate of the plane.

8. A plane flew 1500 miles against the wind in the same amount of time it took it to fly 1800 miles with the wind. If the speed of the wind was 50 mph, find the speed of the plane in still air.

9. Shareen makes a trip at an average speed of 60 mph, and Abdul makes the same trip at an average speed of 50 mph. If Abdul's time is $1\frac{1}{2}$ hours more than Shareen's, how long did it take Shareen to make the trip?

Exercise Set 7.7

Solve the following by using the same method as in Example 1:

1. The numerator and the denominator of a fraction are the same. If 2 is added to the numerator, the new fraction is equivalent to $\frac{14}{10}$. Find the original numerator.

2. The numerator and the denominator of a fraction are the same. If 4 is subtracted from the denominator, the new fraction is equivalent to $\frac{39}{27}$. Find the original fraction.

3. The denominator of a fraction is 3 less than the numerator. If 2 is added to both the numerator and the denominator, the new fraction is equivalent to $\frac{12}{9}$. Find the original fraction.

4. The numerator of a fraction is 5 more than the denominator. If 3 is subtracted from both the numerator

Answers:

Practice Exercises 6–9: 6. 20 mph 7. 300 mph 8. 550 mph 9. 7.5 hours

and the denominator, the result is 6. Find the original fraction.

5. One integer is 3 more than another integer. If $\frac{1}{4}$ of the larger is equal to $\frac{1}{3}$ of the smaller, find the two integers.

6. One integer is 12 less than another integer. If $\frac{1}{7}$ of the larger is equal to $\frac{1}{5}$ of the smaller, find the two integers.

7. Two numbers differ by 6. If $\frac{2}{3}$ of the larger is 5 more than $\frac{1}{2}$ of the smaller, find the two numbers.

8. Two numbers differ by 8. If $\frac{3}{2}$ of the smaller is 6 less than $\frac{3}{4}$ of the larger, find the two numbers.

9. The sum of the reciprocals of two consecutive even integers is $\frac{5}{12}$. Find the two consecutive even integers.

10. Three times the reciprocal of the sum of two consecutive odd integers is $\frac{3}{20}$. Find the two consecutive odd integers.

11. A car costs $\frac{5}{6}$ as much as a van. If together they cost $33,000, how much does the van cost?

12. An electric light bulb uses $\frac{1}{18}$ as much power as an electric toaster. If together they use 1140 watts of power, how many watts does the toaster use?

Solve the following by using the same method as in Example 2. Convert fractions to mixed numbers.

13. Jason can wash and wax his car in four hours. His younger sister can wash and wax the same car in six hours. Working together, how fast can they wash and wax the car?

14. Jon can clean the house in seven hours and Linda can clean the house in five hours. How long would it take if they worked together?

15. Grandma and Grandpa have volunteered to make quilts for HIV-positive babies. If Grandma can make a quilt in 25 days and Grandpa can make a quilt in 35 days, in how many days can they make a quilt working together?

16. It takes Alice 90 minutes to put a futon frame together, and it takes Maya 60 minutes to put the same type of frame together. If they worked together, how long would it take to put a frame together?

17. Working together, it takes two roofers four hours to put a new roof on a portable classroom. If the first

roofer can do the job by himself in six hours, how many hours would it take for the second roofer to do the job by himself?

18. Working together, Rita and Tiffany can cut and trim a lawn in two hours. If it takes Rita five hours to do the lawn by herself, how long would it take Tiffany to do the lawn by herself?

19. With both the cold water and the hot water faucets open, it takes 9 minutes to fill a bathtub. The cold water faucet alone takes 15 minutes to fill the tub. How long would it take to fill the tub with the hot water faucet alone?

20. The cargo hold of a ship has two loading pipes. Used together, the two pipes can fill the cargo hold in six hours. If the larger pipe alone can fill the hold in eight hours, how many hours would it take the smaller pipe to fill the hold by itself?

Solve the following by using the same method as in Example 3.

21. Two buses leave Kansas City traveling in opposite directions, one going east and one going west. The eastbound bus travels at 40 mph and the westbound bus travels at 65 mph. If both buses leave at 10:00 A.M., at what time will they be 525 miles apart?

22. Sailing in opposite directions, an aircraft carrier and a destroyer leave their base in Hawaii at 5:00 A.M. If the destroyer sails at 30 mph and the aircraft carrier sails at 20 mph, at what time will the two ships be 300 miles apart?

23. Rapid City, South Dakota, is 360 miles from Sioux Falls. At 6:00 A.M., a freight train leaves Rapid City for Sioux Falls, and at the same time a passenger

train leaves Sioux Falls for Rapid City. The two trains meet at 9:00 A.M. If the freight train travels $\frac{3}{5}$ as fast as the passenger train, how fast does the passenger train travel?

24. Houston, Texas, and Calgary, Alberta, are about 2100 miles apart. At 2:00 P.M., an airplane leaves Houston for Calgary flying at 250 mph. At the same time, an airplane leaves Calgary for Houston flying at 450 mph. How long will it take before the two airplanes meet?

25. An automobile makes a trip of 270 miles in $2\frac{1}{2}$ hours less than it takes the same auto to make a trip of 420 miles traveling at the same speed. What is the speed of the automobile?

26. A ship leaves port traveling at 15 mph. Two hours later, a speedboat leaves the same port traveling at 40 mph. How long will it take for the speedboat to overtake the ship?

27. An airliner flies against the wind from Washington, D.C., to San Francisco in five and one-half hours. It flies back to Washington, D.C., with the wind in five

hours. If the average speed of the wind is 21 mph, what is the speed of the airliner in still air?

28. A river has a current of 3 mph. A boat goes 40 miles upstream in the same amount of time as it goes 50 miles downstream. Find the speed of the boat in still water.

Solve the following:

29. The denominator of a fraction is twice the numerator. If 3 is added to the numerator and 3 is subtracted from the denominator, the new fraction is $\frac{11}{13}$. Find the original fraction.

30. The same number is added to the numerator and the denominator of the fraction $\frac{4}{5}$. If the new fraction is equivalent to $\frac{33}{36}$, find the number added.

31. The pitcher of a baseball team has $\frac{2}{3}$ as many hits as the catcher, and together they have 85 hits. How many hits does each player have?

32. Two numbers differ by 6. If $\frac{3}{4}$ of the smaller number is $\frac{1}{4}$ less than $\frac{2}{3}$ of the larger, find the two numbers.

33. Hank can load a moving van in 12 hours, and Andy can load a moving van in 15 hours. Working together, how long would it take to load the moving van?

34. In a bicycle race, the first rider crossed the finish line $2\frac{1}{4}$ hours before the last rider. If the winner had an average speed of 26 mph, and the last rider had a average speed of 16 mph, how long did it take the winner to finish the race?

Challenge Exercises:

35. Find a positive number that is equal to 1 less than 12 times its reciprocal.

36. With both hot and cold water faucets on, it takes 5 minutes to fill a sink. However, the stopper is broken, so that water is running out of the sink at the same time that water is running into the sink. If the

leak can empty the sink in 15 minutes, how long will it take the sink to fill $\frac{1}{2}$ full with both faucets on?

37. Dr. Gilbert spent a total of four hours on a trip in the country. He rode out to the country at 15 mph. His bicycle broke down, and he had to walk back at 3 mph. How far out in the country did he get?

Writing Exercises or Group Activities:

If done as a group activity, each group should write two exercises of each of the following types, exchange with another group, and then solve them:

38. Write and solve a "number" word problem.

39. Write and solve a "work" word problem.

40. Write and solve a "distance, time, and rate" word problem.

Chapter 7   Summary

Adding Rational Numbers or Rational Expressions with Common Denominators: [Section 7.1]

- To add rational numbers or rational expressions with common denominators, add the numerators and put the sum over the common denominator. In symbols, $\frac{a}{c} + \frac{b}{c} = \frac{a+b}{c}$. If necessary, reduce the sum to lowest terms.

Least Common Multiple: [Section 7.2]

- The integer a is the **least common multiple** of the integers b and c if a is the smallest positive integer that is a multiple of both b and c. Equivalently, a is the least common multiple of b and c if a is the smallest positive integer that is divisible by both b and c.

Finding the Least Common Multiple (LCM): [Section 7.2]

1. Write each number or expression as the product of its prime factors.
2. Select every factor that appears in any prime factorization. If a factor appears in more than one of the prime factorizations, select the factor with the largest exponent.
3. The LCM is the product of the factors selected in step 2.

Equivalent Fractions: [Section 7.2]
- Two or more fractions are equivalent if they represent the same quantity. Equivalently, $\frac{a}{b} = \frac{c}{d}$ if $a \cdot d = b \cdot c$.

Fundamental Property of Fractions: [Section 7.2]
- If $\frac{a}{b}$ is a fraction and c is any number except 0, then $\frac{a}{b} = \frac{a}{b} \cdot \frac{c}{c} = \frac{ac}{bc}$. In words, we can multiply both the numerator and the denominator of a fraction by any number except 0 and get a fraction equivalent to the original fraction.

Finding the Missing Numerator to Get Equivalent Fractions: [Section 7.2]
- Determine the number or expression that you must multiply the denominator of the given fraction by in order to get the denominator of the new fraction. Multiply both the numerator and the denominator of the given fraction by this number to get an equivalent fraction with the given denominator.

Least Common Denominator (LCD): [Section 7.3]
- The least common denominator of two or more rational numbers or rational expressions is the least common multiple of the denominators of the fractions.

Adding Rational Numbers or Rational Expressions with Unlike Denominators: [Section 7.4]
- To add rational numbers or rational expressions with unlike denominators, change each rational number or rational expression into an equivalent rational number or rational expression with the least common denominator as its denominator. Then add the fractions using the method of adding rational numbers or rational expressions with common denominators.

Simplifying Complex Fractions Using Division: [Section 7.5]
1. If necessary, rewrite the numerator and/or the denominator as single fractions by performing whatever operations are indicated in either.
2. Rewrite the complex fraction as the division of the numerator by the denominator.
3. Perform the division by inverting the fraction on the right and multiplying.
4. If necessary, reduce to lowest terms.

Simplifying Complex Fractions Using the LCD: [Section 7.5]
1. Find the LCD of all fractions that occur in the numerator and/or the denominator.
2. Multiply both the numerator and the denominator by the LCD.
3. Simplify the results.

Solving Equations Involving Rational Numbers or Rational Expressions: [Section 7.6]
1. Determine the value(s) of the variable, if any, that would make any denominator equal to 0.
2. Multiply both sides of the equation by the LCD of all the rational numbers or rational expressions that appear in the equation.
3. Solve the resulting equation.
4. Check for extraneous solutions.

Solving Number Problems: [Section 7.7]
1. Read the problem and assign a variable to the unknown.
2. If there is more than one unknown, represent the other unknowns in terms of the variable assigned in step 1.
3. Write the equation.
4. Solve the equation.
5. Check the solution(s) using the wording in the original problem.

Solving Work Problems: [Section 7.7]
1. Assign a variable to the unknown.
2. Represent the part of the job that can be done in one unit of time by each individual or thing, both working alone and working together.
3. The equation is the sum of the parts done by each individual or thing made equal to the part done while working together.

4. Solve the equation.

5. See if the results make sense when compared to the words of the original problem.

Solving Distance, Rate, and Time Problems: [Section 7.7]

1. Make a chart with columns for the moving things, the distance, the rate, and the time.

2. Fill in one column with known numerical values.

3. Assign a variable and fill in another column with expressions containing variables.

4. Fill in the remaining column from the first two using the appropriate relationships between d, r, and t.

5. Write the equation using the information in the last column filled in.

6. Solve the equation.

7. Check the solution against the wording of the original problem.

Chapter 7   Review Exercises

Find the following sums or differences. Leave the answers reduced to the lowest terms. [Section 7.1]

1. $\dfrac{5}{7} - \dfrac{3}{7}$

2. $\dfrac{3}{8} - \dfrac{-5}{8}$

3. $\dfrac{4}{13} + \dfrac{6}{13} + \dfrac{2}{13}$

4. $\dfrac{9}{11} + \dfrac{5}{11} - \dfrac{7}{11}$

5. $9\dfrac{7}{10} + 6\dfrac{1}{10}$

6. $19\dfrac{11}{12} - 13\dfrac{4}{12}$

7. $11\dfrac{7}{9} - 4\dfrac{2}{9} - \dfrac{5}{9}$

8. $15\dfrac{15}{17} - 8\dfrac{11}{17} + 3\dfrac{4}{17}$

9. $\dfrac{9r}{14x} + \dfrac{2p}{14x}$

10. $\dfrac{4}{y+2} - \dfrac{3}{y+2}$

11. $\dfrac{x^2 - 3x - 8}{x^2 + 7x + 10} + \dfrac{-2x - 6}{x^2 + 7x + 10}$

12. $\dfrac{x^2 + 2x - 5}{x^2 - 3x - 18} - \dfrac{3x + 7}{x^2 - 3x - 18}$

13. $\dfrac{2p^2 - p + 2}{p^2 - 16} + \dfrac{p^2 + 4p - 8}{p^2 - 16}$

14. $\dfrac{x - 3y}{x - y} - \dfrac{5x + 2y}{y - x}$

Answer the following: [Section 7.1]

15. A stock began the day at $32\dfrac{3}{8}$ and rose $3\dfrac{5}{8}$ points. What was the closing price of the stock?

16. A rectangular tablecloth is $66\dfrac{9}{16}$ inches long and $44\dfrac{5}{16}$ inches wide. A border that costs \$0.02 per inch is sewn around the tablecloth. Find the cost of the border to the nearest cent.

17. Find the perimeter of a triangle the length of whose sides are $3\dfrac{2}{9}$ m, $4\dfrac{5}{9}$ m, and $6\dfrac{1}{9}$ m.

18. Find the perimeter of a rectangle whose length is $\dfrac{7}{16}$ km and whose width is $\dfrac{3}{16}$ km.

Find the LCM of the following: [Section 7.2]

19. 6 and 7

20. 25 and 40

21. 16, 24, 32

22. 18, 27, 36

23. a^2y and ay^4

24. $12t^5v^3$ and $15t^2v^7$

25. $14x^2w^3$ and $21x^5w^4$

26. $z - 4$ and $z + 3$

27. $a + 2$ and $a^2 + 4a + 4$

28. $y^2 + 8y + 16$, $y + 4$, $y^2 + y - 12$

Find the missing numerator so that the fractions will be equal. [Section 7.2]

29. $\dfrac{6}{5} = \dfrac{?}{20}$

30. $3 = \dfrac{?}{15}$

31. $\dfrac{4r}{7p^3} = \dfrac{?}{35p^5}$

32. $\dfrac{8x^3}{13tz^4} = \dfrac{?}{52t^3z^7}$

33. $\dfrac{4a}{a + 5} = \dfrac{?}{2a + 10}$

34. $\dfrac{7}{b - 3} = \dfrac{?}{b^2 - 9}$

35. $\dfrac{p-8}{p^2+10p+25}=\dfrac{?}{(p+5)(p-3)(p+5)}$

36. $\dfrac{2q}{y^2-5y}=\dfrac{?}{y(y-5)(y+3)}$

Find the least common denominator of the following. Then write each fraction as an equivalent fraction with the LCD as its denominator. [Section 7.3]

37. $\dfrac{4}{7}$ and $\dfrac{5}{2}$

38. $\dfrac{5}{6}$ and $\dfrac{7}{18}$

39. $\dfrac{8}{9}$ and $\dfrac{11}{12}$

40. $\dfrac{6}{15}, \dfrac{3}{5},$ and $\dfrac{13}{20}$

41. $\dfrac{4}{t^2v^5}$ and $\dfrac{6}{tv^3}$

42. $\dfrac{8b}{9p^3q^4}$ and $\dfrac{11c}{12p^2q^5}$

43. $\dfrac{7}{t-4}$ and $\dfrac{9}{t+2}$

44. $\dfrac{y}{y-11}$ and $\dfrac{y}{y-3}$

45. $\dfrac{4u}{8u+12}$ and $\dfrac{9u}{14u+21}$

46. $\dfrac{2w}{w^2-16}$ and $\dfrac{6w}{w^2+5w+4}$

47. $\dfrac{3a}{a^2+6a+9}$ and $\dfrac{10a}{a^2-2a-15}$

48. $\dfrac{m+1}{m^2-2m-8}$ and $\dfrac{m-1}{m^2-3m-4}$

Find the following sums or differences. Leave the answers reduced to lowest terms. [Section 7.4]

49. $\dfrac{5}{12}+\dfrac{7}{16}$

50. $\dfrac{3}{4}-\dfrac{2}{5}$

51. $\dfrac{17}{5}-4$

52. $\dfrac{3}{14}+\dfrac{5}{7}-\dfrac{4}{21}$

53. $7\dfrac{5}{6}-5\dfrac{1}{3}$

54. $\dfrac{11}{2}-5+3\dfrac{5}{8}$

55. $\dfrac{7}{m}+\dfrac{b}{n}$

56. $\dfrac{10}{9x}-\dfrac{2}{3x}$

57. $\dfrac{y-3}{4y}+\dfrac{y+2}{5y}$

58. $\dfrac{2t-3}{18t^4u^2}+\dfrac{5t+1}{12t^3u^5}$

59. $\dfrac{8}{w-3}-\dfrac{-3}{w}$

60. $\dfrac{6r}{r-5}+\dfrac{2}{5-r}$

61. $\dfrac{v+4}{4v+8}-\dfrac{v-2}{2v-6}$

62. $\dfrac{t^2-5t}{t^2+8t+16}+\dfrac{6}{t+4}$

63. $\dfrac{2w+5}{w^2-25}+\dfrac{6w}{w^2-3w-10}$

64. $\dfrac{z+9}{4z^2+33z+35}-\dfrac{z-12}{z^2+14z+49}$

Answer the following: [Section 7.4]

65. A rectangular computer screen is $8\dfrac{5}{8}$ inches wide and $11\dfrac{1}{4}$ inches high. What is the perimeter of the screen?

66. A carpenter is putting molding around a triangular opening the lengths of whose sides are $8\dfrac{3}{4}$ inches, $7\dfrac{5}{6}$ inches, and $8\dfrac{3}{8}$ inches. If the molding costs $.04 per inch, find the cost of the molding to the nearest cent.

Simplify the following complex fractions using both the division and the LCD methods: [Section 7.5]

67. $\dfrac{\dfrac{4}{5}}{\dfrac{3}{10}}$

68. $\dfrac{\dfrac{5}{18}}{\dfrac{11}{30}}$

69. $\dfrac{\dfrac{8}{9}}{16}$

70. $\dfrac{12}{\dfrac{24}{13}}$

71. $\dfrac{\dfrac{4}{3}-\dfrac{8}{9}}{\dfrac{5}{6}-\dfrac{4}{12}}$

72. $\dfrac{6-\dfrac{10}{7}}{\dfrac{14}{3}-2}$

73. $\dfrac{\dfrac{x}{9}}{\dfrac{y}{27}}$

74. $\dfrac{\dfrac{p}{r}}{s}$

75. $\dfrac{\dfrac{4x}{t^2}}{\dfrac{20x}{t^3}}$

76. $\dfrac{\dfrac{u^4v^2}{w}}{\dfrac{uv^3}{w^2}}$

77. $\dfrac{\dfrac{x^6y^3}{t^4}}{\dfrac{x^2y^2}{t}}$

78. $\dfrac{2w-\dfrac{w}{4}}{6-\dfrac{w}{4}}$

79. $$\dfrac{1 - \dfrac{1}{y} - \dfrac{12}{y}}{1 - \dfrac{6}{y} + \dfrac{8}{y}}$$

80. $$\dfrac{x + \dfrac{25}{x + 10}}{1 - \dfrac{5}{x + 10}}$$

Solve the following equations: [Section 7.6]

81. $\dfrac{y}{4} - \dfrac{y}{5} = \dfrac{3}{20}$

82. $\dfrac{6p}{7} - \dfrac{5p}{14} = 1$

83. $\dfrac{7}{2}t - \dfrac{4}{9}t = \dfrac{2}{3}$

84. $\dfrac{4w}{5} + \dfrac{1}{6} = \dfrac{7w}{3} - \dfrac{13}{10}$

85. $\dfrac{2v - 5}{4} = 3 - v$

86. $\dfrac{z + 6}{7} - \dfrac{3}{14} = \dfrac{5 - 4z}{21}$

87. $\dfrac{5}{4t} - \dfrac{3}{8t} = \dfrac{1}{2}$

88. $\dfrac{12}{m - 2} = 9 + \dfrac{m}{m - 2}$

89. $\dfrac{1}{q + 4} = \dfrac{q}{3q + 2}$

90. $\dfrac{5}{v^2 + 5v + 6} - \dfrac{2}{v^2 - 2v - 8} = \dfrac{3}{v^2 - 16}$

Solve the following: [Section 7.7]

91. The numerator of a fraction is 3 less than the denominator. If 5 is added to the numerator and 5 is subtracted from the denominator, the result is $\dfrac{3}{2}$. Find the original fraction.

92. The difference of an integer and ten times its reciprocal is 3. Find the integer.

93. George can write a chapter for a mathematics textbook in 30 days. Leon can write a chapter in 45 days. How long would it take them to write a chapter together?

94. The Gulf Stream is an ocean current off the eastern coast of the United States. A Coast Guard cutter (ship) can sail 200 miles against the current in the same time that it takes it to sail 260 miles with the current. If the current is 3 mph, how fast can the cutter sail in still water?

95. It is 510 road miles from Denver, Colorado, to Salt Lake City, Utah. A truck leaves Denver for Salt Lake City at 4:00 A.M. At the same time, an automobile leaves Salt Lake City for Denver. If the truck travels at 45 mph and the auto travels at 55 mph, what time will it be when the auto and truck meet? (Give answer to the nearest minute.)

Chapter 7   Test

Find the LCM of the following:

1. 12 and 16

2. $9x^2y^4$ and $15xy^3$

Find the missing numerator so that the fractions are equal.

3. $\dfrac{3t}{4u^3} = \dfrac{?}{20u^5w}$

4. $\dfrac{6p}{q^2 + 4q} = \dfrac{?}{q(q^2 - 16)}$

Find the least common denominator of each of the following. Then write each fraction as an equivalent fraction with the LCD as its denominator.

5. $\dfrac{2}{9}, \dfrac{7}{18},$ and $\dfrac{4}{27}$

6. $\dfrac{7a}{a^2 - 3a - 18}$ and $\dfrac{b}{a + 3}$

Find the following sums or differences. Reduce the answers to the lowest terms.

7. $\dfrac{9}{7} - \dfrac{5}{7} + \dfrac{1}{7}$

8. $\dfrac{r^3}{r^2 + 16} + \dfrac{r - r^3}{r^2 + 16}$

9. $\dfrac{5}{6} - \dfrac{7}{12}$

10. $\dfrac{2}{5} - \dfrac{3}{10} + \dfrac{4}{15}$

11. $\dfrac{17}{4} - 6 + 4\dfrac{5}{6}$

12. $\dfrac{v + 9}{3v} - \dfrac{v - 6}{9v^2}$

13. $\dfrac{5w}{3w - 12} + \dfrac{7w - 1}{5w - 20}$

14. $\dfrac{t + 3}{t^2 - 10t + 25} - \dfrac{t - 4}{2t^2 - 50}$

Find the perimeters of each of the following triangles the lengths of whose sides are given:

15. $2\dfrac{3}{5}$ in., $3\dfrac{7}{10}$ in., $4\dfrac{2}{5}$ in.

16. $\dfrac{8}{2a + 3}$ m, $\dfrac{5}{a - 2}$ m, $\dfrac{6}{2a + 3}$ m

Find the perimeters of the following rectangles the lengths of whose sides are given:

17. $L = \dfrac{6}{x}$ ft, $W = \dfrac{2}{y}$ ft

18. $L = \dfrac{4m}{3m - n}$ yd, $W = \dfrac{3n}{2m + 3n}$ yd

19. A window is $69\frac{1}{8}$ inches wide and $47\frac{3}{5}$ inches high. A decorative molding that costs $0.02 per inch is put around the window. Find the cost of the molding to the nearest cent.

Simplify the following complex fractions:

20. $\dfrac{\dfrac{1}{8}}{\dfrac{5}{12}}$

21. $\dfrac{\dfrac{7}{12} + \dfrac{2}{3}}{\dfrac{9}{16} - \dfrac{6}{24}}$

22. $\dfrac{\dfrac{m^2}{3z}}{\dfrac{m}{4z}}$

23. $\dfrac{8 - \dfrac{2u}{3}}{5u + \dfrac{u}{3}}$

Solve the following equations:

24. $\dfrac{5v}{6} - \dfrac{1}{4} = \dfrac{4}{3} - \dfrac{7v}{12}$

25. $\dfrac{p - 7}{7} + \dfrac{9}{14} = \dfrac{6 - 5p}{2}$

26. $\dfrac{11}{t - 3} + 6 = \dfrac{7t}{t - 3}$

27. $\dfrac{4}{w^2 + 7w + 10} + \dfrac{3}{w^2 - 4} = \dfrac{14}{w^2 + 7w + 10}$

Solve each of the following:

28. Two numbers differ by 4. If $\frac{3}{4}$ of the smaller number is increased by $\frac{1}{2}$ of the larger number, the sum is 2. Find the numbers.

29. Working together, it takes Elena and Eduardo 10 days to do an architectural project. If Elena can do the project by herself in 15 days, how long would it take Eduardo to do the project working by himself?

30. An air cargo plane and an airliner take off at the same time from the same airport. They fly in opposite directions. The air cargo plane flies at 420 mph and the airliner flies at 530 mph. How long is it before the two airplanes are 1900 miles apart?

Ratios, Percents, and Applications

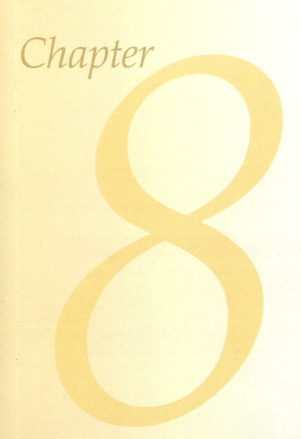

Chapter

8

The idea of ratio is not only very old but very useful in applied mathematics. Ratios have been used to describe the relationships between numbers and lengths for thousands of years. Ratios occur in the fundamental mathematics of arithmetic, in geometry to compare lengths of parts of geometric figures, and in higher mathematics such as algebra, trigonometry, and calculus. Other applications of ratios occur in the study of music. Architects are interested in the ratio of the rise to the width of a roof. Engineers are interested in things like gear ratios. Business executives are interested in earnings ratios. Chemists are interested in the ratios by which molecules combine to make compounds. Physicists are interested in ratios of atoms and forces. Many applied problems that seem impossible at first glance yield to a solution involving ratios.

The idea of percent or per hundred came about because of the need for business to calculate profit and loss and for government to levy taxes. Even though the word percent comes from the Latin phrase *per centum* meaning *per hundred,* the idea of using a fraction with a denominator of 100 did not become popular until well into the sixteenth century in both Europe and the East. As people of the Middle Ages became more interested in trade, money became available for lending (at interest) and governments found a new source of revenue (taxing business). Thus, there was a need for one type of fraction to handle both needs. The symbol for percent, %, is relatively new since it was gradually developed from older symbols over the last century.

Ratios and Rates

O B J E C T I V E S

When you complete this section, you will be able to:

a. Recognize the ratio of two or more numbers.

b. Simplify ratios to lowest terms.

c. Recognize the rate of two measurements.

d. Simplify rates to lowest terms.

e. Recognize and use unit pricing.

Introduction In this section, we will explore a simple but important concept in applied mathematics. It is useful in solving problems that otherwise could be complicated. It is defined as follows:

Ratio

> **DEFINITION Ratio**
>
> A ratio is a comparison of two or more quantities by division. That is, if *a* and *b* are quantities, then *a* : *b* is a ratio of *a* to *b*. If *a, b, c* are quantities, then *a* : *b* : *c* is a ratio of *a* to *b* to *c*.

The quantities in a ratio may be either numbers or measurements. If two quantities have the same unit of measurement, then the ratio is a comparison of numbers. For example, 7 meters compared to 10 meters becomes the ratio 7 to 10. A ratio like 7 to 10 is in the form of a **comparison**. Ratios may be written in **fractional** form as well. For example, 2 : 3 can be written as $\frac{2}{3}$.

There may be more than two quantities in a ratio. Suppose that we have the results of a test. There are 5 As, 8 Bs, 12 Cs, 4 Ds, and 2 Fs on the test. Then the ratio of As to Bs to Cs to Ds to Fs is 5 : 8 : 12 : 4 : 2. Ratios may be written in several ways, but they all mean the same thing. If there are 5 men and 8 women in the math lab, we say that the ratio of men to women is 5 to 8, or 5 : 8, or $\frac{5}{8}$, or five men to eight women. We can summarize this in the form of a table.

Ways to Write Ratios

Comparison	Colon	Fraction	Words
5 to 8	5 : 8	$\frac{5}{8}$	Five to eight.
12 to 5	12 : 5	$\frac{12}{5}$	Twelve to five.
3 to 4 to 7	3 : 4 : 7	———	Three to four to seven.

Note: Ratios with more than two quantities cannot be written in fractional notation.

Ratios in Lowest Terms

As with rational numbers, ratios are in lowest terms when the parts of a ratio are integers that contain no common factors other than the number 1.

Simplifying Ratios Ratios are usually simplified or reduced to lowest terms just as fractions are reduced to lowest terms. Recall from Chapter 6 that we write the numerator and the denominator of the fractions in terms of prime factors and divide by the factors common to both to reduce to lowest terms. We do the same with ratios.

Processes for Simplifying Ratios

a. If the ratio is in fractional form, we divide the numerator and the denominator by the factors common to both. For example,

$$\frac{4}{6} = \frac{2 \cdot 2}{2 \cdot 3} = \frac{2}{3} \text{ or 2 to 3 or } 2 : 3.$$

b. If the ratio is in another form such as $a : b$ or a to b, we divide each part by the factors common to both. For example, $4 : 6$ becomes $2 \cdot 2 : 2 \cdot 3$. Dividing by the common factor of 2, we have $\frac{2 \cdot 2}{2} : \frac{2 \cdot 3}{2}$ or $2 : 3$.

Let us look at a few examples.

Example 1

Simplify the following ratios:

a. $15 : 10 =$ Write each number in terms of prime factors.

$3 \cdot 5 : 2 \cdot 5 =$ Divide both parts by the common factor of 5.

$\dfrac{3 \cdot \cancel{5}}{\cancel{5}} : \dfrac{2 \cdot \cancel{5}}{\cancel{5}} =$ Simplify.

$3 : 2$ Hence, $15 : 10$ is the same as $3 : 2$.

b. $\dfrac{21}{27} =$ Write the numerator and the denominator in terms of prime factors.

$\dfrac{3 \cdot 7}{3 \cdot 3 \cdot 3} =$ Divide by the common factor of 3.

$\dfrac{\cancel{3} \cdot 7}{\cancel{3} \cdot 3 \cdot 3} =$ Multiply remaining factors.

$\dfrac{7}{9}$ Hence, $\dfrac{21}{27} = \dfrac{7}{9}$.

c. 63 to $9 =$ Write each number in terms of prime factors.

$3 \cdot 3 \cdot 7$ to $3 \cdot 3 =$ Divide both parts by the common factors.

$\dfrac{\cancel{3} \cdot \cancel{3} \cdot 7}{\cancel{3} \cdot \cancel{3}}$ to $\dfrac{\cancel{3} \cdot \cancel{3}}{\cancel{3} \cdot \cancel{3}} =$ Simplify.

7 to 1 Hence, 63 to 9 is the same as 7 to 1.

Practice Exercises

Simplify the following ratios:

1. $16 : 24$ **2.** $\dfrac{25}{45}$ **3.** 32 to 40

Often, we need to write English sentences as ratios so that we may perform the necessary mathematical operations.

Example 2

Write each of the following as ratios in lowest terms:

a. At a large university there are 30,000 students. There are 25,000 men and 5000 women. What is the ratio of men to women?

Solution:
We compare the number of men students to the number of women students. As a comparison, the ratio is of the form:

"number of men" to "number of women"   Substitute. (This could be written in any of the three forms.)

25,000 to 5000   Simplify. Divide by the common factor of 1000.

$\dfrac{25{,}000}{1000}$ to $\dfrac{5000}{1000}$ Simplify.

25 to 5   Factor 25 into prime factors.

$5 \cdot 5$ to 5   Divide by the common factor of 5.

$\dfrac{\cancel{5} \cdot 5}{\cancel{5}}$ to $\dfrac{\cancel{5}}{\cancel{5}}$ Simplify.

5 to 1   Hence, the ratio of men students to women students is 5 to 1.

A ratio of 5 men to 1 woman means that there are five times as many men as women.

b. One recipe for homemade cereal calls for 6 cups of rolled oats, 1 cup of sunflower seeds, 2 cups of chopped walnuts, and 4 cups of dried fruit. Write the ratio of the ingredients in the order given.

Solution:
Since they are all measured in cups, we have $6 : 1 : 2 : 4$.

Practice Exercises

Write each of the following as ratios in lowest terms:

4. A citrus grove produced 18,000 boxes of oranges and 12,000 boxes of grapefruit. What is the ratio of the number of boxes of oranges to the number of boxes of grapefruit?

5. A recipe for fruit punch calls for 96 ounces of water, 16 ounces of tea, 12 ounces of frozen lemonade, 64 ounces of cranberry juice, 32 ounces of apple juice, and 16 ounces of orange juice. Write the ratio of the ingredients in the order given.

Unless a special case is called for, a ratio should be simplified so that the parts of the ratio are natural numbers.

Example 3

Write each of the following as a ratio of whole numbers:

a. In a one-liter beaker, we have a solution that is $\frac{2}{5}$ alcohol and $\frac{3}{5}$ water. What is the ratio of alcohol to water?

Solution:
Compare the part that is alcohol to the part that is water. Both of the measurements are in terms of liters, that is $\frac{2}{5}$ liter of alcohol and $\frac{3}{5}$ liter of water.

Therefore, the units are the same and we write

$$\frac{2}{5} \text{ alcohol to } \frac{3}{5} \text{ water, or}$$

$$\frac{2}{5} : \frac{3}{5} \qquad \text{Multiply both parts of the ratio by 5.}$$

$$5\left(\frac{2}{5}\right) : 5\left(\frac{3}{5}\right) = 2 : 3 \qquad \text{Hence, the ratio of alcohol to water is } 2 : 3.$$

This means that for every 2 parts of alcohol there are 3 parts of water.

Note: This could also be done using complex fractions as follows.

$$\frac{\dfrac{2}{5}}{\dfrac{3}{5}} = \qquad \text{Multiply numerator and denominator by 5.}$$

$$\frac{\dfrac{2}{5} \cdot 5}{\dfrac{3}{5} \cdot 5} = \qquad \text{Multiply.}$$

$$\frac{2}{3} \qquad \text{Answer.}$$

b. Recently a research study showed that 2.4 out of every 100 women giving birth at a certain hospital were HIV-positive. Write the ratio of HIV-positive women giving birth to the total number of women giving birth.

Solution:
The number of HIV-positive women to the total was 2.4 : 100. The number 2.4 is not a natural number. We multiple both parts by 10 to make 2.4 a whole number. That is,

2.4 : 100	Multiple both parts by 10.
2.4(10) : 100(10)	Complete multiplication.
24 : 1000	Reduce to lowest terms.
3 : 125	This means that 3 out of every 125 women tested were HIV-positive.

Practice Exercises

Write each of the following as a ratio of whole numbers.

6. Homemade candy is composed of $\frac{1}{6}$ cup of chocolate, $\frac{2}{6}$ cup of nuts, and $\frac{3}{6}$ cup of caramel. What is the ratio of chocolate to caramel?

7. At a certain community college, 56.6 out of 100 faculty members are male. What is the ratio of men to total faculty?

Chemists and other people who are concerned about pollution of the environment by toxic chemicals often use ratios that have very large denominators. These denominators may be thousands, millions, or even billions. Our goal is to have natural numbers in both parts of the ratio.

Example 4

Solve the following.

a. Fish with less than 5 parts of mercury per 10,000,000 parts of flesh are considered safe to eat. Is a fish with .3 parts of mercury per 100,000 parts of flesh safe to eat?

Solution:
We need to know if .3 parts per 100,000 is less than 5 parts per 10,000,000. Therefore, we must have the same number in the second parts of the ratios so that we can compare the first parts.

.3 : 100,000	Multiply both parts by 100 to get 10,000,000 in the second part.
.3(100) : 100,000(100)	Perform multiplication.
30 : 10,000,000	Since the second part is 10,000,000, we can compare this ratio to 5 : 10,000,000.

Is 30 : 10,000,000 less than 5 : 10,000,000?

No!

Conclusion: Do not eat the fish.

Practice Exercise

8. If the level of insecticide to total fluid is over 15 parts per 1,000,000, the fluid cannot be processed for release into streams and rivers. Can fluid that has a level of .12 parts insecticide per 1000 parts of fluid be processed?

Rates   Ratios that have different units of measurement are called **rates**. Rates are usually expressed in lowest terms and use the words "per" or "for every." Since rates are a **Simplifying rates**   special type of ratio, all of the properties that we have discussed for ratios are true for rates as well. Rates are different from ratios in that they are often expressed in a form where the second part is equal to 1 and the ratio appears as a number multiplied by the ratio of the units.

Example 5

Express each ratio with the second part equal to 1.

a. Professor Smart drove her automobile 240 miles in six hours. At what average rate did she drive in miles per hour?

Solution: $\dfrac{240 \text{ miles}}{6 \text{ hours}} =$ Ratio of miles to hours. Reduce to lowest terms, but leave the denominator as 1.

$\dfrac{40 \text{ miles}}{1 \text{ hour}} =$ This is read, "40 miles in 1 hour."

$\dfrac{40 \text{ miles}}{\text{hour}} =$ This is read, "40 miles per hour."

b. There are 160 students in five sections of Integrated Algebra and Arithmetic. What is the average number of students per section?

Solution: $\dfrac{160 \text{ students}}{5 \text{ sections}} =$ Ratio of students to sections. Reduce to lowest terms but leave the denominator as 1.

$\dfrac{32 \text{ students}}{1 \text{ section}} =$ This is read "32 students in 1 section."

$\dfrac{32 \text{ students}}{\text{section}} =$ This is read "32 students per section."

Practice Exercises

Express each ratio with the second part equal to 1.

9. If ten children at a day-care school must share 160 pieces of a construction toy, how many pieces are there per child?

10. A family of five people just won $200,000 in the lottery. What are the average winnings per person?

Unit rate, unit price   A rate that expresses the cost compared to one unit is called a **unit price.** Sometimes expressing the second part of a ratio or rate as 1 may result in the first part of the ratio or rate being a fraction or decimal.

Example 6

Express each of the following as a unit price:

a. Jose Valdez paid $360 to take 12 credits at Grand Valley University. How much did he pay per credit?

Solution: $\dfrac{\$360}{12 \text{ credits}}$ Ratio of cost to credits. Divide 12 into 360 and leave the denominator as 1.

$\dfrac{\$30}{1 \text{ credit}}$ This is read, "$30 per 1 credit."

$\dfrac{\$30}{\text{credit}}$ This is read "$30 per credit."

b. A ten-pound bag of potatoes sells for $1.49. What is the unit price rounded to the nearest cent?

Solution: $\dfrac{\$1.49}{10 \text{ pounds}} =$ Ratio of cost to the number of pounds. Change $1.49 to 149 cents.

$\dfrac{149 \text{ cents}}{10 \text{ pounds}}$ Divide 10 into 149 and write the denominator as 1.

$\dfrac{14.9 \text{ cents}}{1 \text{ pound}} =$ Round to the nearest cent.

$\dfrac{15 \text{ cents}}{\text{pound}}$ Therefore, the potatoes cost 15 cents per pound.

Practice Exercises

Express each of the following as a unit price to the nearest cent:

11. If 6 ounces of yogurt costs $0.49, what is the price per ounce?

12. If 24 hamburger rolls cost $1.79, what is the price per hamburger roll?

Exercise Set 8.1

Simplify the following ratios:

1. 8 to 12

2. 4 to 24

3. 21 : 35

4. 18 : 36

5. 30 to 12

6. 48 to 18

7. 27 : 9

8. 45 : 20

9. $\dfrac{15}{35}$

10. $\dfrac{14}{21}$

11. $\dfrac{48}{36}$

12. $\dfrac{72}{42}$

13. $\dfrac{22}{55}$

14. $\dfrac{34}{68}$

15. $\dfrac{52}{28}$

16. $\dfrac{65}{39}$

Write each of the following as a ratio in simplest form:

17. One year an auto dealer sold 500 cars and 245 trucks. What was the ratio of cars to trucks?

18. Last month the college bookstore sold 624 T-shirts and 156 sweatshirts. What was the ratio of T-shirts to sweatshirts?

19. In a county referendum, 50,000 people voted for higher taxes and 65,000 voted against higher

taxes. What was the ratio of people who voted for the higher taxes to those who voted against higher taxes?

20. On a freight train, there are 72 flatcars and 42 boxcars. What is the ratio of boxcars to flatcars?

21. Gasohol is $\frac{1}{5}$ alcohol and $\frac{4}{5}$ gasoline. What is the ratio of gasoline to alcohol?

22. At a certain temperature, an antifreeze mixture should be $\frac{4}{10}$ antifreeze and $\frac{6}{10}$ water. What is the ratio of antifreeze to water?

23. A recipe calls for 3 cups of water and $\frac{3}{4}$ cup of grits. What is the ratio of water to grits? What is the ratio of grits to water?

24. A biscuit recipe calls for $2\frac{1}{4}$ cups of baking mix and $\frac{2}{3}$ cup of milk. What is the ratio of mix to milk? What is the ratio of milk to mix?

25. In a certain baking mix, there are 140 milligrams of sodium in $\frac{1}{2}$ ounce of mix. What is the ratio of sodium to mix?

26. In a popular cereal, there is .5 gram of soluble fiber and 1.0 gram of insoluble fiber. What is the ratio of soluble to insoluble fiber?

27. At an urban community college, 20.2 out of 100 faculty members have doctorate degrees. What is the ratio of doctorate-holding faculty to total faculty?

28. A taxpayer pays 0.007 of a dollar for every dollar of assessed value of his house for school taxes. What is the ratio of tax to value of the house?

29. Water that has over a level of 1 part of carcinogen (cancer-producing agent) per 1 billion parts of water should not be consumed. If a sample contains .006 parts of carcinogen per million parts of water, is it safe to drink?

30. An acceptable level of emission of lead from an incinerator is 2.788 grams per second. What is the ratio of grams to seconds?

Express each of the following with the second part equal to 1. If necessary, round answers to the nearest cent.

31. Dr. Small drove 350 miles in 7 hours. At what rate did he drive his car?

32. Hank Diaz drove 320 miles and used 8 gallons of gasoline. What is the number of miles per gallon?

33. Four students pooled their money and bought a case of sodas. If there are 24 cans of soda in a case, what is the number of cans per student?

34. A study shows that 36,000 people rode in 20,000 cars. What is the average number of people per car?

Express each of the following as a unit price. If necessary, round answers to the nearest cent.

35. Six bananas cost $0.24. What is the cost per banana?

36. An 8-pound bag of ice costs $1.29. What is the cost per pound?

37. Eight sticks of cheese cost $2.15. What is the cost per stick?

38. A box containing 20 ounces of cereal costs $2.95. What is the cost per ounce?

39. A prescription medicine costs $11.19 for 30 capsules. What is the cost per capsule?

40. A bottle with 100 aspirin tablets costs $2.49. What is the cost per tablet?

41. A hotel charges $250 for a room with four people. What is the cost per person?

42. A family bought a season pass that permits two adults and three children to sit in the end zone at the local college football games. If the pass cost $140, what is the average cost per person?

Challenge Exercises:

43. A can of mixed nuts contained 250 peanuts, 100 cashews, 50 Brazil nuts, 75 almonds, 50 filberts, and 100 pecans. What is the ratio of peanuts to cashews, to Brazil nuts, to almonds, to filberts, to pecans?

44. If $\frac{1}{4}$ cup of cereal weighs 28.35 grams, what is the ratio of grams per cup?

Writing Exercise:

45. Name and give an example of at least five situations that can be described using a ratio.

| Section 8.2 | Proportions |

OBJECTIVES

When you complete this section, you will be able to:

a. Recognize a proportion.

b. Simplify proportions.

c. Solve proportions.

d. Solve application problems using proportions.

Introduction   In the last section, we looked at the general idea of ratio. In this section, we emphasize the idea of a ratio as a fraction. Since some ratios are equal, we make the following definition:

Proportion

DEFINITION   Proportion

A **proportion** is an equation that states that two or more ratios are equal. In symbols, $\frac{a}{b} = \frac{c}{d}$ is a proportion.

We have worked with proportions before. When we studied equivalent or equal fractions, we were studying proportions. Determining if two fractions form a proportion is the same as determining if two fractions are equal. We cross multiply and if the cross products are the same, the two fractions are equal and we have a proportion.

Procedure for Cross Multiplication

If $\frac{a}{b} = \frac{c}{d}$, then $a \cdot d = b \cdot c$ if $b, d \neq 0$. It is also true that if $a \cdot d = b \cdot c$, then $\frac{a}{b} = \frac{c}{d}$.

For two ratios $\frac{a}{b}$ and $\frac{c}{d}$, and , if $a \cdot d = b \cdot c$, then $\frac{a}{b} = \frac{c}{d}$ and we have a proportion.

Note: Suppose we begin with the proportion $\frac{a}{b} = \frac{c}{d}$ and we multiply both sides by the LCD which is bd. We get $bd \cdot \frac{a}{b} = bd \cdot \frac{c}{d}$, which simplifies to $a \cdot d = b \cdot c$ or the same result that we get when we cross multiply.

Example 1

Determine if the following fractions can form a proportion:

a. $\frac{4}{5}$ and $\frac{16}{20}$ Cross multiply.

$\frac{4}{5} \times \frac{16}{20}$ Multiply.

$4(20) \stackrel{?}{=} 5(16)$

$80 = 80$ Therefore, the fractions are equal and can form a proportion. So $\frac{4}{5} = \frac{16}{20}$.

b. $\frac{2}{3}$ and $\frac{9}{12}$

$\frac{2}{3} \times \frac{9}{12}$ Cross multiply.

$2(12) \stackrel{?}{=} 3(9)$

$24 \neq 27$ Therefore, the fractions are not equal and cannot form a proportion. So $\frac{2}{3} \neq \frac{9}{12}$.

Practice Exercises

Determine if the following fractions can form a proportion:

1. $\frac{8}{9}$ and $\frac{30}{36}$

2. $\frac{16}{28}$ and $\frac{4}{7}$

Answers:

Practice Exercises 1–2: 1. no 2. yes

If more practice is needed, do the Additional Practice Exercises.

Additional Practice Exercises

Determine if the following fractions can form a proportion:

a. $\dfrac{3}{4}$ and $\dfrac{15}{30}$

b. $\dfrac{18}{21}$ and $\dfrac{12}{14}$

Simplifying proportions

Proportions may be used to express the relationship between both known and unknown quantities. If one of the quantities of a proportion is unknown, we may solve the proportion for that quantity by cross multiplying and solving the resulting equation. To simplify a proportion means to simplify one or more of the ratios in the proportion.

Solving proportions

Procedure: Solving Proportions

When solving a proportion, perform the following steps:
1. If desired, reduce all fractions to lowest terms.
2. Cross multiply the proportion.
3. Solve the resulting equation.

Example 2

Solve the following proportions for x:

a. $\dfrac{x}{20} = \dfrac{9}{12}$ Reduce $\dfrac{9}{12}$ to lowest terms.

$\dfrac{x}{20} = \dfrac{3}{4}$ Cross multiply.

$4x = 20(3)$ $20(3) = 60$.

$4x = 60$ Divide both sides of the equation by 4.

$x = 15$ Solution.

c. $\dfrac{2x+2}{4} = \dfrac{3x+1}{5}$ Cross multiply.

$5(2x+2) = 4(3x+1)$ Simplify both sides of the equation.

$10x + 10 = 12x + 4$ Subtract 10x from both sides of the equation.

$10x - 10x + 10 = 12x - 10x + 4$ Simplify both sides.

$10 = 2x + 4$ Subtract 4 from both sides of the equation.

$10 - 4 = 2x + 4 - 4$ Simplify both sides of the equation.

$6 = 2x$ Divide both sides of the equation by 2.

$3 = x$ Solution.

b. $\dfrac{9}{t} = \dfrac{3}{14}$ Cross multiply.

$9(14) = 3t$ Multiply 9 and 14.

$126 = 3t$ Divide both sides of the equation by 3.

$42 = t$ Solution.

d. $\dfrac{x+2}{3} = \dfrac{2}{x-3}$ Cross multiply.

$(x+2)(x-3) = 3 \cdot 2$ Multiply.

$x^2 - x - 6 = 6$ Subtract 6 from both sides of the equation.

$x^2 - x - 12 = 0$ Factor.

$(x-4)(x+3) = 0$ Set each factor equal to 0.

$x - 4 = 0, x + 3 = 0$ Solve each linear equation.

$x = 4, x = -3$ Therefore, 4 and −3 are the solutions.

Practice Exercises

Solve the following proportions:

3. $\dfrac{12}{16} = \dfrac{y}{20}$ **4.** $\dfrac{y}{6} = \dfrac{30}{15}$ **5.** $\dfrac{4}{3} = \dfrac{2x+2}{2x-1}$ **6.** $\dfrac{x-4}{2} = \dfrac{8}{x+2}$

If more practice is needed, do the Additional Practice Exercises.

Additional Practice Exercises

Solve the following proportions.

c. $\dfrac{6}{y} = \dfrac{12}{8}$ **d.** $\dfrac{6}{5} = \dfrac{18}{v}$ **e.** $\dfrac{3}{2x+1} = \dfrac{5}{4x-1}$ **f.** $\dfrac{4}{x+5} = \dfrac{x-3}{5}$

Fractions and decimals may occur in a proportion. We treat them the same way as whole numbers and integers.

Example 3

Solve the following proportions:

a. $\dfrac{\frac{1}{2}}{5} = \dfrac{x}{40}$ Cross multiply.

$\dfrac{1}{2}(40) = 5x$ Multiply.

$20 = 5x$ Divide both sides of the equation by 5.

$4 = x$ Solution.

b. $\dfrac{y}{4.2} = \dfrac{8}{3}$ Cross multiply.

$3y = 4.2(8)$ Multiply 4.2 and 8.

$3y = 33.6$ Divide both sides of the equation by 3.

$y = 11.2$ Solution.

Often we have proportions that consist only of variables. They are solved using the same techniques as with constants.

Example 4

Solve each of the following for the indicated variable:

a. $\dfrac{C}{D} = \pi;\ C$

Solution:

$\dfrac{C}{D} = \pi$ Think of π as $\dfrac{\pi}{1}$.

$\dfrac{C}{D} = \dfrac{\pi}{1}$ Cross multiply.

$C = D \cdot \pi$ Solution.

b. $\dfrac{T}{P} = V;\ P$

Solution:

$\dfrac{T}{P} = V$ Think of V as $\dfrac{V}{1}$.

$\dfrac{T}{P} = \dfrac{V}{1}$ Cross multiply.

$T = P \cdot V$ Divide both sides of the equation by V.

$\dfrac{T}{V} = P$ Solution.

Practice Exercises

Solve each of the following proportions:

7. $\dfrac{\frac{2}{3}}{\frac{3}{4}} = \dfrac{8}{x}$

8. $\dfrac{7}{2.5} = \dfrac{z}{15}$

Solve for the indicated variable:

9. $D = \dfrac{W}{V}; W$

10. $i = \dfrac{V}{R}; R$

If more practice is needed, do the Additional Practice Exercises.

Additional Practice Exercises

Solve the following proportions:

g. $\dfrac{y}{12} = \dfrac{\frac{5}{3}}{10}$

h. $\dfrac{3.2}{r} = \dfrac{4}{5}$

Solve for the indicated variable:

i. $\dfrac{A}{L} = W; A$

j. $\dfrac{i}{p} = r; p$

Application problems using proportions   There are many applications of proportions. There is more than one correct way of setting up a proportion from an application problem. For example, all of the following are equivalent proportions: $\frac{a}{b} = \frac{c}{d}, \frac{a}{c} = \frac{b}{d}, \frac{d}{c} = \frac{b}{a}, \frac{d}{b} = \frac{c}{a}$. Note that if we cross multiply each of these we get $ad = bc$.

Example 5

Solve each of the following by using a proportion:

a. An automobile travels 224 miles on 8 gallons of gasoline. How far can it travel on 15 gallons of gasoline?

Solution:
We let x = the number of miles the car can travel on 15 gallons. The key to the solution is the assumption that the rate of consumption in the first ratio will be the same for the second ratio. One method of solution is to set up the first ratio by comparing miles to gallons. Then we set up the second ratio by keeping the same pattern of comparing miles to gallons. In words:

$$\frac{\text{First number of miles}}{\text{First number of galllons}} = \frac{\text{Second number of miles}}{\text{Second number of gallons}} \qquad \text{Substitute.}$$

$$\frac{224 \text{ miles}}{8 \text{ gallons}} = \frac{x \text{ miles}}{15 \text{ gallons}} \qquad \text{Drop units.}$$

$$\frac{224}{8} = \frac{x}{15} \qquad \text{Reduce } \frac{224}{8}.$$

$$\frac{28}{1} = \frac{x}{15} \qquad \text{Cross multiply.}$$

$$28(15) = 1(x) \qquad \text{Multiply.}$$

$$420 = x \qquad \text{Therefore, the car can travel 420 miles on 15 gallons of gasoline.}$$

We say that 224 miles is to 8 gallons as 420 miles is to 15 gallons.

Note: Other proportions that could also be used are $\frac{15}{8} = \frac{x}{224}$, $\frac{224}{x} = \frac{8}{15}$ and $\frac{8}{224} = \frac{15}{x}$. Notice that all of these give the same thing when we cross multiply.

b. The state wildlife department wants to estimate the number of deer in a game preserve. In September, they capture 34 deer, tag them, and release them. Later, they capture a sample of 45 deer and find that 17 are tagged. Estimate the number of deer in the preserve.

Solution:
Let y = the number of deer in the preserve. One method is to assume that the ratio of tagged deer to total deer in the sample is the same as the ratio of tagged deer to total deer in the preserve. We know there are 34 tagged deer in the preserve. In words:

$$\frac{\text{Number of tagged deer in sample}}{\text{Number of deer in sample}} = \frac{\text{Number of tagged deer in preserve}}{\text{Number of deer in preserve}}$$

$\dfrac{17 \text{ tagged}}{45 \text{ total}} = \dfrac{34 \text{ tagged}}{y \text{ total}}$	Drop labels.
$\dfrac{17}{45} = \dfrac{34}{y}$	Cannot simplify. Cross multiply.
$17y = 45(34)$	Multiply 45 and 34.
$17y = 1530$	Divide both sides by 17.
$y = 90$	Therefore, there are about 90 deer in the preserve.

Note: Other proportions that could have been used are $\frac{y}{45} = \frac{34}{17}$, $\frac{17}{34} = \frac{45}{y}$, and $\frac{45}{17} = \frac{y}{34}$. Notice that all of them give the same thing when we cross multiply.

c. The distance between two cities is 150 miles. The scaled distance between the cities on a map is drawn as $1\frac{1}{2}$ inches. If the distance between two other cities is 900 miles, what is the scaled distance between them on the map?

Solution:
Let d = the distance between the two cities on the map. The ratio of miles between the cities to inches on the map will remain the same for all distances. Therefore, we set up the ratio of miles between cities to inches on the map as follows. In words:

$\dfrac{\text{First number of miles}}{\text{First number of inches}} = \dfrac{\text{Second number of miles}}{\text{Second number of inches}}$	Substitute.
$\dfrac{150 \text{ miles}}{1\frac{1}{2} \text{ inches}} = \dfrac{900 \text{ miles}}{d \text{ inches}}$	Drop units.
$\dfrac{150}{1\frac{1}{2}} = \dfrac{900}{d}$	Convert mixed number to improper fraction or decimal.
$\dfrac{150}{\frac{3}{2}} = \dfrac{900}{d}$	Cross multiply.
$150d = \dfrac{3}{2}(900)$	Multiply $\frac{3}{2}$ and 900.
$150d = 1350$	Divide both sides by 150.
$d = 9$	Therefore, the scaled distance between the cities on the map is 9 inches.

Practice Exercises

Solve each of the following:

11. A chef knows that he can serve eight people with 3 pounds of fish fillets. How many pounds of fish fillets does he need to serve 32 people?

12. A manufacturer tested 400 randomly chosen widgets as they came off the production line. If she

found 15 defective widgets, how many defective widgets should she expect to find in 2000 widgets?

13. On a map, $1\frac{1}{4}$ inches represents 100 miles on the surface of Earth. What is the distance on Earth between two locations that are 15 inches apart on the map?

Exercise Set 8.2

Determine if the following fractions can form a proportion:

1. $\frac{4}{5}$ and $\frac{12}{15}$ **2.** $\frac{20}{24}$ and $\frac{25}{30}$ **3.** $\frac{2}{3}$ and $\frac{8}{12}$ **4.** $\frac{5}{6}$ and $\frac{4}{3}$

5. $\frac{4}{9}$ and $\frac{20}{45}$ **6.** $\frac{5}{11}$ and $\frac{7}{9}$ **7.** $\frac{7}{12}$ and $\frac{17\frac{1}{2}}{30}$ **8.** $\frac{3}{8}$ and $\frac{15}{40}$

Solve the following proportions:

9. $\frac{4}{5}=\frac{12}{x}$ **10.** $\frac{y}{24}=\frac{25}{30}$ **11.** $\frac{4}{6}=\frac{t}{18}$

12. $\frac{9}{5}=\frac{y}{35}$ **13.** $\frac{y}{8}=\frac{14}{56}$ **14.** $\frac{22}{33}=\frac{n}{44}$

15. $\frac{13}{6}=\frac{t}{42}$ **16.** $\frac{2}{14}=\frac{8}{y}$ **17.** $\frac{28}{16}=\frac{49}{p}$

18. $\frac{56}{16}=\frac{q}{10}$ **19.** $\frac{9}{w}=\frac{22.5}{1.5}$ **20.** $\frac{y}{.7}=\frac{5}{7}$

21. $\frac{y}{14}=\frac{\frac{3}{7}}{2}$ **22.** $\frac{\frac{6}{5}}{8}=\frac{r}{20}$ **23.** $\frac{3\frac{1}{2}}{x}=\frac{14}{12}$

24. $\frac{9}{m}=\frac{6}{4\frac{2}{3}}$ **25.** $\frac{.6}{t}=\frac{5}{8}$ **26.** $\frac{7}{.5}=\frac{n}{3}$

27. $\frac{14}{3.5}=\frac{w}{5}$ **28.** $\frac{v}{8}=\frac{13}{2.6}$ **29.** $\frac{2}{3}=\frac{8}{x+8}$

30. $\frac{3}{4}=\frac{9}{x+5}$ **31.** $\frac{x+5}{6}=\frac{7}{3}$ **32.** $\frac{x+4}{20}=\frac{3}{4}$

33. $\frac{3}{4}=\frac{x+1}{x+3}$ **34.** $\frac{2}{3}=\frac{x+1}{x+6}$ **35.** $\frac{3}{2x}=\frac{5}{3x+2}$

36. $\frac{4}{3x-3}=\frac{7}{4x+1}$ **37.** $\frac{4x-4}{8}=\frac{2x+1}{5}$ **38.** $\frac{7x-1}{8}=\frac{2x+4}{3}$

39. $\dfrac{x-3}{2} = \dfrac{5}{x+6}$

40. $\dfrac{x+5}{7} = \dfrac{1}{x-1}$

41. $\dfrac{5}{2x+3} = \dfrac{x-5}{3}$

42. $\dfrac{2}{3x+2} = \dfrac{x-3}{17}$

Solve for the indicated variable:

43. $P = \dfrac{W}{t}; \; W$

44. $\dfrac{q}{t} = i; \; t$

45. $\dfrac{PV}{T} = K; \; P$

46. $P = \dfrac{F}{LW}; \; F$

Solve each of the following:

47. A truck travels 440 miles in 11 hours. How far will the truck travel in 7 hours?

48. An airplane uses 350 pounds of fuel to fly 1400 miles. How much fuel will it use to fly 3500 miles?

49. In 8 months, Joel can read 5 novels. How many novels can he read in 12 months?

50. A machine can produce 160 toys in three hours. How long will it take to produce 400 toys?

51. A baseball player got 6 hits in 14 innings of play. How many innings must he play to get 33 hits?

52. Martha can do 32 mathematics exercises in 15 minutes. How long will it take her to do 96 exercises?

53. Two rivers that are 270 miles apart are drawn as 3 inches apart on a map. How many inches apart on the map will two locations that are 675 miles apart on Earth be?

54. On a map, $\frac{3}{4}$ inch represents a distance of 75 miles on Earth's surface. What is the distance on Earth between two cities which are 6 inches apart on the map?

55. The state game commission wants to estimate how many trout there are in a lake. They release 145 tagged trout into the lake. Six months later, they return and capture 160 trout, 58 of which are tagged. Estimate the number of trout in the lake.

56. The wildlife commission wants to determine how many alligators are in a swamp. Conservation officers capture, tag, and release 33 alligators. One year later, they capture 60 alligators and find that 12 are tagged. Approximately how many alligators are in the swamp?

57. A model airplane is built at a scale of 1 : 40. If a wing on the model is six inches long, how long is the wing of the airplane in feet?

58. A model car is built at a scale of 1 : 15. If the bumper of the model car is four inches long, how long is the bumper of the car in feet?

59. If a tile store will install 23 square feet of tile for $92, how much would they charge to install 62 square feet?

60. If a VCR counter registers 660 after playing 22 minutes, what will it register after playing 36 minutes?

Challenge Exercises:

61. One of the professors at Urban University can walk 3 miles in 39 minutes. Assuming that he walks the same pace, how long will it take him to walk 5 miles?

62. A physician can see 40 patients in 7 hours. How many can he see in 35 hours (one week)? At the rate of $60 per patient, how much can he make in 50 weeks (one year)?

63. A 44-pound bag of fertilizer will cover 5000 square feet of lawn. How many pounds of fertilizer are needed to cover 17,500 square feet of lawn? If the fertilizer costs $13.95 a bag, how much will it cost to fertilize the 17,500 square feet?

Writing Exercises:

64. What is the underlying assumption on which all of the preceding proportion exercises are based?

65. Name three types of situations where proportions are used.

66. Explain why $\frac{a}{b} = \frac{c}{d}$ and $\frac{a}{c} = \frac{b}{d}$ are essentially the same proportion.

Writing Exercises or Group Projects:

67. Interview science teachers or students, nursing teachers or students, and other professionals and make a list of five situations where proportions are used in their fields/professions.

68. There is a special ratio called the Golden Ratio that often appears in art, architecture, mathematics, and many other fields. Go to the library, research the topic, and write a one-page report on the Golden Ratio.

Section 8.3	**Percent**

OBJECTIVES

When you complete this section, you will be able to:

a. Recognize percent notation.

b. Convert percent notation to fractions.

c. Convert percent notation to decimals.

d. Convert decimals to percent notation.

e. Convert fractions to percent notation.

f. Add and subtract in percent notation.

Introduction

Throughout recorded history, business people and governments have looked for a method that they could use to calculate profits, taxes, and so on. The most commonly used method is the use of the fraction $\frac{1}{100}$. Over the years the number of hundredths has become known in the Western world as percent. The word *percent* comes from the Latin phrase *per centum*, or per hundred, and has a special symbol, %.

Percent Notation

Percent notation is a special way of writing a fraction with a denominator of 100. For any number n, $n\% = \frac{n}{100}$.

Percent notation

This means that all we have to do to change a number in percent notation into a fraction is write it as a fraction with a denominator of 100 and simplify. For example, $17\% = \frac{17}{100}$ and $25\% = \frac{25}{100} = \frac{1}{4}$. However, when converting something like $5\frac{4}{7}\%$ or .06% to a fraction, the arithmetic becomes very messy. For example,

$$5\frac{4}{7}\% = \frac{5\frac{4}{7}}{100} = \frac{\frac{39}{7}}{100} = \frac{39}{7} \div 100 = \frac{39}{7} \cdot \frac{1}{100} = \frac{39}{700}.$$

An alternate method that leads to nicer arithmetic involves thinking of $x\%$ as $x \cdot 1\%$. By definition, $1\% = \frac{1}{100}$. Consequently, $x\% = x \cdot 1\% = x \cdot \frac{1}{100}$. Since we are converting a percent to a fraction, we replace the % sign with the fraction $\frac{1}{100}$.

Percent to fraction

Converting From Percent To a Fraction

To convert from percent notation to a fraction:

a. Think of $x\%$ as $x \cdot 1\%$.

b. Replace 1% with $\frac{1}{100}$.

c. Multiply.

Note: Rather than thinking of $x\%$ as $x \cdot 1\%$ and then replacing 1% with $\frac{1}{100}$, it is a common practice to replace the % sign with $\frac{1}{100}$, which amounts to the same thing.

In using the preceding technique, we also need to recall how to change decimals into fractions.

Example 1

Convert each of the following percents to fractions:

a. 75%   Think of 75% as $75 \cdot 1\%$ and replace 1% with $\frac{1}{100}$.

$75 \cdot \dfrac{1}{100}$ Multiply. Remember, think of 75 as $\frac{75}{1}$.

$\dfrac{75}{100}$ Write numerator and denominator in prime factorizations and divide by common factors. Multiply the remaining factors.

$\dfrac{3 \cdot \cancel{5} \cdot \cancel{5}}{2 \cdot 2 \cdot \cancel{5} \cdot \cancel{5}}$

$\dfrac{3}{4}$ Hence, $75\% = \frac{3}{4}$.

c. 8.5%

$8.5 \cdot \dfrac{1}{100}$ Think of 8.5% as $8.5 \cdot 1\%$ and replace 1% with $\frac{1}{100}$.

$8.5 \cdot \dfrac{1}{100}$ Change 8.5 into $8\frac{5}{10}$.

$8\dfrac{5}{10} \cdot \dfrac{1}{100}$ Reduce $\frac{5}{10}$ to lowest terms.

$8\dfrac{1}{2} \cdot \dfrac{1}{100}$ Change $8\frac{1}{2}$ into an improper fraction.

$\dfrac{17}{2} \cdot \dfrac{1}{100}$ Multiply

$\dfrac{17}{200}$ Hence, $8.5\% = \frac{17}{200}$.

b. .08%   Think of .08% as $.08 \cdot 1\%$ and replace 1% with $\frac{1}{100}$.

$.08 \cdot \dfrac{1}{100}$ Change .08 into the fraction $\frac{8}{100}$.

$\dfrac{8}{100} \cdot \dfrac{1}{100}$ Reduce $\frac{8}{100}$ to lowest terms.

$\dfrac{2}{25} \cdot \dfrac{1}{100}$ Multiply.

$\dfrac{2}{2500}$ Reduce $\frac{2}{2500}$ to lowest terms.

$\dfrac{1}{1250}$ Hence, $.08\% = \frac{1}{1250}$.

d. $5\dfrac{4}{7}\%$ Think of $5\frac{4}{7}\%$ as $5\frac{4}{7} \cdot 1\%$ and change 1% to $\frac{1}{100}$.

     Rewrite $5\frac{4}{7}$ as an improper fraction.

$\dfrac{39}{7} \cdot \dfrac{1}{100}$ Multiply.

$\dfrac{39}{700}$ Hence, $5\frac{4}{7}\% = \frac{39}{700}$.

Practice Exercises

Convert each of the following percents to fractions:

1. 90%

3. 6.8%

2. .09%

4. $8\dfrac{6}{7}\%$

Additional Practice Exercises

Convert each of the following percents to fractions:

a. 62%

c. 2.3%

b. .75%

d. $5\dfrac{7}{12}\%$

Before converting percents to decimals we need to recall from Chapter 6 the procedure for converting fractions to decimals. Remember that we divide the denominator into the numerator, and if necessary, round the answer to a specified number of decimal places. For example, we write $\frac{5}{8}$ as a decimal as follows:

$$8\overline{)5}$$

Write the decimal point behind the 5 and bring the decimal point up to the quotient. Write zeroes after the 5 as needed.

$$
\begin{array}{r}
.625 \\
8\overline{)5.000} \\
-48 \\
\hline
20 \\
-16 \\
\hline
40 \\
-40 \\
\hline
0
\end{array}
$$

Therefore, $\frac{5}{8} = .625$.

If we were to write $\frac{5}{6}$ as a decimal, the division would never terminate, so it is necessary to round the answer to a specified number of decimal places. If we wanted to write $\frac{5}{6}$ as a decimal rounded to the nearest hundredth, we would proceed as follows:

$$6\overline{)5}$$

Write the decimal point behind the 5 and bring the decimal point up to the quotient. Write zeroes after the 5 as needed.

$$
\begin{array}{r}
.833 \\
6\overline{)5.000} \\
-48 \\
\hline
20 \\
-18 \\
\hline
20 \\
-18 \\
\hline
2
\end{array}
$$

Since we want to round to the nearest hundredth, we need to see the thousandths digit.

Round .833 to the nearest hundredth. (See Section 0.8.)

Therefore, $\frac{5}{6} \approx .83$ to the nearest hundredth.

IMPORTANT: Some conversions to repeating decimals occur so often in applications that they should be memorized to help computations. Examples of these are $.\overline{3} = \frac{1}{3}$ and $.\overline{6} = \frac{2}{3}$, $.\overline{1} = \frac{1}{9}$, $.\overline{2} = \frac{2}{9}$, and so on.

Converting Percent to Decimals   We can also use the definition of percent to change a percent into a decimal. However, the arithmetic is much nicer if we use a method similar to the one used to convert percents to fractions. We use the fact that $1\% = \frac{1}{100} = .01$. So we think of $x\%$ as being $x \cdot 1\%$ and replace 1% with $.01$. Since we are converting a percent to a decimal, we replace the % sign with the decimal $.01$.

Converting From Percent to Decimals

To convert from percent notation to a decimal:

a. Think of $x\%$ as $x \cdot 1\%$.

b. Replace 1% with $.01$.

c. Multiply.

Example 2

Convert the following percents to decimals. If necessary, round answers to the nearest hundredths.

a. 67%   Think of 67% as 67 · 1% and replace 1% with .01.
67 · .01   Multiply.
.67   Hence, 67% = .67.

b. 180%   Think of 180% as 180 · 1% and replace 1% with .01.
180 · .01   Multiply.
1.8   Hence, 180% = 1.8.

c. 12.6%   Think of 12.6% as 12.6 · 1% and replace 1% with .01.
12.6 · .01   Multiply.
.126   Hence, 12.6% = .126.

d. $4\frac{5}{7}$%   Think of $4\frac{5}{7}$% as $4\frac{5}{7}$ · 1% and replace 1% with .01.

$4\frac{5}{7}$ · .01   Change $\frac{5}{7}$ into a decimal.

(4.714285...)(.01)   Multiply.

.04714285   Round to the nearest hundredth.

.05   Hence $4\frac{5}{7}$% = .05 to the nearest hundredth.

Note: Since multiplying by .01 moves the decimal point two places to the left, we could shorten the preceding procedure to the familiar "drop the % sign and move the decimal two places to the left." For example, 45% = .45, 174% = 1.74 and $4\frac{5}{7}$% = 4.714285...% = .04714285... = .05 to the nearest hundredth.

Practice Exercises

Convert the following percents to decimals. If necessary, round to the nearest thousandth.

5. 35%   **6.** 9%   **7.** 55.7%   **8.** $11\frac{8}{17}$%

If more practice is needed, do the Additional Practice Exercises.

Additional Practice Exercises

Convert the following percents to decimals:

e. 47%   **f.** 8.4%   **g.** 44.3%   **h.** $\frac{3}{4}$%

Converting decimals to percent   If we were to write 100% as a decimal, we would have 100% = 100 · .01 = 1. Thus 100% = 1. We also know that 1 is the identity for multiplication because the product of any number and 1 is identical to the number itself. Consequently, to convert a decimal to a percent, we will write the decimal as the product of itself and 1 and then change 1 to 100% and multiply. We summarize as follows:

Converting Decimals to Percents

To convert a decimal into percent notation:

a. Write the decimal as the product of itself and 1.

b. Change 1 to 100%.

c. Multiply.

Answers:

Practice Exercises 5–8: 5. .35 6. .09 7. .557 8. .115   **Additional Practice Exercises e–h:** e. .47 f. .084 g. .443 h. .008

Example 3

Convert the following decimals to percent notation:

a. .17   Write .17 as .17 · 1.

.17 · 1   Change 1 into 100%.

.17 · 100%   Multiply.

17%   Hence, .17 = 17%.

b. .008   Write .008 as .008 · 1.

.008 · 1   Change 1 into 100%.

.008 · 100%   Multiply.

.8%   Hence, .008 = .8%.

c. 2.54   Write 2.54 as 2.54 · 1.

2.54 · 1   Change 1 into 100%.

2.54 · 100%   Multiply.

254%   Hence, 2.54 = 254%.

d. .333. . .   Think of .333. . . as (.333. . .)(1).

(.333. . .)(1)   Change 1 into 100%.

(.333. . .)(100%)   Multiply.

33.333. . .%   Replace .333. . . with fractional equivalent, $\frac{1}{3}$.

$33\frac{1}{3}$%   Hence, .333. . . = $33\frac{1}{3}$%.

e. 8   Think of 8 as 8 · 1.

8 · 1   Change 1 into 100%.

8 · 100%   Multiply.

800%   Hence, 8 = 800%.

Note: Since multiplying by 100 moves the decimal two places to the right, we can shorten the above procedure to "move the decimal two places to the right and add the % sign." For example .62 = 62% and .123 = 12.3%.

Practice Exercises

Convert the following decimals to percent notation.

9. .87

10. .006

11. 2.53

12. 5.333...

13. 9

If more practice is needed, do the Additional Practice Exercises.

Additional Practice Exercises

Convert the following decimals to percent notation.

i. .95

j. .055

k. 1.64

l. 4.666...

m. 4

Fractions to percent   To convert a fraction to a percent, we will use a procedure that is similar to the one used to convert a decimal to a percent. If we change 100% to a fraction, we would have $100\% = 100 \cdot \frac{1}{100} = \frac{100}{100}$. Therefore, 100% as a fraction is 1. Consequently, we will write the fraction as the product of itself and 1 and then change 1 to 100% and multiply.

Converting Fractions to Percents:

To convert a fraction to percent notation:

a. Write the fraction as the product of itself and 1.

b. Change 1 to 100%.

c. Multiply, and if necessary, round to a specified number of decimal places.

Answers:

Example 4

Convert the following fractions to percent notation. If necessary, round to the nearest hundredth of a percent.

a. $\dfrac{5}{8}$ Write as $\dfrac{5}{8}$ as $\dfrac{5}{8} \cdot 1$.

$\dfrac{5}{8} \cdot 1$ Replace 1 with 100%.

$\dfrac{5}{8} \cdot 100\%$ Multiply.

$\dfrac{500}{8}\%$ Divide.

62.5%   Therefore, $\dfrac{5}{8} = 62.5\%$.

c. $\dfrac{4}{9}$ Think of as $\dfrac{4}{9}$ as $\dfrac{4}{9} \cdot 1$.

$\dfrac{4}{9} \cdot 1$ Replace 1 with 100%.

$\dfrac{4}{9} \cdot 100\%$ Multiply.

$\dfrac{400}{9}\%$ Divide.

44.4444... or $44\dfrac{4}{9}\%$ Hence, $\dfrac{4}{9} \approx 44\dfrac{4}{9}\%$.

b. $\dfrac{2}{3}$ Write as $\dfrac{2}{3} \cdot 1$.

$\dfrac{2}{3} \cdot 1$ Replace 1 with 100%.

$\dfrac{2}{3} \cdot 100\%$ Multiply.

$\dfrac{200}{3}\%$ Divide.

66.66...% or $66\dfrac{2}{3}\%$ Hence, $\dfrac{2}{3} = 66\dfrac{2}{3}\%$.

d. $\dfrac{3}{7}$ Think of as $\dfrac{3}{7}$ as $\dfrac{3}{7} \cdot 1$.

$\dfrac{3}{7} \cdot 1$ Replace 1 with 100%.

$\dfrac{3}{7} \cdot 100\%$ Multiply.

$\dfrac{300}{7}\%$ Divide.

42.8571%   Round to the nearest hundredth of a percent.

42.86%   Hence, $\dfrac{3}{7} \approx 42.86\%$.

Note: An alternative approach is to first change the fraction into a decimal and then change the decimal into a percent. For example, $\dfrac{1}{8} = .125 = 12.5\%$.

Practice Exercises

Convert the following fractions to percent notation. If necessary, round to the nearest hundredth of a percent.

14. $\dfrac{7}{8}$

15. $\dfrac{1}{6}$

16. $\dfrac{2}{11}$

17. $\dfrac{5}{13}$

If more practice is needed, do the Additional Practice Exercises.

Additional Practice Exercises

Convert the following fractions to percent notation. If necessary, round to the nearest hundredth of a percent.

n. $\dfrac{3}{8}$

o. $\dfrac{5}{9}$

p. $\dfrac{1}{8}$

q. $\dfrac{5}{6}$

Answers:

Additional Practice Exercises n–q: **n.** 37.5% **o.** 55.56% **p.** 12.5% **q.** 83.3%

Practice Exercises 14–17: **14.** 87.5% **15.** 16.67% **16.** 18.18% **17.** 38.46%

Since percent represents hundredths, adding and subtracting percents is the same as adding and subtracting fractions whose common denominator is 100. In doing so, we add the numerators and put the sum over the common denominator of 100. Since the amount of percent is the numerator, we simply add and subtract the percents as with any other rational numbers. For example, $5\% + 23\% = \frac{5}{100} + \frac{23}{100} = \frac{5+23}{100} = \frac{28}{100} = 28\%$. So $5\% + 23\% = (5 + 23)\% = 28\%$.

Example 5

Perform the indicated operation:

a. $9.3\% - 6\%$

Solution:
$9.3\% - 6\% = 3.3\%$ Subtract percents like subtracting rational numbers.

b. $5\% + 9\frac{1}{2}\%$

Solution:
$5\% + 9\frac{1}{2}\% = 14\frac{1}{2}\%$ Add percents like adding rational numbers.

c. Arthur Gomez paid 14% of his salary in Federal income tax last year. Since he lives in New York, he also paid a 2% income tax to the city. If he made $25,000 last year, what percent of his salary did he pay in income taxes?

Solution:
Ignore the $25,000.   It has nothing to do with the question asked.

$14\% + 2\% = 16\%$ Since both the percents are calculated on the same thing, his salary, we can add them. Hence, he spent 16% on income taxes.

d. Jenny Song pays 25% of her monthly income for housing and 18% for food. How much greater is the percent of her income spent on housing than on food?

Solution:
Percent spent on housing minus the percent spent on food.   Since both percents are calculated on the same thing, the income, we can subtract them.

$25\% - 18\% = 7\%$ Hence, she spent 7% more on housing.

Practice Exercises

Perform the indicated operation:

18. $25\% + 7\frac{3}{4}\%$

19. $48.2\% - 14\%$

Solve the following:

20. The price tag on the nurse's uniform read "25% off." However, the clerk told Francesca that since she worked at the county hospital, the manager would give her another 10% off the original price. What percent did Francesca get off the uniform?

21. When Anne went to pay her bill at the restaurant, she found that a 7% sales tax and a 1.5% restaurant tax had been added to the bill. What percent more did she pay for the sales tax than for the restaurant tax?

Being able to convert between percents, decimals, and fractions is a very important everyday skill.

Exercise Set 8.3

Convert each of the following percents to fractions:

1. 16%

2. 5%

3. .09%

4. .2%

5. 13.6%

6. 45.9%

7. 325%

8. 550%

9. $5\frac{1}{3}\%$

10. $7\frac{3}{4}\%$

11. $12\frac{1}{7}\%$

12. $15\frac{10}{11}\%$

Convert the following percents to decimals. If necessary, round to the nearest thousandths.

13. 42%

14. 15%

15. .6%

16. .9%

17. 37.5%

18. 62.5%

19. 120%

20. 230%

21. 8.08%

22. 4.02%

23. $12\frac{7}{9}\%$

24. $11\frac{5}{14}\%$

Convert the following decimals to percent notation:

25. .2

26. .4

27. .38

28. .19

29. .008

30. .0065

31. 3.21

32. 5.67

33. .555. . .

34. .3999. . .

35. 5

36. 8

37. 10

38. 25

Convert the following fractions to percent notation using the method that works best for each fraction. If necessary, round to the nearest tenth of a percent.

39. $\frac{2}{5}$

40. $\frac{7}{10}$

41. $\frac{9}{20}$

42. $\frac{3}{4}$

43. $\frac{19}{25}$

44. $\frac{13}{50}$

45. $\frac{3}{8}$

46. $\frac{5}{11}$

47. $\frac{4}{3}$

48. $\frac{2}{7}$

49. $\frac{7}{6}$

50. $\frac{5}{9}$

51. $\frac{9}{5}$

52. $\frac{7}{12}$

53. $\frac{3}{2}$

54. $\frac{4}{13}$

Perform the indicated operation.

55. 25% + 14%

56. 15% − 8%

57. 37.5% − 16.6%

58. $8\frac{3}{4}\% - 5\frac{1}{3}\%$

59. 6.23% + 8%

60. 19.61% + 12%

61. 11% − 4.6%

62. 76% − 18.2%

Solve the following:

63. Greg Mendes noticed a 7% sales tax and a 2.5% room tax on his hotel bill. What percent of his hotel bill was tax?

64. The owner of a piece of downtown business property pays a 7% real estate tax on the business. On her home she pays a 2% real estate tax. How much greater is the percentage of taxes that she pays on her business than on her home?

65. Albert Young is a writer. In his home country, he would have to pay 48% of his income in taxes. In his adopted country, he must pay only 21% of his income in taxes. What percent does he save in taxes by living in his adopted country?

66. Janet Sanchez received two bonuses last year. She received 10% of her salary for meeting a sales quota and 5% for signing a big client for the firm. What percent of her salary did she receive in bonuses?

Challenge Exercises: (67–70)

Convert the fraction to percent notation. Round to the nearest one hundredth of a percent.

67. $\dfrac{3}{17}$

Convert the percent to a fraction.

69. $9\dfrac{4}{11}\%$

Convert the decimal to percent notation. Round to the nearest tenth.

68. .8333...

Convert the percent to a decimal.

70. $18\dfrac{2}{3}\%$

Writing Exercises:

71. Why do we have percent notation?

72. Can you multiply in percent notation? Why or why not? Give an example.

73. Can you divide in percent notation? Why or why not? Give an example.

Section 8.4	Applications of Percent

OBJECTIVES

When you complete this section, you will be able to:

a. Translate mathematical expressions involving percent into English.

b. Recognize variations of simple percent problems.

c. Solve percent problems.

d. Solve percent problems with proportions.

Introduction

The applications of percent are numerous. Percent is used in business, education, economics, medicine, and many other areas. In this section, we will begin with simple applications of percent and continue to the more complicated applications in the following section.

Let us begin by familiarizing ourselves with expressions involved in a **simple percent situation**. Simple percent situations are represented by equations that relate percent, the number of which we are taking the percent, and the result of taking the percent. These situations can be represented as _____% of _____ is _____ where we are given two of these values and are asked to find the third. The first blank represents the percent, the second blank represents the base, and the third blank represents the amount or part. Consequently, this method can be thought of as percent · base = amount.

Translation: percent to English

In translating from mathematics to English, we use the following translations:
multiplication symbol " · " translates as "of,"
equal sign "=" translates as "is,"
and the variable translates as "what number," "some number," or "a number."

Example 1

Variations of percent problems

Translate each percent equation into an English sentence:

a. $75\% \cdot 80 = x$

Solution:
75% of 80 is what number?   The symbol " · " becomes "of," "=" becomes "is," and "x" becomes "what number."

b. $150\% \cdot x = 18$

Solution:
150% of what number is 18?

The symbol " · " becomes "of," "x" becomes "what number," "=" becomes "is."

c. $x\% \cdot 40 = 14$

Solution:
What percent of 40 is 14?

x% becomes "what percent," " · " becomes "of," and "=" becomes "is."

Practice Exercises

Translate the following percent equations into an English sentence:

1. $80\% \cdot 60 = x$

2. $75\% \cdot y = 30$

3. $t\% \cdot 500 = 60$

If more practice is needed, do the Additional Practice Exercises.

Additional Practice Exercises

Translate each percent equation into an English sentence.

a. $x = 85\% \cdot 400$

b. $150 = 250\% \cdot y$

c. $t\% \cdot 75 = 45$

Since we cannot multiply or divide in percent notation, we will change the percents into decimals because the arithmetic is usually easier than with fractions. Consequently, _____ % of _____ is _____ becomes _____ · _____ = _____ where the first blank is a decimal.

Example 2

Translate each of the following into an equation and solve. Use y as the variable.

a. 4% of 1500 is what number?

"Of" means multiply, "is" means equals, and "what number" is the variable.

$4\% \cdot 1500 = y$ — Convert 4% to a decimal.

$.04(1500) = y$ — Multiply.

$60 = y$ — Therefore, 4% of 1500 is 60.

b. 90% of what number is 63?

"Of" means multiply, "what number" is the variable, and "is" means equals.

$90\% \cdot y = 63$ — Change 90% to a decimal.

$.90(y) = 63$ — Divide both sides of the equation by .90.

$\dfrac{.90y}{.90} = \dfrac{63}{.90}$ — Divide.

$y = 70$ — Hence, 90% of 70 is 63.

c. What percent of 40 is 8?

"What percent" is the variable, "of" means multiply, and "is" means equals.

$y\% \cdot 40 = 8$ — Since we cannot convert y% to a decimal, think of y% as $y \cdot 1\%$ and change 1% to .01.

$y(.01)(40) = 8$ — Multiply .01 and 40.

$.40y = 8$ — Divide both sides of the equation by .40.

$\dfrac{.40y}{.40} = \dfrac{8}{.40}$ — Simplify both sides.

$y = 20$ — Hence, 20% of 40 is 8.

Answers:

Practice Exercises

Solve the following equations:

4. 10% of 46 is what number?

5. 180% of what number is 45?

6. What percent of 500 is 110?

Additional Practice Exercises

If more practice is needed, do the Additional Practice Exercises.

Solve the following equations:

d. What number is 20% of 150?

e. 45 is 90% of what number?

f. 54 is what percent of 300?

Alternative Method: Solving percent problems using proportions.

Solving percent problems using proportions is often the preferred method. Remember that percent is the number of hundredths. Any percent may be expressed as a ratio with the second part or the denominator as 100. That is, 35% may be written as a ratio 35 : 100 or $\frac{35}{100}$. The 35 is the **number of percent**. The number of which we are taking the percent is called the **base** and follows the word "of" in the English sentence because we are taking a percentage of the base. The result of taking the percent is called the **amount** or **part**.

Percent Proportion

If P represents the number of percent, A represents the amount, and B represents the base, the relationship between these quantities is given by the following proportion:

$$\frac{P}{100} = \frac{A}{B}$$

Solving percent with proportions

Consider the following:

35% of 20 = 7 becomes the proportion $\frac{35}{100} = \frac{7}{20}$. This is read "35 is to 100 as 7 is to 20."

Procedure: Solving Simple Percent Situations using Proportions

The procedure for solving a percent situation with the percent proportion is as follows:

a. Identify the number of percent, the amount, and the base of the second ratio.

b. Substitute the appropriate values into the percent proportion.

c. Solve the proportion for the indicated variable.

Example 3

Solve the following equations using proportions:

a. 75% of 50 is what number?

Solution:
Recall the formula. We identify each part of the proportion and substitute them into the proportion. The number of percent is 75. The variable is placed in the unknown position for the amount. The base of the second ratio is 50 because it follows "of."

$$\frac{P}{100} = \frac{A}{B}$$ Substitute 75 for P and 50 for B.

$$\frac{75}{100} = \frac{A}{50}$$ Reduce $\frac{75}{100}$ to lowest terms.

$$\frac{3}{4} = \frac{A}{50}$$ Cross multiply.

$$3(50) = 4A$$ 3(50) = 150 and divide both sides of the equation by 4.

$$\frac{150}{4} = \frac{4A}{4}$$

$$37.5 = A$$ Therefore, 75% of 50 is 37.5.

Note: It is not necessary to reduce $\frac{75}{100}$ to lowest terms, especially if we are using a calculator. Reducing just makes the numbers smaller and easier to work with.

b. 50% of what number is 90?

Solution:
The number of percent is 50. The amount is 90. The variable is placed in the position for the unknown base of the second ratio.

$$\frac{P}{100} = \frac{A}{B}$$ Substitute 50 for P and 90 for A.

$$\frac{50}{100} = \frac{90}{B}$$ Reduce $\frac{50}{100}$ to lowest terms.

$$\frac{1}{2} = \frac{90}{B}$$ Cross multiply.

$$1(B) = 2(90)$$ Multiply.

$$B = 180$$ Therefore, 50% of 180 is 90.

c. What percent of 60 is 72?

Solution:
The number of percent is unknown. The amount is 72. The base of the second ratio is 60 because it follows "of."

$$\frac{P}{100} = \frac{A}{B}$$ Substitute 72 for A and 60 for B.

$$\frac{P}{100} = \frac{72}{60}$$ Reduce $\frac{72}{60}$ to lowest terms.

$$\frac{P}{100} = \frac{6}{5}$$ Cross multiply.

$$5P = 100(6)$$ 100(6) = 600 and divide both sides of the equation by 5.

$$\frac{5P}{5} = \frac{600}{5}$$ Divide.

$$P = 120$$ Hence, 120% of 60 is 72.

Practice Exercises

Solve the following equations using proportions:

7. 45% of 80 is what number?

8. 15 is 2.5% of what number?

9. 43.4 is what percent of 70?

If more practice is needed, do the Additional Practice Exercises.

Additional Practice Exercises

Solve the following equations using proportions:

g. What number is 3.5% of 400?   **h.** 24% of what number is 12?   **i.** 48 is what percent of 30?

Word problems involving applications of percent are solved in much the same manner as other word problems. We have a strategy.

Solving percent problems   We solve them by rewriting them in the form of _____ % of _____ is _____, translating into an equation or proportion, and solving. We summarize as follows:

> **Procedure: Solving Percent Problems**
>
> 1. Rewrite the problem in the form of _____ % of _____ is _____.
> 2. Translate the statement in step 1 into an equation or proportion.
> 3. Solve the equation or proportion.
> 4. Check your answer.

Example 4

Solve the following and round answers to the nearest percent or hundredth where appropriate:

a. Tanya correctly answered 123 questions out of 150 questions on an arithmetic final exam. What percent of the questions did she answer correctly?

Solution:
Rewrite the problem in the form of _____ % of _____ is _____. The question asks "what percent," so we know that we are looking for the percent. Since she answered 123 out of 150 correctly, this translates as:

What percent of 150 is 123?   *Rewrite as an equation. Let p represent the percent.*

$p\% \cdot 150 = 123$ *Since we cannot change $p\%$ into a decimal, think of $p\%$ as $p \cdot 1\%$ and change 1% to .01.*

$p(.01) \cdot 150 = 123$ *Multiply .01 and 150.*

$1.5p = 123$ *Divide both sides by 1.5.*

$p = 82$ *Therefore, she answered 82% of the questions correctly.*

Check:
Is 82% of 150 equal to 123? (82%)(150) = (.82)(150) = 123. Yes. Therefore, our answer is correct.

Note: This exercise could have been done by using the proportion $\frac{y}{100} = \frac{123}{150}$.

b. This year Harold spent $250 on books. If this amount represents 5% of his student loan, how much was the loan?

Solution:
Rewrite the problem in the form of _____ % of _____ is _____. The problem states that the amount Harold spent for books represents 5% *of his loan* and we are asked to find the amount of the loan. Consequently, this translates as:

5% of what is $250? Let y represent the amount of the loan and write an equation.

$5\% \cdot y = 250$ Change 5% to a decimal.

$.05y = 250$ Divide both sides by .05.

$$\frac{.05y}{.05} = \frac{250}{.05}$$ Divide.

$y = \$5000$ Therefore, Harold's student loan is **$5000**.

Check:
Is 5% of $5000 equal to $250? 5%(5000) $=.05(\$5000) = \$250$. Yes. So, our solution is correct.

Note: This exercise could have been done by using the proportion $\frac{5}{100} = \frac{250}{y}$.

c. A recent study showed that 12.1% of the women giving birth at a certain hospital had taken cocaine within the last 72 hours. If 32,000 women gave birth at this hospital last year, how many of them had taken cocaine within 72 hours of giving birth?

Solution:
Rewrite the problem in the form of _____ % of _____ is _____. The problem states that 12.1% <u>of the women</u> who gave birth had taken cocaine and that 32,000 women had given birth. This translates as:

12.1% of 32,000 is what number? Let z represent the number of women who had taken cocaine and write an equation.

$12.1\% \cdot 32,000 = z$ Convert 12.1% to a decimal.

$.121(32,000) = z$ Multiply.

$z = 3872$ Therefore, 3872 mothers had taken cocaine within 72 hours of giving birth.

Check:
Is 12.1% of 32,000 = 3872? Yes, .121(32,000) = 3872.

Note: This exercise could have been done by using the proportion $\frac{12.1}{100} = \frac{z}{32,000}$.

Practice Exercises

Solve and round answers to the nearest percent or hundredth where appropriate.

10. Out of every 5000 adults in the United States, 4000 do not exercise properly. What percent of the adults do not exercise properly?

11. A score of 70% on a test corresponds to 84 correctly answered test items. How many items were on the test?

12. A family must pay 22% of its income for income tax. How much is the tax if the family earned $65,000 that year?

If more practice is needed, do the Additional Practice Exercises.

Additional Practice Exercises

Solve and round answer to the nearest percent or hundredth.

j. Thirty-five out of 700 lightbulbs failed after 1000 hours of operation. What percent of the bulbs failed?

k. During a recent downsizing, 8% of the employees of a company were laid off. If 160 people were laid off, how many employees did the company have before the downsizing?

l. At a local college, 15% of the employees are administrators or staff for the administration. If 320 people are employed by the college, how many are administrators or staff for the administrators?

Exercise Set 8.4

Translate the following percent equations into English sentences:

1. $x = 25\% \cdot 36$

2. $50\% \cdot 32 = x$

3. $40\% \cdot 60 = v$

4. $v = 90\% \cdot 150$

5. $20\% \cdot y = 8$

6. $35 = 35\% \cdot y$

7. $52 = 65\% \cdot w$

8. $32\% \cdot w = 16$

9. $t\% \cdot 30 = 21$

10. $32 = t\% \cdot 20$

11. $4.8 = z\% \cdot 80$

12. $z\% \cdot 90 = 7.2$

Solve the following by using simple percent sentences:

13. 5% of 200 is what number?

14. What number is 15% of 80?

15. What number is 80% of 35?

16. 60% of 60 is what number?

17. 60 is 75% of what number?

18. 50% of what number is 45?

19. .9% of what number is 54?

20. 18 is 37.5% of what number?

21. 24 is what percent of 80?

22. What percent of 90 is 32.4?

23. What percent of 96 is 12?

24. 30 is what percent of 24?

Solve the following equations using proportions:

25. What number is 55% of 160?

26. 92% of 650 is what number?

27. .5% of 35 is what number?

28. What number is 18.5% of 40?

29. 3.6 is 5% of what number?

30. 12% of what number is 17.4?

31. 100 is 125% of what number?

32. 8% of what number is 800?

33. What percent of 25 is 1?

34. 125 is what percent of 200?

35. 22.5 is what percent of 150?

36. What percent of 800 is 356?

Solve each of the following and round the answers to the nearest percent or nearest hundredth where appropriate:

37. A recent survey indicated that 405 out of 500 women did not consider physical attraction an essential requirement for a spouse. This represents what percent of the women surveyed ?

38. Nicky has read 198 pages of a book that has 660 pages. What percent of the book has she read?

39. Research shows that 63% of all reported traffic accidents involve men. If 1500 traffic accidents are reported, how many involve men?

40. In a University of California study, 58% of 17-year-old girls considered themselves overweight. If 600 girls were studied, how many considered themselves to be overweight? (Only 17% were actually overweight.)

41. By volume a certain wine is 14% alcohol. If a bottle contains 750 milliliters of wine, how many milliliters of alcohol are in the wine bottle?

42. A solution of salt and water is called a saline solution. One saline solution is 8% salt. How much salt is in 50 milliliters of solution?

43. In mixing salad dressing, we use 3 ounces of water and 2 ounces of olive oil. What percent of the salad dressing is olive oil?

44. If we heat metals to liquid form and combine them, we get a mixture called an alloy. If 18 pounds of copper is mixed with 12 pounds of tin, what percent of the alloy is copper?

45. A family paid $2210 in real estate taxes last year. If this is a tax rate of 2.6%, how much is their house worth?

46. Out of 650 greeting cards sold at a certain card store, 585 were bought by women. What percent of the cards were bought by women?

47. The body of a male human is 63% water by weight. If a man's body contains 120 pounds of water, how much does he weigh to the nearest pound?

48. At a certain ranch, 2475 cattle represent 45% of a herd of cattle. How many cattle are in the herd?

Challenge Exercises:

Solve the following:

49. $33\frac{1}{3}$% of 120 is what number?

50. What number is $6\frac{1}{2}$% of 900?

51. Vinegar is 5% acetic acid by volume. How much vinegar would 5 milliliters of acetic acid make?

52. Muriatic acid is a 2% solution of hydrochloric acid. How many gallons of muriatic acid can be made from 1 gallon of hydrochloric acid?

Writing Exercise:

53. Write down five household items that are mixtures and list their ingredients by percent.

Section 8.5 | # Further Applications of Percent

OBJECTIVES

When you complete this section, you will be able to:

a. Compute sales tax.

b. Compute discount.

c. Compute commission.

d. Compute simple interest.

e. Compute compound interest.

f. Calculate percent increase and percent decrease.

Introduction

In the last section, we used the simple percent situation to solve percent problems. In this section, we will apply the simple percent situation to the specific situations of computing sales tax, discounts, and commission. We will also use percent to compute interest and percent increase and decrease.

Sales tax is the amount of money that we pay to the government when we make certain types of purchases. Sales tax is calculated as a fractional part, usually expressed as a percent, of the amount we pay for the goods purchased. This fraction is found by multiplying the purchase price by a percent called the **sales tax rate**.

Computation of sales tax

Calculating sales tax is a special type of simple percent situation and thus takes the form of _____ % of _____ is _____. In this case it becomes

sales tax rate · purchase price = the sales tax.

The total cost of an item is the purchase amount plus the sales tax.

Total Cost = Purchase Amount + Sales Tax

Example 1

Solve each of the following for the missing value:

a. Find the sales tax on a purchase of $400 if the sales tax rate is 4.5%.

Solution:
We need to think of this in the form of
_____ % of _____ is _____, or
sales tax rate · purchase price = the sales tax.
We know the sales tax rate is 4.5% and the purchase price is $400. Consequently:

4.5% of $400 is sales tax	Write as an equation using T as the taxes.
$4.5\% \cdot 400 = T$	Convert 4.5% to a decimal.
$.045 \cdot 400 = T$	Multiply.
$18.00 = T$	Hence, the sales tax on $400 at 4.5% is $18.00.

Note: This exercise could have been done using the proportion $\frac{4.5}{100} = \frac{T}{400}$.

b. Rebecca purchased a refrigerator for $600 and paid $30 in sales tax. What is the tax rate?

Solution:

We need to think of this in the form of

_____ % of _____ is _____, is or

sales tax rate · purchase price = the sales tax.

We know the purchase price is $600 and the sales tax is $30. Consequently:

Sales tax rate · 600 is 30	Write as an equation using r as the sales tax rate.
$r\% \cdot 600 = 30$	Think of $r\%$ as $r \cdot 1\%$ and change 1% to .01.
$r(.01)(600) = 30$	Multiply .01 and 600.
$6r = 30$	Divide both sides by 6.
$r = 5$	Therefore, the sales tax rate is 5%.

Note: This exercise could have been done using the proportion $\frac{r}{100} = \frac{30}{600}$.

c. The sales tax in a certain state is 4%. If Rebecca paid $15 in sales tax, what was the amount of purchase? What was the total cost?

Solution:

We need to think of this in the form of

_____ % of _____ is _____ or

sales tax rate · purchase price = the sales tax.

We know the sales tax rate is 40% and the sales tax is $15. Consequently:

4% of purchase price is $15.	Write as an equation using P as the purchase price.
$4\% \cdot P = 15$	Convert 4% to a decimal.
$.04 \cdot P = 15$	Divide both sides of the equation by .04.
$\frac{.04 \cdot P}{.04} = \frac{15}{.04}$	Divide.
$P = 375$	Hence, the purchase amount is $375.

Note: This exercise could have been done using the proportion $\frac{4}{100} = \frac{15}{P}$.

To find the total cost, recall the relationship.

Total Cost = Purchase Amount + Sales Tax	
Total Cost = $375 + $15	Substitute values and add.
Total Cost = $390	Thus, the cost is $390.

Practice Exercises

Solve each of the following for the missing value:

1. Find the sales tax on an item that had a purchase price of $120 if the sales tax rate is 6.5%.

2. John purchased a bow for $250 and paid $17.50 in sales tax. What was the sales tax rate?

3. The sales tax rate in a certain city is 1.5%. If Juan Carlos paid $6.75 in sales tax, what was the amount of purchase? What was the total cost?

If you need more practice, do the Additional Practice Exercises.

Additional Practice Exercises

Solve each of the following for the missing value:

a. John purchased a stereo for $450 and paid a local city sales tax of 3.5%. Find the sales tax.

b. Find the purchase price if the sales tax is $16 and the sales tax rate is 8%.

c. Julia bought an item for $70. If the sales tax is 7%, what is the amount of the sales tax? What is the total cost?

Answers:

Practice Exercises 1–3: 1. $7.80 2. 7% 3. $450, $456.75   *Additional Practice Exercises a–c:* a. $15.75 b. $200 c. $4.90, $74.90

A **discount** is the amount of money that is taken off (subtracted from) the original price of an item. The amount of the discount is calculated as a percent of the price marked on the goods purchased. This percent is called the **discount rate**.

This is another special type of simple percent situation. If we think of this in terms of _____ % of _____ is _____, it becomes <u>discount rate</u> of <u>original price</u> is <u>the amount of the discount</u>.

The **sale price (S)** of an item is the original price minus the discount.

Sale price = Original price − Discount.

$$S = O - D$$

Example 2

Solve each of the following for the missing value:

a. Find the amount of discount of a bedroom suite priced at $1400 with a 25% discount.

Solution:
We need to think of this in the form of _____ % of _____ is _____, or <u>discount rate</u> · <u>original price</u> = <u>amount of the discount</u>. We know the discount rate is 25% and the original price is $1400. Consequently:

25% of 1400 is the discount Write as an equation using D as the discount.

$25\% \cdot 1400 = D$ Convert 25% to a decimal.

$.25 \cdot 1400 = D$ Multiply.

$350 = D$ Therefore, the discount is $350.

Note: This exercise could have been done using the proportion $\frac{25}{100} = \frac{D}{1400}$.

b. Find the original price of a TV if the discount is $157.50 and the discount rate is 35%.

Solution:
We need to think of this in the form of _____ % of _____ is _____, or <u>discount rate</u> · <u>original price</u> = <u>amount of the discount</u>. We know the amount of the discount is $157.50 and the discount rate is 3%. Consequently:

35% of original price is 157.50 Write as an equation using O as the original price.

$35\% \cdot O = 157.50$ Convert 35% to a decimal.

$.35 \cdot O = 157.50$ Divide both sides of the equation by .35.

$\dfrac{.35 \cdot O}{.35} = \dfrac{157.50}{.35}$ Divide.

$O = 450$ Therefore, the original price is $450 .

Note: This exercise could have been done using the proportion $\frac{35}{100} = \frac{157.50}{O}$.

c. Mark paid $280 for a suit that was discounted $120. What was the original price of the suit? What was the discount rate?

Solution:
First, we solve for the original price. Recall the sale price relationship.

Sale price = Original price − Discount.

$S = O - D$ Substitute values for variables.

$280 = O - 120$ Add $120 to both sides of the equation.

$280 + 120 = O - 120 + 120$ Add.

$400 = O$ Hence, the original price is $400.

Now that we know the original price, we can solve for the discount rate.

We need to think of this in the form of _____ % of _____ is _____ , or discount rate · original price = amount of the discount. We know the original price is $400 and the amount of the discount is $120. Consequently:

The discount rate of 400 is 120.   Write as an equation using r as the discount rate.

$r\% \cdot 400 = 120$ Think of $r\%$ as $r \cdot 1\%$ and change 1% to .01.

$r(.01)(400) = 120$ Multiply .01 and 400.

$4r = 120$ Divide both sides of the equation by 4.

$\dfrac{4r}{4} = \dfrac{120}{4}$ Divide.

$r = 30$ Therefore, the rate of discount is 30%.

Note: This could have been done using the proportion $\dfrac{r}{100} = \dfrac{120}{400}$.

Practice Exercises

Solve each of the following for the missing value:

4. Find the amount of the discount if an item has an original price of $150 and the discount rate is 20%.

5. Jin Ho bought a bed whose original price was $460 with a discount of $69. What was the discount rate?

6. Monica paid $150 for a suit that was discounted $100. What was the original price of the suit? What was the discount rate?

If you need more practice, do the Additional Practice Exercises.

Additional Practice Exercises

Solve each of the following for the missing value:

d. Find the original price if the discount rate is 35% and the amount of the discount is $77.

e. Pamika bought a blouse whose original price was $65 with a discount of 25%. Find the sales price.

f. A sale sign said "45% off." If an item is originally $120, how much is the sale price?

A **commission** is the amount of money that is paid to a person for selling an item. Commission is calculated as a percent of the price paid for the goods purchased. This price is called the **sales**. The amount of the commission is found by multiplying the purchase price by a percent called the **commission rate**.

Computation of commission   This too is an application of the simple percent situation. In this case, _____ % of _____ is _____ becomes commission rate of sales is amount of commission.

Example 3

Solve each of the following for the missing value.

a. Find the amount of the commission on sales of $560 if the commission rate is 16%

Solution:
We need to think of this in the form of _____ % of _____ is _____, or commission rate · sales = amount of the commission. We know the commission rate is 16% and the sales are $560. Consequently:

16% of $560 is commission	Write as an equation using C as the commission.
16% · 560 = C	Convert 16% to a decimal.
.16(560) = C	Multiply.
$89.60 = C	Therefore, the commission is $89.60.

Note: This exercise could have been done using the proportion $\frac{16}{100} = \frac{C}{560}$.

b. Karla sells jewelry and is paid a 30% commission on all her sales. What was the selling price on a bracelet for which she received a commission of $78?

Solution:
We need to think of this in the form of _____ % of _____ is _____, or commission rate · sales = amount of the commission. We know the commission rate is 30% and her commission is $78. Consequently:

30% of sales is $78.	Write as an equation using S for sales.
30% · S = 78	Convert 30% to a decimal.
.3S = 78	Divide both sides of the equation by .3. Remember, .30 = .3.
S = 260	Therefore, the sales is $260.

Note: This exercise could have been done using the proportion $\frac{30}{100} = \frac{78}{S}$.

c. Sam received $287.50 commission on a refrigerator selling for $1150. What was the commission rate?

Solution:
We need to think of this in the form of _____ % of _____ is _____, or commission rate · sales = amount of the commission. We know the sales are $1150 and his commission is $287.50. Consequently:

Commission rate of 1150 is 287.50.	Write as an equation using r for rate.
r% · 1150 = 287.50	Write r% as r · 1% and write 1% as .01.
r(.01)(1150) = 287.50	Multiply .01 and 1150.
11.5r = 287.50	Divide both sides of the equation by 11.5.
$\dfrac{11.5r}{11.5} = \dfrac{287.50}{11.5}$	Divide.
25 = r	Therefore, the commission rate is 25%.

Practice Exercises

Solve each of the following for the missing value:

7. Find the amount of the commission for an item that sold for $175 if the commission rate is 12%.

Answer:

8. Karl works at a department store and receives a 5% commission on all sales. If Karl received a commission of $11.50 for the sale of a suit, what was the selling price of the suit?

9. Marla received a commission of $160 for selling an $800 stove. What was her commission rate?

If you need more practice, do the Additional Practice Exercises.

Additional Practice Exercises

Solve each of the following for the missing value:

g. Find the amount of the commission on a sale of $124 if the commission rate is 8.5%.

h. Sue received a commission of $135 on a computer monitor that sold for $540. What was her rate of commission?

i. Dion received a commission of $420 for selling a refrigerator. If she has a commission rate of 35%, what was the sales amount?

Interest is money that you must pay to someone or to an institution to use their money, or money paid to you by someone who uses your money. The usual way to use someone's money is to get a loan and the usual way for someone to use your money is for you to make a deposit at a bank or savings institution. The amount of money borrowed or deposited is called the **principal.**

Interest is calculated as a part or percentage of the money being loaned or deposited. This amount is found by multiplying the principal by a percent called the **annual rate of interest,** then multiplying the resulting product by the length of time (usually in years) the loan is active. Thus, we have the relationship:

$$\text{Interest} = (\text{annual interest rate})(\text{principal})(\text{time})$$

Computation of simple interest

$$\$ \quad = \quad \% \quad \cdot \quad \$ \quad \cdot \text{ years}$$

Hence, we have the formula for calculating interest.

Interest Calculation Formula

If I = interest, p = principal, r = annual interest rate in percent, and t = time in years, then interest is calculated by the following formula.

$$I = p \cdot r \cdot t$$

Note: In applying this formula, r must be expressed as a decimal.

Example 4

Solve each of the following for the unknown value:

a. $p = \$800$, $r = 18\%$, $t = 1$ year, $I = ?$

Solution:

$I = p \cdot r \cdot t$ Recall the interest calculation formula.
            Substitute values for variables.

$I = 800 \cdot 18\% \cdot 1$ Convert 18% to a decimal.

$I = (800)(.18) \cdot 1$ Multiply.

$I = 144$ Therefore, the interest on $800 at 18% interest rate for one year is $144.

b. $I = \$22.26 \quad p = \$742, \; t = \frac{1}{2}$ year, $r = ?$

Solution:

$I = p \cdot r \cdot t$ Recall the interest calculation formula.
            Substitute values for variables.

$\$22.26 = 742 \cdot r \cdot \dfrac{1}{2}$ Divide 2 into 742.

$22.26 = 371 \cdot r$ Divide both sides of the equation by 371.

$\dfrac{22.26}{371} = \dfrac{371 \cdot r}{371}$ Divide.

$.06 = r$ Convert to percent notation.

$6\% = r$ Therefore, the interest on $742 for $\frac{1}{2}$ year at 6% interest rate is $22.26.

c. Rosa Lee borrowed some money at 12% for three years. If she paid $1800 in interest, how much did she borrow?

Solution:

$I = p \cdot r \cdot t$ Recall the interest calculation formula.
            Substitute values for variables.

$\$1800 = p \cdot 12\% \cdot 3$ Convert 12% to a decimal.

$1800 = p \cdot .12 \cdot 3$ Multiply.

$1800 = .36p$ Divide both sides of the equation by .36.

$\dfrac{1800}{.36} = \dfrac{.36p}{.36}$ Divide.

$5000 = p$ Therefore, she paid $1800 interest on $5000 over the three years.

Practice Exercises

Solve each of the following for the unknown value:

10. $p = \$15,000, \; r = 8\%, \; t = 2$ years, $I = ?$

11. $I = \$833.40, \; r = 12\%, \; t = \frac{3}{4}$ years, $p = ?$

12. Edward deposited $2500 at 16% interest rate. If he was paid $1600 interest, how many years was the money on deposit?

If you need more practice, do the Additional Practice Exercises.

Additional Practice Exercises

j. $p = \$15,000, \; r = 4\%, \; t = 1$ year, $I = ?$

k. $I = \$40.50, \; p = \$900, \; t = \frac{1}{4}$ year, $r = ?$

l. Lilian borrowed $2500 for $\frac{1}{2}$ year at an 8% interest rate. How much interest did she pay?

Answers:

From Example 4 we see that the length of time used in interest calculations is always measured in years. Any other expression of time must be converted to years before a calculation can be made.

In order to make calculations easier, the banking world has set up its own system of time by making a banking month equal to 30 days. See the following table:

Banking Time

30 banking days = 1 banking month

12 banking months = 1 banking year

360 banking days = 1 banking year

We may calculate interest for a certain number of days or months as parts of a year.

Example 5

Calculate the interest for each of the following sets of information:

a. $r = 11\%$, $p = \$1250$, $t = 90$ days

Solution:

First we need to change 90 days into a part of a year. 90 days $= \frac{90}{360}$ years $= \frac{1}{4}$ year.

$$I = p \cdot r \cdot t \qquad \text{Substitute values for variables. Express 11\% as .11.}$$

$$I = 1250(.11)\frac{1}{4} \qquad \text{Multiply.}$$

$$I = \frac{137.5}{4} \qquad \text{Divide.}$$

$$I = 34.375 \qquad \text{Round to nearest cent.}$$

$$I = 34.38 \qquad \text{Hence, the interest on \$1250 at 11\% for 90 days is \$34.38 to the nearest cent.}$$

b. $r = 7.5\%$, $p = \$5500$, $t = 4$ months

Solution:

First we need to change four months into a part of a year.
Four months $= \frac{4}{12}$ year $= \frac{1}{3}$ year.

$$I = p \cdot r \cdot t \qquad \text{Substitute values for variables and express 7.5\% as .075.}$$

$$I = 5500(.075)\frac{1}{3} \qquad \text{Multiply.}$$

$$I = \frac{412.5}{3} \qquad \text{Divide.}$$

$$I = 137.5 \qquad \text{Hence, the interest on \$5500 at 7.5\% for four months is \$137.50.}$$

c. Felicia took out a book loan for \$460 at a 6% interest rate. If the loan is to be repaid in ten months, how much interest will she have to pay?

Solution:

First we need to change ten months into years. Ten months $= \frac{10}{12}$ years $= \frac{5}{6}$ year.

$$I = p \cdot r \cdot t \qquad \text{Substitute values for variables and express 6\% as .06.}$$

$$I = 460(.06)\frac{5}{6} \qquad \text{Multiply.}$$

$$I = \frac{138}{6} \qquad \text{Divide.}$$

$$I = 23 \qquad \text{Hence, Felicia will pay \$23.00 interest on a loan of \$460 at an interest rate of 6\% for ten months.}$$

Practice Exercises

Calculate the interest for each of the following sets of information:

13. $p = \$900$, $r = 18\%$, $t = 30$ days

14. $p = \$2400$, $r = 5\%$, $t = 8$ months

15. Andre put $3200 in an 18-month certificate of deposit that paid at a 9% interest rate. How much interest did he earn?

If you need more practice, do the Additional Practice Exercises.

Additional Practice Exercises

Calculate the interest for each of the following sets of information:

m. $p = \$400$, $r = 12\%$, $t = 6$ months

n. $p = \$6000$, $r = 3\%$, $t = 9$ months

o. Monica got a 60-day loan for $300 at 19.2%. How much interest did she pay?

Until now we have been discussing **simple interest.** Simple interest is interest calculated only on the principal. **Compound interest** is interest computed on previously earned interest in addition to principal. Compound interest is calculated at various lengths of time expressed as parts of a year. Some lengths of time are used so often that they have been given names. See the following chart:

Computation of compound interest

Common Lengths of Time

Part of a Year	Phrase	Number of Times a Year Interest is Computed
1	annually	1
$\frac{1}{2}$	semiannually	2
$\frac{1}{4}$	quarterly	4
$\frac{1}{12}$	monthly	12
$\frac{1}{360}$	daily	360

The **balance** of an account is the amount of money in the account at a particular time. To **post** an amount means to credit the account with the money deposited by adding this amount to the balance.

$$\text{Balance} = \text{Principal} + \text{Interest}$$

Answers:

Practice Exercises 13–15: **13.** $13.50 **14.** $80 **15.** $432   Additional Practice Exercises *m–o*: **m.** $24 **n.** $135 **o.** $9.60

Example 6

Calculate the balance after one year for an account that pays 8% interest compounded semiannually on a beginning balance of $700.00.

Solution:

Interest must be calculated two times, once for the first $\frac{1}{2}$ of the year and once for the second $\frac{1}{2}$ of the year. The principal for the second half of the year will be the original principal plus the interest earned during the first half of the year. We may consider the posting dates as January 1, 2000; July 1, 2000; and January 1, 2001.

Balance:	$700.00	_____	_____
Date:	Jan. 1, 2000	July 1, 2000	Jan. 1, 2001

The first calculation is simple interest.

$$I = p \cdot r \cdot t \qquad \text{Substitute values for variables.}$$

$$I = 700 \cdot 8\% \cdot \frac{1}{2} \qquad \text{Convert 8\% to a decimal or fraction.}$$

$$I = 700 \cdot .08 \cdot \frac{1}{2} \qquad \text{Multiply.}$$

$$I = \frac{56}{2} \qquad \text{Divide.}$$

$$I = 28 \qquad \text{The interest for the first } \tfrac{1}{2} \text{ year is \$28.00.}$$

We post this interest by adding it to the original principal. Therefore, the balance on July 1 is $700 + $28 = $728.

Balance:	$700.00	$728.00	_____
Date:	Jan. 1, 2000	July 1, 2000	Jan. 1, 2001

The second calculation will include the interest from the first calculation, and hence be compound interest. Therefore, the principal for the second half of the year is $728.

$$I = p \cdot r \cdot t \qquad \text{Substitute values for variables.}$$

$$I = 728 \cdot 8\% \cdot \frac{1}{2} \qquad \text{Convert 8\% to a decimal.}$$

$$I = 728 \cdot .08 \cdot \frac{1}{2} \qquad \text{Multiply.}$$

$$I = \frac{58.24}{2} \qquad \text{Divide.}$$

$$I = 29.12 \qquad \text{The interest for the second } \tfrac{1}{2} \text{ year is \$29.12.}$$

We post this interest by adding it to the July 1, 2000 balance to get the January 1, 2001 balance. Therefore, the balance on January 1 is $728 + $29.12 = $757.12.

Balance:	$700.00	$728.00	$757.12
Date:	Jan. 1, 2000	July 1, 2000	Jan. 1, 2001

Hence, the final balance is $757.12, which includes $28 + $29.12 = $57.12 interest.

Note: In the business world, there is a formula that is used for computing compound interest. See Challenge Exercises 51 and 52.

Practice Exercise

16. Calculate the balance after one year for an account that pays 6.5% interest compounded semiannually on a beginning balance of $1300.00.

If you need more practice, do the Additional Practice Exercises.

Additional Practice Exercise

p. Calculate the balance after three years for an account that pays 10% interest compounded annually on a beginning balance of $884.00.

Increase or decrease

Another situation involving percent is where there is an increase or decrease in a quantity. In an increase situation, we start at a value called the **base** or original amount and add a quantity called the **increase**. The increase may be expressed as a number or as a percent of the base. In a decrease situation, we start at an original amount and subtract a quantity called the **decrease**. The decrease may be expressed as a number or as a percent of the base. To find the percent increase or decrease, divide the amount of the increase or decrease by the original amount and convert to a percent.

Procedure: Finding the Percent Increase or Decrease

To find the percent increase or decrease, find $\dfrac{\text{amount of increase or decrease}}{\text{original amount}}$ and convert to a percent.

Example 7

a. In a three-year span, a tree grew 5 feet from a height of 25 feet. What is the percent of growth?

Solution:
Since the tree increased in height, we are asked to find the percent of increase. We know that the original height of the tree is 25 feet and the increase in height is 5 feet. Consequently we find:

$$\frac{\text{Amount of the increase}}{\text{Original amount}}$$ Substitute for amount of increase and the original amount.

$$\frac{5}{25} =$$ Reduce $\frac{5}{25}$ to lowest terms.

$$\frac{1}{5} =$$ Change $\frac{1}{5}$ to a percent.

$$\frac{1}{5} \cdot 1 = \frac{1}{5} \cdot 100\%$$ Multiply.

$$20\%$$ Therefore, the tree has grown by 20% in three years.

Check:
Is 20% of 25 = 5?   Yes, since .20(25) = 5. Therefore, our solution is correct.

Answers:

Practice Exercise

17. In 1999, a certain computer cost $3100. In 2000, the same computer cost 48% less. How much did the computer cost in 2000?

If you need more practice, do the Additional Practice Exercise.

Additional Practice Exercise

q. The price of gasoline rose from $1.12 per gallon to $1.54 per gallon. What was the percent of increase to the nearest percent?

Exercise Set 8.5

Solve the following problems involving sales tax:

1. Find the tax on an item that sells for $90 if the tax rate is 7%.

2. Find the tax on an item that sells for $120 if the tax rate is 5.5%.

3. If the taxes on a microwave oven that sells for $460 are $20.70, find the tax rate.

4. If the taxes on a DVD player that sells for $310 are $15.50, find the tax rate.

5. The sales tax on a TV set is $15.30. If the sales tax rate is 4.5%, what is the purchase amount?

6. The sales tax on a microwave oven is $10.80. If the sales tax rate is 6%, what is the purchase amount?

7. The sales tax rate in a particular state is 7.5%. If you purchase a washing machine for $550, what is the total cost?

8. The sales tax rate in a certain city is 8%. If Olga pays $33.60 in sales tax, what is the total cost?

Solve the following problems involving discounts:

9. If an item has an original price of $180 and the discount rate is 15%, find the discount.

10. If an item has an original price of $240 and the discount rate is 25%, find the discount.

11. A blouse has an original price of $55 and is discounted by $11. What is the rate of the discount?

12. A cleaning kit has an original price of $75 and is discounted by $7.50. Find the rate of the discount.

13. Victor paid $105 for a coat that was discounted $45. What was the original price of the coat? What was the discount rate?

14. Karen paid $165 for an outfit that was discounted $55. What was the original price of the outfit? What was the discount rate?

Solve the following problems involving commission:

15. If the commission rate is 15%, find the commission on sales of $300.

16. If the rate of commission is 30%, find the commission on sales of $750.

17. John works in a department store. There was a shirt that had been on the racks for a long time. John's boss told him that he would give him 10% commission if he could sell it. If John's commission was $12.50, what was the selling price of the shirt?

18. Sing Le worked for 25% commission. If she received a commission of $64.50 on the sale of a washer/dryer combination, what was the selling price?

19. Sharon received $20 commission for selling a wooden worktable that sold for $200. What was the commission rate?

20. Tim received $196 commission for selling a $560 vacuum cleaner. What was the commission rate?

Given the interest formula, $I = r \cdot p \cdot t$, solve each of the following for the unknown value. If necessary round to the nearest cent, percent, or year.

21. $r = 6\%, p = \$400, t = $ two years, $I = $ **22.** $r = 4.5\%, p = \$800, t = $ one year, $I = $

23. $r = 8.5\%, I = \$221, t = \frac{3}{4}, p = $ **24.** $r = 9\%, I = \$22.50, t = \frac{1}{2}, p = $

25. $r = 7\%, I = \$109.20, p = \$520, t = $ **26.** $r = 3\%, I = \$240, P = \$4000, t = $

27. $p = \$1200, I = \$480, t = $ five years, $r = $

28. $p = \$1500, I = \$180, t = $ three years, $r = $

29. James borrowed $3000 at an interest rate of 5% for four years. How much interest did he pay?

30. Irene deposited $2500 in her bank at an interest rate of 6.5% for two years. How much interest did she receive?

31. Blanca received $306 interest at an interest rate of 9% over a five-year period. How much did she deposit?

32. Miki paid $288 interest over $\frac{1}{2}$ year on a loan for $4500. What interest rate did she pay?

Calculate the interest to the nearest cent for each of the following sets of information:

33. $p = \$1000, r = 17\%, t = 60$ days   **34.** $p = \$1800, r = 16\%, t = 120$ days

35. $p = \$4800, r = 19.2\%, t = $ five months   **36.** $p = \$3600, r = 18\frac{1}{2}\%, t = $ nine months

37. Janice borrowed $1600 for eight months at an 18% interest rate. How much interest did she have to pay?

38. Diego deposited $720 in the student depository for seven months. If he received 6% interest on his deposit, how much interest did he earn?

39. Calculate the balance after one year for an account that pays 8% compounded semiannually on a beginning balance of $3500.

40. Calculate the balance after one year for an account that pays 6% compounded semiannually on a beginning balance of $1400.

Solve the following using percent increase and percent decrease. If necessary, round to the nearest tenth.

41. Christopher just went for his yearly physical examination. Last year, he measured 50 inches tall and this year he measured 54 inches tall. What is the percent of increase in his height?

42. In Siberia, houses sometimes sink into the frozen ground because the heat from the house melts the ice. If 20% of a 12-foot tall house sinks, how high is the remaining portion of the house above ground?

43. Through exercise and diet, Willie lost 15 pounds. When he began this weight reduction program, he weighed 180 pounds. What was the percent of weight loss?

44. An airline pilot requests a change in altitude from 28,000 feet to 35,000 feet to avoid bad weather. What was the percent of change?

45. On a very bad day, the value of a mutual fund fell 36%. If the price per share was $14.25 before the fall, how much was it after the fall?

46. A company experienced a 2% increase in sales. If the total sales before the increase were $4,500,000, what were the sales after the increase?

Challenge Exercises:

47. Calculate the balance after one year for an account that pays 5% compounded quarterly on a beginning balance of $6500.

48. In a certain county, the county sales tax is 1.5% and the hotel room tax is 1%. If the state sales tax rate is 5.5%, how much tax must a tourist pay on a hotel room which rents for $120 a day for 5 days?

49. A sewing machine that is marked at $849.99 is on sale for $499.99. What is the percent of discount?

50. Marie received $5400 for selling a house. If the commission rate is 6%, how much did the house sell for?

The formula used for computing compound interest is $A = P(1 + \frac{r}{n})^{nt}$ where A is the accumulated amount, P is the principle (amount invested or borrowed), r is the interest rate written as a decimal, n is the number of times per year that the interest is compounded, and t is the number of years. For example, if $1000 is invested at 6% interest compounded quarterly (four times per year) for 10 years, the accumulated amount would be $A = 1000 (1 + \frac{.06}{4})^{(4)(10)} = 1000(1 + .015)^{40} = 1000(1.015)^{40} = \$1814.02$. We used a calculator to perform the last computation.

51. Find the accumulated amount if $5000 is invested at 5% interest compounded semiannually (twice per year) for 12 years.

52. Find the accumulated amount if $8000 is invested at 9% compounded monthly (twelve times per year) for 15 years.

Writing Exercise:

53. Name at least five business applications of percent not mentioned in the text.

Chapter 8	Summary

Ratio: [Section 8.1]
- A ratio is a comparison of two or more quantities by division. If a and b are quantities, then $a : b$, or $\frac{a}{b}$, is a ratio of a to b. If a, b, c are quantities, then $a : b : c$ is a ratio of a to b to c.

Ratios in Lowest Terms: [Section 8.1]
- As with rational numbers, ratios are in lowest terms when the parts of a ratio are nonzero integers that contain no common factors other than the number 1.

Processes for Simplifying Ratios: [Section 8.1]
- To put a ratio into lowest terms, we do one of the following:
  a. If the ratio is in fractional form, factor the numerator and the denominator into prime factors and divide the numerator and denominator by the factors common to both.
  b. If the ratio is in the form $a : b$ or a to b, factor both sides into prime factors and divide both sides by the factors common to both.

Rates: [Section 8.1]
- **Rates** are ratios with different units of measurement.

Unit Price: [Section 8.1]
- A **unit price** is a rate that expresses the cost compared to one unit.

Proportion: [Section 8.2]
- A proportion is an equation that states that two or more ratios are equal. In symbols, $\frac{a}{b} = \frac{c}{d}$ is a proportion. The fractions $\frac{a}{b}$ and $\frac{c}{d}$ can form a proportion if $a \cdot d = b \cdot c$.

Procedure for Solving Proportions: [Section 8.2]
- When solving a proportion, we perform the following steps:
  1. Simplify all fractional ratios.
  2. Cross multiply the fractional ratios.
  3. Solve the resulting equation.

Percent Notation: [Section 8.3]
- **Percent notation** is a special way of writing a fraction with a denominator of 100. That is, if n is some number, then $\frac{n}{100} = n\%$.

Conversions: [Section 8.3]
- Percent to a Fraction
  a. Think of $x\%$ as $x \cdot 1\%$.
  b. Replace 1% with $\frac{1}{100}$.
  c. Multiply.
- Percent to a Decimal
  a. Think of $x\%$ as $x \cdot 1\%$.
  b. Replace 1% with .01.
  c. Multiply.
- Decimal to Percent
  a. Write the decimal as the product of itself and 1.
  b. Change 1 to 100%.
  c. Multiply.
- Fraction to Percent
  a. Write the fraction as the product of itself and 1.
  b. Change 1 to 100%.
  c. Multiply, and if necessary, round to a specified number of decimal places.

Simple Percent Situation: [Section 8.4]
- **Simple percent** situations are represented by equations that relate percent, the number of which we are taking the percent, and the result of taking the percent. Simple percent situations can be represented as _____ % of _____ is _____. The first blank is called the percent, the second the base, and the third the amount.

Percent Proportion: [Section 8.4]
- If P represents the number of percent, A represents the amount, and B represents the base, the relationship between these quantities is given by the following proportion:

$$\frac{P}{100} = \frac{A}{B}$$

Procedure for Solving Simple Percent Situations Using Proportions: [Section 8.4]
- The procedure for solving a percent situation with the percent proportion is as follows:
  a. Identify the number of percent, the amount, and the base of the second ratio.
  b. Substitute the appropriate values into the simple percent proportion.
  c. Solve the proportion for the indicated variable.

Procedure for Solving Percent Problems: [Section 8.4]
1. Rewrite the problem in the form of _____ % of _____ is _____.
2. Translate the statement in step 1 into an equation or proportion.
3. Solve the equation or proportion.
4. Check your answer.

Sales Tax: [Section 8.5]
- **Sales tax** is the amount of money that we pay to the government when we make certain types of purchases.

Sales Tax Rate: [Section 8.5]
- The **sales tax rate** is a portion of the amount paid for the goods purchased. The sales tax rate is expressed as a percent.

Sales Tax Calculation: [Section 8.5]
- Calculating sales tax is a special type of simple percent situation and thus takes the form of _____ % of _____ is _____. In this case, it becomes sales tax rate · purchase price = the sales tax.

Total Cost = Purchase Amount + Sales Tax: [Section 8.5]

Discount: [Section 8.5]
- A **discount** is the amount of money that is taken off (subtracted from) the original price of an item.

Discount Rate: [Section 8.5]
- The **discount rate** is a portion of the original price of the item. It is expressed in terms of percent.

Discount Calculation: [Section 8.5]
- Calculating discount rates is a special type of simple percent situation and takes the form of _____ % of _____ is _____, it becomes discount rate of original price is the amount of the discount.

Sale Price = Original Price − Discount: [Section 8.5]
- That is, $S = O - D$

Commission: [Section 8.5]
- **Commission** is the amount of money that is paid to a person for selling an item.

Commission Rate: [Section 8.5]
- The **commission rate** is a portion of the amount of sales. It is expressed in terms of percent.

Commission Calculation: [Section 8.5]
- Calculating commission rates is a special type of simple percent situation and takes the form of _____ % of _____ is _____ and becomes commission rate of sales is amount of commission.

Interest: [Section 8.5]
- Interest is money that you must pay to someone or to an institution to use their money, or money paid to you by someone who uses your money.

Principal: [Section 8.5]
- **Principal** is the amount of money borrowed or deposited into an account.

Annual Rate of Interest: [Section 8.5]
- The **annual rate of interest** is a portion of the principal expressed as a percent.

Interest Calculation Formula: [Section 8.5]
- If I = interest, p = principal, r = interest rate in percent, and t = time in years, then interest is calculated by the following formula:

$$I = p \cdot r \cdot t$$

Balance: [Section 8.5]   • The **balance** of an account is the amount of money in the account at a particular time.

$$Balance = Principal + Interest$$

Increase: [Section 8.5]   • Base + Increase = Result

    • Base − Decrease = Result

    • $\dfrac{amount\ of\ increase\ or\ decrease}{original\ amount}$ and convert to a percent.

Chapter 8   Review Exercises

Simplify the following ratios: [Section 8.1]

1. 12 : 20

2. 35 to 30

3. $\dfrac{27}{15}$

4. $\dfrac{18}{24}$

Write each of the following as ratios in lowest terms: [Section 8.1]

5. In a college parking lot, there were 625 compact cars and 350 middle-sized cars. What was the ratio of compact cars to middle-sized cars?

6. A party mix calls for 4 cups of pretzels, 2 cups of peanuts, and 2 cups of raisins. What is the ratio of the ingredients?

Write each of the following as a ratio of whole numbers. [Section 8.1]

7. A recipe for homemade coconut candy calls for $\frac{4}{3}$ cups of fresh coconut, $\frac{2}{3}$ cup of sugar, and $\frac{1}{3}$ cup of margarine. What is the ratio of the ingredients?

8. There are 2.27 kilograms in 5 pounds of sugar. What is the ratio of kilograms to pounds?

Express each ratio with the second part equal to 1. [Section 8.1]

9. A group of 15 students orders five large pizzas that are cut into 12 pieces each. How many pieces should each student get?

10. There are five families in a food co-operative. If they buy 100 pounds of rice, how much does each family receive?

Express each of the following as a unit price. If necessary, round to the nearest cent. [Section 8.1]

11. If 8 ounces of tomato sauce costs $0.36, what is the price per ounce?

12. If 75 square feet of aluminum foil costs $2.49, what is the price per square foot?

Determine if the following fractions can form a proportion: [Section 8.2]

13. $\dfrac{12}{15}$ and $\dfrac{20}{25}$

14. $\dfrac{.8}{.24}$ and $\dfrac{6}{18}$

Solve the following proportions: [Section 8.2]

15. $\dfrac{2}{x}=\dfrac{4}{5}$

16. $\dfrac{8}{5}=\dfrac{y}{60}$

17. $\dfrac{48}{15}=\dfrac{32}{t}$

18. $\dfrac{u}{14}=\dfrac{0}{4}$

19. $\dfrac{y}{2}=\dfrac{8.4}{3}$

20. $\dfrac{11}{3}=\dfrac{x}{.6}$

21. $\dfrac{\frac{2}{3}}{5}=\dfrac{v}{15}$

22. $\dfrac{6}{\frac{4}{5}}=\dfrac{35}{v}$

23. $\dfrac{9}{4\frac{1}{2}}=\dfrac{10}{t}$

24. $\dfrac{w}{15}=\dfrac{3\frac{2}{5}}{17}$

25. $\dfrac{3}{4}=\dfrac{9}{x+5}$

26. $\dfrac{4}{5}=\dfrac{2}{x-1}$

27. $\dfrac{y+5}{2}=\dfrac{y}{-3}$

28. $\dfrac{3(5t-1)}{7}=\dfrac{4t+2}{2}$

29. $\dfrac{x+2}{2}=\dfrac{9}{x-5}$

30. $\dfrac{3}{2x-3}=\dfrac{x+2}{3}$

Solve each of the following: [Section 8.2]

31. A lawn mower uses 2 quarts of gasoline to mow $\frac{1}{3}$ acre of grass. How many acres can be mowed with 9 quarts of gasoline?

32. A machine can fill 3600 cans in 25 minutes. How many cans can be filled in 70 minutes?

33. On a map, two airports are $4\frac{1}{2}$ inches apart. If 90 miles is represented as $\frac{3}{4}$ inch apart on the map, how many miles apart are the two airports?

34. A naturalist group is interested in how many birds are nesting in a conservation area. They capture 210 birds, tag them, and release them. Later 288 birds are captured and 32 of them are tagged. How many birds are in the conservation area?

Convert the following fractions to percent notation. If necessary, round to the nearest tenth of a percent. [Section 8.3]

35. $\dfrac{1}{4}$ **36.** $\dfrac{7}{8}$ **37.** $\dfrac{5}{2}$

38. 4   **39.** $\dfrac{3}{7}$ **40.** $\dfrac{5}{6}$

Convert the following decimals to fractions: [Section 8.3]

41. .4   **42.** 1.09

43. .082   **44.** .333. . .

Convert the following decimals to percent notation: [Section 8.3]

45. .025   **46.** 6.57

47. 2.05   **48.** .0461

Perform the indicated operation. [Section 8.3]

49. 88% − 27.5%   **50.** $32\dfrac{1}{3}\% + 18\%$

Convert each of the following percents to fractions: [Section 8.3]

51. 8%   **52.** 12.5%

53. $9\dfrac{1}{4}\%$ **54.** .04%

Convert the following percents to decimals: [Section 8.3]

55. 18%   **56.** $6\dfrac{3}{4}\%$

57. 2.6%   **58.** .005%

59. 27%   **60.** 9.7%

Solve the following by simple percent sentences: [Section 8.4]

61. 24.5 is 70% of what number?

62. 48 is what percent of 60?

63. 60 is what percent of 40?

64. 26 is 65% of what number?

Solve each of the following and round the answers to the nearest percent or nearest hundredth where appropriate: [Section 8.4]

65. A college football quarterback completed 16 out of 25 passes attempted during a recent game. What percent of his passes were completed?

66. At one point during the Persian Gulf conflict, the United States lost four aircraft, which was said to be a 1% loss rate. How many U.S. aircraft were involved in this war?

67. The federal government spends $77 million a year investigating ways to prevent breast cancer and $648 million for investigating ways to prevent heart disease. What percent of the amount spent on heart disease is the amount spent on breast cancer?

68. On the basis of calories consumed, the average adult eats a diet that is 42% fat. If a person consumes 2000 calories of food per day, how many calories are from fat?

Solve the following problems involving sales tax: [Section 8.5]

69. Find the tax rate if the taxes on a refrigerator that sold for $980 were $63.70.

70. The sales tax on an automobile purchase is $750. If the sales tax rate is 6%, what was the total cost of the automobile including tax?

Solve the following problems involving discounts: [Section 8.5]

71. Find the original price of a dining room suite that was discounted by $138 if the discount rate was 30%.

72. Mark paid $664 for a suit that was discounted 20%. What was the original price of the suit?

Solve the following problems involving commissions: [Section 8.5]

73. If Hans sold a used car for $1800 and received a commission of $396, what was his rate of commission?

74. Jim earned a 2.5% commission for selling a $98,000 house. How much money did he earn for selling the house?

Given the interest formula I = p · r · t, *solve for the missing value.* [Section 8.5]

75. $r = 8\%$, $I = \$120$, $p = ?$, $t = \frac{1}{2}$ year

76. Sophia paid $2880 to borrow $4000 for four years. What was the annual interest rate?

77. Patrick borrowed $560 for 90 days at 16% interest. How much interest did he pay?

78. Calculate the balance after one year for an account that pays 7% interest compounded quarterly on a beginning balance of $2500.

Solve the following exercises. If necessary, round to the nearest tenth. [Section 8.5]

79. The federal government expects an 85% increase on its costs to control pollution from 1990 to the year 2003. If the 1990 costs were $100 billion, what will be the cost in the year 2003?

Chapter 8   Test

Simplify the ratio.

1. $8 : 6 : 10$

Write the ratio in lowest terms.

2. At a certain college, 840 faculty members teach 84,000 students. What is the ratio of faculty to students?

Write as a ratio of whole numbers.

3. A recipe for pancakes calls for 2 cups of pancake mix, $\frac{1}{4}$ cup of cholesterol-free egg product, and $1\frac{1}{4}$ cups of skim milk. What is the ratio of the ingredients?

Express the ratio with the second part equal to 1.

4. A grant of $5,600,000 was given to seven centers for educational research. How much money was given to each center?

Express as a unit price. If necessary, round to the nearest cent.

5. If one-half gallon (64 ounces) of ice cream costs $2.88, what is the price per ounce?

Solve the following proportions:

6. $\frac{15}{27} = \frac{y}{18}$

7. $\frac{3}{4} = \frac{2x+1}{4x-4}$

8. $\frac{x+2}{-2} = \frac{4}{x+8}$

Solve each of the following:

9. If a washing machine can wash three loads of clothes in 75 minutes, how many minutes will it take to wash seven loads of clothes?

10. In Europe, a forester wants to know how many wild hogs there are in a forest. He captures 12 hogs, tags them, and releases them. A few months later, he captures 18 hogs and finds that 4 are tagged. Estimate the number of hogs in the forest.

Convert the fractions to percents. If necessary, round to the nearest tenth of a percent.

11. $\frac{8}{5}$

12. $\frac{4}{7}$

Convert the decimals to percents. If necessary, round to the nearest tenth of a percent.

13. .308

14. 7.666. . .

Convert the percents to fractions.

15. 37.5%

16. $9\frac{1}{4}\%$

Solve each of the following. Round the answers to the nearest percent or nearest hundredth where appropriate.

17. 950 is 102% of what number?

18. If 350 jobs are 8% of the jobs in a company, how many jobs are in the company?

19. A major airline decided to cancel 97 of their 291 flights in a certain state. What percent of their flights in that state did they cancel?

Solve each of the following:

20. Fred borrowed $2400 at an 18.5% interest rate for a period of 90 days. How much interest did he pay?

21. Helene paid $25.50 in sales tax on a dishwasher that costs $425. What was the sales tax rate?

22. An item marked at $350 was discounted 15%. How much was the purchase price?

23. A book sales representative earns 5% commission on all books sold. If he sold 20,000 books at $35 each, what was his commission?

24. In 1991, United Press International announced a 35% cut in management salaries. If an executive was making $45,000 a year, what was his new salary?

25. Find the balance after one year in a savings account that pays 6% interest compounded semiannually if the beginning balance is $12,000.

26. Francine purchased a boat for $18,500. If the sales tax rate is 7%, what was the total cost of the boat?

In this chapter, we continue our study of linear equations with two variables. We know from Chapter 4 that a linear equation with two variables has an infinite number of solutions that are written as ordered pairs. In this chapter, we will be finding the single ordered pair (if there is one) that solves two linear equations with two variables. We will learn three techniques for finding this point, if it exists. These techniques also provide us with another technique for solving application problems with two unknowns.

Systems
of Linear
Equations

Chapter

9

Defining Linear Systems and Solving by Graphing

OBJECTIVES

When you complete this section, you will be able to:

a. Determine whether a given ordered pair is the solution of a linear system with two variables.

b. Solve a linear system with two variables by graphing.

c. Determine whether a linear system with two variables is consistent, inconsistent, or dependent.

Introduction   As mentioned in the introduction, in Chapter 4 we found that a linear equation with two variables has an infinite number of solutions, each of which is an ordered pair. In this section, we will be looking for one ordered pair that is the solution of two linear equations. For example, some solutions of $x + y = 4$ are $(0, 4)$, $(4, 0)$, $(1, 3)$, $(2, 2)$, $(5, -1)$, $(-1, 5)$, and $(-2, 6)$. Some solutions of $x - y = 6$ are $(0, -6)$, $(6, 0)$, $(1, -5)$, $(3, -3)$, $(5, -1)$, $(-2, -8)$, and $(-3, -9)$. By examining the ordered pairs that are solutions of each equation, we see that the ordered pair $(5, -1)$ is a solution of both equations. The equations $x + y = 4$ and $x - y = 6$ are an example of a **system of linear equations.** Therefore, $(5, -1)$ is a solution of this system of equations. Systems of equations are usually written with one equation beneath the other as in the following:

$$\begin{cases} x + y = 4 \\ x - y = 6 \end{cases}$$

System of Linear Equations

A system of linear equations is two or more linear equations with the same variables.

We observed that the ordered pair $(5, -1)$ solved each of the equations $x + y = 4$ and $x - y = 6$. Any ordered pair that solves all the equations in a system of equations is a **solution of the system.**

Solution(s) of a System of Linear Equations

The solution(s) of a system of linear equations with two variables consists of all ordered pairs that solve all the equations of the system.

To determine if an ordered pair is a solution of a system of linear equations, we must show that the ordered pair solves each equation of the system. This means that if we replace x in each equation by the first element of the ordered pair and replace y with the second element of the ordered pair, both equations will result in true statements when both sides are simplified.

Example 1

Determine if the given ordered pair is a solution of the system of linear equations. The ordered pair is in the form (x, y).

a. $(3, -1)$

$$\begin{cases} 2x + y = 5 \\ 3x - y = 10 \end{cases}$$

Solution:
To determine whether the given ordered pair is a solution of the system, we must determine if it solves each equation in the system. Therefore, we substitute 3 for x and -1 for y in each equation and simplify.

$$2x + y = 5 \qquad\qquad 3x - y = 10$$
$$2(3) + (-1) = 5 \qquad 3(3) - (-1) = 10$$
$$6 - 1 = 5 \qquad\qquad 9 + 1 = 10$$
$$5 = 5 \qquad\qquad 10 = 10$$

Since $(3, -1)$ solves both equations of the system, $(3, -1)$ is a solution of the system.

b. $(2, -3)$
$$\begin{cases} 4x + 3y = -1 \\ 2x - 5y = -11 \end{cases}$$

Solution:
Substitute 2 for x and -3 for y in each equation and simplify.

$$4x + 3y = -1 \qquad\qquad 2x - 5y = -11$$
$$4(2) + 3(-3) = -1 \qquad 2(2) - 5(-3) = -11$$
$$8 - 9 = -1 \qquad\qquad 4 + 15 = -11$$
$$-1 = -1 \qquad\qquad 19 \neq -11$$

Since $(2, -3)$ does not solve both of the equations in the system, $(2, -3)$ is not a solution of the system.

Practice Exercises

Determine whether the given ordered pair is a solution of the system of linear equations. The ordered pair is in the form (x, y).

1. $(4, 1)$
$$\begin{cases} 3x + y = 13 \\ 2x + y = 9 \end{cases}$$

2. $(-3, 2)$
$$\begin{cases} 2x + 3y = -12 \\ 4x - 5y = 2 \end{cases}$$

If more practice is needed, do the Additional Practice Exercises.

Additional Practice Exercises

Determine whether the given ordered pair is a solution of the system of linear equations. The ordered pair is in the form (x, y).

a. $(2, -4)$
$$\begin{cases} x - 2y = -10 \\ x + 4y = -14 \end{cases}$$

b. $(2, -1)$
$$\begin{cases} 5x + 3y = 7 \\ 4x - y = 9 \end{cases}$$

In the previous examples and practice exercises, the ordered pair was given and we were asked to determine whether the ordered pair was a solution of the system. Now we will find the ordered pair that is the solution of a system of linear equations. Recall from Chapter 4 that if a point lies on the graph of a line, then the coordinates of the point must solve the equation of the line. Consequently, if the graphs of two lines intersect at a point, then the coordinates of the point of intersection must solve both equations. This leads to the following method of solving a system of linear equations with two variables:

Solving a System of Linear Equations by Graphing

To solve a system of linear equations with two variables, graph each equation of the system. If the graphs of the equations intersect, the coordinates of the point(s) of intersection are the solutions of the system.

Example 2

Find the solution(s) of the following systems of linear equations by graphing:

a. $\begin{cases} x + y = 6 \\ x - y = -2 \end{cases}$

Solution:
To find the solution(s) of the system, we graph both equations on the same coordinate system and find the point(s) of intersection of the graphs. Remember, a quick way to graph the equations is by using the intercepts. To find the x-intercept we let $y = 0$ and to find the y-intercept we let $x = 0$.

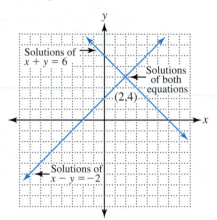

From the graph, we see that the two lines seem to intersect at the point whose coordinates are (2, 4), Therefore, (2, 4) is a possible solution of the system.

Check:
If (2, 4) is the solution of the system, then (2, 4) must solve both equations. Substitute 2 for x and 4 for y into both equations and simplify.

$$
\begin{array}{ll}
x + y = 6 & \quad x - y = -2 \\
2 + 4 = 6 & \quad 2 - 4 = -2 \\
6 = 6 & \quad -2 = -2
\end{array}
$$

Since (2, 4) solves both equations, (2, 4) is the solution of the system.

b. $\begin{cases} 3x + 2y = 4 \\ 2x + 3y = 1 \end{cases}$

Solution:
To find the solution(s) of the system, we graph both equations on the same coordinate system and find the point(s) of intersection of the graphs.

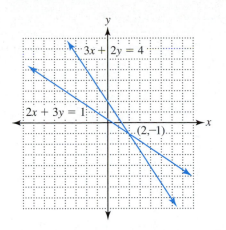

From the graph, we see that the two lines seem to intersect at the point whose coordinates are $(2, -1)$. Therefore, $(2, -1)$ is a possible solution of the system.

Check:

If $(2, -1)$ is the solution of the system, then $(2, -1)$ must solve both equations in the system. Substitute 2 for x and -1 for y into both equations and simplify.

$$3x + 2y = 4 \qquad\qquad 2x + 3y = 1$$
$$3(2) + 2(-1) = 4 \qquad 2(2) + 3(-1) = 1$$
$$6 - 2 = 4 \qquad\qquad 4 - 3 = 1$$
$$4 = 4 \qquad\qquad 1 = 1$$

Since $(2, -1)$ solves both equations, $(2, -1)$ is the solution of the system.

Practice Exercises

Find the solution(s) of the following systems of linear equations with two variables by graphing. Assume each unit on the coordinate system represents 1.

3. $\begin{cases} x + y = 1 \\ x - y = 5 \end{cases}$

4. $\begin{cases} 2x + 3y = 12 \\ 2x + 7 = 8 \end{cases}$

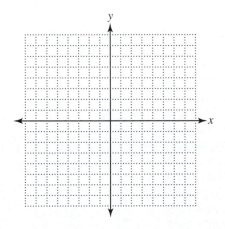

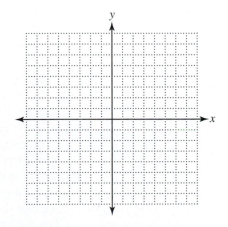

If more practice is needed, do the Additional Practice Exercises.

Additional Practice Exercises

Find the solution(s) of the following linear systems with two variables by graphing. Assume each unit on the coordinate system represents 1.

c. $\begin{cases} x - y = -6 \\ x + 2y = 0 \end{cases}$

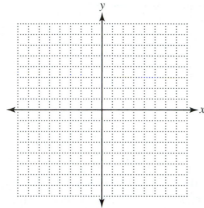

d. $\begin{cases} 3x - 2y = -12 \\ x - 2y = -8 \end{cases}$

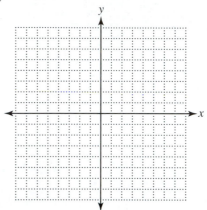

One of the problems with solving systems of equations graphically is that the two lines do not always intersect at a point whose coordinates are integers. In this case, it may not be possible to determine the exact solution of the system by the graphing method as illustrated in the following example:

Example 3

Find the solution(s) of the following linear system with two variables by graphing. Assume each unit on the coordinate system represents 1.

a. $\begin{cases} 6x + 6y = 5 \\ 2x - 3y = 10 \end{cases}$

Solution:

To find the solution(s) of the system, we graph both equations on the same coordinate system and find the point(s) of intersection of the graphs.

From the graph, we see that the x-value of the point of intersection is between -1 and -2 and the y-value is between 2 and 3. The only way to find the exact coordinates of the point of intersection using the graphical method is to approximate the coordinates and then substitute them into both equations. This can be very time-consuming and there is no guarantee that the exact solution will ever be found. For that reason, in this section we will limit ourselves to systems whose graphs intersect at points whose coordinates are integers . We will learn two other methods that are more efficient for solving systems of linear equations in Sections 9.2 and 9.3.

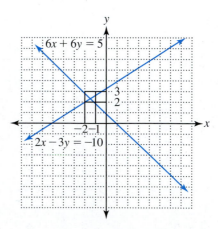

If a system of two linear equations with two variables is graphed on one coordinate system, there are three things that can happen.

1. The lines can intersect in exactly one point as in Examples 2 and the preceding example. If this occurs, the system is said to be **consistent** and the system has one solution, which is the coordinates of the point of intersection.

2. The lines can be parallel. If this occurs, the system is said to be **inconsistent** and the system has no solution, since parallel lines do not intersect.

3. The lines can coincide. If this occurs, the system is said to be **dependent** and there are an infinite number of solutions since the lines intersect at an infinite number of points. Any ordered pair that solves one equation of a dependent system also solves the other. In a dependent system, one equation is a constant multiple of the other.

Examples of inconsistent and dependent systems are given as follows:

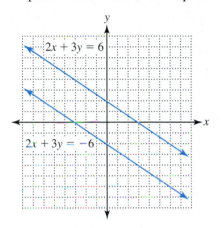

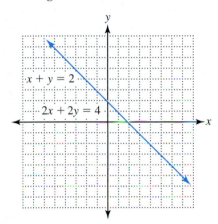

Inconsistent system

No solution

Dependent system

Infinite number of solutions

Example 4

Determine whether the following systems are consistent, inconsistent, or dependent. If the system is consistent, find the solution of the system.

a. $\begin{cases} 2x + y = 4 \\ 4x + 2y = -7 \end{cases}$

Solution:

To determine whether the system is consistent, inconsistent, or dependent, we can draw the graph of the system on one coordinate system.

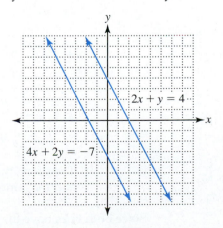

The two lines appear to be parallel. To confirm that they are, we need to find the slope of each line. Remember that parallel lines have equal slopes. To find the slope of each line, write each equation in slope-intercept form.

$$2x + y = 4 \qquad\qquad 4x + 2y = -7$$
$$2x - 2x + y = -2x + 4 \qquad 4x - 4x + 2y = -4x - 7$$
$$y = -2x + 4 \qquad\qquad 2y = -4x - 7$$

Therefore, the slope is -2. $\qquad\qquad y = -2x - \dfrac{7}{2}$

Therefore, the slope is -2.

Since both lines have the same slope, the lines are parallel. Since the lines have different y-intercepts, they are not the same line. Consequently, the system is inconsistent.

Note: It is possible to determine if a system is inconsistent without graphing by writing each equation in slope-intercept form. The lines of an inconsistent system are parallel so they must have the same slope. In order to be distinct lines, rather than the same line, they must have different y-intercepts.

b. $\begin{cases} 4x + 6y = 4 \\ 6x + 9y = 6 \end{cases}$

To determine whether the system is consistent, inconsistent, or dependent, we can draw the graph of the system on one coordinate system.

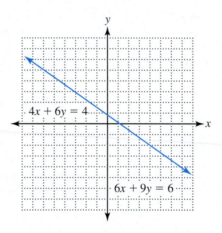

The graphs of the two lines appear to be the same line. To confirm that they are, write each equation in slope-intercept form.

$$4x + 6y = 4 \qquad\qquad 6x + 9y = 6$$
$$4x - 4x + 6y = -4x + 4 \qquad 6x - 6x + 9y = -6x + 6$$
$$6y = -4x + 4 \qquad\qquad 9y = -6x + 6$$
$$\dfrac{6y}{6} = \dfrac{-4x + 4}{6} \qquad\qquad \dfrac{9y}{9} = \dfrac{-6x + 6}{9}$$
$$y = -\dfrac{4}{6}x + \dfrac{4}{6} \qquad\qquad y = -\dfrac{6}{9}x + \dfrac{6}{9}$$
$$y = -\dfrac{2}{3}x + \dfrac{2}{3} \qquad\qquad y = -\dfrac{2}{3}x + \dfrac{2}{3}$$

Therefore, the slope is $-\dfrac{2}{3}$ and the y-intercept is $\dfrac{2}{3}$. $\qquad$ Therefore, the slope is $-\dfrac{2}{3}$ and the y-intercept is $\dfrac{2}{3}$.

Since both lines have the same slope and the same y-intercept, they are the same line. Consequently, the system is dependent. The solutions of a dependent system are usually written as the set of ordered pairs that solve either of the equations. The solutions of this system could be written as $\{(x, y) : 4x + 6y = 4\}$. This is read "the set of all ordered pairs such that $4x + 6y = 4$."

Note: It is possible to determine whether a system is dependent by writing both equations in slope-intercept form. If both lines have the same slope and the same y-intercept, they are the same line. Therefore, the system is dependent.

c. $\begin{cases} 2x - y = 5 \\ 3x + 4y = 24 \end{cases}$

To determine whether the system is consistent, inconsistent, or dependent, we draw the graph of the system on one coordinate system.

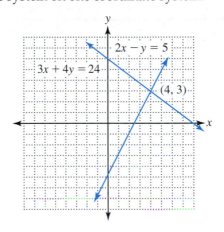

Since the lines intersect at a single point, the system is consistent. The coordinates of the point of intersection are $(4, 3)$. Therefore, $(4, 3)$ is a possible solution of the system.

Check:
The check is left as an exercise for the student.

Note: It is possible to determine if a system is consistent without graphing by writing the equations of the system in slope-intercept form. Since any two lines in the same plane that are not parallel must intersect, the system is consistent if the slopes of the lines are not the same.

Practice Exercises

Determine whether the following systems are consistent, inconsistent, or dependent. If the system is consistent, find the solution of the system.

5. $\begin{cases} 3x + 2y = 12 \\ 6x + 4y = -12 \end{cases}$

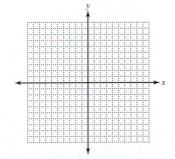

6. $\begin{cases} x + 2y = 6 \\ 2x = 12 - 4y \end{cases}$

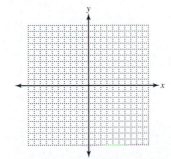

7. $\begin{cases} 5x + 3y = 30 \\ x + y = 8 \end{cases}$

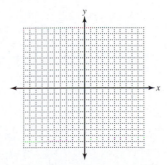

If more practice is needed, do the Additional Practice Exercises.

Additional Practice Exercises

Determine whether the following systems are consistent, inconsistent, or dependent. If the system is consistent, find the solution of the system.

e. $\begin{cases} x + 2y = 6 \\ 2x = -4y - 7 \end{cases}$

f. $\begin{cases} 4x + y = 7 \\ 2y = -8x + 14 \end{cases}$

g. $\begin{cases} x + 2y = 3 \\ y = -2x + 12 \end{cases}$

Exercise Set 9.1

Determine whether the given ordered pair is a solution of the system of linear equations. The ordered pair is in the form (x, y).

1. $(3, 4)$

$\begin{cases} x + 3y = 15 \\ x + 2y = 11 \end{cases}$

2. $(-4, 1)$

$\begin{cases} 2x + y = -7 \\ 3x + y = -11 \end{cases}$

3. $(-2, -3)$

$\begin{cases} 4x + 2y = 5 \\ x + 3y = 1 \end{cases}$

4. $(1, 6)$

$\begin{cases} x - y = -1 \\ 2x + y = 7 \end{cases}$

5. $(-1, 2)$

$\begin{cases} x - 3y = -7 \\ 6x + 2y = -2 \end{cases}$

6. $(5, 1)$

$\begin{cases} 4x - 5y = 13 \\ x + y = 17 \end{cases}$

7. $(-2, 0)$

$\begin{cases} 2x + 3y = 4 \\ x + 2y = 4 \end{cases}$

8. $(0, 3)$

$\begin{cases} 5x + 3y = 9 \\ -4x + 7y = 21 \end{cases}$

Find the solution(s) of the following systems of linear equations by graphing. Assume each unit on the coordinate system represents 1.

9. $\begin{cases} x + y = 1 \\ x - y = -1 \end{cases}$

10. $\begin{cases} x + y = -1 \\ x - y = 3 \end{cases}$

11. $\begin{cases} 3x - 2y = -12 \\ 2x + y = -3 \end{cases}$

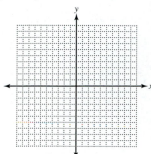

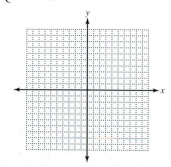

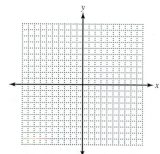

12. $\begin{cases} x - 2y = 2 \\ x + 4y = 8 \end{cases}$

13. $\begin{cases} x - y = -7 \\ x = -3 \end{cases}$

14. $\begin{cases} y = 5 \\ x + y = 4 \end{cases}$

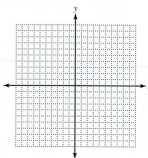

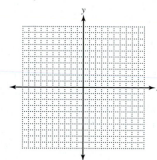

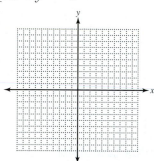

15. $\begin{cases} 2x - y = -4 \\ 2x - 3y = 0 \end{cases}$

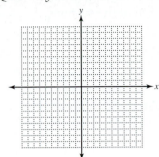

16. $\begin{cases} x - 4y = 8 \\ x + 2y = 2 \end{cases}$

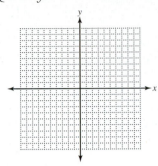

17. $\begin{cases} x - y = 4 \\ x + 5y = 10 \end{cases}$

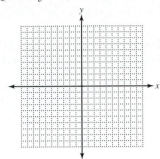

18. $\begin{cases} x - 2y = -12 \\ x + 4y = 6 \end{cases}$

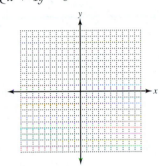

19. $\begin{cases} y = 3 \\ 2x - y = 3 \end{cases}$

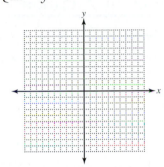

20. $\begin{cases} x = -5 \\ 4x - 5y = -40 \end{cases}$

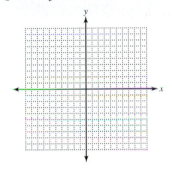

21. $\begin{cases} 6x - 3y = 0 \\ 3x - 2y = 0 \end{cases}$

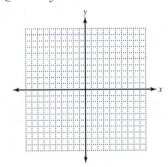

22. $\begin{cases} 4x - 2y = 4 \\ 2x + 3y = -6 \end{cases}$

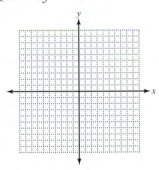

Determine whether the following systems are consistent, inconsistent, or dependent. If the system is consistent, find the solution of the system. Grids are not provided as with 9–22.

23. $\begin{cases} x - 3y = -12 \\ 4x - 12y = -36 \end{cases}$

24. $\begin{cases} 3x + 2y = 6 \\ 6x + 4y = 24 \end{cases}$

25. $\begin{cases} 2x + 3y = 9 \\ 4x + 6y = 12 \end{cases}$

26. $\begin{cases} 3x + y = 6 \\ 6x + 2y = 18 \end{cases}$

27. $\begin{cases} 3x + y = -6 \\ 9x + 3y = 9 \end{cases}$

28. $\begin{cases} -2x - y = 4 \\ 6x + 3y = 18 \end{cases}$

29. $\begin{cases} x + y = -5 \\ 6x - y = 12 \end{cases}$

30. $\begin{cases} 3x - 2y = 12 \\ x + y = -1 \end{cases}$

Challenge Exercises:

Determine whether the given ordered pair is a solution of the system of linear equations. The ordered pair is in the form (x, y).

31. $\left(\dfrac{1}{4}, 2 \right)$

$\begin{cases} 4x + y = 3 \\ 8x + 3y = 8 \end{cases}$

32. $(-24, 12)$

$\begin{cases} 2x - 3y = -84 \\ x + 5y = 36 \end{cases}$

Writing Exercises:

33. If the slopes of two lines are not equal, what type of system is composed of these lines? Why?

34. If the slopes of two lines are equal, what type of system(s) is/are composed of these lines? Why?

35. Why are the coordinates of the point of intersection of two lines the solution of the system?

Critical-Thinking Exercises:

36. Find a system of linear equations whose solution is (2, 3), draw the graph of the system, and explain how you found the equations of the system.

37. Find a system of equations with no solution and describe how you found the equations of the system.

Section 9.2

Solving Systems of Linear Equations Using Elimination by Addition

OBJECTIVES

When you complete this section, you will be able to:

a. Solve consistent systems of linear equations with two variables by eliminating a variable using the addition method.

b. Recognize inconsistent and dependent systems of equations when using the addition method.

Introduction

We found in Section 9.1 that one of the difficulties with solving systems of equations using the graphing method was that the lines did not always intersect at a point whose coordinates were integers. In this section, we will develop another method of solving systems of linear equations with two variables without graphing the system. It can be shown that if two lines intersect at a point, then the sum of the equations of the lines, or any multiples of these equations, is the equation of another line that intersects the two given lines at the same point of intersection.

In Chapter 3, we introduced the Addition Property of Equality, which stated that if $a = b$, then $a + c = b + c$. We will need an extension of this property, which is given as follows:

Addition of Equals

If $a = b$ and $c = d$, then $a + c = b + d$. In words, equals added to equals results in equals.

In the following, we repeat Example 2 from Section 9.1.

Find the solution(s) of the following systems of linear equations by graphing:

a. $\begin{cases} x + y = 6 \\ x - y = -2 \end{cases}$

Solution:

To find the solution(s) of the system, we graph both equations on the same coordinate system and find the point(s) of intersection of the graphs.

From the graph, we see that the two lines seem to intersect at the point whose coordinates are (2, 4), Therefore, (2, 4) is a possible solution of the system.

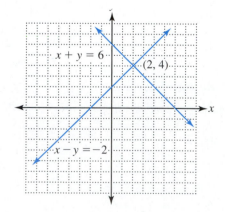

Notice that if we added the two equations, the *y*s would drop out.

$$x + y = 6$$ Add the equations.

$$\underline{x - y = -2}$$ Divide both sides of the equation by 2.

$$2x \quad\;\; = 4$$ *Note:* This is the *x*-value of the point of intersection

$$x = 2$$ of the two lines because it represents the equation of a vertical line that passes through the point of intersection. To find the *y*-value, substitute 2 for *x* in either of the original equations.

Use $x + y = 6$.

$$x + y = 6$$ Substitute 2 for *x*.

$$2 + y = 6$$ Subtract 2 from both sides of the equation.

$$y = 4$$ *Note:* This is the *y*-value of the point of intersection. Therefore, we have solved the system without graphing it.

Summarizing the preceding, the addition of equals property and the fact that the sums of multiples of a system's equations give another equation whose graph passes through the same point of intersection as the original system lead to another method of solving systems of linear equations. Following is an outline of the procedure, which will be made clearer by some subsequent examples:

Solving Systems of Linear Equations Using Elimination by Addition

1. If necessary, rewrite each equation in the form $ax + by = c$ with a, b, and c integers.

2. Choose the variable to be eliminated.

3. If necessary, multiply one or both of the equations of the system by the appropriate constant(s) so the coefficients of the variable to be eliminated are additive inverses.

4. Add the equations.

5. If necessary, solve the equation resulting from step 4 for the remaining variable.

6. Substitute the value found for the variable in step 5 into either of the equations of the original system and solve for the other variable.

7. Check the solution in both equations of the system.

We illustrate this procedure with some examples.

Example 1

Solve the following systems of equations using elimination by addition:

a. $\begin{cases} x + y = 3 \\ x - y = 9 \end{cases}$

Solution:
It is easier to eliminate *y* than to eliminate *x* because we can eliminate *y* simply by adding the two equations.

$$x + y = 3$$

$$\underline{x - y = 9}$$ Add the equations.

$$2x \quad\;\; = 12$$ Divide both sides of the equation by 2.

$$x = 6$$ Substitute $x = 6$ in either equation to find *y*.
Use $x + y = 3$.

$$x + y = 3$$ Substitute 6 for *x*.

$$6 + y = 3$$ Subtract 6 from both sides of the equation.

$$y = -3$$ Therefore, the solution of the system is $(6, -3)$.

Check:

Substitute 6 for x and -3 for y in both equations.

$$
\begin{array}{ll}
x + y = 3 & x - y = 9 \\
6 + (-3) = 3 & 6 - (-3) = 9 \\
3 = 3 & 6 + 3 = 9 \\
& 9 = 9
\end{array}
$$

Since $(6, -3)$ solves both equations, it is the correct solution.

b. $\begin{cases} 2x + 3y = 1 \\ x + 4y = -7 \end{cases}$

Solution:

Notice that neither variable will be eliminated if we add the equations just as they are. In order for a variable to be eliminated, the coefficients of the variables must be additive inverses. It appears easier to make the coefficients of x additive inverses since we would have to multiply only one equation by a constant. Therefore, we will eliminate x by multiplying $x + 4y = -7$ by -2 and then adding the equations.

$$2x + 3y = 1 \qquad \text{Leave } 2x + 3y = 1 \text{ unchanged.}$$
$$x + 4y = -7 \qquad \text{Multiply } x + 4y = -7 \text{ by } -2.$$

$$2x + 3y = 1 \qquad \text{Leave } 2x + 3y = 1 \text{ unchanged.}$$
$$-2(x + 4y) = -2(-7) \qquad \text{Simplify both sides of the equation.}$$

$$
\begin{array}{ll}
\begin{array}{r} 2x + 3y = 1 \\ -2x - 8y = 14 \\ \hline -5y = 15 \end{array} & \text{Add the equations.} \\
& \text{Divide both sides of the equation by } -5. \\
y = -3 & \text{Substitute } -3 \text{ for } y \text{ in either equation to find } x. \\
& \text{Use } x + 4y = -7. \\
x + 4y = -7 & \text{Substitute } -3 \text{ for } y. \\
x + 4(-3) = -7 & \text{Multiply 4 and } -3. \\
x - 12 = -7 & \text{Add 12 to both sides of the equation.} \\
x = 5 & \text{Therefore, the solution is } (5, -3).
\end{array}
$$

Check:

Substitute 5 for x and -3 for y in both equations.

$$
\begin{array}{ll}
2x + 3y = 1 & x + 4y = -7 \\
2(5) + 3(-3) = 1 & 5 + 4(-3) = -7 \\
10 - 9 = 1 & 5 - 12 = -7 \\
1 = 1 & -7 = -7
\end{array}
$$

Since $(5, -3)$ solves both equations, it is the correct solution.

c. $\begin{cases} 3x - 2y = 7 \\ 4x - 3y = 10 \end{cases}$

Solution:

If we add the equations as they are, neither variable would be eliminated. Also, we cannot multiply just one equation by an integer to make the coefficients of either variable additive inverses. Consequently, we must multiply both equations. If we choose to eliminate x, we must make the coefficients of x the least common multiple of 3 and 4, but with opposite signs. The LCM of 3 and 4 is 12. If we choose to eliminate y, we must make the coefficients of y the LCM of 2 and 3, but with opposite signs. The LCM of 2 and 3 is 6. Since 6 is smaller than 12 (It's as good a reason as any!), we choose to eliminate y by making one coefficient 6 and the other -6.

$$3x - 2y = 7 \qquad \text{Multiply } 3x - 2y = 7 \text{ by 3.}$$
$$4x - 3y = 10 \qquad \text{Multiply } 4x - 3y = 10 \text{ by } -2.$$

$$3(3x - 2y) = 3(7) \qquad \text{Simplify both sides of the equation.}$$
$$-2(4x - 3y) = -2(10) \qquad \text{Simplify both sides of the equation.}$$

$$\begin{array}{l} 9x - 6y = 21 \\ \underline{-8x + 6y = -20} \\ \qquad x = 1 \end{array} \qquad \begin{array}{l} \text{Add the equations.} \\ \text{Substitute 1 for } x \text{ in either equation and solve for } y. \\ \text{Use } 3x - 2y = 7. \end{array}$$

$$3x - 2y = 7 \qquad \text{Substitute 1 for } x.$$
$$3(1) - 2y = 7 \qquad \text{Multiply 3 and 1.}$$
$$3 - 2y = 7 \qquad \text{Subtract 3 from both sides of the equation.}$$
$$-2y = 4 \qquad \text{Divide both sides of the equation by } -2.$$
$$y = -2 \qquad \text{Therefore, the solution is } (1, -2).$$

Check: The check is left as an exercise for the student.

Note: Example 1d, which follows, exhibits the advantage of the addition method when the solutions are fractions instead of integers.

d. $\begin{cases} x + 3y = -1 \\ 3x + 6y = -1 \end{cases}$

Solution:
Let us eliminate x by multiplying $x + 3y = -1$ by -3 and adding the equations.

$$x + 3y = -1 \qquad \text{Multiply } x + 3y = -1 \text{ by } -3.$$
$$3x + 6y = -1 \qquad \text{Leave } 3x + 6y = -1 \text{ unchanged.}$$

$$-3(x + 3y) = -3(-1) \qquad \text{Simplify both sides of the equation.}$$
$$3x + 6y = -1 \qquad \text{Leave unchanged.}$$

$$\begin{array}{l} -3x - 9y = 3 \\ \underline{\quad 3x + 6y = -1} \\ \qquad -3y = 2 \end{array} \qquad \begin{array}{l} \text{Divide both sides of the equation by } -3. \\ \text{Add the equations.} \end{array}$$

$$y = -\frac{2}{3} \qquad \begin{array}{l} \text{Substitute } -\frac{2}{3} \text{ for } y \text{ in either equation and} \\ \text{solve for } x. \text{ Use } x + 3y = -1. \end{array}$$

$$x + 3y = -1 \qquad \text{Substitute } -\frac{2}{3} \text{ for } y.$$

$$x + 3\left(-\frac{2}{3}\right) = -1 \qquad \text{Multiply 3 and } -\frac{2}{3}.$$

$$x - 2 = -1 \qquad \text{Add 2 to both sides of the equation.}$$

$$x = 1 \qquad \text{Therefore, } \left(1, -\frac{2}{3}\right) \text{ is the solution.}$$

Check:
Substitute 1 for x and $-\frac{2}{3}$ for y in both equations.

$$x + 3y = -1 \qquad\qquad 3x + 6y = -1$$
$$1 + 3\left(-\frac{2}{3}\right) = -1 \qquad\qquad 3(1) + 6\left(-\frac{2}{3}\right) = -1$$
$$1 - 2 = -1 \qquad\qquad 3 - 4 = -1$$
$$-1 = -1 \qquad\qquad -1 = -1$$

Since $\left(1, -\frac{2}{3}\right)$ solves both equations, $\left(1, -\frac{2}{3}\right)$ is the correct solution.

Practice Exercises

Solve the following systems of equations using elimination by addition:

1. $\begin{cases} x - y = 5 \\ x + y = 1 \end{cases}$

2. $\begin{cases} 3x + 4y = 6 \\ x - 3y = -11 \end{cases}$

3. $\begin{cases} 4x - 3y = -23 \\ 5x - 4y = -30 \end{cases}$

4. $\begin{cases} 12x + 18y = 17 \\ 6x + 10y = 9 \end{cases}$

If more practice is needed, do the Additional Practice Exercises.

Additional Practice Exercises

Solve the following systems of equations:

a. $\begin{cases} x + 2y = 10 \\ x - 2y = -6 \end{cases}$

b. $\begin{cases} x + 3y = -3 \\ 5x + 2y = 24 \end{cases}$

c. $\begin{cases} 4x + 5y = 2 \\ 3x - 2y = -10 \end{cases}$

d. $\begin{cases} 4x - 4y = -1 \\ 12x - 8y = 3 \end{cases}$

When one of the variables has a fractional value and this value is substituted back into one of the equations to find the other variable, it is often necessary to combine fractions. Sometimes it is easier to repeat the elimination procedure to find the remaining variable rather than to substitute a fraction and solve the resulting equation with fractions.

Example 2

Solve the following system of linear equations:

$$\begin{cases} 3x - 4y = 15 \\ 4x + 2y = 7 \end{cases}$$

Solution:
Since we can eliminate y by multiplying $4x + 2y = 7$ by 2 and then adding, we choose to eliminate y.

$3x - 4y = 15$	Leave $3x - 4y = 15$ unchanged.
$4x + 2y = 7$	Multiply $4x + 2y = 7$ by 2.
$3x - 4y = 15$	Leave unchanged.
$2(4x + 2y) = 2(7)$	Simplify both sides of the equation.
$3x - 4y = 15$	
$\underline{8x + 4y = 14}$	Add the equations.
$11x \quad\quad = 29$	Divide both sides of the equation by 11.
$x = \dfrac{29}{11}$	Therefore, the x-value of the solution is $\dfrac{29}{11}$.

We could find y by substituting for x in either equation and solving for y, but that does not look like it would be a whole lot of fun. Instead, let us repeat the elimination procedure and eliminate x this time. Since the LCM of 3 and 4 is 12, we need to make one coefficient of x equal to 12 and the other -12.

$$3x - 4y = 15 \qquad \text{Multiply } 3x - 4y = 15 \text{ by } 4.$$
$$4x + 2y = 7 \qquad \text{Multiply } 4x + 2y = 7 \text{ by } -3.$$

$$4(3x - 4y) = 4(15) \qquad \text{Simplify both sides of the equation.}$$
$$-3(4x + 2y) = -3(7) \qquad \text{Simplify both sides of the equation.}$$

$$12x - 16y = 60$$
$$\underline{-12x - 6y = -21} \qquad \text{Add the equations.}$$
$$-22y = 39 \qquad \text{Divide both sides of the equation by } -22.$$
$$y = -\frac{39}{22} \qquad \text{Therefore, the } y\text{-value of the solution is } -\frac{39}{22}.$$

Therefore, the solution of the system is $\left(\frac{29}{11}, -\frac{39}{22}\right)$.

Check:
Substitute $\frac{29}{11}$ for x and $-\frac{39}{22}$ for y in both equations.

$$3x - 4y = 15 \qquad\qquad 4x + 2y = 7$$
$$3\left(\frac{29}{11}\right) - 4\left(-\frac{39}{22}\right) = 15 \qquad 4\left(\frac{29}{11}\right) + 2\left(-\frac{39}{22}\right) = 7$$
$$\frac{87}{11} + \frac{156}{22} = 15 \qquad\qquad \frac{116}{11} - \frac{78}{22} = 7$$
$$\frac{174}{22} + \frac{156}{22} = 15 \qquad\qquad \frac{232}{22} - \frac{78}{22} = 7$$
$$\frac{330}{22} = 15 \qquad\qquad\qquad \frac{154}{22} = 7$$
$$15 = 15 \qquad\qquad\qquad 7 = 7$$

Since $\left(\frac{29}{11}, -\frac{39}{22}\right)$ solves both equations, it is the correct solution.

Practice Exercises

Solve the following systems of equations:

5. $\begin{cases} 3x + 2y = 9 \\ 2x - 5y = 12 \end{cases}$
    6. $\begin{cases} 4x + y = 6 \\ 3x + 2y = 9 \end{cases}$

Before the elimination by addition method can be used, both equations should be written in the form of $ax + by = c$ where a, b, and c are integers. If necessary, rewrite the equations of the system in this form before attempting to solve the system.

Example 3

Solve the following systems of equations using elimination by addition:

a. $\begin{cases} x = 3y + 22 \\ 5y = -2x - 22 \end{cases}$

Solution:
We first have to rewrite each equation in $ax + by = c$ form.

$$x = 3y + 22 \qquad\qquad 5y = -2x - 22$$
$$x - 3y = 3y - 3y + 22 \qquad 2x + 5y = -2x + 2x - 22$$
$$x - 3y = 22 \qquad\qquad 2x + 5y = -22$$

Now solve the system $x - 3y = 22$ and $2x + 5y = -22$ by eliminating x.

$x - 3y = 22$	Multiply $x - 3y = 22$ by -2.
$2x + 5y = -22$	Leave $2x + 5y = -22$ unchanged.
$-2(x - 3y) = -2(22)$	Simplify both sides of the equation.
$2x + 5y = -22$	Leave unchanged.

$$-2x + 6y = -44$$
$$\underline{2x + 5y = -22} \qquad\quad \text{Add the equations.}$$

$11y = -66$	Divide both sides of the equation by 11.
$y = -6$	Substitute $y = -6$ in either of the original equations and find x. Use $x = 3y + 22$.
$x = 3y + 22$	Substitute -6 for y.
$x = 3(-6) + 22$	Multiply 3 and -6.
$x = -18 + 22$	Add -18 and 22.
$x = 4$	Therefore, the solution of the system is $(4, -6)$.

Check: Substitute 4 for x and -6 for y in both of the original equations. The check is left as an exercise for the student.

b.
$$\begin{cases} \dfrac{1}{5}x + \dfrac{1}{2}y = \dfrac{1}{5} \\[2mm] \dfrac{1}{2}x + \dfrac{1}{3}y = -\dfrac{4}{3} \end{cases}$$

Solution:

The coefficients of x and y should be integers since it is difficult to work with fractions. Multiply each equation by its LCD to eliminate fractions.

$\dfrac{1}{5}x + \dfrac{1}{2}y = \dfrac{1}{5}$	Multiply this equation by the LCD 10.
$\dfrac{1}{2}x + \dfrac{1}{3}y = \dfrac{4}{3}$	Multiply this equation by the LCD 6.
$10\left(\dfrac{1}{5}x + \dfrac{1}{2}y\right) = 10\left(\dfrac{1}{5}\right)$	Simplify both sides of the equation.
$6\left(\dfrac{1}{2}x + \dfrac{1}{3}y\right) = 6\left(-\dfrac{4}{3}\right)$	Simplify both sides of the equation.
$2x + 5y = 2$	To eliminate x, multiply this equation by -3.
$3x + 2y = -8$	To eliminate x, multiply this equation by 2.
$-3(2x + 5y) = -3(2)$	Simplify both sides of the equation.
$2(3x + 2y) = 2(-8)$	Simplify both sides of the equation.

$$-6x - 15y = -6$$
$$\underline{6x + 4y = -16} \qquad\quad \text{Add the equations.}$$

$-11y = -22$	Divide both sides of the equation by -11.
$y = 2$	Substitute $y = 2$ into either of the original equations to find x. Use $\dfrac{1}{5}x + \dfrac{1}{2}y = \dfrac{1}{5}$.

$$\frac{1}{5}x + \frac{1}{2}y = \frac{1}{5}$$ Substitute 2 for y.

$$\frac{1}{5}x + \frac{1}{2}(2) = \frac{1}{5}$$ Multiply $\frac{1}{2}$ and 2.

$$\frac{1}{5}x + 1 = \frac{1}{5}$$ Subtract 1 from both sides of the equation.

$$\frac{1}{5}x + 1 - 1 = \frac{1}{5} - 1$$ Add. Write 1 as $\frac{5}{5}$.

$$\frac{1}{5}x = \frac{1}{5} - \frac{5}{5}$$ Add.

$$\frac{1}{5}x = -\frac{4}{5}$$ Multiply both sides of the equation by 5.

$$5\left(\frac{1}{5}x\right) = 5\left(-\frac{4}{5}\right)$$ Simplify both sides of the equation.

$$x = -4$$ Therefore, the solution of the system is $(-4, 2)$.

Check:
The check is left as an exercise for the student.

Practice Exercises

Solve the following systems of equations:

7. $\begin{cases} 5y = 5 - 3x \\ x = -2y + 1 \end{cases}$

8. $\begin{cases} \dfrac{1}{4}x - \dfrac{1}{2}y = -\dfrac{1}{4} \\ \dfrac{1}{3}x + \dfrac{1}{6}y = \dfrac{4}{3} \end{cases}$

If more practice is needed, do the Additional Practice Exercises.

Additional Practice Exercises

Solve the following systems of equations:

e. $\begin{cases} y = 2x + 2 \\ x = -3y + 13 \end{cases}$

f. $\begin{cases} \dfrac{1}{3}x + \dfrac{1}{5}y = \dfrac{1}{3} \\ x + \dfrac{1}{2}y = \dfrac{1}{2} \end{cases}$

Remember, if a system is inconsistent, it has no solutions. If a system is dependent, it has an infinite number of solutions. How can we tell if a system is inconsistent or dependent if we are using the addition method and do not have the graphs? Study the following examples:

Answers:

Example 4

Solve the following systems of equations:

a. $\begin{cases} 4x - 2y = 7 \\ 2x - y = 4 \end{cases}$

Solution:

Let us eliminate y. Make one coefficient -2 and the other 2.

$$4x - 2y = 7 \qquad \text{Leave } 4x - 2y = 7 \text{ unchanged.}$$
$$2x - y = 4 \qquad \text{Multiply } 2x - y = 4 \text{ by } -2.$$

$$4x - 2y = 7 \qquad \text{Leave } 4x - 2y = 7 \text{ unchanged.}$$
$$-2(2x - y) = -2(4) \qquad \text{Simplify both sides of the equation.}$$

$$\begin{aligned} 4x - 2y &= 7 \\ -4x + 2y &= -8 \qquad \text{Add the equations.} \\ \hline 0 &= -1 \end{aligned}$$

This is a false statement, which means that there is no solution. This is denoted by the symbol $\varnothing$. The lines are parallel and the system is inconsistent.

b. $\begin{cases} 4x + 2y = 6 \\ 6x + 3y = 9 \end{cases}$

Solution: Since the coefficients of y are smaller than the coefficients of x, we will eliminate y. Since the LCM of 2 and 3 is 6, we will make one coefficient of y equal to 6 and the other equal to -6.

$$4x + 2y = 6 \qquad \text{Multiply } 4x + 2y = 6 \text{ by } 3.$$
$$6x + 3y = 9 \qquad \text{Multiply } 6x + 3y = 9 \text{ by } -2.$$

$$3(4x + 2y) = 3(6) \qquad \text{Simplify both sides of the equation.}$$
$$-2(6x + 3y) = -2(9) \qquad \text{Simplify both sides of the equation.}$$

$$\begin{aligned} 12x + 6y &= 18 \\ -12x - 6y &= -18 \qquad \text{Add the equations.} \\ \hline 0 &= 0 \end{aligned}$$

This is a true statement for all values of x and y that solve either equation. Hence, there are an infinite number of solutions, which means the system is dependent. We indicate the solutions as $\{(x, y): 4x + 2y = 6\}$ or $\{(x, y): 6x + 3y = 9\}$

Practice Exercises

Solve the following systems of equations. Indicate whether the system is consistent, inconsistent, or dependent.

9. $\begin{cases} 6x + 2y = 6 \\ y = -3x + 3 \end{cases}$

10. $\begin{cases} 12x + 6y = 9 \\ 4x + 2y = 5 \end{cases}$

If more practice is needed, do the Additional Practice Exercises.

Additional Practice Exercises

Solve the following systems of equations. Indicate whether the system is consistent, inconsistent, or dependent.

g. $\begin{cases} 6x + 3y = 5 \\ 4x + 2y = 7 \end{cases}$

h. $\begin{cases} x = -2y + 6 \\ \dfrac{1}{2}x + y = 3 \end{cases}$

Answers:

Practice Exercises 9–10: 9. dependent 10. inconsistent   **Additional Practice Exercises 8–h:** g. inconsistent h. dependent

Exercise Set 9.2

Solve the following systems of equations using elimination by addition and verify the solution from the graph:

1. $\begin{cases} x - y = 2 \\ x + y = 2 \end{cases}$

2. $\begin{cases} y + x = 2 \\ y - x = 8 \end{cases}$

3. $\begin{cases} x + y = 3 \\ 3x - y = 1 \end{cases}$

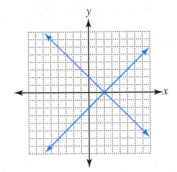
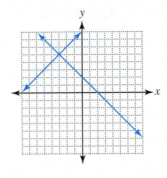
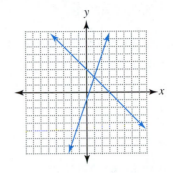

Solve the following systems of equations using elimination by addition.

4. $\begin{cases} x + 3y = -9 \\ -x + 2y = -11 \end{cases}$

5. $\begin{cases} 5x + y = -17 \\ 4x - y = -10 \end{cases}$

6. $\begin{cases} -x + 6y = -30 \\ x - 4y = 20 \end{cases}$

7. $\begin{cases} 5x + y = -1 \\ x - 3y = -13 \end{cases}$

8. $\begin{cases} x - 2y = -4 \\ -2x + 5y = 9 \end{cases}$

9. $\begin{cases} 4x + 5y = 24 \\ -x + 3y = -6 \end{cases}$

10. $\begin{cases} 3x - 5y = -17 \\ 4x + y = -15 \end{cases}$

11. $\begin{cases} 7x - 4y = 4 \\ 5x + y = 26 \end{cases}$

12. $\begin{cases} 5x - 2y = 0 \\ -x + 4y = 0 \end{cases}$

13. $\begin{cases} 3x + 7y = 8 \\ 2x - 4y = 14 \end{cases}$

14. $\begin{cases} 12x - 2y = -54 \\ 13x + 4y = -40 \end{cases}$

15. $\begin{cases} 4x + 5y = -4 \\ 3x + 8y = -20 \end{cases}$

16. $\begin{cases} 10x - 3y = 14 \\ -3x + 9y = 12 \end{cases}$

17. $\begin{cases} 3x - 4y = -14 \\ 5x + 7y = 45 \end{cases}$

18. $\begin{cases} -5x + 4y = -31 \\ 13x - 6y = 85 \end{cases}$

19. $\begin{cases} 3x + 2y = 8 \\ x + 4y = 1 \end{cases}$

20. $\begin{cases} 4x + y = -2 \\ 8x - y = 11 \end{cases}$

21. $\begin{cases} 3x - y = 6 \\ 9x + 2y = -2 \end{cases}$

22. $\begin{cases} 14x + 7y = 6 \\ 7x + 6y = 8 \end{cases}$

23. $\begin{cases} 5x + 6y = 2 \\ 10x + 3y = -2 \end{cases}$

24. $\begin{cases} 7x - 6y = 1 \\ 8x - 12y = 6 \end{cases}$

25. $\begin{cases} 18x + 3y = -1 \\ 4x - 6y = -3 \end{cases}$

26. $\begin{cases} 8x - 18y = -7 \\ 16x + 9y = 11 \end{cases}$

27. $\begin{cases} 13x - 22y = -1 \\ 26x + 33y = 12 \end{cases}$

28. $\begin{cases} 30x + 12y = -3 \\ 15x - 24y = 11 \end{cases}$

29. $\begin{cases} 14x + 14y = 1 \\ 28x - 7y = 23 \end{cases}$

30. $\begin{cases} 20x + 8y = 4 \\ 10x - 12y = 22 \end{cases}$

31. $\begin{cases} x + 2y = 2 \\ 4x = 3y + 36 \end{cases}$

32. $\begin{cases} 5x = -2y - 1 \\ 3x - y = 6 \end{cases}$

33. $\begin{cases} x + 4y + 16 = 0 \\ 3y = 9 - 6x \end{cases}$

34. $\begin{cases} 4x = 8 - 2y \\ 3x = 6y + 21 \end{cases}$

35. $\begin{cases} \dfrac{1}{2}x - \dfrac{1}{4}y = \dfrac{1}{4} \\ \dfrac{2}{3}x + \dfrac{1}{5}y = \dfrac{7}{15} \end{cases}$

36. $\begin{cases} \dfrac{3}{4}x - \dfrac{2}{7}y = -5 \\ \dfrac{5}{8}x + \dfrac{1}{4}y = -\dfrac{3}{4} \end{cases}$

37. $\begin{cases} \dfrac{3}{2}x + \dfrac{1}{4}y = 1 \\ \dfrac{2}{5}x + \dfrac{1}{3}y = \dfrac{4}{3} \end{cases}$
    38. $\begin{cases} \dfrac{1}{2}x - \dfrac{8}{3}y = \dfrac{7}{3} \\ \dfrac{3}{7}x + \dfrac{4}{9}y = \dfrac{88}{63} \end{cases}$
    39. $\begin{cases} .4x + .3y = 4.2 \\ 1.2x + .6y = 12 \end{cases}$

40. $\begin{cases} 2.4x + 3y = 7.2 \\ .8x + 2.4y = 8 \end{cases}$
    41. $\begin{cases} .3x - 1.8y = 2.7 \\ 3.6x + 4y = 6.8 \end{cases}$
    42. $\begin{cases} 6.7x + 3.2y = 2.6 \\ .08x + .5y = 2.34 \end{cases}$

Solve the following systems of equations. Indicate whether the system is consistent, inconsistent, or dependent.

43. $\begin{cases} 4x - 2y = 3 \\ -8x + 4y = -6 \end{cases}$
    44. $\begin{cases} 3x + y = 1 \\ -4x - 2y = -6 \end{cases}$
    45. $\begin{cases} -2x + 5y = 8 \\ x + 4y = 22 \end{cases}$

46. $\begin{cases} x = 2y + 6 \\ 3x = 6y + 9 \end{cases}$
    47. $\begin{cases} 2x - 5y = 4 \\ -8x + 20y = -20 \end{cases}$
    48. $\begin{cases} 5x + y = 4 \\ 3y = 12 - 15x \end{cases}$

Challenge Exercises:

Solve the following systems of equations:

49. $\begin{cases} 6x + 9y = -3 \\ 5x + 37y = -32 \end{cases}$
    50. $\begin{cases} 5x - 7y = -73 \\ -12x + 6y = 132 \end{cases}$
    51. $\begin{cases} 3x - 8y = 11 \\ 12x + 10y = 23 \end{cases}$

52. $\begin{cases} 85 = 24x + 24y \\ 55 = 18x + 16y \end{cases}$
    53. $\begin{cases} \dfrac{5}{6} = \dfrac{1}{4}x - \dfrac{1}{6}y \\ 3\dfrac{1}{5} = 6x + \dfrac{1}{5}y \end{cases}$
    54. $\begin{cases} \dfrac{1}{2}x - 3y = -.21 \\ -\dfrac{1}{3}x + \dfrac{1}{2}y = .11 \end{cases}$

Writing Exercises:

55. If the result of applying the elimination by addition method is the equation $-2 \neq 0$, what does this mean?

56. If the result of applying the elimination by addition method is the equation $4 = 4$, what does this mean?

Section 9.3 # Solving Systems of Linear Equations Using Substitution

OBJECTIVE

When you complete this section, you will be able to:

Solve systems of linear equations with two variables using substitution.

Introduction   In this section, we will learn another method for solving linear systems with two variables. This method is based on the principle of substitution, which allows us to replace a quantity with any other quantity that is equal to it.

To solve a system of linear equations with two variables using substitution, we will solve one equation for one of its variables, if necessary. This expression is then substituted for that variable in the other equation because at the point of intersection these two expressions are equal. This results in an equation with one variable that we can solve using previously learned techniques. We will illustrate the technique with some examples and then formalize the technique.

Example 1

Solve the following system of equations using substitution:

$$\begin{cases} 2x + y = 7 \\ 3x + 2y = 12 \end{cases}$$

Solution:
Solve $2x + y = 7$ for y since it is the easiest variable to solve for in either equation. Substitute the resulting expression for y in the other equation, $3x + 2y = 12$.

$2x + y = 7$	Subtract $2x$ from both sides of the equation.
$2x - 2x + y = 7 - 2x$	Add like terms.
$y = 7 - 2x$	Substitute $7 - 2x$ for y in $3x + 2y = 12$.
$3x + 2(7 - 2x) = 12$	Distribute 2.
$3x + 14 - 4x = 12$	Add like terms.
$-x + 14 = 12$	Subtract 14 from both sides of the equation.
$-x = -2$	Multiply both sides of the equation by -1.
$-1(-x) = -1(-2)$	Perform the multiplications.
$x = 2$	Substitute 2 for x in either of the original equations and solve for y. Use $2x + y = 7$.
$2x + y = 7$	Substitute 2 for x and solve for y.
$2(2) + y = 7$	Perform the multiplication.
$4 + y = 7$	Subtract 4 from both sides of the equation.
$y = 3$	Therefore, $(2,3)$ is the solution of the system.

Check:
Substitute 2 for x and 3 for y in both of the original equations.

$$2x + y = 7 \qquad\qquad 3x + 2y = 12$$
$$2(2) + 3 = 7 \qquad\qquad 3(2) + 2(3) = 12$$
$$4 + 3 = 7 \qquad\qquad 6 + 6 = 12$$
$$7 = 7 \qquad\qquad 12 = 12$$

Since $(2, 3)$ solves both equations, it is the correct solution.

We summarize the procedure in the following box:

> **Solving Systems of Linear Equations Using Substitution**
>
> 1. If necessary, solve one of the equations for one of its variables.
> 2. Substitute the expression found in step 1 into the other equation of the system.
> 3. Solve the resulting equation.
> 4. Substitute the value found in step 3 into either of the original equations of the system and solve for the other variable.
> 5. Check the solution of the system in both of the original equations.

The substitution method is most practical if the coefficient of one of the variables is 1 or -1 since solving for that variable will avoid fractions. If an equation has a variable with a coefficient of 1 or -1, we will solve that equation for that variable. If neither of the variables has a coefficient of 1 in either equation, then solving for either variable will result in fractions that are not easy to work with.

The substitution method gives us another method of solving systems of linear equations. This technique will be useful in later courses when the systems may not be linear.

Example 2

Solve the following systems of equations using substitution:

a. $\begin{cases} 4x + 3y = 10 \\ y = -3x \end{cases}$

Solution:

Since $y = -3x$ is already solved for y, we do not need to solve one of the equations for a variable.

$y = -3x$	Substitute $-3x$ for y in $4x + 3y = 10$.
$4x + 3(-3x) = 10$	Multiply 3 and $-3x$.
$4x - 9x = 10$	Add like terms.
$-5x = 10$	Divide both sides of the equation by -5.
$x = -2$	Substitute -2 for x in either of the original equations and solve for y. Use $y = -3x$.
$y = -3x$	Substitute -2 for x.
$y = -3(-2)$	Multiply.
$y = 6$	Therefore, the solution to the system is $(-2, 6)$.

Check:

Since the check is done in the same manner as solving using the graphical and elimination by addition methods, the check will be omitted.

b. $\begin{cases} x + y = -2 \\ 2x + 7y = -29 \end{cases}$

Solution:

We must solve one of the equations for one of its variables. If possible, we would like to avoid fractions. Since the coefficients of both x and y are one in the equation $x + y = -2$, solving for either x or y would avoid fractions.

$x + y = -2$	Solve for x. Subtract y from both sides of the equation.
$x + y - y = -2 - y$	Add like terms.
$x = -2 - y$	Substitute $-2 - y$ for x in $2x + 7y = -29$.
$2(-2 - y) + 7y = -29$	Distribute 2.
$-4 - 2y + 7y = -29$	Add like terms.
$-4 + 5y = -29$	Add 4 to both sides of the equation.
$5y = -25$	Divide both sides of the equation by 5.
$y = -5$	Substitute -5 for y in either of the original equations and solve for x. Use $x + y = -2$.
$x + y = -2$	Substitute -5 for y.
$x + (-5) = -2$	Add 5 to both sides of the equation.
$x = 3$	Therefore, the solution is $(3, -5)$.

c. $\begin{cases} 2x + 5y = -4 \\ 3x + y = 20 \end{cases}$

Solution:

The only variable with a coefficient of 1 is y in $3x + y = 20$. Therefore, we will solve $3x + y = 20$ for y.

$$3x + y = 20 \qquad \text{Subtract } 3x \text{ from both sides of the equation.}$$
$$3x - 3x + y = 20 - 3x \qquad \text{Add like terms.}$$
$$y = 20 - 3x \qquad \text{Substitute } 20 - 3x \text{ for } y \text{ in } 2x + 5y = -4.$$
$$2x + 5(20 - 3x) = -4 \qquad \text{Distribute 5.}$$
$$2x + 100 - 15x = -4 \qquad \text{Add like terms.}$$
$$-13x + 100 = -4 \qquad \text{Subtract 100 from both sides of the equation.}$$
$$-13x = -104 \qquad \text{Divide both sides of the equation by } -13.$$
$$x = 8 \qquad \text{Substitute 8 for } x \text{ in either of the original}$$
$$\text{equations and solve for } y. \text{ Use } 3x + y = 20.$$
$$3x + y = 20 \qquad \text{Substitute 8 for } x.$$
$$3(8) + y = 20 \qquad \text{Multiply 3 and 8.}$$
$$24 + y = 20 \qquad \text{Subtract 24 from both sides of the equation.}$$
$$y = -4 \qquad \text{Therefore, the solution is } (8, -4).$$

Practice Exercises

Solve the following systems of equations using substitution:

1. $\begin{cases} 2x + 3y = 8 \\ y = -2x \end{cases}$

2. $\begin{cases} 3x + 4y = 13 \\ x + y = 5 \end{cases}$

3. $\begin{cases} x - 3y = -5 \\ 4x + 5y = -3 \end{cases}$

If more practice is needed, do the Additional Practice Exercises.

Additional Practice Exercises

Solve the following systems of equations using substitution:

a. $\begin{cases} x + 4y = 12 \\ x = 2y \end{cases}$

b. $\begin{cases} x - 3y = -26 \\ x + y = 2 \end{cases}$

c. $\begin{cases} 2x + 3y = 1 \\ x + 4y = -7 \end{cases}$

If neither variable has a coefficient of 1 in either equation, the substitution method is made more difficult by the introduction of fractions. Examine both equations carefully and try to determine which variable would be best to solve for. As before, substitute the resulting expression into the other equation for that variable. Actually, solving linear equations of this type is usually easier using the addition method of Section 9.2. It is included here because substitution is often the only method of solving nonlinear systems of equations found in more advanced algebra courses and may be omitted without loss of continuity.

Example 3

Solve the following systems of equations using substitution:

a. $\begin{cases} 3x + 2y = 4 \\ 2x + 3y = 1 \end{cases}$

Solution:
Neither variable has a coefficient of 1 in either equation. There seems to be no particular advantage or disadvantage in solving either equation for either variable. We have to make a choice, so let us solve $3x + 2y = 4$ for y.

$$3x + 2y = 4 \qquad \text{Subtract } 3x \text{ from both sides of the equation.}$$

$$2y = 4 - 3x \qquad \text{Divide both sides of the equation by 2.}$$

$$\frac{2y}{2} = \frac{4 - 3x}{2} \qquad \text{Apply } \frac{a+b}{c} = \frac{a}{c} + \frac{b}{c}.$$

$$y = 2 - \frac{3}{2}x \qquad \text{Substitute } 2 - \frac{3}{2}x \text{ for } y \text{ in } 2x + 3y = 1.$$

$$2x + 3\left(2 - \frac{3}{2}x\right) = 1 \qquad \text{Distribute 3.}$$

$$2x + 6 - \frac{9}{2}x = 1 \qquad \text{Multiply both sides of the equation by the LCD of 2.}$$

$$2\left(2x + 6 - \frac{9}{2}x\right) = 2(1) \qquad \text{Distribute on the left. Multiply on right.}$$

$$4x + 12 - 9x = 2 \qquad \text{Add like terms.}$$

$$-5x + 12 = 2 \qquad \text{Subtract 12 from both sides of the equation.}$$

$$-5x = -10 \qquad \text{Divide both sides of the equation by } -5.$$

$$x = 2 \qquad \text{Substitute 2 for } x \text{ in either of the original equations and solve for } y. \text{ Use } 3x + 2y = 4.$$

$$3x + 2y = 4 \qquad \text{Substitute 2 for } x \text{ and solve for } y.$$

$$3(2) + 2y = 4 \qquad \text{Multiply 3 and 2.}$$

$$6 + 2y = 4 \qquad \text{Subtract 6 from both sides of the equation.}$$

$$2y = -2 \qquad \text{Divide both sides of the equation by 2.}$$

$$y = -1 \qquad \text{Therefore, the solution is } (2, -1).$$

b. $\begin{cases} 15x + 6y = 14 \\ 6x + 14y = 23 \end{cases}$

Solution:

Neither variable has a coefficient of 1 in either equation. Since the coefficient of y in $15x + 6y = 14$ is the smallest coefficient in either equation, we will solve $15x + 6y = 14$ for y. It is as good a reason as any!

$$15x + 6y = 14 \qquad \text{Subtract } 15x \text{ from both sides of the equation.}$$

$$6y = 14 - 15x \qquad \text{Divide both sides of the equation by 6.}$$

$$\frac{6y}{6} = \frac{14 - 15x}{6} \qquad \text{Apply } \frac{a+b}{c} = \frac{a}{c} + \frac{b}{c}.$$

$$y = \frac{14}{6} - \frac{15}{6}x \qquad \text{Reduce to lowest terms.}$$

$$y = \frac{7}{3} - \frac{5}{2}x \qquad \text{Substitute } \frac{7}{3} - \frac{5}{2}x \text{ for } y \text{ in } 6x + 14y = 23 \text{ and solve for } x.$$

$$6x + 14\left(\frac{7}{3} - \frac{5}{2}x\right) = 23 \qquad \text{Distribute 14.}$$

$$6x + \frac{98}{3} - 35x = 23 \qquad \text{Add like terms.}$$

$$-29x + \frac{98}{3} = 23 \qquad \text{Multiply the equation by the LCD 3.}$$

$$3\left(-29x + \frac{98}{3}\right) = 3(23) \qquad \text{Distribute on the left. Multiply on the right.}$$

$$-87x + 98 = 69 \qquad \text{Subtract 98 from both sides of the equation.}$$

$$-87x = -29 \qquad \text{Divide both sides of the equation by } -87.$$

$$x = \frac{29}{87} = \frac{1}{3} \qquad \text{Substitute } \frac{1}{3} \text{ for } x \text{ in either of the original equations and solve for } y. \text{ Use } 15x + 6y = 14.$$

$$15x + 6y = 14 \qquad \text{Substitute } \tfrac{1}{3} \text{ for } x.$$

$$15\left(\frac{1}{3}\right) + 6y = 14 \qquad \text{Multiply 15 and } \tfrac{1}{3}.$$

$$5 + 6y = 14 \qquad \text{Subtract 5 from both sides of the equation.}$$

$$6y = 9 \qquad \text{Divide both sides of the equation by 6.}$$

$$y = \frac{9}{6} = \frac{3}{2} \qquad \text{Therefore, the solution is } \left(\tfrac{1}{3}, \tfrac{3}{2}\right).$$

Practice Exercises

Solve the following systems using the substitution method:

4. $\begin{cases} 2x + 5y = 19 \\ 4x - 3y = -27 \end{cases}$

5. $\begin{cases} 3x + 5y = 8 \\ 3x - 10y = -10 \end{cases}$

If more practice is needed, do the Additional Practice Exercises.

Additional Practice Exercises

Solve the following systems using the substitution method:

d. $\begin{cases} 7x - 3y = 34 \\ 6x + 5y = 14 \end{cases}$

e. $\begin{cases} 12x + 3y = 8 \\ 6x + 6y = 1 \end{cases}$

We recognize inconsistent and dependent systems the same way when using the substitution method as when using the addition method. That is, if we arrive at a false statement, we know the system is inconsistent. If we arrive at a true statement, we know the system is dependent.

Example 4

If possible, solve the following systems of equations. If the system is not consistent, indicate whether it is inconsistent or dependent.

a. $\begin{cases} 2x - y = 7 \\ 4x - 2y = -3 \end{cases}$

Solution:
Since the coefficient of y in $2x - y = 7$ is -1, solve $2x - y = 7$ for y.

$$2x - y = 7 \qquad \text{Subtract } 2x \text{ from both sides of the equation.}$$

$$2x - 2x - y = 7 - 2x \qquad \text{Add like terms.}$$

$$-y = 7 - 2x \qquad \text{Multiply both sides of the equation by } -1.$$

$$-1(-y) = -1(7 - 2x) \qquad \text{Multiply on the left. Distribute on the right.}$$

$$y = -7 + 2x \qquad \text{Substitute } -7 + 2x \text{ for } y \text{ in } 4x - 2y = -3.$$

$$4x - 2(-7 + 2x) = -3 \qquad \text{Distribute } -2.$$

$$4x + 14 - 4x = -3 \qquad \text{Add like terms.}$$

$$14 = -3 \qquad \text{Since this is a false statement, the system is inconsistent and has no solution. Therefore, the solution set is } \varnothing.$$

Answers:

Practice Exercises 4–5: 4. $(-3, 5)$ 5. $\left(\dfrac{2}{3}, \dfrac{6}{5}\right)$ Additional Practice Exercises d–e: d. $(4, -2)$ e. $\left(\dfrac{5}{6}, -\dfrac{2}{3}\right)$

b. $\begin{cases} 3x - y = 4 \\ 2y - 6x = -8 \end{cases}$

Solution: Since the coefficient of y in $3x - y = 4$ is -1, solve $3x - y = 4$ for y.

$3x - y = 4$	Subtract $3x$ from both sides of the equation.
$3x - 3x - y = 4 - 3x$	Add like terms.
$-y = 4 - 3x$	Multiply both sides of the equation by -1.
$-1(-y) = -1(4 - 3x)$	Multiply on the left. Distribute on the right.
$y = -4 + 3x$	Substitute $-4 + 3x$ for y in $2y - 6x = -8$.
$2(-4 + 3x) - 6x = -8$	Distribute 2.
$-8 + 6x - 6x = -8$	Add like terms.
$-8 = -8$	Since this is a true statement, the system is dependent and all solutions are common. Therefore, the solutions may be written as $\{(x, y) : 3x - y = 4\}$.

Practice Exercises

If possible, solve the following systems of equations. If the system is not consistent, indicate whether it is inconsistent or dependent.

6. $\begin{cases} x + 3y = 7 \\ 6y + 2x = -9 \end{cases}$

7. $\begin{cases} 3x + 2y = 6 \\ \dfrac{3}{8}x + \dfrac{1}{4}y = \dfrac{3}{4} \end{cases}$

Exercise Set 9.3

Solve the following system of equations using substitution. Verify your solution using the graph.

1. $\begin{cases} x = 2 \\ 4x + 5y = 3 \end{cases}$

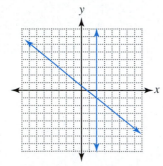

2. $\begin{cases} 3x + y = 5 \\ y = 2 \end{cases}$

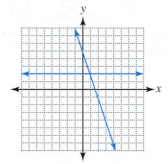

3. $\begin{cases} 3x - y = -2 \\ 5x = y \end{cases}$

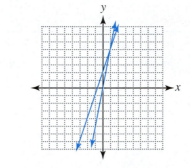

4. $\begin{cases} 2x - y = 9 \\ x = 2y \end{cases}$

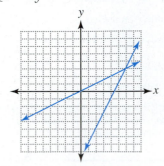

5. $\begin{cases} 4x + y = 5 \\ 3x = -2y \end{cases}$

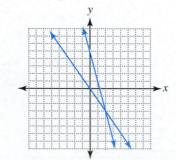

6. $\begin{cases} 3y = -8x \\ 3x - 2y = -25 \end{cases}$

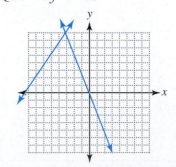

Solve the following system of equations using the substitution method:

7. $\begin{cases} x + y = -5 \\ x - 2y = -2 \end{cases}$

8. $\begin{cases} x + 3y = 1 \\ x + y = -3 \end{cases}$

9. $\begin{cases} 2x - 3y = -6 \\ x - y = -1 \end{cases}$

10. $\begin{cases} 2x + 5y = -7 \\ x - y = -7 \end{cases}$

11. $\begin{cases} x + y = 6 \\ 2x - y = 3 \end{cases}$

12. $\begin{cases} x + y = -10 \\ 4x - 7y = 15 \end{cases}$

13. $\begin{cases} -x + 2y = -12 \\ 2x - 3y = 20 \end{cases}$

14. $\begin{cases} -x + 2y = -9 \\ 4x + 5y = -3 \end{cases}$

15. $\begin{cases} x - 2y = -9 \\ 2x + 3y = -4 \end{cases}$

16. $\begin{cases} x - 3y = 10 \\ 4x + 5y = 6 \end{cases}$

17. $\begin{cases} 2x - y = -4 \\ -x + 3y = -3 \end{cases}$

18. $\begin{cases} x - 3y = -19 \\ 2x + y = -3 \end{cases}$

19. $\begin{cases} 8x - 3y = 10 \\ 4x + 3y = 14 \end{cases}$

20. $\begin{cases} 2x + 3y = 1 \\ 5x + 3y = -7 \end{cases}$

21. $\begin{cases} 5x - 6y = -3 \\ 3x - 5y = 1 \end{cases}$

22. $\begin{cases} 2x - y = -4 \\ 4x - 5y = -51 \end{cases}$

If possible, solve the following systems of equations. If the system is not consistent, indicate whether it is inconsistent or dependent.

23. $\begin{cases} x - 2y = 6 \\ -2x + 4y = 12 \end{cases}$

24. $\begin{cases} x - 3y = -4 \\ -5x + 15y = 6 \end{cases}$

25. $\begin{cases} 8x + 5y = 11 \\ 2x + y = 5 \end{cases}$

26. $\begin{cases} 6x + 2y = 8 \\ 9x + 3y = 12 \end{cases}$

27. $\begin{cases} 3x - 2y = 6 \\ -6x + 4y = 15 \end{cases}$

28. $\begin{cases} 2x + 5y = 8 \\ 3x + 4y = -2 \end{cases}$

29. $\begin{cases} 5x + y = 3 \\ 6x - 7y = -62 \end{cases}$

30. $\begin{cases} -5x + 3y = 15 \\ 15x - 9y = 30 \end{cases}$

Challenge Exercises:

31. $\begin{cases} \dfrac{1}{2}x + y = 1 \\ \dfrac{1}{4}x + y = 0 \end{cases}$

32. $\begin{cases} \dfrac{2}{3}x + y = 1 \\ \dfrac{5}{6}x + \dfrac{2}{5}y = -3 \end{cases}$

33. $\begin{cases} 2.1x + .5y = 16.6 \\ 4x - 5y = -16 \end{cases}$

34. $\begin{cases} 6x - 7y = 14 \\ 4.8x + 3.2y = -6.4 \end{cases}$

35. $\begin{cases} \dfrac{1}{2}x + \dfrac{2}{9}y = 2 \\ .8x + .03y = 3.2 \end{cases}$

36. $\begin{cases} \dfrac{3}{4}x - \dfrac{2}{5}y = 6 \\ 15x - 8y = 120 \end{cases}$

Writing Exercises:

37. When solving a system of linear equations by substitution, how do you determine which equation to solve for which variable?

38. If the result of solving a system of linear equations by substitution is a statement like $8 = 8$, which type of system is it?

39. If the result of solving a system of linear equations by substitution is a statement like $0 \neq 8$, which type of system is it?

| **Section 9.4** | Solving Application Problems Using Systems of Equations |

OBJECTIVE

When you complete this section, you will be able to:

Solve the following types of application problems using systems of linear equations.

a. Numbers

b. Geometry formulas

c. Money

d. Distance, rate, and time

e. Percent

f. Mixture

Introduction   In previous chapters, application problems were done using only one variable. One of the most difficult things in doing application problems that have more than one unknown is representing all the unknowns in terms of the one variable. Using systems of equations, we can assign a different variable to each unknown and eliminate this problem. Some of the examples and exercises in this section are the same as in Chapter 3, except the approach is different. Therefore, when you complete this section you will have a choice of methods to use when solving some types of application problems.

An important thing to remember is that there must be the same number of equations as there are unknowns. Since the systems we have been solving involve two variables, the following application problems will have two unknowns and will require two equations. Following is a general procedure:

Solving Application Problems Using Systems of Equations

1. Represent both of the unknowns in the problem with a variable.
2. From the information given in the problem, write two equations.
3. Solve the system of equations using either the addition or the substitution method.
4. Check the solutions against the wording of the original problem.

Example 1

One number is twice another. If the sum of the two numbers is 51, find the numbers.

Solution:
The unknowns in the problem are the two numbers.
Let x = the smaller number, and
Let y = the larger number. Then,

$$y = 2x \qquad \text{The larger number is twice the smaller.}$$
$$x + y = 51 \qquad \text{The sum of the two numbers is 51.}$$

Since $y = 2x$ is already solved for y, we will use the substitution method.

$$y = 2x \qquad \text{Substitute } 2x \text{ for } y \text{ in } x + y = 51.$$
$$x + (2x) = 51 \qquad \text{Add like terms.}$$
$$3x = 51 \qquad \text{Divide both sides of the equation by 3.}$$
$$x = 17 \qquad \text{Therefore, the smaller number is 17. To find the larger number, substitute 17 for } x \text{ in either of the original equations. Use } y = 2x.$$
$$y = 2x \qquad \text{Substitute 17 for } x.$$
$$y = 2(17) \qquad \text{Multiply 2 and 17.}$$
$$y = 34 \qquad \text{Therefore, the larger number is 34.}$$

Check:
Is one number twice the other? Yes, since 34 is twice 17. Is the sum of the numbers 51? Yes, since $17 + 34 = 51$. Therefore, our solutions are correct.

Practice Exercise

1. One number is three times the other. If the sum of the numbers is 20, find the numbers.

Example 2

A paint contractor paid $372 for 24 gallons of paint to paint the inside and outside of a house. If the paint for the inside costs $12 per gallon and the paint for the outside costs $18 per gallon, how many gallons of each did he buy?

Solution:
We will use a chart as in Chapter 3, except we will use two unknowns instead of one.

The unknowns are the number of gallons that he bought for the inside and the number of gallons that he bought for the outside.

Let x = the number of gallons for the inside, and
Let y = the number of gallons for the outside.

Fill in the table.

Type of Paint	Price per Gallon	Number of Gallons	Total Cost
Inside	12	x	$12x$
Outside	18	y	$18y$
Total		24	372

From the table, we see that $x + y = 24$ (the number of gallons of inside paint plus the number of gallons of outside paint equals the total number of gallons of paint purchased) and $12x + 18y = 372$ (the cost of the inside paint plus the cost of the outside paint equals the total cost of the paint).

Therefore, we need to solve the system. Let's eliminate x.

$$x + y = 24 \qquad \text{Multiply } x + y = 24 \text{ by } -12.$$
$$12x + 18y = 372 \qquad \text{Leave unchanged.}$$

$$-12(x + y) = -12(24) \qquad \text{Distribute on the left. Multiply on the right.}$$
$$12x + 18y = 372 \qquad \text{Leave unchanged.}$$

$$-12x - 12y = -288$$
$$\underline{12x + 18y = 372} \qquad \text{Add the equations.}$$
$$6y = 84 \qquad \text{Divide both sides of the equation by 6.}$$
$$y = 14 \qquad \text{Therefore, he bought 14 gallons of paint for the outside. Substitute 14 for } y \text{ in either of the original equations and solve for } x. \text{ Use } x + y = 24.$$

$$x + y = 24 \qquad \text{Substitute 14 for } y.$$
$$x + 14 = 24 \qquad \text{Subtract 14 from both sides of the equation.}$$
$$x = 10 \qquad \text{Therefore, he bought 10 gallons of paint for the inside.}$$

Check: Did he buy a total of 24 gallons of paint? Yes, since $14 + 10 = 24$. Do 10 gallons of inside paint at $12 per gallon and 14 gallons of outside paint at $18 per gallon cost $372? $10(12) + 14(18) = 120 + 252 = 372$. Yes, so our solutions are correct.

Practice Exercise

2. A landscape company pays $7 each for rose bushes and $5 each for azaleas. If they paid $82 for 14 plants, how many of each did they buy?

Example 3

A collection of 23 coins, made up of nickels and dimes, is worth $1.90. How many of each type of coin are there?

Solution:

Again, we use a table as in Chapter 3.
The two unknowns are the number of nickels and the number of dimes.
Let x = the number of nickels, and
Let y = the number of dimes.

Fill in the table.

Type of Coin	Value of Coin	Number of Coins	Total Value
Nickel	5	x	$5x$
Dime	10	y	$10y$
Total		23	190

From the table, we see that $x + y = 23$ (the number of nickels + the number of dimes = 23) and $5x + 10y = 190$ (value of the nickels + value of the dimes = 190).

We need to solve the system. Let's eliminate x.

$$x + y = 23 \qquad \text{Multiply } x + y = 23 \text{ by } -5.$$
$$5x + 10y = 190 \qquad \text{Leave this equation unchanged.}$$

$$-5(x + y) = -5(23) \qquad \text{Simplify both sides of the equation.}$$
$$5x + 10y = 190 \qquad \text{Leave unchanged.}$$

$$-5x - 5y = -115$$
$$\underline{5x + 10y = 190} \qquad \text{Add the equations.}$$
$$5y = 75 \qquad \text{Divide both sides of the equation by 5.}$$
$$y = 15 \qquad \text{Therefore, there are 15 dimes. Substitute 15 for } y \text{ in either of the original equations and solve for } x. \text{ Use } x + y = 23.$$

$$x + y = 23 \qquad \text{Substitute 15 for } y.$$
$$x + 15 = 23 \qquad \text{Subtract 15 from both sides of the equation.}$$
$$x = 8 \qquad \text{Therefore, there are 8 nickels.}$$

Check:

Is there a total of 23 coins? Yes, since $15 + 8 = 23$. Is the value of 8 nickels and 15 dimes equal to $1.90? $8(5) + 15(10) = 40 + 150 = 190¢ = \$1.90$. Yes, so our solutions are correct.

Practice Exercise

3. A cashier has a total of 30 nickels and quarters in the cash register. If the coins are worth $3.90, how many of each does he have?

Example 4

In four hours, a boat can go 152 kilometers downstream or 104 kilometers upstream. Find the speed of the boat in still water and the speed of the current.

Solution:
The unknowns are the speed of the boat and the speed of the current.
Let x = the speed of the boat in still water, and
Let y = the speed of the current.
Recall that in Section 3.5, we used a chart to do distance, time, and rate problems.

	d	r	t
downstream	152	$x + y$	4
upstream	104	$x - y$	4

Since the rate of the boat is x and the rate of the current is y, the rate downstream is $x + y$ and the rate upstream is $x - y$.

Since d = rt,
$152 = (x + y)4$ (distance downstream) = (rate downstream)(time downstream), and
$104 = (x - y)4$ (distance upstream) = (rate upstream)(time upstream)

Simplify the equations and solve the system.

$$152 = 4x + 4y$$
$$\underline{104 = 4x - 4y}$$ Add the equations.
$$256 = 8x$$ Divide both sides of the equation by 8.
$$32 = x$$ Therefore, the rate of the boat in still water is 32 kilometers per hour. Substitute 32 for x in either original equation and solve for y. Use $152 = (x + y)4$.

$$152 = (x + y)4$$ Substitute 32 for x.
$$152 = (32 + y)4$$ Distribute on the right side.
$$152 = 128 + 4y$$ Subtract 128 from both sides of the equation.
$$24 = 4y$$ Divide both sides of the equation by 4.
$$6 = y$$ Therefore, the speed of the current is 6 kilometers per hour.

Check:
Will the boat go 152 kilometers downstream in four hours? The rate downstream is the rate of the boat plus the rate of the current which is $32 + 6 = 38$ kilometers per hour. In four hours the distance downstream is $4(38) = 152$ kilometers. Will the boat go 104 kilometers upstream in four hours? The rate upstream is the rate of the boat minus the rate of the current, which is $32 - 6 = 26$ kilometers per hour. In four hours, the distance upstream is $4(26) = 104$ kilometers. Therefore, our solutions are correct.

Practice Exercise

4. A car is traveling at an average rate that is 20 miles per hour more than the rate of a bus. Both vehicles leave the same town and travel in opposite directions. Find the rate of each if the distance between them is 300 miles after three hours.

Example 5

Roxanne received an inheritance of $10,000. She invested part of it at 8% and the remainder at 11%. How much did she invest at each rate if the total interest from both investments was $1010 per year?

Solution:
We use a table as before.
The two unknowns are the amount invested at 8% and the amount invested at 11%.
Let x = the amount invested at 8%, and
Let y = the amount invested at 11%.

Fill in the table.

Type of Investment	Interest Rate	Amount Invested	Interest Earned
8%	.08	x	$.08x$
11%	.11	y	$.11y$
Total		10,000	1010

From the table, we see that $x + y = 10,000$ (the total amount invested was $10,000) and $.08x + .11y = 1010$ (the interest earned at 8% + the interest earned at 11% = $1010).

Solve the system.

$$x + y = 10,000 \qquad \text{Leave unchanged.}$$
$$.08x + .11y = 1010 \qquad \text{Multiply by 100 to eliminate the decimals.}$$

$$x + y = 100 \qquad \text{Leave unchanged.}$$
$$100(.08x + .11y) = 100(1010) \qquad \text{Simplify both sides of the equation.}$$

$$x + y = 10,000 \qquad \text{Multiply } x + y = 10,000 \text{ by } -8 \text{ to eliminate } x.$$
$$8x + 11y = 101,000 \qquad \text{Leave unchanged.}$$

$$-8(x + y) = -8(10,000) \qquad \text{Simplify both sides of the equation.}$$
$$8x + 11y = 101,000 \qquad \text{Leave unchanged.}$$

$$-8x - 8y = -80,000$$
$$\underline{8x + 11y = 101,000} \qquad \text{Add the equations.}$$
$$3y = 21,000 \qquad \text{Divide both sides of the equation by 3.}$$
$$y = 7000 \qquad \text{Therefore, there was $7000 invested at 11%.}$$

Substitute 7000 for y in either of the original equations. Use $x + y = 10,000$.

$$x + y = 10,000 \qquad \text{Substitute 7000 for } y.$$
$$x + 7000 = 10,000 \qquad \text{Subtract 7000 from both sides of the equation.}$$
$$x = 3000 \qquad \text{Therefore, there was $3000 invested at 8%.}$$

Check: Is there a total of $10,000 invested? Yes, since $7000 + $3000 = $10,000. Is the total interest earned $1010? The interest earned on $3000 at 8% = .08(3000) = $240. The interest earned on $7000 at 11% is .11(7000) = $770. Since $240 + $770 = $1010, our solutions are correct.

Practice Exercise

5. Jerome received $50,000 from the sale of some property. He invested part of it in a certificate of deposit at 9% and the remainder in bonds at 12%. Find the amount invested in each if the total interest received from the two investments is $5400 per year.

Example 6

A candy store owner mixes some candy worth $2.50 per pound with some worth $3.50 per pound. How many pounds of each would he need if he wants 10 pounds of the mixture worth $2.90 per pound?

Solution:

The two unknowns are the number of pounds of candy worth $2.50 per pound and the number of pounds of candy worth $3.50 per pound. So,

Let x = the number of pounds of candy worth $2.50 per pound, and

Let y = the number of pounds of candy worth $3.50 per pound.

Fill in the table.

Type of Candy	Price per Pound	Number of Pounds	Total Value
$2.50 per pound	2.50	x	$2.50x$
$3.50 per pound	3.50	y	$3.50y$
Mixture	2.90	10	$2.90(10) = 29$

From the table, we see that $x + y = 10$ (the number of pounds of $2.50 candy + the number of pounds of $3.50 candy = the number of pounds in the mixture) and $2.50x + 3.50y = 29$ (the value of the $2.50-per-pound candy) + (the value of the $3.50-per-pound candy) = (the value of the mixture).

Sometimes a diagram of this type of problem, referred to as a mixture problem, is helpful.

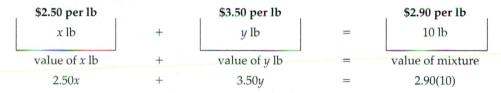

	$2.50 per lb			$3.50 per lb			$2.90 per lb
	x lb	+		y lb	=		10 lb
	value of x lb	+		value of y lb	=		value of mixture
	$2.50x$	+		$3.50y$	=		$2.90(10)$

Solve the system.

$$x + y = 10 \qquad \text{Leave unchanged.}$$
$$2.5x + 3.5y = 29 \qquad \text{Multiply both sides of the equation by 10 to eliminate the decimals.}$$

$$x + y = 10 \qquad \text{Leave unchanged.}$$
$$10(2.5x + 3.5y) = 10(29) \qquad \text{Simplify both sides of the equation.}$$

$$x + y = 10$$
$$25x + 35y = 290$$

Let us use the substitution method. So solve $x + y = 10$ for y.

$$x + y = 10 \qquad \text{Subtract } x \text{ from both sides of the equation.}$$
$$x - x + y = 10 - x \qquad \text{Add like terms.}$$
$$y = 10 - x \qquad \text{Substitute } 10 - x \text{ for } y \text{ in } 25x + 35y = 290.$$

$$25x + 35(10 - x) = 290 \qquad \text{Distribute 35.}$$
$$25x + 350 - 35x = 290 \qquad \text{Add like terms.}$$
$$-10x + 350 = 290 \qquad \text{Subtract 350 from both sides of the equation.}$$
$$-10x = -60 \qquad \text{Divide both sides of the equation by } -10.$$
$$x = 6 \qquad \text{Therefore, he would need 6 pounds of candy worth \$2.50 per pound. Substitute 6 for } x \text{ in either of the original equations and find } y. \text{ Use } x + y = 10.$$

$$6 + y = 10 \qquad \text{Subtract 6 from both sides.}$$
$$y = 4 \qquad \text{Therefore, he needs 4 pounds of candy worth \$3.50 per pound.}$$

Check:
Does the number of pounds of $2.50-per-pound candy + the number of pounds of $3.50-per-pound candy = the number of pounds of candy in the mixture? Yes, since 6 + 4 = 10. Does the value of the $2.50-per-pound candy + the value of the $3.50-per-pound candy = the value of the mixture? The value of the $2.50-per-pound candy is 6($2.50) = $15.00. The value of the $3.50-per-pound candy is 4($3.50) = $14.00. The value of the mixture is 10($2.90) = $29. Since $15.00 + $14.00 = $29.00, the answer is yes. Therefore, our solutions are correct.

Practice Exercise

6. The Smokehouse restaurant specializes in a barbecue that is a mixture of pork and beef. They need 20 pounds of this mixture for a picnic they are catering. If pork barbecue sells for $3.50 per pound and beef barbecue sells for $5.00 per pound, how many pounds of each will they need if the mixture sells for $4.10 per pound?

Example 7

How much of a solution that is 40% alcohol must be added to a solution that is 60% alcohol to obtain 50 liters of a solution that is 55% alcohol?

Solution:
We use a chart as before.

The unknowns are the number of liters of 40% alcohol and the number of liters of 60% alcohol. So,
Let x = the number of liters of 40% alcohol and
let y = the number of liters of 60% alcohol.

Fill in the table.

Type of Solution	Part Pure Alcohol	Volume	Amount Pure Alcohol
40%	.40	x	$.40x$
60%	.60	y	$.60y$
55% (mixture)	.55	50	$.55(50) = 27.5$

From the table, we see that $x + y = 50$ (the number of liters of 40% alcohol + the number of liters of 60% alcohol = the number of liters in the mixture) and $.40x + .60y = 27.5$ (the amount of pure alcohol in the 40% solution + the amount of pure alcohol in the 60% solution = the amount of pure alcohol in the mixture).

Solve the system.

$x + y = 50$	Solve for x.
$.40x + .60y = 27.5$	Leave unchanged.
$x = 50 - y$	Substitute for x in $.40x + .60y = 27.5$.
$.40(50 - y) + .60y = 27.5$	Distribute .40.
$20 - .40y + .60y = 27.5$	Add like terms.
$20 + .20y = 27.5$	Subtract 20 from both sides of the equation.
$.20y = 7.5$	Divide both sides of the equation by .20.
$y = 37.5$	Therefore, 37.5 liters of the 60% solution are needed.
	To find x, substitute 37.5 for y in the equation $x + y = 50$.
$x + 37.5 = 50$	Subtract 37.5 from both sides.
$x = 12.5$	Therefore, 12.5 liters of the 40% solution are needed.

Check: The check is left as an exercise for the student.

Answers:

Practice Exercise

7. How many pounds of an alloy that is 30% tin are to be mixed with an alloy that is 70% tin in order to get 100 pounds of an alloy that is 60% tin?

Exercise Set 9.4

Solve each of the following application problems using a system of linear equations:

1. The difference of two numbers is 8. If 5 times the smaller number is 3 times the larger, find the numbers.

2. One positive number is 3 times another positive number. If the difference of the two numbers is 16, find the numbers.

3. The sum of two numbers is 15. If the difference of twice one of the numbers and the other number is 18, find the numbers.

4. The difference of two numbers is 8. If 5 times the first number is 3 times the second number, find the numbers.

5. An electrical contractor paid $1040 for 36 light fixtures to be installed in a hotel lobby. If the first type of fixture costs $25 each and the second type of fixture costs $32 each, how many of each kind did she buy?

6. A locksmith installed inside and outside locks in a new classroom building. The locksmith paid $1216 for 80 locks. If inside locks cost $14 each and outside locks cost $26 each, how many of each kind of lock did he buy?

7. The managers of an office building have decided to reduce the cost of air-conditioning by installing overhead fans. They paid $5225 for 105 fans, some of which measure 36 inches and the remainder of which measure 54 inches. If the 36-inch fans cost $45 each and the 54-inch fans cost $65 each, how many of each kind of fan did they buy?

8. An accounting firm purchased a supply of computer disks. They paid $725 for 60 boxes of disks. If the double density disks cost $8 a box and the high density disks cost $15 a box, how many of each kind did they buy?

9. A certain vending machine accepts only dimes and quarters. If the person who services the machine collects 52 coins with a total value of $8.95, how many of each kind of coin was in the machine?

10. At the end of his shift, a cashier has only $5 and $20 bills in his drawer. If he has 48 bills worth a total of $900, how many of each kind of bill does he have?

11. Winston paid $7.46 for 36 stamps. If he bought eight fewer 25-cent stamps than 18-cent stamps, how many of each kind did he buy?

12. Gayle paid $17.40 for 42 pens at the campus bookstore. If she bought some pens costing $0.35 each and others costing $0.50 each, how many of each kind did she buy?

13. A freight train and a passenger train leave Memphis, Tennessee, at the same time traveling in opposite directions. The freight train is traveling 15 miles per hour slower than the passenger train. Find the rate of each train if the distance between them is 525 miles after five hours.

14. At 8:00 A.M., two airplanes leave cities that are 2250 miles apart, flying toward each other. The rate of one airplane is 150 miles per hour less than twice the rate of the other. If the planes meet at 11:00 A.M., how fast is each plane flying?

15. In seven hours, a coast guard cutter can sail 126 miles with an ocean current or it can sail 84 miles against the current. Find the speed of the cutter in still water and find the speed of the current.

16. An airplane flies 3000 miles coast to coast in 6 hours with a tailwind pushing it. When it flies back against the wind, it takes $7\frac{1}{2}$ hours. What is the speed of the airplane in still air and what is the speed of the wind?

17. Rochelle received a $20,000 bonus for selling over $1,000,000 worth of life insurance. She put part of the money into a savings account paying 6% interest and the rest of the money into municipal bonds paying 8.5% interest. How much did she invest at each rate if the total interest from both investments was $1650 per year?

Answers:

18. The manager of a supermarket received a $5000 bonus for exceeding the sales quota for her store. She put part of the money into a savings account paying 5% interest and the rest into a certificate of deposit paying 9% interest. How much did she invest at each rate if the total interest from both investments was $370 per year?

19. A college development fund has managed to raise $250,000. They invested part of the money in home mortgages with an expected return of 22% per year. The rest was invested in common stock with an expected return of 28% per year. If the fund earns $61,000 in one year, how much was put into each type of investment?

20. A certain college invests its operating funds in two types of U.S. government securities. The average amount invested is $750,000, which is split between U.S. government guaranteed mortgages at 12% interest and U.S. government treasury bills at 9% interest. If a total of $79,500 is earned per year, how much is invested in each type of security?

21. A specialty store owner mixes peanuts worth $1.80 per pound with cashew nuts worth $5.40 per pound. How many pounds of each type of nut is needed to make a 12-pound mixture worth $3.00 a pound?

22. A coffee company wishes to make a special blend of coffee by mixing coffee beans that cost $3.20 per pound with coffee beans that cost $6.20 per pound. How many pounds of each type of coffee bean are needed to make 40 pounds of coffee that costs $3.95 per pound?

23. A 12% solution of salt is mixed with a 4% solution. How many liters of each is needed to obtain 30 liters of an 8% solution of salt?

24. A 70% solution of alcohol is mixed with a 30% solution. How many gallons of each is needed to obtain 25 gallons of a 60% solution of alcohol?

25. The sum of two numbers is 25. If the difference of these numbers is 5, find the two numbers.

26. One half of a number is equal to three times another number. If the difference of the two numbers is 20, find the numbers.

27. The owners of an exterminator business have replaced their fleet of 27 cars and trucks by purchasing new cars for $12,000 each and new trucks for $9500 each. If they paid a total of $294,000, how many of each type of vehicle did they buy?

28. A building contractor has paid $2088 for 54 doors. If inside doors cost $35 each and outside doors cost $46 each, how many of each kind of door did he buy?

29. The metro rail (rapid transit system) accepts only quarters and silver dollars. If at the end of the day 900 coins worth $375 are collected, how many of each type of coin were in the machines?

30. Mr. and Mrs. Wong had a birthday party at the movie theater for their 8-year-old daughter. They paid $89 for 22 people to attend the movie. If a child's ticket is $3.50 and an adult ticket is $5.50, how many children and how many adults attended the party?

31. In four hours, a speedboat can go 180 kilometers downstream or 140 kilometers upstream. Find the speed of the boat in still water and the speed of the current.

32. Two ships leave port at 5:00 A.M., both traveling in the same direction. At noon, the ships are 56 miles apart and the sum of the distances they have traveled is 224 miles. How fast is each ship traveling?

33. A book salesperson received $3000 for signing six authors last year. She invested part of the money in a mutual fund paying 18% interest and the remainder in a certificate of deposit paying 9% interest. If she earns $414 per year interest from her investments, how much did she put in each investment?

34. A well-known author received a $160,000 royalty check for his new novel. He decides to invest part of the money in bonds at 8% and part of the money in stocks at a 24% expected return. If at the end of the year he expects to earn $25,600, how much did he put in each investment?

35. Pure rice (100%) and a mixture that is 50% rice and 50% beans are mixed. How many cups of each are needed to get 20 cups of a mixture that is 70% rice and 30% beans?

36. A solution that is 15% baking soda is mixed with a solution that is 5% baking soda. How many liters of each is needed to obtain 12 liters of a 10% solution of baking soda?

37. The length of a rectangular sheet of plywood is 4 feet more than the width. If the perimeter is 32 feet, find the length and width.

38. The length of a rectangular picture is 2 inches less than twice the width. If the perimeter is 26 inches, find the length and width of the picture.

39. One side of a triangular sign is 8 cm. The length of the smaller of the remaining two sides is 2 cm less than the length of the longer side. If the perimeter is 20 cm, find the lengths of the remaining two sides.

40. The parallel sides of a trapezoid are 6 m and 9 m in length. The length of the longer of the two nonparallel sides is 1 less than twice the length of the shorter side. If the perimeter is 23 m, find the lengths of the nonparallel sides.

41. Recall from Section 3.7 that two angles are supplementary if the sum of their measures is 180. If two angles are supplementary and the measure of the larger is 30 more than the measure of the smaller, find the measure of each angle.

42. Recall from Section 3.7 that two angles are complementary if the sum of their measures is 90. If two angles are complementary and the measure of the smaller is 14 less than the measure of the larger, find the measure of each angle.

Writing Exercise:

43. Which do you think is easier? Solving application problems using a system of two variables or solving application problems using one variable as we did in Chapters 3 and 7? Why?

Writing Exercises or Group Project:

If done as a group project, each group should write two exercises of each of the following types, exchange with another group, and then solve them.

44. Write and solve a number problem using a system of linear equations.

45. Write and solve a rate, time, and distance problem using a system of linear equations.

46. Write and solve a mixture problem using a system of linear equations.

Section 9.5 | Systems of Linear Inequalities

OBJECTIVE

When you complete this section, you will be able to:

Graphically represent the solutions of a system of linear inequalities.

Introduction   In Section 4.7 we learned to graph linear inequalities with two variables. The procedure is given in the following box:

Graphing Linear Inequalities with Two Variables

1. Graph the equality $ax + by = c$. If the line is part of the solution ($\leq$ or $\geq$), draw a solid line. If the line is not part of the solution ($<$ or $>$), draw a dashed line.

2. Pick a test point not on the line. If the coordinates of the test point satisfy the inequality, then all points on the same side of the line as the test point solve the inequality. If the coordinates of the test point do not satisfy the inequality, then all points on the other side of the line from the test point solve the inequality.

3. Shade the region on the side of the line that contains the solutions of the inequality.

Example 1b from Section 4.7 is repeated as follows to remind you of this procedure:

Example 1

Graph $3x - y \le 6$.

Solution:
Draw the graph of $3x - y = 6$ as a solid line since the points on the line solve the inequality less than or *equal to*. Use the *x*-intercept $(2, 0)$ and the *y*-intercept $(0, -6)$.

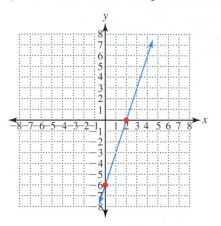

Pick a test point. The origin, $(0, 0)$, is a convenient choice. Substitute $(0, 0)$ into the inequality. $3(0) - 0 \le 6, 0 - 0 \le 6, 0 \le 6$. This is a true statement. Therefore, $(0, 0)$ solves the inequality. Consequently, the coordinates of all points on the same side of the line as $(0, 0)$ solve the inequality. Therefore, shade the region on the same side of the line as $(0, 0)$.

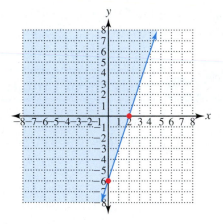

A system of linear inequalities with two variables is much like a system of linear equations with two variables.

System of Linear Inequalities

A system of linear inequalities is two or more linear inequalities with the same variables.

To solve a system of linear inequalities, we graph each inequality. The solutions of the system are the coordinates of all points that solve all the inequalities. Therefore, the region where the solutions overlap (intersect) represents the solutions of the system.

Example 2

Graph the solutions of the following systems of linear inequalities:

a. $\begin{cases} x + y > 6 \\ x - y < -2 \end{cases}$

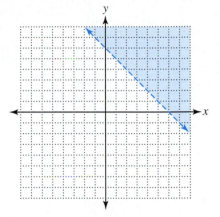

Solution:
The solutions of the system are represented by the region where the solutions of $x + y > 6$ and $x - y < -2$ overlap. Therefore, we draw the graphs of $x + y > 6$ and $x - y < -2$ on the same set of axes. First draw the graph of $x + y > 6$.

Now draw the graph of $x - y < -2$ on the same set of axes as the graph of $x + y > 6$.

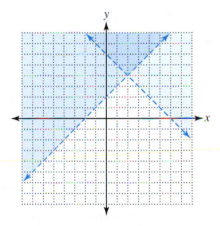

The heavily shaded region where the solutions of $x + y > 6$ and $x - y < -2$ overlap represents the points whose coordinates are the solutions of the system. Since the lines are dashed, the coordinates of the points on the lines are not solutions of the system.

b. $\begin{cases} 2x + 3y \le 12 \\ 2x + y \le 8 \end{cases}$

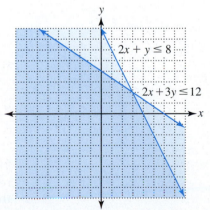

Solution:
The solutions of the system are represented by the region where the solutions of $2x + 3y \le 12$ and $2x + y \le 8$ overlap. Therefore, draw the graphs of $2x + 3y \le 12$ and $2x + y \le 8$ on the same set of axes.

The heavily shaded region where the solutions of $2x + 3y \le 12$ and $2x + y \le 8$ overlap (the intersection of the two sets) represents the points whose coordinates are the solutions of the system. Since the lines are solid, the coordinates of the points on the lines (within the shaded region) are also solutions of the system.

c. $\begin{cases} x + 4y > 8 \\ x \geq 5 \end{cases}$

Solution:

The solutions of the system are represented by the region where the solutions of $x + 4y > 8$ and $x \geq 5$ overlap. Therefore, draw the graphs of $x + 4y > 8$ and $x \geq 5$ on the same set of axes.

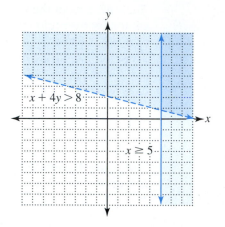

The heavily shaded region where the solutions of $x + 4y > 8$ and $x \geq 5$ overlap represents the points whose coordinates are the solutions of the system. Within the shaded region, the coordinates of the points on the line $x + 4y = 8$ are not solutions of the system but the coordinates of the points on the line $x = 5$ are.

d. $\begin{cases} x \leq -4 \\ y > 3 \end{cases}$

Solution:

The solutions of the system are represented by the region where the solutions of $x \leq -4$ and $y > 3$ overlap. Therefore, draw the graphs of $x \leq -4$ and $y > 3$ on the same set of axes.

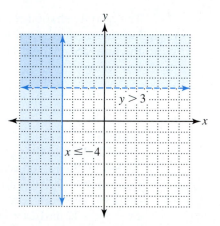

The heavily shaded region where the solutions of $x \leq -4$ and $y > 3$ overlap represents the points whose coordinates are the solutions of the system. Within the shaded region, the coordinates of the points on the line $x = -4$ are solutions of the system but the coordinates of the points on the line $y = 3$ are not.

Practice Exercises

Graph the solution of the following systems of linear inequalities:

1. $\begin{cases} x + y > 4 \\ x - y < 6 \end{cases}$

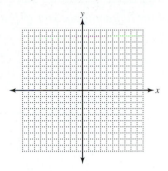

2. $\begin{cases} 2x - y \leq 6 \\ x + 2y \geq 8 \end{cases}$

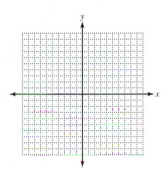

3. $\begin{cases} 3x - 2y \leq 6 \\ y > 5 \end{cases}$

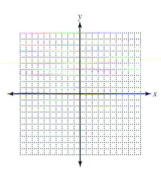

4. $\begin{cases} x \geq 4 \\ y < -3 \end{cases}$

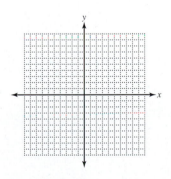

Answers:

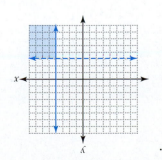

4.

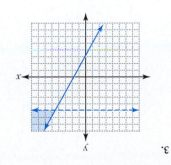

3.

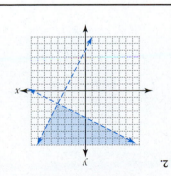

2.

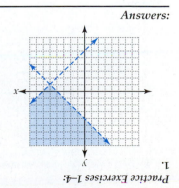

1.

Practice Exercises 1–4:

If more practice is needed, do the Additional Practice Exercises.

Additional Practice Exercises

Graph the solutions of the following systems of linear inequalities:

a. $\begin{cases} x + y < 3 \\ x - y > 5 \end{cases}$

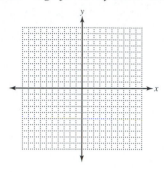

b. $\begin{cases} x + 3y \leq 6 \\ x - 2y \geq 2 \end{cases}$

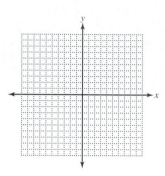

c. $\begin{cases} 3x - 2y \geq 6 \\ x < 6 \end{cases}$

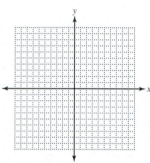

d. $\begin{cases} x > 4 \\ y < -3 \end{cases}$

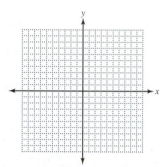

Exercise Set 9.5

Graph the solutions of the following systems of linear inequalities:

1. $\begin{cases} x - y > 2 \\ x + y > 4 \end{cases}$

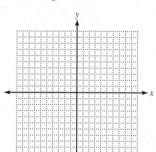

2. $\begin{cases} x - y < -5 \\ 2x - y < -7 \end{cases}$

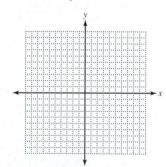

3. $\begin{cases} x + y < -2 \\ x - y > 6 \end{cases}$

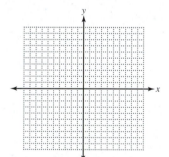

4. $\begin{cases} y - x > 5 \\ x + y < 3 \end{cases}$

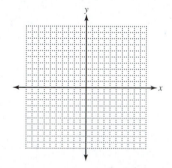

Answers:

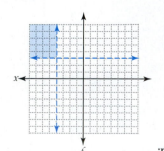

d.

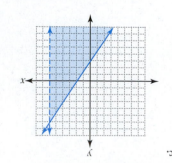

c.

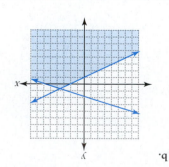

b.

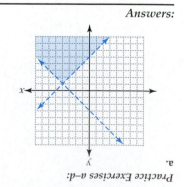

a.

Practice Exercises a–d:

5. $\begin{cases} 2x + y \geq -1 \\ x - y \leq 5 \end{cases}$

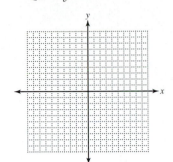

6. $\begin{cases} x + 3y \leq 11 \\ x - 2y \leq 1 \end{cases}$

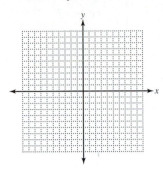

7. $\begin{cases} 2x + 3y < 1 \\ x - 4y \geq 3 \end{cases}$

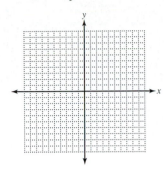

8. $\begin{cases} 4x - y > -14 \\ 5x + 2y \leq 2 \end{cases}$

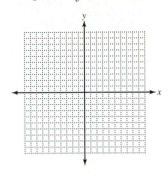

9. $\begin{cases} 2x + 5y \leq 7 \\ 3x + y > -9 \end{cases}$

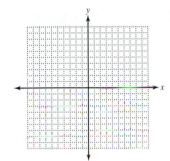

10. $\begin{cases} x - 2y > 3 \\ 2x - 4y \leq 20 \end{cases}$

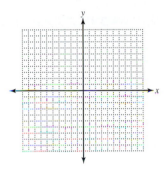

11. $\begin{cases} x + y \geq 4 \\ y \geq 2 \end{cases}$

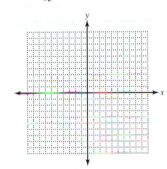

12. $\begin{cases} x - 2y > 0 \\ x < 0 \end{cases}$

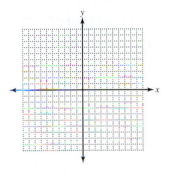

13. $\begin{cases} 2x + y \geq 0 \\ x < 3 \end{cases}$

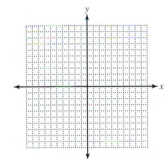

14. $\begin{cases} 5x - 6y > -12 \\ y > 2 \end{cases}$

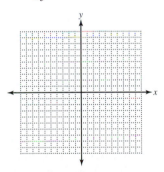

15. $\begin{cases} 5x + 3y < -4 \\ 2y \geq 4 \end{cases}$

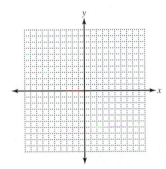

16. $\begin{cases} 3x - y \geq -13 \\ 4x < 12 \end{cases}$

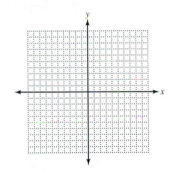

17. $\begin{cases} x \geq -2 \\ y < 4 \end{cases}$

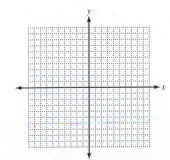

18. $\begin{cases} y \geq -4 \\ x < 3 \end{cases}$

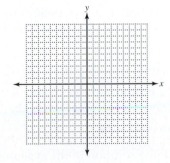

19. $\begin{cases} x < 1 \\ y \geq 0 \end{cases}$

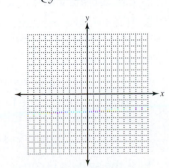

20. $\begin{cases} x \geq -2 \\ y \geq 1 \end{cases}$

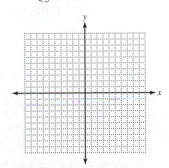

Challenge Exercises:

21. $\begin{cases} 2x + y \geq 1 \\ x - y > -1 \\ x > 2 \end{cases}$

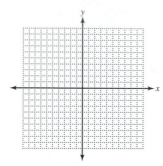

22. $\begin{cases} x + y \geq 1 \\ y - x \geq -5 \\ y > -2 \end{cases}$

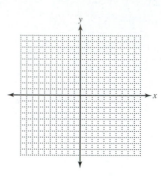

Writing Exercise:

23. Describe the region that is the solution of the following systems of inequalities:

    a. $x > 0$ and $y > 0$ **b.** $x > 0$ and $y < 0$

Chapter 9   Summary

Definition of a System of Linear Equations:
[Section 9.1]

- A **system of linear equations** is two or more linear equations with the same variables.

Solution(s) of a System of Linear Equations:
[Section 9.1]

- The solution(s) of a system of linear equations with two variables consists of all ordered pairs that solve both equations of the system.

Solving a System of Linear Equations by Graphing:
[Section 9.1]

- To solve a system of linear equations with two variables graphically, graph each equation of the system. The coordinates of the point(s) of intersection are the solutions of the system.

Types of Systems of Equations:
[Section 9.1]

- A system is **consistent** if the graphs of the equations of the system intersect at a single point. The coordinates of the point of intersection are the solution(s) of the system.

- A system is **inconsistent** if the graphs of the equations of the system are parallel lines. Therefore, the system has no solution. This is usually indicated by the symbol $\varnothing$.

- A system is **dependent** if the graphs of the equations of the system are the same line. There are an infinite number of solutions that are represented by $\{(x, y): ax + by = c\}$ where $ax + by = c$ is either of the equations.

Solving Linear Systems Using Elimination by Addition:
[Section 9.2]

1. If necessary, rewrite each equation in the form $ax + by = c$ with a, b, and c integers.

2. Choose the variable to be eliminated.

3. If necessary, multiply one or both of the equations of the system by the appropriate constant(s) so the coefficients of the chosen variable will be additive inverses.

4. Add the equations.

5. If necessary, solve the equation resulting from step 4 for the remaining variable.

6. Substitute the value found for the variable in step 5 into either of the equations of the original system and solve for the other variable.

7. Check the solution in both equations.

Recognizing Types of Equations When Using Addition: [Section 9.2]

- If the system is consistent, the solution of the system is found.

- If the system is inconsistent, a false statement will result when the equations are added.

- If the system is dependent, a true statement will result when the equations are added.

Solving Systems of Linear Equations Using Substitution: [Section 9.3]

1. If necessary, solve one of the equations for a chosen variable.
2. Substitute the expression found in step 1 for the chosen variable in the other equation.
3. Solve the resulting equation.
4. Substitute the value found in step 3 into either of the original equations and solve for the remaining variable.
5. Check the solution of the system in both of the original equations.

Recognizing Types of Equations When Using Substitution: [Section 9.3]

- Same as when using the addition method.

Solving Applications Problems Using Systems of Equations: [Section 9.4]

1. Represent both of the unknowns in the problem with a variable.
2. From the information given in the problem, write two equations.
3. Solve the system of equations using either the addition or the substitution method.
4. Check the solutions against the wording of the original problem.

Systems of Linear Inequalities: [Section 9.5]

- To graphically represent the solutions of a system of linear inequalities, graph all the inequalities of the system on the same coordinate axes. The region where the individual solutions overlap represents all points whose coordinates solve the system.

Chapter 9 **Review Exercises**

Determine whether the given ordered pair is a solution of the system of linear equations. The ordered pair is in the form (x, y). [Section 9.1]

1. $(5, -5)$

$$\begin{cases} 7x - 3y = 36 \\ 4x + 2y = 2 \end{cases}$$

2. $\left(\dfrac{2}{3}, -2\right)$

$$\begin{cases} 3x + 4y = -6 \\ 3x - y = 4 \end{cases}$$

3. $(2, 1)$

$$\begin{cases} \dfrac{1}{2}x + 7y = 8 \\ \dfrac{1}{4}x + 3y = \dfrac{7}{2} \end{cases}$$

4. $(-3, 4)$

$$\begin{cases} .3x + .7y = -1.9 \\ x - .5y = 5 \end{cases}$$

Find the solution(s) of the following systems of linear equations by graphing. Assume that each unit on the coordinate system represents 1. [Section 9.1]

5. $\begin{cases} x = -1 \\ 4x + y = -2 \end{cases}$

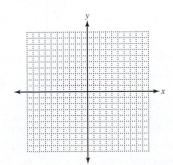

6. $\begin{cases} x + y = 1 \\ 2x - 3y = 12 \end{cases}$

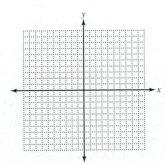

7. $\begin{cases} x - y = -1 \\ 2x - 3y = -9 \end{cases}$

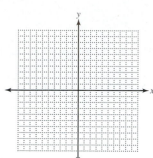

8. $\begin{cases} 8x + 5y = 0 \\ 2x + y = -2 \end{cases}$

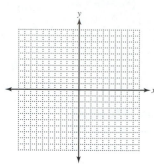

Determine whether the following systems are consistent, inconsistent, or dependent. If the system is consistent, find the solution of the system. Use the method of your choice. [Sections 9.1–9.3]

9. $\begin{cases} 3x - 7y = 4 \\ 12x - 28y = 16 \end{cases}$

10. $\begin{cases} 3x - 5y = 42 \\ x + y = -2 \end{cases}$

11. $\begin{cases} 6x - 5y = 52 \\ -6x + 2y = -10 \end{cases}$

12. $\begin{cases} -6x + 24y = -9 \\ 2x - 8y = 3 \end{cases}$

Solve the following systems of equations using the elimination by addition method. If the system is not consistent, indicate whether it is inconsistent or dependent. [Section 9.2]

13. $\begin{cases} x - y = 12 \\ x + y = 14 \end{cases}$

14. $\begin{cases} 2x - y = -2 \\ 2x - 3y = -30 \end{cases}$

15. $\begin{cases} 4x + 2y = -4 \\ 5x - 3y = -38 \end{cases}$

16. $\begin{cases} 3x - 7y = -10 \\ 8x - 3y = -11 \end{cases}$

17. $\begin{cases} x - 8y = -5 \\ 3x + 12y = -6 \end{cases}$

18. $\begin{cases} 6x + 6y = 7 \\ 9x - 10y = 20 \end{cases}$

19. $\begin{cases} 12x + 11y = 12 \\ 24x - 22y = 4 \end{cases}$

20. $\begin{cases} x = 1 - 7y \\ 2y = 3x + 20 \end{cases}$

21. $\begin{cases} \dfrac{1}{2}x - \dfrac{3}{4}y = -8 \\ \dfrac{1}{4}x - \dfrac{3}{8}y = 2 \end{cases}$

22. $\begin{cases} .5x + .3y = 3 \\ .4x - .7y = 2.3 \end{cases}$

23. $\begin{cases} 3x - 4y = 2 \\ -15x + 20y = -10 \end{cases}$

24. $\begin{cases} 5x - 6y = 4 \\ 30x - 36y = 12 \end{cases}$

Solve the following systems of equations using the substitution method. If the system is not consistent, indicate whether it is inconsistent or dependent. [Section 9.3]

25. $\begin{cases} x = -2y \\ 6x - 5y = -17 \end{cases}$

26. $\begin{cases} x - y = -2 \\ x - 2y = -7 \end{cases}$

27. $\begin{cases} 2x - y = 2 \\ 8x - 7y = -10 \end{cases}$

28. $\begin{cases} 9x + 4y = 18 \\ 5x - 2y = -28 \end{cases}$

29. $\begin{cases} 4x - 3y = 7 \\ 8x - 3y = -13 \end{cases}$

30. $\begin{cases} 5x - y = 3 \\ 10x - 2y = 5 \end{cases}$

Solve the following systems of linear equations by the most appropriate method. If the system is not consistent, indicate whether it is inconsistent or dependent. [Sections 9.1–9.3]

31. $\begin{cases} x + y = -5 \\ 3x - y = -7 \end{cases}$

32. $\begin{cases} 2x = 1 + y \\ 5x - 4y = -8 \end{cases}$

33. $\begin{cases} 6x - 15y = 9 \\ -2x + 5y = -3 \end{cases}$

34. $\begin{cases} 5x - 2y = 0 \\ 3x + 3y = -21 \end{cases}$

35. $\begin{cases} y = -4x \\ 11x + 5y = 9 \end{cases}$

36. $\begin{cases} 8x - 11y = 7 \\ 16x - 22y = 10 \end{cases}$

37. $\begin{cases} 4x + y = 4 \\ x - .5y = -.5 \end{cases}$

38. $\begin{cases} 8x - 5y = 4 \\ \dfrac{1}{2}x - y = -\dfrac{1}{40} \end{cases}$

Solve each of the following using a system of linear equations: [Section 9.4]

39. The difference of two numbers is 7. If the sum of the numbers is 25, find the numbers.

40. A university in the western part of the country changes to a new phone system. They pay a total of $6273 for 425 basic phone sets and 17 executive phone sets. If the cost of a basic phone and an executive phone together is $33, how much did each type of phone set cost?

41. A 9-year-old boy finds that his piggy bank has $5.25 in nickels and dimes. If there are 65 coins in the

piggy bank, how many nickels and how many dimes does he have?

42. A motorcyclist and a bicyclist leave Buffalo, New York, going in opposite directions. The motorcyclist travels four times as fast as the bicyclist. If they are 225 miles apart after three hours, how fast is each traveling?

43. A professional athlete received a $50,000 bonus for signing a contract to skate in an ice show. She in-

vested part of the money in a certificate of deposit that pays 12% interest and part of the money in tax-free bonds paying an average of 7.5% interest. If she earns $4200 interest, how much did she put in each investment?

44. Pure antifreeze (100%) and a solution that is 40% antifreeze are mixed. How many quarts of each are needed to obtain 8 quarts of 70% antifreeze solution?

Graph the solutions of the following systems of linear inequalities: [Section 9.5]

45. $\begin{cases} x + y > 5 \\ x - y < -1 \end{cases}$

46. $\begin{cases} x + y \geq 2 \\ x + 2y \leq 6 \end{cases}$

47. $\begin{cases} 2x - y \leq 3 \\ 2x > 6 \end{cases}$

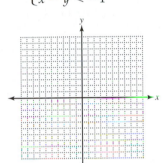
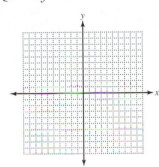
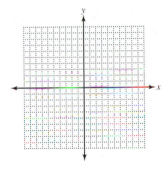

48. $\begin{cases} 3x < 15 \\ y \leq -4 \end{cases}$

49. $\begin{cases} 2x + 3y \leq 4 \\ y + x > 0 \end{cases}$

50. $\begin{cases} y - x \leq 0 \\ x \geq 1 \end{cases}$

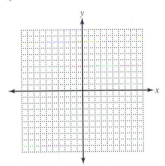
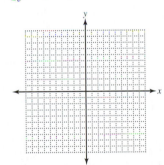
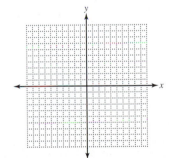

Chapter 9 Test

Determine whether the given ordered pair is a solution of the system of linear equations. The ordered pair is in the form (x, y).

1. $(-3, -1)$
$\begin{cases} 2x + 5y = 1 \\ \frac{1}{3}x - y = 2 \end{cases}$

Find the solution(s) of the following systems of linear equations by graphing. Assume that each unit on the coordinate system represents 1.

2. $\begin{cases} x - y = 8 \\ 2x - y = 12 \end{cases}$

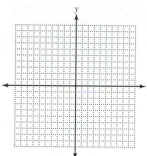

3. $\begin{cases} 4x - 3y = -18 \\ 2x - 3y = -12 \end{cases}$

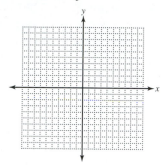

Determine whether the following system is consistent, inconsistent, or dependent. If the system is consistent, find the solution of the system.

4. $\begin{cases} 4x - 6y = 12 \\ 2x - 3y = 6 \end{cases}$

Solve the following systems of equations using the elimination by addition method. If the system is not consistent, indicate whether it is inconsistent or dependent.

5. $\begin{cases} 2x + 3y = 24 \\ 5x + 4y = 46 \end{cases}$

6. $\begin{cases} .9x - .2y = 3.1 \\ 11x - 8y = -1 \end{cases}$

7. $\begin{cases} \dfrac{2}{3}x + \dfrac{5}{8}y = 3 \\ \dfrac{5}{6}x - \dfrac{1}{8}y = -\dfrac{7}{2} \end{cases}$

Solve the following systems of equations using the substitution method. If the system is not consistent, indicate whether it is inconsistent or dependent.

8. $\begin{cases} x + 2y = -6 \\ 3x + 4y = -10 \end{cases}$

9. $\begin{cases} 4x + y = 17 \\ 3x + 2y = 14 \end{cases}$

10. $\begin{cases} 3x + 4y = 14 \\ 2x - 3y = -19 \end{cases}$

Solve the following systems of linear equations by the most appropriate method. If the system is not consistent, indicate whether it is inconsistent or dependent.

11. $\begin{cases} 3x = y - 5 \\ 4x = 3y + 5 \end{cases}$

12. $\begin{cases} 10x + 5y = -2 \\ 5x + 2y = 0 \end{cases}$

13. $\begin{cases} y - x = 1 \\ .8x - .7y = -.4 \end{cases}$

14. $\begin{cases} x = \dfrac{1}{2}y - 6 \\ \dfrac{2x}{7} - 3y = 4 \end{cases}$

Solve each of the following using a system of linear equations:

15. The sum of 3 times a number and 8 is equal to another number. Four times the first number equals the difference of the second number and 6. Find the numbers.

16. The machines at a coin laundry take only dimes and quarters. At the end of the day, one washer has 90 coins with a total value of $13.50. How many of each kind of coin were in the washer?

17. Two cities are 600 miles apart. A car and a bus leave the two cities at the same time, traveling toward each other. The car is traveling 10 miles an hour faster than the bus. If they meet after six hours, how fast was each vehicle traveling?

18. Christopher has just won $16,000 in a video game contest. His parents put part of the money into a mutual fund paying an expected return of 22% and part of the money into tax-free bonds paying 7% interest. If he earns $2920 in interest in a year, how much was put into each investment?

Graph the solutions of the following systems of linear inequalities:

19. $\begin{cases} x - y \geq 5 \\ 2x + 3y > -3 \end{cases}$

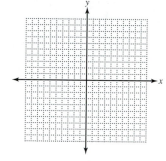

20. $\begin{cases} y - x < 3 \\ x \geq -2 \end{cases}$

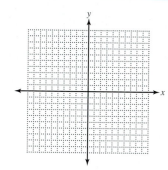

Roots and Radicals

In Section 1.1, we learned to raise numbers to powers. For example, $2^2 = 4$ and $3^3 = 27$. In this chapter we will learn the "reverse" procedure called *finding a root*. For example, we will be asked questions like, "The square of what number is 4?" or "The cube of what number is 27?"

The result of finding a root is not always a rational number. For example, there is no rational number whose square is 3 and there is no rational number whose cube is 4. Thus, we will introduce a new type of number called **irrational numbers** that will be used in solving equations in Chapter 11.

We know how to perform the operations of addition, subtraction, multiplication, division, and raising to powers on rational numbers. In this chapter, we will learn how to perform these same operations on radicals. In addition, we will define the simplest form for a radical and learn to write radicals in this form.

We also know how to solve equations involving rational numbers. In Section 10.5, we will learn to solve equations involving radicals and irrational numbers.

We end the chapter with a very important relationship from geometry involving the sides of a right triangle. Finding one side of a right triangle when given the other two sides often results in an irrational number.

Chapter **10**

Defining and Finding Roots

OBJECTIVES

When you complete this section, you will be able to:

a. Find square roots and higher-order roots of numbers.

b. Determine whether a root is rational or irrational.

c. Find decimal approximations of square and cube roots of irrational numbers using a calculator or a table.

Introduction

The operations of addition and subtraction are inverse operations since one "undoes" the other. Begin with a number, say 3. Add any number to 3. From this sum subtract the same number you previously added. The result will be 3. Multiplication and division (except by 0) are also inverse operations.

Defining roots of numbers

In Section 1.1, we learned to raise numbers to positive integral powers. In this section, we will learn the inverse operation, which, with some restrictions, is finding the roots of numbers.

The inverse of squaring a number is finding the square root. A square root of a number is a number that we must square to get the given number. Since $4^2 = 16$, a square root of 16 is 4. Another square root of 16 is -4 since $(-4)^2 = 16$.

The inverse of cubing a number is finding the cube root. We know that $2^3 = 8$. Therefore, the cube root of 8 is 2. Note that 2 is the only real number whose cube is 8, so 2 is the only real cube root of 8.

The inverse of raising a number to the fourth power is finding the fourth root. We know that $3^4 = 81$. So, a fourth root of 81 is 3. Since $(-3)^4 = 81$, -3 is also a fourth root of 81.

Based on these examples, we make the following definitions:

DEFINITION   Roots of a Number

If the number b is a square root of the number a, then $b^2 = a$.

If the number b is a cube root of the number a, then $b^3 = a$.

If the number b is a fourth root of the number a, then $b^4 = a$.

If the number b is an nth root of the number a, then $b^n = a$.

In words, the square root of a given number is a number whose square equals the given number. The cube root of a given number is a number whose cube equals the given number. The fourth root of a given number is a number whose fourth power equals the given number. In general, the **nth root** of a given number is a number whose nth power equals the given number.

To indicate the root of a number, we use the symbol $\sqrt{}$, called a **radical sign.** In the expression $\sqrt[n]{a}$, read "the nth root of a," n is called the **root index** and indicates which root we are to find. If no root index is given, we assume it to be 2, which means find the square root. The number a is called the **radicand.** The entire expression is called a **radical** and any expression containing a radical is called a **radical expression.**

Using radicals, we restate the definition previously given:

DEFINITION   Roots of Numbers Using Radicals

If $\sqrt{a} = b$, then $b^2 = a$. That is, $\sqrt{a}$ is the number whose square is a.

If $\sqrt[3]{a} = b$, then $b^3 = a$. That is, $\sqrt[3]{a}$ is the number whose cube is a.

If $\sqrt[4]{a} = b$, then $b^4 = a$. That is, $\sqrt[4]{a}$ is the number whose fourth power is a.

In general, if $\sqrt[n]{a} = b$, then $b^n = a$. That is, $\sqrt[n]{a}$ is the number whose nth power is a.

Based on the preceding definition, if $\sqrt{9} = 3$, then $3^2 = 9$. Also, since $5^2 = 25$, $\sqrt{25} = 5$. If $\sqrt[3]{27} = 3$, then $3^3 = 27$. Since $4^3 = 64$, then $\sqrt[3]{64} = 4$. Similar statements can be made for other powers and roots.

Using radicals, $\sqrt{4}$ is read "the square root of 4," $\sqrt[3]{8}$ is read "the cube root of 8," $\sqrt[4]{16}$ is read "the fourth root of 16," and $\sqrt[5]{32}$ is read "the fifth root of 32."

Previously, we noted that 16 has two square roots, 4 and -4. This is usually written as ± 4. Also, 81 has two fourth roots, ± 3. Consequently, without some agreement, this can lead to confusion as to what the symbols represent. If the root index is even, then $\sqrt[n]{a}$ denotes the nonnegative root only and is called the **principal root.** We write the negative even root of a number as $-\sqrt[n]{a}$. We write both the positive and negative even roots as $\pm\sqrt[n]{a}$. Consider the following symbols and their meanings for n even:

Symbol	Meaning
$\sqrt{16}$	4. (The positive square root of 16.)
$-\sqrt{16}$	-4. (The negative square root of 16.)
$\pm\sqrt{16}$	± 4. (The positive and negative square roots of 16.)
$\sqrt[4]{81}$	3. (The positive fourth root of 81.)
$-\sqrt[4]{81}$	-3. (The negative fourth root of 81.)
$\pm\sqrt[4]{81}$	± 3. (The positive and negative fourth roots of 81.)
$\sqrt[n]{a}$	The positive nth root of a.
$-\sqrt[n]{a}$	The negative nth root of a.
$\pm\sqrt[n]{a}$	The positive and negative nth roots of a.

It should also be noted that if the root index is even, the radicand must be positive. For example, $\sqrt{-4}$ does not exist as a real number since there is no real number whose square is -4. Also, $\sqrt[4]{-81}$ does not exist as a real number since there is no real number whose fourth power is -81. These types of numbers are called **imaginary numbers** and are usually studied in more advanced math courses.

If the root index is odd, there is no problem with finding roots of any real number. For example, $\sqrt[3]{27} = 3$ and only 3 since 3 is the only real number whose cube is 27. Also, $\sqrt[3]{-27} = -3$ and only -3 since -3 is the only real number whose cube is -27. We summarize as follows:

Existence of Roots

a. If n is even and $a > 0$, then $\sqrt[n]{a}$ exists as a real number.

b. If n is even and $a < 0$, then $\sqrt[n]{a}$ does not exist as a real number.

c. If n is odd, then $\sqrt[n]{a}$ exists as a real number for all values of a.

Example 1

Find the following roots, if they exist. If the root does not exist, write, "This root does not exist as a real number."

a. $\sqrt{9} = 3$ since 3 is positive and $3^2 = 9$.

b. $-\sqrt{9} = -3$ since -3 is negative and $(-3)^2 = 9$.

c. $\sqrt{-9}$ does not exist as a real number since there is no real number whose square is -9.

d. $\pm\sqrt{9} = \pm 3$ since we want both roots and $(\pm 3)^2 = 9$.

e. $\sqrt[3]{8} = 2$ since $2^3 = 8$.

f. $\sqrt[3]{-8} = -2$ since $(-2)^3 = -8$.

g. $\sqrt[4]{16} = 2$ since 2 is positive and $2^4 = 16$.

Practice Exercises

Find the following roots, if they exist. If the root does not exist, write, "This root does not exist as a real number."

1. $\sqrt{36}$

2. $-\sqrt{36}$

3. $\sqrt{-36}$

4. $\pm\sqrt{36}$

5. $\sqrt[3]{64}$

6. $-\sqrt[3]{64}$

7. $\sqrt[3]{-64}$

8. $-\sqrt[4]{81}$

If more practice is needed, do the Additional Practice Exercises.

Additional Practice Exercises

Find the following roots, if they exist. If the root does not exist, write, "This root does not exist as a real number."

a. $\sqrt{64}$

b. $-\sqrt{64}$

c. $\sqrt{-64}$

d. $\pm\sqrt{64}$

e. $\sqrt[5]{32}$

f. $-\sqrt[5]{32}$

g. $\sqrt[5]{-32}$

h. $-\sqrt[3]{125}$

Determining whether a root is rational or irrational

Only certain numbers have roots that are rational numbers. Remember, a rational number is any number that can be written as the ratio of two integers. As a decimal, a rational number either terminates or repeats.

The numbers 1, 4, 9, 16, 25, 36, . . . , which have rational square roots, are called **perfect squares.** The numbers ± 1, ± 8, ± 27, ± 64, . . . , which have rational cube roots, are called **perfect cubes.** Similarly, we have numbers that have rational fourth, and fifth, roots, and so on. Table 1 in the Appendix gives the squares and cubes of the integers from 1 to 100. Therefore, Table 1 also gives the perfect-square integers from 1 to 10,000 and the perfect-cube integers from 1 to 1,000,000.

If a number is not a perfect square, then it is impossible to represent the square root exactly as a terminating or repeating decimal. For example, we cannot represent $\sqrt{3}$ exactly as a decimal since there is no decimal whose square is 3. However, we may approximate $\sqrt{3}$ as closely as we please by using a sufficient number of decimal places. To the nearest tenth, $\sqrt{3} = 1.7$ and $(1.7)^2 = 2.89$. To the nearest hundredth, $\sqrt{3} = 1.73$ and $(1.73)^2 = 2.9929$, which is closer to 3 than $(1.7)^2$, so 1.73 is a better approximation of $\sqrt{3}$ than 1.7. To the nearest thousandth, $\sqrt{3} = 1.732$ and $(1.732)^2 = 2.999824$, which is closer to 3 than $(1.73)^2$, so 1.732 is a better approximation of $\sqrt{3}$ than 1.73.

Numbers like $\sqrt{3}$ are called **irrational** numbers. Irrational numbers cannot be represented by terminating or repeating decimals since terminating and repeating decimals are rational numbers. Hence, irrational numbers are nonrepeating, nonterminating decimals. In general, any root of a number that is not rational is irrational. Decimal approximations of irrational roots may be found by using a calculator or tables. Table 1 in the Appendix gives decimal approximations of square and cube roots of numbers from 1 to 100 and square roots of 10 times any number from 1 to 100. Calculators afford an easy method for approximating irrational roots.

Using Table 1

Finding decimal approximations of irrational square and cube roots

The headings at the top of Table 1 appear as follows:

n	n^2	n^3	$\sqrt{n}$	$\sqrt[3]{n}$	$\sqrt{10n}$

Under the *n* column are the integers 1–100. The other columns are self-explanatory. For example, if we go down to 8 under the *n* column, we find the row,

n	n^2	n^3	$\sqrt{n}$	$\sqrt[3]{n}$	$\sqrt{10n}$
8	64	512	2.828	2.000	8.944

From this row, we see that $8^2 = 64$, $8^3 = 512$, $\sqrt{8} = 2.828$, $\sqrt[3]{8} = 2.000$, and $\sqrt{80} = 8.944$ to the nearest thousandths.

We may also use the table in another way. Since 64 is in the n^2 column, $8^2 = 64$, which means $\sqrt{64} = 8$. Likewise, since 512 is in the n^3 column, $8^3 = 512$, which means $\sqrt[3]{512} = 8$. In general, any number in the *n* column is the square root of the corresponding number in the n^2 column and the cube root of any number in the n^3 column.

Example 2

Identify each of the following as rational or irrational. If the number is rational, find the root exactly. If the number is irrational, approximate the number to the nearest thousandth by using a calculator or Table 1. The symbol "≈" means **approximately equals.**

a. $\sqrt{12}$ Irrational. $\sqrt{12} \approx 3.464$.

b. $\sqrt[3]{30}$ Irrational. $\sqrt[3]{30} \approx 3.107$.

c. $\sqrt{81}$ Rational. $\sqrt{81} = 9$.

d. $\sqrt{5184}$ Rational. $5184 = 72^2$, so $\sqrt{5184} = 72$.

e. $\sqrt{9.61}$ Rational. $9.61 = 3.1^2$, so $\sqrt{9.61} = 3.1$.

f. $\sqrt[3]{753,571}$ Rational. $753,571 = 91^3$, so $\sqrt[3]{753,571} = 91$.

g. $\sqrt[3]{68}$ Irrational. $\sqrt[3]{68} \approx 4.082$.

h. $\sqrt{670}$ Irrational. $\sqrt{670} \approx 25.884$. (Look under $\sqrt{10n}$ column where $n = 67$.)

Practice Exercises

Identify each of the following as rational or irrational. If the number is rational, find the root exactly. If the number is irrational, approximate the number to the nearest thousandth by using a calculator or Table 1.

9. $\sqrt{19}$ **10.** $\sqrt[3]{63}$ **11.** $\sqrt{121}$ **12.** $\sqrt{3364}$

13. $\sqrt{12.25}$ **14.** $\sqrt[3]{262,144}$ **15.** $\sqrt[3]{86}$ **16.** $\sqrt{440}$

If more practice is needed, do the Additional Practice Exercises.

Additional Practice Exercises

Identify each of the following as rational or irrational. If the number is rational, find the root exactly. If the number is irrational, approximate the number to the nearest thousandth by using a calculator or Table 1.

i. $\sqrt{24}$ **j.** $\sqrt[3]{49}$ **k.** $\sqrt{49}$ **l.** $\sqrt{1156}$

m. $\sqrt{8.41}$ **n.** $\sqrt[3]{2744}$ **o.** $\sqrt[3]{21}$ **p.** $\sqrt{450}$

Answers:

Practice Exercises 9–16: 9. irrational, 4.359 **10.** irrational, 3.979 **11.** rational, 11 **12.** rational, 58 **13.** rational, 3.5 **14.** rational, 64 **15.** irrational, 4.414 **16.** irrational, 20.976   *Additional Practice Exercises i–p:* **i.** irrational, 4.899 **j.** irrational, 3.659 **k.** rational, 7 **l.** rational, 34 **m.** rational, 2.9 **n.** rational, 14 **o.** irrational, 2.759 **p.** irrational, 21.213

For convenience in finding powers and roots we include the following table of powers of numbers that occur often in this chapter:

n	n^2	n^3	n^4	n^5	n^6
2	4	8	16	32	64
3	9	27	81	243	729
4	16	64	256		
5	25	125	625		
6	36	216	1296		
7	49	343			

The definition of a root can also be used to find roots with variable radicands. Since even roots of negative radicands do not exist, all variables are assumed to represent nonnegative numbers. Recall that $(a^m)^n = a^{mn}$.

Example 3

Find the following roots. Assume all variables represent nonnegative numbers.

a. $\sqrt{x^2} = (\ )$ By definition of square root, $(\ )^2 = x^2$. Since $(x)^2 = x^2$, the $(\ )$ represents x.

$\sqrt{x^2} = x$ Therefore, $\sqrt{x^2} = x$. Remember, x must be nonnegative.

Note: If we do not assume that $x \geq 0$, then $\sqrt{x^2} \neq x$ since the radical always denotes the nonnegative square root and x can be positive, zero, or negative. Consequently, $\sqrt{x^2} = |x|$ since $|x|$ is also nonnegative. This situation is usually addressed in intermediate or college algebra and is why we will assume, when finding even roots, that all variables have nonnegative values.

b. $\sqrt{a^4} = (\ )$ By the definition of a square root, $(\ )^2 = a^4$. Since $(a^2)^2 = a^4$, the $(\ )$ represents a^2.

$\sqrt{a^4} = a^2$ Therefore, $\sqrt{a^4} = a^2$.

c. $\sqrt[3]{y^6} = (\ )$ By the definition of a cube root, $(\ )^3 = y^6$. Since $(y^2)^3 = y^6$, the $(\ )$ represents y^2.

$\sqrt[3]{y^6} = y^2$ Therefore, $\sqrt[3]{y^6} = y^2$.

Note: You may have noticed that to find the root of a variable radicand, the exponent on the variable must be divisible by the root index. In fact, the exponent on the root is the exponent on the radicand divided by the index. For example, to find $\sqrt[4]{y^8}$, divide 8 by 4. Since $8 \div 4 = 2$, the exponent on the root is 2. Therefore, $\sqrt[4]{y^8} = y^2$. This works because of the definition of multiplication. If $\sqrt[4]{y^8} = y^n$, then $(y^n)^4 = y^8$, which gives $y^{4n} = y^8$. Consequently $4n = 8$ or $n = \dfrac{8 \text{ (the exponent)}}{4 \text{ (the index)}}$.

Practice Exercises

Find the following roots. Assume all expressions have nonnegative values.

17. $\sqrt{z^2}$ **18.** $\sqrt{c^6}$ **19.** $\sqrt[4]{w^{12}}$

If more practice is needed, do the Additional Practice Exercises.

Additional Practice Exercises

Find the following roots. Assume all variables have nonnegative values.

q. $\sqrt{r^4}$ **r.** $\sqrt{p^8}$ **s.** $\sqrt[5]{x^{15}}$

Answers:

Practice Exercises 17–19: 17. z 18. c^3 19. w^3 Additional Practice Exercises q–s: q. r^2 r. p^4 s. x^3

Often the radicands of radical expressions are not monomials. However, the same procedures apply.

Example 4

Find the following roots. Assume all expressions have nonnegative values.

a. $\sqrt{(a-2)^2} =$ We need the expression whose square is $(a-2)^2$, which is $a-2$.

$a-2$ Therefore, $\sqrt{(a-2)^2} = a-2$.

b. $\sqrt{(2x-3y)^2} =$ We need the expression whose square is $(2x-3y)^2$, which is $2x-3y$.

$2x-3y$ Therefore, $\sqrt{(2x-3y)^2} = 2x-3y$.

c. $\sqrt{x^2+6x+9} =$ Rewrite x^2+6x+9 as $(x+3)^2$.

$\sqrt{(x+3)^2} =$ We need the expression whose square is $(x+3)^2$, which is $x+3$.

$x+3$ Therefore, $\sqrt{x^2+6x+9} = x+3$.

Practice Exercises

Find the following roots. Assume all radicands have nonnegative values.

20. $(x+4)^2$ **21.** $\sqrt{(4x-5y)^2}$ **22.** $\sqrt{a^2+4a+4}$

Exercise Set 10.1

Find the following roots, if they exist. If the root does not exist, write, "This root does not exist as a real number."

1. $\sqrt{25}$ **2.** $\sqrt{-25}$ **3.** $-\sqrt{25}$ **4.** $\pm\sqrt{25}$

5. $-\sqrt{100}$ **6.** $\sqrt{100}$ **7.** $\sqrt{-100}$ **8.** $\pm\sqrt{100}$

9. $\sqrt{625}$ **10.** $-\sqrt{289}$ **11.** $\sqrt{169}$ **12.** $\pm\sqrt{256}$

13. $\sqrt[3]{27}$ **14.** $-\sqrt[3]{27}$ **15.** $\sqrt[3]{-27}$ **16.** $-\sqrt[3]{216}$

17. $\sqrt[3]{-216}$ **18.** $\sqrt[3]{216}$ **19.** $-\sqrt[3]{-216}$ **20.** $\sqrt[4]{625}$

21. $\sqrt[4]{-625}$ **22.** $\sqrt[5]{243}$ **23.** $-\sqrt[5]{243}$ **24.** $\sqrt[5]{-243}$

Identify each of the following as rational or irrational. If the number is rational, find the root exactly. If the number is irrational, approximate the number to the nearest thousandth by using a calculator or Table 1.

25. $\sqrt{24}$ **26.** $\sqrt{32}$ **27.** $\sqrt{144}$ **28.** $\sqrt{400}$

29. $\sqrt[3]{50}$ **30.** $\sqrt[3]{21}$ **31.** $\sqrt[3]{64}$ **32.** $\sqrt[3]{27}$

33. $\sqrt{1.69}$ **34.** $\sqrt{.64}$ **35.** $\sqrt{810}$ **36.** $\sqrt{640}$

37. $\sqrt[3]{.125}$ **38.** $\sqrt[3]{.216}$ **39.** $\sqrt[3]{100}$ **40.** $\sqrt[3]{80}$

41. $\sqrt{2116}$ **42.** $\sqrt{3844}$ **43.** $\sqrt[3]{9261}$ **44.** $\sqrt[3]{24,389}$

Find the following roots. Assume all variables represent nonnegative numbers.

45. $\sqrt{b^2}$ **46.** $\sqrt{c^2}$ **47.** $\sqrt{n^4}$ **48.** $\sqrt{m^4}$

49. $\sqrt{r^8}$ **50.** $\sqrt{a^{10}}$ **51.** $\sqrt[3]{n^3}$ **52.** $\sqrt[3]{m^3}$

Answers:

53. $\sqrt[3]{r^6}$

54. $\sqrt[3]{a^{12}}$

55. $\sqrt[3]{c^3}$

56. $\sqrt[4]{d^8}$

57. $\sqrt[5]{s^{10}}$

58. $\sqrt[5]{y^{15}}$

59. $\sqrt[6]{a^6}$

60. $\sqrt[6]{d^{12}}$

Find the following roots. Assume all expressions have nonnegative values.

61. $\sqrt{(x-5)^2}$

62. $\sqrt{(r+1)^2}$

63. $\sqrt{(2a+b)^2}$

64. $\sqrt{(3x-2y)^2}$

65. $\sqrt{x^2 + 10x + 25}$

66. $\sqrt{a^2 - 8a + 16}$

67. $\sqrt{4a^2 - 12ab + 9b^2}$

68. $\sqrt{9a^2 - 30ab + 25b^2}$

Writing Exercise:

69. Why is $\sqrt{-16}$ not a real number?

70. In order for $\sqrt[n]{x}$ to exist as a real number when n is even, x must be nonnegative. However, if n is odd, then $\sqrt[n]{x}$ exists for all values of x whether positive or negative. Explain.

Group Project:

71. Interview some science instructors, engineering instructors, and so on, and make a list of a total of five situations where radical expressions are used in areas other than mathematics.

Section 10.2	Simplifying Radicals

OBJECTIVES

When you complete this section, you will be able to:

a. Simplify square roots using the product rule.

b. Simplify higher-order roots using the product rule.

c. Use the alternative rule to simplify radicals.

Introduction   Fractions are not considered to be in their simplest form unless they are reduced to their lowest terms. Likewise, radical expressions are not considered to be in their simplest form unless they satisfy certain conditions. In order to simplify radicals, we need to develop some properties of radicals.

OPTIONAL   **CALCULATOR EXPLORATION ACTIVITY**

Use your calculator to evaluate columns A and B below and round your answer to the nearest ten thousandth. Compare the results obtained in each line of the columns.

Column A	**Column B**
a. $\sqrt{2} \cdot \sqrt{3} = $ _____	$\sqrt{6} = $ _____
b. $\sqrt{3} \cdot \sqrt{6} = $ _____	$\sqrt{18} = $ _____
c. $\sqrt{5} \cdot \sqrt{6} = $ _____	$\sqrt{30} = $ _____
d. $\sqrt{20} \cdot \sqrt{30} = $ _____	$\sqrt{600} = $ _____

1. Based on your answers to the corresponding lines in columns A and B, $\sqrt{a} \cdot \sqrt{b} = $ _____.

Consider the following. If we find each root and then multiply, we have $\sqrt{9} \cdot \sqrt{16} = 3 \cdot 4 = 12$. If we multiply the radicands and then find the root, we have $\sqrt{9} \cdot \sqrt{16} = \sqrt{9 \cdot 16} = \sqrt{144} = 12$. Note that we arrived at the same answer using either method. This suggests the following rule for multiplying radicals:

Multiplying radicals

Products of Radicals

If $\sqrt[n]{a}$ and $\sqrt[n]{b}$ are both real numbers, then $\sqrt[n]{a} \cdot \sqrt[n]{b} = \sqrt[n]{a \cdot b}$.

Simplifying radicals

In this section, we will be using this rule in the form $\sqrt[n]{ab} = \sqrt[n]{a} \cdot \sqrt[n]{b}$. For example, $\sqrt{4 \cdot 3} = \sqrt{4} \cdot \sqrt{3} = 2\sqrt{3}$.

A radical is in **simplified form** if the radicand contains no factor that can be written to a power greater than or equal to the index. In other words, the radicand of a square root cannot contain a factor that is a perfect square, the radicand of a cube root cannot contain a factor that is a perfect cube, and so on. Consequently, the first step in simplifying a *square* root is to write the radicand as the product of the largest possible perfect square and a number that has no perfect-square factors.

Recall from Section 5.1 that any number is divisible by all of its factors. Consequently, if a radicand has a perfect-square factor, it must be divisible by an integer that is a perfect square. If so, by the definition of division, the number of times that the perfect square divides into the radicand is the other factor of the radicand. Remember, the integers that are perfect squares are 1, 4, 9, 16, 25, 36, ... Since 1 is a factor of every number, we ignore 1 when finding perfect-square factors of a radicand.

Example 1

Write each of the following as the product of the largest possible perfect square and another number that has no perfect-square factors:

a. 27

Solution:
See if 27 is divisible by a perfect-square integer. Is 27 divisible by 4? No. Is 27 divisible by 9? Yes, $27 \div 9 = 3$. Since 3 has no perfect-square factors, 9 is the largest perfect-square factor of 27. Therefore, $27 = 9 \cdot 3$.

b. 150

Solution:
See if 150 is divisible by a perfect-square integer. Is 150 divisible by 4? No. Is 150 divisible by 9? No. Is 150 divisible by 16? No. Is 150 divisible by 25? Yes, $150 \div 25 = 6$. Since 6 has no perfect-square factors, 25 is the largest perfect-square factor of 150. Therefore, $150 = 25 \cdot 6$.

c. 80

Solution:
See if 80 is divisible by a perfect-square integer. Is 80 divisible by 4? Yes, $80 \div 4 = 20$. Does 20 have a perfect-square factor? Yes, $20 \div 4 = 5$. Consequently, 4 is not the largest perfect-square factor of 80. Keep going. Is 80 divisible by 9? No. Is 80 divisible by 16? Yes, $80 \div 16 = 5$. Since 5 has no perfect-square factors, 16 is the largest perfect-square factor of 80. Therefore, $80 = 16 \cdot 5$.

Note: We can determine the greatest perfect-square factor of a number by using a procedure similar to the one we used to write a number in terms of its prime factors. The procedure is as follows:

Determining the Greatest Perfect-Square Factor

1. Divide the number by any perfect-square factor.
2. If possible, divide the resulting quotient by any perfect-square factor.
3. Continue dividing the quotients by perfect-square factors until you arrive at a quotient with no perfect-square factors.
4. The greatest perfect-square factor is the product of all the perfect-square divisors and the last quotient is the other factor.

Using the procedure on Example 1c would result in the following solution:

Four is a perfect-square factor of 80, so divide 4 into 80.

$$\frac{20}{4\overline{)80}} \qquad \text{20 has a perfect-square factor of 4, so divide 4 into 20.}$$

$$\frac{5}{4\overline{)20}} \qquad \text{5 has no perfect-square factors.}$$

Therefore, $80 = 4 \cdot 4 \cdot 5 = 16 \cdot 5$ where 16 is the greatest perfect-square factor of 80.

d. 72

Solution:
We will use the preceding procedure. Four is a perfect-square factor of 72, so divide 4 into 72.

$$\frac{18}{4\overline{)72}} \qquad \text{9 is a perfect-square factor of 18, so divide 9 into 18.}$$

$$\frac{2}{9\overline{)18}} \qquad \text{2 has no perfect-square factors.}$$

Therefore, $72 = 4 \cdot 9 \cdot 2 = 36 \cdot 2$ where 36 is the greatest perfect-square factor of 72.

Practice Exercises

Write each of the following as the product of the largest possible perfect square and another number that has no perfect-square factors:

1. 40   **2.** 125   **3.** 96   **4.** 108

The procedure for simplifying *square roots* is outlined as follows:

Simplifying Square Roots

1. Write the radicand as the product of the largest possible perfect square and a number that has no perfect-square factors.
2. Apply the product rule in the form $\sqrt{a \cdot b} = \sqrt{a} \cdot \sqrt{b}$ where a is a perfect square.
3. Find the square root of the perfect-square radicand.

Example 2

Simplify the following radical expressions:

a. $\sqrt{18} =$ The largest perfect-square factor of 18 is 9. Rewrite 18 as $9 \cdot 2$.

$\sqrt{9 \cdot 2} =$ Apply $\sqrt{ab} = \sqrt{a} \cdot \sqrt{b}$.

$\sqrt{9} \cdot \sqrt{2} =$ $\sqrt{9} = 3$.

$3\sqrt{2}$ Therefore, $\sqrt{18} = 3\sqrt{2}$.

b. $\sqrt{72} =$ The largest perfect-square factor of 72 is 36. Rewrite 72 as $36 \cdot 2$.

$\sqrt{36 \cdot 2} =$ Apply $\sqrt{ab} = \sqrt{a} \cdot \sqrt{b}$.

$\sqrt{36} \cdot \sqrt{2} =$ $\sqrt{36} = 6$.

$6\sqrt{2}$ Therefore, $\sqrt{72} = 6\sqrt{2}$.

Note: Even though 4 is a perfect-square factor of 72, we should not write $\sqrt{72} = \sqrt{4 \cdot 18}$ because 18 has 9 as a perfect-square factor. However, we could proceed as follows:

$$\sqrt{72} = \sqrt{4 \cdot 18} = \sqrt{4} \cdot \sqrt{18} = 2\sqrt{18} = 2\sqrt{9 \cdot 2} = 2\sqrt{9} \cdot \sqrt{2} = 2 \cdot 3\sqrt{2} = 6\sqrt{2}$$

and arrive at the same answer but with more steps. That is why it is preferable to find the largest perfect-square factor.

c. $2\sqrt{80} =$ The largest perfect-square factor of 80 is 16. Rewrite 80 as $16 \cdot 5$.

$2\sqrt{16 \cdot 5} =$ Apply $\sqrt{ab} = \sqrt{a} \cdot \sqrt{b}$.

$2\sqrt{16} \cdot \sqrt{5} =$ $\sqrt{16} = 4$.

$2 \cdot 4\sqrt{5} =$ Multiply 2 and 4.

$8\sqrt{5}$ Therefore, $2\sqrt{80} = 8\sqrt{5}$.

OPTIONAL   **ALTERNATIVE METHOD**

Sometimes finding the greatest perfect-square factor of the radicand, especially if the radicand is a large number, can be troublesome. Another method, using the prime factorization of the radicand, is often useful under these circumstances. A perfect-square integer can be thought of as an integer that can be expressed as the product of two identical factors. For example, $49 = 7 \cdot 7$ and $121 = 11 \cdot 11$. The square root of a perfect square is one of these two identical factors. For example, $\sqrt{49} = \sqrt{7 \cdot 7} = 7$, which is one of the two identical factors of 49. We can use this idea to simplify a radical by writing the radicand as the product of its prime factors without using exponents. Each factor that appears twice will have a square root equal to one of the two factors. The remaining factors stay underneath the radical sign. We illustrate with some examples.

Example 3

Simplify the following radicals using prime factorizations:

a. $\sqrt{72} =$ Write 72 as the product of its prime factors.

$\sqrt{2 \cdot 2 \cdot 2 \cdot 3 \cdot 3} =$ $\sqrt{2 \cdot 2} = 2$ and $\sqrt{3 \cdot 3} = 3$.

$2 \cdot 3\sqrt{2} =$ $2 \cdot 3 = 6$.

$6\sqrt{2}$ Therefore, $\sqrt{72} = 6\sqrt{2}$.

b. $\sqrt{375} =$ Write 375 as the product of its prime factors.

$\sqrt{5 \cdot 5 \cdot 5 \cdot 3} =$ $\sqrt{5 \cdot 5} = 5$.

$5\sqrt{3 \cdot 5} =$ $3 \cdot 5 = 15$.

$5\sqrt{15}$ Therefore, $\sqrt{375} = 5\sqrt{15}$.

Practice Exercises

Simplify the following square roots:

5. $\sqrt{8}$ **6.** $\sqrt{50}$ **7.** $2\sqrt{54}$ **8.** $5\sqrt{108}$

If more practice is needed, do the Additional Practice Exercises.

Additional Practice Exercises

Simplify the following square roots.

a. $\sqrt{20}$ **b.** $\sqrt{162}$ **c.** $3\sqrt{45}$ **d.** $5\sqrt{44}$

We can apply either procedure for simplifying radicals whose radicands contain variables. Remember, to find the root of a variable expression, the exponent on the variable must be divisible by the root index. For example, $\sqrt{x^8} = x^{\frac{8}{2}} = x^4$. In general $\sqrt{x^n} = x^{\frac{n}{2}}$. Consequently, variable factors that are perfect squares must have even exponents. Therefore, to find the square root, we rewrite the radicand as the product of the variable with the largest possible even exponent and the variable to the first power.

Example 4

Simplify the following roots. Assume all variables have nonnegative values.

a. $\sqrt{x^3} =$ The largest perfect-square factor of x^3 is x^2. Rewrite x^3 as $x^2 \cdot x$.

$\sqrt{x^2 \cdot x} =$ Apply $\sqrt{ab} = \sqrt{a} \cdot \sqrt{b}$.

$\sqrt{x^2} \cdot \sqrt{x} =$ $\sqrt{x^2} = x$.

$x\sqrt{x}$ Therefore, $\sqrt{x^3} = x\sqrt{x}$.

b. $a^2\sqrt{a^7} =$ The largest perfect-square factor of a^7 is a^6. Rewrite a^7 as $a^6 \cdot a$.

$a^2\sqrt{a^6 \cdot a} =$ Apply $\sqrt{ab} = \sqrt{a} \cdot \sqrt{b}$.

$a^2\sqrt{a^6} \cdot \sqrt{a} =$ $\sqrt{a^6} = a^3$.

$a^2 \cdot a^3\sqrt{a} =$ Multiply $a^2 \cdot a^3$.

$a^5\sqrt{a}$ Therefore, $a^2\sqrt{a^7} = a^5\sqrt{a}$.

c. $\sqrt{c^5d^3} =$ The largest perfect-square factor of c^5 is c^4, and the largest perfect-square factor of d^3 is d^2. Rewrite c^5d^3 as $c^4d^2 \cdot cd$.

$\sqrt{c^4d^2 \cdot cd} =$ Apply $\sqrt{ab} = \sqrt{a} \cdot \sqrt{b}$.

$\sqrt{c^4d^2} \cdot \sqrt{cd} =$ Apply $\sqrt{ab} = \sqrt{a} \cdot \sqrt{b}$ to $\sqrt{c^4d^2}$.

$\sqrt{c^4} \cdot \sqrt{d^2} \cdot \sqrt{cd} =$ $\sqrt{c^4} = c^2$ and $\sqrt{d^2} = d$.

$c^2d\sqrt{cd}$ Therefore, $\sqrt{c^5d^3} = c^2d\sqrt{cd}$.

Note: If we used the alternative method of simplifying square roots, Example 4c would appear as follows:

Answers:
Practice Exercises 5–8: 5. $2\sqrt{2}$ 6. $5\sqrt{2}$ 7. $6\sqrt{6}$ 8. $30\sqrt{3}$ **Additional Practice Exercises a–d:** a. $2\sqrt{5}$ b. $9\sqrt{2}$
c. $9\sqrt{5}$ d. $10\sqrt{11}$

$$\sqrt{c^5 d^3} =$$ Write $c^5 d^3$ as the product of its prime factors.

$$\sqrt{c \cdot c \cdot c \cdot c \cdot c \cdot d \cdot d \cdot d} =$$ $\sqrt{c \cdot c} = c$ and $\sqrt{d \cdot d} = d$.

$$c \cdot c \cdot d \sqrt{c \cdot d} =$$ $c \cdot c = c^2$.

$$c^2 d \sqrt{cd} =$$ Therefore, $\sqrt{c^5 d^3} = c^2 d \sqrt{cd}$.

d. $3n\sqrt{28n^9} =$ The largest perfect-square factor of 28 is 4 and the largest perfect-square factor of n^9 is n^8. Rewrite $28n^9$ as $4n^8 \cdot 7n$.

$$3n\sqrt{4n^8 \cdot 7n} =$$ Apply $\sqrt{ab} = \sqrt{a} \cdot \sqrt{b}$.

$$3n\sqrt{4n^8} \cdot \sqrt{7n} =$$ Apply $\sqrt{ab} = \sqrt{a} \cdot \sqrt{b}$ to $\sqrt{4n^8}$.

$$3n\sqrt{4} \cdot \sqrt{n^8} \cdot \sqrt{7n} =$$ $\sqrt{4} = 2$ and $\sqrt{n^8} = n^4$.

$$3n \cdot 2n^4\sqrt{7n} =$$ Multiply $3n \cdot 2n^4$.

$$6n^5\sqrt{7n}$$ Therefore, $3n\sqrt{28n^9} = 6n^5\sqrt{7n}$.

Practice Exercises

Simplify the following square roots:

9. $\sqrt{a^5}$ **10.** $c^2\sqrt{c^{11}}$ **11.** $\sqrt{r^7 s^3}$ **12.** $4b\sqrt{27b^7}$

Simplifying higher-order roots   To simplify cube roots, we rewrite the radicand as the product of a perfect cube and some number that has no perfect-cube factors. Recall that the integers that are perfect cubes are 1, 8, 27, 64, 125, ... To determine if the radicand has a perfect-cube factor, see if the radicand is divisible by any integer that is a perfect cube. For example, 32 has the perfect-cube factor 8 since $32 \div 8 = 4$. Consequently, $32 = 8 \cdot 4$.

Remember: in finding cube roots of variables, the exponent on the radicand must be divisible by the index 3. For example, $\sqrt[3]{x^9} = x^{\frac{9}{3}} = x^3$. In general, $\sqrt[3]{x^n} = x^{\frac{n}{3}}$. So variable factors must be written as the product of the variable to the largest possible power divisible by 3 and the variable to a power smaller than 3.

Example 5

Simplify the following radical expressions:

a. $\sqrt[3]{24} =$ The largest perfect-cube factor of 24 is 8. Rewrite 24 as $8 \cdot 3$.

$$\sqrt[3]{8 \cdot 3} =$$ Apply the product rule.

$$\sqrt[3]{8} \cdot \sqrt[3]{3} =$$ $\sqrt[3]{8} = 2$.

$$2\sqrt[3]{3}$$ Therefore, $\sqrt[3]{24} = 2\sqrt[3]{3}$.

b. $5\sqrt[3]{108} =$ The largest perfect-cube factor of 108 is 27. Rewrite 108 as $27 \cdot 4$.

$$5\sqrt[3]{27 \cdot 4} =$$ Apply the product rule.

$$5\sqrt[3]{27} \cdot \sqrt[3]{4} =$$ $\sqrt[3]{27} = 3$.

$$5 \cdot 3\sqrt[3]{4} =$$ Multiply 5 and 3.

$$15\sqrt[3]{4}$$ Therefore, $5\sqrt[3]{108} = 15\sqrt[3]{4}$.

c. $\sqrt[3]{x^5} =$ The largest perfect-cube factor of x^5 is x^3. Rewrite x^5 as $x^3 \cdot x^2$.

$$\sqrt[3]{x^3 \cdot x^2} =$$ Apply the product rule.

$$\sqrt[3]{x^3} \cdot \sqrt[3]{x^2} =$$ $\sqrt[3]{x^3} = x$.

$$x\sqrt[3]{x^2}$$ Therefore, $\sqrt[3]{x^5} = x\sqrt[3]{x^2}$.

d. $\sqrt[3]{40x^{10}} =$ The largest perfect-cube factor of 40 is 8 and the largest perfect-cube factor of x^{10} is x^9. Rewrite $40x^{10}$ as $8x^9 \cdot 5x$.

$\sqrt[3]{8x^9 \cdot 5x} =$ Apply the product rule.

$\sqrt[3]{8x^9} \cdot \sqrt[3]{5x} =$ Apply the product rule to $\sqrt[3]{8x^9}$.

$\sqrt[3]{8} \cdot \sqrt[3]{x^9} \cdot \sqrt[3]{5x} =$ $\sqrt[3]{8} = 2$ and $\sqrt[3]{x^9} = x^3$.

$2x^3\sqrt[3]{5x}$ Therefore, $\sqrt[3]{40x^{10}} = 2x^3\sqrt[3]{5x}$.

| OPTIONAL | **ALTERNATIVE METHOD** |

The alternative method of simplifying radicands is easily extended to roots other than square roots. For example, a perfect cube can be thought of as a number that can be written as the product of three identical factors, such as, $8 = 2 \cdot 2 \cdot 2$. Therefore, the cube root of a perfect cube is one of its three identical factors. For example, $\sqrt[3]{8} = \sqrt[3]{2 \cdot 2 \cdot 2} = 2$, which is one of the three identical factors of 8. Using the alternative method, Example 5d would appear as follows:

$\sqrt[3]{40x^{10}}$ Write $40x^{10}$ as the product of its prime factors.

$\sqrt[3]{2 \cdot 2 \cdot 2 \cdot 5 \cdot x \cdot x \cdot x \cdot x \cdot x \cdot x \cdot x \cdot x \cdot x \cdot x}$ $\sqrt[3]{2 \cdot 2 \cdot 2} = 2$ and $\sqrt[3]{x \cdot x \cdot x} = x$.

$2 \cdot x \cdot x \cdot x\sqrt[3]{5x}$ $x \cdot x \cdot x = x^3$.

$2x^3\sqrt[3]{5x}$ Therefore, $\sqrt[3]{40x^{10}} = 2x^3\sqrt[3]{5x}$.

Practice Exercises

Simplify the following radical expressions:

13. $\sqrt[3]{56}$ **14.** $6\sqrt[3]{128}$ **15.** $\sqrt[3]{a^8}$ **16.** $\sqrt[3]{81c^{13}}$

If more practice is needed, do the Additional Practice Exercises.

Additional Practice Exercises

Simplify the following roots:

e. $\sqrt[3]{72}$ **f.** $4\sqrt[3]{48}$ **g.** $\sqrt[3]{r^4}$ **h.** $\sqrt[3]{250n^{14}}$

Exercise Set 10.2

Simplify the following:

1. $\sqrt{12}$ **2.** $\sqrt{28}$ **3.** $\sqrt{98}$ **4.** $\sqrt{48}$

5. $\sqrt{128}$ **6.** $\sqrt{180}$ **7.** $5\sqrt{54}$ **8.** $3\sqrt{63}$

9. $6\sqrt{80}$ **10.** $4\sqrt{50}$ **11.** $5\sqrt{112}$ **12.** $3\sqrt{245}$

13. $7\sqrt{243}$ **14.** $6\sqrt{242}$ **15.** $\sqrt{a^3}$ **16.** $\sqrt{d^5}$

17. $\sqrt{x^2y^4}$ **18.** $\sqrt{a^6b^2}$ **19.** $\sqrt{x^6y^8z^{10}}$ **20.** $\sqrt{p^4q^8r^8}$

21. $c\sqrt{c^3d^6}$ **22.** $x\sqrt{x^7y^4}$ **23.** $rs^2\sqrt{r^9s^5}$ **24.** $a^2b^3\sqrt{a^{11}b^3}$

---

25. $\sqrt{20x^6}$

26. $\sqrt{45y^4}$

27. $3\sqrt{72x^5}$

28. $6\sqrt{75c^5d^3}$

29. $5x\sqrt{150x^7}$

30. $4a^2b\sqrt{216a^6b^9}$

31. $8a^3c\sqrt{99a^{13}c^4}$

32. $11r^3s^2\sqrt{405r^{11}s^7}$

33. $\sqrt[3]{32}$

34. $\sqrt[3]{54}$

35. $4\sqrt[3]{40}$

36. $7\sqrt[3]{81}$

37. $9\sqrt[3]{56}$

38. $7\sqrt[3]{320}$

39. $\sqrt[3]{x^7}$

40. $\sqrt[3]{b^{11}}$

41. $\sqrt[3]{x^6y^5}$

42. $\sqrt[3]{m^{13}n^9}$

43. $\sqrt[3]{128z^8}$

44. $\sqrt[3]{48h^{14}}$

45. $2\sqrt[3]{24}$

46. $4\sqrt[3]{250}$

Functions that contain radicals are referred to as root functions. Evaluate the following:

47. If $f(x) = \sqrt{3x}$, find the following: **a.** $f(1)$, **b.** $f(3)$, **c.** $f(4)$.

48. If $f(x) = \sqrt{6x}$, find the following: **a.** $f(0)$, **b.** $f(2)$, **c.** $f(6)$.

49. If $f(x) = \sqrt{3x - 5}$, find the following: **a.** $f(2)$, **b.** $f(3)$, **c.** $f(5)$.

50. If $f(x) = \sqrt{2x + 3}$, find the following: **a.** $f(2)$, **b.** $f(3)$, **c.** $f(-1)$.

Challenge Exercises: (51–57)

Simplify the following:

51. $\sqrt[4]{162}$

52. $\sqrt[4]{x^9}$

53. $\sqrt[4]{a^6b^7}$

54. $\sqrt[5]{64}$

55. $\sqrt[5]{972}$

56. $\sqrt[5]{x^{12}}$

57. $\sqrt[5]{a^9b^{16}}$

Writing Exercises:

58. When simplifying square-root radicals, we wrote the radicand as the product of a perfect square and another number that had no perfect-square factors. When simplifying cube roots, we wrote the radicand as the product of a perfect cube and another number with no perfect-cube factors. Using this same line of reasoning, how would you rewrite the radicand of a fourth root before simplifying? A fifth root?

59. To find the root of a variable expression, the exponent must be divisible by the root index. If the root index is 4, what exponents would you want on the factor of the radicand whose fourth root you will find? Fifth root?

Critical-Thinking Exercises:

60. Write a procedure for finding the largest perfect-cube factor similar to that for finding the largest perfect-square factor, following Example 1c.

61. Use $a = 6$ and $b = 8$ to answer the following:
   a. Does $\sqrt{a^2 + b^2} = \sqrt{a^2} + \sqrt{b^2} = a + b$?
   b. Does $\sqrt{(a + b)^2} = a + b$?

| **Section 10.3** | **Products and Quotients of Radicals** |

OBJECTIVES

When you complete this section, you will be able to:

a. Find the products of radicals and simplify the results.

b. Find the quotients of radicals and simplify the results.

c. Simplify radical expressions using both the product and quotient rules.

Introduction   We know how to perform the operations of addition, subtraction, multiplication, and division on rational numbers, but how do we perform these operations on radical expressions that are irrational numbers? In this section, we will multiply and divide radicals. In Section 10.4 we will learn how to add and subtract radical expressions.

We learned how to multiply radicals in Section 10.2. For reference, the product rule for radicals is repeated below.

Multiplying radicals

Products of Radicals
If $\sqrt[n]{a}$ and $\sqrt[n]{b}$ are both real numbers, then $\sqrt[n]{a} \cdot \sqrt[n]{b} = \sqrt[n]{a \cdot b}$.

In section 10.2 we used this rule in the form $\sqrt[n]{ab} = \sqrt[n]{a} \cdot \sqrt[n]{b}$. In this section, we will be using this rule in the form $\sqrt[n]{a} \cdot \sqrt[n]{b} = \sqrt[n]{ab}$.

Example 1

Find the following products of radicals. Assume all variables have nonnegative values.

a. $\sqrt{3} \cdot \sqrt{5} =$ Apply $\sqrt{a} \cdot \sqrt{b} = \sqrt{ab}$.
$\sqrt{3 \cdot 5} =$ Multiply 3 and 5.
$\sqrt{15}$ Therefore, $\sqrt{3} \cdot \sqrt{5} = \sqrt{15}$.

b. $\sqrt{2} \cdot \sqrt{8} =$ Apply $\sqrt{a} \cdot \sqrt{b} = \sqrt{ab}$.
$\sqrt{2 \cdot 8} =$ Multiply 2 and 8.
$\sqrt{16} =$ Find the $\sqrt{16}$.
4 Therefore, $\sqrt{2} \cdot \sqrt{8} = 4$.

c. $\sqrt{7} \cdot \sqrt{x} =$ Apply $\sqrt{a} \cdot \sqrt{b} = \sqrt{ab}$.
$\sqrt{7x}$ Therefore, $\sqrt{7} \cdot \sqrt{x} = \sqrt{7x}$.

d. $\sqrt{x} \cdot \sqrt{x} =$ Apply $\sqrt{a} \cdot \sqrt{b} = \sqrt{ab}$.
$\sqrt{x \cdot x} =$ Multiply x and x.
$\sqrt{x^2} =$ Find the $\sqrt{x^2}$.
x Therefore, $\sqrt{x} \cdot \sqrt{x} = x$.

Practice Exercises

Find the following products of radicals. Assume all variables have nonnegative values.

1. $\sqrt{3} \cdot \sqrt{7}$ **2.** $\sqrt{3} \cdot \sqrt{27}$ **3.** $\sqrt{5} \cdot \sqrt{a}$ **4.** $\sqrt{r} \cdot \sqrt{r}$

If more practice is needed, do the Additional Practice Exercises.

Additional Practice Exercises

Find the following products of radicals. Assume all variables have nonnegative values.

a. $\sqrt{6} \cdot \sqrt{5}$ **b.** $\sqrt{2} \cdot \sqrt{32}$ **c.** $\sqrt{7} \cdot \sqrt{p}$ **d.** $\sqrt{q} \cdot \sqrt{q}$

It is often necessary to simplify the products of radicals. If the radical expressions have coefficients, multiply the coefficients and then multiply the radicals in much the same manner that terms containing variables are multiplied. See the following:

Monomials	**Radicals**	
$2x \cdot 3x^2$	$2\sqrt{3} \cdot 3\sqrt{5}$	Regroup.
$(2 \cdot 3)(x \cdot x^2)$	$(2 \cdot 3)\left(\sqrt{3} \cdot \sqrt{5}\right)$	Multiply.
$6x^3$	$6\sqrt{15}$	Product.

Answers:

Practice Exercises 1–4: **1.** $\sqrt{21}$ **2.** 9 **3.** $\sqrt{5a}$ **4.** r *Additional Practice Exercises a–d:* **a.** $\sqrt{30}$ **b.** 8 **c.** $\sqrt{7p}$ **d.** q

Example 2

Find the following products. Leave answers in simplified form. Remember, you may use the alternative method to simplify the radicals.

a. $\sqrt{2} \cdot \sqrt{6} =$ Apply the product rule in the form $\sqrt{a} \cdot \sqrt{b} = \sqrt{ab}$.

$\sqrt{2 \cdot 6} =$ Multiply.

$\sqrt{12} =$ 12 has a perfect-square factor of 4. Rewrite 12 as $4 \cdot 3$.

$\sqrt{4 \cdot 3} =$ Apply the product rule in the form $\sqrt{ab} = \sqrt{a} \cdot \sqrt{b}$.

$\sqrt{4} \cdot \sqrt{3} =$ $\sqrt{4} = 2$.

$2\sqrt{3}$ Therefore, $\sqrt{2} \cdot \sqrt{6} = 2\sqrt{3}$.

b. $5\sqrt{3} \cdot 3\sqrt{15} =$ Regroup and multiply 5 and 3, and $\sqrt{3}$ and $\sqrt{15}$.

$(5 \cdot 3)\left(\sqrt{3} \cdot \sqrt{15}\right) =$ Multiply.

$15\sqrt{45} =$ 45 has a perfect-square factor of 9. Rewrite 45 as $9 \cdot 5$.

$15\sqrt{9 \cdot 5} =$ Apply the product rule in the form $\sqrt{ab} = \sqrt{a} \cdot \sqrt{b}$.

$15\sqrt{9} \cdot \sqrt{5} =$ $\sqrt{9} = 3$.

$15 \cdot 3\sqrt{5} =$ Multiply 15 and 3.

$45\sqrt{5}$ Therefore, $5\sqrt{3} \cdot 3\sqrt{15} = 45\sqrt{5}$.

c. $\sqrt{x^3} \cdot \sqrt{x^5} =$ Apply the product rule in the form $\sqrt{a} \cdot \sqrt{b} = \sqrt{ab}$.

$\sqrt{x^3 \cdot x^5} =$ $x^3 \cdot x^5 = x^{3+5} = x^8$.

$\sqrt{x^8} =$ Find the root.

x^4 Therefore, $\sqrt{x^3} \cdot \sqrt{x^5} = x^4$.

d. $2\sqrt{5a^2} \cdot 3\sqrt{10a^5} =$ Regroup and multiply 2 and 3 and $\sqrt{5a^2}$ and $\sqrt{10a^5}$.

$(2 \cdot 3)\left(\sqrt{5a^2} \cdot \sqrt{10a^5}\right) =$ Multiply.

$2 \cdot 3\sqrt{5a^2 \cdot 10a^5} =$ Multiply.

$6\sqrt{50a^7} =$ The largest perfect-square factor of 50 is 25, and the largest perfect-square factor of a^7 is a^6. Rewrite $50a^7$ as $25a^6 \cdot 2a$.

$6\sqrt{25a^6 \cdot 2a} =$ Apply the product rule in the form $\sqrt{ab} = \sqrt{a} \cdot \sqrt{b}$.

$6\sqrt{25a^6} \cdot \sqrt{2a} =$ Using the product rule, $\sqrt{25a^6} = 5a^3$.

$6 \cdot 5a^3\sqrt{2a} =$ Multiply 6 and 5.

$30a^3\sqrt{2a}$ Therefore, $2\sqrt{5a^2} \cdot 3\sqrt{10a^5} = 30a^3\sqrt{2a}$.

Practice Exercises

Find the following products. Leave answers in simplified form. Assume all variables have nonnegative values.

5. $\sqrt{3} \cdot \sqrt{21}$ **6.** $4\sqrt{6} \cdot 7\sqrt{15}$ **7.** $\sqrt{a^5} \cdot \sqrt{a^5}$ **8.** $4\sqrt{10c} \cdot 2\sqrt{6c^4}$

If more practice is needed, do the Additional Practice Exercises.

Additional Practice Exercises

Find the following products of radicals. Leave answers in simplified form. Assume all variables have nonnegative values.

e. $\sqrt{6} \cdot \sqrt{15}$ **f.** $4\sqrt{14} \cdot 2\sqrt{6}$ **g.** $\sqrt{a} \cdot \sqrt{a^3}$ **h.** $5\sqrt{10m^2} \cdot \sqrt{15m^5}$

Answers:

The product rule for radicals can be used to develop a very important property of radicals. In Example 1d, we found that the product $\sqrt{x} \cdot \sqrt{x} = x$. By the definition of an exponent, $\sqrt{x} \cdot \sqrt{x} = \left(\sqrt{x}\right)^2$. Consequently, $\left(\sqrt{x}\right)^2 = x$. Also, $\sqrt[3]{x} \cdot \sqrt[3]{x} \cdot \sqrt[3]{x} = \sqrt[3]{x^3} = x$. But $\sqrt[3]{x} \cdot \sqrt[3]{x} \cdot \sqrt[3]{x} = \left(\sqrt[3]{x}\right)^3$ also. Consequently, $\left(\sqrt[3]{x}\right)^3 = x$. This leads to the following generalization:

Powers of Roots

If $\sqrt[n]{a}$ is a real number, then $\left(\sqrt[n]{a}\right)^n = a$.

Example 3

Find the following. Assume all variables have nonnegative values.

a. $\left(\sqrt{2}\right)^2 = 2$

b. $\left(\sqrt[3]{a}\right)^3 = a$

c. $\left(\sqrt[4]{x}\right)^4 = x$

d. $\left(\sqrt[5]{9}\right)^5 = 9$

e. $\left(\sqrt[3]{3x^2y}\right)^3 = 3x^2y$

Practice Exercises

Find the following. Assume all variables have nonnegative values.

9. $\left(\sqrt{5}\right)^2$

10. $\left(\sqrt[3]{r}\right)^3$

11. $\left(\sqrt[4]{y}\right)^4$

12. $\left(\sqrt[5]{16}\right)^5$

13. $\left(\sqrt[4]{6ab}\right)^4$

Dividing radicals   Look for a pattern for quotients of radicals in the following Calculator Exploration Activity.

OPTIONAL   **CALCULATOR EXPLORATION ACTIVITY**

Use your calculator to evaluate columns A and B below and round your answer off to the nearest ten-thousandth. Compare the results obtained in each line.

Column A	Column B
a. $\dfrac{\sqrt{10}}{\sqrt{2}} = $ ____	$\sqrt{5} = $ ____
b. $\dfrac{\sqrt{18}}{\sqrt{6}} = $ ____	$\sqrt{3} = $ ____
c. $\dfrac{\sqrt{48}}{\sqrt{8}} = $ ____	$\sqrt{6} = $ ____
d. $\dfrac{\sqrt{42}}{\sqrt{6}} = $ ____	$\sqrt{7} = $ ____

1. Based on your answers to the corresponding lines of columns A and B, $\dfrac{\sqrt{a}}{\sqrt{b}} = $ ____ .

A rule similar to the product rule can be established for quotients. Consider the following. Simplify $\dfrac{\sqrt{36}}{\sqrt{4}}$ by first evaluating the numerator and the denominator: $\dfrac{\sqrt{36}}{\sqrt{4}} = \dfrac{6}{2} = 3$. If we divide the radicands, we get $\dfrac{\sqrt{36}}{\sqrt{4}} = \sqrt{\dfrac{36}{4}} = \sqrt{9} = 3$. Note that we arrived at the same answer. Consequently, $\dfrac{\sqrt{36}}{\sqrt{4}} = \sqrt{\dfrac{36}{4}}$. This suggests the following rule for quotients of radicals:

Quotients of Radicals

If $\sqrt[n]{a}$ and $\sqrt[n]{b}$ and both exist and $b \neq 0$, then $\dfrac{\sqrt[n]{a}}{\sqrt[n]{b}} = \sqrt[n]{\dfrac{a}{b}}$.

Like the product rule, this rule may be applied in either order as $\dfrac{\sqrt[n]{a}}{\sqrt[n]{b}} = \sqrt[n]{\dfrac{a}{b}}$ or as $\sqrt[n]{\dfrac{a}{b}} = \dfrac{\sqrt[n]{a}}{\sqrt[n]{b}}$.

Example 4

Find the following quotients. Assume all variables have positive values.

a. $\dfrac{\sqrt{28}}{\sqrt{7}} =$ Apply $\dfrac{\sqrt{a}}{\sqrt{b}} = \sqrt{\dfrac{a}{b}}$.

$\sqrt{\dfrac{28}{7}} =$ Divide.

$\sqrt{4} =$ Simplify.

2 Therefore, $\dfrac{\sqrt{28}}{\sqrt{7}} = 2$.

b. $\sqrt{\dfrac{4}{9}} =$ Apply $\dfrac{\sqrt{a}}{\sqrt{b}} = \sqrt{\dfrac{a}{b}}$.

$\dfrac{\sqrt{4}}{\sqrt{9}} =$ Simplify the numerator and the denominator.

$\dfrac{2}{3}$ Therefore, $\sqrt{\dfrac{4}{9}} = \dfrac{2}{3}$.

c. $\dfrac{\sqrt{32x^5}}{\sqrt{2x}} =$ Apply $\dfrac{\sqrt{a}}{\sqrt{b}} = \sqrt{\dfrac{a}{b}}$.

$\sqrt{\dfrac{32x^5}{2x}} =$ Divide.

$\sqrt{16x^4} =$ Simplify.

$4x^2$ Therefore, $\dfrac{\sqrt{32x^5}}{\sqrt{2x}} = 4x^2$.

d. $\dfrac{\sqrt{81x^4y^3}}{\sqrt{3xy^2}} =$ Apply $\dfrac{\sqrt{a}}{\sqrt{b}} = \sqrt{\dfrac{a}{b}}$.

$\sqrt{\dfrac{81x^4y^3}{3xy^2}} =$ Divide.

$\sqrt{27x^3y}$ The largest perfect-square factor of 27 is 9, and the largest perfect-square factor of x^3 is x^2. Rewrite $27x^3y$ as $9x^2 \cdot 3xy$.

$\sqrt{9x^2 \cdot 3xy} =$ Apply $\sqrt{ab} = \sqrt{a} \cdot \sqrt{b}$.

$\sqrt{9x^2} \cdot \sqrt{3xy} =$ $\sqrt{9x^2} = 3x$.

$3x\sqrt{3xy} =$ Therefore, $\dfrac{\sqrt{81x^4y^3}}{\sqrt{3xy^2}} = 3x\sqrt{3xy}$.

e. $\dfrac{8\sqrt{756a^8b^5}}{2\sqrt{7a^4b^2}} =$ Divide coefficient into coefficient and radical into radical.

$\dfrac{8}{2} \cdot \dfrac{\sqrt{756a^8b^5}}{\sqrt{7a^4b^2}} =$ Divide coefficients and apply $\dfrac{\sqrt{a}}{\sqrt{b}} = \sqrt{\dfrac{a}{b}}$.

$4\sqrt{\dfrac{756a^8b^5}{7a^4b^2}} =$ Divide under the radical sign.

$4\sqrt{108a^4b^3} =$ Rewrite the radicand with a perfect-square factor.

$4\sqrt{36a^4b^2 \cdot 3b} =$ Apply $\sqrt{ab} = \sqrt{a} \cdot \sqrt{b}$.

$4\sqrt{36a^4b^2} \cdot \sqrt{3b} =$ Find the square roots.

$4 \cdot 6a^2b\sqrt{3b} =$ Multiply.

$24a^2b\sqrt{3b}$ Answer.

Therefore, $\dfrac{8\sqrt{756a^8b^5}}{2\sqrt{7a^4b^2}} = 24a^2b\sqrt{3b}$.

Practice Exercises

Find the following quotients. Assume all variables have positive values.

14. $\dfrac{\sqrt{75}}{\sqrt{3}}$

15. $\sqrt{\dfrac{16}{25}}$

16. $\dfrac{\sqrt{48x^3}}{\sqrt{12x}}$

17. $\dfrac{\sqrt{16a^5b^4}}{\sqrt{2a^2b^2}}$

18. $\dfrac{12\sqrt{480a^7b^6}}{3\sqrt{6a^3b^3}}$

If more practice is needed, do the Additional Practice Exercises.

Additional Practice Exercises

Find the following quotients:

i. $\dfrac{\sqrt{45}}{\sqrt{5}}$

j. $\sqrt{\dfrac{64}{25}}$

k. $\dfrac{\sqrt{54x^7}}{\sqrt{6x^3}}$

l. $\dfrac{\sqrt{32m^4n^3}}{\sqrt{4m^2n^2}}$

m. $\dfrac{14\sqrt{315x^{11}y^8}}{2\sqrt{5x^6y^5}}$

Simplifying expressions using both the product and quotient rules   We can combine the product and quotient rules to simplify radical expressions involving both multiplication and division.

Example 5

Simplify the following:

a. $\sqrt{\dfrac{2}{3}} \cdot \sqrt{\dfrac{10}{3}} =$ Apply the product rule.

$\sqrt{\dfrac{2}{3} \cdot \dfrac{10}{3}} =$ Multiply the fractions.

$\sqrt{\dfrac{20}{9}} =$ Apply the quotient rule.

$\dfrac{\sqrt{20}}{\sqrt{9}} =$ Rewrite 20 as $4 \cdot 5$ and $\sqrt{9} = 3$.

$\dfrac{\sqrt{4 \cdot 5}}{3} =$ Apply the product rule.

$\dfrac{\sqrt{4} \cdot \sqrt{5}}{3} =$ $\sqrt{4} = 2$.

$\dfrac{2\sqrt{5}}{3}$ Therefore, $\sqrt{\dfrac{2}{3}} \cdot \sqrt{\dfrac{10}{3}} = \dfrac{2\sqrt{5}}{3}$.

Note: We could also have applied the quotient rule first and then the product rule and done Example 5a as follows:

$$\sqrt{\dfrac{2}{3}} \cdot \sqrt{\dfrac{10}{3}} = \dfrac{\sqrt{2}}{\sqrt{3}} \cdot \dfrac{\sqrt{10}}{\sqrt{3}} = \dfrac{\sqrt{2 \cdot 10}}{\sqrt{3 \cdot 3}} = \dfrac{\sqrt{20}}{\sqrt{9}} = \dfrac{\sqrt{4 \cdot 5}}{3} = \dfrac{\sqrt{4} \cdot \sqrt{5}}{3} = \dfrac{2\sqrt{5}}{3}.$$

b. $\sqrt{\dfrac{x}{5}} \cdot \sqrt{\dfrac{x^3}{125}} =$ Apply the product rule.

$\sqrt{\dfrac{x}{5} \cdot \dfrac{x^3}{125}} =$ Multiply the rational expressions.

$\sqrt{\dfrac{x^4}{625}} =$ Apply the quotient rule.

$\dfrac{\sqrt{x^4}}{\sqrt{625}} =$ $\sqrt{x^4} = x^2$ and $\sqrt{625} = 25$.

$\dfrac{x^2}{25}$ Therefore, $\sqrt{\dfrac{x}{5}} \cdot \sqrt{\dfrac{x^3}{125}} = \dfrac{x^2}{25}$.

Practice Exercises

Simplify the following:

19. $\sqrt{\dfrac{3}{5}} \cdot \sqrt{\dfrac{6}{5}}$ **20.** $\sqrt{\dfrac{2}{a}} \cdot \sqrt{\dfrac{8}{a}}$

Exercise Set 10.3

Find the following products or powers of roots. Leave all answers in simplified form. Assume all variables have nonnegative values.

1. $\sqrt{6} \cdot \sqrt{5}$ **2.** $\sqrt{6} \cdot \sqrt{7}$ **3.** $\sqrt{7} \cdot \sqrt{5}$ **4.** $\sqrt{3} \cdot \sqrt{13}$

5. $\sqrt{2} \cdot \sqrt{18}$ **6.** $\sqrt{2} \cdot \sqrt{50}$ **7.** $\sqrt{3} \cdot \sqrt{75}$ **8.** $\sqrt{27} \cdot \sqrt{3}$

9. $\sqrt{5} \cdot \sqrt{x}$ **10.** $\sqrt{11} \cdot \sqrt{c}$ **11.** $\sqrt{15} \cdot \sqrt{y}$ **12.** $\sqrt{13} \cdot \sqrt{k}$

13 $\sqrt{x} \cdot \sqrt{y}$ **14.** $\sqrt{a} \cdot \sqrt{b}$ **15.** $\sqrt{k} \cdot \sqrt{k}$ **16.** $\sqrt{m} \cdot \sqrt{m}$

17 $\sqrt{x} \cdot \sqrt{x^3}$ **18.** $\sqrt{s^3} \cdot \sqrt{s}$ **19.** $\left(\sqrt{w}\right)^2$ **20.** $\left(\sqrt{z}\right)^2$

21. $\left(\sqrt{x^3}\right)^2$ **22.** $\left(\sqrt{b^5}\right)^2$ **23.** $\left(\sqrt[3]{w}\right)^3$ **24.** $\left(\sqrt[4]{5x}\right)^4$

25. $\left(\sqrt[7]{2ab}\right)^7$ **26.** $\left(\sqrt[5]{4x^2}\right)^5$ **27.** $\left(\sqrt[8]{3x^2y}\right)^8$ **28.** $\sqrt{3} \cdot \sqrt{6}$

29. $\sqrt{6} \cdot \sqrt{8}$ **30.** $\sqrt{3} \cdot \sqrt{21}$ **31.** $\sqrt{5} \cdot \sqrt{15}$ **32.** $2\sqrt{7} \cdot 3\sqrt{14}$

33. $4\sqrt{6} \cdot 2\sqrt{15}$ **34.** $5\sqrt{10} \cdot 3\sqrt{14}$ **35.** $2\sqrt{6} \cdot 5\sqrt{21}$ **36.** $4\sqrt{10} \cdot \sqrt{20}$

37. $\sqrt{a^3} \cdot \sqrt{a}$ **38.** $\sqrt{r^3} \cdot \sqrt{r^5}$ **39.** $\sqrt{y^3} \cdot \sqrt{y^2}$ **40.** $\sqrt{n^5} \cdot \sqrt{n^2}$

41. $\sqrt{m^7} \cdot \sqrt{m^4}$ **42.** $\sqrt{x^2y^3} \cdot \sqrt{x^4y^3}$ **43.** $x\sqrt{x^5y^2} \cdot y\sqrt{xy^3}$ **44.** $c\sqrt{c^5d^3} \cdot c\sqrt{c^5d^2}$

45. $\sqrt{a^5b} \cdot \sqrt{a^2b^2}$ **46.** $2\sqrt{5a} \cdot 3\sqrt{10a^3}$ **47.** $4\sqrt{6c^3} \cdot 3\sqrt{10c^5}$ **48.** $6\sqrt{15c^2} \cdot 2\sqrt{10c^5}$

Find the following quotients. Assume all variables have positive values. Leave answers in simplest form.

49. $\sqrt{\dfrac{1}{4}}$ **50.** $\sqrt{\dfrac{9}{25}}$ **51.** $\sqrt{\dfrac{49}{64}}$ **52.** $\sqrt{\dfrac{81}{64}}$

Answers:

53. $\dfrac{\sqrt{32}}{\sqrt{8}}$

54. $\dfrac{\sqrt{125}}{\sqrt{5}}$

55. $\dfrac{\sqrt{98}}{\sqrt{2}}$

56. $\dfrac{\sqrt{27}}{\sqrt{3}}$

57. $\dfrac{6\sqrt{48}}{2\sqrt{6}}$

58. $\dfrac{8\sqrt{54}}{4\sqrt{3}}$

59. $\dfrac{9\sqrt{160}}{3\sqrt{8}}$

60. $\dfrac{10\sqrt{280}}{2\sqrt{10}}$

61. $\dfrac{\sqrt{x^3}}{\sqrt{x}}$

62. $\dfrac{\sqrt{a^5}}{\sqrt{a^3}}$

63. $\dfrac{\sqrt{c^5 d^4}}{\sqrt{cd^3}}$

64. $\dfrac{\sqrt{m^6 n^5}}{\sqrt{m^3 n^3}}$

65. $\dfrac{\sqrt{18x^3}}{\sqrt{2x}}$

66. $\dfrac{\sqrt{20r^5}}{\sqrt{5r^3}}$

67. $\dfrac{8\sqrt{45a^5}}{2\sqrt{5a}}$

68. $\dfrac{6\sqrt{48n^7}}{3\sqrt{3n^3}}$

69. $\dfrac{12\sqrt{72c^5}}{4\sqrt{6c^2}}$

70. $\dfrac{15\sqrt{48a^7}}{5\sqrt{2a^2}}$

71. $\dfrac{36\sqrt{96x^6 y^{11}}}{4\sqrt{3x^2 y^4}}$

72. $\dfrac{54\sqrt{240r^9 s^{10}}}{9\sqrt{5r^6 s^4}}$

Simplify the following:

73. $\sqrt{\dfrac{3}{7}} \cdot \sqrt{\dfrac{8}{7}}$

74. $\sqrt{\dfrac{8}{5}} \cdot \sqrt{\dfrac{6}{5}}$

75. $\sqrt{\dfrac{a^3}{2}} \cdot \sqrt{\dfrac{a^5}{2}}$

76. $\sqrt{\dfrac{c^2}{6}} \cdot \sqrt{\dfrac{c^5}{6}}$

77. $\sqrt{\dfrac{3x^5}{2}} \cdot \sqrt{\dfrac{15x^5}{8}}$

78. $\sqrt{\dfrac{5y}{3}} \cdot \sqrt{\dfrac{10}{27}}$

Writing Exercise:

79. Compare simplifying $(3x)(2x)$ with simplifying $(3\sqrt{2})(2\sqrt{2})$.

Section 10.4 | **Addition, Subtraction, and Mixed Operations with Radicals**

OBJECTIVES

When you complete this section, you will be able to:

a. Add and subtract like radicals.

b. Simplify radicals and then add and/or subtract like radicals.

c. Simplify radical expressions that contain mixed operations.

Introduction We have found that $\sqrt[n]{a} \cdot \sqrt[n]{b} = \sqrt[n]{ab}$ and $\dfrac{\sqrt[n]{a}}{\sqrt[n]{b}} = \sqrt[n]{\dfrac{a}{b}}$, provided that $\sqrt[n]{a}$ and $\sqrt[n]{b}$ are real numbers. Does $\sqrt[n]{a} + \sqrt[n]{b} = \sqrt[n]{a+b}$?

OPTIONAL | **CALCULATOR EXPLORATION ACTIVITY I**

Use your calculator to evaluate columns A and B below and round your answers off to the nearest ten-thousandth. Compare the results of the corresponding lines.

Column A	Column B
a. $\sqrt{2} + \sqrt{3} =$ _____	$\sqrt{5} =$ _____
b. $\sqrt{5} + \sqrt{12} =$ _____	$\sqrt{17} =$ _____
c. $\sqrt{23} + \sqrt{19} =$ _____	$\sqrt{42} =$ _____
d. $\sqrt{31} + \sqrt{43} =$ _____	$\sqrt{74} =$ _____

1. Based on your answers to the corresponding lines in columns A and B, does $\sqrt{a} + \sqrt{b} = \sqrt{a+b}$?

Answer:

Calculator Activity I: **1.** no

CALCULATOR EXPLORATION ACTIVITY II

Use your calculator to evaluate columns A and B below. Round your answers to the nearest ten-thousandth. Compare the results of the corresponding lines.

Column A	Column B
a. $2\sqrt{2} + 3\sqrt{2} = $ _____	$5\sqrt{2} = $ _____
b. $4\sqrt{6} + 5\sqrt{6} = $ _____	$9\sqrt{6} = $ _____
c. $-4\sqrt{3} + 7\sqrt{3} = $ _____	$3\sqrt{3} = $ _____
d. $-3\sqrt{5} - 6\sqrt{5} = $ _____	$-9\sqrt{5} = $ _____

2. By looking at the results of the corresponding lines, state a rule for adding radical expressions with the same root indices and the same radicands. Following is justification for this rule:

Addition and subtraction of radicals is done in much the same manner as addition and subtraction of variables. The distributive property was used to add like terms and will be used to add like radicals. Remember that like terms have the same variables and the same exponents on the variables. Similarly, **like radical expressions** have the same radicand and the same root indices. Study the following comparisons of addition of terms and addition of radical expressions:

$$3x + 5x \qquad\qquad\qquad 3\sqrt{2} + 5\sqrt{2}$$

These are like terms, so we can apply the distributive property and add.

$$3x + 5x = (3 + 5)x = 8x$$

These are like radical expressions, so we can apply the distributive property and add.

$$3\sqrt{2} + 5\sqrt{2} = (3 + 5)\sqrt{2} = 8\sqrt{2}$$

$$3x + 5y \qquad\qquad\qquad 3\sqrt{2} + 5\sqrt{3}$$

These are not like terms because the variables are different. Consequently, we cannot add these terms.

These are not like radical expressions because the radicands are different. Consequently, we cannot add these radicals.

$$3x^2 + 5x^3 \qquad\qquad\qquad 3\sqrt{2} + 5\sqrt[3]{2}$$

These are not like terms because the exponents on the variables are different. Consequently, we cannot add these terms.

These are not like radical expressions because the root indices are different. Consequently, we cannot add these radicals.

Instead of applying the distributive property each time we combined like terms, we observed that you add the coefficients and leave the variable portion unchanged. Similarly, we have the following rule for adding like radical expressions:

Adding Like Radical Expressions

To add like radical expressions, add the coefficients and leave the radical portion unchanged.

Note: Since $a - b = a + (-b)$, we do not need a separate rule for differences.

Answer:

Calculator Activity II: 2. Add the coefficients and leave the radical portions unchanged.

Example 1

Find the following sums of radicals:

a. $7\sqrt{5} + 2\sqrt{5} =$ Add the coefficients and leave the radicals unchanged.

$9\sqrt{5}$ Therefore, $7\sqrt{5} + 2\sqrt{5} = 9\sqrt{5}$.

b. $4\sqrt{x} - 8\sqrt{x} =$ Add the coefficients and leave the radicals unchanged.

$-4\sqrt{x}$ Therefore, $4\sqrt{x} - 8\sqrt{x} = -4\sqrt{x}$.

c. $6\sqrt[3]{4} + \sqrt[3]{4} =$ Add the coefficients and leave the radicals unchanged.

$7\sqrt[3]{4}$ Therefore, $6\sqrt[3]{4} + \sqrt[3]{4} = 7\sqrt[3]{4}$.

d. $6x\sqrt[4]{3} - 2x\sqrt[4]{3} =$ Add the coefficients and leave the radicals unchanged.

$4x\sqrt[4]{3}$ Therefore, $6x\sqrt[4]{3} - 2x\sqrt[4]{3} = 4x\sqrt[4]{3}$.

e. $3a^2\sqrt{2a} - 7a^2\sqrt{2a} =$ Add the coefficients and leave the radicals unchanged.

$-4a^2\sqrt{2a}$ Therefore, $3a^2\sqrt{2a} - 7a^2\sqrt{2a} = -4a^2\sqrt{2a}$.

> **Be Careful** *A very common error is to add radicands when finding the sums of radicals. The following illustrates that this cannot be done:*
>
> $\sqrt{9} + \sqrt{16} \neq \sqrt{9 + 16}$
> $3 + 4 \neq \sqrt{25}$
> $7 \neq 5$
>
> ***Never add radicands!***

Practice Exercises

Find the following sums of radicals:

1. $5\sqrt{6} + 2\sqrt{6}$

2. $3\sqrt{c} - 9\sqrt{c}$

3. $5\sqrt[4]{5} + \sqrt[4]{5}$

4. $7y\sqrt[3]{7} - 12y\sqrt[3]{7}$

5. $7y^3\sqrt{6y} + 8y^3\sqrt{6y}$

If more practice is needed, do the Additional Practice Exercises.

Additional Practice Exercises

Find the following sums of radicals:

a. $4\sqrt{5} + 6\sqrt{5}$

b. $3\sqrt{y} - 7\sqrt{y}$

c. $9\sqrt[3]{3x} - \sqrt[3]{3x}$

d. $z\sqrt{5} + 8z\sqrt{5}$

e. $5a^2\sqrt{7a} + 3a^2\sqrt{7a}$

It is often necessary to simplify radicands before they can be added.

Example 2

Find the following sums. Assume all variables represent nonnegative quantities.

a. $\sqrt{12} + \sqrt{75} =$ Rewrite 12 as $4 \cdot 3$ and 75 as $25 \cdot 3$.

$\sqrt{4 \cdot 3} + \sqrt{25 \cdot 3} =$ Apply the product rule.

$\sqrt{4} \cdot \sqrt{3} + \sqrt{25} \cdot \sqrt{3} =$ Find $\sqrt{4}$ and $\sqrt{25}$.

$2\sqrt{3} + 5\sqrt{3} =$ Add like radicals.

$7\sqrt{3}$ Therefore, $\sqrt{12} + \sqrt{75} = 7\sqrt{3}$.

b. $3\sqrt{80} - 2\sqrt{245} =$ Rewrite 80 as $16 \cdot 5$ and 245 as $49 \cdot 5$.

$\quad 3\sqrt{16 \cdot 5} - 2\sqrt{49 \cdot 5} =$ Apply the product rule.

$\quad 3\sqrt{16} \cdot \sqrt{5} - 2\sqrt{49} \cdot \sqrt{5} =$ Find $\sqrt{16}$ and $\sqrt{49}$.

$\quad 3 \cdot 4\sqrt{5} - 2 \cdot 7\sqrt{5} =$ Multiply.

$\quad 12\sqrt{5} - 14\sqrt{5} =$ Add like radicals.

$\quad -2\sqrt{5}$ Therefore, $3\sqrt{80} - 2\sqrt{245} = -2\sqrt{5}$.

c. $\sqrt{48x^3} + \sqrt{12x^3} =$ Rewrite $48x^3$ as $16x^2 \cdot 3x$ and $12x^3$ as $4x^2 \cdot 3x$.

$\quad \sqrt{16x^2 \cdot 3x} + \sqrt{4x^2 \cdot 3x} =$ Apply the product rule.

$\quad \sqrt{16x^2} \cdot \sqrt{3x} + \sqrt{4x^2} \cdot \sqrt{3x} =$ Find $\sqrt{16x^2}$ and $\sqrt{4x^2}$.

$\quad 4x\sqrt{3x} + 2x\sqrt{3x} =$ Add like radicals.

$\quad 6x\sqrt{3x}$ Therefore, $\sqrt{48x^3} + \sqrt{12x^3} = 6x\sqrt{3x}$.

Practice Exercises

Find the following sums. Assume all variables represent nonnegative quantities.

6. $\sqrt{45} + \sqrt{20}$ **7.** $4\sqrt{24} - 6\sqrt{54}$ **8.** $\sqrt{50x^5} - \sqrt{18x^5}$

If more practice is needed, do the Additional Practice Exercises.

Additional Practice Exercises

Find the following sums of radicals. Assume all variables represent nonnegative quantities.

f. $\sqrt{72} + \sqrt{32}$ **g.** $5\sqrt{48} - 2\sqrt{75}$ **h.** $\sqrt{28a^5} - \sqrt{63a^5}$

Mixed operations with radicals   Now that we know how to perform the operations of addition, subtraction, multiplication, and division with radicals, it is possible for us to simplify radical expressions that contain more than one operation. The order of operations applies to radical expressions just as it does to any other numbers. All answers should be left in simplest form as described in the following box:

Simplest Form of Radicals

A radical expression is in its simplest form if:

1. All rational roots have been found.
2. No factors that can be removed may be left in a radicand. For example, no square root may have a radicand with a perfect-square factor.
3. All possible products, quotients, sums, and differences have been found.
4. There are no radicals in the denominator of a fraction. (This will be discussed in Section 10.5.)

Answers:

Practice 6–8: 6. $5\sqrt{5}$ 7. $-10\sqrt{6}$ 8. $2x^2\sqrt{2x}$ **Additional Practice Exercises f–h:** f. $10\sqrt{2}$ g. $10\sqrt{3}$ h. $-a^2\sqrt{7a}$

Example 3

Perform the following operations:

a. $\sqrt{2} \cdot \sqrt{10} + \sqrt{3} \cdot \sqrt{15} =$ Apply the product rule.

$\sqrt{2 \cdot 10} + \sqrt{3 \cdot 15} =$ Find the products.

$\sqrt{20} + \sqrt{45} =$ Rewrite 20 as $4 \cdot 5$ and 45 as $9 \cdot 5$.

$\sqrt{4 \cdot 5} + \sqrt{9 \cdot 5} =$ Apply the product rule.

$\sqrt{4} \cdot \sqrt{5} + \sqrt{9} \cdot \sqrt{5} =$ Find $\sqrt{9}$ and $\sqrt{4}$.

$2\sqrt{5} + 3\sqrt{5} =$ Add like radicals.

$5\sqrt{5}$ Therefore, $\sqrt{2} \cdot \sqrt{10} + \sqrt{3} \cdot \sqrt{15} = 5\sqrt{5}$.

b. $\dfrac{\sqrt{54}}{\sqrt{3}} + \sqrt{32} =$ Apply the quotient rule and rewrite 32 as $16 \cdot 2$.

$\sqrt{\dfrac{54}{3}} + \sqrt{16 \cdot 2} =$ Divide 54 by 3. Apply the product rule to $\sqrt{16 \cdot 2}$.

$\sqrt{18} + \sqrt{16} \cdot \sqrt{2} =$ Rewrite 18 as $9 \cdot 2$ and find $\sqrt{16}$.

$\sqrt{9 \cdot 2} + 4\sqrt{2} =$ Apply the product rule to $\sqrt{9 \cdot 2}$.

$\sqrt{9} \cdot \sqrt{2} + 4\sqrt{2} =$ Find $\sqrt{9}$.

$3\sqrt{2} + 4\sqrt{2} =$ Add like radicals.

$7\sqrt{2}$ Therefore, $\dfrac{\sqrt{54}}{\sqrt{3}} + \sqrt{32} = 7\sqrt{2}$.

Note: It is possible to perform operations on radical expressions using calculators and to find decimal approximations for the results. For example, the decimal approximation for the result of Example 3b, accurate to the nearest thousandth, is 9.899.

Practice Exercises

Perform the following operations:

9. $2\sqrt{3} \cdot \sqrt{6} + 4\sqrt{7} \cdot \sqrt{14}$

10. $\sqrt{63} + \dfrac{\sqrt{140}}{\sqrt{5}}$

If more practice is needed, do the Additional Practice Exercises.

Additional Practice Exercises

Perform the following operations with radicals:

i. $\sqrt{5} \cdot \sqrt{15} + 2\sqrt{7} \cdot \sqrt{21}$

j. $\dfrac{\sqrt{72}}{\sqrt{6}} + \sqrt{75}$

Products involving sums   Products involving sums and differences of radicals are found in much the same way as products of polynomials.

Example 4

Find the following products. Assume all variables have nonnegative values.

a. $\sqrt{3}(\sqrt{3} + \sqrt{15}) =$ Apply the distributive property.

$\sqrt{3} \cdot \sqrt{3} + \sqrt{3} \cdot \sqrt{15} =$ Apply the product rule.

$\sqrt{3 \cdot 3} + \sqrt{3 \cdot 15} =$ Find the products.

$\sqrt{9} + \sqrt{45} =$ Find $\sqrt{9}$ and simplify $\sqrt{45}$.

$3 + 3\sqrt{5}$ Therefore, $\sqrt{3}(\sqrt{3} + \sqrt{15}) = 3 + 3\sqrt{5}$.

b. $2\sqrt{6}(3 + 5\sqrt{5}) =$ Apply the distributive property.

$2\sqrt{6} \cdot 3 + 2\sqrt{6} \cdot 5\sqrt{5} =$ Apply the product rule.

$2 \cdot 3\sqrt{6} + 2 \cdot 5\sqrt{6 \cdot 5} =$ Find the products.

$6\sqrt{6} + 10\sqrt{30}$ Therefore, $2\sqrt{6}(3 + 5\sqrt{5}) = 6\sqrt{6} + 10\sqrt{30}$.

c. $(2 + \sqrt{3})(\sqrt{5} - \sqrt{6}) =$ Multiply using FOIL.

$2\sqrt{5} - 2\sqrt{6} + \sqrt{3} \cdot \sqrt{5} - \sqrt{3} \cdot \sqrt{6} =$ Apply the product rule.

$2\sqrt{5} - 2\sqrt{6} + \sqrt{15} - \sqrt{18} =$ Simplify $\sqrt{18}$.

$2\sqrt{5} - 2\sqrt{6} + \sqrt{15} - 3\sqrt{2}$ Therefore, $(2 + \sqrt{3})(\sqrt{5} - \sqrt{6}) = 2\sqrt{5} - 2\sqrt{6} + \sqrt{15} - 3\sqrt{2}$.

d. $(3\sqrt{x} + \sqrt{y})(2\sqrt{x} - 5\sqrt{y}) =$ Multiply using FOIL.

$3\sqrt{x} \cdot 2\sqrt{x} - 3\sqrt{x} \cdot 5\sqrt{y} + \sqrt{y} \cdot 2\sqrt{x} - \sqrt{y} \cdot 5\sqrt{y} =$ Apply the product rule.

$6x - 15\sqrt{xy} + 2\sqrt{xy} - 5y =$ Add like radicals.

$6x - 13\sqrt{xy} - 5y$ Therefore, $(3\sqrt{x} + \sqrt{y})(2\sqrt{x} - 5\sqrt{y}) = 6x - 13\sqrt{xy} - 5y$.

e. $(5 + \sqrt{3})^2 =$ Apply $(a + b)^2 = a^2 + 2ab + b^2$.

$5^2 + 2 \cdot 5\sqrt{3} + (\sqrt{3})^2 =$ Simplify.

$25 + 10\sqrt{3} + 3 =$ Add 25 and 3.

$28 + 10\sqrt{3}$ Therefore, $(5 + \sqrt{3})^2 = 28 + 10\sqrt{3}$.

Practice Exercises

Find the following products. Assume all variables represent nonnegative quantities.

11. $\sqrt{7}(\sqrt{3} + \sqrt{7})$ **12.** $2\sqrt{11}(3 - 3\sqrt{6})$ **13.** $(4 - \sqrt{3})(\sqrt{2} - \sqrt{11})$

14. $(2\sqrt{a} + 3\sqrt{b})(\sqrt{a} - 3\sqrt{b})$ **15.** $(3 + \sqrt{5})^2$

If more practice is needed, do the Additional Practice Exercises.

Additional Practice Exercises

Find the following products:

k. $\sqrt{5}(\sqrt{6} - \sqrt{2})$ **l.** $7\sqrt{13}(3\sqrt{2} - 4)$ **m.** $(\sqrt{3} - 5)(\sqrt{7} - \sqrt{5})$

n. $(3\sqrt{n} + 2\sqrt{m})(5\sqrt{n} - 3\sqrt{m})$ **o.** $(4 - \sqrt{3})^2$

Answers:

Additional Practice Exercises k–o: **k.** $\sqrt{30} - \sqrt{10}$ **l.** $21\sqrt{26} - 28\sqrt{13}$ **m.** $\sqrt{21} - \sqrt{15} - 5\sqrt{7} + 5\sqrt{5}$ **n.** $15n + \sqrt{mn} - 6m$ **o.** $19 - 8\sqrt{3}$

Practice Exercises 11–15: **11.** $\sqrt{21} + 7$ **12.** $6\sqrt{11} - 6\sqrt{66}$ **13.** $4\sqrt{2} - 4\sqrt{11} - \sqrt{6} + \sqrt{33}$ **14.** $2a - 3\sqrt{ab} - 9b$ **15.** $14 + 6\sqrt{5}$

One form of products of sums and differences of radicals is of particular interest. Recall that the product of binomials of the form $(a + b)(a - b) = a^2 + ab - ab - b^2 = a^2 - b^2$. For example, $(x + 3)(x - 3) = x^2 - 3^2 = x^2 - 9$. If square roots are involved, expressions of the form $a + b$ and $a - b$ are called **conjugates**. For example, $3 + \sqrt{2}$ and $3 - \sqrt{2}$ are conjugates and their product is $(3 + \sqrt{2})(3 - \sqrt{2}) = 3^2 - (\sqrt{2})^2 = 9 - 2 = 7$. (Remember that $(\sqrt{x})^2 = x$ if $x \geq 0$.)

Example 5

Find the following products using the fact that $(a + b)(a - b) = a^2 - b^2$:

a. $(2 + \sqrt{3})(2 - \sqrt{3}) =$ Apply $(a + b)(a - b) = a^2 - b^2$.

  $2^2 - (\sqrt{3})^2 =$ $2^2 = 4$ and $(\sqrt{3})^2 = 3$.

  $4 - 3 =$ Add 4 and -3.

  1 Therefore, $(2 + \sqrt{3})(2 - \sqrt{3}) = 1$.

b. $(\sqrt{5} - 3\sqrt{3})(\sqrt{5} + 3\sqrt{3}) =$ Apply $(a + b)(a - b) = a^2 - b^2$.

  $(\sqrt{5})^2 - (3\sqrt{3})^2 =$ $(\sqrt{5})^2 = 5$ and $(3\sqrt{3})^2 = 3^2(\sqrt{3})^2 = 9 \cdot 3$.

  $5 - 9 \cdot 3 =$ $9 \cdot 3 = 27$.

  $5 - 27 =$ Add 5 and -27.

  -22 Therefore, $(\sqrt{5} - 3\sqrt{3})(\sqrt{5} + 3\sqrt{3}) = -22$.

Note: The product of conjugates always results in a rational number. This will be useful in the next section on rationalizing the denominator.

Practice Exercises

Find the following products:

16. $(4 + \sqrt{10})(4 - \sqrt{10})$

17. $(3\sqrt{5} + 2\sqrt{6})(3\sqrt{5} - 2\sqrt{6})$

If more practice is needed, do the Additional Practice Exercises.

Additional Practice Exercises

Find the following products:

p. $(\sqrt{7} - 3)(\sqrt{7} + 3)$

q. $(5\sqrt{7} - \sqrt{3})(5\sqrt{7} + \sqrt{3})$

In Section 11.3, it will be necessary to simplify and reduce expressions of the form $\dfrac{a + b\sqrt{c}}{d}$ when solving quadratic equations. It will be helpful if we practice this procedure at this time. Remember, when reducing fractions you must divide the numerator and the denominator by the greatest factor common to both the numerator and the denominator.

Example 6

Simplify the following:

a. $\dfrac{4 + 2\sqrt{5}}{6} =$ Remove the common factor of 2 from the numerator.

$\dfrac{2(2 + \sqrt{5})}{6} =$ Divide the numerator and the denominator by 2.

$\dfrac{2 + \sqrt{5}}{3}$ Therefore, $\dfrac{4 + 2\sqrt{5}}{6} = \dfrac{2 + \sqrt{5}}{3}$.

b. $\dfrac{6 + 4\sqrt{18}}{6} =$ Simplify $\sqrt{18}$.

$\dfrac{6 + 4 \cdot 3\sqrt{2}}{6} =$ Multiply 4 and 3.

$\dfrac{6 + 12\sqrt{2}}{6} =$ Remove the common factor of 6 from the numerator.

$\dfrac{6(1 + 2\sqrt{2})}{6} =$ Divide the numerator and the denominator by 6.

$1 + 2\sqrt{2}$ Therefore, $\dfrac{6 + 4\sqrt{18}}{6} = 1 + 2\sqrt{2}$.

Practice Exercises

Simplify the following:

18. $\dfrac{8 - 4\sqrt{2}}{2}$

19. $\dfrac{10 + 3\sqrt{75}}{5}$

If more practice is needed, do the Additional Practice Exercises.

Additional Practice Exercises

Simplify the following:

r. $\dfrac{12 + 6\sqrt{3}}{3}$

s. $\dfrac{4 - 3\sqrt{8}}{4}$

Exercise Set 10.4

Find the following sums of radicals. Assume all variables have nonnegative values.

1. $2\sqrt{7} + 5\sqrt{7}$

2. $3\sqrt{3} + 8\sqrt{3}$

3. $9\sqrt{6} - 15\sqrt{6}$

4. $2\sqrt{10} - 13\sqrt{10}$

5. $7\sqrt{a} + 2\sqrt{a}$

6. $5\sqrt{y} + 7\sqrt{y}$

7. $-5\sqrt{n} + 2\sqrt{n}$

8. $-8\sqrt{m} + 4\sqrt{m}$

9. $6x\sqrt[3]{9} - 3x\sqrt[3]{9}$

10. $4y\sqrt[3]{3} - y\sqrt[3]{3}$

11. $6x^2\sqrt[4]{5x} + 12x^2\sqrt[4]{5x}$

12. $3y^3\sqrt[4]{8y} - 9y^3\sqrt[4]{8y}$

Simplify the following radicals and then find the sums. Assume all variables have nonnegative values.

13. $\sqrt{27} + \sqrt{12}$

14. $\sqrt{8} + \sqrt{32}$

15. $\sqrt{48} - \sqrt{75}$

16. $\sqrt{80} - \sqrt{20}$

17. $\sqrt{80y} - \sqrt{125y}$

18. $\sqrt{27x} + \sqrt{75x}$

19. $\sqrt{24n^2} - \sqrt{96n^2}$

20. $\sqrt{48x^2} + \sqrt{75x^2}$

21. $\sqrt{18y^5} + \sqrt{72y^5}$

22. $\sqrt{48a^7} + \sqrt{108a^7}$

23. $12\sqrt{5} + \sqrt{45}$

24. $14\sqrt{6} - \sqrt{24}$

25. $\sqrt{80} - 4\sqrt{45}$

26. $\sqrt{20} - 2\sqrt{180}$

27. $3\sqrt{96} - 2\sqrt{54}$

28. $5\sqrt{63} + 2\sqrt{28}$

29. $6\sqrt{48a^3} - 2\sqrt{75a^3}$

30. $4\sqrt{98y^5} - 7\sqrt{128y^5}$

31. $\sqrt{150} - \sqrt{54} + \sqrt{24}$

32. $\sqrt{20} + \sqrt{125} - \sqrt{80}$

33. $2\sqrt{8} - 3\sqrt{48} + 2\sqrt{98} - \sqrt{75}$

34. $3\sqrt{216} - \sqrt{147} - 4\sqrt{96} - \sqrt{108}$

35. $4\sqrt{72x^2y} - 2x\sqrt{128y} + 5\sqrt{32x^2y}$

36. $3\sqrt{24a^3b^2} - 2a\sqrt{150ab^2} + 4ab\sqrt{96a}$

Simplify the following:

37. $\sqrt{3} \cdot \sqrt{15} + \sqrt{8} \cdot \sqrt{10}$

38. $\sqrt{6} \cdot \sqrt{8} + \sqrt{5} \cdot \sqrt{15}$

39. $3\sqrt{3} \cdot \sqrt{18} - 4\sqrt{18} \cdot \sqrt{12}$

40. $6\sqrt{8} \cdot \sqrt{10} - 5\sqrt{12} \cdot \sqrt{15}$

41. $\dfrac{\sqrt{40}}{\sqrt{5}} + \sqrt{50}$

42. $\dfrac{\sqrt{60}}{\sqrt{5}} + \sqrt{48}$

43. $\dfrac{\sqrt{540}}{\sqrt{3}} - 4\sqrt{125}$

44. $\dfrac{\sqrt{288}}{\sqrt{6}} - 6\sqrt{108}$

Simplify the following products. Assume all variables have nonnegative values.

45. $\sqrt{2}(3 + \sqrt{2})$

46. $\sqrt{5}(4 - \sqrt{5})$

47. $\sqrt{3}(\sqrt{3} - \sqrt{15})$

48. $\sqrt{6}(\sqrt{6} + \sqrt{2})$

49. $\sqrt{5}(\sqrt{3} + 2\sqrt{15})$

50. $\sqrt{7}(\sqrt{5} - 3\sqrt{14})$

51. $4\sqrt{3}(2\sqrt{3} - 4\sqrt{6})$

52. $6\sqrt{2}(3\sqrt{2} + 2\sqrt{10})$

53. $(3 + \sqrt{5})(4 - \sqrt{2})$

54. $(3 + \sqrt{7})(7 - \sqrt{3})$

55. $(3 + \sqrt{x})(2 + \sqrt{x})$

56. $(5 - \sqrt{a})(2 + \sqrt{a})$

57. $(4 - 2\sqrt{6})(2 - 5\sqrt{6})$

58. $(9 - 2\sqrt{2})(1 + 4\sqrt{2})$

59. $(2 + 3\sqrt{3})(3 + 5\sqrt{2})$

60. $(7 - 3\sqrt{5})(2 - 2\sqrt{10})$

61. $(\sqrt{2} + \sqrt{3})(\sqrt{3} + \sqrt{5})$

62. $(\sqrt{5} + \sqrt{2})(\sqrt{2} + \sqrt{7})$

63. $(\sqrt{x} + \sqrt{y})(\sqrt{x} - 2\sqrt{y})$

64. $(2\sqrt{a} + \sqrt{b})(\sqrt{a} + \sqrt{b})$

65. $(\sqrt{3} + 2\sqrt{2})(\sqrt{3} - 4\sqrt{2})$

66. $(\sqrt{5} - 4\sqrt{6})(\sqrt{5} + 2\sqrt{6})$

67. $(5\sqrt{2} - 3\sqrt{5})(2\sqrt{2} + 6\sqrt{5})$

68. $(2\sqrt{2} + 3\sqrt{3})(3\sqrt{2} - 4\sqrt{3})$

69. $(4\sqrt{2} + 2\sqrt{5})(3\sqrt{7} - 3\sqrt{3})$

70. $(8\sqrt{2} - 2\sqrt{3})(2\sqrt{5} + 3\sqrt{10})$

71. $(2\sqrt{a} + 3\sqrt{b})(4\sqrt{a} - \sqrt{b})$

72. $(\sqrt{m} - 4\sqrt{n})(2\sqrt{m} - 3\sqrt{n})$

73. $(4 + \sqrt{6})^2$

74. $(1 - \sqrt{2})^2$

75. $(6 + \sqrt{3})^2$

76. $(5 + \sqrt{2})^2$

77. $(2 + 2\sqrt{3})^2$

78. $(3 + 2\sqrt{5})^2$

Find the following products of conjugates. Assume all variables have nonnegative values.

79. $\left(2 + \sqrt{3}\right)\left(2 - \sqrt{3}\right)$

80. $\left(3 + \sqrt{5}\right)\left(3 - \sqrt{5}\right)$

81. $\left(4 + 2\sqrt{3}\right)\left(4 - 2\sqrt{3}\right)$

82. $\left(5 + 3\sqrt{3}\right)\left(5 - 3\sqrt{3}\right)$

83. $\left(3 + \sqrt{x}\right)\left(3 - \sqrt{x}\right)$

84. $\left(5 + \sqrt{y}\right)\left(5 - \sqrt{y}\right)$

85. $\left(\sqrt{2} + 4\right)\left(\sqrt{2} - 4\right)$

86. $\left(\sqrt{7} - 6\right)\left(\sqrt{7} + 6\right)$

87. $\left(5\sqrt{2} + 5\right)\left(5\sqrt{2} - 5\right)$

88. $\left(6\sqrt{3} - 2\right)\left(6\sqrt{3} + 2\right)$

89. $\left(\sqrt{3} + \sqrt{2}\right)\left(\sqrt{3} - \sqrt{2}\right)$

90. $\left(\sqrt{5} - \sqrt{3}\right)\left(\sqrt{5} + \sqrt{3}\right)$

91. $\left(\sqrt{x} + \sqrt{y}\right)\left(\sqrt{x} - \sqrt{y}\right)$

92. $\left(\sqrt{a} - \sqrt{b}\right)\left(\sqrt{a} + \sqrt{b}\right)$

93. $\left(3\sqrt{7} + \sqrt{13}\right)\left(3\sqrt{7} - \sqrt{13}\right)$

94. $\left(4\sqrt{5} - 3\sqrt{2}\right)\left(4\sqrt{5} + 3\sqrt{2}\right)$

95. $\left(3\sqrt{5} - 2\sqrt{6}\right)\left(3\sqrt{5} + 2\sqrt{6}\right)$

96. $\left(5\sqrt{2} - 2\sqrt{7}\right)\left(5\sqrt{2} + 2\sqrt{7}\right)$

Simplify the following:

97. $\dfrac{6 + 3\sqrt{2}}{3}$

98. $\dfrac{8 + 4\sqrt{5}}{4}$

99. $\dfrac{10 + 20\sqrt{6}}{5}$

100. $\dfrac{12 - 18\sqrt{2}}{3}$

101. $\dfrac{3 + 2\sqrt{45}}{3}$

102. $\dfrac{12 - 3\sqrt{20}}{6}$

103. $\dfrac{9 - 2\sqrt{54}}{3}$

104. $\dfrac{10 - 3\sqrt{50}}{5}$

105. $\dfrac{4 + 6\sqrt{3}}{8}$

106. $\dfrac{6 - 3\sqrt{5}}{9}$

107. $\dfrac{8 - 4\sqrt{6}}{6}$

108. $\dfrac{12 + 6\sqrt{7}}{9}$

Challenge Exercises: (109–116)

Perform the following operations on radical expressions whose indices are greater than 2:

109. $\sqrt[3]{16} + \sqrt[3]{54}$

110. $\sqrt[3]{24} + \sqrt[3]{81}$

111. $\sqrt[3]{250x^4y^5} + \sqrt[3]{128x^4y^5}$

112. $\sqrt[3]{108a^6b^4} + \sqrt[3]{32a^6b^4}$

113. $\sqrt[4]{48} + \sqrt[4]{243}$

114. $\sqrt[4]{64} - \sqrt[4]{324}$

115. $\left(\sqrt[3]{2} + \sqrt[3]{4}\right)\left(2\sqrt[3]{2} - 3\sqrt[3]{4}\right)$

116. $\left(\sqrt[3]{9} - 2\sqrt[3]{3}\right)\left(2\sqrt[3]{9} + 3\sqrt[3]{3}\right)$

Writing Exercises:

117. Compare finding the product $(2x + 3y)(3x - 4y)$ with finding the product $\left(2\sqrt{2} + 3\sqrt{5}\right)\left(3\sqrt{2} - 4\sqrt{5}\right)$. How are they alike and how are they different?

118. Compare adding $3x + 2x$ with adding $3\sqrt{2} + 2\sqrt{2}$.

119. Why is the product of conjugates always a rational number instead of an irrational number?

120. Why is $\sqrt{a} + \sqrt{b} \neq \sqrt{a + b}$? Use examples if necessary.

| Section 10.5 | Rationalizing the Denominator |

OBJECTIVES

When you complete this section, you will be able to:

a. Rationalize the denominator when the denominator contains a single radical that is a square root or a cube root.

b. Rationalize the denominator when the denominator contains an expression of the form $a \pm b$ where a and/or b is a square root.

Introduction   In Section 10.2 we discussed certain conditions that must be satisfied in order for a radical to be in simplest form. We did not discuss the condition that the denominator of a fraction cannot contain a radical that is an irrational number. The process of making the denominator a rational number is called **rationalizing the denominator**. One of the reasons for rationalizing the denominator is in finding decimal approximations. Consider the following two methods of approximating $\frac{2}{\sqrt{3}}$:

Method 1

$\dfrac{2}{\sqrt{3}}$ Replace $\sqrt{3}$ with 1.732.

$\dfrac{2}{1.732}$ Divide.

1.155   Therefore, $\dfrac{2}{\sqrt{3}} \approx 1.155$.

Method 2

$\dfrac{2}{\sqrt{3}}$ Multiply by $\dfrac{\sqrt{3}}{\sqrt{3}}$.

$\dfrac{2}{\sqrt{3}} \cdot \dfrac{\sqrt{3}}{\sqrt{3}}$ Simplify.

$\dfrac{2\sqrt{3}}{\sqrt{9}}$

Replace $\sqrt{3}$ with 1.732 and find $\sqrt{9} = 3$.

$\dfrac{2(1.732)}{3}$ Simplify.

Method 2 is a little longer, but the arithmetic is much easier. With the use of calculators, ease in approximating radical expressions is no longer an important reason for rationalizing the denominator. In more advanced math courses, there are other reasons for rationalizing the denominator (and the numerator) that make it an important skill to possess.

Rationalizing denominators
of the form $\sqrt{a}$ The technique used in rationalizing the denominator is to multiply the fraction by an appropriate radical expression divided by itself (hence, its value is 1), so that the denominator of the fraction becomes a radical expression whose value is rational. This means that if the denominator of the original fraction is a square root, we will multiply this fraction by a square root divided by itself (thus, its value is 1), so that the radicand in the denominator is a perfect square. If the radicand of the original fraction contains variables, we must multiply the fraction by a radical expression divided by itself that will make the exponent(s) of the variable(s) in the radicand of the denominator divisible by two. Cube roots will be discussed later in this section.

Example 1

Rationalize the denominators of the following fractions. Assume all variables have positive values.

a. $\dfrac{3}{\sqrt{5}} =$ Since $\sqrt{5} \cdot \sqrt{5} = \sqrt{25} = 5$, multiply by $\dfrac{\sqrt{5}}{\sqrt{5}}$, which equals 1.

$\dfrac{3}{\sqrt{5}} \cdot \dfrac{\sqrt{5}}{\sqrt{5}} =$ Multiply.

$\dfrac{3\sqrt{5}}{\sqrt{25}} =$ $\sqrt{25} = 5$.

$\dfrac{3\sqrt{5}}{5}$ Therefore, $\dfrac{3}{\sqrt{5}} = \dfrac{3\sqrt{5}}{5}$.

b. $\dfrac{8}{\sqrt{6}} =$ Since, $\sqrt{6} \cdot \sqrt{6} = \sqrt{36} = 6$, multiply by $\dfrac{\sqrt{6}}{\sqrt{6}}$, which equals 1.

$\dfrac{8}{\sqrt{6}} \cdot \dfrac{\sqrt{6}}{\sqrt{6}} =$ Multiply.

$\dfrac{8\sqrt{6}}{\sqrt{36}} =$ $\sqrt{36} = 6$.

$\dfrac{8\sqrt{6}}{6} =$ Reduce to lowest terms.

$\dfrac{4\sqrt{6}}{3}$ Therefore, $\dfrac{8}{\sqrt{6}} = \dfrac{4\sqrt{6}}{3}$.

c. $\dfrac{\sqrt{3}}{\sqrt{8}} =$ Since $\sqrt{8} \cdot \sqrt{2} = \sqrt{16} = 4$, multiply by $\dfrac{\sqrt{2}}{\sqrt{2}}$, which equals 1.

$\dfrac{\sqrt{3}}{\sqrt{8}} \cdot \dfrac{\sqrt{2}}{\sqrt{2}} =$ Multiply.

$\dfrac{\sqrt{6}}{\sqrt{16}} =$ $\sqrt{16} = 4$.

$\dfrac{\sqrt{6}}{4} =$ Therefore, $\dfrac{\sqrt{3}}{\sqrt{8}} = \dfrac{\sqrt{6}}{4}$.

Note: One method of determining the radicand of the radical expression by which you multiply the original fraction is to find the smallest perfect square that is divisible by the radicand of the original fraction. For example, in Example 1c the smallest perfect square that is divisible by 8 is 16 and $16 \div 8 = 2$. Therefore, the radicand of the expression that we multiply by is 2. As another example, given $\dfrac{1}{\sqrt{12}}$, the smallest perfect square divisible by 12 is 36 and $36 \div 12 = 3$. Therefore, we would multiply by $\dfrac{1}{\sqrt{12}}$ by $\dfrac{\sqrt{3}}{\sqrt{3}}$.

d. $\dfrac{3}{\sqrt{x}} =$ Since $\sqrt{x} \cdot \sqrt{x} = \sqrt{x^2} = x$, multiply by $\dfrac{\sqrt{x}}{\sqrt{x}}$, which equals 1.

$\dfrac{3}{\sqrt{x}} \cdot \dfrac{\sqrt{x}}{\sqrt{x}} =$ Multiply.

$\dfrac{3\sqrt{x}}{\sqrt{x^2}} =$ $\sqrt{x^2} = x$. Remember $x \geq 0$.

$\dfrac{3\sqrt{x}}{x}$ Therefore, $\dfrac{3}{\sqrt{x}} = \dfrac{3\sqrt{x}}{x}$.

OPTIONAL	**ALTERNATIVE METHOD**

In Section 10.2, we gave an alternative method of simplifying radicals in that we wrote the radicand in terms of prime factors. Each time a factor appeared two times, the square root was one of the two equal factors. For example, $\sqrt{2 \cdot 2} = 2$, $\sqrt{2 \cdot 2 \cdot 2 \cdot 2} = 2 \cdot 2 = 4$, and $\sqrt{x \cdot x} = x$. We illustrate this technique by reworking Examples 1a and 1c.

Example 2

Rationalize the following denominators using prime factorization. Assume all variables have positive values.

a. $\dfrac{3}{\sqrt{5}} =$ Since $\sqrt{5 \cdot 5} = 5$, we need another factor of 5 in the radicand of the denominator. So, multiply by $\dfrac{\sqrt{5}}{\sqrt{5}}$.

$\dfrac{3}{\sqrt{5}} \cdot \dfrac{\sqrt{5}}{\sqrt{5}} =$ Multiply.

$\dfrac{3\sqrt{5}}{\sqrt{5 \cdot 5}} =$ $\sqrt{5 \cdot 5} = 5$.

$\dfrac{3\sqrt{5}}{5}$ Therefore, $\dfrac{3}{\sqrt{5}} = \dfrac{3\sqrt{5}}{5}$.

b. $\dfrac{\sqrt{3}}{\sqrt{8}} =$ Write 8 in terms of prime factors.

$\dfrac{\sqrt{3}}{\sqrt{2 \cdot 2 \cdot 2}} =$ Since we need an even number of factors of 2 in the radicand of the denominator, multiply by $\dfrac{\sqrt{2}}{\sqrt{2}}$.

$\dfrac{\sqrt{3}}{\sqrt{2 \cdot 2 \cdot 2}} \cdot \dfrac{\sqrt{2}}{\sqrt{2}} =$ Multiply.

$\dfrac{\sqrt{6}}{\sqrt{2 \cdot 2 \cdot 2 \cdot 2}} =$ $\sqrt{2 \cdot 2 \cdot 2 \cdot 2} = 2 \cdot 2 = 4$.

$\dfrac{\sqrt{6}}{4}$ Therefore, $\dfrac{\sqrt{3}}{\sqrt{8}} = \dfrac{\sqrt{6}}{4}$.

Practice Exercises

Rationalize the denominators of the following fractions. Assume all variables have positive values.

1. $\dfrac{5}{\sqrt{2}}$ **2.** $\dfrac{8}{\sqrt{10}}$ **3.** $\dfrac{\sqrt{5}}{\sqrt{12}}$ **4.** $\dfrac{b}{\sqrt{a}}$

If more practice is needed, do the Additional Practice Exercises.

Additional Practice Exercises

Rationalize the following denominators. Assume all variables have positive values.

a. $\dfrac{7}{\sqrt{5}}$ **b.** $\dfrac{10}{\sqrt{18}}$ **c.** $\dfrac{\sqrt{3}}{\sqrt{7}}$ **d.** $\dfrac{c}{\sqrt{d}}$

If the radicand is a fraction, apply the quotient rule for radicals. This results in a fraction with a radical in the denominator that is rationalized using the technique of Example 1.

Example 3

Simplify the following. Assume all variables have positive values.

a. $\sqrt{\dfrac{7}{2}} =$ Rewrite using the quotient rule.

$\dfrac{\sqrt{7}}{\sqrt{2}} =$ Since $\sqrt{2} \cdot \sqrt{2} = \sqrt{4} = 2$, multiply by $\dfrac{\sqrt{2}}{\sqrt{2}}$.

$\dfrac{\sqrt{7}}{\sqrt{2}} \cdot \dfrac{\sqrt{2}}{\sqrt{2}} =$ Multiply.

$\dfrac{\sqrt{14}}{\sqrt{4}} =$ $\sqrt{4} = 2$.

$\dfrac{\sqrt{14}}{2}$ Therefore, $\sqrt{\dfrac{7}{2}} = \dfrac{\sqrt{14}}{2}$.

b. $\sqrt{\dfrac{a}{b^3}} =$ Rewrite using the quotient rule.

$\dfrac{\sqrt{a}}{\sqrt{b^3}} =$ Since $\sqrt{b^3} \cdot \sqrt{b} = \sqrt{b^4} = b^2$, multiply by $\dfrac{\sqrt{b}}{\sqrt{b}}$.

$\dfrac{\sqrt{a}}{\sqrt{b^3}} \cdot \dfrac{\sqrt{b}}{\sqrt{b}} =$ Multiply.

$\dfrac{\sqrt{ab}}{\sqrt{b^4}} =$ $\sqrt{b^4} = b^2$.

$\dfrac{\sqrt{ab}}{b^2}$ Therefore, $\sqrt{\dfrac{a}{b^3}} = \dfrac{\sqrt{ab}}{b^2}$.

c. $\sqrt{\dfrac{2}{3x}} =$ Rewrite using the quotient rule.

$\dfrac{\sqrt{2}}{\sqrt{3x}} =$ Since $\sqrt{3x} \cdot \sqrt{3x} = \sqrt{9x^2} = 3x$, multiply by $\dfrac{\sqrt{3x}}{\sqrt{3x}}$.

$\dfrac{\sqrt{2}}{\sqrt{3x}} \cdot \dfrac{\sqrt{3x}}{\sqrt{3x}} =$ Multiply.

$\dfrac{\sqrt{6x}}{\sqrt{9x^2}} =$ $\sqrt{9x^2} = 3x$.

$\dfrac{\sqrt{6x}}{3x}$ Therefore, $\sqrt{\dfrac{2}{3x}} = \dfrac{\sqrt{6x}}{3x}$.

Practice Exercises

Simplify the following. Assume all variables have positive values.

5. $\sqrt{\dfrac{5}{3}}$ **6.** $\sqrt{\dfrac{c}{d^3}}$ **7.** $\sqrt{\dfrac{5}{2y}}$

Rationalizing denominators that are cube roots   If the denominator contains a cube root instead of a square root, the technique is essentially the same. We multiply the original fraction by the cube root of a quantity divided by itself such that the resulting radicand in the denominator is a perfect cube.

Answers:

Practice Exercises 5–7: 5. $\dfrac{\sqrt{15}}{3}$ 6. $\dfrac{\sqrt{cd}}{d^2}$ 7. $\dfrac{\sqrt{10y}}{2y}$

Example 4

Rationalize the denominators of the following:

a. $\dfrac{3}{\sqrt[3]{2}} =$ Since $\sqrt[3]{2} \cdot \sqrt[3]{4} = \sqrt[3]{8} = 2$, multiply by $\dfrac{\sqrt[3]{4}}{\sqrt[3]{4}}$.

$\dfrac{3}{\sqrt[3]{2}} \cdot \dfrac{\sqrt[3]{4}}{\sqrt[3]{4}} =$ Multiply.

$\dfrac{3\sqrt[3]{4}}{\sqrt[3]{8}} =$ $\sqrt[3]{8} = 2$.

$\dfrac{3\sqrt[3]{4}}{2}$ Therefore, $\dfrac{3}{\sqrt[3]{2}} = \dfrac{3\sqrt[3]{4}}{2}$.

b. $\dfrac{\sqrt[3]{a}}{\sqrt[3]{b}} =$ Since $\sqrt[3]{b} \cdot \sqrt[3]{b^2} = \sqrt[3]{b^3} = b$, multiply by $\dfrac{\sqrt[3]{b^2}}{\sqrt[3]{b^2}}$.

$\dfrac{\sqrt[3]{a}}{\sqrt[3]{b}} \cdot \dfrac{\sqrt[3]{b^2}}{\sqrt[3]{b^2}} =$ Multiply.

$\dfrac{\sqrt[3]{ab^2}}{\sqrt[3]{b^3}} =$ $\sqrt[3]{b^3} = b$.

$\dfrac{\sqrt[3]{ab^2}}{b}$ Therefore, $\dfrac{\sqrt[3]{a}}{\sqrt[3]{b}} = \dfrac{\sqrt[3]{ab^2}}{b}$.

Note: We can also determine the radicand of the radical expression by which we multiply the original fraction by finding the smallest perfect cube divisible by the radicand. For example, in Example 4a the smallest perfect cube that is divisible by 2 is 8 and $8 \div 2 = 4$. Therefore, the radicand of the radical expression that we multiply by is 4.

c. $\sqrt[3]{\dfrac{5}{9a^2}} =$ Rewrite $\sqrt[3]{\dfrac{5}{9a^2}}$ as $\dfrac{\sqrt[3]{5}}{\sqrt[3]{9a^2}}$.

$\dfrac{\sqrt[3]{5}}{\sqrt[3]{9a^2}} =$ Since $\sqrt[3]{9a^2} \cdot \sqrt[3]{3a} = \sqrt[3]{27a^3} = 3a$, multiply by $\dfrac{\sqrt[3]{3a}}{\sqrt[3]{3a}}$.

$\dfrac{\sqrt[3]{5}}{\sqrt[3]{9a^2}} \cdot \dfrac{\sqrt[3]{3a}}{\sqrt[3]{3a}} =$ Multiply.

$\dfrac{\sqrt[3]{15a}}{\sqrt[3]{27a^3}} =$ $\sqrt[3]{27a^3} = 3a$.

$\dfrac{\sqrt[3]{15a}}{3a}$ Therefore, $\sqrt[3]{\dfrac{5}{9a^2}} = \dfrac{\sqrt[3]{15a}}{3a}$.

OPTIONAL   ALTERNATIVE METHOD

The method using prime factorization also works very well for radicands containing cube roots. Each time a factor appears three times, the cube root is one of the three equal factors. For example, $\sqrt[3]{a \cdot a \cdot a} = a$, and $\sqrt[3]{2 \cdot 2 \cdot 2 \cdot 2 \cdot y \cdot y \cdot y \cdot y \cdot y} = 2y\sqrt[3]{2y^2}$. We illustrate this technique by reworking Examples 4a and 4c.

Example 5

Rationalize the following denominators using prime factorization. Assume all variables have positive values.

a. $\dfrac{3}{\sqrt[3]{2}} =$ Since $\sqrt[3]{2 \cdot 2 \cdot 2} = 2$, we need two more factors of 2 in the radicand of the denominator. So, multiply by $\dfrac{\sqrt[3]{2 \cdot 2}}{\sqrt[3]{2 \cdot 2}}$.

$\dfrac{3}{\sqrt[3]{2}} \cdot \dfrac{\sqrt[3]{2 \cdot 2}}{\sqrt[3]{2 \cdot 2}} =$ Multiply.

$\dfrac{3\sqrt[3]{2 \cdot 2}}{\sqrt[3]{2 \cdot 2 \cdot 2}} =$ $\sqrt[3]{2 \cdot 2 \cdot 2} = 2$.

$\dfrac{3\sqrt[3]{4}}{2}$ Therefore, $\dfrac{3}{\sqrt[3]{2}} = \dfrac{3\sqrt[3]{4}}{2}$.

b. $\sqrt[3]{\dfrac{5}{9a^2}} =$ Rewrite $\sqrt[3]{\dfrac{5}{9a^2}}$ as $\dfrac{\sqrt[3]{5}}{\sqrt[3]{9a^2}}$.

$\dfrac{\sqrt[3]{5}}{\sqrt[3]{9a^2}} =$ Write $9a^2$ in terms of prime factors.

$\dfrac{\sqrt[3]{5}}{\sqrt[3]{3 \cdot 3 \cdot a \cdot a}} =$ Since $\sqrt[3]{3 \cdot 3 \cdot 3} = 3$ and $\sqrt[3]{a \cdot a \cdot a} = a$, we need another factor of 3 and a in the radicand of the denominator. So, multiply by $\dfrac{\sqrt[3]{3a}}{\sqrt[3]{3a}}$.

$\dfrac{\sqrt[3]{5}}{\sqrt[3]{3 \cdot 3 \cdot a \cdot a}} \cdot \dfrac{\sqrt[3]{3a}}{\sqrt[3]{3a}} =$ Multiply.

$\dfrac{\sqrt[3]{15a}}{\sqrt[3]{3 \cdot 3 \cdot 3 \cdot a \cdot a \cdot a}} =$ $\sqrt[3]{3 \cdot 3 \cdot 3 \cdot a \cdot a \cdot a} = 3a$.

$\dfrac{\sqrt[3]{15a}}{3a}$ Therefore, $\sqrt[3]{\dfrac{5}{9a^2}} = \dfrac{\sqrt[3]{15a}}{3a}$.

Practice Exercises

Rationalize the denominators of the following:

8. $\dfrac{6}{\sqrt[3]{3}}$

9. $\sqrt[3]{\dfrac{m}{n}}$

10. $\dfrac{4}{\sqrt[3]{4y^2}}$

If more practice is needed, do the Additional Practice Exercises.

Additional Practice Exercises

Rationalize the denominators of the following:

e. $\dfrac{5}{\sqrt[3]{9}}$

f. $\sqrt[3]{\dfrac{3}{x^2}}$

g. $\dfrac{8}{\sqrt[3]{25c}}$

Rationalizing the denominators of the form $a \pm b$ where a and/or b are square roots

Recall from Section 10.4 that the product of conjugates always results in a rational number. Consequently, if the denominator contains an expression of the form $a + b$ where a and/or b is a square root, multiply the original fraction by the conjugate of the denominator divided by itself.

Example 6

Rationalize the denominators of the following:

a. $\dfrac{3}{4 + \sqrt{5}} =$

Since the conjugate of $4 + \sqrt{5}$ is $4 - \sqrt{5}$, multiply by $\dfrac{4 - \sqrt{5}}{4 - \sqrt{5}}$.

$\dfrac{3}{4 + \sqrt{5}} \cdot \dfrac{4 - \sqrt{5}}{4 - \sqrt{5}} =$ Multiply the fractions.

$\dfrac{3(4 - \sqrt{5})}{(4 + \sqrt{5})(4 - \sqrt{5})} =$ Perform the multiplications.

$\dfrac{12 - 3\sqrt{5}}{4^2 - (\sqrt{5})^2} =$ Simplify the denominator.

$\dfrac{12 - 3\sqrt{5}}{16 - 5} =$ Add 16 and -5.

$\dfrac{12 - 3\sqrt{5}}{11}$ Therefore, $\dfrac{3}{4 + \sqrt{5}} = \dfrac{12 - 3\sqrt{5}}{11}$.

b. $\dfrac{2 + \sqrt{2}}{4 - \sqrt{2}} =$

Since the conjugate of $4 - \sqrt{2}$ is $4 + \sqrt{2}$, multiply by $\dfrac{4 + \sqrt{2}}{4 + \sqrt{2}}$.

$\dfrac{2 + \sqrt{2}}{4 - \sqrt{2}} \cdot \dfrac{4 + \sqrt{2}}{4 + \sqrt{2}} =$ Multiply the fractions.

$\dfrac{(2 + \sqrt{2})(4 + \sqrt{2})}{(4 - \sqrt{2})(4 + \sqrt{2})} =$ Perform the multiplications.

$\dfrac{8 + 2\sqrt{2} + 4\sqrt{2} + \sqrt{4}}{4^2 - (\sqrt{2})^2} =$ Simplify.

$\dfrac{8 + 6\sqrt{2} + 2}{16 - 2} =$ Continue simplifying.

$\dfrac{10 + 6\sqrt{2}}{14} =$ Factor the numerator.

$\dfrac{2(5 + 3\sqrt{2})}{14} =$ Reduce to lowest terms.

$\dfrac{5 + 3\sqrt{2}}{7}$ Therefore, $\dfrac{2 + \sqrt{2}}{4 - \sqrt{2}} = \dfrac{5 + 3\sqrt{2}}{7}$.

Answers:

Practice Exercises 8–10: 8. $2\sqrt[3]{9}$ 9. $\dfrac{\sqrt[3]{mn^2}}{n}$ 10. $\dfrac{\sqrt[3]{2y}}{2y}$ **Additional Practice Exercises e–g:** e. $\dfrac{5\sqrt[3]{3}}{3}$ f. $\dfrac{\sqrt[3]{3x}}{x}$ g. $\dfrac{8\sqrt[3]{5c^2}}{5c}$

c. $\dfrac{1 + \sqrt{y}}{\sqrt{x} - \sqrt{y}} =$ Since the conjugate of $\sqrt{x} - \sqrt{y}$ is $\sqrt{x} + \sqrt{y}$,

multiply by $\dfrac{\sqrt{x} + \sqrt{y}}{\sqrt{x} + \sqrt{y}}$.

$\dfrac{1 + \sqrt{y}}{\sqrt{x} - \sqrt{y}} \cdot \dfrac{\sqrt{x} + \sqrt{y}}{\sqrt{x} + \sqrt{y}} =$ Multiply the fractions.

$\dfrac{(1 + \sqrt{y})(\sqrt{x} + \sqrt{y})}{(\sqrt{x} - \sqrt{y})(\sqrt{x} + \sqrt{y})} =$ Perform the multiplications.

$\dfrac{\sqrt{x} + \sqrt{y} + \sqrt{xy} + \sqrt{y^2}}{(\sqrt{x})^2 - (\sqrt{y})^2} =$ Simplify.

$\dfrac{\sqrt{x} + \sqrt{y} + \sqrt{xy} + y}{x - y}$ Therefore, $\dfrac{1 + \sqrt{y}}{\sqrt{x} - \sqrt{y}} = \dfrac{\sqrt{x} + \sqrt{y} + \sqrt{xy} + y}{x - y}$.

Practice Exercises

Rationalize the denominators of the following. Assume all variables have nonnegative values.

11. $\dfrac{7}{6 + \sqrt{6}}$ 12. $\dfrac{4 - \sqrt{2}}{3 - \sqrt{2}}$ 13. $\dfrac{\sqrt{3} + \sqrt{5}}{\sqrt{3} - \sqrt{5}}$

If more practice is needed, do the Additional Practice Exercises.

Additional Practice Exercises

Rationalize the denominators of the following:

h. $\dfrac{7}{3 + \sqrt{7}}$ i. $\dfrac{3 + \sqrt{3}}{5 - \sqrt{3}}$ j. $\dfrac{\sqrt{5} + \sqrt{2}}{\sqrt{5} - \sqrt{2}}$

Exercise Set 10.5

Rationalize the following denominators. Assume all variables have positive values.

1. $\dfrac{3}{\sqrt{7}}$ 2. $\dfrac{2}{\sqrt{3}}$ 3. $\dfrac{5}{\sqrt{5}}$ 4. $\dfrac{6}{\sqrt{6}}$

5. $\dfrac{6}{\sqrt{2}}$ 6. $\dfrac{12}{\sqrt{3}}$ 7. $\dfrac{20}{\sqrt{10}}$ 8. $\dfrac{30}{\sqrt{15}}$

9. $\dfrac{4}{\sqrt{x}}$ 10. $\dfrac{6}{\sqrt{y}}$ 11. $\dfrac{m}{\sqrt{n}}$ 12. $\dfrac{r}{\sqrt{s}}$

13. $\dfrac{\sqrt{5}}{\sqrt{2}}$ 14. $\dfrac{\sqrt{7}}{\sqrt{5}}$ 15. $\dfrac{\sqrt{3}}{\sqrt{8}}$ 16. $\dfrac{\sqrt{11}}{\sqrt{18}}$

17. $\dfrac{\sqrt{5}}{\sqrt{32}}$ 18. $\dfrac{\sqrt{7}}{\sqrt{27}}$ 19. $\dfrac{\sqrt{a}}{\sqrt{b}}$ 20. $\dfrac{\sqrt{c}}{\sqrt{d}}$

21. $\dfrac{\sqrt{d}}{\sqrt{e}}$

Answers:

Practice Exercises 11–13: 11. $\dfrac{42 - 7\sqrt{6}}{30}$ 12. $\dfrac{10 + \sqrt{2}}{7}$ 13. $-4 - \sqrt{15}$ Additional Practice Exercises h–j: h. $\dfrac{21 - 7\sqrt{7}}{2}$ i. $\dfrac{9 + 4\sqrt{3}}{11}$ j. $\dfrac{7 + 2\sqrt{10}}{3}$

Simplify the following. Assume all variables have nonnegative values.

22. $\sqrt{\dfrac{3}{2}}$

23. $\sqrt{\dfrac{5}{3}}$

24. $\sqrt{\dfrac{7}{8}}$

25. $\sqrt{\dfrac{7}{12}}$

26. $\sqrt{\dfrac{u}{v}}$

27. $\sqrt{\dfrac{p}{q}}$

28. $\sqrt{\dfrac{3}{y^3}}$

29. $\sqrt{\dfrac{7}{x^5}}$

30. $\sqrt{\dfrac{3}{2x}}$

31. $\sqrt{\dfrac{7}{5y}}$

32. $\sqrt{\dfrac{4}{3b}}$

33. $\sqrt{\dfrac{9}{7a}}$

Rationalize the denominators of the following. Assume all variables have positive values.

34. $\dfrac{5}{\sqrt[3]{4}}$

35. $\dfrac{2}{\sqrt[3]{9}}$

36. $\dfrac{8}{\sqrt[3]{2}}$

37. $\dfrac{10}{\sqrt[3]{25}}$

38. $\dfrac{\sqrt[3]{x}}{\sqrt[3]{y}}$

39. $\dfrac{\sqrt[3]{r}}{\sqrt[3]{s}}$

40. $\dfrac{\sqrt[3]{a^2}}{\sqrt[3]{b^2}}$

41. $\dfrac{\sqrt[3]{x}}{\sqrt[3]{y^2}}$

42. $\sqrt[3]{\dfrac{2}{3a}}$

43. $\sqrt[3]{\dfrac{5}{4b}}$

44. $\sqrt[3]{\dfrac{5}{9x^2}}$

45. $\sqrt[3]{\dfrac{7}{2d^2}}$

Rationalize the denominators of the following:

46. $\dfrac{23}{5+\sqrt{2}}$

47. $\dfrac{-1}{2+\sqrt{5}}$

48. $\dfrac{19}{5-\sqrt{6}}$

49. $\dfrac{7}{3-\sqrt{3}}$

50. $\dfrac{x}{5-\sqrt{x}}$

51. $\dfrac{a}{7+\sqrt{a}}$

52. $\dfrac{3+\sqrt{2}}{2+\sqrt{2}}$

53. $\dfrac{4+\sqrt{3}}{3-\sqrt{3}}$

54. $\dfrac{\sqrt{6}-5}{\sqrt{6}+3}$

55. $\dfrac{\sqrt{5}-6}{\sqrt{5}+4}$

56. $\dfrac{3+\sqrt{2}}{2+\sqrt{3}}$

57. $\dfrac{5+\sqrt{3}}{4-\sqrt{5}}$

58. $\dfrac{2-\sqrt{b}}{3+\sqrt{b}}$

59. $\dfrac{7+\sqrt{a}}{2-\sqrt{a}}$

60. $\dfrac{\sqrt{3}+\sqrt{2}}{\sqrt{3}-\sqrt{2}}$

61. $\dfrac{\sqrt{5}-\sqrt{3}}{\sqrt{5}+\sqrt{3}}$

62. $\dfrac{\sqrt{a}+\sqrt{b}}{\sqrt{a}-\sqrt{b}}$

63. $\dfrac{\sqrt{x}+\sqrt{y}}{\sqrt{x}-\sqrt{y}}$

64. $\dfrac{4+\sqrt{2}}{\sqrt{3}}$

65. $\dfrac{5-\sqrt{6}}{\sqrt{2}}$

Challenge Exercises: (66–73)

Rationalize the denominators of the following:

66. $\dfrac{4-2\sqrt{3}}{3\sqrt{2}-2\sqrt{5}}$

67. $\dfrac{2\sqrt{5}+3\sqrt{6}}{3\sqrt{2}-3\sqrt{7}}$

68. $\dfrac{2}{\sqrt[4]{2}}$

69. $\dfrac{3}{\sqrt[4]{9}}$

70. $\dfrac{x^2}{\sqrt[4]{x^2}}$

71. $\dfrac{y^2}{\sqrt[4]{y^3}}$

72. $\dfrac{2}{\sqrt[5]{2}}$

73. $\dfrac{6}{\sqrt[3]{8}}$

Writing Exercises:

74. In your own words, describe the procedures used in rationalizing the denominator.

75. What is wrong with the following? $\dfrac{9}{16} = \dfrac{3}{4}$

76. If we were to rationalize the denominator of $\dfrac{1}{\sqrt[3]{a}+\sqrt[3]{b}}$, we would not multiply by $\dfrac{\sqrt[3]{a}-\sqrt[3]{b}}{\sqrt[3]{a}-\sqrt[3]{b}}$. Why not?

Solving Equations with Radicals

O B J E C T I V E S

When you complete this section, you will be able to:

a. Solve equations that contain one radical (square root).

b. Solve equations that have two radicals (square roots).

Introduction

Solving equations that contain radicals that are square roots depends upon two concepts. The first, presented in Section 10.1, is $\left(\sqrt{x}\right)^2 = x$ For example, $\left(\sqrt{3}\right)^2 = 3$, $\left(\sqrt{x+2}\right)^2 = x + 2$, and so on. The second, presented in Section 2.4, is $(a + b)^2 = a^2 + 2ab + b^2$. For example, $(x + 5)^2 = x^2 + 2 \cdot 5x + 5^2 = x^2 + 10x + 25$, and $\left(2 + \sqrt{x}\right)^2 = 2^2 + 2 \cdot 2\sqrt{x} + \left(\sqrt{x}\right)^2 = 4 + 4\sqrt{x} + x$.

Solving equations with one radical

The key to solving equations that contain one radical is to isolate the radical on one side of the equation. After this is accomplished, square both sides of the equation. However, there may be a problem in doing so. When both sides of an equation are raised to a power, *the resulting equation may have solutions that do not solve the original equation*. Consider the following:

Begin with the equation $x = 4$.

$x = 4$	The solution is obviously 4. Square both sides.
$x^2 = 4^2$	$4^2 = 16$.
$x^2 = 16$	Subtract 16 from both sides.
$x^2 - 16 = 0$	Factor $x^2 - 16$.
$(x + 4)(x - 4) = 0$	Set each factor equal to 0.
$x + 4 = 0, x - 4 = 0$	Solve each equation.
$x = -4, x = 4$	Therefore, the solutions of $x^2 = 4^2$ are 4 and -4.

You will note that the equation that resulted from squaring both sides of the original equation not only has the solution of the original equation (4), but also has another solution (-4). This leads to the following observation:

Observation

If both sides of an equation are raised to a power, the resulting equation contains all the solutions of the original equation and perhaps contains some solutions that are not solutions of the original equation. Any solution of the resulting equation that does not solve the original equation is called an **extraneous** solution and is discarded.

This means that when we solve an equation by raising both sides to a power, we will get all the solutions of the original equation, but we may get some solutions that will not solve the original equation (extraneous solutions). Consequently, we must check all solutions of the resulting equation in the original equation. This difficulty results because the symbol $\sqrt{}$ denotes the positive square root only. Following is the procedure for solving equations that contain one square root:

Solving Equations Containing One Square Root

If an equation contains one square root, it can be solved by:

1. Isolating the square root on one side of the equation.

2. Squaring both sides of the equation.

3. Solving the resulting equation.

4. Checking all solutions in the original equation.

Example 1

Solve the following equations:

a. $\sqrt{x} = 4$ Since $\left(\sqrt{x}\right)^2 = x$, square both sides of the equation.

$\left(\sqrt{x}\right)^2 = 4^2$ Simplify both sides.

$x = 16$ Therefore, 16 is a possible solution.

CHECK:

$\sqrt{x} = 4$ Substitute 16 for x.

$\sqrt{16} = 4$ Remember, the radical sign denotes the positive root only.

$4 = 4$ Therefore, 16 is the correct solution.

b. $\sqrt{2x + 3} - 5 = 0$ Isolate the radical by adding 5 to both sides of the equation.

$\sqrt{2x + 3} - 5 + 5 = 0 + 5$ Simplify both sides.

$\sqrt{2x + 3} = 5$ Square both sides.

$\left(\sqrt{2x + 3}\right)^2 = 5^2$ Simplify both sides.

$2x + 3 = 25$ Subtract 3 from both sides.

$2x = 22$ Divide both sides by 2.

$x = 11$ Therefore, 11 is a possible solution.

CHECK:

$\sqrt{2x + 3} - 5 = 0$ Substitute 11 for x.

$\sqrt{2(11) + 3} - 5 = 0$ Simplify.

$\sqrt{22 + 3} - 5 = 0$ Continue simplifying.

$\sqrt{25} - 5 = 0$ $\sqrt{25} = 5$.

$5 - 5 = 0$ Simplify.

$0 = 0$ Therefore, 11 is the correct solution.

c. $\sqrt{3x - 3} + 6 = 0$ Isolate the radical by subtracting 6 from both sides of the equation.

$\sqrt{3x - 3} = -6$ Square both sides.

$\left(\sqrt{3x - 3}\right)^2 = (-6)^2$ Simplify both sides.

$3x - 3 = 36$ Add 3 to both sides.

$3x = 39$ Divide both sides by 3.

$x = 13$ Therefore, 13 is a possible solution.

CHECK:

$\sqrt{3x - 3} + 6 = 0$ Substitute 13 for x.

$\sqrt{3(13) - 3} + 6 = 0$ Simplify.

$\sqrt{39 - 3} + 6 = 0$ Continue simplifying.

$\sqrt{36} + 6 = 0$ $\sqrt{36} = 6$.

$6 + 6 = 0$ Simplify.

$12 \neq 0$ Therefore, 13 is not a solution.

Since 13 is the only possible solution, $\sqrt{3x - 3} + 6 = 0$ has no solution. Remember, this is usually denoted by $\varnothing$, which is called the empty or null set. This means that 13 is an extraneous solution.

d.

$$\sqrt{3x + 10} = x + 2$$ Square both sides of the equation.

$$\left(\sqrt{3x + 10}\right)^2 = (x + 2)^2$$ Simplify both sides.

$$3x + 10 = x^2 + 2 \cdot 2x + 4$$ Simplify the right side.

$$3x + 10 = x^2 + 4x + 4$$ Subtract 3x and 10 from both sides.

$$3x - 3x + 10 - 10 = x^2 + 4x - 3x + 4 - 10$$ Simplify both sides.

$$0 = x^2 + x - 6$$ Factor the right side.

$$0 = (x + 3)(x - 2)$$ Set each factor equal to 0.

$$x + 3 = 0, x - 2 = 0$$ Solve each *equation.*

$$x = -3, x = 2$$ Therefore, the possible solutions are -3 and 2.

CHECK:

$$x = -3$$

$$\sqrt{3x + 10} = x + 2$$ Substitute -3 for x.

$$\sqrt{3(-3) + 10} = (-3) + 2$$ Simplify.

$$\sqrt{-9 + 10} = -1$$ Continue simplifying.

$$\sqrt{1} = -1$$ $\sqrt{1} = 1$.

$$1 \neq -1$$ Therefore, -3 is not a solution.

$$x = 2$$

$$\sqrt{3x + 10} = x + 2$$ Substitute 2 for x.

$$\sqrt{3(2) + 10} = (2) + 2$$ Simplify.

$$\sqrt{6 + 10} = 4$$ Continue simplifying.

$$\sqrt{16} = 4$$ $\sqrt{16} = 4$.

$$4 = 4$$ Therefore, 2 is a solution.

Since -3 did not check, 2 is the only solution of the equation and -3 is an extraneous solution. Also, in the check, we could tell that -3 was not going to check when we arrived at the line $\sqrt{-9 + 10} = -1$, because the left side of the equation is positive (the radical denotes the positive root only) and the right is negative. Consequently, they cannot be equal.

Practice Exercises

Solve the following equations:

1. $\sqrt{a} - 6 = 0$ **2.** $\sqrt{4x - 8} - 4 = 0$

3. $\sqrt{2x - 5} + 7 = 0$ **4.** $\sqrt{x - 1} = x - 3$

If more practice is needed, do the Additional Practice Exercises.

Additional Practice Exercises

Solve the following equations:

a. $\sqrt{b} + 5 = 0$ **b.** $\sqrt{3x + 6} = 6$

c. $\sqrt{4 - 2x} + 8 = 0$ **d.** $\sqrt{3x + 3} = 2x - 1$

Answers:

Practice Exercises 1–4: 1. $a = 36$ 2. $x = 6$ 3. $\varnothing$ 4. $x = 5$ **Additional Practice Exercises a–d:** a. $\varnothing$ b. $x = 10$ c. $\varnothing$ d. $x = 2$

Solving equations with two radicals   If the equation contains more than one radical, the procedure for solving is more complicated and is outlined as follows:

> **Solving Equations Containing Two Square Roots**
>
> If an equation contains two square roots, it can be solved by:
> 1. Isolating one of the radicals on one side of the equation.
> 2. Squaring both sides of the equation and, if necessary, simplifying each side.
> 3. Isolating the radical and squaring both sides of the equation again if the equation still contains a radical.
> 4. Checking all solutions in the original equation.

Example 2

Solve the following equations:

a. $\sqrt{4x-3} - \sqrt{2x+7} = 0$ Isolate $\sqrt{4x-3}$ by adding $\sqrt{2x+7}$ to both sides of the equation.

$$\sqrt{4x-3} - \sqrt{2x+7} + \sqrt{2x+7} = 0 + \sqrt{2x+7} \quad \text{Simplify.}$$
$$\sqrt{4x-3} = \sqrt{2x+7} \quad \text{Square both sides.}$$
$$\left(\sqrt{4x-3}\right)^2 = \left(\sqrt{2x+7}\right)^2 \quad \text{Simplify.}$$
$$4x - 3 = 2x + 7 \quad \text{Subtract } 2x \text{ from both sides.}$$
$$2x - 3 = 7 \quad \text{Add 3 to both sides.}$$
$$2x = 10 \quad \text{Divide both sides by 2.}$$
$$x = 5 \quad \text{Therefore, 5 is a possible solution.}$$

CHECK:

$$\sqrt{4x-3} - \sqrt{2x+7} = 0 \quad \text{Substitute 5 for } x.$$
$$\sqrt{4(5)-3} - \sqrt{2(5)+7} = 0 \quad \text{Simplify.}$$
$$\sqrt{20-3} - \sqrt{10+7} = 0 \quad \text{Continue simplifying.}$$
$$\sqrt{17} - \sqrt{17} = 0 \quad \text{Add like radicals.}$$
$$0 = 0 \quad \text{Therefore, 5 is the solution.}$$

b. $\sqrt{x+4} = 3 - \sqrt{x-2}$ Since one radical is isolated, square both sides of the equation.

$$\left(\sqrt{x+4}\right)^2 = \left(3 - \sqrt{x-2}\right)^2 \quad \text{Simplify both sides.}$$
$$x + 4 = 3^2 - 2 \cdot 3\sqrt{x-2} + \left(\sqrt{x-2}\right)^2 \quad \text{Simplify the right side.}$$
$$x + 4 = 9 - 6\sqrt{x-2} + x - 2 \quad \text{Continue simplifying the right side.}$$
$$x + 4 = x + 7 - 6\sqrt{x-2} \quad \text{Subtract } x \text{ from both sides.}$$
$$4 = 7 - 6\sqrt{x-2} \quad \text{Subtract 7 from both sides.}$$
$$-3 = -6\sqrt{x-2} \quad \text{Divide both sides by } -3.$$
$$1 = 2\sqrt{x-2} \quad \text{Square both sides.}$$
$$1^2 = \left(2\sqrt{x-2}\right)^2 \quad \text{Simplify both sides.}$$
$$1 = 4(x-2) \quad \text{Simplify the right side.}$$
$$1 = 4x - 8 \quad \text{Add 8 to both sides.}$$
$$9 = 4x \quad \text{Divide both sides by 4.}$$
$$\frac{9}{4} = x \quad \text{Therefore, } \frac{9}{4} \text{ is a possible solution.}$$

CHECK:

$$\sqrt{x + 4} = 3 - \sqrt{x - 2}$$ Substitute $\frac{9}{4}$ for x.

$$\sqrt{\frac{9}{4} + 4} = 3 - \sqrt{\frac{9}{4} - 2}$$ Get common denominators.

$$\sqrt{\frac{9}{4} + \frac{16}{4}} = 3 - \sqrt{\frac{9}{4} - \frac{8}{4}}$$ Add the fractions.

$$\sqrt{\frac{25}{4}} = 3 - \sqrt{\frac{1}{4}}$$ Find the square roots.

$$\frac{5}{2} = 3 - \frac{1}{2}$$ Common denominator on the right.

$$\frac{5}{2} = \frac{6}{2} - \frac{1}{2}$$ Add the fractions.

$$\frac{5}{2} = \frac{5}{2}$$ Therefore, $\frac{9}{4}$ is the correct solution.

Practice Exercises

Solve the following equations.

5. $\sqrt{3x + 2} - \sqrt{x + 4} = 0$ **6.** $\sqrt{12 + x} = 6 - \sqrt{x}$

If more practice is needed, do the Additional Practice Exercises.

Additional Practice Exercises

Solve the following equations:

e. $\sqrt{5x + 2} + \sqrt{x - 6} = 0$ **f.** $\sqrt{x + 4} - \sqrt{x - 4} = 4$

Radical equations are frequently used in real-world situations.

Example 3

a. The distance a car will skid when the brakes are applied is approximated by the formula $s = k\sqrt{d}$ where s is the speed of the car in miles per hour, d is the distance the car will skid in feet, and k is a constant determined by road conditions, weight of the car, type of tires, and so on. If $k = 4.21$, find the following:

1. If the length of the skid mark is 196 feet, find the speed of the car.

     Solution:

     $s = k\sqrt{d}$ Substitute 4.21 for k and 196 for d.

     $s = 4.21\sqrt{196}$ $\sqrt{196} = 14$.

     $s = 4.21(14)$ Multiply.

     $s = 58.94$ Therefore, the speed of the car was 58.94 miles per hour.

2. If the car is traveling at a speed of 63.15 miles per hour, how far will it take it to skid to a stop?

Answers:

Practice Exercises 5–6: **5.** $x = 1$ **6.** $x = 4$ **Additional Practice Exercises e–f:** **e.** $\emptyset$ **f.** $\emptyset$

Solution:

$$s = k\sqrt{d}$$ Substitute 63.15 for s and 4.21 for k.

$$63.15 = 4.21\sqrt{d}$$ Divide both sides by 4.21.

$$15 = \sqrt{d}$$ Square both sides of the equation.

$$225 = d$$ Therefore, it would take the car 225 feet to stop.

b. Under certain conditions, the distance one can see to the horizon is given by $s = \sqrt{1.75h}$ where s is the distance in miles and h is the elevation of the viewer in feet.

1. How far can a person see to the horizon if his or her elevation is 1372 feet?

   Solution:

   $$s = \sqrt{1.75h}$$ Substitute 1372 for h.

   $$s = \sqrt{(1.75)(1372)}$$ Multiply.

   $$s = \sqrt{2401}$$ $\sqrt{2401} = 49$.

   $$s = 49$$ Therefore, you would be able to see 49 miles to the horizon.

2. What would the elevation of the viewer have to be in order to see 70 miles to the horizon?

   Solution:

   $$s = \sqrt{1.75h}$$ Substitute 70 for s.

   $$70 = \sqrt{1.75h}$$ Square both sides.

   $$4900 = 1.75h$$ Divide both sides by 1.75.

   $$2800 = h$$ Therefore, the elevation would have to be 2800 feet.

Practice Exercises

Answer the following:

7. The number of amperes in an electrical system is calculated by the formula amperes $= \sqrt{\dfrac{\text{watts}}{\text{ohms}}}$.

   a. How many amperes are used if there are 8000 watts and 20 ohms?

   b. How many watts of power are used by an appliance that uses 8 amperes when the resistance is 35 ohms?

8. The amount of time it takes an object to fall s feet is given by the formula $t = \dfrac{\sqrt{s}}{4}$ where t is the time in seconds and s is distance in feet.

   a. How long will it take an object to reach the ground if it is dropped from a height of 256 feet?

   b. If an object is dropped from the top of a tall building and it takes 6 seconds for it to hit the ground, how tall is the building?

Exercise Set 10.6

Solve the following equations:

1. $\sqrt{a} = 2$

2. $\sqrt{c} = 9$

3. $\sqrt{z} - 7 = 0$

4. $\sqrt{n} - 5 = 0$

5. $\sqrt{a} + 4 = 0$

6. $\sqrt{b} + 2 = 0$

7. $2\sqrt{x} = 8$

8. $3\sqrt{y} = 9$

9. $\sqrt{y} - 4 = 4$

10. $\sqrt{x} + 2 = 6$

11. $\sqrt{x} + 3 = 8$

12. $\sqrt{y} - 4 = 2$

13. $\sqrt{5h - 21} = 3$

14. $\sqrt{2d + 39} = 7$

15. $\sqrt{7r - 26} - 4 = 0$

16. $\sqrt{7d - 24} - 2 = 0$

17. $\sqrt{8r - 15} = r$

18. $\sqrt{4d + 32} = d$

19. $\sqrt{2m + 24} - m = 0$

20. $\sqrt{6h - 8} - h = 0$

21. $\sqrt{w - 3} = w - 5$

22. $\sqrt{2t + 9} = t - 3$

23. $\sqrt{4k + 25} = k - 5$

24. $\sqrt{4f + 25} = f - 5$

25. $\sqrt{w - 4} - w + 6 = 0$

26. $\sqrt{x - 2} - x + 4 = 0$

27. $\sqrt{y - 4} - y + 4 = 0$

28. $\sqrt{3n + 16} - n + 4 = 0$

29. $\sqrt{4f - 20} = \sqrt{f - 2}$

30. $\sqrt{10n - 3} = \sqrt{4n + 3}$

31. $\sqrt{5n - 21} = \sqrt{n - 1}$

32. $\sqrt{6f - 4} = \sqrt{4f + 8}$

33. $\sqrt{3a - 7} - \sqrt{a - 1} = 0$

34. $\sqrt{6r - 26} - \sqrt{r - 1} = 0$

35. $\sqrt{4 - x} = 2 - \sqrt{x}$

36. $\sqrt{9 - 11x} = 3 - \sqrt{x}$

37. $\sqrt{a - 4} = 4 - \sqrt{a + 4}$

38. $\sqrt{b + 8} = 6 - \sqrt{b - 4}$

39. $\sqrt{c + 10} + \sqrt{c + 2} = 2$

40. $\sqrt{n + 7} + \sqrt{n - 4} = 1$

41. $\sqrt{x + 5} + \sqrt{x - 3} = 4$

42. $\sqrt{m - 4} + \sqrt{m + 5} = 3$

Answer the following:

43. Using the formula $s = k\sqrt{d}$ from Example 3a with $k = 3.7$, find the following:

   a. Find how fast a car was traveling if the length of the skid mark is 324 feet.

   b. Find the length of the skid mark for a car traveling 74 miles per hour.

44. The distance a person can see to the horizon on a particular day is given by $s = \sqrt{1.8h}$ where s is the distance in miles and h is the elevation of the person in feet. Answer the following:

   a. How far can a person see to the horizon from a height of 200 feet?

   b. How high would a person have to be in order to see a distance of 5 miles?

45. Using the formula $\text{amperes} = \sqrt{\dfrac{\text{watts}}{\text{ohms}}}$ given in Practice Exercise 7, find the following:

   a. Find the number of amperes if the number of watts is 4000 and the number of ohms is 10.

   b. Find the number of watts used by an electrical system that draws 15 amperes when the resistance is 120 ohms.

46. Using the formula $t = \dfrac{\sqrt{s}}{4}$ given in Practice Exercise 8, find the following:

   a. Find how long it takes an object to hit the ground if it is dropped from a height of 3600 feet.

   b. Find the height of a cliff if it takes 16 seconds for a rock dropped from the top of the cliff to reach the ground below.

47. The equation for the length of time it takes a pendulum to make one swing is $t = 2\pi\sqrt{\dfrac{L}{32}}$ where t is the time in seconds and L is the length of the pendulum in feet. Find the following:

   a. Find how long it takes a pendulum that is 2 feet long to make one swing.

   b. If a pendulum makes one swing every 1.5 seconds, find the length of the pendulum to the nearest one hundredth of a foot.

48. If air resistance is neglected, the velocity (v) of an object after it has fallen a distance (d) is given by $v = 8\sqrt{d}$ where v is the velocity in feet per second and d is the distance in feet. Find the following:

   a. Find the velocity of an object that has fallen 289 feet.

   b. If an object hits the ground with a velocity of 120 feet per second, how far has it fallen?

Challenge Exercises: (49–58)

Solve the following equations:

49. $\sqrt{2x - 3} - \sqrt{2x - 11} = 2$

50. $\sqrt{2x + 20} + \sqrt{2x + 5} = 3$

51. $\sqrt[3]{x} = 2$

52. $\sqrt[3]{x} = -3$

53. $\sqrt[3]{x + 3} = -2$

54. $\sqrt[3]{x - 5} = 1$

55. $\sqrt[4]{x} = 2$

56. $\sqrt[4]{x + 2} = 1$

57. $\sqrt[4]{2x - 1} = 3$

58. $\sqrt[4]{2x + 4} = 2$

Writing Exercise:

59. What is wrong with the following "solution" of the equation?

$$\sqrt{3x - 9} = 5 + \sqrt{2x + 6}$$
$$\left(\sqrt{3x - 9}\right)^2 = \left(5 + \sqrt{2x + 6}\right)^2$$
$$3x - 9 = 25 + 2x + 6$$
$$3x - 9 = 2x + 31$$
$$x - 9 = 31$$
$$x = 40$$

Section 10.7	Pythagorean Theorem

OBJECTIVES

When you complete this section, you will be able to:

a. Find the unknown side of a right triangle.

b. Find the length of a diagonal of a rectangle or square.

c. Find the length of the missing side of a rectangle when given the length of a diagonal and the length of one side.

d. Find the length of a diagonal of a square.

e. Find the length of each side of a square when given the length of a diagonal.

f. Solve application problems involving right triangles.

Introduction

One place where radicals are used in real-world situations is in finding an unknown side of a right triangle. The procedure for finding an unknown side of a right triangle is attributed to the Greek mathematician and philosopher Pythagoras and for that reason is called the Pythagorean theorem.

There is some doubt as to whether Pythagoras himself actually discovered the relationship between the sides of a right triangle or whether it was discovered by one of his followers.

A triangle one of whose angles is a right (90°) angle is called a **right triangle.** The sides that form the right angle are called the **legs** of the right triangle and the side opposite the right angle is called the **hypotenuse.** See the figure that follows:

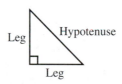

Conclusions drawn from observations are never accepted as fact until they are proven mathematically. Many proofs of the Pythagorean theorem exist, but none are included in this text. Once a fact has been proven, it is called a theorem.

The Pythagorean theorem probably arose from observing that if squares were constructed on each of the sides of a right triangle, the sum of the areas of the squares on the legs was equal to the area of the square on the hypotenuse. See the following figure where the length of the legs are 3 and 4 and the length of the hypotenuse is 5.

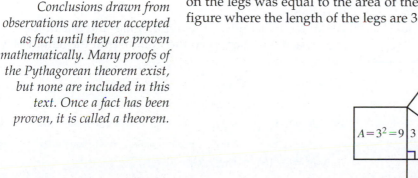

From the figure, we see that $3^2 + 4^2 = 5^2$, because $9 + 16 = 25$. Thus the sum of the areas of the squares on the legs is equal to the area of the square on the hypotenuse. The Pythagorean theorem is no longer stated in terms of areas of squares, but is stated as follows:

Pythagorean Theorem

If the legs of a right triangle have lengths of a and b and the hypotenuse has length of c, then $a^2 + b^2 = c^2$.

In words, the sum of the squares of the lengths of the legs of a right triangle is equal to the square of the length of the hypotenuse.

Before finding an unknown side of a right triangle, we need one other concept that will be discussed in detail in Section 11.1. Consider an equation like $x^2 = 16$. This asks the question, "The square of what number is 16?" This is the definition of square root. Consequently, the number(s) whose square(s) equal 16 are the square roots of 16, which are 4 and -4. Algebraically we would write this as follows: If $x^2 = 16$, then $\sqrt{x^2} = \pm\sqrt{16}$ so $x = 4$ or $x = -4$. Likewise, if $x^2 = 3$, then $\sqrt{x^2} = \pm\sqrt{3}$ so $x = \sqrt{3}$ or $-\sqrt{3}$ and if $x^2 = 15$, then $\sqrt{x^2} = \pm\sqrt{15}$ so $x = \sqrt{15}$ or $-\sqrt{15}$. Since the lengths of sides of triangles are positive, we will be limited to taking the *positive square roots only*. Since a and b represent the legs, we may substitute the lengths of a leg for either a or b interchangeably.

Example 1

Find the length of the unknown side of the following right triangles:

a.

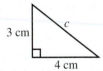

Solution:

The lengths of the legs are 3 cm and 4 cm. We need to find the length of the hypotenuse. Use the Pythagorean theorem.

$a^2 + b^2 = c^2$	Substitute for a and b.
$3^2 + 4^2 = c^2$	Square 3 and 4.
$9 + 16 = c^2$	Add 9 and 16.
$25 = c^2$	Take the positive square root of each side of the equation.
$\sqrt{25} = \sqrt{c^2}$	Find $\sqrt{25}$ and $\sqrt{c^2}$.
$5 = c$	Therefore, the length of the hypotenuse is 5 cm.

b.

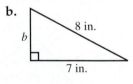

Solution:

We know the length of the hypotenuse and one of the legs. Use the Pythagorean theorem to find the length of the other leg.

$a^2 + b^2 = c^2$	Substitute for a and c.
$7^2 + b^2 = 8^2$	Square 7 and 8.
$49 + b^2 = 64$	Subtract 49 from both sides of the equation.
$b^2 = 15$	Take the positive square root of each side.
$\sqrt{b^2} = \sqrt{15}$	Find $\sqrt{b^2}$.
$b = \sqrt{15}$	Therefore, the length of the other leg is $\sqrt{15}$ inches.

c.

We know the length of the hypotenuse and one leg. Use the Pythagorean theorem to find the length of the other leg.

$$a^2 + b^2 = c^2 \qquad \text{Substitute for } b \text{ and } c.$$
$$a^2 + \left(\sqrt{7}\right)^2 = 8^2 \qquad \text{Square } \sqrt{7} \text{ and } 8.$$
$$a^2 + 7 = 64 \qquad \text{Subtract 7 from both sides of the equation.}$$
$$a^2 = 57 \qquad \text{Take the positive square root of each side.}$$
$$\sqrt{a^2} = \sqrt{57} \qquad \text{Find } \sqrt{a^2}.$$
$$a = \sqrt{57} \qquad \text{Therefore, the length of the other leg is } \sqrt{57} \text{ feet.}$$

d. Find the length of the other leg of a right triangle, one of whose legs is $2\sqrt{3}$ feet and whose hypotenuse is $4\sqrt{5}$ feet.

Solution:
Use the Pythagorean theorem. Substitute for a or b and c.

$$a^2 + b^2 = c^2 \qquad \text{Substitute for } b \text{ and } c.$$
$$a^2 + \left(2\sqrt{3}\right)^2 = \left(4\sqrt{5}\right)^2 \qquad \text{Apply } (ab)^n = a^n b^n.$$
$$a^2 + 2^2 \cdot \left(\sqrt{3}\right)^2 = 4^2 \cdot \left(\sqrt{5}\right)^2 \qquad \text{Square.}$$
$$a^2 + 4 \cdot 3 = 16 \cdot 5 \qquad \text{Multiply.}$$
$$a^2 + 12 = 80 \qquad \text{Subtract 12 from both sides of the equation.}$$
$$a^2 = 68 \qquad \text{Take the positive square root of each side.}$$
$$\sqrt{a^2} = \sqrt{68} \qquad \text{Find } \sqrt{a^2} \text{ and simplify } \sqrt{68}.$$
$$a = 2\sqrt{17} \qquad \text{Therefore, the unknown leg is } 2\sqrt{17} \text{ feet.}$$

Practice Exercises

Find the length of the unknown side of each of the following right triangles:

1.

6 in.   *c*   8 in.

2.

a   8 m   4 m

3.

3 yd $\sqrt{3}$ yd   *a*

4. Find the length of the other leg of a right triangle, one of whose legs is $4\sqrt{3}$ inches and whose hypotenuse is $4\sqrt{6}$ inches.

If more practice is needed, do the Additional Practice Exercises.

Additional Practice Exercises

Find the length of the unknown side of each of the following right triangles:

a.

12 m   *c*   5 m

b.

b   10 cm   12 cm

c.

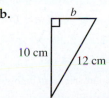
$\sqrt{7}$ in.   7 in.   *b*

d. The hypotenuse of a right triangle is $4\sqrt{3}$ inches and one leg is $3\sqrt{2}$ inches. Find the length of the other leg.

A diagonal of a square or a rectangle divides the square or rectangle into two right triangles. See the following figure.

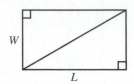

The sides of the rectangle become the legs of the right triangle and the diagonal of the rectangle becomes the hypotenuse of the right triangle. Consequently, if we know the lengths of the sides of the rectangle, we can find the length of the diagonal. Also, if we know the length of the diagonal and the length of one of the sides of the rectangle, we can find the length of the other side.

Example 2

a. Find the length of the diagonal of the following rectangle:

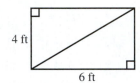

Solution:
The lengths of the legs of the right triangle are 6 feet and 4 feet. The diagonal is the hypotenuse of a right triangle.

$$a^2 + b^2 = c^2 \qquad \text{Substitute for } a \text{ and } b.$$
$$6^2 + 4^2 = c^2 \qquad \text{Square 6 and 4.}$$
$$36 + 16 = c^2 \qquad \text{Add 36 and 16.}$$
$$52 = c^2 \qquad \text{Take the positive square root of each side of the equation.}$$
$$\sqrt{52} = \sqrt{c^2} \qquad \text{Simplify } \sqrt{52} \text{ and find } \sqrt{c^2}.$$
$$2\sqrt{13} = c \qquad \text{Therefore, the diagonal of the rectangle is } 2\sqrt{13} \text{ feet.}$$

b. The length of a diagonal of a rectangle is $\sqrt{26}$ meters and the width of the rectangle is $2\sqrt{2}$ meters. Find the length.

Solution:
Draw and label a figure.

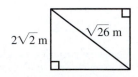

The hypotenuse is $\sqrt{26}$ meters long and one leg is $2\sqrt{2}$ meters long.

$$a^2 + b^2 = c^2 \qquad \text{Substitute for } a \text{ (or } b\text{) and } c.$$
$$\left(2\sqrt{2}\right)^2 + b^2 = \left(\sqrt{26}\right)^2 \qquad \text{Square } 2\sqrt{2} \text{ and } \sqrt{26}.$$
$$2^2\left(\sqrt{2}\right)^2 + b^2 = 26 \qquad \text{Square 2 and } \sqrt{2}.$$
$$4 \cdot 2 + b^2 = 26 \qquad \text{Multiply 4 and 2.}$$
$$8 + b^2 = 26 \qquad \text{Subtract 8 from both sides of the equation.}$$
$$b^2 = 18 \qquad \text{Take the positive square root of each side.}$$
$$\sqrt{b^2} = \sqrt{18} \qquad \text{Simplify both sides.}$$
$$b = \sqrt{18} = 3\sqrt{2} \qquad \text{Therefore, the length is } 3\sqrt{2} \text{ meters.}$$

c. The diagonal of a square is $4\sqrt{2}$ inches long. Find the length of the sides of the square.

Solution:
Draw and label a figure. Remember, all four sides of a square are equal in length. Therefore, let each side have length of x.

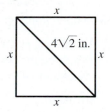

The legs are each x inches long and the hypotenuse is 4 inches long.

$a^2 + b^2 = c^2$	Substitute x for a and b and $4\sqrt{2}$ for c.
$x^2 + x^2 = \left(4\sqrt{2}\right)^2$	Add x^2 and x^2 and square $4\sqrt{2}$.
$2x^2 = 4^2 \cdot \left(\sqrt{2}\right)^2$	Square 4 and $\sqrt{2}$.
$2x^2 = 16 \cdot 2$	Multiply 16 and 2.
$2x^2 = 32$	Divide both sides of the equation by 2.
$x^2 = 16$	Take the positive square root of each side.
$\sqrt{x^2} = \sqrt{16}$	Simplify both sides.
$x = \sqrt{16} = 4$	Therefore, each side of the square is 4 inches long.

Practice Exercises

5. Find the length of the diagonal of the following rectangle:

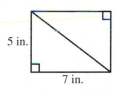

5 in.

7 in.

6. The diagonal of a rectangle is $2\sqrt{11}$ centimeters and the length is $2\sqrt{7}$ centimeters. Find the width.

7. The diagonal of a square is $3\sqrt{2}$ yards. Find the length of the sides of the square.

If more practice is needed, do the Additional Practice Exercises.

Additional Practice Exercises

e. Find the length of the diagonal of the following rectangle:

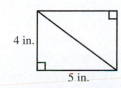

4 in.

5 in.

f. The diagonal of a rectangle is $\sqrt{30}$ feet and the width is $\sqrt{14}$ feet. Find the length.

g. The diagonal of a square is $4\sqrt{3}$ meters. Find the length of each side of the square.

Answers:

Practice Exercises 5–7: 5. $\sqrt{74}$ in. 6. 4 cm 7. 3 yd   **Additional Practice Exercises e–g:** e. $\sqrt{41}$ in. f. 4 ft 8. $2\sqrt{6}$ m

Applications of right triangles are found in many real-world situations. Knowledge of how to find the unknown side of a right triangle is important in solving many types of problems.

Example 3

Solve the following:

a. A building contractor wants to be sure that the walls of a building meet at right angles. If he marks a point 6 feet from the corner of one wall and a point 8 feet from the corner of the other wall, what should the distance be between these two points?

Solution:
If the walls meet at right angles, the distances measured along each wall will be the legs of a right triangle and the distance between the points will be the hypotenuse. See the following figure:

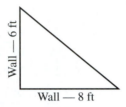

Use the Pythagorean theorem to find the distance between the points.

$a^2 + b^2 = c^2$ Substitute for a and b.

$6^2 + 8^2 = c^2$ Square 6 and 8.

$36 + 64 = c^2$ Add 36 and 64.

$100 = c^2$ Take the positive square root of each side of the equation.

$10 = c$ If the walls meet at right angles, there will be 10 feet between the two points.

b. A house is 48 feet wide. The roof is supported by rafters that are 7 feet above the level of the walls at the center of the house. Find the length of each rafter. See the following figure:

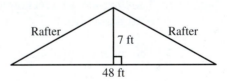

Solution:
The length of each rafter is the length of the hypotenuse of a right triangle. Since the rafters meet at the center of the house, the point below where the rafters meet is 24 feet from each wall. See the following figure:

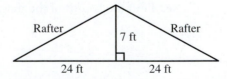

Thus, we have two right triangles whose legs are 24 feet and 7 feet. Use the Pythagorean theorem to find the hypotenuse.

$a^2 + b^2 = c^2$ Substitute for a and b.

$24^2 + 7^2 = c^2$ Square 24 and 7.

$576 + 49 = c^2$ Add 576 and 49.

$625 = c^2$ Take the positive square root of each side.

$25 = c$ Therefore, each rafter is 25 feet long.

Practice Exercises

Solve the following.

8. Television screens are measured diagonally. For example, if a television is advertised as having a 17-inch screen, it means the screen measures 17 inches diagonally. If a television is advertised as having a 25-inch rectangular screen and the screen is 20 inches long, how wide is the screen?

9. The range for a particular brand of marine radio is 20 miles. Bill and Don are in separate boats that are equipped with these radios. Their boats are traveling on courses that are at right angles to each other. If Bill is 8 miles from port and Don is 15 miles from the same port, can they contact each other by radio? Why or why not?

Exercise Set 10.7

Find the value of x in each of the following right triangles:

1.
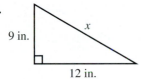
9 in. — x — 12 in.

2.
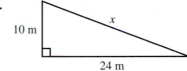
10 m — x — 24 m

3.
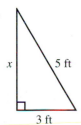
x — 5 ft — 3 ft

4.

x — 13 ft — 5 ft

5.
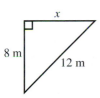
x — 8 m — 12 m

6.

x — 7 in. — 5 in.

7.
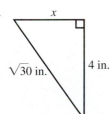
x — $\sqrt{30}$ in. — 4 in.

8.
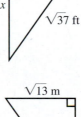
5 ft — x — $\sqrt{37}$ ft

9.
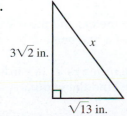
$3\sqrt{2}$ in. — x — $\sqrt{13}$ in.

10.
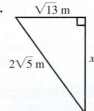
$\sqrt{13}$ m — $2\sqrt{5}$ m — x

Find the length of the unknown side of each of the following right triangles:

11. The lengths of the legs are 6 feet and 9 feet.

12. The lengths of the legs are 7 inches and 12 inches.

13. One leg is 4 meters long and the hypotenuse is 6 meters long.

14. One leg is 10 centimeters long and the hypotenuse is 12 centimeters long.

15. The lengths of the legs are $\sqrt{38}$ inches and $\sqrt{14}$ inches.

16. The lengths of the legs are $\sqrt{10}$ yards and $\sqrt{22}$ yards.

17. One leg is $2\sqrt{6}$ centimeters long and the hypotenuse is $4\sqrt{2}$ centimeters long.

18. One leg is $2\sqrt{7}$ inches long and the hypotenuse is $4\sqrt{3}$ inches long.

Find the lengths of the diagonals of the following rectangles or squares:

19.
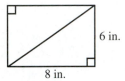
6 in.

8 in.

20.
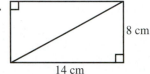
8 cm

14 cm

21.

6 ft

6 ft

22.
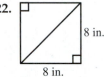
8 in.

8 in.

23.
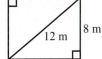
$4\sqrt{6}$ mm

$4\sqrt{6}$ mm

24.
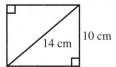
$3\sqrt{5}$ in.

$3\sqrt{5}$ in.

Find the length of the unknown side(s) of each of the following rectangles:

25.
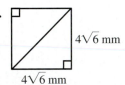
12 m   8 m

26.
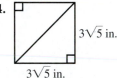
14 cm   10 cm

27.

$3\sqrt{6}$ m

$2\sqrt{6}$ m

28.

$4\sqrt{2}$ m

$3\sqrt{2}$ m

Solve the following:

29. The length of a rectangle is 5 centimeters and the width is 8 centimeters. Find the length of a diagonal.

30. The length of a rectangle is 9 meters and the width is 6 meters. Find the length of a diagonal.

31. The length of a diagonal of a rectangle is 12 inches and the length is $4\sqrt{5}$ inches. Find the width.

32. The length of a diagonal of a rectangle is 14 yards and the length is $4\sqrt{6}$ yards. Find the width.

33. The length of each side of a square is 7 feet. Find the length of a diagonal.

34. The length of each side of a square is 5 feet. Find the length of a diagonal.

35. The length of each side of a square is $5\sqrt{2}$ centimeters. Find the length of a diagonal.

36. The length of each side of a square is $2\sqrt{7}$ feet. Find the length of a diagonal.

37. The length of a diagonal of a square is 8 inches. Find the length of each side of the square.

38. The length of a diagonal of a square is 10 yards. Find the length of each side of the square.

Solve the following:

39. In Example 3a, a building contractor wanted to be sure that the walls of a building met at right angles. Suppose he marks a point on one wall 8 feet from the corner and a point 15 feet from the corner on the other wall. If the walls are perpendicular, how far is it between the two points?

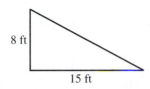

8 ft

15 ft

40. In Example 3b, we found the length of some rafters. How long would the rafters be for a house 48 feet wide if the rafters meet at the center of the house at a point 10 feet above the level of the walls?

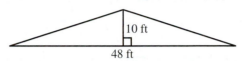

10 ft

48 ft

41. Practice Exercise 8 dealt with the measurement of television screens. If a television has a 13-inch screen that is 10 inches long, how wide is it?

42. Practice Exercise 9 dealt with a particular brand of marine radio. Suppose a marine radio has a range of 25 miles. Bill and Don are traveling in separate boats equipped with these radios on courses that are at right angles to each other, as in Practice Exercise 9. If Bill is 12 miles from port and Don is 16 miles from the same port, can they contact each other by radio? Why or why not?

43. When hurricanes threaten, a common practice is to tape windows to keep the glass from shattering. If a rectangular window is 40 inches long and 30 inches wide, how long is a piece of tape that goes diagonally across the window?

44. Suppose a plate glass window 72 inches long and 60 inches wide is taped as in Exercise 43. How long is a piece of tape that goes diagonally across this window?

45. A newly transplanted tree is supported by three ropes that are attached to the tree at a point 24 feet above the ground. These ropes are tied to pegs that are driven into the ground 7 feet from the base of the tree. What is the total length of the three pieces of rope?

46. A guy wire is attached to a telephone pole at a point 30 feet above the ground. How long is the wire if the other end is attached at a point on the ground 10 feet from the base of the pole?

47. An advertising sign is in the shape of a rectangle. A stripe is painted diagonally across the sign. If the length of the stripe is 8 feet and the sign is 6 feet long, find the width of the sign.

48. Two pieces of wood are nailed together at right angles. To give added support, a third piece of wood is nailed to each of these at points 3 feet from the corner. Find the length of this piece of wood.

Writing Exercise:

49. Write and solve an application problem involving the Pythagorean theorem.

Group Project:

50. Make a list of five everyday situations where right triangles have a practical application.

Chapter 10   Summary

Roots of a Number: [Section 10.1]
- A number b is an nth root of a number a if $b^n = a$. Written symbolically, $\sqrt[n]{a} = b$ if $b^n = a$. If n is even, then a must be nonnegative. If n is odd, then a may have any real number value.

Radical: [Section 10.1]
- A **radical** is an algebraic expression of the form $\sqrt[n]{a}$. The symbol $\sqrt{\ }$, is the radical sign, n is the **root index**, and a is the **radicand.**

Perfect nth Powers: [Section 10.1]
- The perfect squares are $\{1, 4, 9, 16, 25, \ldots\}$.
- The perfect cubes are $\{\pm 1, \pm 8, \pm 27, \pm 64, \pm 125, \ldots\}$.
- The perfect fourth powers are $\{1, 16, 81, 256, \ldots\}$.
- And so on.
- Numbers that are perfect nth powers have rational nth roots.

Irrational Numbers: [Section 10.1]	• An **irrational** number is any real number that is not rational. Written as a decimal, irrational numbers neither terminate nor repeat. Examples are $\sqrt{2}$, $\sqrt[3]{5}$, and $\sqrt[5]{8}$.
Products of Radicals: [Section 10.2]	• If $\sqrt[n]{a}$ and $\sqrt[n]{b}$ both exist, then $\sqrt[n]{a} \cdot \sqrt[n]{b} = \sqrt[n]{a \cdot b}$. This rule is also used in the form of $\sqrt[n]{ab} = \sqrt[n]{a} \cdot \sqrt[n]{b}$.
Radicals to Powers: [Section 10.2]	• In general, if $\sqrt[n]{a}$ exists, then $\left(\sqrt[n]{a}\right)^n = a$.
Quotients of Radicals: [Section 10.3]	• If $\sqrt[n]{a}$ and $\sqrt[n]{b}$ both exist and $b \neq 0$, then $\dfrac{\sqrt[n]{a}}{\sqrt[n]{b}} = \sqrt[n]{\dfrac{a}{b}}$. This rule is also used in the form $\sqrt[n]{\dfrac{a}{b}} = \dfrac{\sqrt[n]{a}}{\sqrt[n]{b}}$.
Like Radical Expressions: [Section 10.4]	• **Like radical expressions** must have the same root index and the same radicand.
Adding Like Radical Expressions: [Section 10.4]	• To add like radicals, add the coefficients and leave the radical portion unchanged. Never add the radicands. It is often necessary to simplify the radical expressions before they can be added.

Simplest Form of Radicals:
[Sections 10.3 and 10.4]

1. Find all rational roots.

2. Remove all possible factors from the radicand. For example, no square root may have a radicand with a perfect-square factor.

3. Find all possible products, quotients, sums, and differences.

4. Rationalize all the denominators. This also means no radicand may contain a fraction.

Rationalizing the Denominator:
[Section 10.5]

1. If the denominator is in the form of $\sqrt{a}$, multiply the numerator and the denominator of the original fraction by the same appropriate radical so that the denominator of the original fraction becomes a radical expression whose value is rational.

2. If the denominator is of the form $a + b$ where a and/or b is a square root, multiply the numerator and the denominator of the fraction by the conjugate of the denominator. The conjugate is found by changing the middle sign of the expression $a + b$ where a and/or b is a square root.

Roots of Equations Containing Radicals: [Section 10.6]

• If both sides of an equation are raised to a power, the resulting equation contains all the solutions of the original equation and perhaps contains some solutions that do not solve the original equation.

Extraneous Solutions: [Section 10.6]

• Any solution of an equation resulting from raising both sides of an equation to a power that is not a solution of the original equation is called an **extraneous** solution and is discarded.

Solving Equations Containing Square Roots: [Section 10.6]

1. If the equation contains only one square root, isolate the radical on one side of the equation. Then square both sides of the equation and solve the resulting equation. Check all solutions in the original equation.

2. If the equation contains two square roots, isolate one radical on one side of the equation. Then square both sides of the equation and simplify the resulting equation. Isolate the remaining radical on one side of the equation and square both sides again. Solve the resulting equation and check all solutions in the original equation.

Pythagorean theorem: [Section 10.7]

• It the lengths of the legs of a right triangle are a and b and the length of the hypotenuse is c, then $a^2 + b^2 = c^2$.

Chapter 10   Review Exercises

Find the following roots, if they exist. If the root does not exist, write, "This root does not exist as a real number." Assume all variables have nonnegative values.
[Sections 10.1 and 10.2]

1. $\sqrt{16}$ **2.** $\sqrt{-16}$ **3.** $-\sqrt{16}$ **4.** $\pm\sqrt{16}$

5. $\sqrt[3]{125}$ **6.** $\sqrt[3]{-125}$ **7.** $-\sqrt[3]{125}$ **8.** $-\sqrt[3]{-125}$

9. $\sqrt[4]{81}$ **10.** $\sqrt[4]{-81}$ **11.** $-\sqrt[4]{81}$ **12.** $\sqrt{b^4}$

13. $\sqrt{n^5}$ **14.** $\sqrt[3]{m^5}$ **15.** $\sqrt[5]{z^8}$ **16.** $\sqrt{m^5 n^6}$

17. $\sqrt[3]{a^4 b^6}$ **18.** $\sqrt[4]{x^6 y^7}$

Identify each of the following as rational or irrational. If the number is rational, find the root exactly. If the number is irrational, approximate the number to the nearest thousandth by using a calculator or Table 1. [Section 10.1]

19. $\sqrt{32}$ **20.** $\sqrt{169}$ **21.** $\sqrt[3]{63}$

22. $\sqrt[3]{1331}$ **23.** $\sqrt{1.44}$ **24.** $\sqrt[3]{.027}$

Simplify the following radicals: [Section 10.2]

25. $\sqrt{63}$ **26.** $\sqrt{44}$ **27.** $4\sqrt{125}$

28. $5\sqrt{108}$ **29.** $\sqrt[3]{54}$ **30.** $6\sqrt[3]{32}$

31. $\sqrt{162x^7}$ **32.** $\sqrt{200a^8}$ **33.** $3\sqrt{40z^5 y^3}$

34. $\sqrt[3]{128a^4}$ **35.** $5\sqrt[3]{108n^3 m^7}$ **36.** $\sqrt[4]{162z^5}$

Find the following products or powers of roots. Leave answers in simplified form. Assume all variables have nonnegative values. [Section 10.3]

37. $\left(\sqrt{r}\right)^2$ **38.** $\left(\sqrt{x^3}\right)^2$ **39.** $\left(\sqrt{5}\right)^2$ **40.** $\sqrt{5}\cdot\sqrt{7}$

41. $\sqrt{6}\cdot\sqrt{r}$ **42.** $\sqrt{m}\cdot\sqrt{n}$ **43.** $\sqrt{7}\cdot\sqrt{7}$

44. $\sqrt{5}\cdot\sqrt{35}$ **45.** $\sqrt{45}\cdot\sqrt{5}$ **46.** $4\sqrt{6}\cdot 2\sqrt{24}$

47. $5\sqrt{75}\cdot 3\sqrt{6}$ **48.** $7\sqrt{18}\cdot 3\sqrt{6}$ **49.** $\sqrt{c^3}\cdot\sqrt{c}$

50. $\sqrt{s^5}\cdot\sqrt{s^3}$ **51.** $\sqrt{v}\cdot\sqrt{v^4}$ **52.** $2\sqrt{30n^3}\cdot 3\sqrt{10n^3}$

53. $6\sqrt{5v^4}\cdot 2\sqrt{15v^3}$

Find the following quotients. Leave all answers in simplified form. Assume all variables have positive values. [Section 10.3]

54. $\sqrt{\dfrac{25}{9}}$ **55.** $\sqrt{\dfrac{81x^2}{4y^4}}$ **56.** $\sqrt{\dfrac{13}{25}}$

57. $\dfrac{\sqrt{48}}{\sqrt{3}}$ **58.** $\dfrac{6\sqrt{72}}{3\sqrt{8}}$ **59.** $\dfrac{\sqrt{x^7}}{\sqrt{x^3}}$

60. $\dfrac{\sqrt{c^5 d^4}}{\sqrt{c^2 d^2}}$ **61.** $\dfrac{10\sqrt{63a^5}}{5\sqrt{7a}}$ **62.** $\dfrac{6a^3\sqrt{140n^5}}{2a\sqrt{5n^2}}$

Simplify the following. Leave answers in simplified form. Assume all variables have positive values. [Section 10.3]

63. $\sqrt{\dfrac{3}{2}}\cdot\sqrt{\dfrac{5}{8}}$ **64.** $\sqrt{\dfrac{n^2}{m^3}}\cdot\sqrt{\dfrac{n^5}{m}}$ **65.** $\sqrt{\dfrac{7a}{18b}}\cdot\sqrt{\dfrac{3a}{2b}}$

Find the following sums. Leave answers in simplified form. Assume all variables have nonnegative values. [Section 10.4]

66. $7\sqrt{5}-9\sqrt{5}$ **67.** $6\sqrt{a}+12\sqrt{a}$ **68.** $4a\sqrt[3]{x}-6a\sqrt[3]{x}$

69. $\sqrt{3}+\sqrt{75}$ **70.** $\sqrt{48}-\sqrt{3}$ **71.** $4\sqrt{27}-5\sqrt{12}$

72. $-4\sqrt{28}-5\sqrt{63}$ **73.** $2a\sqrt{20}-5a\sqrt{45}$

74. $4\sqrt{63a^3}+6\sqrt{64a^3}$

Simplify the following: [Section 10.4]

75. $\sqrt{6}\cdot\sqrt{2}+\sqrt{6}\cdot\sqrt{8}$ **76.** $\dfrac{\sqrt{300}}{\sqrt{2}}-\sqrt{96}$

77. $\sqrt{5}\left(3-\sqrt{30}\right)$ **78.** $\sqrt{3}\left(\sqrt{3}+2\sqrt{6}\right)$

79. $\left(4+\sqrt{5}\right)\left(5-\sqrt{5}\right)$ **80.** $\left(2\sqrt{3}+\sqrt{5}\right)\left(\sqrt{3}-4\sqrt{5}\right)$

81. $\left(3\sqrt{3}+\sqrt{6}\right)\left(3\sqrt{3}-\sqrt{6}\right)$ **82.** $\left(2-\sqrt{7}\right)^2$

Reduce the following to lowest terms: [Section 10.4]

83. $\dfrac{6-2\sqrt{2}}{12}$ **84.** $\dfrac{8+4\sqrt{12}}{4}$

Rationalize the following denominators. Leave answers in simplest form. Assume all variables have positive values. [Section 10.5]

85. $\dfrac{6}{\sqrt{7}}$ **86.** $\dfrac{8}{\sqrt{18}}$ **87.** $\sqrt{\dfrac{5}{3}}$ **88.** $\dfrac{a}{\sqrt{b}}$

89. $\sqrt{\dfrac{m}{n}}$ **90.** $\dfrac{5}{\sqrt{5n}}$ **91.** $\dfrac{4}{\sqrt[3]{9}}$ **92.** $\dfrac{n}{\sqrt[3]{n^2}}$

93. $\dfrac{4a}{\sqrt[3]{2a}}$ **94.** $\dfrac{4}{2-\sqrt{5}}$ **95.** $\dfrac{\sqrt{7}}{\sqrt{7}+2}$ **96.** $\dfrac{6}{\sqrt{3}-\sqrt{5}}$

97. $\dfrac{\sqrt{12}}{3-\sqrt{6}}$ **98.** $\dfrac{2+\sqrt{5}}{2-\sqrt{5}}$ **99.** $\dfrac{5+\sqrt{3}}{\sqrt{5}-\sqrt{3}}$

Solve the following equations: [Section 10.6]

100. $\sqrt{a}=9$ **101.** $3\sqrt{n}-18=0$

102. $\sqrt{x+4}=5$ **103.** $2\sqrt{2x+4}+6=2$

104. $\sqrt{3n-5}=\sqrt{5n+9}$ **105.** $4\sqrt{s}=\sqrt{15s-5}$

106. $\sqrt{7x - 10} = x$ **107.** $\sqrt{4x - 12} = x - 3$

108. $\sqrt{n + 2} + \sqrt{n - 6} = 4$

Find the value of x in each of the following: [Section 10.7]

109.
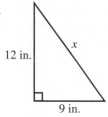

110.
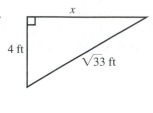

111.
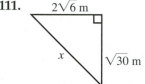

112.
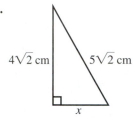

113.
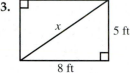

114.
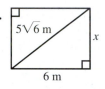

115. The lengths of the legs of a right triangle are 9 inches and 10 inches. Find the length of the hypotenuse.

116. The length of the hypotenuse of a right triangle is 12 meters and the length of one leg is 4 meters. Find the length of the other leg.

117. A rectangle has a length of $6\sqrt{3}$ inches and a width of $2\sqrt{5}$ inches. Find the length of a diagonal.

118. A rectangle has a diagonal length of 9 yards and a width of $3\sqrt{2}$ yards. Find the length.

119. Each side of a square is 10 inches long. Find the length of a diagonal.

120. The length of a diagonal of a square is $3\sqrt{10}$ feet. Find the length of each side of the square.

121. A television antenna is supported by two guy wires. The wires are attached to the antenna at a point 15 feet above the ground. The other ends of the wires are anchored to the ground at a point 5 feet from the base of the antenna. Find the length of each of the wires.

122. A tree fell during a storm with the broken portion still resting on the stump, forming a right triangle. The point where the tree broke is 15 feet above the ground. The top of the tree is on the ground at a point 36 feet from the base of the tree. How high was the tree before it broke?

Chapter 10   Test

Find the following roots, if they exist. If the root does not exist, write, "The root does not exist as a real number."

1. $\sqrt{81}$ **2.** $\sqrt[3]{-64}$ **3.** $\sqrt{x^8}$ **4.** $\sqrt[3]{c^9 d^6}$

Simplify the following:

5. $\sqrt{96}$ **6.** $3\sqrt{98x^5}$ **7.** $4\sqrt[3]{40}$ **8.** $3\sqrt[3]{64y^7}$

Find the following products and/or quotients. Leave answers in simplified form.

9. $2\sqrt{35} \cdot 3\sqrt{5}$ **10.** $3\sqrt{15a^3} \cdot 4\sqrt{10a^4}$ **11.** $\dfrac{\sqrt{96}}{\sqrt{6}}$

12. $\dfrac{8\sqrt{56c^7}}{\sqrt{7c^2}}$ **13.** $\sqrt{\dfrac{21c^3}{5}} \cdot \sqrt{\dfrac{3c}{5}}$

Find the following sums:

14. $6\sqrt{6} - 12\sqrt{6}$ **15.** $3\sqrt{80} + 5\sqrt{20}$ **16.** $\sqrt{180a^3} + 4\sqrt{45a^3}$

Simplify the following. Leave answers in simplified form. Assume all variables have nonnegative values.

17. $\sqrt{3}(\sqrt{6} + 4\sqrt{18})$ **18.** $(\sqrt{5} + 2\sqrt{6})(3\sqrt{5} - \sqrt{6})$

19. $(4 + 2\sqrt{7})(4 - 2\sqrt{7})$ **20.** $(5 - \sqrt{x})^2$

Rationalize the denominators of the following:

21. $\dfrac{8}{\sqrt{12}}$ **22.** $\dfrac{8}{\sqrt[3]{16}}$ **23.** $\sqrt{\dfrac{m}{n}}$ **24.** $\dfrac{12}{7 - \sqrt{5}}$

25. $\dfrac{5 + \sqrt{6}}{2 - \sqrt{6}}$

Solve the following equations. If there is no solution, write $\varnothing$.

26. $2\sqrt{b} - 10 = 0$ **27.** $\sqrt{7z + 58} - 11 = 0$

28. $\sqrt{x + 27} - \sqrt{x + 6} = 3$

29. Find the value of x in the following right triangle:

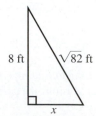

30. Two cars approach the same intersection at right angles to each other. If one car is 45 feet from the intersection and the other is 60 feet from the intersection, find the distance between the two cars.

Q uadratic equations were introduced in Section 5.8. Recall that a quadratic equation is any equation that may be put in the form of $ax^2 + bx + c = 0$ with $a \neq 0$. We solved quadratic equations by factoring, and the solutions were rational numbers.

Solving Quadratic Equations

However, if asked to solve the equation $x^2 + x - 5 = 0$ by factoring, we could not since $x^2 + x - 5$ cannot be factored. In this chapter, we will learn two methods of solving this type of equation. One method is called *completing the square* and involves creating a perfect-square trinomial, which is the square of a binomial. The procedure for squaring a binomial was discussed in Section 2.4. You may want to review this procedure prior to doing Section 11.2. The second method is using the *quadratic formula*. The quadratic formula is derived by using the completing the square method. The solutions of quadratic equations are often irrational numbers, which were discussed in Chapter 10. Section 11.4 contains types of equations that we have previously solved (linear, quadratic, radical, rational), except that now many of the radical and rational equations will result in quadratic equations after simplification.

Quadratic equations also allow us to solve a wide variety of application problems. We end the chapter with a section on application problems that can be solved only by the use of quadratic equations.

Chapter 11

| Section 11.1 | Solving Incomplete Quadratic Equations |

O B J E C T I V E S

When you complete this section, you will be able to:

a. Solve quadratic equations of the form $ax^2 + bx = 0$.

b. Solve quadratic equations of the form $x^2 = k$ by extraction of roots.

c. Solve quadratic equations of the form $(ax + b)^2 = c$ by extraction of roots.

Introduction   In Chapter 5, we solved quadratic equations by factoring. Recall that after the equation was written in $ax^2 + bx + c = 0$ form, we factored $ax^2 + bx + c$ and set each factor equal to 0. We then solved the resulting linear equations. For example, solve $x^2 + 2x - 15 = 0$.

$$x^2 + 2x - 15 = 0 \qquad \text{Factor } x^2 + 2x - 15.$$
$$(x + 5)(x - 3) = 0 \qquad \text{Set each factor equal to 0.}$$
$$x + 5 = 0, x - 3 = 0 \qquad \text{Solve each equation.}$$
$$x = -5, \quad x = 3 \qquad \text{Therefore, the solutions are } -5 \text{ and } 3.$$

This method, however, cannot be used to solve all quadratic equations since not all quadratic polynomials are factorable. For example, $x^2 + x + 3 = 0$ cannot be solved by factoring since $x^2 + x + 3$ cannot be factored. In this section, we will learn a technique that will be used in Section 11.2 to develop a procedure that will allow us to solve any quadratic equation.

Solving quadratic equations   A quadratic equation in the form of $ax^2 + bx + c = 0$, in which either b or $c = 0$,
of the form $ax^2 + bx = 0$ is called an **incomplete quadratic equation**. If $c = 0$, the equation is of the form $ax^2 + bx = 0$. This type can be solved by factoring and was discussed in Section 5.8. We will do a couple of examples to refresh your memory.

Example 1

Solve the following equations:

a. $2x^2 + 3x = 0$ Remove the common factor x.

$x(2x + 3) = 0$ Set each factor equal to 0.

$x = 0$, or $2x + 3 = 0$ $x = 0$ is one solution. Solve $2x + 3 = 0$.

$\qquad 2x = -3$

$\qquad x = -\dfrac{3}{2}$ Therefore, the solutions are 0 and $-\dfrac{3}{2}$.

CHECK: $x = 0$ $x = -\dfrac{3}{2}$

Substitute 0 for x.   Substitute $-\dfrac{3}{2}$ for x.

$2x^2 + 3x = 0$ $2x^2 + 3x = 0$

$2(0)^2 + 3(0) = 0$ $2\left(-\dfrac{3}{2}\right)^2 + 3\left(-\dfrac{3}{2}\right) = 0$

$2(0) + 0 = 0$ $2\left(\dfrac{9}{4}\right) - \dfrac{9}{2} = 0$

$0 + 0 = 0$ $\dfrac{9}{2} - \dfrac{9}{2} = 0$

$0 = 0$ $0 = 0$

Therefore, $x = 0$ and $x = -\dfrac{3}{2}$ are the correct solutions.

b. $4x^2 = 8x$ Subtract $8x$ from both sides of the equation.

$4x^2 - 8x = 0$ Remove the common factor of $4x$.

$4x(x - 2) = 0$ Set each factor equal to 0.

$4x = 0$, or $x - 2 = 0$ Solve each equation.

$x = 0 \qquad x = 2$ Therefore, the solutions are 0 and 2.

CHECK:

$x = 0 \qquad x = 2$

Substitute 0 for x.   Substitute 2 for x.

$4x^2 = 8x$

$4(0)^2 - 8(0) = 0$ $4(2)^2 = 8(2)$

$\qquad 4(0) - 0 = 0$ $4(4) = 16$

$\qquad 0 - 0 = 0$ $16 = 16$

$\qquad 0 = 0$

Therefore, $x = 0$ and $x = 2$ are the correct solutions.

Practice Exercises

Solve the following equations:

1. $3x^2 + 9x = 0$

2. $5x^2 = 2x$

Solving quadratic equations that can be put in the form $x^2 = c$

Incomplete quadratic equations with $b = 0$ have the form $ax^2 + c = 0$. We also solved some special cases of this type of equation in Section 5.8 by factoring. For example, $x^2 - 9 = 0$ can be solved by factoring. However, $x^2 - 3 = 0$ cannot be solved by factoring. The technique we will use in this section to solve equations of the form $ax^2 + c = 0$ is known as **extraction of roots** and was used in Section 10.7 on the Pythagorean theorem.

We begin with the equation $x^2 - 9 = 0$. This equation can be solved by factoring as follows:

$$x^2 - 9 = 0$$
$$(x + 3)(x - 3) = 0$$
$$x + 3 = 0, x - 3 = 0$$
$$x = -3, \quad x = 3$$

Therefore, the solutions are -3 and 3.

However, $x^2 - 9 = 0$ can be written as $x^2 = 9$. This equation is asking the question, "The square of what number is equal to 9?" This is precisely the definition of square root. Thus, we can solve the equation $x^2 = 9$ by extracting the two square roots of 9. Since the two square roots of 9 are 3 and -3, the solutions of $x^2 = 9$ are 3 and -3. This procedure is summarized as follows:

Solving Quadratic Equations by Extraction of Roots

To solve quadratic equations of the form $ax^2 + c = 0$:

1. Solve the equation for x^2. This results in an equation of the form $x^2 = k$ where k is a constant.

2. The solutions of $x^2 = k$ are both the positive and negative square roots of k written as $x = \pm\sqrt{k}$ (read "x equals the positive and negative square roots of k"). That is, $x = \sqrt{k}$, or $x = -\sqrt{k}$.

3. Write the answers in simplified form.

Note: If k is a negative number, then $\sqrt{k}$ is not a real number. Therefore, the equation has no real-number solutions. As mentioned in Section 10.1, the square roots of negative numbers result in **imaginary numbers** and are usually discussed in intermediate algebra. Consequently, when we encounter such situations in this text, we will write, "There are no real number solutions," and we will indicate the solution set as $\varnothing$.

Example 2

Solve the following equations:

a. $x^2 = 16$ Take the positive and negative square roots of 16.

$x = \pm 4$ Therefore, the solutions are 4 and -4.

CHECK: $x = 4$ $x = -4$

Substitute 4 for x.   Substitute -4 for x.

$x^2 = 16$ $x^2 = 16$

$4^2 = 16$ $(-4)^2 = 16$

$16 = 16$ $16 = 16$

Therefore, ± 4 are the correct solutions.

b. $x^2 - 27 = 0$ Add 27 to both sides of the equation.

$x^2 = 27$ Take the positive and negative square roots of 27.

$x = \pm\sqrt{27}$ Rewrite 27 as $9 \cdot 3$.

$x = \pm\sqrt{9 \cdot 3}$ Apply the product rule for radicals.

$x = \pm\sqrt{9} \cdot \sqrt{3}$ $\sqrt{9} = 3$.

$x = \pm 3\sqrt{3}$ Therefore, the solutions are $3\sqrt{3}$ and $-3\sqrt{3}$.

CHECK: $x = 3\sqrt{3}$ $x = -3\sqrt{3}$

Substitute $3\sqrt{3}$ for x.   Substitute $-3\sqrt{3}$ for x.

$x^2 - 27 = 0$ $x^2 - 27 = 0$

$(3\sqrt{3})^2 - 27 = 0$ $(-3\sqrt{3})^2 - 27 = 0$

$3^2(\sqrt{3})^2 - 27 = 0$ $(-3)^2(\sqrt{3})^2 - 27 = 0$

$9 \cdot 3 - 27 = 0$ $9 \cdot 3 - 27 = 0$

$27 - 27 = 0$ $27 - 27 = 0$

$0 = 0$ $0 = 0$

Therefore, $\pm 3\sqrt{3}$ are the correct solutions.

c. $2x^2 - 6 = 0$ Add 6 to both sides of the equation.

$2x^2 = 6$ Divide both sides of the equation by 2.

$x^2 = 3$ Take the positive and negative square roots of 3.

$x = \pm\sqrt{3}$ Therefore, the solutions are $\sqrt{3}$ and $-\sqrt{3}$.

CHECK:

$x = \sqrt{3}$ $x = -\sqrt{3}$

Substitute $\sqrt{3}$ for x.   Substitute $-\sqrt{3}$ for x.

$2x^2 - 6 = 0$ $2x^2 - 6 = 0$

$2(\sqrt{3})^2 - 6 = 0$ $2(-\sqrt{3})^2 - 6 = 0$

$2(3) - 6 = 0$ $2(3) - 6 = 0$

$6 - 6 = 0$ $6 - 6 = 0$

$0 = 0$ $0 = 0$

Therefore, $\pm\sqrt{3}$ are the correct solutions.

d. $3x^2 - 7 = 0$ Add 7 to both sides of the equation.

$3x^2 = 7$ Divide both sides by 3.

$x^2 = \dfrac{7}{3}$ Take the positive and negative square roots of $\dfrac{7}{3}$.

$x = \pm\sqrt{\dfrac{7}{3}}$ Apply the quotient rule for radicals.

$x = \pm\dfrac{\sqrt{7}}{\sqrt{3}}$ Rationalize the denominator.

$x = \pm\dfrac{\sqrt{7}}{\sqrt{3}} \cdot \dfrac{\sqrt{3}}{\sqrt{3}}$ Apply the product rule for radicals.

$x = \pm\dfrac{\sqrt{21}}{\sqrt{9}}$ $\sqrt{9} = 3$.

$x = \pm\dfrac{\sqrt{21}}{3}$ Therefore, the solutions are $\dfrac{\sqrt{21}}{3}$ and $-\dfrac{\sqrt{21}}{3}$.

CHECK:
The check is left as an exercise for the student.

e. $x^2 + 4 = 0$ Subtract 4 from both sides of the equation.

$x^2 = -4$ Take the positive and negative square roots of -4.

$x = \pm\sqrt{-4}$ Since $\sqrt{-4}$ does not exist as a real number, this equation has no real number solutions. So the solution set is $\varnothing$.

> **Be Careful** Do not confuse solving equations like $x^2 = 9$ with finding $\sqrt{9}$. Remember, the symbol $\sqrt{}$ means find the positive square root only. Hence, $\sqrt{9} = 3$ only. There is no radical sign in $x^2 = 9$, so we are not limited to just the positive square root. We are asked to find the number(s) whose square is 9. Consequently, we find both the positive and negative square roots of 9.

Practice Exercises

Solve the following equations:

3. $x^2 = 49$

4. $x^2 - 8 = 0$

5. $4x^2 - 24 = 0$

6. $5x^2 - 2 = 0$

7. $x^2 + 16 = 0$

If more practice is needed, do the Additional Practice Exercises.

Additional Practice Exercises

Solve the following equations:

a. $x^2 = 81$

b. $x^2 - 18 = 0$

c. $6x^2 - 42 = 0$

d. $7x^2 - 5 = 0$

e. $2x^2 + 18 = 0$

Solving quadratic equations of the form $(ax^2 + b)^2 = c$ The method of extraction of roots can be extended to solve equations like $(2x + 3)^2 = 9$. We illustrate this with some examples.

Example 3

Solve the following equations. If there are no real-number solutions, indicate the solution set as $\varnothing$.

a. $(x + 3)^2 = 16$ Take the positive and negative square roots of 16.

$x + 3 = \pm 4$ Rewrite as two equations.

$x + 3 = 4, x + 3 = -4$ Solve each equation.

$x = 1, \quad x = -7$ Therefore, the solutions are 1 and −7.

CHECK: $x = 1$ $x = -7$

Substitute 1 for x.   Substitute −7 for x.

$(x + 3)^2 = 16$ $(x + 3)^2 = 16$

$(1 + 3)^2 = 16$ $(-7 + 3)^2 = 16$

$4^2 = 16$ $(-4)^2 = 16$

$16 = 16$ $16 = 16$

Therefore, 1 and −7 are the correct solutions.

b. $(x + 3)^2 = 5$ Take the positive and negative square roots of 5.

$x + 3 = \pm\sqrt{5}$ Rewrite as two equations.

$x + 3 = \sqrt{5}, x + 3 = -\sqrt{5}$ Solve each equation.

$x = -3 + \sqrt{5}, x = -3 - \sqrt{5}$ Therefore, the solutions are $-3 + \sqrt{5}$ and $-3 - \sqrt{5}$.

Note: The solutions of Example 3b are usually written as $-3 \pm \sqrt{5}$.

CHECK: $x = -3 + \sqrt{5}$ $x = -3 - \sqrt{5}$

Substitute $-3 + \sqrt{5}$ for x.   Substitute $-3 - \sqrt{5}$ for x.

$(x + 3)^2 = 5$ $(x + 3)^2 = 5$

$(-3 + \sqrt{5} + 3)^2 = 5$ $(-3 - \sqrt{5} + 3)^2 = 5$

$(\sqrt{5})^2 = 5$ $(-\sqrt{5})^2 = 5$

$5 = 5$ $5 = 5$

Therefore, $-3 + \sqrt{5}$ and $-3 - \sqrt{5}$ are the correct solutions.

c. $(2x + 4)^2 = 25$ Take the positive and negative square roots of 25.

$2x + 4 = \pm 5$ Rewrite as two equations.

$2x + 4 = 5, 2x + 4 = -5$ Solve each equation.

$2x = 1, 2x = -9$

$x = \dfrac{1}{2}, \quad x = -\dfrac{9}{2}$ Therefore, the solutions are $\dfrac{1}{2}$ and $-\dfrac{9}{2}$.

CHECK: The check is left as an exercise for the student.

d. $(3x - 1)^2 = 7$

$3x - 1 = \pm\sqrt{7}$

$3x - 1 = \sqrt{7}, 3x - 1 = -\sqrt{7}$

$3x = 1 + \sqrt{7}, 3x = 1 - \sqrt{7}$

$x = \dfrac{1 + \sqrt{7}}{3}, \quad x = \dfrac{1 - \sqrt{7}}{3}$

Take the positive and negative square roots of 7.
Rewrite as two equations.

Solve each equation.

Therefore, the solutions are $\dfrac{1 + \sqrt{7}}{3}$ and $\dfrac{1 - \sqrt{7}}{3}$.

e. $(2a - 3)^2 + 12 = 0$

$(2a - 3)^2 = -12$

$2a - 3 = \pm\sqrt{12}$

Subtract 12 from both sides of the equation.
Take the positive and negative square roots of -12.
Since $\sqrt{-12}$ does not exist as a real number, this equation has no real-number solutions. So the solution set is $\varnothing$.

Note: The solutions to Example 3d are often written as $\dfrac{1 \pm \sqrt{7}}{3}$.

CHECK: $x = \dfrac{1 + \sqrt{7}}{3}$ $x = \dfrac{1 - \sqrt{7}}{3}$

Substitute $\dfrac{1 + \sqrt{7}}{3}$ for x.   Substitute $\dfrac{1 - \sqrt{7}}{3}$ for x.

$(3x - 1)^2 = 7$ $(3x - 1)^2 = 7$

$\left(3\left(\dfrac{1 + \sqrt{7}}{3}\right) - 1\right)^2 = 7$ $\left(3\left(\dfrac{1 - \sqrt{7}}{3}\right) - 1\right)^2 = 7$

$(1 + \sqrt{7} - 1)^2 = 7$ $(1 - \sqrt{7} - 1)^2 = 7$

$(\sqrt{7})^2 = 7$ $(-\sqrt{7})^2 = 7$

$7 = 7$ $7 = 7$

Therefore, $\dfrac{1 \pm \sqrt{7}}{3}$ are the correct solutions.

Practice Exercises

Solve the following equations. If there are no real-number solutions, indicate the solution set as $\varnothing$.

8. $(x + 5)^2 = 36$ **9.** $(x - 4)^2 = 8$ **10.** $(3x - 2)^2 = 4$

11. $(5x - 3)^2 = 10$ **12.** $(3x - 1)^2 + 16 = 0$

If more practice is needed, do the Additional Practice Exercises.

If more practice is needed, do the Additional Practice Exercises.

Additional Practice Exercises

Solve the following equations:

f. $(x - 6)^2 = 64$ **g.** $(x + 4)^2 = 2$ **h.** $(4x + 3)^2 = 9$

i. $(5x - 1)^2 = 11$ **j.** $(x + 5)^2 + 25 = 0$

Quadratic equations are also used in the solution of application problems.

Answers:

Additional Practice Exercises f–j: f. $x = 14, -2$ g. $x = -4 \pm \sqrt{2}$ h. $x = 0, -\dfrac{3}{2}$ i. $x = \dfrac{1 \pm \sqrt{11}}{5}$ j. $\varnothing$

Practice Exercises 8–12: 8. $x = 1, -11$ 9. $x = 4 \pm 2\sqrt{2}$ 10. $x = 0, \dfrac{4}{3}$ 11. $x = \dfrac{3 \pm \sqrt{10}}{5}$ 12. $\varnothing$

Example 4

a. The square of 3 less than a number is 36. Find the number(s).

Solution:
Let x = the number. Then,

$x - 3$ = 3 less than the number. Therefore, the equation is:

$$(x - 3)^2 = 36$$
$$x - 3 = \pm 6$$
$$x - 3 = 6, x - 3 = -6$$
$$x = 9, x = -3$$

Take the positive and negative square roots of 36.
Rewrite as two equations.
Solve each equation. Therefore, the numbers are 9 and −3.

Check:
Three less than 9 is $9 - 3 = 6$ and $6^2 = 36$. Therefore, 9 is a solution. Three less than −3 is $-3 - 3 = -6$ and $(-6)^2 = 36$. Therefore, −3 is also a solution.

b. Find the length of each side of a square whose area is 49 square inches.

Solution:
The formula for the area of a square is $A = s^2$ where s is the length of a side of the square.

$$A = s^2$$
$$49 = s^2$$
$$\pm 7 = s$$
$$7 = s$$

Substitute 49 for A.
Take the positive and negative square roots of 49.
Since the length of a side of a square must be positive, the only solution is 7 inches.

Practice Exercises

13. The square of the sum of a number and 4 is 64. Find the number(s).

14. Find the length of each side of a square whose area is 169 square centimeters.

Exercise Set 11.1

Solve each of the following equations. If there are no real-number solutions, indicate the solution set as $\varnothing$.

1. $4x^2 + 8x = 0$

2. $2x^2 + 10x = 0$

3. $7a^2 = -21a$

4. $5r^2 = -45r$

5. $9z^2 = 6z$

6. $12t^2 = 30t$

7. $x(x + 2) = 4x$

8. $n(n - 4) = 2n$

9. $r^2 = 25$

10. $y^2 = 100$

11. $a^2 = -36$

12. $p^2 = -64$

13. $x^2 - 121 = 0$

14. $x^2 - 144 = 0$

15. $x^2 - 7 = 0$

16. $x^2 - 11 = 0$

17. $x^2 + 5 = 0$

18. $t^2 + 7 = 0$

19. $x^2 - 48 = 0$

20. $x^2 - 32 = 0$

21. $w^2 - 28 = 0$

22. $m^2 - 45 = 0$

23. $2x^2 - 8 = 0$

24. $3x^2 - 27 = 0$

25. $4x^2 - 20 = 0$

26. $5x^2 - 30 = 0$

27. $2m^2 - 40 = 0$

28. $3r^2 - 54 = 0$

29. $5q^2 - 40 = 0$

30. $7b^2 - 56 = 0$

31. $7x^2 + 28 = 0$

32. $3a^2 + 30 = 0$

33. $2x^2 - 7 = 0$

34. $3x^2 - 5 = 0$

35. $5a^2 - 11 = 0$

36. $7r^2 - 5 = 0$

37. $3z^2 - 4 = 0$

38. $6a^2 - 25 = 0$

39. $(x + 4)^2 = 25$

40. $(a - 2)^2 = 100$

41. $(x + 5)^2 = 44$

42. $(n - 7)^2 = 54$

43. $(3c + 4)^2 = 49$

44. $(2a - 5)^2 = 64$

45. $(2r - 3)^2 = 121$

46. $(3s - 6)^2 = 81$

47. $(x - 3)^2 = 7$

48. $(x + 4)^2 = 3$

49. $(m + 5)^2 = 11$

50. $(n - 9)^2 = 13$

51. $(2a + 4)^2 = 2$

52. $(3s - 2)^2 = 5$

53. $(s - 2)^2 = 98$

54. $(n - 5)^2 = 75$

55. $(3x + 2)^2 = 52$

56. $(2a - 6)^2 = 72$

Answers:

Solve the following application problems. Remember from Section 5.8 that if the solution of an application problem requires a quadratic equation, not all solutions of the equation will always satisfy the conditions of the problem. Always check the solution(s) against the wording of the problem.

57. The square of 3 more than a number is 25. Find the number(s).

58. The square of 2 less than three times a number is 16. Find the number(s).

Exercises 59 through 62 require the use of the formula for the area of a square. The formula for the area of a square is $A = s^2$ where s represents the length of each side.

59. Find the length of each side of a square if the area is 64 square inches.

60. Find the length of each side of a square if the area is 121 square meters.

61. If the length of each side of a square is doubled and then increased by 4, the area would be 100 square centimeters. Find the length of each side of the square.

62. If the length of each side of a square is tripled and then decreased by 3, the area would be 36 square feet. Find the length of each side of the square.

Challenge Exercises: (63–66)

Solve the following equations. If there are no real-number solutions, indicate the solution set as $\varnothing$.

63. $3(x + 5)^2 = 7$ **64.** $2(a - 3)^2 = 5$ **65.** $5(2x + 3)^2 - 11 = 0$ **66.** $7(4x - 1)^2 - 15 = 0$

Writing Exercises:

67. Compare and contrast the meanings of $\sqrt{49}$ and $x^2 = 49$. How are they similar and how are they different?

68. Explain in detail why the equation $(x + 2)^2 = -9$ has no real-number solution.

Section 11.2   # Solving Quadratic Equations by Completing the Square

OBJECTIVES

When you complete this section, you will be able to:

a. Determine the number to be added to a polynomial of the form $x^2 + bx$ in order to form a perfect-square trinomial.

b. Solve quadratic equations by completing the square.

Introduction   Recall from Section 5.6 that a perfect-square trinomial is a trinomial that is the square of a binomial. For example, $(x + 3)^2 = (x)^2 + 2(x)(3) + 3^2 = x^2 + 6x + 9$. Since $x^2 + 6x + 9 = (x + 3)^2$, it is a perfect-square trinomial. By examining $x^2 + 6x + 9 = (x + 3)^2$ and the procedure for squaring a binomial, we make the following observations about a binomial whose square gives a perfect-square trinomial.

1. The first term of the binomial is the square root of the first term of the trinomial.

2. The second term of the binomial is the square root of the last term of the trinomial.

3. The sign of the binomial is the same as the sign of the middle term of the trinomial.

Using the above example, we have:

$$x^2 + 6x + 9$$
$$\left(\sqrt{x^2} + \sqrt{9}\right)^2$$

That is, $x^2 + 6x + 9 = (\sqrt{x^2} + \sqrt{9})^2 = (x + 3)^2$.

Completing the square

Suppose we know the first two terms of a trinomial whose squared term has a coefficient of 1, and we want to find the constant term necessary for the trinomial to be a perfect-square trinomial. For example, what constant term is necessary for $x^2 + 6x +$ _____ to be a perfect-square trinomial, and what is the binomial whose square gives this perfect-square trinomial?

If we represent the last term of the preceding trinomial with a^2, we have $x^2 + 6x + a^2$. We need to find a. The first term of the binomial whose square gives $x^2 + 6x + a^2$ is $\sqrt{x^2} = x$. The second term of the binomial is $\sqrt{a^2} = a$. Since the sign of the second term of the trinomial is $+$, we have:

$$x^2 + 6x + a^2 = (x + a)^2$$

Expand $(x + a)^2$ and we have:

$$x^2 + 6x + a^2 = x^2 + 2ax + a^2$$

Since two polynomials are equal if and only if their corresponding terms are equal, we conclude that:

$$6 = 2a$$
$$\frac{6}{2} = a$$
$$3 = a$$

Since the last term of the perfect-square trinomial is a^2 and $a = 3$, the number needed in order for $x^2 + 6x +$ _____ to be a perfect-square trinomial is $3^2 = 9$. Hence, $x^2 + 6x + 9$ is a perfect-square trinomial and $x^2 + 6x + 9 = (x + 3)^2$.

The procedure for finding the number that will make a polynomial of the form $x^2 + bx$ a perfect-square trinomial is called **completing the square**. We do not want to go through all these steps each time we need to complete the square. The key to the procedure lies in the steps that we repeat as follows:

$$6 = 2a$$
$$\frac{6}{2} = a$$
$$3 = a$$

We then squared 3 to get the last term of the perfect-square trinomial. Since 6 is the coefficient of the first degree term of the trinomial, we found the last term of the perfect-square trinomial by taking half the coefficient of the first-degree term and squaring it. We generalize this discussion to the following procedure for completing the square:

Completing the Square

To find the number needed to make a polynomial of the form $x^2 + bx$ a perfect-square trinomial, take $\frac{1}{2}$ of the coefficient of the first degree term and square it. Since the square of a negative number is positive, we can ignore the sign of the first-degree term.

Example 1

Find the number needed to make each of the following a perfect-square trinomial. Then write each trinomial as the square of a binomial.

a. $x^2 + 8x +$ _____ $=$ Take $\frac{1}{2}$ of 8 and square it. $\frac{1}{2}(8) = 4$ and $4^2 = 16$.
$x^2 + 8x + 16 =$ Therefore, the number needed is 16.
$(x + 4)^2 =$ Write as the square of a binomial.

b. $a^2 - 10a +$ _____ $=$ Take $\frac{1}{2}$ of 10 and square it. $\frac{1}{2}(10) = 5$ and $5^2 = 25$.
$a^2 - 10a + 25 =$ Therefore, the number needed is 25.
$(a - 5)^2$ Write as the square of a binomial.

c. $b^2 + 3b +$ _____ $=$ Take $\frac{1}{2}$ of 3 and square it. $\frac{1}{2}(3) = \frac{3}{2}$ and $\left(\frac{3}{2}\right)^2 = \frac{9}{4}$.
$b^2 + 3b + \frac{9}{4} =$ Therefore, the number needed is $\frac{9}{4}$.
$\left(b + \frac{3}{2}\right)^2$ Write as the square of a binomial.

d. $r^2 - \frac{7}{3}r +$ _____ $=$ Take $\frac{1}{2}$ of $\frac{7}{3}$ and square it. $\frac{1}{2}\left(\frac{7}{3}\right) = \frac{7}{6}$ and $\left(\frac{7}{6}\right)^2 = \frac{49}{36}$.
$r^2 - \frac{7}{3}r + \frac{49}{36} =$ Therefore, the number needed is $\frac{49}{36}$.
$\left(r - \frac{7}{6}\right)^2$ Write as the square of a binomial.

Practice Exercises

Find the number needed to make each of the following a perfect-square trinomial. Then write each trinomial as the square of a binomial.

1. $x^2 + 2x + $ _____

2. $n^2 - 14n + $ _____

3. $m^2 + 7m + $ _____

4. $y^2 - \dfrac{5}{2}y + $ _____

If more practice is needed, do the Additional Practice Exercises.

Additional Practice Exercises

Find the number needed to make each of the following a perfect-square trinomial. Then write each trinomial as the square of a binomial.

a. $x^2 + 12x + $ _____

b. $r^2 - 4r + $ _____

c. $c^2 + 9c + $ _____

d. $r^2 - \dfrac{3}{5}r + $ _____

Solving quadratic equations by completing the square   By combining the techniques of completing the square and the extracting of roots, we can solve any quadratic equation using the procedure outlined in the following box:

Solving Quadratic Equations by Completing the Square

To solve a quadratic equation by completing the square:

1. If the coefficient of the x^2 term is not 1, divide by the coefficient.
2. Write the equation in the form $x^2 + bx = c$.
3. Add the constant needed to make $x^2 + bx$ a perfect square to both sides of the equation.
4. Write the perfect-square trinomial as the square of a binomial.
5. Solve by extraction of roots.
6. Check all solutions in the original equation.

Although the equations in Example 2 could be solved by factoring, we will solve them by completing the square for illustrative purposes.

Example 2

Solve the following equations by completing the square:

a. $x^2 - 2x - 8 = 0$ Add 8 to both sides of the equation.

$x^2 - 2x = 8$ Find the number that must be added to $x^2 - 2x$ to make it a perfect-square trinomial, and add it to both sides of the equation.

$x^2 - 2x + $ _____ $= 8 + $ _____   Take $\dfrac{1}{2}$ of 2, and square it. $\dfrac{1}{2}(2) = 1$ and $1^2 = 1$. Therefore, add

$x^2 - 2x + 1 = 8 + 1$ 1 to both sides of the equation.

$(x - 1)^2 = 9$ Write $x^2 - 2x + 1$ as the square of a binomial.
    Solve using extraction of roots.

$x - 1 = \pm 3$ Write as two equations.

$x - 1 = 3, x - 1 = -3$ Solve each equation.

$x = 4, \quad x = -2$ Therefore, the solutions are 4 and -2.

Answers:

CHECK: $x = 4$ $\quad$ $x = -2$

Substitute 4 for x. $\quad$ Substitute -2 for x.

$x^2 - 2x - 8 = 0$ $\qquad$ $x^2 - 2x - 8 = 0$

$4^2 - 2(4) - 8 = 0$ $\qquad$ $(-2)^2 - 2(-2) - 8 = 0$

$16 - 8 - 8 = 0$ $\qquad$ $4 + 4 - 8 = 0$

$0 = 0$ $\qquad\qquad$ $0 = 0$

Therefore, 4 and -2 are the correct solutions.

b. $a^2 + 5a + 4 = 0$ $\qquad$ Subtract 4 from both sides of the equation.

$a^2 + 5a = -4$ $\qquad$ Find the number that must be added to $a^2 + 5a$ to make it a perfect-square trinomial, and add it to both sides of the equation.

$a^2 + 5a + \underline{\hspace{1cm}} = -4 + \underline{\hspace{1cm}}$ $\qquad$ Take $\dfrac{1}{2}$ of 5 and square it. $\dfrac{1}{2}(5) = \dfrac{5}{2}$ and $\left(\dfrac{5}{2}\right)^2 = \dfrac{25}{4}$. Therefore, add $\dfrac{25}{4}$ to both sides of the equation.

$a^2 + 5a + \dfrac{25}{4} = -4 + \dfrac{25}{4}$ $\qquad$ Write the left side of the equation as the square of a binomial and write the right side with the LCD of 4.

$\left(a + \dfrac{5}{2}\right)^2 = -\dfrac{16}{4} + \dfrac{25}{4}$ $\qquad$ Simplify the right side.

$\left(a + \dfrac{5}{2}\right)^2 = \dfrac{9}{4}$ $\qquad$ Solve by using extraction of roots.

$a + \dfrac{5}{2} = \pm\dfrac{3}{2}$ $\qquad$ Write as two equations.

$a + \dfrac{5}{2} = \dfrac{3}{2}, a + \dfrac{5}{2} = -\dfrac{3}{2}$ $\qquad$ Solve each equation.

$a = \dfrac{3}{2} - \dfrac{5}{2}, a = -\dfrac{3}{2} - \dfrac{5}{2}$

$a = -\dfrac{2}{2} = -1, a = -\dfrac{8}{2} = -4$ $\qquad$ Therefore, the solutions are -1 and -4.

CHECK: To check, substitute -1 and -4 in the original equation. This is left as an exercise for the student.

Practice Exercises

Solve the following equations by completing the square:

5. $x^2 - 8x + 12 = 0$ $\qquad\qquad\qquad$ **6.** $r^2 + 3r - 18 = 0$

If more practice is needed, do the Additional Practice Exercises.

Additional Practice Exercises

Solve the following equations by completing the square:

e. $x^2 - 6x + 8 = 0$ $\qquad\qquad\qquad$ **f.** $b^2 + 3b + 2 = 0$

As previously indicated, the equations in Example 2 could have been solved by factoring. The real advantage to solving equations by completing the square is in solving equations that cannot be solved by factoring.

Answers:

Practice Exercises 5–6: 5. $x = 2, 6$ 6. $r = 3, -6$ $\quad$ **Additional Practice Exercises e–f:** e. $x = 2, 4$ f. $b = -2, -1$

Example 3

Solve the following equations by completing the square:

a. $x^2 - 4x + 2 = 0$ Subtract 2 from both sides of the equation.

$x^2 - 4x = -2$ Find the number that must be added to $x^2 - 4x$ to make it a perfect-square trinomial and add it to both sides of the equation.

$x^2 - 4x + \underline{\hphantom{xx}} = -2 + \underline{\hphantom{xx}}$ Take $\frac{1}{2}$ of 4 and square it. $\frac{1}{2}(4) = 2$ and $2^2 = 4$. Therefore, add 4 to both sides.

$x^2 - 4x + 4 = -2 + 4$ Write the left side as the square of a binomial.

$(x - 2)^2 = 2$ Solve using extraction of roots.

$x - 2 = \pm\sqrt{2}$ Write as two equations.

$x - 2 = \sqrt{2}, \ x - 2 = -\sqrt{2}$ Solve each equation.

$x = 2 + \sqrt{2}, \ x = 2 - \sqrt{2}$ Therefore, the solutions are $2 \pm \sqrt{2}$.

CHECK: $x = 2 + \sqrt{2}$ $x = 2 - \sqrt{2}$

Substitute $2 + \sqrt{2}$ for x.   Substitute $2 - \sqrt{2}$ for x.

$x^2 - 4x + 2 = 0$ $x^2 - 4x + 2 = 0$

$(2 + \sqrt{2})^2 - 4(2 + \sqrt{2}) + 2 = 0$ $(2 - \sqrt{2})^2 - 4(2 - \sqrt{2}) + 2 = 0$

$4 + 4\sqrt{2} + 2 - 8 - 4\sqrt{2} + 2 = 0$ $4 - 4\sqrt{2} + 2 - 8 + 4\sqrt{2} + 2 = 0$

$0 = 0$ $0 = 0$

Therefore, the correct solutions are $2 \pm \sqrt{2}$.

Note: Radical expressions like $a + \sqrt{b}$ and $a - \sqrt{b}$ are called **conjugate pairs.** Irrational solutions to quadratic equations always occur as conjugate pairs.

b. $a^2 = -7a - 1$ We need the equation to be in $a^2 + ba = c$ form, so add $7a$ to both sides of the equation. Find the number that must be added to $a^2 + 7a$ to make it a perfect-square trinomial, and add it to both sides of the equation.

$a^2 + 7a = -1$

$a^2 + 7a + \underline{\hphantom{xx}} = -1 + \underline{\hphantom{xx}}$ Take $\frac{1}{2}$ of 7 and square it. $\frac{1}{2}(7) = \frac{7}{2}$ and $\left(\frac{7}{2}\right)^2 = \frac{49}{4}$. Therefore, add $\frac{49}{4}$ to both sides.

$a^2 + 7a + \dfrac{49}{4} = -1 + \dfrac{49}{4}$ Write the left side as the square of a binomial, and write the right side with the LCD of 4.

$\left(a + \dfrac{7}{2}\right)^2 = -\dfrac{4}{4} + \dfrac{49}{4}$ Simplify the right side.

$\left(a + \dfrac{7}{2}\right)^2 = \dfrac{45}{4}$ Solve using extraction of roots.

$a + \dfrac{7}{2} = \pm\sqrt{\dfrac{45}{4}}$ Apply the quotient rule, and write 45 as $9 \cdot 5$.

$a + \dfrac{7}{2} = \pm\dfrac{\sqrt{9 \cdot 5}}{\sqrt{4}}$ Apply the product rule to $\sqrt{9 \cdot 5}$ and $\sqrt{4} = 2$.

$a + \dfrac{7}{2} = \pm\dfrac{\sqrt{9} \cdot \sqrt{5}}{2}$ $\sqrt{9} = 3$.

$a + \dfrac{7}{2} = \pm\dfrac{3\sqrt{5}}{2}$ Write as two equations.

$a + \dfrac{7}{2} = \dfrac{3\sqrt{5}}{2}, \ a + \dfrac{7}{2} = -\dfrac{3\sqrt{5}}{2}$ Solve each equation.

$a = -\dfrac{7}{2} + \dfrac{3\sqrt{5}}{2}, \ a = -\dfrac{7}{2} - \dfrac{3\sqrt{5}}{2}$ Add the fractions.

$a = \dfrac{-7 + 3\sqrt{5}}{2}, \ a = \dfrac{-7 - 3\sqrt{5}}{2}$ Therefore, the solutions are $x = \dfrac{-7 \pm 3\sqrt{5}}{2}$.

CHECK: The check is complicated and is left as a challenge exercise for the student.

c. $x^2 + 8x + 20 = 0$ Subtract 20 from both sides of the equation.

$x^2 + 8x +$ _____ $= -20 +$ _____   Take $\frac{1}{2}$ of 8 and square it. $\frac{1}{2}(8) = 4$ and $4^2 = 16$. Therefore, add 16 to both sides.

$x^2 + 8x + 16 = -20 + 16$ Write the left side as the square of a binomial. Simplify the right side.

$(x + 4)^2 = -4$ Solve using extraction of roots.

$x + 4 = \pm\sqrt{-4}$ Since $\sqrt{-4}$ is not a real number, this equation has no real-number solutions. Hence the solution set is $\varnothing$.

Practice Exercises

Solve the following equations by completing the square. If there is no real solution, indicate the solution set as $\varnothing$.

7. $x^2 - 4x - 6 = 0$ **8.** $b^2 - 5 = -5b$ **9.** $x^2 - 10x + 34 = 0$

If more practice is needed, do the Additional Practice Exercises.

Additional Practice Exercises

Solve the following equations by completing the square:

g. $r^2 = -6r + 4$ **h.** $x^2 - 4 = -5x$ **i.** $a^2 - 6a + 13 = 0$

The procedure for completing the square works only if the coefficient of the second-degree term is 1. If the coefficient of the second-degree term is not 1, we must divide both sides of the equation by that coefficient in order to make it 1. We illustrate with some examples.

Example 4

Solve the following equations by completing the square:

a. $2x^2 - 12 = 5x$ Rewrite in the form $ax^2 + bx = c$.

$2x^2 - 5x = 12$ Divide both sides of the equation by 2 to get x^2.

$\dfrac{2x^2 - 5x}{2} = \dfrac{12}{2}$ Simplify both sides.

$x^2 - \dfrac{5}{2}x = 6$ Find the number that must be added to $x^2 - \dfrac{5}{2}x$ to make it a perfect-square trinomial, and add it to both sides of the equation.

$x^2 - \dfrac{5}{2}x +$ _____ $= 6 +$ _____   Take $\frac{1}{2}$ of $\frac{5}{2}$ and square it. $\frac{1}{2} \cdot \frac{5}{2} = \frac{5}{4}$ and $\left(\frac{5}{4}\right)^2 = \frac{25}{16}$. Add $\frac{25}{16}$ to both sides.

$x^2 - \dfrac{5}{2}x + \dfrac{25}{16} = 6 + \dfrac{25}{16}$ Write the left side as the square of a binomial, and write the right side with the LCD of 16.

$\left(x - \dfrac{5}{4}\right)^2 = \dfrac{96}{16} + \dfrac{25}{16}$ Simplify the right side.

$\left(x - \dfrac{5}{4}\right)^2 = \dfrac{121}{16}$ Solve using extraction of roots.

$x - \dfrac{5}{4} = \pm\dfrac{11}{4}$ Write as two equations.

$x - \dfrac{5}{4} = \dfrac{11}{4}, x - \dfrac{5}{4} = -\dfrac{11}{4}$ Solve each equation.

$x = \dfrac{11}{4} + \dfrac{5}{4}, x = -\dfrac{11}{4} + \dfrac{5}{4}$ Add the fractions.

$x = \dfrac{16}{4} = 4, x = -\dfrac{6}{4} = -\dfrac{3}{2}$ Therefore, the solutions are 4 and $-\dfrac{3}{2}$.

CHECK: The check is left as an exercise for the student.

Answers:

Practice Exercises 7–9: 7. $x = 2 \pm \sqrt{10}$ 8. $b = \dfrac{-5 \pm 3\sqrt{5}}{2}$ 9. $\varnothing$ **Additional Practice Exercises g–i:** g. $-3 \pm \sqrt{13}$ h. $x = \dfrac{-5 \pm \sqrt{41}}{2}$ i. $\varnothing$

b. $3x^2 + 6x + 1 = 0$ Subtract 1 from both sides of the equation.

$3x^2 + 6x = -1$ Divide both sides by 3 to get x^2.

$\dfrac{3x^2 + 6x}{3} = \dfrac{-1}{3}$ Simplify the left side.

$x^2 + 2x = -\dfrac{1}{3}$ Find the number that must be added to $x^2 + 2x$ to make it a perfect-square trinomial and add it to both sides.

$x^2 + 2x + \underline{\hspace{1cm}} = -\dfrac{1}{3} + \underline{\hspace{1cm}}$ Take $\dfrac{1}{2}$ of 2 and square it. $\dfrac{1}{2}(2) = 1$ and $1^2 = 1$. Add 1 to both sides.

$x^2 + 2x + 1 = -\dfrac{1}{3} + 1$ Write the left side as a binomial squared, and write the right side with the LCD of 3.

$(x + 1)^2 = -\dfrac{1}{3} + \dfrac{3}{3}$ Simplify the right side.

$(x + 1)^2 = \dfrac{2}{3}$ Solve using extraction of roots.

$x + 1 = \pm\sqrt{\dfrac{2}{3}}$ Apply the quotient rule for radicals.

$x + 1 = \pm\dfrac{\sqrt{2}}{\sqrt{3}}$ Rationalize the denominator.

$x + 1 = \pm\dfrac{\sqrt{2}}{\sqrt{3}} \cdot \dfrac{\sqrt{3}}{\sqrt{3}}$ Simplify the right side.

$x + 1 = \pm\dfrac{\sqrt{6}}{3}$ Rewrite as two equations.

$x + 1 = \dfrac{\sqrt{6}}{3}, x + 1 = -\dfrac{\sqrt{6}}{3}$ Solve each equation.

$x = -1 + \dfrac{\sqrt{6}}{3}, x = -1 - \dfrac{\sqrt{6}}{3}$ Write the right side with the LCD of 3.

$x = -\dfrac{3}{3} + \dfrac{\sqrt{6}}{3}, x = -\dfrac{3}{3} - \dfrac{\sqrt{6}}{3}$ Add the fractions.

$x = \dfrac{-3 + \sqrt{6}}{3}, x = \dfrac{-3 - \sqrt{6}}{3}$ Therefore, the solutions are $\dfrac{-3 \pm \sqrt{6}}{3}$.

CHECK:
The check is complicated and is left as a challenge exercise for the student.

Practice Exercises

Solve the following equations by completing the square. If there is no real solution, indicate the solution set as $\varnothing$.

10. $3x^2 + 2 = 5x$ **11.** $2x^2 + 6x = -3$

If more practice is needed, do the Additional Practice Exercises.

Additional Practice Exercises

Solve the following equations by completing the square. If there is no real solution, indicate the solution set as $\varnothing$.

j. $-3x + 5 = 2x^2$ **k.** $3x^2 + 12x = 5$

As you have no doubt noticed, solving quadratic equations by completing the square can be very tedious and sometimes difficult. In the next section, we will use completing the square to develop another method that is usually easier.

Exercise Set 11.2

Find the number needed to make each of the following a perfect-square trinomial. Then, write each trinomial as the square of a binomial.

1. $a^2 + 6a +$ _____

2. $r^2 + 14r +$ _____

3. $x^2 - 18x +$ _____

4. $a^2 - 20a +$ _____

5. $n^2 + 5n +$ _____

6. $m^2 + 11m +$ _____

7. $b^2 - 7b +$ _____

8. $r^2 - 9r +$ _____

9. $x^2 + \frac{5}{2}x +$ _____

10. $c^2 + \frac{5}{3}c +$ _____

11. $a^2 - \frac{3}{7}a +$ _____

12. $n^2 - \frac{1}{2}n +$ _____

Solve the following equations by completing the square. Most of Exercises 13 through 32 have rational solutions or no solutions. If there is no real solution, indicate the solution set as $\varnothing$.

13. $r^2 - 8r + 15 = 0$

14. $x^2 - 6x + 5 = 0$

15. $x^2 = -8x - 12$

16. $x^2 = 6x - 8$

17. $z^2 - 4z = 12$

18. $b^2 - 2b = 24$

19. $x^2 - 10 = -3x$

20. $a^2 + 6 = -9a$

21. $n^2 + 9n = -20$

22. $n^2 - 5n = 14$

23. $x^2 - \frac{7}{2}x + 3 = 0$

24. $x^2 - \frac{13}{2}x + 10 = 0$

25. $x^2 - \frac{7}{3}x = 2$

26. $y^2 - \frac{24}{5}y = 1$

27. $2x^2 = 3x + 20$

28. $2y^2 = 15y - 18$

29. $3x^2 + 6 = 11x$

30. $3x^2 - 24 = 14x$

31. $a^2 - 4a + 6 = 0$

32. $n^2 + 6n + 10 = 0$

Solve the following by completing the square. Most of Exercises 33 through 52 have irrational solutions or no real solutions. If there is no real solution, indicate the solution set as $\varnothing$.

33. $x^2 + 4x - 1 = 0$

34. $a^2 + 6a - 4 = 0$

35. $r^2 - 8r = -5$

36. $z^2 - 10z = -6$

37. $a^2 - 2 = 14a$

38. $t^2 - 4 = 3t$

39. $x^2 + 8x + 3 = 0$

40. $y^2 + 6y + 2 = 0$

41. $x^2 - 10x = 5$

42. $z^2 - 2z = 6$

43. $t^2 - 3 = -5t$

44. $z^2 - 1 = -7z$

45. $2x^2 = 4x + 3$

46. $2x^2 = -6x - 5$

47. $2a^2 - 5a + 4 = 0$

48. $2x^2 - 7x + 6 = 0$

49. $3x^2 + 6x = 1$

50. $3s^2 + 8s = 3$

51. $2x^2 + 3x + 5 = 0$

52. $3x^2 - 5x + 8 = 0$

Challenge Exercises: (53–56)

Solve the following by completing the square:

53. $x^2 + \frac{1}{3}x - \frac{2}{9} = 0$

54. $x^2 + \frac{5}{2}x + \frac{21}{16} = 0$

55. Check Example 3b.

56. Check Example 4b.

Writing Exercises:

57. Why do some quadratic equations have no real-number solutions?

58. What is the advantage to solving quadratic equations by completing the square?

Critical-Thinking Exercises:

59. Write a quadratic equation that has no real solutions, and explain how you arrived at that equation.

Solving Quadratic Equations by the Quadratic Formula

O B J E C T I V E S

When you complete this section, you will be able to:

Solve quadratic equations by using the quadratic formula.

Introduction

Thus far, we have learned two methods of solving quadratic equations: factoring and completing the square. The disadvantage of factoring is that not all quadratic equations can be solved by factoring. The disadvantage of completing the square is that it is complicated, and there are many places where errors can be made. In this section, we will learn a third method of solving quadratic equations that also allows us to solve *any* quadratic equation. This method involves using the **quadratic formula** that we will now develop.

Developing the quadratic formula

Let us begin with the general form of a quadratic equation and solve for x by completing the square. This will result in a formula that gives the solutions of the equation in terms of its coefficients.

$$ax^2 + bx + c = 0$$ Subtract c from both sides of the equation

$$ax^2 + bx = -c$$ Divide both sides by a to get x^2.

$$\frac{ax^2 + bx}{a} = -\frac{c}{a}$$ Simplify the left side.

$$x^2 + \frac{b}{a}x = -\frac{c}{a}$$ Find the expression needed to make the left side a perfect-square trinomial, and add it to both sides.

$$x^2 + \frac{b}{a}x + \underline{} = \underline{} + -\frac{c}{a}$$ $\frac{1}{2} \cdot \frac{b}{a} = \frac{b}{2a}$ and $\left(\frac{b}{2a}\right)^2 = \frac{b^2}{4a^2}$. Therefore, add $\frac{b^2}{4a^2}$ to both sides.

$$x^2 + \frac{b}{a}x + \frac{b^2}{4a^2} = \frac{b^2}{4a^2} - \frac{c}{a}$$ Write the left side as the square of a binomial, and write the right side with the LCD of $4a^2$.

$$\left(x + \frac{b}{2a}\right)^2 = \frac{b^2}{4a^2} - \frac{c}{a} \cdot \frac{4a}{4a}$$ Simplify the right side.

$$\left(x + \frac{b}{2a}\right)^2 = \frac{b^2}{4a^2} - \frac{4ac}{4a^2}$$ Add the fractions on the right side.

$$\left(x + \frac{b}{2a}\right)^2 = \frac{b^2 - 4ac}{4a^2}$$ Solve using extraction of roots.

$$x + \frac{b}{2a} = \pm\sqrt{\frac{b^2 - 4ac}{4a^2}}$$ Apply the quotient rule for radicals.

$$x + \frac{b}{2a} = \pm\frac{\sqrt{b^2 - 4ac}}{\sqrt{4a^2}}$$ $\sqrt{4a^2} = 2a$ if $a > 0$.

$$x + \frac{b}{2a} = \pm\frac{\sqrt{b^2 - 4ac}}{2a}$$ Subtract $\frac{b}{2a}$ from both sides.

$$x = -\frac{b}{2a} \pm \frac{\sqrt{b^2 - 4ac}}{2a}$$ Add the fractions.

$$x = \frac{-b \pm \sqrt{b^2 - 4ac}}{2a}$$ Therefore, the solutions of $ax^2 + bx + c = 0$ are $\frac{-b + \sqrt{b^2 - 4ac}}{2a}$ and $\frac{-b - \sqrt{b^2 - 4ac}}{2a}$.

Note: The letters a, b, and c in the quadratic formula come from the equation written in the form $ax^2 + bx + c = 0$. This means that the solutions of a quadratic equation are determined by its coefficients.

The Quadratic Formula

The solutions of any equation written in the form $ax^2 + bx + c = 0$ are
$$x = \frac{-b \pm \sqrt{b^2 - 4ac}}{2a}.$$

The first task necessary in using the quadratic formula is to find the values of a, b, and c. This is done by rewriting the given equation, if necessary, in the form of $ax^2 + bx + c = 0$ and comparing the coefficients.

Example 1

Find the values of a, b, and c in each of the following:

a. $3x^2 + 2x - 5 = 0$ Compare with $ax^2 + bx + c = 0$.

$a = 3, b = 2, c = -5$

b. $x^2 - 3x = 5$ Subtract 5 from both sides of the equation.

$x^2 - 3x - 5 = 0$ Compare with $ax^2 + bx + c = 0$.

$a = 1, b = -3, c = -5$

c. $x^2 = x - 7$ Subtract x from and add 7 to both sides of the equation.

$x^2 - x + 7 = 0$ Compare with $ax^2 + bx + c = 0$.

$a = 1, b = -1, c = 7$

d. $3 = x^2 + 7x$ Subtract 3 from both sides of the equation.

$0 = x^2 + 7x - 3$ Rewrite in the form $ax^2 + bx + c = 0$.

$x^2 + 7x - 3 = 0$ Compare with $ax^2 + bx + c = 0$.

$a = 1, b = 7, c = -3$

Practice Exercises

Find the values of a, b, and c in each of the following:

1. $4x^2 + 5x - 6 = 0$ **2.** $x^2 - x = 9$

3. $6x + 2 = -4x^2$ **4.** $x^2 + 5 = -4x$

To solve quadratic equations using the quadratic formula, we use the following procedure:

Solving Quadratic Equations with the Quadratic Formula

1. If necessary, rewrite the equation in the form $ax^2 + bx + c = 0$.
2. Find a, b, and c.
3. Write the quadratic formula.
4. Substitute the values for a, b, and c into the formula.
5. Simplify the results.
6. Check the solution(s) in the original equation.

In Example 2, the equations could be solved by factoring, but we solved them using the quadratic formula for illustrative purposes.

Answers:

Example 2

Solve the following equations using the quadratic formula. If there is no real solution, indicate the solution set as $\varnothing$.

a. $x^2 - 2x - 8 = 0$ Find a, b, and c.

$a = 1$, $b = -2$, $c = -8$ Write the quadratic formula.

$x = \dfrac{-b \pm \sqrt{b^2 - 4ac}}{2a}$ Substitute for a, b, and c.

$x = \dfrac{-(-2) \pm \sqrt{(-2)^2 - 4(1)(-8)}}{2(1)}$ Simplify. Watch the signs!

$x = \dfrac{2 \pm \sqrt{4 + 32}}{2}$ Add 4 and 32.

$x = \dfrac{2 \pm \sqrt{36}}{2}$ $\sqrt{36} = 6$.

$x = \dfrac{2 \pm 6}{2}$ Write as two separate fractions.

$x = \dfrac{2 + 6}{2}, x = \dfrac{2 - 6}{2}$ Simplify each fraction.

$x = \dfrac{8}{2}, \qquad x = \dfrac{-4}{2}$ Continue simplifying.

$x = 4, \qquad x = -2$ Therefore, the solutions are 4 and -2.

Note: These equations are checked as was done in Section 11.2. Consequently, the checks are omitted.

b. $6x^2 + x = 12$ Rewrite in the form $ax^2 + bx + c = 0$.

$6x^2 + x - 12 = 0$ Find a, b, and c.

$a = 6$, $b = 1$, $c = -12$ Write the quadratic formula.

$x = \dfrac{-b \pm \sqrt{b^2 - 4ac}}{2a}$ Substitute for a, b, and c.

$x = \dfrac{-1 \pm \sqrt{1^2 - 4(6)(-12)}}{2(6)}$ Simplify.

$x = \dfrac{-1 \pm \sqrt{1 + 288}}{12}$ Continue simplifying.

$x = \dfrac{-1 \pm \sqrt{289}}{12}$ $\sqrt{289} = 17$.

$x = \dfrac{-1 \pm 17}{12}$ Rewrite as two separate fractions.

$x = \dfrac{-1 + 17}{12}, x = \dfrac{-1 - 17}{12}$ Simplify each fraction.

$x = \dfrac{16}{12}, \qquad x = \dfrac{-18}{12}$ Reduce each to lowest terms.

$x = \dfrac{4}{3}, \qquad x = -\dfrac{3}{2}$ Therefore, the solutions are $\dfrac{4}{3}$ and $-\dfrac{3}{2}$.

Practice Exercises

Solve the following equations using the quadratic formula. If there is no real solution, indicate the solution set as $\varnothing$.

5. $x^2 + 7x + 10 = 0$ **6.** $6x^2 - x - 2 = 0$

If more practice is needed, do the Additional Practice Exercises.

Additional Practice Exercises

Solve the following equations using the quadratic formula:

a. $x^2 + x - 12 = 0$ **b.** $8x^2 + 10x - 3 = 0$

The real advantage to solving quadratic equations using the quadratic formula is in solving equations that do not factor. Although this type of equation can be solved by using completing the square, the quadratic formula is usually easier.

Answers:

b. $x = \dfrac{1}{4}, -\dfrac{3}{2}$

Practice Exercises 5–6: **5.** $x = -2, -5$ **6.** $x = \dfrac{2}{3}, -\dfrac{1}{2}$ **Additional Practice Exercises *a–b*:** **a.** $x = 3, -4$

Example 3

Solve the following equations using the quadratic formula. If there is no real solution, indicate the solution set as ∅.

a. $x^2 + 3x - 1 = 0$ Find a, b, and c.

$a = 1, b = 3, c = -1$ Write the quadratic formula.

$x = \dfrac{-b \pm \sqrt{b^2 - 4ac}}{2a}$ Substitute for a, b, and c.

$x = \dfrac{-3 \pm \sqrt{3^2 - 4(1)(-1)}}{2(1)}$ Simplify.

$x = \dfrac{-3 \pm \sqrt{9 + 4}}{2}$ Continue simplifying.

$x = \dfrac{-3 \pm \sqrt{13}}{2}$ Therefore, the solutions are $\dfrac{-3 + \sqrt{13}}{2}$ and $\dfrac{-3 - \sqrt{13}}{2}$.

b. $x(x + 6) = 4$ Distribute x.

$x^2 + 6x = 4$ Subtract 4 from both sides of the equation to put in $ax^2 + bx + c = 0$ form.

$x^2 + 6x - 4 = 0$ Find a, b, and c.

$a = 1, b = 6, c = -4$ Write the quadratic formula.

$x = \dfrac{-b \pm \sqrt{b^2 - 4ac}}{2a}$ Substitute for a, b, and c.

$x = \dfrac{-6 \pm \sqrt{6^2 - 4(1)(-4)}}{2(1)}$ Simplify.

$x = \dfrac{-6 \pm \sqrt{36 + 16}}{2}$ Continue simplifying.

$x = \dfrac{-6 \pm \sqrt{52}}{2}$ Rewrite 52 as $4 \cdot 13$.

$x = \dfrac{-6 \pm \sqrt{4 \cdot 13}}{2}$ Apply the product rule for radicals.

$x = \dfrac{-6 \pm 2\sqrt{13}}{2}$ Factor 2 from the numerator.

$x = \dfrac{2(-3 \pm \sqrt{13})}{2}$ Divide by the common factor of 2.

$x = -3 \pm \sqrt{13}$ Therefore, the solutions are $-3 + \sqrt{13}$ and $-3 - \sqrt{13}$.

c. $2x^2 = 7x - 1$ Subtract $7x$ from and add 1 to both sides of the equation to put in $ax^2 + bx + c = 0$ form.

$2x^2 - 7x + 1 = 0$ Find a, b, and c.

$a = 2, b = -7, c = 1$ Write the quadratic formula.

$x = \dfrac{-b \pm \sqrt{b^2 - 4ac}}{2a}$ Substitute for a, b, and c.

$x = \dfrac{-(-7) \pm \sqrt{(-7)^2 - 4(2)(1)}}{2(2)}$ Simplify.

$x = \dfrac{7 \pm \sqrt{49 - 8}}{4}$ Continue simplifying.

$x = \dfrac{7 \pm \sqrt{41}}{4}$ Therefore, the solutions are $\dfrac{7 + \sqrt{41}}{4}$ and $\dfrac{7 - \sqrt{41}}{4}$.

d. $2x^2 + 6 = -2x$ Add $2x$ to both sides of the equation and write in $ax^2 + bx + c = 0$ form.

$2x^2 + 2x + 6 = 0$ Find a, b, and c.

$a = 2, b = 2, c = 6$ Write the quadratic formula.

$x = \dfrac{-b \pm \sqrt{b^2 - 4ac}}{2a}$ Substitute for a, b, and c.

$x = \dfrac{-2 \pm \sqrt{2^2 - 4(2)(6)}}{2(2)}$ Simplify.

$x = \dfrac{-2 \pm \sqrt{4 - 48}}{4}$ Continue simplifying.

$x = \dfrac{-2 \pm \sqrt{-44}}{4}$ Since $\sqrt{-44}$ does not exist as a real number, this equation has no real solutions. Hence the solution set is ∅.

Practice Exercises

Solve the following equations using the quadratic formula. If there is no real solution, indicate the solution set as ∅.

7. $x^2 + 6x = -6$ **8.** $x^2 = 4x + 23$ **9.** $3x^2 - 9x + 3 = 0$ **10.** $x(3x + 2) = -6$

If more practice is needed, do the Additional Practice Exercises.

Additional Practice Exercises

Solve the following equations using the quadratic formula. If there is no solution, indicate the solution set as $\varnothing$.

c. $x^2 - 4x = -1$ **d.** $x(x + 6) = 3$ **e.** $2x^2 - 6x - 1 = 0$ **f.** $2x^2 + 5 = -2x$

Given a choice of which method to use in solving a quadratic equation, we recommend the following procedure:

> ### Solving Quadratic Equations
>
> 1. Try to solve by factoring.
> 2. If $b = 0$, use extraction of roots.
> 3. If the equation cannot be solved by factoring, or if it is difficult to factor, then solve it by using the quadratic formula.

You might ask why we learned to solve quadratic equations by completing the square if we do not use it. Remember, we needed to know how to solve quadratic equations by completing the square in order to derive the quadratic formula. Though completing the square is rarely used in solving quadratic equations, this procedure has many important applications including graphing of second-degree equations.

Exercise Set 11.3

Solve the following using the quadratic formula. If there is no real solution, indicate the solution set as $\varnothing$. Most of the solutions to Exercises 1–20 are rational numbers.

1. $x^2 - 6x + 5 = 0$ **2.** $r^2 - 8r + 15 = 0$ **3.** $x^2 = 6x - 8$ **4.** $x^2 = -8x - 12$

5. $b^2 - 2b = 24$ **6.** $z^2 - 4z = 12$ **7.** $a^2 + 6 = -7a$ **8.** $x^2 - 10 = -3x$

9. $n^2 - 5n = 14$ **10.** $n^2 + 9n = -20$ **11.** $2y^2 - 15y + 18 = 0$ **12.** $2x^2 - 3x - 20 = 0$

13. $3x^2 - 14x = 24$ **14.** $3x^2 - 11x = 6$ **15.** $n^2 = -6n - 10$ **16.** $a^2 = -4a - 6$

17. $8x^2 - 14x - 15 = 0$ **18.** $9x^2 - 6x - 8 = 0$ **19.** $15a^2 = 29a + 2$ **20.** $6x^2 = 25x - 25$

Solve the following quadratic equations using the quadratic formula. If there is no real solution, indicate the solution set as $\varnothing$. Most of Exercises 21 through 48 have irrational solutions.

21. $x^2 + 3x - 1 = 0$ **22.** $x^2 + 5x - 2 = 0$ **23.** $a^2 = -a + 3$ **24.** $b^2 - 5b = -3$

25. $2c^2 - 3c = 4$ **26.** $2n^2 - 5n + 1 = 0$ **27.** $5r^2 - r - 3 = 0$ **28.** $4y^2 = 5y + 2$

29. $3q^2 + 3 = -2q$ **30.** $2m^2 + 2 = -2m$ **31.** $x^2 + 7x + 1 = 0$ **32.** $2x^2 + 9x = 1$

33. $2a^2 = 9a + 3$ **34.** $2d^2 = 5d + 4$ **35.** $x^2 - 2x = 2$ **36.** $a^2 + 4a = -2$

37. $x^2 - 10x = -20$ **38.** $x^2 + 12x = -34$ **39.** $b^2 = 8b - 8$ **40.** $w^2 + 8w = 11$

41. $2x^2 - 4x - 5 = 0$ **42.** $2x^2 - 6x = -3$ **43.** $4r^2 + 10r + 5 = 0$ **44.** $4x^2 - 12x + 7 = 0$

45. $3x^2 = -2x + 4$ **46.** $3x^2 = -8x + 2$ **47.** $5z^2 + 2z + 10 = 0$ **48.** $6a^2 - 4a + 2 = 0$

Answers:

Solve the following application problems:

49. Find two consecutive odd integers the sum of whose squares is 34.

50. Find two consecutive even integers the sum of whose squares is 52.

Exercises 51 through 54 require the use of the formula for the area of a rectangle. The formula for the area of a rectangle is A = LW.

51. The length of a rectangle is 6 inches more than the width. If the area is 55 square inches, find the length and width.

53. The length of a rectangle is 1 yard less than twice the width. If the area is 28 square yards, find the length and width.

52. The width of a rectangle is 3 meters less than the length. If the area is 70 square meters, find the length and width.

54. The length of a rectangle is 2 centimeters more than three times the width. If the area is 33 square centimeters, find the length and width.

A function of the form $f(x) = ax^2 + bx + c$ *is a quadratic function.*

55. If $f(x) = x^2 + 2x - 3$, find the following: **a.** $f(0)$, **b.** $f(2)$, **c.** $f(-3)$

56. If $f(x) = x^2 - 4x + 5$, find the following: **a.** $f(0)$, **b.** $f(3), f(-1)$.

57. If $f(x) = 3x^2 - 2x + 1$, find the following: **a.** $f(0)$, **b.** $f(1)$, **c.** $f(-2)$.

58. If $f(x) = 2x^2 + 4x - 3$, find the following: **a.** $f(0)$, **b.** $f(2), f(-2)$.

Challenge Exercises: (59–62)

Solve the following quadratic equations using the quadratic formula. If there is no real solution, indicate the solution set as $\varnothing$.

59. $x^2 - \dfrac{7}{2}x + 3 = 0$

60. $x^2 - \dfrac{13}{2}x + 10 = 0$

61. $x^2 - \dfrac{7}{3}x - 2 = 0$

62. $y^2 - \dfrac{24}{5}y - 1 = 0$

Writing Exercises:

63. What are the advantages of using the quadratic formula instead of completing the square?

64. In the quadratic formula, $x = \dfrac{-b \pm \sqrt{b^2 - 4ac}}{2a}$, the expression $b^2 - 4ac$ is called the *discriminant*. For what values (types of numbers) of the discriminant will a quadratic equation have:

a. No real solutions? Why?

b. Two irrational solutions? Why?

c. Two rational solutions? Why?

d. Exactly one solution? Why?

65. Given a quadratic equation in the form $ax^2 + bx + c = 0$, if b is an even integer, then $\sqrt{b^2 - 4ac}$ can always be simplified. Why?

Section 11.4	Solving Mixed Equations

OBJECTIVES

When you complete this section, you will be able to:
Solve the following:

a. Linear equations.

b. Quadratic equations.

c. Equations containing rational numbers and/or rational expressions, which result in linear or quadratic equations when simplified.

d. Equations containing radicals, which result in linear or quadratic equations when simplified.

Introduction   Previously we learned to solve linear, quadratic, rational, and radical equations. The techniques that we use to solve each type of equation differ greatly. In this section, we will not learn anything new since we are going to review the techniques needed to solve the different types of equations. We have learned many arithmetic and algebraic techniques since we first solved equations, so the initial appearance of the equation may be different than you have seen before. However, all the equations that we have solved, or will solve, will ultimately be quadratic or linear after simplification. Some of these techniques are needed in Section 11.5 when solving application problems. First, we review the techniques needed to solve linear equations.

Solving Linear Equations

1. If necessary, simplify both sides of the equation as much as possible.
2. If necessary, use the Addition Property of Equality to get all the terms with variables on one side of the equation and all the constant terms on the other.
3. If necessary, use the Multiplication Property of Equality to eliminate any coefficient on the variable.
4. Check the answer in the original equation.

Example 1

Solve the following equations. All checks are left as exercises for the student.

a. $5(2x - 3) + 2(3x - 7) + 1 = 4 - 6(x - 2)$ Apply the distributive property.

$10x - 15 + 6x - 14 + 1 = 4 - 6x + 12$ Add like terms.

$16x - 28 = 16 - 6x$ Add $6x$ to both sides of the equation.

$22x - 28 = 16$ Add 28 to both sides.

$22x = 44$ Divide both sides by 22.

$x = 2$ Therefore, $x = 2$ is the solution.

b. $(x + 4)^2 - x^2 = 48$ Square $x + 4$.

$x^2 + 8x + 16 - x^2 = 48$ Add like terms.

$8x + 16 = 48$ Subtract 16 from both sides of the equation.

$8x = 32$ Divide both sides by 8.

$x = 4$ Therefore, $x = 4$ is the solution.

c. $(2x + 3)(x - 2) - 18 = (x - 6)(2x - 1) + 6$ Apply FOIL on each side of the equation.

$2x^2 - x - 6 - 18 = 2x^2 - 13x + 6 + 6$ Add like terms on each side.

$2x^2 - x - 24 = 2x^2 - 13x + 12$ Subtract $2x^2$ from both sides.

$-x - 24 = -13x + 12$ Add $13x$ to both sides.

$12x - 24 = 12$ Add 24 to both sides.

$12x = 36$ Divide both sides by 12.

$x = 3$ Therefore, the solution is $x = 3$.

Practice Exercises

Solve the following equations:

1. $3(2x + 1) - 4(3x - 8) = 2(4x + 5) - 3$

2. $(a + 3)^2 - a^2 = 33$

3. $(b - 3)(2b + 5) + 4 = (2b - 5)(b + 4) + 4$

In Section 11.3, we solved quadratic equations by using the quadratic formula only. However, near the end of the section, we outlined a procedure for solving quadratic equations if the technique for solving was not specified. We repeat the procedure in the following box:

Solving Quadratic Equations

To solve a quadratic equation in the form $ax^2 + bx + c = 0$:

1. Try to solve by factoring.

2. If $b = 0$, use extraction of roots.

3. If the equation cannot be solved by factoring or if it is difficult to factor, then solve it by using the quadratic formula.

Example 2

Solve the following quadratic equations using any method. The checks are left as exercises for the student.

a. $(2x + 3)(2x + 5) = 35$ Multiply on the left side of the equation.

$4x^2 + 16x + 15 = 35$ Subtract 35 from both sides.

$4x^2 + 16x - 20 = 0$ Remove the common factor of 4.

$4(x^2 + 4x - 5) = 0$ Factor $x^2 + 4x - 5$.

$4(x + 5)(x - 1) = 0$ Set each factor equal to 0 ($4 \neq 0$).

$x + 5 = 0, x - 1 = 0$ Solve each linear equation.

$x = -5, \quad x = 1$ Therefore, the solutions are -5 and 1.

b. $x^2 + (2x - 4)^2 = 100$ Square $2x - 4$.

$x^2 + 4x^2 - 16x + 16 = 100$ Add x^2 and $4x^2$.

$5x^2 - 16x + 16 = 100$ Subtract 100 from both sides of the equation.

$5x^2 - 16x - 84 = 0$ Factor the left side.

$(5x + 14)(x - 6) = 0$ Set each factor equal to 0.

$5x + 14 = 0, x - 6 = 0$ Solve each equation.

$5x = -14, \quad x = 6$

$x = -\dfrac{14}{5}$ Therefore, the solutions are 6 and $-\dfrac{14}{5}$.

c. $2x^2 - 12x + 12 = -3$ Add 3 to both sides of the equation.

$2x^2 - 12x + 15 = 0$ First, try factoring.

$(2x - 1)(x - 15)$ Incorrect.

$(2x - 15)(x - 1)$ Incorrect.

$(2x - 3)(x - 5)$ Incorrect.

$(2x - 5)(x - 3)$ These are the only possibilities and none work. Therefore, use the quadratic formula. Identify a, b, and c.

$a = 2, b = -12, c = 15$

$x = \dfrac{-b \pm \sqrt{b^2 - 4ac}}{2a}$ Substitute for a, b, and c.

$x = \dfrac{-(-12) \pm \sqrt{(-12)^2 - 4(2)(15)}}{2(2)}$ Simplify.

$x = \dfrac{12 \pm \sqrt{144 - 120}}{4}$ Continue simplifying.

$x = \dfrac{12 \pm \sqrt{24}}{4}$ $\sqrt{24} = 2\sqrt{6}$.

$x = \dfrac{12 \pm 2\sqrt{6}}{4}$ Factor the numerator.

$x = \dfrac{2(6 \pm \sqrt{6})}{4}$ Reduce.

$x = \dfrac{6 \pm \sqrt{6}}{2}$ Therefore, the solutions are $\dfrac{6 + \sqrt{6}}{2}$ and $\dfrac{6 - \sqrt{6}}{2}$.

d. $6x^2 - 72 = 0$ Since $b = 0$, use extraction of roots. Add 72 to both sides of the equation.

$6x^2 = 72$ Divide both sides by 6.

$x^2 = 12$ Take the square roots of both sides.

$x = \pm\sqrt{12}$ $\sqrt{12} = 2\sqrt{3}$.

$x = \pm 2\sqrt{3}$ Therefore, the solutions are $2\sqrt{3}$ and $-2\sqrt{3}$.

Practice Exercises

4. $(2x + 2)(2x + 5) = 28$ **5.** $x^2 + (2x + 2)^2 = 169$ **6.** $x^2 - 2x - 10 = 0$ **7.** $5x^2 - 40 = 0$

In Section 5.7 we solved equations that contained rational numbers or expressions. The procedure for solving these types of equations is outlined in the following box:

Solving Equations Containing Rational Expressions

1. Multiply both sides of the equation by the LCD in order to eliminate all rational expressions from the equation.

2. Solve the resulting equation using the appropriate techniques, depending upon whether it is linear or quadratic.

3. At the minimum, check the solution(s) of the equation in step 2 into the original equation to see if it or they make any denominator equal to 0. If so, that solution of the equation in step 2 is not a solution of the original equation.

Example 3

Solve the following equations containing rational expressions:

a. $\dfrac{1}{x} + \dfrac{1}{x + 2} = \dfrac{12}{35}$ Multiply both sides by the LCD $35x(x + 2)$.

$$35x(x + 2)\left(\frac{1}{x} + \frac{1}{x + 2}\right) = 35x(x + 2)\left(\frac{12}{35}\right)$$

Simplify both sides.

$$35x(x + 2)\frac{1}{x} + 35x(x + 2)\frac{1}{x + 2} = 12x(x + 2)$$

Continue simplifying.

$35(x + 2) + 35x = 12x^2 + 24x$ Multiply 35 and $x + 2$.

$35x + 70 + 35x = 12x^2 + 24x$ Add like terms.

$70x + 70 = 12x^2 + 24x$ Subtract $70x$ from both sides of the equation.

$70 = 12x^2 - 46x$ Subtract 70 from both sides.

$0 = 12x^2 - 46x - 70$ Remove the common factor of 2.

$0 = 2(6x^2 - 23x - 35)$ Factor $6x^2 - 23x - 35$.

$0 = 2(6x + 7)(x - 5)$ Set each factor equal to 0 $(2 \neq 0)$.

$6x + 7 = 0, x - 5 = 0$ Solve each equation.

$6x = -7, \qquad x = 5$

$x = -\dfrac{7}{6}$ Therefore, the solutions are 5 and $-\dfrac{7}{6}$.

Neither solution makes any denominator equal to 0, so both solutions may be solutions of the original equation. The only way to verify that both are solutions is to check both in the original equation.

b. $\dfrac{240}{x + 2} = \dfrac{240}{x} - 20$ Multiply both sides of the equation by the LCD of $x(x + 2)$.

$$x(x + 2)\left(\frac{240}{x + 2}\right) = x(x + 2)\left(\frac{240}{x} - 20\right)$$ Simplify both sides.

$240x = x(x + 2)\dfrac{240}{x} - 20x(x + 2)$ Multiply on the right.

$240x = 240(x + 2) - 20x^2 - 40x$ Multiply 240 and $x + 2$.

$240x = 240x + 480 - 20x^2 - 40x$ Add like terms.

$240x = -20x^2 + 200x + 480$ Write in $ax^2 + bx + c = 0$ form.

$20x^2 + 40x - 480 = 0$ Remove the common factor of 20.

$20(x^2 + 2x - 24) = 0$ Factor $x^2 + 2x - 24$.

$20(x + 6)(x - 4) = 0$ Set each factor equal to 0 $(20 \neq 0)$.

$x + 6 = 0, x - 4 = 0$ Solve each equation.

$x = -6, \qquad x = 4$ Therefore, the solutions are -6 and 4.

Neither solution makes any denominator equal to 0, so both solutions may be solutions of the original equation. The only way to verify that both are solutions is to check both against the original equation.

Answers:

Practice Exercises 4–7: 4. $x = -\dfrac{7}{2}$, 1   5. $x = -\dfrac{33}{5}$, 5   6. $x = 1 \mp \sqrt{11}$ 7. $x = \pm 2\sqrt{2}$

c. $\dfrac{1}{x-3} + \dfrac{1}{3} = \dfrac{6}{x^2-9}$ Factor $x^2 - 9$.

$\dfrac{1}{x-3} + \dfrac{1}{3} = \dfrac{6}{(x+3)(x-3)}$ Multiply both sides of the equation by the LCD $3(x+3)(x-3)$.

$3(x+3)(x-3)\left(\dfrac{1}{x-3} + \dfrac{1}{3}\right) = 3(x+3)(x-3)\left(\dfrac{6}{(x+3)(x-3)}\right)$ Simplify.

$3(x+3)(x-3)\dfrac{1}{x-3} + 3(x+3)(x-3)\dfrac{1}{3} = 3 \cdot 6$ Continue simplifying.

$3(x+3) + (x+3)(x-3) = 18$ Continue simplifying.

$3x + 9 + x^2 - 9 = 18$ Continue simplifying.

$x^2 + 3x = 18$ Subtract 18 from both sides.

$x^2 + 3x - 18 = 0$ Factor the left side.

$(x+6)(x-3) = 0$ Set each factor equal to 0.

$x+6 = 0, \; x-3 = 0$ Solve each equation.

$x = -6, \quad x = 3$ Therefore, the solutions are -6 and 3. However, 3 makes the denominators of $\dfrac{1}{x-3}$ and $\dfrac{6}{x^2-9}$ equal to 0. Therefore, the only solution (which can verified by substituting into the original equation) is -6.

Practice Exercises

Solve the following:

8. $\dfrac{1}{x} + \dfrac{1}{x+2} = \dfrac{5}{12}$

9. $\dfrac{120}{x+1} = \dfrac{120}{x} - 10$

10. $\dfrac{x}{x+2} + \dfrac{3}{4x} = \dfrac{3x+1}{2x+4}$

In Section 10.6, we learned to solve equations that contain radicals. We repeat the procedures in the following boxes. Remember, squaring both sides of an equation can introduce extraneous solutions, so all answers must be checked in the original equation.

Solving Equations Containing One Square Root

If an equation contains one square root, it can be solved by:

1. Isolating the square root on one side of the equation.
2. Squaring both sides of the equation.
3. Solving the resulting equation.
4. Checking all solutions in the original equation.

Solving Equations Containing Two Square Roots

If an equation contains two square roots, it can be solved by:

1. Isolating one of the radicals on one side of the equation.
2. Squaring both sides of the equation, and if necessary, simplifying each side by combining like terms.
3. Isolating the radical on one side of the equation, if the equation still contains a radical, and squaring both sides of the equation again; then solving the resulting equation.
4. Checking all solutions in the original equation.

Answers:

Example 4

Solve the following:

a. $\sqrt{3x + 3} = 2x - 1$ Square both sides of the equation.

$(\sqrt{3x + 3})^2 = (2x - 1)^2$ Simplify both sides.

$3x + 3 = 4x^2 - 4x + 1$ Subtract $3x$ and 3 from both sides.

$0 = 4x^2 - 7x - 2$ Factor.

$0 = (4x + 1)(x - 2)$ Set each factor equal to 0.

$4x + 1 = 0, x - 2 = 0$ Solve each equation.

$4x = -1, \quad x = 2$

$x = -\dfrac{1}{4}$ Therefore, the possible solutions are $-\dfrac{1}{4}$ and 2.

CHECK: $x = -\dfrac{1}{4}$ $x = 2$

$\sqrt{3\left(-\dfrac{1}{4}\right) + 3} = 2\left(-\dfrac{1}{4}\right) - 1$ $\sqrt{3(2) + 3} = 2(2) - 1$

$\sqrt{-\dfrac{3}{4} + 3} = -\dfrac{2}{4} - 1$ $\sqrt{6 + 3} = 4 - 1$

$\sqrt{\dfrac{9}{4}} = -\dfrac{1}{2} - 1$ $\sqrt{9} = 3$

$\dfrac{3}{2} \neq -\dfrac{3}{2}$ $3 = 3$

Therefore, 2 is the only solution.

b. $\sqrt{2x} - \sqrt{3x + 10} = -2$ Add $\sqrt{3x + 10}$ to both sides of the equation.

$\sqrt{2x} = \sqrt{3x + 10} - 2$ Square both sides.

$(\sqrt{2x})^2 = (\sqrt{3x + 10} - 2)^2$ Simplify both sides.

$2x = 3x + 10 - 4\sqrt{3x + 10} + 4$ Simplify the right side.

$2x = 3x - 4\sqrt{3x + 10} + 14$ Subtract $3x$ from both sides.

$-x = -4\sqrt{3x + 10} + 14$ Subtract 14 from both sides.

$-x - 14 = -4\sqrt{3x + 10}$ Multiply both sides by -1.

$x + 14 = 4\sqrt{3x + 10}$ Square both sides.

$x^2 + 28x + 196 = 16(3x + 10)$ Multiply on the right.

$x^2 + 28x + 196 = 48x + 160$ Subtract $48x$ and 160 from both sides.

$x^2 - 20x + 36 = 0$ Factor the left side.

$(x - 18)(x - 2) = 0$ Set each factor equal to 0.

$x - 18 = 0, x - 2 = 0$ Solve each equation.

$x = 18, \quad x = 2$ Therefore, the possible solutions are 18 and 2.

CHECK: $x = 18$ $x = 2$

$\sqrt{2 \cdot 18} - \sqrt{3 \cdot 18 + 10} = -2$ $\sqrt{2 \cdot 2} - \sqrt{3 \cdot 2 + 10} = -2$

$\sqrt{36} - \sqrt{54 + 10} = -2$ $\sqrt{4} - \sqrt{6 + 10} = -2$

$6 - \sqrt{64} = -2$ $2 - \sqrt{16} = -2$

$6 - 8 = -2$ $2 - 4 = -2$

$-2 = -2$ $-2 = -2$

Therefore, 18 and 2 are both solutions.

Practice Exercises

Solve the following:

11. $\sqrt{3x + 1} = 2x - 6$

12. $\sqrt{3x + 4} - \sqrt{x + 2} = 2$

Exercise Set 11.4

Solve the following equations. Each equation is linear or will be linear when simplified.

1. $3(x + 2) - 12 = 4(2x - 6) - 2(3x + 2)$

2. $4(2x - 1) + 3 = 3(3x - 2) + 2(x - 5)$

3. $(b + 1)^2 = b^2 + 9$

4. $(c + 2)^2 = (c + 1)^2 + 11$

5. $(n + 2)^2 - (n - 2)^2 = 50$

6. $(m + 2)^2 - (m - 2)^2 = 20$

7. $(2z + 3)(z - 4) - 8 = (z - 6)(2z + 5)$

8. $(3x + 2)(x - 1) = (x + 5)(3x - 2) - 6$

Solve the following quadratic equations using any method:

9. $(2x + 3)(2x + 5) = 63$

10. $(2x + 2)(2x + 4) = 80$

11. $(a + 2)(a + 5) - a(a + 3) = 22$

12. $(x + 4)(x + 8) - x(x + 4) = 48$

13. $b^2 + (b + 3)^2 = 225$

14. $x^2 + (2x - 1)^2 = 289$

15. $6c^2 + 8c - 5 = -8$

16. $4d^2 + 6d + 2 = 5$

17. $6n + 7 = 5 - 3n^2$

18. $-10m + 8 = 4 - m^2$

19. $8a^2 - 192 = 0$

20. $3b^2 - 96 = 0$

Solve the following equations containing rational expressions:

21. $\dfrac{1}{x} + \dfrac{1}{x + 2} = \dfrac{3}{4}$

22. $\dfrac{1}{x} + \dfrac{1}{x + 1} = \dfrac{7}{12}$

23. $\dfrac{60}{x + 2} = \dfrac{60}{x} - 5$

24. $\dfrac{120}{x + 2} = \dfrac{120}{x} - 5$

25. $\dfrac{1}{p - 4} + \dfrac{1}{4} = \dfrac{8}{p^2 - 16}$

26. $\dfrac{1}{y - 5} + \dfrac{1}{5} = \dfrac{10}{y^2 - 25}$

27. $\dfrac{6}{2y + 5} = \dfrac{2}{y + 5} + \dfrac{1}{5}$

28. $\dfrac{6}{2x + 3} = \dfrac{2}{x - 6} + \dfrac{4}{3}$

Solve the following equations containing radicals:

29. $\sqrt{2a + 5} = 3a - 3$

30. $\sqrt{3b + 1} = 5b - 3$

31. $\sqrt{8m + 1} - 2m + 1 = 0$

32. $\sqrt{6x + 1} - 2x + 3 = 0$

33. $\sqrt{4x + 1} = \sqrt{x + 2} + 1$

34. $\sqrt{3x + 7} = \sqrt{2x + 3} + 1$

35. $\sqrt{2x + 1} - \sqrt{3x + 4} = -1$

36. $\sqrt{2x - 1} - \sqrt{4x + 5} = -2$

Challenge Exercises: (37–42)

Solve the following:

37. $\dfrac{3}{2n + 5} + \dfrac{2}{n + 1} = \dfrac{8}{3n + 2}$

38. $\dfrac{4}{x + 2} - \dfrac{6}{2x + 4} = \dfrac{2}{x + 7}$

39. $\sqrt{3x + 1} - \sqrt{x - 1} = \sqrt{2x - 6}$

40. $\sqrt{4x + 1} - \sqrt{2x - 3} = \sqrt{x - 2}$

41. $(3x - 2)(x - 4) - (2x + 3)(x + 2) = 4$

42. $(2x + 1)^2 - (x - 5)^2 = x - 14$

Answers:

Section 11.5	Applications Involving Quadratic Equations

OBJECTIVES

When you complete this section, you will be able to:
Solve the following types of application problems involving quadratic equations:

a. Numbers.

b. Geometric figures.

c. Pythagorean theorem.

d. Distance, rate, and time.

e. Work.

f. Applications from science.

Introduction   Application problems that resulted in quadratic equations were first introduced in Section 5.8. In that section, we were limited to integer problems and areas of squares and rectangles. In later chapters, we learned to solve other types of equations. Consequently, in this section, we will concentrate on application problems that involve solving these other types of equations and equations from the sciences.

As in Section 5.8, care must be taken when solving application problems that involve quadratic equations. Quite often, a number may solve the equation but not satisfy the physical conditions of the problem. For example, the length of the side of a rectangle cannot be negative. Consequently, it is very important that all solutions of the equation be checked against the wording of the original problem. Also, as in Section 5.8, most of the equations in this section can be solved by factoring, though you have the option of using the quadratic formula if you wish.

Example 1

The sum of the reciprocals of two consecutive even integers is $\frac{7}{24}$. Find the integers.

Number problems   **Solution:**
Let $x =$ the smaller of the two consecutive even integers. Then,
$x + 2 =$ the larger of the two consecutive even integers.

$\frac{1}{x} =$ the reciprocal of the smaller integer.

$\frac{1}{x + 2} =$ the reciprocal of the larger integer.

Since the sum of the reciprocals is $\frac{7}{24}$, the equation is:

$$\frac{1}{x} + \frac{1}{x + 2} = \frac{7}{24}$$ Multiply both sides of the equation by the LCD $24x(x + 2)$.

$$24x(x + 2)\left(\frac{1}{x} + \frac{1}{x + 2}\right) = 24x(x + 2)\left(\frac{7}{24}\right)$$ Simplify both sides.

$$24x(x + 2)\left(\frac{1}{x}\right) + 24x(x + 2)\left(\frac{1}{x + 2}\right) = x(x + 2)(7)$$ Multiply.

$$24(x + 2) + 24x = 7x(x + 2)$$ Distribute.

$$24x + 48 + 24x = 7x^2 + 14x$$ Add like terms.

$$48x + 48 = 7x^2 + 14x$$ Subtract $48x$ and 48 from both sides.

$$0 = 7x^2 - 34x - 48$$ Factor $7x^2 - 34x - 48$.

$$0 = (7x + 8)(x - 6)$$ Set each factor equal to 0.

$$7x + 8 = 0, \ x - 6 = 0$$ Solve each equation.

$$7x = -8, \quad x = 6$$

$$x = -\frac{8}{7}$$ Since $-\frac{8}{7}$ is not an integer, the only answer is 6.

Have we answered the question that was asked? No, we need the other integer.

If $x = 6$, then $x + 2 = 8$. Therefore, the two consecutive even integers are 6 and 8.

Check:
Is the sum of the reciprocals of 6 and 8 equal to $\frac{7}{24}$? Since $\frac{1}{6} + \frac{1}{8} = \frac{4}{24} + \frac{3}{24} = \frac{7}{24}$, our solutions are correct.

Practice Exercise

1. The product of the reciprocals of two consecutive integers is $\frac{1}{12}$. Find the integers.

Example 2

Geometric problems   *A rectangular picture, 4 inches by 6 inches, is enclosed by a frame of uniform width. If the area of the picture and the frame is 80 square inches, find the width of the frame.*

Solution:
Let x = the width of the frame.

Draw and label a figure.

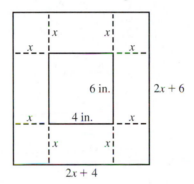

Since there are x inches of frame to the left and the right of the picture and the picture is 4 inches wide, the outside dimension is $2x + 4$ inches. Since there are x inches of the frame above and below the picture and the picture is 6 inches high, the outside dimension is $2x + 6$ inches.

The formula for finding the area of a rectangle is $A = LW$. Since the area of the picture and the frame is 80 square inches, substitute 80 for A, $2x + 6$ for L, and $2x + 4$ for W into the formula.

$A = LW$	Substitute for A, L, and W.
$80 = (2x + 6)(2x + 4)$	Multiply the right side of the equation.
$80 = 4x^2 + 20x + 24$	Subtract 80 from both sides.
$0 = 4x^2 + 20x - 56$	Remove the common factor of 4.
$0 = 4(x^2 + 5x - 14)$	Factor $x^2 + 5x - 14$.
$0 = 4(x + 7)(x - 2)$	Set each factor, except 4, equal to 0.
$4 \neq 0, x + 7 = 0, x - 2 = 0$	Solve each equation.
$x = -7, \quad \mathbf{x = 2}$	Since a frame cannot have a width of -7 inches, 2 inches is the only answer.

Since x is the width of the frame, we have answered the question that was asked.

Check:
If the width of the frame is 2 inches, is the area of the picture and the frame 80 square inches? If the width of the frame is 2 inches, the outside dimensions are $4 + 4 = 8$ inches and $4 + 6 = 10$ inches. The area of a rectangle that is 8 inches by 10 inches is $(8)(10) = 80$ square inches. Therefore, our solutions are correct.

Practice Exercise

2. The length of a rectangular flower garden is 2 feet more than its width. The flower garden is surrounded by a uniform border of mulch that is 2 feet wide. If the area of the flower garden and the border is 80 square feet, find the length and width of the flower bed.

Answers:

Example 3

Pythagorean Theorem   *The longer side of a rectangular mural is 3 feet less than three times the shorter side. If the mural is 13 feet across diagonally, find the lengths of the sides.*

Solution:
Let x = the length of the shorter side. Then,
$3x - 3$ = the length of the longer side.

Draw and label a figure.

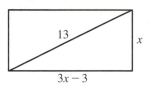

The diagonal divides the rectangle into two right triangles. Remember, the Pythagorean theorem states that in a right triangle, $a^2 + b^2 = c^2$, where a and b are the lengths of the legs and c is the length of the hypotenuse.

$a^2 + b^2 = c^2$	Substitute for a, b, and c.
$x^2 + (3x - 3)^2 = 13^2$	Raise to powers.
$x^2 + 9x^2 - 18x + 9 = 169$	Add like terms.
$10x^2 - 18x + 9 = 169$	Subtract 169 from both sides of the equation.
$10x^2 - 18x - 160 = 0$	Remove the common factor of 2.
$2(5x^2 - 9x - 80) = 0$	Factor the trinomial.
$2(5x + 16)(x - 5) = 0$	Set each factor, except 2, equal to 0.
$5x + 16 = 0, x - 5 = 0$	Solve each equation.
$5x = -16,\quad \mathbf{x = 5}$	
$x = -\dfrac{16}{5}$	Since a side of a rectangle cannot be $-\dfrac{16}{5}$, the only answer is 5 feet.

Have we answered the question asked? No, we also need the length.

Since $x = 5$ feet is the length of the shorter side, and $3x - 3$ = the length of the longer side, the length of the longer side = $3(5) - 3 = 15 - 3 = 12$ feet.

Check: Do the lengths of the sides satisfy the Pythagorean Theorem? Since $5^2 + 12^2 = 25 + 144 = 169 = 13^2$, our solutions are correct.

Practice Exercise

3. The distance diagonally across a rectangular vacant lot is 10 feet more than the length of the longer side. If the shorter side is 30 feet long, find the length of the longer side.

Example 4

The average speed of a car is 10 miles per hour more than the average speed of a bus. The time required for the bus to travel 200 miles is one hour more than the time required by the car. Find how long it takes the car to travel 200 miles.

Solution:
Remember from Section 7.7 that we use a chart to solve distance, rate, and time problems.

Let x = the time for the car to travel 200 miles. Then, $x + 1$ = the time for the bus to travel 200 miles.

	d	r	t
bus			
car			

Since both distances equal 200 miles, fill in the distance column also.

Since $r = \frac{d}{t}$, the rate of the bus is $\frac{200}{x+1}$ and the rate of the car is $\frac{200}{x}$. Put these in the rate column.

	d	r	t
bus	200		$x+1$
car	200		x

	d	r	t
bus	200	$\frac{200}{x+1}$	$x+1$
car	200	$\frac{200}{x}$	x

The last column filled in was the rate column, so write an equation involving rates. Since the rate of the car is 10 miles per hour faster than the rate of the bus, (the rate of the car) = (the rate of the bus) + 10. Consequently, the equation is:

$$\frac{200}{x} = \frac{200}{x+1} + 10$$

Multiply both sides by the LCD $x(x+1)$.

$$x(x+1)\left(\frac{200}{x}\right) = x(x+1)\left(\frac{200}{x+1} + 10\right)$$

Simplify both sides of the equation.

$$(x+1)(200) = x(x+1)\left(\frac{200}{x+1}\right) + x(x+1)(10)$$

Continue simplifying.

$$200x + 200 = 200x + 10x^2 + 10x$$

Subtract 200x and 200 from both sides.

$$0 = 10x^2 + 10x - 200$$

Remove the common factor of 10.

$$0 = 10(x^2 + x - 20)$$

Factor $x^2 + x - 20$.

$$0 = 10(x+5)(x-4)$$

Set each factor, except 10, equal to 0.

$$x + 5 = 0, x - 4 = 0$$

Solve each equation.

$$x = -5, \quad x = 4 \text{ hours}$$

Since time cannot be negative, 4 is the only answer. Therefore, it takes the car four hours to go 200 miles.

Since x represents the time of the car, we have answered the question that was asked.

Check:
The rate of the car is $\frac{200}{x} = \frac{200}{4} = 50$ miles per hour. So in four hours the car can travel 50(4) = 200 miles. The time necessary for the bus to travel 200 miles is $x + 1 = 4 + 1 =$ five hours. The rate of the bus is $\frac{200}{x+1}$. Therefore, the rate of the bus is $\frac{200}{4+1} = \frac{200}{5} = 40$ miles per hour. So in five hours, the bus can travel 5(40) = 200 miles. Therefore, our answers are correct.

Practice Exercise

4. The average speed of a passenger train is 25 miles per hour more than the average speed of a car. The time required for the car to travel 300 miles is two hours more than the time required for the train. Find the average speed of the car.

Work problems   In Section 7.7, we did work problems that resulted in linear equations, but they often result in quadratic equations as well.

Example 5

It would take Ramon 4 hours longer to plant a garden than it would take his wife, Aurea. If they can plant the garden in $4\frac{4}{5}$ hours working together, how long would it take each to plant the garden alone?

Solution:

Remember from Section 7.7 that we used a chart to do work problems.

Let $x =$ the number of hours for Aurea to plant the garden working alone. Then, $x + 4 =$ the number of hours for Ramon to plant the garden working alone.

	Time to do the job alone	rate	time worked	fractional part of the total job done
Aurea	x	$\frac{1}{x}$	$\frac{24}{5}$	$\frac{1}{x} \cdot \frac{24}{5} = \frac{24}{5x}$
Ramon	$x + 4$	$\frac{1}{x+4}$	$\frac{24}{5}$	$\frac{1}{x+4} \cdot \frac{24}{5} = \frac{24}{5(x+4)}$

The relationship is:

(the fractional part Aurea does) + (the part fractional Ramon does) = 1 (The entire job).

Hence, the equation is:

$$\frac{24}{5x} + \frac{24}{5(x+4)} = 1$$

Multiply both sides of the equation by the LCD $5x(x+4)$.

$$5x(x+4)\left(\frac{24}{5x} + \frac{24}{5(x+4)}\right) = 5x(x+4)(1)$$

Simplify both sides.

$$5x(x+4)\left(\frac{24}{5x}\right) + 5x(x+4)\left(\frac{24}{5(x+4)}\right) = 5x^2 + 20x$$

Continue simplifying.

$$24(x+4) + 24x = 5x^2 + 20x$$

Distribute.

$$24x + 96 + 24x = 5x^2 + 20x$$

Add like terms.

$$48x + 96 = 5x^2 + 20x$$

Subtract $48x$ and 96 from both sides.

$$0 = 5x^2 - 28x - 96$$

Factor $5x^2 - 28x - 96$.

$$0 = (5x + 12)(x - 8)$$

Set each factor equal to 0.

$$5x + 12 = 0, x - 8 = 0$$

Solve each equation.

$$5x = -12, \quad x = 8$$

$$x = -\frac{12}{5}$$

Since the garden cannot be planted in $-\frac{12}{5}$ hours, 8 is the only answer.

Have we answered the question that was asked? No, we have found the number of hours for Aurea to plant the garden when working alone. We need to find the number of hours for Ramon as well. It would take Ramon $x + 4 = 8 + 4 = 12$ hours to plant the garden alone.

Check:

Since Aurea can plant the garden in 8 hours, she can plant $\frac{1}{8}$ of the garden in 1 hour. Ramon can plant the garden in 12 hours alone, so he can plant $\frac{1}{12}$ of the garden in 1 hour. Together they can plant $\frac{1}{8} + \frac{1}{12} = \frac{3}{24} + \frac{2}{24} = \frac{5}{24}$ of the garden in 1 hour. So in $4\frac{4}{5}$ hours they can plant $4\frac{4}{5} \cdot \frac{5}{24} = \frac{24}{5} \cdot \frac{5}{24} = 1$ garden.

Practice Exercise

5. Terri and Tommy run a flea market. It takes Terri 2 hours longer to put the merchandise out than it does Tommy. If they can put the merchandise out together in $1\frac{7}{8}$ hours, how long would it take each working alone?

Example 6

Science problems

A particle moves along a straight line according to the formula $s = 2t^2 + t - 6$, where s represents the distance in inches from the beginning point and t represents the time in seconds. Find the time(s), to the nearest tenth of a second, when the particle is 3 inches from the beginning point.

Solution:
Since s represents the distance from the beginning point, we need to find t when $s = 3$.

$s = 2t^2 + t - 6$	Substitute 3 for s.
$3 = 2t^2 + t - 6$	Subtract 3 from both sides of the equation.
$0 = 2t^2 + t - 9$	Solve using the quadratic formula.
$t = \dfrac{-1 \pm \sqrt{1^2 - 4(2)(-9)}}{2(2)}$	Simplify the radical.
$t = \dfrac{-1 \pm \sqrt{1 + 72}}{4}$	Add 1 and 72.
$t = \dfrac{-1 \pm \sqrt{73}}{4}$	Write as two separate answers.
$t = \dfrac{-1 + \sqrt{73}}{4}, \quad t = \dfrac{-1 - \sqrt{73}}{4}$	$\sqrt{73} \approx 8.5440$.
$t = \dfrac{-1 + 8.5440}{4}, t = \dfrac{-1 - 8.5440}{4}$	Simplify.
$t = 1.886, \quad\quad t = -2.386$	Round to the nearest tenth.
$\mathbf{t = 1.9}, \quad\quad\; t = -2.4$	Since we cannot have negative time, the only answer is 1.9 seconds.

Check:
Substitute 3 for s and 1.9 for t in the formula $s = 2t^2 + t - 6$.

$$3 \approx 2(1.9)^2 + 1.9 - 6$$
$$3 \approx 2(3.61) + 1.9 - 6$$
$$3 \approx 7.22 + 1.9 - 6$$
$$3 \approx 3.12$$

These are not exactly equal due to the round-off error. Since they are approximately equal, our answer is correct.

Practice Exercise

6. A ball is thrown upward with an initial velocity of 32 feet per second from the top of a building that is 64 feet high. The distance, s, above the ground t seconds after the ball is thrown is given by the formula $s = -16t^2 + 32t + 64$, where s is measured in feet. Find the time(s) when the ball is 32 feet above the ground.

Answers:

Exercise Set 11.5

Solve the following:

1. The sum of the reciprocals of two consecutive integers is $\frac{9}{20}$. Find the integers.

2. The sum of the reciprocals of two consecutive odd integers is $\frac{12}{35}$. Find the integers.

3. One integer is 3 more than another. If the sum of the squares of the integers is 117, find the integers.

4. The larger of two integers is 1 less than twice the smaller. If the difference of the squares of the integers is 56, find the integers.

5. The product of the reciprocals of two consecutive odd integers is $\frac{1}{35}$. Find the integers.

6. The product of the reciprocals of two consecutive even integers is $\frac{1}{48}$. Find the integers.

7. The length of a small rectangular field is 10 yards more than its width. A road 5 yards wide completely encircles the field. If the area of the road and the field together is 4200 square yards, find the dimensions of the field.

8. A rectangular slab whose length is two feet more than its width is poured for a utility shed. The shed is placed on the slab so that 1 foot of the slab is exposed on all sides of the shed. If the area of the floor of the shed is 120 square feet, find the dimensions of the slab.

9. A rectangular sheet of paper is 3 inches longer than it is wide. Material is printed on the page with a 1-inch margin at the top and bottom of the page and a $\frac{1}{2}$-inch margin on the left and right. If the area of the printed material is 63 square inches, find the length and width of the page.

10. The length of a rectangular picture is 2 more than its width. It is to be mounted with one inch of matting on the sides and $1\frac{1}{2}$ inches at the top and bottom. If the area of the picture and the matting is 130 square inches, find the length and width of the picture.

11. The sides of a right triangle are three consecutive integers. Find the lengths of the sides.

12. The sides of a right triangle are three consecutive even integers. Find the lengths of the sides.

13. A guy wire is attached to a vertical pole at a point 24 feet above the ground. The length of the wire is 6 feet more than twice the distance from the pole to the point where the wire is attached to the ground. Find the length of the wire.

14. A telephone pole is supported by two guy wires. The distance from the pole to the points on the ground where the wires are attached is 15 feet. The length of each wire is 5 feet more than the distance from the ground to the point on the pole where the wires are attached. Find the total length of the wires.

15. Two cars are approaching the same intersection at right angles to each other. The distance from the intersection to one of the cars is 10 feet less than twice the distance from the intersection to the other car. If the distance between the cars is 170 feet, find the distance from each car to the intersection.

16. Two ships are approaching the same port and are traveling on courses that are at right angles to each other. One ship is 8 miles from port. The distance of the other ship from port is 2 miles less than the distance between the ships. If the maximum range of their radios is 20 miles, can the ships communicate with each other?

17. The average rate of a bus is 15 miles per hour more than the average rate of a truck. It takes the truck one hour longer to go 180 miles than it takes the bus. How long does it take the bus to travel 180 miles?

18. The time required for a train to travel 200 miles is two hours more than the time required for a car to travel 180 miles. The average rate of the car is 20 miles per hour more than the average rate of the train. Find the rate of the train.

19. The time required for a bus to travel 360 miles is three hours more than the time required for a motorcycle to travel 300 miles. The average rate of the motorcycle is 15 miles per hour more than the average rate of the bus. Find the average rate of the bus.

20. The time required for a truck to travel 400 miles is three hours more than the time required for a car to travel 300 miles. The average rate of the car is 10 miles per hour more than the average rate of the truck. Find the average rate of the truck.

21. Billy and Jody are commercial fishermen. Working alone, it takes Billy 2 hours longer to run the hoop nets than it takes Jody working alone. Together,

they can run the hoop nets in $2\frac{2}{5}$ hours. How long does it take each when working alone?

22. Using a riding lawn mower, Fran can mow the grass at the campground in 4 hours less time than it takes Donnie using a push-type mower. Together, they can mow the grass in $2\frac{2}{3}$ hours. How long does it take each when working alone?

23. It takes Buster 1 hour longer to stack a load of hay than it takes Ronnie. Together, they can stack a load of hay in $1\frac{1}{5}$ hours. How long does it take each when working alone?

24. It takes John, a novice mechanic, 5 hours longer to repair a transmission than it takes Ramon, an experienced mechanic. Together, they can repair the transmission in $3\frac{1}{3}$ hours. How long does it take each when working alone?

25. A particle is moving along a straight line according to the formula $s = t^2 - 2t + 3$, where s is the distance in inches and t is the time in seconds. Find the time(s) when the particle is 18 inches from the beginning point.

26. A particle is moving along a straight line according to the formula $s = t^2 - 6t + 12$, where s is the dis-

tance in inches and t is the time in seconds. Find the time(s) when the particle is 4 inches from the beginning point.

27. The distance a free-falling object falls is given by $s = 16t^2$, where s is in feet and t is in seconds. Find the time(s) to the nearest tenth of a second it takes a free-falling object to fall 48 feet.

28. The distance a free-falling object falls is given by $s = 16t^2$, where s is in feet and t is in seconds. Find the time(s) to the nearest tenth of a second it takes a free-falling object to fall 80 feet.

29. An object is thrown upward with an initial velocity of 48 feet per second from a height of 80 feet. The distance, s, above the ground after t seconds is given by $s = -16t^2 + 48t + 80$, where s is measured in feet. Find the time(s) to the nearest tenth of a second when the object is 40 feet above the ground.

30. An object is thrown upward with an initial velocity of 52 feet per second from a height of 100 feet. The distance, s, above the ground after t seconds is given by $s = -16t^2 + 52t + 100$, where s is measured in feet. Find the time(s) to the nearest tenth of a second when the object is 120 feet above the ground.

Writing Exercises:

31. Write an application problem that results in a quadratic equation and involves integers.

32. Write an application problem that results in a quadratic equation and involves geometric figures.

Chapter 11   Summary

Solving Incomplete Quadratic Equations: [Section 11.1]

1. Solve incomplete quadratic equations that can be put in the form of $ax^2 + bx = 0$ by factoring.

2. Solve incomplete quadratics that can be put in the form $ax^2 + c = 0$ by extraction of roots. This involves the following:

   a. Solve the equation for x^2. This gives an equation of the form $x^2 = k$.

   b. Find the positive and negative square roots of k.

   c. Write the solutions in simplest form.

3. Solve equations of the form $(ax + b)^2 = c$ by doing the following:

   a. Set $ax + b = \pm\sqrt{c}$.

   b. Rewrite as two equations.

   c. Solve each equation.

Completing the Square: [Section 11.2]

• To determine the number to be added to $x^2 + bx$ in order to get a perfect-square trinomial, find one-half of b and square it. The resulting trinomial is the square of a binomial whose first term is the square root of the first term of the trinomial, whose second term is the square root of the last term of the trinomial, and whose sign is the same as the sign of the middle term of the trinomial.

Solving Quadratic Equations by Completing the Square: [Section 11.2]

1. Write the equation in the form $x^2 + bx = c$. This may require several steps, depending upon the form of the given equation.

2. Determine the number needed to make $x^2 + bx$ a perfect-square trinomial, and add that number to both sides of the equation.

3. Write one side of the equation as the square of a binomial, and add the numbers on the other side of the equation.

4. Solve the equation resulting from step 3 by extraction of roots.

5. Solve each of the resulting equations.

6. Check each solution in the original equation.

Solving Quadratic Equations Using the Quadratic Formula: [Section 11.3]

1. If necessary, rewrite the equation in the form of $ax^2 + bx + c = 0$.

2. Find the values of a, b, and c.

3. Substitute the values of a, b, and c into the quadratic formula, which is
$$x = \frac{-b \pm \sqrt{b^2 - 4ac}}{2a}.$$

4. Simplify the results.

5. Check the solutions in the original equation.

Solving Linear Equations: [Section 11.4]

1. If necessary, simplify both sides of the equation as much as possible.

2. If necessary, use the Addition Property of Equality to get all the terms with variables on one side of the equation and all the constants on the other.

3. If necessary, use the Multiplication Property of Equality to eliminate any coefficient on the variable.

4. Check the answer in the original equation.

Solving Equations Containing Rational Expressions: [Section 11.4]

1. Multiply both sides of the equation by the LCD in order to eliminate all rational expressions from the equation.

2. Solve the resulting equation using the appropriate techniques depending upon whether it is linear or quadratic.

3. At the minimum, insert the solution(s) of the equation from step 2 into the original equation to see if it or they make any denominator equal to 0. If so, that solution of the equation in step 2 is not a solution of the original equation.

Solving Equations Containing Radicals: [Section 11.4]

- If an equation contains one square root, it can be solved by doing the following:

  1. Isolate the square root on one side of the equation.

  2. Square both sides of the equation.

  3. Solve the resulting equation.

  4. Check all solutions in the original equation.

- If an equation contains two square roots, it can be solved by doing the following:

  1. Isolate one of the radicals on one side of the equation.

  2. Square both sides of the equation, and if necessary, simplify each side by combining like terms.

  3. Isolate the radical on one side of the equation, if the equation still contains a radical, and square both sides of the equation again. Solve the resulting equation.

  4. Check all solutions in the original equation.

Solving Application Problems Involving Quadratic Equations: [Section 11.5]

- Solve the application problems following the same procedure you used earlier. Be sure you answered the question(s) asked. Check the solution(s) against the wording of the problem. Be aware that some solutions of the equation may not satisfy the conditions of the original problem.

Chapter 11   Review Exercises

Solve the following incomplete quadratic equations. If there is no real solution, write $\varnothing$. [Section 11.1]

1. $6x^2 + 18x = 0$

2. $5x^2 - 80 = 0$

3. $5a^2 = -13a$

4. $2c^2 = 13c$

5. $n(n + 3) = 5n$

6. $a(a - 5) = 8a$

7. $r^2 = 144$

8. $c^2 = 400$

9. $x^2 - 28 = 0$

10. $c^2 - 45 = 0$

11. $b^2 + 63 = 0$

12. $k^2 + 48 = 0$

13. $3x^2 = 75$

14. $2q^2 = 98$

15. $4x^2 - 80 = 0$

16. $5w^2 - 90 = 0$

17. $(x + 6)^2 = 49$

18. $(a - 3)^2 = 64$

19. $(3x + 4)^2 = 80$

20. $(2x - 5)^2 = 125$

Solve the following quadratic equations by completing the square. If there is no real solution, write $\varnothing$. [Section 11.2]

21. $x^2 + 10x + 24 = 0$

22. $x^2 - 10x + 16 = 0$

23. $c^2 = 5c + 14$

24. $r^2 - 18 = -3r$

25. $2w^2 + 6 = -13w$

26. $3n^2 = 4n + 4$

27. $y^2 + 8y - 2 = 0$

28. $c^2 - 10c + 4 = 0$

29. $a^2 + 7a = 3$

30. $n^2 - 9n + 2 = 0$

31. $x^2 + 3x + 5 = 0$

32. $y^2 + 7y + 9 = 0$

33. $2x^2 + 12x - 5 = 0$

34. $2x^2 - 14x + 3 = 0$

Solve the following quadratic equations using the quadratic formula. If there is no real solution, write $\varnothing$. [Section 11.3]

35. $a^2 + 3a + 2 = 0$

36. $b^2 + 5b + 4 = 0$

37. $a^2 = 8a - 12$

38. $b^2 - 3b = 10$

39. $2x^2 + x - 10 = 0$

40. $3t^2 + t - 2 = 0$

41. $6v^2 = 8v + 3$

42. $6m^2 = -5m + 4$

43. $x^2 + 6x - 4 = 0$

44. $q^2 + 8q - 2 = 0$

45. $2x^2 + 1 = 8x$

46. $4n^2 = -6n + 5$

47. $6r^2 + 2r = 3$

48. $6x^2 = 4x + 1$

49. $3d^2 + 2d + 4 = 0$

50. $4z^2 - 3z + 5 = 0$

51. $2x^2 + 5x - 2 = 0$

52. $3x^2 + 7x = -1$

Solve the following equations: [Section 11.4]

53. $3(4x - 2) - 2(2x + 5) = 4(x - 6) + 4$

54. $2(5x - 6) - 3(4x + 2) = 5(x + 2) - 7$

55. $(x + 2)(x + 3) = x(x + 1) + 26$

56. $(2x + 5)(2x + 6) = 56$

57. $\dfrac{1}{x} + \dfrac{1}{x + 1} = \dfrac{11}{30}$

58. $\dfrac{8}{3x + 1} - \dfrac{5}{4x - 2} = \dfrac{3}{2x + 4}$

59. $\sqrt{5x + 1} = 4x - 8$

60. $\sqrt{x + 2} + \sqrt{4x + 1} = 5$

Solve the following application problems:

61. One integer is 3 more than the other. The sum of the reciprocals of the integers is $\frac{7}{10}$. Find the integers.

62. One integer is 4 more than the other. The sum of the squares of the integers is 40. Find the integers.

63. The sum of the squares of two consecutive integers is 61. Find the integers.

64. One integer is twice the other. The product of the reciprocals of the integers is $\frac{1}{32}$. Find the integers.

65. A rectangular room is 4 feet longer than it is wide. A rectangular rug is placed on the floor, and there is a uniform border of exposed floor 1 foot wide all the way around the rug. If the area of the rug is 140 square feet, find the dimensions of the room.

66. A rectangular piece of plywood is 3 feet longer than it is wide. The whole piece of plywood is to be painted with a black rectangle surrounded by a uniform border of red that is 1 foot wide. If the area of the red border is 30 square feet, find the length and width of the piece of plywood.

67. Two airplanes are approaching the same airport and are flying on courses that are perpendicular to each other. One airplane is 1 mile less than twice as far from the airport as the other. If the distance between the airplanes is 17 miles, how far is each from the airport?

68. A diver's flag is rectangular in shape and is red with a black stripe going diagonally across it. If the width of the flag is 5 inches less than the length and the length of the black stripe is 25 inches, find the length and width of the flag.

69. The average rate of a recreational vehicle is 20 miles per hour less than the average rate of a car. It takes the recreational vehicle 1 hour more to travel 200 miles than it takes the car to travel 240 miles. Find the average rate of the recreational vehicle.

70. It takes a bus two hours longer to travel 200 miles than it takes a motorcycle to travel 120 miles. The

average rate of the motorcycle is 10 miles per hour more than the average rate of the bus. Find how long it takes the motorcycle to travel 120 miles.

71. Horace is planning to build a house. The lot he has chosen has to be built up with fill dirt. It would take Frank 4 hours longer to spread the dirt than it would take his worker, Ahmad. If they can spread the dirt in $3\frac{3}{4}$ hours working together, how long would it take each when working alone?

72. After Horace built his house in Exercise 71, he and Ahmad painted it. It would take Horace two days longer to paint the house than it would take Ahmad. Together, they could paint the house in $3\frac{3}{7}$ days. How long would it take each when working alone?

Chapter 11   Test

Solve the following incomplete quadratic equations. If there is no real solution, write $\varnothing$.

1. $6x^2 - 12x = 0$

2. $4x^2 = 20x$

3. $x^2 - 121 = 0$

4. $3x^2 = 81$

5. $(x - 6)^2 = 36$

6. $(3x + 2)^2 = 18$

Find the number needed to make each of the following a perfect-square trinomial. Then write each as the square of a binomial.

7. $a^2 - 20a +$ _____

8. $b^2 + \frac{7}{3}b +$ _____

Solve the following quadratic equations by completing the square. If there is no real solution, write $\varnothing$.

9. $x^2 - 6x - 16 = 0$

10. $2x^2 - 6x - 5 = 0$

Solve the following quadratic equations using the quadratic formula. If there is no real solution, write $\varnothing$.

11. $a^2 - 5a + 6 = 0$

12. $2c^2 - 3c = 20$

13. $r^2 - 2r + 6 = 0$

14. $3w^2 = -8w + 3$

Solve the following quadratic equations using the method that seems most appropriate. If there is no real solution, write $\varnothing$.

15. $x^2 = 2x + 35$

16. $a^2 + \frac{7}{2}a = 3$

17. $3d^2 + 8 = 5d$

18. $x^2 - 8x - 109 = 0$

Solve the following equations:

19. $5(x - 2) - 3(2x + 1) = -3(x - 5) - 8$

20. $(2x + 4)(2x + 7) = 88$

21. $\dfrac{1}{x} + \dfrac{1}{x + 2} = \dfrac{7}{24}$

22. $\sqrt{5x} - \sqrt{3x + 1} = 1$

Solve the following application problems:

23. Find two consecutive even integers the sum of whose squares is 164.

24. The length of a picture is 2 inches more than the width. The picture is in a frame that is $1\frac{1}{2}$ inches wide. If the area of the picture and the frame is 99 square inches, find the length and width of the picture.

25. In mountainous country, it takes a recreational vehicle (RV) one hour longer to travel 120 miles than it takes a van. If the rate of the van is 10 miles per hour more than the rate of the RV, find how long it takes the van to travel 120 miles.

Answers for Odd-Numbered Exercises

Exercise Set 0.1

1. twelve   **3.** nine hundred five   **5.** seven thousand, one hundred forty-nine   **7.** eight thousand, nine hundred two   **9.** thirty thousand, two hundred nine   **11.** four million, seventy-eight thousand, seventy-four   **13.** twenty-nine million, seven hundred fifty-six thousand, eleven   **15.** 407   **17.** 14,073   **19.** 902,460   **21.** hundred   **23.** million   **25.** one   **27.** $7 \cdot 10 = 70$ **29.** $0 \cdot 10,000 = 0$ **31.** $4 \cdot 1,000,000 = 4,000,000$ **33.** $300 + 90 + 7$ **35.** $3,000 + 30 + 3$ **37.** $400,000 + 9,000 + 100 + 30 + 5$ **39.** $5000 + 200 + 80$ ft.   **41.** 八十六

Exercise Set 0.2

1. 68   **3.** 868   **5.** 930   **7.** 1010   **9.** 9741   **11.** 12,922   **13.** 79,962   **15.** 10,021   **17.** 87   **19.** 298   **21.** 2112   **23.** 1724   **25.** 442   **27.** 20,773   **29.** 6529   **29.** 28 ft   **C1.** 454,739   **C3.** 178,751   **31.** 13_{five}

Exercise Set 0.3

1. 25   **3.** 61   **5.** 33   **7.** 22   **9.** 211   **11.** 351   **13.** 1051   **15.** 1247   **17.** 38   **19.** 336   **21.** 453   **23.** 4775   **25.** 876   **27.** 42   **29.** 38   **31.** 108   **33.** 370   **35.** 1873   **37.** $1270   **39.** $3000   **41.** 400 mph   **C1.** 22,917   **C3.** 22,685   **C5.** 4411   **43.** $220,000

Exercise Set 0.4

1. 68   **3.** 230   **5.** 158   **7.** 1794   **9.** 520   **11.** 1026   **13.** 11,860   **15.** 8772   **17.** 103,632   **19.** 217,136   **21.** 690 chairs   **23.** $33,600   **25.** 96 windows   **27.** 525 entries   **C1.** 2340 pages   **C3.** 180,000 trees   **29.** no   **31.** $520

Exercise Set 0.5

1. 7 R 8   **3.** 7 R 2   **5.** 9 R 3   **7.** 9 R 6   **9.** 18   **11.** 118 R 4   **13.** 21 R 4   **15.** 14 R 41   **17.** 22 R 216   **19.** 16   **21.** 29   **23.** 19   **25.** 16 children   **27.** $30   **29.** 24 bunches   **31.** $5   **33.** five days   **C1.** 1314   **C3.** 3045 full-time students   **35.** 10 in. by 8 in.

Exercise Set 0.6

1. $\dfrac{9}{12}$ **3.** $\dfrac{1}{3}$ **5.** $\dfrac{36}{60}$

7.

9.

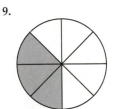

11. $\dfrac{1}{15}$ **13.** $\dfrac{8}{45}$ **15.** $\dfrac{12}{35}$ **17.** $\dfrac{35}{204}$ **19.** 2   **21.** 2   **23.** 6   **25.** $\dfrac{12}{35}$ **27.** $\dfrac{9}{10}$ **29.** $\dfrac{56}{33}$ **31.** $\dfrac{2}{15}$ **33.** $\dfrac{28}{3}$ **35.** $\dfrac{5}{7}$ **37.** $\dfrac{15}{17}$ **39.** $\dfrac{2}{5}$ **41.** $\dfrac{5}{13}$ **43.** $\dfrac{6}{7}$ **45.** $\dfrac{9}{13}$ **47.** 5 cups   **49.** $\dfrac{5}{8}$ **51.** $\dfrac{49}{12}$ ft   **53.** 18 pieces   **55.** $\dfrac{4}{15}$ hr

Exercise Set 0.7

1. six tenths   **3.** eighty-four hundredths   **5.** seven hundred sixty-three thousandths   **7.** nineteen and forty-five hundredths   **9.** sixty-eight and six hundred seven thousandths   **11.** tenths   **13.** hundredths   **15.** tenths

17. 52.0 **19.** 18043 **21.** 298.20 **23.** .66499
25. 1.1022 **27.** 1.4 **29.** .71 **31.** 6.06 **33.** 119.497
35. 1190.61 **37.** 206.27 **39.** .6 **41.** .58 **43.** 2.57
45. 35.26 **47.** 113.956 **49.** 19.1631 **51.** 32.34
53. 220.5 **55.** 1158.353 **57.** seven hundred six and
$\frac{45}{100}$ dollars **59.** three hundred five and $\frac{2}{100}$ dollars
61. $8.29 million **63.** .8 ft **65.** 298.1 mi **67.** $.07
69. $3.03 **71.** 28.971 gal **C1.** $881.29 **73.** 105.9 cents

Exercise Set 0.8

1. 3 **3.** .034 **5.** 56.7 **7.** 14.08 **9.** .936
11. 23.4 **13.** 12.8 **15.** 42.84 **17.** 6.9 **19.** .267
21. .3885 **23.** 3.17781 **25.** 4 **27.** 3 **29.** .82
31. 8.4 **33.** 65 **35.** .007 **37.** 5.0 **39.** .7
41. .08 **43.** .7 **45.** 22.67 **47.** 1476.21 **49.** 36.57
51. 96 km/hr **53.** 4.17 pages per day **55.** $32.67
57. $1,142.86 **C1.** 241.05 **C3.** 0.12 **59.** $504

Exercise Set 0.9

1. 24 in. **3.** 3 yd **5.** 26,400 ft **7.** 3520 yd **9.** 5 ft
11. 8 yd **13.** 8 yd **15.** 765 ft **17.** 9240 yd
19. 90 in. **21.** 300 m **23.** 4 dm **25.** 8000 cm
27. 1400 m **29.** 700 mm **31.** 4.58 dam **33.** 10 km
35. .84 m **37.** 9 m **39.** .0162 hm **41.** 15 yd
43. 35,376 ft **45.** .00528 km **47.** .0001 mm

Chapter 0 Review Exercises

1. six thousand fifty-one **2.** thirty thousand, four hundred
fifty-nine **3.** 37,204 **4.** ten thousand **5.** 400,000
6. 300,000 + 90,000 + 6,000 + 70 + 1 **7.** < **8.** >
9. 99 **10.** 85 **11.** 628 **12.** 6156 **13.** 729,351
14. 14,383 **15.** 42 points **16.** 41 ft **17.** 45,000 trees
18. 20 ft **19.** 688 **20.** 79 **21.** 75,156 **22.** 182 mi
23. 136 books **24.** 34 cousins **25.** $506 **26.** 455
27. 2964 **28.** 12,852 **29.** 20,130 **30.** 44,820
31. 360 test tubes **32.** 195 **33.** $157 **34.** 6 R 5
35. 89 R 2 **36.** 37 **37.** 791 **38.** 33 **39.** 454 R 8
40. $4,000 **41.** 35 classes **42.** $\frac{15}{8}$ **43.** $\frac{54}{77}$ **44.** $\frac{7}{11}$
45. $\frac{4}{9}$ **46.** 1.129 **47.** 197.149 **48.** 2.20 **49.** 146.81
50. 503.74 **51.** 467.942 **52.** 3.81 pounds **53.** $34.57
54. .088 **55.** .585 **56.** .9585 **57.** .034 **58.** 38.646
59. 4 **60.** 38.69 **61.** 3 **62.** 31.13 **63.** 30.80
64. $330.48 **65.** 29.58 ml **66.** $18.30 **67.** 60 in.
68. 3.6 ft **69.** 7.5 ft **70.** 3700 mm **71.** 1,800,000 cm
72. 890 dm **73.** 9 in. **74.** 3.2 mi **75.** 9.215 m
76. 43,600 dm

Chapter 0 Test

1. ten thousand **2.** 80,000 + 2,000 + 60 + 3 **3.** 1402
4. 96,021 **5.** 20,183 **6.** 533 cards **7.** 5981
8. 517,370 **9.** 86 seeds **10.** 5992 **11.** 53,935
12. 646 pencils **13.** 95 **14.** 37.15 **15.** $\frac{18}{5}$ **16.** $\frac{5}{12}$
17. 80.967 **18.** 538.034 **19.** 14.654 **20.** 6.2 pounds
21. 3.8775 **22.** 1.90938 **23.** 3.66 **24.** 65.00
25. 2.10 **26.** 936.67 **27.** 14 yd **28.** 66 in.
29. 76,000 cm **30.** 90 mm

Exercise Set 1.1

1. 8 squared or 8 to the second power, 64 **3.** 5 cubed or 5
to the third power, 125 **5.** 9 to the fourth power, 6561
7. 2 to the fifth power, 32 **9.** 10 to the sixth power,
1,000,000 **11.** four-fifths squared or four-fifths to the second
power, $\frac{16}{25}$ **13.** 1.3 cubed or 1.3 to the third power, 2.197
15. $a \cdot a \cdot a \cdot a$ **17.** $a \cdot a \cdot b \cdot b \cdot b$ **19.** $5 \cdot x \cdot x \cdot x$
21. $4a \cdot 4a \cdot 4a \cdot 4a$ **23.** p^4 **25.** $b^4 d^3$ **27.** $3x^4$
29. $(4r)^6$ **31.** 9 **33.** 2 **35.** 19 **37.** 56 **39.** 46
41. 6 **43.** 12 **45.** 288 **47.** 2304 **49.** $\frac{1}{18}$ **51.** 56
53. 24 **55.** 77 **57.** 16 **59.** 21 **61.** 5 **63.** 3
65. 32 **67.** 4 **69.** 10 **71.** 24 **73.** 4 **75.** 61
77. 2 **79.** 3 **81.** 4 **C1.** 2187 **C3.** 53.5824
C5. 43.7235 **C7.** 430.204708 **83.** 18 **85.** 32
87. 12 **89.** 180 **91.** 48 **93.** 8 **95.** 2 **97.** $\frac{4}{75}$
99. $\frac{4}{25}$ **C9.** 13.44 **C11.** 181.8432 **C13.** 99.5328
101. $654 **103.** $5160 **105.** $396,000 **107.** $7.38
109. $13,514 **111.** 70 **113.** 1096

Exercise Set 1.2

1. 14 m **3.** 33 in. **5.** 72 in. **7.** 42 m **9.** 180 ft
11. 28 ft **13.** 4.8 yd **15.** 56 cm **17.** 26.2 m
19. 30 mm **21.** 22 in. **23.** 14π m **25.** 34.5 ft
27. $\frac{44}{3}$ or $14\frac{2}{3}$ in. **29.** 40 ft **31.** 32.28 ft **33.** 208 in.
35. 69 ft **37.** $990 **39.** 314 ft **41.** 50.24 in.
43. 219.8 in.

Exercise Set 1.3

1. 54 m^2 **3.** 1400 ft^2 **5.** 45 ft^2 **7.** 1.44 yd^2
9. $\frac{25}{49}$ cm^2 **11.** 19.32 m^2 **13.** 30 mm^2 **15.** 15 in.^{2}
17. 49π m^2 **19.** 95.0 ft^2 **21.** $\frac{154}{9}$ or $17\frac{1}{9}$ in.^{2}
23. 112 mm^2 **25.** 1.56 mi^2 **27.** 7.13 cm^2 **29.** 42 mi^2
31. 107.8 in.^{2} **33.** 52 ft^2 **35.** 7.74 cm^2 **37.** 30,000
plants, $88,500 **39.** 18,000 yd^2 **41.** 260 pieces of tile,
$673.40 **43.** $1,050

Exercise Set 1.4

1. $V = 72$ ft^3, $SA = 108$ ft^2 **3.** $V = 270.4$ m^3, $SA = 254.8$ m^2
5. $V = 125$ mi^3, $SA = 150$ mi^2 **7.** $V = \frac{27}{125}$ cm^3, $SA = \frac{54}{25}$ cm^2
9. $V = 300\pi$ cm^3, $SA = 170\pi$ cm^2 **11.** $V = 105.84\pi$ m^3,
$SA = 85.68\pi$ m^2 **13.** $V = 42.9$ m^3 **15.** $V = \frac{1100}{147}$ or
$\frac{71}{147}$ in.^{3} **17.** 250π in.^{3} **19.** 108π in.^{3} **21.** 14.22 yd^3
23. $31,752\pi$ in.^{3} **25.** $720 **27.** $454.80
29. 1,272,000 in.^{2} **31.** $576 **33.** $8.40 **35.** 50.24 in.^{2}
37. $1256

Exercise Set 1.5

1. -8 **3.** losing 10 lbs **5.** 250 mi west

7.
$-6\ -5\ -4\ -3\ -2\ -1\ 0\ 1\ 2\ 3\ 4\ 5\ 6$

9.

$$\underset{-10\,-9\,-8\,-7\,-6\,-5\,-4\,-3\,-2\,-1\ 0\ 1\ 2\ 3\ 4}{\leftarrow\!\!+\!\!\bullet\!\!+\!\!+\!\!+\!\!+\!\!+\!\!+\!\!+\!\!+\!\!+\!\!\bullet\!\!+\!\!+\!\!\bullet\!\!+\!\!\bullet\!\!+\!\!+\!\!+\!\!\bullet\!\!\rightarrow}$$

11. 0 or any negative integer **13.** 0 **15.** sometimes
17. always **19.** always **21.** -17 **23.** 15
25. -8 **27.** -8 **29.** 15 **31.** $>$ **33.** $<$ **35.** $>$
37. $>$ **39.** $=$ **41.** $<$ **43.** $>$ **45.** $=$ **47.** 5
49. 78

Exercise Set 1.6

1. 18 **3.** 11 **5.** -7 **7.** 9 **9.** -11 **11.** -11
13. 48 **15.** 11 **17.** -6 **19.** 29 **21.** 1.8 **23.** -4.7
25. 7 **27.** -5 **29.** -25 **31.** -70 **33.** 3

35. -11 **37.** -17 **39.** -5 **41.** $\dfrac{4}{11}$ **43.** $-\dfrac{4}{17}$

45. $-\dfrac{7}{11}$ **C1.** -18.59 **C3.** -942.2249 **C5.** -1086.78

47. commutative Property of Addition **49.** additive inverse
51. commutative Property of Addition **53.** additive inverse
55. commutative Property of Addition **57.** associative
property of addition **59.** $6 + (-5)$ **61.** $-4 + (12 + 9)$
or $(9 + 12) + (-4)$ **63.** 0 **65.** $7(8 + 5)$ **67.** $-11 + 8$
$= -3$ **69.** $-6 + (-4) + 10 = 0$ **71.** $8 + 11 = 19$
73. $-12 + 7 = -5$ **75.** $49 + 35 = 84$ **77.** $-18 + 36 = 18$
79. $92° + (-25°), 67°$ **81.** $12° + (-19°), -7°$
83. $-12° + 24°, 12°$ **85.** $-21° + (-7°), -28°$
87. -68 ft $+ 39$ ft, -29 ft **89.** -65 ft $+ (-43$ ft$), -108$ ft
91. 3rd parking level below ground level **93.** 12th floor

95. $143 **97.** $376 **99.** $39\dfrac{1}{4}$ **101.** $946 **103.** a. yes,

b. Total net gain was 11 yards **105.** 18 **107.** 504

Exercise Set 1.7

1. 6 **3.** -6 **5.** 12 **7.** 4 **9.** -8 **11.** -1.7
13. -9 **15.** -12 **17.** 25 **19.** 2.2 **21.** 4 **23.** 12
25. 1 **27.** -14 **29.** 39 **31.** 2 **33.** 17 **35.** -26

37. -1 **39.** $-\dfrac{23}{20}$ **41.** $-\dfrac{7}{17}$ **43.** $-\dfrac{29}{24}$ **45.** $-\dfrac{6}{13}$

C1. 16,138.3227 **C3.** -8.93 **47.** $11x$ **49.** $6y$
51. $7ab$ **53.** $-7x^2$ **55.** $4z + 3$ **57.** $-6y - 3$
59. $-11x$ **61.** $-5r$ **63.** $-4xy + 8x$ **65.** $-8x^2 - 2y^2$
67. $12z + x - 12$ **69.** $7b^2 - 10b + 12$ **71.** $m^2r + 6r$
73. $7 - (-5) = 12$ **75.** $-10a - 6a = -16a$
77. $8 - 5 = 3$ **79.** $-5m - 8m = -13m$ **81.** $6 - 9 = -3$
83. $9 - (-13) = 22$ **85.** $26r - 32r = -6r$ **87.** $14 +$
$[6 - (-5)] = 25$ **89.** $[-4ab - (-6ab)] + 15ab = 17ab$
91. $[9 - (-4)] - (-5 + 6) = 12$ **93.** $[7y - (-3y)] +$
$[-5y - (-2y)] = 7y$ **95.** -62 ft $- (-18$ ft$), -44$ ft
97. -15 ft $- (-34$ ft$), 19$ ft **99.** $350 - (-145) = 495$ ft
101. $12,000 - (-5000) = $17,000 **103.** $-125 - 250 = -375$ ft
105. $2500 - (-187) = 2687$ ft **107.** $-6 - 17 = -23$ psi

109. $+$157 **111.** 100 **113.** 100 **115.** $\dfrac{2}{5}x^2y + \dfrac{5}{7}xy^2$

117. $-\dfrac{4}{13}m^3n^2 - \dfrac{7}{9}mn^2$ **119.** $-\dfrac{11}{23}x^3y - \dfrac{9}{29}xy^2$

Exercise Set 1.8

1. monomial **3.** trinomial **5.** binomial **7.** trinomial
9. none of these **11.** monomial **13.** none of these
15. $3; x; 1$ **17.** $1; x; 2$ **19.** $-5; x; 1$ **21.** $4; x; 3$
23. $-1; x; 3$ **25.** $10; x$ and $y; 6$ **27.** $-13; x$ and $z; 6$
29. $-5x + 4; 1$ **31.** $2x^2 + 3x - 4; 2$ **33.** $-x^4 + x^3 - 4x^2 +$
$3x; 4$ **35.** $3x^3 + x; 3$ **37.** 0 **39.** 26 **41.** -1

43. 45 **45.** 19 **C1.** 4.674 **47.** 100 ft **49.** 27 units
51. $9800 **53.** 2500 sq ft **55.** $9x - 3y$ **57.** $2x + 2$
59. $5x^2 - 5x + 14$ **61.** $5m^2c^2 + 4mc - 9c^2 + 3$ **63.** $-5r^2 +$
$6r - 10$ **65.** $rp^2 - 5rp - r^3$ **67.** $h^2 - 9$ **69.** $11x^2 +$
$6x + 3$ **71.** $-3x^2 - 10x + 4$ **73.** $8x + 5$ **75.** $2x + 5$
77. $3x^2 + x - 2$ **79.** $12z - 11$ **81.** $x^2 - 11x + 6$
83. $4x - 18$ **85.** $8x$ cm **87.** $(8x + 1)$ in. **89.** $16x$ in.
91. $(7x + 9)$ ft **93.** $(3x - 2)$ ft **95.** $(x - 2)$ ft

97. $(5x - 2)$ m **99.** $\dfrac{6}{7}x - \dfrac{1}{11}y$ **101.** $\dfrac{2}{13}r - \dfrac{16}{17}m$

103. $\dfrac{1}{3}a^2 - \dfrac{2}{5}a + \dfrac{7}{15}$ **105.** $\dfrac{-4}{23}y^2 - \dfrac{1}{19}y + \dfrac{4}{13}$

Chapter 1 Review Exercises

1. 9^5 **2.** $(-2)^4$ **3.** $3^3 \cdot 5^4$ **4.** $12 \cdot 9^3 \cdot 2^2$ **5.** $3^3 \cdot a^3$
6. $7^3r^4s^2$ **7.** 324 **8.** 80 **9.** 900 **10.** 50,000
11. 54 **12.** 36 **13.** 90 **14.** 384 **15.** 169 sq in.
16. 216 cm^3 **17.** 2 **18.** 20 **19.** 9 **20.** 23
21. 10 **22.** 4 **23.** 31 **24.** 27 **25.** 12 **26.** 72
27. 270 **28.** 20 in. **29.** 36 ft **30.** 21 ft **31.** 31.4 m
32. 21 in.^{2} **33.** 81 ft^2 **34.** 18 ft^2 **35.** 78.5 m^2
36. 84 yd^2 **37.** 81 cm^2 **38.** $V = 240$ ft^3, $SA = 236$ ft^2
39. $V = 216$ in.^{3}, $SA = 216$ in.^{2} **40.** $V = 175\pi$ m^3, $SA =$
120π in.^{2} **41.** 32π m^3 **42.** 1000 ft **43.** 942 ft
44. $17.50 **45.** $1820 **46.** 2704 in.^{2} **47.** 64 in.^{2}
48. 70,650 ft^2 **49.** 120 in.^{2} **50.** 48 in.^{3} $SA = 124$ in.^{2}
51. a. 35.325 in.^{3} **b.** 61.23 in.^{2} **52.** 24π in.^{3} **53.** $-$$50
54. 7 ft **55.** -23 **56.** 35 **57.** -7 **58.** -32
59. 64 **60.** 86 **61.** $<$ **62.** $<$ **63.** $>$ **64.** $=$
65. $<$ **66.** $>$ **67.** 6 **68.** 15 **69.** -7 **70.** 13
71. 23 **72.** -7 **73.** -95 **74.** -30 **75.** commutative of addition **76.** associative of addition **77.** commutative of addition **78.** commutative of addition
79. additive inverse **80.** additive Identity **81.** $-9 + 6$
$= -3$ **82.** $15 - (-8) = 23$ **83.** $-5 + 16 = 11$
84. $3 - (-6) = 9$ **85.** $-9 + 8 = -1$ **86.** $-3 - (-7) = 4$
87. $(-5 + 13) + [18 + (-8)] = 18$ **88.** $[-9 - (-2)] +$
$(-5 - 9) = -21$ **89.** $[14 + (-8)] - [4 - (-5)] = -3$
90. $5(-6 + 10) = 20$ **91.** $-3x$ **92.** $2x + y$
93. $-5w^3 - 6u^2 + 3$ **94.** $-5x^2y + 6xy^2 + 4$ **95.** $9; x; 2$
96. -13, a, sixth **97.** $-9; x$ and $y; 9$ **98.** $4x^3 + 6x$,
binomial, third **99.** $7y^2 - 8y + 2$; trinomial; 2 **100.** $5x^7$,
monomial, seventh **101.** $-7x^5 + 9x^4 + 2x$; trinomial; 5
102. $14a^6 + 2a^3 - 5a^2 + 19$, none of these, sixth **103.** not a
polynomial **104.** -3 **105.** 18 **106.** 26 **107.** 253
108. $10x - 5$ **109.** $5a + 11b$ **110.** $5u^4 - 5u^3 + 2u$
111. $2a^2 - 14a + 5$ **112.** $z^4 + z^3 + 2z + 2$
113. $5x^4 - 4x^2 - 1$ **114.** $z^2 + 6z - 8$ **115.** $11x^2 - 11x + 4$
116. $3x^3 + 6x^2 - 4x - 11$ **117.** $-2x^2 + 13x - 5$
118. $-2u^4 + 3u^3 - 3u^2 - 3u + 8$ **119.** $-2x^3 - 13x^2 + 2x + 5$
120. $(9x - 6)$ ft **121.** $(x + 11)$ ft **122.** $(4x - 3)$ ft
123. $2x - 9$ dollars

Chapter 1 Test

1. $6^4 \cdot 5^2$ **2.** $4^2 \cdot x^4 \cdot y^2$ **3.** $3x^5$ **4.** $(4a)^3$ **5.** 33
6. 23 **7.** 70 **8.** 2 **9.** 256 **10.** 4 **11.** $P = 33$ cm,
$A = 42$ cm^2 **12.** $P = 26$ ft, $A = 36$ ft^2 **13.** 40 m^2
14. 38 ft^2 **15.** $C = 43.96$ m, $A = 153.86$ m^2 **16.** $V =$
90 in.^{3}, $SA = 126$ in.^{2} **17.** $V = 1538.6$ m^3, $SA = 747.32$ m^2
18. $36 **19.** $72 **20.** 4500 lbs **21.** $<$ **22.** $<$
23. -11 **24.** -7 **25.** -1 **26.** -44
27. commutative of addition **28.** additive inverse

29. $-7 + 11 = 4$ **30.** 3   **31.** -15 **32.** $-x^2 - 11x + 5$
33. $-x + 2$ **34.** $x + 8$ meters   **35. a.** 4; **b.** 3; **c.** $4x^3 + 3x^2 + 6x$; **d.** trinomial   **36.** 16   **37.** $5xy - 4yz$
38. $5x^2 - 3x - 4$ **39.** $6n^3 - 2n^2 + 7n - 2$ **40.** No.
$-x$ is positive if x is negative because the opposite of a negative number is a positive number.

Exercise Set 2.1

1. -24 **3.** -24 **5.** 24   **7.** -21 **9.** 44   **11.** 3.5
13. -9.12 **15.** 15   **17.** -48 **19.** -90 **21.** -120
23. 120   **25.** -81 **27.** 121   **29.** -81 **31.** -72

33. -108 **35.** 125   **37.** $-\dfrac{8}{15}$ **39.** $-\dfrac{56}{75}$ **41.** $\dfrac{9}{40}$

43. $-\dfrac{45}{64}$ **C1.** -49.651 **C3.** 6029.6236   **45.** -6

47. 10   **49.** 24   **51.** 40   **53.** -256 **55.** -128
57. -9 **59.** 33   **61.** 48   **63.** 10   **65.** -28
67. associative of multiplication   **69.** commutative of multiplication   **71.** identity for multiplication
73. commutative of multiplication   **75.** inverse for multiplication   **77.** multiplication by 0   **79.** $[3(-9)](-7)$
81. $(-7)(3)$ **83.** $4(-9) + 4(2)$ **85.** $(-9 + 2)4$ **87.** -2
89. 1   **91.** 0   **93.** 14   **95.** -11 **97.** $-12°$
99. -70 ft   **101.** -6 points

Exercise Set 2.2

1. r^8 **3.** 4^7 **5.** x^9 **7.** $20x^9$ **9.** $21a^8$ **11.** x^8y^7
13. $8a^{10}b^{11}$ **15.** a^5b^5 **17.** $16y^4$ **19.** $16x^2$ **21.** $-25x^2$
23. $27c^3$ **25.** $-27b^3$ **27.** y^{18} **29.** 4^{15} **31.** x^7y^7
33. 4^8x^8 **35.** $c^{24}d^{18}$ **37.** $4^{20}a^{12}$ **39.** $-4^{20}b^{25}$
41. -6^6b^9 **43.** $108a^{17}b^{12}$ **45.** $4^{20}b^{28}$ **47.** $a^{14}b^{16}c^{22}$
49. $16a^2$ **51.** $3x^3$ **53.** $89x^6$ **55.** $-24a^4$ **57.** $-2m^4n^6$
59. a^{8x} **61.** a^{5b-3} **63.** $27^xa^{5y}b^{3z}$ **65.** $2^cx^{ac}y^{bc}$ **67.** b^{6n-3}

Exercise Set 2.3

1. $-7a^3x^3$ **3.** $42f^7k^2b^5$ **5.** $24p^5q^5r^8$ **C1.** $-53.7654x^9y^{13}$
C3. $-34.56x^{11}y^9$ **7.** $2x + 6$ **9.** $-8x + 12$ **11.** $6x^2 - 18x + 15$ **13.** $-12b^2 + 4b - 8$ **15.** $4b^3 - 8b^2 + 10b$
17. $2h^5 + 16h^4 - 10h^3$ **19.** $2b^6 + 14b^5 - 12b^4$ **21.** $-6r^6 + 8r^5 - 4r^4 + 6r^3$ **23.** $10x^4y^3 - 20x^3y^4$ **25.** $-12p^6q^4 + 4p^4q^5 - 4p^2q^4$ **C5.** $22.79x^5 - 33.54x^4 - 30.96x^3$ **27.** $7x + 6$
29. $12a - 16$ **31.** $2y - 4$ **33.** $-11a - 6b$ **35.** $-14x - 7$
37. $24c^2 - c - 6$ **39.** $7r^2 + 16r - 13$ **41.** $-14a^2 + 21a + 4$
43. $11a^2 + 4a + 14$ **45.** $a^2 + 9a + 20$ **47.** $z^2 - 3z - 40$
49. $x^2 - 16$ **51.** $4x^2 - y^2$ **53.** $x^2 - xy - 2y^2$ **55.** $20w^2 - 17w - 24$ **57.** $6z^2 - 26z + 24$ **59.** $6a^2 + 5ab - 4b^2$
61. $x^3 + 2x - 3$ **63.** $2x^3 + 2x^2 - 9x + 9$ **65.** $x^3 + 8$
67. $6x^3 - x^2 - 18x + 9$ **C7.** $37.63x^2 - 15x - 17.28$
69. $12a^4 + 5a^2b^2 - 2b^4$ **71.** $4a^4 + 8a^3 - 4a^2 + 12a - 15$
73. $a^4 - a^3 - 7a^2 + a + 6$ **75.** $x^{2n} - x^n - 6$ **77.** $2x^2 + 10x + 12$ **79.** $6x^2 - 25x + 25$ **81.** $2a^3 - 6a^2 - 8a + 24$
83. $8y^3 + 8y^2 - 14y + 4$ **85.** $2x^2$ **87.** $4x^2 + 24x$
89. $2x^2 + x - 6$ **91.** $3x^3 - x^2 + 4x + 4$ **93.** $4y^2$
95. $9z^2$ **97.** $(2y^2 - 8y)$ ft^2 **99.** $(9x^2 + 9x - 10)$ ft^2
101. $(12.5x^2 + 37.5x)$ dollars

Exercise Set 2.4

1. $x^2 + 5x + 6$ **3.** $a^2 - a - 12$ **5.** $c^2 - 7c + 10$
7. $a^2 + 7ab + 12b^2$ **9.** $r^2 - 2rs - 15s^2$ **11.** $6a^2 + 13ab + 6b^2$ **13.** $6x^2 - xy - 15y^2$ **15.** $6x^2 - 19xy + 10y^2$
C1. $25.42x^2 + 35.54xy - 32.76y^2$ **17.** $x^2 - 9$ **19.** $p^2 - q^2$

21. $a^2 - 16b^2$ **23.** $36a^2 - b^2$ **25.** $16a^2 - 25$
27. $4x^2 - 25y^2$ **C3.** $51.84a^2 - 44.89$ **29.** $x^2 + 4x + 4$
31. $a^2 - 8a + 16$ **33.** $r^2 + 2rs + s^2$ **35.** $t^2 - 2ts + s^2$
37. $4x^2 + 12x + 9$ **39.** $9a^2 - 6a + 1$ **41.** $4x^2 + 20xy + 25y^2$
43. $16x^2 - 24xy + 9y^2$ **45.** $25t^4 - 10t^2w + w^2$
C5. $22.09x^2 + 34.78xy + 13.69y^2$ **47.** $6a^2 + 11ab - 10b^2$
49. $16a^2 - 9b^2$ **51.** $9a^2 + 30a + 25$ **53.** $8 - 18b + 9b^2$
55. $36 - y^2$ **57.** $16 + 24w + 9w^2$ **59.** $16c^2 - 49d^2$
61. $(3 + x)^2 = 9 + 6x + x^2$ **63.** $x^2 + 3$ **65.** $(x - 5)^2 = 25$
67. $x^2 - 3x - 40$ **69.** $12a^2 + 21a - 45$ **71.** $x^2 + 16x + 64$
73. $9a^2 - 30ab + 25b^2$ **75.** $p^3 + q^3$ **77.** $x^3 + 27$
79. $b^3 - 64$ **81.** $27a^3 + 8b^3$

Exercise Set 2.5

1. -4 **3.** -6 **5.** 6   **7.** undefined   **9.** -8

11. $-\dfrac{9}{20}$ **13.** $-\dfrac{12}{25}$ **15.** 5   **17.** 8   **19.** 16   **21.** 10

23. -5 **25.** -4 **27.** $-7°$ per hour   **29.** $-\dfrac{3}{50}$ or

$-.06$ ft per foot   **31.** -9 ft per sec   **33.** -2 **35.** 22
37. -16 **39.** -33 **41.** 4   **43.** 12   **45.** -22
47. 11   **49.** -6 **51.** 1   **53.** -5 **55.** -16 **57.** 27

59. -4 **61.** 2   **63.** -2 **65.** $\dfrac{17}{2}$ **67.** $-\dfrac{14}{29}$

C1. 60.96   **C3.** 5.8   **C5.** 9948.2   **C7.** -2665.18
C9. 2.35   **69.** 10   **71.** -89 **73.** -7 **75.** 36
77. 90   **79.** -324 **81.** 27   **83.** 60   **85.** -26

87. -92 **89.** $\dfrac{5}{3}$ **91.** $-\dfrac{5}{18}$ **93.** $\dfrac{43}{42}$ **95.** $6.49 profit

97. $300 loss   **99.** $1398 profit

Exercise Set 2.6

1. 2^4 **3.** c^4 **5.** z^4 **7.** $\dfrac{1}{2^4}$ **9.** $\dfrac{1}{a^6}$ **11.** 5^3 **13.** y^5

15. $\dfrac{1}{x^3}$ **17.** $\dfrac{1}{5^3}$ **19.** 1   **21.** 3   **23.** 1   **25.** 2

27. -1 **29.** $\dfrac{5}{a^2}$ **31.** $\dfrac{-9}{c^6}$ **33.** $\dfrac{1}{5^2a^2}$ **35.** 2^2 **37.** $\dfrac{1}{5^2}$

39. $\dfrac{1}{x^5}$ **41.** $\dfrac{1}{z^2}$ **43.** $\dfrac{1}{3^4}$ **45.** $\dfrac{1}{y^9}$ **47.** 6^7 **49.** r^7

51. 6   **53.** $\dfrac{1}{2^4}$ **55.** 8^{18} **57.** r^{20} **59.** $\dfrac{1}{6^{15}}$ **61.** $\dfrac{1}{y^{21}}$

63. $\dfrac{1}{2^4}$ **65.** $\dfrac{1}{z^{15}}$ **67.** 9^{10} **69.** $-\dfrac{108x^3}{y^5}$ **71.** $-\dfrac{2m^{14}}{n^{16}}$

73. $\dfrac{8b^{18}}{a^{21}}$ **75.** x^{2a} **77.** a^{m+1}

Exercise Set 2.7

1. $\dfrac{4}{9}$ **3.** $\dfrac{27}{125}$ **5.** $\dfrac{125}{8}$ **7.** $\dfrac{4}{9}$ **9.** $\dfrac{2401}{81}$ **11.** $\dfrac{a^3}{b^3}$

13. $\dfrac{q^4}{p^4}$ **15.** $\dfrac{8x^3}{y^3}$ **17.** $\dfrac{27q^3}{p^3}$ **19.** $\dfrac{a^{12}}{b^{20}}$ **21.** $\dfrac{y^{12}}{x^{24}}$

23. $\dfrac{1}{64x^6}$ **25.** $\dfrac{y^{10}}{4}$ **27.** $\dfrac{25}{a^4b^{10}}$ **29.** $\dfrac{z^4}{4w^6}$ **31.** $\dfrac{w^9}{z^{18}}$

33. $\dfrac{z^{30}}{x^{20}}$ **35.** $24w^3$ **37.** $\dfrac{3}{10x}$ **39.** x^2 **41.** $\dfrac{1}{a^{15}}$

43. a^{14} **45.** $\dfrac{16}{y^8}$ **47.** x^4 **49.** $\dfrac{1}{4x^4}$ **51.** $\dfrac{g^4}{16}$ **53.** $\dfrac{1}{x^{15}}$

55. $\dfrac{1}{q}$ **57.** $\dfrac{1}{a^{12}}$ **59.** $\dfrac{8}{x^{12}}$ **61.** $81a^3$ **63.** $\dfrac{216a^7}{b^{19}}$

65. $a^{mq}b^{nq}$

Exercise Set 2.8

1. $5x$ **3.** $-3z^2$ **5.** $2x^2$ **7.** $-\dfrac{5}{a^3}$ **9.** $-6n^2$ **11.** 2

13. x^2y^2 **15.** $-\dfrac{s^4}{r^2}$ **17.** $3x^2y^2$ **19.** $-\dfrac{4y^3}{x^3}$ **21.** $\dfrac{xy^2}{z^2}$

23. $-\dfrac{4y^3}{x^2z^2}$ **25.** $\dfrac{2n^2}{p^3}$ **27.** $9x^2y^2$ **29.** $-\dfrac{5}{y^2}$ **31.** $x+3$

33. $a-1$ **35.** $3+\dfrac{7}{y}$ **37.** $2z-1$ **39.** $3y+\dfrac{2}{y}$

41. $4a^3b-2ab^4$ **43.** $\dfrac{2n}{m}+\dfrac{3m^2}{n}$ **45.** $2x^2+3x-1$

47. y^3-y^2-y **49.** $3m^3-m^2+\dfrac{3}{5}m$ **51.** $3y^2-4y+\dfrac{2}{y}$

53. $3m-2m^2n+4m^3n^2$ **55.** $3b-\dfrac{3}{4b}$ **57.** $\dfrac{4n}{m^2}+\dfrac{3m}{n^2}$

59. $4r^2-2+\dfrac{3}{r^2}$ **61.** $\dfrac{8}{pq}-\dfrac{3p^2}{4q^2}+5q^2$ **63.** $2mn^2$

65. $\dfrac{2n}{m^3}$ **67.** $2a^4-3a^2$ **69.** $3y^2-4-\dfrac{2}{y^2}$ **71.** $2x^2y^2$

73. $3y^3-5y+4$

Exercise Set 2.9

1. $35{,}000$ **3.** $.0095$ **5.** $4{,}790{,}000$ **7.** $.00000924$
9. 100 **11.** $.00000001$ **13.** 7.6×10^3 **15.** 3.5×10^{-4}
17. 8.57×10^5 **19.** 4.98×10^{-7} **21.** 6×10^5
23. 1×10^{-8} **25.** $300{,}000{,}000$ **27.** 1376 **29.** $.0131976$
31. $.000000264388$ **33.** 2000 **35.** $.00000015$
37. $1{,}200{,}000$ **39.** 50 **41.** $25{,}000$ **43.** $9{,}000{,}000$
45. $126{,}500{,}000$ **47.** $.252$ **49.** $.00000000000009$
51. $23{,}000$ **53.** $.000025$ **55.** $22{,}000{,}000$ **57.** $.18$
59. $.06$ **61.** 8 **63.** 1.5×10^8 km   **65.** 1.36×10^4 kg
per cubic meter   **67.** 1.3×10^{-6} meters   **69.** $299{,}792{,}500$
meters per second   **71.** $19{,}300$ kg per cubic meter
73. $.000000000000001673$ grams   **75.** 4.2 light-years
77. 1.5 moles   **79.** approximately $115{,}068$ years. Yes, since
this is considerably longer than our life span.

Chapter 2 Review Exercises

1. -8 **2.** 15 **3.** -72 **4.** -120 **5.** 144 **6.** 64
7. -15 **8.** -12 **9.** -72 **10.** -96 **11.** -25
12. 19 **13.** -42 **14.** commutative of multiplication
15. associative of multiplication   **16.** commutative of multiplication   **17.** distributive   **18.** multiplicative inverse
19. identity for multiplication   **20.** w^{13} **21.** 4^{14}
22. $-40x^{10}$ **23.** $-30a^6b^{12}$ **24.** x^7y^7 **25.** $125d^3$
26. $16a^4b^4$ **27.** x^9y^9 **28.** $-2048x^8$ **29.** x^{20} **30.** 4^{24}
31. $a^{16}b^{24}$ **32.** $4^4a^{12}b^{28}$ or $256a^{12}b^{28}$ **33.** $a^{37}b^{19}$
34. $72m^{13}n^{19}$ **35.** $15x^{10}y^5$ **36.** $18y^3-12y$
37. $-20x^4+25x^2$ **38.** $21c^5-49c^4+14c^3$
39. $-6x^6y^4+16x^3y^8-8x^5y^3$ **40.** $12x+13$
41. $3x+9$ **42.** $5a^2-23a+13$ **43.** $-b^2+19b+13$
44. $x^2+7x+10$ **45.** $12m^2-9m-30$
46. $20m^2-9mn-18n^2$ **47.** x^2-81 **48.** $25x^2-36$
49. $4s^2-25t^2$ **50.** $m^2+2mn+n^2$ **51.** $9a^2-12a+4$
52. $36a^2-24ab+4b^2$ **53.** $a^3-a^2-16a+16$ **54.** $6m^3-$
$11m^2-18m+20$ **55.** $12x^3-26x^2y+18xy^2-4y^3$

56. $-4x+19$ **57.** $6m-21$ **58.** x^3-64 **59.** $27x^3+$
125 **60.** $15a^2-11a-14$ **61.** $10x^2$ **62.** $9z^2+48z+$
64 **63.** $36x^2-48x+16$ **64.** $9z^2+12z+4$ **65.** -8
66. 2 **67.** -9 **68.** -4 **69.** 2 **70.** 2 **71.** 4
72. -2 **73.** -12 **74.** 8 **75.** -61 **76.** 3
77. -20 **78.** -40 **79.** 16 **80.** -2 **81.** -6
82. -3 **83.** -1 **84.** 35 **85.** 432 **86.** 72
87. -41 **88.** -10 **89.** $-\dfrac{13}{12}$ **90.** $-\dfrac{34}{57}$ **91.** 2^5
92. m^2 **93.** $\dfrac{1}{3^4}$ **94.** $\dfrac{1}{x^6}$ **95.** $\dfrac{1}{5^2}$ **96.** 4 **97.** $\dfrac{1}{5}$
98. z^2 **99.** $\dfrac{1}{x^6}$ **100.** $\dfrac{1}{5^6}$ **101.** a^3 **102.** 7^{10}
103. $\dfrac{9}{25}$ **104.** $\dfrac{4}{9}$ **105.** $\dfrac{1}{3^{10}}$ **106.** r^{20} **107.** $\dfrac{x^{12}}{y^{12}}$
108. $\dfrac{x^6}{64}$ **109.** $\dfrac{1}{5^3a^{12}}$ **110.** $\dfrac{x^{12}}{8y^{18}}$ **111.** $\dfrac{n^6}{m^8}$
112. m^{32} **113.** $8n^3$ **114.** mn^4 **115.** $\dfrac{6}{m^2}$ **116.** $-5x^2$
117. $\dfrac{3}{m^3n^6}$ **118.** m^7 **119.** $x-2$ **120.** $5y^2+3$
121. $7a-4$ **122.** $\dfrac{4m^2}{n^2}+\dfrac{2n^2}{m}$ **123.** $4x^2-3x-7$
124. $\dfrac{1}{mn^2}-\dfrac{4n^2}{m^4}+\dfrac{3}{n}$ **125.** $-\dfrac{3m^2}{n^2}$ **126.** $3a^3-2a$
127. $2x^2-3x-\dfrac{1}{3}$ **128.** $7x^2-\dfrac{10}{x}$ **129.** $4x-2-\dfrac{2}{x}$
130. $460{,}000$ **131.** $3{,}060{,}000$ **132.** $.0000703$
133. 0.007123 **134.** 9.7×10^{10} **135.** 4.67×10^4
136. 4.78×10^{-6} **137.** 3.07×10^{-4} **138.** $.01395$
139. $.00000205842$ **140.** $.03$ **141.** $.00000003$
142. $.00552$ **143.** $.00000000001598$ **144.** $50{,}000{,}000$
145. 1500

Chapter 2 Test

1. 60 **2.** -72 **3.** $-6x^5y^5$ **4.** $48x^6y^3$ **5.** $6x^3-8x^2$
6. $2p^3q^4-4p^4q^2-2p^2q$ **7.** $3x+8$ **8.** $6p^2+5p-4$
9. $6a^2-11ab+4b^2$ **10.** $3y^3+5y^2+13y-5$
11. $25x^2-9y^2$ **12.** $4a^2-28a+49$ **13.** -4 **14.** 3
15. $-\dfrac{24n^7}{m^3}$ **16.** 7 **17.** $-\dfrac{2z^3}{x^3}$ **18.** $\dfrac{x^{15}}{27y^{12}}$ **19.** $\dfrac{2a^3}{b^{18}}$
20. $\dfrac{2b^6}{a^{10}}$ **21.** $2m^3n-8mn^3+5m^2$ **22.** $4x-5+\dfrac{3}{x}$
23. -14 **24.** -3 **25.** -5 **26.** 18 **27.** $5x^2-6x-10$
28. a. 4.79×10^9 **b.** 7.49×10^{-8} **29.** 1075 **30.** 4.2×10^{10}

Exercise Set 3.1

1. $t=8$ **3.** $r=-6$ **5.** $a=8$ **7.** $x=10.5$
9. $m=-1.2$ **11.** $w=9$ **13.** $x=7$ **15.** $z=3$
17. $t=-1$ **19.** $x=8.1$ **21.** $r=4$ **23.** $x=0$
25. $x=-2$ **27.** $x=6$ **29.** $x=2$ **31.** $x=-2.6$
33. $y=-7.7$ **35.** $t=12$ **37.** $x=4$ **39.** $x=10$
41. $x=-.77$ **43.** Some number minus five is nineteen.
A number decreased by five is nineteen.   **45.** The sum of
thirteen and some number is seven. Some number more than
thirteen is seven.   **47.** Seventeen is some number decreased
by six. Seventeen equals the difference of a number and six.
49. $x-2=9$, $x=11$ **51.** $x-14=8$, $x=22$
53. $x-2=15$, $x=17$ **55.** $5+x=13$, $x=8$

57. $x - 8 = -10, x = -2$ **59.** $x - 2 = -8, x = -6$
61. $x + 275 = 550, x = 275$ miles   **63.** $x + 75 = 160, x = \$85$
65. $x - 8 = 24, x = 32$ points   **67.** $632 + x = 800, x = 168$
points   **69.** $x + 46 = 326, x = \$280$ **71.** $x - 23 = 118$,
141 lbs   **73.** $x + 18{,}000 = 40{,}000, \$22{,}000$ **75.** 4 in.

Exercise Set 3.2

1. $u = 8$ **3.** $n = -12$ **5.** $x = -3$ **7.** $p = 4$ **9.** $u = .8$
11. $x = 50$ **13.** $t = 110$ **15.** $n = 16$ **17.** $q = -18$
19. $z = -49$ **21.** $w = 9$ **23.** $m = -75$ **25.** $x = 9$
27. $a = -2$ **29.** $y = 7$ **31.** $x = 3$ **33.** $x = -4$
35. $x = -3$ **37.** $u = 9$ **39.** $x = 3$ **41.** Sixty-two is
equal to four times a number. Four times a number is sixty-
two.   **43.** Negative thirty-nine is three times a number.
Three times a number is negative thirty-nine.   **45.** Two and
seven-tenths times a number is equal to five and six tenths.
Five and six-tenths is two and seven-tenths times a number.
47. $5x = 35, x = 7$ **49.** $4x = -28, x = -7$ **51.** $.4x = 2$,
$x = 5$ **53.** $\dfrac{x}{4} = 5, x = 20$ **55.** $\dfrac{x}{-3} = 8, x = -24$
57. 8 ft   **59.** 30 yd   **61.** 150 calories   **63.** $8.50
65. 115 people   **67.** 4 miles   **69.** 287 employees
71. $17,500   **73.** $500

Exercise Set 3.3

1. $x = 2$ **3.** $x = -2$ **5.** $a = 5$ **7.** $x = 3$ **9.** $u = 2$
11. $v = 1$ **13.** $n = 0$ **15.** $p = -2$ **17.** $v = 2$
19. $k = 1$ **21.** $a = \dfrac{9}{17}$ **23.** $y = 4$ **25.** $n = -8$ **27.** $r = 5$
29. $u = 13$ **31.** $w = 2$ **33.** $y = \dfrac{5}{2}$ or 2.5   **35.** $x = 10$
37. $n = 1$ **39.** $p = 2$ **41.** $n = -2$ **43.** $u = 1$
45. $b = 1$ **47.** $a = -9$ **49.** $v = \dfrac{13}{2}$ or 6.5   **51.** identity,
all real numbers   **53.** identity, all real numbers
55. contradiction, no solutions or $\varnothing$ **57.** $x = 2$
59. Four times a number decreased by three is equal to seven.
Seven is the difference of four times a number and three.
61. Six times a number decreased by nine is three times that
number. Three times a number is equal to six times the number minus nine.   **63.** Two and eight-tenths more than seven
times a number is six and three-tenths. Six and three-tenths is
equal to the sum of seven times a number and two and eight-
tenths.   **65.** Twelve times the difference of a number and
one is nine. The product of twelve and a number decreased
by one is nine.   **67.** $3x - 8 = 13, x = 7$ **69.** $5x - 6 = 9$,
$x = 3$ **71.** $0 = 9(x + 3), x = -3$ **73.** $4(3x - 2) = 16, x = 2$
75. 4 hours   **77.** 5 hours   **79.** $160   **81.** $9.25
83. 121 copies   **85.** $85.50   **87.** 100 minutes
89. 12 oz salt, 96 oz of water   **91.** 550 yd   **93.** 8 ft
95. 5 in.

Exercise Set 3.4

1.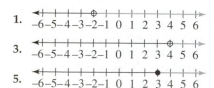
$\begin{array}{c}\longleftrightarrow\!\!+\!\!+\!\!+\!\!+\!\!+\!\!+\!\!+\!\!+\!\!+\!\!+\!\!+\!\!+\!\!\longrightarrow\\ {\scriptstyle -6\;-5\;-4\;-3\;-2\;-1\;\;0\;\;1\;\;2\;\;3\;\;4\;\;5\;\;6}\end{array}$

3. $\begin{array}{c}\longleftrightarrow\!\!+\!\!+\!\!+\!\!+\!\!+\!\!+\!\!+\!\!+\!\!+\!\!+\!\!+\!\!+\!\!\longrightarrow\\ {\scriptstyle -6\;-5\;-4\;-3\;-2\;-1\;\;0\;\;1\;\;2\;\;3\;\;4\;\;5\;\;6}\end{array}$

5. $\begin{array}{c}\longleftrightarrow\!\!+\!\!+\!\!+\!\!+\!\!+\!\!+\!\!+\!\!+\!\!+\!\!+\!\!+\!\!+\!\!\longrightarrow\\ {\scriptstyle -6\;-5\;-4\;-3\;-2\;-1\;\;0\;\;1\;\;2\;\;3\;\;4\;\;5\;\;6}\end{array}$

7. $\begin{array}{c}\longleftrightarrow\!\!+\!\!+\!\!+\!\!+\!\!+\!\!+\!\!+\!\!+\!\!+\!\!+\!\!+\!\!+\!\!\longrightarrow\\ {\scriptstyle -6\;-5\;-4\;-3\;-2\;-1\;\;0\;\;1\;\;2\;\;3\;\;4\;\;5\;\;6}\end{array}$

9. $\begin{array}{c}\longleftrightarrow\!\!+\!\!+\!\!+\!\!+\!\!+\!\!+\!\!+\!\!+\!\!+\!\!+\!\!+\!\!+\!\!\longrightarrow\\ {\scriptstyle -6\;-5\;-4\;-3\;-2\;-1\;\;0\;\;1\;\;2\;\;3\;\;4\;\;5\;\;6}\end{array}$

11. $\begin{array}{c}\longleftrightarrow\!\!+\!\!+\!\!+\!\!+\!\!+\!\!+\!\!+\!\!+\!\!+\!\!+\!\!+\!\!+\!\!\longrightarrow\\ {\scriptstyle -6\;-5\;-4\;-3\;-2\;-1\;\;0\;\;1\;\;2\;\;3\;\;4\;\;5\;\;6}\end{array}$

13. $q < -2$ $\begin{array}{c}\longleftrightarrow\!\!+\!\!+\!\!+\!\!+\!\!+\!\!+\!\!+\!\!+\!\!+\!\!+\!\!+\!\!+\!\!\longrightarrow\\ {\scriptstyle -6\;-5\;-4\;-3\;-2\;-1\;\;0\;\;1\;\;2\;\;3\;\;4\;\;5\;\;6}\end{array}$

15. $p \le 2$ $\begin{array}{c}\longleftrightarrow\!\!+\!\!+\!\!+\!\!+\!\!+\!\!+\!\!+\!\!+\!\!+\!\!+\!\!+\!\!+\!\!\longrightarrow\\ {\scriptstyle -6\;-5\;-4\;-3\;-2\;-1\;\;0\;\;1\;\;2\;\;3\;\;4\;\;5\;\;6}\end{array}$

17. $p < -2$ $\begin{array}{c}\longleftrightarrow\!\!+\!\!+\!\!+\!\!+\!\!+\!\!+\!\!+\!\!+\!\!+\!\!+\!\!+\!\!+\!\!\longrightarrow\\ {\scriptstyle -6\;-5\;-4\;-3\;-2\;-1\;\;0\;\;1\;\;2\;\;3\;\;4\;\;5\;\;6}\end{array}$

19. $s \ge 1$ $\begin{array}{c}\longleftrightarrow\!\!+\!\!+\!\!+\!\!+\!\!+\!\!+\!\!+\!\!+\!\!+\!\!+\!\!+\!\!+\!\!\longrightarrow\\ {\scriptstyle -6\;-5\;-4\;-3\;-2\;-1\;\;0\;\;1\;\;2\;\;3\;\;4\;\;5\;\;6}\end{array}$

21. $x > 2$ $\begin{array}{c}\longleftrightarrow\!\!+\!\!+\!\!+\!\!+\!\!+\!\!+\!\!+\!\!+\!\!+\!\!+\!\!+\!\!+\!\!\longrightarrow\\ {\scriptstyle -6\;-5\;-4\;-3\;-2\;-1\;\;0\;\;1\;\;2\;\;3\;\;4\;\;5\;\;6}\end{array}$

23. $u > 41$ $\begin{array}{c}\longleftrightarrow\!\!+\!\!+\!\!+\!\!+\!\!+\!\!+\!\!+\!\!+\!\!+\!\!+\!\!+\!\!+\!\!\longrightarrow\\ {\scriptstyle 34\;35\;36\;37\;38\;39\;40\;41\;42\;43\;44\;45\;46}\end{array}$

25. $t \le -3$ $\begin{array}{c}\longleftrightarrow\!\!+\!\!+\!\!+\!\!+\!\!+\!\!+\!\!+\!\!+\!\!+\!\!+\!\!+\!\!+\!\!\longrightarrow\\ {\scriptstyle -6\;-5\;-4\;-3\;-2\;-1\;\;0\;\;1\;\;2\;\;3\;\;4\;\;5\;\;6}\end{array}$

27. $y < 5$ **29.** $u < -3$ **31.** $a > 7$ **33.** $x > -15$
35. $r < 2$ **37.** $q > 12$ **39.** $a \ge 3$ **41.** $p \ge 5$
43. $w > -8$ **45.** $a \le 1.2$ **47.** $t > 30$ **49.** $s > -1$
51. $p \le 3$ **53.** $w \ge -\dfrac{5}{7}$ **55.** $m > 4$ **57.** $x \le -1$
59. $u \le -4$ **61.** $u > -1$ **63.** $n > 0$ **65.** $z \le 1$
67. $r > -5$ **69.** $4 < x < 10$ **71.** $-3 \le y \le 4$
73. $-2 \le x < 3$ **75.** $-1 \le y < 1$ **77.** $-1 < t < 5$
79. $y \ge 5$ **81.** Five times a number is greater than or equal
to ten. Five times a number is at least ten.   **83.** Three times
a number decreased by five is greater than two. Two is less
than the difference of three times a number and five.
85. Twenty-eight is less than or equal to the product of four
and the difference of a number and two. Four times a number
reduced by two is at least twenty-eight.   **87.** $6x < 18, x < 3$
89. $19 - 4x > 18, x < \dfrac{1}{4}$ **91.** $3(2x + 7) \ge -3, x \ge -4$
93. $4x - 6 < 18, x < 6$ **95.** $4(x - 3) \le 8, x \le 5$
97. $2(x - 1) + 3 \le 5, x \le 2$

Exercise Set 3.5

1. 3 ft, 5 ft   **3.** 523 mi   **5.** Richard = $257, Patricia = $300
7. $229 and $365   **9.** $12 per day   **11.** 1500
13. $34,596.15   **15.** 11, 13   **17.** 24, 25, 26   **19.** $-6, -5, -4$
21. 18, 20, 22   **23.** 6, 8, 10   **25.** seven hrs
27. eight hrs   **29.** nine hrs   **31.** six hrs   **33.** five hrs
35. 9 ft, 12 ft, 15 ft, 18 ft

Exercise Set 3.6:

1. $(400 - x)$ miles   **3.** $(24 - x)$ dimes   **5.** $150 - x$
camelias   **7.** 16 at $25, 20 at $32   **9.** 80 of 36," 25 of 54 in.
11. 22 at 18¢, 14 at 25¢   **13.** 12 trucks, 15 cars   **15.** four
$5 bills, forty-four $20 bills   **17.** 25 dimes, 30 nickels
19. 20 dimes, 35 quarters   **21.** $10,000 at 6%, $26,000 at 10%
23. $40,000   **25.** 4 oz   **27.** $3.00 per pound   **29.** 8 lb
peanuts, 4 lb cashews   **31.** 18 ml   **33.** 30 tons
35. 3.53 liters   **37.** 40 gallons   **39.** 25 at $.49, 15 at $1.19
41. 6.25 gallons   **43.** 12 nickels, 15 dimes, 8 quarters

Exercise Set 3.7

1. 3 ft   **3.** $L = 75$ yards, $W = 35$ yards   **5.** $L = 22$ ft, $W = 8$ ft   **7.** 4 in., 6 in., 8 in.   **9.** 8 ft, 8 ft, 12 ft
11. 30°, 60°   **13.** 25°, 75°, 80°   **15.** 50°, 40°   **17.** 40°, 140°   **19.** $x = 20$, $m \angle A = m \angle B = 78°$ **21.** $x = 15$, $m \angle A = m \angle B = 57°$ **23.** $x = 20$, $m \angle A = 70°$, $m \angle B = 110°$ **25.** 65°

Exercise Set 3.8

1. $A = 8$ sq mi   **3.** $h = 4$ ft   **5.** 1.5 hours   **7.** $r = 4$ in.
9. $P = 42$ in.   **11.** $A = 24$ **13.** $V = 15$ **15.** $P = 60$
17. $W = 2624$ **19.** $P = 12$ **21.** $C = K - 273$

23. $t = \dfrac{P}{k}$ **25.** $m = \dfrac{F}{a}$ **27.** $t = \dfrac{v}{g}$ **29.** $V = \dfrac{kT}{P}$

31. $m = \dfrac{E}{c^2}$ **33.** $r = \dfrac{I}{pt}$ **35.** $a^2 = c^2 - b^2$ **37.** $g = \dfrac{2d}{t^2}$

39. $M = DV$ **41.** $y = -2x + 3$ **43.** $y = \dfrac{-3x + 6}{2}$

45. $y = \dfrac{-6x + 7}{-3}$ **47.** $B = \dfrac{2A - hb}{h}$ or $\dfrac{2A}{h} - b$

49. $q_1 = \dfrac{Fr^2}{kq^2}$

Chapter 3 Review Exercises

1. $x = 4$ **2.** $x = \dfrac{1}{6}$ **3.** $v = 2$ **4.** $z = -3$ **5.** $z = 40$

6. $w = -9$ **7.** $u = -3$ **8.** $b = 7$ **9.** $a = -12$
10. $z = 5$ **11.** The difference of twice a number and 5 is 14. Five less than twice a number is 14.   **12.** The sum of 5 and some number is -2. Five increased by some number is -2.
13. $x + 12 = -2$, $x = -14$ **14.** $x - 4 = 6$, $x = 10$

15. \$140   **16.** \$280   **17.** $-\dfrac{7}{2}$ **18.** $\dfrac{1}{.4}$ **19.** $x = -8$

20. $y = -5$ **21.** $t = 15$ **22.** $x = -12$ **23.** Fourteen times a number is negative twenty-eight.   **24.** Two thirds of some number is -6 **25.** $4x = -12$, $x = -3$

26. $\dfrac{x}{-2} = 3$, $x = -6$ **27.** \$156   **28.** 6 m   **29.** $w = -14$

30. $z = -10$ **31.** \$165   **32.** 24   **33.** 75 teachers

34. \$180,000   **35.** 9 cm   **36.** 8 in.   **37.** $t = -2$
38. $u = -\dfrac{25}{17}$ **39.** $r = 3$ **40.** $s = \dfrac{26}{3}$ **41.** $v = 1$

42. $p = 13$ **43.** $q = 5$ **44.** $a = 4$ **45.** contradiction, no solutions or $\varnothing$ **46.** Identity, all real numbers
47. The difference of 3 times a number and 2 is 5.   **48.** Four times the difference of twice a number and 9 is 22.
49. $4x + 3 = 11$, $x = 2$ **50.** $54 = 6(3x - 9)$, $x = 6$ **51.** \$45
52. \$35

53. $-6\,-5\,-4\,-3\,-2\,-1\ 0\ 1\ 2\ 3\ 4\ 5\ 6$

54. $-6\,-5\,-4\,-3\,-2\,-1\ 0\ 1\ 2\ 3\ 4\ 5\ 6$

55. $-6\,-5\,-4\,-3\,-2\,-1\ 0\ 1\ 2\ 3\ 4\ 5\ 6$

56. $-6\,-5\,-4\,-3\,-2\,-1\ 0\ 1\ 2\ 3\ 4\ 5\ 6$
57.
$p < 15$ **58.** $q \le 12$ **59.** $x > 8$ **60.** $y \le 2.2$
61. $z > -12$ **62.** $s > -20$ **63.** Nineteen is less than or equal to the sum of a number and nine. Nine more than some number is at least nineteen.   **64.** $11 \ge x + 6$

65. $t \ge -7$ **66.** $u > -6$ **67.** $x < 2$ **68.** $y \ge 3$
69. $z > 5$ **70.** $r < 1$ **71.** $u \le 6$ **72.** $v > 1$
73. $s \ge -1$ **74.** $t \le -5$ **75.** $-4 < x < 1$
76. $5 \le x \le 7$ **77.** $-2 \le x \le 2$ **78.** $-3 \le x < 3$
79. 138 women   **80.** 7 ft   **81.** \$31.82   **82.** 10, 11, 12
83. 5, 7, 9   **84.** 30 light, 20 heavy   **85.** 20 nickels, 12 dimes   **86.** \$3500 at 6%, \$3000 at 5%   **87.** 200 ml at 25%, 300 ml at 50%   **88.** 8 lbs   **89.** $L = 100$ ft, $W = 75$ ft
90. 40°, 95°   **91.** 37°, 53°   **92.** $x = 25$ **93.** 18.84
94. $\dfrac{c - b}{a}$

Chapter 3 Test

1. $u = -13$ **2.** $v = -3$ **3.** $k = -6$ **4.** $m = -.675$

5. $v = -10$ **6.** $t = \dfrac{37}{8}$ or 4.625   **7.** $y = 5$ **8.** $a = -2.2$

9. $p > 4$ **10.** $q \ge 8$ **11.** $u \ge 17$ **12.** $x > 7$
13. $y \ge -6$ **14.** $z > -14$ **15.** $-2 < x < 3$
16. Four more than the quotient of a number and three is 6.
17. Two times the difference of a number and five is equal to three times the number.   **18.** $3(x - 6) = x + 1$ **19.** 8 ft
20. 10 in.   **21.** \$5250   **22.** \$36   **23.** 90 ml   **24.** 2532

25. $-1, 0, 1$ **26.** 15 incandescent, 10 flourescent   **27.** $\dfrac{1}{2}$ hr

28. $W = 31$ cm, $L = 40$ cm   **29.** $W = \dfrac{P - 2L}{2}$ or $W = \dfrac{P}{2} - L$
30. $B = \dfrac{2A - bh}{h}$ or $\dfrac{2A}{h} - b$

Section 4.1

1. yes   **3.** no   **5.** yes   **7.** no   **9.** yes   **11.** yes

13. no   **15.** yes   **17.** $(0, 3), (-\dfrac{3}{2}, 0), (3, 9), (2, 7)$

19. $(0, -4), (-2, 0), (-3, 2), (2, -8)$ **21.** $(0, -4), (6, 0),$ $(-6, -8), (-3, -6)$ **23.** $(0, 4), (5, 0), (-10, 12), (-5, 8)$
25. $(0, -2), (-3, 0), (9, -8), (3, -4)$ **27.** $(0, 4), (-3, 0),$ $(6, 12), (-12, -12)$ **29.** $(4, 0), (4, 3), (4, -2), (4, -4)$
31. $(2, 4), (3, 4), (-4, 4), (-1, 4)$

33.

35.

37.

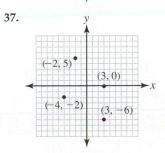

39. $A(3, 1)$ $B(-4, 2)$ $C(-2, -3)$ $D(4, -5)$ **41.** $A(3, 0)$ $B(0, 4)$ $C(-5, 0)$ $D(0, -6)$ **43.** $A(3, 4)$ $B(0, 6)$ $C(-4, -7)$ $D(4, 0)$
45. $(1, 40)$, $(2, 80)$, $(5, 200)$. After eight hours, the motorcycle will have traveled 320 miles. **47.** $(2, 12)$, $(4, 24)$, $(5, 30)$. When the length is 9 m, the area is 54 m². **49.** $(1, 60)$, $(3, 180)$, $(7, 420)$. Four candy bars cost 240 cents. **51.** $(2, 10)$, $(5, 25)$, $(8, 40)$. The value of 10 five-dollar bills is $50. **53.** $(1, 68)$, $(3, 4)$, $(6, -92)$. After 2 seconds, the velocity of the ball is 36 feet per second. **55.** $(1, 6)$, $(3, 54)$, $(5, 150)$. After 2 seconds, the particle is 24 inches from the starting point. **57.** $(1, 334)$, $(2, 436)$, $(4, 544)$. After 3 seconds, the object is 506 feet above the ground. **59.** $(1, 200)$, $(2, 400)$, $(3, 800)$. After 4 seconds, there are 1600 bacteria present.

Section 4.2

1. $(0, -3)$ $(2, 1)$ $(-2, -7)$ **3.** $(0, 5)$ $(5, 0)$ $(2, 3)$

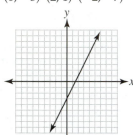
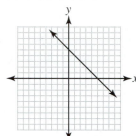

5. $(0, 6)$ $(3, 0)$ $(1, 4)$

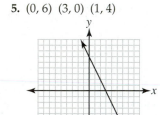

7.
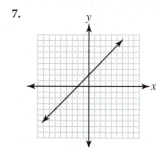

9.
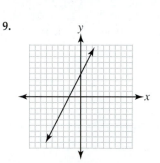

11.
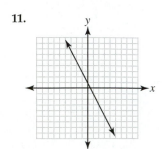

13.
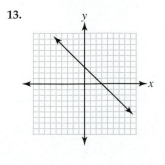

15.
17.
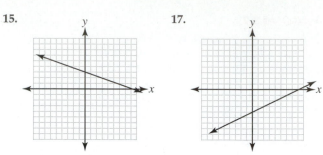

19.
21.
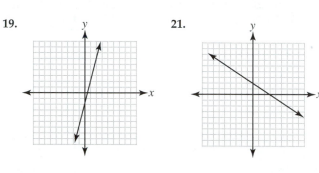

23.
25.
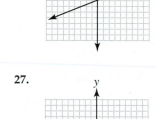
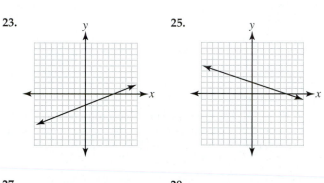

27.
29.
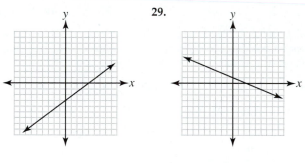

31.
33.
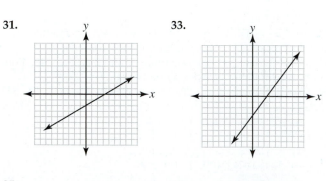

35.
37.
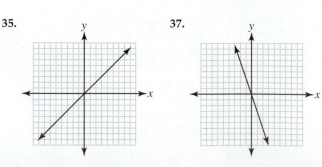

39.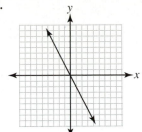

41.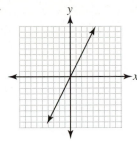

63. $x - y = 4$

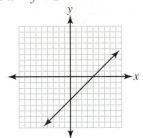

65. $3x - 4y = 12$

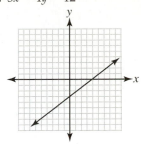

43.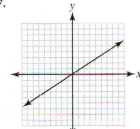

45.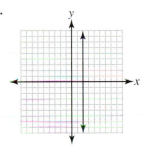

67. $y = 3x + 2$

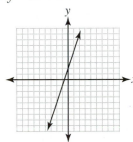

69. a. $(0, 0), (2, -64), (4, -128)$

b.

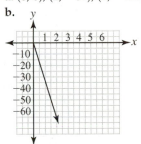

c. 3 seconds

47.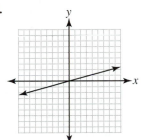

49.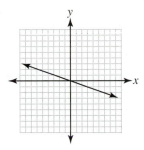

Section 4.3

1. $(-4, 0)$ $(0, 3)$ **3.** $(-4, 0)$ $(0, -3)$ **5.** $(6, 0)$ no y-intercept

7.

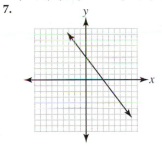

9.

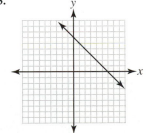

51.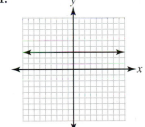

53.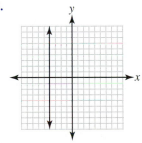

11.

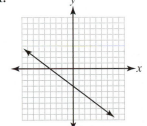

13.

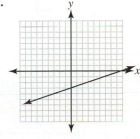

55.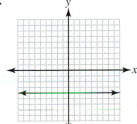

57.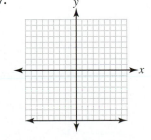

15.

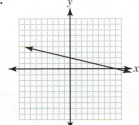

17.

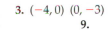

59.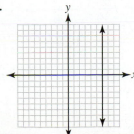

61.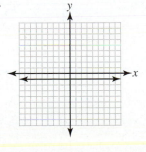

19.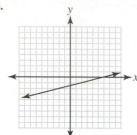

21.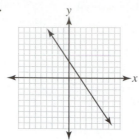

43.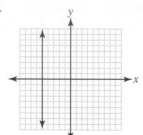

45.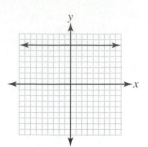

23.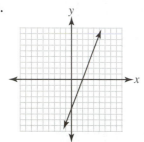

25.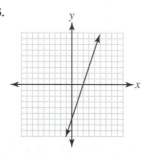

47.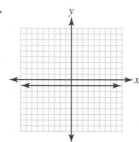

27.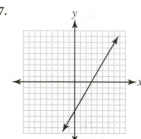

29.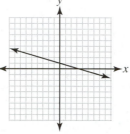

49. $2x + 2y = 32$ **51.** $6x + 9y = 420$ **53.** $2x + y = 27$
55. $a = -2, b = 4$ **57.** $(-3, 0)\ (-1, 0)\ (1, 0)\ (0, -3)$
59. $(-3, 0)\ (0, -1)\ (0, 1)\ (0, 3)$

Section 4.4

1. $\dfrac{1}{2}$ **3.** -2 **5.** -2 **7.** $\dfrac{4}{-6} = -\dfrac{2}{3}$ **9.** -1 **11.** $\dfrac{3}{4}$

13. $\dfrac{4}{8} = \dfrac{1}{2}$ **15.** 0 **17.** undefined   **19.** 0 **21.** 1

23. $-\dfrac{1}{2}$ **25.** $\dfrac{3}{4}$ **27.** undefined   **29.** 0

31.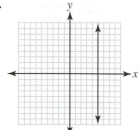

33.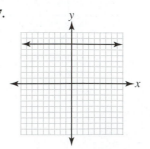

31.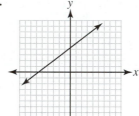

33.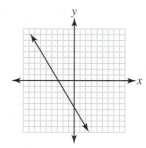

35.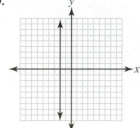

37.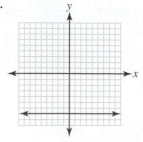

35.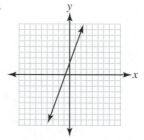

37.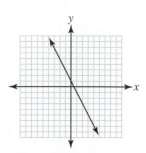

39.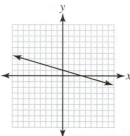

41.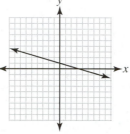

39.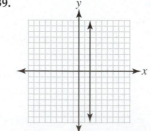

41.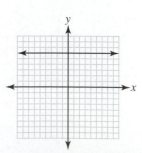

43. undefined   **45.** undefined   **47.** 0   **49.** 0

51. a. $\dfrac{30}{500} = \dfrac{3}{50}$ **b.** 6%   **53.** $\dfrac{24}{9} = \dfrac{8}{3}$

Section 4.5

1. $-2, (0.9)$

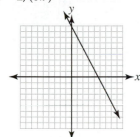

3. $3, (0, -7)$

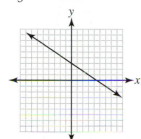

5. $-\dfrac{2}{3}, (0, 3)$

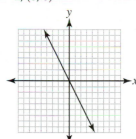

7. $\dfrac{5}{3}, (0, -5)$

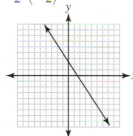

9. $-2, (0, 0)$

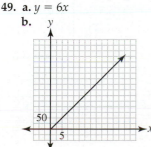

11. $-\dfrac{3}{2}, \left(0, \dfrac{5}{2}\right)$

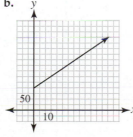

13. undefined   **15.** undefined   **17.** 0   **19.** 0

21. $y = 2x + 4$ **23.** $y = \dfrac{3}{5}x - 1$ **25.** $y = -\dfrac{2}{3}x + \dfrac{4}{5}$

27. parallel   **29.** perpendicular   **31.** perpendicular
33. neither   **35.** parallel   **37.** perpendicular
39. parallel   **41.** perpendicular   **43.** parallel
45. parallel   **47.** perpendicular

49. a. $y = 6x$
b. *y*

c. same

51. a. $y = 2x + 100$
b. *y*

c. same

Section 4.6

1. $2x - y = 5$ **3.** $2x + y = 0$ **5.** $x - 4y = -23$
7. $x + 3y = -10$ **9.** $4x - 3y = 17$ **11.** $2x + 3y = -19$
13. $x = 3$ **15.** $y = 3$ **17.** $x = 5$ **19.** $y = 4$
21. $2x + y = 5$ **23.** $x - y = -4$ **25.** $2x + y = -7$
27. $2x + 3y = 12$ **29.** $x = 5$ **31.** $y = -4$
33. $x + 2y = 1$ **35.** $x + 2y = -11$ **37.** $4x - y = 17$
39. $3x - y = -18$ **41.** $x = 6$ **43.** $y = -4$
45. a. $y = .15x + 25$ **b.** \$100 **c.** 250 miles   **47.** $y = .8x + 1.2$
b. \$7.60 **c.** 5 miles   **49. a.** $y = -5000x + 50,000$
b. \$10,000 **c.** 10 years   **51. a.** $y = .05x + 200$
b. \$12,700 **c.** \$175,000

Section 4.7

1. no   **3.** no   **5.** yes

7.

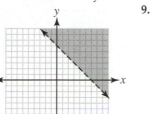

9.

11.

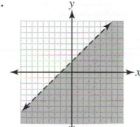

13.

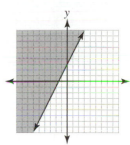

15.

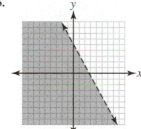

17.

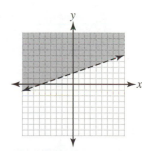

19.

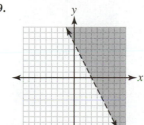

21.

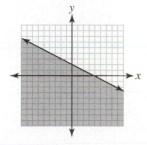

23.

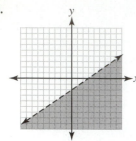

25.

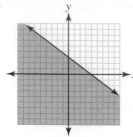

43. $x - 2y \geq 8$

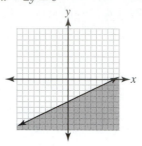

45. $y < 3x + 5$

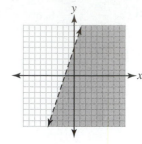

27.

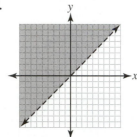

29.

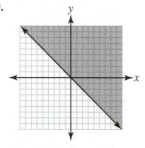

47. $20x + 25y \leq 150$ **49.** $3x + 2y \geq 500$
51. a. $x + y \leq 200$ **53. a.** $2x + 3y \leq 84$
b.

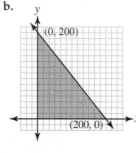

b.

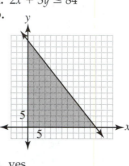

c. no   **c.** yes

31.

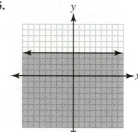

33.

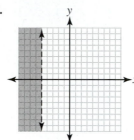

Section 4.8

1. yes, $D = \{-6, -3, 5, 7\}, R = \{2, -2, -7, 9\}$ **3.** yes,
$D = \{-4, -1, 3, 5\}, R = \{2, 3, 1\}$ **5.** no, $D = \{-8, -6, 5\}$,
$R = \{2, 3, -3, 1\}$ **7.** yes   **9.** no   **11.** yes   **13.** no
15. yes   **17.** yes   **19.** no   **21.** no   **23.** no
25. yes   **27.** yes   **29.** no   **31.** yes   **33.** 5, (0, 5)
35. $3a + 5, (a, 3a + 5)$ **37.** $-1, (-1, -1)$ **39.** 16, (1, 16)
41. $16a^2, (a, 16a^2)$ **43.** $-6, (-2, -6)$ **45.** 7, $(-2, 7)$
47. $|2z - 3|, (z, |2z - 3|)$ **49. a.** \$34, **b.** \$64, **c.** The cost
of driving 250 miles is \$52.   **51. a.** \$1675 **b.** \$1175 **c.** Her
salary for \$20,000 of sales is \$2175.   **53.** no   **55.** yes

35.

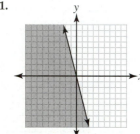

37.

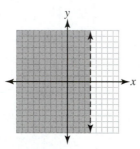

Chapter 4 Review Exercises

1. yes   **2.** no   **3.** yes   **4.** yes   **5.** yes   **6.** no

7. $(0, 3), (1, 0), (3, -6), (2, -3)$ **8.** $(0, 2), (-\frac{1}{2}, 0), (-5, -18)$,

$(-2, -6)$ **9.** $(0, 6), (4, 0), (-4, 12), (2, 3)$ **10.** $(0, -5)$,
$(3, 0), (6, 5), (6, 5)$ **11.** $(2, 120), (4, 240), (7, 420)$, In three
hours, the motorcycle will have traveled 180 miles.
12. $(2, 50), (5, 125), (7, 175)$. The accountant is paid \$100 for
4 hours work.

39.

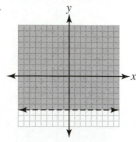

41. $2x + 4y < 12$

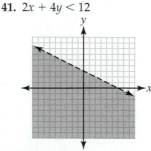

13.

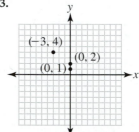

14.

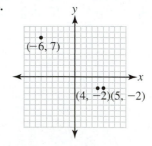

15.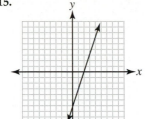

16.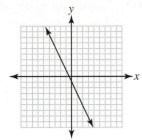

35.

36.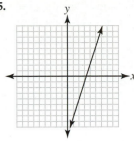

17.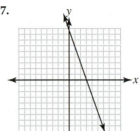

18.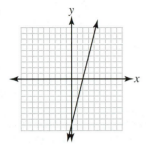

37.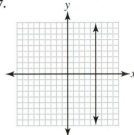

38.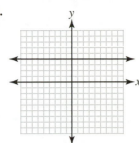

19.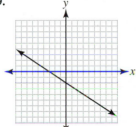

20.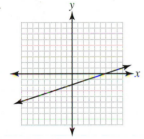

39. $\frac{20}{30} = \frac{2}{3}$ **40.** $-\frac{2}{25}$ **41.** $-\frac{1}{2}, (0, 4)$ **42.** $\frac{5}{2}, \left(0, -\frac{15}{2}\right)$

43. undefined, none **44.** 0, (0, −3) **45.** parallel
46. perpendicular **47.** neither **48.** perpendicular
49. parallel **50.** neither **51.** parallel

52. perpendicular **53.** $y = -3x + \frac{5}{6}$ **54.** $y = \frac{6}{5}x + 4$

55. $y = -3$ **56.** $3x - y = 0$ **57.** $2x + y = 1$
58. $5x + 2y = 18$ **59.** $4x - 3y = -32$ **60.** $x = -7$
61. $y = -5$ **62.** $y = 2$ **63.** $x = -8$ **64.** $x + 2y = 5$
65. $x - y = 4$ **66.** $3x + 2y = 12$ **67.** $3x - y = 12$
68. $y = 5$ **69.** $x = 2$ **70.** $3x + 5y = -2$ **71.** $3x + y = -9$ **72.** $2x + 3y = 2$ **73.** $x + 4y = -5$ **74.** $x = 2$
75. $y = -1$ **76.** $y = 3$ **77. a.** $y = 1.5x + 250$ **b.** \$700
c. 225 sandwiches

21.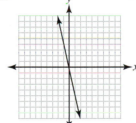

22.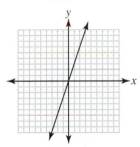

78.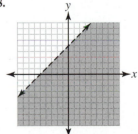

79.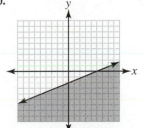

23.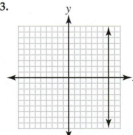

24.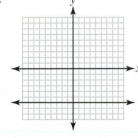

80.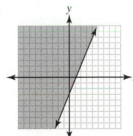

81.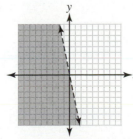

25. $2x + 4y = 16$ **26.** $5x + 7y = 145$ **27.** 2 **28.** $\frac{2}{3}$
29. $\frac{3}{-6} = -\frac{1}{2}$ **30.** $\frac{5}{2}$ **31.** undefined **32.** 0

33.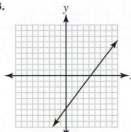

34.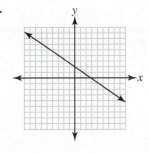

82.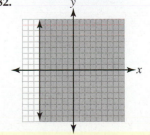

83.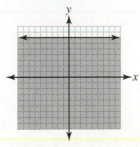

84. $2x - 3y \geq 8$ **85.** $18x + 15y \leq 180$ **86.** no, $D =$ $\{-3, -1, 0\}$, $R = \{2, 4, 7, 5\}$ **87.** yes, $D = \{-4, -1, 3, 6\}$, $R = \{2, 4, 5\}$ **88.** yes   **89.** no   **90.** no   **91.** yes
92. yes   **93.** yes   **94.** no   **95.** yes   **96. a.** $f(2) = 11$, $(2, 11)$ **b.** $f(-3) = 21$, $(-3, 21)$ **c.** $f(a) = 2a^2 + 3$, $(a, 2a^2 + 3)$
97. a. $g(-2) = -18$, $(-2, -18)$ **b.** $g(0) = 2$, $(0, 2)$ $g(b) = b^3 - 3b^2 + 2$, $(b, b^3 - 3b^2 + 2)$ **98. a.** $22,800 **b.** $12,000 **c.** After 8 years the machinery is worth $15,600.

Chapter 4 Test

1. yes   **2.** $(0, -6)$, $(-8, 0)$, $(-4, -3)$, $(-12, 3)$
3. $(1, 2)$, $(2, 4)$, $(4, 8)$, Ten feet of pipe costs $20.
4.

5.

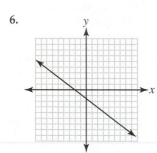

6.   **7.**

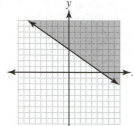

8. $9x + 20y = 480$
9.   **10.**

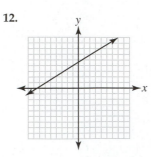

11.   **12.**

13.

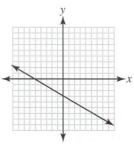

14. -1 **15.** 0 **16.** $\dfrac{1}{4}$ **17.** $\dfrac{5}{2}$ **18.** Undefined
19. perpendicular   **20.** parallel   **21.** $5x + 2y = 16$
22. $x - 2y = 8$ **23.** $x - 3y = -11$ **24.** $y = -6$
25. The line with slope of -2 slants downward as you go left to right. It drops two units vertically for each horizontal change of one unit to the right. The line with slope of 2 slants upward as you go left to right. It rises two units vertically for each horizontal change of one unit to the right.   **26.** no, $D = \{-4, 8, 6\}$, $R = \{7, -2, -4, 3\}$ **27.** no   **28.** yes
29. a. $f(2) = 0$, $(2, 0)$ **b.** $f(-1) = 6$, $(-1, 6)$ **30. a.** $4,500
b. It costs $7000 to operate the business 10 days.

Section 5.1

1. composite   **3.** prime   **5.** composite   **7.** composite
9. prime   **11.** 2, 3, 5   **13.** 3   **15.** 2, 3   **17.** none
19. 5   **21.** 2, 3   **23.** 2, 3, 5   **25.** $2 \cdot 7$ **27.** $2 \cdot 23$
29. $2^2 \cdot 7$ **31.** prime   **33.** $3^2 \cdot 5$ **35.** $2^2 \cdot 5^2$
37. $2 \cdot 3 \cdot 7$ **39.** $2 \cdot 7 \cdot 11$ **41.** $2^2 \cdot 3^2 \cdot 5$ **43.** $3^2 \cdot 5 \cdot 7$
45. $2^2 \cdot 3^3 \cdot 7$ **47.** 1, 2, 3, 4, 6, 9, 12, 18, 27, 36, 54, 108
49. $2^3 \cdot 3^3 \cdot 7$ **51.** 3   **53.** 5   **55.** 1   **57.** 3   **59.** 10
61. 90   **63.** 1   **65.** 8   **67.** 8   **69.** 24   **71.** mn
73. cd^2 **75.** a^3b^2 **77.** 1   **79.** a^2bc^2 **81.** pr^3
83. $6pq$ **85.** $9s^2t^3$ **87.** $3abc^2$ **89.** $4a^2b^3c^2$ **91.** 3
93. 2 nickels and 5 dimes   **95.** four Yankees and eight Braves   **97.** 3 redfish and 8 trout   **99.** 48
101. $18x^4y^7(a - 3)^4$

Exercise Set 5.2

1. $c(d + f)$ **3.** $r(s - t)$ **5.** $3(x + 3)$ **7.** $4(2x - 3)$
9. $x(x + 1)$ **11.** $r^2(r - 1)$ **13.** $7a^2(a + 2)$ **15.** $9p^3(2 - p^2)$
17. $5x(3x^2 + 1)$ **19.** $11r^2(2r^3 - 1)$ **21.** $4c^2(3c + 2)$
23. $6z^2(3 - 2z^4)$ **25.** $x^2y^2(y + 1)$ **27.** $u^3v^2(u - v)$
29. $cd^2(c^2d^2 + 1)$ **31.** $x^2y^2(1 - xy^2)$ **33.** $6a^2b^2(3b^2 + 2a)$
35. $3xy^2(3y - 4x)$ **37.** $3xy^3(3xy + 1)$ **39.** $3(3x^2 - 4x + 2)$
41. $5x^2(3x^2 - 2x - 4)$ **43.** $6(a^4 + 2a^2 - 4)$
45. $11x(2x^2 - 3x - 1)$ **47.** $4cd^2(4c^2 - 6cd^2 + 9)$
49. $7xy^2(2x^2y - 3xy^2 - 1)$ **51.** $-3(m - 2n)$
53. $-8(2c - d)$ **55.** $-7(2x - 1)$ **57.** $-4(x^2 - 2x + 4)$
59. $-5x(2x^2 + 3x - 5)$ **61.** $(m + n)(a + b)$
63. $(c + 2)(a - b)$ **65.** $(t - 3)(t + 6)$ **67.** $(a - 6)(a - 7)$
69. $2x + 3$ and $3x - 5$ **71.** $180x^6y^4(2y - 3x^2)$
73. $3(a + b)^2(a + b + 3)$ **75.** $(x + y)(z + w)$
77. $(n - 4)(m + 3)$ **79.** $(b + 5)(a + 3)$ **81.** $(x - y)(x + 5)$
83. $(b - 3)(a + 1)$ **85.** $(m + n)(x - y)$ **87.** $(n - 3)(m - 6)$
89. $(d - e)(c - 3)$ **91.** $(x + 3)(2x - y)$ **93.** $(s + 4)(r - 1)$
95. $(d - 3)(c - 1)$ **97.** $(2b + 5)(2a + 3)$
99. $(2z - w)(4x - y)$ **101.** $(3c + 2d)(2a - 3b)$
103. $(2x + 3)(4x - y)$

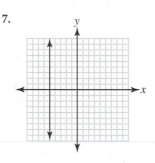
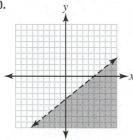

Exercise Set 5.3

1. $(x + 1)(x + 2)$ **3.** $(x - 3)(x - 1)$ **5.** $(a + 2)(a + 3)$
7. $(b - 4)(b - 2)$ **9.** $(y - 5)(y + 1)$ **11.** $(a - 5)(a + 7)$
13. $(x + 2)(x + 8)$ **15.** $(r - 8)(r + 9)$ **17.** $(a - 9)(a - 3)$
19. prime   **21.** $(x - 7)(x - 5)$ **23.** $(y + 7)(y - 6)$
25. $(z + 3)(z + 12)$ **27.** $(a + b)(a + 3b)$ **29.** $(r - 7s)(r - s)$
31. $(c - 2d)(c + 9d)$ **33.** $(r - 11s)(r + 4s)$
35. $(y + 8z)(y - 5z)$ **37.** $(a - 9b)(a + 6b)$
39. $(r - 12s)(r - 3s)$ **41.** $3(a + 2)(a + 3)$
43. $5(z - 3)(z - 2)$ **45.** $x(x - 7)(x + 4)$
47. $y^2(y - 15)(y - 5)$ **49.** $3a(a^2 - a - 4)$
51. $xy(x - 8)(x + 10)$ **53.** $y^2(x + 4)(x + 5)$
55. $4s(r - 2)(r + 7)$ **57.** $3b(a^2 - a - 28)$ **59.** $(x - 12)(x + 9)$
61. $(y + 8)(y + 16)$ **63.** $(4 + x)(6 - x)$ **65.** $(6 + a)(3 + a)$
67. $(8 + b)(2 - b)$ **69.** $7, -7, 8, -8, 13, -13$
71. $(x + 4), (x + 8)$ **73.** $x, (x + 10), (x - 4)$

Exercise Set 5.4

1. $(a + 1)(2a + 5)$ **3.** $(a + 3)(2a - 1)$ **5.** $(7m + 1)(m - 5)$
7. prime   **9.** prime   **11.** $(11x + 5y)(x + y)$ **13.** prime
15. $(5x - 4)(x + 2)$ **17.** $(2x + 1)(2x + 3)$
19. $(2x - 7)(5x + 1)$ **21.** $(2x + 3)(x + 4)$
23. $(3a - 8)(a - 2)$ **25.** $(5c - 4)(c + 3)$ **27.** $(2b - 3)(b + 8)$
29. $(2x + 5)(3x + 1)$ **31.** $(2x - 5)(4x - 1)$
33. $(4a - 3)(6a + 1)$ **35.** $(2n + 1)(8n - 5)$
37. $(2x + 3)(3x + 5)$ **39.** $(3c - 2)(4c - 3)$
41. $(3a + 5)(6a - 7)$ **43.** $(2w + 3)(4w - 9)$ **45.** prime
47. prime   **49.** $(2a + 5b)(3a + 7b)$ **51.** $(3c - 4d)(6c + 5d)$
53. $(3a + 4b)(4a - 3b)$ **55.** $(2f - 3g)(6f - 7g)$
57. $3(2a - 3)(a + 2)$ **59.** $y(4x - 1)(x - 3)$
61. $3x(x - 1)(2x - 5)$ **63.** $3x(2x - 3)(3x + 4)$
65. $(4x + 5)(9x - 4)$
67. $[(x + 2) - 3][(x + 2) + 2] = (x - 1)(x + 4)$
69. $(3 + 2a)(2 - a)$ **71.** $(3 - 2b)(4 + 3b)$
73. $2(3 + 4n)(3 - n)$ **75.** $7, 11$

Exercise Set 5.5

1. $(m + n)(m - n)$ **3.** $(x + 2)(x - 2)$ **5.** $(r + 8)(r - 8)$
7. prime   **9.** $(c + 5d)(c - 5d)$ **11.** $(a + 10b)(a - 10b)$
13. $(2x + 5y)(2x - 5y)$ **15.** $(7p - 9q)(7p + 9q)$ **17.** prime
19. $(11a + 7b)(11a - 7b)$ **21.** $(a^2 + b^2)(a + b)(a - b)$
23. $(x^2 + 1)(x + 1)(x - 1)$ **25.** $(4x^2 + 9)(2x + 3)(2x - 3)$
27. $(x + y + 3)(x + y - 3)$ **29.** $(6 + x - y)(6 - x + y)$
31. $(2x - 2y + 3)(2x - 2y - 3)$ **33.** $3(x + 5)(x - 5)$
35. $4(2x + 3y)(2x - 3y)$ **37.** $3(3x + 4)(3x - 4)$
39. $9(x + 3)(x - 3)$ **41.** $3(x^2 + 4)(x + 2)(x - 2)$
43. $(3x + 7)(3x - 7)$ **45.** $(14x - 25)(14x + 25)$
47. $(16c - 19d)(16c + 19d)$ **49.** $2(13x - 12)(13x + 12)$
51. $(x^4 + y^4)(x^2 + y^2)(x + y)(x - y)$ **53.** $(9x + 18 + 4y)(9x + 18 - 4y)$ **55.** $(x^n + y^n)(x^n - y^n)$ **57.** $(x + y)(x^2 - xy + y^2)$
59. $(c + 2)(c^2 - 2c + 4)$ **61.** $(n - 4)(n^2 + 4n + 16)$
63. $(m + 1)(m^2 - m + 1)$ **65.** $(2x + y)(4x^2 - 2xy + y^2)$
67. $(3a - 2)(9a^2 + 6a + 4)$ **69.** $(2x + 5y)(4x^2 - 10xy + 25y^2)$

Exercise Set 5.6

1. $(x + 2)^2$ **3.** $(x + 1)^2$ **5.** $(x + 6)^2$ **7.** $(y + 10)^2$
9. $(5x - 1)^2$ **11.** $(a - b)^2$ **13.** $(2a + b)^2$ **15.** $(4c - 3)^2$
17. $(3x + 5)^2$ **19.** $(4x - y)^2$ **21.** $(5x + 4y)^2$ **23.** prime
25. $2(2x - 3)^2$ **27.** $2(x + 7)^2$ **29.** $4(x + 10)^2$
31. $x^2(5x - 3)^2$ **33.** $2xy(5x + 4y)^2$ **35.** $x - 7$
37. $(9x - 5y)^2$ **39.** $25(2a^2 + 3b^2)^2$ **41.** 49 **43.** 25

Exercise Set 5.7

1. $8(2 - x^2)$ **3.** $(c + 3)(c + 9)$ **5.** $3a(a + 3)(a - 6)$
7. $(x + 10)(x - 10)$ **9.** prime   **11.** $(a + b)(x + 2)$
13. $6(a - 2)(a - 6)$ **15.** $(r + 9)(r - 8)$ **17.** $m^2n^2(m + n^2)$
19. $3a^2(2a - 5)(2a + 5)$ **21.** $(3a + 5b)^2$
23. $8(r + 4)(2r - 3)$ **25.** $(ab + 8)(ab - 8)$
27. $(5n + 2)(n + 4)$ **29.** prime   **31.** $(y + z)(x + w)$
33. $(a - b + 6)(a - b - 6)$ **35.** $(2y - 3)(5x - 2)$
37. $(a^2 + 2b)(a^2 - 2b)$ **39.** prime
41. $(4b^2 + 1)(2b + 1)(2b - 1)$ **43.** $(6x + 1)(x + 5)$
45. $(2x + 3z)(3x - y)$ **47.** $(b + 3)(a - 1)$
49. $2u(8u - v)(u + 3v)$ **51.** $ab(2a - 5ab + 7b)$
53. $(8c + d)(c + 4d)$ **55.** $3(3a + b)(x - 1)$
57. $(3x - 5)(2x + 3)$ **59.** $(3a + 2b)(4c - 3d)$
61. $36(x + 2y)(x - 2y)$ **63.** $(x - 10)(x + 12)$
65. $3a^2b^3(4a + b)(3a - 4b)$ **67.** $(2z + 5)(4z^2 - 10z + 25)$
69. $(3c - 2d)(9c^2 + 6cd + 4d^2)$ **71.** $3(2c - d)(4c^2 + 2cd + d^2)$
73. $(3x - 8)(4x + 3)$ **75.** $(9a^2 + 25b^2)(3a + 5b)(3a - 5b)$
77. $(x + 5 - y)(x + 5 + y)$

Exercise Set 5.8

1. $6, -4$ **3.** $5, 9$ **5.** $0, \dfrac{2}{5}$ **7.** $0, -\dfrac{11}{7}$ **9.** $3, -2, 5$
11. $-7, \dfrac{5}{3}, \dfrac{1}{4}$ **13.** 2 **15.** $-2, 3$ **17.** $\dfrac{3}{4}, 5$ **19.** $0, 2$
21. $0, \dfrac{2}{3}$ **23.** $2, -2$ **25.** $5, -5$ **27.** $3, 4$ **29.** $-\dfrac{2}{3}, 4$
31. 6 **33.** $-\dfrac{5}{2}$ **35.** $12, 14$ **37.** $-6, -4$
39. -4 or -8 **41.** 3 **43.** $w = 4$ ft, $l = 8$ ft
45. $w = 8$ in., $l = 11$ in.   **47.** $b = 6$ mi, $h = 2$ mi
49. a. 7 sec, **b.** $3, 4$ sec   **51.** 4 sec   **53.** 7 positive integers
55. 6 logs   **57.** 9 offices   **59. a.** -8, **b.** 0 **c.** -5
61. $0, 4, 9$ **63.** $0, -6$

Chapter 5 Review Exercises

1. composite   **2.** prime   **3.** $2^2 \cdot 3 \cdot 5$ **4.** $2^2 \cdot 7 \cdot 11$
5. 3 **6.** 2 **7.** 14 **8.** 6 **9.** m^2n **10.** b^2
11. $6r^2s^2t^2$ **12.** $5ab^2$ **13.** $a(x + y)$ **14.** $3(2x^2 + 1)$
15. $6xy(3x - 4)$ **16.** $r^4s^3(1 + r^2s)$ **17.** $6(3x^2 - 4x + 2)$
18. $2ab^3(4ab - 6a^2b^2 + 9)$ **19.** $(s - 2)(r - 5)$ **20.** $(n - p)$
$(m - q)$ **21.** $-5x(3x^2 - x + 5)$ **22.** $(a + b)(x + y)$
23. $(s - 8)(r + 3)$ **24.** $(a - 3)(b - 6)$ **25.** $(x - 7)(y - 4)$
26. $(b - 1)(3a - 1)$ **27.** $(2x + 5)(3x - 2)$ **28.** $(x + 11)$
$(x + 1)$ **29.** $(x - 5)(x - 7)$ **30.** $(x + 7)(x - 1)$ **31.** $(m - 5)$
$(m + 2)$ **32.** $(x - 8)(x - 3)$ **33.** $(r + 9s)(r - 8s)$
34. $(a + 27b)(a - 2b)$ **35.** $(m - 21n)(m + 2n)$
36. $2(x + 7)(x + 5)$ **37.** $y(y + 11z)(y - 4z)$ **38.** $(x - 1)$
$(5x - 2)$ **39.** $(z + 1)(3z - 5)$ **40.** $(x + 4)(5x - 2)$
41. $(3a + 2)(a + 8)$ **42.** prime   **43.** $(9a + 5)(2a - 1)$
44. $(5x + 2y)(7x + 3y)$ **45.** $(5a + 6b)(4a - 3b)$
46. $2(2a - 7)(5a + 1)$ **47.** $3c(3c - 2d)(4c - 3d)$
48. $(g + h)(g - h)$ **49.** $(r + 11)(r - 11)$ **50.** $(5x - 4)$
$(5x + 4)$ **51.** $(7a + 8b)(7a - 8b)$ **52.** $6(x + 2)(x - 2)$
53. $16(2c - d)(2c + d)$ **54.** $(m + n + 9)(m + n - 9)$
55. $(8 + c - d)(8 - c + d)$ **56.** prime   **57.** $(m^2 + n^2)$
$(m + n)(m - n)$ **58.** $(m + n)(m^2 - mn + n^2)$ **59.** $(r - 1)$
$(r^2 + r + 1)$ **60.** $(a + 5)(a^2 - 5a + 25)$ **61.** $(4x + 3y)$
$(16x^2 - 12xy + 9y^2)$ **62.** $(r + 9)^2$ **63.** $(x - 12)^2$
64. prime   **65.** $(7a - 1)^2$ **66.** $(7a + 2b)^2$
67. $(5c + 3d)^2$ **68.** $2(9p^2 + 12p + 16)$ **69.** $4(x - 5)^2$
70. $13a(2a + 1)$ **71.** $(9x + y)(9x - y)$ **72.** $2(x - 8)(x + 5)$

73. $(x - 9)(x + 7)$ **74.** $11a^2b(5 - ab)$ **75.** $3(3m - 4)(m - 4)$
76. $(x - 9)(3y - 4)$ **77.** $(2a + 2b + 7)(2a + 2b - 7)$
78. $4(2x^2 + 7x - 14)$ **79.** $(7a - 2b)^2$ **80.** $3(x + 5)(2x - 3)$
81. $2(y - 5)(3x + 1)$ **82.** $3y(3x - 5)(3x + 5)$
83. $(11a + 13)(11a - 13)$ **84.** $(9x - 4y)^2$ **85.** $9(x^2 + 9)$
86. $4(x + 3)(2x - y)$ **87.** $(r + 10)(r - 4)$ **88.** prime
89. $(y^2 + 25)(y + 5)(y - 5)$ **90.** $3xy^2(x - 9y)(x + 5y)$
91. $(8x - 7)(x - 1)$ **92.** $x = 0, -5$ **93.** $0, 6$

94. $x = \dfrac{9}{2}, -\dfrac{9}{2}$ **95.** $\dfrac{8}{3}, -\dfrac{8}{3}$ **96.** $x = 7, -5$ **97.** $7, -3$

98. $x = 4, -\dfrac{1}{2}$ **99.** $\dfrac{2}{3}, -4$ **100.** $W = 5$ cm, $L = 5$ cm

101. -9 and -7 or 7 and 9 **102.** 6 yds   **103.** -7 or 4

Chapter 5 Test

1. a. composite **b.** prime   **2.** $2^2 \cdot 3^2 \cdot 7$ **3.** 24
4. $12x^2y$ **5.** $x^2(x + 1)$ **6.** $(m - 9)^2$
7. $(m - 8n)(m + 5n)$ **8.** $(7a + 3b)(7a - 3b)$
9. $7x^2y(2x - 4y + 5x^2y^2)$ **10.** $(3y - 4)^2$ **11.** prime
12. $(s + 5)(r - 7)$ **13.** $(x - y - 10)(x - y + 10)$
14. $3(q - 5)(q + 7)$ **15.** $(a - 6)(3a + 5)$
16. $(2a - 5b)(4a - 7b)$ **17.** $3(3x - 2y)(3x + 2y)$
18. $(3b - 2)(2a + 5)$ **19.** $5(a + 5b)^2$
20. $(x^2 + 25)(x + 5)(x - 5)$ **21.** $(4a - 3b)(5a + 2b)$

22. $(x - 6)$ and $(x + 4)$ **23.** $0, -\dfrac{2}{3}$ **24.** $\dfrac{2}{3}, \dfrac{5}{2}$

25. $w = 4$ in., $l = 8$ in.

Exercise Set 6.1

1. $\dfrac{9}{12}$ **3.** $\dfrac{1}{3}$ **5.** $\dfrac{34}{60}$ **7.** $\dfrac{7}{52}$ **9.** $\dfrac{7}{10}, \dfrac{3}{10}$

11. $\dfrac{47}{50}, \dfrac{3}{50}$ **13.** [grid figure]   **15.** [circle figure]

17. $\dfrac{16}{12}, 1\dfrac{4}{12}$ **19.** $\dfrac{5}{4}, 1\dfrac{1}{4}$ **21.** $\dfrac{36}{36}$ **23.** $\dfrac{3}{4}$ **25.** $\dfrac{4}{6}$
27. $\dfrac{5}{13}$ **29.** $\dfrac{7}{20}$ **31.** $\dfrac{357}{400}$ **33.** $\dfrac{3837}{5000}$ **35.** $\dfrac{481}{450}$

37. $\dfrac{115}{12}, 9\dfrac{7}{12}$ **39.** [triangles figure]

41.

43. [figure]

45. 10 students   **47.** 7 slices   **49.** 12 buildings   **51.** .25
53. .88   **55.** .56   **57.** .67   **59.** 2.25   **61.** 4.4

63. 1.32   **65.** 2.21   **67.** $2\dfrac{3}{5}$ **69.** $5\dfrac{1}{7}$ **71.** $3\dfrac{8}{13}$

73. $2\dfrac{5}{29}$ **75.** $\dfrac{22}{5}$ **77.** $\dfrac{49}{9}$ **79.** $\dfrac{77}{12}$ **81.** $\dfrac{77}{6}$ **83.** $\dfrac{343}{23}$

85. [number line: -4 to 4]

87. [number line: -4 to 4]

89. [number line: -4 to 4]

Exercise Set 6.2

1. $\dfrac{1}{6}$ **3.** $\dfrac{3}{2}$ **5.** $\dfrac{10}{21}$ **7.** $\dfrac{2}{3}$ **9.** $\dfrac{3}{2}$ **11.** $\dfrac{4}{3}$ **13.** $\dfrac{2}{3}$

15. $\dfrac{5}{4}$ **17.** $\dfrac{21}{11}$ **19.** $\dfrac{2}{7}$ **21.** $\dfrac{b^2}{a^2}$ **23.** $-r$ **25.** $\dfrac{1}{ab^2}$

27. $4x^3y$ **29.** $-2xy^2$ **31.** $-\dfrac{7}{4m^4n}$ **33.** $\dfrac{a^2c^2}{b^3}$

35. $\dfrac{16n^2}{9p}$ **37.** $-\dfrac{3}{5rs}$ **39.** $-2x^2y$

Exercise Set 6.3

1. $x \neq -5$ **3.** $r \neq 8$ **5.** $x \neq -4$ **7.** $x \neq \dfrac{4}{3}$

9. $x \neq -2y$ or $y \neq -\dfrac{x}{2}$ **11.** $x \neq 7$ or -3 **13.** $x \neq -2$ or 5

15. $r \neq -\dfrac{3}{2}$ or 5 **17.** $\dfrac{4}{9}$ **19.** $-\dfrac{5}{6}$ **21.** $\dfrac{2}{3}$ **23.** $\dfrac{2}{3}$

25. $\dfrac{2}{5}$ **27.** $-\dfrac{3}{4}$ **29.** $\dfrac{x + 2}{x - 3}$ **31.** $\dfrac{a - 2}{a + 3}$ **33.** $\dfrac{1}{x + 2}$

35. $\dfrac{x - 3}{x - 5}$ **37.** $\dfrac{2x - 3y}{3x - 5y}$ **39.** $\dfrac{2a - 3b}{3a - 2b}$ **41.** $\dfrac{c - d}{x + y}$

43. $\dfrac{y - 4}{y - 6}$ **45.** $-(x - y)$ **47.** $-(c - 2d)$ **49.** $-(2s - 3r)$

51. -1 **53.** -1 **55.** $-a - 5$ **57.** $-x - 2$

59. $\dfrac{-1}{x - 5}$ **61. a.** $-\dfrac{1}{2}$ **b.** $-\dfrac{1}{3}$ **c.** undefined

Exercise Set 6.4

1. $\dfrac{6}{35}$ **3.** $-\dfrac{45}{88}$ **5.** 14   **7.** $\dfrac{4}{3}$ **9.** $\dfrac{4}{3}$ **11.** 4

13. $-\dfrac{10}{9}$ **15.** $\dfrac{3}{2}$ **17.** $\dfrac{11}{6}$ **19.** -8 **21.** $\dfrac{48}{5}$

23. $-\dfrac{25}{4}$ **25.** 68   **27.** 39   **29.** $-\dfrac{6}{5}$ **31.** 15   **33.** 3

35. 27   **37.** $-\dfrac{7}{8}$ **39.** $-\dfrac{1}{2}$ **41.** 13   **43.** 45   **45.** $\dfrac{s^2}{r^3}$

47. xy^2 **49.** $\dfrac{1}{m^2}$ **51.** $\dfrac{a^2x}{b^2y^2}$ **53.** $\dfrac{x}{y^2}$ **55.** $\dfrac{1}{rs}$

57. $\dfrac{3xy^2}{2}$ **59.** $\dfrac{10y^2z^2}{3x^3w}$ **61.** $\dfrac{4xy^2}{9}$ **63.** $\dfrac{8c}{3b^3}$

65. 104 sq. yds   **67.** 1452 yds   **69.** $\dfrac{2}{9}$ of a beaker

71. $\dfrac{9}{4}$ ft   **73.** $\dfrac{75}{8}$ sq. cm   **75.** 38 sq. ft   **77.** $\dfrac{63}{160}$ sq. m

79. $\dfrac{169}{24}$ sq. ft   **81.** 11 cm^3 **83.** 72 yd^3 **85.** 10 sq. in.

Exercise Set 6.5

1. $\dfrac{4a(x+y)}{3}$ 3. $\dfrac{xy^2(r+s)}{8}$ 5. $\dfrac{6}{35}$ 7. $\dfrac{2}{5}$

9. $\dfrac{4(a+1)}{3(a+4)}$ 11. $(2x-5)(3x+4)$ 13. $-\dfrac{9}{7}$ 15. $-\dfrac{3}{2}$

17. $-\dfrac{4y^2}{3x}$ 19. $\dfrac{1}{(x-1)(x+4)}$ 21. $\dfrac{x-2}{x+3}$ 23. $\dfrac{x-1}{x+4}$

25. $\dfrac{x+2}{x-2}$ 27. 1 29. $\dfrac{x+2}{x-3}$ 31. $-b-5$

33. $-(3x-1)$ or $1-3x$ 35. $\dfrac{(d-5)(a+2)}{(b-4)(d+1)}$ 37. $\dfrac{c-d}{m+n}$

39. $2a-b$

Exercise Set 6.6

1. $\dfrac{7}{16}$ 3. $-\dfrac{3}{5}$ 5. $\dfrac{6}{5}$ 7. $\dfrac{20}{11}$ 9. $-\dfrac{8}{5}$ 11. $\dfrac{4}{3}$

13. 9 15. -27 17. $-\dfrac{56}{5}$ 19. $\dfrac{2}{9}$ 21. $\dfrac{7}{45}$ 23. $\dfrac{9}{5}$

25. $-\dfrac{16}{9}$ 27. $\dfrac{28}{27}$ 29. $-\dfrac{9}{20}$ 31. $\dfrac{21}{2}$ 33. $-\dfrac{5}{6}$

35. $\dfrac{4}{5}$ 37. $\$16$ 39. 14 lots   41. $\dfrac{1}{6}$ of a bag

43. $\dfrac{1}{a^2b^2}$ 45. $\dfrac{2}{3ab}$ 47. $\dfrac{a^2}{b^3c^2d^2}$ 49. $\dfrac{32x^6}{45}$ 51. $\dfrac{15}{28}$

53. $\dfrac{4a}{9}$ 55. $\dfrac{8}{5}$ 57. $\dfrac{1}{4}$ 59. $-\dfrac{2}{3}$ 61. $\dfrac{1}{(h+7)^2}$

63. $\dfrac{d-4}{2d(3d+4)}$ 65. $\dfrac{(m-5)(m+6)}{6}$ 67. $\dfrac{x-8}{x+8}$

69. $\dfrac{a-5}{a-6}$ 71. $\dfrac{x-2}{x+3}$ 73. $\dfrac{b-4}{c-3}$

Exercise Set 6.7

1. $x-5$ 3. $x+3$ 5. $x+2$ 7. $2x+5$

9. $2x-5$ 11. $x+5+\dfrac{1}{x-3}$ 13. $3x+1+\dfrac{9}{4x-3}$

15. $3x-4+\dfrac{12}{2x+7}$ 17. $2x^2-6x+3$ 19. $2x^2+4x-6$

21. $3x^2-2x+1+\dfrac{4}{2x+5}$ 23. $3x^2+15x+2$

25. $2x^2+2x+3+\dfrac{27}{2x-3}$ 27. $4x^2-3x+2+\dfrac{4}{4x+3}$

29. x^2+2x+4 31. x^3+2x^2+4x+8

33. $x^2-3x+9+\dfrac{-18}{x+3}$ 35. 18 37. -5

Chapter 6 Review Exercises

1. $\dfrac{3}{9}$ 2. $\dfrac{18}{45}$ 3.

4.
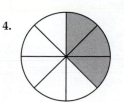

5. $\dfrac{21}{15}, 1\dfrac{6}{15}$ 6. $\dfrac{13}{4}, 3\dfrac{1}{4}$ 7. $\dfrac{7}{10}$ 8. $\dfrac{6}{8}$ or $\dfrac{3}{4}$ 9. $\dfrac{113}{120}$

10. $\dfrac{197}{250}$ 11. $\dfrac{387}{500}$ 12. $\dfrac{12}{52}$ or $\dfrac{3}{13}$ 13. $\dfrac{28}{45}, \dfrac{17}{45}$

14. $\dfrac{225}{250}$ or $\dfrac{9}{10}$, $\dfrac{25}{250}$ or $\dfrac{1}{10}$ 15. $\dfrac{573}{600}$ 16. $\dfrac{750}{1000}$ or $\dfrac{3}{4}$

17. $8\dfrac{12}{25}, \dfrac{212}{25}$ 18. $3\dfrac{3}{10}, \dfrac{33}{10}$ 19. 12 females   20. 9

21. .31   22. .78   23. .85   24. .63   25. 2.2

26. 4.25   27. 1.42   28. 2.09   29. $2\dfrac{1}{8}$ 30. $5\dfrac{5}{6}$

31. $2\dfrac{11}{21}$ 32. $2\dfrac{5}{11}$ 33. $\dfrac{39}{7}$ 34. $\dfrac{61}{9}$ 35. $\dfrac{61}{18}$

36. $\dfrac{189}{23}$ 37. $\dfrac{7}{5}$ 38. $\dfrac{15}{7}$

39.

40.

41. $\dfrac{6}{7}$ 42. $\dfrac{8}{9}$ 43. $\dfrac{30}{31}$ 44. $\dfrac{7}{8}$ 45. $\dfrac{x}{y^2}$ 46. $\dfrac{1}{m^2n}$

47. r^2s^2 48. $4xy^2$ 49. $-\dfrac{14p^3}{9q^2}$ 50. $-\dfrac{1}{3m^2n^3}$

51. $x \neq -9$ 52. $x \neq -7,-2$ 53. $x \neq 5y$ or $y \neq \dfrac{x}{5}$

54. $a \neq -6, 2$ 55. $\dfrac{7}{5}$ 56. $\dfrac{4}{7}$ 57. $\dfrac{a-3}{a+2}$ 58. $\dfrac{1}{3x+4}$

59. $\dfrac{5x-3}{2x-1}$ 60. $\dfrac{1}{4c+7d}$ 61. -1 62. $\dfrac{-1}{2x+5}$

63. $\dfrac{3}{2}$ 64. $-\dfrac{9}{4}$ 65. $-\dfrac{7}{12}$ 66. 16   67. 76   68. $\dfrac{3}{2}$

69. $\$44.20$ 70. $\dfrac{7}{2}$ ft   71. 3 sq. ft   72. $\dfrac{42}{5}$ in.^{2}

73. $\dfrac{x^2}{y}$ 74. $\dfrac{10mn^2p^2}{3q^3}$ 75. $\dfrac{9n^2}{8m^2(m-n)}$ 76. $\dfrac{7}{8}$

77. $\dfrac{3(5x+4)}{4}$ 78. $-\dfrac{5}{6}$ 79. $\dfrac{x-4}{x+5}$ 80. $\dfrac{4x-5}{3x+2}$

81. $-y-2$ 82. 1   83. $-\dfrac{8}{5}$ 84. -256 85. $\dfrac{4}{57}$

86. $-\dfrac{42}{5}$ 87. $-\dfrac{32}{9}$ 88. $\$360$ 89. $32\dfrac{1}{2}$ mpg

90. $\dfrac{14b^3y^2}{9ax^2}$ 91. $\dfrac{40}{27}$ 92. $\dfrac{3}{4}$ 93. $-\dfrac{3}{2}$ 94. $\dfrac{1}{3z(z-4)}$

95. $\dfrac{(2p+3)(6p-7)}{20}$ 96. $\dfrac{5p+3q}{2p+5q}$ 97. $4x+5$

98. $3x-7+\dfrac{-2}{5x+2}$ 99. $2a^2-3a+1+\dfrac{3}{3a+4}$

100. $4z-1+\dfrac{9}{4z-3}$ 101. $4x^2+5x-6$

102. $3x^2+9x+3+\dfrac{2}{x-3}$

Chapter 6 Test

1. $\dfrac{23}{84}, \dfrac{61}{84}$ **2.** $2\dfrac{5}{12}, \dfrac{29}{12}$ **3.** $\dfrac{9}{20}$ **4.** $.31$ **5.** $2\dfrac{3}{13}$

6. $\dfrac{46}{7}$ **7.** **8.** $\dfrac{6}{7}$

9. $a \neq 4$ **10.** $b \neq -2$ or 4 **11.** $\dfrac{21a^2}{8b^2}$ **12.** $\dfrac{3x-2}{2x-3}$

13. $10\dfrac{1}{2}$ gal **14.** $\$4.50$ **15.** $\dfrac{21}{20}$ **16.** $\dfrac{25}{2}$ **17.** $\dfrac{2}{21}$

18. $\dfrac{15b^2y^2}{4a^3x^3}$ **19.** $-\dfrac{13}{21}$ **20.** $\dfrac{a+2}{a-4}$ **21.** $\dfrac{y(4y-5)}{(2y-7)(2y-3)}$

22. $4x + 3 + \dfrac{3}{3x-5}$

Exercise Set 7.1

1. $\dfrac{5}{7}$ **3.** $\dfrac{2}{3}$ **5.** $\dfrac{9}{16}$ **7.** $\dfrac{1}{4}$ **9.** $\dfrac{15}{17}$ **11.** $-\dfrac{2}{11}$

13. 1 **15.** $\dfrac{3}{5}$ **17.** $\dfrac{1}{3}$ **19.** $\dfrac{1}{3}$ **21.** $\dfrac{89}{7}$ **23.** $\dfrac{16}{3}$

25. $\dfrac{61}{9}$ **27.** $\dfrac{a+c}{b}$ **29.** $\dfrac{u-5}{v}$ **31.** $\dfrac{11r+4m}{21u}$

33. $\dfrac{5}{2y}$ **35.** $-\dfrac{x}{7y^2}$ **37.** $\dfrac{9}{n+5}$ **39.** $\dfrac{2}{t-8}$ **41.** $\dfrac{3c}{c+2}$

43. $\dfrac{v-11}{v-6}$ **45.** $\dfrac{6a}{a+2}$ **47.** 3 **49.** $\dfrac{x+2}{x-4}$ **51.** $\dfrac{t-1}{t+7}$

53. $\dfrac{u+5}{u+3}$ **55.** $\dfrac{x-1}{x-6}$ **57.** $\dfrac{x-3}{x+5}$ **59.** $-\dfrac{1}{7}$

61. $-\dfrac{1}{17}$ **63.** $\dfrac{8}{4-t}$ or $\dfrac{-8}{t-4}$ **65.** $\dfrac{-4y}{v-u}$ or $\dfrac{4y}{u-v}$

67. $\dfrac{3m-n-3}{m-n}$ **69.** $\dfrac{-a-5b}{3a-b}$ **71.** $10\dfrac{1}{4}$ ft **73.** 22 in.

75. $\dfrac{23}{24a}$ ft **77.** $\dfrac{7a+7b}{2a+3b}$ dm **79.** $\dfrac{x-4}{x-5}$ cm **81.** $\dfrac{20}{x}$ m

83. $\dfrac{6a+2b}{2a-3b}$ ft **85.** $\dfrac{4x+6}{2x-1}$ m **87.** $\dfrac{1}{x}$ **89.** $\dfrac{t-3}{t-1}$

Exercise Set 7.2

1. 15 **3.** 99 **5.** 150 **7.** 180 **9.** 770 **11.** 252
13. x^3y^3 **15.** u^6v^4 **17.** $a^3b^6c^4$ **19.** $x^4y^6z^3$ **21.** $9u^3v^5$
23. $72y^5z^4$ **25.** $45m^5np^9$ **27.** $60r^4s^8t^3$ **29.** $108u^4v^6$
31. $84r^4s^9t^6$ **33.** $(y+4)(y+1)$ **35.** $w(w+3)(w-6)$
37. $(v-5)(v+3)$ **39.** $(x-4)(x+3)(x+5)$
41. $(z+1)^2(z-5)$ **43.** $(t-3)(t+3)(t-5)(t-4)$
45. 132 **47.** $6(y+5)(y-5)^2$ **49.** 21 **51.** 35
53. 5 **55.** 60 **57.** 243 **59.** cf **61.** 24 **63.** $35tz^2$
65. $99rs^3$ **67.** $63pq^2$ **69.** $108y^2z^5$ **71.** $27tx^3y^2$
73. $5y$ **75.** $9t$ **77.** $16c$ **79.** $11r+33$
81. $5v^2+10v$ **83.** y^2+2y-8 **85.** w^2-4w+3
87. $n^2-12n+27$ **89.** $(x+2)t$ or $xt+2t$

Exercise Set 7.3

1. $14, \dfrac{7}{14}, \dfrac{2}{14}$ **3.** $15, \dfrac{10}{15}, \dfrac{9}{15}$ **5.** $30, \dfrac{25}{30}, \dfrac{24}{30}$

7. $15, \dfrac{6}{15}, \dfrac{7}{15}$ **9.** $72, \dfrac{27}{72}, \dfrac{32}{72}$ **11.** $30, \dfrac{3}{30}, \dfrac{2}{30}$

13. $120, \dfrac{33}{120}, \dfrac{52}{120}$ **15.** $336, \dfrac{189}{336}, \dfrac{176}{336}$

17. $120, \dfrac{20}{120}, \dfrac{45}{120}, \dfrac{108}{120}$ **19.** $72, \dfrac{66}{72}, \dfrac{20}{72}, \dfrac{51}{72}$

21. $mn, \dfrac{2m}{mn}, \dfrac{3n}{mn}$ **23.** $bd, \dfrac{ad}{bd}, \dfrac{bc}{bd}$ **25.** $abc, \dfrac{6c}{abc}, \dfrac{5a}{abc}$

27. $mstu, \dfrac{rm}{mstu}, \dfrac{ns}{mstu}$ **29.** $swv^2, \dfrac{sx}{swv^2}, \dfrac{vwy}{swv^2}$

31. $a^3b^5, \dfrac{2at}{a^3b^5}, \dfrac{3b^2z}{a^3b^5}$ **33.** $x^2y^5z^4, \dfrac{5ay^4}{x^2y^5z^4}, \dfrac{7dxz}{x^2y^5z^4}$

35. $24u^2v^4, \dfrac{28}{24u^2v^4}, \dfrac{9uv}{24u^2v^4}$ **37.** $60p^6q^8, \dfrac{16}{60p^6q^8}, \dfrac{27p^2q^3}{60p^6q^8}$

39. $14x^2y^2, \dfrac{2ry}{14x^2v^2}, \dfrac{tx}{14x^2y^2}$ **41.** $50w^6z^7, \dfrac{15a}{50w^6z^7}, \dfrac{12bw^4z}{50w^6z^7}$

43. $(u+8)(u-10), \dfrac{3u-30}{(u+8)(u-10)}, \dfrac{9u+72}{(u+8)(u-10)}$

45. $(x+7)(x-5), \dfrac{x^2-5x}{(x+7)(x-5)}, \dfrac{x^2+7x}{(x+7)(x-5)}$

47. $12(t-3), \dfrac{4t}{12(t-3)}, \dfrac{15t}{12(t-3)}$

49. $45(y+2), \dfrac{30y}{45(y+2)}, \dfrac{24y}{45(y+2)}$

51. $(a+1)(a-1)(a+4), \dfrac{3a^2+12a}{(a+1)(a-1)(a+4)}, \dfrac{a^2+a}{(a+1)(a-1)(a+4)}$

53. $(v+1)(v-3)(v-2), \dfrac{v^2-7v+10}{(v+1)(v-3)(v-2)}, \dfrac{v^2+4v+3}{(v+1)(v-3)(v-2)}$

55. $(m+4)^2(m+1), \dfrac{m^2+2m+1}{(m+4)^2(m+1)}, \dfrac{m^2-16}{(m+4)^2(m+1)}$

57. $x^2(x+3)(x-2)(x+4), \dfrac{x^3+9x^2+20x}{x^2(x+3)(x-2)(x+4)}, \dfrac{x^2-5x+6}{x^2(x+3)(x-2)(x+4)}$

Exercise Set 7.4

1. $\dfrac{19}{21}$ **3.** $\dfrac{7}{10}$ **5.** $\dfrac{1}{2}$ **7.** $-\dfrac{5}{24}$ **9.** $\dfrac{23}{9}$ **11.** $-\dfrac{11}{4}$

13. $\dfrac{13}{12}$ **15.** $-\dfrac{1}{36}$ **17.** $-\dfrac{5}{36}$ **19.** $\dfrac{11}{36}$ **21.** $\dfrac{53}{8}$

23. $\dfrac{283}{24}$ **25.** $\dfrac{205}{36}$ **27.** $\dfrac{49}{16}$ **29.** $-\dfrac{19}{20}$ **31.** $\dfrac{a^2+2b}{ab}$

33. $\dfrac{1}{24u}$ **35.** $\dfrac{31z}{20x}$ **37.** $\dfrac{m+38}{6m}$

39. $\dfrac{45p^2q-10pq+9p+3}{30p^3q^2}$ **41.** $\dfrac{-2k+8}{k(k+2)}$

43. $\dfrac{9w+13}{(w-3)(w+7)}$ **45.** $\dfrac{7a+3}{a-3}$ **47.** 1 **49.** $\dfrac{x-13}{6(x+3)}$

51. $\dfrac{8t-24}{(t-4)^2}$ **53.** $\dfrac{u+8}{u-3}$ **55.** $\dfrac{h^2+15h}{(h+5)(h-5)}$

57. $\dfrac{2x^2-3x+19}{(x-3)(x+4)(x+2)}$ **59.** $\dfrac{v^2+16v+51}{(v+4)(v-4)(v+3)}$

61. $\dfrac{2z^2-z+9}{(z+3)^2(z-2)}$ **63.** $\dfrac{2x^2-5x-12}{x(x+6)(x-6)(x+3)}$

65. 17 miles **67.** $\$7.10$ **69.** $\dfrac{3a+1}{a^2}$ ft

71. $\dfrac{2yz+3xz+4xy}{xyz}$ in. **73.** $\dfrac{3a+3b+10c+10d}{(a+b)(c+d)}$ ft

75. $\dfrac{8x + 6y}{xy}$ ft   77. $\dfrac{4x^2 + 2y^2}{(x + y)(2x - y)}$ cm

79. $\dfrac{4a^2 - 2a + 44}{(a + 4)(a - 6)(a + 2)}$ in.

Exercise Set 7.5

1. $\dfrac{4}{5}$ 3. $\dfrac{14}{9}$ 5. $\dfrac{9}{2}$ 7. 2   9. $\dfrac{1}{16}$ 11. $\dfrac{3}{35}$ 13. 4

15. $\dfrac{4}{11}$ 17. $\dfrac{2}{13}$ 19. $\dfrac{230}{71}$ 21. $-\dfrac{5}{3}$ 23. $\dfrac{187}{30}$

25. $\dfrac{88}{15}$ 27. $\dfrac{3m}{2n}$ 29. $\dfrac{5f}{3d}$ 31. $\dfrac{xt}{yr}$ 33. $\dfrac{g}{4h}$ 35. $\dfrac{ac}{b}$

37. $\dfrac{a}{xb}$ 39. $\dfrac{uv}{w}$ 41. $\dfrac{1}{b}$ 43. $\dfrac{u^4}{w^4v^2}$ 45. $\dfrac{4}{2z - z^2}$

47. $\dfrac{2 + x}{2 - x}$ 49. $\dfrac{t^2 - t^3}{1 - t^3} = \dfrac{t^2}{t^2 + t + 1}$ 51. $\dfrac{v + 2}{v - 3}$

53. $\dfrac{1}{u + 6}$ 55. $\dfrac{x - 6}{x(x + 3)}$ 57. $\dfrac{5r + 44}{(r + 8)^2}$ 59. $\dfrac{t^2 - 7t + 9}{11t - 80}$

61. $\dfrac{2a^2 + a + 6}{9a^3 + 54a^2 - 3a}$

Exercise Set 7.6

1. $x = 2$ 3. $t = 5$ 5. $u = 6$ 7. $m = -\dfrac{15}{2}$ 9. $x = 2$

11. $t = 5$ 13. $x = 2$ 15. $y = \dfrac{1}{3}$ 17. $x = -2$

19. $v = 1$ 21. $q = 1$ 23. $m = 24$ 25. $r = \dfrac{2}{3}$

27. $u = 1$ 29. $w = -2$ 31. $x = 4$ 33. $t = 0$

35. $a = 11$ 37. $c = 2 \text{ or } -5$ 39. $y = \dfrac{1}{3} \text{ or } y = -2$

41. $m = 0$ 43. $y = -3$ 45. $x = 3, x \neq 2$ (makes the

denominator 0)   47. $t = 1$ 49. $w = -\dfrac{35}{4}$

51. $b \neq -1 \text{ or } 2$ 53. $x = \dfrac{9}{4}$

Exercise Set 7.7

1. 5   3. $\dfrac{10}{7}$ 5. 9, 12   7. 6, 12   9. 4, 6

11. \$18,000   13. $\dfrac{12}{5}$ hr $= 2\dfrac{2}{5}$ = 2 hr 24 min

15. $\dfrac{175}{12} = 14\dfrac{7}{12}$ days   17. 12 hrs   19. $22\dfrac{1}{2}$ min

21. 3 P.M.   23. 75 mi/hr   25. 60 mi/hr

27. 441 mi/hr   29. $\dfrac{8}{16}$ 31. catcher = 51, pitcher = 34

33. $\dfrac{20}{3}$ hr $= 6\dfrac{2}{3}$ hr = 6 hr 40 min   35. 3   37. 10 mi

Chapter 7 Review Exercises

1. $\dfrac{2}{7}$ 2. 1   3. $\dfrac{12}{13}$ 4. $\dfrac{7}{11}$ 5. $15\dfrac{4}{5}$ 6. $\dfrac{79}{12}$ or $6\dfrac{7}{12}$

7. 7   8. $\dfrac{178}{17}$ or $10\dfrac{8}{17}$ 9. $\dfrac{9r + 2p}{14x}$ 10. $\dfrac{1}{y + 2}$

11. $\dfrac{x - 7}{x + 5}$ 12. $\dfrac{x - 4}{x - 6}$ 13. $\dfrac{3p^2 + 3p - 6}{p^2 - 16}$

14. $\dfrac{6x - y}{x - y}$ 15. 36   16. \$4.44   17. $13\dfrac{8}{9}$ m

18. $1\dfrac{1}{4}$ km   19. 42   20. 200   21. 96   22. 108

23. a^2y^4 24. $60t^5v^7$ 25. $42x^5w^4$ 26. $(z - 4)(z + 3)$

27. $(a + 2)^2$ 28. $(y + 4)^2(y - 3)$ 29. 24   30. 45

31. $20rp^2$ 32. $32x^3t^2z^3$ 33. $8a$ 34. $7b + 21$

35. $p^2 - 11p + 24$ 36. $2qy + 6q$ 37. $14, \dfrac{8}{14}, \dfrac{35}{14}$

38. $18, \dfrac{15}{18}, \dfrac{7}{18}$ 39. $36, \dfrac{32}{36}, \dfrac{33}{36}$ 40. $60, \dfrac{24}{60}, \dfrac{36}{60}, \dfrac{39}{60}$

41. $t^2v^5, \dfrac{4}{t^2v^5}, \dfrac{6tv^2}{t^2v^5}$ 42. $36p^3q^5, \dfrac{32bq}{36p^3q^5}, \dfrac{33cp}{36p^3q^5}$

43. $(t - 4)(t + 2), \dfrac{7t + 14}{(t - 4)(t + 2)}, \dfrac{9t - 36}{(t - 4)(t + 2)}$

44. $(y - 11)(y - 3), \dfrac{y^2 - 3y}{(y - 11)(y - 3)}, \dfrac{y^2 - 11y}{(y - 11)(y - 3)}$

45. $28(2u + 3), \dfrac{28u}{28(2u + 3)}, \dfrac{36u}{28(2u + 3)}$

46. $(w + 4)(w - 4)(w + 1), \dfrac{2w^2 + 2w}{(w + 4)(w - 4)(w + 1)},$

$\dfrac{6w^2 - 24w}{(w + 4)(w - 4)(w + 1)}$

47. $(a + 3)^2(a - 5), \dfrac{3a^2 - 15a}{(a + 3)^2(a - 5)}, \dfrac{10a^2 + 30a}{(a + 3)^2(a - 5)}$

48. $(m - 4)(m + 2)(m + 1), \dfrac{m^2 + 2m + 1}{(m - 4)(m + 2)(m + 1)},$

$\dfrac{m^2 + m - 2}{(m - 4)(m + 2)(m + 1)}$ 49. $\dfrac{41}{48}$ 50. $\dfrac{7}{20}$ 51. $-\dfrac{3}{5}$

52. $\dfrac{31}{42}$ 53. $\dfrac{5}{2}$ 54. $\dfrac{33}{8}$ 55. $\dfrac{7n + bm}{mn}$ 56. $\dfrac{4}{9x}$

57. $\dfrac{9y - 7}{20y}$ 58. $\dfrac{4tu^3 - 6u^3 + 15t^2 + 3t}{36t^4u^5}$ 59. $\dfrac{11w - 9}{w(w - 3)}$

60. $\dfrac{6r - 2}{r - 5}$ 61. $\dfrac{-v^2 + v - 4}{4(v + 2)(v - 3)}$ 62. $\dfrac{t^2 + t + 24}{(t + 4)^2}$

63. $\dfrac{8w^2 + 39w + 10}{(w + 5)(w - 5)(w + 2)}$ 64. $\dfrac{-3z^2 + 59z + 123}{(z + 7)^2(4z + 5)}$

65. $39\dfrac{3}{4}$ in.   66. \$1.00   67. $\dfrac{8}{3}$ 68. $\dfrac{25}{33}$ 69. $\dfrac{1}{18}$

70. $\dfrac{13}{2}$ 71. $\dfrac{8}{9}$ 72. $\dfrac{12}{7}$ 73. $\dfrac{3x}{y}$ 74. $\dfrac{ps}{r}$ 75. $\dfrac{t}{5}$

76. $\dfrac{u^3w}{v}$ 77. $\dfrac{x^4y}{t^3}$ 78. $\dfrac{7w}{24 - w}$ 79. $\dfrac{y - 13}{y + 2}$

80. $x + 5$ 81. $y = 3$ 82. $p = 2$ 83. $t = \dfrac{12}{55}$

84. $w = \dfrac{22}{23}$ 85. $v = \dfrac{17}{6}$ 86. $z = -\dfrac{17}{14}$ 87. $t = \dfrac{7}{4}$

88. $m = 3$ 89. $q = -2 \text{ or } 1$ 90. $v = -\dfrac{122}{29}$ 91. $\dfrac{16}{19}$

92. 5 or -2 93. 18 days   94. 23 mph   95. 9:06 A.M.

Chapter 7 Test

1. 48   **2.** $45x^2y^4$ **3.** $15tu^2w$ **4.** $6p(q-4)$

5. $54, \dfrac{12}{54}, \dfrac{21}{54}, \dfrac{8}{54}$

6. $(a+3)(a-6)\dfrac{7a}{(a+3)(a-6)}, \dfrac{b(a-6)}{(a+3)(a-6)}$ **7.** $\dfrac{5}{7}$

8. $\dfrac{r}{r^2+16}$ **9.** $\dfrac{1}{4}$ **10.** $\dfrac{11}{30}$ **11.** $\dfrac{37}{12}$ or $3\dfrac{1}{12}$

12. $\dfrac{3v^2+26v+6}{9v^2}$ **13.** $\dfrac{46w-3}{15(w-4)}$ **14.** $\dfrac{t^2+25t+10}{2(t+5)(t-5)^2}$

15. $10\dfrac{7}{10}$ in.   **16.** $\dfrac{24a-13}{(2a+3)(a-2)}$ **17.** $\dfrac{4x+12y}{xy}$ ft

18. $\dfrac{16m^2+42mn-6n^2}{(3m-n)(2m+3n)}$ yd   **19.** $4.67   **20.** $\dfrac{3}{10}$ **21.** 4

22. $\dfrac{4m}{3}$ **23.** $\dfrac{12-u}{8u}$ **24.** $v=\dfrac{19}{17}$ **25.** $p=\dfrac{47}{37}$

26. $t=-7$ **27.** $w=5$ **28.** 0 and 4   **29.** 30 days
30. 2 hr

Exercise Set 8.1

1. 2 to 3   **3.** $3:5$ **5.** 5 to 2   **7.** $3:1$ **9.** $\dfrac{3}{7}$ **11.** $\dfrac{4}{3}$

13. $\dfrac{2}{5}$ **15.** $\dfrac{13}{7}$ **17.** $100:49$ **19.** $\dfrac{10}{13}$ **21.** $\dfrac{4}{1}$

23. $\dfrac{4}{1}, \dfrac{1}{4}$ **25.** 280 mg sodium : 1 oz mix   **27.** $\dfrac{101}{500}$

29. no   **31.** 50 mi per hr   **33.** 6 cans per student
35. $.04   **37.** $.27   **39.** $.37   **41.** $62.50
43. $10:4:2:3:2:4$

Exercise Set 8.2

1. yes   **3.** yes   **5.** yes   **7.** yes   **9.** $x=15$
11. $t=12$ **13.** $v=2$ **15.** $t=91$ **17.** $p=28$
19. $w=.6$ **21.** $y=3$ **23.** $x=3$ **25.** $t=.96$
27. $w=20$ **29.** $x=4$ **31.** $x=9$ **33.** $x=5$

35. $x=6$ **37.** $x=7$ **39.** $x=4,-7$ **41.** $x=-\dfrac{5}{2},6$

43. $W=Pt$ **45.** $P=\dfrac{KT}{V}$ **47.** 280 miles   **49.** 7.5 novels

51. 77 innings   **53.** $7\dfrac{1}{2}$ in.   **55.** 400 trout   **57.** 20 ft

59. $248   **61.** 65 min   **63.** 154 lb, $48.83

Exercise Set 8.3

1. $\dfrac{4}{25}$ **3.** $\dfrac{9}{10000}$ **5.** $\dfrac{17}{125}$ **7.** $\dfrac{13}{4}$ **9.** $\dfrac{4}{75}$ **11.** $\dfrac{17}{140}$

13. .42   **15.** .006.   **17.** .375   **19.** 1.20   **21.** .0808
23. .128 . . .   **25.** 20%   **27.** 38%   **29.** .8%   **31.** 321%

33. $55\dfrac{5}{9}\%$ **35.** 500%   **37.** 1000%   **39.** 40%

41. 45%   **43.** 76%   **45.** 37.5%   **47.** $133\dfrac{1}{3}\%$

49. $116\dfrac{2}{3}\%$ **51.** 180%   **53.** 150%   **55.** 39%

57. 20.9%   **59.** 14.23%   **61.** 6.4%   **63.** 9.5%

65. 27%   **67.** 17.65%   **69.** $\dfrac{103}{1100}$

Exercise Set 8.4

1. What number is 25% of 36?   **3.** 40% of 60 is what number?
5. 20% of what number is 8?   **7.** 52 is 65% of what number?
9. What percent of 30 is 21?   **11.** 4.8 is what percent of 80?
13. 10   **15.** 28   **17.** 80   **19.** 6000   **21.** 30%
23. 12.5%   **25.** 88   **27.** .175   **29.** 72   **31.** 80
33. 4%   **35.** 15%   **37.** 81%   **39.** 945 accidents
41. 105 ml   **43.** 40%   **45.** $85,000   **47.** 190 lb
49. 40   **51.** 100 ml

Exercise Set 8.5

1. $6.30   **3.** 4.5%   **5.** $340   **7.** $591.25   **9.** $27
11. 20%   **13.** $150, 30%   **15.** $45   **17.** $125
19. 10%   **21.** $48   **23.** $3466.67   **25.** $t=3$
27. $r=8\%$ **29.** $600   **31.** $680   **33.** $28.33
35. $384   **37.** $192   **39.** $3785.60   **41.** 8%

43. $8\dfrac{1}{3}\%$ nearest tenth 8.3%   **45.** $9.12   **47.** $6831.15

49. 41.2%   **51.** $9043.63

Chapter 8 Review Exercises

1. $3:5$ **2.** 7 to 6   **3.** $9:5$ **4.** $\dfrac{3}{4}$ **5.** $25:14$

6. $2:1:1$ **7.** $4:2:1$ **8.** $\dfrac{227}{500}$ **9.** 4 pieces

10. 20 pounds   **11.** $.05   **12.** 3 cents   **13.** yes
14. no   **15.** $x=2.5$ **16.** $y=96$ **17.** $t=10$
18. $u=0$ **19.** $y=5.6$ **20.** $x=2.2$ **21.** $v=2$

22. $v=\dfrac{14}{3}$ **23.** 5   **24.** $w=3$ **25.** $x=7$

26. $x=3.5$ **27.** $y=-3$ **28.** $t=10$ **29.** $x=-4,7$

30. $x=\dfrac{5}{2},-3$ **31.** $1\dfrac{1}{2}$ acres   **32.** 10,080   **33.** 540 mi

34. 1890   **35.** 25%   **36.** 87.5%   **37.** 250%

38. 400%   **39.** 42.9%   **40.** 83.3%   **41.** $\dfrac{2}{5}$ **42.** $\dfrac{109}{100}$

43. $\dfrac{41}{500}$ **44.** $\dfrac{1}{3}$ **45.** 2.5%   **46.** 657%   **47.** 205%

48. 4.61%   **49.** 60.5%   **50.** $50\dfrac{1}{3}\%$ **51.** $\dfrac{2}{25}$ **52.** $\dfrac{1}{8}$

53. $\dfrac{37}{400}$ **54.** $\dfrac{1}{2500}$ **55.** .18   **56.** .0675   **57.** .026

58. .00005   **59.** .27   **60.** .097   **61.** 35   **62.** 80%
63. 150%   **64.** 40   **65.** 64%   **66.** 400   **67.** 12%
nearest percent   **68.** 840   **69.** 6.5%   **70.** $12,500
71. $460   **72.** $830   **73.** 22%   **74.** $2,450
75. $3,000   **76.** 18%   **77.** $22.40   **78.** $2,679.44
79. $185 billion

Chapter 8 Test

1. $4:3:5$ **2.** faculty to students, $1:100$ **3.** $8:1:5$

4. $800,000:1$ **5.** $\dfrac{\$.05}{\text{oz}}$ **6.** $y=10$ **7.** $x=4$

8. $x=-4,-6$ **9.** 175 min   **10.** 54 hogs   **11.** 160%

12. 57.1%   **13.** 30.8%   **14.** 767%   **15.** $\dfrac{3}{8}$ **16.** $\dfrac{37}{400}$

17. 931.37　**18.** 4375　**19.** $33\frac{1}{3}\%$　**20.** \$111　**21.** 6%

22. \$297.50　**23.** \$35,000　**24.** \$29,250　**25.** \$12,730.80

26. \$19,795

Exercise Set 9.1

1. yes　**3.** no　**5.** yes　**7.** no　**9.** $(0, 1)$　**11.** $(-2, 3)$
13. $(-3, 4)$　**15.** $(-3, -2)$　**17.** $(5, 1)$　**19.** $(3, 3)$
21. $(0, 0)$　**23.** inconsistent　**25.** inconsistent
27. inconsistent　**29.** consistent $(1, -6)$　**31.** yes

Exercise Set 9.2

1. $(2, 0)$　**3.** $(1, 2)$　**5.** $(-3, -2)$　**7.** $(-1, 4)$　**9.** $(6, 0)$
11. $(4, 6)$　**13.** $(5, -1)$　**15.** $(4, -4)$　**17.** $(2, 5)$
19. $\left(3, -\frac{1}{2}\right)$　**21.** $\left(\frac{2}{3}, -4\right)$　**23.** $\left(-\frac{2}{5}, \frac{2}{3}\right)$
25. $\left(-\frac{1}{8}, \frac{5}{12}\right)$　**27.** $\left(\frac{3}{13}, \frac{2}{11}\right)$　**29.** $\left(\frac{47}{70}, -\frac{3}{5}\right)$
31. $\left(\frac{78}{11}, -\frac{28}{11}\right)$　**33.** $(4, -5)$　**35.** $\left(\frac{5}{8}, \frac{1}{4}\right)$　**37.** $(0, 4)$
39. $(9, 2)$　**41.** $(3, -1)$　**43.** dependent　**45.** $(6, 4)$
47. inconsistent　**49.** $(1, -1)$　**51.** $\left(2\frac{1}{3}, -\frac{1}{2}\right)$
53. $\left(\frac{2}{3}, -4\right)$

Exercise Set 9.3

1. $(2, -1)$　**3.** $(1, 5)$　**5.** $(2, -3)$　**7.** $(-4, -1)$
9. $(3, 4)$　**11.** $(3, 3)$　**13.** $(4, -4)$　**15.** $(-5, 2)$
17. $(-3, -2)$　**19.** $(2, 2)$　**21.** $(-3, -2)$
23. inconsistent　**25.** $(7, -9)$　**27.** inconsistent
29. consistent, $(-1, 8)$　**31.** $(4, -1)$　**33.** $(6, 8)$
35. $(4, 0)$

Exercise Set 9.4

1. 20, 12　**3.** 11, 4　**5.** 16 light fixtures at \$25 each, 20
light fixtures at \$32 each　**7.** 80 36-in. fans, 25 54-in. fans
9. 27 dimes, 25 quarters　**11.** 22 18¢ stamps, 14 25¢ stamps
13. freight train: 45 mph; passenger train: 60 mph
15. speed of cutter: 15 mph; speed of current: 3 mph
17. \$2000 at 6%, \$18,000 at 8.5%　**19.** \$150,000 at 22%,
\$100,000 at 28%　**21.** 8 lb of peanuts, 4 lb of cashews
23. 12% solution: 15 L; 4% solution: 15 L　**25.** 10, 15
27. 15 cars, 12 trucks　**29.** 700 quarters, 200 silver dollars
31. 40 mph still water, 5 mph current　**33.** \$1600 at 18%,
\$1400 at 9%　**35.** 8 cups rice, 12 cups mixture
37. $w = 6$ ft, $l = 10$ ft　**39.** 5 cm, 7 cm　**41.** 75°, 105°

Exercise Set 9.5

1.
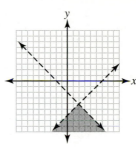

3.
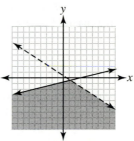

5.
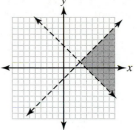

7.
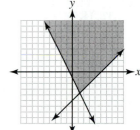

9.
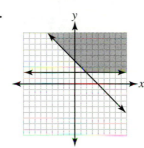

11.
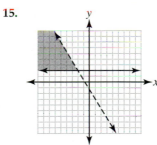

13.
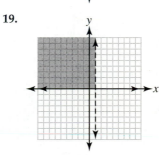

15.

17.

19.

21.
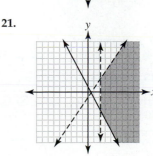

Chapter 9 Review Exercises

1. no   **2.** yes   **3.** yes   **4.** no   **5.** $(-1, 2)$ **6.** $(3, -2)$
7. $(6, 7)$ **8.** $(-5, 8)$ **9.** dependent   **10.** consistent;
$(4, -6)$ **11.** consistent; $(-3, -14)$ **12.** dependent
13. $(13, 1)$ **14.** $(6, 14)$ **15.** $(-4, 6)$ **16.** $(-1, 1)$
17. $\left(-3, \frac{1}{4}\right)$ **18.** $\left(\frac{5}{3}, \frac{1}{-2}\right)$ **19.** $\left(\frac{7}{12}, \frac{5}{11}\right)$ **20.** $(-6, 1)$

21. inconsistent   **22.** $(5.93617, 0.106382)$ or $\left(\frac{279}{47}, \frac{5}{47}\right)$

23. dependent   **24.** inconsistent   **25.** $(-2, 1)$

26. $(3, 5)$ **27.** $(4, 6)$ **28.** $(-2, 9)$ **29.** $(-5, -9)$
30. inconsistent   **31.** $(-3, -2)$ **32.** $(4, 7)$
33. dependent   **34.** $(-2, -5)$ **35.** $(-1, 4)$

36. inconsistent   **37.** $\left(\frac{1}{2}, 2\right)$ **38.** $\left(\frac{3}{4}, \frac{2}{5}\right)$ **39.** $(16, 9)$

40. basic $= \$14$, executive $= \$19$ **41.** 25 nickels, 40 dimes
42. motorcycle $= 60$ mph, bicycle $= 15$ mph   **43.** $\$10,000$ at
12%, $\$40,000$ at 7.5%   **44.** 4 quarts of each
45.   **46.**

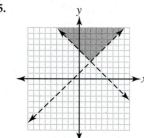
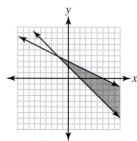

47.   **48.**

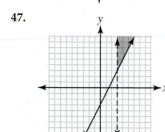
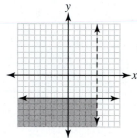

49.   **50.**

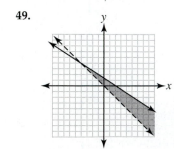
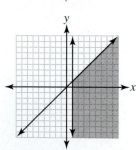

Chapter 9 Test

1. no   **2.** $(4, -4)$ **3.** $(-3, 2)$ **4.** dependent   **5.** $(6, 4)$
6. $(5, 7)$ **7.** $(-3, 8)$ **8.** $(2, -4)$ **9.** $(4, 1)$ **10.** $(-2, 5)$

11. $(-4, -7)$ **12.** $(\frac{4}{5}, -2)$ **13.** $(3, 4)$ **14.** $(-7, -2)$

15. 2 and 14   **16.** 60 dimes, 30 quarters   **17.** car $= 55$
mph, bus $= 45$ mph   **18.** $\$12,000$ at 22% and $\$4000$ at 7%

19.   **20.**

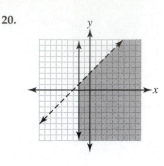

Section 10.1

1. 5   **3.** -5 **5.** -10 **7.** Does not exist as a real number   **9.** 25   **11.** 13   **13.** 3   **15.** -3 **17.** -6
19. 6   **21.** Does not exist as a real number   **23.** -3
25. irrational, 4.899   **27.** rational, 12   **29.** irrational, 3.684
31. rational, 4   **33.** rational, 1.3   **35.** irrational, 28.460
37. rational, .5   **39.** irrational, 4.642   **41.** rational, 46
43. rational, 21   **45.** b **47.** n^2 **49.** r^4 **51.** n
53. r^2 **55.** c **57.** s^2 **59.** a **61.** $x - 5$
63. $2a + b$ **65.** $x + 5$ **67.** $2a - 3b$

Section 10.2

1. $2\sqrt{3}$ **3.** $7\sqrt{2}$ **5.** $8\sqrt{2}$ **7.** $15\sqrt{6}$ **9.** $24\sqrt{5}$
11. $20\sqrt{7}$ **13.** $63\sqrt{3}$ **15.** $a\sqrt{a}$ **17.** xy^2 **19.** $x^3y^4z^5$
21. $c^2d^3\sqrt{c}$ **23.** $r^5s^4\sqrt{rs}$ **25.** $2x^3\sqrt{5}$ **27.** $18x^2\sqrt{2x}$
29. $25x^4\sqrt{6x}$ **31.** $24a^9c^3\sqrt{11a}$ **33.** $2\sqrt[3]{4}$ **35.** $8\sqrt[3]{5}$
37. $18\sqrt[3]{7}$ **39.** $x^2\sqrt[3]{x}$ **41.** $x^2y\sqrt[3]{y^2}$ **43.** $4z^2\sqrt[3]{2z^2}$
45. $4\sqrt[3]{3}$ **47. a.** $f(1) = \sqrt{3}$ **b.** $f(3) = 3$ **c.** $f(4) = 2\sqrt{3}$
49. a. $f(2) = 1$ **b.** $f(3) = 2$ **c.** $f(5) = \sqrt{10}$ **51.** $3\sqrt[4]{2}$
53. $ab\sqrt[4]{a^2b^3}$ **55.** $3\sqrt[5]{4}$ **57.** $ab^3\sqrt[5]{a^4b}$

Section 10.3

1. $\sqrt{30}$ **3.** $\sqrt{35}$ **5.** 6   **7.** 15   **9.** $\sqrt{5x}$ **11.** $\sqrt{15y}$
13. $\sqrt{xy}$ **15.** k **17.** x^2 **19.** w **21.** x^3 **23.** w
25. $2ab$ **27.** $3x^2y$ **29.** $4\sqrt{3}$ **31.** $5\sqrt{3}$ **33.** $24\sqrt{10}$
35. $30\sqrt{14}$ **37.** a^2 **39.** $y^2\sqrt{y}$ **41.** $m^5\sqrt{m}$
43. $x^4y^3\sqrt{y}$ **45.** $a^3b\sqrt{ab}$ **47.** $24c^4\sqrt{15}$ **49.** $\frac{1}{2}$
51. $\frac{7}{8}$ **53.** 2   **55.** 7   **57.** $6\sqrt{2}$ **59.** $6\sqrt{5}$
61. x **63.** $c^2\sqrt{d}$ **65.** $3x$ **67.** $12a^2$ **69.** $6c\sqrt{3c}$
71. $36x^2y^3\sqrt{2y}$ **73.** $\frac{2\sqrt{6}}{7}$ **75.** $\frac{a^4}{2}$ **77.** $\frac{3x^5\sqrt{5}}{4}$

Section 10.4

1. $7\sqrt{7}$ **3.** $-6\sqrt{6}$ **5.** $9\sqrt{a}$ **7.** $-3\sqrt{n}$ **9.** $3x\sqrt[3]{9}$
11. $18x^2\sqrt[4]{5x}$ **13.** $5\sqrt{3}$ **15.** $-\sqrt{3}$ **17.** $-\sqrt{5y}$
19. $-2n\sqrt{6}$ **21.** $9y^2\sqrt{2y}$ **23.** $15\sqrt{5}$ **25.** $-8\sqrt{5}$
27. $6\sqrt{6}$ **29.** $14a\sqrt{3a}$ **31.** $4\sqrt{6}$ **33.** $18\sqrt{2} - 17\sqrt{3}$
35. $28x\sqrt{2y}$ **37.** $7\sqrt{5}$ **39.** $-15\sqrt{6}$ **41.** $7\sqrt{2}$
43. $-14\sqrt{5}$ **45.** $3\sqrt{2} + 2$ **47.** $3 - 3\sqrt{5}$
49. $\sqrt{15} + 10\sqrt{3}$ **51.** $24 - 48\sqrt{2}$
53. $12 - 3\sqrt{2} + 4\sqrt{5} - \sqrt{10}$ **55.** $6 + 5\sqrt{x} + x$
57. $68 - 24\sqrt{6}$ **59.** $6 + 10\sqrt{2} + 9\sqrt{3} + 15\sqrt{6}$
61. $\sqrt{6} + \sqrt{10} + 3 + \sqrt{15}$ **63.** $x - \sqrt{xy} - 2y$

65. $-13 - 2\sqrt{6}$ **67.** $-70 + 24\sqrt{10}$
69. $12\sqrt{14} - 12\sqrt{6} + 6\sqrt{35} - 6\sqrt{15}$ **71.** $8a + 10\sqrt{ab} - 3b$
73. $22 + 8\sqrt{6}$ **75.** $39 + 12\sqrt{3}$ **77.** $16 + 8\sqrt{3}$ **79.** 1
81. 4   **83.** $9 - x$ **85.** -14 **87.** 25   **89.** 1
91. $x - y$ **93.** 50   **95.** 21   **97.** $2 + \sqrt{2}$ **99.** $2 + 4\sqrt{6}$

101. $1 + 2\sqrt{5}$ **103.** $3 - 2\sqrt{6}$ **105.** $\dfrac{2 + 3\sqrt{3}}{4}$

107. $\dfrac{4 - 2\sqrt{6}}{3}$ **109.** $5\sqrt[3]{2}$ **111.** $9xy\sqrt[3]{2xy^2}$ **113.** $5\sqrt[4]{3}$

115. $2\sqrt[3]{4} - 2 - 6\sqrt[3]{2}$

Section 10.5

1. $\dfrac{3\sqrt{7}}{7}$ **3.** $\sqrt{5}$ **5.** $3\sqrt{2}$ **7.** $2\sqrt{10}$ **9.** $\dfrac{4\sqrt{x}}{x}$

11. $\dfrac{m\sqrt{n}}{n}$ **13.** $\dfrac{\sqrt{10}}{2}$ **15.** $\dfrac{\sqrt{6}}{4}$ **17.** $\dfrac{\sqrt{10}}{8}$ **19.** $\dfrac{\sqrt{ab}}{b}$

21. $\dfrac{\sqrt{de}}{e}$ **23.** $\dfrac{\sqrt{15}}{3}$ **25.** $\dfrac{\sqrt{21}}{6}$ **27.** $\dfrac{\sqrt{pq}}{q}$ **29.** $\dfrac{\sqrt{7x}}{x^3}$

31. $\dfrac{\sqrt{35y}}{5y}$ **33.** $\dfrac{3\sqrt{7a}}{7a}$ **35.** $\dfrac{2\sqrt[3]{3}}{3}$ **37.** $2\sqrt[3]{5}$

39. $\dfrac{\sqrt[3]{rs^2}}{s}$ **41.** $\dfrac{\sqrt[3]{xy}}{y}$ **43.** $\dfrac{\sqrt[3]{10b^2}}{2b}$ **45.** $\dfrac{\sqrt[3]{28d}}{2d}$

47. $2 - \sqrt{5}$ **49.** $\dfrac{21 + 7\sqrt{3}}{6}$ **51.** $\dfrac{7a - a\sqrt{a}}{49 - a}$

53. $\dfrac{15 + 7\sqrt{3}}{6}$ **55.** $\dfrac{29 - 10\sqrt{5}}{-11}$

57. $\dfrac{20 + 5\sqrt{5} + 4\sqrt{3} + \sqrt{15}}{11}$ **59.** $\dfrac{14 + 9\sqrt{a} + a}{4 - a}$

61. $4 - \sqrt{15}$ **63.** $\dfrac{x + 2\sqrt{xy} + y}{x - y}$ **65.** $\dfrac{5\sqrt{2} - 2\sqrt{3}}{2}$

67. $-\dfrac{2\sqrt{10} + 2\sqrt{35} + 6\sqrt{3} + 3\sqrt{42}}{15}$ **69.** $\sqrt[4]{9}$ **71.** $y\sqrt[4]{y}$

73. $3\sqrt[5]{4}$

Section 10.6

1. 4   **3.** 49   **5.** $\varnothing$ **7.** 16   **9.** 64   **11.** 25   **13.** 6
15. 6   **17.** 3, 5   **19.** 6   **21.** 7   **23.** 14   **25.** 8
27. 4, 5   **29.** 6   **31.** 5   **33.** 3   **35.** 0, 4   **37.** 5
39. $\varnothing$ **41.** 4   **43. a.** 66.6 mph **b.** 400 ft   **45. a.** 20 amps
b. 27,000 watts   **47. a.** $\dfrac{\pi}{2}$ or 1.57 seconds **b.** 1.82 ft.   **49.** 6
51. $x = 8$ **53.** $x = -11$ **55.** $x = 16$ **57.** $x = 41$

Section 10.7

1. 15 in.   **3.** 4 ft   **5.** $4\sqrt{5}$ m   **7.** $\sqrt{14}$ in.   **9.** $\sqrt{31}$ in.
11. $3\sqrt{13}$ ft   **13.** $2\sqrt{5}$ m   **15.** $2\sqrt{13}$ in.   **17.** $2\sqrt{2}$ cm
19. 10 in.   **21.** $6\sqrt{2}$ ft   **23.** $8\sqrt{3}$ mm   **25.** $4\sqrt{5}$ m
27. $\sqrt{30}$ m   **29.** $\sqrt{89}$ cm   **31.** 8 in.   **33.** $7\sqrt{2}$ ft
35. 10 cm   **37.** $4\sqrt{2}$ in.   **39.** 17 ft   **41.** $\sqrt{69}$ in.
43. 50 in.   **45.** 75 ft   **47.** $2\sqrt{7}$ ft

Chapter 10 Review Exercises

1. 4   **2.** does not exist as a real number   **3.** -4 **4.** ± 4
5. 5   **6.** -5 **7.** -5 **8.** 5   **9.** 3   **10.** does not
exist as a real number   **11.** -3 **12.** b^2 **13.** $n^2\sqrt{n}$
14. $m^3\sqrt{m^2}$ **15.** $z^5\sqrt{z^3}$ **16.** $m^2n^3\sqrt{m}$ **17.** $ab^2\sqrt[3]{a}$
18. $xy\sqrt[4]{x^2y^3}$ **19.** irrational, 5.657   **20.** rational, 13
21. irrational, 3.979   **22.** rational, 11   **23.** rational, 1.2
24. rational, .3   **25.** $3\sqrt{7}$ **26.** $2\sqrt{11}$ **27.** $20\sqrt{5}$
28. $30\sqrt{3}$ **29.** $3\sqrt[3]{2}$ **30.** $12\sqrt[3]{4}$ **31.** $9x^3\sqrt{2x}$
32. $10a\sqrt[4]{2}$ **33.** $6z^2y\sqrt{10zy}$ **34.** $4a\sqrt[3]{2a}$
35. $15nm^2\sqrt[3]{4m}$ **36.** $3z\sqrt[4]{2z}$ **37.** r **38.** x^3 **39.** 5
40. $\sqrt{35}$ **41.** $\sqrt{6r}$ **42.** $\sqrt{mn}$ **43.** 7   **44.** $5\sqrt{7}$
45. 15   **46.** 96   **47.** $225\sqrt{2}$ **48.** $126\sqrt{3}$ **49.** c^2
50. s^4 **51.** $v^2\sqrt{v}$ **52.** $60n^3\sqrt{3}$ **53.** $60v^3\sqrt{3v}$ **54.** $\dfrac{5}{3}$
55. $\dfrac{9x}{2y^2}$ **56.** $\dfrac{\sqrt{13}}{5}$ **57.** 4   **58.** 6   **59.** x^2 **60.** $cd\sqrt{c}$
61. $6a^2$ **62.** $6a^2n\sqrt{7n}$ **63.** $\dfrac{\sqrt{15}}{4}$ **64.** $\dfrac{n^3\sqrt{n}}{m^2}$
65. $\dfrac{a\sqrt{21}}{6b}$ **66.** $-2\sqrt{5}$ **67.** $18\sqrt{a}$ **68.** $-2a\sqrt[3]{x}$
69. $6\sqrt{3}$ **70.** $3\sqrt{3}$ **71.** $2\sqrt{3}$ **72.** $-23\sqrt{7}$
73. $-11a\sqrt{5}$ **74.** $12a\sqrt{7a} + 48a\sqrt{a}$ **75.** $6\sqrt{3}$
76. $\sqrt{6}$ **77.** $3\sqrt{5} - 5\sqrt{6}$ **78.** $3 + 6\sqrt{2}$ **79.** $15 + \sqrt{5}$
80. $-14 - 7\sqrt{15}$ **81.** 21   **82.** $11 - 4\sqrt{7}$ **83.** $\dfrac{3 - \sqrt{2}}{6}$
84. $2 + 2\sqrt{3}$ **85.** $\dfrac{6\sqrt{7}}{7}$ **86.** $\dfrac{4\sqrt{2}}{3}$ **87.** $\dfrac{\sqrt{15}}{3}$
88. $\dfrac{a\sqrt{b}}{b}$ **89.** $\dfrac{\sqrt{mn}}{n}$ **90.** $\dfrac{\sqrt{5n}}{n}$ **91.** $\dfrac{4\sqrt[3]{3}}{3}$ **92.** $\sqrt[3]{n}$
93. $2a\sqrt[3]{4a^2}$ **94.** $-8 - 4\sqrt{5}$ **95.** $\dfrac{7 - 2\sqrt{7}}{3}$
96. $-3\sqrt{3} - 3\sqrt{5}$ **97.** $2\sqrt{2} + 2\sqrt{3}$ **98.** $-9 - 4\sqrt{5}$
99. $\dfrac{5\sqrt{5} + 5\sqrt{3} + \sqrt{15} + 3}{2}$ **100.** 81   **101.** 36
102. 21   **103.** $\varnothing$ **104.** $\varnothing$ **105.** $\varnothing$ **106.** $x = 2, 5$
107. 3, 7   **108.** 7   **109.** 15 in.   **110.** $x = \sqrt{17}$ ft
111. $3\sqrt{6}$ m   **112.** $x = 3\sqrt{2}$ cm   **113.** $\sqrt{89}$ ft
114. $x = \sqrt{114}$ m   **115.** $\sqrt{181}$ in.   **116.** $8\sqrt{2}$ m
117. $8\sqrt{2}$ in.   **118.** $8\sqrt{7}$ yds.   **119.** $10\sqrt{2}$ in.
120. $3\sqrt{5}$ ft   **121.** $5\sqrt{10}$ ft   **122.** 54 ft.

Chapter 10 Test

1. 9   **2.** -4 **3.** x^4 **4.** c^3d^2 **5.** $4\sqrt{6}$ **6.** $21x^2\sqrt{2x}$
7. $8\sqrt[3]{5}$ **8.** $12y^2\sqrt[3]{y}$ **9.** $30\sqrt{7}$ **10.** $60a^3\sqrt{6a}$ **11.** 4
12. $16c^2\sqrt{2c}$ **13.** $\dfrac{3c^2\sqrt{7}}{5}$ **14.** $-6\sqrt{6}$ **15.** $22\sqrt{5}$

16. $18a\sqrt{5a}$ **17.** $3\sqrt{2} + 12\sqrt{6}$ **18.** $3 + 5\sqrt{30}$
19. -12 **20.** $25 - 10\sqrt{x} + x$ **21.** $\dfrac{4\sqrt{3}}{3}$ **22.** $2\sqrt[3]{4}$
23. $\dfrac{\sqrt{mn}}{n}$ **24.** $\dfrac{21 + 3\sqrt{5}}{11}$ **25.** $-\dfrac{16 + 7\sqrt{6}}{2}$ **26.** 25
27. 9 **28.** -2 **29.** $3\sqrt{2}$ ft **30.** 75 ft

Exercise Set 11.1

1. $0, -2$ **3.** $0, -3$ **5.** $0, \dfrac{2}{3}$ **7.** $0, 2$ **9.** ± 5 **11.** $\varnothing$
13. ± 11 **15.** $\pm\sqrt{7}$ **17.** $\varnothing$ **19.** $\pm 4\sqrt{3}$
21. $\pm 2\sqrt{7}$ **23.** ± 2 **25.** $\pm\sqrt{5}$ **27.** $\pm 2\sqrt{5}$
29. $\pm 2\sqrt{2}$ **31.** $\varnothing$ **33.** $\pm\dfrac{\sqrt{14}}{2}$ **35.** $\pm\dfrac{\sqrt{55}}{5}$
37. $\pm\dfrac{2\sqrt{3}}{3}$ **39.** $1, -9$
41. $-5 \pm 2\sqrt{11}$ **43.** $1, -\dfrac{11}{3}$ **45.** $-4, 7$
47. $3 \pm \sqrt{7}$ **49.** $-5 \pm \sqrt{11}$ **51.** $\dfrac{-4 \pm \sqrt{2}}{2}$
53. $2 \pm 7\sqrt{2}$ **55.** $\dfrac{-2 \pm 2\sqrt{13}}{3}$
57. 2 and -8 **59.** 8 in. **61.** 3 cm **63.** $\dfrac{-15 \pm \sqrt{21}}{3}$
65. $\dfrac{-15 \pm \sqrt{55}}{10}$

Exercise Set 11.2

1. $9, (a + 3)^2$ **3.** $81, (x - 9)^2$ **5.** $\dfrac{25}{4}, \left(x + \dfrac{5}{2}\right)^2$
7. $\dfrac{49}{4}, \left(b - \dfrac{7}{2}\right)^2$ **9.** $\dfrac{25}{16}, \left(x + \dfrac{5}{4}\right)^2$ **11.** $\dfrac{9}{196}, \left(a - \dfrac{3}{14}\right)^2$
13. $3, 5$ **15.** $-2, -6$ **17.** $-2, 6$ **19.** $2, -5$
21. $-4, -5$ **23.** $2, \dfrac{3}{2}$ **25.** $3, \dfrac{2}{3}$ **27.** $4, \dfrac{5}{2}$ **29.** $3, \dfrac{2}{3}$
31. $\varnothing$ **33.** $-2 \pm \sqrt{5}$ **35.** $4 \pm \sqrt{11}$ **37.** $7 \pm \sqrt{51}$
39. $-4 \pm \sqrt{13}$ **41.** $5 \pm \sqrt{30}$ **43.** $\dfrac{-5 \pm \sqrt{37}}{2}$
45. $\dfrac{2 \pm \sqrt{10}}{2}$ **47.** $\varnothing$ **49.** $\dfrac{-3 \pm 2\sqrt{3}}{3}$ **51.** $\varnothing$
53. $\dfrac{1}{3}, -\dfrac{2}{3}$

Exercise Set 11.3

1. $1, 5$ **3.** $2, 4$ **5.** $-4, 6$ **7.** $-1, -6$ **9.** $-2, 7$
11. $6, \dfrac{3}{2}$ **13.** $6, -\dfrac{4}{3}$ **15.** $\varnothing$ **17.** $\dfrac{5}{2}, -\dfrac{3}{4}$ **19.** $2, -\dfrac{1}{15}$
21. $\dfrac{-3 \pm \sqrt{13}}{2}$ **23.** $\dfrac{-1 \pm \sqrt{13}}{2}$ **25.** $\dfrac{3 \pm \sqrt{41}}{4}$
27. $\dfrac{1 \pm \sqrt{61}}{10}$ **29.** $\varnothing$ **31.** $\dfrac{-7 \pm 3\sqrt{5}}{2}$ **33.** $\dfrac{9 \pm \sqrt{105}}{4}$
35. $1 \pm \sqrt{3}$ **37.** $5 \pm \sqrt{5}$ **39.** $4 \pm 2\sqrt{2}$ **41.** $\dfrac{2 \pm \sqrt{14}}{2}$
43. $\dfrac{-5 \pm \sqrt{5}}{4}$ **45.** $\dfrac{-1 \pm \sqrt{13}}{3}$ **47.** $\varnothing$ **49.** 3 and 5 or
-3 and -5 **51.** $W = 5$ in., $L = 11$ in. **53.** $W = 4$ yds,
$L = 7$ yds **55. a.** $f(0) = -3$ **b.** $f(2) = 5$ **c.** $f(-3) = 0$
57. a. $f(0) = 1$ **b.** $f(1) = 2$ **c.** $f(-2) = 17$ **59.** $2, \dfrac{3}{2}$ **61.** $3, -\dfrac{2}{3}$

Section 11.4

1. -22 **3.** 4 **5.** $\dfrac{25}{4}$ **7.** -5 **9.** $2, -6$ **11.** 3
13. $9, -12$ **15.** $\varnothing$ **17.** $\dfrac{-3 \pm \sqrt{3}}{3}$ **19.** $\pm 2\sqrt{6}$
21. $-\dfrac{4}{3}, 2$ **23.** $-6, 4$ **25.** -8 **27.** $-\dfrac{15}{2}, 5$ **29.** 2
31. 3 **33.** 2 **35.** $0, 4$ **37.** $-\dfrac{7}{5}, 2$ **39.** 5
41. $x = \dfrac{21 \pm \sqrt{449}}{2}$

Section 11.5

1. $4, 5$ **3.** 6 and 9 or -6 and -9 **5.** 5 and 7 or -5 and
-7 **7.** $w = 50$ yd, $l = 60$ yd **9.** $w = 8$ in., $l = 11$ in.
11. $3, 4,$ and 5 **13.** 26 ft **15.** 80 ft and 150 ft
17. three hrs **19.** 45 mph **21.** Jody = four hrs,
Billy = 6 hrs **23.** Ronnie = two hrs, Buster = three hrs
25. 5 sec **27.** 1.7 sec **29.** 3.7 sec

Chapter 11 Review Exercises

1. $0, -3$ **2.** $x = \pm 4$ **3.** $0, -\dfrac{13}{5}$ **4.** $c = 0, \dfrac{13}{2}$
5. $0, 2$ **6.** $a = 0, 13$ **7.** ± 12 **8.** $c = \pm 20$
9. $\pm 2\sqrt{7}$ **10.** $c = \pm 3\sqrt{5}$ **11.** $\varnothing$ **12.** $\varnothing$ **13.** ± 5
14. $q = \pm 7$ **15.** $\pm 2\sqrt{5}$ **16.** $w = \pm 3\sqrt{2}$ **17.** $1, -13$
18. $-5, 11$ **19.** $\dfrac{-4 \pm 4\sqrt{5}}{3}$ **20.** $\dfrac{5 \pm 5\sqrt{5}}{2}$ **21.** $-4, -6$
22. $x = 2, 8$ **23.** $-2, 7$ **24.** $3, -6$ **25.** $-6, -\dfrac{1}{2}$
26. $2, -\dfrac{2}{3}$ **27.** $-4 \pm 3\sqrt{2}$ **28.** $5 \pm \sqrt{21}$
29. $\dfrac{-7 \pm \sqrt{61}}{2}$ **30.** $\dfrac{9 \pm \sqrt{73}}{2}$ **31.** $\varnothing$ **32.** $\dfrac{-7 \pm \sqrt{13}}{2}$
33. $\dfrac{-6 \pm \sqrt{46}}{2}$ **34.** $\dfrac{7 \pm \sqrt{43}}{2}$ **35.** $-1, -2$
36. $b = -1, -4$ **37.** $2, 6$ **38.** $5, -2$ **39.** $2, -\dfrac{5}{2}$
40. $t = \dfrac{2}{3}, -1$ **41.** $\dfrac{4 \pm \sqrt{34}}{6}$ **42.** $\dfrac{1}{2}, -\dfrac{4}{3}$
43. $-3 \pm \sqrt{13}$ **44.** $-4 \pm 3\sqrt{2}$ **45.** $\dfrac{4 \pm \sqrt{14}}{2}$
46. $\dfrac{-3 \pm \sqrt{29}}{4}$ **47.** $\dfrac{-1 \pm \sqrt{19}}{6}$ **48.** $\dfrac{2 \pm \sqrt{10}}{6}$ **49.** $\varnothing$
50. $\varnothing$ **51.** $\dfrac{-5 \pm \sqrt{41}}{4}$ **52.** $\dfrac{7 \pm \sqrt{37}}{6}$ **53.** -1
54. -3 **55.** 5 **56.** $1, -\dfrac{13}{2}$ **57.** $-\dfrac{6}{11}, 5$ **58.** $3, 13$
59. 3 **60.** 2 **61.** 2 and 5 **62.** 2 and 6 or -6 and -2
63. 5 and 6 or -5 and -6 **64.** $x = 4$ and 8 or -4 and -8
65. $w = 12$ ft and $l = 16$ ft **66.** $L = 10$ ft, $W = 7$ ft
67. 8 mi and 15 mi **68.** $L = 20$ in., $W = 15$ in.
69. 40 mph **70.** 2 hrs **71.** Ahmad = six hrs and
Frank = ten hrs **72.** Ahmad = 6 days, Horace = 8 days

Chapter 11 Test

1. $0, 2$ **2.** $0, 5$ **3.** ± 11 **4.** $\pm 3\sqrt{3}$ **5.** $0, 12$

6. $\dfrac{-2 \pm 3\sqrt{2}}{3}$ **7.** $100, (a - 10)^2$ **8.** $\dfrac{49}{36}$ **9.** $-2, 8$

10. $\dfrac{3 \pm \sqrt{19}}{2}$ **11.** $2, 3$ **12.** $4, -\dfrac{5}{2}$ **13.** $\varnothing$

14. $-3, \dfrac{1}{3}$ **15.** $-5, 7$ **16.** $\dfrac{-7 \pm \sqrt{97}}{4}$ **17.** $\varnothing$

18. $4 \pm 5\sqrt{5}$ **19.** 10 **20.** $-\dfrac{15}{2}, 2$ **21.** $-\dfrac{8}{7}, 6$ **22.** 5

23. 8 and 10 or -8 and -10 **24.** $w = 6$ in., $1 = 8$ in.

25. three hrs

Index

Table 1: Powers and Roots

n	n^2	n^3	$\sqrt{n}$	$\sqrt[3]{n}$	$\sqrt{10n}$
1	1	1	1.000	1.000	3.162
2	4	8	1.414	1.260	4.472
3	9	27	1.732	1.442	5.477
4	16	64	2.000	1.587	6.325
5	25	125	2.236	1.710	7.071
6	36	216	2.449	1.817	7.746
7	49	343	2.646	1.913	8.367
8	64	512	2.828	2.000	8.944
9	81	729	3.000	2.080	9.487
10	100	1,000	3.162	2.154	10.000
11	121	1,331	3.317	2.224	10.488
12	144	1,728	3.464	2.289	10.954
13	169	2,197	3.606	2.351	11.402
14	196	2,744	3.742	2.410	11.832
15	225	3,375	3.873	2.466	12.247
16	256	4,096	4.000	2.520	12.649
17	289	4,913	4.123	2.571	13.038
18	324	5,832	4.243	2.621	13.416
19	361	6,859	4.359	2.688	13.784
20	400	8,000	4.472	2.714	14.142
21	441	9,261	4.583	2.759	14.491
22	484	10,648	4.690	2.802	14.832
23	529	12,167	4.796	2.844	15.166
24	576	13,824	4.899	2.884	15.492
25	625	15,625	5.000	2.924	15.811
26	676	17,576	5.099	2.962	16.125
27	729	19,683	5.196	3.000	16.432
28	784	21,952	5.292	3.037	16.733
29	841	24,389	5.385	3.072	17.029
30	900	27,000	5.477	3.107	17.321
31	961	29,791	5.568	3.141	17.607
32	1,024	32,768	5.657	3.175	17.889
33	1,089	35,937	5.745	3.208	18.166
34	1,156	39,304	5.831	3.240	18.439
35	1,225	42,875	5.916	3.271	18.708
36	1,296	46,656	6.000	3.302	18.974
37	1,369	50,653	6.083	3.332	19.235
38	1,444	54,872	6.164	3.362	19.494
39	1,521	59,319	6.245	3.391	19.748
40	1,600	64,000	6.325	3.420	20.000
41	1,681	68,921	6.403	3.448	20.248
42	1,764	74,088	6.481	3.476	20.494
43	2,849	79,507	6.557	3.503	20.736
44	2,936	85,184	6.633	3.530	20.976
45	2,025	91,125	6.708	3.557	21.213
46	2,116	97,336	6.782	3.583	21.148
47	2,209	103,823	6.856	3.609	21.679
48	2,304	110,592	6.928	3.534	21.909
49	2,401	117,649	7.000	3.659	22.136
50	2,500	125,000	7.071	3.684	22.361

Table 1: Powers and Roots

n	n^2	n^3	$\sqrt{n}$	$\sqrt[3]{n}$	$\sqrt{10n}$
51	2,601	132,651	7.141	3.708	22.583
52	2,704	140,608	7.211	3.733	22.804
53	2,809	148,877	7.280	3.756	23.022
54	2,916	157,464	7.348	3.780	23.238
55	3,025	166,375	7.416	3.803	23.452
56	3,136	175,616	7.483	3.826	23.664
57	3,249	185,193	7.550	3.849	13.875
58	3,364	195,112	7.616	3.871	24.083
59	3,481	205,379	7.681	3.893	24.290
60	3,600	216,000	7.746	3.915	24.495
61	3,721	226,981	7.810	3.936	24.698
62	3,844	238,328	7.874	3.958	24.900
63	3,969	250,047	7.937	3.979	25.100
64	4,096	262,144	8.000	4.000	25.298
65	4,225	274,625	8.062	4.021	25.495
66	4,356	287,496	8.124	4.041	25.690
67	4,489	300,763	8.185	4.062	25.884
68	4,624	314,432	8.246	4.082	26.077
69	4,761	328,509	8.307	4.102	26.268
70	4,900	343,000	8.367	4.121	26.458
71	5,041	357,911	8.426	4.141	26.646
72	5,184	373,248	8.485	4.160	26.833
73	5,329	389,017	8.544	4.179	27.019
74	5,476	405,224	8.602	4.198	27.203
75	5,625	421,875	8.660	4.217	27.386
76	5,776	438,976	8.718	4.236	27.568
77	5,929	456,533	8.775	4.254	27.749
78	6,084	474,552	8.832	4.273	27.928
79	6,241	493,039	8.888	4.291	28.107
80	6,400	512,000	8.944	4.309	28.284
81	6,561	531,441	9.000	4.327	28.460
82	6,724	551,368	9.055	4.344	28.636
83	6,889	571,787	9.110	4.362	28.810
84	7,056	592,704	9.165	4.380	28.983
85	7,225	614,125	9.220	4.397	29.155
86	7,396	636,056	9.274	4.414	29.326
87	7,569	658,503	9.327	4.431	29.496
88	7,744	981,472	9.381	4.448	29.665
89	7,921	704,969	9.434	4.465	29.833
90	8,100	729,000	9.487	4.481	30.000
91	8,281	753,571	9.539	4.498	30.166
92	8,464	778,688	9.592	4.514	30.332
93	8,649	804,357	9.644	4.531	30.496
94	8,836	830,584	9.695	4.547	30.659
95	9,025	857,375	9.747	4.563	30.882
96	9,216	884,736	9.798	4.579	30.984
97	9,409	912,673	9.849	4.595	31.145
98	9,604	941,192	9.899	4.610	31.305
99	9,801	970,299	9.950	4.626	31.464
100	10,000	1,000,000	10.000	4.642	21.623